KT-574-661

ADVANCED ENGINEERING MATHEMATICS

FOURTH EDITION

ADVANCED ENGINEERING MATHEMATICS

ERWIN KREYSZIG
Professor of Mathematics
Ohio State University
Columbus, Ohio

JOHN WILEY & SONS
New York Chichester Brisbane Toronto

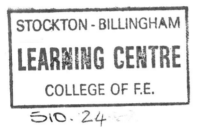

STOCKTON - BILLINGHAM
LEARNING CENTRE
COLLEGE OF F.E.
510.24

Copyright © 1962, 1967, 1972, 1979, by John Wiley & Sons, Inc.

All rights reserved. Published simultaneously in Canada.

Reproduction or translation of any part of this work beyond that permitted by Sections 107 and 108 of the 1976 United States Copyright Act without the permission of the copyright owner is unlawful. Requests for permission or further information should be addressed to the Permissions Department, John Wiley & Sons.

Library of Congress Cataloging in Publication Data:

Kreyszig, Erwin.
Advanced engineering mathematics.

Bibliography: p.
Includes index.
1. Mathematical physics. 2. Engineering mathematics. I. Title.
QA401.K7 1979 510'.2'462 78-5073

Printed in the United States of America

10 9 8 7 6 5 4

PREFACE

Purpose of the book. This book is intended to introduce students of engineering and physics to those areas of mathematics which, from a modern point of view, seem to be the most important in connection with practical problems. Topics are chosen according to the frequency of occurrence in applications. New ideas of modern mathematical training, as expressed in recent symposia on engineering education, were taken into account. The book should suit those institutions that have offered extended mathematical training for a long time as well as those that intend to follow the general trend of broadening the program of instruction in mathematics.

A course in elementary calculus is the sole prerequisite.

The material included in the book has formed the basis of various lecture courses given to undergraduate and graduate students of engineering, physics, and mathematics in this country, in Canada, and in Europe.

Changes in the fourth edition

The present edition differs essentially from the first, second and third editions in the following respects.

Problem sets. The problems have been changed. More applications are included in the problems.

Modeling is given an even greater emphasis by further applications in various sections.

Ordinary linear differential equations. A new section on *differential operators* is included in Chapter 2. *Phase plane methods, stability* and the famous *Van der Pol equation* are considered in a new chapter (Chap. 3).

Systems of differential equations. An elementary approach (without matrices) has been added. The approach by means of matrices is extended and is now presented in a new section.

Laplace transformation. *Convolution* is included. The entire chapter has been rewritten to make it more compact. In this new version, the two shifting theorems are included in the same section, differential equations appear earlier, convolution is introduced and applied, and partial fractions are discussed later and with less emphasis.

Partial differential equations. This chapter includes a new section on the application of the Laplace transformation to partial differential equations.

Matrices. The material in this chapter has been reorganized and rewritten in part, in accordance with the modern trend in linear algebra. The new version also includes more applications.

Complex analysis has been streamlined as follows. Three short sections on complex numbers (instead of the single long section in the previous editions) provide an easier introduction to this topic. For similar reasons, the chapter on conformal mapping now includes more examples at the beginning. The chapter on sequences and series has been rewritten, so that in classes on complex analysis only the first two sections of the chapter are required and, in this way, much time can be saved without loss of continuity. Power series, Taylor series and Laurent series now appear in the same chapter, that is, they are presented closer together than in previous editions.

Numerical analysis. The present edition includes a new section on *splines,* which are of increasing importance to the engineer.

Probability and statistics. The problem sets have been extended by including more applications.

References in Appendix 1 have been updated.

Content and arrangement. The arrangement of the subject matter in major parts can be seen from the diagram.

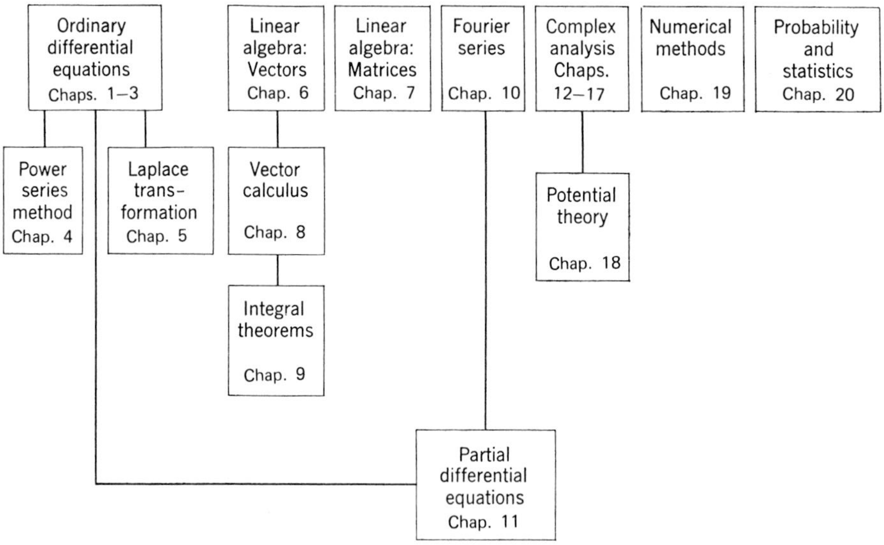

Much space is devoted to ordinary differential equations, linear algebra and vector analysis, and complex analysis, probably the three most important fields for engineers. But also the length of the other chapters on Fourier series, partial differential equations, numerical methods, etc., is such that the presentation may serve as a text for courses of the usual type.

To facilitate the use of parts of the book, the chapters are kept as independent of one another as possible.

The chapters are subdivided into relatively short sections. Each section includes typical examples and problems illustrating the concepts, methods, and results and their engineering applications.

Historical notes, references to original literature, and about 400 figures are included in the text.

References. A list of some books for reference and further study can be found at the end of the book, starting on page A-1. Some formulas for special functions are included in Appendix 3, starting on page A-54.

Problems and answers. The book contains more than 3500 carefully selected problems, which range from simple routine exercises to practical applications of considerable complexity. Answers to odd-numbered problems are included at the end of the book, starting on page A-6.

Tables of functions are included in Appendix 4, starting on page A-61.

Suggestions for a sequence of courses. The material may be taken in sequence and is then suitable for four consecutive semester courses, meeting 3–5 hours a week as follows:

First semester.	Ordinary differential equations (Chaps. 1–5).
Second semester.	Linear algebra and vector analysis (Chaps. 6–9).
Third semester.	Fourier series and partial differential equations (Chaps. 10, 11), Numerical methods (Chap. 19).
Fourth semester.	Complex analysis (Chaps. 12–18).

A course on **engineering statistics** (3–5 hours a week; Chap. 20) may be taught during any of those four semesters or afterwards.

Independent one-semester courses. The book is also suitable for various independent one-semester courses meeting 3 hours a week; for example:

Introduction to ordinary differential equations (Chaps. 1, 2).
Laplace transformation (Chap. 5).
Vector algebra and calculus (Chaps. 6, 8).
Matrices and systems of linear equations (Chap. 9).
Fourier series and partial differential equations (Chaps. 10, 11).
Complex analysis (Chaps. 12–17).
Numerical analysis (Chap. 19).

Shorter courses. Sections that may be omitted in a shorter course are indicated at the beginning of each chapter.

Principles for selection of topics. Which topics should be contained in a book of the present type, and how should these topics be arranged and presented?

To find some answer to these basic questions we may take a look at the historical development of engineering mathematics. This development shows the following two interesting features:

1. Mathematics has become more and more important to engineering science, and it is easy to conjecture that this trend will also continue in the future. In fact, problems in modern engineering are so complex that most of them cannot be solved solely by using physical intuition and past experience. The empirical approach has been successful in the solution of many problems in the past, but fails as soon as high speeds, large forces, high temperatures or other abnormal conditions are involved, and the situation becomes still more critical by the fact that various modern materials (plastics, alloys, etc.) have unusual physical properties. Experimental work has become complicated, time-consuming, and expensive. Here mathematics offers aid in planning constructions and experiments, in evaluating experimental data, and in reducing the work and cost of finding solutions.

2. Mathematical methods which were developed for purely theoretical reasons suddenly became of great importance in engineering mathematics. Examples are the theory of matrices, conformal mapping, and the theory of differential equations having periodic solutions.

What are the reflections of this situation on the teaching of engineering mathematics? Since the engineer will need more and more mathematics, should we try to include more and more topics in our courses, devoting less and less time to each topic? Or should we concentrate on a few carefully selected basic ideas of general practical importance which are especially suitable for teaching the student mathematical thinking and developing his own creative ability?

Sixty or eighty years ago no one was able to predict that conformal mapping or matrices would ever be of importance in the mathematical part of engineering work. Similarly, it is difficult to conjecture which new mathematical theories will have applications to engineering twenty or thirty years from now. But no matter what happens in that respect, if a student has a good training in the fundamentals of mathematics he will meet the future needs because he will be able to get acquainted with new methods by his own further study.

It follows that the most important objective and purpose in engineering mathematics seems to be that the student becomes familiar with mathematical thinking. He should learn to recognize the guiding principles and ideas "behind the scenes," which are more important than formal manipulations. He should get the impression that mathematics is not a collection of tricks and recipes but a systematic science of practical importance, resting on a relatively small number of basic concepts and involving powerful unifying methods. He should soon convince himself of the necessity for applying mathematical procedures to engineering problems, and he will find that the theory and its applications are related to each other like a tree and its fruits.

The student will see that the application of mathematics to an engineering problem consists essentially of three phases:

1. Translation of the given physical information into a mathematical form ***(modeling).*** *In this way we obtain a mathematical model of the physical situation. This model may be a differential equation, a system of linear equations, or some other mathematical expression.*

2. Treatment of the model by mathematical methods. This will lead to the solution of the given problem in mathematical form.

3. Interpretation of the mathematical result in physical terms.

All three steps seem to be of equal importance, and the presentation in this book is such that it will help the student to develop skill in carrying out all these steps. In this connection, preference has been given to applications which are of a general nature.

In some considerations it will be unavoidable to rely on results whose proofs are beyond the level of a book of the present type. In any case these points are marked distinctly, because hiding difficulties and oversimplifying matters is no real help to the student in preparing him for his professional work.

These are some of the guiding principles I used in selecting and presenting the material in this book. I made the choice with greatest care, using past and present teaching and research experience and resisting the temptation to include "everything which is important" in engineering mathematics.

Particular efforts have been made in presenting the topics as simply, clearly, and accurately as possible; this refers also to the choice of the notations. In each chapter the level increases gradually, avoiding jumps and accumulations of difficult theoretical considerations.

The end of a proof is marked by ■. This sign is also used at the end of some of the definitions and at the end of a set of examples when followed by further text.

Acknowledgment. I am indebted to many of my former teachers, colleagues, and students for advice and help in preparing this book. Various parts of the manuscript were distributed to my classes in mimeographed form and returned to me with suggestions for improvement. Discussions with engineers and mathematicians (as well as written comments) were of great help to me; I want to mention particularly Professors S. Bergman(†), S. L. Campbell, J. T. Cargo, P. L. Chambré, A. Cronheim, J. Delany, J. W. Dettman, R. G. Helsel, W. N. Huff, E. C. Klipple, V. Komkow, H. Kuhn, G. Lamb, H. B. Mann, I. Marx, W. D. Munroe, H.-W. Pu, T. Rado(†) P. V. Reichelderfer, J. T. Scheick, H. A. Smith, J. P. Spencer, J. Todd, H. J. Weiss, A. Wilansky, who are all in this country, Professor H. S. M. Coxeter of Toronto, and Professors B. Baule(†), H. Behnke, H. Florian, H. Graf, F. Hohenberg, K. Klotter, M. Pinl, F. Reutter, C. Schmieden, H. Unger, A. Walther(†), H. Wielandt, all in Europe. I can offer here only an inadequate acknowledgment of my appreciation.

Finally, I thank John Wiley and Sons for their effective cooperation and their great care in preparing this edition.

Suggestions of many readers were evaluated in preparing the present edition. Any further comment and suggestion for improvement of the book will be gratefully received.

ERWIN KREYSZIG

CONTENTS

ADVANCED ENGINEERING MATHEMATICS

CHAPTER 1

ORDINARY DIFFERENTIAL EQUATIONS OF THE FIRST ORDER

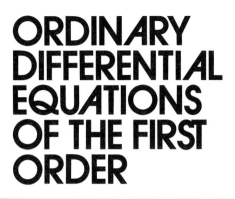

Differential equations are of fundamental importance in engineering mathematics because many physical laws and relations appear mathematically in the form of such equations. We shall consider various physical and geometrical problems which lead to differential equations and the most important standard methods for solving such equations. These methods will in general involve integration.

We shall pay particular attention to the derivation of differential equations from given physical situations. This transition from the physical problem to a corresponding "mathematical model" is called **modeling.** It is of great practical importance to the engineer and physicist, and will be illustrated by typical examples.

The first five chapters of the book are devoted to ordinary differential equations. In the present chapter we shall start with the simplest of these equations, the so-called differential equations of the first order.

Differential equations are particularly convenient for modern computers. Corresponding **numerical methods** for obtaining approximate solutions of first-order differential equations are included in Sec. 19.7, which is independent of the other sections in Chap. 19. Numerical methods for second-order differential equations are considered in Sec. 19.8.

Prerequisite for this chapter: integral calculus.
Sections which may be omitted in a shorter course: 1.8–1.12.
References: Appendix 1, Part B.
Answers to Problems: Appendix 2.

1.1 Basic Concepts and Ideas

In this section we shall define and explain the basic concepts that are essential in connection with differential equations and illustrate these concepts by examples. Then we shall consider two simple practical problems taken from physics and geometry. This will give us a first idea of the nature and purpose of differential equations and their application.

By an **ordinary differential equation** we mean a relation which involves one or several derivatives of an unspecified function y of x; the relation may also involve y itself, given functions of x, and constants.

For example,

$$y' = \cos x, \tag{1}$$

$$y'' + 4y = 0, \tag{2}$$

$$x^2 y''' y' + 2e^x y'' = (x^2 + 2)y^2 \tag{3}$$

are ordinary differential equations.

The term *ordinary* distinguishes it from a *partial differential equation*, which involves partial derivatives of an unspecified function of two or more independent variables. For example,

$$\frac{\partial^2 u}{\partial x^2} + \frac{\partial^2 u}{\partial y^2} = 0$$

is a partial differential equation. In the present chapter we shall consider only ordinary differential equations.

An ordinary differential equation is said to be of **order** n if the nth derivative of y with respect to x is the highest derivative of y in that equation.

The notion of the order of a differential equation leads to a useful classification of the equations into equations of first order, second order, etc. Thus (1) is a first-order equation, (2) is of the second order, and (3) is of the third order.

In the present chapter we shall consider first-order equations. Equations of second and higher order will be discussed in Chaps. 2–5.

A function

$$y = g(x) \tag{4}$$

is called a **solution** of a given first-order differential equation on some interval, say, $a < x < b$ (perhaps infinite) if $g(x)$ is defined and differentiable throughout that interval and is such that the equation becomes an identity when y and y' are replaced by g and g', respectively.

For example, the function

$$y = g(x) = e^{2x}$$

is a solution of the first-order differential equation

$$y' = 2y$$

for all x, because by differentiation we obtain

$$g' = 2e^{2x},$$

and by inserting g and g' we see that the equation reduces to the identity

$$2e^{2x} = 2e^{2x}.$$

Sometimes a solution of a differential equation will appear as an implicit function, that is, implicitly given in the form

$$G(x, y) = 0,$$

and is then called an *implicit solution*, in contrast to the *explicit solution* (4).

For example,

$$x^2 + y^2 = 1 \qquad (y > 0)$$

is an implicit solution of the differential equation

$$yy' = -x$$

on the interval $-1 < x < 1$, as the student may verify.

The principal task of the theory of ordinary differential equations is to find all the solutions of a given differential equation and investigate their properties. We shall discuss various standard methods which were developed for that purpose.

A differential equation may have many solutions. Let us illustrate this fact by the following examples.

Example 1

Each of the functions

$$y = \sin x, \qquad y = \sin x + 3, \qquad y = \sin x - \tfrac{4}{5}$$

is a solution of the differential equation (1),

$$y' = \cos x,$$

and we know from calculus that every solution of the equation is of the form

$$y = \sin x + c \tag{5}$$

where c is a constant. If we regard c as arbitrary, then (5) represents the totality of all solutions of the differential equation. (See Fig. 1.)

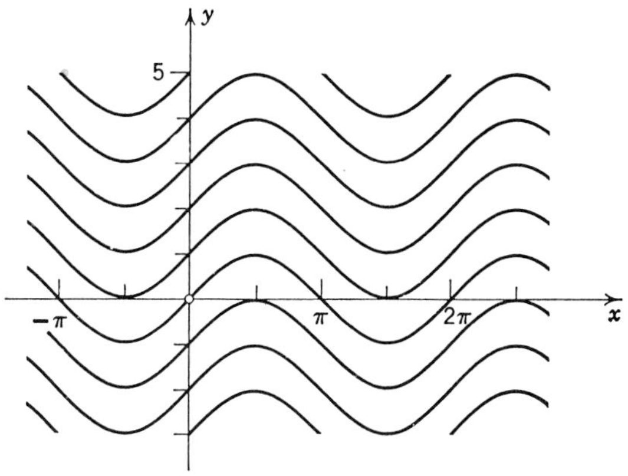

Fig. 1. Solutions of $y' = \cos x$

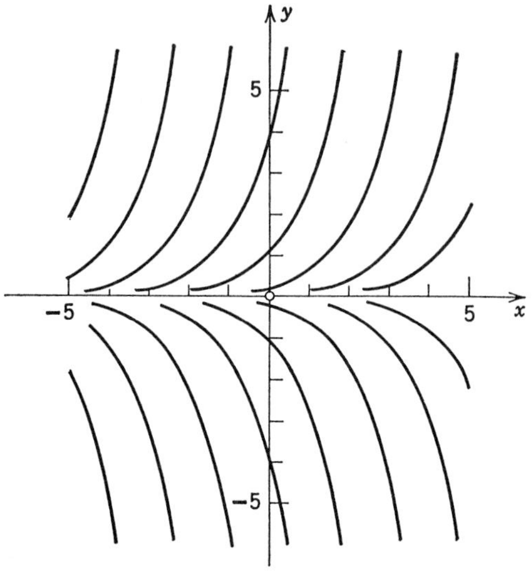

Fig. 2. Solutions of $y' = y$

Example 2

The student may verify that each of the functions

$$y = e^x, \qquad y = 2e^x, \qquad y = -\tfrac{6}{5}e^x \tag{6}$$

is a solution of the differential equation

$$y' = y$$

for all x. We shall see later that every solution of this equation is of the form

$$y = ce^x \tag{7}$$

where c is a constant. Formula (7), with arbitrary c, thus represents the totality of all solutions of the differential equation. (See Fig. 2.) ∎

Our examples illustrate that a differential equation may (and, in general, will) have more than one solution, even infinitely many solutions, which can be represented by a single formula involving an arbitrary constant c. It is customary to call such a function which contains an arbitrary[1] constant a **general solution** of the corresponding differential equation of the first order. If we assign a definite value to that constant, then the solution so obtained is called a **particular solution.**

Thus (7) is a general solution of the equation $y' = y$, and (6) are particular solutions.

In the following sections we shall develop various methods for obtaining general solutions of first-order equations. For a given equation, a general solution obtained by such a method is unique, except for notation, and will then be called ***the*** general solution of that equation.

[1]The range of the constant may have to be restricted in some cases to avoid imaginary expressions or other degeneracies.

We mention that in some cases there may be further solutions of a given equation which cannot be obtained by assigning a definite value to the arbitrary constant in the general solution; such a solution is then called a **singular solution** of the equation.

For example, the equation

$$y'^2 - xy' + y = 0 \tag{8}$$

has the general solution

$$y = cx - c^2$$

representing a family of straight lines where each line corresponds to a definite value of c. A further solution is

$$y = \frac{x^2}{4}$$

as the reader may readily verify by substitution. This is a singular solution of (8), since we cannot obtain it by assigning a definite value to c in the general solution. Obviously, each particular solution represents a tangent to the parabola represented by the singular solution (Fig. 3).

Singular solutions will rarely occur in engineering problems.

It should be noted that in some books the notion *general solution* means a formula which includes *all* solutions of an equation, that is, both the particular and the singular ones. We do not adopt this definition for two reasons. First of all, it is frequently quite difficult to prove that a formula includes *all* solutions; hence that definition of a general solution is rather useless in practice. Furthermore, we shall see later that a large and very important class of equations (the so-called *linear* differential equations) does not have singular solutions, and our definition of a general solution can be easily generalized to equations of higher order so that the resulting notion includes all solutions of a differential equation which is linear.

We shall see that the conditions under which a given differential equation has solutions are fairly general. But we should note that there are simple equations which do not have solutions at all, and others which do not have a general solution. For example, the equation $y'^2 = -1$ does not have a solution for real y. (Why?) The equation $|y'| + |y| = 0$ does not have a general solution, because its only solution is $y \equiv 0$.

Differential equations are of great importance in engineering, because many

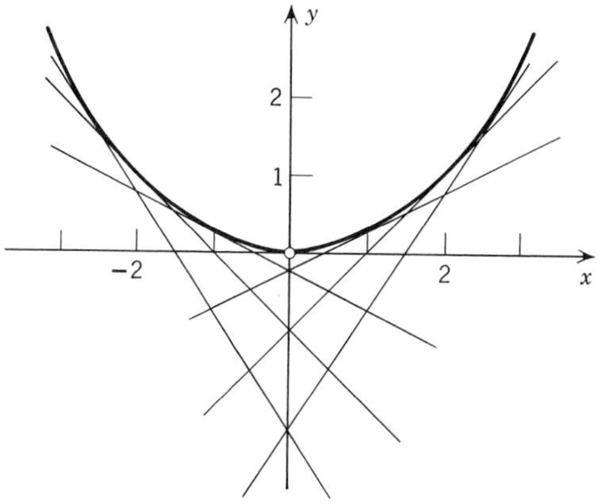

Fig. 3. Singular solution (representing a parabola) and particular solutions of the equation (8)

physical laws and relations appear mathematically in the form of differential equations.

Let us consider a simple physical example which will illustrate the typical steps of **modeling,** that is, the steps which lead from the physical situation (*physical system*) to a mathematical formulation (*mathematical model*) and solution, and to the physical interpretation of the result. This may be the easiest way of obtaining a first idea of the nature and purpose of differential equations and their applications.

Example 3. Radioactivity, exponential decay

Experiments show that a radioactive substance decomposes at a rate proportional to the amount present. Starting with a given amount of substance, say, 2 grams, at a certain time, say, $t = 0$, what can be said about the amount available at a later time?

1st Step (Mathematical description of the physical process by a differential equation). We denote by $y(t)$ the amount of substance still present at time t. The rate of change is dy/dt. According to the physical law governing the process of radiation, dy/dt is proportional to y. Thus,

$$\frac{dy}{dt} = ky. \tag{9}$$

Here k is a definite physical constant whose numerical value is known for various radioactive substances. (For example, in the case of radium, $k \approx -1.4 \cdot 10^{-11}\ \text{sec}^{-1}$.) Clearly, since the amount of substance is positive and decreases with time, dy/dt is negative, and so is k. We see that the physical process under consideration is described mathematically by an ordinary differential equation of the first order. Hence this equation is the mathematical model of that physical process. Whenever a physical law involves a rate of change of a function, such as velocity, acceleration, etc., it will lead to a differential equation. For this reason differential equations occur frequently in physics and engineering.

2nd Step (Solving the differential equation). At this early stage of our discussion no systematic method for solving (9) is at our disposal. However, (9) tells us that if there is a solution $y(t)$, its derivative must be proportional to y. From calculus we remember that exponential functions have this property. In fact, the function e^{kt} or more generally

$$y(t) = ce^{kt} \tag{10}$$

where c is any constant, is a solution of (9) for all t, as can readily be verified by substituting (10) into (9). Since (10) involves an arbitrary constant, it is a general solution of the first order equation (9). [We shall see later (in Sec. 1.3) that (10) includes *all* solutions of (9), that is, (9) does not have singular solutions.]

3rd Step (Determination of a particular solution). It is clear that our physical process has a unique behavior. Hence we can expect that by using further given information we shall be able to select a definite numerical value of c in (10) so that the resulting particular solution will describe the process properly. The amount of substance $y(t)$ still present at time t will depend on the initial amount of substance given. This amount is 2 grams at $t = 0$. Hence we have to specify the value of c so that $y = 2$ when $t = 0$. This condition is called an **initial condition,** since it refers to the initial state of the physical system. By inserting this condition

$$y(0) = 2 \tag{11}$$

in (10) we obtain

$$y(0) = ce^0 = 2 \qquad \text{or} \qquad c = 2.$$

If we use this value of c, then (10) becomes

$$y(t) = 2e^{kt}. \tag{12}$$

This particular solution of (9) characterizes the amount of substance still present at any time $t \geqq 0$. The physical constant k is negative, and $y(t)$ decreases, as is shown in Fig. 4.

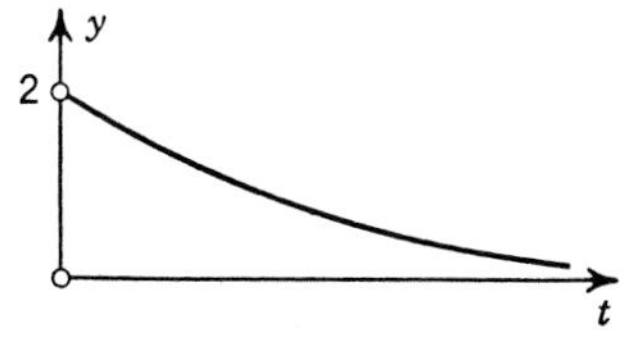

Fig. 4. Radioactivity (exponential decay)

4th Step (Checking). From (12) we have

$$\frac{dy}{dt} = 2ke^{kt} = ky \qquad \text{and} \qquad y(0) = 2e^0 = 2.$$

The function (12) satisfies the equation (9) as well as the initial condition (11).

The student should never forget to carry out the important final step which shows him whether the function is (or is not) the solution of the problem.

Let us illustrate that geometrical problems also lead to differential equations.

Example 4. A geometrical application

Find the curve through the point (1, 1) in the xy-plane having at each of its points the slope $-y/x$. Clearly the function representing the desired curve must be a solution of the differential equation

$$y' = -\frac{y}{x}. \tag{13}$$

We shall soon learn how to solve such an equation. For the time being the student may verify that

$$y = \frac{c}{x}$$

is a solution of (13) for every value of the constant c. Some of the corresponding curves are shown in Fig. 5. Since we are looking for the curve which passes through (1, 1), we must have $y = 1$ when $x = 1$. This yields $c = 1$. Hence the solution of our problem is

$$y = \frac{1}{x}.$$

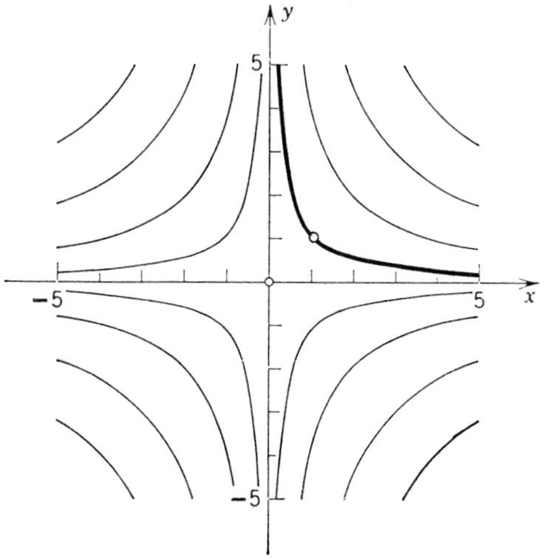

Fig. 5. Solutions of $y' = -y/x$

Problems for Sec. 1.1

In each case state the order of the differential equation and verify that the given function is a solution.

1. $y' + 3y = 0, \quad y = ce^{-3x}$

2. $y'' = e^x, \quad y = e^x + ax + b$

3. $y'' + 9y = 0, \quad y = A \cos 3x + B \sin 3x$

4. $y' + y \tan x = 0, \quad y = c \cos x$

5. $xy' + y = \sin x, \quad y = (c - \cos x)/x$

6. $y'' + 4y' + 5y = 0, \quad y = e^{-2x}(A \cos x + B \sin x)$

In each case verify that the given function is a solution of the corresponding equation and graph the corresponding curves for some values of the constant c.

7. $2y' + y = 0, \quad y = ce^{-0.5x}$

8. $y' + 2xy = 0, \quad y = ce^{-x^2}$ (*"bell-shaped curves"*)

9. $y' = -x/y, \quad x^2 + y^2 = c$

10. $y'^2 = 4y, \quad y = 0, \quad y = (x + c)^2$

Solve the following differential equations.

11. $y' = e^{-x}$ **12.** $y''' = 1$ **13.** $3y'' = \cos 2x$

Find a first-order differential equation involving both y and y' for which the given function is a solution.

14. $y = \cos x$ **15.** $y = xe^{-x}$ **16.** $y = 5x^3$

In each case find $y = \int f(x)\, dx$ and determine the constant of integration c so that y satisfies the given condition.

17. $f = x^2, \quad y = -5$ when $x = 2$

18. $f = xe^{-x^2}, \quad y = 1$ when $x = 0$

19. $f = \cos^2 x, \quad y = \pi/2$ when $x = \pi/2$

In each case verify that the given function is a solution of the corresponding differential equation and determine c so that the resulting particular solution satisfies the given condition. Graph this solution.

20. $y' + y = 1, \quad y = 1 + ce^{-x}, \quad y = 3$ when $x = 0$

21. $xy' = 3y, \quad y = cx^3, \quad y = 1$ when $x = -2$

22. $y' = -2xy, \quad y = ce^{-x^2}, \quad y = 3$ when $x = 0$

23. **(Falling body)** Experiments show that if a body falls in vacuum due to the action of gravity, then its acceleration is constant (equal to $g = 9.80$ meters/sec^2 = 32.1 ft/sec^2; this is called the *acceleration of gravity*). State this law as a differential equation for $s(t)$, the distance fallen as a function of t. Show that by solving this equation one can obtain the familiar law

$$s(t) = \frac{g}{2}t^2.$$

(In practice, this also applies to the free fall in air if we can neglect the air resistance, for instance, if we drop a stone or an iron ball from a position not very high above the ground.)

24. If in Prob. 23 the body starts at $t = 0$ from initial position $s = s_0$ with initial velocity $v = v_0$, show that then the solution is

$$s(t) = \frac{g}{2}t^2 + v_0 t + s_0.$$

25. The makers of a certain car advertise that the car will accelerate from 0 to 100 mph in 30 sec. If the acceleration is constant, how far will the car go in this time?

26. An airplane taking off from a landing field has a run of 2 kilometers. If the plane starts with speed 10 meters/sec, moves with constant acceleration and makes the run in 50 sec, with what speed does it take off?

27. A rocket is shot straight up. During the initial stages of flight it has acceleration $7t$ meters/sec^2. The engine cuts out at $t = 10$ sec. How high will the rocket go? (Neglect air resistance.)

28. Linear accelerators are used in physics for accelerating charged particles. Suppose that an alpha particle enters an accelerator and undergoes a constant acceleration which increases its speed from 10^3 meters/sec to 10^4 meters/sec in 10^{-3} sec. Find the acceleration a and the distance traveled during this period of 10^{-3} sec.

29. **(Exponential decay; atmospheric pressure)** Observations show that the rate of change of the atmospheric pressure p with altitude h is proportional to the pressure. Assuming that the pressure at 6000 meters (about 18,000 ft) is 1/2 of its value p_0 at sea level, find the formula for the pressure at any height. *Hint.* Remember from calculus that if $y = e^{kx}$, then $y' = ke^{kx} = ky$.

30. **(Exponential growth; a population model)** If relatively small populations (of humans, animals, bacteria, etc.) are left undisturbed, they often grow according to *Malthus's law,* which states that the time rate of growth is proportional to the population present. Formulate this as a differential equation. Show that the solution is $y(t) = y_0 e^{kt}$. For the United States, observed values, in millions, are as follows.

t	0	30	60	90	120
Year	1800	1830	1860	1890	1920
Population	5.3	12.9	31.4	62.9	106.5

Use the first two columns for determining y_0 and k. Then calculate values for 1860, 1890 and 1920 and compare them with the observed values. Comment.

1.2 Geometrical Considerations. Isoclines

We shall now start with a systematic treatment of first-order differential equations. Any such equation may be written in the *implicit form*

$$F(x, y, y') = 0. \tag{1}$$

Not always, but in many cases we shall be able to write a first-order differential equation in the **explicit form**

$$y' = f(x, y). \tag{2}$$

Before we discuss various standard methods for solving equations of the form (2) we shall demonstrate that (2) has a very simple geometric interpretation. This will immediately lead to a useful graphical method for obtaining a rough picture of the particular solutions of (2), without actually solving the equation.

We assume that the function f is defined in some region of the xy-plane, so that to each point in that region it assigns one (and only one) value. The solutions of (2) can be plotted as curves in the xy-plane. We do not know the solutions, but we see from (2) that a solution passing through a point (x_0, y_0) must have the slope $f(x_0, y_0)$ at this point. This suggests the following method. We first graph some of the curves in the xy-plane along which $f(x, y)$ is constant. These curves are called *curves of constant slope,* or **isoclines** of (2). Hence every isocline of (2) is given by

$$f(x, y) = k = const$$

(where the constant k differs from isocline to isocline). Along the isocline $f(x, y) = k$ we draw a number of parallel short line segments **(lineal elements)** with slope k, which is the slope of solution curves of (2) at any point of that isocline. This we do for all isoclines which we graphed before. In this way we obtain a field of lineal elements, called the **direction field** of (2). With the help of the lineal elements we can now easily graph approximation curves to the (unknown) solution curves of the given equation (2) and thus obtain a qualitatively correct picture of these solution curves.

Let us illustrate the method by a simple example.

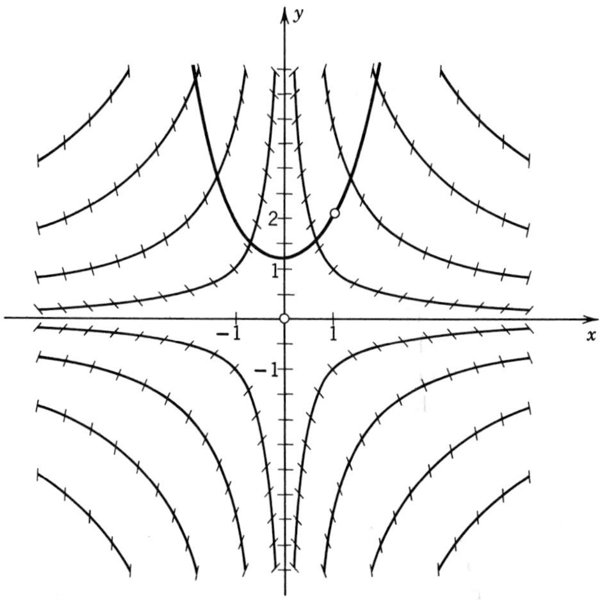

Fig. 6. Direction field of the differential equation (3)

Example 1. Isoclines, direction field

Graph the direction field of the first-order differential equation

$$y' = xy \tag{3}$$

and an approximation to the solution curve through the point (1, 2). The isoclines are the equilateral hyperbolas $xy = k$ together with the two coordinate axes. We graph some of them. Then we draw lineal elements by sliding a triangle along a fixed ruler. The result is shown in Fig. 6, which also shows an approximation to the solution curve passing through the point (1, 2).

In the next section we shall see that (3) can easily be solved, so that our example merely illustrates the technical details of the method of direction field. This method is particularly helpful in engineering problems involving a first-order differential equation whose solution cannot be expressed in terms of known functions or is a very complicated expression.

Problems for Sec. 1.2

In each case draw a good direction field. Plot several approximate solution curves. Then solve the equation analytically and compare, to get a feeling for the accuracy of the present method.

1. $y' = x$ **2.** $y' = x^2$ **3.** $y' = y$

Draw good direction fields for the following differential equations and plot several approximate solution curves.

4. $y' = -xy$ **5.** $y' = -x/y$ **6.** $y' = x^2 + y^2$

7. $y' = x + y$ **8.** $4yy' + x = 0$ **9.** $y' = \sin y$

Using direction fields, plot an approximate solution curve of the given differential equation satisfying the given condition.

10. $y' = 2x,\quad y(2) = 4$ **11.** $yy' + 9x = 0,\quad y(1) = 3$

12. $y' + 2y^2 = 0,\quad y(4) = 1$ **13.** $y' + 4x^3y = 0,\quad y(0) = 5$

14. (Approximate integration) Using direction fields, find approximate values of the following integrals (the last two of which are not elementary).

$$\int_0^1 x^2\,dx, \qquad \int_0^\pi \frac{\sin x}{x}\,dx, \qquad \int_0^1 e^{-x^2}\,dx.$$

15. (Skydiver) Graph the direction field of the differential equation

$$\frac{dv}{dt} = 10 - 0.4v^2$$

and draw from it the following conclusions. The isoclines are horizontal straight lines. The isocline $v = 5$ is at the same time a solution curve. All solution curves in the upper half-plane ($v > 0$) seem to approach the line $v = 5$ as $t \to \infty$; they are monotone increasing if $0 < v(0) < 5$ and monotone decreasing if $v(0) > 5$. Verify that the equation is satisfied by

$$v(t) = 5\frac{1 + ce^{-4t}}{1 - ce^{-4t}}$$

where c is arbitrary. (We shall see in the next section that the equation governs the motion of a parachutist under the assumption that the air resistance is proportional to the square of the velocity $v(t)$. Hence our result shows that $v = 5$ is a limiting speed which is attained, practically speaking, after a sufficiently long time.)

16. **(A pursuit problem, tractrix)** In a pursuit problem, an object or *target* moves along a given curve and a second object or *pursuer* follows or *pursues* the target, that is, moves in the direction of the target at all times. Assume the target to be a ship S that moves along a straight line and the pursuer to be a destroyer D that moves so that the distance a from D to S is constant, say, $a = 1$ (nautical mile). Show that (Fig. 7)

$$y' = -y/\sqrt{a^2 - y^2}.$$

Draw a direction field (using $a = 1$) and plot an approximate solution curve corresponding to $y(0) = a = 1$; see Fig. 7. (Such a curve is called a *tractrix,* from Latin *trahere* = to pull.)

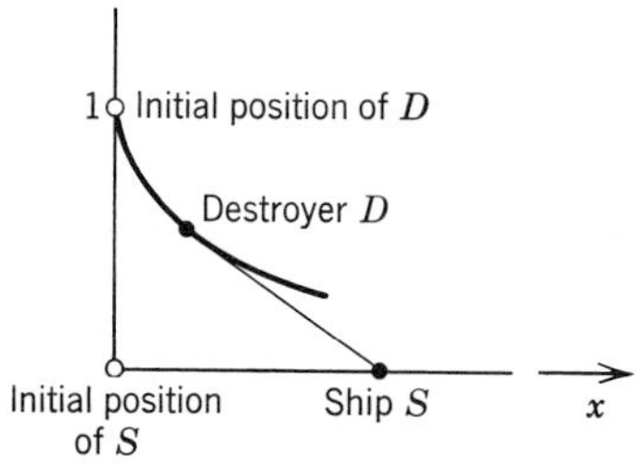

Fig. 7. Tractrix and notations in Prob. 16

17. **(Spirals)** Show that the isoclines of

$$y' = \frac{y - x}{y + x}$$

are straight lines through the origin and that the direction field suggests the solution curves to be spirals.

18. Draw a direction field of $y' = y - x$. Notice that the direction field shows the following. The isocline $y - x = 1$ is at the same time a solution curve and the other solution curves seem to approach that isocline as $x \to -\infty$. Confirm this by verifying that $y = x + 1 + ce^x$ is a solution of the equation.

19. Graph a direction field of

$$y' = ay - by^2$$

where $a = 3$ and $b = 1$, and draw from it the following conclusions. The isoclines $y = 0$ and $y = 3$ are also solution curves. All solution curves in the horizontal strip between $y = 0$ and $y = 3$ are s-shaped and monotone increasing. Also $y(x) \to 3$ as $x \to \infty$ and $y(x) \to 0$ as $x \to -\infty$ for any such curve. What about solution curves above the line $y = 3$? In the lower half-plane $y < 0$?

20. **(A population model)** *Malthus's law* states that the time rate of change of a population $p(t)$ is proportional to $p(t)$. This holds for many populations as long as they are not too large. A more refined model is the *logistic law* given by

$$\frac{dp}{dt} = ap - bp^2.$$

For the United States, Verhulst predicted in 1845 that $a = 0.03$ and $b = 1.6 \cdot 10^{-4}$, where $p(t)$ is measured in millions. Graph a direction field for

$1850 \leqq t \leqq 1950$ and $0 \leqq p \leqq 200$. Plot the solution curve passing through the point (1900, 76) and compare the accuracy with some actual values

1850	1870	1890	1910	1930
23	39	63	92	123

Show that the direction field indicates that $p(t) \to a/b$ as $t \to \infty$.

1.3 Separable Equations

Many first-order differential equations can be reduced to the form

$$g(y)y' = f(x) \tag{1}$$

by algebraic manipulations. Since $y' = dy/dx$, we find it convenient to write

$$g(y)\,dy = f(x)\,dx, \tag{2}$$

but we keep in mind that this is merely another way of writing (1). Such an equation is called an *equation with separable variables*, or a **separable equation,** because in (2) the variables x and y are *separated* so that x appears only on the right and y appears only on the left. By integrating on both sides of (2) we obtain

$$\int g(y)\,dy = \int f(x)\,dx + c. \tag{3}$$

If we assume that f and g are continuous functions, the integrals in (3) will exist, and by evaluating these integrals we obtain the general solution of (1).

Example 1
Solve the differential equation

$$9yy' + 4x = 0.$$

By separating variables we have

$$9y\,dy = -4x\,dx.$$

By integrating on both sides we obtain the general solution

$$\tfrac{9}{2}y^2 = -2x^2 + \tilde{c} \quad \text{or} \quad \frac{x^2}{9} + \frac{y^2}{4} = c \qquad \left(c = \frac{\tilde{c}}{18}\right).$$

The solution represents a family of ellipses.

Example 2
Solve the differential equation

$$y' = -2xy.$$

By separating variables we have

$$\frac{dy}{y} = -2x\,dx \qquad (y \neq 0).$$

Integration yields

$$\ln|y| = -x^2 + \tilde{c}. \tag{4}$$

In fact, the result on the left may be verified by differentiation as follows. When $y > 0$, then $(\ln y)' = y'/y$. When $y < 0$, then $-y > 0$ and $(\ln(-y))' = -y'/(-y) = y'/y$. Now $y = |y|$ when $y > 0$ and $-y = |y|$ when $y < 0$. Hence we may combine both formulas, finding $(\ln|y|)' = y'/y$.
It is of great importance that the constant of integration is introduced immediately when the integration is carried out. From (4) we obtain

$$|y| = e^{-x^2+\tilde{c}}$$

or by noting that $e^{a+b} = e^a e^b$ and setting $e^{\tilde{c}} = c$ when $y > 0$ and $e^{\tilde{c}} = -c$ when $y < 0$,

$$y = ce^{-x^2}.$$

This solution represents a family of so-called **bell-shaped curves** (Fig. 8). We may also admit $c = 0$, which gives the solution $y \equiv 0$.

Example 3
Solve the differential equation

$$y' = 1 + y^2.$$

By separating variables and integrating we obtain

$$\frac{dy}{1+y^2} = dx, \qquad \arctan y = x + c, \qquad y = \tan(x + c). \qquad \blacksquare$$

In many engineering applications we are not interested in the general solution of a given differential equation but only in the particular solution $y(x)$ satisfying a given **initial condition,** say, the condition that at some point x_0 the solution $y(x)$ has a prescribed value y_0, briefly

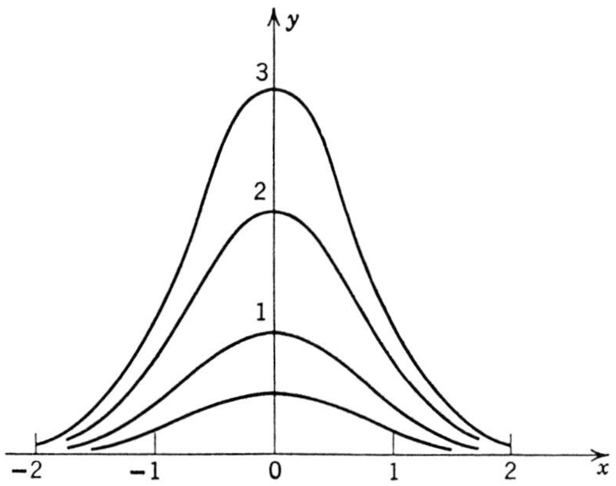

Fig. 8. Solutions of $y' = -2xy$ ("bell-shaped curves") in the upper half plane ($y > 0$)

(5) $$y(x_0) = y_0.$$

A first-order differential equation together with an initial condition is called an **initial value problem.** To solve such a problem we must find the particular solution of the equation which satisfies the given initial condition.

The term "initial value problem" is suggested by the fact that the independent variable often is time, so that (5) prescribes the initial situation at some instant and the solution of the problem shows what happens later.

Example 4

Solve the initial value problem

$$(x^2 + 1)y' + y^2 + 1 = 0, \qquad y(0) = 1.$$

1st Step. Separating variables, we find

$$\frac{dy}{1 + y^2} = -\frac{dx}{1 + x^2}.$$

By integration we obtain

$$\text{arc tan } y = -\text{arc tan } x + c$$

or

$$\text{arc tan } y + \text{arc tan } x = c.$$

Taking the tangent on both sides, we have

(6) $$\tan(\text{arc tan } y + \text{arc tan } x) = \tan c.$$

Now the addition formula for the tangent is

$$\tan(a + b) = \frac{\tan a + \tan b}{1 - \tan a \tan b}.$$

For $a = \text{arc tan } y$ and $b = \text{arc tan } x$ this becomes

$$\tan(\text{arc tan } y + \text{arc tan } x) = \frac{y + x}{1 - xy}.$$

Consequently, (6) may be written

(7) $$\frac{y + x}{1 - xy} = \tan c.$$

2nd Step. We determine c from the initial condition. Setting $x = 0$ and $y = 1$ in (7), we have $1 = \tan c$, so that

$$\frac{y + x}{1 - xy} = 1 \qquad \text{or} \qquad y = \frac{1 - x}{1 + x}.$$

3rd Step. Check the result.

Example 5 (Newton's law of cooling[2])

A copper ball is heated to a temperature of 100° C. Then at time $t = 0$ it is placed in water which is maintained at a temperature of 30° C. At the end of 3 minutes the temperature of the ball is reduced to 70° C. Find the time at which the temperature of the ball is reduced to 31° C.

Physical information. Experiments show that the time rate of change of the temperature T of the ball is proportional to the difference between T and the temperature of the surrounding medium (*Newton's law of cooling*).

Experiments also show that heat flows so rapidly in copper that at any time the temperature is practically the same at all points of the ball.

1st Step. The mathematical formulation of Newton's law of cooling in our case is

$$\frac{dT}{dt} = -k(T - 30) \tag{8}$$

where we denoted the constant of proportionality by $-k$ in order that $k > 0$.

2nd Step. The general solution of (8) is obtained by separating variables; we find

$$T(t) = ce^{-kt} + 30.$$

3rd Step. The given initial condition is $T(0) = 100$. The particular solution satisfying this condition is

$$T(t) = 70e^{-kt} + 30.$$

4th Step. k can be determined from the given information $T(3) = 70$. We obtain

$$T(3) = 70e^{-3k} + 30 = 70 \qquad \text{or} \qquad k = \tfrac{1}{3}\ln\tfrac{7}{4} = 0.1865.$$

Using this value of k we see that the temperature $T(t)$ of the ball is

$$T(t) = 70e^{-0.1865t} + 30.$$

It follows that the value $T = 31°$ C is reached when

$$70e^{-0.1865t} = 1 \qquad \text{or} \qquad 0.1865t = \ln 70, \qquad t = \frac{\ln 70}{0.1865} = 22.78,$$

that is, after approximately 23 minutes.

5th Step. Check the result.

Example 6. Flow of water through orifices (Torricelli's law[3])

A cylindrical tank 1.50 meter high stands on its circular base of diameter 1.00 meter and is initially filled with water. At the bottom of the tank there is a hole of diameter 1.00 cm, which is opened at some instant, so that the water starts draining under the influence of gravity (Fig. 9). Find the height $h(t)$ of the water in the tank at any time t. Find the times at which the tank is one-half full, one-quarter full, and empty.

1st Step. We set up the mathematical model (the differential equation) of the problem. The volume of water which flows out during a short interval of time Δt is

$$\Delta V = Av\Delta t$$

[2]Sir ISAAC NEWTON (1642–1727), English physicist and mathematician. Newton and the German mathematician and philosopher, GOTTFRIED WILHELM LEIBNIZ (1646–1716), invented (independently) the differential and integral calculus. Newton discovered many basic physical laws and introduced the method of investigating physical problems by means of calculus. His work is of greatest importance to both mathematics and physics. Sometimes the Latin spelling LEIBNITZ is used; this spelling appeared in Leibniz's scientific publications.

[3]EVANGELISTA TORRICELLI (1608–1647), Italian physicist. The "contraction factor" 0.6 was introduced by J. C. BORDA in 1766 because of the fact that the cross section of the outflowing stream of liquid is somewhat smaller than that of the orifice.

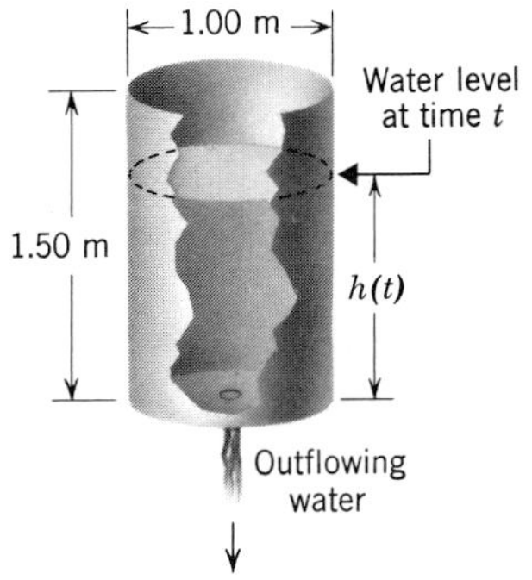

Fig. 9. Tank in Example 6

where $A = 0.500^2\pi \text{ cm}^2$ is the cross-sectional area of the outlet and v is the velocity of the outflowing water. Torricelli's law states that the velocity with which water issues from an orifice is

$$v = 0.600\sqrt{2gh}$$

where $g = 980 \text{ cm/sec}^2$ (about 32.17 ft/sec^2) is the acceleration of gravity at the surface of the earth and h is the instantaneous height of the water above the orifice. This formula looks reasonable. Indeed, $\sqrt{2gh}$ is the speed a body acquires if it falls a distance h and the air resistance is so small that it can be neglected. The factor 0.600 is introduced since the cross section of the outflowing stream is somewhat smaller than that of the orifice.

ΔV must equal the change ΔV^* of the volume of water in the tank. Clearly,

$$\Delta V^* = -B\Delta h$$

where $B(h)$ is the cross-sectional area of the container at height $h(t)$ and Δh is the decrease of the height $h(t)$ of the water. The minus sign appears since the volume of water in the tank decreases. Since $\Delta V = \Delta V^*$, we have

$$Av\Delta t = 0.600A\sqrt{2gh} = -B\Delta h.$$

Division by Δt yields

$$\frac{\Delta h}{\Delta t} = -\frac{0.600A\sqrt{2gh}}{B}.$$

Letting $\Delta t \to 0$, we obtain the differential equation

$$\frac{dh}{dt} = -0.600A\sqrt{2g}\frac{\sqrt{h}}{B(h)}. \tag{9}$$

In our case, $A = 0.500^2\pi \text{ cm}^2$ (see above) and the cross-sectional area of the tank does not vary with h but is constant, namely, $B = 50.0^2\pi \text{ cm}^2$. Since $A/B = 1.00 \cdot 10^{-4}$ and $\sqrt{2g} = \sqrt{2 \cdot 980} = 44.3$ $[\text{cm}^{1/2}/\text{sec}]$, we thus obtain

$$\frac{dh}{dt} = -0.600\frac{A}{B}\sqrt{2gh} = -0.00266h^{1/2}.$$

The initial condition is

$$h(0) = 150 \text{ cm},$$

where $t = 0$ is the instant at which the hole is opened.

2nd Step. We solve the differential equation. Separation of variables yields

$$h^{-1/2}\,dh = -0.00266\,dt.$$

By integration,

$$2h^{1/2} = -0.00266t + \tilde{c}.$$

Dividing and squaring, we have, writing $\tilde{c}/2 = c$,

$$h(t) = (c - 0.00133t)^2.$$

By the initial condition, $h(0) = c^2 = 150$. Hence the particular solution corresponding to our problem is

$$h(t) = (12.25 - 0.00133t)^2.$$

3rd Step. To answer the remaining questions, we express t in terms of h:

$$12.25 - 0.00133t = \sqrt{h}, \qquad \text{hence} \qquad t = \frac{12.25 - \sqrt{h}}{0.00133}.$$

This shows that the tank is one-half full at

$$t = \frac{12.25 - \sqrt{75.0}}{0.00133} = 2.70 \cdot 10^3 \text{ sec} = 45.0 \text{ min},$$

one-quarter full at $t = 76.8$ min and empty at $t = 154$ min.

4th Step. Check the result.

Example 7. Velocity of escape from the earth

Find the minimum initial velocity of a projectile which is fired in radial direction from the earth and is supposed to escape from the earth. Neglect the air resistance and the gravitational pull of other celestial bodies.

1st Step. According to Newton's law of gravitation, the gravitational force, and therefore also the acceleration a of the projectile, is proportional to $1/r^2$, where r is the distance from the center of the earth to the projectile. Thus

$$a(r) = \frac{dv}{dt} = \frac{k}{r^2}$$

where v is the velocity and t is the time. Since v is decreasing, $a < 0$ and thus $k < 0$. Let R be the radius of the earth. When $r = R$ then $a = -g$, the acceleration of gravity at the surface of the earth. Note that the minus sign occurs because g is positive and the attraction of gravity acts in the negative direction (the direction towards the center of the earth). Thus

$$-g = a(R) = \frac{k}{R^2} \qquad \text{and} \qquad a(r) = -\frac{gR^2}{r^2}.$$

Now $v = dr/dt$, and differentiation yields

$$a = \frac{dv}{dt} = \frac{dv}{dr}\frac{dr}{dt} = \frac{dv}{dr}v.$$

Hence the differential equation for the velocity is

$$\frac{dv}{dr}v = -\frac{gR^2}{r^2}.$$

2nd Step. Separating variables and integrating, we obtain

$$v\,dv = -gR^2\frac{dr}{r^2} \qquad \text{and} \qquad \frac{v^2}{2} = \frac{gR^2}{r} + c. \tag{10}$$

3rd Step. On the earth's surface, $r = R$ and $v = v_0$, the initial velocity. For these values of r and

v, formula (10) becomes

$$\frac{v_0^2}{2} = \frac{gR^2}{R} + c, \qquad \text{and} \qquad c = \frac{v_0^2}{2} - gR.$$

By inserting this expression for c into (10) we have

$$v^2 = \frac{2gR^2}{r} + v_0^2 - 2gR. \tag{11}$$

4th Step. If $v^2 = 0$, then $v = 0$, the projectile stops, and it is clear from the physical situation that the velocity will change from positive to negative, so that the projectile will return to the earth. Consequently, we have to choose v_0 so large that this cannot happen. If we choose

$$v_0 = \sqrt{2gR}, \tag{12}$$

then in (11) the expression $v_0^2 - 2gR$ is zero and v^2 is never zero for any r. However, if we choose a smaller value for the initial velocity v_0, then $v = 0$ for a certain r. The expression in (12) is called the *velocity of escape* from the earth. Since $R = 6372$ km $= 3960$ miles and $g = 9.80$ meters/sec^2 $= 0.00980$ km/sec^2 $= 32.17$ ft/sec^2 $= 0.00609$ miles/sec^2, we have

$$v_0 = \sqrt{2gR} \approx 11.2 \text{ km/sec} \approx 6.95 \text{ miles/sec}.$$

5th Step. Check the result.

Example 8. Skydiver

Suppose that a skydiver falls from rest toward the earth and the parachute opens at an instant, call it $t = 0$, when the skydiver's speed is $v(0) = v_0 = 10.0$ meters/sec. Find the speed $v(t)$ of the skydiver at any later time t. Does $v(t)$ increase indefinitely?

Physical assumptions and laws. Suppose that the weight of the man plus the equipment is $W = 712$ nt (read "newtons"; about 160 lb), the air resistance U is proportional to v^2, say, $U = bv^2$ nt, where the constant of proportionality b depends mainly on the parachute, and we assume that $b = 30.0$ nt·sec^2/meter^2 $= 30.0$ kg/meter.

1st Step. We set up the mathematical model (the differential equation) of the problem. **Newton's second law** is

$$\text{Mass} \times \text{Acceleration} = \text{Force}$$

where "force" means the resultant of the forces acting on the skydiver at any instant. These forces are the weight W and the air resistance U. The weight is $W = mg$, where $g = 9.80$ meters/sec^2 (about 32.2 ft/sec^2) is the acceleration of gravity at the earth's surface. Hence the mass of the man plus the equipment is $m = W/g = 72.7$ kg (about 5.00 slugs). The air resistance U acts upward (against the direction of the motion), so that the resultant is

$$W - U = mg - bv^2.$$

The acceleration a is the time derivative of v, that is, $a = dv/dt$. Hence by Newton's second law,

$$m\frac{dv}{dt} = mg - bv^2.$$

This is the differential equation of our problem. By assumption, $v = v_0 = 10$ meters/sec when $t = 0$. Hence we are looking for the solution $v(t)$ satisfying the *initial condition*

$$v(0) = v_0 = 10.$$

2nd Step. We solve the differential equation. Division by m gives

$$\frac{dv}{dt} = -\frac{b}{m}(v^2 - k^2) \qquad\qquad k^2 = \frac{mg}{b}.$$

Separating variables, we have

$$\frac{dv}{v^2 - k^2} = -\frac{b}{m}\,dt. \tag{13}$$

On the left,

$$\frac{1}{v^2 - k^2} = \frac{1}{2k}\left(\frac{1}{v - k} - \frac{1}{v + k}\right),$$

as can be readily verified. This is a representation in terms of partial fractions which is convenient for integration:

$$\int \frac{dv}{v^2 - k^2} = \frac{1}{2k}[\ln (v - k) - \ln (v + k)] = \frac{1}{2k} \ln \frac{v - k}{v + k}.$$

From (13) we thus obtain

$$\frac{1}{2k} \ln \frac{v - k}{v + k} = -\frac{b}{m}t + \tilde{c}.$$

Multiplying by $2k$ and taking exponents, we get

$$\frac{v - k}{v + k} = ce^{-pt} \qquad p = \frac{2kb}{m}, \tag{14*}$$

where $c = e^{2k\tilde{c}}$. Solving this for v, we have

$$v(t) = k\frac{1 + ce^{-pt}}{1 - ce^{-pt}}. \tag{14}$$

We see that $v(t) \to k$ as $t \to \infty$; that is, $v(t)$ does not increase indefinitely but approaches a limit, k. It is interesting to note that this limit is independent of the initial condition $v(0) = v_0$.

3rd Step. We determine c in (14) such that we obtain the particular solution satisfying the initial condition $v(0) = v_0$. From (14*) with $t = 0$ we immediately have

$$c = \frac{v_0 - k}{v_0 + k}.$$

With this c, formula (14) represents the solution we are looking for.

4th Step. Note that we did not yet make use of the given numerical values. Very often it is a good principle to hunt for general formulas first and to substitute numerical data later. In this way one often gets a better idea of what is going on and, moreover, saves work if one is interested in various solutions corresponding to different sets of data.

In the present case we now calculate (see before)

$$k^2 = \frac{mg}{b} = \frac{W}{b} = \frac{712}{30.0} = 23.7 \text{ [meters}^2/\text{sec}^2].$$

Hence $k = 4.87$ meters/sec. This is the limiting speed. Practically speaking, this is the speed of the skydiver after a sufficiently long time.

To $v(0) = v_0 = 10.0$ meters/sec there corresponds

$$c = \frac{v_0 - k}{v_0 + k} = 0.345.$$

Finally,

$$p = \frac{2kb}{m} = \frac{2 \cdot 4.87 \cdot 30.0}{72.7} = 4.02 \text{ [sec}^{-1}].$$

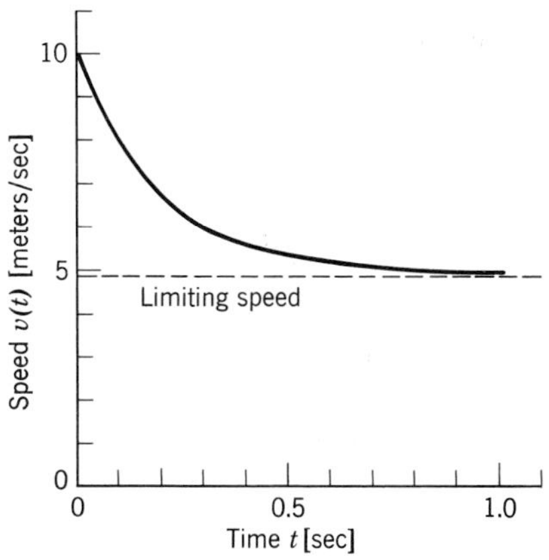

Fig. 10. Speed $v(t)$ of the skydiver in Example 8

This altogether yields the result (Fig. 10)

$$v(t) = 4.87\frac{1 + 0.345e^{-4.02t}}{1 - 0.345e^{-4.02t}}. \tag{15}$$

5th Step. Check the result.

Problems for Sec. 1.3

1. Using (3), show that the solution of (1) satisfying (5) is obtained from

$$\int_{y_0}^{y} g(y^*)\, dy^* = \int_{x_0}^{x} f(x^*)\, dx^*. \tag{16}$$

Use this formula for confirming the result in Example 4.

2. Once in a while a student asks whether the derivation of (3) from (2) is correct since one seems to integrate with respect to different variables on both sides. What would you answer?

3. Why is it important to introduce the constant of integration immediately when the integration is performed?

Find the general solution of the following equations (where a, b, n are constants).

4. $y' = -xy$

5. $yy' + x = 0$

6. $y' = -\dfrac{x - a}{y - b}$

7. $y' = \dfrac{y}{x \ln x}$

8. $y' = ny/x$

9. $y' + ay + b = 0 \quad (a \neq 0)$

10. $yy' = 2x \exp(y^2)$

11. $xy' = 2\sqrt{y - 1}$

12. $y' \sin 2x = y \cos 2x$

13. $y' + 2y \tanh x = 0$

Solve the following initial value problems (where L, R, α are constants and the dot denotes differentiation with respect to time t).

14. $y' = 3y/x, \quad y(1) = 2$

15. $L\dot{I} + RI = 0, \quad I(0) = I_0$

16. $(x + 1)y' = 2y, \quad y(0) = 1$

17. $y' = \sec y, \quad y(0) = 0$

18. $v\, dv/dx = g = const, \quad v(x_0) = v_0$

19. $y' = -2xy, \quad y(0) = y_0$

20. $2yy' = \sin^2 \alpha x, \quad y(0) = 0$

21. $(x \ln x)\, dy = y\, dx, \quad y(3) = 4$

22. $y' = -6y \tan 2x, \quad y(0) = -2$

23. $dr \sin \theta = 2r \cos \theta\, d\theta, \quad r(\pi/2) = 2$

Physical applications

24. Show that a ball thrown vertically upward with initial velocity v_0 takes twice as much time to return as to reach the highest point. Find the velocity upon return. (Air resistance is assumed negligible.)

25. **(Velocity of escape)** At the earth's surface the velocity of escape is 11.2 km/sec; cf. Example 7. If the projectile is carried by a rocket and is separated from the rocket at a distance of 1000 km from the earth's surface, what would be the minimum velocity at this point sufficient for escape from the earth?

26. If in Example 7 the initial velocity is half the velocity of escape, what maximum height above the earth's surface would the projectile reach? First guess, then calculate.

27. **(Boyle–Mariotte's law for ideal gases[4])** Experiments show that for a gas at low pressure p (and constant temperature) the rate of change of the volume $V(p)$ equals $-V/p$. Solve the corresponding differential equation.

28. **(Exponential growth)** If in a culture of bacteria the rate of growth is proportional to the population $p(t)$ present at time t and the population doubles in 1 day, how much can be expected after 1 week at the same rate of growth?

29. **(Exponential decay) Lambert's law of absorption[5]** states that the absorption of light in a very thin transparent layer is proportional to the thickness of the layer and to the amount incident on that layer. Formulate this in terms of a differential equation and solve it.

30. **(Exponential decay, half-life)** Experiments show that most radioactive substances disintegrate at a rate proportional to the amount of substance instantaneously present. The time in which 50% of a given amount disappears is called the *half-life* of the substance. For radium $_{88}Ra^{226}$ the half-life is $H = 1600$ years. (See CRC, *Handbook of Chemistry and Physics,* 54th ed., 1973–74, page B-504.) What percentage will disappear in 1 year? In 10 years?

31. Show that the half-life H of a radioactive substance can be determined from two measurements $y_1 = y(t_1)$ and $y_2 = y(t_2)$ of the amounts present at times t_1 and t_2 by the formula

$$H = \frac{(t_2 - t_1) \ln 2}{\ln (y_1/y_2)}.$$

32. Experiments show that the rate of inversion of cane sugar in dilute solution is proportional to the concentration $y(t)$ of unaltered sugar. Assume that the concentration is $1/100$ at $t = 0$ and is $1/300$ at $t = 4$ hours. Find $y(t)$.

33. **(Flywheel)** A flywheel of moment of inertia I is rotating with a relatively small constant angular speed ω_0 (radians/sec). At some instant, call it $t = 0$, the power is shut off and the motion starts slowing down because of friction. Assuming the friction torque to be proportional to $\sqrt{\omega}$, where ω is the instantaneous angular speed, and using **Newton's second law** *in torsional form*

$$\text{Moment of inertia} \times \text{Angular acceleration} = \text{Torque},$$

[4] ROBERT BOYLE (1627–1691), English physicist; EDMÉ MARIOTTE (1620–1684), French physicist.

[5] JOHANN HEINRICH LAMBERT (1728–1777), German physicist and mathematician.

find $\omega(t)$ and the instant t_1 at which the wheel is rotating with $\omega_0/2$ and t_2 when it comes to rest.

34. (Evaporation) Experiments show that a wet porous substance in the open air loses its moisture at a rate proportional to the moisture content. If a sheet hung in the wind loses half its moisture during the first hour, when will it be practically dry, say, when will it have lost 99.9% of its moisture, weather conditions remaining the same?

35. (Mothball) Suppose that a mothball loses volume by evaporation at a rate proportional to its instantaneous area. If the diameter of the ball decreases from 2 cm to 1 cm in 3 months, how long will it take until the ball has practically gone, say, until its diameter is 1 mm?

36. Two liquids are boiling in a vessel. It is found that the ratio of the quantities of each passing off as vapor at any instant is proportional to the ratio of the quantities x and y still in the liquid state. Show that

$$\frac{dy}{dx} = k\frac{y}{x}$$

and solve this equation.

37. (Newton's law of cooling) A thermometer, reading 15° C, is brought into a room whose temperature is 23° C. One minute later the thermometer reading is 19° C. How long does it take until the reading is practically 23° C, say, 22.9° C?

38. The time to empty the tank in Example 6 is greater than twice the time at which the tank is one-half full. Is this physically understandable?

39. A conical tank (Fig. 11) of circular cross section whose angle at the apex is 60° has an outlet of cross-sectional area of 0.5 cm². The tank contains water. At time $t = 0$ the outlet is opened and the water flows out. Determine the time when the tank will be empty, assuming that the initial height of water is $h(0) = 1$ meter. *Hint.* You can use (9). Why?

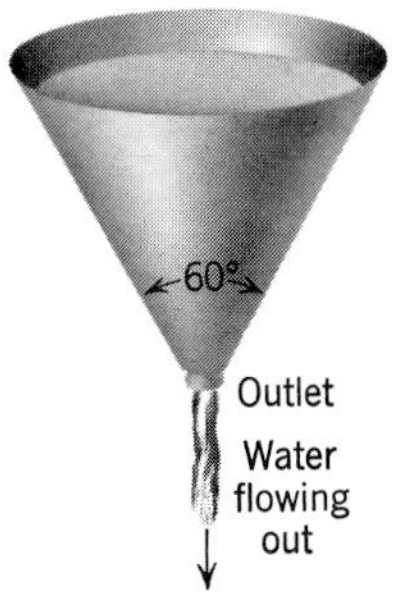

Fig. 11. Conical tank in Prob. 39

40. Suppose that the tank in Prob. 39 is hemispherical, of radius R, is initially full of water, and has an outlet of 5 cm² cross-sectional area at the bottom. The outlet is opened at some instant. Find the time it takes to empty the tank (a) for any given R, (b) for $R = 1$ meter.

41. (Skydiver) (a) To what height of free fall does the limiting speed (4.87 meters/sec) in Example 8 correspond? (b) How does the equation and solution in Example 8 change if we assume the air resistance to be proportional to v (instead of v^2), say, $U = \tilde{b}v$ where $\tilde{b} = 7.3$ kg/sec? Is this model still reasonable?

42. (Friction) If a body moves (slides) on a surface, it experiences a force in the direction against the motion. This force is called *friction*. *Coulomb's law of kinetic friction without lubrication*[6] is

$$|F| = \mu\,|N|$$

where F is the friction, N is the normal force (force which holds the two surfaces together; see Fig. 12), and the constant of proportionality μ is called the *coefficient of kinetic friction*. Using Newton's second law (mass $\times$ acceleration $=$ force), set up the differential equation in Fig. 12. Find the velocity when the body reaches the bottom of the slide, assuming that the body weighs 45 nt (about 10 lb), $\mu = 0.10$, $\alpha = 30°$, the length of the slide is 10 meters, the air resistance is negligible, and the initial velocity is zero.

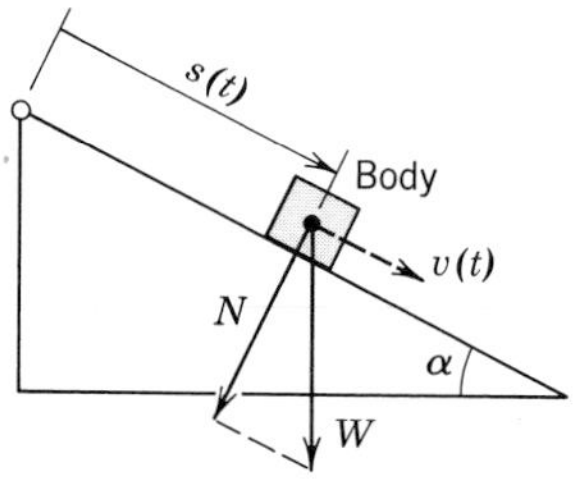

Fig. 12. Problem 42

43. Assume the body in Prob. 42 to be such that the air resistance is no longer negligible but causes a force of magnitude $0.5v$. Find $v(t)$, the distance $s(t)$ and $s(2.22)$, where $t = 2.22$ sec is the time to cover the whole distance of 10 meters if the air resistance is neglected.

44. **(Rope)** If we wind a rope around a rough cylinder which is fixed on the ground, we need only a small force at one end to resist a large force at the other. Let S be the force in the rope. Experiments show that the change ΔS of S in a small portion of the rope is proportional to S and to the small angle $\Delta\phi$ in Fig. 13 on the next page. Show that the differential equation for S is

$$\frac{dS}{d\phi} = \mu S.$$

If $\mu = 0.2$ radian^{-1}, how many times must the rope be snubbed around the cylinder in order that a man holding one end of the rope can resist a force one thousand times greater than he can exert?

45. **(Law of mass action)** The *law of mass action* states that, if the temperature is kept constant, the velocity of a chemical reaction is proportional to the product of the concentrations of the substances which are reacting. In a bimolecular reaction

$$A + B \rightarrow M$$

a moles per liter of a substance A and b moles of a substance B are combined. If y is the number of moles per liter which have reacted after time t, the rate of reaction is given by

$$\frac{dy}{dt} = k(a - y)(b - y).$$

Solve this equation, assuming that $a \neq b$.

[6]CHARLES AUGUSTIN DE COULOMB (1736–1806), French physicist.

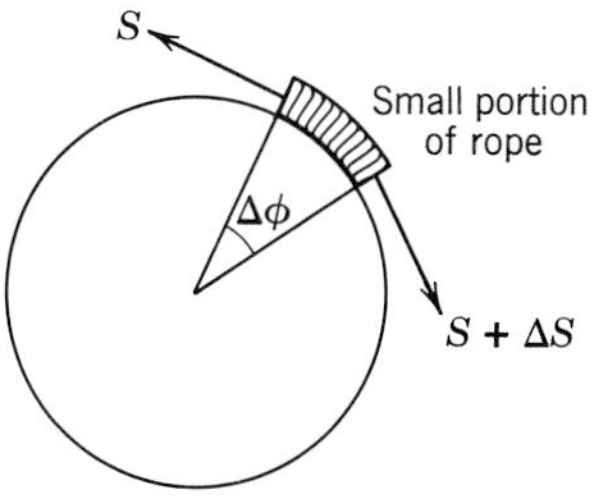

Fig. 13. Problem 44

Geometrical applications

46. **(Lemniscate)** Solve the initial value problem

$$\frac{dr}{d\theta} + \frac{a^2}{r}\sin 2\theta = 0, \qquad r^2(0) = a^2,$$

where a is a constant and r and θ are polar coordinates. The solution curve is called a *lemniscate*.

In each case find all curves in the xy-plane having the indicated property.

47. The slope of the tangent is proportional to x/y, the constant of proportionality being positive.

48. For every tangent, the segment between the coordinate axes is bisected by the point of tangency.

49. The normals all pass through the origin.

50. The normal at each point (x, y) intersects the x-axis at $(x + \frac{1}{2}, 0)$.

51. The tangents all pass through the origin.

52. The slope of the tangent at any point (x, y) is $1/2$ the slope of the line from the origin to the point.

53. The distance from (x, y) to the origin is the same as the distance from (x, y) to the point of intersection of the normal at (x, y) and the x-axis.

54. The y-axis bisects the segment of the normal at (x, y) whose endpoints are (x, y) and the point of intersection of the normal and the x-axis.

55. The distances from (x, y) to the points of intersection between the x-axis and the tangent and the normal are equal.

1.4 Equations Reducible to Separable Form

Certain first-order differential equations are not separable but can be made separable by a simple change of variables. This holds for equations of the form[7]

$$y' = g\left(\frac{y}{x}\right) \tag{1}$$

[7] These equations are sometimes called **homogeneous equations.** We shall not use this terminology but reserve the term "homogeneous" for a much more important purpose (cf. Sec. 1.7).

where g is any given function of y/x, for example $(y/x)^3$, $\sin(y/x)$, etc. The form of the equation suggests that we set

$$\frac{y}{x} = u,$$

remembering that y and u are functions of x. Then $y = ux$, and by differentiation,

$$(2) \qquad y' = u + u'x.$$

By inserting this into (1) and noting that $g(y/x) = g(u)$ we have

$$u + u'x = g(u).$$

Now we may separate the variables u and x, finding

$$\frac{du}{g(u) - u} = \frac{dx}{x}.$$

If we integrate and then replace u by y/x, we obtain the general solution of (1).

Example 1
Solve

$$2xyy' - y^2 + x^2 = 0.$$

Dividing by x^2, we have

$$2\frac{y}{x}y' - \left(\frac{y}{x}\right)^2 + 1 = 0.$$

If we set $u = y/x$ and use (2), the equation becomes

$$2u(u + u'x) - u^2 + 1 = 0 \qquad \text{or} \qquad 2xuu' + u^2 + 1 = 0.$$

Separating variables, we find

$$\frac{2u\,du}{1 + u^2} = -\frac{dx}{x}.$$

By integration

$$\ln(1 + u^2) = -\ln|x| + c^* \qquad \text{or} \qquad 1 + u^2 = \frac{c}{x}.$$

Replacing u by y/x, we finally obtain

$$x^2 + y^2 = cx \qquad \text{or} \qquad \left(x - \frac{c}{2}\right)^2 + y^2 = \frac{c^2}{4}.$$

Sometimes the form of a given differential equation suggests other simple substitutions, as the following example illustrates.

Example 2
Solve the first-order differential equation

$$(2x - 4y + 5)y' + x - 2y + 3 = 0.$$

We set $x - 2y = v$. Then $y' = \frac{1}{2}(1 - v')$ and the equation takes the form

$$(2v + 5)v' = 4v + 11.$$

Separating variables and integrating, we find

$$\left(1 - \frac{1}{4v + 11}\right) dv = 2\,dx \qquad \text{and} \qquad v - \tfrac{1}{4}\ln|4v + 11| = 2x + c^*.$$

Since $v = x - 2y$, this may be written

$$4x + 8y + \ln|4x - 8y + 11| = c.$$

Further simple substitutions are illustrated by the equations in Probs. 11–15.

Problems for Sec. 1.4

In each case find the general solution.

1. $xy' = x + y$

2. $xy' - 2y = 3x$

3. $x^2y' = x^2 - xy + y^2$

4. $(x^2 + 1)y(xy' - y) = x^3$

5. $xy' = y + x^2 \sec(y/x)$

6. $xy' - y - x^2 \tan(y/x) = 0$

Solve the following initial value problems.

7. $y' = \dfrac{y - x}{y + x}, \quad y(1) = 1$

8. $y' = \dfrac{y + x}{y - x}, \quad y(0) = 2$

9. $xy' = y + (y - x)^3, \quad y(1) = 1.5$

10. $xyy' = 2y^2 + 4x^2, \quad y(2) = 4$

Using the indicated transformations, find the general solution of the following equations.

11. $y' = (y - x)^2 \quad (y - x = v)$

12. $y' = \cot(y + x) - 1 \quad (y + x = v)$

13. $xy' = e^{-xy} - y \quad (xy = v)$

14. $e^y y' = k(x + e^y) - 1 \quad (x + e^y = v)$

15. $y' = \dfrac{y - x}{y - x - 1} \quad (y - x = v)$

16. Consider $y' = f(ax + by + k)$, where f is continuous. If $b = 0$, the solution is immediate. (Why?) If $b \neq 0$, show that one obtains a separable equation by using $u(x) = ax + by + k$ as a new dependent variable.

17. Show that a straight line through the origin intersects all solution curves of a given differential equation $y' = g(y/x)$ at the same angle.

18. Find the curve which passes through the point $(x, y) = (\sqrt{e}, \sqrt{e})$ and has the slope $x/y + y/x$.

19. Find the curve $y(x)$ which passes through $(1, 1/2)$ and is such that at each point (x, y) the intercept of the tangent on the y-axis is equal to $2xy^2$.

20. The positions of four battle ships on the ocean are such that they form the vertices of a square of length l. At some instant each ship fires a missile which directs its motion steadily toward the missile on its right. Assuming that the four missiles fly horizontally and with the same speed, find the path of each.

1.5 Exact Differential Equations

A first-order differential equation of the form

$$M(x, y)\,dx + N(x, y)\,dy = 0 \tag{1}$$

is said to be **exact** if the left-hand side is the total or exact differential

$$du = \frac{\partial u}{\partial x}dx + \frac{\partial u}{\partial y}dy \tag{2}$$

of some function $u(x, y)$. Then the differential equation (1) can be written

$$du = 0.$$

By integration we immediately obtain the general solution of (1) in the form

$$u(x, y) = c. \tag{3}$$

Comparing (1) and (2), we see that (1) is exact, if there is some function $u(x, y)$ such that

$$\text{(a)}\quad \frac{\partial u}{\partial x} = M, \qquad \text{(b)}\quad \frac{\partial u}{\partial y} = N. \tag{4}$$

Suppose that M and N are defined and have continuous first partial derivatives in a region in the xy-plane whose boundary is a closed curve having no self-intersections. Then from (4),

$$\frac{\partial M}{\partial y} = \frac{\partial^2 u}{\partial x\,\partial y}, \qquad \frac{\partial N}{\partial x} = \frac{\partial^2 u}{\partial y\,\partial x}.$$

By the assumption of continuity the second derivatives are equal. Thus

$$\frac{\partial M}{\partial y} = \frac{\partial N}{\partial x}. \tag{5}$$

This condition is not only necessary but also sufficient[8] for $M\,dx + N\,dy$ to be a total differential.

If (1) is exact, the function $u(x, y)$ can be found by guessing or in the following systematic way. From (4a) we have by integration with respect to x

$$u = \int M\,dx + k(y); \tag{6}$$

[8] We shall prove this fact at another occasion (in Sec. 9.12); the proof can also be found in some books on elementary calculus; cf. Ref. [A14] in Appendix 1.

in this integration, y is to be regarded as a constant, and $k(y)$ plays the role of a "constant" of integration. To determine $k(y)$, we derive $\partial u/\partial y$ from (6), use (4b) to get dk/dy, and integrate.

Formula (6) was obtained from (4a). Instead of (4a) we may equally well use (4b). Then instead of (6) we first have

$$u = \int N\,dy + l(x) \tag{6*}$$

To determine $l(x)$ we derive $\partial u/\partial x$ from (6*), use (4a) to get dl/dx, and integrate.

Example 1
Solve

$$xy' + y + 4 = 0.$$

We write this equation in the form (1), that is,

$$(y + 4)\,dx + x\,dy = 0.$$

We now see that $M = y + 4$ and $N = x$. Hence (5) holds, so that the equation is exact. From (6*) we have

$$u = \int N\,dy + l(x) = \int x\,dy + l(x) = xy + l(x).$$

To determine $l(x)$, we differentiate this formula with respect to x and use (4a), finding

$$\frac{\partial u}{\partial x} = y + \frac{dl}{dx} = M = y + 4.$$

Thus $dl/dx = 4$ and $l = 4x + c^*$. Hence we obtain the general solution of our equation in the form

$$u = xy + l(x) = xy + 4x + c^* = const.$$

Division by x gives the explicit form

$$y = \frac{c}{x} - 4.$$

Intelligent students, who know that thinking is important in mathematics, write the given equation in the form

$$y\,dx + x\,dy = -4\,dx,$$

recognize the left side as the total differential of xy, and integrate, finding $xy = -4x + c$, which is equivalent to our previous result.

Example 2
Solve

$$2x \sin 3y\,dx + 3x^2 \cos 3y\,dy = 0.$$

The equation is exact, and from (6) we obtain

$$u = \int 2x \sin 3y\,dx + k(y) = x^2 \sin 3y + k(y).$$

Therefore $\partial u/\partial y = 3x^2 \cos 3y + dk/dy$, and $dk/dy = 0$ or $k = c^* = const.$ The general solution is $u = const$ or $x^2 \sin 3y = c$.

Problems for Sec. 1.5

In each case graph some *level curves* $u(x, y) = c = const$ and find the total differential of u.

1. $u = 4x^2 + 9y^2$

2. $u = y/x$

3. $u = e^{xy}$

4. $u = \sin(y^2 - x)$

Find the exact differential equations that have the following general solutions.

5. $x^3y + \cos 2x = c$

6. $e^x \sin y = c$

7. $xe^{-2xy} = c$

8. $\tan x = c \tan y$

Show that the following differential equations are exact and solve them.

9. $y^2\,dx + 2xy\,dy = 0$

10. $\cos x\,dx + y\,dy = 0$

11. $x^{-1}\,dy - x^{-2}y\,dx = 0$

12. $3re^{3\theta}\,d\theta + e^{3\theta}\,dr = 0$

13. $2x \ln y\,dx + y^{-1}x^2\,dy = 0$

14. $(\cot y + x^2)\,dx = x \operatorname{cosec}^2 y\,dy$

15. If a differential equation is exact, we can find $u(x, y) = c$ from (6) by the method explained in the text. However, if the equation is not exact, that method will not work. To see this, consider $x\,dy - y\,dx = 0$. Show that this equation is not exact. Obtain u from (6), calculate $\partial u/\partial y$ from the result and show that (4b) yields a contradiction. Solve the equation by some other method which has been discussed before.

Are the following equations exact? Find the general solution.

16. $ye^{xy}\,dx + (1 + xe^{xy})\,dy = 0$

17. $x\,dy + 3y^2\,dx = 0$

18. $xy(dx + dy) = 0$

19. $\cosh x \sin y\,dy = \sinh x \cos y\,dx$

20. $(1 + x^2)\,dy + (1 + y^2)\,dx = 0$

21. $x^{-1} \cos 2y\,dx = 2 \ln x \sin 2y\,dy$

22. If an equation is separable, show that it is exact. Is the converse true?

It will sometimes be possible to solve a differential equation by several methods. To illustrate this, solve the following equations (*a*) by the present method, (*b*) by separating variables, (*c*) by inspection.

23. $b^2x\,dx + a^2y\,dy = 0$

24. $4x^{-5}y\,dx = x^{-4}\,dy$

25. $2x\,dx + x^{-2}(x\,dy - y\,dx) = 0$

26. $\cos x \cos y\,dx = \sin x \sin y\,dy$

27. If $F(x, y)\,dx + G(x, y)\,dy = 0$ is exact, show that the following equation is also exact.

$$[F(x, y) + f(x)]\,dx + [G(x, y) + g(y)]\,dy = 0$$

28. Under what conditions is $(ax + by)\,dx + (kx + ly)\,dy = 0$ exact? (Here, a, b, k, l are constants.) Solve the exact equation.

29. Under what conditions is $[f(x) + g(y)]\,dx + [h(x) + p(y)]\,dy = 0$ exact?

30. Under what conditions is $f(x, y)\,dx + g(x)h(y)\,dy = 0$ exact?

Solve the following initial value problems.

31. $3x^2y^4\,dx + 4x^3y^3\,dy = 0, \quad y(1) = 2$

32. $4x\,dx + 9y\,dy = 0, \quad y(3) = 0$

33. $(y - 1)\,dx + (x - 3)\,dy = 0, \quad y(0) = 2/3$

34. $\cos \pi x \cos 2\pi y\,dx = 2 \sin \pi x \sin 2\pi y\,dy, \quad y(3/2) = 1/2$

35. $[e^x \cos y + 2(x - y)]\,dx = [e^x \sin y + 2(x - y)]\,dy, \qquad y(0) = \pi$

36. $\cos 2x \cosh 2y\,dx + \sin 2x \sinh 2y\,dy = 0, \qquad y(\pi/4) = 0$

Find and graph the curve having the given slope y' and passing through the given point (x, y).

37. $y' = \dfrac{x + 2}{y + 1}, \quad (-3, -1)$

38. $y' = -\dfrac{y - 2}{x - 2}, \quad (0, 0)$

39. $y' = \dfrac{1 - x}{1 + y}, \quad (1, 0)$

40. $y' = \dfrac{x}{4 - 4y}, \quad (0, 2)$

1.6 Integrating Factors

Sometimes a given differential equation

$$P(x, y)\,dx + Q(x, y)\,dy = 0 \tag{1}$$

is not exact but can be made exact by multiplying it by a suitable function $F(x, y)$ $(\not\equiv 0)$. This function is then called an **integrating factor** of (1). With some experience, integrating factors can be found by inspection. For this purpose the student should keep in mind the differentials listed in Example 2 below. In some important special cases, integrating factors can be determined in a systematic way which will be illustrated in the next section.

Example 1

Solve

$$x\,dy - y\,dx = 0.$$

The differential equation is not exact. An integrating factor is $F = 1/x^2$, and we obtain

$$F(x)(x\,dy - y\,dx) = \frac{x\,dy - y\,dx}{x^2} = d\left(\frac{y}{x}\right) = 0, \qquad y = cx.$$

Theorem 1

If (1) *is not exact and has a general solution* $u(x, y) = c$, *then there exists an integrating factor of* (1) (*even infinitely many such factors*).

Proof. By differentiating $u(x, y) = c$ we have

$$du = \frac{\partial u}{\partial x}dx + \frac{\partial u}{\partial y}dy = 0.$$

Comparing this and (1), we see that

$$\frac{\partial u}{\partial x} : \frac{\partial u}{\partial y} = P : Q$$

must hold identically. Hence there exists a function $F(x, y)$ such that

$$\frac{\partial u}{\partial x} = FP, \qquad \frac{\partial u}{\partial y} = FQ.$$

Using these expressions for $\partial u/\partial x$ and $\partial u/\partial y$, we have

$$du = FP\,dx + FQ\,dy = F(P\,dx + Q\,dy). \tag{2}$$

This shows that F is an integrating factor of (1). Furthermore, multiplication of (2) by a function $H(u)$ yields the expression

$$H(u)F(P\,dx + Q\,dy) = H(u)\,du$$

which is exact. Hence $H(u)F(x, y)$ is another integrating factor, and since H is arbitrary, there are infinitely many integrating factors of (1). ∎

Example 2

Find integrating factors of $x\,dy - y\,dx = 0$; cf. Example 1. Since

$$d\left(\frac{x}{y}\right) = \frac{y\,dx - x\,dy}{y^2}, \qquad d\left(\ln\frac{y}{x}\right) = \frac{x\,dy - y\,dx}{xy}, \qquad d\left(\arctan\frac{y}{x}\right) = \frac{x\,dy - y\,dx}{x^2 + y^2}$$

the functions $1/y^2$, $1/xy$, and $1/(x^2 + y^2)$ are such factors. The corresponding solutions

$$\frac{x}{y} = c, \qquad \ln\frac{y}{x} = c, \qquad \arctan\frac{y}{x} = c$$

are essentially the same, because each represents a family of straight lines through the origin.

Problems for Sec. 1.6

Show that the given function is an integrating factor and solve:

1. $2\cos \pi y\,dx = \pi \sin \pi y\,dy, \quad e^{2x}$

2. $y\cos x\,dx + 3\sin x\,dy = 0, \quad y^2$

3. $3(y + 1)\,dx = 2x\,dy, \quad (y + 1)/x^4$

4. $2\sinh x\,dx + \cosh x\,dy = 0, \quad e^y \cosh x$

5. $(\sin xy \cos xy + xy)\,dx + x^2\,dy = 0, \quad \sec^2 xy$

6. $(a + 1)y\,dx + (b + 1)x\,dy = 0, \quad x^a y^b$

In each case find an integrating factor F and solve:

7. $3y\,dx + 2x\,dy = 0$

8. $2\,dx - e^{y-x}\,dy = 0$

9. $\sin y\,dx + \cos y\,dy = 0$

10. $x\cosh y\,dy - \sinh y\,dx = 0$

11. $2\cos x \cos y\,dx - \sin x \sin y\,dy = 0$

12. $(y + 1)\,dx - (x + 1)\,dy = 0$

Find integrating factors and solve the following initial value problems.

13. $2y\,dx + x\,dy = 0, \quad y(2) = -1$

14. $2y\,dx - x\,dy = 0, \quad y(1) = 2$

15. $(2y + xy)\,dx + 2x\,dy = 0, \quad y(3) = \sqrt{2}$

16. $dx + \frac{1}{2}\sec x \cos y\,dy = 0, \quad y(0) = 0$

In each case show that the given function is an integrating factor and find further integrating factors.

17. $y\,dx + 2x\,dy = 0, \quad F = y$

18. $2y\sinh x\,dx + \cosh x\,dy = 0, \quad F = \cosh x$

In each case find conditions such that F is an integrating factor of (1). *Hint.* Assume $F(P\,dx + Q\,dy) = 0$ to be exact and apply (5) in Sec. 1.5.

19. $F = x^a$ **20.** $F = y^b$ **21.** $F = e^y$ **22.** $F = x^a y^b$

23. Using the result of Prob. 19, solve Probs. 10, 13 and 14.

24. Checking of solutions is always important. In connection with the present method it is particularly essential since one may have to exclude the function $y(x)$ given by $F(x, y) = 0$. To see this, consider $(xy)^{-1}\,dy - x^{-2}\,dx = 0$; show that an integrating factor is $F = y$ and leads to $d(y/x) = 0$, hence $y = cx$, where c is arbitrary, but $F = y = 0$ is not a solution of the original equation.

25. It is interesting to note that if $F(x, y)$ is an integrating factor, $1/F(x, y) = 0$ may sometimes yield an additional solution of the corresponding equation. To see this, consider $x\,dy - y\,dx = 0$; show that an integrating factor is $F = 1/x^2$ and leads to $y = cx$. Show that $1/F = x^2 = 0$ yields the additional solution $x = 0$ of the given equation which is not included in $y = cx$.

1.7 Linear First-Order Differential Equations

A first-order differential equation is said to be **linear** if it can be written in the form

$$y' + f(x)y = r(x). \tag{1}$$

The characteristic feature of this equation is that it is linear in y and y', whereas f and r may be *any* given functions of x.

If $r(x) \equiv 0$, the equation is said to be **homogeneous;** otherwise, it is said to be **nonhomogeneous.** (Here, $r \equiv 0$ means $r = 0$ for all x in the domain of r.)

Let us find a formula for the general solution of (1) in some interval I, assuming that f and r are continuous in I. For the homogeneous equation

$$y' + f(x)y = 0 \tag{2}$$

this is very simple. By separating variables we have

$$\frac{dy}{y} = -f(x)\,dx \qquad \text{and thus} \qquad \ln|y| = -\int f(x)\,dx + c^*$$

or

$$y(x) = ce^{-\int f(x)\,dx} \qquad (c = \pm e^{c^*} \quad \text{when} \quad y \gtrless 0); \tag{3}$$

here we may also take $c = 0$ and obtain the *trivial solution* $y \equiv 0$.

To solve the nonhomogeneous equation (1), let us write it in the form

$$(fy - r)\,dx + dy = 0$$

and show that we can find an integrating factor $F(x)$ *depending only on* x. If such a factor exists,

$$F(x)(fy - r)\,dx + F(x)\,dy = 0$$

must be exact. Now for this equation the condition (5) of Sec. 1.5 assumes the form

$$\frac{\partial}{\partial y}[F(fy - r)] = \frac{dF}{dx} \qquad \text{or} \qquad Ff = \frac{dF}{dx}.$$

By separating variables we have $f\,dx = dF/F$. Integration gives

$$\ln|F| = \int f(x)\,dx.$$

Hence

$$F(x) = e^{h(x)} \qquad \text{where} \qquad h(x) = \int f(x)\,dx.$$

This shows that $F(x)$ is an integrating factor of (1). Let us now multiply (1) by this factor, finding

$$e^h(y' + fy) = e^h r.$$

Since $h' = f$, this may be written

$$\frac{d}{dx}(ye^h) = e^h r.$$

Now we can integrate on both sides and obtain

$$ye^h = \int e^h r\,dx + c.$$

By dividing both sides by e^h we have the desired formula

(4) $$y(x) = e^{-h}\left[\int e^h r\,dx + c\right], \qquad h = \int f(x)\,dx,$$

which represents the general solution of (1) in the form of an integral. (The choice of the values of the constants of integration is immaterial; cf. Prob. 2 below.)

Example 1
Solve the linear differential equation

$$y' - y = e^{2x}.$$

Here

$$f = -1, \qquad r = e^{2x}, \qquad h = \int f\,dx = -x$$

and from (4) we obtain the general solution

$$y(x) = e^{x}\left[\int e^{-x}e^{2x}\,dx + c\right] = e^{x}[e^{x} + c] = ce^{x} + e^{2x}.$$

Alternatively, we may multiply the given equation by $e^{h} = e^{-x}$,

$$(y' - y)e^{-x} = (ye^{-x})' = e^{2x}e^{-x} = e^{x}$$

and integrate on both sides, obtaining the same result as before:

$$ye^{-x} = e^{x} + c \qquad \text{hence} \qquad y = e^{2x} + ce^{x}.$$

Example 2
Solve

$$xy' + y + 4 = 0.$$

We write this equation in the form (1), that is,

$$y' + \frac{1}{x}y = -\frac{4}{x}.$$

Hence, $f = \frac{1}{x}$, $r = -\frac{4}{x}$ and, therefore,

$$h = \int f\,dx = \ln|x|, \qquad e^{h} = x, \qquad e^{-h} = \frac{1}{x}.$$

From this and (4) we obtain the general solution

$$y(x) = \frac{1}{x}\left[\int x\left(-\frac{4}{x}\right)dx + c\right] = \frac{c}{x} - 4,$$

in agreement with Example 1 of Sec. 1.5.

Of course, in simple cases such as Examples 2, 3 and 5 (below) we may solve the equation without using (4). For instance, in Example 2 the student may write the given equation in the form $(xy)' + 4 = 0$ and can now obtain the solution by integration.

Example 3
Solve the linear equation

$$xy' + y = \sin x.$$

We can write the equation in the form

$$d(xy) = \sin x\,dx$$

and integrate on both sides, finding

$$xy = -\cos x + c \qquad \text{or} \qquad y = \frac{1}{x}(c - \cos x).$$

Let us now illustrate the procedure of solving initial value problems.

Example 4
Solve the initial value problem

$$y' + y \tan x = \sin 2x, \qquad y(0) = 1.$$

Here $f = \tan x$, $r = \sin 2x = 2 \sin x \cos x$, and

$$\int f\,dx = \int \tan x\,dx = \ln |\sec x|.$$

From this we see that in (4),

$$e^h = \sec x, \qquad e^{-h} = \cos x, \qquad e^h r = 2 \sin x$$

and the general solution of our equation is

$$y(x) = \cos x \left[2 \int \sin x\,dx + c\right] = c \cos x - 2 \cos^2 x.$$

According to the initial condition, $y = 1$ when $x = 0$, that is

$$1 = c - 2 \qquad \text{or} \qquad c = 3,$$

and the solution of our initial value problem is

$$y = 3 \cos x - 2 \cos^2 x.$$

Example 5
Solve the initial value problem

$$x^2y' + 2xy - x + 1 = 0, \qquad y(1) = 0.$$

The equation may be written as $d(x^2y) = (x - 1)\,dx$. By integration we obtain

$$x^2y = \tfrac{1}{2}x^2 - x + c \qquad \text{or} \qquad y(x) = \frac{1}{2} - \frac{1}{x} + \frac{c}{x^2}.$$

From this and the initial condition we have $y(1) = \frac{1}{2} - 1 + c = 0$ or $c = \frac{1}{2}$. Thus

$$y(x) = \frac{1}{2} - \frac{1}{x} + \frac{1}{2x^2}.$$ ∎

Very often the solutions of differential equations in engineering applications are not elementary functions. For instance, the integral in (4) may not be elementary. If it is not a tabulated function,[9] we may develop the integrand in a power series and integrate term by term, or we may apply a method of numerical integration (cf. Sec. 19.6) or a numerical method for solving first-order differential equations (cf. Sec. 19.7).

In applications the independent variable x will often be the time; the function $r(x)$ on the right-hand side of (1) may represent a force, and the solution $y(x)$ a displacement, a current, or some other variable physical quantity. In engineer-

[9] Whether tables of a certain function were computed can be found in the extremely useful book Ref. [A6] listed in Appendix 1.

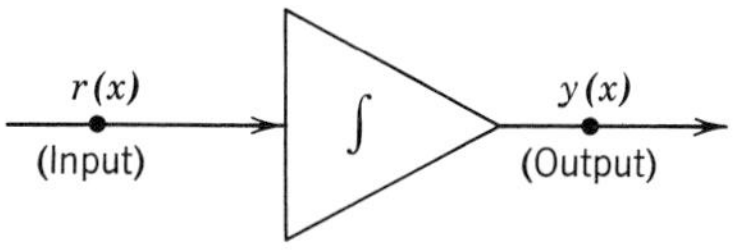

Fig. 14. Simple block diagram showing a single block

ing mathematics $r(x)$ is frequently called the **input,** and $y(x)$ is called the **output** or *response to the input* (and the initial conditions). For instance, in electrical engineering the differential equation may govern the behavior of an electric circuit and the output $y(x)$ is obtained as the solution of that equation corresponding to the input $r(x)$. In the field of analogue computers the physical system may be an integrating unit; then the corresponding differential equation is $y' = r(x)$, and the output $y(x)$ corresponding to an input $r(x)$ and a fixed initial condition, say, $y(x_0) = 0$, is the integral of $r(x)$:

$$y(x) = \int_{x_0}^{x} r(x^*)\, dx^*.$$

The sytem may be represented by a "block diagram" as shown in Fig. 14. Such a diagram displays functional relationships rather than physical details of systems. More complicated block diagrams consist of two or more blocks and their interrelationships. Various examples will occur in our further considerations.

Problems for Sec. 1.7

1. Show that $e^{-\ln x} = 1/x$ (but not $-x$) and $e^{-\ln(\operatorname{cosec} x)} = \sin x$.

2. Show that the choice of the value of the constant of integration in $\int f\,dx$ [cf. (4)] is immaterial (so that we may choose it to be zero).

Find the general solutions of the following differential equations.

3. $y' - y = 1 - x$ **4.** $y' + y = 1$ **5.** $xy' + y = 2x$

6. $y' + xy = 2x$ **7.** $y' - 4y = 2x - 4x^2$ **8.** $y' + 2y = \cos x$

9. $y' + 3y = e^{2x} + 6$ **10.** $y' \tan x = y - 1$ **11.** $y' = 2y/x + x^2e^x$

12. $(x^2 - 1)y' = xy - x$ **13.** $xy' + 2y = e^{x^2}$ **14.** $x^2y' + 2xy = \sinh 3x$

Solve the following initial value problems.

15. $y' - y = e^x, \quad y(1) = 0$

16. $y' - y \cot x = 2x - x^2 \cot x, \quad y(\tfrac{1}{2}\pi) = \tfrac{1}{4}\pi^2 + 1$

17. $y' - (1 + 3x^{-1})y = x + 2, \quad y(1) = e - 1$

18. $y' - x^3y = -4x^3, \quad y(0) = 6$

Prove and illustrate with examples that the linear differential equations (1) and (2) have the following important properties. [In the next chapter (Secs. 2.3 and 2.9) we shall see that linear differential equations of higher order enjoy similar properties. This fact is quite important since it will enable us to obtain new solutions from given ones.]

19. The homogeneous equation (2) has the *"trivial solution"* $y \equiv 0$.

20. If y_1 is a solution of (2), then $y = cy_1$ (c any constant) is a solution of (2).

21. If y_1 and y_2 are solutions of (2), then their sum $y = y_1 + y_2$ is a solution of (2).

22. If y_1 is a solution of (1), then $y = cy_1$ is a solution of $y' + fy = cr$.

23. If y_1 is a solution of (1) and y_2 is a solution of (2), then $y = y_1 + y_2$ is a solution of (1).

24. The difference $y = y_1 - y_2$ of two solutions of (1) is a solution of (2).

25. If $f(x)$ and $r(x)$ in (1) are constant, say, $f(x) = f_0$ and $r(x) = r_0$, show that separating variables is simpler than the method explained in the text and leads to the same result, namely, (4).

26. Newton's law of cooling leads to the differential equation (cf. Sec. 1.3)

$$\frac{dT}{dt} = -k(T - T_1)$$

where $T(t)$ is the temperature of a body placed in a medium which is kept at a constant temperature T_1. Solve the equation by the method developed in the present section, assuming the initial temperature of the body to be $T(0) = T_0$.

27. (Motion of a boat) Two men are riding in a motorboat, the combined weight of the men and the boat being 4900 nt (about 1100 lb). Suppose that the motor exerts a constant force of 200 nt (about 45 lb) and the resistance R of the water is proportional to the speed v, say, $R = kv$ nt, where $k = 10$ nt · sec/meter. Set up the differential equation for $v(t)$, using Newton's second law (mas × acceleration = force). Find $v(t)$ satisfying $v(0) = 0$. Find the maximum speed v_∞ at which the boat will travel (practically after a sufficiently long time). If the boat starts from rest, how long will it take to reach $0.9v_\infty$ and what distance does the boat travel during that time?

28. (Air resistance) An extended object falling downward is known to experience a resistive force of the air (called *drag*). We assume the magnitude of this force to be proportional to the speed v. Using Newton's second law, show that

$$m\dot{v} = -kv - mg$$

where $\dot{v} = dv/dt$ and g ($= 9.80$ meters/sec²) is the acceleration of gravity at the surface of the earth. Solve this equation (a) by the present method, (b) by separating variables, assuming that $v(0) = v_0$. (c) Integrating the result, find $y(t)$, the distance at time t measured from the starting point $y_0 = y(0)$.

29. (Atomic waste disposal) The Atomic Energy Commission put atomic waste into sealed containers and dumped them in the ocean. It is important that the containers do not break when they hit the bottom of the ocean. Assume that this is the case as long as the speed is less than 12 meters/sec. Show that Newton's second law yields the equation of motion (Fig. 15)

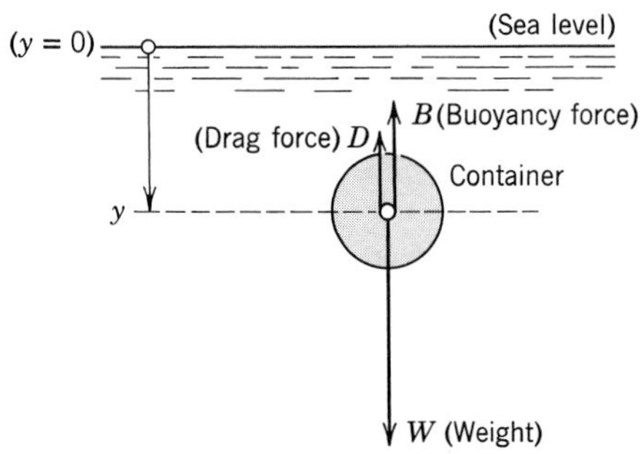

Fig. 15. System and forces in Prob. 29

$$m\frac{dv}{dt} = W - B - kv, \qquad v(0) = 0$$

where the drag force $D = -kv$ is assumed to be proportional to the velocity v. (This is experimentally confirmed for velocities which are not too large.) Solve the equation to obtain $v(t)$. Integrate to obtain $y(t)$ such that $y(0) = 0$. Determine the critical time t_{crit} when the container reaches the critical speed $v_{\text{crit}} = 12$ meters/sec, assuming that $W = 2254$ nt (about 500 lb), $B = 2090$ nt (about 460 lb), $k = 0.637$ kg/sec. Show that the container will break if it is dumped at points where the ocean is deeper than 105 meters, approximately.

30. The speed $v(t)$ in Prob. 29 approaches a limiting value as $t \to \infty$. How can this limit be obtained from the differential equation without actually solving it? What would happen if there were no drag force?

1.8 Variation of Parameters

There is another interesting way of obtaining the general solution of the linear differential equation

$$y' + f(x)y = r(x). \tag{1}$$

We have seen that a solution of the corresponding homogeneous equation is

$$v(x) = e^{-\int f(x)\,dx}. \tag{2}$$

Using this function $v(x)$, let us try to determine a function $u(x)$ such that

$$y(x) = u(x)v(x) \tag{3}$$

is the general solution of (1). This attempt is suggested by the form of the general solution $cv(x)$ of the homogeneous equation and consists in replacing the parameter c by a variable $u(x)$. Therefore, this approach, credited to Lagrange,[10] is called the **method of variation of parameters.** It can be generalized to equations of higher order where it is of great importance. Our present simple case is a good occasion for getting acquainted with this basic method.

By substituting (3) into (1) we obtain

$$u'v + u(v' + fv) = r.$$

Since v is a solution of the homogeneous equation, this reduces to

$$u'v = r \qquad \text{or} \qquad u' = \frac{r}{v}.$$

[10] JOSEPH LOUIS LAGRANGE (1736–1813), a great French mathematician, spent 20 years of his life in Prussia and then returned to Paris. His important major work was in the calculus of variations, celestial and general mechanics, differential equations, and algebra.

Integration yields

$$u = \int \frac{r}{v}\,dx + c.$$

We thus obtain the result

$$y = uv = v\left(\int \frac{r}{v}\,dx + c\right), \tag{4}$$

and from (2) we see that this is identical with (4) of the previous section.

Example 1
Illustrate the method of variation of parameters by solving

$$y' - \frac{2y}{x} = x^2 \cos 3x.$$

The homogeneous equation is $y' - 2y/x = 0$. A solution is x^2. Hence we have to substitute $y = ux^2$ and $y' = u'x^2 + 2ux$ into the given equation. We obtain

$$u'x^2 + 2ux - 2ux = x^2 \cos 3x, \qquad u' = \cos 3x, \qquad u = \tfrac{1}{3}\sin 3x + c.$$

This yields

$$y = ux^2 = (\tfrac{1}{3}\sin 3x + c)x^2.$$

The reader may solve the equation by (4), Sec. 1.7.

Problems for Sec. 1.8

Solve the following equations by the method of variation of parameters.

1. $y' + y = 2$

2. $y' + 2y = x^2$

3. $y' - y = 3e^x$

4. $(x + 4)y' + 3y = 3$

Taking x as the dependent variable, solve:

5. $(2x + y^4)y' = y$

6. $y' = 1/(\cos y - 2x)$

Reduction of nonlinear equations to linear form. The following problems illustrate that *certain nonlinear first-order differential equations may be reduced to linear form by a suitable change of the dependent variable.*

7. (Bernoulli equation) The differential equation

$$y' + f(x)y = g(x)y^a \qquad (a \text{ any real number})$$

is called the **Bernoulli**[11] **equation.** For $a = 0$ and $a = 1$ the equation is linear, and otherwise it is nonlinear. Set $[y(x)]^{1-a} = u(x)$ and show that the equation assumes the linear form

$$u' + (1 - a)f(x)u = (1 - a)g(x).$$

[11] JACOB BERNOULLI (1654–1705), Swiss mathematician, who contributed to the theory of elasticity and to the theory of mathematical probability. The method for solving Bernoulli's equation was discovered by Leibniz in 1696.

Solve the following Bernoulli equations.

8. $y' + x^{-1}y = xy^2$

9. $y' + y = y^2$

10. $3y' + y = (1 - 2x)y^4$

11. $y' = x^3y^2 + xy$

Applying suitable substitutions, reduce to linear form and solve:

12. $y' - 1 = e^{-y} \sin x$

13. $2xyy' + (x - 1)y^2 = x^2e^x$ (put $y^2 = xz$)

14. Show that the transformation $x = \phi(t)$ (where ϕ has a continuous derivative) transforms (1) into a linear differential equation involving y and dy/dt.

15. Show that the transformation $y = a(x)u + b(x)$ with differentiable $a \not\equiv 0$ and b transforms (1) into a linear differential equation involving u and du/dx.

16. Show that in (4), one term is independent of the function on the right-hand side of (1) and the other term is independent of an initial condition which we may want to impose.

17. Suppose that f and r in (1) are continuous. Why can you conclude from (4) that an initial value problem consisting of (1) and an initial condition $y(x_0) = y_0$ has a unique solution?

18. If the input r satisfies $|r(x)| \leqq M$ for all $x \geqq 0$, show that the output (the solution) of the initial value problem

$$y' + y = r(x), \quad y(0) = 0$$

satisfies $|y(x)| < M$ for all $x > 0$.

19. **(Direction field)** Show that the direction field of (1) has the following interesting property. The lineal elements on any vertical line $x = x_0$ all pass through a single point with coordinates

$$\xi = x_0 + \frac{1}{f(x_0)}, \qquad \eta = \frac{r(x_0)}{f(x_0)}.$$

20. Applying the result in Prob. 19, find an approximate solution curve of the initial value problem $y' = 2xy + 1$, $y(1) = 1$.

1.9 Electric Circuits

Linear first-order differential equations have various applications in physics and engineering. To illustrate this, let us consider some standard examples in connection with electric circuits. These and similar considerations are important, because they help *all* students to learn how to set up **models**, that is, express physical situations in terms of mathematical relations. The transition from the physical system to a corresponding mathematical model is always the first step in engineering mathematics, and this important step requires experience and training which can only be gained by considering typical examples from various fields.

The simplest electric circuit is a series circuit in which we have a source of electric energy (*electromotive force*) such as a generator or a battery, and a resistor, which uses energy, for example an electric light bulb (Fig. 16). If we close the switch, a current I will flow through the resistor, and this will cause a

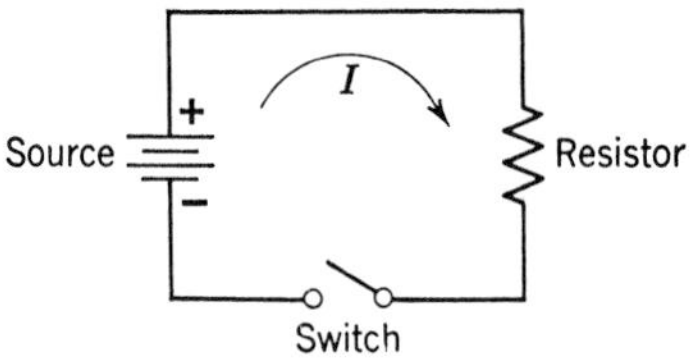

Fig. 16. Circuit

voltage drop, that is, the electric potential at the two ends of the resistor will be different; this potential difference or voltage drop can be measured by a voltmeter. Experiments show that the following law holds.

The voltage drop E_R across a resistor is proportional to the instantaneous current I, say,

$$E_R = RI \tag{1}$$ **(Ohm's law)**

where the constant of proportionality R is called the **resistance** of the resistor. The current I is measured in *amperes,* the resistance R in *ohms,* and the voltage E_R in *volts.*[12]

The other two important elements in more complicated circuits are *inductors* and *capacitors.* An inductor opposes a change in current, having an inertia effect in electricity similar to that of mass in mechanics; we shall consider this analogy later (Sec. 2.14). Experiments yield the following law.

The voltage drop E_L across an inductor is proportional to the instantaneous time rate of change of the current I, say,

$$E_L = L\frac{dI}{dt} \tag{2}$$

where the constant of proportionality L is called the **inductance** of the inductor and is measured in *henrys;* the time t is measured in seconds.

A capacitor is an element which stores energy. Experiments yield the following law.

The voltage drop E_C across a capacitor is proportional to the instantaneous electric charge Q on the capacitor, say,

$$E_C = \frac{1}{C}Q \tag{3*}$$

where C is called the **capacitance** and is measured in *farads;* the charge Q is measured in *coulombs.* Since

$$I(t) = \frac{dQ}{dt} \tag{3'}$$

[12]These and the subsequent units are named after ANDRÉ MARIE AMPÈRE (1775–1836), French physicist, GEORG SIMON OHM (1789–1854), German physicist, ALLESSANDRO VOLTA (1745–1827), Italian physicist, JOSEPH HENRY (1797–1878), American physicist, MICHAEL FARADAY (1791–1867), English physicist, and CHARLES AUGUSTIN DE COULOMB (1736–1806), French physicist.

this may be written

$$E_C = \frac{1}{C}\int_{t_0}^{t} I(t^*)\,dt^*. \tag{3}$$

The current $I(t)$ in a circuit may be determined by solving the equation (or equations) resulting from the application of the following physical law.

Kirchhoff's voltage law (KVL)[13]

The algebraic sum of all the instantaneous voltage drops around any closed loop is zero, or the voltage impressed on a closed loop is equal to the sum of the voltage drops in the rest of the loop.

Example 1. RL-circuit

For the "RL-circuit" in Fig. 17 we obtain from Kirchhoff's voltage law and (1), (2)

$$L\frac{dI}{dt} + RI = E(t). \tag{4}$$

Case A (Constant electromotive force). If $E = E_0 = const$, then (4) in Sec. 1.7 yields the general solution

$$\begin{aligned} I(t) &= e^{-\alpha t}\left[\frac{E_0}{L}\int e^{\alpha t}\,dt + c\right] \qquad (\alpha = R/L) \\ &= \frac{E_0}{R} + ce^{-(R/L)t}. \end{aligned} \tag{5}$$

The last term approaches zero as t tends to infinity, so that $I(t)$ tends to the limiting value E_0/R; after a sufficiently long time, I will practically be constant, its value being independent of c, hence independent of an initial condition which we may want to impose. The particular solution corresponding to the initial condition $I(0) = 0$ is (Fig. 18)

$$I(t) = \frac{E_0}{R}(1 - e^{-(R/L)t}) = \frac{E_0}{R}(1 - e^{-t/\tau_L}) \tag{5*}$$

where $\tau_L = L/R$ is called the **inductive time constant** of the circuit (cf. Probs. 5 and 7).

Case B (Periodic electromotive force). If $E(t) = E_0 \sin \omega t$, then, by (4) in Sec. 1.7, the general solution of (4) in this current section is

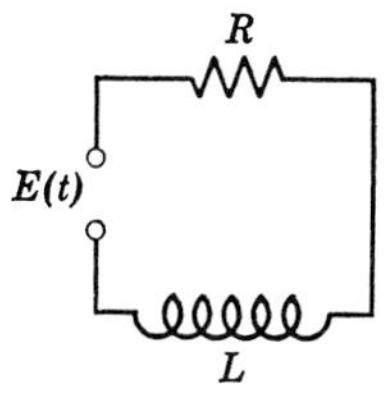

Fig. 17. *RL*-circuit

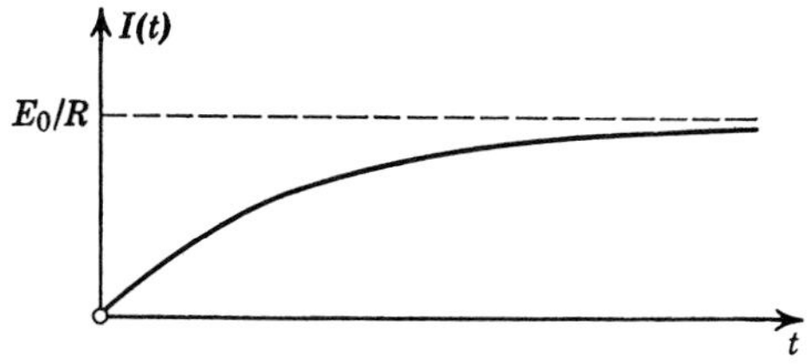

Fig. 18. Current in an *RL*-circuit due to a constant electromotive force

[13] GUSTAV ROBERT KIRCHHOFF (1824–1887), German physicist. Later we shall also need

Kirchoff's current law (KCL)

At any point of a circuit, the sum of the inflowing currents is equal to the sum of the outflowing currents.

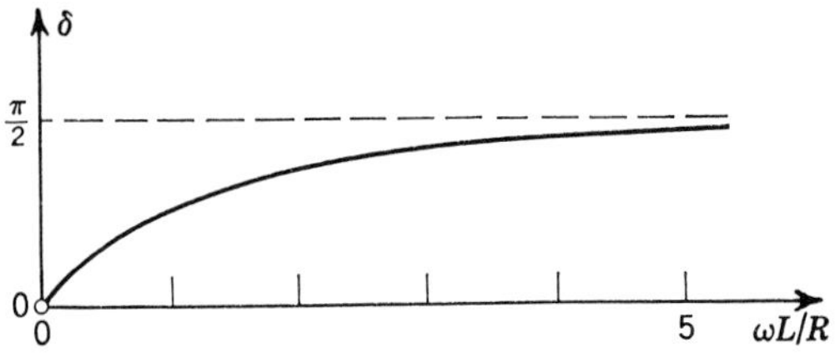

Fig. 19. Phase angle δ in (6) as a function of $\omega L/R$

$$I(t) = e^{-\alpha t}\left[\frac{E_0}{L}\int e^{\alpha t}\sin \omega t\, dt + c\right] \qquad (\alpha = R/L).$$

Integration by parts yields

$$I(t) = ce^{-(R/L)t} + \frac{E_0}{R^2 + \omega^2 L^2}(R \sin \omega t - \omega L \cos \omega t).$$

This may be written [cf. (14) in Appendix 3]

$$I(t) = ce^{-(R/L)t} + \frac{E_0}{\sqrt{R^2 + \omega^2 L^2}}\sin(\omega t - \delta), \qquad \delta = \arctan\frac{\omega L}{R}. \tag{6}$$

The exponential term will approach zero as t approaches infinity. This means that after a sufficiently long time the current $I(t)$ executes practically harmonic oscillations. (Cf. Fig. 20.) Figure 19 shows the phase angle δ as a function of $\omega L/R$. If $L = 0$, then $\delta = 0$, and the oscillations of $I(t)$ are in phase with those of $E(t)$. ∎

An electrical (or dynamical) system is said to be in the **steady state** when the variables describing its behavior are periodic functions of the time or constant, and it is said to be in the **transient state** (or *unsteady state*) when it is not in the steady state. The corresponding variables are called *steady-state functions* and *transient functions,* respectively.

In Example 1, Case A, the function E_0/R is the steady-state function or **steady-state solution** of (4), and in Case B the steady-state solution is represented by the last term in (6). Before the circuit (practically) reaches the steady state it is in the transient state. It is clear that such an interim or transient period occurs because inductors and capacitors store energy, and the corresponding inductor currents and capacitor voltages cannot be changed instantly. Similar situations arise in various physical systems. For example, if a radio receiver having heater-type vacuum tubes is turned on, a transient interval of time must

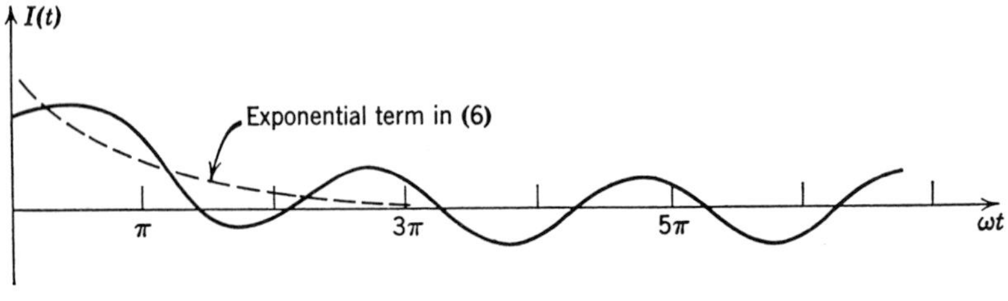

Fig. 20. Current in an RL-circuit due to a sinusoidal electromotive force, as given by (6) (with $\delta = \pi/4$)

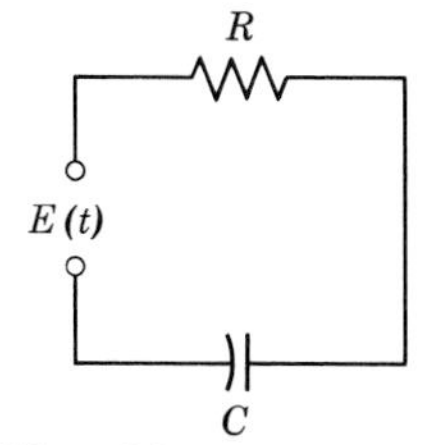

Fig. 21. *RC*-circuit

elapse during which the tubes change from "cold" to "hot." Practically, a transient state of our present system will last only for a short time.

Example 2. RC-circuit

By applying Kirchhoff's voltage law and (1), (3) to the RC-circuit in Fig. 21 we obtain the equation

$$RI + \frac{1}{C}\int I\,dt = E(t). \tag{7}$$

To get rid of the integral we differentiate the equation with respect to t, finding

$$R\frac{dI}{dt} + \frac{1}{C}I = \frac{dE}{dt}. \tag{8}$$

According to (4) in Sec. 1.7 this differential equation has the general solution

$$I(t) = e^{-t/RC}\left(\frac{1}{R}\int e^{t/RC}\frac{dE}{dt}\,dt + c\right). \tag{9}$$

Case A (Constant electromotive force). If E is constant, then $\frac{dE}{dt} = 0$, and (9) assumes the simple form (Fig. 22)

$$I(t) = ce^{-t/RC} = ce^{-t/\tau_C} \tag{10}$$

where $\tau_C = RC$ is called the **capacitive time constant** of the circuit.

Case B (Sinusoidal electromotive force). If $E(t) = E_0 \sin \omega t$, then

$$\frac{dE}{dt} = \omega E_0 \cos \omega t.$$

By inserting this in (9) and integrating by parts we find

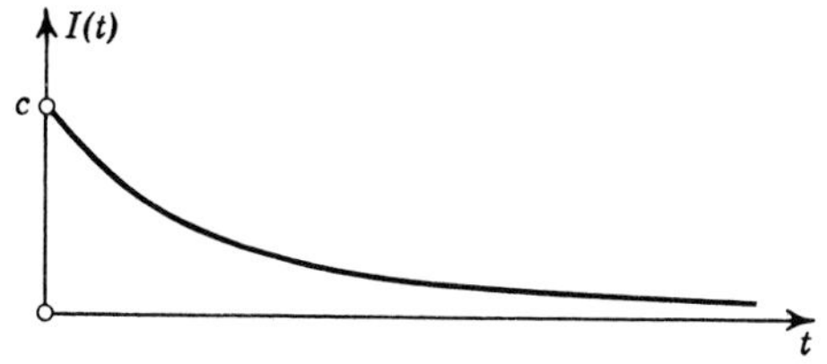

Fig. 22. Current in an *RC*-circuit due to a constant electromotive force

$$I(t) = ce^{-t/RC} + \frac{\omega E_0 C}{1 + (\omega RC)^2} (\cos \omega t + \omega RC \sin \omega t)$$

(11)

$$= ce^{-t/RC} + \frac{\omega E_0 C}{\sqrt{1 + (\omega RC)^2}} \sin (\omega t - \delta),$$

where $\tan \delta = -1/\omega RC$. The first term decreases steadily as t increases, and the last term represents the steady-state current, which is sinusoidal. The graph of $I(t)$ is similar to that in Fig. 20. ▌

More complicated circuits and the analogy between electrical and mechanical vibrations will be considered in connection with second-order differential equations in Sec. 2.14.

These more complicated circuits will be called *RLC-circuits* since they contain resistors, inductors and capacitors. Their mathematical model will be a second-order differential equation obtained from Kirchhoff's laws.

Electrical networks will be considered in Secs. 3.1, 5.7, 7.5 and 7.11.

Problems for Sec. 1.9

1. Derive and check (5).

2. In (5) the solution approaches E_0/R as $t \to \infty$. Could this be seen directly from (4) with $E(t) = E_0$, without solving (4)?

3. Derive and check (6).

4. Derive the steady-state solution in (6) by substituting $I_p = A \cos \omega t + B \sin \omega t$ into (4) with $E(t) = E_0 \sin \omega t$ and determining A and B by equating the cosine and sine terms in the resulting equation. (Note that in this way one avoids integration by parts.)

5. Show that the inductive time constant $\tau_L = L/R$ is the time t at which the current (5*) reaches about 63% of its final value. Find the time at which the current reaches about 99% of its final value, assuming that $L = 1$ henry and $R = 500$ ohms.

6. In Example 1, case A, let $R = 20$ ohms, $L = 0.03$ millihenry ($3 \cdot 10^{-5}$ henry) and $I(0) = 0$. Find the time when the current reaches 99.9% of its final value.

7. In Example 1, case A, let $R = 100$ ohms, $L = 2.5$ henrys, $E_0 = 110$ volts and $I(0) = 0$. Find the time constant and the time necessary for the current to rise from 0 to 0.6 ampere.

8. If $L = 10$ henrys, what R should we choose in order that (5*) reach 99% of its final value at $t = 1$ sec?

9. For what initial condition does (6) yield the steady-state solution?

10. Obtain from (6) the particular solution satisfying the initial condition $I(0) = 0$.

11. Verify that (11) is a solution of (8) with $E = E_0 \sin \omega t$.

12. (Discharge of a capacitor) Show that (7) can also be written

$$R\frac{dQ}{dt} + \frac{1}{C}Q = E(t).$$

Solve the equation with $E(t) = 0$, assuming $Q(0) = Q_0$. Find the time when the capacitor has lost 99% of its initial charge.

13. Write the general solution of

$$R\frac{dQ}{dt} + \frac{1}{C}Q = E(t)$$

in the form of an integral and derive from it formula (9) by differentiation and subsequent integration by parts.

14. In Prob. 13, let $R = 20$ ohms, $C = 0.01$ farad and $E(t)$ be exponentially decaying, say, $E(t) = 60e^{-2t}$ volts. Assuming $Q(0) = 0$, find and graph $Q(t)$. Also determine the time when $Q(t)$ reaches a maximum and that maximum charge.

15. A capacitor ($C = 0.1$ farad) in series with a resistor ($R = 200$ ohms) is charged from a source ($E_0 = 12$ volts); see Fig. 21 with $E(t) = E_0$. Find the voltage $V(t)$ on the capacitor, assuming that at $t = 0$ the capacitor is completely uncharged.

16. Find the current $I(t)$ in the RC-circuit shown in Fig. 21, assuming that $E = 100$ volts, $C = 0.25$ farad, R is variable according to $R = (100 - t)$ ohms when $0 \leqq t \leqq 100$ sec, $R = 0$ when $t > 100$ sec, and $I(0) = 1$ ampere.

17. Obtain from (11) the particular solution satisfying the initial condition $I(0) = 0$.

18. Suppose that in Fig. 23 we have $C = 0.1$ farad and $R = 100$ ohms, and the initial charge on the capacitor is 2 coulombs. At $t = 0$ the switch is closed and the capacitor starts discharging. Find the current $I(t)$, the charge $Q(t)$ and the voltage $V(t)$ on the capacitor.

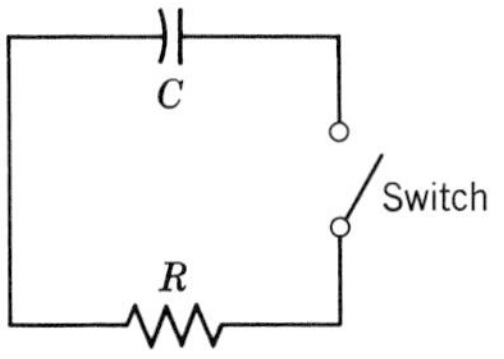

Fig. 23. Problem 18

19. Periodic electromotive forces other than pure sine or cosine functions occur quite frequently in applications. An example is shown in Fig. 24, where the discontinuities (jumps) are convenient mathematical approximations of very abrupt changes of $E(t)$ from 0 to 1 and conversely. Assume that this $E(t)$ is applied to an RL-circuit with $R = 1$ ohm and $L = 100$ henrys. Assuming $I(0) = 0$, find the current $I(t)$.

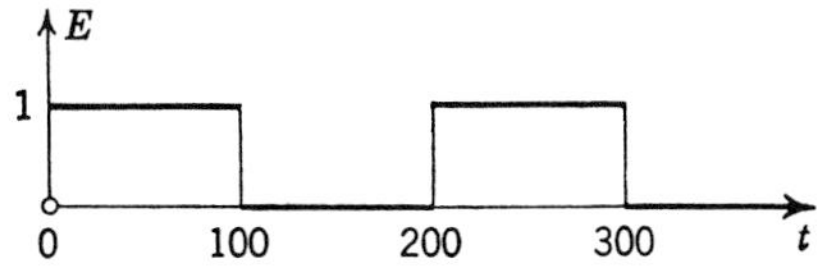

Fig. 24. Problem 19

20. It is worth noting that periodicity of the input does not entail periodicity of the output since, in general, there will be a transient state. But there are obvious exceptions. To illustrate this, show that $I(t)$ in Prob. 19 satisfies $I(t + 200) = I(t)$ for all $t \geqq 0$ if and only if we choose the initial condition $I(0) = (e - 1)/(e^2 - 1)$.

1.10 Families of Curves. Orthogonal Trajectories

If for each fixed real value of c the equation

$$F(x, y, c) = 0 \tag{1}$$

represents a curve in the xy-plane and if for variable c it represents infinitely many curves, then the totality of these curves is called a **one-parameter family of curves,** and c is called the *parameter* of the family.

Example 1

The equation

$$F(x, y, c) = x + y + c = 0 \tag{2}$$

represents a family of parallel straight lines; each line corresponds to precisely one value of c. The equation

$$F(x, y, c) = x^2 + y^2 - c^2 = 0 \tag{3}$$

represents a family of concentric circles of radius c with center at the origin. ▮

The general solution of a first-order differential equation involves a parameter c and thus represents a family of curves. This yields a possibility for representing many one-parameter families of curves by such differential equations. The practical use of such representations will become obvious from our further considerations.

Example 2

By differentiating (2) we see that

$$y' + 1 = 0$$

is the differential equation of that family of straight lines. Similarly, the differential equation of the family (3) is

$$y' = -x/y.$$ ▮

If the equation obtained by differentiating (1) still contains the parameter c, then we have to eliminate c from this equation by using (1).

Example 3

The differential equation of the family of parabolas

$$y = cx^2 \tag{4}$$

is obtained by differentiating (4),

$$y' = 2cx, \tag{5}$$

and by eliminating c from (5). From (4) we have $c = y/x^2$, and by substituting this into (5) we find the desired result

$$y' = 2y/x. \tag{6}$$

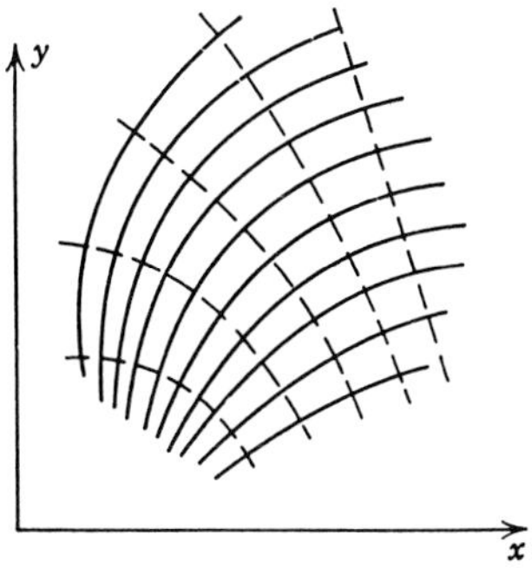

Fig. 25. Curves and their orthogonal trajectories

Note that we may also proceed as follows. We solve (4) for c, finding $c = y/x^2$, and differentiate this equation with respect to x, obtaining

$$0 = \frac{y'}{x^2} - \frac{2y}{x^3}. \qquad \text{Hence} \qquad y' = \frac{2y}{x},$$

as before. ∎

In many engineering applications, a family of curves is given, and it is required to find another family whose curves intersect each of the given curves at right angles.[14] Then the curves of the two families are said to be *mutually orthogonal,* they form an *orthogonal net,* and the curves of the family to be obtained are called the **orthogonal trajectories** of the given curves (and conversely); cf. Fig. 25.

Let us mention some familiar examples. The meridians on the earth's surface are the orthogonal trajectories of the parallels. On a map the curves of steepest descent are the orthogonal trajectories of the contour lines. In electrostatics the equipotential lines and the lines of electric force are orthogonal trajectories of each other. An illustrative example is shown in Fig. 26. We shall see later that

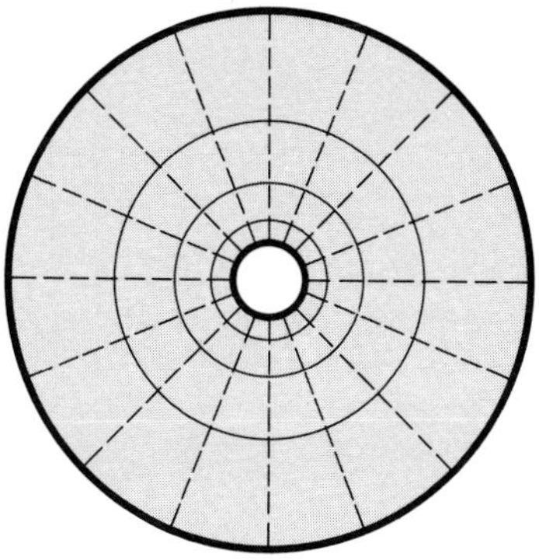

Fig. 26. Equipotential lines and lines of electric force (dashed) between two concentric cylinders

[14]Remember that the angle of intersection of two curves is defined to be the angle between the tangents of the curves at the point of intersection.

orthogonal trajectories play an important role in various fields of physics, for example, in hydrodynamics and heat conduction.

Given a family $F(x, y, c) = 0$ which can be represented by a differential equation

$$y' = f(x, y) \tag{7}$$

we may find the corresponding orthogonal trajectories as follows. From (7) we see that a curve of the given family which passes through a point (x_0, y_0) has the slope $f(x_0, y_0)$ at this point. The slope of the orthogonal trajectory through (x_0, y_0) at this point should be the negative reciprocal of $f(x_0, y_0)$, that is, $-1/f(x_0, y_0)$, because this is the condition for the tangents of the two curves at (x_0, y_0) to be perpendicular. Consequently, the differential equation of the orthogonal trajectories is

$$y' = -\frac{1}{f(x, y)}, \tag{8}$$

and the trajectories are obtained by solving this equation.

It can be shown that, under relatively general conditions, a family of curves has orthogonal trajectories, but we shall not discuss this question here.

Example 4. Orthogonal trajectories

Find the orthogonal trajectories of the parabolas in Example 3. From (6) we see that the differential equation of the orthogonal trajectories is

$$y' = -\frac{1}{2y/x} = -\frac{x}{2y}.$$

By separating variables and integrating we find that the orthogonal trajectories are the ellipses (Fig. 27)

$$\frac{x^2}{2} + y^2 = c^*.$$

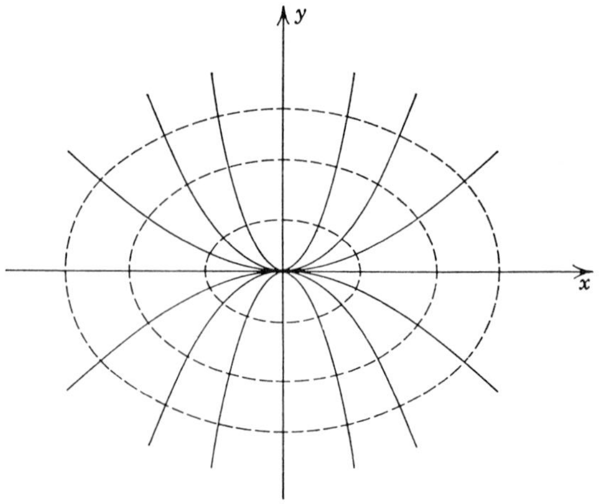

Fig. 27. Parabolas and their orthogonal trajectories in Example 4

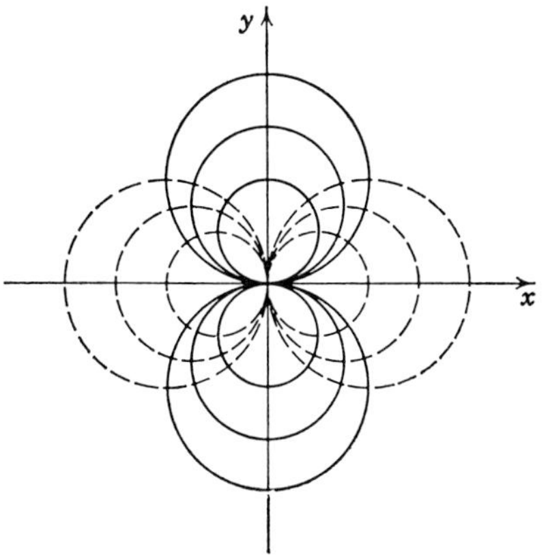

Fig. 28. Circles and their orthogonal trajectories (dashed) in Example 5

Example 5. Orthogonal trajectories
Find the orthogonal trajectories of the circles

$$x^2 + (y - c)^2 = c^2. \tag{9}$$

We first determine the differential equation of the given family. By differentiating (9) with respect to x we obtain

$$2x + 2(y - c)y' = 0. \tag{10}$$

We must eliminate c. Solving (9) for c, we have

$$c = \frac{x^2 + y^2}{2y}.$$

By inserting this into (10) and simplifying we get

$$x + \frac{y^2 - x^2}{2y}y' = 0 \qquad \text{or} \qquad y' = \frac{2xy}{x^2 - y^2}.$$

From this and (8) we see that the differential equation of the orthogonal trajectories is

$$y' = -\frac{x^2 - y^2}{2xy} \qquad \text{or} \qquad 2xyy' - y^2 + x^2 = 0.$$

The orthogonal trajectories obtained by solving this equation (cf. Example 1 in Sec. 1.4) are the circles (Fig. 28)

$$(x - \tilde{c})^2 + y^2 = \tilde{c}^2.$$

Problems for Sec. 1.10

What families of curves are represented by the following equations? Sketch some of the curves.

1. $3y - x + c = 0$ **2.** $(x - c)^2 + y^2 = 1$ **3.** $xy = c$

4. $cx^2 + y^2 = 1$ **5.** $y + 2(x - c)^2 = 0$ **6.** $y^2 - (x - c)^3 = 0$

Represent the following families of curves in the form (1). Sketch some of the curves.

7. All nonvertical straight lines through the point $(4, -1)$.

8. The catenaries obtained by translating the catenary $y = \cosh x$ in the direction of the straight line $y = -x$.

9. All ellipses with foci -1 and 1 on the y-axis.

Represent the following families of curves by differential equations.

10. $xy = c$ **11.** $y = cx^3$ **12.** $y = e^{cx}$

13. $y = ce^x$ **14.** $y = \sin cx$ **15.** $c^2x^2 + y^2 = c^2$

Using differential equations, find the orthogonal trajectories of the following curves. Graph some of the curves and the trajectories.

16. $y = 2x + c$ **17.** $y = -\frac{1}{2}x^2 + c$ **18.** $y = \ln|x| + c$

19. $y = cx^3$ **20.** $xy = c$ **21.** $y = ce^x$

22. $x^2 + 2y^2 = c$ **23.** $y^2 - x^2 = c$ **24.** $y = c\sqrt{x}$

25. $y = ce^{x^2}$ **26.** $y = ce^{-x^2}$ **27.** $y = cx^{3/2}$

Find the orthogonal trajectories of the following curves. (Here, n is a positive integer.)

28. $x^2 + ny^2 = c$ **29.** $e^x \sin y = c$ **30.** $x^2 + y^2 = cx$

31. Show that (8) may be written in the form

$$\frac{dx}{dy} = -f(x, y).$$

Using this result, find the orthogonal trajectories of the curves $y = \sqrt{x + c}$.

32. Show that the orthogonal trajectories of a given family $g(x, y) = c$ can be obtained from the differential equation

$$\frac{dy}{dx} = \frac{\partial g/\partial y}{\partial g/\partial x}.$$

Using this equation, find the orthogonal trajectories of the curves $ye^{2x} = c$.

33. Show that if $u(x, y) = c = const$ represents a family of curves, a representation of its orthogonal trajectories of the form $v(x, y) = c^* = const$ can be obtained from

$$\frac{\partial u}{\partial x} = \frac{\partial v}{\partial y}, \quad \frac{\partial u}{\partial y} = -\frac{\partial v}{\partial x}.$$

(These so-called *Cauchy-Riemann differential equations* are basic in complex analysis and will be considered in Chap. 12.)

34. Using the result in Prob. 33, find the orthogonal trajectories of the curves $e^x \cos y = c$.

35. (Electric field) If an electrical current is flowing in a wire along the z-axis, the resulting electric lines of force in the xy-plane are the straight lines $y = cx$, and the equipotential lines are the orthogonal trajectories. Find the differential equation of these trajectories and solve it.

36. Experiments show that the lines of electric force of two opposite charges of the same strength at $(-1, 0)$ and $(1, 0)$ are the circles through $(-1, 0)$ and $(1, 0)$. Show that these circles have the representation $x^2 + (y - c)^2 = 1 + c^2$. Show that the equipotential lines (orthogonal trajectories) are the circles $(x + c^*)^2 + y^2 = c^{*2} - 1$, which are dashed in Fig. 29.

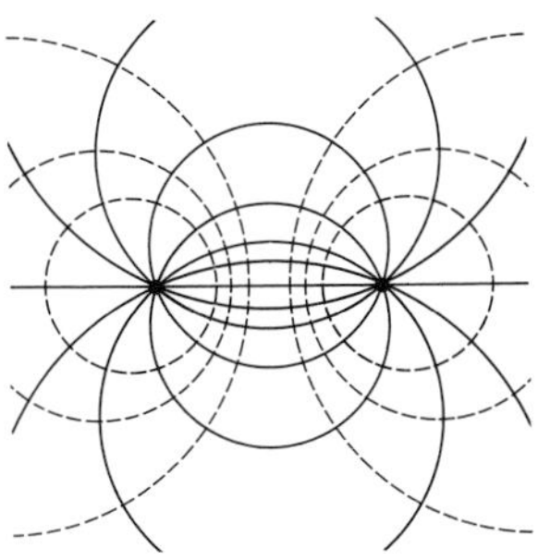

Fig. 29. Electric field in Problem 36

37. Starting from the equipotential lines in Prob. 36, regain the lines of electric force by the method discussed in this section.

38. (Temperature field) Curves of constant temperature $T(x, y) = const$ in a temperature field are called *isotherms*. Their orthogonal trajectories are the curves along which heat will flow (in regions that are free of sources or sinks of heat and are filled with a homogeneous medium). If the isotherms are given by $y^2 + 2xy - x^2 = c$, what are the curves of heat flow?

39. (Fluid flow) Another area in which orthogonal trajectories play a role is fluid flow. Here the path of a particle of fluid is called a *streamline* and the orthogonal trajectories of the streamlines are called *equipotential lines* (for reasons to be discussed in some other context, in Sec. 18.2). Suppose that the streamlines are $xy = c$. Show that the walls of the channel in Fig. 30 are streamlines, so that the flow may be regarded as a flow around a corner. Find and graph the equipotential lines.

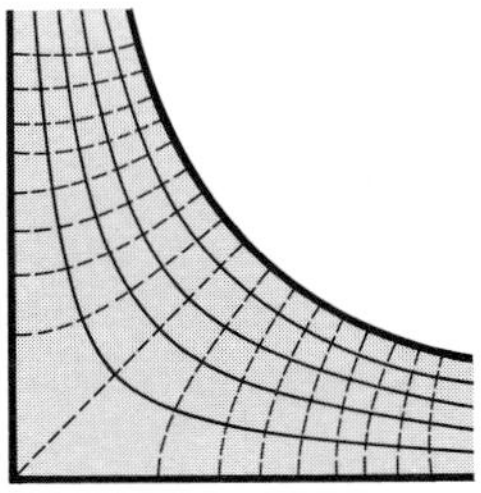

Fig. 30. Flow around a corner in Problem 39

40. Isogonal trajectories of a given family of curves are curves which intersect the given curves at a constant angle θ. Show that at each point the slopes m_1 and m_2 of the tangents to the corresponding curves satisfy the relation

$$\frac{m_2 - m_1}{1 + m_1 m_2} = \tan\theta = const.$$

Using this formula, determine the curves which intersect the circles $x^2 + y^2 = c$ at an angle of $45°$.

1.11 Picard's Iteration Method

There are various differential equations of the first order which cannot be solved by one of the standard methods discussed before or by another elementary method for obtaining exact solutions.[15] Such equations occur quite frequently in engineering problems. Then one may apply an approximation method for obtaining approximate solutions, for instance, a numerical method such as those to be discussed in Sec. 19.7. In practical problems, approximate solutions will often be sufficient.

In this section we shall consider an approximation method, which is called **Picard's iteration method**[16] and gives approximate solutions of an initial value problem which is of the form

$$y' = f(x, y), \qquad y(x_0) = y_0 \tag{1}$$

and is assumed to have a unique solution on some interval containing x_0. Picard's method is of great theoretical value, as we shall see in the next section. Its practical value is limited because it involves integrations which may sometimes be complicated.

The basic idea of Picard's method is very simple. By integration we see that (1) may be written in the form

$$y(x) = y_0 + \int_{x_0}^{x} f[t, y(t)]\, dt, \tag{2}$$

where t denotes the variable of integration. In fact, when $x = x_0$ the integral is zero and $y = y_0$, so that (2) satisfies the initial condition in (1); furthermore, by differentiating (2) we obtain the differential equation in (1).

To find approximations to the solution $y(x)$ of (2) we proceed as follows. We substitute the crude approximation $y = y_0 = const$ on the right; this yields the presumably better approximation

$$y_1(x) = y_0 + \int_{x_0}^{x} f(t, y_0)\, dt.$$

Then we substitute $y_1(x)$ in the same way to get

$$y_2(x) = y_0 + \int_{x_0}^{x} f[t, y_1(t)]\, dt,$$

etc. The nth step of this iteration yields

[15] Reference [B12] in Appendix 1 includes more than 1500 important differential equations and their solutions, arranged in systematic order and accompanied by numerous references to original literature.

[16] EMILE PICARD (1856–1941), French mathematician who made important contributions to the theory of complex analytic functions and differential equations.

$$y_n(x) = y_0 + \int_{x_0}^{x} f[t, y_{n-1}(t)]\, dt. \tag{3}$$

In this way we obtain a sequence of approximations

$$y_1(x), \qquad y_2(x), \ \cdots, \qquad y_n(x), \ \cdots,$$

and we shall see in the next section that the conditions under which this sequence converges to the solution $y(x)$ of (1) are relatively general.

An **iteration method** is a method that yields a sequence of approximations to an (unknown) function, say, $y_1, y_2, \cdots$, where the nth approximation, y_n, is obtained in the nth step by using one (or several) of the previous approximations, and the operation performed in each step is the same. In the simplest case, y_n is obtained from y_{n-1}; denoting the operation by T, we may write

$$y_n = T(y_{n-1}).$$

Picard's method is of this type, because (3) may be written

$$y_n(x) = T(y_{n-1}(x)) = y_0 + \int_{x_0}^{x} f(t, y_{n-1}(t))\, dt.$$

To illustrate the method, let us apply it to an equation which we can readily solve exactly, so that we may compare the approximations with the exact solution.

Example 1. Picard iteration

Find approximate solutions to the initial value problem

$$y' = 1 + y^2, \qquad y(0) = 0.$$

In this case, $x_0 = 0$, $y_0 = 0$, $f(x, y) = 1 + y^2$, and (3) becomes

$$y_n(x) = \int_0^x [1 + y_{n-1}^2(t)]\, dt = x + \int_0^x y_{n-1}^2(t)\, dt.$$

Starting from $y_0 = 0$, we thus obtain (cf. Fig. 31)

$$y_1(x) = x + \int_0^x 0 \cdot dt = x$$

$$y_2(x) = x + \int_0^x t^2\, dt = x + \tfrac{1}{3}x^3$$

$$y_3(x) = x + \int_0^x \left(t + \frac{t^3}{3}\right)^2 dt = x + \tfrac{1}{3}x^3 + \tfrac{2}{15}x^5 + \tfrac{1}{63}x^7$$

etc. Of course, we can obtain the exact solution of our present problem by separating variables (Example 3 in Sec. 1.3), finding

$$y(x) = \tan x = x + \tfrac{1}{3}x^3 + \tfrac{2}{15}x^5 + \tfrac{17}{315}x^7 + \cdots \qquad \left(-\frac{\pi}{2} < x < \frac{\pi}{2}\right). \tag{4}$$

The first three terms of $y_3(x)$ and the series in (4) are the same. The series in (4) converges for

T005456

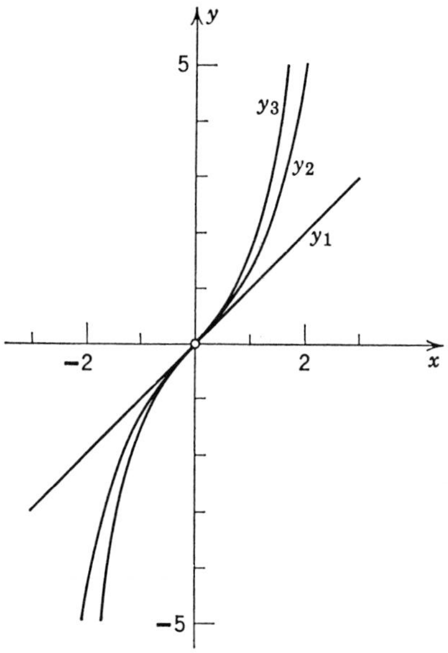

Fig. 31. Approximate solutions in Example 1

$|x| < \pi/2$, and all we may expect is that our sequence $y_1, y_2, \cdots$ converges to a function which is the solution of our problem for $|x| < \pi/2$. This illustrates that the problem of convergence is of practical importance.

Problems for Sec. 1.11

Apply Picard's method to the following initial value problems. Determine also the exact solution. Compare.

1. $y' = 2y,\ y(0) = 1$

2. $y' = xy,\ y(0) = 1$

3. $y' = y^2 + 4,\ y(0) = 0$

4. $y' = x + y,\ y(0) = -1$

5. $y' = x + y,\ y(0) = 1$

6. $y' = xy + 2x - x^3,\ y(0) = 0$

7. Apply Picard's method to $y' = y$, $y(0) = 1$, and show that the successive approximations tend to $y = e^x$, the exact solution.

8. Apply Picard's method to $y' = 2xy$, $y(0) = 1$. Graph y_1, y_2, y_3 and the exact solution for $0 \leqq x \leqq 2$.

9. In Prob. 8, compute the values $y_1(1)$, $y_2(1)$, $y_3(1)$ and compare them with the exact value $y(1) = e = 2.718 \cdots$.

10. Show that if f in (1) does not depend on y, then the approximations obtained by Picard's method are identical with the exact solution. Why?

1.12 Existence and Uniqueness of Solutions

The initial value problem

$$|y'| + |y| = 0, \qquad y(0) = 1$$

has no solution because $y \equiv 0$ is the only solution of the differential equation. (Why?) The initial value problem

$$y' = x, \qquad y(0) = 1$$

has precisely one solution, namely, $y = \frac{1}{2}x^2 + 1$. The initial value problem

$$xy' = y - 1, \qquad y(0) = 1$$

has infinitely many solutions, namely, $y = 1 + cx$ where c is arbitrary. From these examples we see that an initial value problem

$$y' = f(x, y), \qquad y(x_0) = y_0 \tag{1}$$

may have none, precisely one, or more than one solution. This leads to the following two fundamental questions.

Problem of existence. *Under what conditions does an initial value problem of the form (1) have at least one solution?*

Problem of uniqueness. *Under what conditions does that problem have a unique solution, that is, only one solution?*

Theorems which state such conditions are called **existence theorems** and **uniqueness theorems,** respectively.

Of course, our three examples are so simple that we can find the answer to those two questions by inspection, without using any theorems. However, it is clear that in more complicated cases—for example, when the equation cannot be solved by elementary methods—existence and uniqueness theorems will be of great importance. As a matter of fact, a more advanced course in differential equations consists mainly of considerations concerning the existence, uniqueness, and general behavior of solutions of various types of differential equations, and many of these considerations have far-reaching practical consequences. For example, results about the existence and uniqueness of periodic solutions of certain differential equations, which were obtained by Poincaré for entirely theoretical reasons about 90 years ago, are nowadays the base of many practical investigations in nonlinear mechanics.

Uniqueness is of importance, for instance, if we attempt to predict the future behavior of a physical system governed by an initial value problem. Our model may be complicated, so that we have to apply a numerical method for obtaining an approximate solution. But before doing so, we should make sure that the model will yield a unique solution.

We want to mention that in many branches of mathematics existence and uniqueness theorems are of similar importance. Let us illustrate this by a familiar example in linear algebra. The three systems of linear equations

$$\text{(a)}\quad \begin{aligned} x + y &= 1 \\ x + y &= 0 \end{aligned} \qquad \text{(b)}\quad \begin{aligned} x + y &= 1 \\ x - y &= 0 \end{aligned} \qquad \text{(c)}\quad \begin{aligned} x + y &= 1 \\ 2x + 2y &= 2 \end{aligned}$$

have, respectively, none, precisely one, and infinitely many solutions. This is immediately clear, and we need no theorems. However, if we have a system of many equations in many unknowns, there is no immediate answer to the question of whether the system has a solution at all and, if so, how many solutions there are. Then the need for existence and uniqueness theorems becomes obvious.

In connection with differential equations, a student who is exclusively interested in applications and does not like theoretical considerations (an attitude which will prevent him from having good success in the applied area) may be inclined to reason in the following manner. The physical problem corresponding to a certain differential equation has a unique solution, and the same must therefore be true for the differential equation. In simpler cases he may be right. However, he should keep in mind that the differential equation is merely an abstraction of the reality obtained by disregarding certain physical facts which seem to be of minor influence, and in complicated physical situations there may be no way of judging a priori the importance of various factors. In such a case there will then be no a priori guarantee that the differential equation is a faithful model, that is, leads to a faithful picture of the reality. This is part of the art of modeling: to develop a feeling for the relative importance of various factors in a given physical, chemical, biological or other system.

In the case of an initial value problem of the form

(1) $$y' = f(x, y), \qquad y(x_0) = y_0$$

there are simple conditions for the existence and uniqueness of the solution. If f is continuous in some region of the xy-plane containing the point (x_0, y_0), then the problem (1) has at least one solution. If, moreover, the partial derivative $\partial f/\partial y$ exists and is continuous in that region, then the problem (1) has precisely one solution. This solution can then be obtained by Picard's iteration method. Let us formulate these three statements in a precise manner.

Existence theorem

If $f(x, y)$ is continuous at all points (x, y) in some rectangle (Fig. 32)

$$R: \quad |x - x_0| < a, \qquad |y - y_0| < b$$

and bounded[17] *in R, say,*

[17] A function $f(x, y)$ is said to be **bounded** when (x, y) varies in a region in the xy-plane, if there is a number K such that $|f| \leqq K$ when (x, y) is in that region. For example, $f = x^2 + y^2$ is bounded, with $K = 2$ if $|x| < 1$ and $|y| < 1$. The function $f = \tan(x + y)$ is not bounded for $|x + y| < \pi/2$.

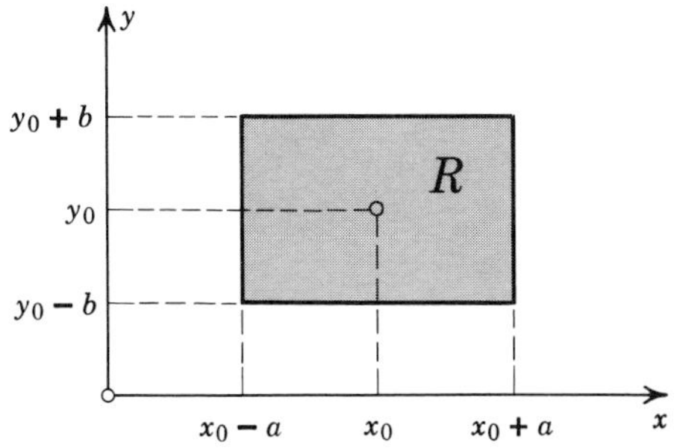

Fig. 32. Rectangle R in the existence and uniqueness theorems

$$(2) \qquad |f(x, y)| \leqq K \qquad \textit{for all } (x, y) \textit{ in } R,$$

then the initial value problem (1) *has at least one solution* $y(x)$, *which is defined at least for all* x *in the interval* $|x - x_0| < \alpha$ *where* α *is the smaller of the two numbers* a *and* b/K.

Uniqueness theorem

If $f(x, y)$ *and* $\partial f/\partial y$ *are continuous for all* (x, y) *in that rectangle* R *and bounded, say,*

$$(3) \qquad (a) \quad |f| \leqq K, \qquad (b) \quad \left|\frac{\partial f}{\partial y}\right| \leqq M \qquad \textit{for all } (x, y) \textit{ in } R,$$

then the initial value problem (1) *has only one solution* $y(x)$, *which is defined at least for all* x *in that interval* $|x - x_0| < \alpha$. *This solution can then be obtained by Picard's iteration method, that is, the sequence* $y_0, y_1, \cdots, y_n, \cdots$ *where*

$$y_n(x) = y_0 + \int_{x_0}^{x} f[t, y_{n-1}(t)]\, dt, \qquad n = 1, 2, \cdots,$$

converges to that solution $y(x)$.

Since the proofs of these theorems require familiarity with uniformly convergent series and other concepts to be considered later in this book, we shall not present these proofs at this time but refer the student to Ref. [B11] in Appendix 1. However, we want to include some remarks and examples which may be helpful for a good understanding of the two theorems.

Since $y' = f(x, y)$, the condition (2) implies that $|y'| \leqq K$, that is, the slope of any solution curve $y(x)$ in R is at least $-K$ and at most K. Hence a solution curve which passes through the point (x_0, y_0) must lie in the shaded region in Fig. 33 bounded by the lines l_1 and l_2 whose slopes are $-K$ and K, respectively. Depending on the form of R, two different cases may arise. In the first case, shown in Fig. 33a, we have $b/K \geqq a$ and therefore $\alpha = a$ in the existence theorem, which then asserts that the solution exists for all x between $x_0 - a$ and $x_0 + a$. In the second case, shown in Fig. 33b, we have $b/K < a$. Therefore $\alpha = b/K$, and all we can conclude from the theorems is that the solution exists

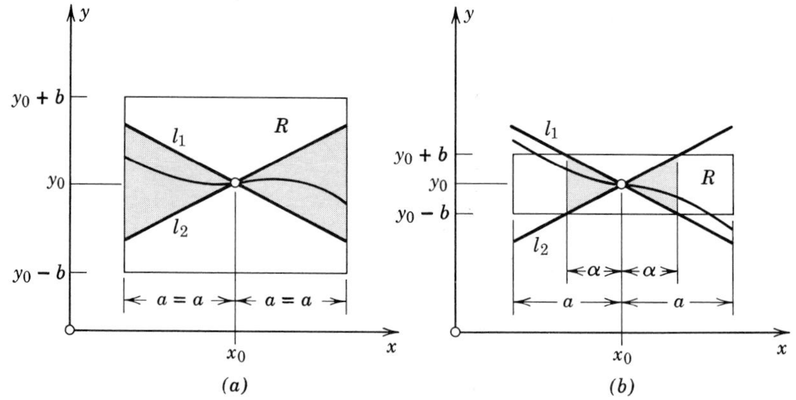

Fig. 33. The condition (2) of the existence theorem. (*a*) First case. (*b*) Second case

for all x between $x_0 - b/K$ and $x_0 + b/K$; for larger or smaller x's the solution curve may leave the rectangle R, and since nothing is assumed about f outside R, nothing can be concluded about the solution for those corresponding values of x.

Example 1

Consider the problem

$$y' = 1 + y^2, \qquad y(0) = 0$$

(cf. Example 1 in Sec. 1.11) and take R: $|x| < 5$, $|y| < 3$. Then

$$a = 5,\ b = 3,\ |f| = |1 + y^2| \leqq K = 10,\ |\partial f/\partial y| = 2\,|y| \leqq M = 6,\ \alpha = b/K = 0.3 < a.$$

In fact, the solution $y = \tan x$ of the problem is discontinuous at $x = \pm\pi/2$, and there is no continuous solution valid in the entire interval $|x| < 5$ from which we started. ∎

The conditions in the two theorems are sufficient conditions rather than necessary ones, and can be lessened. For example, by the mean value theorem of differential calculus we have

$$f(x, y_2) - f(x, y_1) = (y_2 - y_1)\left.\frac{\partial f}{\partial y}\right|_{y=\tilde{y}}$$

where (x, y_1) and (x, y_2) are assumed to be in R, and $\tilde{y}$ is a suitable value between y_1 and y_2. From this and (3*b*) it follows that

$$(4) \qquad |f(x, y_2) - f(x, y_1)| \leqq M\,|y_2 - y_1|,$$

and it can be shown that (3*b*) may be replaced by the weaker condition (4) which is known as a **Lipschitz condition.**[18] However, continuity of $f(x, y)$ is not enough to guarantee the *uniqueness* of the solution. This may be illustrated by the following example.

[18]RUDOLF LIPSCHITZ (1831–1903), German mathematician, who contributed also to algebra and differential geometry.

Example 2. Non-uniqueness

The initial value problem

$$y' = \sqrt{|y|}, \qquad y(0) = 0$$

has the two solutions

$$y \equiv 0 \qquad \text{and} \qquad y^* = \begin{cases} x^2/4 & \text{when } x \geqq 0 \\ -x^2/4 & \text{when } x \leqq 0 \end{cases}$$

although $f(x, y) = \sqrt{|y|}$ is continuous for all y. The Lipschitz condition (4) is violated in any region which includes the line $y = 0$, because for $y_1 = 0$ and positive y_2 we have

$$\text{(5)} \qquad \frac{|f(x, y_2) - f(x, y_1)|}{|y_2 - y_1|} = \frac{\sqrt{y_2}}{y_2} = \frac{1}{\sqrt{y_2}}, \qquad (\sqrt{y_2} > 0)$$

and this can be made as large as we please by choosing y_2 sufficiently small, whereas (4) requires that the quotient on the left-hand side of (5) does not exceed a fixed constant M. ∎

Numerical methods for first-order differential equations are presented in Sec. 19.7, which is independent of the other sections in Chap. 19 and may be studied now.

Problems for Sec. 1.12

1. Show that the differential equation $|y'| + |y| = -1$ has no solution.
2. Determine the largest set M in the xy-plane such that through each point of M there passes one and only one solution curve of $x\,dy = y\,dx$.
3. Find all solutions of the initial value problem $xy' = 4y$, $y(0) = 0$. Graph some of them.
4. Consider the differential equation $xy' = 2y$ and find all initial conditions $y(x_0) = y_0$ such that the resulting initial value problem has (a) no solution, (b) more than one solution, (c) precisely one solution.
5. How many solutions does the initial value problem $xy' = 3y$, $y(0) = 1$ have? Explain.
6. Work Prob. 4, for the differential equation $(x^2 - x)y' = (2x - 1)y$.
7. Show that if $y' = f(x, y)$ satisfies the assumptions of the present theorems in a rectangle R and y_1 and y_2 are two solutions of the equation whose curves lie in R, then these curves cannot have a point in common (unless they are identical).
8. If the assumptions of the present theorems are satisfied not merely in a rectangle R but even in a vertical strip given by $|x - x_0| \leqq a$, show that then the solution of (1) exists for all x in the interval $|x - x_0| \leqq a$.
9. It is worth noting that the solution of an initial value problem may exist in an interval larger than that in the theorem, the latter being $|x - x_0| < \alpha$ with α defined in the theorem. This is illustrated by Example 1 in the text. Consider that example. Find α in terms of general a and b. What is the maximum possible α that we can achieve by a suitable choice of a and b?

Find the largest α for which the present theorems guarantee the existence of the solution.

10. $y' = y^2$, $y(1) = 1$
11. $y' - y^2 = 4$, $y(0) = 0$

12. Show that $f(x, y) = |\sin y| + x$ satisfies a Lipschitz condition (4), with $M = 1$, on the whole xy-plane, but $\partial f/\partial y$ does not exist when $y = 0$.

13. Does $f(x, y) = |x| + |y|$ satisfy a Lipschitz condition in the xy-plane? Does $\partial f/\partial y$ exist?

14. Write the *linear* differential equation $y' + g(x)y = r(x)$ in the form (1). If g and r are continuous for all x such that $|x - x_0| \leqq a$, show that $f(x, y)$ in this equation satisfies a Lipschitz condition. [By our present theorems, this implies the uniqueness of solutions, which, of course, also follows from (4) in Sec. 1.7.]

15. Find all solutions of the initial value problem $y' = 2\sqrt{y}$, $y(1) = 0$. Which of them do we obtain by Picard's method if we start from $y_0 = 0$? Does $2\sqrt{y}$ satisfy a Lipschitz condition?

CHAPTER

2 ORDINARY LINEAR DIFFERENTIAL EQUATIONS

The ordinary differential equations may be divided into two large classes, the so-called linear equations and the nonlinear equations. While nonlinear equations (of the second and higher orders) are rather difficult, linear equations are much simpler in many respects, because various properties of their solutions can be characterized in a general way, and standard methods are available for solving many of these equations. In the present chapter we shall consider linear differential equations and their applications. These equations play an important role in engineering mathematics, for example, in connection with mechanical vibrations and electric circuits and networks.

Much space will be devoted to linear equations of the second order. This seems to be justified by the following facts. First of all, second-order equations are the most important ones from the practical point of view. Furthermore, the necessary theoretical considerations will become easier if we first concentrate upon second-order equations. Once the student understands what is needed to handle second-order equations, he will easily become familiar with the generalizations of the concepts, methods, and results to differential equations of the third and higher orders.

Important engineering applications in connection with mechanical and electrical oscillations will be considered in Secs. 2.6 and 2.13–2.15.

Numerical methods for solving second-order differential equations are included in Sec. 19.8, which is independent of the other sections in Chap. 19.

(Legendre's, Bessel's, and the hypergeometric equations will be considered in Chap. 4.)

Prerequisite for this chapter: Chap. 1, in particular Secs. 1.7 and 1.8.
Sections which may be omitted in a shorter course: 2.8–2.10, 2.15, 2.16.
References: Appendix 1, Part B.
Answers to problems: Appendix 2.

2.1 Homogeneous Linear Equations of the Second Order

The student has already met with linear differential equations of the first order (Secs. 1.7–1.9), and we shall now define and consider linear differential equations of the second order.

A second-order differential equation is said to be **linear** if it can be written in the form

$$y'' + f(x)y' + g(x)y = r(x). \tag{1}$$

The characteristic feature of this equation is that it is linear in the unknown function y and its derivatives, whereas f, g, and r may be any given functions of x. We shall see that (1) has somewhat simpler properties when the function $r(x)$ on the right is identically zero. Then (1) becomes simply

$$y'' + f(x)y' + g(x)y = 0 \tag{2}$$

and is said to be **homogeneous.** If $r(x) \not\equiv 0$, then (1) is said to be **nonhomogeneous.**

For example,

$$y'' + 4y = e^{-x} \sin x$$

is a nonhomogeneous linear differential equation, whereas

$$x^2y'' + xy' + (x^2 - 1)y = 0$$

is a homogeneous linear differential equation.

Any differential equation of the second order which cannot be written in the form (1) is said to be **nonlinear.** For example, nonlinear equations are

$$y''y + y' = 0$$

and

$$y'' = \sqrt{y'^2 + 1}.$$

Linear differential equations of the second order play a basic role in many engineering problems. We shall see that some of these equations are very simple because their solutions are elementary functions. Others are more complicated, their solutions being important higher functions, such as Bessel and hypergeometric functions.

The functions f and g in (1) and (2) are called the **coefficients** of the equations.

In all our considerations x will be assumed to vary in some arbitrary fixed range, for example, in some finite interval or on the entire x-axis. All the assumptions and statements will refer to such a fixed range, which need not be specified in each case.

A function

$$y = \phi(x)$$

is called a **solution** of a (linear or nonlinear) differential equation of the second order *on some interval* (perhaps infinite), if $\phi(x)$ is defined and twice differentiable throughout that interval and is such that the equation becomes an identity when we replace the unspecified function y and its derivatives by ϕ and its corresponding derivatives.

This definition is analogous to that in the case of first-order differential equations in Sec. 1.1.

Example 1

The functions $y = \cos x$ and $y = \sin x$ are solutions of the homogeneous linear differential equation

$$y'' + y = 0$$

for all x, since

$$(\cos x)'' + \cos x = -\cos x + \cos x = 0$$

and similarly for $y = \sin x$. We can even go an important step further. If we multiply the first solution by a constant, for example, by 3, the resulting function $y = 3 \cos x$ is also a solution because

$$(3 \cos x)'' + 3 \cos x = 3[(\cos x)'' + \cos x] = 0.$$

It is clear that instead of 3 we may choose any other constant, for example, -5 or $\frac{2}{9}$. We may even multiply $\cos x$ and $\sin x$ by different constants, say, by 2 and -8, respectively, and add the resulting functions, having

$$y = 2 \cos x - 8 \sin x,$$

and this function is also a solution of our homogeneous equation for all x, because

$$(2 \cos x - 8 \sin x)'' + 2 \cos x - 8 \sin x = 2[(\cos x)'' + \cos x] - 8[(\sin x)'' + \sin x] = 0. \quad ∎$$

This example illustrates the very important fact that we may obtain new solutions of the homogeneous linear equation (2) from known solutions by multiplication by constants and by addition. Of course, this is a great practical and theoretical advantage, because this property will enable us to generate more complicated solutions from simple solutions. We may characterize this basic property as follows.

Fundamental Theorem 1

If a solution of the ***homogeneous linear*** *differential equation* (2) *on some interval J is multiplied by any constant, the resulting function is also a solution of* (2) *on J. The sum of two solutions of* (2) *on J is also a solution of* (2) *on that interval.*[1]

Proof. We assume that $\phi(x)$ is a solution of (2) on some interval J and show that then $y = c\phi(x)$ is also a solution of (2) on J. If we substitute $y = c\phi(x)$ into (2), the left-hand side of (2) becomes

$$(c\phi)'' + f(c\phi)' + gc\phi = c[\phi'' + f\phi' + g\phi].$$

Since ϕ satisfies (2), the expression in the brackets is zero, and the first statement of the theorem is proved. The proof of the last statement is quite simple, too, and is left to the reader. ∎

The student should know this highly important theorem as well as he knows his multiplication table, but he should not forget that *the theorem* **does not hold**

[1] This theorem is sometimes called the **superposition principle** or *linearity principle*.

for nonhomogeneous linear equations or nonlinear equations, as may be illustrated by the following two examples.

Example 2. A nonhomogeneous linear equation
Substitution shows that the functions

$$y = 1 + \cos x \qquad \text{and} \qquad y = 1 + \sin x$$

are solutions of the nonhomogeneous linear equation

$$y'' + y = 1,$$

but the functions

$$2(1 + \cos x) \qquad \text{and} \qquad (1 + \cos x) + (1 + \sin x)$$

are not solutions of that equation.

Example 3. A nonlinear differential equation
Substitution shows that the functions

$$y = x^2 \qquad \text{and} \qquad y = 1$$

are solutions of the nonlinear differential equation

$$y''y - xy' = 0,$$

but the functions

$$-x^2 \qquad \text{and} \qquad x^2 + 1$$

are not solutions of that equation.

Problems for Sec. 2.1

Important general properties of homogeneous and nonhomogeneous linear differential equations. Prove the following statements, which refer to any fixed interval J, and illustrate them with examples. Here we assume $r(x) \not\equiv 0$ in (1).

1. $y \equiv 0$ is a solution of (2) (called the "*trivial solution*") but not of (1).

2. The sum of two solutions of (2) is a solution of (2).

3. A "linear combination" $y = c_1y_1 + c_2y_2$ of solutions y_1 and y_2 of (2) is a solution of (2).

4. A multiple $y = cy_1$ of a solution of (1) is **not** a solution of (1), unless $c = 1$.

5. The sum of two solutions of (1) is **not** a solution of (1).

6. The difference $y = y_1 - y_2$ of two solutions of (1) is a solution of (2).

7. The sum $y = y_1 + y_2$ of a solution y_1 of (1) and y_2 of (2) is a solution of (1).

8. (Second-order differential equations reducible to the first order) Certain second-order equations can be reduced to the first order. This holds for equations $F(x, y', y'') = 0$, which do not contain y explicitly. Show that by setting $y' = z$ we obtain a first-order equation in z and from a solution z a solution y of the given equation by integration.

Reduce to first order and solve:

9. $y'' + y' = x + 1$ **10.** $2xy'' = 3y'$ **11.** $y'' = y' \tanh x$

12. $y'' = 1 + y'^2$ **13.** $xy'' + 2y' = 0$ **14.** $xy'' + y' = y'^2$

15. A particle moves on a straight line so that its acceleration equals its velocity. At time $t = 0$ its distance from the origin is 1 meter and its velocity is 2 meters/sec. Find its position and velocity at $t = 5$ sec.

16. A particle moves on a straight line so that the product of its velocity and acceleration is constant, say, 1 meter^2/sec^3. If at $t = 0$ the distance from the origin is 2 meters and the velocity is 2 meters/sec, what are distance and velocity at $t = 6$?

17. Find the curve $y(x)$ through the origin for which $y'' = y'$ and the tangent at the origin is $y = x$.

18. **(Hanging cable)** It can be shown that the curve $y(x)$ of an inextensible flexible homogeneous cable hanging between two fixed points is obtained by solving $y'' = k\sqrt{1 + y'^2}$, where the constant k depends on the weight. This curve is called a *catenary* (from Latin *catena* = the chain). Find and graph $y(x)$, assuming $k = 1$ and those fixed points are $(-1, 0)$ and $(1, 0)$ in a vertical xy-plane.

19. **(Reduction to first order)** Another type of equations reducible to first order is $F(y, y', y'') = 0$, in which the independent variable x does not appear explicitly. Using the chain rule, show that $y'' = (dz/dy)z$, where $z = y'$, so that we obtain a first-order equation with y as the independent variable.

Reduce to first order and solve:

20. $yy'' + 3y'^2 = 0$ **21.** $y'' + e^{2y}y'^3 = 0$ **22.** $yy'' + (y + 1)y'^2 = 0$

23. Find the curve $y(x)$ through the origin for which $y'' = 12\sqrt{y}$ and the tangent at the origin is the x-axis.

24. Find the curve $y(x)$ through $(0, 0)$ and $(1, 1)$ for which $2y''y = y'^2$.

25. Given $\ddot{y} + a\dot{y}^2 + f(y) = 0$, where a is a constant. Show that the substitutions $\dot{y} = z$, $u = z^2$ lead to the linear first-order equation

$$\frac{du}{dy} + 2au + 2f(y) = 0.$$

2.2 Homogeneous Second-Order Equations with Constant Coefficients

We shall now consider linear differential equations of the second order whose coefficients are constant. These equations have important engineering applications, especially in connection with mechanical and electrical vibrations, as we shall see in Secs. 2.6, 2.13, and 2.14.

We start with homogeneous equations, that is, equations of the form

$$y'' + ay' + by = 0 \tag{1}$$

where a and b are constants. *We assume that a and b are real* and the range of x considered is the entire x-axis.

How to solve equation (1)? We remember that the solution of the *first-order* homogeneous linear equation with constant coefficients

$$y' + ky = 0$$

is an exponential function, namely,

$$y = ce^{-kx}.$$

We thus conjecture that

$$y = e^{\lambda x} \tag{2}$$

may be a solution of (1) if λ is properly chosen. Substituting (2) and its derivatives

$$y' = \lambda e^{\lambda x} \qquad \text{and} \qquad y'' = \lambda^2 e^{\lambda x}$$

into our equation (1), we obtain

$$(\lambda^2 + a\lambda + b)e^{\lambda x} = 0.$$

So (2) is a solution of (1), if λ is a solution of the quadratic eqution

$$\lambda^2 + a\lambda + b = 0. \tag{3}$$

This equation is called the **characteristic equation** (or *auxiliary equation*) of (1). Its roots are

$$\lambda_1 = \tfrac{1}{2}(-a + \sqrt{a^2 - 4b}), \qquad \lambda_2 = \tfrac{1}{2}(-a - \sqrt{a^2 - 4b}). \tag{4}$$

Our derivation shows that the functions

$$y_1 = e^{\lambda_1 x} \qquad \text{and} \qquad y_2 = e^{\lambda_2 x} \tag{5}$$

are solutions of (1). The student may check this result by substituting (5) into (1).

From elementary algebra we know that, since a and b are real, the characteristic equation may have

(Case I) *two distinct real roots,*

(Case II) *two complex conjugate roots, or*

(Case III) *a real double root.*

These cases will be discussed separately. For the time being let us illustrate each case by a simple example.

Example 1. Distinct real roots

Find solutions of the equation

$$y'' + y' - 2y = 0.$$

The characteristic equation is

$$\lambda^2 + \lambda - 2 = 0.$$

The roots are 1 and -2, and we obtain the two solutions

$$y_1 = e^x \qquad \text{and} \qquad y_2 = e^{-2x}.$$

Example 2. Complex conjugate roots
Find solutions of the equation

$$y'' + y = 0.$$

The characteristic equation is

$$\lambda^2 + 1 = 0.$$

The roots are i $(= \sqrt{-1})$ and $-i$, and we obtain the two solutions

$$y_1 = e^{ix} \qquad \text{and} \qquad y_2 = e^{-ix}.$$

In Sec. 2.4 we shall see that it is quite simple to obtain real solutions from such complex solutions.

Example 3. A real double root
Find solutions of the equation

$$y'' - 2y' + y = 0.$$

The characteristic equation

$$\lambda^2 - 2\lambda + 1 = 0$$

has the real double root 1, and we obtain at first only one solution

$$y_1 = e^x.$$

The case of a real double root will also be discussed in Sec. 2.4.

Problems for Sec. 2.2

Find solutions of the following equations.

1. $y'' - y = 0$ **2.** $y'' + 4y' + 3y = 0$ **3.** $y'' + y' - 2y = 0$
4. $y'' + 2y' = 0$ **5.** $y'' - y' = 0$ **6.** $y'' - 3y' + 2y = 0$
7. $y'' - 9y = 0$ **8.** $y'' + 6y' + 8y = 0$ **9.** $y'' + 4y' + 5y = 0$

10. Show that Cases I, II, and III occur if and only if $a^2 > 4b$, $a^2 < 4b$ and $a^2 = 4b$, respectively.

11. Show that $y = \cosh x$ and $y = \sinh x$ are solutions of $y'' - y = 0$. How can these be obtained from the answer to Prob. 1?

12. Show that $y = \cosh 3x$ and $y = \sinh 3x$ are solutions of $y'' - 9y = 0$ and how they can be obtained from the answer to Prob. 7.

13. Find solutions of $y'' + 4y' = 0$ (*a*) by the present method, (*b*) by reduction to the first order.

14. Show that a and b in (1) can be expressed in terms of λ_1 and λ_2 by the formulas $a = -\lambda_1 - \lambda_2$, $b = \lambda_1\lambda_2$.

Find a differential equation of the form (1) for which the following functions are solutions.

15. e^x, e^{-3x} **16.** e^{2ix}, e^{-2ix} **17.** $1, e^{2x}$
18. e^{kx}, e^{lx} **19.** $e^{(-1+2i)x}, e^{(-1-2i)x}$ **20.** $e^{-(\alpha+i\omega)x}, e^{-(\alpha-i\omega)x}$

2.3 General Solution. Basis. Initial Value Problem

In our further discussion of linear equations with constant coefficients, it will be necessary to know about some important general properties of solutions of linear differential equations. For this purpose we first introduce the following two concepts.

A solution of a differential equation of the second order (linear or not) is called a **general solution** if it contains two arbitrary[2] independent constants. Here *independence* means that the solution cannot be reduced to a form containing only one arbitrary constant or none. If we assign definite values to those two constants, then the solution so obtained is called a **particular solution.**

These notions are obvious generalizations of those in the case of first-order differential equations (Sec. 1.1).

It will be shown later (in Secs. 2.8 and 2.11) that if a differential equation of the second order is linear (homogeneous or not), then a general solution includes *all* solutions of the equation; any solution containing no arbitrary constants can be obtained by assigning definite values to the arbitrary constants in that general solution (cf. also the general remark in Sec. 1.1, p. 5).

In the present section we consider the homogeneous linear equation

$$y'' + f(x)y' + g(x)y = 0, \tag{1}$$

and we want to show that a general solution of such an equation can be readily obtained if two suitable solutions y_1 and y_2 are known.

Clearly, if $y_1(x)$ and $y_2(x)$ are solutions of (1) on some interval I, then, by Fundamental Theorem 1 in Sec. 2.1,

$$y(x) = c_1y_1(x) + c_2y_2(x) \tag{2}$$

is a solution of (1) on I; here, c_1 and c_2 are arbitrary constants. Hence, by definition, y is a general solution of (1) on I, provided y cannot be reduced to an expression containing fewer than two arbitrary constants.

Under what conditions will such a reduction be possible? To answer this question, we first introduce the following basic concepts, which are important in other considerations, too.

Two functions $y_1(x)$ and $y_2(x)$ are said to be **linearly dependent** *on an open interval* I where both functions are defined, if they are proportional on I, that is, if[3]

$$\text{(a) } y_1 = ky_2 \qquad \text{or} \qquad \text{(b) } y_2 = ly_1 \tag{3}$$

[2]The range of the constants may have to be restricted in some cases to avoid imaginary expressions or other degeneracies.

[3]Note that if (3a) holds for a $k \neq 0$, we may divide by k, finding $y_2 = y_1/k$, so that in this case (3a) implies (3b) (with $l = 1/k$). However, if $y_1 \equiv 0$, then (3a) holds for $k = 0$, but does not imply (3b); in fact, in this case (3b) will not hold unless $y_2 \equiv 0$, too. (An interval I is said to be **open** if its two endpoints are not regarded as points belonging to I. It is said to be **closed** if the endpoints are regarded as points of I.)

holds for all x on I; here k and l are numbers, zero or not. If the functions are not proportional on I, they are said to be **linearly independent** *on* I.

Note that these concepts of linear dependence and independence of functions refer to an infinite set (the points of I), but not merely to a single point.

If at least one of the functions y_1 and y_2 is identically zero on I, then the functions are linearly dependent on I. In any other case the functions are linearly dependent on I if and only if the quotient y_1/y_2 is constant on I. Hence if y_1/y_2 depends on x on I, then y_1 and y_2 are linearly independent on I.

In fact, if $y_1 \equiv 0$, then (3a) holds with $k = 0$; and if $y_2 \equiv 0$, then (3b) holds with $l = 0$. This proves the first statement, and the other statements are immediate consequences of (3), too. Of course, in the last statement, we admit that y_1/y_2 may become infinite, namely, for those x for which $y_2 = 0$ (or even undefined at a point if y_1 and y_2 have a common zero).

Example 1

The functions

$$y_1 = 3x \qquad \text{and} \qquad y_2 = x$$

are linearly dependent on any interval, because $y_1/y_2 = 3 = \textit{const}$. The functions

$$y_1 = x^2 \qquad \text{and} \qquad y_2 = x$$

are linearly independent on any interval because $y_1/y_2 = x \neq \textit{const}$, and so are the functions

$$y_1 = x \qquad \text{and} \qquad y_2 = x + 1.$$

∎

If the solutions y_1 and y_2 in (2) are linearly dependent on I, then (3a) or (3b) holds and we see that we may then reduce (2) to one of the forms

$$y = Ay_2 \qquad (A = c_1k + c_2)$$

and

$$y = By_1 \qquad (B = c_1 + lc_2)$$

containing only one arbitrary constant. Hence in this case (2) is certainly not a general solution of (1) on I.

On the other hand, if y_1 and y_2 are linearly independent on I, then they are not proportional, and no such reduction can be performed. Hence, in this case, (2) is a general solution of (1) on I.

Two linearly independent solutions of (1) on I are called a **basis**[4] or a **fundamental system** of solutions of (1) on I. Using this concept, we may formulate our result as follows.

Theorem 1 (General solution, basis)

The solution

$$y(x) = c_1y_1(x) + c_2y_2(x) \qquad (c_1,\ c_2\ \textit{arbitrary})$$

[4]Readers familiar with the notion of a vector space (Sec. 6.4) will notice that, by Fundamental Theorem 1 in Sec. 2.1, the solutions of (1) on I form a vector space, and a basis y_1, y_2 is a basis of that vector space in the sense of linear algebra.

is a general solution of the differential equation (1) *on an interval I of the x-axis if and only if the functions* y_1 *and* y_2 *constitute a basis of solutions of* (1) *on I.*

y_1 *and* y_2 *constitute such a basis if and only if their quotient*[5] y_1/y_2 *is not constant on I but depends on x.*

Example 2

The functions

$$y_1 = e^x \quad \text{and} \quad y_2 = e^{-2x}$$

considered in Example 1, Sec. 2.2, are solutions of

$$y'' + y' - 2y = 0.$$

Since y_1/y_2 is not constant, these solutions constitute a basis, and the corresponding general solution for all x is

$$y = c_1y_1 + c_2y_2 = c_1e^x + c_2e^{-2x}.$$

Example 3

The functions

$$y_1 = e^x \quad \text{and} \quad y_2 = 3e^x$$

are solutions of the equation in Example 2, Sec. 2.2, but since their quotient is constant, they are linearly dependent, so that they do not form a basis. ∎

In the case of a *first-order* equation, a general solution contained *one* arbitrary constant and we needed *one* condition for obtaining a particular solution. In the present case we have *two* arbitrary constants and need *two* conditions. In many applications these conditions are of the type

$$y(x_0) = K, \qquad y'(x_0) = L, \tag{4}$$

where $x = x_0$ is a given point and K and L are given numbers. Hence we are looking for a particular solution of (1) which at x_0 has the value K and the value of the derivative L. The conditions in (4) are called **initial conditions.** Equation (1) and conditions (4) together constitute what is known as an **initial value problem.**

Let us illustrate the matter by a typical example.

Example 4. Initial value problem

Solve

$$y'' + y' - 2y = 0, \qquad y(0) = 4, \qquad y'(0) = 1.$$

A general solution of the equation is (cf. Example 2)

$$y(x) = c_1e^x + c_2e^{-2x}.$$

Hence $y(0) = c_1 + c_2 = 4$ by the first initial condition. Differentiation gives

$$y'(x) = c_1e^x - 2c_2e^{-2x}.$$

[5] Of course, here we exclude the trivial case $y_2 \equiv 0$ in which y_1 and y_2 do not constitute a basis, and we also admit that y_1/y_2 may become infinite (namely, for those values of x for which $y_2 = 0$) or even undefined if y_1 and y_2 are both zero at some x.

Hence $y'(0) = c_1 - 2c_2 = 1$ by the second initial condition. Together, $c_1 = 3$, $c_2 = 1$. The solution is

$$y(x) = 3e^x + e^{-2x}.$$

We mention that some applications lead to another type of conditions, namely,

$$y(A) = k_1, \qquad y(B) = k_2. \tag{5}$$

These are called *boundary conditions* since they refer to the endpoints A, B (*boundary points* A, B) of an interval I. The task is to determine a solution of (1) on I satisfying (5). Equation (1) and conditions (5) constitute what is called a **boundary value problem.**

For instance, a boundary value problem is

$$y'' + y = 0, \qquad y(0) = 3, \qquad y(\pi) = -3.$$

A solution is

$$y = 3 \cos x + c_2 \sin x.$$

Here c_2 is still arbitrary. This is a surprise. Of course, the reason is that $\sin x$ is zero at 0 and π. The reader may conclude and prove (Prob. 33) that the solution of a boundary value problem (1), (5) is unique if and only if no solution $y \not\equiv 0$ of (1) satisfies $y(A) = y(B) = 0$.

We are now prepared for continuing our discussion of differential equations with constant coefficients, and shall, therefore, postpone further theoretical considerations until Sec. 2.8.

Problems for Sec. 2.3

Are the following functions linearly dependent or independent on the given interval?

1. e^x, e^{-x}, any interval

2. $e^{\lambda_1 x}$, $e^{\lambda_2 x}$, $\lambda_1 \neq \lambda_2$, any interval

3. x, x^2 $(-1 < x < 1)$

4. x, $x + 1$ $(0 < x < 1)$

5. $x^2 - 2$, $4 - 2x^2$ $(0 < x < 1)$

6. $2 \sin x$, $\sin 2x$ $(0 < x < 2\pi)$

7. $\sin 2x$, $\cos 2x$ $(0 < x < \pi)$

8. $\sin 2x$, $\cos (2x + \pi/2)$ $(x > 0)$

9. $|x|x$, x^2 $(0 < x < 1)$

10. $|x|x$, x^2 $(-1 < x < 1)$

11. $\ln x$, $\ln x^2$ $(x > 1)$

12. $\ln 2x$, $2 \ln x$ $(x > 1)$

13. $e^{\lambda x}$, $xe^{\lambda x}$, any interval

14. $\sin 2x$, $\sin x \cos x$, any interval

Prove and illustrate with simple examples and graphs:

15. If y and w are linearly independent on an interval I, then y and w are linearly independent on any interval containing I on which y and w are defined.

16. y and w in Prob. 15 may be linearly dependent on a subinterval of I.

17. If y and w are linearly dependent on an interval I, they are linearly dependent on any subinterval of I.

18. y and w in Prob. 17 may be linearly independent on an interval containing I.

Find a general solution.

19. $y'' - 25y = 0$

20. $y'' - 3y' + 2y = 0$

21. $y'' - 3y' = 0$

22. $4y'' + y = 0$

23. $y'' + 2y' - 15y = 0$

24. $y'' + y' - 6y = 0$

Solve the following initial value problems.

25. $y'' - 4y' + 3y = 0$, $y(0) = -1$, $y'(0) = 1$

26. $y'' + 2y' = 0$, $y(0) = 1$, $y'(0) = 2$

27. $y'' - y = 0$, $y(0) = 3$, $y'(0) = 0$

28. $y'' + y' - 2y = 0$, $y(0) = 4$, $y'(0) = -8$

Solve the following boundary value problems.

29. $y'' - 4y = 0$, $y(-2) = y(2) = \cosh 4$

30. $y'' - 2y' = 0$, $y(0) = 0$, $y(\frac{1}{2}) = e - 1$

31. $3y'' - 8y' - 3y = 0$, $y(-3) = e$, $y(3) = 1/e$

32. $y'' - y' - 2y = 0$, $y(0) = 3$, $y(1) = 2e^2 + 1/e$

33. Show that the solution of a boundary value problem (1), (5) is unique if and only if no solution $y \not\equiv 0$ of (1) satisfies $y(A) = y(B) = 0$.

34. Give a detailed proof of the fact that if y_1 and y_2 are linearly independent on an interval I, then $c_1y_1 + c_2y_2$ with arbitrary c_1 and c_2 cannot be reduced to a form containing fewer than two arbitrary constants.

35. If y_1, y_2 is a basis for (1) on an interval I, show that $y_3 = y_1 + y_2$, $y_4 = y_1 - y_2$ is a basis for (1) on I. Give examples.

36. If y_1, y_2 is a basis for (1) on an interval I, show that $y_3 = \alpha y_1 + \beta y_2$, $y_4 = \gamma y_1 + \delta y_2$ is a basis for (1) on I if and only if $\alpha\delta \neq \beta\gamma$.

Find a differential equation $y'' + ay' + by = 0$ for which the given functions are a basis. Test linear independence by means of Theorem 1.

37. $e^{i\omega x}$, $e^{-i\omega x}$ $(\omega \neq 0)$

38. 1, e^{-4x}

39. $e^{(-3+i)x}$, $e^{(-3-i)x}$

40. $e^{(2+i)x}$, $e^{(2-i)x}$

2.4 Real Roots, Complex Roots, Double Root of the Characteristic Equation

From Sec. 2.2 we recall that a homogeneous second-order equation with real constant coefficients a, b is of the form

$$y'' + ay' + by = 0. \qquad (1)$$

A function

$$y = e^{\lambda x} \qquad (2)$$

is a solution of (1) if λ is a root of the characteristic equation

$$\lambda^2 + a\lambda + b = 0. \tag{3}$$

These roots are

$$\lambda_1 = \tfrac{1}{2}(-a + \sqrt{a^2 - 4b}), \qquad \lambda_2 = \tfrac{1}{2}(-a - \sqrt{a^2 - 4b}). \tag{4}$$

Since a and b are real, the characteristic equation may have

(Case I) *two distinct real roots,*

(Case II) *two complex conjugate roots, or*

(Case III) *a real double root.*

This is the point we reached in Sec. 2.2. We can now discuss these cases separately.

Case I. Two distinct real roots. A basis on any interval is

$$y_1 = e^{\lambda_1 x}, \qquad y_2 = e^{\lambda_2 x}.$$

Indeed, linear independence follows from the fact that y_1/y_2 is not constant on any interval. The corresponding general solution is

$$y = c_1 e^{\lambda_1 x} + c_2 e^{\lambda_2 x}. \tag{5}$$

Example 1. General solution in the case of distinct real roots
Solve

$$y'' + y' - 2y = 0.$$

The characteristic equation $\lambda^2 + \lambda - 2 = 0$ has the roots 1 and -2, so that we obtain the general solution

$$y = c_1 e^x + c_2 e^{-2x}.$$

Case II. Complex roots. Complex roots must be conjugate, say,

$$\lambda_1 = p + iq, \qquad \lambda_2 = p - iq$$

where p and q are real and $q \neq 0$. This follows from the assumption that a and b are real, as we know from algebra. Hence we first obtain the basis

$$y_1 = e^{(p+iq)x}, \qquad y_2 = e^{(p-iq)x}$$

consisting of two *complex* functions. It is of great practical interest to obtain real solutions from these complex solutions.

This we can do, once and forever, by applying the Euler formulas[6]

[6] Students not familiar with these important formulas from elementary calculus will find a proof in Sec. 12.7. Of course, once we obtained (6) we may verify directly that it satisfies (1) in the present case.

$$e^{i\theta} = \cos\theta + i\sin\theta, \qquad e^{-i\theta} = \cos\theta - i\sin\theta,$$

taking $\theta = qx$. In fact, from the first Euler formula we obtain

$$y_1 = e^{(p+iq)x} = e^{px}e^{iqx} = e^{px}(\cos qx + i\sin qx),$$

and from the second Euler formula,

$$y_2 = e^{(p-iq)x} = e^{px}e^{-iqx} = e^{px}(\cos qx - i\sin qx).$$

From these two expressions for y_1 and y_2 we now derive by addition and subtraction

$$\tfrac{1}{2}(y_1 + y_2) = e^{px}\cos qx,$$

$$\frac{1}{2i}(y_1 - y_2) = e^{px}\sin qx.$$

The two functions on the right are real-valued, and from Fundamental Theorem 1 in Sec. 2.1 we conclude that they are solutions of the differential equation (1). Since their quotient is not constant on any interval, they are linearly independent on any interval. Hence in the case of complex roots a basis on any interval is

$$e^{px}\cos qx, \qquad e^{px}\sin qx.$$

The corresponding general solution is

$$y(x) = e^{px}(A\cos qx + B\sin qx) \tag{6}$$

where A and B are arbitrary constants.

Example 2. General solution in the case of complex conjugate roots

Find a general solution of the equation

$$y'' - 2y' + 10y = 0.$$

The characteristic equation

$$\lambda^2 - 2\lambda + 10 = 0$$

has the roots

$$\lambda_1 = p + iq = 1 + 3i \qquad \text{and} \qquad \lambda_2 = p - iq = 1 - 3i.$$

Thus $p = 1$, $q = 3$. This yields the basis

$$e^x\cos 3x, \qquad e^x\sin 3x.$$

The corresponding general solution is

$$y = e^x(A\cos 3x + B\sin 3x).$$

Example 3. An initial value problem
Solve the initial value problem

$$y'' - 2y' + 10y = 0, \qquad y(0) = 4, \qquad y'(0) = 1.$$

The equation is the same as in the previous example. Hence we can take the general solution just obtained and its derivative

$$y'(x) = e^{x}(A \cos 3x + B \sin 3x - 3A \sin 3x + 3B \cos 3x).$$

From y, y' and the initial conditions,

$$y(0) = A = 4$$
$$y'(0) = A + 3B = 1.$$

Hence $A = 4$, $B = -1$, and the answer is

$$y = e^{x}(4 \cos 3x - \sin 3x).$$

Example 4
A general solution of the equation

$$y'' + \omega^2 y = 0 \qquad (\omega \text{ constant, not zero})$$

is

$$y = A \cos \omega x + B \sin \omega x.$$

Case III. Double root. This is sometimes called the **critical case.** From (4) we see that it arises if and only if the discriminant of the characteristic equation (3) is zero, that is,

$$a^2 - 4b = 0. \qquad \text{Then} \qquad b = \tfrac{1}{4}a^2.$$

Hence (1) now takes the form

$$y'' + ay' + \tfrac{1}{4}a^2 y = 0. \tag{7}$$

The double root is $\lambda = -a/2$, and we have at first only one solution

$$y_1 = e^{\lambda x} \qquad \left(\lambda = -\frac{a}{2}\right).$$

To find another solution y_2 we may apply the method of variation of parameters (cf. Sec. 1.8), starting from

$$y_2(x) = u(x)y_1(x) \qquad \text{where} \qquad y_1(x) = e^{-ax/2}.$$

We have to determine u. We substitute y_2 and its derivatives into (7) and collect terms, obtaining

$$u(y_1'' + ay_1' + \tfrac{1}{4}a^2 y_1) + u'(2y_1' + ay_1) + u''y_1 = 0.$$

Since y_1 is a solution, the expression in the first parentheses is zero. Since

$$2y_1' = 2\left(-\frac{a}{2}\right)e^{-ax/2} = -ay_1,$$

the expression in the second parentheses is zero, and our equation reduces to $u''y_1 = 0$. Hence $u'' = 0$. A solution is $u = x$. Consequently,

$$y_2(x) = xe^{\lambda x} \qquad \left(\lambda = -\frac{a}{2}\right).$$

y_1 and y_2 are linearly independent on any interval since their quotient is $y_2/y_1 = x$, which is not constant. Hence in the case of a double root a basis of solutions of (1) on any interval is

$$e^{\lambda x}, \qquad xe^{\lambda x} \qquad \left(\lambda = -\frac{a}{2}\right).$$

The corresponding general solution is

$$\textbf{(8)} \qquad y = (c_1 + c_2x)e^{\lambda x} \qquad \left(\lambda = -\frac{a}{2}\right).$$

Note that if λ is a *simple* root of (3), then (8) is **not** a solution of (1).

Example 5. General solution in the case of a double root
Solve

$$y'' + 8y' + 16y = 0.$$

The characteristic equation has the double root $\lambda = -4$. Hence a basis is

$$e^{-4x} \quad \text{and} \quad xe^{-4x}$$

and the corresponding general solution is

$$y = (c_1 + c_2x)e^{-4x}.$$

Example 6. An initial value problem
Solve the initial value problem

$$y'' - 4y' + 4y = 0, \qquad y(0) = 3, \qquad y'(0) = 1.$$

A general solution of the differential equation is

$$y(x) = (c_1 + c_2x)e^{2x}.$$

By differentiation we obtain

$$y'(x) = c_2e^{2x} + 2(c_1 + c_2x)e^{2x}.$$

From this and the initial conditions it follows that

$$y(0) = c_1 = 3, \qquad y'(0) = c_2 + 2c_1 = 1.$$

Hence $c_1 = 3$, $c_2 = -5$, and the answer is

$$y = (3 - 5x)e^{2x}.$$ ∎

This completes the discussion of all three cases, and we may sum up our result in tabular form:

Case	Roots of (3)	Basis of (1)	General solution of (1)
I	Distinct real λ_1, λ_2	$e^{\lambda_1 x}, e^{\lambda_2 x}$	$y = c_1 e^{\lambda_1 x} + c_2 e^{\lambda_2 x}$
II	Complex conjugate $\lambda_1 = p + iq, \lambda_2 = p - iq$	$e^{px} \cos qx$ $e^{px} \sin qx$	$y = e^{px}(A \cos qx + B \sin qx)$
III	Real double root $\lambda\ (= -a/2)$	$e^{\lambda x}, xe^{\lambda x}$	$y = (c_1 + c_2 x)e^{\lambda x}$

Problems for Sec. 2.4

Verify that the following functions are solutions of the given differential equations and obtain from them a real-valued general solution of the form (6).

1. $y = c_1 e^{i\omega x} + c_2 e^{-i\omega x}, \quad y'' + \omega^2 y = 0$

2. $y = c_1 e^{4ix} + c_2 e^{-4ix}, \quad y'' + 16y = 0$

3. $y = c_1 e^{-(2+2i)x} + c_2 e^{-(2-2i)x}, \quad y'' + 4y' + 8y = 0$

4. $y = c_1 e^{-(\alpha+i)x} + c_2 e^{-(\alpha-i)x}, \quad y'' + 2\alpha y' + (\alpha^2 + 1)y = 0$

5. $y = c_1 e^{(1+3i)x} + c_2 e^{(1-3i)x}, \quad y'' - 2y' + 10y = 0$

State whether the given equation corresponds to Case I, Case II, or Case III and find a general solution involving real-valued functions.

6. $y'' - 2y' - 3y = 0$

7. $y'' + \pi^2 y = 0$

8. $y'' - 2y' + y = 0$

9. $y'' + 6y' + 9y = 0$

10. $y'' - 6y' + 25y = 0$

11. $2y'' + 3y' - 2y = 0$

12. $y'' + 2y' + (\pi^2 + 1)y = 0$

13. $4y'' - 4y' + y = 0$

14. $y'' + 2ky' + k^2 y = 0$

15. $8y'' - 2y' - y = 0$

16. Assuming that (6) is a solution of (1), express a and b in terms of p and q.

Find a differential equation of the form (1) such that y is a general solution.

17. $y = e^{-x}(A \cos x + B \sin x)$

18. $y = e^{-3x}(A \cos \sqrt{3}x + B \sin \sqrt{3}x)$

19. $y = (c_1 + c_2 x)e^{4x}$

20. $y = (c_1 + c_2 x)e^{-x/2}$

21. $y = A \cosh 3x + B \sinh 3x$

22. $y = e^{-x/2}(A \cos 2x + B \sin 2x)$

23. $y = c_1 + c_2 e^{4x}$

24. $y = (c_1 + c_2 x)e^{kx}$

25. Apply the method of variation of parameters to $y'' - 2y' + y = 0$, using $y_1 = e^x$.

26. Apply the method of variation of parameters to $y'' + 10y' + 25y = 0$, using $y_1 = e^{-5x}$.

27. Verify directly that in the case of a double root, $xe^{\lambda x}$ with $\lambda = -a/2$ is a solution of (1).

28. Verify that if (1) does not have a double root, then $xe^{\lambda x}$ is **not** a solution of (1).

Find a differential equation (1) for which the given y is a general solution and determine the constants so that the given initial conditions are satisfied.

29. $y = e^{-2x}(A \cos x + B \sin x), \quad y(0) = 0, \quad y'(0) = -3$

30. $y = (c_1 + c_2 x)e^{2x}, \quad y(-1) = 0, \quad y'(-1) = 1/e^2$

31. $y = e^x(A \cos \pi x + B \sin \pi x), \quad y(0) = 1, \quad y'(0) = 1 - \pi$

32. $y = c_1 e^{-2x} + c_2 e^{-3x}, \quad y(0) = 2, \ y'(0) = -3$

Solve the following initial value problems.

33. $y'' - 4y = 0, \quad y(0) = 2, \quad y'(0) = 4$

34. $y'' + 4y = 0, \quad y(0) = 0, \quad y'(0) = 6$

35. $y'' - 6y' + 9y = 0, \quad y(0) = 2, \quad y'(0) = 8$

36. $y'' + 2y' + y = 0, \quad y(0) = 1, \quad y'(0) = -2$

37. $y'' + 4y' + 5y = 0, \quad y(0) = 1, \quad y'(0) = -3$

38. $4y'' + 4y' + y = 0, \quad y(0) = -2, \quad y'(0) = 1$

39. $y'' - 3y' + 2y = 0, \quad y(0) = -1, \quad y'(0) = 0$

40. Consider Example 3 and start from another general solution, say,

$$y(x) = c_1 e^{(1+3i)x} + c_2 e^{(1-3i)x}.$$

Show that from the initial conditions we obtain $c_1 = 2 + i/2$, $c_2 = 2 - i/2$ and the resulting particular solution may be simplified so that it becomes identical with that in Example 3. This illustrates that the choice of a general solution is not essential.

2.5 Differential Operators

By an *operator* we mean a transformation which transforms a function into another function. Operators and corresponding techniques ("*operational methods*") play an increasing role in engineering mathematics.

Differentiation suggests an operator as follows. Let D denote differentiation with respect to x, that is, write

$$Dy = y'.$$

D is an operator; it transforms y (assumed differentiable) into its derivative y'. For example,

$$D(x^2) = 2x, \qquad D(\sin x) = \cos x.$$

Applying D twice, we obtain the second derivative $D(Dy) = Dy' = y''$. We simply write $D(Dy) = D^2y$, so that

$$Dy = y', \qquad D^2y = y'', \qquad D^3y = y''', \qquad \cdots.$$

More generally,

$$L = P(D) = D^2 + aD + b \tag{1}$$

is called a *second-order differential operator*. Here a and b are constant. P suggests "polynomial." L suggests "linear" (explained below). When L is applied to a function y (assumed twice differentiable), it produces

$$L[y] = (D^2 + aD + b)y = y'' + ay' + by. \tag{2}$$

L is a *linear* operator. By definition this means that

$$L[\alpha y + \beta w] = \alpha L[y] + \beta L[w]$$

for any constants α and β and any functions y and w to which L is applicable.

The homogeneous linear second-order differential equation $y'' + ay' + by = 0$ may now be simply written

$$L[y] = P(D)[y] = 0. \tag{3}$$

For example,

$$L[y] = (D^2 + D - 6)y = y'' + y' - 6y = 0. \tag{4}$$

Since

$$D[e^{\lambda x}] = \lambda e^{\lambda x}, \qquad D^2[e^{\lambda x}] = \lambda^2 e^{\lambda x},$$

we have from (2) and (3)

$$P(D)[e^{\lambda x}] = (\lambda^2 + a\lambda + b)e^{\lambda x} = P(\lambda)e^{\lambda x} = 0. \tag{5}$$

This confirms our result of the last section that $e^{\lambda x}$ is a solution of (3) if and only if λ is a solution of the characteristic equation $P(\lambda) = 0$. If $P(\lambda)$ has two different roots, we obtain a basis. If $P(\lambda)$ has a double root, we need a second independent solution. Differentiating (5) with respect to λ and interchanging differentiation with respect to λ and x, we obtain

$$P(D)[xe^{\lambda x}] = P'(\lambda)e^{\lambda x} + P(\lambda)xe^{\lambda x} = 0$$

where $P' = dP/d\lambda$. For a double root, $P(\lambda) = P'(\lambda) = 0$. This suggests that the desired second solution is $xe^{\lambda x}$, a result which is confirmed by substitution.

$P(\lambda)$ is a polynomial in λ, in the ordinary sense of algebra. Replacing λ by D, we have the "operator polynomial" $P(D)$. The point of this "operational calculus" is that $P(D)$ can be treated just like an algebraic quantity. In particular, we can factor it. To illustrate this, consider

$$(D + 3)(D - 2) = D^2 + D - 6$$

appearing in (4). By definition, $(D - 2)y = y' - 2y$. Hence

$$\begin{aligned}(D + 3)(D - 2)y &= (D + 3)[y' - 2y] \\ &= y'' - 2y' + 3y' - 6y \\ &= y'' + y' - 6y,\end{aligned}$$

which shows that the above factorization is "permissible," that is, yields the same result. A solution of $(D + 3)y = 0$ is $y = e^{-3x}$. A solution of $(D - 2)y = 0$ is $y = e^{2x}$. This gives the basis obtained before. The result is not unexpected because we factored the operator polynomial $P(D)$ in the same way in which we had factored the characteristic polynomial $P(\lambda) = \lambda^2 + \lambda - 6$.

Without going into details at this time, we want to mention that operational methods can also be used for operators $M = D^2 + fD + g$ with *variable* coefficients $f(x)$ and $g(x)$ but are more difficult and require more care. For example, $xD \neq Dx$ because

$$xDy = xy' \qquad \text{but} \qquad Dxy = (xy)' = y + xy'.$$

If operational calculus were limited to the simple situations illustrated in this section, it would perhaps not be worth mentioning. Actually, the power of the linear operator approach comes out in more complicated engineering problems, as we shall see in Chap. 5.

Problems for Sec. 2.5

In each case apply the given operator to each of the given functions.

1. $D - 4$; $4x^2 - 3x$, $6e^{4x}$, $\cos x - \sin x$

2. $D^2 + 3D$; $\cosh 3x$, $e^{-x} + e^{2x}$, $10 - e^{-3x}$

3. $(D - 2)(D + 1)$; e^{2x}, xe^{2x}, e^{-x}, xe^{-x}

4. $(D + 3)^2$; $x^3 + \sin \pi x$, e^{-3x}, xe^{-3x}

5. Prove that the differential operator D is linear.

6. Illustrate the linearity of L in (2) by taking $\alpha = 3$, $\beta = -2$, $y = e^{-2x}$ and $w = \sin x$.

7. Prove that the operator L in (2) is linear.

Find a general solution of the following equations.

8. $(D^2 + 4D + 4)y = 0$

9. $(D^2 - D - 2)y = 0$

10. $(4D^2 + 12D + 13)y = 0$

11. $(D^2 + 2D + 4\pi^2 - 1)y = 0$

12. $(4D^2 + 20D + 25)y = 0$

13. $(6D^2 - D - 1)y = 0$

14. Verify by direct calculation that $D^2[e^{-kx}f(x)] = e^{-kx}(D - k)^2 f(x)$. Choosing $e^{-kx}f(x) = c_1 + c_2x$, show that $f(x)$ is a general solution of (3) in the case of a double root.

15. If $(D^2 + aD + 6)y = 0$ is such that the characteristic equation has distinct roots λ and μ, show that a particular solution is

$$y = \frac{1}{\mu - \lambda}(e^{\mu x} - e^{\lambda x}).$$

Obtain from this a solution of the form $xe^{\lambda x}$ by letting $\mu \to \lambda$ and applying l'Hospital's rule to y, considered as a function of μ.

2.6 Free Oscillations

Linear homogeneous differential equations with constant coefficients have important engineering applications. In this section we shall consider such an application, which is fundamental. It is taken from mechanics, but we shall see later that it has a complete analogue in electric circuits. In our discussion we shall gain some more experience in **modeling,** that is, setting up mathematical models of physical systems. We shall also see that the mathematical distinction between three cases (I, II, III) in Sec. 2.4 is of physical significance because these cases correspond to three different types of motion.

We take an ordinary spring which resists compression as well as extension and suspend it vertically from a fixed support (Fig. 34). At the lower end of the spring we attach a body of mass m. We assume m to be so large that we may disregard the mass of the spring. If we pull the body down a certain distance and then release it, it undergoes a motion. We assume that the body moves strictly vertically.

We want to determine the motion of our mechanical system. For this purpose we consider the forces[7] acting on the body during the motion. This will lead to a differential equation, and by solving this equation we shall then obtain the displacement as a function of the time.

We may choose the *downward direction* as the positive direction and regard forces which act downward as positive and upward forces as negative.

The most obvious force acting on the body is the *attraction of gravity.*

$$F_1 = mg \tag{1}$$

where m is the mass of the body and g ($= 980\ \text{cm/sec}^2$) is the acceleration of gravity.

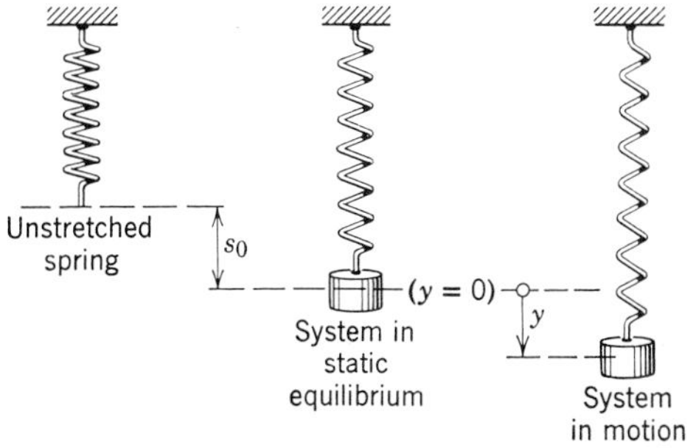

Fig. 34. Mechanical system under consideration

[7]The most important systems of units are shown in the table below. The Engineering System is on its way out. Since 1960, the Mks system is also known as the *International System of Units* (abbreviated *SI System*), and the abbreviations s (instead of sec) and N (instead of nt) are also used.

System of units	Length	Mass	Time	Force
Cgs system	centimeter (cm)	gram (gm)	second (sec)	dyne
Mks system	meter (m)	kilogram (kg)	second (sec)	newton (nt)
Engineering system	foot (ft)	slug	second (sec)	pound (lb)

1 ft = 30.4800 cm = 0.304800 meter, 1 slug = 14,594 g = 14.594 kg,
1 lb = 444,822 dynes = 4.44822 nt

For further details see, e.g., D. Halliday and R. Resnick, *Physics*. New York: Wiley, 1966. See also AN American National Standard, ASTM/IEEE Standard Metric Practice, Institute of Electrical and Electronics Engineers, Inc., 345 East 47th Street, New York, N.Y. 10017.

The next force to be considered is the *spring force* exerted by the spring if the spring is stretched. Experiments show that this force is proportional to the stretch, say,

$$F = ks \qquad \textbf{(Hooke's}^{8}\textbf{ law)}$$

where s is the stretch. The constant of proportionality k is called the *spring modulus*.

If $s = 1$, then $F = k$. The stiffer the spring is, the larger is k.

When the body is at rest we describe its position as the *static equilibrium position*. Clearly in this position the spring is stretched by an amount s_0 such that the resultant of the corresponding spring force and the gravitational force (1) is zero. Hence this spring force acts upward and its magnitude ks_0 is equal to that of F_1; that is,

$$ks_0 = mg. \tag{2}$$

Let $y = y(t)$ denote the displacement of the body from the static equilibrium position, with the positive direction downward (Fig. 34).

From Hooke's law it follows that the spring force corresponding to a displacement y is

$$F_2 = -ks_0 - ky, \tag{3}$$

the resultant of the spring force $-ks_0$ when the body is in static equilibrium position and the additional spring force $-ky$ caused by the displacement. Note that the sign of the last term in (3) is properly chosen, because when y is positive, $-ky$ is negative and, according to our assumption, represents an upward force, while for negative y the force $-ky$ represents a downward force.

The resultant of the forces F_1 and F_2 given by (1) and (3) is

$$F_1 + F_2 = mg - ks_0 - ky,$$

and because of (2) this becomes

$$F_1 + F_2 = -ky. \tag{4}$$

Undamped System. If the damping of the system is so small that it can be disregarded, then (4) is the resultant of all the forces acting on the body. The differential equation will now be obtained by the use of **Newton's second law**

$$\text{Mass} \times \text{Acceleration} = \text{Force}$$

where *force* means the resultant of the forces acting on the body at any instant. In our case, the acceleration is $\ddot{y} = d^2y/dt^2$ and that resultant is given by (4). Thus

$$m\ddot{y} = -ky.$$

[8]ROBERT HOOKE (1635–1703), English physicist.

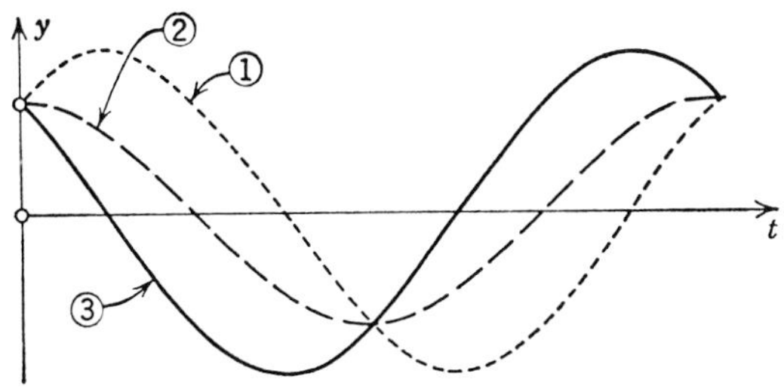

Fig. 35. Harmonic oscillations

Hence the motion of our system is governed by the linear differential equation with constant coefficients

(5) $$m\ddot{y} + ky = 0.$$

The general solution is

(6) $$y(t) = A \cos \omega_0 t + B \sin \omega_0 t \qquad \text{where } \omega_0 = \sqrt{k/m}.$$

The corresponding motion is called a **harmonic oscillation.** Figure 35 shows typical forms of (6) corresponding to various initial conditions.

By applying the addition formula for the cosine, the student may verify that (6) can be written [cf. also (13) in Appendix 3]

(6*) $$y(t) = C \cos (\omega_0 t - \delta) \qquad \left(C = \sqrt{A^2 + B^2}, \quad \tan \delta = \frac{B}{A}\right).$$

Since the period of the trigonometric functions in (6) is $2\pi/\omega_0$, the body executes $\omega_0/2\pi$ cycles per second. The quantity $\omega_0/2\pi$ is called the **frequency** of the oscillation and is measured in cycles per second. A new name for cycles/sec is[9] hertz (Hz).

Damped System. If we connect the mass to a dashpot (Fig. 36), then we have to take the corresponding viscous damping into account. The corresponding damping force has the direction opposite to the instantaneous motion, and we shall assume that it is proportional to the velocity $\dot{y} = dy/dt$ of the body. This is generally a good approximation, at least for small velocities. Thus the damping force is of the form

$$F_3 = -c\dot{y}.$$

Let us show that the *damping constant* c is positive. If $\dot{y}$ is positive, the body moves downward (in the positive y-direction) and $-c\dot{y}$ must be an upward force, that is, by agreement, $-c\dot{y} < 0$, which implies $c > 0$. For negative $\dot{y}$ the

[9] H. HERTZ (1857–1894), German physicist.

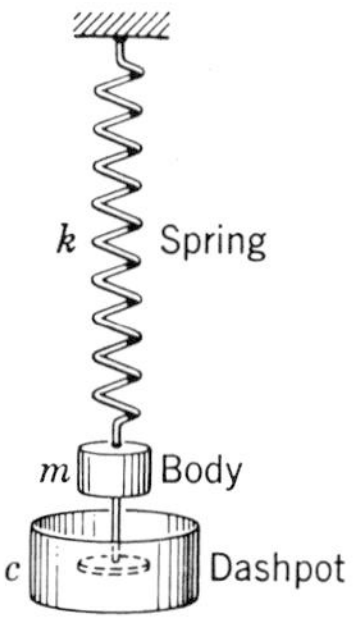

Fig. 36. Damped system

body moves upward and $-c\dot{y}$ must represent a downward force, that is, $-c\dot{y} > 0$ which implies $c > 0$.

The resultant of the forces acting on the body is now [cf. (4)]

$$F_1 + F_2 + F_3 = -ky - c\dot{y}.$$

Hence, by Newton's second law,

$$m\ddot{y} = -ky - c\dot{y},$$

and we see that the motion of the damped mechanical system is governed by the linear differential equation with constant coefficients

$$m\ddot{y} + c\dot{y} + ky = 0. \tag{7}$$

The corresponding characteristic equation is

$$\lambda^2 + \frac{c}{m}\lambda + \frac{k}{m} = 0.$$

The roots are

$$\lambda_{1,2} = -\frac{c}{2m} \pm \frac{1}{2m}\sqrt{c^2 - 4mk}.$$

Using the abbreviated notations

$$\alpha = \frac{c}{2m} \quad \text{and} \quad \beta = \frac{1}{2m}\sqrt{c^2 - 4mk}, \tag{8}$$

we can write

$$\lambda_1 = -\alpha + \beta \quad \text{and} \quad \lambda_2 = -\alpha - \beta.$$

The form of the solution of (7) will depend on the damping, and, as in Sec. 2.2, we may now distinguish between the following three cases:

Case I. $c^2 > 4mk$. *Distinct real roots λ_1, λ_2.* ***(Overdamping.)***

Case II. $c^2 < 4mk$. *Complex conjugate roots.* ***(Underdamping.)***

Case III. $c^2 = 4mk$. *A real double root.* ***(Critical damping.)***

Let us discuss these three cases separately.

Case I. Overdamping. When the damping constant c is so large that $c^2 > 4mk$, then λ_1 and λ_2 are distinct real roots, the general solution of (7) is

$$y(t) = c_1 e^{-(\alpha-\beta)t} + c_2 e^{-(\alpha+\beta)t}, \tag{9}$$

and we see that in this case the body does not oscillate. For $t > 0$ both exponents in (9) are negative because $\alpha > 0$, $\beta > 0$, and $\beta^2 = \alpha^2 - k/m < \alpha^2$. Hence both terms in (9) approach zero as t approaches infinity. Practically speaking, after a sufficiently long time the mass will be at rest in the static equilibrium position ($y = 0$). This is physically understandable since the damping takes energy from the system and there is no external force that keeps the motion going. Figure 37 shows (9) for some typical initial conditions.

Case II. Underdamping. This is the most interesting case. If the damping constant c is so small that $c^2 < 4mk$, then β in (8) is pure imaginary, say

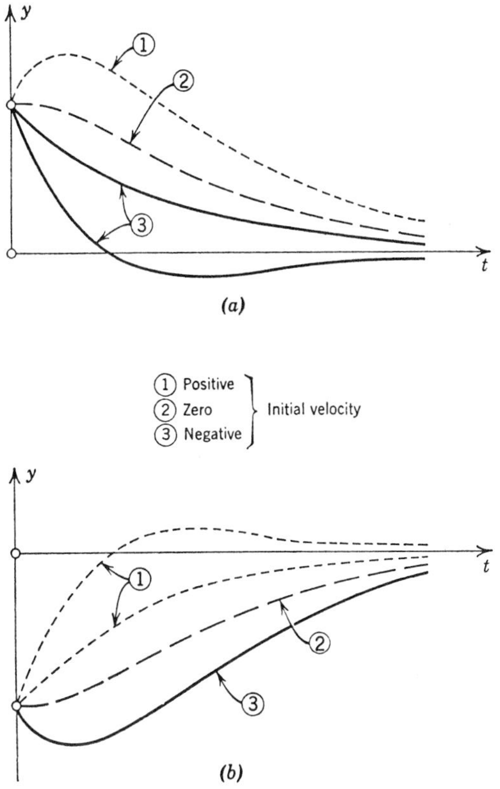

Fig. 37. Typical motions in the overdamped case
(*a*) Positive initial displacement
(*b*) Negative initial displacement

(10) $$\beta = i\omega^* \quad \text{where} \quad \omega^* = \frac{1}{2m}\sqrt{4mk - c^2} \quad (>0).$$

(We write ω^* to reserve ω for Sec. 2.13.) The roots of the characteristic equation are complex conjugate,

$$\lambda_1 = -\alpha + i\omega^*, \qquad \lambda_2 = -\alpha - i\omega^*,$$

and the general solution is

(11) $$y(t) = e^{-\alpha t}(A \cos \omega^* t + B \sin \omega^* t) = Ce^{-\alpha t} \cos (\omega^* t - \delta)$$

where $C^2 = A^2 + B^2$ and $\tan \delta = B/A$ [as in (6*)].

This solution represents damped oscillations. Since $\cos (\omega^* t - \delta)$ varies between -1 and 1, the curve of the solution lies between the curves $y = Ce^{-\alpha t}$ and $y = -Ce^{-\alpha t}$ in Fig. 38, touching these curves when $\omega^* t - \delta$ is an integral multiple of π.

The frequency is $\omega^*/2\pi$ cycles per second. From (10) we see that the smaller c (>0) is, the larger is ω^* and the more rapid the oscillations become. As c approaches zero, ω^* approaches the value $\omega_0 = \sqrt{k/m}$ corresponding to the harmonic oscillation (6).

Case III. Critical damping. If $c^2 = 4mk$, then $\beta = 0$, $\lambda_1 = \lambda_2 = -\alpha$, and the general solution is

(12) $$y(t) = (c_1 t + c_2)e^{-\alpha t}.$$

Since the exponential function is never zero and $c_1 t + c_2$ can have at most one positive zero, it follows that the motion can have at most one passage through the equilibrium position ($y = 0$). If the initial conditions are such that c_1 and c_2 have the same sign, there is no such passage at all. This is similar to Case I. Figure 39 shows typical forms of (12).

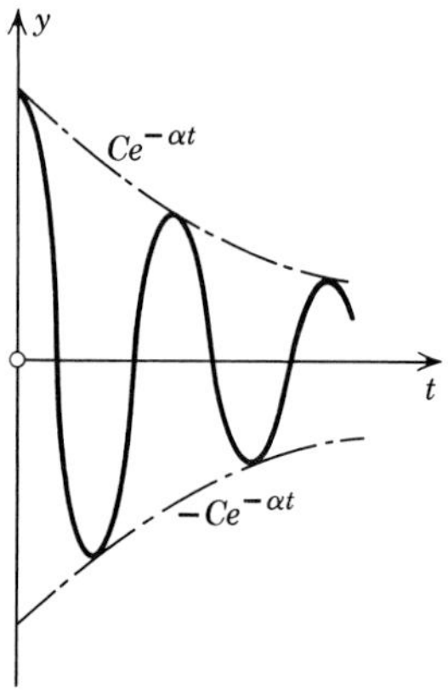

Fig. 38. Damped oscillation in Case II

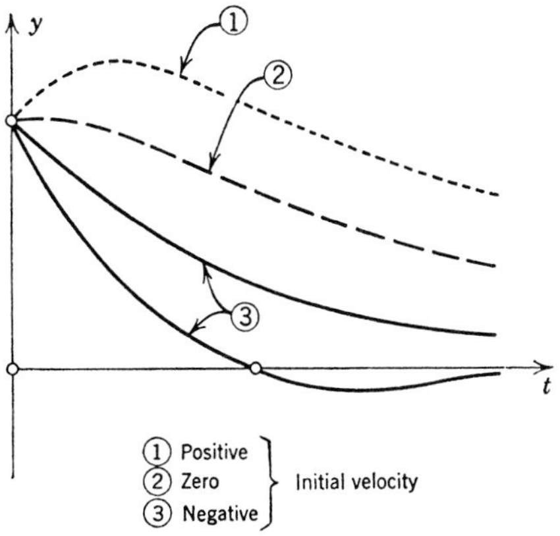

Fig. 39. Critical damping

Problems for Sec. 2.6

Harmonic oscillations (Undamped motion)

1. How does the frequency of a harmonic oscillation depend on the initial conditions? If a body of 20 nt is replaced by a body of 40 nt, the spring remaining the same, how does the frequency change?
2. Consider (6), assuming that $m = 10$ kg and $k = 10$ nt/meter. Determine A and B such that the initial displacement is 0.01 meter and the initial velocity is (*a*) -0.01 meter/sec $= -1$ cm/sec, (*b*) 0, (*c*) 0.01 meter/sec $= 1$ cm/sec. Represent these functions in the form (6*) and graph their curves.
3. Graph the harmonic oscillation (6) for various values of the spring modulus, say, $k = 1, 4, 9, 16$, assuming that the mass is 1 and the motion starts from $y = 1$ with initial velocity 0. What are the frequencies?
4. A spring is such that a 20.0 nt weight (about 4.5 lb) stretches it 9.8 cm. The lower end of the spring, with the 20.0 nt weight attached, is pulled down 5.0 cm and given an upward velocity of 30.0 cm/sec. Find the position of the weight at all later times, assuming zero damping.
5. Determine the harmonic oscillation (6) that starts with initial displacement y_0 and initial velocity v_0 and represent it in the form (6*).
6. If a spring is such that a weight of 20 nt (about 4.5 lb) would stretch it 2 cm, what would the frequency of the corresponding harmonic oscillation be? The period?
7. Show that the frequency of a harmonic oscillation of a body on a spring is $(\sqrt{g/s_0})/2\pi$, so that the period is $2\pi\sqrt{s_0/g}$, where s_0 is the elongation shown in Fig. 34.
8. If a body hangs on a spring of modulus $k_1 = 10$, which in turn hangs on a spring of modulus $k_2 = 20$, what is the modulus k of this combination of springs?
9. **(Pendulum)** Determine the frequency of oscillation of the pendulum of length l in Fig. 40 on the next page. Neglect air resistance and the weight of the rod. Assume that θ is small enough that $\sin\theta \approx \theta$.
10. A clock has a 1 meter pendulum. The clock ticks once for each time that the pendulum completes a swing, returning to its original position. How many times does the clock tick in 1 minute?
11. Suppose that the system in Fig. 41 consists of a pendulum as in Prob. 9 and two springs with constants k_1 and k_2 attached to the vibrating body and two vertical

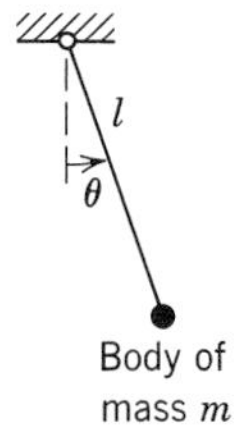

Fig. 40. Pendulum in Problem 9

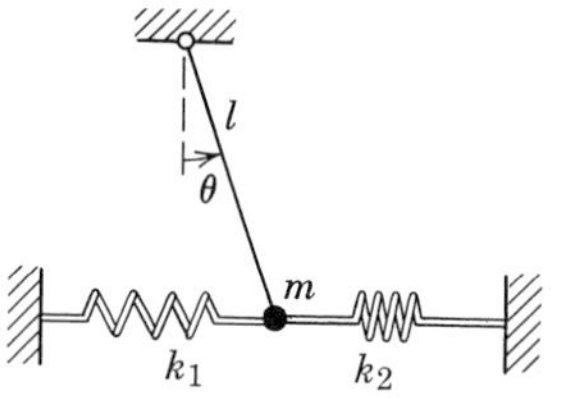

Fig. 41. System in Problem 11

walls such that $\theta = 0$ remains the position of static equilibrium and $\theta(t)$ remains small during the motion. Find the period T.

12. A cylindrical buoy 60 cm in diameter stands in water with its axis vertical (Fig. 42). When depressed slightly and released, it is found that the period of vibration is 2 sec. Determine the weight of the buoy. *Hint.* By *Archimedes' principle,* the buoyancy force equals the weight of the water displaced by the body (assumed partly or totally submerged).

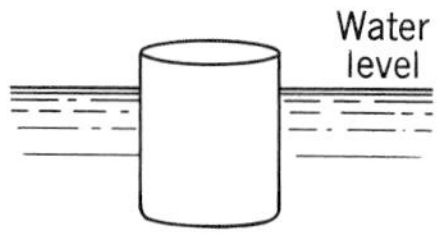

Fig. 42. Buoy in Problem 12

13. If 1 liter of water is allowed to vibrate up and down in a U-shaped tube 2 cm in diameter (Fig. 43), what is the frequency? Neglect friction.

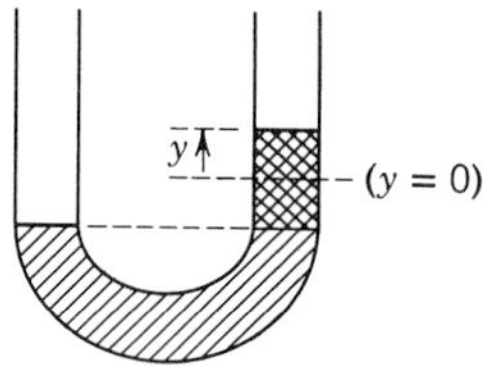

Fig. 43. Tube in Problem 13

14. **(Damping)** Determine and graph the motion $y(t)$ of the mechanical system described by (7) for increasing damping, say, $c = 0, 1, 2, 10$. Assume $m = 1$, $k = 1$, initial displacement 1 and initial velocity 0.

Overdamped motion

15. Show that for (9) to satisfy initial conditions $y(0) = y_0$ and $v(0) = v_0$ we must have

$$c_1 = [(1 + \alpha/\beta)y_0 + v_0/\beta]/2 \text{ and } c_2 = [(1 - \alpha/\beta)y_0 - v_0/\beta]/2.$$

16. Show that in the overdamped case, the body can pass through $y = 0$ at most once (Fig. 37).

17. In Prob. 16 find conditions for c_1 and c_2 such that the body does not pass through $y = 0$ at all.

18. Show that an overdamped motion with zero initial displacement cannot pass through $y = 0$.

Underdamped motion (Damped oscillation)

19. Show that the damped oscillation satisfying the initial conditions $y(0) = y_0$, $v(0) = v_0$ is

$$y = e^{-\alpha t}(y_0 \cos \omega^* t + \omega^{*-1}(v_0 + \alpha y_0) \sin \omega^* t).$$

20. Find and graph the three damped oscillations of the form

$$y = e^{-t}(A \cos t + B \sin t) = Ce^{-t} \cos (t - \delta)$$

starting from $y = 1$ with initial velocity -1, 0, 1, respectively.

21. Show that the frequency $\omega^*/2\pi$ of the underdamped motion decreases as the damping increases.

22. Show that for small damping,

$$\omega^* \approx \omega_0 \left(1 - \frac{c^2}{8mk}\right).$$

23. Determine the values of t corresponding to the maxima and minima of the oscillation $y(t) = e^{-t} \sin t$. Check your result by graphing $y(t)$.

24. Show that the maxima and minima of an underdamped motion occur at equidistant values of t, the distance between two consecutive maxima being $2\pi/\omega^*$.

25. **(Logarithmic decrement)** Prove that the ratio of two consecutive maximum amplitudes of a damped oscillation (11) is constant, the natural logarithm of this ratio being $\Delta = 2\pi\alpha/\omega^*$. ($\Delta$ is called the *logarithmic decrement* of the oscillation.) Find Δ in the case of $y = e^{-t} \cos t$ and determine the values of t corresponding to the maxima and minima.

26. Consider an underdamped motion of a body of mass $m = 0.5$ kg. If the time between two consecutive maxima is 1 sec and the maximum amplitude decreases to 1/2 its initial value after 10 cycles, what is the damping constant of the system?

Critical damping

27. Find the critical motion (12) which starts from y_0 with initial velocity v_0.

28. Under what conditions does (12) have a maximum or minimum at some instant $t > 0$?

29. Represent the maximum or minimum amplitude in Prob. 28 in terms of the initial values y_0 and v_0.

30. **(Buckling of a rod)** Equations $y'' + \gamma^2 y = 0$ also have applications in which the independent variable is a space variable (instead of time t). Figure 44 shows a famous example, the buckling of a rod under a constant vertical force F which is applied at the upper end of the rod. It is shown in mechanics that the curve $y(x)$ which the rod assumes under the load is a solution of the second-order

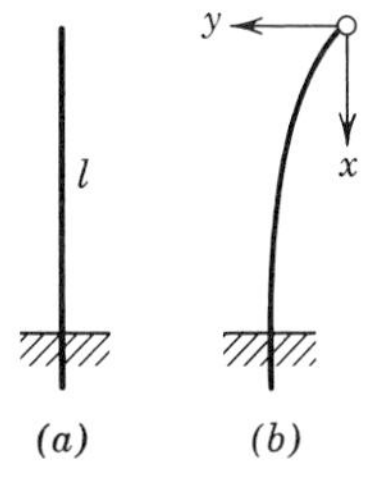

Fig. 44. Buckling of a thin rod

differential equation

$$EIy'' = -Fy.$$

E is the modulus of elasticity of the material of the rod. I is the moment of inertia of the cross section (assumed circular) about an axis through the center of the circle. We assume E and I to be constant, the lower end of the rod to be clamped and the upper end free. Show that with respect to the coordinates in Fig. 44*b* this leads to the conditions

$$y(0) = 0, \; y'(l) = 0$$

where l is the length of the rod and y is assumed to be small. A solution for any F is $y = 0$ (Fig. 44*a*), but one knows that if F exceeds a critical value, the equilibrium in Fig. 44*a* is no longer stable, that is, if slightly displaced, the rod will not return but become curved. Show that the critical force is $F_{\text{crit}} = (\pi/2l)^2 EI$ and the corresponding $y(x)$ in Fig. 44*b* is a portion of a sine curve.

2.7 Cauchy Equation

The so-called **Cauchy equation** or **Euler equation**[10]

(1) $$x^2y'' + axy' + by = 0 \qquad (a,\ b \text{ constant})$$

can also be solved by purely algebraic manipulations. By substituting

(2) $$y = x^m$$

and its derivatives into the equation (1) we find

$$x^2m(m-1)x^{m-2} + axmx^{m-1} + bx^m = 0.$$

By omitting the common power x^m, which is not zero when $x \neq 0$, we obtain the auxiliary equation

(3) $$m^2 + (a-1)m + b = 0.$$

If the roots m_1 and m_2 of this equation are different, then the two functions

$$y_1(x) = x^{m_1} \qquad \text{and} \qquad y_2(x) = x^{m_2}$$

constitute a basis of solutions of (1) for all x for which these functions are defined. The corresponding general solution is

(4) $$y = c_1x^{m_1} + c_2x^{m_2} \qquad (c_1,\ c_2 \text{ arbitrary}).$$

[10] LEONHARD EULER (1707–1783) was an enormously creative Swiss mathematician. He contributed to almost all branches of mathematics and its application to physical problems. His important books on algebra and calculus include numerous results of his own research work. The great French mathematician, AUGUSTIN LOUIS CAUCHY (1789–1857), is the father of modern analysis. He exercised a great influence on the theory of infinite series, complex analysis, and differential equations.

Example 1
Solve

$$x^2y'' - 1.5xy' - 1.5y = 0.$$

The auxiliary equation is

$$m^2 - 2.5m - 1.5 = 0.$$

The roots are $m_1 = -0.5$ and $m_2 = 3$. Hence a basis of real solutions for all positive x is

$$y_1 = \frac{1}{\sqrt{x}}, \qquad y_2 = x^3$$

and the corresponding general solution for all those x is

$$y = \frac{c_1}{\sqrt{x}} + c_2x^3.$$

∎

The auxiliary equation (3) has a *double root* $m_1 = m_2$ if and only if

$$b = \tfrac{1}{4}(1 - a)^2, \qquad \text{and then} \qquad m_1 = m_2 = \frac{1 - a}{2}.$$

In this *critical case* we may obtain a second solution by applying the method of variation of parameters. The procedure is similar to that in Sec. 2.4 and yields (cf. Prob. 1)

$$y_2 = uy_1 = (\ln x)y_1.$$

Thus, writing m in place of m_1, we see that

$$\text{(5)} \qquad y_1 = x^m \quad \text{and} \quad y_2 = x^m \ln x \qquad \left(m = \frac{1 - a}{2}\right)$$

are solutions of (1) in the case of a double root m of (3). Since these solutions are linearly independent, they constitute a basis of real solutions of (1) for all positive x, and the corresponding general solution is

$$\text{(6)} \qquad y = (c_1 + c_2 \ln x)x^m, \qquad (c_1, c_2 \text{ arbitrary}).$$

Example 2. General solution in the case of a double root
Solve

$$x^2y'' - 3xy' + 4y = 0.$$

The auxiliary equation has the double root $m = 2$. Hence a basis of real solutions for all positive x is x^2, $x^2 \ln x$, and the corresponding general solution is

$$y = (c_1 + c_2 \ln x)x^2.$$

Problems for Sec. 2.7

1. Derive y_2 in (5) by the method of variation of parameters.
2. Verify directly that in the case of a double root of (3) the function y_2 in (5) is a solution of (1).
3. Show that if the roots m_1 and m_2 of (3) are different, the functions $x^{m_1} \ln x$ and $x^{m_2} \ln x$ are **not** solutions of (1).

Find a general solution of the following differential equations.

4. $x^2y'' + xy' - y = 0$

5. $x^2y'' - 3xy' + 3y = 0$

6. $x^2y'' + xy' - 4y = 0$

7. $x^2y'' - xy' + y = 0$

8. $(x^2D^2 - xD + 0.75)y = 0$

9. $(xD^2 - 3D)y = 0$

10. $(x^2D^2 + 0.25)y = 0$

11. $(xD^2 + D)y = 0$

Reduce to the form (1) and solve:

12. $(z + 1)^2y'' + 5(z + 1)y' + 3y = 0$

13. $(2z - 3)^2y'' + 7(2z - 3)y' + 4y = 0$

Solve the following initial value problems.

14. $x^2y'' + xy' - 0.25y = 0, \quad y(1) = 2, \quad y'(1) = 1$

15. $x^2y'' - 3xy' + 4y = 0, \quad y(1) = 1, \quad y'(1) = 1$

16. $(x^2D^2 + xD - 2.25)y = 0, \quad y(1) = 2, \quad y'(1) = 0$

17. $(x^2D^2 - 3xD + 3)y = 0, \quad y(1) = 0, \quad y'(1) = -2$

18. Show that by setting $x = e^t$ $(x > 0)$ the Cauchy equation (1) can be transformed into the equation $\ddot{y} + (a - 1)\dot{y} + by = 0$, whose coefficients are constant.

19. Transform the equation in Prob. 18 back into (1).

20. Show that if we apply the transformation in Prob. 18, then (2) yields an expression of the form (2), Sec. 2.2, and (6) yields an expression of the form (8), Sec. 2.4, except for notation.

2.8 Existence and Uniqueness of Solutions

An **initial value problem** *for a second-order differential equation* consists of the equation and two **initial conditions,** one for the solution $y(x)$ and one for its derivative $y'(x)$. We shall now consider initial value problems of the form

$$y'' + f(x)y' + g(x)y = 0, \qquad y(x_0) = K_1, \qquad y'(x_0) = K_2 \tag{1}$$

involving a second-order homogeneous linear differential equation; K_1 and K_2 are given constants. The important result will be that continuity of the coefficients f and g is sufficient for the existence and uniqueness of the solution of the problem (1).

Existence and uniqueness theorem

If $f(x)$ and $g(x)$ are continuous functions on an open interval I and x_0 is in I, then the initial value problem (1) *has a unique solution $y(x)$ on I.*

The *proof of existence* uses the same prerequisites as those of the existence theorem in Sec. 1.12 and will not be presented here; it can be found in Ref. [B11] in Appendix 1.

Proof of uniqueness.[11] Assuming that (1) has two solutions $y_1(x)$ and $y_2(x)$ on

[11] This proof was suggested by my colleague, Prof. A. D. Ziebur.

the interval I, we show that their difference

$$y(x) = y_1(x) - y_2(x)$$

is identically zero on I; then $y_1 \equiv y_2$ on I, which implies uniqueness.

Since the equation in (1) is homogeneous and linear, y is a solution of that equation on I, and since y_1 and y_2 satisfy the same initial conditions, y satisfies the conditions

$$y(x_0) = 0, \qquad y'(x_0) = 0. \tag{2}$$

We consider the function

$$z(x) = y(x)^2 + y'(x)^2$$

and its derivative

$$z' = 2yy' + 2y'y''.$$

From the differential equation we have

$$y'' = -fy' - gy.$$

By substituting this in the expression for z' we obtain

$$z' = 2yy' - 2fy'^2 - 2gyy'. \tag{3}$$

Now, since y and y' are real,

$$(y \pm y')^2 = y^2 \pm 2yy' + y'^2 \geqq 0,$$

and from this we immediately obtain the two inequalities

$$\text{(a) } 2yy' \leqq y^2 + y'^2 = z, \qquad \text{(b) } -2yy' \leqq y^2 + y'^2 = z. \tag{4}$$

For the last term in (3) we now have

$$-2gyy' \leqq |-2gyy'| = |g||2yy'| \leqq |g|z.$$

Using this result and applying (4a) to the term $2yy'$ in (3), we find

$$z' \leqq z + 2|f|y'^2 + |g|z.$$

Since $y'^2 \leqq y^2 + y'^2 = z$, we obtain from this

$$z' \leqq (1 + 2|f| + |g|)z$$

or, denoting the function in parentheses by h,

$$z' \leqq hz \qquad \text{for all } x \text{ on } I. \tag{5a}$$

Similarly, from (3) and (4) it follows that

$$\begin{aligned} -z' &= -2yy' + 2fy'^2 + 2gyy' \\ &\leqq z + 2|f|z + |g|z = hz. \end{aligned} \tag{5b}$$

The inequalities (5a) and (5b) are equivalent to the inequalities

$$z' - hz \leqq 0, \qquad z' + hz \geqq 0. \tag{6}$$

Integrating factors for the two expressions on the left are

$$F_1 = e^{-\int h(x)\,dx} \qquad \text{and} \qquad F_2 = e^{\int h(x)\,dx}.$$

The integrals in the exponents exist since h is continuous. Since F_1 and F_2 are positive, we thus obtain from (6)

$$F_1(z' - hz) = (F_1 z)' \leqq 0 \qquad \text{and} \qquad F_2(z' + hz) = (F_2 z)' \geqq 0,$$

which means that $F_1 z$ is nonincreasing and $F_2 z$ is nondecreasing on I. Since $z(x_0) = 0$ by (2), we thus obtain when $x \leqq x_0$

$$F_1 z \geqq (F_1 z)_{x_0} = 0, \qquad F_2 z \leqq (F_2 z)_{x_0} = 0$$

and similarly, when $x \geqq x_0$,

$$F_1 z \leqq 0, \qquad F_2 z \geqq 0.$$

Dividing by F_1 and F_2 and noting that these functions are positive, we altogether have

$$z \leqq 0, \qquad z \geqq 0 \qquad \text{for all } x \text{ on } I.$$

This implies $z = y^2 + y'^2 \equiv 0$ on I. Hence $y \equiv 0$ or $y_1 \equiv y_2$ on I, and the proof is complete. ∎

The remaining part of the present section will be devoted to some important consequences of this theorem.

To prepare for this consideration let us return to the concepts of linear dependence and independence of functions introduced in Sec. 2.3. Let $y_1(x)$ and $y_2(x)$ be given functions defined on an open interval I and consider the relation

$$k_1 y_1(x) + k_2 y_2(x) = 0$$

on I, where k_1 and k_2 are constants to be determined. This relation certainly holds for $k_1 = 0, k_2 = 0$. It may happen that we can find a $k_1 \neq 0$ such that the relation holds on I. Then we may divide by k_1 and solve for y_1:

$$y_1 = -\frac{k_2}{k_1} y_2.$$

Similarly, if that relation holds for a $k_2 \neq 0$ on I we may solve for y_2:

$$y_2 = -\frac{k_1}{k_2}y_1.$$

We see that in both cases y_1 and y_2 are linearly dependent on I.

However, if $k_1 = 0, k_2 = 0$ is the only pair of numbers for which that relation holds on I, we cannot solve for y_1 or y_2, and the functions are linearly independent on I.

Consequently, *two functions $y_1(x)$ and $y_2(x)$ are linearly dependent on an interval I where both functions are defined if and only if we can find constants k_1, k_2, not both zero such that*

(7) $$k_1y_1(x) + k_2y_2(x) = 0$$

for all x on I.

This result will be used in the subsequent proof.

In our further consideration we shall need another criterion for testing linear dependence and independence of two solutions $y_1(x)$ and $y_2(x)$ of a differential equation of the form

(8) $$y'' + f(x)y' + g(x)y = 0.$$

This criterion uses the determinant

$$W(y_1, y_2) = \begin{vmatrix} y_1 & y_2 \\ y_1' & y_2' \end{vmatrix} = y_1y_2' - y_2y_1',$$

which is called the *Wronski determinant*[12] or briefly the **Wronskian** of y_1 and y_2. The criterion can be stated as follows.

Theorem 2 (Linear dependence and independence of solutions) *Suppose that the coefficients $f(x)$ and $g(x)$ of the differential equation* (8) *are continuous on an open interval I. Then any two solutions of* (8) *on I are linearly dependent on I if and only if their Wronskian W is zero for some $x = x_0$ in I.* (*If $W = 0$ for $x = x_0$, then $W \equiv 0$ on I.*)

Proof. **(a)** Let $y_1(x)$ and $y_2(x)$ be linearly dependent solutions of (8) on I. Then there are constants k_1 and k_2, not both zero, such that

$$k_1y_1 + k_2y_2 = 0$$

holds on I. By differentiation we obtain

[12] Introduced by HÖNE (1778–1853), who changed his name to Wronski. Second-order determinants should be familiar to the reader from elementary calculus; otherwise he may consult Sec. 7.9, which is independent of the other sections in Chap. 7.

$$k_1y_1' + k_2y_2' = 0.$$

Suppose that $k_1 \neq 0$. Then by multiplying the first equation by y_2'/k_1 and the last by $-y_2/k_1$ and adding the resulting equations, we obtain

$$y_1y_2' - y_2y_1' = 0, \qquad \text{that is,} \qquad W(y_1, y_2) = 0.$$

If $k_1 = 0$, then $k_2 \neq 0$, and the argument is similar.

(b) Conversely, suppose that $W(y_1, y_2) = 0$ for some $x = x_0$ in I. Consider the system of linear equations

$$\text{(9)} \qquad \begin{aligned} k_1y_1(x_0) + k_2y_2(x_0) &= 0 \\ k_1y_1'(x_0) + k_2y_2'(x_0) &= 0 \end{aligned}$$

in the unknowns k_1, k_2. Since the system is homogeneous and the determinant of the system is just the Wronskian $W[y_1(x_0), y_2(x_0)]$ and $W = 0$, the system has a solution k_1, k_2 where k_1 and k_2 are not both zero (cf. Sec. 7.9). Using these numbers k_1, k_2, we introduce the function

$$y(x) = k_1y_1(x) + k_2y_2(x).$$

By Fundamental Theorem 1 in Sec. 2.1 the function $y(x)$ is a solution of (8) on I. From (9) we see that it satisfies the initial conditions $y(x_0) = 0$, $y'(x_0) = 0$. Now another solution of (8) satisfying the same initial conditions is $y^* \equiv 0$. Since f and g are continuous, the uniqueness theorem is applicable and we conclude that $y^* \equiv y$, that is,

$$k_1y_1 + k_2y_2 \equiv 0$$

on I. Since k_1 and k_2 are not both zero, this means that y_1 and y_2 are linearly dependent on I. Furthermore, from this and the first part of the proof it follows that $W(y_1, y_2) \equiv 0$ on I. ∎

Example 1

The functions $y_1 = \cos \omega x$ and $y_2 = \sin \omega x$ are solutions of the equation

$$y'' + \omega^2 y = 0 \qquad (\omega \neq 0).$$

A simple calculation shows that their Wronskian has the value ω. Hence these functions constitute a basis of solutions for all x. ∎

Further examples are included in Secs. 2.3 and 2.4. The following example illustrates the fact that *the assumption of continuity of the coefficients in Theorem 2 cannot be omitted.*

Example 2

The functions

$$y_1 = x^3 \qquad \text{and} \qquad y_2 = \begin{cases} x^3 & \text{when } x \geqq 0 \\ -x^3 & \text{when } x < 0 \end{cases}$$

are solutions of the differential equation

$$y'' - \frac{3}{x}y' + \frac{3}{x^2}y = 0 \qquad (10)$$

for all x, as can be readily verified by substituting each of the functions and its derivatives into (10). The reader may show that the functions are linearly independent on the x-axis, although their Wronskian is identically zero. This does not contradict Theorem 2, because the coefficients of (10) are not continuous at $x = 0$. ∎

Of course, Example 2 also illustrates the fact that *if u_1 and u_2 are any differentiable functions of x, the condition $W(u_1, u_2) = 0$ is only necessary for linear dependence of u_1 and u_2* [as follows from part **(a)** of the proof of Theorem 2], *but not sufficient.*

Let us now show that *an equation of the form* (8) *whose coefficients are continuous on an open interval I has a basis of solutions on I, and, in fact, infinitely many such bases.*

Proof. From the existence theorem it follows that (8) has a solution $y_1(x)$ satisfying the initial conditions

$$y_1(x_0) = 1, \qquad y_1'(x_0) = 0,$$

and a solution $y_2(x)$ satisfying the initial conditions

$$y_2(x_0) = 0, \qquad y_2'(x_0) = 1,$$

where x_0 is any fixed point in I. At x_0 we have $W(y_1, y_2) = 1$. Hence, by Theorem 2, the two solutions are linearly independent on I and form a basis of (8) on I. Another basis is ay_1, by_2 where $a \neq 0$, $b \neq 0$ are any constants, so that there are infinitely many such bases. This completes the proof. ∎

We shall now prove the important fact that a general solution of an equation of the form (8) with continuous coefficients includes the totality of all solutions of (8).

Theorem 3 (General solution)

Let

$$y(x) = c_1 y_1(x) + c_2 y_2(x) \qquad (11)$$

be a general solution of the differential equation (8) *on an open interval I where the coefficients f and g are continuous. Let $Y(x)$ be any solution of* (8) *on I containing no arbitrary constants. Then $Y(x)$ is obtained from* (11) *by assigning suitable values to the constants c_1 and c_2.*

Proof. Let $Y(x)$ be any solution of (8) on I containing no arbitrary constants, and let $x = x_0$ be any fixed point in I. We first show that we can find values of c_1 and c_2 in (11) such that

$$y(x_0) = Y(x_0), \qquad y'(x_0) = Y'(x_0)$$

written out

$$\text{(12)} \qquad \begin{aligned} c_1y_1(x_0) + c_2y_2(x_0) &= Y(x_0), \\ c_1y_1'(x_0) + c_2y_2'(x_0) &= Y'(x_0). \end{aligned}$$

In fact, this is a system of linear equations in the unknowns c_1, c_2. Its determinant is the Wronskian of y_1 and y_2 at $x = x_0$. Since (11) is a general solution, y_1 and y_2 are linearly independent on I, and from Theorem 2 it follows that their Wronskian is not zero. Hence the system has a unique solution $c_1 = c_1^*$, $c_2 = c_2^*$ which can be obtained by Cramer's rule (Sec. 7.9). By using these constants we obtain from (11) the particular solution

$$y^*(x) = c_1^*y_1(x) + c_2^*y_2(x).$$

From (12) we see that

$$\text{(13)} \qquad y^*(x_0) = Y(x_0), \qquad y^{*\prime}(x_0) = Y'(x_0).$$

From this and the uniqueness theorem we conclude that y^* and Y must be identical on I, and the proof is complete. ∎

Problems for Sec. 2.8

Using the Wronskian and Theorem 2, show linear independence of the following pairs of functions which occurred as bases in Secs. 2.4 and 2.7.

1. $e^{\lambda_1 x}$, $e^{\lambda_2 x}$ **2.** $e^{(p+iq)x}$, $e^{(p-iq)x}$ **3.** $e^{px} \cos qx$, $e^{px} \sin qx$
4. $e^{\lambda x}$, $xe^{\lambda x}$ **5.** x^{m_1}, x^{m_2} **6.** x^m, $x^m \ln x$

In each case find a second-order homogeneous linear differential equation for which the given functions are solutions. Show linear independence by Theorem 2.

7. $\cos \omega x$, $\sin \omega x$ **8.** e^x, xe^x **9.** x^2, $x^{1/2}$
10. 1, e^{-x} **11.** 1, x^2 **12.** x^2, $x^2 \ln x$
13. $e^{-x} \cos x$, $e^{-x} \sin x$ **14.** $\cosh x$, $\sinh x$ **15.** $e^{i\beta x}$, $e^{-i\beta x}$

Using (7), find out whether the following functions are linearly dependent or independent.

16. x, x^3 **17.** $\ln x$, $\ln x^3$ $(x > 0)$ **18.** e^x, 1
19. $x + 1$, $x - 1$ **20.** $\sqrt{x}$, $x^{3/2}$ $(x > 0)$ **21.** $|x|x^2$, x^3 $(x < 0)$

22. Show that y_1, y_2 in Example 2 constitute a basis of solutions of (10), but $W(y_1, y_2) \equiv 0$.

23. Show that $W(x^2, x|x|) = 0$ but the functions x^2, $x|x|$ are linearly independent on the interval $-1 < x < 1$. Does this contradict Theorem 2?

24. Suppose that y_1, y_2 constitute a basis of solutions of a differential equation satisfying the assumptions of Theorem 2. Show that $z_1 = a_{11}y_1 + a_{12}y_2$, $z_2 = a_{21}y_1 + a_{22}y_2$ is a basis of that equation if and only if the determinant of the coefficients a_{jk} is not zero.

25. Suppose that (8) has continuous coefficients on I. Show that two solutions of (8) on I which are zero at the same point in I cannot form a basis of solutions of (8) on I. Give examples.

2.9 Homogeneous Linear Equations of Arbitrary Order

A differential equation of nth order is said to be **linear** if it can be written in the form

$$(1) \qquad y^{(n)} + f_{n-1}(x)y^{(n-1)} + \cdots + f_1(x)y' + f_0(x)y = r(x)$$

where the function r on the right-hand side and the *coefficients* $f_0, f_1, \cdots, f_{n-1}$ are any given functions of x, and $y^{(n)}$ is the nth derivative of y with respect to x, etc. Any differential equation of order n which cannot be written in the form (1) is said to be **nonlinear.**

If $r(x) \equiv 0$, equation (1) becomes

$$(2) \qquad y^{(n)} + f_{n-1}(x)y^{(n-1)} + \cdots + f_1(x)y' + f_0(x)y = 0$$

and is said to be **homogeneous.** If $r(x)$ is not identically zero, the equation is said to be **nonhomogeneous.**

Using operator notations (Sec. 2.5), we can write (1) in the form

$$(1^*) \qquad \begin{aligned} L[y] &= P(D)[y] \\ &= [D^n + f_{n-1}(x)D^{n-1} + \cdots + f_1(x)D + f_0(x)]y = r(x) \end{aligned}$$

and similarly for (2).

A function $y = \phi(x)$ is called a **solution** of a (linear or nonlinear) differential equation of nth order *on an interval I*, if $\phi(x)$ is defined and n times differentiable on I and is such that the equation becomes an identity when we replace the unspecified function y and its derivatives in the equation by ϕ and its corresponding derivatives.

We shall now see that our considerations in Secs. 2.3 and 2.8 can be readily extended to linear homogeneous equations of arbitrary order n.

Existence and uniqueness theorem

If $f_0(x), \cdots, f_{n-1}(x)$ in (2) are continuous functions on an open interval I, then the initial value problem consisting of the equation (2) and the n ***initial conditions***

$$y(x_0) = K_1, \quad y'(x_0) = K_2, \cdots, y^{(n-1)}(x_0) = K_n$$

has a unique solution $y(x)$ on I; here x_0 is any fixed point in I, and $K_1, \cdots, K_n$ are given numbers.

The proof of existence can be found in Ref. [B11] in Appendix 1, and uniqueness can be proved by a slight generalization of the proof in Sec. 2.8.

To extend the considerations of Sec. 2.3 we need the concepts of linear dependence and independence of n ($\geqq 2$) functions, say, $y_1(x), \cdots, y_n(x)$. These functions are said to be **linearly dependent** *on some interval* I where they are defined, if (at least) one of them can be represented on I as a **"linear combination"** of the other $n - 1$ functions, that is, as a sum of those functions, each multiplied by a constant (zero or not). If none of the functions can be represented in this way, they are said to be **linearly independent** *on* I.

This definition includes that in Sec. 2.3 for two functions as a particular case. n linearly dependent functions on I are also called a **linearly dependent set** *of functions on* I. Similarly for linearly independent functions.

For example, the functions $y_1 = x$, $y_2 = x^2$, $y_3 = x^3$ are linearly independent on any interval. The functions $y_1 = x$, $y_2 = 3x$, $y_3 = x^2$ are linearly dependent on any interval, because $y_2 = 3y_1 + 0y_3$.

n functions $y_1(x), \cdots, y_n(x)$ are linearly dependent on an interval I where they are defined if and only if we can find constants $k_1, \cdots, k_n$, not all zero, such that the relation

$$k_1y_1(x) + \cdots + k_ny_n(x) = 0 \tag{3}$$

holds for all x on I.

In fact, if (3) holds for some $k_1 \neq 0$, we may divide by k_1 and express y_1 as a linear combination of $y_2, \cdots, y_n$:

$$y_1 = -\frac{1}{k_1}(k_2y_2 + \cdots + k_ny_n),$$

which proves linear dependence. Similarly, if (3) holds for some $k_i \neq 0$, we may express y_i as a linear combination of the other functions. On the other hand, if $k_1 = 0, \cdots, k_n = 0$ is the only set of constants for which (3) holds on I, then we cannot solve for any of the functions in terms of the others, and the functions are linearly independent. This proves the statement. ∎

A solution of a differential equation of the order n (linear or not) is called a **general solution** if it contains n arbitrary[13] independent constants. Here *independence* means that the solution cannot be reduced to a form containing less than n arbitrary constants. If we assign definite values to the n constants, then the solution so obtained is called a **particular solution** of that equation.

A set of n linearly independent solutions $y_1(x), \cdots, y_n(x)$ of the linear homogeneous equation (2) on I is called a **basis** or a **fundamental system** of solutions of (2) on I.

If $y_1, \cdots, y_n$ is such a basis, then

$$y(x) = c_1y_1(x) + \cdots + c_ny_n(x) \qquad (c_1, \cdots, c_n \text{ arbitrary}) \tag{4}$$

[13]Cf. footnote 2 in Sec. 2.3.

is a general solution of (2) on I. In fact, since (2) is linear and homogeneous, (4) is a solution of (2), and since $y_1, \cdots, y_n$ are linearly independent functions on I, none of them can be expressed as a linear combination of the others, that is, (4) cannot be reduced to a form containing less than n arbitrary constants.

The test for linear dependence and independence of solutions (Theorem 2 in Sec. 2.8) can be generalized to nth order equations as follows.

Theorem 2 (Linear dependence and independence of solutions)
Suppose that the coefficients $f_0(x), \cdots, f_{n-1}(x)$ of (2) are continuous on an open interval I. Then n solutions $y_1, \cdots, y_n$ of (2) on I are linearly dependent on I if and only if their **Wronskian**

$$(5) \qquad W(y_1, \cdots, y_n) = \begin{vmatrix} y_1 & y_2 & \cdots & y_n \\ y_1' & y_2' & \cdots & y_n' \\ \cdot & \cdot & \cdots & \cdot \\ y_1^{(n-1)} & y_2^{(n-1)} & \cdots & y_n^{(n-1)} \end{vmatrix}$$

is zero for some $x = x_0$ in I. (If $W = 0$ at $x = x_0$, then $W \equiv 0$ on I.)

The idea of the proof is similar to that of Theorem 2 in Sec. 2.8.

A general solution of an equation of the form (2) with continuous coefficients includes the totality of all solutions of (2). In fact, the following generalization of Theorem 3 in Sec. 2.8 holds true.

Theorem 3 (General solution)
Let (4) be a general solution of (2) on an open interval I where $f_0(x), \cdots, f_{n-1}(x)$ are continuous, and let $Y(x)$ be any solution of (2) on I involving no arbitrary constants. Then $Y(x)$ is obtained from (4) by assigning suitable values to the arbitrary constants $c_1, \cdots, c_n$.

The idea of the proof is similar to that of Theorem 3 in Sec. 2.8.

Example 1
The functions $y_1 = e^{-x}$, $y_2 = e^x$, and $y_3 = e^{2x}$ are solutions of the equation

$$(6) \qquad y''' - 2y'' - y' + 2y = 0.$$

The Wronskian is

$$W(e^{-x}, e^x, e^{2x}) = \begin{vmatrix} e^{-x} & e^x & e^{2x} \\ -e^{-x} & e^x & 2e^{2x} \\ e^{-x} & e^x & 4e^{2x} \end{vmatrix} = 6e^{2x} \neq 0,$$

which shows that the functions constitute a basis of solutions of (6) for all x, and the corresponding general solution is

$$y = c_1e^{-x} + c_2e^x + c_3e^{2x}.$$

Problems for Sec. 2.9

Prove:

1. If $y \equiv 0$ is a function of a set of functions on an interval I, the set is linearly dependent on I.
2. If one of the functions of a set of functions on I is a constant multiple of another function of the set on I, the set is linearly dependent on I.
3. If a set of p functions on I is linearly dependent on I, a set of n ($\geqq p$) functions on I containing that set is linearly dependent on I.
4. If a set of functions is linearly dependent on I, it is linearly dependent on any subinterval of I.
5. If a set of functions is linearly independent on I, it is linearly independent on any interval J containing I on which all the functions are defined.

Are the following functions linearly dependent or independent on the positive x-axis?

6. $1, x, x^2$
7. $\cosh^2 x, \sinh^2 x, 1$
8. $\cos x, \sin x, 1$
9. $\ln x, \ln x^2, \ln x^3$
10. e^x, xe^x, x^2e^x
11. $1, x, 2x$
12. $x, x - 1, x + 1$
13. $1, x^2, x^4, x^6$
14. $e^{2x}, e^{-2x}, 0$

Using Theorem 2, show that the given functions form a basis of solutions of the corresponding differential equation.

15. $e^{-x}, e^x, xe^x, y''' - y'' - y' + y = 0$
16. $e^{-x}, xe^{-x}, x^2e^{-x}, y''' + 3y'' + 3y' + y = 0$
17. $x, x^2, x^3, x^3y''' - 3x^2y'' + 6xy' - 6y = 0$
18. $x^{1/2}, x^{-1/2}, 1, x^2y''' + 3xy'' + 0.75y' = 0$
19. $e^x, xe^x, x^2e^x, x^3e^x, y^{\text{IV}} - 4y''' + 6y'' - 4y' + y = 0$
20. $e^x, e^{-x}, \cos x, \sin x, y^{\text{IV}} - y = 0$

2.10 Homogeneous Linear Equations of Arbitrary Order with Constant Coefficients

The method in Sec. 2.2 may easily be extended to a linear homogeneous equation of any order n with constant coefficients

$$y^{(n)} + a_{n-1}y^{(n-1)} + \cdots + a_1y' + a_0y = 0. \tag{1}$$

Substituting $y = e^{\lambda x}$ and its derivatives into (1), we obtain the *characteristic equation*

$$\lambda^n + a_{n-1}\lambda^{n-1} + \cdots + a_1\lambda + a_0 = 0. \tag{2}$$

If this equation has n distinct roots $\lambda_1, \cdots, \lambda_n$, then the n solutions

$$y_1 = e^{\lambda_1 x}, \qquad \cdots, \qquad y_n = e^{\lambda_n x} \tag{3}$$

constitute a basis for all x, and the corresponding general solution of (1) is

$$y = c_1 e^{\lambda_1 x} + \cdots + c_n e^{\lambda_n x}. \tag{4}$$

If a *double root* occurs, say, $\lambda_1 = \lambda_2$, then $y_1 = y_2$ in (3) and we take y_1 and $y_2 = xy_1$ as two linearly independent solutions corresponding to that root; this is as in Sec. 2.4.

If a *triple root* occurs, say, $\lambda_1 = \lambda_2 = \lambda_3$, then $y_1 = y_2 = y_3$ in (3) and three linearly independent solutions corresponding to that root are

$$y_1, \qquad xy_1, \qquad x^2y_1; \tag{5}$$

these solutions can be obtained by the method of variation of parameters. More generally, if λ is a root of order m, then m corresponding linearly independent solutions are

$$e^{\lambda x}, \quad xe^{\lambda x}, \quad \cdots, \quad x^{m-1}e^{\lambda x}. \tag{6}$$

Indeed, linear independence is almost obvious. Substitution shows that these are solutions. The corresponding formulas become a little simpler if we use operator notation (Sec. 2.5), writing (1) in the form

$$L[y] = P(D)[y] = [D^n + a_{n-1}D^{n-1} + \cdots + a_0]y = 0.$$

By assumption,

$$P(D) = Q(D)(D - \lambda)^m \tag{7}$$

where $Q(D)$ is a differential operator of order $n - m$ and corresponds to other roots which the characteristic equation of (1) may have. Now for the functions $x^k e^{\lambda x}$ ($k = 0, \cdots, m - 1$) in (6) we obtain

$$(D - \lambda)(x^k e^{\lambda x}) = kx^{k-1}e^{\lambda x} + \lambda x^k e^{\lambda x} - \lambda x^k e^{\lambda x} = kx^{k-1}e^{\lambda x}.$$

Hence by applying $D - \lambda$ to this we have

$$(D - \lambda)^2(x^k e^{\lambda x}) = k(k - 1)x^{k-2}e^{\lambda x}$$

and so on. Finally

$$(D - \lambda)^k(x^k e^{\lambda x}) = k!e^{\lambda x},$$

so that

$$(D - \lambda)^{k+1}(x^k e^{\lambda x}) = k!(D - \lambda)e^{\lambda x} = 0.$$

Since $k + 1 \leqq m$ in (6), the desired result follows:

$$(D - \lambda)^m(x^k e^{\lambda x}) = 0 \qquad\qquad k = 0, \cdots, m - 1.$$

Because of (7) this shows that the functions in (6) are solutions of (1) in the case of a root of order m. ∎

The whole situation is rather simple and may be illustrated by the following two examples.

Example 1
Solve the differential equation

$$y''' - 2y'' - y' + 2y = 0.$$

The roots of the characteristic equation

$$\lambda^3 - 2\lambda^2 - \lambda + 2 = 0$$

are -1, 1, and 2, and the corresponding general solution (4) is

$$y = c_1 e^{-x} + c_2 e^{x} + c_3 e^{2x}.$$

Example 2
Solve the differential equation

$$y^{(V)} - 3y^{(IV)} + 3y''' - y'' = 0.$$

The characteristic equation

$$\lambda^5 - 3\lambda^4 + 3\lambda^3 - \lambda^2 = 0$$

has the roots $\lambda_1 = \lambda_2 = 0$ and $\lambda_3 = \lambda_4 = \lambda_5 = 1$, and the answer is

$$y = c_1 + c_2 x + (c_3 + c_4 x + c_5 x^2)e^{x}. \tag{8}$$

Problems for Sec. 2.10

1. Show that (8) may also be obtained by setting $y'' = z$, finding a general solution z of the new third-order equation and integrating z twice.

Find a general solution.

2. $y''' - 3y'' + 3y' - y = 0$
3. $y^{IV} + 3y'' - 4y = 0$
4. $y^{IV} - y = 0$
5. $y''' + y' = 0$
6. $y^{IV} + 4y''' + 6y'' + 4y' + y = 0$
7. $y^{IV} - 5y'' + 4y = 0$

Find an equation for which the given functions form a basis.

8. $\cosh x$, $\sinh x$, $\cos 2x$, $\sin 2x$
9. e^{x}, xe^{x}, x^2e^{x}
10. e^{-x}, xe^{-x}, e^{x}, xe^{x}
11. 1, x, e^{3x}, xe^{3x}
12. $\sin x$, $\sin 2x$, $\cos x$, $\cos 2x$
13. e^{-2x}, e^{-x}, e^{x}, e^{2x}, 1

Reduction of order. It may sometimes be relatively easy to find one solution of a given differential equation. Then the order of the equation can be reduced, that is, further solutions of the given equation can be obtained from a differential equation of lower order. The details are as follows.

14. If $y_1(x)$ is a solution of $y'' + f(x)y' + g(x)y = 0$, show that another solution is $y_2(x) = u(x)y_1(x)$, where $u(x)$ is obtained by solving the first-order equation in $z = u'$:

$$y_1u'' + (2y_1' + fy_1)u' = 0.$$

15. If $y_1(x)$ is a solution of $y''' + f(x)y'' + g(x)y' + h(x)y = 0$, show that another solution is $y_2(x) = u(x)y_1(x)$, where $u(x)$ is obtained by solving the second-order equation in $z = u'$:

$$y_1u''' + (3y_1' + fy_1)u'' + (3y_1'' + 2fy_1' + gy_1)u' = 0.$$

Using the method of reduction of order, solve:

16. $y''' - 3y'' - y' + 3y = 0, \quad y_1 = e^x$

17. $x^3y''' - 3xy' + 3y = 0, \quad y_1 = 1/x$

18. $y'' - 4xy' + (4x^2 - 2)y = 0, \quad y_1 = e^{x^2}$

19. $x^3y''' - 3x^2y'' + (6 - x^2)xy' - (6 - x^2)y = 0, \quad y_1 = x$

20. **(Cauchy equation)** The *Cauchy equation* or *Euler equation of the third order* is

$$x^3y''' + ax^2y'' + bxy' + cy = 0.$$

Generalizing the method of Sec. 2.7, show that $y = x^m$ is a solution of the equation if and only if m is a root of the auxiliary equation

$$m^3 + (a - 3)m^2 + (b - a + 2)m + c = 0.$$

Solve $x^3y''' - 3x^2y'' + 6xy' - 6y = 0$.

2.11 Nonhomogeneous Linear Equations

So far we have considered *homogeneous* linear equations. We shall now discuss linear equations which are *nonhomogeneous*. For the sake of simplicity we concentrate on second-order equations

$$y'' + f(x)y' + g(x)y = r(x), \tag{1}$$

but it will be obvious that the results in this section can easily be generalized to linear equations of any order.

We shall first show that a general solution $y(x)$ of (1) can be obtained in a simple way if a general solution $y_h(x)$ of the corresponding homogeneous equation

$$y'' + f(x)y' + g(x)y = 0, \tag{2}$$

is known. We claim that we then obtain $y(x)$ by adding to y_h any solution $\tilde{y}$ of (1) involving no arbitrary constants, that is,

$$y(x) = y_h(x) + \tilde{y}(x). \tag{3}$$

In fact, since y_h involves two arbitrary constants, so does y and will therefore be a general solution of (1) as defined in Sec. 2.3, provided y is a solution of (1) at all. To show this, we substitute (3) into (1). Then the left-hand side of (1) becomes

$$(y_h + \tilde{y})'' + f(y_h + \tilde{y})' + g(y_h + \tilde{y}).$$

This may be written

$$(y_h'' + fy_h' + gy_h) + \tilde{y}'' + f\tilde{y}' + g\tilde{y}.$$

Since y_h is a solution of (2), the expression in the parentheses is zero. Sincy $\tilde{y}$ satisfies (1), the sum of the other terms equals $r(x)$, and the statement is proved.

The remaining practical question is how to obtain a solution $\tilde{y}$ of (1). Corresponding methods will be discussed in the next section and in Secs. 2.15 and 2.16.

At present let us prove that a general solution of an equation (1) with continuous functions f, g, and r represents the totality of all solutions of (1) as follows.

Theorem 1 (General solution)
Suppose that $f(x)$, $g(x)$, and $r(x)$ in (1) *are continuous*[14] *functions on an open interval I. Let $Y(x)$ be any solution of* (1) *on I containing no arbitrary constants. Then $Y(x)$ is obtained from* (3) *by assigning suitable values to the two arbitrary constants contained in the general solution $y_h(x)$ of* (2). *In* (3), *the function $\tilde{y}(x)$ is any solution of* (1) *on I containing no arbitrary constants.*

Proof. We set $Y - \tilde{y} = y^*$. Then

$$y^{*\prime\prime} + fy^{*\prime} + gy^* = (Y'' + fY' + gY) - (\tilde{y}'' + f\tilde{y}' + g\tilde{y}) = r - r = 0,$$

that is, y^* is a solution of (2). Now y^* does not contain arbitrary constants. Hence, by Theorem 3, Sec. 2.8 (with Y replaced by y^*), it can be obtained from y_h by assigning suitable values to the arbitrary constants in y_h. From this, since $Y = y^* + \tilde{y}$, the statement follows. ∎

This theorem shows that any solution of (1) containing no arbitrary constants is a particular solution of (1) in the sense of our definition in Sec. 2.3, and we may now formulate our results as follows.

Theorem 2 (General solution)
A general solution $y(x)$ of the linear nonhomogeneous differential equation (1) *is the sum of a general solution $y_h(x)$ of the corresponding homogeneous equation* (2) *and an arbitrary particular solution $y_p(x)$ of* (1):

$$y(x) = y_h(x) + y_p(x). \tag{4}$$

Problems for Sec. 2.11

1. Show that $y_{p1} = \sin 2x$ and $y_{p2} = \sin 2x - \sin x$ are particular solutions of $y'' + y = -3 \sin 2x$. Find corresponding general solutions (4) and show that one of them can be transformed into the other, so that the choice of y_p is immaterial.

[14] This suffices for the existence of y_h on I, as we know, and for the existence of $\tilde{y}$ on I, as will become obvious in Sec. 2.16.

2. Show that two general solutions (4) of (1) which correspond to different y_p can be transformed into each other.
3. Show that if y_1 is a solution of (1) with $r = r_1$ and y_2 is a solution of (1) with $r = r_2$, then $y = y_1 + y_2$ is a solution of (1) with $r = r_1 + r_2$.
4. Show that $y_1 = e^x$ is a solution of $y'' + y = 2e^x$ and $y_2 = x \sin x$ is a solution of $y'' + y = 2 \cos x$. Using Prob. 3, find a general solution of the equation $y'' + y = 2e^x + 2 \cos x$.
5. Extend Theorems 1 and 2 to an equation of any order n.

In each case verify that $y_p(x)$ is a solution of the given differential equation and find a general solution.

6. $y'' + y = 2e^x, \quad y_p = e^x$
7. $y'' - y = 2e^x, \quad y_p = xe^x$
8. $y'' - 3y' + 2y = 2x^2 - 6x + 2, \quad y_p = x^2$
9. $y'' + 4y = -12 \sin 2x, \quad y_p = 3x \cos 2x$
10. $y'' + 9y = 18x, \quad y_p = 2x$
11. $(D^2 - 1)y = 2 \cos x, \quad y_p = -\cos x$
12. $(D^2 + D)y = 2 \sin x, \quad y_p = -x \cos x$
13. $(4x^2D^2 + 1)y = (1 - x^2) \cos 0.5x, \quad y_p = \cos 0.5x$
14. $(x^2D^2 - 3xD + 3)y = 3 \ln x - 4, \quad y_p = \ln x$
15. $(D^2 + 6D + 9)y = e^{-3x}/(x^2 + 1), \quad y_p = (x \arctan x - 0.5 \ln (x^2 + 1))\, e^{-3x}$

2.12 A Method for Solving Nonhomogeneous Linear Equations

For obtaining a general solution of a given nonhomogeneous linear differential equation we need a particular solution $y_p(x)$ of that equation. This was shown in the last section. How can we find such a function y_p?

We shall consider a general method for finding y_p in Sec. 2.16. In the present section we discuss a much simpler special method, the so-called **method of undetermined coefficients,** which is basic in engineering problems, for instance in connection with forced mechanical or electrical vibrations. This method is applicable to equations

$$y'' + ay' + by = r(x) \tag{1}$$

when $r(x)$ is such that the form of a particular solution $y_p(x)$ of (1) may be guessed;[15] for example, r may be a single power of x, a polynomial, an exponential function, a sine or cosine, or a sum of such functions. The method consists in assuming for y_p an expression similar to that of $r(x)$ containing unknown coefficients which are to be determined by inserting y_p and its derivatives into (1). This suggests the name of the method.

[15] The student will see in Sec. 2.16 that our present method actually results from the general method.

Let us begin with some examples, so that we get a feeling for what is going on.

Example 1
Solve the nonhomogeneous equation

$$y'' - 4y' + 3y = 10e^{-2x}. \tag{2}$$

The derivatives of e^{-2x} are e^{-2x} times some constants, so let us try

$$y_p = ke^{-2x}.$$

Substitution of y_p, y_p', y_p'' into (2) gives

$$4ke^{-2x} - 4(-2ke^{-2x}) + 3ke^{-2x} = 10e^{-2x}.$$

Hence $4k + 8k + 3k = 10$, and $k = 2/3$. A basis of the homogeneous equation is e^x, e^{3x}. Theorem 2 in the preceding section thus yields the answer

$$y = y_h + y_p = c_1e^x + c_2e^{3x} + \tfrac{2}{3}e^{-2x}.$$

Example 2
Solve the nonhomogeneous equation

$$y'' + 4y = 8x^2. \tag{3}$$

We try $y_p = Kx^2$, but this attempt will fail. We have $y_p'' = 2K$. By substitution in (3) we obtain

$$2K + 4Kx^2 = 8x^2.$$

For this to hold for all x, the coefficients of each power of x on both sides (x^2 and x^0) must be the same:

$$4K = 8 \qquad \text{and} \qquad 2K = 0,$$

which has no solution. Let us try

$$y_p = Kx^2 + Lx + M. \qquad \text{Then} \qquad y_p'' = 2K.$$

Substitution in (3) yields

$$2K + 4(Kx^2 + Lx + M) = 8x^2.$$

Equating the coefficients of x^2, x, and x^0 on both sides, we have $4K = 8$, $4L = 0$, $2K + 4M = 0$. Thus $K = 2$, $L = 0$, $M = -1$. Hence $y_p = 2x^2 - 1$, and a general solution of (3) is

$$y = A\cos 2x + B\sin 2x + 2x^2 - 1.$$

Can you see the reason why our first attempt failed?

Example 3
Solve the nonhomogeneous equation

$$y'' - y' - 2y = 10\cos x. \tag{4}$$

The student may try $y_p = K\cos x$ and see that this attempt fails. Since differentiations of the cosine give sine *and* cosine functions, we try

$$y_p = K\cos x + M\sin x.$$

Then

$$y_p' = -K\sin x + M\cos x,$$

$$y_p'' = -K\cos x - M\sin x.$$

By inserting this into (4) we obtain

$$(-3K - M)\cos x + (K - 3M)\sin x = 10\cos x.$$

By equating the coefficients of $\cos x$ and $\sin x$ on both sides we have

$$-3K - M = 10, \qquad K - 3M = 0.$$

Hence $K = -3$, $M = -1$, and $y_p = -3\cos x - \sin x$. A general solution of the corresponding homogeneous equation is

$$y_h = c_1e^{-x} + c_2e^{2x}.$$

Altogether,

$$y = y_h + y_p = c_1e^{-x} + c_2e^{2x} - 3\cos x - \sin x.$$

∎

Our examples suggest the following rule.

Rule

If $r(x)$ in (1) is one of the functions in column 1 of Table 2.1, choose for y_p a linear combination with undetermined coefficients of $r(x)$ and its linearly independent derivatives (cf. column 2) and determine the coefficients by substitution.

If $r(x)$ is a sum of functions in column 1, choose for y_p the sum of the functions in the corresponding lines.

If a term in $r(x)$ is a solution of the homogeneous equation corresponding to (1), modify your choice according to the following

Modification rule

Multiply the expression in the appropriate line of column 2 by x or x^2 depending on whether the number in column 3 is a simple or double root of the characteristic equation of the homogeneous equation corresponding to (1). Proof below.

Let us consider two more examples, one in which we do not need the Modification rule and one in which we do.

Example 4

Find a particular solution of the equation

$$y'' - 3y' + 2y = 4x + e^{3x}.$$

We start from

$$y_p = K_1x + K_0 + Ce^{3x}.$$

Table 2.1
Method of Undetermined Coefficients

Term in $r(x)$	Choice for y_p	
ke^{px}	Ce^{px}	p
kx^n $(n = 0, 1, \ldots)$	$K_nx^n + K_{n-1}x^{n-1} + \cdots + K_1x + K_0$	0
$k\cos qx$	$K\cos qx + M\sin qx$	iq
$k\sin qx$	$K\cos qx + M\sin qx$	iq

By substitution we find $K_1 = 2$, $K_0 = 3$, $C = \frac{1}{2}$, and therefore

$$y_p = 2x + 3 + \tfrac{1}{2}e^{3x}.$$

Example 5

Solve the nonhomogeneous equation

$$y'' - 2y' + y = (D - 1)^2 y = e^x + x. \tag{5}$$

By Table 2.1, the term x indicates a particular solution choice

$$K_1 x + K_0.$$

Since 1 is a double root of the characteristic equation $(\lambda - 1)^2 = 0$, by the Modification rule the term e^x calls for the particular solution

$$Cx^2e^x \qquad \text{(instead of } Ce^x\text{).}$$

Together,

$$y_p = K_1x + K_0 + Cx^2e^x.$$

Substituting this into (5), we obtain

$$y_p'' - 2y_p' + y_p = 2Ce^x + K_1x - 2K_1 + K_0 = e^x + x.$$

Hence $C = \frac{1}{2}$, $K_1 = 1$, $K_0 = 2$, and a general solution of (5) is

$$y = y_h + y_p = (c_1x + c_2)e^x + \tfrac{1}{2}x^2e^x + x + 2.$$

∎

Our rules are simple. It is essential that they work in the cases listed in Table 2.1. Of minor importance is a proof. We give it for completeness, using operators (Sec. 2.5) as a convenient tool. The reader may skip it, or he may enjoy it as an application of operator methods.

For line 1 of Table 2.1, equation (1) is of the form

$$(D - \lambda_1)(D - \lambda_2)y = ke^{px}. \tag{6}$$

Applying $D - p$ on both sides, we obtain

$$(D - p)(D - \lambda_1)(D - \lambda_2)y = k(D - p)[e^{px}] = 0. \tag{7}$$

Every solution of (6) satisfies (7) as we see directly from the form of (6) and (7). Hence every solution of (6) can be obtained from a general solution of (7). Equation (7) is a homogeneous third-order equation. The form of its general solutions depends on whether the roots are different or equal. General solutions of (7) are

$$\text{(8)}\quad \begin{array}{lll} \text{(a)} & y = (c_1e^{\lambda_1 x} + c_2e^{\lambda_2 x}) + Ce^{px} & \text{if } p \neq \lambda_1 \neq \lambda_2 \neq p \\ & y = (c_1e^{\lambda x} + c_2xe^{\lambda x}) + Ce^{px} & \text{if } \lambda_1 = \lambda_2 = \lambda \neq p \\ \text{(b)} & y = (c_1e^{\lambda_1 x} + c_2e^{px}) + Cxe^{px} & \text{if } \lambda_1 \neq \lambda_2 = p \\ \text{(c)} & y = (c_1e^{px} + c_2xe^{px}) + Cx^2e^{px} & \text{if } \lambda_1 = \lambda_2 = p. \end{array}$$

The functions in the parentheses are general solutions of the homogeneous equation corresponding to (6). Hence the last term with suitably determined C is a particular solution y_p of (6). This shows that (8a) proves the Rule, (8b) the Modification rule for a simple root and (8c) for a double root when $r(x) = ke^{px}$. For lines 2 and 3 of Table 2.1 the idea of proof is the same; instead of $D - p$ we then have to apply D^{n+1} in line 2, which gives $D^{n+1}[x^n] = 0$, and $D^2 + q^2$ in line 3, which gives $(D^2 + q^2)[\cos qx] = 0$ and $(D^2 + q^2)[\sin qx] = 0$. ▮

Problems for Sec. 2.12

1. Give the details of the proof of the modification rule in the case of line 2 in Table 2.1.

2. Perform the same task as in Prob. 1, for line 3 in Table 2.1.

Find a particular solution of the following differential equations.

3. $y'' + y = x^2 + x$

4. $y'' + 2y' + y = 2x^2$

5. $y'' + 5y' + 6y = 9x^4 - x$

6. $y'' - y' - 2y = \sin x$

7. $y'' + y' - 6y = 52 \cos 2x$

8. $y'' + y' - 2y = 3e^x$

9. $y'' + y = 2 \sin x$

10. $y'' - 2y' + 2y = 2e^x \cos x$

11. $y''' + 2y'' - y' - 2y = 1 - 4x^3$

12. $y^{IV} - 5y'' + 4y = 10 \cos x$

Find a general solution of the following differential equations.

13. $y'' + y = -x - x^2$

14. $y'' - y = e^x$

15. $y'' + 4y = e^{-x}$

16. $y'' - y' - 2y = 4 \sin x$

17. $y'' + y = \sin x$

18. $y'' + y' - 2y = e^x$

19. $y'' - 4y' + 3y = e^{3x}$

20. $y'' - 4y' + 9y = 10e^{2x} - 12 \cos 3x$

The method of undetermined coefficients is useful for certain first-order equations, too, and simpler than the usual method (Sec. 1.7). Using both methods, solve

21. $y' + y = x^4$

22. $y' - 2y = \sin 4x$

Solve the following initial value problems.

23. $y'' + 25y = 5x, \quad y(0) = 5, \quad y'(0) = -4.8$

24. $y'' - y = x, \quad y(0) = 1, \quad y'(0) = 0$

25. $y'' - 2y' + y = 2x^2 - 8x + 4, \quad y(0) = 3, \quad y'(0) = 3$

26. $y'' - y' - 2y = 10 \sin x, \quad y(0) = 1, \quad y'(0) = -3$

27. $y'' - y' - 2y = 3e^{2x}, \quad y(0) = 0, \quad y'(0) = -2$

28. $y'' + 2y' + 2y = -2 \cos 2x - 4 \sin 2x, \quad y(0) = 1, \quad y'(0) = 1$

29. $y'' + 4y' + 8y = 4 \cos x + 7 \sin x, \quad y(0) = 1, \quad y'(0) = -1$

30. $y'' + 2y' + 10y = 4.5 \cos x - \sin x, \quad y(0) = 0.5, \quad y'(0) = 3$

2.13 Forced Oscillations. Resonance

In Sec. 2.6 we considered free oscillations of a body on a spring as shown in Fig. 45. These motions are governed by the homogeneous equation

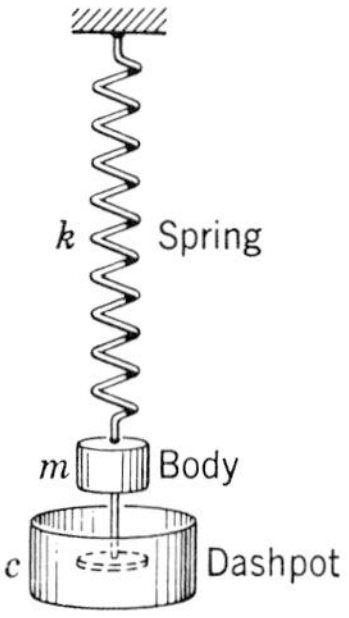

Fig. 45. Mass on a spring

$$m\ddot{y} + c\dot{y} + ky = 0 \tag{1}$$

where m is the mass of the body, c is the damping constant, and k is the spring modulus. We shall now extend our consideration, assuming that a variable force $r(t)$ acts on the system. In this way we shall become familiar with further interesting facts which are fundamental in engineering mathematics, in particular with resonance.

We remember that the equation (1) was obtained by considering the forces acting on the body and using Newton's second law. From this it is immediately clear that the differential equation corresponding to the present situation is obtained from (1) by adding the force $r(t)$; this yields

$$m\ddot{y} + c\dot{y} + ky = r(t).$$

$r(t)$ is called the **input** or **driving force,** and a corresponding solution is called an **output** or a **response** *of the system to the driving force.* (Cf. also Sec. 1.7.) The resulting motion is called a **forced motion,** in contrast to the **free motion** corresponding to (1), which is a motion in the absence of an external force $r(t)$.

Of particular interest are periodic inputs. We shall consider a sinusoidal input, say,

$$r(t) = F_0 \cos \omega t \qquad (F_0 > 0,\ \omega > 0).$$

More complicated periodic inputs will be considered later in Sec. 10.7.

The differential equation now under consideration is

$$m\ddot{y} + c\dot{y} + ky = F_0 \cos \omega t. \tag{2}$$

We want to determine and discuss its general solution, which represents the general output of our vibrating system. Since a general solution of the corresponding homogeneous equation (1) is known from Sec. 2.6, we must now determine a particular solution $y_p(t)$ of (2).

This can be done by the method of undetermined coefficients (Sec. 2.12), starting from

$$y_p(t) = a \cos \omega t + b \sin \omega t. \tag{3}$$

By differentiating this function we have

$$\dot{y}_p = -\omega a \sin \omega t + \omega b \cos \omega t, \qquad \ddot{y}_p = -\omega^2 a \cos \omega t - \omega^2 b \sin \omega t.$$

Substituting these expressions into (2) and collecting the cosine and the sine terms, we obtain

$$[(k - m\omega^2)a + \omega cb] \cos \omega t + [-\omega ca + (k - m\omega^2)b] \sin \omega t = F_0 \cos \omega t.$$

By equating the coefficients of the cosine and sine terms on both sides we have

$$\text{(4)} \qquad \begin{aligned} (k - m\omega^2)a + \quad \omega cb \quad &= F_0 \\ -\omega ca \quad + (k - m\omega^2)b &= 0. \end{aligned}$$

This is a system of two linear algebraic equations in the two unknowns a and b. The solution is obtained in the usual way by elimination or Cramer's rule (if necessary, see Sec. 7.9). We find

$$a = F_0 \frac{k - m\omega^2}{(k - m\omega^2)^2 + \omega^2 c^2}, \qquad b = F_0 \frac{\omega c}{(k - m\omega^2)^2 + \omega^2 c^2}$$

provided the denominator is not zero. If we set $\sqrt{k/m} = \omega_0 \ (>0)$ as in Sec. 2.6, this becomes

$$\textbf{(5)} \quad a = F_0 \frac{m(\omega_0^2 - \omega^2)}{m^2(\omega_0^2 - \omega^2)^2 + \omega^2 c^2}, \qquad b = F_0 \frac{\omega c}{m^2(\omega_0^2 - \omega^2)^2 + \omega^2 c^2}.$$

We thus obtain the general solution

$$\text{(6)} \qquad y(t) = y_h(t) + y_p(t),$$

where y_h is a general solution of (1) and y_p is given by (3) with coefficients (5).

We shall now discuss the behavior of the mechanical system, distinguishing between the two cases $c = 0$ (no damping) and $c > 0$ (damping). These cases will correspond to two different types of output.

Case 1. Undamped forced oscillations. If there is no damping, then $c = 0$. We first assume that $\omega^2 \neq \omega_0^2$. This is essential. We then obtain from (3) and (5)

$$\text{(7)} \qquad y_p(t) = \frac{F_0}{m(\omega_0^2 - \omega^2)} \cos \omega t = \frac{F_0}{k[1 - (\omega/\omega_0)^2]} \cos \omega t.$$

From this and (6*) in Sec. 2.6 we have the general solution

$$\textbf{(8)} \qquad y(t) = C \cos (\omega_0 t - \delta) + \frac{F_0}{m(\omega_0^2 - \omega^2)} \cos \omega t.$$

This output represents a superposition of two harmonic oscillations; the frequen-

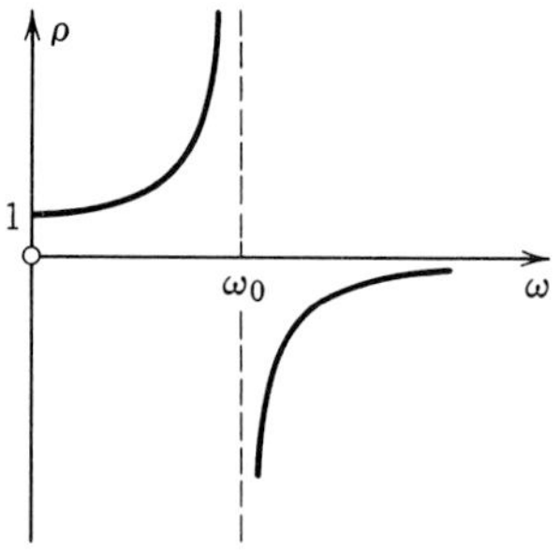

Fig. 46. Resonance factor $\rho(\omega)$

*cies are the **"natural frequency"** $\omega_0/2\pi$ [cycles/sec] of the system (that is, the frequency of the free undamped motion) and the frequency $\omega/2\pi$ of the input.*

From (7) we see that the maximum amplitude of y_p is

$$a_0 = \frac{F_0}{k}\rho \qquad \text{where} \qquad \rho = \frac{1}{1-(\omega/\omega_0)^2}. \tag{9}$$

It depends on ω and ω_0. As $\omega \to \omega_0$, the quantities ρ and a_0 tend to infinity. This phenomenon of excitation of large oscillations by matching input and natural frequencies ($\omega = \omega_0$) is known as **resonance,** and is of basic importance in the study of vibrating systems (cf. below). The quantity ρ is called the *resonance factor* (Fig. 46). From (9) we see that ρ/k is the ratio of the amplitudes of the function y_p and the input.

In the case of resonance, equation (2) becomes

$$\ddot{y} + \omega_0^2 y = \frac{F_0}{m}\cos\omega_0 t. \tag{10}$$

From the modification rule in the preceding section we conclude that a particular solution of (10) has the form

$$y_p(t) = t(a\cos\omega_0 t + b\sin\omega_0 t).$$

By substituting this into (10) we find $a = 0$, $b = F_0/2m\omega_0$, and (Fig. 47)

$$y_p(t) = \frac{F_0}{2m\omega_0}t\sin\omega_0 t. \tag{11}$$

We see that y_p becomes larger and larger. In practice, this means that systems with very little damping may undergo large vibrations which can destroy the system; we shall return to this practical aspect of resonance later in this section.

Another interesting and highly important type of oscillation is obtained when ω is close to ω_0. Take, for example, the particular solution [cf. (8)]

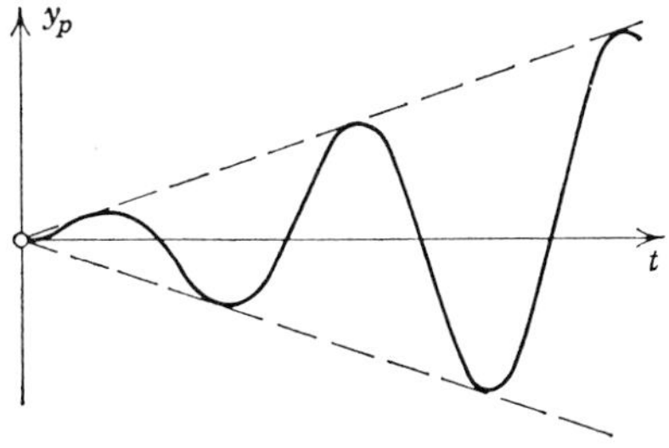

Fig. 47. Particular solution in the case of resonance

(12) $$y(t) = \frac{F_0}{m(\omega_0^2 - \omega^2)}(\cos \omega t - \cos \omega_0 t) \qquad (\omega \neq \omega_0)$$

corresponding to the initial conditions $y(0) = 0, \dot{y}(0) = 0$. This may be written [cf. (12) in Appendix 3]

$$y(t) = \frac{2F_0}{m(\omega_0^2 - \omega^2)} \sin \frac{\omega_0 + \omega}{2} t \sin \frac{\omega_0 - \omega}{2} t.$$

Since ω is close to ω_0 the difference $\omega_0 - \omega$ is small, so that the period of the last sine function is large, and we obtain an oscillation of the type shown in Fig. 48.

Case 2. Damped forced oscillations. If there is damping, then $c > 0$, and we know from Sec. 2.6 that the general solution y_h of (1) approaches zero as t approaches infinity (practically, after a sufficiently long time); that is, the general solution (6) of (2) now represents the **transient solution** and tends to the **steady-state solution** y_p. *Hence, after a sufficiently long time, the output corresponding to a purely sinusoidal input will practically be a harmonic oscillation whose frequency is that of the input. This is what happens in practice, because no physical system is completely undamped.*

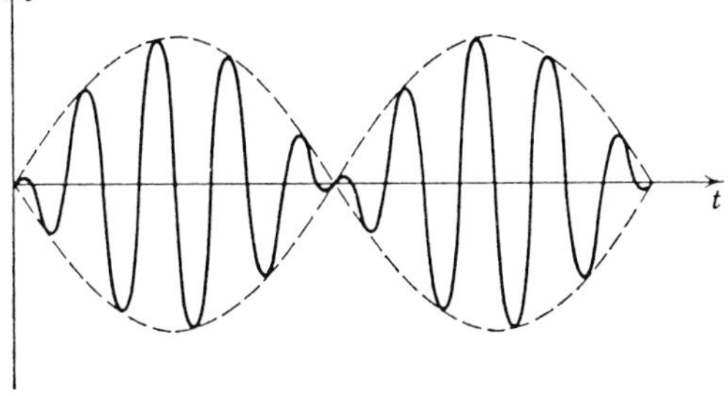

Fig. 48. Forced undamped oscillation when the difference of the input and natural frequencies is small **("beats")**

While in the undamped case the amplitude y_p approaches infinity as ω approaches ω_0, this will not happen in the present case. *The amplitude will always be finite,* but may have a maximum for some ω, depending on c. This may be called **practical resonance.** It is of great importance because it shows that some input may excite oscillations with such a large amplitude that the system may be destroyed. Such cases happened in practice, in particular in earlier times when less was known about resonance. Machines, cars, ships, airplanes and bridges are vibrating mechanical systems, and it is sometimes rather difficult to find constructions which are completely free of undesired resonance effects.

To investigate the amplitude of y_p as a function of ω, we write (3) in the form

$$y_p(t) = C^* \cos(\omega t - \eta) \tag{13}$$

where, according to (5),

$$\text{(14)} \qquad \begin{aligned} C^*(\omega) &= \sqrt{a^2 + b^2} = \frac{F_0}{\sqrt{m^2(\omega_0{}^2 - \omega^2)^2 + \omega^2 c^2}}, \\ \tan \eta &= \frac{b}{a} = \frac{\omega c}{m(\omega_0{}^2 - \omega^2)}. \end{aligned}$$

Let us determine the maximum of $C^*(\omega)$. By equating $dC^*/d\omega$ to zero we find

$$[-2m^2(\omega_0{}^2 - \omega^2) + c^2]\,\omega = 0.$$

The expression in brackets is zero for

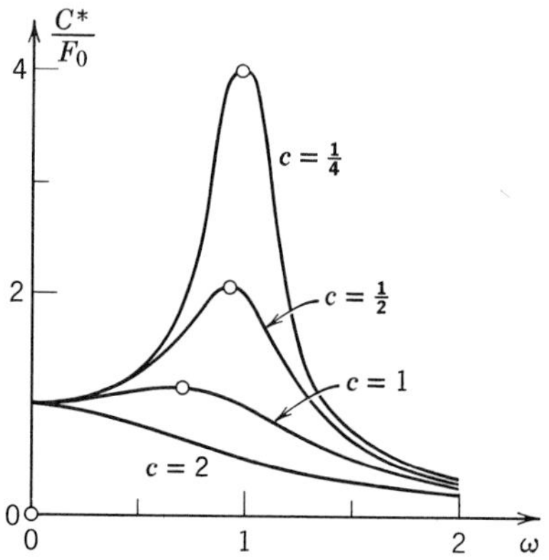

Fig. 49. Amplification C^*/F_0 as a function of ω for $m = 1$, $k = 1$, and various values of the damping constant c

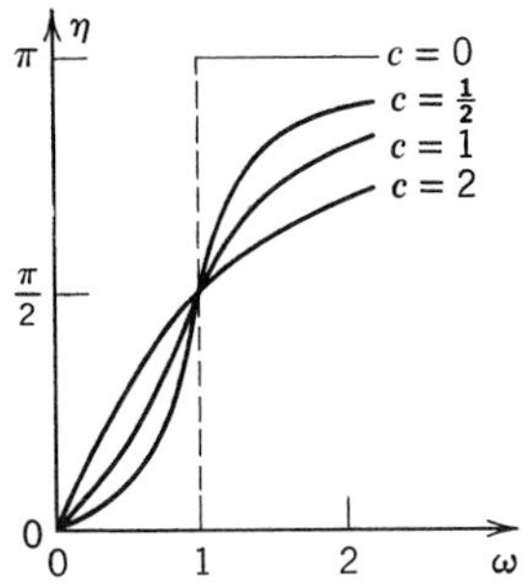

Fig. 50. Phase lag η as a function of ω for $m = 1$, $k = 1$, and various values of the damping constant c

$$c^2 = 2m^2(\omega_0^2 - \omega^2). \tag{15}$$

For sufficiently large damping ($c^2 > 2m^2\omega_0^2 = 2mk$) equation (15) has no real solution, and C^* decreases in a monotone way as ω increases (Fig. 49). If $c^2 \leqq 2mk$, equation (15) has a real solution $\omega = \omega_{max}$, which increases as c decreases and approaches ω_0 as c approaches zero. The amplitude $C^*(\omega)$ has a maximum at $\omega = \omega_{max}$, and by inserting $\omega = \omega_{max}$ into (14) we find that

$$C^*(\omega_{max}) = \frac{2mF_0}{c\sqrt{4m^2\omega_0^2 - c^2}}. \tag{16}$$

We see that $C^*(\omega_{max})$ is finite when $c > 0$. Since $dC^*(\omega_{max})/dc < 0$ when $c^2 < 2mk$, the value of $C^*(\omega_{max})$ increases as c ($\leqq \sqrt{2mk}$) decreases and approaches infinity as c approaches zero, in agreement with our result in Case 1. Figure 49 shows the **amplification** C^*/F_0 (ratio of the amplitudes of output and input) as a function of ω for $m = 1$, $k = 1$, and various values of the damping constant c.

The angle η in (14) is called the **phase angle** or **phase lag** (Fig. 50) because it measures the lag of the output with respect to the input. If $\omega < \omega_0$, then $\eta < \pi/2$; if $\omega = \omega_0$, then $\eta = \pi/2$, and if $\omega > \omega_0$, then $\eta > \pi/2$.

Problems for Sec. 2.13

In each case find the steady-state oscillations.

1. $\ddot{y} + y = 3\cos 2t$

2. $\ddot{y} + 4y = \sin t$

3. $(D^2 + 16)y = \cos t - \sin t$

4. $\ddot{y} + 3\dot{y} + 2y = 10\sin t$

5. $(2D^2 + 2D + 3)y = 87\cos 3t - 5\sin t$

6. $(D^2 + D + 1)y = 2\cos t + 26\cos 2t$

In each case find the transient solution.

7. $\ddot{y} + 3\dot{y} + 2y = 10 \sin t$
8. $\ddot{y} + 4\dot{y} + 3y = 65 \cos 2t$
9. $\ddot{y} + 2\dot{y} + 2y = 85 \cos 3t$
10. $\ddot{y} + 4\dot{y} + 5y = 37.7 \sin 4t$
11. $(D^2 + 2D + 5)y = -\sin t$
12. $(D^2 + 2D + 1)y = 50 \sin 3t$

Solve the following initial value problems.

13. $\ddot{y} + 25y = 24 \sin t, \quad y(0) = 1, \quad \dot{y}(0) = 1$
14. $\ddot{y} + y = -9 \sin 2t, \quad y(0) = 1, \quad \dot{y}(0) = 0$
15. $\ddot{y} + y = \cos \omega t, \quad y(0) = y_0, \quad \dot{y}(0) = v_0, \quad \omega^2 \neq 1$
16. $\ddot{y} + \omega_0^2 y = \cos \omega t, \quad y(0) = y_0, \quad \dot{y}(0) = v_0, \quad \omega^2 \neq \omega_0^2$

17. For what initial conditions does the solution of Prob. 15 represent an oscillation whose frequency equals the frequency of the input? Can we find initial conditions such that the solution of Prob. 16 represents an oscillation whose frequency equals the natural frequency of the system?
18. Solve the initial value problem $\ddot{y} + y = \cos \omega t, \omega^2 \neq 1, y(0) = 0, \dot{y}(0) = 0$. Graph the maximum amplitude as a function of ω. Show that the solution can be written

$$y(t) = \frac{2}{1 - \omega^2} \sin [\tfrac{1}{2}(1 + \omega)t] \sin [\tfrac{1}{2}(1 - \omega)t].$$

Plot good graphs of $y(t)$ with $\omega = 0.5, 0.9, 1.1, 2$.

19. In (12) let ω approach ω_0. Show that this leads to a solution of the form (11).
20. Find a particular solution of the differential equation

$$m\ddot{y} + c\dot{y} + ky = F_0 \sin \omega t \qquad (c \neq 0).$$

Solve the following initial value problems.

21. $\ddot{y} + 10\dot{y} + 29y = 10 \cos t + 28 \sin t, \quad y(0) = 1, \quad \dot{y}(0) = -4$
22. $0.5\ddot{y} + 2\dot{y} + 2.5y = \cos 5t + \sin 5t, \quad y(0) = -0.1, \quad \dot{y}(0) = 0$
23. $\ddot{y} + 2\dot{y} + 2y = \sin 2t - 2 \cos 2t, \quad y(0) = 0, \quad \dot{y}(0) = 0$
24. $\ddot{y} + 4\dot{y} + 20y = 23 \sin t - 15 \cos t, \quad y(0) = 0, \quad \dot{y}(0) = -1$

25. **(Gun barrel)** Solve

$$\ddot{y} + y = \begin{cases} 1 - t^2/\pi^2 & \text{if } 0 \leqq t \leqq \pi \\ 0 & \text{if } t > \pi \end{cases} \qquad y(0) = \dot{y}(0) = 0.$$

This may be interpreted as an undamped system on which a force F acts during some interval of time (see the figure), for instance, the force acting on a gun barrel when a shell is fired, the barrel being braked by heavy springs (and then damped by a dashpot which we disregard for simplicity). *Hint.* At $t = \pi$ both y and $\dot{y}$ must be continuous.

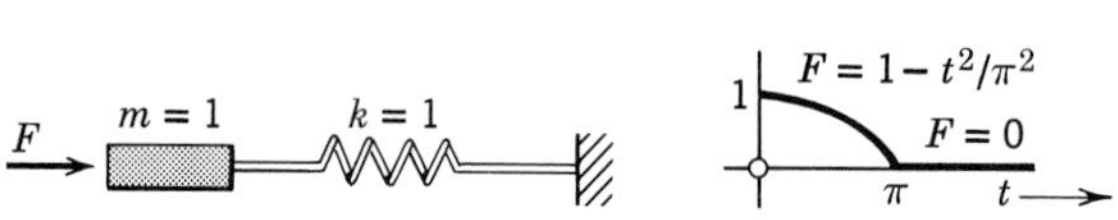

Problem 25

2.14 Electric Circuits

The last section was devoted to the study of a mechanical system which is of great practical interest. We shall now consider a similarly important electrical system, which may be regarded as a basic building block in electrical networks. This consideration will also provide a striking example of the important fact that entirely different physical systems may correspond to the same differential equation, and thus illustrate the role which mathematics plays in unifying various phenomena of entirely different physical nature. We shall obtain a correspondence between mechanical and electrical systems which is not merely qualitative but strictly quantitative in the sense that to a given mechanical system we can construct an electric circuit whose current will give the exact values of the displacement in the mechanical system when suitable scale factors are introduced. The practical importance of such an analogy between mechanical and electrical systems is almost obvious. The analogy may be used for constructing an "electrical model" of a given mechanical system; in many cases this will be an essential simplification, because electric circuits are easy to assemble and currents and voltages are easy to measure, whereas the construction of a mechanical model may be complicated and expensive, and the measurement of displacements will be time-consuming and inaccurate. (For similar ideas in connection with analog computers, see the simple book by Truitt and Rogers [G18] listed in Appendix 1.)

We shall consider the "*RLC*-circuit" in Fig. 51. Simpler circuits were considered in Sec. 1.9, and it will be assumed that the student is familiar with those simpler cases, because they will immediately provide the starting point for obtaining the differential equation of our present circuit. We remember that the differential equations of those circuits were obtained by equating the sum of the voltage drops across the elements of the circuits to the impressed electromotive force. This was done by using Kirchhoff's voltage law, which therefore plays a role in circuits similar to that of Newton's second law (Sec. 2.6) in mechanical systems.

To obtain the differential equation of the circuit in Fig. 51, we may use the equation of an *RC*-circuit [cf. (7), Sec. 1.9] and add to it the voltage drop $L\dot{I}$ across the inductor in our present circuit. This yields

$$L\dot{I} + RI + \frac{1}{C}\int I\,dt = E(t) = E_0 \sin \omega t. \tag{1'}$$

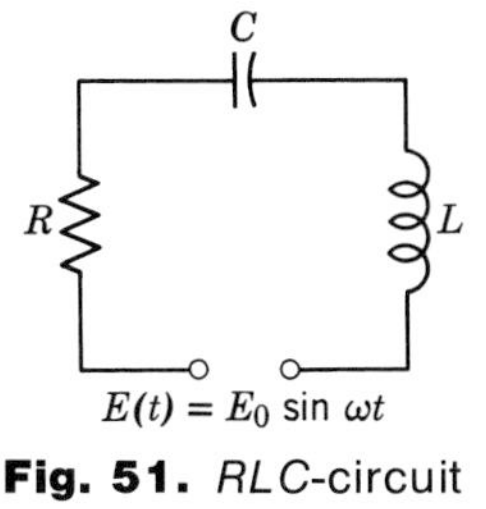

Fig. 51. *RLC*-circuit

From Sec. 1.9 we remember that $I = \dot{Q} = dQ/dt$, that is, the current is the time rate of change of the charge. Hence $\dot{I} = \ddot{Q}$ and $\int I\,dt = Q$ in (1′), so that (1′) can be written

$$L\ddot{Q} + R\dot{Q} + \frac{1}{C}Q = E_0 \sin \omega t. \tag{1''}$$

This is the differential equation for the charge Q on the capacitor in our RLC-circuit. Now in most practical problems, not the charge $Q(t)$ but the current $I(t)$ is the physical quantity of interest. Consequently, more important than (1″) is the differential equation for the current $I(t)$ in the RLC-circuit. We obtain this equation by differentiating (1′) with respect to t. The result is

$$L\ddot{I} + R\dot{I} + \frac{1}{C}I = E_0\omega \cos \omega t. \tag{1}$$

This is of the same form as (2), *Sec.* 2.13. *Hence our RLC-circuit is the electrical analogue of the mechanical system in Sec.* 2.13. *The corresponding analogy of electrical and mechanical quantities is shown in Table* 2.2.

To obtain a particular solution of (1) we may proceed as in Sec. 2.13. By substituting

$$I_p(t) = a \cos \omega t + b \sin \omega t \tag{2}$$

into (1) we obtain

$$a = \frac{-E_0 S}{R^2 + S^2}, \qquad b = \frac{E_0 R}{R^2 + S^2} \tag{3}$$

where S is the so-called **reactance**, given by the expression

$$S = \omega L - \frac{1}{\omega C}. \tag{4}$$

In any practical case, $R \neq 0$ so that the denominator in (3) is not zero. The result is that (2), with a and b given by (3), is a particular solution of (1).

Table 2.2
Analogy of Electrical and Mechanical Quantities in (1), this Section, and (2), Sec. 2.13

Electrical System	Mechanical System
Inductance L	Mass m
Resistance R	Damping constant c
Reciprocal of capacitance $1/C$	Spring modulus k
Derivative $E_0\omega \cos \omega t$ of electromotive force	Driving force $F_0 \cos \omega t$
Current $I(t)$	Displacement $y(t)$

Using (3), we may write I_p in the form

$$I_p(t) = I_0 \sin(\omega t - \theta) \tag{5}$$

where [cf. (14) in Appendix 3]

$$I_0 = \sqrt{a^2 + b^2} = \frac{E_0}{\sqrt{R^2 + S^2}}, \qquad \tan\theta = -\frac{a}{b} = \frac{S}{R}.$$

The quantity $\sqrt{R^2 + S^2}$ is called the **impedance.** Our formula shows that the impedance equals the ratio E_0/I_0, which is somewhat analogous to $E/I = R$ (Ohm's law).

The characteristic equation of the homogeneous differential equation corresponding to (1) is

$$\lambda^2 + \frac{R}{L}\lambda + \frac{1}{LC} = 0$$

and has the roots $\lambda_1 = -\alpha + \beta$ and $\lambda_2 = -\alpha - \beta$ where

$$\alpha = \frac{R}{2L}, \qquad \beta = \frac{1}{2L}\sqrt{R^2 - \frac{4L}{C}}.$$

As in Sec. 2.13 we conclude that if $R > 0$ (which, of course, is true in any practical case) the general solution $I_h(t)$ of the homogeneous equation approaches zero as t approaches infinity (practically: after a sufficiently long time). Hence, the transient current $I = I_h + I_p$ tends to the steady-state current I_p, and *after a sufficiently long time the output will practically be a harmonic oscillation, which is given by* (5) *and whose frequency is that of the input.*

Example 1. *RLC*-circuit

Find the current $I(t)$ in an RLC-circuit with $R = 100$ ohms, $L = 0.1$ henry, $C = 10^{-3}$ farad, which is connected to a source of voltage $E(t) = 155 \sin 377t$ (hence 60 Hz = 60 cycles/sec), assuming zero charge and current when $t = 0$.

Equation (1) is

$$0.1\ddot{I} + 100\dot{I} + 1000I = 155 \cdot 377 \cos 377t.$$

We calculate the reactance $S = 37.7 - 1/0.377 = 35.0$ and the steady-state current

$$I_p(t) = a \cos 377t + b \sin 377t$$

where

$$a = \frac{-155 \cdot 35.0}{100^2 + 35^2} = -0.483, \qquad b = \frac{155 \cdot 100}{100^2 + 35^2} = 1.381.$$

Then we solve the characteristic equation

$$0.1\lambda^2 + 100\lambda + 1000 = 0.$$

The roots are $\lambda_1 = -10$ and $\lambda_2 = -990$. Hence the general solution is

$$I(t) = c_1 e^{-10t} + c_2 e^{-990t} - 0.483 \cos 377t + 1.381 \sin 377t. \tag{6}$$

We determine c_1 and c_2 from the initial conditions $Q(0) = 0$ and $I(0) = 0$. By the second condition,

$$I(0) = c_1 + c_2 - 0.483 = 0. \tag{7}$$

How to use $Q(0) = 0$? Solving (1′) algebraically for $\dot{I}$, we have

$$\dot{I}(t) = \frac{1}{L}\left[E(t) - RI(t) - \frac{1}{C}Q(t)\right] \tag{8}$$

since $\int I\,dt = Q$. Here $E(0) = 0$, $I(0) = 0$ and $Q(0) = 0$, so that $\dot{I}(0) = 0$. Differentiating (6), we thus obtain

$$\dot{I}(0) = -10c_1 - 990c_2 + 1.381 \cdot 377 = 0. \tag{9}$$

The solution of (7) and (9) is $c_1 = -0.043$, $c_2 = 0.526$. From (6) we thus have the answer

$$I(t) = -0.043e^{-10t} + 0.526e^{-990t} - 0.483 \cos 377t + 1.381 \sin 377t.$$

The first two terms will die out rapidly, and after a very short time the current will practically execute harmonic oscillations of frequency 60 Hz = 60 cycles/sec, which is the frequency of the impressed voltage. Note that by (5) we can write the steady-state current in the form

$$I_p(t) = 1.463 \sin (377t - 0.34).$$

Problems for Sec. 2.14

1. Derive (3) in two ways, namely, (*a*) directly by substituting (2) into (1), (*b*) from (5) in Sec. 2.13, using Table 2.2 and taking $E_0\omega$ instead of F_0.
2. Find the natural frequency (= frequency of free oscillations) of an *LC*-circuit (*a*) directly, (*b*) from Sec. 2.6 by means of Table 2.2.
3. What are the conditions for an *RLC*-circuit to be overdamped (Case I), underdamped (Case II) and critically damped (Case III)? In particular, what is the critical resistance R_{crit} (the analogue of the critical damping constant $c_{crit} = 2\sqrt{mk}$)?
4. It was claimed in the text that if $R > 0$, then the transient current approaches $I_p(t)$ as $t \to \infty$. How can this be proved?
5. **(Tuning)** In tuning a radio to a station we turn a knob on the radio which changes C (or perhaps L) in an *RLC*-circuit (Fig. 51) so that the amplitude of the steady-state current becomes maximum. For what C will this be the case?

Find the steady-state current in the *RLC*-circuit in Fig. 52 where

6. $R = 20$ ohms, $L = 10$ henrys, $C = 0.05$ farad, $E = 50 \sin t$ volts
7. $R = 240$ ohms, $L = 40$ henrys, $C = 10^{-3}$ farad, $E = 369 \sin 10t$ volts
8. $R = 40$ ohms, $L = 10$ henrys, $C = 0.02$ farad, $E = 800 \cos 5t$ volts

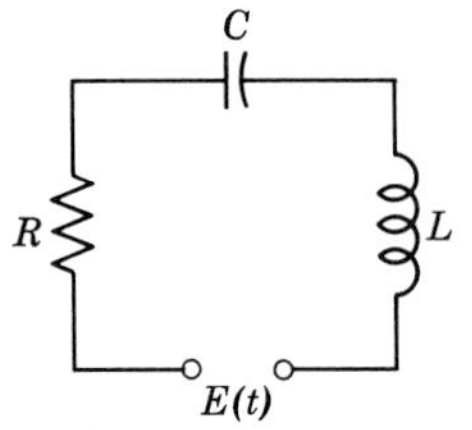

Fig. 52. *RLC*-circuit

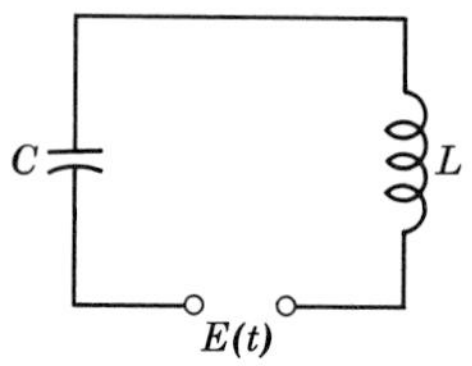

Fig. 53. *LC*-circuit

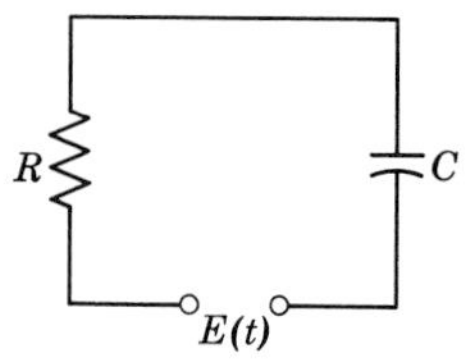

Fig. 54. *RC*-circuit

Find the transient current in the RLC-circuit in Fig. 52 where

9. $R = 200$ ohms, $L = 100$ henrys, $C = 5 \cdot 10^{-3}$ farad, $E = 2500 \sin t$ volts

10. $R = 20$ ohms, $L = 5$ henrys, $C = 0.01$ farad, $E = 850 \sin 4t$ volts

11. $R = 16$ ohms, $L = 8$ henrys, $C = 0.125$ farad, $E = 300 \cos 2t$ volts

Find the current $I(t)$ in the LC-circuit in Fig. 53, assuming zero initial current and charge, and

12. $L = 10$ henrys, $C = 4 \cdot 10^{-3}$ farad, $E = 250$ volts

13. $L = 1$ henry, $C = 0.25$ farad, $E = 90 \cos t$ volts

14. $L = 2$ henrys, $C = 5 \cdot 10^{-3}$ farad, $E = 210 \sin 4t$ volts

Find the current $I(t)$ in the RLC-circuit in Fig. 52, assuming zero initial current and charge, and

15. $R = 80$ ohms, $L = 20$ henrys, $C = 0.01$ farad, $E = 100$ volts

16. $R = 160$ ohms, $L = 20$ henrys, $C = 2 \cdot 10^{-3}$ farad, $E = 481 \sin 10t$ volts

17. $R = 6$ ohms, $L = 1$ henry, $C = 0.04$ farad, $E = 24 \cos 5t$ volts

18. Show that if $E(t)$ in Fig. 53 has a jump of magnitude J at $t = a$, then $\dot{I}(t)$ has a jump of magnitude J/L at $t = a$ while $I(t)$ is continuous at $t = a$.

Using the result of Prob. 18, find the current $I(t)$ in the LC-circuit in Fig. 53, assuming $L = 1$ henry, $C = 1$ farad, zero initial current and charge, and

19. $E = t$ when $0 < t < a$ and $E = 0$ when $t > a$

20. $E = 1$ when $0 < t < a$ and $E = 0$ when $t > a$

21. $E = 1 - e^{-t}$ when $0 < t < \pi$ and $E = 0$ when $t > \pi$

22. Show that if the initial charge in the capacitor in Fig. 54 is $Q(0)$, the initial current in the RC-circuit is

$$I(0) = \frac{E(0)}{R} - \frac{Q(0)}{RC}.$$

23. Show that if $E(t)$ in Fig. 54 has a jump of magnitude J at $t = a$, then the current $I(t)$ in the circuit has a jump of magnitude J/R at $t = a$.

Using the result of Prob. 23, find the current $I(t)$ in the RC-circuit in Fig. 54, assuming $R = 1$ ohm, $C = 1$ farad, zero initial charge on the capacitor, and

24. $E = t$ when $0 < t < a$ and $E = a$ when $t > a$

25. $E = t$ when $0 < t < a$ and $E = 0$ when $t > a$

2.15 Complex Method for Obtaining Particular Solutions

Given an equation of the form (1) in Sec. 2.14, for example,

$$\ddot{I} + \dot{I} + 2I = 6 \cos t, \qquad (1)$$

we know that we can obtain a particular solution $I_p(t)$ by the method of undetermined coefficients, that is, by substituting

$$I_p(t) = a \cos t + b \sin t$$

into (1) and determining a and b. The result will be

$$I_p(t) = 3 \cos t + 3 \sin t.$$

Engineers often prefer a simple and elegant complex method for obtaining $I_p(t)$. In this method we note that $6 \cos t$ in (1) is the real part of

$$6e^{it} = 6(\cos t + i \sin t)$$

(cf. the Euler formula in Sec. 2.4) and, instead of (1), we consider the differential equation

$$\ddot{I} + \dot{I} + 2I = 6e^{it} \qquad (i = \sqrt{-1}). \tag{2}$$

We determine a complex particular solution of the form

$$I_p^*(t) = Ke^{it}. \tag{3}$$

By substituting this function and its derivatives

$$\dot{I}_p^* = iKe^{it}, \qquad \ddot{I}_p^* = -Ke^{it}$$

into the equation (2) we have

$$(-1 + i + 2)Ke^{it} = 6e^{it}.$$

Solving for K, we obtain

$$K = \frac{6}{1+i} = \frac{6(1-i)}{(1+i)(1-i)} = 3 - 3i.$$

Hence

$$I_p^*(t) = (3 - 3i)e^{it} = (3 - 3i)(\cos t + i \sin t)$$

is a solution of (2). The real part of I_p^* is

$$I_p(t) = 3 \cos t + 3 \sin t$$

and this function is a solution of the real part of the differential equation (2), that is, of the given differential equation (1). In fact, the function is identical with that obtained above. This illustrates the practical procedure of the complex method.

Our equation (1) is a particular case of the equation [cf. (1), Sec. 2.14]

(4) $$L\ddot{I} + R\dot{I} + \frac{1}{C}I = E_0\omega \cos \omega t$$

which we shall now consider, assuming that $R \neq 0$. The corresponding complex equation is

(5) $$L\ddot{I} + R\dot{I} + \frac{1}{C}I = E_0\omega e^{i\omega t}.$$

The function on the right suggests a particular solution of the form

(6) $$I_p^*(t) = Ke^{i\omega t} \qquad (i = \sqrt{-1}).$$

By substituting this function and its derivatives

$$\dot{I}_p^* = i\omega Ke^{i\omega t} \qquad \ddot{I}_p^* = -\omega^2 Ke^{i\omega t}$$

into (5) we immediately have

$$\left(-\omega^2 L + i\omega R + \frac{1}{C}\right)Ke^{i\omega t} = E_0\omega e^{i\omega t}.$$

Solving this equation for K, we obtain

(7) $$K = \frac{E_0\omega}{\dfrac{1}{C} - \omega^2 L + i\omega R} = \frac{E_0}{iZ}$$

where Z is the so-called **complex impedance,** given by

(8) $$Z = R + i\left(\omega L - \frac{1}{\omega C}\right).$$

It follows that (6) with K given by (7) is a solution of (5).

We see that the imaginary part of Z is the reactance S defined by (4), Sec. 2.14, and $|Z|$ is the impedance defined in connection with (5), Sec. 2.14. Thus (cf. Fig. 55)

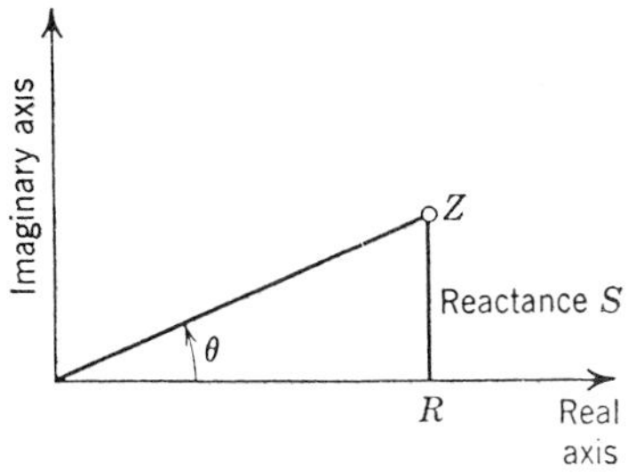

Fig. 55. Complex impedance Z

$$Z = |Z|e^{i\theta} \quad \text{where} \quad \tan\theta = \frac{S}{R}.$$

Consequently, formula (6) with K according to (7) can be written in the form

$$I_p^*(t) = \frac{E_0}{iZ}e^{i\omega t} = -i\frac{E_0}{|Z|}e^{i(\omega t-\theta)}.$$

The real part is

$$(9) \qquad I_p(t) = \frac{E_0}{|Z|}\sin(\omega t - \theta) = \frac{E_0}{\sqrt{R^2+S^2}}\sin(\omega t - \theta),$$

and this is a solution of the real part of the differential equation (5), that is, $I_p(t)$ is a solution of (4). We see that this function is identical with (5) in the previous section. Since

$$\sin(\omega t - \theta) = \sin\omega t\cos\theta - \cos\omega t\sin\theta$$

and furthermore

$$\cos\theta = \frac{\operatorname{Re} Z}{|Z|} = \frac{R}{\sqrt{R^2+S^2}}, \qquad \sin\theta = \frac{\operatorname{Im} Z}{|Z|} = \frac{S}{\sqrt{R^2+S^2}},$$

the solution can be written in the form

$$(10) \qquad I_p(t) = \frac{E_0}{R^2+S^2}[R\sin\omega t - S\cos\omega t].$$

This is identical with (2) in the previous section, where a and b are given by (3) in that section.

Problems for Sec. 2.15

Using the complex method, find the steady-state current $I_p(t)$ in the circuit governed by (4), where

1. $R = 20,\ L = 10,\ C = 0.05,\ E_0 = 5,\ \omega = 2$

2. $R = 100,\ L = 100,\ C = 10^{-3},\ E_0 = 3,\ \omega = 4$

3. $R = 50,\ L = 25,\ C = 0.01,\ E_0 = 500,\ \omega = 3$

4. $R = 10,\ L = 20,\ C = 0.1,\ E_0 = 4,\ \omega = 2$

5. $R = 25,\ L = 15,\ C = 0.05,\ E_0 = 100,\ \omega = 4$

6. Find the complex impedance Z and the reactance S in Prob. 5.

Using the complex method, find the steady-state output of the following equations.

7. $\ddot{y} + 3\dot{y} + 16y = 24\cos 4t$

8. $\ddot{y} + \dot{y} + 4y = 8\sin 2t$

9. $\ddot{y} + 5\dot{y} + \frac{1}{8}y = 25\cos 10t$

10. $\ddot{y} + \dot{y} + 9y = -3\sin 3t$

2.16 General Method for Solving Nonhomogeneous Equations

The method of undetermined coefficients in Sec. 2.12 is a simple procedure for obtaining particular solutions of nonhomogeneous linear differential equations, and we have seen in Secs. 2.13–2.15 that it has important applications in connection with vibrations. However, it is limited to certain simpler types of differential equations. The method to be discussed in this section will be completely general but more complicated than that method.

We consider linear differential equations of the form

$$y'' + f(x)y' + g(x)y = r(x) \tag{1}$$

assuming that f, g, and r are continuous on an open interval I. We shall obtain a particular solution of (1) by using the method of **variation of parameters**[16] as follows.

We know that the corresponding homogeneous differential equation

$$y'' + f(x)y' + g(x)y = 0 \tag{2}$$

has a general solution $y_h(x)$ on I, which is of the form

$$y_h(x) = c_1 y_1(x) + c_2 y_2(x).$$

The method consists in replacing c_1 and c_2 by functions $u(x)$ and $v(x)$ to be determined so that the resulting function

$$y_p(x) = u(x)y_1(x) + v(x)y_2(x) \tag{3}$$

is a particular solution of (1) on I. By differentiating (3) we obtain

$$y_p' = u'y_1 + uy_1' + v'y_2 + vy_2'.$$

We shall see that we can determine u and v such that

$$u'y_1 + v'y_2 = 0. \tag{4}$$

This reduces the expression for y_p' to the form

$$y_p' = uy_1' + vy_2'. \tag{5}$$

By differentiating this function we have

$$y_p'' = u'y_1' + uy_1'' + v'y_2' + vy_2''. \tag{6}$$

[16] The student may wish to review the simpler application of the method in Sec. 1.8.

By substituting (3), (5), and (6) into (1) and collecting terms containing u and terms containing v we readily obtain

$$u(y_1'' + fy_1' + gy_1) + v(y_2'' + fy_2' + gy_2) + u'y_1' + v'y_2' = r.$$

Since y_1 and y_2 are solutions of the homogeneous equation (2), this reduces to

$$u'y_1' + v'y_2' = r.$$

Equation (4) is

$$u'y_1 + v'y_2 = 0.$$

This is a system of two linear algebraic equations for the unknown functions u' and v'. The solution is obtained by Cramer's rule (cf. Sec. 7.9). We find

$$u' = -\frac{y_2 r}{W}, \qquad v' = \frac{y_1 r}{W}, \tag{7}$$

where

$$W = y_1 y_2' - y_1' y_2$$

is the Wronskian of y_1 and y_2 (cf. Sec. 2.8). Clearly, $W \neq 0$ since y_1, y_2 constitute a basis of solutions. Integration of (7) gives

$$u = -\int \frac{y_2 r}{W} dx, \qquad v = \int \frac{y_1 r}{W} dx.$$

These integrals exist because $r(x)$ is continuous. Substituting these expressions for u and v into (3), we obtain the desired solution of (1),

$$y_p(x) = -y_1 \int \frac{y_2 r}{W} dx + y_2 \int \frac{y_1 r}{W} dx. \tag{8}$$

Note that if the constants of integration in (8) are left arbitrary, then (8) represents a general solution of (1).

Example 1

Solve the differential equation

$$y'' + y = \sec x. \tag{9}$$

The method of undetermined coefficients (Sec. 2.12) cannot be applied. The functions

$$y_1 = \cos x, \qquad y_2 = \sin x$$

constitute a basis of solutions of the homogeneous equation. Their Wronskian is

$$W(y_1, y_2) = \cos x \cos x - (-\sin x) \sin x = 1.$$

From (8) we thus obtain the following particular solution of (9):

$$y_p = -\cos x \int \sin x \sec x \, dx + \sin x \int \cos x \sec x \, dx$$

$$= \cos x \ln |\cos x| + x \sin x.$$

The corresponding general solution of the differential equation (9) is

$$y = y_h + y_p = [c_1 + \ln |\cos x|] \cos x + (c_2 + x) \sin x.$$

Problems for Sec. 2.16

Find a general solution of the following equations.

1. $y'' + y = \csc x + x$

2. $y'' + 9y = \sec 3x$

3. $y'' - 4y' + 4y = e^{2x}/x$

4. $y'' + 6y' + 9y = e^{-3x}/(x^2 + 1)$

5. $y'' + 2y' + y = e^{-x} \cos x$

6. $y'' + 2y' + y = e^{-x} \ln x$

7. $(D^2 - 2D + 1)y = x^{3/2}e^x$

8. $(D^2 - 2D + 1)y = e^x/x^3$

9. $(D^2 + 2D + 2)y = e^{-x}/\cos^3 x$

10. $(D^2 - 4D + 5)y = e^{2x}/\sin x$

11. $(x^2D^2 + xD - 1)y = 4$

12. $(x^2D^2 - 2)y = 3x^2$

13. $(x^2D^2 - 2xD + 2)y = x^4$

14. $(x^2D^2 + xD - 0.25)y = 1/x$

15. $(x^2D^2 - 2xD + 2)y = 1/x^2$

16. $(x^2D^2 - 2xD + 2)y = x^3 \cos x$

17. $x^2y'' - 4xy' + 6y = 1/x^4$

18. $x^2y'' - 4xy' + 6y = x^4 \sin x$

19. $xy'' - y' = 2x^2e^x$

20. $x^2y'' + xy' - y = x^3e^x$

CHAPTER

3 SYSTEMS OF DIFFERENTIAL EQUATIONS, PHASE PLANE, STABILITY

Section 3.1 is devoted to systems of differential equations. This is an elementary introduction, without employing matrices. (A matrix approach to systems will be presented in Sec. 7.15.)

In Sec. 3.2 we consider differential equations in the phase plane.

Stability is a concept of increasing importance in modern engineering mathematics, for instance, in connection with feedback and controls. It is suggested by physics, where it means that, roughly speaking, small changes of a physical system at some instant cause only small changes in the behavior of the system at all later times. We consider stability in Sec. 3.3, in connection with differential equations, using the phase plane.

Prerequisite for this chapter: Chaps. 1 and 2.
References: Appendix 1, Part B.
Answers to problems: Appendix 2.

3.1 Systems of Differential Equations

Systems of differential equations have important applications. For instance, they arise quite frequently as models of mechanical or electrical systems which are combinations of the simple systems discussed in Secs. 2.6, 2.13, and 2.14.

The present section contains an elementary[1] approach to systems of differential equations, and we shall consider the method of **solution by elimination.** In this method, unknown functions and their derivatives are successively eliminated until one arrives at a single higher order differential equation containing only one unknown function and its derivatives. This equation is solved, and then the other unknown functions are found in turn. Let us explain the details in terms of a typical example.

Example 1

Consider the system

$$(1)\qquad \begin{aligned} &\text{(a)}\quad \dot{x} = 4x - 2y \\ &\text{(b)}\quad \dot{y} = x + y \end{aligned}$$

[1]That is, without the use of vectors and matrices. The vector approach is included in the chapter on matrices (in Sec. 7.15). Of course, readers familiar with vectors and matrices may immediately study Sec. 7.15.

consisting of two first-order linear differential equations. x and y are unknown functions of t. By a *solution* of (1) we mean a pair of functions x, y which when substituted into (1) reduce these equations to an identity. We solve the system by elimination, as follows. From (1a) we have

$$y = -\tfrac{1}{2}\dot{x} + 2x. \tag{2}$$

We differentiate (1a). In the resulting equation we first substitute $\dot{y}$ as given by (1b) and y as given by (2). This yields

$$\begin{aligned}\ddot{x} &= 4\dot{x} - 2\dot{y}\\ &= 4\dot{x} - 2(x + y)\\ &= 4\dot{x} - 2x - 2(-\tfrac{1}{2}\dot{x} + 2x).\end{aligned}$$

Ordering terms and simplifying, we have

$$\ddot{x} - 5\dot{x} + 6x = 0.$$

A general solution is

$$x = c_1e^{3t} + c_2e^{2t}. \tag{3a}$$

By differentiation,

$$\dot{x} = 3c_1e^{3t} + 2c_2e^{2t}.$$

Hence from (2),

$$y = \tfrac{1}{2}c_1e^{3t} + c_2e^{2t}. \tag{3b}$$

∎

More generally, the method of elimination applies to systems of the form

$$\text{(4)}\qquad \begin{aligned}&\text{(a)} \quad \dot{x} = a_1x + b_1y + f_1(t)\\ &\text{(b)} \quad \dot{y} = a_2x + b_2y + f_2(t)\end{aligned}$$

where f_1 and f_2 are given functions and a_1, b_1, a_2, b_2 are constants. If $b_1 = a_2 = 0$, the system breaks up into two separate first-order linear differential equations which can be solved by the method of Sec. 1.7. If one of these two constants, say b_1, is not zero, we may proceed in a fashion similar to that in Example 1. That is, we first obtain from (4a)

$$b_1y = \dot{x} - a_1x - f_1. \tag{5}$$

We now differentiate (4a). In the resulting equation we substitute $\dot{y}$ as given by (4b) and then b_1y as given by (5). This yields

$$\begin{aligned}\ddot{x} &= a_1\dot{x} + b_1\dot{y} + \dot{f}_1\\ &= a_1\dot{x} + b_1(a_2x + b_2y + f_2) + \dot{f}_1\\ &= a_1\dot{x} + b_1a_2x + b_2(\dot{x} - a_1x - f_1) + b_1f_2 + \dot{f}_1.\end{aligned}$$

Collecting terms, we have

$$\ddot{x} - (a_1 + b_2)\dot{x} + (a_1b_2 - b_1a_2)x = r(t)$$

where

$$r(t) = b_1 f_2(t) - b_2 f_1(t) + \dot{f}_1(t).$$

By solving this equation we get x. Then y is obtained from (5).

Can the method of elimination be extended to include more general systems of practical interest? The answer is yes, and we discuss two such extensions in terms of typical examples.

Example 2. Mechanical system

Figure 56 shows a mechanical system consisting of two masses on two springs. Set up a mathematical model and solve it; that is, find the displacements $y_1(t)$ and $y_2(t)$ of the masses from their positions of static equilibrium ($y_1 = 0$ and $y_2 = 0$) under the assumptions made in Sec. 2.6 (vertical motion, no damping, masses of springs neglected).

Solution. As in Sec. 2.6, the differential equations governing our mechanical system are obtained from Newton's second law:

$$\ddot{y}_1 = -3y_1 + 2(y_2 - y_1)$$
$$\ddot{y}_2 = -2(y_2 - y_1).$$

This can be written

$$\text{(6)} \qquad \begin{aligned} &\text{(a)} \quad \ddot{y}_1 = -5y_1 + 2y_2 \\ &\text{(b)} \quad \ddot{y}_2 = \ \ 2y_1 - 2y_2. \end{aligned}$$

From (6a) we have

$$\text{(7)} \qquad y_2 = \tfrac{1}{2}\ddot{y}_1 + \tfrac{5}{2}y_1.$$

In Example 1 we differentiated once. Here we differentiate (6a) twice. In the resulting equation we substitute $\ddot{y}_2$ as given by (6b) and y_2 as given by (7). This yields

$$\begin{aligned} y_1^{(iv)} &= -5\ddot{y}_1 + 2\ddot{y}_2 \\ &= -5\ddot{y}_1 + 2(2y_1 - 2y_2) \\ &= -5\ddot{y}_1 + 4y_1 - 4(\tfrac{1}{2}\ddot{y}_1 + \tfrac{5}{2}y_1). \end{aligned}$$

Ordering terms, we have

$$\text{(8)} \qquad y_1^{(iv)} + 7\ddot{y}_1 + 6y_1 = 0.$$

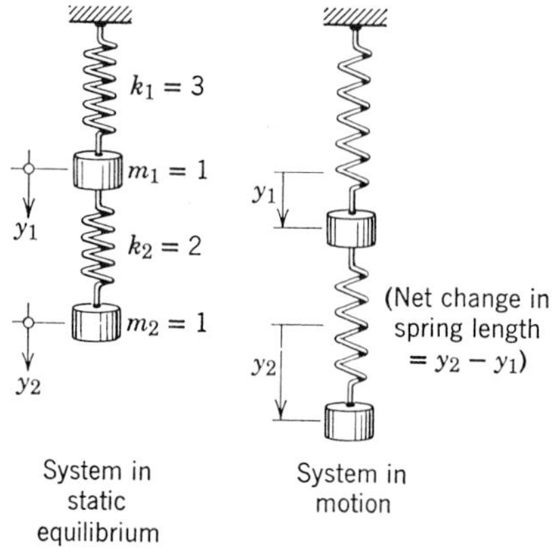

Fig. 56. Example 2

The characteristic equation $\lambda^4 + 7\lambda^2 + 6 = 0$ is quadratic in $p = \lambda^2$. The roots are $p_1 = -1$ and $p_2 = -6$. Hence λ has the four values i, $-i$, $\sqrt{6}\,i$, $-\sqrt{6}\,i$. We thus obtain the general solution of (8),

$$y_1 = a_1 \cos t + b_1 \sin t + a_2 \cos \sqrt{6}t + b_2 \sin \sqrt{6}t. \tag{9a}$$

Differentiating twice, we have

$$\ddot{y}_1 = -a_1 \cos t - b_1 \sin t - 6a_2 \cos \sqrt{6}t - 6b_2 \sin \sqrt{6}t.$$

Using this and (9a), we obtain from (7)

$$y_2 = 2a_1 \cos t + 2b_1 \sin t - \tfrac{1}{2}a_2 \cos \sqrt{6}t - \tfrac{1}{2}b_2 \sin \sqrt{6}t. \tag{9b}$$

This solution (9) of (6) represents harmonic oscillations of the two masses of our mechanical system.

Example 3. Electrical network

Find the currents $I_1(t)$ and $I_2(t)$ in the network shown in Fig. 57, assuming that all charges and currents are zero when the switch is closed at $t = 0$.

Solution. The mathematical model of the network is obtained from Kirchhoff's voltage law, as in Sec. 2.14. The left loop yields

$$\dot{I}_1 + 4(I_1 - I_2) = 12.$$

For the right loop we obtain

$$6I_2 + 4(I_2 - I_1) + 4\int I_2\,dt = 0.$$

Rewriting the first equation and differentiating the second, we see that the currents I_1 and I_2 in the network are governed by the system

$$(10)\qquad \begin{aligned} &\text{(a)} & \dot{I}_1 + 4I_1 - 4I_2 &= 12 \\ &\text{(b)} & -4\dot{I}_1 + 10\dot{I}_2 + 4I_2 &= 0. \end{aligned}$$

Note that the second equation involves the derivatives of both unknown functions. We solve (10). From (10a) we have

$$I_2 = \tfrac{1}{4}\dot{I}_1 + I_1 - 3. \tag{11}$$

By differentiation,

$$\dot{I}_2 = \tfrac{1}{4}\ddot{I}_1 + \dot{I}_1.$$

Substituting this and (11) into (10b), we obtain

$$-4\dot{I}_1 + 10(\tfrac{1}{4}\ddot{I}_1 + \dot{I}_1) + 4(\tfrac{1}{4}\dot{I}_1 + I_1 - 3) = 0.$$

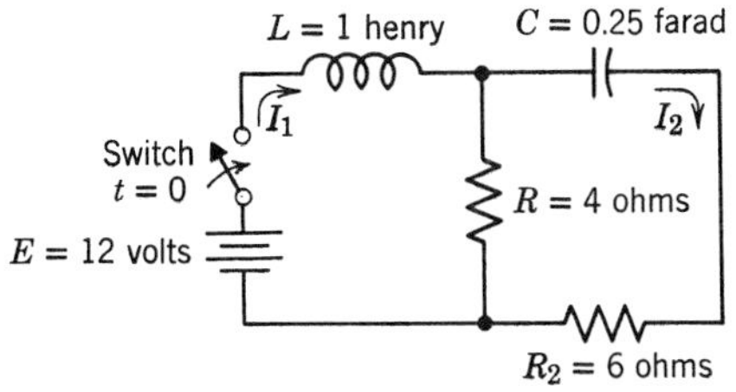

Fig. 57. Example 3

Simplification yields

$$\ddot{I}_1 + \tfrac{14}{5}\dot{I}_1 + \tfrac{8}{5}I_1 = \tfrac{24}{5}.$$

A general solution of this equation is

$$I_1 = c_1e^{-2t} + c_2e^{-0.8t} + 3. \tag{12a}$$

Its derivative is given by

$$\dot{I}_1 = -2c_1e^{-2t} - 0.8c_2e^{-0.8t}.$$

We can now calculate I_2 from this and (11), finding

$$I_2 = \tfrac{1}{2}c_1e^{-2t} + \tfrac{4}{5}c_2e^{-0.8t}. \tag{12b}$$

The initial conditions are $I_1(0) = 0$ and $I_2(0) = 0$. Hence by (12),

$$I_1(0) = c_1 + c_2 + 3 = 0$$

$$I_2(0) = \tfrac{1}{2}c_1 + \tfrac{4}{5}c_2 = 0.$$

This gives $c_1 = -8$ and $c_2 = 5$. Hence the solution of our physical problem is

$$I_1 = -8e^{-2t} + 5e^{-0.8t} + 3$$

$$I_2 = -4e^{-2t} + 4e^{-0.8t}.$$

We see that I_1 approaches the value of 3 amperes, whereas I_2 approaches 0 as $t \to \infty$. This had to be expected. Why?

Problems for Sec. 3.1

Solve the following systems of differential equations (where $D = d/dt$).

1. $\dot{x} = y$, $\dot{y} = x$

2. $Dx = x + y$, $Dy = 4x + y$

3. $\dot{x} + \dot{y} = \cos t$, $\dot{x} - \dot{y} = \sin t$

4. $(D + 4)x + 6y = 0$, $(D - 1)y - x = 0$

5. $\ddot{x} = y + 1$, $\ddot{y} = x + t$

6. $(D - 2)x + 2Dy = 2 - 4e^{2t}$, $(2D - 3)x + (3D - 1)y = 0$

Find the solution satisfying the given conditions.

7. $Dx = 3x + 4y$, $Dy = 4x - 3y$, $x(0) = 1,\ y(0) = 3$

8. $\dot{x} = x + 2y$, $\dot{y} = -8x + 11y$, $x(0) = 1,\ y(0) = 1$

9. $(D - 2)x - 3y = 2e^{2t}$, $-x + (D - 4)y = 3e^{2t}$, $x(0) = -2/3,\ y(0) = 1/3$

10. $\dot{x} = 5x + 4y - 5t^2 + 6t + 25$, $\dot{y} = x + 2y - t^2 + 2t + 4$, $x(0) = 0,\ y(0) = 0$

11. The solution (9) in Example 2 is an interesting superposition of two harmonic oscillations. Show that special cases are

$$(i) \quad y_1 = \sin t, \qquad y_2 = 2\sin t$$

$$(ii) \quad y_1 = \sin\sqrt{6}t, \qquad y_2 = -\tfrac{1}{2}\sin\sqrt{6}t.$$

Show that in (i) at each instant the masses are moving both upward or both downward, so that, at any given time, the springs are both compressed or both extended. Graph (i), choosing separate coordinate systems such that the axes are parallel and the origin of the ty_2-system is two units vertically below the origin of the ty_1-system. Discuss and graph (ii) in a similar fashion. Compare. Is it plausible that (ii) has a higher frequency than (i)?

12. Set up a model for the mechanical system in the figure. Solve it under the physical assumptions made in Example 2. In particular, find, graph and compare the solutions corresponding to zero initial velocities and initial displacements

$$(i) \qquad y_1(0) = 1, \quad y_2(0) = 1$$
$$(ii) \qquad y_1(0) = 1, \quad y_2(0) = -1.$$

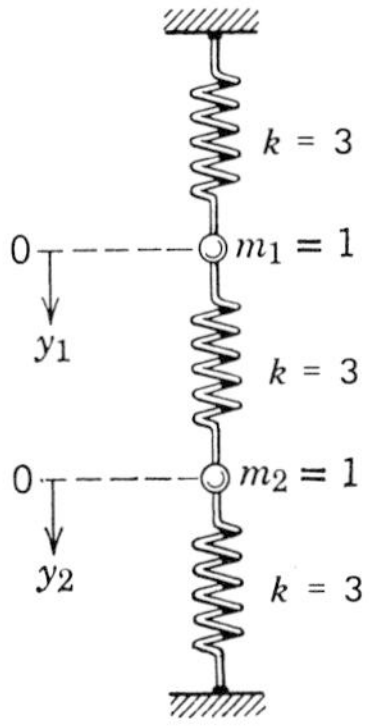

Problem 12 (System in static equilibrium)

13. Solve the model for the network in Fig. 57, assuming that $E = 100$ volts, the other data being as before. Compare with Example 3 and comment.

Set up a model (system of differential equations) and solve it. Determine also the steady-state currents.

14.

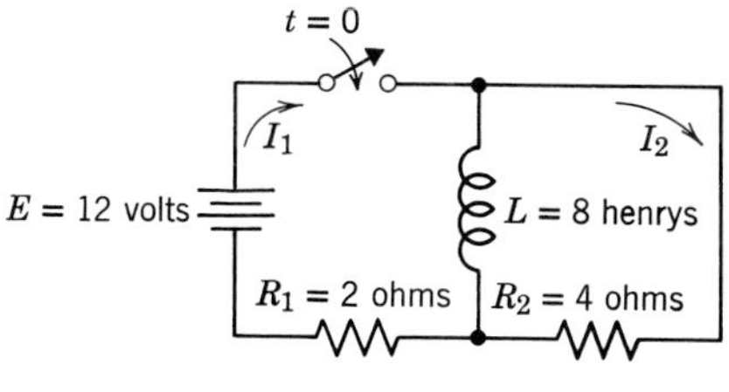

15.

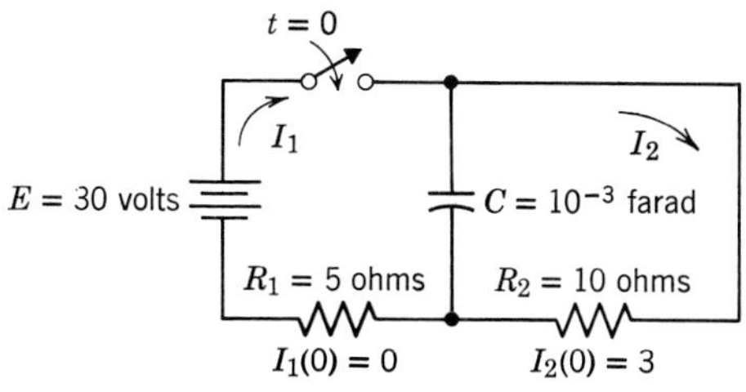

3.2 Phase Plane

A physical system is said to be *autonomous* if its differential equation does not contain the independent variable (time t, say) explicitly. Hence if the differential equation is of second order, it is of the form

$$F(y, \dot{y}, \ddot{y}) = 0. \tag{1}$$

Here $\dot{y} = dy/dt = v$ is the velocity. By the chain rule,

$$\dot{v} = \frac{dv}{dt} = \frac{dv}{dy}\frac{dy}{dt} = \frac{dv}{dy}v. \tag{2}$$

We thus obtain a first-order differential equation for v as a function of the variable y, which now becomes the *independent* variable. (Cf. also the problem set of Sec. 2.1.) Solutions of this new differential equation represent curves in the yv-plane. The yv-plane is called the **phase plane.** This term is suggested by mechanics, where "phase plane" means the yp-plane, $p = mv$ being the momentum and m the mass of a moving particle.

Example 1. Harmonic oscillations

The autonomous vibrating system governed by (cf. Sec. 2.6)

$$\ddot{y} + y = 0, \qquad y(0) = 0, \qquad \dot{y}(0) = 1$$

has the solution (Fig. 58*a*)

$$y = \sin t.$$

Using (2), we obtain

$$\frac{dv}{dy}v + y = 0, \qquad v(0) = 1.$$

Hence

$$v\,dv + y\,dy = 0, \qquad \text{and} \qquad v^2 + y^2 = 1.$$

This shows that in the phase plane a harmonic oscillation appears as a circle (Fig. 58*b*). Note that to each t there corresponds a point on the circle with coordinates $y(t)$, $v(t)$. If we wish, we can label the points as shown in Fig. 58*b* for $t = 0$ and $t = \pi$ and indicate the sense of increasing t by an arrow. ▮

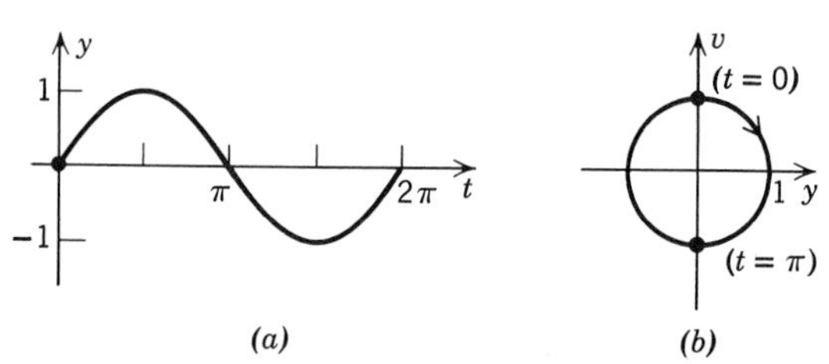

Fig. 58. Graphs of a harmonic oscillation (*a*) in the *ty*-plane, (*b*) in the phase plane

The new equation obtained by the use of (2) is of first order. Hence we may use a direction field (Sec. 1.2) in the phase plane for discussing the general behavior of solutions $v(y)$ without actually solving that equation. Engineers make frequent use of this possibility, particularly in connection with nonlinear differential equations. Let us illustrate this with a very important example.

Example 2. Self-sustained oscillations, van der Pol equation

There are physical systems such that for small oscillations, energy is fed into the system, whereas for large oscillations, energy is taken from the system. In other words, large oscillations will be damped, whereas for small oscillations there is "negative damping" (feeding of energy into the system). For physical reasons we expect such a system to approach a periodic behavior, which will thus appear as a closed curve in the phase plane, called a **limit cycle.** A differential equation describing such vibrations is the famous *van der Pol equation*

$$\ddot{y} - \mu(1 - y^2)\dot{y} + y = 0 \qquad (\mu > 0). \tag{3}$$

It occurs in the study of electrical circuits with vacuum tubes. For $\mu = 0$ this is $\ddot{y} + y = 0$ and we obtain harmonic oscillations. Let $\mu > 0$. The damping term has the coefficient $-\mu(1 - y^2)$. This is negative for small oscillations, namely, $y^2 < 1$, so that we have "negative damping," is zero for $y^2 = 1$ (no damping) and is positive if $y^2 > 1$ (positive damping, loss of energy). If μ is small, we expect a limit cycle which is almost a circle because then our equation differs but little from $\ddot{y} + y = 0$. If μ is large, the limit cycle will probably look different.

Using $\ddot{y} = v\,dv/dy$, we have from (3)

$$\frac{dv}{dy}v - \mu(1 - y^2)v + y = 0.$$

The isoclines are $dv/dy = k = const$, that is,

$$\frac{dv}{dy} = \mu(1 - y^2) - \frac{y}{v} = k. \tag{4}$$

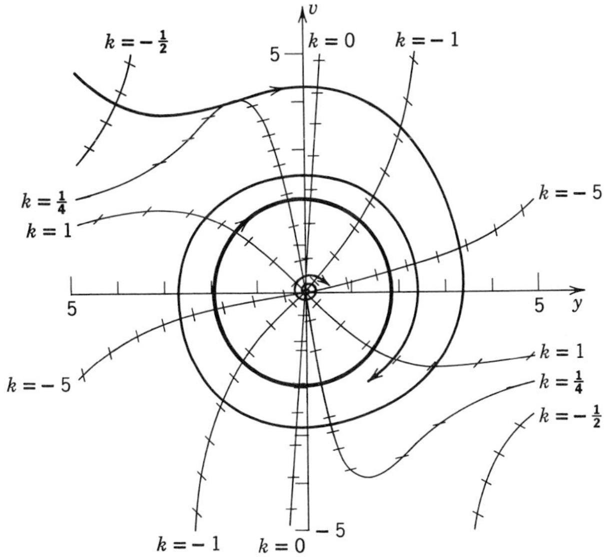

Fig. 59. Lineal element diagram for the van der Pol equation with $\mu = 0.1$ in the phase plane, showing also the limit cycle and two solution curves

This yields

$$v = \frac{y}{\mu(1 - y^2) - k}.$$

These curves are a bit complicated. Figure 59 shows some of them for a small $\mu = 0.1$ as well as the limit cycle (almost a circle) and two solution curves approaching the limit cycle, one from the outside and one from the inside. The latter is a narrow spiral and only the initial part of it is shown in the figure. For larger μ the situation changes and the limit cycle no longer resembles a circle. Figure 60 illustrates this for $\mu = 1$. Note that the approach of solution curves $v(y)$ to the limit cycle is much more rapid than for $\mu = 0.1$.

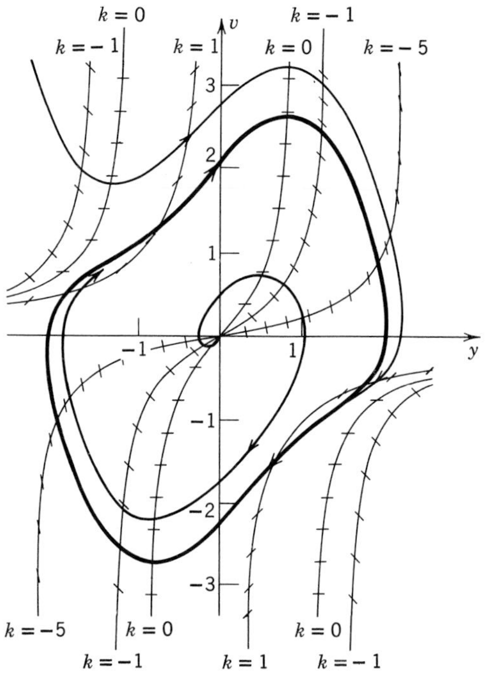

Fig. 60. Lineal element diagram for the van der Pol equation with $\mu = 1$ in the phase plane, showing also the limit cycle and two solution curves approaching it

Problems for Sec. 3.2

1. In Example 1, a general solution is $y = A \cos t + B \sin t$. What is the radius of the corresponding circle in the phase plane?
2. Consider the damped oscillation $y(t) = e^{-t} \sin t$. Plot some points of the corresponding curve in the phase plane to obtain the impression that this must be a spiral.

Solve the following equations and indicate the type of corresponding curves in the phase plane.

3. $4\ddot{y} + y = 0$
4. $\ddot{y} + \omega^2 y = 0$
5. $\ddot{y} - y = 0$
6. $\ddot{y} - k^2 y = 0$

Reduce to first order, solve and graph some of the solution curves in the phase plane.

7. $\ddot{y} = \dot{y}$ **8.** $\ddot{y} + 3\dot{y} = 0$

9. $y\ddot{y} + \dot{y}^2 = 0$ **10.** $\ddot{y} + \dot{y}^2 = 0$

11. (Pendulum) Show that the motion of an undamped pendulum of length l (Fig. 40 in Sec. 2.6) is governed by the nonlinear equation

$$\ddot{\theta} + \frac{g}{l}\sin\theta = 0$$

where $g = 9.80 \text{ m/sec}^2$ ($= 32.17 \text{ ft/sec}^2$) is the acceleration of gravity at the surface of the earth. Graph the phase plane, assuming $l = 2$ m and the curve corresponding to the solution $\theta(t)$ for which $\theta(0) = 0$, $\dot{\theta}(0) = 1$ m/sec.

3.3 Critical Points. Stability

The phase plane can give information about the general behavior of solutions of differential equations without actually solving the equations. The more complicated the equations are, the more important this approach becomes.

In this section we shall see that systems of differential equations can also be studied in the phase plane. This will lead, in a rather natural way, to stability considerations. Stability concepts are suggested by physics, where *stability* means, roughly speaking, that a small change (small disturbance) of a physical system at some instant changes the behavior of the system only slightly at all future times.

We first note that a differential equation

$$\ddot{y} = G(y, \dot{y})$$

can be written as a system

$$\dot{y} = v$$

$$\dot{v} = G(y, v)$$

and a solution $y(t)$, $v(t)$ of this system represents a curve in the yv-plane, the phase plane (Sec. 3.2).

For our present more general discussion it is convenient to change our notation, replacing y by x and v by y. Then the phase plane is the xy-plane. And our system is $\dot{x} = y$, $\dot{y} = G(x, y)$. More generally, we consider systems of the form

$$\dot{x} = F(x, y)$$
(1)
$$\dot{y} = G(x, y)$$

where y is any variable, not necessarily $\dot{x}$. A solution $x(t)$, $y(t)$ of (1) represents a curve C in the xy-plane (or a point as a degenerate case). This curve is called a

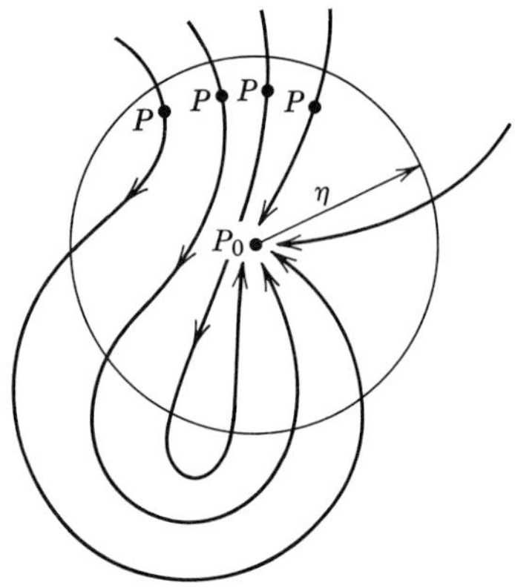

Fig. 61. Attractive critical point P_0 of (1)

solution curve or **path** (sometimes a *trajectory*) of (1). The sense of increasing t is called the *positive sense* on C and can be marked by an arrow head. This defines an *orientation* on C. If t is time and C the path of a moving body, the positive sense is the sense in which the body moves along C as time progresses.

From (1) we see that the slope of a path passing through a point P: (x, y) is

$$\frac{dy}{dx} = \frac{dy/dt}{dx/dt} = \frac{G(x, y)}{F(x, y)}. \tag{2}$$

Note that (2) gives no information about the orientation of a path. Note further that we must have $F(x, y) \neq 0$ at P. If $F(x, y) = 0$ but $G(x, y) \neq 0$ at P, we can take $dx/dy = F(x, y)/G(x, y)$ instead of (2) and conclude from $dx/dy = 0$ that the tangent of C at P is vertical. However, what can we do if both F and G are zero at some point? This problem is the main topic of the present section and will lead to interesting results of practical importance.

A point P_0: (x_0, y_0) at which both F and G are zero is called a **critical point** of (1).

There are different types of critical points P_0 depending on the behavior of paths near such a point. In discussing this we assume that P_0 is *isolated,* that is, P_0 is the only critical point of (1) in some circular disk about P_0. For our purpose the following concepts will be fundamental.

A critical point P_0 of (1) is called **attractive** if, roughly speaking, P_0 eventually attracts all paths of (1) which at some instant are sufficiently close to P_0; precisely: if there is a disk D of radius $\eta > 0$ about P_0 such that every path of (1) which has a point in D (marked P in Fig. 61) approaches P_0 as $t \to \infty$.

A critical point P_0 of (1) is called **stable**[2] if, roughly speaking, all paths of (1) which at some instant are sufficiently close to P_0 remain close to P_0 at all future times; precisely: if for every disk D_ϵ of radius $\epsilon > 0$ about P_0 there is a disk D_δ of radius $\delta > 0$ about P_0 such that every path of (1) which has a point P_1 (corresponding to $t = t_1$, say) in D_δ has all its points corresponding to $t \geqq t_1$ in D_ϵ. See Fig. 62.

A critical point P_0 of (1) is called **asymptotically stable** if P_0 is both stable and attractive.

[2] More precisely: *stable in the sense of Liapunov.* There exist other definitions of stability, but Liapunov's definition—the only we shall consider—is probably the most useful one.

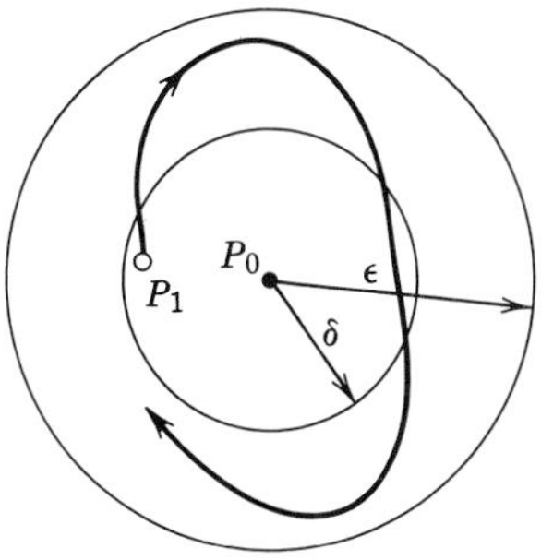

Fig. 62. Stable critical point P_0 of (1). (The path initiating at P_1 stays in the disk of radius ε.)

A critical point of (1) which is not stable is called **unstable.**

Clearly, a stable critical point P_0 need not be asymptotically stable since P_0 may not be attractive. Indeed, stability does not imply that paths of (1) in a neighborhood of P_0 must approach P_0. On the other hand, an attractive critical point P_0 need not be stable. This is less obvious, perhaps even surprising. It is true, since paths may be such that they first come arbitrarily close to P_0 but then move away from P_0 before they approach P_0 in the limit. (Formally, for every $\delta > 0$ and every M there is a path having distance less than δ from P_0 at some t_1, reaching a distance greater than M from P_0 at some $t_2 > t_1$ and approaching P_0 as $t \to \infty$.) The paths to the left of P_0 in Fig. 61 are supposed to illustrate the situation. Corresponding examples of differential equations are complicated and hardly encountered in practice. (Such examples are given in [B1] and [B9] listed in Appendix 1.)

So far we were concerned with the stability or instability of critical points. We now proceed to a further classification. We shall see that there are different types of critical points depending on the shape of the paths in a neighborhood of such a point.

A **proper node** is a critical point P_0 such that every path approaches P_0 in a definite direction as $t \to +\infty$ or $t \to -\infty$ and, given any direction, there is a path that approaches P_0 in this direction.

For example, the system

$$\dot{x} = x, \qquad \dot{y} = y \tag{3}$$

has an unstable proper node at (0, 0) because a general solution is (Fig. 63)

$$x = Ae^t, \quad y = Be^t,$$

so that the paths are rays directed away from (0, 0). Note that $y = (B/A)x$ when $A \neq 0$ and $x = (A/B)y$ when $B \neq 0$.

Similarly, the system

$$\dot{x} = -x, \qquad \dot{y} = -y \tag{4}$$

has a stable and attractive proper node at the origin (Fig. 64).

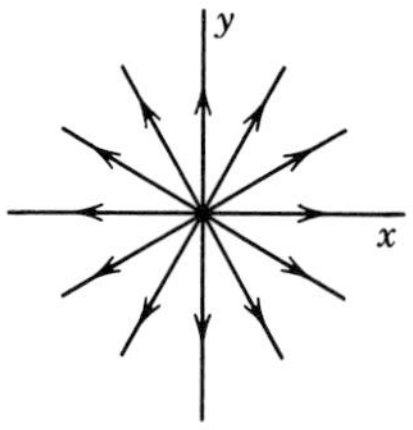

Fig. 63. Unstable proper node

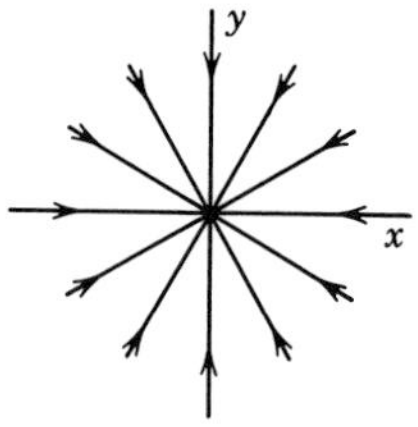

Fig. 64. Stable and attractive proper node

An **improper node** is a critical point P_0 such that every path, with the possible exception of one pair of paths, has the same limiting direction at P_0.

For example, the system

(5) $$\dot{x} = x, \qquad \dot{y} = 2y$$

has an unstable improper node at the origin (Fig. 65). The system

(6) $$\dot{x} = -x, \qquad \dot{y} = -2y$$

has a stable and attractive improper node at the origin (Fig. 66).

The system

(7) $$\dot{x} = x, \qquad \dot{y} = x + y$$

has an unstable improper node at (0, 0) such that *all* paths have the same limiting direction as $t \to -\infty$ (Fig. 67). Indeed, the reader may show that a general solution is

$$x = Ae^t, \qquad y = (At + B)e^t.$$

For $A = 0$ this gives vertical rays ($x = 0$), and for $A \neq 0$ we obtain

$$\frac{\dot{y}}{\dot{x}} = \frac{(At + B + A)e^t}{Ae^t} \to -\infty \qquad \text{as } t \to -\infty,$$

that is, the paths approach the origin in vertical direction as $t \to -\infty$. Similarly, the system

(8) $$\dot{x} = -x, \qquad \dot{y} = x - y$$

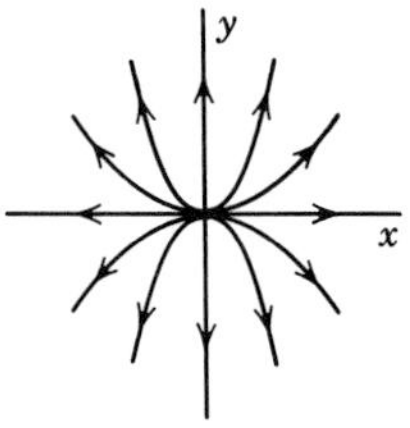

Fig. 65. Unstable improper node

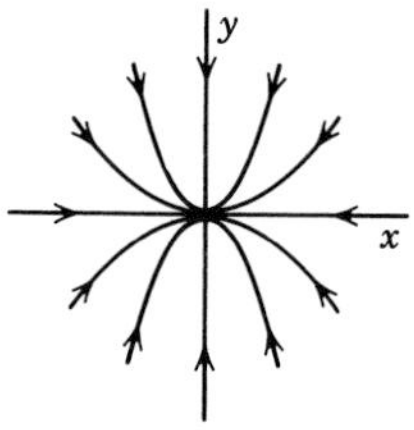

Fig. 66. Stable and attractive improper node

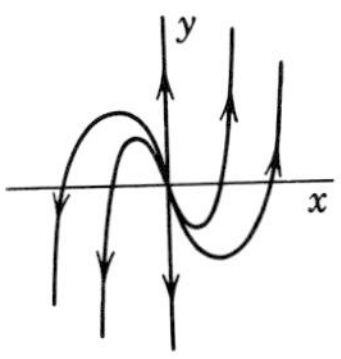

Fig. 67. Unstable improper node

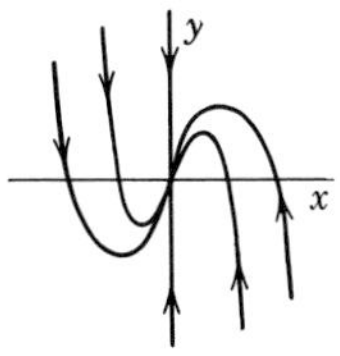

Fig. 68. Stable and attractive improper node

has a stable attractive improper node at (0, 0) such that *all* the paths have the same limiting direction as $t \to +\infty$ (Fig. 68).

A **saddle point** is a critical point P_0 such that finitely many paths approach P_0 as $t \to +\infty$ and finitely many paths approach P_0 as $t \to -\infty$. A saddle point is unstable.

For example, the system

$$\dot{x} = x, \qquad \dot{y} = -y \tag{9}$$

has a saddle point at the origin (Fig. 69). Two paths approach the origin as $t \to +\infty$, and two paths approach it as $t \to -\infty$. The curved paths are branches of hyperbolas.

A **center** is a critical point P_0 such that the paths form closed curves containing P_0 in the region enclosed by any path.

For example, the system

$$\dot{x} = y, \qquad \dot{y} = -x \tag{10}$$

has a center at the origin (Fig. 70).

A **spiral point** or **vortex point** is a critical point P_0 such that the paths form spirals about P_0 with P_0 as the asymptotic point.

For example, the system

$$\dot{x} = -x + y, \qquad \dot{y} = -x - y \tag{11}$$

has a stable spiral point at the origin (Fig. 71). Indeed, the reader may verify that a general solution is

$$x = e^{-t}(A\cos t + B\sin t), \qquad y = e^{-t}(B\cos t - A\sin t).$$

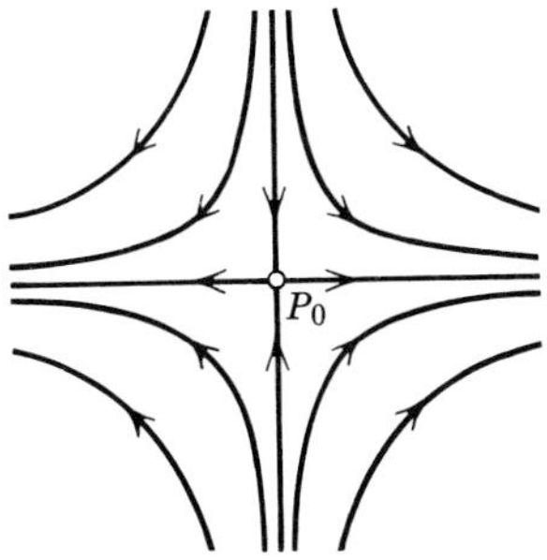

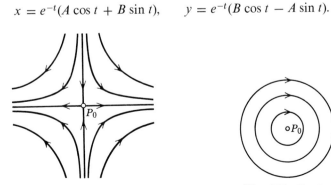

Fig. 69. Saddle point

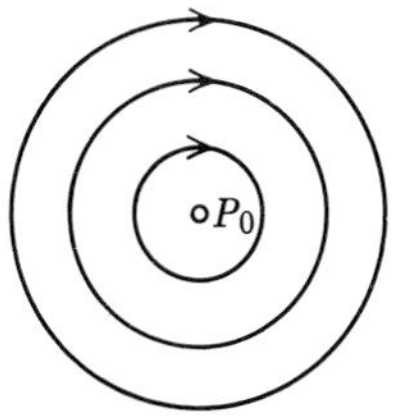

Fig. 70. Center

Fig. 71. Stable and attractive spiral point

Setting $C^2 = A^2 + B^2$, $\cos \alpha = A/C$ and $\sin \alpha = B/C$, we may write this

$$x = Ce^{-t} \cos (t - \alpha), \qquad y = -Ce^{-t} \sin (t - \alpha).$$

Using polar coordinates r, θ defined by

$$x = r \cos \theta, \qquad y = r \sin \theta,$$

we have by comparison $r = Ce^{-t}$ and $\theta = \alpha - t$. Hence $t = \alpha - \theta$ and

$$r(\theta) = C_0 e^{\theta} \qquad (C_0 = Ce^{-\alpha})$$

which represents a spiral.

The systems we considered as examples are of the form

$$\text{(12)} \qquad \begin{aligned} &\text{(a)} \quad \dot{x} = ax + by \\ &\text{(b)} \quad \dot{y} = cx + dy \end{aligned}$$

where $ad - bc \neq 0$. The solutions must be exponential functions (perhaps multiplied by t). To see this, one may obtain from (12) a second-order differential equation for x by differentiating (12a), substituting $\dot{y}$ from (12b) and then by from (12a). The result is

$$\text{(13)} \qquad \ddot{x} - (a + d)\dot{x} + (ad - bc)x = 0.$$

One could now discuss the form of solutions in terms of the roots of the characteristic equation

$$\lambda^2 - (a + d)\lambda + ad - bc = 0,$$

with x obtained from (13) and y from x and (12). Such a discussion would show that our figures illustrate all possible types of critical points of (12). (The interested reader wishing to pursue this subject further will find details in [B4] listed in Appendix 1.)

It is worth noting that we may consider a damped linear oscillator (mass on a spring; cf. Sec. 2.6) in the light of our new terminology. The differential equation is $m\ddot{x} + c\dot{x} + kx = 0$ or

$$\ddot{x} + \frac{c}{m}\dot{x} + \frac{k}{m}x = 0$$

where $m > 0$, $c \geqq 0$, $k > 0$. This is of the form (13) and can be written as a system

$$\dot{x} = v, \qquad \dot{v} = -\frac{c}{m}v - \frac{k}{m}x.$$

If $c = 0$ (no damping), the paths are circles about the origin, which is a center. If $c > 0$ (damping), the paths are spirals about the origin, which is an asymptotically stable spiral point.

Our discussion concerned the *linear* system (12), but we mention that the *nonlinear* system (1) in general shows the same behavior near a critical point. More precisely, the following holds true. Assume (0, 0) to be a critical point P_0 of (1), that is,

$$F(0, 0) = 0, \qquad G(0, 0) = 0.$$

Suppose that F and G are continuous and have continuous partial derivatives in a neighborhood of P_0. Writing the linear terms in F and G explicitly, we have from (1)

$$\text{(14)} \qquad \begin{aligned} \dot{x} &= ax + by + F_1(x, y) \\ \dot{y} &= cx + dy + G_1(x, y) \end{aligned}$$

It can be shown that if $ad - bc \neq 0$ and P_0 is *isolated,* that is, there is a disk about P_0 containing no further critical points of (1), then the paths of the linear system (12) are good approximations to the paths of the nonlinear system (1) near P_0, except for a few cases which are not very essential. An illustrative example is as follows.

Example 1. Pendulum

Figure 40 in Sec. 2.6 shows a pendulum of mass m and length l. The displacement from the equilibrium position is $s = l\theta$. Hence the acceleration is $l\ddot{\theta}$. The restoring force is the tangential component $mg \sin \theta$ of the weight mg. Here we neglect the mass of the rod. We also neglect damping. Then by Newton's second law,

$$ml\ddot{\theta} + mg \sin \theta = 0.$$

Dividing by ml and setting $\theta = x$, $\dot{\theta} = y$ (to have our previous notation), we obtain

$$\text{(15)} \qquad \dot{x} = y, \qquad \dot{y} = -\frac{g}{l}\sin x.$$

This is a nonlinear system with a critical point at (0, 0). The exact solution is not an elementary function. We write the system in the form (14):

$$\begin{aligned} \dot{x} &= y \\ \dot{y} &= -\frac{g}{l}x + \frac{g}{l}(x - \sin x). \end{aligned}$$

Hence $a = 0$, $b = 1$, $c = -g/l$, $d = 0$ in (14), so that $ad - bc \neq 0$. Also the above differentiability condition holds. Hence the critical point of (15) is of the same type as that of the corresponding linear system

$$\text{(16)} \qquad \dot{x} = y, \qquad \dot{y} = -\frac{g}{l}x.$$

That is, it is a center. This is understandable since $\sin x \approx x$ for small $|x|$, so that (16) is a good approximation to (15) in the case of small vibrations. These are nearly harmonic, because (16) implies $\ddot{y} + \omega^2 y = 0$, where $\omega = \sqrt{g/l}$. We mention that this is no longer true for large vibrations. In that case, the period depends on the initial displacement, as can be shown.

Problems for Sec. 3.3

1. Show that in (3) for every choice of A and B one obtains a ray which is directed away from the origin (so that this critical point is unstable).
2. Solve (4) and verify that (0, 0) is a stable proper node.
3. Solve (5). What curves are the paths? What are the directions of approach of the paths at the origin?
4. Solve (6). What curves are the paths?
5. In (6) let t represent time. Introduce $\tau = -t$ as a new independent variable, solve and compare with the solution of (5). What does that transformation mean in terms of mechanics?
6. Derive the solution of (7).
7. Solve (8).
8. Solve (9) and show that the curved paths are branches of hyperbolas.
9. Solve (10).
10. Derive (13) from (12).

CHAPTER

4 POWER SERIES SOLUTIONS OF DIFFERENTIAL EQUATIONS. ORTHOGONAL FUNCTIONS

In Chap. 2 we have seen that a homogeneous linear differential equation whose coefficients are *constant* can be solved by algebraic methods, and the solutions are elementary functions known from calculus. However, if those coefficients are not constant but depend on x, the situation is more complicated and the solutions may be nonelementary functions. Bessel's equation, Legendre's equation and the hypergeometric equation are of this type. Since these and other equations and their solutions play an important role in engineering mathematics, we shall now consider a method for solving such equations. The solutions will appear in the form of power series. For this reason the method is called the *power series method.* We shall also consider some basic properties of the solutions. This will help the student to get acquainted with these "higher transcendental functions" and with some standard procedures used in connection with special functions, in particular for the purpose of establishing properties of the functions and relations between them, some of which are basic in numerical computations.

The last three sections (Secs. 4.7–4.9) are devoted to the orthogonality of Legendre polynomials, Bessel functions, and other sets of solutions of certain boundary value problems, which are known as *Sturm–Liouville problems.*

Prerequisite for this chapter: Chap. 2.
Sections which may be omitted in a shorter course: 4.2, 4.6, 4.9.
References: Appendix 1, Part B.
Answers to problems: Appendix 2.

4.1 The Power Series Method

We shall now consider solving differential equations by the so-called *power series method* which yields solutions in the form of power series. It is a very efficient standard procedure in connection with linear differential equations whose coefficients are variable.

We first remember that a **power series**[1] (in powers of $x - a$) is an infinite

[1]The term "power series" alone usually refers to a series of the form (1), but does not include series of negative powers of x such as $c_0 + c_1x^{-1} + c_2x^{-2} + \cdots$ or series involving fractional powers of x. Note that in (1) we write, for convenience, $(x - a)^0 = 1$, even when $x = a$.

series of the form

$$\sum_{m=0}^{\infty} c_m(x-a)^m = c_0 + c_1(x-a) + c_2(x-a)^2 + \cdots, \tag{1}$$

where $c_0, c_1, \cdots$ are constants, called the **coefficients** of the series, a is a constant, called the **center,** and x is a variable.

If in particular $a = 0$, we obtain a *power series in powers of* x

$$\sum_{m=0}^{\infty} c_m x^m = c_0 + c_1 x + c_2 x^2 + c_3 x^3 + \cdots. \tag{2}$$

We shall assume in this section that all variables and constants are real.

Familiar examples of power series are the Maclaurin series

$$\frac{1}{1-x} = \sum_{m=0}^{\infty} x^m = 1 + x + x^2 + \cdots \qquad (|x| < 1, \textit{ geometric series}),$$

$$e^x = \sum_{m=0}^{\infty} \frac{x^m}{m!} = 1 + x + \frac{x^2}{2!} + \frac{x^3}{3!} + \cdots,$$

$$\cos x = \sum_{m=0}^{\infty} \frac{(-1)^m x^{2m}}{(2m)!} = 1 - \frac{x^2}{2!} + \frac{x^4}{4!} - + \cdots,$$

$$\sin x = \sum_{m=0}^{\infty} \frac{(-1)^m x^{2m+1}}{(2m+1)!} = x - \frac{x^3}{3!} + \frac{x^5}{5!} - + \cdots.$$

The basic idea of the power series method for solving differential equations is very simple and natural. We shall describe the practical procedure and illustrate it by simple examples, postponing the mathematical justification of the method to the next section.

A differential equation being given, we first represent all given functions in the equation by power series in powers of x (or in powers of $x - a$, if solutions in the form of power series in powers of $x - a$ are wanted). In many practical cases the given functions will be polynomials, and in such a case, nothing need be done in this first step. Then we assume a solution in the form of a power series, say,

$$y = c_0 + c_1 x + c_2 x^2 + c_3 x^3 + \cdots = \sum_{m=0}^{\infty} c_m x^m \tag{3}$$

and insert this series and the series obtained by termwise differentiation,

(4)

$$\text{(a)} \quad y' = c_1 + 2c_2 x + 3c_3 x^2 + \cdots = \sum_{m=1}^{\infty} m c_m x^{m-1}$$

$$\text{(b)} \quad y'' = 2c_2 + 3 \cdot 2c_3 x + 4 \cdot 3c_4 x^2 + \cdots = \sum_{m=2}^{\infty} m(m-1) c_m x^{m-2}$$

etc., into the equation. Then we collect like powers of x and equate the sum of the coefficients of each occurring power of x to zero, starting with the constant terms, the terms containing x, the terms containing x^2, etc. This gives relations from which we can determine the unknown coefficients in (3) successively.

Let us illustrate the procedure for some simple equations which can also be solved by elementary methods.

Example 1

Solve

$$y' - y = 0.$$

In the first step, we insert (3) and (4a) into the equation:

$$(c_1 + 2c_2x + 3c_3x^2 + \cdots) - (c_0 + c_1x + c_2x^2 + \cdots) = 0.$$

Then we collect like powers of x, finding

$$(c_1 - c_0) + (2c_2 - c_1)x + (3c_3 - c_2)x^2 + \cdots = 0.$$

Equating the coefficient of each power of x to zero, we have

$$c_1 - c_0 = 0, \qquad 2c_2 - c_1 = 0, \qquad 3c_3 - c_2 = 0, \quad \cdots .$$

Solving these equations, we may express $c_1, c_2, \cdots$ in terms of c_0 which remains arbitrary:

$$c_1 = c_0, \qquad c_2 = \frac{c_1}{2} = \frac{c_0}{2!}, \qquad c_3 = \frac{c_2}{3} = \frac{c_0}{3!}, \quad \cdots .$$

With these values (3) becomes

$$y = c_0 + c_0x + \frac{c_0}{2!}x^2 + \frac{c_0}{3!}x^3 + \cdots,$$

and we see that we have obtained the familiar general solution

$$y = c_0\left(1 + x + \frac{x^2}{2!} + \frac{x^3}{3!} + \cdots\right) = c_0e^x.$$

Of course, if we write the given equation in the form $y' = y$ and insert (3) and (4a) as before, we obtain

$$\begin{aligned} &c_1 + 2c_2x + 3c_3x^2 + \cdots \\ &= c_0 + \ c_1x + \ c_2x^2 + \cdots \end{aligned}$$

and see that we must have

$$c_1 = c_0, \quad 2c_2 = c_1, \quad 3c_3 = c_2, \quad \cdots,$$

etc., as before.

Example 2

Solve

$$y'' + y = 0.$$

By inserting (3) and (4b) into the equation we obtain

$$(2c_2 + 3 \cdot 2c_3x + 4 \cdot 3c_4x^2 + \cdots) + (c_0 + c_1x + c_2x^2 + \cdots) = 0.$$

Collecting like powers of x, we find

$$(2c_2 + c_0) + (3 \cdot 2c_3 + c_1)x + (4 \cdot 3c_4 + c_2)x^2 + \cdots = 0.$$

Equating the coefficient of each power of x to zero, we have

$$\begin{aligned} 2c_2 + c_0 &= 0 && \text{coefficient of } x^0 \\ 3 \cdot 2c_3 + c_1 &= 0 && \text{coefficient of } x^1 \\ 4 \cdot 3c_4 + c_2 &= 0 && \text{coefficient of } x^2 \end{aligned}$$

etc. Solving these equations, we see that $c_2, c_4, \cdots$ may be expressed in terms of c_0, and $c_3, c_5, \cdots$ may be expressed in terms of c_1:

$$c_2 = -\frac{c_0}{2!}, \qquad c_3 = -\frac{c_1}{3!}, \qquad c_4 = -\frac{c_2}{4 \cdot 3} = \frac{c_0}{4!}, \quad \cdots;$$

c_0 and c_1 are arbitrary. With these values (3) becomes

$$y = c_0 + c_1 x - \frac{c_0}{2!}x^2 - \frac{c_1}{3!}x^3 + \frac{c_0}{4!}x^4 + \frac{c_1}{5!}x^5 + \cdots.$$

This can be written

$$y = c_0\left(1 - \frac{x^2}{2!} + \frac{x^4}{4!} - + \cdots\right) + c_1\left(x - \frac{x^3}{3!} + \frac{x^5}{5!} - + \cdots\right)$$

and we recognize the familiar general solution

$$y = c_0 \cos x + c_1 \sin x.$$

Example 3
Solve

$$(x + 1)y' - (x + 2)y = 0.$$

By inserting (3) and (4a) into the equation we obtain

$$(x + 1)(c_1 + 2c_2 x + 3c_3 x^2 + \cdots) - (x + 2)(c_0 + c_1 x + c_2 x^2 + \cdots) = 0.$$

By performing the indicated multiplications we get

$$\begin{aligned} & c_1 x + 2c_2 x^2 + 3c_3 x^3 + 4c_4 x^4 + 5c_5 x^5 + \cdots + sc_s x^s + \cdots \\ + \; & c_1 + 2c_2 x + 3c_3 x^2 + 4c_4 x^3 + 5c_5 x^4 + 6c_6 x^5 + \cdots + (s+1)c_{s+1} x^s + \cdots \\ - \; & c_0 x - c_1 x^2 - c_2 x^3 - c_3 x^4 - c_4 x^5 - \cdots - c_{s-1} x^s - \cdots \\ - \; & 2c_0 - 2c_1 x - 2c_2 x^2 - 2c_3 x^3 - 2c_4 x^4 - 2c_5 x^5 - \cdots - 2c_s x^s - \cdots = 0. \end{aligned}$$

Since this must be an identity in x, the sum of the coefficients of each power of x must be zero; we thus obtain

(5) $\quad$ (a) $c_1 - 2c_0 = 0 \quad$ (b) $2c_2 - c_1 - c_0 = 0, \quad$ etc.,

and in general

(6) $$sc_s + (s + 1)c_{s+1} - c_{s-1} - 2c_s = 0.$$

By solving (5a) for c_1 and (6) for c_{s+1} we have

(7) $\quad$ (a) $c_1 = 2c_0, \quad$ (b) $c_{s+1} = \dfrac{1}{s+1}[c_{s-1} + (2 - s)c_s], \qquad s = 1, 2, \cdots.$

Formula (7b) is called a **recurrence relation** or **recursion formula.** From it we can now determine c_2, c_3, $\cdots$ successively. If we wish, we may arrange the calculations in (7b) in tabular form:

s	c_{s-1}	$(2-s)c_s$	Sum	$s+1$	$c_{s+1} = \dfrac{\text{Sum}}{s+1}$	c_{s+1} in terms of c_0
						$c_1 = 2c_0$
1	c_0	c_1	$c_0 + c_1$	2	$\frac{c_0}{2} + \frac{c_1}{2}$	$c_2 = \frac{3}{2}c_0$
2	c_1	0	c_1	3	$\frac{c_1}{3}$	$c_3 = \frac{2}{3}c_0$
3	c_2	$-c_3$	$c_2 - c_3$	4	$\frac{c_2}{4} - \frac{c_3}{4}$	$c_4 = \frac{5}{24}c_0$
.	. .	. .	. . .	.	. . .	. . .

With these values (3) becomes

$$y = c_0(1 + 2x + \tfrac{3}{2}x^2 + \tfrac{2}{3}x^3 + \tfrac{5}{24}x^4 + \cdots).$$

The reader may verify that the explicitly written terms are the first few terms of the Maclaurin series of

$$y = c_0(1 + x)e^x,$$

the general solution of the equation obtained by separating variables.

Problems for Sec. 4.1

Apply the power series method to the following differential equations.

1. $y' = 3y$ **2.** $y' + 2y = 0$ **3.** $y' = ky$ **4.** $y' = 2xy$
5. $(1 - x)y' = y$ **6.** $y' = xy$ **7.** $(1 - x)y' = y$ **8.** $(1 - x^2)y' = y$
9. $y'' = y$ **10.** $y'' + 9y = 0$ **11.** $y'' = 4y$ **12.** $y'' = y'$

(More problems of this type are included at the end of the next section.)

4.2 Theoretical Basis of the Power Series Method

We have seen that the power series method yields solutions of differential equations in the form of power series. The solution y of a given equation is assumed in the form of a power series with undetermined coefficients, and the coefficients are determined successively by inserting that series and the series for the derivatives of y into the given equation.

The practical usefulness of the method becomes obvious if we remember that power series may be used for computing numerical values of solutions. Furthermore, many general properties of the solutions can be derived from their power series; this will be seen in our further discussion.

The power series method involves various operations on power series, for example, differentiation, addition, and multiplication of power series. Therefore, to justify the method we have to consider the theoretical basis of these operations. This will be done in the present section. The discussion will involve some concepts and facts which may be already known to the student from elementary calculus, as well as other concepts which are not considered in elementary classes. [Corresponding proofs and further details (which will not be used in this chapter) can be found in Secs. 16.1 and 16.2.]

A **power series** is an infinite series of the form

$$\sum_{m=0}^{\infty} c_m(x-a)^m = c_0 + c_1(x-a) + c_2(x-a)^2 + \cdots, \tag{1}$$

and we assume that the variable x, the *center* a, and the *coefficients* $c_0, c_1, \cdots$ are real, as in the previous section.

The expression

$$s_n(x) = c_0 + c_1(x-a) + c_2(x-a)^2 + \cdots + c_n(x-a)^n \tag{2}$$

(n a positive integer) is called the nth **partial sum** of the series (1). Clearly, if we omit the terms of s_n from (1), the remaining expression is

$$R_n(x) = c_{n+1}(x-a)^{n+1} + c_{n+2}(x-a)^{n+2} + \cdots. \tag{3}$$

This expression is called the **remainder** *of* (1) *after the term* $c_n(x-a)^n$.

For example, in the case of the geometric series

$$1 + x + x^2 + \cdots + x^n + \cdots$$

we have

$$s_1 = 1 + x, \qquad R_1 = x^2 + x^3 + x^4 + \cdots,$$

$$s_2 = 1 + x + x^2, \qquad R_2 = x^3 + x^4 + x^5 + \cdots, \text{ etc.}$$

In this way we have now associated with (1) the sequence of the partial sums $s_1(x), s_2(x), \cdots$. It may happen that for some $x = x_0$ this sequence converges, say,

$$\lim_{n\to\infty} s_n(x_0) = s(x_0).$$

Then we say that the series (1) **converges,** or *is convergent, at* $x = x_0$; the number $s(x_0)$ is called the **value** or *sum* of (1) at x_0, and we write

$$s(x_0) = \sum_{m=0}^{\infty} c_m(x_0 - a)^m.$$

If that sequence is divergent at $x = x_0$, then the series (1) is said to **diverge,** or to *be divergent,* at $x = x_0$.

We remember that a *sequence* $s_1, s_2, \cdots$ is said to *converge* to a number s, or to *be convergent* with the *limit* s, if for each given positive number ϵ (no matter how small, but not zero) we can find a number N such that

$$(4) \qquad |s_n - s| < \epsilon \qquad \text{for each } n > N.$$

Geometrically speaking, (4) means that s_n with $n > N$ lies between $s - \epsilon$ and $s + \epsilon$ (Fig. 72). Of course, N will in general depend on the choice of ϵ.

Now in our case, $s = s_n + R_n$ or $R_n = s - s_n$. Hence

$$|s_n - s| = |R_n|$$

in (4), and convergence at $x = x_0$ means that we can make $|R_n(x_0)|$ as small as we please, by taking n large enough. In other words, in the case of convergence, $s_n(x_0)$ is an approximation to $s(x_0)$, and the error $|R_n(x_0)|$ of the approximation can be made smaller than any preassigned positive number ϵ by taking n sufficiently large.

If we choose $x = x_0 = a$ in (1), the series reduces to the single term c_0 because the other terms are zero. This shows that the series (1) converges at $x = a$. In some cases this may be the only value of x for which (1) converges. If there are other values of x for which the series converges, these values form an interval, the **convergence interval,** having the midpoint $x = a$. This interval may be finite, as in Fig. 73. Then the series converges for all x in the interior of the interval, that is, for all x for which

$$(5) \qquad |x - a| < R$$

and diverges when $|x - a| > R$. The interval may also be infinite, that is, the series may converge for all x.

The quantity R in Fig. 73 is called the **radius of convergence** of (1); it is the distance of each endpoint of the convergence interval from the center a. If the series converges for all x, then we set $R = \infty$ (and $1/R = 0$).

The radius of convergence can be determined from the coefficients of the series by means of each of the formulas

$$\textbf{(6)} \qquad \text{(a)}\ \frac{1}{R} = \lim_{m\to\infty} \sqrt[m]{|c_m|} \qquad \text{(b)}\ \frac{1}{R} = \lim_{m\to\infty} \left|\frac{c_{m+1}}{c_m}\right|$$

provided the limits in (6) exist. (Proofs of these facts can be found in Sec. 16.1.)

For each x for which (1) converges, it has a certain value $s(x)$ depending on

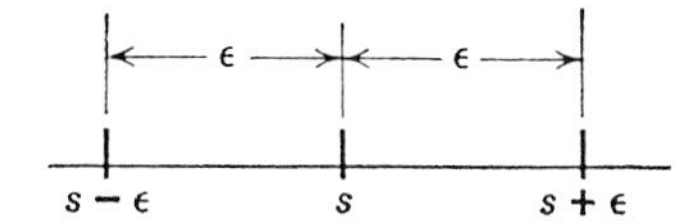

Fig. 72. Convergence of a sequence

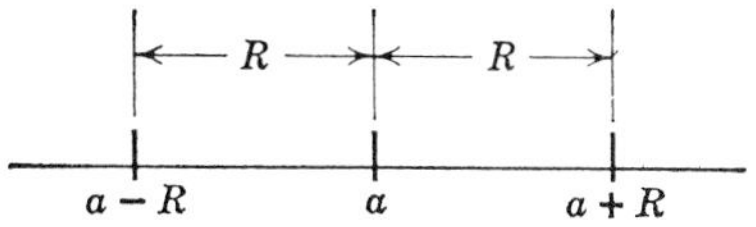

Fig. 73. Convergence interval (5) of a power series with center a

x; if the radius of convergence R of (1) is not zero, we write

$$s(x) = \sum_{m=0}^{\infty} c_m(x-a)^m \qquad (|x-a| < R)$$

and we say that the series (1) *represents* the function $s(x)$ in the interval of convergence.

Let us consider three typical examples illustrating the three possibilities that the radius of convergence R may be positive and finite, or infinite, or zero.

Example 1. Geometric series

In the case of the *geometric series* we have

$$\frac{1}{1-x} = \sum_{m=0}^{\infty} x^m = 1 + x + x^2 + \cdots \qquad (|x| < 1).$$

In fact, $c_m = 1$ for all m, and from (6) we obtain $R = 1$, that is, the geometric series converges and represents $1/(1-x)$ when $|x| < 1$.

Example 2

In the case of the series

$$e^x = \sum_{m=0}^{\infty} \frac{x^m}{m!} = 1 + x + \frac{x^2}{2!} + \cdots$$

we have $c_m = 1/m!$. Hence in (6b),

$$\frac{c_{m+1}}{c_m} = \frac{1/(m+1)!}{1/m!} = \frac{1}{m+1} \to 0 \qquad \text{as } m \to \infty,$$

which means $R = \infty$; the series converges for all x.

Example 3

In the case of the series

$$\sum_{m=0}^{\infty} m!x^m = 1 + x + 2x^2 + 6x^3 + \cdots$$

we have $c_m = m!$, and in (6b),

$$\frac{c_{m+1}}{c_m} = \frac{(m+1)!}{m!} = m + 1 \to \infty \qquad \text{as } m \to \infty.$$

Thus $R = 0$, and the series converges only at the center $x = 0$. Such a series is useless.

Operations on Power Series

We shall now consider the operations on power series which are used in connection with the power series method. The following three operations are permissible, in the sense explained in each case. We also list a condition concerning the vanishing of all coefficients, which is a basic tool of the power series method.

Termwise differentiation

A power series may be differentiated term by term. More precisely: if

$$y(x) = \sum_{m=0}^{\infty} c_m(x - a)^m$$

converges for $|x - a| < R$, where $R > 0$, then the series obtained by differentiating term by term also converges for those x and represents the derivative y' of y, that is

$$y'(x) = \sum_{m=1}^{\infty} mc_m(x - a)^{m-1}.$$

(Proof in Sec. 16.2, Theorems 3 and 5.)

Termwise addition

Two power series may be added term by term. More precisely: if the series

$$\sum_{m=0}^{\infty} b_m(x - a)^m \quad \text{and} \quad \sum_{m=0}^{\infty} c_m(x - a)^m \tag{7}$$

have positive radii of convergence and their sums are $f(x)$ and $g(x)$, then the series

$$\sum_{m=0}^{\infty} (b_m + c_m)(x - a)^m$$

converges and represents $f(x) + g(x)$ for each x which lies in the interior of the convergence interval of each of the given series. (Proof in Sec. 16.1.)

Termwise multiplication

Two power series may be multiplied term by term. More precisely: Suppose that the series (7) have positive radii of convergence and let $f(x)$ and $g(x)$ be their sums. Then the series obtained by multiplying each term of the first series by each term of the second series and collecting like powers of $x - a$, that is,

$$b_0c_0 + (b_0c_1 + b_1c_0)(x - a) + \cdots = \sum_{n=0}^{\infty} (b_0c_n + b_1c_{n-1} + \cdots + b_nc_0)(x - a)^n$$

converges and represents $f(x)g(x)$ for each x in the interior of the convergence interval of each of the given series. (Proof in Sec. 16.1, Theorem 3.)

Vanishing of all coefficients

If a power series has a positive radius of convergence and a sum which is identically zero throughout its interval of convergence, then each coefficient of the series is zero. (Proof in Sec. 16.2, Theorem 2.)

These properties of power series form the theoretical basis of the power series method. The remaining question is whether a given differential equation has solutions representable by power series at all. We may answer this question by using the following concept.

A function $f(x)$ is said to be **analytic** *at a point* $x = a$ if it can be represented by a power series in powers of $x - a$ with radius of convergence $R > 0$.

Using this notion, we may formulate a basic criterion of the desired type as follows.

Theorem 1 (Existence of power series solutions)
If the functions f, g and r in the differential equation

$$y'' + f(x)y' + g(x)y = r(x) \tag{8}$$

are analytic at $x = a$, *then every solution* $y(x)$ *of* (8) *is analytic at* $x = a$ *and can thus be represented by a power series in powers of* $x - a$ *with radius of convergence*[2] $R > 0$.

The proof requires advanced methods of complex analysis and can be found in Ref. [B11] in Appendix 1. In applying this theorem it is important to write the linear equation in the form (8), with 1 as the coefficient of y''.

Problems for Sec. 4.2

Determine the radius of convergence of the following series.

1. $\sum_{m=0}^{\infty} \frac{x^m}{3^m}$
2. $\sum_{m=0}^{\infty} \frac{(-1)^m x^m}{m!}$
3. $\sum_{m=0}^{\infty} x^{2m}$
4. $\sum_{m=0}^{\infty} (-1)^m x^{3m}$
5. $\sum_{m=0}^{\infty} \frac{(-1)^m x^{2m}}{(2m)!}$
6. $\sum_{m=0}^{\infty} \frac{(-1)^m x^{2m+1}}{(2m+1)!}$
7. $\sum_{m=1}^{\infty} \frac{x^m}{m}$
8. $\sum_{m=0}^{\infty} \frac{x^{2m+1}}{2m+1}$
9. $\sum_{m=0}^{\infty} \frac{x^{2m+1}}{(2m+1)m!}$
10. $\sum_{m=1}^{\infty} \frac{(4m)!}{(m!)^4} x^m$
11. $\sum_{m=2}^{\infty} \frac{m(m-1)}{3^m} x^m$
12. $\sum_{m=0}^{\infty} \frac{(-1)^m}{k^m} x^{2m}$
13. $\sum_{m=0}^{\infty} (-1)^m (x-1)^m$
14. $\sum_{m=0}^{\infty} \frac{1}{2^m} (x-3)^{2m}$
15. $\sum_{m=0}^{\infty} \frac{m}{3^m} (x-2)^m$

16. **(Shift of summation index)** In the power series method it will sometimes be helpful to make simple shifts of summation indices. To get used to this, show that

$$\sum_{m=2}^{\infty} m(m-1)c_m x^{m-2} = \sum_{p=1}^{\infty} (p+1)pc_{p+1}x^{p-1} = \sum_{s=0}^{\infty} (s+2)(s+1)c_{s+2}x^s.$$

What is the relation between m, p and s?

[2] R is at least equal to the distance between the point $x = a$ and that point (or those points) closest to $x = a$ at which one of the functions f, g, r, *as functions of a complex variable*, is not analytic. (Note that that point may not lie on the x-axis, but somewhere in the complex plane.)

Apply the power series method to the following differential equations.

17. $xy' - (x + 2)y = -2x^2 - 2x$
18. $xy' = (x + 1)y$
19. $x(x + 1)y' - (2x + 1)y = 0$
20. $xy' - 3y = 3$
21. $y'' - 3y' + 2y = 0$
22. $y'' - y = x$
23. $(1 - x^2)y'' - 2xy' + 2y = 0$
24. $y'' - 4xy' + (4x^2 - 2)y = 0$
25. $y'' - xy' + y = 0$
26. $(1 - x^2)y'' - 2xy' + 6y = 0$

27. Show that $y' = (y/x) + 1$ cannot be solved for y as a power series in x. Solve this equation for y as a power series in powers of $x - 1$. (*Hint.* Introduce $t = x - 1$ as a new independent variable and solve the resulting equation for y as a power series in t.) Compare the result with that obtained by the appropriate elementary method.

Solve for y as a power series in powers of $x - 1$:

28. $y' = 2y$
29. $y'' - y = 0$
30. $xy' + y = 0$

4.3 Legendre's Equation. Legendre Polynomials

Legendre's differential equation[3]

$$(1 - x^2)y'' - 2xy' + n(n + 1)y = 0 \tag{1}$$

arises in numerous physical problems, particularly in boundary value problems for spheres. The parameter n in (1) is a given real number. Any solution of (1) is called a **Legendre function.**

Dividing (1) by $1 - x^2$, we obtain the standard form (8), Sec. 4.2, and we see that the coefficients of the resulting equation are analytic at $x = 0$, so that we may apply the power series method. Substituting

$$y = \sum_{m=0}^{\infty} c_m x^m \tag{2}$$

and its derivatives into (1) and denoting the constant $n(n + 1)$ by k we obtain

$$(1 - x^2) \sum_{m=2}^{\infty} m(m - 1)c_m x^{m-2} - 2x \sum_{m=1}^{\infty} mc_m x^{m-1} + k \sum_{m=0}^{\infty} c_m x^m = 0.$$

By writing the first expression as two separate series we have

$$\sum_{m=2}^{\infty} m(m - 1)c_m x^{m-2} - \sum_{m=2}^{\infty} m(m - 1)c_m x^m - 2 \sum_{m=1}^{\infty} mc_m x^m + k \sum_{m=0}^{\infty} c_m x^m = 0, \tag{1*}$$

[3]ADRIEN MARIE LEGENDRE (1752–1833), French mathematician, who made important contributions to the theory of numbers and elliptic functions.

written out

$$\begin{aligned}
&2\cdot 1c_2 + 3\cdot 2c_3x + 4\cdot 3c_4x^2 + \cdots + (s+2)(s+1)c_{s+2}x^s + \cdots \\
&\qquad\qquad - 2\cdot 1c_2x^2 - \cdots \qquad\qquad - s(s-1)c_sx^s - \cdots \\
&\qquad - 2\cdot 1c_1x - 2\cdot 2c_2x^2 - \cdots \qquad\qquad - 2sc_sx^s - \cdots \\
&+ kc_0 \quad + kc_1x \quad + kc_2x^2 + \cdots \qquad\qquad + kc_sx^s + \cdots = 0.
\end{aligned}$$

Since this must be an identity in x if (2) is to be a solution of (1), the sum of the coefficients of each power of x must be zero; remembering that $k = n(n+1)$ we thus have

$$\text{(3a)} \qquad \begin{aligned} 2c_2 + n(n+1)c_0 &= 0, &&\text{coefficients of } x^0 \\ 6c_3 + [-2 + n(n+1)]c_1 &= 0, &&\text{coefficients of } x^1 \end{aligned}$$

and in general, when $s = 2, 3, \cdots,$

$$\text{(3b)} \qquad (s+2)(s+1)c_{s+2} + [-s(s-1) - 2s + n(n+1)]c_s = 0.$$

The expression in brackets $[\cdots]$ can be written

$$(n-s)(n+s+1).$$

We thus obtain from (3)

$$\textbf{(4)} \qquad c_{s+2} = -\frac{(n-s)(n+s+1)}{(s+2)(s+1)}c_s \qquad (s = 0, 1, \cdots).$$

This is called a **recurrence relation** or **recursion formula.** It gives each coefficient in terms of the second one preceding it, except for c_0 and c_1, which are left as arbitrary constants. We find successively

$c_2 = -\frac{n(n+1)}{2!}c_0$	$c_3 = -\frac{(n-1)(n+2)}{3!}c_1$
$c_4 = -\frac{(n-2)(n+3)}{4\cdot 3}c_2$	$c_5 = -\frac{(n-3)(n+4)}{5\cdot 4}c_3$
$= \frac{(n-2)n(n+1)(n+3)}{4!}c_0$	$= \frac{(n-3)(n-1)(n+2)(n+4)}{5!}c_1$

etc. By inserting these values for the coefficients into (2) we obtain

$$\text{(5)} \qquad y(x) = c_0y_1(x) + c_1y_2(x)$$

where

$$(6) \quad y_1(x) = 1 - \frac{n(n+1)}{2!}x^2 + \frac{(n-2)n(n+1)(n+3)}{4!}x^4 - + \cdots$$

and

$$(7) \quad y_2(x) = x - \frac{(n-1)(n+2)}{3!}x^3 + \frac{(n-3)(n-1)(n+2)(n+4)}{5!}x^5 - + \cdots$$

These series converge for $|x| < 1$. Since (6) contains even powers of x only, while (7) contains odd powers of x only, the ratio y_1/y_2 is not a constant, and y_1 and y_2 are linearly independent solutions. Hence (5) is a general solution of (1) on the interval $-1 < x < 1$.

Legendre polynomials

In many applications the parameter n in Legendre's equation will be a nonnegative integer. Then the right-hand side of (4) is zero when $s = n$, and, therefore, $c_{n+2} = 0$, $c_{n+4} = 0$, $c_{n+6} = 0$, $\cdots$. Hence, if n is even, $y_1(x)$ reduces to a polynomial of degree n. If n is odd, the same is true with respect to $y_2(x)$. These polynomials, multiplied by some constants, are called **Legendre polynomials.** Since they are of great practical importance, let us consider them in more detail. For this purpose we write (4) in the form

$$(8) \qquad c_s = -\frac{(s+2)(s+1)}{(n-s)(n+s+1)}c_{s+2} \qquad (s \leqq n-2)$$

and may then express all the nonvanishing coefficients in terms of the coefficient c_n of the highest power of x of the polynomial. The coefficient c_n is then arbitrary. It is customary to choose $c_n = 1$ when $n = 0$ and

$$(9) \qquad c_n = \frac{(2n)!}{2^n(n!)^2} = \frac{1 \cdot 3 \cdot 5 \cdots (2n-1)}{n!}, \qquad n = 1, 2, \cdots,$$

the reason being that for this choice of c_n all those polynomials will have the value 1 when $x = 1$; this follows from (14) in Prob. 8. We then obtain from (8) and (9)

$$(9^*) \qquad \begin{aligned} c_{n-2} &= -\frac{n(n-1)}{2(2n-1)}c_n = -\frac{n(n-1)(2n)!}{2(2n-1)2^n(n!)^2} \\ &= -\frac{n(n-1)2n(2n-1)(2n-2)!}{2(2n-1)2^n n(n-1)!\, n(n-1)(n-2)!}, \end{aligned}$$

that is,

$$c_{n-2} = -\frac{(2n-2)!}{2^n(n-1)!\,(n-2)!}.$$

Similarly,

$$c_{n-4} = -\frac{(n-2)(n-3)}{4(2n-3)}c_{n-2} = \frac{(2n-4)!}{2^n 2!\,(n-2)!\,(n-4)!}$$

etc., and in general, when $n - 2m \geqq 0$,

$$c_{n-2m} = (-1)^m \frac{(2n-2m)!}{2^n m!\,(n-m)!\,(n-2m)!}. \tag{10}$$

The resulting solution of Legendre's equation is called the **Legendre polynomial** *of degree n* and is denoted by $P_n(x)$; from (10) we obtain

$$\begin{aligned} P_n(x) &= \sum_{m=0}^{M} (-1)^m \frac{(2n-2m)!}{2^n m!\,(n-m)!\,(n-2m)!} x^{n-2m} \\ &= \frac{(2n)!}{2^n (n!)^2} x^n - \frac{(2n-2)!}{2^n 1!\,(n-1)!\,(n-2)!} x^{n-2} + - \cdots, \end{aligned} \tag{11}$$

where $M = n/2$ or $(n-1)/2$, whichever is an integer. In particular (Fig. 74),

$$\begin{aligned} &P_0(x) = 1, && P_1(x) = x, \\ &P_2(x) = \tfrac{1}{2}(3x^2 - 1), && P_3(x) = \tfrac{1}{2}(5x^3 - 3x), \\ &P_4(x) = \tfrac{1}{8}(35x^4 - 30x^2 + 3), && P_5(x) = \tfrac{1}{8}(63x^5 - 70x^3 + 15x), \end{aligned} \tag{11'}$$

etc.

The so-called "orthogonality" of the Legendre polynomials will be considered in Sec. 4.9.

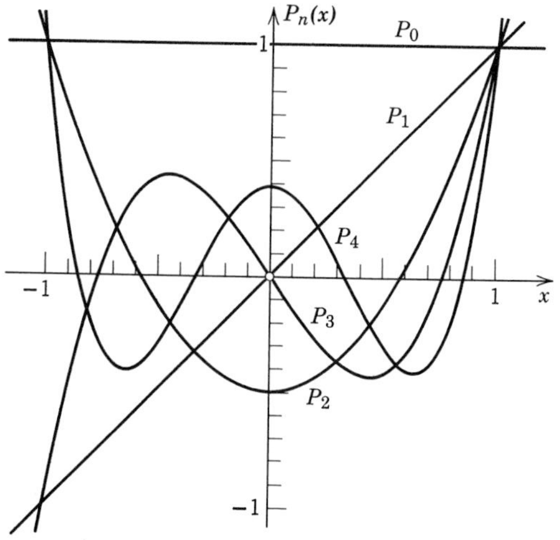

Fig. 74. Legendre polynomials

Problems for Sec. 4.3

1. Derive (11′) from (11). Find and graph $P_6(x)$.
2. Verify by substitution that $P_0, \cdots, P_5$ given in (11′) satisfy the Legendre equation (1).
3. Show that we can get (3) from (1*) more quickly if we write $m - 2 = s$ in the first sum in (1*) and $m = s$ in the other sums, obtaining

$$\sum_{s=0}^{\infty} \{(s+2)(s+1)c_{s+2} - [s(s-1) - 2s + k]c_s\}x^s = 0.$$

4. Show that for any n for which the series (6) or (7) do not reduce to a polynomial, these series have radius of convergence 1.
5. Show that the equation in Prob. 23, Sec. 4.2, is a special Legendre equation and obtain a general solution from (5), this section.
6. Prove

$$(12) \qquad P_n(x) = \frac{1}{2^n n!} \frac{d^n}{dx^n}[(x^2 - 1)^n] \qquad \textbf{(Rodrigues' formula)}$$

by applying the binomial theorem to $(x^2 - 1)^n$, differentiating the result n times term by term, and comparing with (11).

7. Using (12) and integrating n times by parts, show that

$$(13) \qquad \int_{-1}^{1} P_n^2(x)\,dx = \frac{2}{2n+1} \qquad (n = 0, 1, \cdots).$$

8. **(Generating function)** Show that

$$(14) \qquad \frac{1}{\sqrt{1 - 2xu + u^2}} = \sum_{n=0}^{\infty} P_n(x)u^n.$$

Hint. Start from the binomial expansion of $1/\sqrt{1-v}$, set $v = 2xu - u^2$, multiply the powers of $2xu - u^2$ out, collect all the terms involving u^n, and verify that the sum of these terms is $P_n(x)u^n$.

The function on the left-hand side of (14) is called a *generating function* of the Legendre polynomials.

9. Let A_1 and A_2 be two points in space (Fig. 75, $r_2 > 0$). Using (14), show that

$$\frac{1}{r} = \frac{1}{\sqrt{r_1^2 + r_2^2 - 2r_1 r_2 \cos\theta}} = \frac{1}{r_2}\left[P_0 + P_1(\cos\theta)\frac{r_1}{r_2} + P_2(\cos\theta)\left(\frac{r_1}{r_2}\right)^2 + \cdots\right].$$

This formula has applications in potential theory.

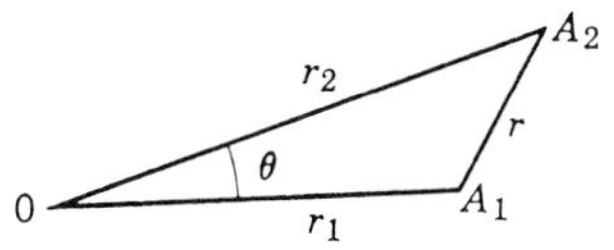

Fig. 75. Problem 9

10. (Associated Legendre functions) Consider

$$(1 - x^2)y'' - 2xy' + \left[n(n + 1) - \frac{m^2}{1 - x^2}\right]y = 0.$$

Substituting $y(x) = (1 - x^2)^{m/2}u(x)$, show that u satisfies

$$(1 - x^2)u'' - 2(m + 1)xu' + [n(n + 1) - m(m + 1)]u = 0.$$

Starting from (1) and differentiating it m times, show that a solution u of the present equation is

$$u = \frac{d^m P_n}{dx^m}.$$

The corresponding $y(x)$ is denoted by $P_n{}^m(x)$ and is called an *associated Legendre function.* Thus

$$P_n{}^m(x) = (1 - x^2)^{m/2}\frac{d^m P_n}{dx^m}.$$

It plays a role in physics (in particular, quantum physics).

4.4 Extended Power Series Method. Indicial Equation

Several second-order differential equations, which are highly important in many applications, have coefficients which are not analytic at $x = 0$ (definition in Sec. 4.2), but are such that the following theorem can be applied.

Theorem 1

Any differential equation of the form

(1) $$y'' + \frac{a(x)}{x}y' + \frac{b(x)}{x^2}y = 0,$$

where the functions $a(x)$ and $b(x)$ are analytic at $x = 0$, has at least one solution which can be represented in the form

(2) $$y(x) = x^r \sum_{m=0}^{\infty} c_m x^m = x^r(c_0 + c_1 x + c_2 x^2 + \cdots) \qquad (c_0 \neq 0)$$

where the exponent r may be any (real or complex) number (and is chosen so that $c_0 \neq 0$).[4]

The point is that in (2) we have a power series times a single power of x whose exponent r is not restricted to be a nonnegative integer. (The latter restriction would make the whole expression a power series, by definition; see footnote 1 in Sec. 4.1.)

[4]In this theorem, the variable x may be replaced by $x - a$ where a is any number.

The proof of the theorem requires advanced methods of complex analysis and can be found in Ref. [B11], Appendix 1. The following method for solving (1) is called the **Frobenius' method.**[5]

To solve (1), we write it in the somewhat more convenient form

$$x^2y'' + xa(x)y' + b(x)y = 0. \tag{1'}$$

We first expand $a(x)$ and $b(x)$ in power series,

$$a(x) = a_0 + a_1x + a_2x^2 + \cdots, \qquad b(x) = b_0 + b_1x + b_2x^2 + \cdots.$$

Then we differentiate (2) term by term, finding

$$y'(x) = \sum_{m=0}^{\infty} (m + r)c_m x^{m+r-1} = x^{r-1}[rc_0 + (r + 1)c_1x + \cdots],$$

$$y''(x) = \sum_{m=0}^{\infty} (m + r)(m + r - 1)c_m x^{m+r-2}$$

$$= x^{r-2}[r(r - 1)c_0 + (r + 1)rc_1x + \cdots].$$

By inserting all these series into (1′) we readily obtain

$$x^r[r(r - 1)c_0 + \cdots] + (a_0 + a_1x + \cdots)x^r(rc_0 + \cdots)$$
$$+ (b_0 + b_1x + \cdots)x^r(c_0 + c_1x + \cdots) = 0.$$

We now equate the sum of the coefficients of each power of x to zero, as before. This yields a system of equations involving the unknown coefficients c_m. The smallest power is x^r, and the corresponding equation is

$$[r(r - 1) + a_0r + b_0]c_0 = 0.$$

Since by assumption $c_0 \neq 0$, we obtain

$$r^2 + (a_0 - 1)r + b_0 = 0. \tag{3}$$

This important quadratic equation is called the **indicial equation** of the differential equation (1). We shall see that our method will yield a basis of solutions; one of the solutions will always be of the form (2), but for the form of the other solution there will be three possible cases depending on the roots of the indicial equation as follows.

Case 1. *Distinct roots which do not differ by an integer integer* (1, 2, 3, $\cdots$).[6]
Case 2. *Double root.*
Case 3. *Distinct roots which differ by an integer.*

Let us discuss these cases separately.

[5] GEORG FROBENIUS (1849–1917), German mathematician, who also made important contributions to the theory of matrices and groups.

[6] Note that this case includes complex conjugate roots r_1 and $r_2 = \bar{r}_1$, because $r_1 - r_2 = r_1 - \bar{r}_1 = 2i \operatorname{Im} r_1 \neq \pm 1, \pm 2, \cdots$.

Case 1. Distinct roots of the indicial equation not differing by an integer. This is the simplest case. Let r_1 and r_2 be the roots of (3). If we insert $r = r_1$ into the aforementioned system of equations and determine the coefficients c_1, c_2, $\cdots$ successively, as before, then we obtain a solution

$$y_1(x) = x^{r_1}(c_0 + c_1 x + c_2 x^2 + \cdots). \tag{4}$$

It can be shown (cf. Ref. [B11]) that in the present case the differential equation has another independent solution $y_2(x)$ of a similar form, namely

$$y_2(x) = x^{r_2}(c_0{}^* + c_1{}^* x + c_2{}^* x^2 + \cdots), \tag{5}$$

which can be obtained by inserting $r = r_2$ into that system of equations and determining the coefficients $c_0{}^*$, $c_1{}^*$, $\cdots$ successively. Linear independence of y_1 and y_2 follows from the fact that y_1/y_2 is not constant because $r_1 - r_2$ is not an integer. Hence y_1 and y_2 form a basis of (1) on the interval of convergence of both series.

Example 1. Illustration of Case 1

We consider the differential equation

$$x^2y'' + (x^2 + \tfrac{5}{36})y = 0.$$

By substituting (2) into this equation we obtain

$$\sum_{m=0}^{\infty} (m + r)(m + r - 1)c_m x^{m+r} + \sum_{m=0}^{\infty} c_m x^{m+r+2} + \frac{5}{36} \sum_{m=0}^{\infty} c_m x^{m+r} = 0. \tag{6}$$

By equating the sum of the coefficients of x^r to zero we obtain the indicial equation

$$r(r - 1) + \tfrac{5}{36} = 0.$$

The roots are $r_1 = \frac{5}{6}$ and $r_2 = \frac{1}{6}$. By equating the sum of the coefficients of x^{r+s} in (6) to zero, we find

$$\text{(7)}\quad \begin{aligned} &\text{(a)} && [(r + 1)r + \tfrac{5}{36}]c_1 = 0 && (s = 1),\\ &\text{(b)} && (s + r)(s + r - 1)c_s + c_{s-2} + \tfrac{5}{36}c_s = 0 && (s = 2, 3, \cdots). \end{aligned}$$

Let us first determine a solution $y_1(x)$ corresponding to $r_1 = \frac{5}{6}$. For this value of r, equation (7a) yields $c_1 = 0$. Equation (7b) may be written

$$s(s + \tfrac{2}{3})c_s + c_{s-2} = 0. \tag{8}$$

From $c_1 = 0$ it follows that $c_3 = 0$, $c_5 = 0$, $\cdots$, successively. If we set $s = 2p$ in (8), we obtain

$$c_{2p} = -\frac{3}{4} \cdot \frac{c_{2p-2}}{p(3p + 1)} \qquad (p = 1, 2, \cdots).$$

From this formula we see that the coefficients are

$$c_2 = -\frac{3}{4} \cdot \frac{c_0}{4}$$

$$c_4 = -\frac{3}{4} \cdot \frac{c_2}{2 \cdot 7} = \left(\frac{3}{4}\right)^2 \frac{c_0}{2!\, 4 \cdot 7}$$

$$c_6 = -\frac{3}{4} \cdot \frac{c_4}{3 \cdot 10} = -\left(\frac{3}{4}\right)^3 \frac{c_0}{3!\, 4 \cdot 7 \cdot 10} \qquad \text{etc.}$$

This yields the solution

$$
\text{(9)}\qquad
\begin{aligned}
y_1(x) &= c_0 \sum_{p=0}^{\infty} (-1)^p \left(\frac{3}{4}\right)^p \frac{x^{2p+5/6}}{p!\,1\cdot 4\cdots(3p+1)} \\
&= c_0 x^{5/6}\left(1 - \frac{3}{16}x^2 + \frac{9}{896}x^4 - + \cdots\right).
\end{aligned}
$$

Another solution $y_2(x)$ corresponding to the root $r_2 = \frac{1}{6}$ may be obtained in a similar fashion. Denoting the coefficients of this solution by $c_0{}^*, c_1{}^*, \cdots$, instead of (7) we now have

$$
\text{(10)}\qquad
\begin{aligned}
&\text{(a)}\quad [(r_2+1)r_2 + \tfrac{5}{36}]c_1{}^* = 0 \\
&\text{(b)}\quad (s+r_2)(s+r_2-1)c_s{}^* + c_{s-2}^* + \tfrac{5}{36}c_s{}^* = 0 \qquad (s = 2, 3, \cdots).
\end{aligned}
$$

From (10a) we see that $c_1{}^* = 0$, and (10b) may be written

$$
\text{(11)}\qquad s(s - \tfrac{2}{3})c_s{}^* + c_{s-2}^* = 0 \qquad (s = 2, 3, \cdots).
$$

Since $c_1{}^* = 0$, if follows that $c_3{}^* = 0, c_5{}^* = 0, \cdots$, successively. If we set $s = 2p$ in (11), we obtain

$$
c_{2p}{}^* = -\frac{3}{4}\cdot\frac{c_{2p-2}^*}{p(3p-1)} \qquad (p = 1, 2, \cdots).
$$

From this formula we find

$$
c_2{}^* = -\frac{3}{4}\cdot\frac{c_0{}^*}{2}
$$

$$
c_4{}^* = -\frac{3}{4}\cdot\frac{c_2{}^*}{2\cdot 5} = \left(\frac{3}{4}\right)^2\frac{c_0{}^*}{2!\,2\cdot 5}
$$

etc., and the corresponding solution is

$$
\text{(12)}\qquad
\begin{aligned}
y_2(x) &= c_0{}^* + c_0{}^* \sum_{p=1}^{\infty} (-1)^p \left(\frac{3}{4}\right)^p \frac{x^{2p+1/6}}{p!\,2\cdot 5\cdots(3p-1)} \\
&= c_0{}^* x^{1/6}\left(1 - \frac{3}{8}x^2 + - \cdots\right).
\end{aligned}
$$

The series (9) and (12) converge for all x, and y_1 and y_2 are linearly independent. Hence these functions constitute a basis of solutions for all values of x. (In the next section we shall see that our equation is closely related to Bessel's equation with the parameter $\nu = \frac{1}{3}$.) ∎

Case 2. Double root of the indicial equation. The indicial equation (3) has a double root r if, and only if, $(a_0 - 1)^2 - 4b = 0$, and then $r = (1 - a_0)/2$. We may determine a first solution

$$
\textbf{(13)}\qquad y_1(x) = x^r(c_0 + c_1x + c_2x^2 + \cdots) \qquad \left(r = \frac{1-a_0}{2}\right)
$$

as before. To find another solution we apply the method of variation of parameters, that is, we replace the constant c in the solution cy_1 by a function $u(x)$ to be determined so that

$$
\text{(14)}\qquad y_2(x) = u(x)y_1(x)
$$

is a solution of (1). By inserting (14) and the derivatives

$$
y_2{}' = u'y_1 + uy_1{}', \qquad y_2{}'' = u''y_1 + 2u'y_1{}' + uy_1{}''
$$

into the differential equation (1′) we obtain

$$x^2(u''y_1 + 2u'y_1' + uy_1'') + xa(u'y_1 + uy_1') + buy_1 = 0.$$

Since y_1 is a solution of (1′), the sum of the terms involving u is zero, and the last equation reduces to

$$x^2y_1u'' + 2x^2y_1'u' + xay_1u' = 0.$$

By dividing by x^2y_1 and inserting the power series for a we obtain

$$u'' + \left(2\frac{y_1'}{y_1} + \frac{a_0}{x} + \cdots\right)u' = 0.$$

Here and in the following the dots designate terms which are constant or involve positive powers of x. Now from (13) it follows that

$$\begin{aligned}\frac{y_1'}{y_1} &= \frac{x^{r-1}[rc_0 + (r+1)c_1x + \cdots]}{x^r[c_0 + c_1x + \cdots]}\\ &= \frac{1}{x}\left(\frac{rc_0 + (r+1)c_1x + \cdots}{c_0 + c_1x + \cdots}\right) = \frac{r}{x} + \cdots.\end{aligned}$$

Hence the last equation can be written

$$u'' + \left(\frac{2r + a_0}{x} + \cdots\right)u' = 0. \tag{15}$$

Since $r = (1 - a_0)/2$ the term $(2r + a_0)/x$ equals $1/x$, and by dividing by u' we thus have

$$\frac{u''}{u'} = -\frac{1}{x} + \cdots.$$

By integration we obtain

$$\ln u' = -\ln x + \cdots \qquad \text{or} \qquad u' = \frac{1}{x}e^{(\cdots)}.$$

Expanding the exponential function in powers of x and integrating once more, we see that the expression for u will be of the form

$$u = \ln x + k_1x + k_2x^2 + \cdots.$$

By inserting this into (14) we have the following result.

Form of the solutions in the case of a double root of the indicial equation. *If the indicial equation* (3) *has a double root r, then linearly independent solutions of* (1) *are of the form* (13) *and*

$$y_2(x) = y_1(x)\ln x + x^r\sum_{m=1}^{\infty} A_mx^m. \tag{16}$$

The simplest illustration of this case is given by the Cauchy equation (Sec. 2.7) in the critical case (cf. also Example 5, below).

Example 2. Illustration of Case 2

Solve the differential equation

$$x(x - 1)y'' + (3x - 1)y' + y = 0. \tag{17}$$

Writing this equation in the standard form (1), we see that it satisfies the assumptions in Theorem 1. By inserting (2) and its derivatives into (17) we obtain

$$\begin{aligned}&\sum_{m=0}^{\infty} (m + r)(m + r - 1)c_m x^{m+r} - \sum_{m=0}^{\infty} (m + r)(m + r - 1)c_m x^{m+r-1}\\ &\qquad + 3 \sum_{m=0}^{\infty} (m + r)c_m x^{m+r} - \sum_{m=0}^{\infty} (m + r)c_m x^{m+r-1} + \sum_{m=0}^{\infty} c_m x^{m+r} = 0.\end{aligned} \tag{18}$$

The smallest power is x^{r-1}; by equating the sum of its coefficients to zero we have

$$[-r(r - 1) - r]c_0 = 0 \qquad \text{or} \qquad r^2 = 0.$$

Hence this indicial equation has the double root $r = 0$. We insert this value into (18) and equate the sum of the coefficients of the power x^s to zero, finding

$$s(s - 1)c_s - (s + 1)sc_{s+1} + 3sc_s - (s + 1)c_{s+1} + c_s = 0$$

or $c_{s+1} = c_s$. Hence $c_0 = c_1 = c_2 = \cdots$, and by choosing $c_0 = 1$ we obtain the solution

$$y_1(x) = \sum_{m=0}^{\infty} x^m = \frac{1}{1 - x}.$$

To derive a second solution we substitute $y_2 = uy_1$ and its derivatives into (17), finding

$$x(x - 1)(u''y_1 + 2u'y_1' + uy_1'') + (3x - 1)(u'y_1 + uy_1') + uy_1 = 0.$$

Since y_1 is a solution of (17), this reduces to

$$x(x - 1)(u''y_1 + 2u'y_1') + (3x - 1)u'y_1 = 0.$$

Inserting the expressions for y_1 and y_1' and simplifying, we obtain the differential equation

$$xu'' + u' = 0 \qquad \text{or} \qquad \frac{u''}{u'} = -\frac{1}{x}.$$

By integrating twice we finally obtain $u = \ln x$, and a second solution for all positive x is

$$y_2 = uy_1 = \frac{\ln x}{1 - x}.$$

Case 3. Roots of the indicial equation differing by an integer. If the roots r_1 and r_2 of the indicial equation (3) differ by an integer, say, $r_1 = r$ and $r_2 = r - p$, where p is a positive integer, then we may always determine one solution as before, namely, the solution

$$y_1(x) = x^{r_1}(c_0 + c_1x + c_2x^2 + \cdots)$$

corresponding to the root r_1. However, for the root r_2 there may arise difficulties, and it may not be possible to determine a solution y_2 by the method used in

Case 1 (cf. Example 3, below). To determine y_2, we may proceed as in Case 2. The first steps are literally the same and yield equation (15). We determine $2r + a_0$ in (15). From elementary algebra we know that the coefficient $a_0 - 1$ in (3) equals $-(r_1 + r_2)$. In our case, $r_1 = r$, $r_2 = r - p$, and, therefore, $a_0 - 1 = p - 2r$. Hence in (15) we have $2r + a_0 = p + 1$. We thus obtain

$$\frac{u''}{u'} = -\left(\frac{p+1}{x} + \cdots\right).$$

The further steps are as in Case 2. Integrating, we find

$$\ln u' = -(p+1)\ln x + \cdots \qquad \text{or} \qquad u' = x^{-(p+1)}e^{(\cdots)}$$

where the dots stand for some series of nonnegative powers of x. By expanding the exponential function as before we obtain a series of the form

$$u' = \frac{1}{x^{p+1}} + \frac{k_1}{x^p} + \cdots + \frac{k_p}{x} + k_{p+1} + k_{p+2}x + \cdots.$$

By integrating once more we have

$$u = -\frac{1}{px^p} - \cdots + k_p \ln x + k_{p+1}x + \cdots.$$

Multiplying this expression by the series

$$y_1 = x^{r_1}(c_0 + c_1x + \cdots)$$

and remembering that $r_1 - p = r_2$, we see that $y_2 = uy_1$ is of the form

$$\textbf{(19)} \qquad y_2(x) = k_p y_1(x) \ln x + x^{r_2} \sum_{m=0}^{\infty} C_m x^m \qquad (x > 0).$$

We may sum up our result as follows.

Form of the solutions when the roots of the indicial equation differ by an integer. *If the roots of the indicial equation* (3) *differ by an integer, then to the root with the larger real part* (*say,* r_1) *there corresponds a solution of* (1) *of the form* (4), *and to the other root* (r_2) *there corresponds a solution of the form* (19). *The solutions form a basis of* (1).

While for a double root of (3) the second solution always contains a logarithmic term, the coefficient k_p in (19) may be zero and in the present case the logarithmic term may be missing (cf. Example 4).[7]

[7] A general theory of convergence of those series will not be presented here, but in each individual case convergence may be tested in the usual way.

Example 3. Illustration of Case 3

We consider

$$(x^2 - 1)x^2y'' - (x^2 + 1)xy' + (x^2 + 1)y = 0.$$

By substituting (2) and its derivatives into the differential equation we first have

$$(x^2 - 1) \sum_{m=0}^{\infty} (m + r)(m + r - 1)c_m x^{m+r} - (x^2 + 1) \sum_{m=0}^{\infty} (m + r)c_m x^{m+r} + (x^2 + 1) \sum_{m=0}^{\infty} c_m x^{m+r} = 0.$$

Performing the indicated multiplications and simplifying, we obtain

$$(20) \qquad \sum_{m=0}^{\infty} (m + r - 1)^2 c_m x^{m+r+2} - \sum_{m=0}^{\infty} (m + r + 1)(m + r - 1)c_m x^{m+r} = 0.$$

By equating the coefficient of x^r to zero we see that the indicial equation is

$$(r + 1)(r - 1) = 0.$$

The roots $r_1 = 1$ and $r_2 = -1$ differ by an integer. By equating the coefficient of x^{r+1} in (20) to zero we have

$$-(r + 2)rc_1 = 0.$$

For $r = r_1$ as well as for $r = r_2$ this implies $c_1 = 0$. By equating the sum of the coefficients of x^{s+r+2} in (20) to zero we find

$$(21) \qquad (s + r - 1)^2 c_s = (s + r + 3)(s + r + 1)c_{s+2} \qquad (s = 0, 1, \cdots).$$

Inserting $r = r_1 = 1$ and solving for c_{s+2}, we see that

$$c_{s+2} = \frac{s^2}{(s + 4)(s + 2)} c_s \qquad (s = 0, 1, \cdots).$$

Since $c_1 = 0$, it follows that $c_3 = 0$, $c_5 = 0$, etc. For $s = 0$ we obtain $c_2 = 0$, and from this by taking $s = 2, 4, \cdots$, we have $c_4 = 0$, $c_6 = 0$, etc. Hence to the larger root $r_1 = 1$ there corresponds the solution

$$y_1 = c_0 x.$$

Let us determine another solution. If we insert the smaller root $r = r_2 = -1$ into (21), we get

$$(s - 2)^2 c_s = (s + 2)sc_{s+2} \qquad (s = 0, 1, \cdots).$$

With $s = 0$ this becomes $4c_0 = 0$, which implies $c_0 = 0$. This contradicts our initial assumption $c_0 \neq 0$. In fact, we shall now see that our equation does not have a second independent solution $y_2(x)$ of the form (2) corresponding to the smaller root r_2. From (19) we see that $y_2(x)$ must be of the form

$$y_2(x) = kx \ln x + \frac{1}{x} \sum_{m=0}^{\infty} C_m x^m.$$

Substituting this and the derivatives into the differential equation, we obtain

$$(x^2 - 1)x^2\left(\frac{k}{x} + \sum_{m=0}^{\infty} (m - 1)(m - 2)C_m x^{m-3}\right) - (x^2 + 1)x\left(k \ln x + k + \sum_{m=0}^{\infty} (m - 1)C_m x^{m-2}\right)$$

$$+ (x^2 + 1)\left(kx \ln x + \sum_{m=0}^{\infty} C_m x^{m-1}\right) = 0.$$

The logarithmic terms drop out, and simplification yields

$$-2kx + \sum_{m=0}^{\infty} (m - 2)^2 C_m x^{m+1} - \sum_{m=0}^{\infty} m(m - 2)C_m x^{m-1} = 0.$$

By equating the sum of the coefficients of x^s to zero we obtain $C_1 = 0$ $(s = 0)$, $-2k + 4C_0 = 0$ $(s = 1)$, and

$$(s - 3)^2 C_{s-1} - (s^2 - 1)C_{s+1} = 0 \qquad (s = 2, 3, \cdots).$$

Solving for C_{s+1}, we have

$$C_{s+1} = \frac{(s - 3)^2}{s^2 - 1} C_{s-1} \qquad (s = 2, 3, \cdots).$$

Since $C_1 = 0$, we see that $C_3 = 0$, $C_5 = 0$, etc. $s = 3$ yields $C_4 = 0$, hence $C_6 = 0$, $C_8 = 0$, etc. $k = 2C_0$ and C_2 remain arbitrary. Consequently,

$$y_2(x) = 2C_0 x \ln x + \frac{1}{x}(C_0 + C_2 x^2).$$

Since the last term is $C_2 y_1/c_0$, we may take $C_2 = 0$. With $C_0 = \frac{1}{2}$,

$$y_2(x) = x \ln x + \frac{1}{2x}.$$

In Cases 2 and 3 it may sometimes be simpler to obtain a second independent solution by the method of variation of parameters.

Let us illustrate this for the equation under consideration. Since x is a solution, we may set

$$y_2(x) = xu(x).$$

Substituting this into the equation and simplifying, we find

$$(x^3 - x)u'' + (x^2 - 3)u' = 0.$$

Using partial fractions, we obtain

$$\frac{u''}{u'} = \frac{3 - x^2}{x^3 - x} = -\frac{3}{x} + \frac{1}{x + 1} + \frac{1}{x - 1}.$$

Integrating on both sides, we have

$$\ln u' = -3 \ln x + \ln (x + 1) + \ln (x - 1) = \ln \frac{x^2 - 1}{x^3}.$$

Taking exponentials and integrating again, we find

$$u = \ln x + 1/(2x^2).$$

This yields the same expression as before, namely,

$$y_2(x) = xu(x) = x \ln x + \frac{1}{2x}.$$

Example 4. Case 3, second solution without logarithmic term

We consider the differential equation

$$x^2 y'' + xy' + (x^2 - \tfrac{1}{4})y = 0.$$

Substituting (2) and the derivatives into this equation, we obtain

$$\sum_{m=0}^{\infty} [(m + r)(m + r - 1) + (m + r) - \tfrac{1}{4}]c_m x^{m+r} + \sum_{m=0}^{\infty} c_m x^{m+r+2} = 0.$$

By equating the coefficient of x^r to zero we get the indicial equation

$$r(r - 1) + r - \tfrac{1}{4} = 0 \qquad \text{or} \qquad r^2 = \tfrac{1}{4}.$$

The roots $r_1 = \frac{1}{2}$ and $r_2 = -\frac{1}{2}$ differ by an integer. By equating the sum of the coefficients of x^{s+r} to zero we find

$$\text{(22)}\quad \begin{aligned} &\text{(a)} && [(r+1)r + (r+1) - \tfrac{1}{4}]c_1 = 0 && (s = 1) \\ &\text{(b)} && [(s+r)(s+r-1) + s + r - \tfrac{1}{4}]c_s + c_{s-2} = 0 && (s = 2, 3, \cdots). \end{aligned}$$

For $r = r_1 = \frac{1}{2}$, equation (22a) yields $c_1 = 0$, and (22b) becomes

$$\text{(23)}\qquad (s+1)sc_s + c_{s-2} = 0.$$

From this and $c_1 = 0$ we obtain $c_3 = 0$, $c_5 = 0$, etc. Solving (23) for c_s and setting $s = 2p$, we get

$$c_{2p} = -\frac{c_{2p-2}}{2p(2p+1)} \qquad (p = 1, 2, \cdots).$$

Hence the nonzero coefficients are

$$c_2 = -\frac{c_0}{3!}, \qquad c_4 = -\frac{c_2}{4\cdot 5} = \frac{c_0}{5!}, \qquad c_6 = -\frac{c_0}{7!}, \qquad \text{etc.,}$$

and we obtain the solution

$$\text{(24)}\qquad y_1(x) = c_0 x^{1/2} \sum_{m=0}^{\infty} \frac{(-1)^m x^{2m}}{(2m+1)!} = c_0 x^{-1/2} \sum_{m=0}^{\infty} \frac{(-1)^m x^{2m+1}}{(2m+1)!} = c_0 \frac{\sin x}{\sqrt{x}}.$$

From (19) we see that a second independent solution is of the form

$$y_2(x) = ky_1(x)\ln x + x^{-1/2} \sum_{m=0}^{\infty} C_m x^m.$$

If we substitute this and the derivatives into the differential equation, we see that the three expressions involving $\ln x$ and the expressions ky_1 and $-ky_1$ drop out. Simplifying the remaining equation, we thus obtain

$$2kxy_1' + \sum_{m=0}^{\infty} m(m-1)C_m x^{m-1/2} + \sum_{m=0}^{\infty} C_m x^{m+3/2} = 0.$$

From (24) we find $2kxy' = -kc_0 x^{1/2} + \cdots$. Since there is no further term involving $x^{1/2}$ and $c_0 \neq 0$, we must have $k = 0$. The sum of the coefficients of the power $x^{s-1/2}$ is

$$s(s-1)C_s + C_{s-2} \qquad (s = 2, 3, \cdots).$$

Equating this to zero and solving for C_s, we have

$$C_s = -\frac{C_{s-2}}{s(s-1)} \qquad (s = 2, 3, \cdots).$$

We thus obtain

$$C_2 = -\frac{C_0}{2!}, \qquad C_4 = -\frac{C_2}{4\cdot 3} = \frac{C_0}{4!}, \qquad C_6 = -\frac{C_0}{6!}, \qquad \text{etc.,}$$

$$C_3 = -\frac{C_1}{3!}, \qquad C_5 = -\frac{C_3}{5\cdot 4} = \frac{C_1}{5!}, \qquad C_7 = -\frac{C_1}{7!}, \qquad \text{etc.,}$$

We may take $C_1 = 0$, because the odd powers would yield $C_1 y_1/c_0$. Then

$$\text{(25)}\qquad y_2(x) = C_0 x^{-1/2} \sum_{m=0}^{\infty} \frac{(-1)^m x^{2m}}{(2m)!} = C_0 \frac{\cos x}{\sqrt{x}}.$$

$y_1(x)$ and $y_2(x)$ constitute a basis of solutions for all $x \neq 0$.

Example 5. Cauchy equation

In Sec. 2.7 we have seen that the Cauchy equation

$$x^2y'' + a_0xy' + b_0y = 0 \qquad (a_0,\ b_0 \text{ constant}) \tag{26}$$

can be solved by substituting $y = x^r$ into (26). This yields the auxiliary equation

$$r^2 + (a_0 - 1)r + b_0 = 0.$$

Obviously, this is the indicial equation. If $r_1 \neq r_2$, then $y_1 = x^{r_1}, y_2 = x^{r_2}$ form a basis of solutions; this corresponds to our present Cases 1 and 3. If $r_1 = r_2 = r$, then the solutions

$$y_1 = x^r, \qquad y_2 = y_1 \ln x = x^r \ln x$$

form a basis for all positive x, as follows by substituting (16) in (26), in agreement with the result in Sec. 2.7.

Problems for Sec. 4.4

Find a basis of solutions of the following differential equations. Try to identify the series obtained by the Frobenius method as expansions of known functions.

1. $x^2y'' + 4xy' + (x^2 + 2)y = 0$

2. $2x(1 - x)y'' + (1 + x)y' - y = 0$

3. $y'' + xy' + (1 - 2x^{-2})y = 0$

4. $xy'' + 2y' + 4xy = 0$

5. $xy'' + 3y' + 4x^3y = 0$

6. $xy'' - y = 0$

7. $x^2y'' - 5xy' + 9y = 0$

8. $xy'' + 2y' + xy = 0$

9. $xy'' + y' - xy = 0$

10. $(x - 1)xy'' + (4x - 2)y' + 2y = 0$

11. $x^2y'' + xy' - 4y = 0$

12. $2x^2y'' - xy' - 2y = 0$

13. $x^2y'' + 6xy' + (6 - 4x^2)y = 0$

14. $x^2y'' - 2y = 0$

15. $xy'' + (1 - 2x)y' + (x - 1)y = 0$

16. $xy'' + (2 - 2x)y' + (x - 2)y = 0$

17. (Gauss's[8] hypergeometric equation) The differential equation

$$x(1 - x)y'' + [c - (a + b + 1)x]y' - aby = 0 \tag{27}$$

(where a, b, c are constants) is known as *Gauss's hypergeometric equation*. Show that the corresponding indicial equation has the roots $r_1 = 0$ and $r_2 = 1 - c$. Show that for $r_1 = 0$ the Frobenius method yields

$$\begin{aligned} y_1(x) = 1 &+ \frac{ab}{1!\,c}x + \frac{a(a+1)b(b+1)}{2!\,c(c+1)}x^2 \\ &+ \frac{a(a+1)(a+2)b(b+1)(b+2)}{3!\,c(c+1)(c+2)}x^3 + \cdots \end{aligned} \tag{28}$$

$$(c \neq 0, -1, -2, \cdots).$$

[8]CARL FRIEDRICH GAUSS (1777–1855), great German mathematician, whose work was of basic importance in algebra, number theory, differential equations, differential geometry, non-Euclidean geometry, complex analysis, numerical analysis, and theoretical mechanics. He also paved the way for a general and systematic use of complex numbers.

This series is called the **hypergeometric series.** Its sum $y_1(x)$ is usually denoted by $F(a, b, c; x)$ and is called the **hypergeometric function.**

18. Prove that the series (28) converges for $|x| < 1$.

19. Show that when $a = b = c = 1$, the series (28) reduces to the geometric series.

20. For what values of a and b does (28) reduce to a polynomial?

21. Show that to $r_2 = 1 - c$ in Prob. 17 there corresponds the solution

$$(29)\quad \begin{aligned} y_2(x) = x^{1-c}\Bigg(1 &+ \frac{(a-c+1)(b-c+1)}{1!\,(-c+2)}x \\ &+ \frac{(a-c+1)(a-c+2)(b-c+1)(b-c+2)}{2!\,(-c+2)(-c+3)}x^2 + \cdots\Bigg) \end{aligned}$$

$$(c \neq 2, 3, 4, \cdots).$$

22. Show that in Prob. 21,

$$y_2(x) = x^{1-c}F(a - c + 1, b - c + 1, 2 - c; x).$$

23. Show that if $c \neq 0, \pm 1, \pm 2, \cdots$, the functions (28) and (29) constitute a basis of solutions of (27).

24. Show that

$$\frac{dF(a, b, c; x)}{dx} = \frac{ab}{c}F(a + 1, b + 1, c + 1; x),$$

$$\frac{d^2F(a, b, c; x)}{dx^2} = \frac{a(a+1)b(b+1)}{c(c+1)}F(a + 2, b + 2, c + 2; x), \qquad \text{etc.}$$

Many elementary functions are special cases of $F(a, b, c; x)$. Prove

25. $\dfrac{1}{1-x} = F(1, 1, 1; x) = F(1, b, b; x) = F(a, 1, a; x)$

26. $(1 + x)^n = F(-n, b, b; -x), \quad (1 - x)^n = 1 - nxF(1 - n, 1, 2; x)$

27. $\arctan x = xF(\tfrac{1}{2}, 1, \tfrac{3}{2}; -x^2), \quad \arcsin x = xF(\tfrac{1}{2}, \tfrac{1}{2}, \tfrac{3}{2}; x^2)$

28. $\ln(1 + x) = xF(1, 1, 2; -x), \quad \ln\dfrac{1+x}{1-x} = 2xF(\tfrac{1}{2}, 1, \tfrac{3}{2}; x^2)$

29. Consider the differential equation

$$(30)\qquad (t^2 + At + B)\ddot{y} + (Ct + D)\dot{y} + Ky = 0$$

where A, B, C, D, K are constants and $t^2 + At + B$ has distinct zeros t_1 and t_2. Show that by introducing the new independent variable $x = (t - t_1)/(t_2 - t_1)$ the equation (30) becomes the hypergeometric equation where $Ct_1 + D = -c(t_2 - t_1)$, $C = a + b + 1$, $K = ab$.

Solve, in terms of hypergeometric functions, the following equations.

30. $2x(1 - x)y'' + (1 - 5x)y' - y = 0$

31. $2x(1 - x)y'' - (1 + 5x)y' - y = 0$

32. $5x(1 - x)y'' + (4 - x)y' + y = 0$

33. $4(t^2 - 3t + 2)\ddot{y} - 2\dot{y} + y = 0$

34. $3t(1 + t)\ddot{y} + t\dot{y} - y = 0$

35. $2(t^2 - 5t + 6)\ddot{y} + (2t - 3)\dot{y} - 8y = 0$

4.5 Bessel's Equation. Bessel Functions of the First Kind

One of the most important differential equations in applied mathematics is **Bessel's differential equation**[9]

$$x^2y'' + xy' + (x^2 - \nu^2)y = 0 \tag{1}$$

where the parameter ν is a given number. We assume that ν is real and nonnegative. This equation is of the type characterized in Theorem 1, Sec. 4.4. Hence it has a solution of the form

$$y(x) = \sum_{m=0}^{\infty} c_m x^{m+r} \qquad (c_0 \neq 0). \tag{2}$$

Substituting this expression and its derivatives into Bessel's equation, we have

$$\sum_{m=0}^{\infty} (m + r)(m + r - 1)c_m x^{m+r} + \sum_{m=0}^{\infty} (m + r)c_m x^{m+r} + \sum_{m=0}^{\infty} c_m x^{m+r+2} - \nu^2 \sum_{m=0}^{\infty} c_m x^{m+r} = 0.$$

By equating the sum of the coefficients of x^{s+r} to zero we find

$$\text{(3)}\quad \begin{aligned} &\text{(a)} && r(r - 1)c_0 + rc_0 - \nu^2 c_0 = 0 && (s = 0) \\ &\text{(b)} && (r + 1)rc_1 + (r + 1)c_1 - \nu^2 c_1 = 0 && (s = 1) \\ &\text{(c)} && (s + r)(s + r - 1)c_s + (s + r)c_s + c_{s-2} - \nu^2 c_s = 0 && (s = 2, 3, \cdots). \end{aligned}$$

From (3a) we obtain the indicial equation

$$(r + \nu)(r - \nu) = 0. \tag{4}$$

The roots are $r_1 = \nu$ ($\geqq 0$) and $r_2 = -\nu$.

Let us first determine a solution corresponding to the root r_1. For this value of r, equation (3b) yields $c_1 = 0$. Equation (3c) may be written

$$(s + r + \nu)(s + r - \nu)c_s + c_{s-2} = 0,$$

and for $r = \nu$ this takes the form

$$(s + 2\nu)sc_s + c_{s-2} = 0. \tag{5}$$

[9] FRIEDRICH WILHELM BESSEL (1784–1846), German astronomer and mathematician.

Since $c_1 = 0$ and $\nu \geqq 0$, it follows that $c_3 = 0, c_5 = 0, \cdots$, successively. If we set $s = 2m$ in (5), we obtain

$$c_{2m} = -\frac{1}{2^2 m(\nu + m)} c_{2m-2}, \qquad m = 1, 2, \cdots \tag{6}$$

and may determine the coefficients $c_2, c_4, \cdots$ successively.

c_0 is arbitrary. It is customary to put

$$c_0 = \frac{1}{2^\nu \Gamma(\nu + 1)} \tag{7}$$

where $\Gamma(\nu + 1)$ is the **gamma function.** Some basic formulas for this important function are included in Appendix 3. For the present purpose it suffices to know that $\Gamma(\alpha)$ is defined by the integral

$$\Gamma(\alpha) = \int_0^\infty e^{-t} t^{\alpha-1}\, dt \qquad (\alpha > 0). \tag{8}$$

By integration by parts we obtain

$$\Gamma(\alpha + 1) = \int_0^\infty e^{-t} t^\alpha\, dt = -e^{-t} t^\alpha \Big|_0^\infty + \alpha \int_0^\infty e^{-t} t^{\alpha-1}\, dt.$$

The first expression on the right is zero, and the integral on the right is $\Gamma(\alpha)$. This yields the basic relation

$$\Gamma(\alpha + 1) = \alpha\Gamma(\alpha). \tag{9}$$

Since

$$\Gamma(1) = \int_0^\infty e^{-t}\, dt = 1,$$

we conclude from (9) that

$$\Gamma(2) = \Gamma(1) = 1!, \qquad \Gamma(3) = 2\Gamma(2) = 2!, \quad \cdots$$

and in general

$$\Gamma(k + 1) = k! \qquad (k = 0, 1, \cdots). \tag{10}$$

This shows that the gamma function may be regarded as a generalization of the factorial function known from elementary calculus.

Returning to our problem, we obtain from (6), (7), and (9)

$$c_2 = -\frac{c_0}{2^2(\nu + 1)} = -\frac{1}{2^{2+\nu} 1!\, \Gamma(\nu + 2)}$$

$$c_4 = -\frac{c_2}{2^2 2(\nu + 2)} = \frac{1}{2^{4+\nu} 2!\, \Gamma(\nu + 3)}$$

and so on; in general,

$$c_{2m} = \frac{(-1)^m}{2^{2m+\nu} m!\, \Gamma(\nu + m + 1)}. \tag{11}$$

By inserting these coefficients in (2) and remembering that $c_1 = 0$, $c_3 = 0$, $\cdots$ we obtain a particular solution of Bessel's equation which is denoted by $J_\nu(x)$:

$$J_\nu(x) = x^\nu \sum_{m=0}^{\infty} \frac{(-1)^m x^{2m}}{2^{2m+\nu} m!\, \Gamma(\nu + m + 1)}. \tag{12}$$

This solution of (1) is known as the **Bessel function of the first kind** *of order ν*. The ratio test shows that the series converges for all x.

Integer values *of ν are frequently denoted by n.* Thus, for $n \geqq 0$,

$$J_n(x) = x^n \sum_{m=0}^{\infty} \frac{(-1)^m x^{2m}}{2^{2m+n} m!\, (n + m)!}. \tag{13}$$

Replacing ν by $-\nu$ in (12), we have

$$J_{-\nu}(x) = x^{-\nu} \sum_{m=0}^{\infty} \frac{(-1)^m x^{2m}}{2^{2m-\nu} m!\, \Gamma(m - \nu + 1)}. \tag{14}$$

Since Bessel's equation involves ν^2, the functions J_ν and $J_{-\nu}$ are solutions of the equation for the same ν. If ν is not an integer, they are linearly independent, because the first term in (12) and the first term in (14) are finite nonzero multiples of x^ν and $x^{-\nu}$, respectively. This yields the following result.

Theorem 1 (General solution)

If ν is not an integer, a general solution of Bessel's equation for all $x \neq 0$ is

$$y(x) = a_1 J_\nu(x) + a_2 J_{-\nu}(x). \tag{15}$$

However, if ν is an integer, then (15) is not a general solution. This follows from

Theorem 2 (Linear dependence)

For integer $\nu = n$ the Bessel functions $J_n(x)$ and $J_{-n}(x)$ are linearly dependent, because

$$J_{-n}(x) = (-1)^n J_n(x) \qquad (n = 1, 2, \cdots). \tag{16}$$

Proof. We use (14) and let ν approach a positive integer n. Then the gamma functions in the coefficients of the first n terms become infinite (cf. Fig. 405 in Appendix 3), the coefficients become zero, and the summation starts with

$m = n$. Since in this case $\Gamma(m - n + 1) = (m - n)!$ [cf. (10)], we obtain

$$J_{-n}(x) = \sum_{m=n}^{\infty} \frac{(-1)^m x^{2m-n}}{2^{2m-n} m!\,(m-n)!} = \sum_{s=0}^{\infty} \frac{(-1)^{n+s} x^{2s+n}}{2^{2s+n}(n+s)!\,s!}$$

where $m = n + s$ and $s = m - n$. From (13) we see that the last series represents $(-1)^n J_n(x)$. This completes the proof. ∎

A general solution of the Bessel equation with integer $\nu = n$ will be obtained in the next section. Numerical values of J_0 and J_1 are included in Appendix 4.

Problems for Sec. 4.5

1. Show that the series in (12) converge for all x.
2. Show that $J_n(x)$ is an even function for even n and an odd function for odd n.
3. Show that (cf. Fig. 76)

$$J_0(x) = 1 - \frac{x^2}{2^2(1!)^2} + \frac{x^4}{2^4(2!)^2} - \frac{x^6}{2^6(3!)^2} + - \cdots \tag{17}$$

and

$$J_1(x) = \frac{x}{2} - \frac{x^3}{2^3 1!\,2!} + \frac{x^5}{2^5 2!\,3!} - \frac{x^7}{2^7 3!\,4!} + - \cdots. \tag{18}$$

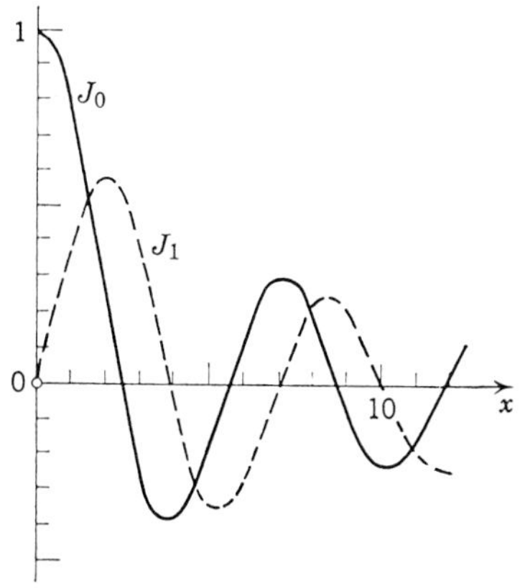

Fig. 76. Bessel functions of the first kind

4. Show that for small $|x|$ we have $J_0(x) \approx 1 - 0.25x^2$. Using this formula, compute $J_0(x)$ for $x = 0.1, 0.2, \cdots, 1.0$ and determine the relative error by comparing with Table A2 in Appendix 4.
5. How many terms of (17) are needed to compute $J_0(1)$ with an error less than 1 unit of the fifth decimal place? How many terms would be needed to compute ln 2 from the Maclaurin series of $\ln(1 + x)$ with the same degree of accuracy?
6. It can be shown that for large x,

$$(19) \qquad \begin{aligned} J_{2n}(x) &\approx (-1)^n (\pi x)^{-1/2}(\cos x + \sin x) \\ J_{2n+1}(x) &\approx (-1)^{n+1}(\pi x)^{-1/2}(\cos x - \sin x). \end{aligned}$$

Using (19), graph $J_0(x)$ for large x, compute approximate values of the first five positive zeros of $J_0(x)$, and compare them with the more accurate values 2.405, 5.520, 8.654, 11.792, 14.931.

Relations between Bessel functions. Bessel functions are important in engineering applications. (An illustration will be given in Sec. 11.10.) In using them, it is necessary to know that they satisfy various relations. (Ref. [B18], the standard book on Bessel functions, and Ref. [A5] in Appendix 1 contain an incredibly large number of formulas.) Some important relations are included in the following problems. Formulas (22) and (23) are useful for computing tables.

7. Using (12) and (9), show that

$$[x^{\nu}J_{\nu}(x)]' = x^{\nu}J_{\nu-1}(x). \tag{20}$$

Using (12), show that

$$[x^{-\nu}J_{\nu}(x)]' = -x^{-\nu}J_{\nu+1}(x). \tag{21}$$

8. Derive from (20) and (21) the **recurrence relation**

$$J_{\nu-1}(x) + J_{\nu+1}(x) = \frac{2\nu}{x}J_{\nu}(x). \tag{22}$$

9. Derive from (20) and (21) the **recurrence relation**

$$J_{\nu-1}(x) - J_{\nu+1}(x) = 2J_{\nu}'(x). \tag{23}$$

10. Derive the Bessel equation (1) from (20) and (21).

11. Using (20) and (21) and Rolle's theorem, show that between two consecutive zeros of $J_0(x)$ there is precisely one zero of $J_1(x)$.

12. Show that between any two consecutive positive zeros of $J_n(x)$ there is precisely one zero of $J_{n+1}(x)$.

13. Using Table A2 in Appendix 4 and (22), compute $J_2(x)$ for the values $x = 0, 0.1, 0.2, \cdots, 1.0$.

Integrals involving Bessel functions can often be evaluated or at least simplified by the use of (20)–(23). Show that

14. $\int x^{\nu}J_{\nu-1}(x)\,dx = x^{\nu}J_{\nu}(x) + c$

15. $\int x^{-\nu}J_{\nu+1}(x)\,dx = -x^{-\nu}J_{\nu}(x) + c$

16. $\int J_{\nu+1}(x)\,dx = \int J_{\nu-1}(x)\,dx - 2J_{\nu}(x)$

Using the formulas in Probs. 14–16 and, if necessary, integration by parts, evaluate

17. $\int J_3(x)\,dx$

18. $\int J_5(x)\,dx$

19. $\int x^{-3}J_4(x)\,dx$

20. $\int x^3J_0(x)\,dx$

(In other cases, one may be left with the integral of $J_0(x)$, which cannot be evaluated in finite form, but has been tabulated; cf. Ref. [B18] in Appendix 1.)

Elementary functions. For certain values of ν the Bessel function J_{ν} can be expressed in terms of cosine and sine:

21. Using (12) and the relation $\Gamma(\frac{1}{2}) = \sqrt{\pi}$, show that

$$J_{1/2}(x) = \sqrt{\frac{2}{\pi x}}\sin x, \qquad J_{-1/2}(x) = \sqrt{\frac{2}{\pi x}}\cos x. \tag{24}$$

22. Using (22) and (24), show that

$$J_{3/2}(x) = \sqrt{\frac{2}{\pi x}}\left(\frac{\sin x}{x} - \cos x\right), \qquad J_{-3/2}(x) = -\sqrt{\frac{2}{\pi x}}\left(\frac{\cos x}{x} + \sin x\right).$$

23. Conclude from (22) and (24) that the *Bessel functions* $J_\nu(x)$ *of order* $\nu = \pm\frac{1}{2}, \pm\frac{3}{2}, \pm\frac{5}{2}, \ldots$ *are elementary functions.*

24. (Elimination of first derivative) Consider the differential equation

$$y'' + f(x)y' + g(x)y = 0.$$

Substitute $y(x) = u(x)v(x)$ and determine v such that the resulting second-order differential equation does not involve u'.

25. Show that for the Bessel equation the substitution in Prob. 24 is $y = ux^{-1/2}$ and gives

$$x^2u'' + (x^2 + \tfrac{1}{4} - \nu^2)u = 0. \tag{25}$$

Using this result, find a general solution of (1) with $\nu = \frac{1}{2}$.

4.6 Bessel Functions of the Second Kind

For integer $\nu = n$ the Bessel functions $J_n(x)$ and $J_{-n}(x)$ are linearly dependent (Sec. 4.5), so that they do not form a basis of solutions. We shall now obtain a second independent solution, starting with the case $n = 0$. In this case Bessel's equation may be written

$$xy'' + y' + xy = 0, \tag{1}$$

the indicial equation has the double root $r = 0$, and we see from (16) in Sec. 4.4, that the desired solution must have the form

$$y_2(x) = J_0(x) \ln x + \sum_{m=1}^{\infty} A_m x^m. \tag{2}$$

We substitute y_2 and its derivatives

$$y_2' = J_0' \ln x + \frac{J_0}{x} + \sum_{m=1}^{\infty} mA_m x^{m-1}$$

$$y_2'' = J_0'' \ln x + \frac{2J_0'}{x} - \frac{J_0}{x^2} + \sum_{m=1}^{\infty} m(m-1)A_m x^{m-2}$$

into (1). Then the logarithmic terms disappear because J_0 is a solution of (1), the other two terms containing J_0 cancel, and we find

$$2J_0' + \sum_{m=1}^{\infty} m(m-1)A_m x^{m-1} + \sum_{m=1}^{\infty} mA_m x^{m-1} + \sum_{m=1}^{\infty} A_m x^{m+1} = 0.$$

From (12) in Sec. 4.5 we obtain the power series of J_0' in the form

$$J_0'(x) = \sum_{m=1}^{\infty} \frac{(-1)^m 2m x^{2m-1}}{2^{2m}(m!)^2} = \sum_{m=1}^{\infty} \frac{(-1)^m x^{2m-1}}{2^{2m-1} m!\,(m-1)!}.$$

By inserting this series we have

$$\sum_{m=1}^{\infty} \frac{(-1)^m x^{2m-1}}{2^{2m-2} m!\,(m-1)!} + \sum_{m=1}^{\infty} m^2 A_m x^{m-1} + \sum_{m=1}^{\infty} A_m x^{m+1} = 0.$$

We first show that the A_m with odd subscripts are all zero. The coefficient of the power x^0 is A_1, and so, $A_1 = 0$. By equating the sum of the coefficients of the power x^{2s} to zero we have

$$(2s+1)^2 A_{2s+1} + A_{2s-1} = 0, \qquad s = 1, 2, \cdots.$$

Since $A_1 = 0$, we thus obtain $A_3 = 0, A_5 = 0, \cdots$, successively. We now equate the sum of the coefficients of x^{2s+1} to zero. For $s = 0$ this gives

$$-1 + 4A_2 = 0 \qquad \text{or} \qquad A_2 = \tfrac{1}{4}.$$

For the other values of s we obtain

$$\frac{(-1)^{s+1}}{2^{2s}(s+1)!\,s!} + (2s+2)^2 A_{2s+2} + A_{2s} = 0 \qquad (s = 1, 2, \cdots).$$

For $s = 1$ this yields

$$\tfrac{1}{8} + 16A_4 + A_2 = 0 \qquad \text{or} \qquad A_4 = -\tfrac{3}{128}$$

and in general

$$\text{(3)} \qquad A_{2m} = \frac{(-1)^{m-1}}{2^{2m}(m!)^2}\left(1 + \frac{1}{2} + \frac{1}{3} + \cdots + \frac{1}{m}\right), \qquad m = 1, 2, \cdots.$$

Using the short notation

$$\text{(4)} \qquad h_m = 1 + \frac{1}{2} + \cdots + \frac{1}{m}$$

and inserting (3) and $A_1 = A_3 = \cdots = 0$ into (2), we obtain the result

$$(5)\qquad \begin{aligned} y_2(x) &= J_0(x)\ln x + \sum_{m=1}^{\infty} \frac{(-1)^{m-1}h_m}{2^{2m}(m!)^2}x^{2m} \\ &= J_0(x)\ln x + \tfrac{1}{4}x^2 - \tfrac{3}{128}x^4 + - \cdots . \end{aligned}$$

Since J_0 and y_2 are linearly independent functions, they form a basis of (1). Of course, another basis is obtained if we replace y_2 by an independent particular solution of the form $a(y_2 + bJ_0)$ where a ($\neq 0$) and b are constants. It is customary to choose $a = 2/\pi$ and $b = \gamma - \ln 2$, where the number $\gamma = 0.577\,215\,664\,90 \cdots$ is the so-called **Euler constant,** which is defined as the limit of

$$1 + \frac{1}{2} + \cdots + \frac{1}{s} - \ln s$$

as s approaches infinity. The standard particular solution thus obtained is known as the **Bessel function of the second kind** *of order zero* (Fig. 77) or **Neumann's function** *of order zero* and is denoted by $Y_0(x)$. Thus

$$(6)\qquad Y_0(x) = \frac{2}{\pi}\left[J_0(x)\left(\ln\frac{x}{2} + \gamma\right) + \sum_{m=1}^{\infty} \frac{(-1)^{m-1}h_m}{2^{2m}(m!)^2}x^{2m}\right],$$

where h_m is defined by (4).

If $\nu = n = 1, 2, \cdots$, a second solution can be obtained by similar manipulations, starting from (19), Sec. 4.4. It turns out that also in these cases the solution contains a logarithmic term.

The situation is not yet completely satisfactory, because the second solution is defined differently, depending on whether the order ν is an integer or not. To provide uniformity of formalism and numerical tabulation, it is desirable to adopt a form of the second solution which is valid for all values of the order. This is the reason for introducing a standard second solution $Y_\nu(x)$ defined for

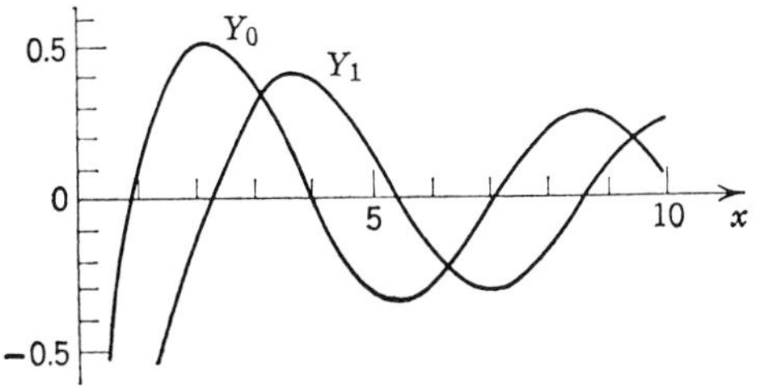

Fig. 77. Bessel functions of the second kind. (A small corresponding table is included in Appendix 4.)

all ν by the formula

$$\text{(7)}\qquad \begin{aligned} &\text{(a)} \quad Y_\nu(x) = \frac{1}{\sin \nu\pi}\,[J_\nu(x) \cos \nu\pi - J_{-\nu}(x)] \\ &\text{(b)} \quad Y_n(x) = \lim_{\nu \to n} Y_\nu(x). \end{aligned}$$

This function is known as the **Bessel function of the second kind** *of order* ν or **Neumann's function**[10] *of order* ν.

For noninteger order ν, the function $Y_\nu(x)$ is evidently a solution of Bessel's equation because $J_\nu(x)$ and $J_{-\nu}(x)$ are solutions of that equation. Since for those ν the solutions J_ν and $J_{-\nu}$ are linearly independent and Y_ν involves $J_{-\nu}$, the functions J_ν and Y_ν are linearly independent. Furthermore, it can be shown that the limit in (7b) exists and Y_n is a solution of Bessel's equation for integer order; cf. Ref. [B18] in Appendix 1. We shall see that the series development of $Y_n(x)$ contains a logarithmic term. Hence $J_n(x)$ and $Y_n(x)$ are linearly independent solutions of Bessel's equation. The series development of $Y_n(x)$ can be obtained if we insert the series (12) and (14), Sec. 4.5, for $J_\nu(x)$ and $J_{-\nu}(x)$ into (7a) and then let ν approach n; for details see Ref. [B18]; the result is

$$\text{(8)}\qquad \begin{aligned} Y_n(x) = \frac{2}{\pi} J_n(x)\left(\ln\frac{x}{2} + \gamma\right) &= \frac{x^n}{\pi} \sum_{m=0}^{\infty} \frac{(-1)^{m-1}(h_m + h_{m+n})}{2^{2m+n} m!\,(m+n)!} x^{2m} \\ &\quad - \frac{x^{-n}}{\pi} \sum_{m=0}^{n-1} \frac{(n-m-1)!}{2^{2m-n} m!} x^{2m} \end{aligned}$$

where $x > 0$, $n = 0, 1, \cdots$, and

$$h_0 = 0, \qquad h_s = 1 + \frac{1}{2} + \frac{1}{3} + \cdots + \frac{1}{s} \qquad (s = 1, 2, \cdots),$$

and when $n = 0$ the last sum in (8) is to be replaced by 0. For $n = 0$ the representation (8) takes the form (6). Furthermore, it can be shown that

$$Y_{-n}(x) = (-1)^n Y_n(x).$$

We may formulate our result, as follows.

Theorem 1 (General solution)
A general solution of Bessel's equation for all values of ν *is*

$$\text{(9)}\qquad y(x) = C_1 J_\nu(x) + C_2 Y_\nu(x).$$

[10]CARL NEUMANN (1832–1925), German mathematician and physicist. The solutions $Y_\nu(x)$ are sometimes denoted by $N_\nu(x)$; in Ref. [B18] they are called Weber's functions. Euler's constant in (6) is often denoted by C or $\ln \gamma$ (instead of γ).

We finally mention that there is a practical need for solutions of Bessel's equation which are complex for real values of x. For this reason the solutions

$$\text{(10)} \qquad \begin{aligned} H_\nu^{(1)}(x) &= J_\nu(x) + iY_\nu(x) \\ H_\nu^{(2)}(x) &= J_\nu(x) - iY_\nu(x) \end{aligned}$$

are frequently used and have been tabulated (cf. Ref. [A6] in Appendix 1). These linearly independent functions are called **Bessel functions of the third kind** *of order* ν or *first and second* **Hankel functions**[11] *of order* ν.

Problems for Sec. 4.6

1. Show that for small $x > 0$ we have $Y_0(x) \approx 2\pi^{-1}(\ln \frac{1}{2}x + \gamma)$. Using this formula, compute an approximate value of the smallest positive zero of $Y_0(x)$ and compare it with the more accurate value 0.9.

2. It can be shown that, for large x,

$$\text{(11)} \qquad Y_n(x) \approx \sqrt{2/\pi x}\, \sin\left(x - \tfrac{1}{2}n\pi - \tfrac{1}{4}\pi\right).$$

Using (11) and Table A2 in Appendix 4, graph $Y_0(x)$ and $Y_1(x)$ for $0 < x \leqq 15$. Using (11), compute approximate values of the first three positive zeros of $Y_0(x)$ and compare these values with the more accurate values 0.89, 3.96, and 7.09.

Differential equations reducible to Bessel's equation. There are various differential equations which can be reduced to Bessel's equation. To illustrate this, use the indicated substitutions and find a general solution of the given equations in terms of Bessel functions.

3. $x^2y'' + xy' + (x^2 - 4)y = 0$

4. $xy'' + y' + xy = 0$

5. $xy'' + y' + \frac{1}{4}y = 0 \qquad (\sqrt{x} = z)$

6. $4x^2y'' + 4xy' + (x - n^2)y = 0 \qquad (\sqrt{x} = z)$

7. $x^2y'' + xy' + (\lambda^2x^2 - n^2)y = 0 \qquad (\lambda x = z)$

8. $x^2y'' + xy' + 4(x^4 - n^2)y = 0 \qquad (x^2 = z)$

9. $xy'' - y' + xy = 0 \qquad (y = xu)$

10. $xy'' + (1 + 2n)y' + xy = 0 \qquad (y = x^{-n}u)$

11. $x^2y'' + (1 - 2n)xy' + n^2(x^{2n} + 1 - n^2)y = 0 \qquad (y = x^nu,\ x^n = z)$

12. $x^2y'' + (x^2 + \frac{1}{4})y = 0 \qquad (y = u\sqrt{x})$

13. $y'' + xy = 0 \qquad (y = u\sqrt{x},\ \frac{2}{3}x^{3/2} = z)$

Show that the given functions are solutions of the corresponding equations.

14. $y'' + k^2xy = 0, \qquad y = \sqrt{x}\,J_{1/3}(2kx^{3/2}/3), \qquad y = \sqrt{x}\,Y_{1/3}(2kx^{3/2}/3)$

15. $y'' + k^2x^2y = 0, \qquad y = \sqrt{x}\,J_{1/4}(kx^2/2), \qquad y = \sqrt{x}\,Y_{1/4}(kx^2/2)$

16. $y'' + k^2x^4y = 0, \qquad y = \sqrt{x}\,J_{1/6}(kx^3/3), \qquad y = \sqrt{x}\,Y_{1/6}(kx^3/3)$

17. (Modified Bessel functions) The function $I_\nu(x) = i^{-\nu}J_\nu(ix)$, $i = \sqrt{-1}$, is called the *modified Bessel function of the first kind of order* ν. Show that $I_\nu(x)$ is a solution of the differential equation

$$\text{(12)} \qquad x^2y'' + xy' - (x^2 + \nu^2)y = 0$$

[11] HERMANN HANKEL (1839–1873), German mathematician.

and has the representation

$$I_\nu(x) = \sum_{m=0}^{\infty} \frac{x^{2m+\nu}}{2^{2m+\nu} m!\, \Gamma(m + \nu + 1)}. \tag{13}$$

18. Show that $I_\nu(x)$ is real for all real x (and real ν), $I_\nu(x) \neq 0$ for all real $x \neq 0$, and $I_{-n}(x) = I_n(x)$, where n is any integer.

19. (Modified Bessel functions) Show that another solution of the differential equation (12) is the so-called *modified Bessel function of the third kind* (sometimes: *of the second kind*)

$$K_\nu(x) = \frac{\pi}{2 \sin \nu\pi} [I_{-\nu}(x) - I_\nu(x)]. \tag{14}$$

20. Show that the Hankel functions (10) constitute a basis of solutions of Bessel's equation for any ν.

4.7 Orthogonal Sets of Functions

Legendre polynomials (Sec. 4.3) and Bessel functions enjoy a property which is called **orthogonality** and is of general importance in engineering mathematics. In the present section we shall introduce the corresponding concepts and notations. In Sec. 4.8 we shall consider certain boundary value problems ("Sturm–Liouville problems") whose solutions form orthogonal sets of functions, and in Sec. 4.9 we shall apply the results of Sec. 4.8 to Legendre polynomials and Bessel functions. This is our program for the remaining portion of this chapter.

So let us first define orthogonality of functions. Let $g_m(x)$ and $g_n(x)$ be two real-valued functions which are defined on an interval $a \leqq x \leqq b$ and are such that the integral of the product $g_m(x)g_n(x)$ over that interval exists. We shall denote this integral by (g_m, g_n). This is a simple standard notation which is widely used. Thus

$$(g_m, g_n) = \int_a^b g_m(x) g_n(x)\, dx. \tag{1}$$

The functions g_m and g_n are said to be **orthogonal** *on the interval* $a \leqq x \leqq b$ if the integral (1) is zero, that is,

$$(g_m, g_n) = \int_a^b g_m(x) g_n(x)\, dx = 0 \qquad (m \neq n). \tag{2}$$

A set of real-valued functions $g_1(x)$, $g_2(x)$, $g_3(x)$, $\cdots$ is called an **orthogonal set** *of functions on an interval* $a \leqq x \leqq b$ if these functions are defined on that interval and if all the integrals (g_m, g_n) exist and are zero for all pairs of distinct functions in the set.

The nonnegative square root of (g_m, g_m) is called the **norm** of $g_m(x)$ and is generally denoted by $\|g_m\|$; thus

$$\text{(3)} \qquad \|g_m\| = \sqrt{(g_m, g_m)} = \sqrt{\int_a^b g_m^2(x)\,dx}.$$

Throughout our discussion we shall make the following

General assumption

All occurring functions are bounded and are such that the occurring integrals exist, and the norms are not zero.

Clearly, an orthogonal set $g_1, g_2, \cdots$ on an interval $a \leqq x \leqq b$ whose functions have norm 1 satisfies the relations

$$\text{(4)} \qquad (g_m, g_n) = \int_a^b g_m(x) g_n(x)\,dx = \begin{cases} 0 & \text{when } m \neq n \\ 1 & \text{when } m = n \end{cases} \qquad \begin{pmatrix} m = 1, 2, \cdots \\ n = 1, 2, \cdots \end{pmatrix}.$$

Such a set is called an **orthonormal set** *of functions on the interval* $a \leqq x \leqq b$.

Obviously, from an *orthogonal* set we may obtain an *orthonormal* set by dividing each function by its norm on the interval under consideration.

Example 1

The functions $g_m(x) = \sin mx$, $m = 1, 2, \cdots$ form an orthogonal set on the interval $-\pi \leqq x \leqq \pi$, because [cf. (11) in Appendix 3]

$$\text{(5)} \qquad \begin{aligned} (g_m, g_n) &= \int_{-\pi}^{\pi} \sin mx \sin nx\,dx \\ &= \frac{1}{2}\int_{-\pi}^{\pi} \cos(m-n)x\,dx - \frac{1}{2}\int_{-\pi}^{\pi} \cos(m+n)x\,dx = 0 \end{aligned} \qquad (m \neq n).$$

The norm $\|g_m\|$ equals $\sqrt{\pi}$, because

$$\|g_m\|^2 = \int_{-\pi}^{\pi} \sin^2 mx\,dx = \pi \qquad (m = 1, 2, \cdots).$$

Hence the corresponding orthonormal set consists of the functions

$$\frac{\sin x}{\sqrt{\pi}}, \quad \frac{\sin 2x}{\sqrt{\pi}}, \quad \frac{\sin 3x}{\sqrt{\pi}}, \quad \cdots.$$

Example 2

The functions

$$1, \quad \cos x, \quad \sin x, \quad \cos 2x, \quad \sin 2x, \quad \cdots$$

form an orthogonal set on the interval $-\pi \leqq x \leqq \pi$. This follows from (5), the analogous relation for the cosine functions, and

$$\int_{-\pi}^{\pi} \cos mx \sin nx\,dx = \frac{1}{2}\int_{-\pi}^{\pi} \sin(m+n)x\,dx - \frac{1}{2}\int_{-\pi}^{\pi} \sin(m-n)x\,dx = 0$$

for all $m, n = 0, 1, \cdots$. The corresponding orthonormal set is

$$\frac{1}{\sqrt{2\pi}}, \quad \frac{\cos x}{\sqrt{\pi}}, \quad \frac{\sin x}{\sqrt{\pi}}, \quad \frac{\cos 2x}{\sqrt{\pi}}, \quad \frac{\sin 2x}{\sqrt{\pi}}, \quad \cdots$$ ∎

Orthogonal sets yield important types of series developments in a relatively simple fashion. In fact, let $g_1(x), g_2(x), \cdots$ be any orthogonal set of functions on an interval $a \leqq x \leqq b$ and let $f(x)$ be a given function which can be represented in terms of the g_j's by a convergent series

(6) $$f(x) = \sum_{n=1}^{\infty} c_n g_n(x) = c_1 g_1(x) + c_2 g_2(x) + \cdots.$$

This series is called a **generalized Fourier series** of $f(x)$ and its coefficients $c_1, c_2, \cdots$ are called the **Fourier constants** *of* $f(x)$ *with respect to that orthogonal set of functions.*

Because of the orthogonality, these constants can be determined in a very simple fashion. In fact, multiplying both sides of (6) by $g_m(x)$ (m fixed), integrating over $a \leqq x \leqq b$, and assuming that term-by-term integration is permissible,[12] we have

$$(f, g_m) = \int_a^b f g_m \, dx = \sum_{n=1}^{\infty} c_n (g_n, g_m) = \sum_{n=1}^{\infty} c_n \int_a^b g_n g_m \, dx.$$

The integral for which $n = m$ equals $(g_m, g_m) = \|g_m\|^2$, whereas all the other integrals on the right are zero, because of the orthogonality. Hence

(7′) $$(f, g_m) = c_m \|g_m\|^2,$$

and the desired formula for the Fourier constants is

(7) $$c_m = \frac{(f, g_m)}{\|g_m\|^2} = \frac{1}{\|g_m\|^2} \int_a^b f(x) g_m(x) \, dx \qquad (m = 1, 2, \cdots).$$

Example 3. Fourier series

In the case of the orthogonal set in Example 2, representation (6) may be written

(8) $$f(x) = a_0 + \sum_{n=1}^{\infty} (a_n \cos nx + b_n \sin nx)$$

and (7) becomes

(9) $$\begin{aligned} a_0 &= \frac{1}{2\pi} \int_{-\pi}^{\pi} f(x) \, dx \\ a_n &= \frac{1}{\pi} \int_{-\pi}^{\pi} f(x) \cos nx \, dx \\ b_n &= \frac{1}{\pi} \int_{-\pi}^{\pi} f(x) \sin nx \, dx \end{aligned} \qquad (n = 1, 2, \cdots)$$

[12] This is justified, for instance, in the case of uniform convergence (cf. Theorem 3 in Sec. 16.6).

If the series in (8) converges and represents $f(x)$, it is called the **Fourier series** of $f(x)$. Its coefficients are called the **Fourier coefficients** of $f(x)$ and formulas (9) are called the **Euler formulas** for these coefficients. Since Fourier series are particularly important in engineering mathematics, we shall devote a whole chapter to them (Chap. 10) and present some basic applications in Chap. 11 in connection with partial differential equations. ∎

Some important sets of real functions $g_1, g_2, \cdots$ occurring in applications are not orthogonal but have the property that for some nonnegative function $p(x)$,

$$\text{(10)} \qquad \int_a^b p(x) g_m(x) g_n(x)\,dx = 0 \qquad \text{when} \quad m \neq n.$$

Such a set is then said to be *orthogonal with respect to the* **weight function** $p(x)$ *on the interval* $a \leqq x \leqq b$. The *norm* of g_m is now defined as

$$\text{(11)} \qquad \|g_m\| = \sqrt{\int_a^b p(x) g_m{}^2(x)\,dx}$$

and if the norm of each function g_m is 1, the set is said to be *orthonormal* on that interval with respect to $p(x)$.

If we set $h_m = \sqrt{p}\, g_m$, then (10) becomes

$$\int_a^b h_m(x) h_n(x)\,dx = 0 \qquad (m \neq n),$$

that is, the functions h_m form an orthogonal set in the usual sense. Important examples will be considered in the following sections.

If $g_1(x), g_2(x), \cdots$ is an orthogonal set on an interval $a \leqq x \leqq b$ with respect to a weight function p, and if a given function $f(x)$ can be represented by a **generalized Fourier series** [cf. (6)]

$$\text{(12)} \qquad f(x) = c_1 g_1(x) + c_2 g_2(x) + \cdots,$$

then the Fourier constants $c_1, c_2, \cdots$ of $f(x)$ with respect to that set can be determined as before; the only difference is that we now have to multiply that series by pg_m (instead of g_m) before integrating. The other steps are as before and give

$$\text{(13)} \qquad c_m = \frac{1}{\|g_m\|^2} \int_a^b p(x) f(x) g_m(x)\,dx \qquad (m = 1, 2, \cdots),$$

where the norm is now defined by (11).

Problems for Sec. 4.7

In each case show that the given set is orthogonal on the given interval I and determine the corresponding orthonormal set.

1. $1, \cos x, \cos 2x, \cos 3x, \cdots, \quad I: 0 \leqq x \leqq 2\pi$

2. $\sin x, \sin 2x, \sin 3x, \cdots, \quad I: 0 \leqq x \leqq \pi$

3. $\sin \pi x, \sin 2\pi x, \sin 3\pi x, \cdots,$ $\quad I: -1 \leqq x \leqq 1$

4. $1, \cos 2x, \cos 4x, \cos 6x, \cdots,$ $\quad I: 0 \leqq x \leqq \pi$

5. $1, \cos \frac{2n\pi}{T} x \, (n = 1, 2, \cdots),$ $\quad I: 0 \leqq x \leqq T$

6. $\sin \frac{2n\pi}{T} x \, (n = 1, 2, \cdots),$ $\quad I: -T/2 \leqq x \leqq T/2$

7. $P_0(x), P_1(x), P_2(x), P_3(x),$ $\quad I: -1 \leqq x \leqq 1$ (cf. Sec. 4.3)

8. Determine constants $a_0, b_0, \cdots, c_2$ so that the functions $g_1 = a_0$, $g_2 = b_0 + b_1 x$, $g_3 = c_0 + c_1 x + c_2 x^2$ form an orthonormal set on the interval $-1 \leqq x \leqq 1$. Compare the result with that of Prob. 7.

9. Show that if the functions $g_1(x), g_2(x), \cdots$ form an orthogonal set on an interval $a \leqq x \leqq b$, then the functions $g_1(ct + k), g_2(ct + k), \cdots, c > 0$, form an orthogonal set on the interval $(a - k)/c \leqq t \leqq (b - k)/c$.

10. Using the result of Prob. 9, derive the orthogonality property of the set in Prob. 5 from that in Prob. 1.

4.8 Sturm-Liouville Problem

In engineering mathematics, various important orthogonal sets of functions arise as solutions of linear second-order differential equations of the form

$$[r(x)y']' + [q(x) + \lambda p(x)]y = 0 \tag{1}$$

on some given interval $a \leqq x \leqq b$, satisfying conditions of the form

(2)

(a) $k_1 y(a) + k_2 y'(a) = 0$ $\quad$ (k_1 and k_2 not both zero)

(b) $l_1 y(b) + l_2 y'(b) = 0$ $\quad$ (l_1 and l_2 not both zero).

Here λ is a parameter, and k_1, k_2, l_1, l_2 are given real constants.

In connection with (1) and (2) one uses the following terms. Equation (1) is called a **Sturm-Liouville equation.**[13] We shall see that Legendre's, Bessel's and other equations can be written in the form (1). Conditions (2) are called **boundary conditions** since they refer to the endpoints (boundary points) $x = a$ and $x = b$ of that interval. A differential equation together with boundary conditions constitutes what is known as a **boundary value problem.** Our boundary value problem (1), (2) is called a **Sturm-Liouville problem.**

We see directly from (1) and (2) that for every number λ, the problem has the trivial solution $y \equiv 0$, that is, $y(x) = 0$ for all x in that interval. Solutions $y \not\equiv 0$—if they exist—are called *characteristic functions* or **eigenfunctions** of the problem, and the values of λ for which such solutions exist are called *characteristic values* or **eigenvalues** of the problem.

[13]JACQUES CHARLES FRANCOIS STURM (1803-1855), Swiss mathematician. JOSEPH LIOUVILLE (1809-1882), French mathematician, known by his important research work in complex analysis.

Example 1

Find the eigenvalues and eigenfunctions of the Sturm-Liouville problem

(3) (a) $y'' + \lambda y = 0$ (b) $y(0) = 0, \ y(\pi) = 0.$

For negative $\lambda = -\nu^2$ a general solution of the equation is

$$y(x) = c_1 e^{\nu x} + c_2 e^{-\nu x}.$$

From (3b) we obtain $c_1 = c_2 = 0$ and $y \equiv 0$, which is not an eigenfunction. For $\lambda = 0$ the situation is similar. For positive $\lambda = \nu^2$ a general solution is

$$y(x) = A \cos \nu x + B \sin \nu x.$$

From the first boundary condition we obtain $y(0) = A = 0$. The second boundary condition then yields

$$y(\pi) = B \sin \nu\pi = 0 \qquad \text{or} \qquad \nu = 0, \pm 1, \pm 2, \cdots.$$

For $\nu = 0$ we have $y \equiv 0$. For $\lambda = \nu^2 = 1, 4, 9, 16, \cdots$, taking $B = 1$, we obtain

$$y(x) = \sin \nu x \qquad \nu = 1, 2, \cdots.$$

Hence the eigenvalues of the problem are $\lambda = \nu^2$ where $\nu = 1, 2, \cdots$, and corresponding eigenfunctions are $y(x) = \sin \nu x$ where $\nu = 1, 2, \cdots$. ▮

It can be shown that under rather general conditions on the functions p, q, and r in (1) the Sturm–Liouville problem (1), (2) has infinitely many eigenvalues; the corresponding rather complicated theory can be found in Ref. [B11]; cf. Appendix 1.

Furthermore, the eigenfunctions of a Sturm–Liouville problem have the following orthogonality property.

Theorem 1 (Orthogonality of eigenfunctions)

Let the functions p, q, r and r' in the Sturm–Liouville equation (1) *be real-valued and continuous on the interval $a \leqq x \leqq b$. Let $y_m(x)$ and $y_n(x)$ be eigenfunctions of the Sturm–Liouville problem* (1), (2) *corresponding to distinct eigenvalues λ_m and λ_n, respectively. Then y_m and y_n are orthogonal on that interval with respect to the weight function p.*

If $r(a) = 0$, then (2a) *can be dropped from the problem. If $r(b) = 0$, then* (2b) *can be dropped. If $r(a) = r(b)$, then* (2) *can be replaced by*

(4) $$y(a) = y(b), \qquad y'(a) = y'(b).$$

Proof. By assumption, y_m satisfies

$$(ry_m')' + (q + \lambda_m p)y_m = 0,$$

and y_n satisfies

$$(ry_n')' + (q + \lambda_n p)y_n = 0.$$

Multiplying the first equation by y_n and the second by $-y_m$ and adding, we get

$$\begin{aligned}(\lambda_m - \lambda_n)py_m y_n &= y_m(ry_n')' - y_n(ry_m')' \\ &= [(ry_n')y_m - (ry_m')y_n]'\end{aligned}$$

where the last equality can be readily verified by carrying out the indicated differentiation in the last expression. That expression is continuous since r and r' are continuous by hypothesis, and y_m and y_n are solutions of (1). Integrating over x from a to b, we thus obtain

$$(5) \qquad (\lambda_m - \lambda_n)\int_a^b p y_m y_n \, dx = \Big[r(y_n' y_m - y_m' y_n) \Big]_a^b .$$

The expression on the right equals

$$(6) \qquad \begin{aligned} &r(b)[y_n'(b)y_m(b) - y_m'(b)y_n(b)] \\ -&r(a)[y_n'(a)y_m(a) - y_m'(a)y_n(a)]. \end{aligned}$$

Case 1. If $r(a) = 0$ and $r(b) = 0$, then the expression in (6) is zero. Hence the expression on the left-hand side of (5) must be zero, and since λ_m and λ_n are distinct, we obtain the desired orthogonality

$$(7) \qquad \int_a^b p(x)y_m(x)y_n(x)\,dx = 0 \qquad (m \neq n)$$

without the use of the boundary conditions (2).

Case 2. Let $r(b) = 0$, but $r(a) \neq 0$. Then the first line in (6) is zero. We consider the remaining expression in (6). From (2a) we have

$$k_1 y_n(a) + k_2 y_n'(a) = 0,$$
$$k_1 y_m(a) + k_2 y_m'(a) = 0.$$

Let $k_2 \neq 0$. Then by multiplying the first equation by $y_m(a)$ and the last by $-y_n(a)$ and adding, we have

$$k_2[y_n'(a)y_m(a) - y_m'(a)y_n(a)] = 0.$$

Since $k_2 \neq 0$, the expression in brackets must be zero. This expression is identical with that in the last line of (6). Hence (6) is zero, and from (5) we obtain (7). If $k_2 = 0$, then by assumption $k_1 \neq 0$, and the argument of proof is similar.

Case 3. If $r(a) = 0$, but $r(b) \neq 0$, the proof is similar to that in Case 2, but instead of (2a) we now have to use (2b).

Case 4. If $r(a) \neq 0$ and $r(b) \neq 0$, we have to use both boundary conditions (2) and proceed as in Cases 2 and 3.

Case 5. Let $r(a) = r(b)$. Then (6) takes the form

$$r(b)[y_n'(b)y_m(b) - y_m'(b)y_n(b) - y_n'(a)y_m(a) + y_m'(a)y_n(a)].$$

We may use (2) as before and conclude that the expression in brackets is zero. However, we immediately see that this would also follow from (4), so that we may replace (2) by (4). Hence (5) yields (7), as before. This completes the proof of Theorem 1. ∎

Example 2

The differential equation in Example 1 is of the form (1) where $r = 1$, $q = 0$, and $p = 1$. From Theorem 1 it follows that the eigenfunctions are orthogonal on the interval $0 \leqq x \leqq \pi$.

Example 3. Fourier series

The reader may show that the functions

$$1, \quad \cos x, \quad \sin x, \quad \cos 2x, \quad \sin 2x, \quad \cdots$$

occurring in the Fourier series in Example 3, Sec. 4.7, are the eigenfunctions of the Sturm–Liouville problem

$$y'' + \lambda y = 0, \qquad y(\pi) = y(-\pi), \qquad y'(\pi) = y'(-\pi).$$

Hence from Theorem 1 it follows that these functions form an orthogonal set on the interval $-\pi \leqq x \leqq \pi$. Note that the boundary conditions of our present problem are of the form (4).

Further examples will be included in the next section. ∎

A generalized Fourier series (cf. Sec. 4.7) in which the orthogonal set is a set of eigenfunctions is called an **eigenfunction expansion.**

The eigenvalues of Sturm–Liouville problems have the following interesting property.

Theorem 2 (Real eigenvalues)

If the Sturm–Liouville problem (1), (2) *satisfies the condition stated in Theorem* 1 *and p is positive in the whole interval* $a \leqq x \leqq b$ *(or negative everywhere in that interval), then all the eigenvalues of the problem are real.*

Proof. Let $\lambda = \alpha + i\beta$ be an eigenvalue of the problem and let

$$y(x) = u(x) + iv(x)$$

be a corresponding eigenfunction; here α, β, u, and v are real. Substituting this into (1) we have

$$(ru' + irv')' + (q + \alpha p + i\beta p)(u + iv) = 0.$$

This complex equation is equivalent to the following pair of equations for the real and imaginary parts:

$$(ru')' + (q + \alpha p)u - \beta pv = 0,$$

$$(rv')' + (q + \alpha p)v + \beta pu = 0.$$

Multiplying the first equation by v and the second by $-u$ and adding, we obtain

$$-\beta(u^2 + v^2)p = u(rv')' - v(ru')'$$

$$= [(rv')u - (ru')v]'.$$

Integrating over x from a to b, we find

$$-\beta \int_a^b (u^2 + v^2)p\, dx = \Big[r(uv' - u'v)\Big]_a^b.$$

From the boundary conditions it follows that the expression on the right is zero; this can be shown in a fashion similar to that in the proof of Theorem 1. Since y is an eigenfunction, $u^2 + v^2 \not\equiv 0$. Since y and p are continuous and $p > 0$ (or $p < 0$) for all x between a and b, the integral on the left is not zero. Hence, $\beta = 0$, which means that $\lambda = \alpha$ is real. This completes the proof. ∎

Theorem 2 is illustrated by Examples 2 and 3, and further examples are included in the next section and the corresponding problem set.

Problems for Sec. 4.8

1. Carry out the proof of Theorem 1 for the special Sturm–Liouville problem in Example 1.
2. Carry out the details of the proof of Theorem 1 in Cases 3 and 4.
3. Show that if $y = y_0$ is an eigenfunction of (1), (2) corresponding to some eigenvalue $\lambda = \lambda_0$, then $y = \alpha y_0$ ($\alpha \neq 0$, arbitrary) is an eigenfunction of (1), (2) corresponding to λ_0. (Note that this property can be used to obtain eigenfunctions whose norm is 1.)

Find the eigenvalues and eigenfunctions of the following Sturm–Liouville problems.

4. $y'' + \lambda y = 0, \quad y(0) = 0, \ y(l) = 0$
5. $y'' + \lambda y = 0, \quad y(0) = 0, \ y'(l) = 0$
6. $y'' + \lambda y = 0, \quad y'(0) = 0, \ y(l) = 0$
7. $y'' + \lambda y = 0, \quad y'(0) = 0, \ y'(l) = 0$
8. $y'' + \lambda y = 0, \quad y(0) = y(2\pi), \ y'(0) = y'(2\pi)$
9. $(xy')' + \lambda x^{-1} y = 0, \quad y(1) = 0, \ y(e) = 0$
10. $(xy')' + \lambda x^{-1} y = 0, \quad y(1) = 0, \ y'(e) = 0$
11. $(e^{2x} y')' + e^{2x}(\lambda + 1) y = 0, \quad y(0) = 0, \ y(\pi) = 0$
12. $(x^{-1} y')' + (\lambda + 1) x^{-3} y = 0, \quad y(1) = 0, \ y(e) = 0$

13. Show that the eigenvalues of the Sturm–Liouville problem $y'' + \lambda y = 0$, $y(0) = 0$, $y(1) + y'(1) = 0$ are obtained as solutions of the equation $\sin k + k \cos k = 0$, where $k = \sqrt{\lambda}$. How many positive solutions does this equation have?

Find a Sturm–Liouville problem whose eigenfunctions are

14. $1, \cos \frac{2n\pi}{T} x, \sin \frac{2n\pi}{T} x, \quad (n = 1, 2, \cdots)$
15. $1, \cos x, \cos 2x, \cos 3x, \cdots$

4.9 Orthogonality of Legendre Polynomials and Bessel Functions

Example 1. Legendre polynomials

Legendre's equation (1) in Sec. 4.3 can be written

$$[(1 - x^2) y']' + \lambda y = 0, \qquad \lambda = n(n + 1).$$

Hence it is a Sturm–Liouville equation (1), Sec. 4.8, with $r = 1 - x^2$, $q = 0$, and $p = 1$. Since $r = 0$ when $x = \pm 1$, no boundary conditions are needed to form a Sturm–Liouville problem on the interval $-1 \leqq x \leqq 1$. We know that, for $n = 0, 1, \cdots$, the Legendre polynomials $P_n(x)$ are solutions of the problem. Hence these are eigenfunctions, and from Theorem 1 in Sec. 4.8 it follows that they are orthogonal on that interval, that is

$$\text{(1)} \qquad \int_{-1}^{1} P_m(x)P_n(x)\,dx = 0 \qquad (m \neq n).$$

The norm is (cf. Prob. 7 in Sec. 4.3)

$$\text{(2)} \qquad \|P_m\| = \sqrt{\int_{-1}^{1} P_m{}^2(x)\,dx} = \sqrt{\frac{2}{2m+1}} \qquad m = 0, 1, \cdots.$$

Example 2. Bessel functions

The orthogonality of the Bessel functions J_n is quite important in engineering applications, for example in connection with vibrations of circular membranes (to be considered in Sec. 11.10). $J_n(s)$ satisfies the Bessel equation (Sec. 4.5)

$$s^2\ddot{J}_n + s\dot{J}_n + (s^2 - n^2)J_n = 0$$

where the dots denote derivatives with respect to s. We assume that n is a nonnegative integer. Setting $s = \lambda x$ where λ is a constant, not zero, we have $dx/ds = 1/\lambda$, and, by the chain rule,

$$\dot{J}_n = J_n'/\lambda, \qquad \ddot{J}_n = J_n''/\lambda^2,$$

where primes denote derivatives with respect to x. By substitution,

$$x^2J_n''(\lambda x) + xJ_n'(\lambda x) + (\lambda^2x^2 - n^2)J_n(\lambda x) = 0.$$

Dividing by x, we may write this equation in the form

$$\text{(3)} \qquad [xJ_n'(\lambda x)]' + \left(-\frac{n^2}{x} + \lambda^2 x\right)J_n(\lambda x) = 0.$$

We see that for each fixed n this is a Sturm–Liouville equation (1), Sec. 4.8, with the parameter written as λ^2 instead of λ, and

$$p(x) = x, \qquad q(x) = -n^2/x, \qquad r(x) = x.$$

Since $r(x) = 0$ at $x = 0$, it follows from Theorem 1 in the last section that those solutions of (3) on a given interval $0 \leqq x \leqq R$ which satisfy the boundary condition

$$\text{(4)} \qquad J_n(\lambda R) = 0$$

form an orthogonal set on that interval with respect to the weight function $p(x) = x$. (Note that for $n \neq 0$ the function q is discontinuous at $x = 0$, but this does not affect the proof of that theorem.) It can be shown (Ref. [B18]) that $J_n(s)$ has infinitely many real zeros; let $\alpha_{1n} < \alpha_{2n} < \alpha_{3n} \cdots$ denote the positive zeros of $J_n(s)$. Then (4) holds for

$$\text{(5)} \qquad \lambda R = \alpha_{mn} \quad \text{or} \quad \lambda = \lambda_{mn} = \frac{\alpha_{mn}}{R} \qquad (m = 1, 2, \cdots),$$

and we obtain the following result.

Theorem 1 (Orthogonality of Bessel functions)

For each fixed $n = 0, 1, \cdots$ the Bessel functions $J_n(\lambda_{1n}x), J_n(\lambda_{2n}x), J_n(\lambda_{3n}x), \cdots$, with λ_{mn} given by (5), form an orthogonal set on the interval $0 \leqq x \leqq R$ with respect to the weight function $p(x) = x$, that is

$$\text{(6)} \qquad \int_0^R xJ_n(\lambda_{mn}x)J_n(\lambda_{kn}x)\,dx = 0, \qquad (k \neq m).$$

Hence we have obtained infinitely many orthogonal sets, each corresponding to one of the fixed values of n.

Since $p(x) = x$, we see from (11)–(13) in Sec. 4.7 that if a given function $f(x)$ has an eigenfunction expansion in terms of one of those orthogonal sets, the expansion is

$$f(x) = c_1 J_n(\lambda_{1n} x) + c_2 J_n(\lambda_{2n} x) + \cdots . \tag{7}$$

This is called a **Fourier-Bessel series.** We shall prove that

$$\|J_n(\lambda_{mn} x)\|^2 = \int_0^R x J_n{}^2(\lambda_{mn} x)\, dx = \frac{R^2}{2} J_{n+1}^2(\lambda_{mn} R), \tag{8}$$

where $\lambda_{mn} = \alpha_{mn}/R$. By (13) in Sec. 4.7 this implies that the Fourier constants c_m of f in (7) are

$$c_m = \frac{2}{R^2 J_{n+1}^2(\alpha_{mn})} \int_0^R x f(x) J_n(\lambda_{mn} x)\, dx, \qquad m = 1, 2, \cdots . \tag{9}$$

We prove (8). Multiplying (3) by $2xJ_n{}'(\lambda x)$, we obtain an equation which can be written

$$\{[xJ_n{}'(\lambda x)]^2\}' + (\lambda^2 x^2 - n^2)\{J_n{}^2(\lambda x)\}' = 0.$$

Integrating over x from 0 to R we find

$$[xJ_n{}'(\lambda x)]^2 \Big|_0^R = -\int_0^R (\lambda^2 x^2 - n^2)\{J_n{}^2(\lambda x)\}'\, dx. \tag{10}$$

From (21) in Prob. 7, Sec. 4.5, writing s and n instead of x and ν, and denoting the derivative with respect to s by a dot, we have

$$-ns^{-n-1}J_n(s) + s^{-n}\dot{J}_n(s) = -s^{-n}J_{n+1}(s).$$

Multiplying by s^{n+1} and setting $s = \lambda x$, we obtain

$$\lambda x J_n{}'(\lambda x)\frac{1}{\lambda} = nJ_n(\lambda x) - \lambda x J_{n+1}(\lambda x),$$

where the prime denotes the derivative with respect to x. Hence the left-hand side of (10) equals

$$\Big[[nJ_n(\lambda x) - \lambda x J_{n+1}(\lambda x)]^2\Big]_{x=0}^R .$$

If $\lambda = \lambda_{mn}$ then $J_n(\lambda R) = 0$, and since $J_n(0) = 0$ $(n = 1, 2, \cdots)$, that left-hand side becomes

$$\lambda_{mn}{}^2 R^2 J_{n+1}^2(\lambda_{mn} R). \tag{11}$$

If we integrate by parts, the right-hand side of (10) becomes

$$-\Big[(\lambda^2 x^2 - n^2)J_n{}^2(\lambda x)\Big]_0^R + 2\lambda^2 \int_0^R x J_n{}^2(\lambda x)\, dx.$$

When $\lambda = \lambda_{mn}$, the first expression is zero at $x = R$. It is also zero at $x = 0$ because $\lambda^2 x^2 - n^2 = 0$ when $n = x = 0$, and $J_n(\lambda x) = 0$ when $x = 0$ and $n = 1, 2, \cdots$. From this and (11) we readily obtain (8). This completes the proof. ∎

Problems for Sec. 4.9

Legendre polynomials

Represent the following polynomials in terms of Legendre polynomials.

1. $1, x, x^2, x^3, x^4$

2. $3x^2 + x$

3. $5x^3 - 3x^2 - x - 1$

4. $35x^4 + 15x^3 - 30x^2 - 15x + 3$

In each case, obtain the first few terms of the expansion of $f(x)$ in terms of Legendre polynomials and graph the first three partial sums.

5. $f(x) = \begin{cases} 0 & \text{if } -1 < x < 0 \\ x & \text{if } \ \ 0 < x < 1 \end{cases}$ **6.** $f(x) = \begin{cases} 0 & \text{if } -1 < x < 0 \\ 1 & \text{if } \ \ 0 < x < 1 \end{cases}$

7. $f(x) = |x| \quad \text{if } -1 < x < 1$

8. Show that the functions $P_n(\cos\theta)$, $n = 0, 1, \cdots$, form an orthogonal set on the interval $0 \leqq \theta \leqq \pi$ with respect to the weight function $\sin\theta$.

Hermite polynomials[14]

The functions

$$He_0 = 1, \qquad He_n(x) = (-1)^n e^{x^2/2} \frac{d^n}{dx^n}(e^{-x^2/2}), \qquad n = 1, 2, \cdots$$

are called *Hermite polynomials.*

Remark. As is true for many special functions, the literature contains more than one notation, and one sometimes defines as Hermite polynomials the functions

$$H_0^* = 1, \qquad H_n^*(x) = (-1)^n e^{x^2} \frac{d^n e^{-x^2}}{dx^n}.$$

This differs from our definition, which is preferably used in applications.

9. Show that

$$He_1(x) = x,\ He_2(x) = x^2 - 1,\ He_3(x) = x^3 - 3x,\ He_4(x) = x^4 - 6x^2 + 3.$$

10. Show that the Hermite polynomials are related to the coefficients of the Maclaurin series

$$e^{tx - t^2/2} = \sum_{n=0}^{\infty} a_n(x) t^n$$

by the formula $He_n(x) = n! a_n(x)$. *Hint.* Note that $tx - t^2/2 = x^2/2 - (x - t)^2/2$. (The exponential function on the left is called the *generating function* of the He_n.)

11. Show that the Hermite polynomials satisfy the relation

$$He_{n+1}(x) = xHe_n(x) - He_n{}'(x).$$

12. Differentiating the generating function in Prob. 10 with respect to x, show that

$$He_n{}'(x) = nHe_{n-1}(x).$$

Using this and the formula in Prob. 11 (with n replaced by $n - 1$), prove that $He_n(x)$ satisfies the differential equation

$$y'' - xy' + ny = 0.$$

13. Using the equation in Prob. 12, show that $w = e^{-x^2/4} He_n(x)$ is a solution of **Weber's**[15] **equation**

$$w'' + (n + \tfrac{1}{2} - \tfrac{1}{4}x^2)w = 0 \qquad (n = 0, 1, \cdots).$$

[14]CHARLES HERMITE (1822–1901), French mathematician, is known by his work in algebra and number theory.

[15]HEINRICH WEBER (1842–1913), German mathematician.

14. Show that the Hermite polynomials are orthogonal on the x-axis $-\infty < x < \infty$ with respect to the weight function $p(x) = e^{-x^2/2}$. *Hint.* Use the formula in Prob. 12 and integrate by parts.

Laguerre polynomials[16]

The functions

$$L_0 = 1, \qquad L_n(x) = \frac{e^x}{n!}\frac{d^n(x^n e^{-x})}{dx^n}, \qquad n = 1, 2, \cdots$$

are called *Laguerre polynomials.*

15. Show that

$$L_1(x) = 1 - x, \quad L_2(x) = 1 - 2x + x^2/2, \quad L_3(x) = 1 - 3x + 3x^2/2 - x^3/6.$$

16. Show that

$$L_n(x) = \sum_{m=0}^{n} \frac{(-1)^m}{m!}\binom{n}{m}x^m = 1 - nx + \frac{n(n-1)}{4}x^2 - + \cdots + \frac{(-1)^n}{n!}x^n.$$

17. $L_n(x)$ satisfies Laguerre's differential equation

$$xy'' + (1 - x)y' + ny = 0.$$

Verify this fact for $n = 0, 1, 2, 3$.

18. Verify by direct integration that L_0, $L_1(x)$, $L_2(x)$ are orthogonal on the positive axis $0 \leqq x < \infty$ with respect to the weight function $p(x) = e^{-x}$.

19. Prove that the set of all Laguerre polynomials is orthogonal on $0 \leqq x < \infty$ with respect to the weight function $p(x) = e^{-x}$. *Hint.* Consider the integral of $e^{-x}L_m(x)L_n(x)$ from 0 to ∞ with $m < n$. Since the highest power in L_m is x^m, conclude that it suffices to show that the integral of $e^{-x}x^kL_n(x)$ $(k < n)$ from 0 to ∞ is zero. Prove this by repeated integration by parts.

Tchebichef polynomials[17]

The functions

$$T_n(x) = \cos(n \arccos x), \qquad U_n(x) = \frac{\sin[(n+1)\arccos x]}{\sqrt{1-x^2}} \qquad n = 0, 1, \cdots$$

are called *Tchebichef polynomials of the first and second kind,* respectively.

20. Show that

$$T_0 = 1, \quad T_1(x) = x, \quad T_2(x) = 2x^2 - 1, \quad T_3(x) = 4x^3 - 3x,$$

$$U_0 = 1, \quad U_1(x) = 2x, \quad U_2(x) = 4x^2 - 1, \quad U_3(x) = 8x^3 - 4x.$$

21. Show that $T_n(x)$ is a solution of the differential equation

$$(1 - x^2)T_n'' - xT_n' + n^2T_n = 0.$$

[16]EDMOND LAGUERRE (1834–1886), French mathematician who did research work in geometry and the theory of infinite series.

[17]PAFNUTI TCHEBICHEF (1821–1894), Russian mathematician, is known by his work in approximation theory and the theory of numbers. Another transliteration of the name is CHEBYSHEV.

22. Show that the Tchebichef polynomials $T_n(x)$ are orthogonal on the interval $-1 \leqq x \leqq 1$ with respect to the weight function $p(x) = 1/\sqrt{1 - x^2}$. *Hint.* To evaluate the integral, set $\arccos x = \theta$.

Bessel functions

23. Graph $J_0(\lambda_{10}x)$, $J_0(\lambda_{20}x)$, $J_0(\lambda_{30}x)$, and $J_0(\lambda_{40}x)$ for $R = 1$ in the interval $0 \leqq x \leqq 1$. (Use Table A2 in Appendix 4.)

24. Graph $J_1(\alpha_{11}x)$, $J_1(\alpha_{21}x)$, $J_1(\alpha_{31}x)$, and $J_1(\alpha_{41}x)$ in the interval $0 \leqq x \leqq 1$. (Use Table A2 in Appendix 4.)

Develop the following functions $f(x)$ $(0 < x < R)$ in a Fourier-Bessel series of the form

$$f(x) = c_1 J_0(\lambda_{10}x) + c_2 J_0(\lambda_{20}x) + c_3 J_0(\lambda_{30}x) + \cdots$$

and graph the first few partial sums.

25. $f(x) = 1$. *Hint.* Use (20), Sec. 4.5.

26. $f(x) = \begin{cases} 1 & \text{if} \quad 0 < x < R/2 \\ 0 & \text{if} \quad R/2 < x < R \end{cases}$

27. $f(x) = \begin{cases} k & \text{if} \quad 0 < x < a \\ 0 & \text{if} \quad a < x < R \end{cases}$

28. $f(x) = \begin{cases} 0 & \text{if} \quad 0 < x < R/2 \\ k & \text{if} \quad R/2 < x < R \end{cases}$

29. $f(x) = 1 - x^2$ $(R = 1)$. *Hint.* Use (20), Sec. 4.5, and integration by parts.

30. $f(x) = R^2 - x^2$

31. $f(x) = x^2$

32. $f(x) = x^4$

33. Show that $f(x) = x^n$ $(0 < x < 1,\ n = 0, 1, \cdots)$ can be represented by the Fourier-Bessel series

$$x^n = \frac{2J_n(\alpha_{1n}x)}{\alpha_{1n}J_{n+1}(\alpha_{1n})} + \frac{2J_n(\alpha_{2n}x)}{\alpha_{2n}J_{n+1}(\alpha_{2n})} + \cdots.$$

34. Find a representation of $f(x) = x^n$ $(0 < x < R,\ n = 0, 1, \cdots)$ similar to that in Prob. 33.

35. Represent $f(x) = x^3$ $(0 < x < 2)$ by a Fourier-Bessel series involving J_3.

CHAPTER

5 LAPLACE TRANSFORMATION

The Laplace transformation is a method for solving differential equations and corresponding initial and boundary value problems. The process of solution consists of three main steps:

1st step. The given "hard" problem is transformed into a "simple" equation (*subsidiary equation*).

2nd step. The subsidiary equation is solved by purely algebraic manipulations.

3rd step. The solution of the subsidiary equation is transformed back to obtain the solution of the given problem.

In this way the Laplace transformation reduces the problem of solving a differential equation to an algebraic problem. The third step is made easier by tables, whose role is similar to that of integral tables in integration. (These tables are also useful in the first step.) One is included at the end of the chapter.

The method is widely used in engineering mathematics, where it has numerous applications. It is particularly useful in problems where the (mechanical or electrical) driving force has discontinuities, for instance, acts for a short time only, or is periodic but is not merely a sine or cosine function. Another advantage is that it solves problems directly. Indeed, initial value problems are solved without first determining a general solution. Similarly, nonhomogeneous equations are solved without first solving the corresponding homogeneous equation.

In this chapter we consider the Laplace transformation from a practical point of view and illustrate its use by important engineering problems.

The application of the Laplace transformation to *ordinary* differential equations is considered in this chapter.

Partial differential equations can also be treated by the Laplace transformation. An introduction to this field is given in Sec. 11.13, at the end of the chapter on partial differential equations.

Prerequisite for this chapter: Chap. 2.
References: Appendix 1, Part B.
Answers to problems: Appendix 2.

5.1 Laplace Transform. Inverse Transform. Linearity

Let $f(t)$ be a given function which is defined for all $t \geqq 0$. We multiply $f(t)$ by e^{-st} and integrate with respect to t from zero to infinity. Then, if the resulting integral exists, it is a function of s, say, $F(s)$:

$$F(s) = \int_0^\infty e^{-st} f(t)\, dt.$$

The function $F(s)$ is called the **Laplace transform**[1] of the original function $f(t)$, and will be denoted by $\mathscr{L}(f)$. Thus

$$\textbf{(1)} \qquad F(s) = \mathscr{L}(f) = \int_0^\infty e^{-st} f(t)\, dt.$$

The operation just described, which yields $F(s)$ from a given $f(t)$, is called the **Laplace transformation.**

Furthermore the original function $f(t)$ in (1) is called the *inverse transform* or **inverse** of $F(s)$ and will be denoted by $\mathscr{L}^{-1}(F)$; that is, we shall write

$$f(t) = \mathscr{L}^{-1}(F).$$

Notation

Original functions are denoted by lower case letters and their transforms by the same letters in capital, so that $F(s)$ denotes the transform of $f(t)$, and $Y(s)$ denotes the transform of $y(t)$, and so on.

Example 1

Let $f(t) = 1$ when $t \geqq 0$. Find $F(s)$. From (1) we obtain by integration

$$\mathscr{L}(f) = \mathscr{L}(1) = \int_0^\infty e^{-st}\, dt = -\frac{1}{s} e^{-st} \Big|_0^\infty ;$$

hence, when $s > 0$,

$$\mathscr{L}(1) = \frac{1}{s}.$$

Our notation in the first line on the right is convenient, but we should say a word about it. The interval of integration in (1) is infinite. Such an integral is called an *improper integral* and, by definition, is evaluated according to the rule

$$\int_0^\infty e^{-st} f(t)\, dt = \lim_{T\to\infty} \int_0^T e^{-st} f(t)\, dt.$$

Hence our convenient notation means

$$\int_0^\infty e^{-st}\, dt = \lim_{T\to\infty} \left[-\frac{1}{s} e^{-st} \right]_0^T = \lim_{T\to\infty} \left[-\frac{1}{s} e^{-sT} + \frac{1}{s} e^{0} \right] = \frac{1}{s} \qquad (s > 0).$$

We shall use that notation throughout this chapter.

Example 2

Let $f(t) = e^{at}$ when $t \geqq 0$, where a is a constant. Find $\mathscr{L}(f)$. Again by (1),

$$\mathscr{L}(e^{at}) = \int_0^\infty e^{-st} e^{at}\, dt = \frac{1}{a-s} e^{-(s-a)t} \Big|_0^\infty ;$$

[1]PIERRE SIMON DE LAPLACE (1749–1827), great French mathematician, who developed the foundation of potential theory and made important contributions to celestial mechanics and probability theory.

hence, when $s - a > 0$,

$$\mathcal{L}(e^{at}) = \frac{1}{s - a}.$$

Must we go on in this fashion and obtain the transform of one function after another directly from the definition? The answer is no. And the reason is that the Laplace transformation has many general properties which are helpful for that purpose. A very important property is that the Laplace transformation is a linear operation, just as differentiation and integration. By this we mean the following.

Theorem 1 (Linearity of the Laplace transformation)
The Laplace transformation is a linear operation, that is, for any functions $f(t)$ and $g(t)$ whose Laplace transform exists and any constants a and b we have

$$\mathcal{L}\{af(t) + bg(t)\} = a\mathcal{L}\{f(t)\} + b\mathcal{L}\{g(t)\}.$$

Proof. By the definition

$$\begin{aligned}\mathcal{L}\{af(t) + bg(t)\} &= \int_0^\infty e^{-st}[af(t) + bg(t)]\,dt \\ &= a\int_0^\infty e^{-st}f(t)\,dt + b\int_0^\infty e^{-st}g(t)\,dt \\ &= a\mathcal{L}\{f(t)\} + b\mathcal{L}\{g(t)\}.\end{aligned}$$

Example 3
Let $f(t) = \cosh at = (e^{at} + e^{-at})/2$. Find $\mathcal{L}(f)$. From Theorem 1 and Example 2 we obtain

$$\mathcal{L}(\cosh at) = \frac{1}{2}\mathcal{L}(e^{at}) + \frac{1}{2}\mathcal{L}(e^{-at}) = \frac{1}{2}\left(\frac{1}{s - a} + \frac{1}{s + a}\right);$$

that is, when $s > a\ (\geqq 0)$,

$$\mathcal{L}(\cosh at) = \frac{s}{s^2 - a^2}.$$

A short list of some important elementary functions and their Laplace transforms is given in Table 5.1, and a more extensive list in Sec. 5.8. Cf. also the references in Appendix 1, Part B.

Once we know the transforms in Table 5.1, nearly all the transforms we shall need can be obtained through the use of some simple general theorems which we consider in the subsequent sections.

Formulas 1, 2, and 3 in Table 5.1 are special cases of formula 4. Formula 4 follows from formula 5 and $\Gamma(n + 1) = n!$, where n is a nonnegative integer [cf. (10) in Sec. 4.5]. Formula 5 can be proved by starting from the definition

$$\mathcal{L}(t^a) = \int_0^\infty e^{-st}t^a\,dt,$$

Table 5.1
Some Elementary Functions $f(t)$ and Their Laplace Transforms $\mathscr{L}(f)$

	$f(t)$	$\mathscr{L}(f)$		$f(t)$	$\mathscr{L}(f)$
1	1	$1/s$	6	e^{at}	$\dfrac{1}{s-a}$
2	t	$1/s^2$	7	$\cos \omega t$	$\dfrac{s}{s^2+\omega^2}$
3	t^2	$2!/s^3$	8	$\sin \omega t$	$\dfrac{\omega}{s^2+\omega^2}$
4	t^n $(n = 1, 2, \cdots)$	$\dfrac{n!}{s^{n+1}}$	9	$\cosh at$	$\dfrac{s}{s^2-a^2}$
5	t^a (a positive)	$\dfrac{\Gamma(a+1)}{s^{a+1}}$	10	$\sinh at$	$\dfrac{a}{s^2-a^2}$

setting $st = x$, and using (8) in Sec. 4.5; then

$$\mathscr{L}(t^a) = \int_0^\infty e^{-x}\left(\frac{x}{s}\right)^a \frac{dx}{s} = \frac{1}{s^{a+1}}\int_0^\infty e^{-x}x^a\,dx = \frac{\Gamma(a+1)}{s^{a+1}} \qquad (s > 0).$$

Formula 6 was proved in Example 2. To prove the formulas 7 and 8 we set $a = i\omega$ in formula 6. Then

$$\mathscr{L}(e^{i\omega t}) = \frac{1}{s-i\omega} = \frac{s+i\omega}{(s-i\omega)(s+i\omega)} = \frac{s+i\omega}{s^2+\omega^2} = \frac{s}{s^2+\omega^2} + i\frac{\omega}{s^2+\omega^2}.$$

On the other hand, by Theorem 1,

$$\mathscr{L}(e^{i\omega t}) = \mathscr{L}(\cos \omega t + i \sin \omega t) = \mathscr{L}(\cos \omega t) + i\mathscr{L}(\sin \omega t).$$

Equating the real and imaginary parts of these two equations, we obtain the formulas 7 and 8. Formula 9 was proved in Example 3, and formula 10 can be proved in a similar manner. ∎

In conclusion of this introductory section we should say something about the existence of the Laplace transform. Roughly and intuitively speaking, the situation is as follows. For a fixed s the integral in (1) will exist if the whole integrand $e^{-st}f(t)$ goes to zero fast enough as $t \to \infty$, say, at least like an exponential function with a negative exponent. This motivates the inequality (2) in the subsequent existence theorem. $f(t)$ need not be continuous. This is of practical importance since discontinuous inputs (driving forces) are just those for which the Laplace transformation becomes particularly useful. It suffices to require that $f(t)$ is piecewise continuous on every finite interval in the range $t \geqq 0$.

By definition, a function $f(t)$ is **piecewise continuous** on a finite interval $a \leqq t \leqq b$ if $f(t)$ is defined on that interval and is such that the interval can be subdivided into finitely many intervals, in each of which $f(t)$ is continuous and has finite limits as t approaches either endpoint of the interval of subdivision from the interior.

It follows from the definition that finite jumps are the only discontinuities which a piecewise continuous function may have; these are known as *ordinary discontinuities*. Figure 78 shows an example. Furthermore, it is clear that the class of piecewise continuous functions includes every continuous function.

Theorem 2 (Existence theorem for Laplace transforms)
Let $f(t)$ be a function which is piecewise continuous on every finite interval in the range $t \geqq 0$ and satisfies

$$|f(t)| \leqq Me^{\gamma t} \qquad \text{for all } t \geqq 0 \tag{2}$$

and for some constants γ and M. Then the Laplace transform of $f(t)$ exists for all $s > \gamma$.

Proof. Since $f(t)$ is piecewise continuous, $e^{-st}f(t)$ is integrable over any finite interval on the t-axis, and from (2),

$$|\mathcal{L}(f)| = \left| \int_0^\infty e^{-st} f(t)\, dt \right| \leqq \int_0^\infty e^{-st} |f(t)|\, dt \leqq \int_0^\infty e^{-st} Me^{\gamma t}\, dt$$
$$= M \int_0^\infty e^{-(s-\gamma)t}\, dt = \frac{M}{s-\gamma} \qquad (s > \gamma).$$

This completes the proof. ∎

The conditions in Theorem 2 are sufficient for most applications, and it is simple to find whether a given function satisfies an inequality of the form (2). For example,

$$\cosh t < e^t, \qquad t^n < n!\, e^t \quad (n = 0, 1, \cdots) \qquad \text{for all } t > 0, \tag{3}$$

and any function which is bounded in absolute value for all $t \geqq 0$, such as the sine and cosine functions of a real variable, satisfies that condition. An example of a function which does not satisfy a relation of the form (2) is the exponential function e^{t^2}, because, no matter how large we choose the numbers M and γ in (2),

$$e^{t^2} > Me^{\gamma t} \qquad \text{for all } t > t_0$$

where t_0 is a sufficiently large number, depending on M and γ.

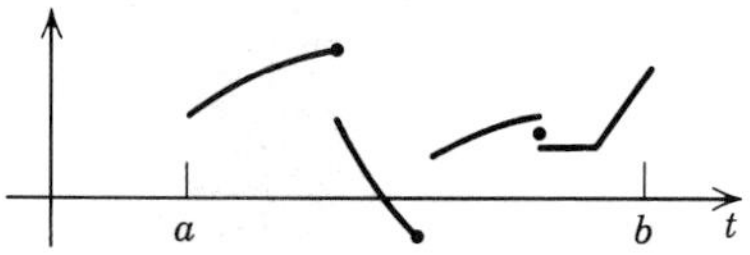

Fig. 78. Example of a piecewise continuous function $f(t)$. (The dots mark the function values at the jumps.)

It should be noted that the conditions in Theorem 2 are sufficient rather than necessary. For example, the function $1/\sqrt{t}$ is infinite at $t = 0$, but its transform exists; in fact, from the definition and $\Gamma(\frac{1}{2}) = \sqrt{\pi}$ [cf. (30) in Appendix 3] we obtain

$$\mathscr{L}(t^{-1/2}) = \int_0^\infty e^{-st}t^{-1/2}\,dt = \frac{1}{\sqrt{s}}\int_0^\infty e^{-x}x^{-1/2}\,dx = \frac{1}{\sqrt{s}}\Gamma(\tfrac{1}{2}) = \sqrt{\frac{\pi}{s}}.$$

If the Laplace transform of a given function exists, it is uniquely determined. Conversely, it can be shown that if two functions (both defined on the positive real axis) have the same transform, these functions cannot differ over an interval of positive length, although they may differ at various isolated points (cf. Ref. [B19] in Appendix 1). Since this is of no importance in applications, we may say that the inverse of a given transform is essentially unique. In particular, if two *continuous* functions have the same transform, they are completely identical. Of course, this *is* of practical importance. Why? (Remember the introduction to the chapter.)

Problems for Sec. 5.1

Find the Laplace transforms of the following functions, where k, a, b, c, θ are constants.

1. 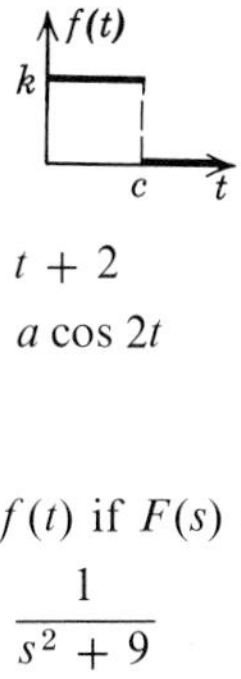

2. 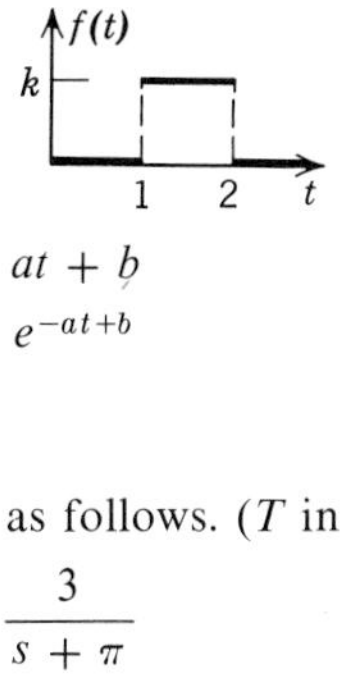

3. 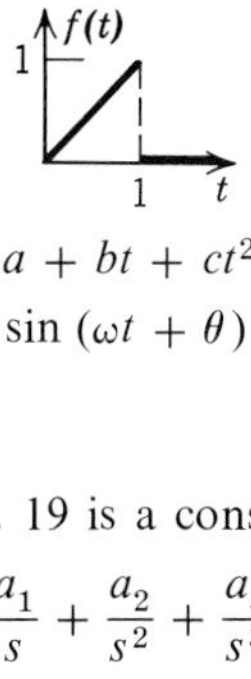

4. 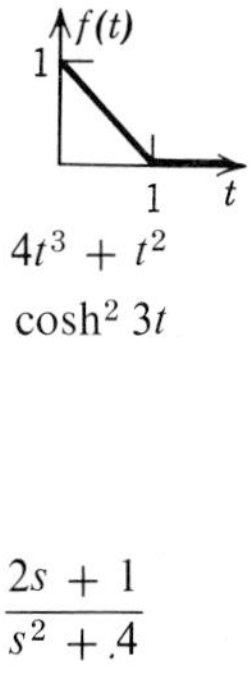

5. $t + 2$ **6.** $at + b$ **7.** $a + bt + ct^2$ **8.** $4t^3 + t^2$

9. $a \cos 2t$ **10.** e^{-at+b} **11.** $\sin(\omega t + \theta)$ **12.** $\cosh^2 3t$

Find $f(t)$ if $F(s) = \mathscr{L}(f)$ is as follows. (T in Prob. 19 is a constant.)

13. $\dfrac{1}{s^2 + 9}$ **14.** $\dfrac{3}{s + \pi}$ **15.** $\dfrac{a_1}{s} + \dfrac{a_2}{s^2} + \dfrac{a_3}{s^3}$ **16.** $\dfrac{2s + 1}{s^2 + 4}$

17. $\dfrac{4(s + 1)}{s^2 - 16}$ **18.** $\dfrac{2}{s} + \dfrac{1}{s + 2}$ **19.** $\dfrac{2n\pi T}{T^2s^2 + (2n\pi)^2}$ **20.** $\dfrac{s}{s^2 + n^2\pi^2}$

21. $\dfrac{1}{(s + 1)(s + 2)}$ **22.** $\dfrac{3}{s^2 + 3s}$ **23.** $\dfrac{3}{4}s^{-5/2}$ **24.** $s^{-3/2}$

25. Obtain the answer to Prob. 4 from Probs. 1 and 3.

26. Prove (3).

27. Obtain formula 10 in Table 5.1 from formula 6.

28. Obtain formula 6 in Table 5.1 from formulas 9 and 10.

29. Derive formulas 7 and 8 in Table 5.1 by integration by parts.

30. Using $\cosh x = \cos ix$ and $\sinh x = -i \sin ix$, $i = \sqrt{-1}$, obtain formulas 9 and 10 in Table 5.1 from formulas 7 and 8.

5.2 Laplace Transforms of Derivatives and Integrals

Probably the most important property of the Laplace transformation is the linearity (Theorem 1 in the previous section). Next in the order of importance comes the fact that, roughly speaking, differentiation of a function $f(t)$ corresponds simply to multiplication of the transform $F(s)$ by s. This permits replacing operations of calculus by simple algebraic operations on transforms.

Furthermore, since integration is the inverse operation of differentiation, we expect it to correspond to division of transforms by s. This is in fact so. Accordingly, our program for the section is as follows. Theorem 1 concerns the differentiation of $f(t)$, Theorem 2 the extension to higher derivatives and Theorem 3 the integration of $f(t)$. We include examples as well as a first application to a differential equation.

Theorem 1 [Differentiation of $f(t)$]

Suppose that $f(t)$ is continuous for all $t \geqq 0$, satisfies (2), p. 204, for some γ and M, and has a derivative $f'(t)$ which is piecewise continuous on every finite interval in the range $t \geqq 0$. Then the Laplace transform of the derivative $f'(t)$ exists when $s > \gamma$, and ∎

$$\mathscr{L}(f') = s\mathscr{L}(f) - f(0) \qquad (s > \gamma). \tag{1}$$

Proof. We first consider the case when $f'(t)$ is continuous for all $t \geqq 0$. Then, by definition and by integrating by parts,

$$\mathscr{L}(f') = \int_0^\infty e^{-st} f'(t)\, dt = [e^{-st} f(t)]_0^\infty + s\int_0^\infty e^{-st} f(t)\, dt.$$

Since f satisfies (2), Sec. 5.1, the integrated portion on the right is zero at the upper limit when $s > \gamma$, and at the lower limit it is $-f(0)$. The last integral is $\mathscr{L}(f)$, the existence for $s > \gamma$ being a consequence of Theorem 2 in Sec. 5.1. This proves that the expression on the right exists when $s > \gamma$, and is equal to $-f(0) + s\mathscr{L}(f)$. Consequently, $\mathscr{L}(f')$ exists when $s > \gamma$, and (1) holds.

If $f'(t)$ is merely piecewise continuous, the proof is quite similar; in this case, the range of integration in the original integral must be broken up into parts such that f' is continuous in each such part. ∎

Remark. *This theorem may be extended to piecewise continuous functions $f(t)$, but in place of (1) we then obtain the formula (1*) in Prob. 12 at the end of the current section.*

By applying (1) to the second-order derivative $f''(t)$ we obtain

$$\begin{aligned}\mathscr{L}(f'') &= s\mathscr{L}(f') - f'(0)\\ &= s[s\mathscr{L}(f) - f(0)] - f'(0);\end{aligned}$$

that is,

$$\mathscr{L}(f'') = s^2\mathscr{L}(f) - sf(0) - f'(0). \tag{2}$$

Similarly,

$$\mathscr{L}(f''') = s^3\mathscr{L}(f) - s^2f(0) - sf'(0) - f''(0), \tag{3}$$

etc. By induction we thus obtain the following extension of Theorem 1.

Theorem 2 (Derivative of any order *n*)

Let $f(t)$ and its derivatives $f'(t), f''(t), \cdots, f^{(n-1)}(t)$ be continuous functions for all $t \geqq 0$, satisfying (2), p. 204, for some γ and M, and let the derivative $f^{(n)}(t)$ be piecewise continuous on every finite interval in the range $t \geqq 0$. Then the Laplace transform of $f^{(n)}(t)$ exists when $s > \gamma$, and is given by the formula

$$\mathscr{L}(f^{(n)}) = s^n\mathscr{L}(f) - s^{n-1}f(0) - s^{n-2}f'(0) - \cdots - f^{(n-1)}(0). \tag{4}$$

Example 1

Let $f(t) = t^2$. Find $\mathscr{L}(f)$. Since $f(0) = 0, f'(0) = 0, f''(t) = 2$, and $\mathscr{L}(2) = 2\mathscr{L}(1) = 2/s$, we obtain from (2)

$$\mathscr{L}(f'') = \mathscr{L}(2) = \frac{2}{s} = s^2\mathscr{L}(f), \quad \text{hence} \quad \mathscr{L}(t^2) = \frac{2}{s^3},$$

in agreement with Table 5.1. The example is typical: it illustrates that in general there are several ways for obtaining the transforms of given functions.

Example 2

Let $f(t) = \sin^2 t$. Find $\mathscr{L}(f)$. We have $f(0) = 0$,

$$f'(t) = 2 \sin t \cos t = \sin 2t$$

and (1) gives

$$\mathscr{L}(\sin 2t) = \frac{2}{s^2 + 4} = s\mathscr{L}(f) \quad \text{or} \quad \mathscr{L}(\sin^2 t) = \frac{2}{s(s^2 + 4)}.$$

Example 3

Let $f(t) = t \sin \omega t$. Find $\mathscr{L}(f)$. We have $f(0) = 0$ and

$$\begin{aligned} f'(t) &= \sin \omega t + \omega t \cos \omega t, \qquad f'(0) = 0, \\ f''(t) &= 2\omega \cos \omega t - \omega^2 t \sin \omega t \\ &= 2\omega \cos \omega t - \omega^2 f(t), \end{aligned}$$

so that by (2),

$$\mathscr{L}(f'') = 2\omega\mathscr{L}(\cos \omega t) - \omega^2\mathscr{L}(f) = s^2\mathscr{L}(f).$$

Using the formula for the Laplace transform of $\cos \omega t$, we thus obtain

$$(s^2 + \omega^2)\mathscr{L}(f) = 2\omega\mathscr{L}(\cos \omega t) = \frac{2\omega s}{s^2 + \omega^2}.$$

Hence the result is

$$\mathscr{L}(t \sin \omega t) = \frac{2\omega s}{(s^2 + \omega^2)^2}.$$

Example 4. A differential equation
Solve the initial value problem

$$y'' + 4y' + 3y = 0, \qquad y(0) = 3, \quad y'(0) = 1.$$

Let $Y(s) = \mathscr{L}(y)$ be the Laplace transform of the (unknown) solution $y(t)$. Then by Theorems 1 and 2 and the initial conditions,

$$\mathscr{L}(y') = sY - y(0) \qquad\qquad = sY - 3,$$

$$\mathscr{L}(y'') = s^2Y - sy(0) - y'(0) = s^2Y - 3s - 1.$$

We substitute this into the Laplace transform of the given differential equation, finding

$$s^2Y + 4sY + 3Y = 3s + 1 + 4 \cdot 3.$$

The equation for the transform $Y(s)$ of the unknown function $y(t)$ is called the **subsidiary equation** of the given differential equation. In our present example it can be written

$$(s + 3)(s + 1)Y = 3s + 13.$$

Solving algebraically for Y and using partial fractions, we obtain

$$Y = \frac{3s + 13}{(s + 3)(s + 1)} = \frac{-2}{s + 3} + \frac{5}{s + 1}.$$

Now from Table 5.1 we see that

$$\mathscr{L}^{-1}\left\{\frac{1}{s + 3}\right\} = e^{-3t}, \qquad \mathscr{L}^{-1}\left\{\frac{1}{s + 1}\right\} = e^{-t}.$$

Using the linearity (Theorem 1 in Sec. 5.1), we see that the solution of our problem is

$$y(t) = -2e^{-3t} + 5e^{-t}.$$

In our approach we assume that the unknown solution $y(t)$ has a transform $Y(s)$ and Theorems 1 and 2 are applicable. Once the solution has been found, these assumptions should be justified. Practically speaking, we find it simpler and more natural to check by substitution whether $y(t)$ satisfies the given equation and initial conditions. This is the case.

Let us summarize our approach:

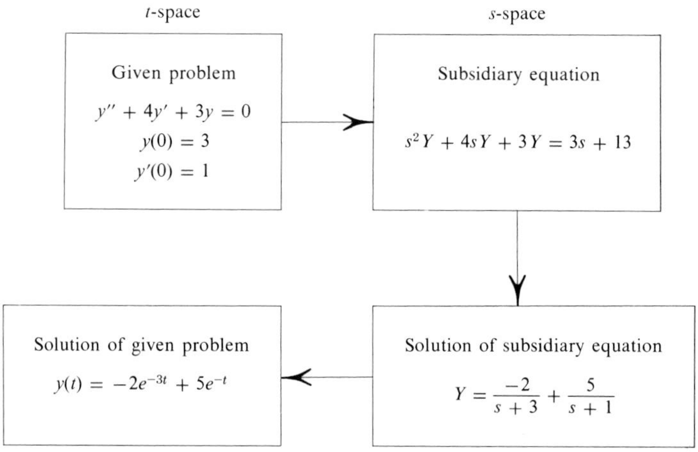

Did we gain anything, compared with the method in Secs. 2.2–2.4? The initial conditions were automatically accounted for. So we gained something but not much, since the example is too simple. However, it shows the typical steps of the new method. Applications illustrating substantial gains will be considered after we know further properties and techniques which make the Laplace transformation powerful and flexible. We start this program in the next section with so-called shifting processes and the unit step function. ∎

We conclude the present section with integration of $f(t)$, the inverse operation of differentiation; we expect it to correspond to division of the transform by s, since division is the inverse operation of multiplication.

Theorem 3 [Integration of *f(t)*]

If $f(t)$ is piecewise continuous and satisfies an inequality of the form (2), *Sec.* 5.1, *then*

$$\mathscr{L}\left\{\int_0^t f(\tau)\,d\tau\right\} = \frac{1}{s}\mathscr{L}\{f(t)\} \qquad (s > 0,\ s > \gamma). \tag{5}$$

Proof. Suppose that $f(t)$ is piecewise continuous and satisfies (2), Sec. 5.1, for some γ and M. Clearly, if (2) holds for some negative γ, it also holds for positive γ, and we may assume that γ is positive. Then the integral

$$g(t) = \int_0^t f(\tau)\,d\tau$$

is continuous, and by using (2) in Sec. 5.1 we obtain

$$|g(t)| \leqq \int_0^t |f(\tau)|\,d\tau \leqq M\int_0^t e^{\gamma\tau}\,d\tau = \frac{M}{\gamma}(e^{\gamma t} - 1) \qquad (\gamma > 0).$$

Furthermore, $g'(t) = f(t)$, except for points at which $f(t)$ is discontinuous. Hence $g'(t)$ is piecewise continuous on each finite interval, and, according to Theorem 1,

$$\mathscr{L}\{f(t)\} = \mathscr{L}\{g'(t)\} = s\mathscr{L}\{g(t)\} - g(0) \qquad (s > \gamma).$$

Clearly, $g(0) = 0$, and, therefore,

$$\mathscr{L}(g) = \frac{1}{s}\mathscr{L}(f).$$

This completes the proof. ∎

Formula (5) has a useful companion, which we obtain by writing $\mathscr{L}\{f(t)\} = F(s)$, interchanging the two sides and taking the inverse transform on both sides. Then

$$\mathscr{L}^{-1}\left\{\frac{1}{s}F(s)\right\} = \int_0^t f(\tau)\,d\tau. \tag{6}$$

Example 5

Let $\mathscr{L}(f) = \dfrac{1}{s^2(s^2 + \omega^2)}$. Find $f(t)$. From Table 5.1 in Sec. 5.1 we have

$$\mathscr{L}^{-1}\left(\frac{1}{s^2 + \omega^2}\right) = \frac{1}{\omega}\sin \omega t.$$

From this and Theorem 3 it follows that

$$\mathcal{L}^{-1}\left\{\frac{1}{s}\left(\frac{1}{s^2+\omega^2}\right)\right\} = \frac{1}{\omega}\int_0^t \sin\omega\tau\, d\tau = \frac{1}{\omega^2}(1-\cos\omega t).$$

Applying Theorem 3 once more, we obtain the desired answer

$$\mathcal{L}^{-1}\left\{\frac{1}{s^2}\left(\frac{1}{s^2+\omega^2}\right)\right\} = \frac{1}{\omega^2}\int_0^t (1-\cos\omega\tau)\, d\tau = \frac{1}{\omega^2}\left(t-\frac{\sin\omega t}{\omega}\right).$$

Problems for Sec. 5.2

Using Theorem 1, derive

1. $\mathcal{L}(\sin t)$ from $\mathcal{L}(\cos t)$

2. $\mathcal{L}(\sinh 2t)$ from $\mathcal{L}(\cosh 2t)$

Using Theorems 1 and 2, show that

3. $\mathcal{L}(t\cos\omega t) = \dfrac{s^2-\omega^2}{(s^2+\omega^2)^2}$

4. $\mathcal{L}(t\cosh at) = \dfrac{s^2+a^2}{(s^2-a^2)^2}$

5. $\mathcal{L}(t\sinh at) = \dfrac{2as}{(s^2-a^2)^2}$

6. $\mathcal{L}(te^{at}) = \dfrac{1}{(s-a)^2}$

Using Example 3 and Problem 3, show that

7. $\mathcal{L}^{-1}\left(\dfrac{1}{(s^2+\omega^2)^2}\right) = \dfrac{1}{2\omega^3}(\sin\omega t - \omega t\cos\omega t)$

8. $\mathcal{L}^{-1}\left(\dfrac{s^2}{(s^2+\omega^2)^2}\right) = \dfrac{1}{2\omega}(\sin\omega t + \omega t\cos\omega t)$

9. Check the result in Example 5 by noting that

$$\mathcal{L}(f) = \frac{1}{\omega^2}\left(\frac{1}{s^2}-\frac{1}{s^2+\omega^2}\right).$$

10. Find $\mathcal{L}(\cos^2 t)$ (a) by the use of the result of Example 2, (b) by the method used in Example 2, (c) by expressing $\cos^2 t$ in terms of $\cos 2t$.

11. Carry out the details of the proof of Theorem 1, assuming that $f'(t)$ has finite jumps at $t_1, t_2, \cdots, t_m$, where $0 < t_1 < t_2 < \cdots < t_m$.

12. (Extension of Theorem 1) The following extension of Theorem 1 is of practical interest in applications. Show that if $f(t)$ is continuous, except for an ordinary discontinuity (finite jump) at $t = a$ (> 0), the other conditions remaining the same as in Theorem 1, then

$$\text{(1*)}\qquad \mathcal{L}(f') = s\mathcal{L}(f) - f(0) - [f(a+0) - f(a-0)]e^{-as}.$$

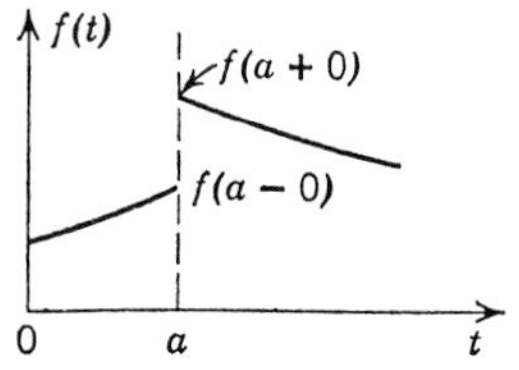

Problem 12

Graph the following functions. Using (1*), find their Laplace transform.

13. $f(t) = 1$ when $1 < t < 2$, $f(t) = 0$ otherwise

14. $f(t) = t$ when $0 < t < 1$, $f(t) = 0$ otherwise

15. $f(t) = t$ when $0 < t < 1$, $f(t) = 1$ when $1 < t < 2$, $f(t) = 0$ otherwise

16. $f(t) = t - 1$ when $1 < t < 2$, $f(t) = 0$ otherwise

Using Theorem 3, find $f(t)$ if $\mathscr{L}(f)$ equals

17. $\dfrac{1}{s(s-2)}$ **18.** $\dfrac{1}{s(s^2+9)}$ **19.** $\dfrac{1}{s(s^2-1)}$ **20.** $\dfrac{1}{s^2(s+1)}$

21. $\dfrac{1}{s^2}\left(\dfrac{s-1}{s+1}\right)$ **22.** $\dfrac{1}{s^2}\left(\dfrac{s-2}{s^2+4}\right)$ **23.** $\dfrac{54}{s^3(s-3)}$ **24.** $\dfrac{2s-\pi}{s^3(s-\pi)}$

Solve the following initial value problems by means of the Laplace transformation.

25. $y'' + 9y = 0, \quad y(0) = 0, \quad y'(0) = 2$

26. $y'' + y' - 2y = 0, \quad y(0) = 0, \; y'(0) = 3$

27. $y'' - 2y' - 3y = 0, \quad y(0) = 1, \; y'(0) = 7$

28. $4y'' + y = 0, \quad y(0) = 1, \; y'(0) = -2$

29. $y'' + 2y' - 8y = 0, \quad y(0) = 1, \; y'(0) = 8$

30. **(Subsidiary equation)** Show that the subsidiary equation of the differential equation

$$y'' + \omega^2 y = r(t) \qquad (\omega \text{ constant})$$

has the solution

$$Y(s) = \frac{sy(0) + y'(0)}{s^2 + \omega^2} + \frac{R(s)}{s^2 + \omega^2}$$

where $R(s)$ is the Laplace transform of $r(t)$. Note that the first term on the right is completely determined by given initial conditions, say, $y(0) = k_1, y'(0) = k_2$, and the second term is independent of those conditions.

5.3 Shifting on the *s*-Axis, Shifting on the *t*-Axis, Unit Step Function

What is the state we have reached, and what is our next goal? We know that the Laplace transformation is linear (Theorem 1, Sec. 5.1), that differentiation of $f(t)$ roughly corresponds to the multiplication of $\mathscr{L}(f)$ by s (Theorems 1 and 2, Sec. 5.2) and that this property is essential in solving differential equations. A first illustration of the technique is given in Example 4, Sec. 5.2. But there the solution may easily be found by the usual methods.

To provide applications such that the Laplace transformation can show its real power, we first have to derive some further general properties of the transformation. Two very important properties concern the shifting on the s-axis and the shifting on the t-axis, as expressed in the two shifting theorems (Theorems 1 and 2 of this section).

Theorem 1 (First shifting theorem; shifting on the *s*-axis)

If $f(t)$ has the transform $F(s)$ where $s > \gamma$, then $e^{at}f(t)$ has the transform

$F(s-a)$ where $s-a>\gamma$; thus, if

$$\mathcal{L}\{f(t)\} = F(s), \qquad \text{then} \qquad \mathcal{L}\{e^{at}f(t)\} = F(s-a).$$

That is, if we replace s by $s-a$ in the transform ("shifting on the s-axis", Fig. 79), this corresponds to the multiplication of the original function by e^{at}.

Proof. By definition,

$$F(s) = \int_0^\infty e^{-st} f(t)\,dt$$

and, therefore,

$$F(s-a) = \int_0^\infty e^{-(s-a)t} f(t)\,dt = \int_0^\infty e^{-st}[e^{at}f(t)]\,dt = \mathcal{L}\{e^{at}f(t)\}. \qquad \blacksquare$$

Example 1

By applying Theorem 1 to the formulas 4, 7, and 8 in Table 5.1 we obtain the following results.

$f(t)$	$\mathcal{L}(f)$
$e^{at}t^n$	$\dfrac{n!}{(s-a)^{n+1}}$
$e^{at}\cos\omega t$	$\dfrac{s-a}{(s-a)^2+\omega^2}$
$e^{at}\sin\omega t$	$\dfrac{\omega}{(s-a)^2+\omega^2}$

Example 2. Damped free vibrations

A small body of mass $m = 2$ is attached at the lower end of an elastic spring whose upper end is fixed, the spring modulus being $k = 10$. Let $y(t)$ be the displacement of the body from the position of static equilibrium. Determine the free vibrations of the body, starting from the initial position $y(0) = 2$ with the initial velocity $y'(0) = -4$, assuming that there is damping proportional to the velocity, the damping constant being $c = 4$.

The motion is described by the solution $y(t)$ of the initial value problem

$$y'' + 2y' + 5y = 0, \qquad y(0) = 2, \qquad y'(0) = -4,$$

cf. (7), Sec. 2.6. Using (1) and (2) in Sec. 5.2, we obtain the subsidiary equation

$$s^2Y - 2s + 4 + 2(sY - 2) + 5Y = 0.$$

The solution is

$$Y(s) = \frac{2s}{(s+1)^2+2^2} = 2\frac{s+1}{(s+1)^2+2^2} - \frac{2}{(s+1)^2+2^2}.$$

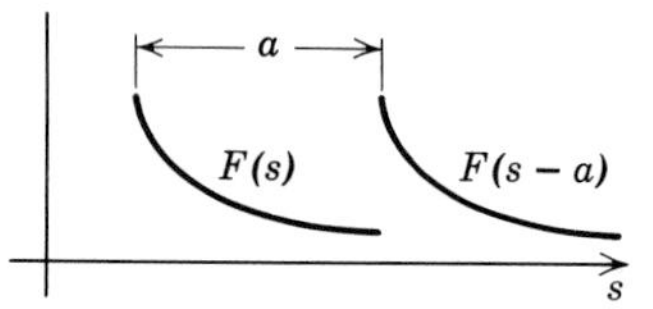

Fig. 79. First shifting theorem, shifting on the s-axis

Now

$$\mathcal{L}^{-1}\left(\frac{s}{s^2+2^2}\right) = \cos 2t, \qquad \mathcal{L}^{-1}\left(\frac{2}{s^2+2^2}\right) = \sin 2t.$$

From this and Theorem 1 we obtain the expected type of solution

$$y(t) = \mathcal{L}^{-1}(Y) = e^{-t}(2\cos 2t - \sin 2t).$$

The first shifting theorem (Theorem 1) concerns shifting on the s-axis: the replacement of s in $F(s)$ by $s - a$ corresponds to the multiplication of the original function $f(t)$ by e^{at}. We shall now state the second shifting theorem (Theorem 2) which concerns shifting on the t-axis: the replacement of t in $f(t)$ by $t - a$ corresponds roughly to the multiplication of the transform $F(s)$ by e^{-as}; the precise formulation of this is as follows.

Theorem 2 (Second shifting theorem; shifting on the *t*-axis)
If $F(s)$ is the transform of $f(t)$, then $e^{-as}F(s)$ ($a > 0$, arbitrary) is the transform of the function

$$\tilde{f}(t) = \begin{cases} 0 & \text{if } t < a \\ f(t-a) & \text{if } t > a. \end{cases}$$

Remark. We can write the last formula in a different way as follows. By definition, the **unit step function** $u_a(t)$ is (see Fig. 80)

(1)
$$u_a(t) = \begin{cases} 0 & \text{if } t < a \\ 1 & \text{if } t > a \end{cases} \qquad (a \geqq 0).$$

We see that $\tilde{f}(t)$ can now be written $f(t-a)u_a(t)$. Figure 81 shows an example. And Theorem 2 asserts that

(2)
$$e^{-as}F(s) = \mathcal{L}\{f(t-a)u_a(t)\}.$$

Equivalently, by taking the inverse transform on both sides,

(2*)
$$\mathcal{L}^{-1}\{e^{-as}F(s)\} = f(t-a)u_a(t).$$

Proof of Theorem 2. From the definition we have

$$e^{-as}F(s) = e^{-as}\int_0^\infty e^{-s\tau}f(\tau)\,d\tau = \int_0^\infty e^{-s(\tau+a)}f(\tau)\,d\tau.$$

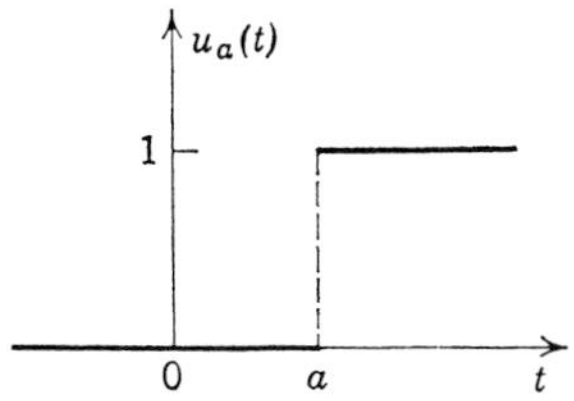

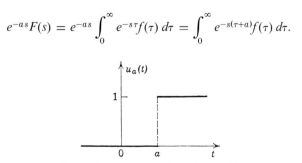

Fig. 80. Unit step function $u_a(t)$

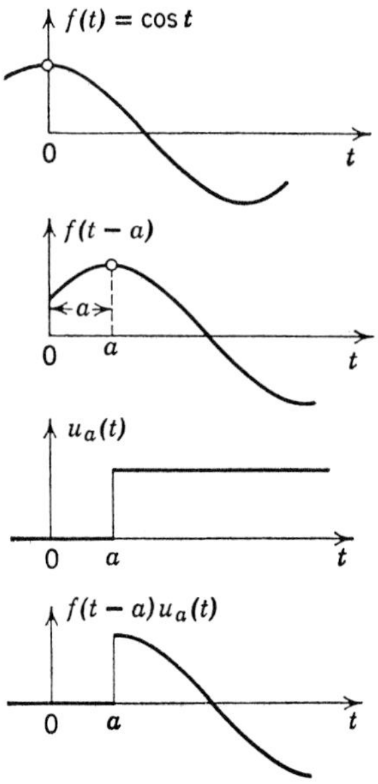

Fig. 81. $f(t-a)u_a(t)$ where $f(t) = \cos t$

Substituting $\tau + a = t$ in the integral, we obtain

$$e^{-as}F(s) = \int_a^\infty e^{-st}f(t-a)\,dt.$$

We can write this as an integral from 0 to ∞ if we make sure that the integrand is zero for all t from 0 to a. We may easily accomplish this by multiplying the present integrand by the step function $u_a(t)$:

$$e^{-as}F(s) = \int_0^\infty e^{-st}f(t-a)u_a(t)\,dt = \mathscr{L}\{f(t-a)u_a(t)\}.$$

This is (2) and the proof is complete. ∎

Before we show applications of the theorem we want to point out that the unit step function $u_a(t)$ is very important. In fact, we shall see that it is the basic building block of certain functions whose knowledge greatly increases the usefulness of the Laplace transformation.

For later use we state that the transform of $u_a(t)$ is

$$\mathscr{L}\{u_a(t)\} = \frac{e^{-as}}{s} \qquad (s > 0). \tag{3}$$

This follows directly from the definition because

$$\mathscr{L}\{u_a(t)\} = \int_0^\infty e^{-st}u_a(t)\,dt = \int_0^a e^{-st}0\,dt + \int_a^\infty e^{-st}1\,dt = -\frac{1}{s}e^{-st}\Big|_a^\infty.$$

It is fair to say that we have already reached the stage where we can attack problems for which the Laplace transformation is preferable to the usual classical method, as Examples 5–7 will illustrate.

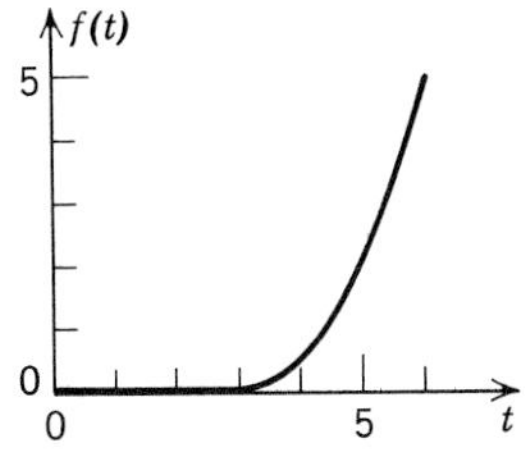

Fig. 82. Example 3

Example 3

Find the inverse transform of e^{-3s}/s^3. Since $\mathscr{L}^{-1}(1/s^3) = t^2/2$ (cf. Table 5.1 in Sec. 5.1), Theorem 2 gives (Fig. 82)

$$\mathscr{L}^{-1}(e^{-3s}/s^3) = \tfrac{1}{2}(t-3)^2 u_3(t).$$

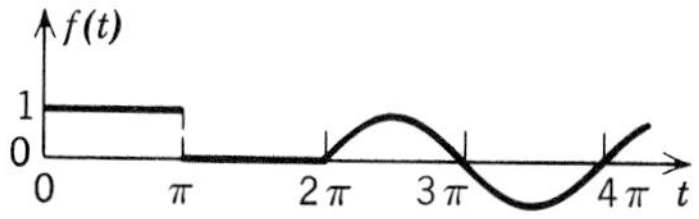

Fig. 83. Example 4

Example 4

Find the transform of the function (Fig. 83)

$$f(t) = \begin{cases} 1 & \text{if} \quad 0 < t < \pi \\ 0 & \text{if} \quad \pi < t < 2\pi \\ \sin t & \text{if} \quad t > 2\pi. \end{cases}$$

We can represent $f(t)$ in terms of unit step functions:

$$f(t) = u_0(t) - u_\pi(t) + u_{2\pi}(t) \sin t.$$

Here $\sin t = \sin(t - 2\pi)$ because of the periodicity. From (3) and (2) and Table 5.1 (Sec. 5.1) we thus obtain

$$\mathscr{L}(f) = \frac{1}{s} - \frac{e^{-\pi s}}{s} + \frac{e^{-2\pi s}}{s^2+1}.$$

Example 5. Response of an *RC*-circuit to a single square wave

Find the current $i(t)$ in the circuit in Fig. 84 if a single square wave with voltage V_0 is applied. The circuit is assumed to be quiescent before the square wave is applied.

The equation of the circuit is (cf. Sec. 1.9)

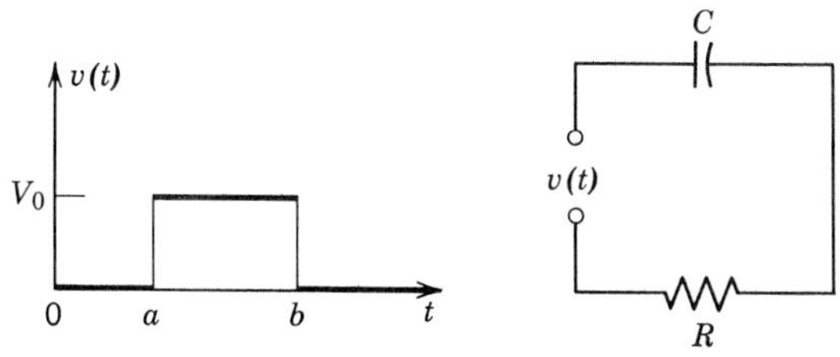

Fig. 84. Example 5

$$Ri(t) + \frac{q(t)}{C} = Ri(t) + \frac{1}{C}\int_0^t i(\tau)\,d\tau = v(t)$$

where $v(t)$ can be represented in terms of two unit step functions:

$$v(t) = V_0[u_a(t) - u_b(t)].$$

Using Theorem 3 in Sec. 5.2 and formula (3), we obtain the subsidiary equation

$$RI(s) + \frac{I(s)}{sC} = \frac{V_0}{s}[e^{-as} - e^{-bs}].$$

The solution of this equation may be written

$$I(s) = F(s)(e^{-as} - e^{-bs}) \qquad \text{where} \qquad F(s) = \frac{V_0/R}{s + 1/RC}.$$

From Table 5.1 in Sec. 5.1 we have

$$\mathcal{L}^{-1}(F) = \frac{V_0}{R}e^{-t/RC}.$$

Hence Theorem 2 yields the solution (Fig. 85)

$$\begin{aligned} i(t) = \mathcal{L}^{-1}(I) &= \mathcal{L}^{-1}\{e^{-as}F(s)\} - \mathcal{L}^{-1}\{e^{-bs}F(s)\} \\ &= \frac{V_0}{R}[e^{-(t-a)/RC}u_a(t) - e^{-(t-b)/RC}u_b(t)]; \end{aligned}$$

that is, $i = 0$ when $t < a$, and

$$i(t) = \begin{cases} K_1 e^{-t/RC} & \text{when } a < t < b \\ (K_1 - K_2)e^{-t/RC} & \text{when } t > b \end{cases}$$

where $K_1 = V_0 e^{a/RC}/R$ and $K_2 = V_0 e^{b/RC}/R$.

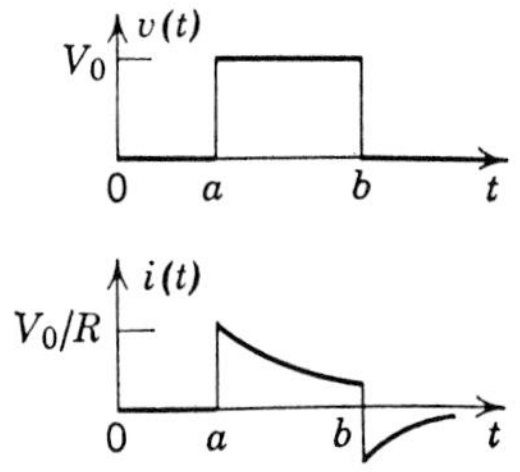

Fig. 85. Voltage and current in Example 5

Example 6. Response of an undamped system to a single square wave
Solve the initial value problem

$$y'' + 2y = r(t), \qquad y(0) = 0, \quad y'(0) = 0$$

where $r(t) = 1$ if $0 < t < 1$ and 0 otherwise (Fig. 86).

From Theorem 2 in Sec. 5.2, the initial conditions and (3) we obtain the subsidiary equation

$$s^2Y + 2Y = \frac{1}{s} - \frac{e^{-s}}{s}.$$

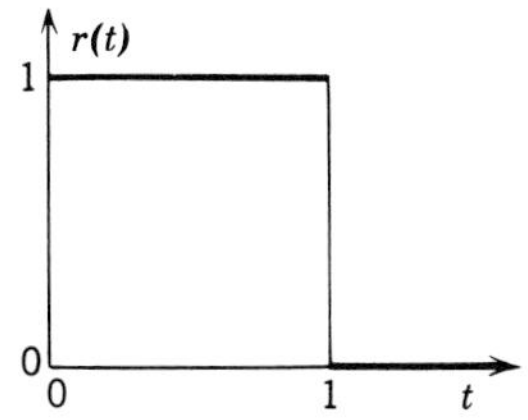

Fig. 86. Input in Examples 6 and 7

The solution is

$$Y(s) = \frac{1}{s(s^2+2)} - \frac{e^{-s}}{s(s^2+2)}.$$

On the right we use partial fractions:

$$\frac{1}{s(s^2+2)} = \frac{1}{2}\left(\frac{1}{s} - \frac{s}{s^2+2}\right).$$

Hence by Table 5.1 (Sec. 5.1) and Theorem 2,

$$y(t) = \tfrac{1}{2}[1 - \cos\sqrt{2}t] - \tfrac{1}{2}[1 - \cos\sqrt{2}(t-1)]u_1(t),$$

that is,

$$y(t) = \begin{cases} \tfrac{1}{2} - \tfrac{1}{2}\cos\sqrt{2}t & \text{if } 0 \leqq t < 1 \\ \tfrac{1}{2}\cos\sqrt{2}(t-1) - \tfrac{1}{2}\cos\sqrt{2}t & \text{if } \quad t > 1. \end{cases}$$

We see that $y(t)$ represents a composite of harmonic oscillations.

Example 7. Response of a damped vibrating system to a single square wave
Determine the response of the damped vibrating system corresponding to

$$y'' + 3y' + 2y = r(t), \qquad y(0) = 0, \quad y'(0) = 0$$

with $r(t)$ as in Example 6 (Fig. 86).

By the same method as before we obtain the subsidiary equation

$$s^2Y + 3sY + 2Y = \frac{1}{s}(1 - e^{-s}).$$

Solving for Y, we have

$$Y(s) = F(s)(1 - e^{-s}) \qquad \text{where} \qquad F(s) = \frac{1}{s(s+1)(s+2)}.$$

In terms of partial fractions,

$$F(s) = \frac{\frac{1}{2}}{s} - \frac{1}{s+1} + \frac{\frac{1}{2}}{s+2}.$$

Hence by Table 5.1 (Sec. 5.1),

$$f(t) = \mathcal{L}^{-1}(F) = \tfrac{1}{2} - e^{-t} + \tfrac{1}{2}e^{-2t}.$$

Therefore, by Theorem 2, we have

$$\mathcal{L}^{-1}\{e^{-s}F(s)\} = f(t-1)u_1(t) = \begin{cases} 0 & (0 \leqq t < 1) \\ \tfrac{1}{2} - e^{-(t-1)} + \tfrac{1}{2}e^{-2(t-1)} & (t > 1) \end{cases}$$

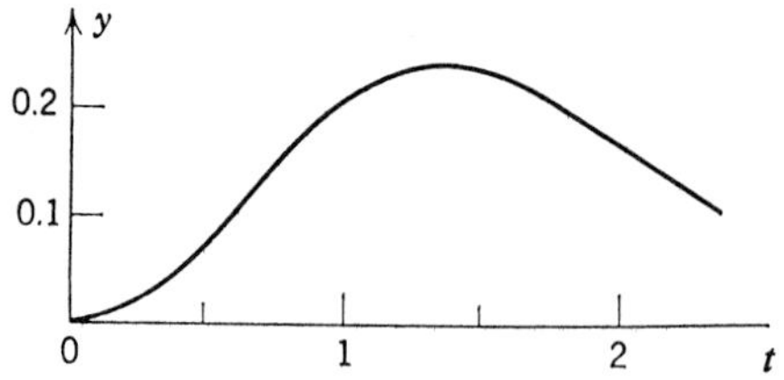

Fig. 87. Output in Example 7

This yields the solution (cf. Fig. 87)

$$y(t) = \mathscr{L}^{-1}(Y) = f(t) - f(t-1)u_1(t) = \begin{cases} \frac{1}{2} - e^{-t} + \frac{1}{2}e^{-2t} & (0 \leqq t < 1) \\ K_1e^{-t} - K_2e^{-2t} & (t > 1) \end{cases}$$

where $K_1 = e - 1$ and $K_2 = (e^2 - 1)/2$.

Problems for Sec. 5.3

Find $\mathscr{L}(f)$ if $f(t)$ equals

1. $e^{-2t} \sin n\pi t$

2. $e^{-t} \sin (\omega t + \theta)$

3. $e^{-at}(A \cos \beta t + B \sin \beta t)$

4. $e^{3t}(a_1 t + a_2 t^2)$

Representing the hyperbolic functions in terms of exponential functions and applying the first shifting theorem, show that

5. $\mathscr{L}(\cosh at \cos at) = \dfrac{s^3}{s^4 + 4a^4}$

6. $\mathscr{L}(\cosh at \sin at) = \dfrac{a(s^2 + 2a^2)}{s^4 + 4a^4}$

7. $\mathscr{L}(\sinh at \cos at) = \dfrac{a(s^2 - 2a^2)}{s^4 + 4a^4}$

8. $\mathscr{L}(\sinh at \sin at) = \dfrac{2a^2 s}{s^4 + 4a^4}$

Find $f(t)$ if $\mathscr{L}(f)$ equals

9. $\dfrac{s + 3}{(s + 1)^2 + 1}$

10. $\dfrac{2s + 1}{s^2 + 4s + 13}$

11. $\dfrac{1}{(s - 2)^3} + \dfrac{1}{(s - 2)^5}$

12. $\dfrac{bs + c}{(s - a)^2 + \omega^2}$

Solve the following initial value problems by the Laplace transformation.

13. $y'' + 2y' + 5y = 0, \quad y(0) = 2, \quad y'(0) = -4$

14. $y'' - 4y' + 4y = 0, \quad y(0) = 0, \quad y'(0) = 2$

15. $y'' + 2y' + 10y = 0, \quad y(0) = 0, \quad y'(0) = 3$

16. $y'' + 4y' + 4y = 0, \quad y(0) = 2, \quad y'(0) = -3$

Represent the following functions in terms of unit step functions and find their Laplace transforms.

17.

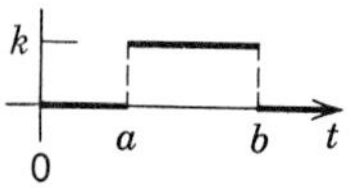

18.

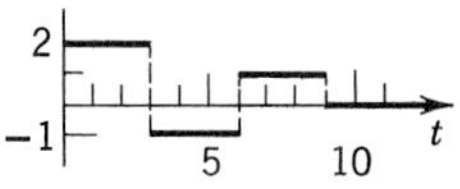

19.

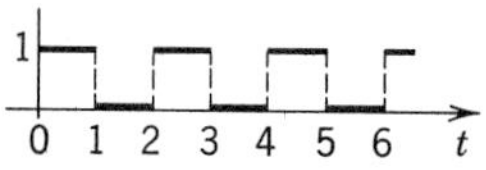

20.

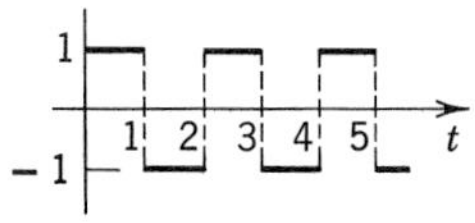

21.

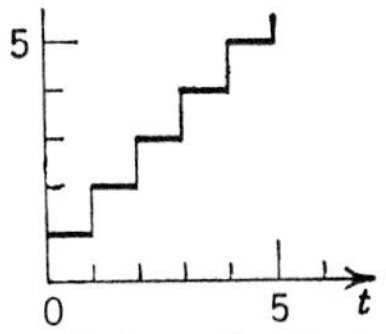

(Staircase function)

22.

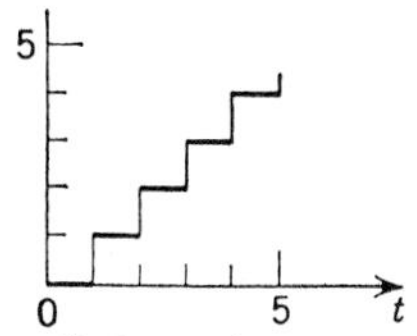

(Staircase function)

Using Theorem 3 in Sec. 5.2, find the Laplace transforms of the following functions.

23.

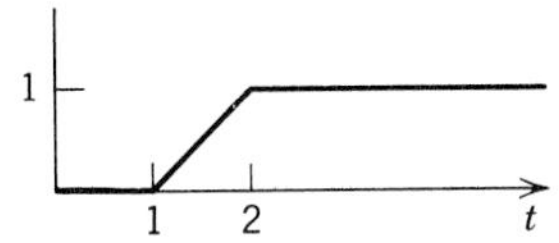

24.

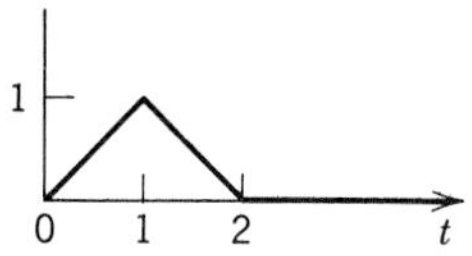

25. A capacitor of capacitance C is charged so that its potential is V_0. At $t = 0$ the switch in the figure is closed and the capacitor starts to discharge through the resistor of resistance R. Using the Laplace transformation, find the charge $q(t)$ on the capacitor.

26. Find the current $i(t)$ in the circuit in the figure, assuming that no current flows when $t \leqq 0$ and the switch is closed at $t = 0$.

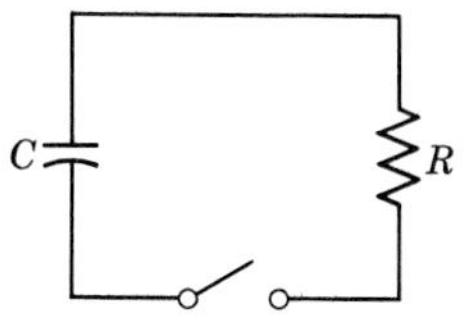

Problem 25

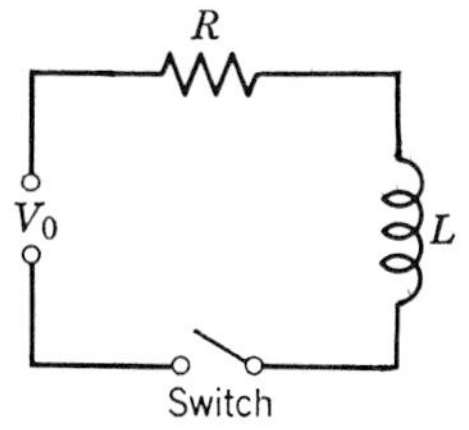

Problem 26

Find and graph $f(t)$ if $\mathscr{L}(f)$ equals

27. $(e^{-3s} - e^{-s})/s$

28. e^{-s}/s^2

29. $(e^{-s} + e^{-2s} - 3e^{-3s} + e^{-6s})/s^2$

30. $(2e^{-as} - 1)/s$

Graph the following functions and find their Laplace transforms.

31. $(t - 1)u_1(t)$

32. $tu_1(t)$

33. $u_\pi(t) \cos t$

34. $u_{2\pi/\omega}(t) \sin \omega t$

35. $e^{kt}u_a(t)$

36. $t^2u_1(t)$

Graph the following functions, which are assumed to be zero outside the given intervals, and find their Laplace transforms.

37. $K \sin \omega t \quad (0 < t < \pi/\omega)$

38. $K \sin t \quad (2\pi < t < 4\pi)$

39. $1 - e^{-t} \quad (0 < t < \pi)$

40. $K \cos \omega t \quad (0 < t < \pi/\omega)$

41. $t^2 \quad (0 < t < 1)$

42. $t^2 \quad (0 < t < 3)$

Find and graph $f(t)$ if $F(s) = \mathscr{L}(f)$ equals

43. e^{-2s}/s

44. e^{-s}/s^3

45. $e^{-2s}/(s - 2)$

46. $e^{-s}/(s^2 + \pi^2)$

47. $e^{-s}/(s - 3)$

48. $(1 - e^{-\pi s})/(s^2 + 4)$

49. $e^{-\pi s}/(s^2 + 2s + 2)$

50. $s(1 + e^{-\pi s})/(s^2 + 1)$

Find the current $i(t)$ in the LC-circuit shown in the figure, assuming $L = 1$ henry, $C = 1$ farad, zero initial current and charge on the capacitor and $v(t)$ as follows.

51. $v = t$ when $0 < t < a$ and 0 otherwise

52. $v = t$ when $0 < t < a$ and $v = a$ when $t > a$

53. $v = 1 - e^{-t}$ when $0 < t < \pi$ and 0 otherwise

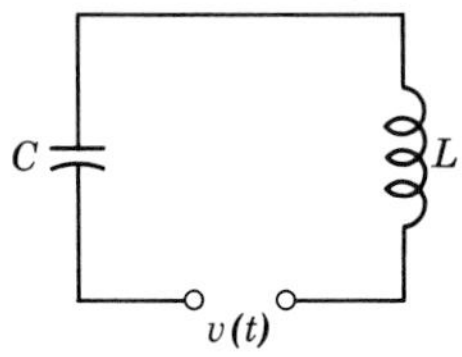

Problems 51–53. *LC*-circuit

54. If $\mathcal{L}\{f(t)\} = F(s)$, show that $\mathcal{L}\{f(at)\} = F(s/a)/a$. Using this formula, derive the transform of $\sin \omega t$ from that of $\sin t$.

55. Show that (2) implies

$$e^{-as}\mathcal{L}\{f(t + a)\} = \mathcal{L}\{f(t)u_a(t)\}, \tag{2**}$$

which is sometimes more convenient in applications.

5.4 Differentiation and Integration of Transforms

The Laplace transformation has surprisingly many general properties which we can use for obtaining transforms or inverse transforms. Corresponding methods are direct integration (Sec. 5.1), use of the linearity (Sec. 5.1), shifting (Sec. 5.3) and differentiation or integration of original functions $f(t)$ (Sec. 5.2). But this is not all: in this section we consider differentiation and integration of transforms $F(s)$ and find out the corresponding operations for original functions $f(t)$. And in the next section we shall consider multiplication of transforms and find out the corresponding operation for original functions.

It can be shown that if $f(t)$ satisfies the conditions of the existence theorem in Sec. 5.1, then the derivative of the corresponding transform

$$F(s) = \mathcal{L}(f) = \int_0^\infty e^{-st} f(t)\, dt$$

with respect to s can be obtained by differentiating under the integral sign with respect to s (proof in Ref. [A3]); thus

$$F'(s) = -\int_0^\infty e^{-st}[tf(t)]\, dt.$$

Consequently, if $\mathcal{L}(f) = F(s)$, then

$$\mathcal{L}\{tf(t)\} = -F'(s); \tag{1}$$

differentiation of the transform of a function corresponds to the multiplication of the function by $-t$.

This property of the Laplace transformation enables us to obtain new transforms from given ones.

Example 1

We shall derive the following three formulas:

	$\mathscr{L}(f)$	$f(t)$
(2)	$\dfrac{1}{(s^2+\beta^2)^2}$	$\dfrac{1}{2\beta^3}(\sin \beta t - \beta t \cos \beta t)$
(3)	$\dfrac{s}{(s^2+\beta^2)^2}$	$\dfrac{t}{2\beta}\sin \beta t$
(4)	$\dfrac{s^2}{(s^2+\beta^2)^2}$	$\dfrac{1}{2\beta}(\sin \beta t + \beta t \cos \beta t)$

From (1) and formula 8 (with $\omega = \beta$) in Table 5.1, Sec. 5.1, we obtain

$$\mathscr{L}(t \sin \beta t) = \frac{2\beta s}{(s^2+\beta^2)^2}.$$

By dividing by 2β we obtain (3). From (1) and formula 9 (with $\omega = \beta$) in Table 5.1, Sec. 5.1, we find

$$\mathscr{L}(t \cos \beta t) = -\frac{(s^2+\beta^2) - 2s^2}{(s^2+\beta^2)^2} = \frac{s^2-\beta^2}{(s^2+\beta^2)^2} = \frac{1}{s^2+\beta^2} - \frac{2\beta^2}{(s^2+\beta^2)^2} \tag{5}$$

where the last equality may be readily verified by direct calculation. Hence

$$t \cos \beta t = \mathscr{L}^{-1}\left\{\frac{1}{s^2+\beta^2} - \frac{2\beta^2}{(s^2+\beta^2)^2}\right\} = \frac{1}{\beta}\sin \beta t - 2\beta^2 \mathscr{L}^{-1}\left\{\frac{1}{(s^2+\beta^2)^2}\right\},$$

and we see that $\mathscr{L}^{-1}\{\cdots\}$ on the right equals $f(t)$ in (2). Finally, by subtracting and adding β^2 in the numerator we have

$$\frac{s^2}{(s^2+\beta^2)^2} = \frac{s^2-\beta^2}{(s^2+\beta^2)^2} + \frac{\beta^2}{(s^2+\beta^2)^2}.$$

From (5) and (2) we see that the two terms on the right are the inverse transforms of $t \cos \beta t$ and $(\sin \beta t - \beta t \cos \beta t)/2\beta$, respectively. This yields (4).

Similarly, if $f(t)$ satisfies the conditions of the existence theorem in Sec. 5.1 and the limit of $f(t)/t$, as t approaches 0 from the right, exists, then

$$\mathscr{L}\left\{\frac{f(t)}{t}\right\} = \int_s^\infty F(\tilde{s})\, d\tilde{s} \qquad (s > \gamma); \tag{6}$$

in this manner, *integration of the transform of a function* $f(t)$ *corresponds to the division of* $f(t)$ *by* t.

In fact, from the definition it follows that

$$\int_s^\infty F(\tilde{s})\, d\tilde{s} = \int_s^\infty \left[\int_0^\infty e^{-\tilde{s}t} f(t)\, dt\right] d\tilde{s},$$

and it can be shown (cf. Ref. [A3]) that under the above assumptions we may reverse the order of integration, that is,

$$\int_s^\infty F(\tilde{s})\,d\tilde{s} = \int_0^\infty \left[\int_s^\infty e^{-\tilde{s}t} f(t)\,d\tilde{s}\right] dt = \int_0^\infty f(t)\left[\int_s^\infty e^{-\tilde{s}t}\,d\tilde{s}\right] dt.$$

The integral over $\tilde{s}$ on the right equals e^{-st}/t when $s > \gamma$, and, therefore,

$$\int_s^\infty F(\tilde{s})\,d\tilde{s} = \int_0^\infty e^{-st}\frac{f(t)}{t}\,dt = \mathcal{L}\left\{\frac{f(t)}{t}\right\} \qquad (s > \gamma).$$

Example 2

Find the inverse transform of the function $\ln\left(1 + \frac{\omega^2}{s^2}\right)$. By differentiation we first have

$$-\frac{d}{ds}\ln\left(1 + \frac{\omega^2}{s^2}\right) = \frac{2\omega^2}{s(s^2+\omega^2)} = \frac{2}{s} - 2\frac{s}{s^2+\omega^2},$$

where the last equality can be readily verified by direct calculation. This is our present $F(s)$. From Table 5.1 in Sec. 5.1 we obtain

$$f(t) = \mathcal{L}^{-1}(F) = \mathcal{L}^{-1}\left\{\frac{2}{s} - 2\frac{s}{s^2+\omega^2}\right\} = 2 - 2\cos\omega t.$$

This function satisfies the conditions under which (6) holds. Therefore,

$$\mathcal{L}^{-1}\left\{\ln\left(1 + \frac{\omega^2}{s^2}\right)\right\} = \int_s^\infty F(\tilde{s})\,d\tilde{s} = \frac{f(t)}{t}.$$

Our result is

$$\mathcal{L}^{-1}\left\{\ln\left(1 + \frac{\omega^2}{s^2}\right)\right\} = \frac{2}{t}(1 - \cos\omega t).$$

Problems for Sec. 5.4

Using (1), find the Laplace transform of

1. te^t **2.** t^2e^{2t} **3.** $t\sin 3t$

4. t^2e^{-t} **5.** $t^2\cos t$ **6.** $te^{-t}\cos t$

7. $te^{-t}\cosh 2t$ **8.** $t^2\sinh 2t$ **9.** $te^{-2t}\sin t$

Using (6), find $f(t)$, if $\mathcal{L}(f)$ equals

10. $\dfrac{1}{(s-4)^2}$ **11.** $\dfrac{2s}{(s^2+1)^2}$ **12.** $\dfrac{2}{(s-a)^3}$

13. $\operatorname{arc\,cot}(s+1)$ **14.** $\operatorname{arc\,cot}\dfrac{s}{\omega}$ **15.** $\ln\dfrac{s}{s-1}$

16. $\ln\dfrac{s+a}{s-a}$ **17.** $\ln\dfrac{s+a}{s+b}$ **18.** $\ln\dfrac{s^2+1}{(s-1)^2}$

19. (Laguerre polynomials) The *Laguerre polynomial of order n* is defined by

$$l_0(t) = 1, \qquad l_n(t) = \frac{e^t}{n!}\frac{d^n}{dt^n}(t^ne^{-t}) \qquad n = 1, 2, \cdots.$$

(L_n is the usual notation, but we write l_n to conform with the use of small letters for functions and capitals for transforms.) Show that $l_n(t)$ is a polynomial of order

n in t. Using a shifting theorem (Sec. 5.3) and the formula for the transform of the nth derivative (Sec. 5.2), show that

$$\mathcal{L}\{l_n(t)\} = \frac{(s-1)^n}{s^{n+1}}.$$

20. Laguerre's differential equation is

$$ty'' + (1-t)y' + ny = 0.$$

Using (1) and Prob. 19, find a solution. (Two observations are interesting. (*i*) For $Y = \mathcal{L}\{y\}$ we still get a differential equation, but one that can easily be solved. (*ii*) We do not get a general solution y. The equation has a second independent solution whose Laplace transform does not exist, as can be shown.)

5.5 Convolution

Another important general property of the Laplace transformation has to do with products of transforms. Indeed, if we know the inverse transforms $f(t)$ of $F(s)$ and $g(t)$ of $G(s)$, we can calculate from $f(t)$ and $g(t)$ the inverse transform $h(t)$ of the product $H(s) = F(s)G(s)$. This function $h(t)$ is written $(f * g)(t)$, which is standard notation, and is called the **convolution** of f and g. Its form is stated in the following theorem, whose practical importance is obvious since the situation and task just described occur quite often in applications.

Theorem 1 (Convolution theorem)
If $f(t)$ and $g(t)$ are the inverse transforms of $F(s)$ and $G(s)$, respectively, and satisfy the hypothesis of the existence theorem (Sec. 5.1), then the inverse transform $h(t)$ of the product $H(s) = F(s)G(s)$ is the convolution of $f(t)$ and $g(t)$, written $(f * g)(t)$ and defined by

$$h(t) = (f * g)(t) = \int_0^t f(\tau)g(t-\tau)\,d\tau. \tag{1}$$

Proof. By the definition of $G(s)$ and the first shifting theorem, for each fixed τ ($\tau \geqq 0$) we have

$$e^{-s\tau}G(s) = \int_0^\infty e^{-st}g(t-\tau)u_\tau(t)\,dt = \int_\tau^\infty e^{-st}g(t-\tau)\,dt$$

where $s > \gamma$. Hence by the definition of $F(s)$ we obtain

$$\begin{aligned} F(s)G(s) &= \int_0^\infty e^{-s\tau}f(\tau)G(s)\,d\tau \\ &= \int_0^\infty f(\tau)\int_\tau^\infty e^{-st}g(t-\tau)\,dt\,d\tau \end{aligned}$$

where $s > \gamma$. Here we integrate over t from τ to ∞ and then over τ from 0 to ∞; this corresponds to the shaded wedge-shaped region extending to infinity in

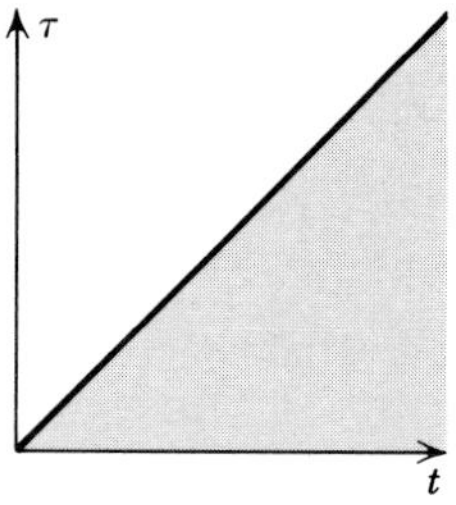

Fig. 88. Region of integration in the $t\tau$-plane in the proof of Theorem 1

the $t\tau$-plane shown in Fig. 88. Our assumptions on f and g are such that the order of integration can be reversed. (A proof requiring the knowledge of uniform convergence is included in [B3] listed in Appendix 1.) We then integrate first over τ from 0 to t (cf. Fig. 88) and then over t from 0 to ∞; thus

$$F(s)G(s) = \int_0^\infty e^{-st} \int_0^t f(\tau)g(t-\tau)\,d\tau\,dt$$

$$= \int_0^\infty e^{-st} h(t)\,dt = \mathscr{L}(h)$$

where h is given by (1). This completes the proof. ∎

Using the definition, the reader may show that the convolution $f * g$ has the properties

$$f * g = g * f \qquad \text{(commutative law)}$$

$$f * (g_1 + g_2) = f * g_1 + f * g_2 \qquad \text{(distributive law)}$$

$$(f * g) * v = f * (g * v) \qquad \text{(associative law)}$$

$$f * 0 = 0 * f = 0,$$

just as the multiplication of numbers. But $1 * g \neq g$ in general. For example if $g(t) = t$, then

$$(1 * g)(t) = \int_0^t 1 \cdot (t - \tau)\,d\tau = \frac{t^2}{2}.$$

Another unusual property is that $(f * f)(t) \geqq 0$ may not hold.

We shall now illustrate that convolution is useful for obtaining inverse transforms and solving differential equations.

Example 1

Let $H(s) = 1/s^2(s-a)$. Find $h(t)$. From Table 5.1 (Sec. 5.1) we know that

$$\mathscr{L}^{-1}\left\{\frac{1}{s^2}\right\} = t, \qquad \mathscr{L}^{-1}\left\{\frac{1}{s-a}\right\} = e^{at}.$$

We use Theorem 1 and integrate by parts, finding

$$\begin{aligned} h(t) = t * e^{at} &= \int_0^t \tau e^{a(t-\tau)}\,d\tau \\ &= e^{at}\int_0^t \tau e^{-a\tau}\,d\tau \\ &= \frac{1}{a^2}(e^{at} - at - 1). \end{aligned}$$

Example 2. Forced oscillations, resonance

Solve the initial value problem

$$my'' + ky = F_0 \sin pt, \qquad y(0) = 0, \quad y'(0) = 0.$$

From Sec. 2.13 we know that this equation governs the forced oscillations of a body of mass m attached at the lower end of an elastic spring whose upper end is fixed (Fig. 89). k is the spring modulus, and $F_0 \sin pt$ is the driving force or input. Setting $\omega_0 = \sqrt{k/m}$, we may write the equation in the form

$$y'' + \omega_0{}^2 y = K \sin pt \qquad \left(K = \frac{F_0}{m}\right).$$

The subsidiary equation is

$$s^2 Y + \omega_0{}^2 Y = K\frac{p}{s^2 + p^2}. \tag{2}$$

The solution is

$$Y(s) = \frac{Kp}{(s^2 + \omega_0{}^2)(s^2 + p^2)}. \tag{3}$$

Now

$$\mathcal{L}^{-1}\left\{\frac{1}{s^2 + \omega_0{}^2}\right\} = \frac{1}{\omega_0}\sin \omega_0 t$$

$$\mathcal{L}^{-1}\left\{\frac{1}{s^2 + p^2}\right\} = \frac{1}{p}\sin pt.$$

Hence by the convolution theorem,

$$y(t) = \frac{Kp}{\omega_0 p}\sin \omega_0 t * \sin pt = \frac{K}{\omega_0}\int_0^t \sin \omega_0 \tau \sin (pt - p\tau)\,d\tau. \tag{4}$$

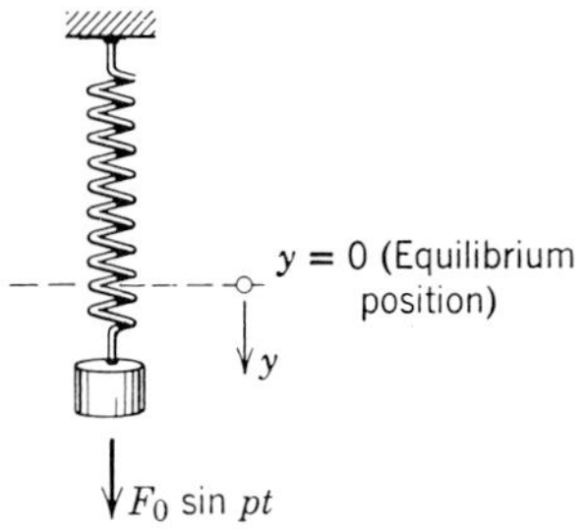

Fig. 89. Vibrating system. Forced oscillations

The integrand equals [cf. (11) in Appendix 3]

$$\tfrac{1}{2}[-\cos(pt + (\omega_0 - p)\tau) + \cos(pt - (\omega_0 + p)\tau)]. \tag{5}$$

No resonance. If $p^2 \neq \omega_0^2$, we obtain by integration with respect to τ

$$\begin{aligned} y(t) &= \frac{K}{2\omega_0}\left[-\frac{\sin(pt + (\omega_0 - p)\tau)}{\omega_0 - p} + \frac{\sin(pt - (\omega_0 + p)\tau)}{-\omega_0 - p}\right]_0^t \\ &= \frac{K}{2\omega_0}\left[\frac{\sin pt - \sin \omega_0 t}{\omega_0 - p} + \frac{\sin pt + \sin \omega_0 t}{\omega_0 + p}\right]. \end{aligned}$$

Forming the common denominator and simplifying, we finally obtain the result

$$y(t) = \frac{K}{p^2 - \omega_0^2}\left[\frac{p}{\omega_0}\sin \omega_0 t - \sin pt\right].$$

This represents a superposition of two harmonic oscillations whose frequencies are the natural frequency of the freely vibrating system and the frequency of the driving force.

Resonance. If $p = \omega_0$, then (5) is simply

$$\tfrac{1}{2}[-\cos \omega_0 t + \cos(\omega_0 t - 2\omega_0 \tau)].$$

Integration with respect to τ yields

$$\begin{aligned} y(t) &= \frac{K}{2\omega_0}\left[-\tau \cos \omega_0 t - \frac{1}{2\omega_0}\sin(\omega_0 t - 2\omega_0\tau)\right]_0^t \\ &= \frac{K}{2\omega_0^2}[\sin \omega_0 t - \omega_0 t \cos \omega_0 t]. \end{aligned}$$

This shows that we now have resonance as studied in Sec. 2.13.

Problems for Sec. 5.5

Find the following convolutions. [*Hint.* In Probs. 6–9 use (11) in Appendix 3.]

1. $1 * 1$ **2.** $1 * \sin t$ **3.** $e^t * e^t$

4. $e^{at} * e^{bt}$ $(a \neq b)$ **5.** $t * e^{at}$ **6.** $\sin t * \sin t$

7. $\sin t * \sin 2t$ **8.** $\cos t * \cos t$ **9.** $\sin t * \cos t$

Find $h(t)$ by the convolution theorem if $H(s) = \mathscr{L}\{h(t)\}$ equals

10. $\dfrac{1}{(s-a)^2}$ **11.** $\dfrac{1}{(s-a)^3}$ **12.** $\dfrac{1}{(s-a)(s-b)}$ $(a \neq b)$

13. $\dfrac{1}{s(s^2+\omega^2)}$ **14.** $\dfrac{1}{s^2(s^2+\omega^2)}$ **15.** $\dfrac{1}{(s^2+1)^2}$

16. $\dfrac{s}{(s^2+\omega^2)^2}$ **17.** $\dfrac{s^2}{(s^2+1)^2}$ **18.** $\dfrac{s}{(s^2+a^2)(s^2+b^2)}$ $(a^2 \neq b^2)$

19. Prove the commutative law $f * g = g * f$.

20. Prove the associative law $(f * g) * v = f * (g * v)$.

21. Prove the distributive law $f * (g_1 + g_2) = f * g_1 + f * g_2$.

22. Using the convolution theorem, prove by induction that

$$\mathscr{L}^{-1}\left\{\frac{1}{(s-a)^n}\right\} = \frac{1}{(n-1)!}t^{n-1}e^{at}.$$

23. How can we obtain Theorem 3 in Sec. 5.2 from the convolution theorem?

24. Give an example showing that $(f * f)(t) \geqq 0$ may not hold. (Take $f(t) = \cos t$.)

Applying convolution, solve the following initial value problems.

25. $y'' + y = \sin 3t, \quad y(0) = 0, \quad y'(0) = 0$

26. $y'' + y = \sin t, \quad y(0) = 0, \quad y'(0) = 0$

27. $y'' + 2y = r(t), \quad r(t) = 1$ if $0 < t < 1$ and 0 if $t > 1$, $y(0) = 0, \quad y'(0) = 0$

Integral equations. An *integral equation* is an equation in which the unknown function, call it $y(t)$, occurs inside an integral. The following problems illustrate that certain types of integral equations can be solved by the convolution theorem. Solve

28. $y(t) = t + \int_0^t y(\tau) \sin(t - \tau)\, d\tau$

29. $y(t) = \sin 2t + \int_0^t y(\tau) \sin 2(t - \tau)\, d\tau$

30. $y(t) = \sin t + \int_0^t y(\tau) \sin(t - \tau)\, d\tau$

5.6 Partial Fractions

Most of the applications considered so far in this chapter led to a subsidiary equation whose solution $Y(s)$ was of the form

$$Y(s) = \frac{F(s)}{G(s)}$$

where $F(s)$ and $G(s)$ are polynomials in s. In such a case, $y(t) = \mathscr{L}^{-1}(Y)$, the solution of the problem, can be determined by first expressing $Y(s)$ in terms of partial fractions. Actually, this is what we did in some simpler cases, ad hoc and without difficulty. In the present section we approach partial fractions more systematically, making the following assumption.

Assumption

$F(s)$ and $G(s)$ have real coefficients and no common factors. The degree of $F(s)$ is lower than that of $G(s)$.

The form of partial fractions depends on the type of factor. We first note that the corresponding inverse transforms are already known to us, as follows. To an unrepeated factor $s - a$ of $G(s)$ there corresponds a partial fraction $A/(s - a)$ and (cf. Table 5.1, Sec. 5.1)

$$\mathscr{L}^{-1}\left\{\frac{A}{s - a}\right\} = Ae^{at}.$$

To a repeated factor $(s - a)^m$ there corresponds

$$\frac{A_m}{(s-a)^m} + \frac{A_{m-1}}{(s-a)^{m-1}} + \cdots + \frac{A_1}{s-a}$$

and (cf. Example 1, Sec. 5.3)

$$\mathcal{L}^{-1}\left\{\frac{A_k}{(s-a)^k}\right\} = A_k \frac{t^{k-1}}{(k-1)!} e^{at}$$

where $k = 1, \cdots, m$. In the case of a complex root $a = \alpha + i\beta$ of $G(s) = 0$, complex fractions can be avoided by noting that $\bar{a} = \alpha - i\beta$ is also a root of $G(s) = 0$, since $G(s)$ has real coefficients. Consequently, using $(s - a)(s - \bar{a}) = (s - \alpha)^2 + \beta^2$, we may set up fractions of the form

$$\frac{As + B}{(s-\alpha)^2 + \beta^2}$$

if $s - a$ and $s - \bar{a}$ are unrepeated factors, and

$$\frac{As + B}{[(s-\alpha)^2 + \beta^2]^2} + \frac{Cs + D}{(s-\alpha)^2 + \beta^2}$$

if $s - a$ and $s - \bar{a}$ are repeated factors. Then we need (cf. Example 1, Sec. 5.3)

$$\begin{aligned}\mathcal{L}^{-1}\left\{\frac{As + B}{(s-a)(s-\bar{a})}\right\} &= \mathcal{L}^{-1}\left\{\frac{A(s-\alpha) + \alpha A + B}{(s-\alpha)^2 + \beta^2}\right\} \\ &= e^{\alpha t}\left[A \cos \beta t + \frac{\alpha A + B}{\beta} \sin \beta t\right]\end{aligned}$$

and (cf. Example 1, Sec. 5.4)

$$\mathcal{L}^{-1}\left\{\frac{A(s-\alpha) + \alpha A + B}{[(s-\alpha)^2 + \beta^2]^2}\right\} = e^{\alpha t}\left[\frac{At}{2\beta} \sin \beta t + \frac{\alpha A + B}{2\beta^3}(\sin \beta t - \beta t \cos \beta t)\right].$$

[We do not consider complex $(s - a)^3$, $(s - a)^4$, $\cdots$, which are of minor practical importance.] The remaining task is the determination of the constants A, B, etc., for which some methods are presented in elementary calculus. The student may now immediately consider Examples 1–4 (below) and verify the given results in his own way, that is, by applying his favorite method of determining those constants.

For the sake of completeness we shall now present a systematic discussion which shows how to obtain those constants from quantities directly related to $Y(s) = F(s)/G(s)$.

Notation

(1a), (2a), $\cdots$, (5a) (below) are representations of $Y(s) = F(s)/G(s)$, and in each case, $W(s)$ denotes the sum of the partial fractions corresponding to all the (unrepeated or repeated) linear factors of $G(s)$ which are not under consideration.

Case 1. Unrepeated factors s − a. *In this case,*

$$\text{(1a)} \qquad Y(s) = \frac{F(s)}{G(s)} = \frac{A}{s-a} + W(s).$$

The inverse transform is

$$\textbf{(1b)} \qquad \mathscr{L}^{-1}(Y) = Ae^{at} + \mathscr{L}^{-1}(W)$$

where A is given by either of the two expressions

$$\textbf{(1c)} \qquad A = Q_a(a) \qquad \text{or} \qquad A = \frac{F(a)}{G'(a)}$$

and $Q_a(s)$ is the function which is left after removing the factor $s - a$ from $G(s)$ in $Y(s)$, that is

$$\text{(1d)} \qquad Q_a(s) = \frac{(s-a)F(s)}{G(s)}.$$

The subscript a of Q_a emphasizes the important fact that Q_a depends on a and changes when we proceed from one linear factor to another.

Case 2. Repeated factor (s − a)^m. *In this case,*

$$\text{(2a)} \qquad Y(s) = \frac{F(s)}{G(s)} = \frac{A_m}{(s-a)^m} + \frac{A_{m-1}}{(s-a)^{m-1}} + \cdots + \frac{A_1}{s-a} + W(s).$$

The inverse transform of this expression is

$$\textbf{(2b)} \qquad \mathscr{L}^{-1}(Y) = e^{at}\left(A_m \frac{t^{m-1}}{(m-1)!} + A_{m-1}\frac{t^{m-2}}{(m-2)!} + \cdots + A_2\frac{t}{1!} + A_1\right) + \mathscr{L}^{-1}(W)$$

where the constants $A_1, \cdots, A_m$ are given by the formula

$$\textbf{(2c)} \qquad A_m = Q_a(a), \quad A_k = \frac{1}{(m-k)!}\left.\frac{d^{m-k}Q_a(s)}{ds^{m-k}}\right|_{s=a} \qquad (k = 1, \cdots, m-1);$$

here,

$$\text{(2d)} \qquad Q_a(s) = \frac{(s-a)^m F(s)}{G(s)}.$$

Complex Factors. The previous cases include real as well as complex numbers a, but if a is complex, other formulas are preferable from the practical point of view. Let

$$a = \alpha + i\beta \qquad \text{and} \qquad \bar{a} = \alpha - i\beta \qquad (\alpha, \beta \text{ real}, \beta \neq 0).$$

Suppose that $s = a$ is a root of $G(s) = 0$. Then since $G(s)$ has real coefficients, $s = \bar{a}$ is also a root of $G(s) = 0$.

Case 3. Unrepeated complex factor s − a. *Writing the partial fraction corresponding to $s - a$ and $s - \bar{a}$ explicitly and noting that*

$$(s - a)(s - \bar{a}) = (s - \alpha)^2 + \beta^2,$$

we have

(3a) $$Y(s) = \frac{F(s)}{G(s)} = \frac{As + B}{(s - \alpha)^2 + \beta^2} + W(s) \qquad (A, B \text{ real}).$$

The inverse transform is

(3b) $$\mathscr{L}^{-1}(Y) = \frac{1}{\beta} e^{\alpha t}(T_a \cos \beta t + S_a \sin \beta t) + \mathscr{L}^{-1}(W)$$

where S_a and T_a are the real and imaginary parts of

(3c) $$R_a(a) = S_a + iT_a;$$

here

(3d) $$R_a(s) = [(s - \alpha)^2 + \beta^2]\frac{F(s)}{G(s)}.$$

Case 4. Repeated complex factor (s − a)². *Writing the partial fractions corresponding to $(s - a)^2$ and $(s - \bar{a})^2$ explicitly, we have*

(4a) $$Y(s) = \frac{F(s)}{G(s)} = \frac{As + B}{[(s - \alpha)^2 + \beta^2]^2} + \frac{Cs + D}{(s - \alpha)^2 + \beta^2} + W(s)$$

where A, B, C, D are real. The inverse transform is

(4b) $$\mathscr{L}^{-1}(Y) = \frac{1}{2\beta^3} e^{\alpha t}[(T_a - \beta S_a^* - \beta S_a t) \cos \beta t + (S_a + \beta T_a^* + \beta T_a t) \sin \beta t] + \mathscr{L}^{-1}(W)$$

where

(4c) $$R_a(a) = S_a + iT_a, \qquad R_a'(a) = S_a^* + iT_a^*$$

and

(4d) $$R_a(s) = [(s - \alpha)^2 + \beta^2]^2\frac{F(s)}{G(s)}.$$

Proofs of the formulas

Proof in Case 1. Multiplying (1a) by $s - a$ and using (1d), we have

$$Q_a(s) = A + (s - a)W(s).$$

Letting $s \to a$ and noting that $W(s)$ does not contain the factor $s - a$, we obtain $A = Q_a(a)$, the first formula in (1c). If we write (1d) in the form

$$Q_a(s) = \frac{F(s)}{G(s)/(s - a)}$$

and let $s \to a$, then $F(s) \to F(a)$ while the denominator appears as an indeterminate of the form 0/0; evaluating it in the usual way, we obtain

$$\lim_{s \to a} \frac{G(s)}{s - a} = \lim_{s \to a} \frac{G'(s)}{(s - a)'} = G'(a).$$

This proves the second formula in (1c). The inverse transform of $A/(s - a)$ is Ae^{at}, and the proof is complete. ∎

Proof in Case 2. Multiplying (2a) by $(s - a)^m$ and using (2d), we have

$$Q_a(s) = A_m + (s - a)A_{m-1} + (s - a)^2 A_{m-2} + \cdots + (s - a)^m W(s).$$

Hence $Q_a(a) = A_m$. Differentiation yields

$$Q_a{}'(s) = A_{m-1} + \text{further terms containing factors } s - a.$$

Hence $Q_a{}'(a) = A_{m-1}$ because those terms are zero when $s = a$. Differentiating again we find $Q_a{}''(a) = 2!\,A_{m-2}$, etc. This proves (2c). The inverse transform of $A_k/(s - a)^k$ is $A_k e^{at} t^{k-1}/(k - 1)!$. This yields (2b). ∎

Proof in Case 3. Multiplying (3a) by $(s - \alpha)^2 + \beta^2$, using (3d), and letting $s \to a$, we obtain

$$R_a(a) = aA + B = (\alpha + i\beta)A + B.$$

The real and imaginary parts on both sides must be equal. Using (3c), we thus have

$$S_a = \alpha A + B, \qquad T_a = \beta A.$$

Hence in (3a),

$$\frac{As + B}{(s - \alpha)^2 + \beta^2} = \frac{A(s - \alpha) + \alpha A + B}{(s - \alpha)^2 + \beta^2} = \frac{1}{\beta} \frac{T_a(s - \alpha) + \beta S_a}{(s - \alpha)^2 + \beta^2}.$$

From this and Example 1 in Sec. 5.3 we obtain (3b). ∎

Proof in Case 4. Introducing the notation

$$p(s) = (s - a)(s - \bar{a}) = (s - \alpha)^2 + \beta^2,$$

multiplying (4a) by p^2, and using (4d), we have

$$R_a(s) = As + B + (Cs + D)p(s) + p^2(s)W(s).$$

We now let $s \to a$. Since $p(a) = 0$, we see that then we simply obtain

$$R_a(a) = aA + B = (\alpha + i\beta)A + B. \tag{5}$$

Differentiating $R_a(s)$ with respect to s, we get

$$R_a'(s) = A + (Cs + D)p'(s) + \text{terms containing a factor } p(s).$$

Since $p(a) = 0$ and $p'(a) = 2(a - \alpha) = 2i\beta$, it follows that

$$R_a'(a) = A + 2i\beta[C(\alpha + i\beta) + D]. \tag{6}$$

In (5) and (6) the real and imaginary parts on both sides must be equal. Using (4c) we thus obtain

$$S_a = \alpha A + B, \qquad T_a = \beta A \tag{7}$$

and furthermore

$$S_a^* = A - 2\beta^2 C, \qquad T_a^* = 2\alpha\beta C + 2\beta D,$$

which yields

$$C = (T_a - \beta S_a^*)/2\beta^3, \qquad \alpha C + D = T_a^*/2\beta. \tag{8}$$

From Example 1 in Sec. 5.3 we see that the partial fraction

$$\frac{Cs + D}{(s - \alpha)^2 + \beta^2} = \frac{C(s - \alpha) + \alpha C + D}{(s - \alpha)^2 + \beta^2}$$

in (4a) has the inverse transform

$$e^{\alpha t}[\beta C \cos \beta t + (D + \alpha C) \sin \beta t]/\beta.$$

The other partial fraction

$$\frac{As + B}{[(s - \alpha)^2 + \beta^2]^2} = \frac{A(s - \alpha) + \alpha A + B}{[(s - \alpha)^2 + \beta^2]^2}$$

has the inverse transform

$$e^{\alpha t}[(A\beta^2 t + \alpha A + B) \sin \beta t - (\alpha A + B)\beta t \cos \beta t]/2\beta^3,$$

as follows from Example 1 in Sec. 5.4. By inserting (7) and (8) and simplifying we obtain (4b), and the proof is complete. ▮

Example 1

Find the inverse transformation of

$$Y(s) = \frac{F(s)}{G(s)} = \frac{s+1}{s^3 + s^2 - 6s}.$$

The denominator has the three distinct linear factors s, $s - 2$, and $s + 3$, corresponding to the roots $a_1 = 0$, $a_2 = 2$, and $a_3 = -3$. Hence

$$Y(s) = \frac{s+1}{s(s-2)(s+3)} = \frac{A_1}{s} + \frac{A_2}{s-2} + \frac{A_3}{s+3},$$

and $F(s) = s + 1$, $G'(s) = 3s^2 + 2s - 6$. Using (1c) we thus obtain

$$A_1 = \frac{F(0)}{G'(0)} = -\frac{1}{6}, \qquad A_2 = \frac{F(2)}{G'(2)} = \frac{3}{10}, \qquad A_3 = \frac{F(-3)}{G'(-3)} = -\frac{2}{15}.$$

From this and (1b) we see that the inverse transform is

$$\mathscr{L}^{-1}(Y) = -\tfrac{1}{6} + \tfrac{3}{10}e^{2t} - \tfrac{2}{15}e^{-3t}.$$

Example 2

Solve the initial value problem

$$y'' - 3y' + 2y = 4t + e^{3t}, \qquad y(0) = 1, \qquad y'(0) = -1.$$

From Table 5.1, Sec. 5.1, and (1) and (2) in Sec. 5.2 we see that the subsidiary equation is

$$s^2Y - s + 1 - 3(sY - 1) + 2Y = \frac{4}{s^2} + \frac{1}{s-3}.$$

Solving for Y and representing Y in terms of partial fractions, we have

$$Y(s) = \frac{F(s)}{G(s)} = \frac{s^4 - 7s^3 + 13s^2 + 4s - 12}{s^2(s-3)(s^2 - 3s + 2)}$$

$$= \frac{A_2}{s^2} + \frac{A_1}{s} + \frac{B}{s-3} + \frac{C}{s-2} + \frac{D}{s-1}.$$

We first determine the constants A_1 and A_2 in the partial fractions corresponding to the root $a = 0$. In (2d) and (2c),

$$Q_0(s) = \frac{F(s)}{(s-3)(s^2 - 3s + 2)}, \qquad A_2 = Q_0(0) = 2, \qquad A_1 = Q_0'(0) = 3.$$

Using (1c) we obtain the remaining coefficients

$$B = \left.\frac{F(s)}{s^2(s^2 - 3s + 2)}\right|_{s=3} = \tfrac{1}{2}, \qquad C = \left.\frac{F(s)}{s^2(s-3)(s-1)}\right|_{s=2} = -2,$$

$$D = \left.\frac{F(s)}{s^2(s-3)(s-2)}\right|_{s=1} = -\tfrac{1}{2}.$$

From (1b) and (2b) we see that the solution is

$$y(t) = \mathscr{L}^{-1}(Y) = 2t + 3 + \tfrac{1}{2}e^{3t} - 2e^{2t} - \tfrac{1}{2}e^{t}.$$

The student may solve the problem by the classical method and convince himself that the present approach is much simpler and much more rapid.

Example 3. Forced oscillations, resonance

Solve the initial value problem in Example 2 of Sec. 5.5 by the method of partial fractions.

No resonance. If $p^2 \neq \omega_0{}^2$, the solution $Y(s)$ of the subsidiary equation (2), Sec. 5.5, and its representation in terms of partial fractions are

$$Y(s) = \frac{Kp}{(s^2 + \omega_0{}^2)(s^2 + p^2)} = \frac{As + B}{s^2 + \omega_0{}^2} + \frac{Ms + N}{s^2 + p^2}. \tag{9}$$

We first determine A and B in the fraction corresponding to $a_1 = \alpha_1 + i\beta_1 = i\omega_0$. In (3c),

$$R_{a_1}(a_1) = S_{a_1} + iT_{a_1} = \frac{Kp}{s^2 + p^2}\bigg|_{s=a_1} = \frac{Kp}{p^2 - \omega_0{}^2}$$

so that $T_{a_1} = 0$. From this and (3b) we see that the inverse transform of the first fraction on the right-hand side of (9) is

$$\frac{Kp}{\omega_0(p^2 - \omega_0{}^2)} \sin \omega_0 t.$$

We determine M and N in the last fraction, which corresponds to $a_2 = \alpha_2 + i\beta_2 = ip$. In (3c) we now have

$$R_{a_2}(a_2) = S_{a_2} + iT_{a_2} = \frac{Kp}{s^2 + \omega_0{}^2}\bigg|_{s=a_2} = \frac{Kp}{\omega_0{}^2 - p^2}.$$

Hence $T_{a_2} = 0$. From (3b) we see that the inverse transform of that fraction is

$$-\frac{K}{p^2 - \omega_0{}^2} \sin pt.$$

By adding these two transforms we obtain the desired solution

$$y(t) = \mathscr{L}^{-1}(Y) = \frac{K}{p^2 - \omega_0{}^2}\left(\frac{p}{\omega_0} \sin \omega_0 t - \sin pt\right).$$

Resonance. If $p = \omega_0$, the solution of the subsidiary equation is

$$Y(s) = \frac{K\omega_0}{(s^2 + \omega_0{}^2)^2}.$$

The denominator has the double roots $a = \alpha + i\beta = i\omega_0$ and $\bar{a} = -i\omega_0$, that is, $\alpha = 0$ and $\beta = \omega_0$. In (4c) and (4d),

$$R_a(s) = K\omega_0, \qquad S_a = K\omega_0, \qquad T_a = 0, \qquad R_a{}'(s) = 0,$$

and (4b) yields the solution

$$y(t) = \mathscr{L}^{-1}(Y) = \frac{K}{2\omega_0{}^2}(\sin \omega_0 t - \omega_0 t \cos \omega_0 t).$$

Our present results agree with those in Example 2 of Sec. 5.5. ∎

Systems of differential equations

The Laplace transformation may also be used for solving systems of differential equations. We explain the method in terms of a typical example.

Example 4. Vibration of two masses

The mechanical system in Fig. 90 consists of two masses on three springs and is governed by the differential equations

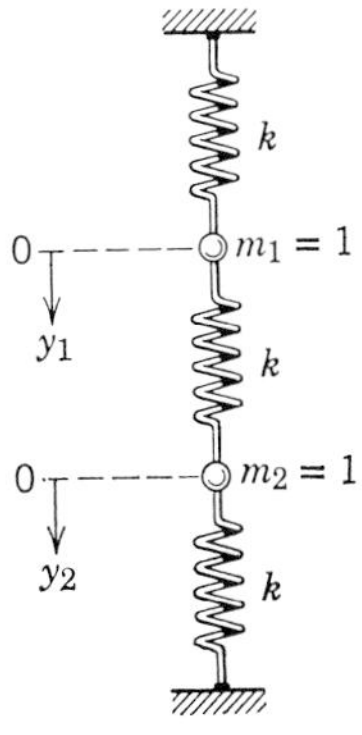

Fig. 90. Example 4

$$(10)\qquad \begin{aligned} y_1'' &= -ky_1 + k(y_2 - y_1) \\ y_2'' &= -k(y_2 - y_1) - ky_2 \end{aligned}$$

where k is the spring modulus of each of the three springs, y_1 and y_2 are the displacements of the masses from their position of static equilibrium; the masses of the springs and the damping are neglected. The derivation of (10) is similar to that of the differential equation in Sec. 2.6. Cf. also Sec. 3.1.

We shall determine the solution corresponding to the initial conditions $y_1(0) = 1$, $y_2(0) = 1$, $y_1'(0) = \sqrt{3k}$, $y_2'(0) = -\sqrt{3k}$. Let $Y_1 = \mathscr{L}(y_1)$ and $Y_2 = \mathscr{L}(y_2)$. Then from (2) in Sec. 5.2 and the initial conditions we obtain the subsidiary equations

$$\begin{aligned} s^2 Y_1 - s - \sqrt{3k} &= -kY_1 + k(Y_2 - Y_1) \\ s^2 Y_2 - s + \sqrt{3k} &= -k(Y_2 - Y_1) - kY_2. \end{aligned}$$

This system of linear algebraic equations in the unknowns Y_1 and Y_2 may be written

$$\begin{aligned} (s^2 + 2k)Y_1 - \quad kY_2 \quad &= s + \sqrt{3k} \\ -kY_1 \quad + (s^2 + 2k)Y_2 &= s - \sqrt{3k}. \end{aligned}$$

Cramer's rule (Sec. 7.9) or elimination yields the solution

$$Y_1 = \frac{(s + \sqrt{3k})(s^2 + 2k) + k(s - \sqrt{3k})}{(s^2 + 2k)^2 - k^2}$$

$$Y_2 = \frac{(s^2 + 2k)(s - \sqrt{3k}) + k(s + \sqrt{3k})}{(s^2 + 2k)^2 - k^2}.$$

The representations in terms of partial fractions are

$$Y_1 = \frac{s}{s^2 + k} + \frac{\sqrt{3k}}{s^2 + 3k}, \qquad Y_2 = \frac{s}{s^2 + k} - \frac{\sqrt{3k}}{s^2 + 3k}.$$

Hence the solution of our initial value problem is

$$\begin{aligned} y_1(t) &= \mathscr{L}^{-1}(Y_1) = \cos\sqrt{k}\,t + \sin\sqrt{3k}\,t \\ y_2(t) &= \mathscr{L}^{-1}(Y_2) = \cos\sqrt{k}\,t - \sin\sqrt{3k}\,t. \end{aligned}$$

Problems for Sec. 5.6

1. Derive (10) in Example 4.

Using partial fractions, find $f(t)$ if $\mathcal{L}(f)$ equals

2. $\dfrac{3s-2}{s^2-s}$ **3.** $\dfrac{1}{(s-a)(s-b)}$ **4.** $\dfrac{s^2+9s-9}{s^3-9s}$

5. $\dfrac{4s+4}{s^2+16}$ **6.** $\dfrac{s}{s^2+2s+2}$ **7.** $\dfrac{s}{(s+1)^2}$

8. $\dfrac{s^3+6s^2+14s}{(s+2)^4}$ **9.** $\dfrac{2s^2-3s}{(s-2)(s-1)^2}$ **10.** $\dfrac{s^2+2s}{(s^2+2s+2)^2}$

11. $\dfrac{s^2-6s+7}{(s^2-4s+5)^2}$ **12.** $\dfrac{3s^2-6s+7}{(s^2-2s+5)^2}$ **13.** $\dfrac{s^3-3s^2+6s-4}{(s^2-2s+2)^2}$

14. Solve Probs. 3 and 10 using convolution formulas.

Show that

15. $\mathcal{L}^{-1}\left\{\dfrac{s^3}{s^4+4a^4}\right\} = \cosh at \cos at$

16. $\mathcal{L}^{-1}\left\{\dfrac{s}{s^4+4a^4}\right\} = \dfrac{1}{2a^2}\sinh at \sin at$

17. $\mathcal{L}^{-1}\left\{\dfrac{s^2}{s^4+4a^4}\right\} = \dfrac{1}{2a}(\cosh at \sin at + \sinh at \cos at)$

18. $\mathcal{L}^{-1}\left\{\dfrac{1}{s^4+4a^4}\right\} = \dfrac{1}{4a^3}(\cosh at \sin at - \sinh at \cos at)$

Solve the following initial value problems by means of the Laplace transformation.

19. $y'' - 3y' + 2y = 6e^{-t}, \quad y(0) = 3, \quad y'(0) = 3$

20. $y'' + 2y' - 3y = 10 \sinh 2t, \quad y(0) = 0, \quad y'(0) = 4$

21. $y'' - 4y = 8t^2 - 4, \quad y(0) = 5, \quad y'(0) = 10$

22. $y'' - 2y' + 5y = 8 \sin t - 4 \cos t, \quad y(0) = 1, \quad y'(0) = 3$

23. $y'' + 2y' + y = 2 \cos t, \quad y(0) = 3, \quad y'(0) = 0$

24. $y'' + y = -2 \sin t, \quad y(0) = 0, \quad y'(0) = 1$

Systems of differential equations. Solve the following initial value problems by means of the Laplace transformation.

25. $y_1' = -y_2, \quad y_2' = y_1, \quad y_1(0) = 1, \quad y_2(0) = 0$

26. $y_1' + y_2 = 2 \cos t, \; y_1 + y_2' = 0, \quad y_1(0) = 0, \; y_2(0) = 1$

27. $y_1'' = y_1 + 3y_2, \quad y_2'' = 4y_1 - 4e^t,$
$y_1(0) = 2, \quad y_1'(0) = 3, \quad y_2(0) = 1, \quad y_2'(0) = 2$

28. $y_1'' + y_2 = -5 \cos 2t, \quad y_2'' + y_1 = 5 \cos 2t,$
$y_1(0) = 1, \quad y_1'(0) = 1, \quad y_2(0) = -1, \quad y_2'(0) = 1$

29. $y_1' + y_2' = 2 \sinh t, \; y_2' + y_3' = e^t, \; y_3' + y_1' = 2e^t + e^{-t},$
$y_1(0) = 1, \; y_2(0) = 1, \; y_3(0) = 0$

30. $-2y_1' + y_2' + y_3' = 0, \; y_1' + y_2' = 4t + 2, \; y_2' + y_3 = t^2 + 2,$
$y_1(0) = y_2(0) = y_3(0) = 0$

5.7 Periodic Functions. Further Applications

Periodic functions appear in many practical problems, and in most cases they are more complicated than single cosine or sine functions. This justifies the topic of the present section, which is a systematic approach to the transformation of periodic functions. The text and the problem set also include further applications.

Let $f(t)$ be a function which is defined for all positive t and has the period p (>0), that is,

$$f(t+p) = f(t) \qquad \text{for all } t > 0.$$

If $f(t)$ is piecewise continuous over an interval of length p, then its Laplace transform exists, and we can write the integral from zero to infinity as the series of integrals over successive periods:

$$\mathscr{L}(f) = \int_0^{\infty} e^{-st} f(t)\, dt = \int_0^{p} e^{-st} f\, dt + \int_p^{2p} e^{-st} f\, dt + \int_{2p}^{3p} e^{-st} f\, dt + \cdots .$$

If we substitute $t = \tau + p$ in the second integral, $t = \tau + 2p$ in the third integral, $\cdots$, $t = \tau + (n-1)p$ in the nth integral, $\cdots$, then the new limits are 0 and p. Since

$$f(\tau + p) = f(\tau), \qquad f(\tau + 2p) = f(\tau),$$

etc., we thus obtain

$$\mathscr{L}(f) = \int_0^{p} e^{-s\tau} f(\tau)\, d\tau + \int_0^{p} e^{-s(\tau+p)} f(\tau)\, d\tau + \int_0^{p} e^{-s(\tau+2p)} f(\tau)\, d\tau + \cdots .$$

The factors which do not depend on τ can be taken out from under the integral signs; this gives

$$\mathscr{L}(f) = [1 + e^{-sp} + e^{-2sp} + \cdots] \int_0^{p} e^{-s\tau} f(\tau)\, d\tau.$$

The series in brackets $[\cdots]$ is a geometric series whose sum is $1/(1 - e^{-ps})$. This establishes the following result.

Theorem 1 (Transform of periodic functions)
The Laplace transform of a piecewise continuous periodic function $f(t)$ with period p is

$$\mathscr{L}(f) = \frac{1}{1 - e^{-ps}} \int_0^{p} e^{-st} f(t)\, dt \qquad (s > 0). \tag{1}$$

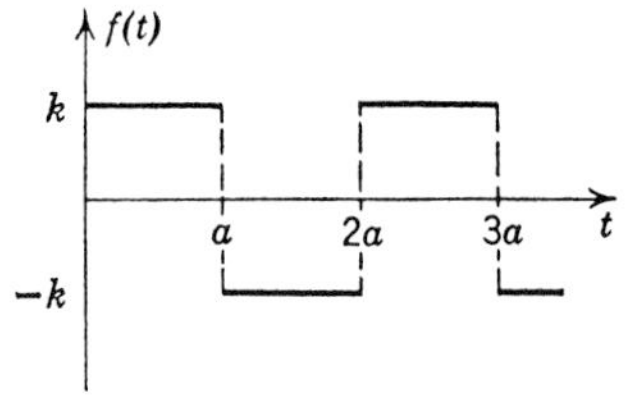

Fig. 91. Example 1

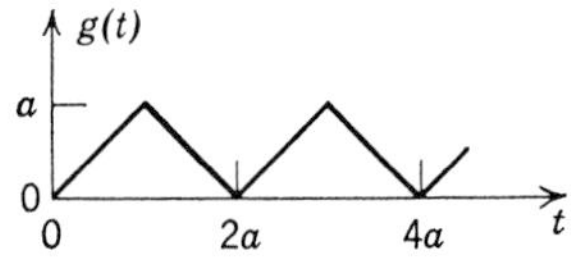

Fig. 92. Example 2

Example 1. Periodic square wave

Find the transform of the square wave shown in Fig. 91. Here $p = 2a$, and from (1) we obtain by direct integration and simplification

$$\begin{aligned}\mathcal{L}(f) &= \frac{1}{1 - e^{-2as}}\left(\int_0^a ke^{-st}\,dt + \int_a^{2a} (-k)e^{-st}\,dt\right) \\ &= \frac{k}{s}\,\frac{1 - 2e^{-as} + e^{-2as}}{(1 + e^{-as})(1 - e^{-as})} = \frac{k}{s}\left(\frac{1 - e^{-as}}{1 + e^{-as}}\right) \\ &= \frac{k}{s}\,\frac{e^{-as/2}(e^{as/2} - e^{-as/2})}{e^{-as/2}(e^{as/2} + e^{-as/2})}.\end{aligned}$$

Hence the result is

$$\mathcal{L}(f) = \frac{k}{s}\tanh\frac{as}{2}.$$

We can also obtain a less elegant but often more useful form of the result if we write

$$\mathcal{L}(f) = \frac{k}{s}\left(\frac{1 - e^{-as}}{1 + e^{-as}}\right) = \frac{k}{s}\left(1 - \frac{2e^{-as}}{1 + e^{-as}}\right) = \frac{k}{s}\left(1 - \frac{2}{e^{as} + 1}\right).$$

Example 2. Periodic triangular wave

Find the transform of the periodic function shown in Fig. 92. We see that $g(t)$ is the integral of the function $f(t)$ with $k = 1$ in Example 1. Hence, by Theorem 3 in Sec. 5.2,

$$\mathcal{L}(g) = \frac{1}{s}\mathcal{L}(f) = \frac{1}{s^2}\tanh\frac{as}{2}.$$

Example 3. Half-wave rectifier

Find the transform of the following function $f(t)$ with period $p = 2\pi/\omega$:

$$f(t) = \begin{cases} \sin \omega t & \text{when} \quad 0 < t < \pi/\omega, \\ 0 & \text{when} \quad \pi/\omega < t < 2\pi/\omega. \end{cases}$$

Note that this function is the half-wave rectification of $\sin \omega t$ (Fig. 93). From (1) we obtain

$$\mathcal{L}(f) = \frac{1}{1 - e^{-2\pi s/\omega}}\int_0^{\pi/\omega} e^{-st}\sin \omega t\,dt.$$

Integrating by parts or noting that the integral is the imaginary part of the integral

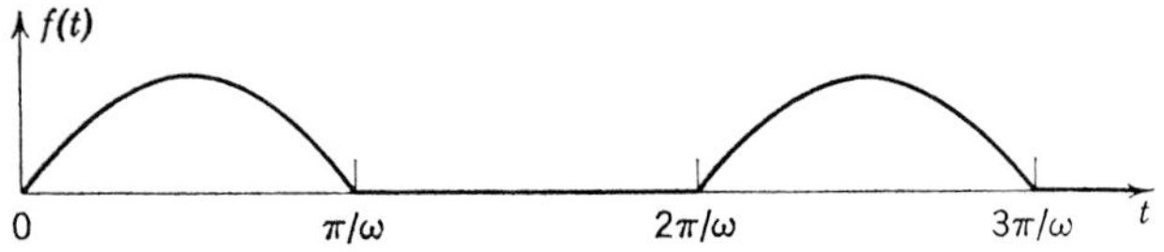

Fig. 93. Half-wave rectification of sin ωt

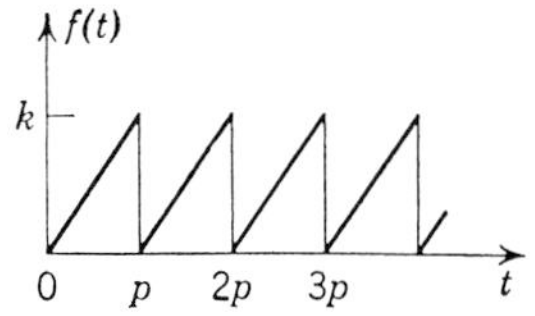

Fig. 94. Saw-tooth wave

$$\int_0^{\pi/\omega} e^{(-s+i\omega)t}\, dt = \frac{1}{-s+i\omega} e^{(-s+i\omega)t}\Big|_0^{\pi/\omega} = \frac{-s-i\omega}{s^2+\omega^2}(-e^{-s\pi/\omega}-1)$$

we obtain the result

$$\mathscr{L}(f) = \frac{\omega(1+e^{-\pi s/\omega})}{(s^2+\omega^2)(1-e^{-2\pi s/\omega})} = \frac{\omega}{(s^2+\omega^2)(1-e^{-\pi s/\omega})}.$$

Example 4. Saw-tooth wave

Find the Laplace transform of the function (Fig. 94)

$$f(t) = \frac{k}{p}t \qquad \text{when } 0 < t < p, \qquad f(t+p) = f(t).$$

Integration by parts yields

$$\int_0^p e^{-st}t\, dt = -\frac{t}{s}e^{-st}\Big|_0^p + \frac{1}{s}\int_0^p e^{-st}\, dt$$

$$= -\frac{p}{s}e^{-sp} - \frac{1}{s^2}(e^{-sp}-1),$$

and from (1) we thus obtain the result

$$\mathscr{L}(f) = \frac{k}{ps^2} - \frac{ke^{-ps}}{s(1-e^{-ps})} \qquad (s > 0).$$

Example 5. Staircase function

Find the Laplace transform of the staircase function (Fig. 95)

$$g(t) = kn \qquad [np < t < (n+1)p, \qquad n = 0, 1, 2, \cdots].$$

Since $g(t)$ is the difference of the functions $h(t) = kt/p$ (whose transform is k/ps^2) and $f(t)$ in Example 4, we obtain

$$\mathscr{L}(g) = \mathscr{L}(h) - \mathscr{L}(f) = \frac{ke^{-ps}}{s(1-e^{-ps})} \qquad (s > 0).$$

The next example illustrates the application of the Laplace transformation to a linear nonhomogeneous differential equation which has a periodic function on the right-hand side.

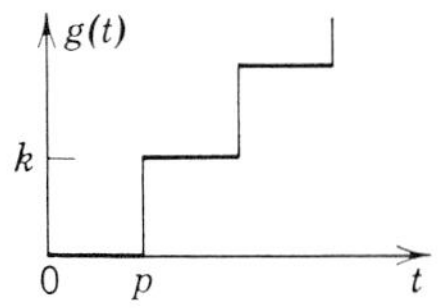

Fig. 95. Staircase function

Example 6. Forced oscillations
Solve the differential equation

$$y'' + 2y' + 10y = r(t) \tag{2}$$

where (Fig. 96)

$$r(t) = \begin{cases} 1 & (0 < t < \pi) \\ -1 & (\pi < t < 2\pi) \end{cases} \qquad r(t + 2\pi) = r(t).$$

The subsidiary equation of (2) is

$$s^2 Y - sy(0) - y'(0) + 2[sY - y(0)] + 10Y = R(s),$$

where $R(s)$ denotes the transform of $r(t)$. By solving for Y we obtain

$$Y(s) = \frac{(s + 2)y(0) + y'(0)}{s^2 + 2s + 10} + \frac{R(s)}{s^2 + 2s + 10}. \tag{3}$$

The roots of the denominator are

$$a = -1 + 3i \qquad \text{and} \qquad \bar{a} = -1 - 3i.$$

The inverse transform of the first term in (3) is

$$\begin{aligned} \mathcal{L}^{-1}\left\{\frac{(s + 2)y(0) + y'(0)}{s^2 + 2s + 10}\right\} &= \mathcal{L}^{-1}\left\{\frac{(s + 1)y(0)}{(s + 1)^2 + 9} + \frac{y(0) + y'(0)}{(s + 1)^2 + 9}\right\} \\ &= e^{-t}\left(y(0)\cos 3t + \frac{y(0) + y'(0)}{3}\sin 3t\right). \end{aligned} \tag{4}$$

This is the solution of the homogeneous equation corresponding to (2). We consider the last term in (3) and determine $R(s)$. We can represent $r(t)$ in terms of step functions (Sec. 5.3):

$$r(t) = u_0(t) - 2u_\pi(t) + 2u_{2\pi}(t) - 2u_{3\pi}(t) + - \cdots.$$

From (3) in Sec. 5.3 we obtain the transform

$$R(s) = \mathcal{L}(r) = \frac{1}{s}[1 - 2e^{-\pi s} + 2e^{-2\pi s} - 2e^{-3\pi s} + - \cdots].$$

Hence

$$\frac{R(s)}{s^2 + 2s + 10} = \frac{1}{s[(s + 1)^2 + 9]}\{1 - 2e^{-\pi s} + - \cdots\}.$$

Now we know that

$$\mathcal{L}^{-1}\left\{\frac{1}{(s + 1)^2 + 9}\right\} = \frac{1}{3}e^{-t}\sin 3t.$$

Therefore, according to Theorem 3 in Sec. 5.2,

$$\mathcal{L}^{-1}\left\{\frac{1}{s}\left(\frac{1}{(s + 1)^2 + 9}\right)\right\} = \frac{1}{3}\int_0^t e^{-\tau}\sin 3\tau \, d\tau = \frac{1}{10}[1 - h(t)] \tag{5}$$

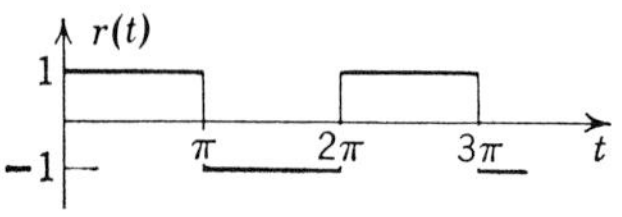

Fig. 96. Input $r(t)$ in Example 6

where

$$h(t) = e^{-t}(\cos 3t + \tfrac{1}{3}\sin 3t). \tag{6}$$

Denoting the function on the right-hand side of (5) by $f(t)$, that is,

$$f(t) = \frac{1}{10} - \frac{h(t)}{10},$$

and applying the second shifting theorem (Sec. 5.3), we thus obtain

$$\mathscr{L}^{-1}\left\{\frac{R(s)}{s^2 + 2s + 10}\right\} = f(t) - 2f(t-\pi)u_\pi(t) + 2f(t-2\pi)u_{2\pi}(t) - + \cdots. \tag{7}$$

Now in the terms on the right,

$$h(t-\pi) = e^{-(t-\pi)}[\cos 3(t-\pi) + \tfrac{1}{3}\sin 3(t-\pi)] = -h(t)e^{\pi}$$

and similarly,

$$h(t-2\pi) = h(t)e^{2\pi}, \qquad h(t-3\pi) = -h(t)e^{3\pi},$$

etc. It follows that in (7),

$$f(t-\pi) = \frac{1}{10} - \frac{h(t-\pi)}{10} = \frac{1}{10} + \frac{h(t)}{10}e^{\pi}$$

$$f(t-2\pi) = \frac{1}{10} - \frac{h(t-2\pi)}{10} = \frac{1}{10} - \frac{h(t)}{10}e^{2\pi}$$

etc., and the right-hand side of (7) equals

$$\frac{1}{10} - \frac{h(t)}{10} \qquad \text{when } 0 < t < \pi,$$

$$-\frac{1}{10} - \frac{h(t)}{10} - \frac{h(t)}{5}e^{\pi} \qquad \text{when } \pi < t < 2\pi,$$

$$\frac{1}{10} - \frac{h(t)}{10} - \frac{h(t)}{5}(e^{\pi} + e^{2\pi}) \qquad \text{when } 2\pi < t < 3\pi,$$

$$\frac{(-1)^n}{10} + \frac{h(t)}{10} - \frac{h(t)}{5}(1 + e^{\pi} + \cdots + e^{n\pi}) \qquad \text{when } n\pi < t < (n+1)\pi.$$

Summing the finite geometric progression in the parentheses, we see that the expression in the last line may be written

$$\frac{(-1)^n}{10} + \frac{h(t)}{10} - \frac{h(t)}{5}\left(\frac{e^{(n+1)\pi} - 1}{e^{\pi} - 1}\right).$$

If we use formula (6), this becomes

$$\begin{aligned} &\left[\frac{1}{10} + \frac{1}{5(e^{\pi}-1)}\right]e^{-t}\left(\cos 3t + \frac{1}{3}\sin 3t\right) \\ &\qquad + \frac{(-1)^n}{10} - \frac{1}{5(e^{\pi}-1)}e^{-[t-(n+1)\pi]}\left(\cos 3t + \frac{1}{3}\sin 3t\right), \end{aligned} \tag{8}$$

and the solution of (2) on the interval $n\pi < t < (n+1)\pi$ is the sum of (8) and the right-hand side of (4).

Clearly the function in (4) and the function in the first line of (8) approach zero as t approaches infinity. The function in the last line of (8) is a periodic function of t with period 2π and thus

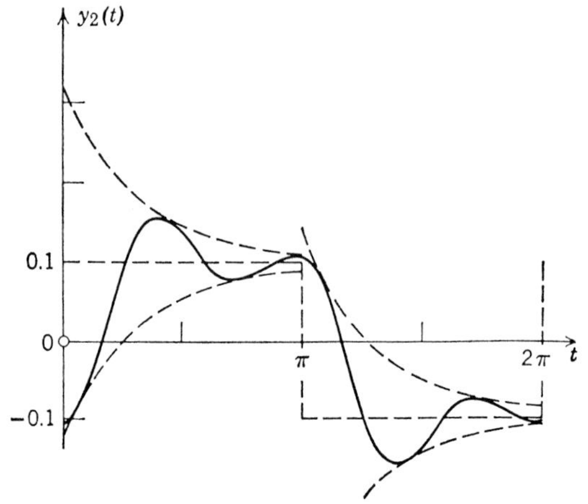

Fig. 97. Steady-state output (9) in Example 6

represents the *steady-state solution* of our problem. To see this more distinctly we denote that function by $y_2(t)$ and set $t - n\pi = \tau$. Then $t = \tau + n\pi$, and τ ranges from 0 to π as t ranges from $n\pi$ to $(n+1)\pi$. Also

$$\cos 3t = \cos(3\tau + 3n\pi) = \cos 3\tau \cos 3n\pi = (-1)^n \cos 3\tau,$$

$$\sin 3t = \sin(3\tau + 3n\pi) = \sin 3\tau \cos 3n\pi = (-1)^n \sin 3\tau,$$

and the function y_2 becomes

$$\begin{aligned} y_2 &= (-1)^n\left[\frac{1}{10} - \frac{e^{-\tau+\pi}}{5(e^{\pi}-1)}\left(\cos 3\tau + \frac{1}{3}\sin 3\tau\right)\right] \\ &\approx (-1)^n[0.1 - 0.22e^{-\tau}\cos(3\tau - 0.32)] \end{aligned} \tag{9}$$

where $0 < \tau < \pi$ and $t = \tau + n\pi$, $n = 0, 1, \cdots$. Figure 97 shows this function; the graph for $0 < t < \pi$ is obtained from (9) with $n = 0$ and $t = \tau$, and for $\pi < t < 2\pi$ it is obtained from (9) with $n = 1$ and $t = \tau + \pi$.

We have the surprising result that the output y_2 tends to oscillate more rapidly than the input $r(t)$. The reason for this unexpected behavior of our system will become obvious from our later consideration of Fourier series in Sec. 10.7.

Problems for Sec. 5.7

1. Apply Theorem 1 to the function $f(t) = 1$, which is periodic with any period p.

2. Show that the solution y_2 in Example 6 is continuous at 0 and π.

3. Find the Laplace transform of the half-wave rectification of $-\sin \omega t$ (Fig. 98).

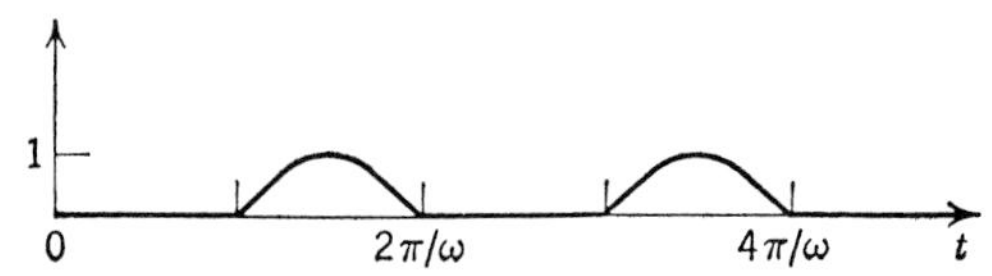

Fig. 98. Problem 3

4. Solve Prob. 3 by applying Theorem 2, Sec. 5.3, to the result of Example 3.

5. Find the Laplace transform of the full-wave rectification of $\sin \omega t$ (Fig. 99).

6. Check the answer to Prob. 5 by using the results of Example 3 and Prob. 3.

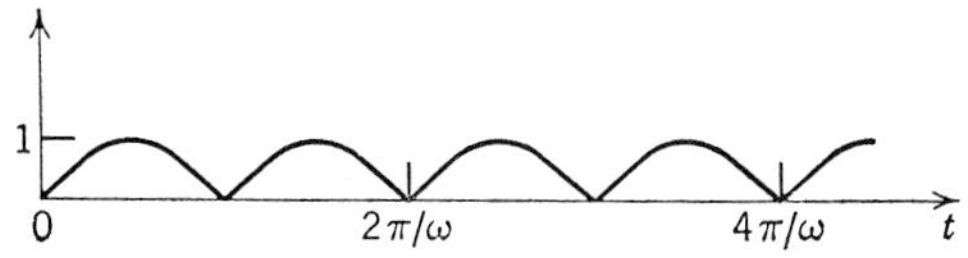

Fig. 99. Problem 5

7. Find the Laplace transform of $|\cos \omega t|$.

Graph the following functions which are assumed to have the period 2π and find their transforms.

8. $f(t) = 2\pi - t \quad (0 < t < 2\pi)$
9. $f(t) = e^t \quad (0 < t < 2\pi)$
10. $f(t) = t^2 \quad (0 < t < 2\pi)$
11. $f(t) = K \sin(t/2) \quad (0 < t < 2\pi)$
12. $f(t) = 1$ when $0 < t < \pi$, $f(t) = -1$ when $\pi < t < 2\pi$
13. $f(t) = t$ when $0 < t < \pi$, $f(t) = \pi - t$ when $\pi < t < 2\pi$
14. $f(t) = 0$ when $0 < t < \pi$, $f(t) = t - \pi$ when $\pi < t < 2\pi$
15. $f(t) = t$ when $0 < t < \pi$, $f(t) = 0$ when $\pi < t < 2\pi$

16. A capacitor ($C = 1$ farad) is charged to the potential $V_0 = 100$ volts and discharged starting at $t = 0$ by closing the switch in Fig. 100. Find the current in the circuit and the charge in the capacitor.

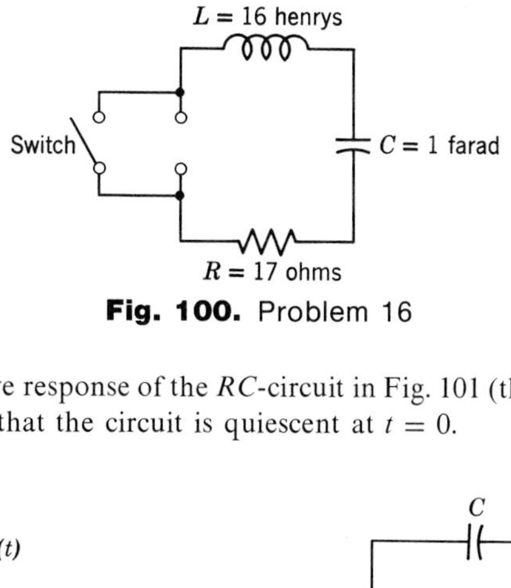

Fig. 100. Problem 16

17. Find the ramp-wave response of the RC-circuit in Fig. 101 (the current $i(t)$ in the circuit), assuming that the circuit is quiescent at $t = 0$.

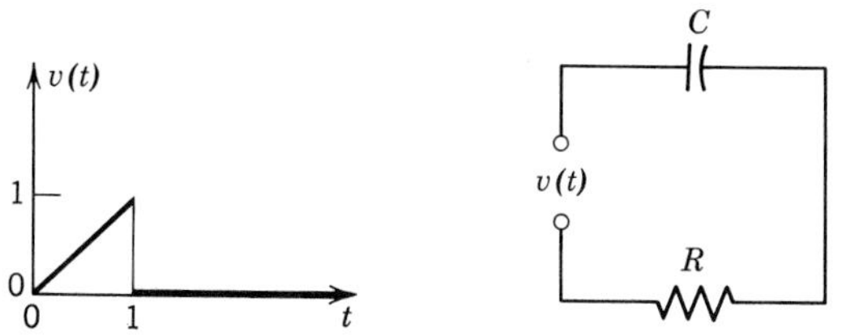

Fig. 101. Problem 17

18. Solve Prob. 17 without the use of the Laplace transformation. Explain the reason for the jump of $i(t)$ at $t = 1$.
19. Find the current $i(t)$ in the RC-circuit in Fig. 101 when $v(t) = \sin \omega t$ $(0 < t < \pi/\omega)$, $v(t) = 0$ $(t > \pi/\omega)$, and $i(0) = 0$.

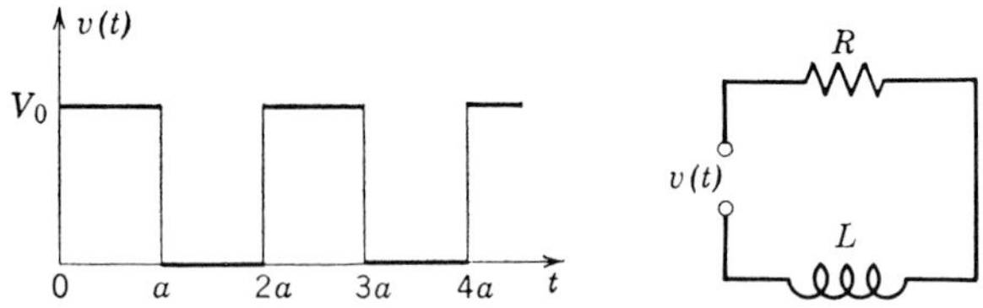

Fig. 102. Problem 20

20. Find the steady-state current in the circuit in Fig. 102.

21. Using the Laplace transformation, show that the current $i(t)$ in the RLC-circuit in Fig. 103 (constant electromotive force V_0, zero initial current and charge) is

$$i(t) = \begin{cases} (K/\omega^*)e^{-\alpha t} \sin \omega^* t & \text{if } \omega^{*2} > 0 \\ Kte^{-\alpha t} & \text{if } \omega^{*2} = 0 \\ (K/\beta)e^{-\alpha t} \sinh \beta t & \text{if } \omega^{*2} = -\beta^2 < 0; \end{cases}$$

here $K = V_0/L$, $\alpha = R/2L$, $\omega^{*2} = (1/LC) - \alpha^2$.

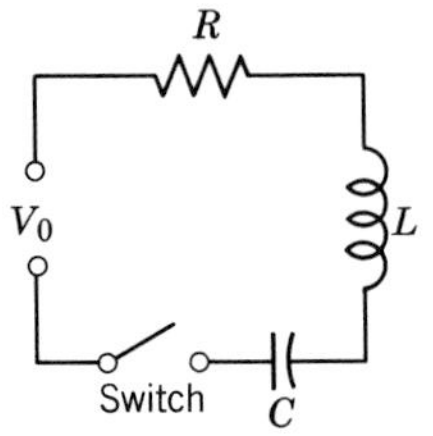

Fig. 103. Problem 21

22. Find the current in the circuit in Prob. 21, assuming that the electromotive force applied at $t = 0$ is $V_0 \sin pt$, and the current and charge at $t = 0$ are zero.

23. Find the current in the RLC-circuit in Fig. 103 if a battery of electromotive force V_0 is connected to the circuit at $t = 0$ and short-circuited at $t = a$. Assume that the initial current and charge are zero, and ω^{*2}, as defined in Prob. 21, is positive.

24. Find the current $i(t)$ in the circuit in Fig. 104, assuming that $i(0) = 0$.

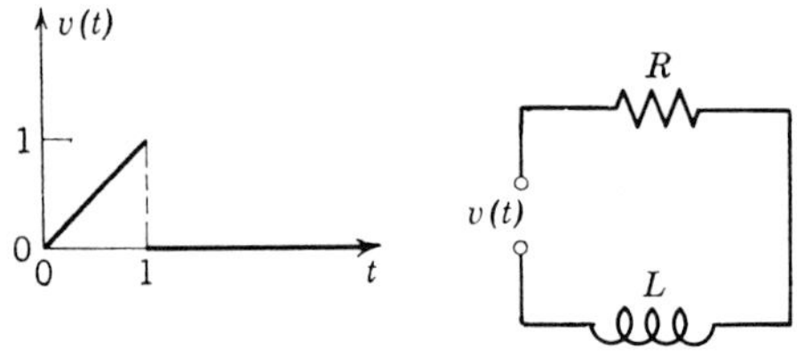

(a) (b)

Fig. 104. Problem 24

25. Steady current is flowing in the circuit in Fig. 105 with the switch closed. At $t = 0$ the switch is opened. Find the current $i(t)$.

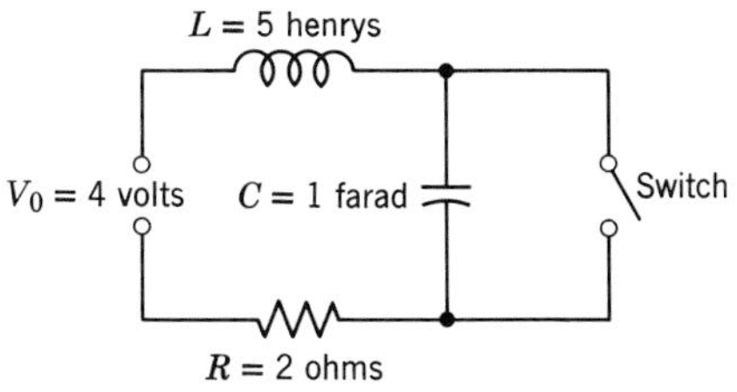

Fig. 105. Problem 25

26. Find the steady-state current in the circuit in Fig. 104*b*, if $v(t) = t$ when $0 < t < 1$ and $v(t + 1) = v(t)$ as shown in Fig. 106.

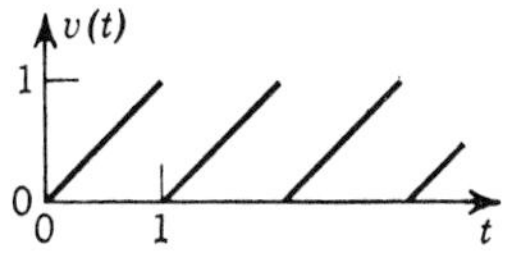

Fig. 106. Problem 26

27. The circuits in Fig. 107 are coupled by mutual inductance M, and at $t = 0$ the currents $i_1(t)$ and $i_2(t)$ are zero. Applying Kirchhoff's voltage law (Sec. 1.9), show that the differential equations for the currents are

$$L_1 i_1' + R_1 i_1 = M i_2' + V_0 u_0(t)$$

$$L_2 i_2' + R_2 i_2 = M i_1'$$

where $u_0(t)$ is the unit step function. Setting $\mathscr{L}(i_1) = I_1(s)$ and $\mathscr{L}(i_2) = I_2(s)$, show that the subsidiary equations are

$$L_1 s I_1 + R_1 I_1 = M s I_2 + \frac{V_0}{s}$$

$$L_2 s I_2 + R_2 I_2 = M s I_1.$$

Assuming that $A \equiv L_1 L_2 - M^2 > 0$, show that the expression for I_2, obtained by solving these algebraic equations, may be written

$$I_2 = \frac{K}{(s + \alpha)^2 + \omega^{*2}}$$

where $K = A^{-1} V_0 M, \alpha = (2A)^{-1}(R_1 L_2 + R_2 L_1), \omega^{*2} = A^{-1} R_1 R_2 - \alpha^2$. Show that $i_2(t) = \mathscr{L}^{-1}(I_2)$ is of the same form as $i(t)$ in Prob. 21, where the constants K, α, and ω^* are now those defined in the present problem.

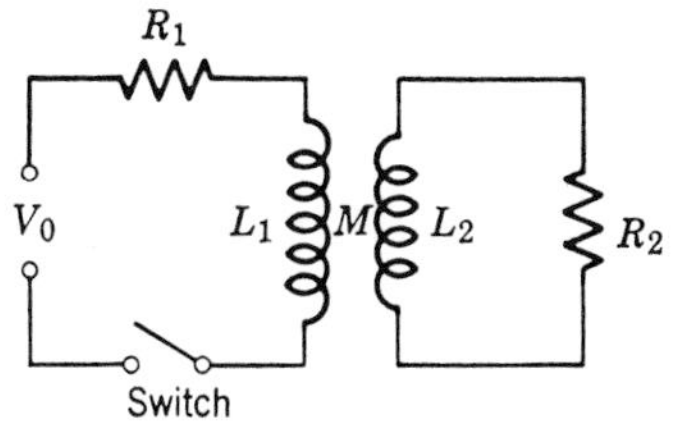

Fig. 107. Problem 27

28. Find $i_1(t)$ in Prob. 27, assuming that $\omega^{*2} > 0$.

29. Show that if $A = 0$ in Prob. 27, then $i_2(t) = K_0 e^{-at}$ where $K_0 = BV_0M$, $a = BR_1R_2$, $B = (R_1L_2 + R_2L_1)^{-1}$.

30. Suppose that in the first circuit in Fig. 107 the switch is closed and a steady current V_0/R_1 is flowing in it. At $t = 0$ the switch is opened. Find the secondary current $i_2(t)$. *Hint.* Note that $i_1(0) = V_0/R_1$ and $i_1 = 0$ when $t > 0$.

31. Figure 108 shows a system for automatic control of pressure. $y(t)$ is the displacement, where $y = 0$ corresponds to the equilibrium position due to a given constant pressure. We make the following assumptions. The damping of the system is proportional to the velocity y'. For $t < 0$, the system is at rest. At $t = 0$ the pressure is suddenly increased in the form of a unit step function. Show that the corresponding differential equation is

$$my'' + cy' + ky = Pu_0(t)$$

(m = effective mass of the moving parts, c = damping constant, k = spring modulus, P = force due to the increase of pressure at $t = 0$). Using the Laplace transformation, show that if $c^2 < 4mk$, then (Fig. 109)

$$y(t) = \frac{P}{k}[1 - e^{-\alpha t}\sqrt{1 + (\alpha/\omega^*)^2}\cos(\omega^* t + \theta)]$$

where $\alpha = c/2m$, $\omega^* = \sqrt{(k/m) - \alpha^2}\,(>0)$, $\tan\theta = -\alpha/\omega^*$.

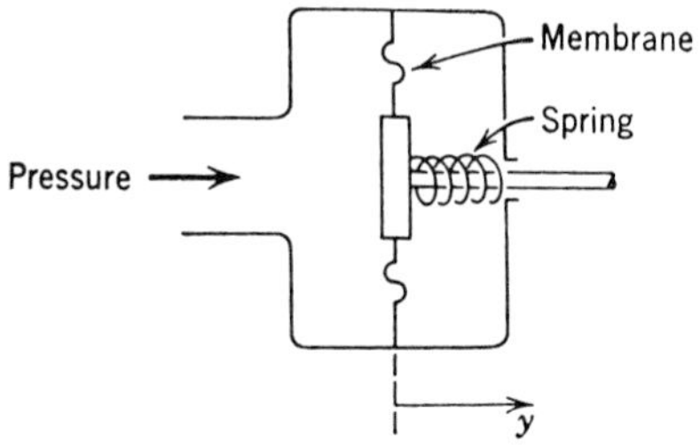

Fig. 108. Problem 31

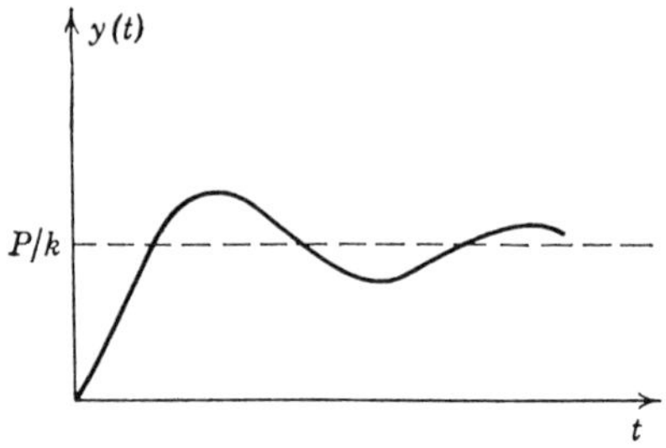

Fig. 109. Displacement $y(t)$ in Problem 31

32. Solve Prob. 31 when $c^2 > 4mk$.

33. Solve Prob. 31 when $c^2 = 4mk$.

34. Two flywheels (moments of inertia M_1 and M_2) are connected by an elastic shaft (moment of inertia negligible) and are rotating with constant angular velocity ω. At $t = 0$ a constant retarding couple P is applied to the first wheel. Find the subsequent angular velocity $v(t)$ of the other wheel.

35. Same conditions as in Prob. 34, but the retarding couple is applied during the interval $0 < t < 1$ only. Find $v(t)$.

5.8 Table of Some Laplace Transforms

Remark. For a more extensive list of Laplace transforms and their inverse transforms, see Ref. [B7] in Appendix 1.

	$F(s) = \mathcal{L}\{f(t)\}$	$f(t)$
1	$1/s$	1
2	$1/s^2$	t
3	$1/s^n, \quad (n = 1, 2, \cdots)$	$t^{n-1}/(n-1)!$
4	$1/\sqrt{s}$	$1/\sqrt{\pi t}$
5	$1/s^{3/2}$	$2\sqrt{t/\pi}$
6	$1/s^a \quad (a > 0)$	$t^{a-1}/\Gamma(a)$
7	$\dfrac{1}{s-a}$	e^{at}
8	$\dfrac{1}{(s-a)^2}$	te^{at}
9	$\dfrac{1}{(s-a)^n} \quad (n = 1, 2, \cdots)$	$\dfrac{1}{(n-1)!} t^{n-1}e^{at}$
10	$\dfrac{1}{(s-a)^k} \quad (k > 0)$	$\dfrac{1}{\Gamma(k)} t^{k-1}e^{at}$
11	$\dfrac{1}{(s-a)(s-b)} \quad (a \neq b)$	$\dfrac{1}{(a-b)}(e^{at} - e^{bt})$
12	$\dfrac{s}{(s-a)(s-b)} \quad (a \neq b)$	$\dfrac{1}{(a-b)}(ae^{at} - be^{bt})$
13	$\dfrac{1}{s^2+\omega^2}$	$\dfrac{1}{\omega}\sin \omega t$
14	$\dfrac{s}{s^2+\omega^2}$	$\cos \omega t$
15	$\dfrac{1}{s^2-a^2}$	$\dfrac{1}{a}\sinh at$
16	$\dfrac{s}{s^2-a^2}$	$\cosh at$
17	$\dfrac{1}{(s-a)^2+\omega^2}$	$\dfrac{1}{\omega}e^{at}\sin \omega t$
18	$\dfrac{s-a}{(s-a)^2+\omega^2}$	$e^{at}\cos \omega t$
19	$\dfrac{1}{s(s^2+\omega^2)}$	$\dfrac{1}{\omega^2}(1 - \cos \omega t)$
20	$\dfrac{1}{s^2(s^2+\omega^2)}$	$\dfrac{1}{\omega^3}(\omega t - \sin \omega t)$
21	$\dfrac{1}{(s^2+\omega^2)^2}$	$\dfrac{1}{2\omega^3}(\sin \omega t - \omega t \cos \omega t)$
22	$\dfrac{s}{(s^2+\omega^2)^2}$	$\dfrac{t}{2\omega}\sin \omega t$
23	$\dfrac{s^2}{(s^2+\omega^2)^2}$	$\dfrac{1}{2\omega}(\sin \omega t + \omega t \cos \omega t)$
24	$\dfrac{s}{(s^2+a^2)(s^2+b^2)} \quad (a^2 \neq b^2)$	$\dfrac{1}{b^2-a^2}(\cos at - \cos bt)$

Table of Some Laplace Transforms (*continued*)

	$F(s) = \mathcal{L}\{f(t)\}$	$f(t)$
25	$\dfrac{1}{s^4 + 4a^4}$	$\dfrac{1}{4a^3}(\sin at \cosh at - \cos at \sinh at)$
26	$\dfrac{s}{s^4 + 4a^4}$	$\dfrac{1}{2a^2}\sin at \sinh at$
27	$\dfrac{1}{s^4 - a^4}$	$\dfrac{1}{2a^3}(\sinh at - \sin at)$
28	$\dfrac{s}{s^4 - a^4}$	$\dfrac{1}{2a^2}(\cosh at - \cos at)$
29	$\sqrt{s-a} - \sqrt{s-b}$	$\dfrac{1}{2\sqrt{\pi t^3}}(e^{bt} - e^{at})$
30	$\dfrac{1}{\sqrt{s+a}\sqrt{s+b}}$	$e^{-(a+b)t/2} I_0\left(\dfrac{a-b}{2}t\right)$ (cf. Sec. 4.6)
31	$\dfrac{1}{\sqrt{s^2 + a^2}}$	$J_0(at)$ (cf. Sec. 4.5)
32	$\dfrac{s}{(s-a)^{3/2}}$	$\dfrac{1}{\sqrt{\pi t}} e^{at}(1 + 2at)$
33	$\dfrac{1}{(s^2 - a^2)^k}$ $(k > 0)$	$\dfrac{\sqrt{\pi}}{\Gamma(k)}\left(\dfrac{t}{2a}\right)^{k-1/2} I_{k-1/2}(at)$ (cf. Sec. 4.6)
34	$\dfrac{1}{s} e^{-k/s}$	$J_0(2\sqrt{kt})$ (cf. Sec. 4.5)
35	$\dfrac{1}{\sqrt{s}} e^{-k/s}$	$\dfrac{1}{\sqrt{\pi t}}\cos 2\sqrt{kt}$
36	$\dfrac{1}{s^{3/2}} e^{k/s}$	$\dfrac{1}{\sqrt{\pi k}}\sinh 2\sqrt{kt}$
37	$e^{-k\sqrt{s}}$ $(k > 0)$	$\dfrac{k}{2\sqrt{\pi t^3}} e^{-k^2/4t}$
38	$\dfrac{1}{s}\ln s$	$-\ln t - \gamma$ $(\gamma \approx 0.5772;$ cf. Sec. 4.6)
39	$\ln \dfrac{s-a}{s-b}$	$\dfrac{1}{t}(e^{bt} - e^{at})$
40	$\ln \dfrac{s^2 + \omega^2}{s^2}$	$\dfrac{2}{t}(1 - \cos \omega t)$
41	$\ln \dfrac{s^2 - a^2}{s^2}$	$\dfrac{2}{t}(1 - \cosh at)$
42	$\text{arc tan}\, \dfrac{\omega}{s}$	$\dfrac{1}{t}\sin \omega t$
43	$\dfrac{1}{s}\,\text{arc cot}\, s$	$\text{Si}(t)$ (cf. Appendix 3)

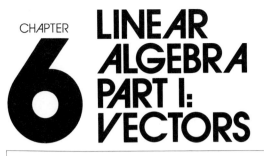

CHAPTER 6 LINEAR ALGEBRA PART I: VECTORS

Two main factors have affected the development of engineering mathematics during the past two decades. These factors are the widespread application of automatic computers to engineering problems and the increasing use of linear algebra and linear analysis, for instance, in handling large problems in systems analysis.

The next four chapters (6–9) are devoted to linear algebra (Chaps. 6, 7) and linear analysis (Chaps. 8, 9). The first two of these chapters include the theory and application of vectors and vector spaces (Secs. 6.1–6.4), dot products and inner product spaces (Sec. 6.5, 6.6), vector products and scalar triple products (Secs. 6.7–6.9), matrices and linear equations (Secs. 7.1–7.8), determinants and Cramer's rule (Secs. 7.9–7.11), quadratic and Hermitian forms (Sec. 7.12) and eigenvalues and eigenvectors (Secs. 7.13, 7.14). Section 7.15 is devoted to the use of matrices in connection with systems of differential equations. Chapter 8 contains vector differential calculus (scalar and vector fields, curves, velocity, directional derivative, gradient, divergence, curl), and Chap. 9 is devoted to vector integral calculus (line, surface, and triple integrals and their transformation by means of the integral theorems of Green, Gauss, and Stokes).

In the present chapter we shall consider basic concepts and methods of vector algebra and its applications to physical and geometrical problems. The usefulness of vectors in engineering mathematics results from the fact that many physical quantities—for example, forces and velocities—may be represented by vectors, and in several respects the rules of vector calculation are as simple as the rules governing the system of real numbers.

It is true that any problem that can be solved by the use of vectors can also be treated by nonvectorial methods, but vector analysis is a shorthand which simplifies many calculations considerably. Furthermore, it is a way of visualizing physical and geometrical quantities and relations between them. For all those reasons extensive use is made of vector notation in modern engineering literature.

Prerequisite for this chapter. In the last two sections, determinants of second and third order will be needed.

Section which may be omitted in a shorter course: Sec. 6.6.

References: Appendix 1, Part C.

Answers to problems: Appendix 2.

6.1 Scalars and Vectors

In physics and geometry there are quantities each of which is completely specified when its magnitude—that is, its size or number of units according to

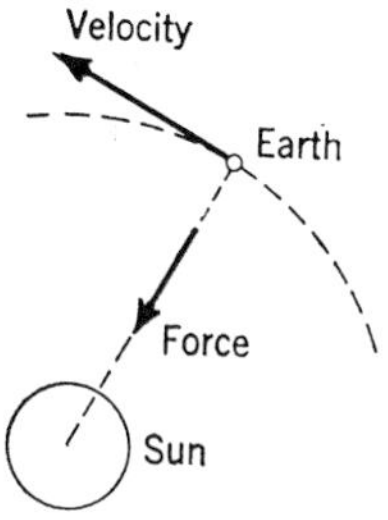

Fig. 110. Force and velocity

some scale—is given. Examples are the mass of a body, the charge of an electron, the specific heat of water, the resistance of a resistor, the diameter of a circle, the area of a triangle, and the volume of a cube. Each of these quantities is described by a single number (after a suitable choice of units of measure). Such a quantity is called a **scalar.**[1]

However, there are other physical and geometrical quantities which cannot be described by a single number, because they require for their complete characterization the specification of a direction as well as a magnitude.

For example, forces in mechanics are quantities of this type. We know that we may represent a force graphically by an arrow, or *directed line segment,* which indicates the direction of the force and whose length is equal to the magnitude of the force according to some suitable scale. Figure 110 shows the force of attraction for the earth's motion around the sun. The instantaneous velocity of the earth may also be represented by an arrow of suitable length and direction, and this illustrates that a velocity is also a quantity which is characterized by a magnitude and a direction.

Figure 111 shows a translation (displacement without rotation) of a triangle in the plane. This motion can be characterized by its magnitude (distance travelled by each point of the triangle) and its direction. We may represent the translation graphically by a directed line segment whose *initial point* is the original position P of a point of the triangle and whose *terminal point* is the new position Q of that point after the translation. If we do this for each point of the triangle, we obtain a family of directed line segments which have the *same length* and the *same direction* (that is, are parallel and are directed in the same sense). We may say that each of these directed line segments "carries" a point of the triangle from its original position to its new position.

This situation suggests the following definition of a vector.

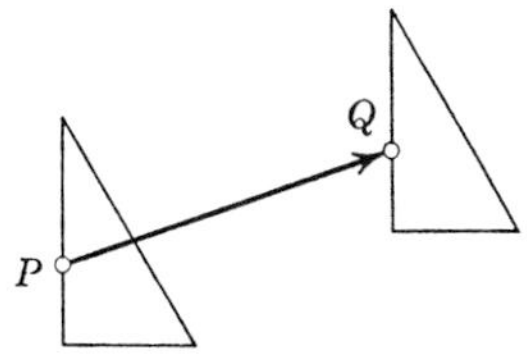

Fig. 111. Translation

[1]Clearly, if a number characterizes a scalar which has a physical or geometrical meaning, this number should be independent of the choice of coordinates; since this point will not matter in the present chapter, we may discuss it later (in Sec. 8.1).

Definition of a vector

A directed line segment is called a **vector.** Its length is called the **length** of the vector and its direction is called the **direction** of the vector. Two vectors are equal if and only if they have the same length and the same direction.

The length of a vector is also called the **Euclidean norm** or **magnitude** of the vector. ▮

Vectors will be denoted by lowercase boldfaced letters,[2] such as **a, b, v,** and the length of a vector **a** by $|\mathbf{a}|$.

Of course, if we want, we may combine the two parts of this definition and define a vector to be the collection of all directed line segments having a given length and a given direction.

From the definition we see that a vector may be arbitrarily translated (that is, displaced without rotation) or, what amounts to the same thing, its initial point may be chosen in an arbitrary fashion. Clearly, if we choose a certain point as the initial point of a given vector, the terminal point of the vector is uniquely determined.

If two vectors **a** and **b** are equal, we write

$$\mathbf{a} = \mathbf{b},$$

and if they are different, we may write

$$\mathbf{a} \neq \mathbf{b}.$$

Any vector may be represented graphically as an arrow of suitable length and direction, as is illustrated in Fig. 112.

A vector of length 1 is called a **unit vector.**

For the sake of completeness, we mention that in physics and geometry there are situations where we want to impose restrictions on the position of the initial point of a vector. For example, as is known from mechanics, a force acting on a rigid body may be equally well applied to any point of the body on its line of action. This suggests the concept of a **sliding vector,** that is, a vector whose initial point can be any point on a straight line which is parallel to the vector. A force acting on an elastic body is a vector whose initial point cannot be changed at all. In fact, if we choose another point of application of the force, the effect of the force will in general be different. This suggests the notion of a **bound vector,** that is, a vector having a certain fixed initial point (*point of application*). Since these concepts occur in the literature, the student should know about them, but they will not be of particular importance in our further considerations.

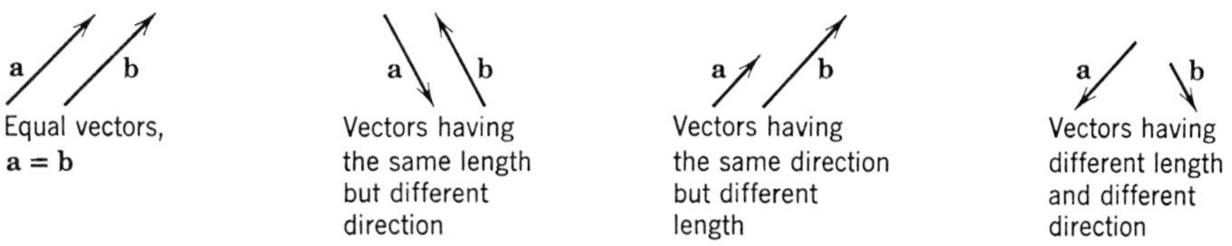

Fig. 112. Vectors

[2]This is customary in printed work; in handwritten work one may characterize vectors by arrows, for example $\vec{a}$ (in place of **a**), $\vec{b}$, etc.

6.2 Components of a Vector

A point in three-dimensional space is a geometric object, but if we introduce a coordinate system, we may describe it by (or even regard it as) an ordered triple of numbers (called its coordinates). Similarly, in the last section we defined a vector in a geometrical fashion (using the notion of a directed line segment), but if we use a coordinate system, we may describe vectors in algebraic terms.

In fact, let us introduce a coordinate system in space whose axes are three mutually perpendicular straight lines. On all three axes we choose the same scale. Then the three *unit points* on the axes, whose coordinates are $(1, 0, 0)$, $(0, 1, 0)$, and $(0, 0, 1)$, have the same distance from the *origin*, the point of intersection of the axes. The rectangular coordinate system thus obtained is called a **Cartesian coordinate system** in space (cf. Fig. 113).

We consider now a vector **a** obtained by directing a line segment PQ such that P is the initial point and Q is the terminal point (Fig. 114). Let (x_1, y_1, z_1) and (x_2, y_2, z_2) be the coordinates of P and Q, respectively. Then the numbers

$$(1) \qquad a_1 = x_2 - x_1, \qquad a_2 = y_2 - y_1, \qquad a_3 = z_2 - z_1$$

are called the **components** of the vector **a** with respect to that Cartesian coordinate system.

By definition, the length $|\mathbf{a}|$ of the vector **a** is the distance $\overline{PQ}$, and from (1) and the theorem of Pythagoras it follows that

$$(2) \qquad |\mathbf{a}| = \sqrt{a_1^2 + a_2^2 + a_3^2}.$$

Example 1. Components and length of a vector

The vector **a** with initial point P: $(3, 1, 4)$ and terminal point Q: $(1, -2, 4)$ has the components

$$a_1 = 1 - 3 = -2, \qquad a_2 = -2 - 1 = -3, \qquad a_3 = 4 - 4 = 0$$

and the length

$$|\mathbf{a}| = \sqrt{(-2)^2 + (-3)^2 + 0^2} = \sqrt{13}.$$

If we choose $(-1, 5, 8)$ as the initial point of **a**, then the corresponding terminal point is $(-3, 2, 8)$.

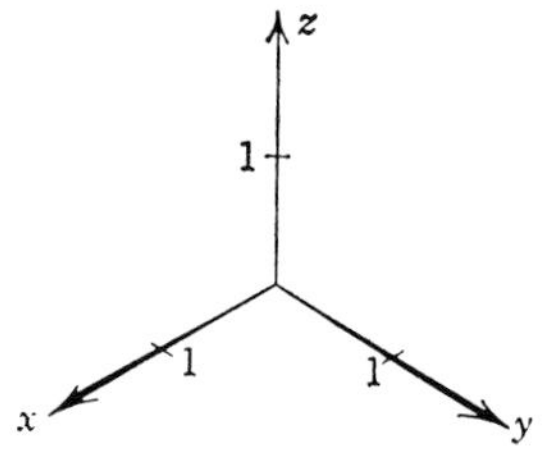

Fig. 113. Cartesian coordinate system

Observe from (1) that if we choose the initial point of a vector to be the origin, then its components are equal to the coordinates of the terminal point and the vector is then called the **position vector** of the terminal point (with respect to our coordinate system) and is usually denoted by **r.** Cf. Fig. 115.

From (1) we also see immediately that *the components* a_1, a_2, a_3 *of the vector* **a** *are independent of the choice of the initial point of* **a,** because if we translate **a,** then corresponding coordinates of P and Q are altered by the same amount. *Hence, a fixed Cartesian coordinate system being given, each vector is uniquely determined by the ordered triple of its components with respect to that coordinate system.*

We now introduce the *null vector or* **zero vector 0,** defined as the vector with components 0, 0, 0. Then any ordered triple of real numbers, including the triple 0, 0, 0, can be chosen as the components of a vector.

The coordinate system being fixed, the correspondence between the ordered triples of real numbers and the vectors in space is one-to-one, that is, to each such triple there corresponds a vector in space that has those three numbers as its components with respect to that coordinate system, and conversely.

It follows that two vectors **a** *and* **b** *are equal if and only if corresponding components of these vectors are equal. Consequently, a vector equation*

$$\mathbf{a} = \mathbf{b}$$

is equivalent to the three equations

$$a_1 = b_1, \qquad a_2 = b_2, \qquad a_3 = b_3$$

for the components a_1, a_2, a_3 *and* b_1, b_2, b_3 *of the vectors with respect to a given Cartesian coordinate system.*

There is yet another consequence of that one-to-one correspondence between all ordered triples of real numbers and all vectors as defined in Sec. 6.1 in a geometrical fashion. That correspondence implies that a vector could equally well be *defined* as an ordered triple of real numbers (called the *components* of the vector). Starting from this definition, we would then obtain the geometrical interpretation of a vector which we used as a definition in Sec. 6.1.

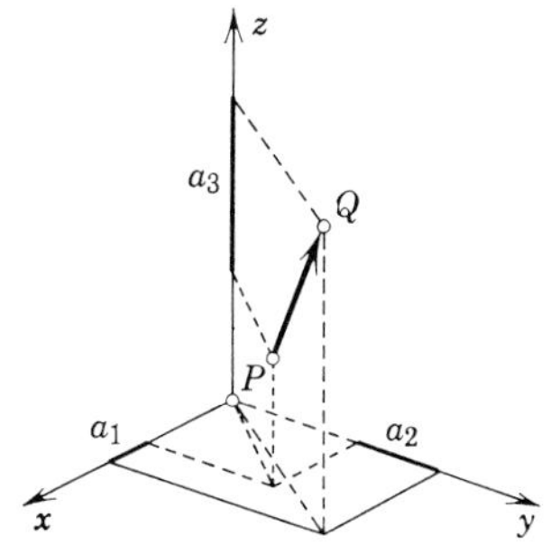

Fig. 114. Components of a vector

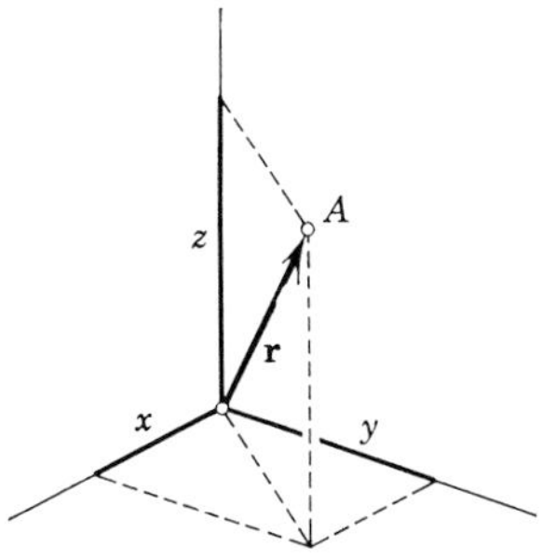

Fig. 115. Position vector **r** of a point A having the coordinates x, y, z

Clearly, if a vector is a physical or geometric object, its length and direction should be independent of the particular choice of a coordinate system, and this requirement implies a certain transformation behavior of the components under the transition to another coordinate system. We shall discuss this problem later (in Sec. 8.9).

In the next chapter we shall also generalize the concept of a vector to include ordered n-tuples of numbers, where n is any fixed positive integer.

Problems for Sec. 6.2

Find the components of the vector $\mathbf{v}$ with given initial point P: (x_1, y_1, z_1) and terminal point Q: (x_2, y_2, z_2). Find $|\mathbf{v}|$. Graph $\mathbf{v}$.

1. P: $(1, 2, 0)$, Q: $(3, 2, 1)$

2. P: $(0, 1, -1)$, Q: $(0, 4, 2)$

3. P: $(-1, 2, -2)$, Q: $(0, 3, -1)$

4. P: $(1, 1, 1)$, Q: $(3, 3, 3)$

5. P: $(-2, -3, -1)$, Q: $(0, 0, 0)$

6. P: $(2, 4, 6)$, Q: $(1, 2, 3)$

7. P: $(0, 2, 0)$, Q: $(0, 1, 0)$

8. P: $(0, 0, 0)$, Q: $(-1, 2, 0)$

9. P: $(3, 8, 2)$, Q: $(3, 8, 2)$

10. P: $(3, 8, 4)$, Q: $(3, 9, 4)$

In each case, the components v_1, v_2, v_3 of a vector $\mathbf{v}$ and a particular initial point P are given. Find the corresponding terminal point.

11. $-1, 2, 5$; P: $(8, 2, 0)$

12. $3, 8, 0$; P: $(-4, 1, 2)$

13. $\frac{1}{2}, 2, -1$; P: $(-2, \frac{1}{2}, 0)$

14. $1, 0, 0$; P: $(0, 0, 0)$

15. $4, -1, 6$; P: $(-4, 1, -6)$

16. $0, 0, 0$; P: $(1, 4, 7)$

17. $3, 8, -9$; P: $(0, 0, 0)$

18. $1, 2, 3$; P: $(1, 2, 3)$

19. $-1, -2, 1$; P: $(-2, 0, -3)$

20. $8, -6, -1$; P: $(-8, 0, 4)$

6.3 Addition of Vectors, Multiplication by Scalars

Since we want to do calculations with vectors, we should now introduce algebraic operations for vectors. We shall define two such operations, which are called addition of vectors and multiplication of vectors by scalars.

Experiments show that the resultant of two forces can be determined by the familiar parallelogram law (Fig. 116). This suggests and motivates the following

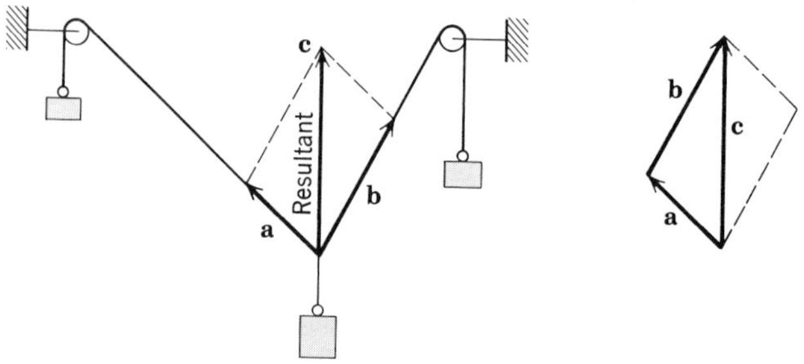

Fig. 116. Resultant of two forces (parallelogram law)

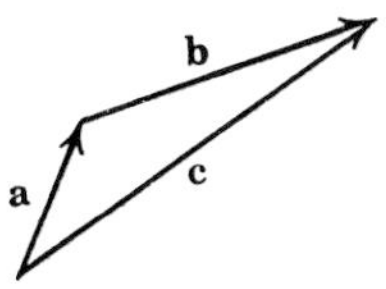

Fig. 117. Vector addition

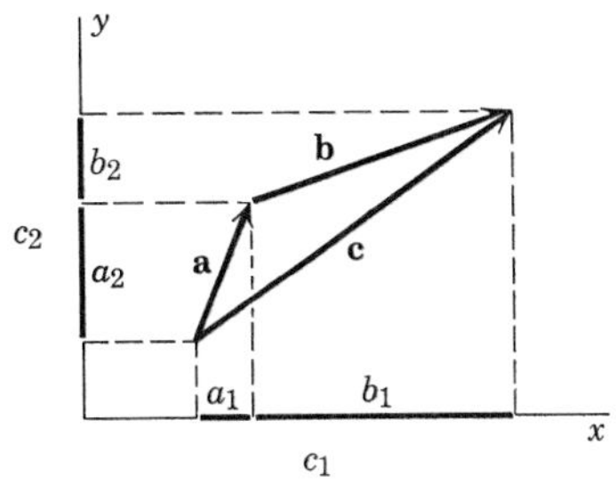

Fig. 118. Vector addition in terms of components

Definition of vector addition

Given two vectors **a** and **b**, put the initial point of **b** at the terminal point of **a**; then the **sum** of **a** and **b** is defined as the vector **c** drawn from the initial point of **a** to the terminal point of **b** (Fig. 117), and we write

$$\mathbf{c} = \mathbf{a} + \mathbf{b}.$$ ∎

From the definition we conclude the following. If in some fixed coordinate system, **a** has the components a_1, a_2, a_3 and **b** has the components b_1, b_2, b_3, then the components c_1, c_2, c_3 of the sum vector $\mathbf{c} = \mathbf{a} + \mathbf{b}$ are obtained by the addition of corresponding components of **a** and **b**; thus

$$c_1 = a_1 + b_1, \qquad c_2 = a_2 + b_2, \qquad c_3 = a_3 + b_3. \tag{1}$$

Figure 118 shows this for the plane, and in space the situation is similar.

From the definition or from (1) we see that vector addition has the following properties. For any vectors (cf. Figs. 119, 120)

(2)

(a) $\mathbf{a} + \mathbf{b} = \mathbf{b} + \mathbf{a}$ (*commutativity*)

(b) $(\mathbf{u} + \mathbf{v}) + \mathbf{w} = \mathbf{u} + (\mathbf{v} + \mathbf{w})$ (*associativity*)

(c) $\mathbf{a} + \mathbf{0} = \mathbf{0} + \mathbf{a} = \mathbf{a}$

(d) $\mathbf{a} + (-\mathbf{a}) = \mathbf{0}$

where $-\mathbf{a}$ denotes the vector having the length $|\mathbf{a}|$ and the direction opposite to that of **a**.

In (2b) we may simply write $\mathbf{u} + \mathbf{v} + \mathbf{w}$, and similarly for sums of more than

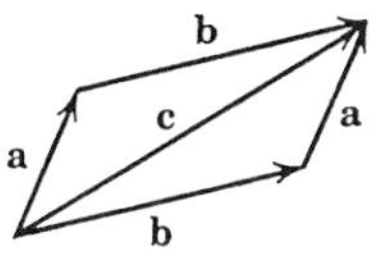

Fig. 119. Commutativity of vector addition

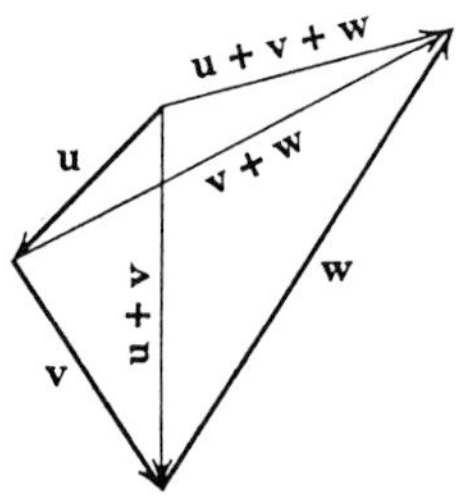

Fig. 120. Associativity of vector addition

three vectors. Instead of $\mathbf{a} + \mathbf{a}$ we also write $2\mathbf{a}$, and so on. This (and the notation $-\mathbf{a}$ used before) suggests that we define the second algebraic operation for vectors as follows.

Multiplication of vectors by scalars (numbers)

Let $\mathbf{a}$ be any vector and q any real number. Then the vector $q\mathbf{a}$ is defined as follows.

The length of $q\mathbf{a}$ is $|q||\mathbf{a}|$.
If $\mathbf{a} \neq \mathbf{0}$ and $q > 0$, then $q\mathbf{a}$ has the direction of $\mathbf{a}$.
If $\mathbf{a} \neq \mathbf{0}$ and $q < 0$, then $q\mathbf{a}$ has the direction opposite to $\mathbf{a}$.
If $\mathbf{a} = \mathbf{0}$ or $q = 0$ (or both), then $q\mathbf{a} = \mathbf{0}$. ▮

Simple examples are shown in Fig. 121.

Clearly, if $\mathbf{a}$ has the components a_1, a_2, a_3, then $q\mathbf{a}$ has the components qa_1, qa_2, qa_3 (with respect to the same coordinate system). Furthermore, from the definitions it follows that for any vectors and real numbers,

$$\text{(3)} \quad \begin{aligned} &\text{(a)} \quad q(\mathbf{a} + \mathbf{b}) = q\mathbf{a} + q\mathbf{b} \\ &\text{(b)} \quad (c + k)\mathbf{a} = c\mathbf{a} + k\mathbf{a} \\ &\text{(c)} \quad c(k\mathbf{a}) = (ck)\mathbf{a} \qquad \text{(written } ck\mathbf{a}\text{)} \\ &\text{(d)} \quad 1\mathbf{a} = \mathbf{a}. \end{aligned}$$

The reader may prove that (2) and (3) imply that for any $\mathbf{a}$,

$$\text{(4)} \quad \begin{aligned} &\text{(a)} \quad 0\mathbf{a} = \mathbf{0} \\ &\text{(b)} \quad (-1)\mathbf{a} = -\mathbf{a}. \end{aligned}$$

Instead of $\mathbf{b} + (-\mathbf{a})$ we simply write $\mathbf{b} - \mathbf{a}$ (Fig. 122).

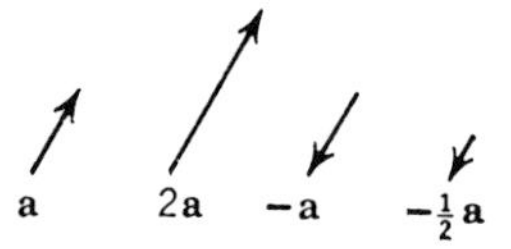

Fig. 121. Multiplication of vectors by numbers

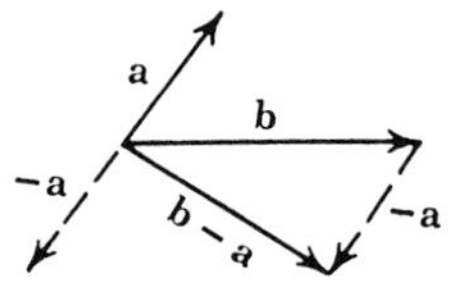

Fig. 122. Difference of vectors

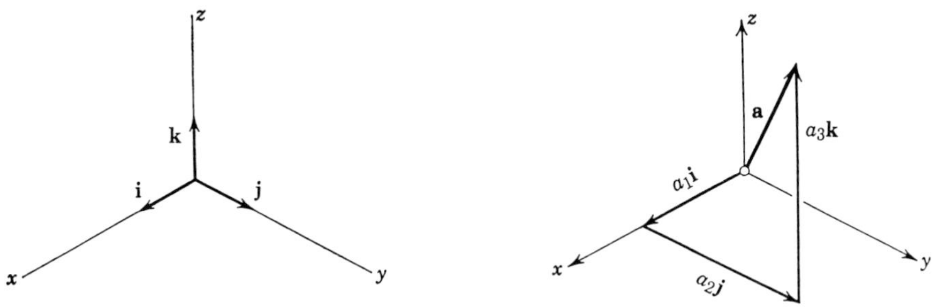

Fig. 123. The unit vectors **i, j, k** and the representation (5)

A Cartesian coordinate system being given, we may now represent a vector **a** with components a_1, a_2, a_3 as the sum of three vectors parallel to the coordinate axes. For this purpose we associate with that coordinate system three unit vectors **i, j, k** which have the positive directions of the three coordinate axes. Then (Fig. 123)

$$\mathbf{a} = a_1\mathbf{i} + a_2\mathbf{j} + a_3\mathbf{k}. \tag{5}$$

Figure 123 shows the vectors **i, j, k** when the origin is chosen as their common initial point. These vectors are mutually perpendicular. Instead of *perpendicular* the term **orthogonal** is also used, and we say that **i, j, k** form a *triple of orthogonal unit vectors.* This triple is also known as the *fundamental orthogonal triad* associated with that coordinate system.

Example 1

With respect to a given coordinate system, let

$$\mathbf{a} = 4\mathbf{i} + \mathbf{k} \quad \text{and} \quad \mathbf{b} = 2\mathbf{i} - 4\mathbf{j} + 3\mathbf{k}.$$

Then

$$2\mathbf{b} = 4\mathbf{i} - 8\mathbf{j} + 6\mathbf{k}, \quad -\mathbf{b} = -2\mathbf{i} + 4\mathbf{j} - 3\mathbf{k}, \quad 1.5\mathbf{a} - \mathbf{b} = 4\mathbf{i} + 4\mathbf{j} - 1.5\mathbf{k}.$$

Problems for Sec. 6.3

Let $\mathbf{a} = \mathbf{i} + 2\mathbf{j} - 3\mathbf{k}$, $\mathbf{b} = 2\mathbf{i} + \mathbf{j} + 4\mathbf{k}$, $\mathbf{c} = -5\mathbf{j}$. Find

1. $-2\mathbf{a}, \frac{1}{2}\mathbf{a}, 3\mathbf{a}$
2. $\mathbf{a} + \mathbf{b}, \mathbf{b} + \mathbf{a}$
3. $\mathbf{a} - \mathbf{b}, \mathbf{b} - \mathbf{a}, \mathbf{b} - \mathbf{c}$
4. $|\mathbf{a} + \mathbf{b}|, |\mathbf{a}| + |\mathbf{b}|$
5. $|\mathbf{a} - \mathbf{c}|, |\mathbf{a}| - |\mathbf{c}|$
6. $\mathbf{a}/|\mathbf{a}|, \mathbf{b}/|\mathbf{b}|, \mathbf{c}/|\mathbf{c}|$
7. $3\mathbf{a} - 3\mathbf{b}, 3(\mathbf{a} - \mathbf{b})$
8. $|-\mathbf{a} + 2\mathbf{b} - 3\mathbf{c}|$
9. $(\mathbf{a} + \mathbf{b}) + \mathbf{c}, \mathbf{a} + (\mathbf{b} + \mathbf{c})$

10. What laws do Probs. 2, 7 and 9 illustrate?

Find the resultant of the following forces.

11. $\mathbf{p} = \mathbf{i} + 3\mathbf{j} - \mathbf{k}, \quad \mathbf{q} = 5\mathbf{i} - 2\mathbf{k}, \quad \mathbf{u} = -\mathbf{j} + 3\mathbf{k}$

12. $\mathbf{p} = 3\mathbf{i} + 4\mathbf{j} + 2\mathbf{k}, \quad \mathbf{q} = -2\mathbf{i} - 5\mathbf{j} - 3\mathbf{k}, \quad \mathbf{u} = \mathbf{k}$

13. Determine a force **p** such that the forces **p**, $\mathbf{q} = \mathbf{i} + \mathbf{j} - 2\mathbf{k}$, and $\mathbf{u} = -\mathbf{i} + 2\mathbf{j}$ are in equilibrium.

14. Let **a, b, c** be the edge vectors of a parallelepiped. Find the vectors corresponding to the four diagonals.

Using vectors, prove the following statements.

15. The diagonals of a parallelogram bisect each other.

16. The four diagonals of a cube have the same length.

17. The line which joins one vertex of a parallelogram to the midpoint of an opposite side divides the diagonal in the ratio 1:2.

18. The sum of the vectors drawn from the center of a regular polygon to its vertices is the zero vector.

19. The medians of a triangle meet at a point P which divides each median in the ratio 1:2.

20. The four diagonals of a parallelepiped meet and bisect each other.

6.4 Vector Spaces. Linear Dependence and Independence

The set V of all vectors with the two algebraic operations of vector addition and multiplication of vectors by scalars defined on V suggests and motivates an algebraic structure which is called a *vector space* (or *linear space*) and is important because it includes many sets (sets of vectors, matrices, functions, operators, etc.) which occur quite often in mathematics and its applications. In fact, the concept of a vector space is basic in *functional analysis* (abstract modern analysis), which has applications to differential equations, numerical analysis, and other fields of practical interest to the engineer. (Cf. Ref. [A10] in Appendix 1.) Let us state the definition:

Definition of a real vector space

A nonempty set V of elements $\mathbf{a}, \mathbf{b}, \cdots$ is called a *real vector space* (or *real linear space*) if in V there are defined two algebraic operations (called *vector addition* and *multiplication by scalars*) as follows.

I. Vector addition associates with every pair of elements $\mathbf{a}$ and $\mathbf{b}$ of V a unique element of V, called the *sum* of $\mathbf{a}$ and $\mathbf{b}$ and denoted by $\mathbf{a} + \mathbf{b}$, such that the following axioms are satisfied.

I.1 *Commutativity.* For any two elements $\mathbf{a}$ and $\mathbf{b}$ of V,

$$\mathbf{a} + \mathbf{b} = \mathbf{b} + \mathbf{a}.$$

I.2 *Associativity.* For any three elements $\mathbf{u}, \mathbf{v}, \mathbf{w}$ of V,

$$(\mathbf{u} + \mathbf{v}) + \mathbf{w} = \mathbf{u} + (\mathbf{v} + \mathbf{w}) \qquad \text{(written } \mathbf{u} + \mathbf{v} + \mathbf{w}\text{)}.$$

I.3 There is a unique element in V, called the *zero element* and denoted by $\mathbf{0}$, such that for every $\mathbf{a}$ in V,

$$\mathbf{a} + \mathbf{0} = \mathbf{a}.$$

I.4 For every $\mathbf{a}$ in V there is a unique element in V which is denoted by $-\mathbf{a}$ and is such that

$$\mathbf{a} + (-\mathbf{a}) = \mathbf{0}.$$

II. Multiplication by scalars. The real numbers are called *scalars*. Multiplication by scalars associates with every $\mathbf{a}$ in V and every scalar q a unique element of V, called the *product* of q and $\mathbf{a}$ and denoted by $q\mathbf{a}$ (or $\mathbf{a}q$) such that the following axioms are satisfied.

II.1 *Distributivity.* For every scalar q and elements $\mathbf{a}$ and $\mathbf{b}$ in V,

$$q(\mathbf{a} + \mathbf{b}) = q\mathbf{a} + q\mathbf{b}.$$

II.2 *Distributivity.* For all scalars c and k and every $\mathbf{a}$ in V,

$$(c + k)\mathbf{a} = c\mathbf{a} + k\mathbf{a}.$$

II.3 *Associativity.* For all scalars c and k and every $\mathbf{a}$ in V,

$$c(k\mathbf{a}) = (ck)\mathbf{a} \qquad \text{(written } ck\mathbf{a}\text{)}.$$

II.4 For every $\mathbf{a}$ in V,

$$1\mathbf{a} = \mathbf{a}.$$

The elements of V are called **vectors.**[3] ∎

A *complex vector space* is obtained if, instead of real numbers, we take complex numbers as scalars.

We do not want to sidetrack our attention from our actual present goal, which is the study of vector algebra, so we shall confine ourselves to illustrating the generality of the notion of a vector space with some typical examples included in the problem set below, and we shall discuss further examples in connection with matrices in the next chapter. Note that we obtained the **abstract concept** of a vector space by the principle of defining it in terms of some of the most important properties of our **concrete model,** the set of all vectors in three-dimensional space. This is the principle by which many abstract mathematical concepts are derived from concrete models. Of course, in each case the selection of properties to be stated in the form of axioms needs experience.

We shall now discuss some important concepts related to the concept of a vector space.

Let $\mathbf{a}_{(1)}, \cdots, \mathbf{a}_{(m)}$ be any vectors in V. Then an expression of the form

$$q_1\mathbf{a}_{(1)} + \cdots + q_m\mathbf{a}_{(m)} \qquad (q_1, \cdots, q_m \text{ any scalars})$$

[3] Regardless of what they actually are, but this convention causes no confusion because in any specific case the nature of those elements is clear from the context.

is called a **linear combination** of those vectors. Clearly, this is a vector in V, by the axioms of a vector space. The set S of all these linear combinations is called the **span** of $\mathbf{a}_{(1)}, \cdots, \mathbf{a}_{(m)}$ and is denoted by span $(\mathbf{a}_{(1)}, \cdots, \mathbf{a}_{(m)})$, and we say that S is **spanned** or *generated* by those m vectors.

It is not difficult to see that the span of a set of vectors is a vector space. (Proof?)

Using the notion of linear combination, we can now introduce the basic concepts of linear dependence and independence of vectors.

A set of m vectors $\mathbf{a}_{(1)}, \cdots, \mathbf{a}_{(m)}$ is called a **linearly dependent set** if at least one of the vectors can be represented as a linear combination of the others (with scalars which may be zero or not). The set is called a **linearly independent set** if none of the vectors can be represented in that fashion.

In this definition, $m \geqq 2$. If $m = 1$, so that that set consists of a single vector $\mathbf{a}$, we call that set *linearly dependent* if $\mathbf{a} = \mathbf{0}$ and *linearly independent* if $\mathbf{a} \neq \mathbf{0}$.

Example 1. Linearly dependent and independent sets of vectors

The vectors $\mathbf{a} = \mathbf{i} + 2\mathbf{j} + \mathbf{k}$, $\mathbf{b} = 3\mathbf{k}$, $\mathbf{c} = 2\mathbf{i} + 4\mathbf{j}$ constitute a linearly dependent set because $6\mathbf{a} - 2\mathbf{b} - 3\mathbf{c} = \mathbf{0}$; we have $\mathbf{a} = \frac{1}{3}\mathbf{b} + \frac{1}{2}\mathbf{c}$. The vectors $\mathbf{i}, \mathbf{j}, \mathbf{k}$ defined in the last section constitute a linearly independent set. ∎

If a vector space V is such that it contains a linearly independent set B of n vectors, whereas any set of $n + 1$ or more vectors in V is linearly dependent, then V is said to have **dimension** n (or to be *n-dimensional*), and B is called a **basis** of V. Then every vector $\mathbf{v}$ of V can be represented as a linear combination of the n vectors of the basis in a unique fashion.

For example, the set of all vectors in space as defined in Sec. 6.1 forms a vector space. It is three-dimensional because a basis is $\mathbf{i}, \mathbf{j}, \mathbf{k}$ (cf. Sec. 6.3).

More generally, if we consider any set M of vectors $\mathbf{a}_{(1)}, \cdots, \mathbf{a}_{(m)}$ in V and if their span has dimension r, this means that M has a subset of r linearly independent vectors. Then the given vectors are linear combinations of those r vectors, which form a basis of the span of M.

There is an important criterion for linear dependence and independence. To obtain it, we consider the vector equation

$$k_1\mathbf{a}_{(1)} + k_2\mathbf{a}_{(2)} + \cdots + k_m\mathbf{a}_{(m)} = \mathbf{0}, \tag{1}$$

where $\mathbf{a}_{(1)}, \cdots, \mathbf{a}_{(m)}$ are given vectors and $k_1, \cdots, k_m$ are scalars and we first assume that $m \geqq 2$. This equation certainly holds for $k_1 = k_2 = \cdots = k_m = 0$. If we can find scalars $k_1, \cdots, k_m$, not all zero, such that (1) holds, then we can divide by a scalar $k_i \neq 0$ and represent the corresponding vector $\mathbf{a}_{(i)}$ as a linear combination of the others. (For example, if (1) holds with a $k_1 \neq 0$, we obtain from (1)

$$\mathbf{a}_{(1)} = l_2\mathbf{a}_{(2)} + \cdots + l_m\mathbf{a}_{(m)} \qquad \text{where} \qquad l_j = -k_j/k_1.)$$

Hence, in this case the vectors constitute a linearly dependent set. However, if

$k_1 = k_2 = \cdots = k_m = 0$ is the only set of scalars for which (1) holds, we cannot solve (1) for any of the vectors, and the vectors constitute a linearly independent set. If $m = 1$, then (1) has the form $k_1\mathbf{a}_{(1)} = \mathbf{0}$. This holds for a $k_1 \neq 0$ if and only if $\mathbf{a}_{(1)} = \mathbf{0}$, which means linear dependence, by definition. This proves the following criterion.

Theorem 1 (Linear dependence)
A set of vectors $\mathbf{a}_{(1)}, \cdots, \mathbf{a}_{(m)}$ *is linearly dependent if and only if* (1) *holds for a set of scalars* $k_1, \cdots, k_m$, *not all zero.*

This necessary and sufficient condition for linear dependence is often taken as a *definition* of linear dependence.

If one of those vectors is the zero vector, that set is linearly dependent, because if, say $\mathbf{a}_{(1)} = \mathbf{0}$, then (1) holds for any k_1 and $k_2 = \cdots = k_m = 0$.

Two vectors in three-dimensional space which form a linearly dependent set are **collinear;** that is, if we let their initial points coincide, they lie in the same line (Fig. 124). If three such vectors **a, b, c** form a linearly dependent set, they

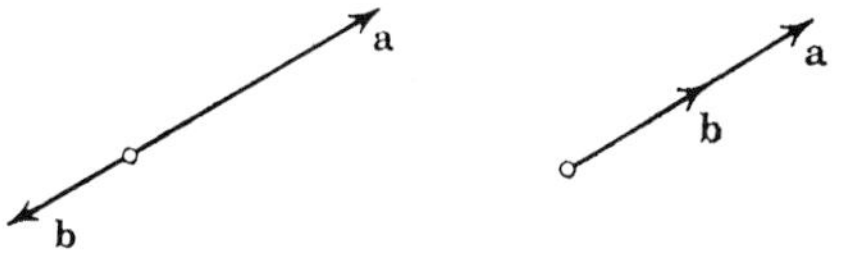

Fig. 124. Collinear vectors

are collinear or **coplanar;** that is, if we let their initial points coincide, they lie in the same plane (Fig. 125). In fact, linear dependence implies that one of the vectors is a linear combination of the other two vectors, for instance, $\mathbf{a} = k\mathbf{b} + l\mathbf{c}$, and thus lies in the plane determined by **b** and **c** (or in a straight line if **b** and **c** are collinear). A set of four or more vectors in three-dimensional space is linearly dependent, because all vectors in that space form a vector space of dimension three.

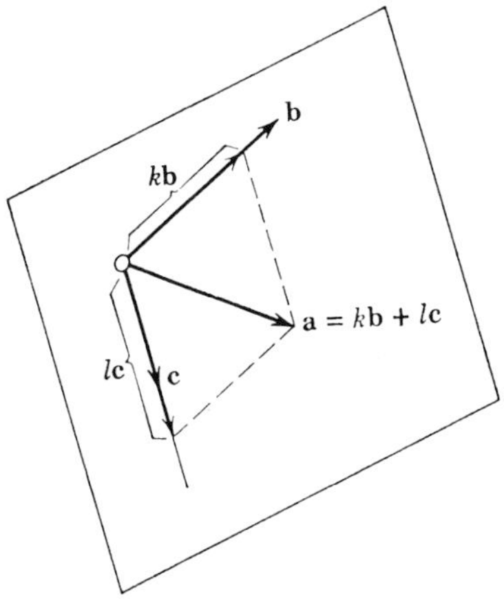

Fig. 125. Coplanar vectors

Problems for Sec. 6.4

Show that each of the following sets constitutes a vector space. Determine its dimension. Find a basis.

1. The set of all vectors in space whose first component is zero.
2. The set of all vectors of the form $c(\mathbf{i} + \mathbf{j} + 2\mathbf{k})$, where c is any scalar.
3. The set of all vectors of the form $b\mathbf{i} + c(\mathbf{j} + \mathbf{k})$, where b and c are any scalars.
4. The set of all ordered pairs of numbers, with the usual addition defined by $(a_1, a_2) + (b_1, b_2) = (a_1 + b_1, a_2 + b_2)$ and multiplication by scalars defined by $c(a_1, a_2) = (ca_1, ca_2)$.
5. The set of all ordered n-tuples of numbers $(a_1, \cdots, a_n)$ with addition defined by $(a_1, \cdots, a_n) + (b_1, \cdots, b_n) = (a_1 + b_1, \cdots, a_n + b_n)$ and multiplication by scalars defined by $c(a_1, \cdots, a_n) = (ca_1, \cdots, ca_n)$.
6. All functions of the form $y(x) = c_1 e^x + c_2 e^{-2x}$ (c_1 and c_2 arbitrary) with the usual addition and multiplication by numbers.
7. All functions of the form $y(x) = a \cos x + b \sin x$ (a and b arbitrary) with the usual addition and multiplication by numbers.
8. All polynomials in x of degree not exceeding 2, with the usual addition and multiplication by numbers.

Describe span $\{\mathbf{u}, \mathbf{v}\}$, where

9. $\mathbf{u} = (1, 0, 0)$, $\mathbf{v} = (1, 2, 0)$
10. $\mathbf{u} = (1, 1, 1)$, $\mathbf{v} = (0, 0, 2)$

11. Show that the representation of a vector $\mathbf{a}$ in a vector space V as a linear combination of basis vectors $\mathbf{e}_{(1)}, \cdots, \mathbf{e}_{(n)}$ is unique.
12. If a set $\{\mathbf{a}_{(1)}, \cdots, \mathbf{a}_{(m)}\}$ is linearly independent, show that every (nonempty) subset of that set is linearly independent.
13. **(Subspace)** A nonempty subset W of a vector space V is called a *subspace* of V if W is itself a vector space with respect to the algebraic operations defined in V. Give examples of one- and two-dimensional subspaces of the space of all vectors in three-dimensional space.
14. Which of the following sets constitute a subspace of the vector space of all vectors of the form $\mathbf{v} = (v_1, v_2, v_3)$? (*a*) All $\mathbf{v}$ with $v_1 = v_2$ and $v_3 = 0$. (*b*) All $\mathbf{v}$ with $v_1 = v_2 + 1$. (*c*) All $\mathbf{v}$ with positive v_1, v_2, v_3. (*d*) All $\mathbf{v}$ with $v_1 - v_2 + v_3 = k = const.$
15. Show that a subset W of a vector space V is a subspace of V if and only if for any $\mathbf{a}$ and $\mathbf{b}$ in W and any scalars c and q the vector $c\mathbf{a} + q\mathbf{b}$ is in W.

6.5 Inner Product (Dot Product)

The **inner product, dot product,** or **scalar product** of two vectors $\mathbf{a}$ and $\mathbf{b}$ in three-dimensional space is written $\mathbf{a} \cdot \mathbf{b}$ and is defined as

$$(1) \qquad \begin{aligned} \mathbf{a} \cdot \mathbf{b} &= |\mathbf{a}|\,|\mathbf{b}| \cos \gamma \qquad &(\text{when } \mathbf{a} \neq \mathbf{0}, \mathbf{b} \neq \mathbf{0}), \\ \mathbf{a} \cdot \mathbf{b} &= 0 \qquad &(\text{when } \mathbf{a} = \mathbf{0} \text{ or } \mathbf{b} = \mathbf{0}); \end{aligned}$$

here γ ($0 \leqq \gamma \leqq \pi$) is the angle between $\mathbf{a}$ and $\mathbf{b}$ (computed when the vectors have their initial point coinciding). (Cf. Fig. 126.)

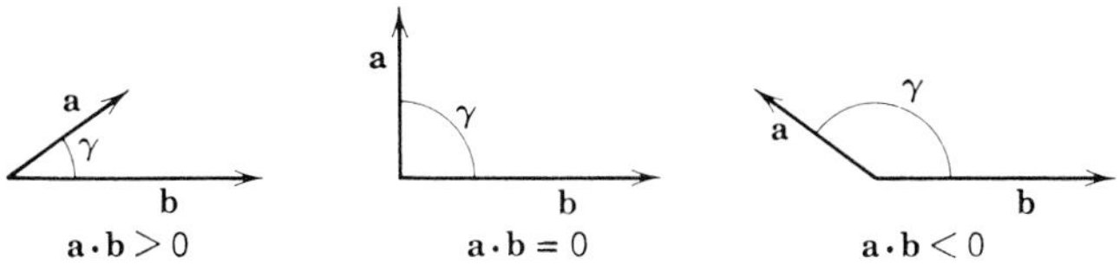

Fig. 126. Angle between vectors

The value of the inner product is a scalar (a real number), and this motivates the term "scalar product." Since the cosine in (1) may be positive, zero or negative, the same is true for the inner product (cf. Fig. 126). Since for γ between 0 and π the cosine is 0 if and only if $\gamma = \pi/2$, we have the following important result.

Theorem 1 (Orthogonality)

Two nonzero vectors are orthogonal (perpendicular) if and only if their inner product (their scalar product) is zero.

Setting $\mathbf{b} = \mathbf{a}$ in (1), we have $\mathbf{a} \cdot \mathbf{a} = |\mathbf{a}|^2$, and this shows that the length (Euclidean norm) of a vector can be written in terms of the inner product,

$$|\mathbf{a}| = \sqrt{\mathbf{a} \cdot \mathbf{a}} \qquad (\geqq 0). \tag{2}$$

From this and (1) we obtain the useful formula

$$\cos \gamma = \frac{\mathbf{a} \cdot \mathbf{b}}{|\mathbf{a}|\,|\mathbf{b}|} = \frac{\mathbf{a} \cdot \mathbf{b}}{\sqrt{\mathbf{a} \cdot \mathbf{a}}\,\sqrt{\mathbf{b} \cdot \mathbf{b}}}. \tag{3}$$

From the definition we also see that the inner product has the properties

(4)

$$\text{(a)} \quad [q_1\mathbf{a} + q_2\mathbf{b}] \cdot \mathbf{c} = q_1\mathbf{a} \cdot \mathbf{c} + q_2\mathbf{b} \cdot \mathbf{c} \qquad \textit{(Linearity)}$$

$$\text{(b)} \quad \mathbf{a} \cdot \mathbf{b} = \mathbf{b} \cdot \mathbf{a} \qquad \textit{(Symmetry)}$$

$$\text{(c)} \quad \left.\begin{matrix} \mathbf{a} \cdot \mathbf{a} \geqq 0 \\ \mathbf{a} \cdot \mathbf{a} = 0 \quad \text{if and only if} \quad \mathbf{a} = \mathbf{0} \end{matrix}\right\} \qquad \textit{(Positive-definiteness).}$$

Hence *dot multiplication is commutative* [cf. (4b)] *and is distributive with respect to vector addition;* in fact, from (4a) with $q_1 = 1$ and $q_2 = 1$ we have

$$(\mathbf{a} + \mathbf{b}) \cdot \mathbf{c} = \mathbf{a} \cdot \mathbf{c} + \mathbf{b} \cdot \mathbf{c} \qquad \textit{(Distributivity).} \tag{4a*}$$

Furthermore, from (1) and $|\cos \gamma| \leqq 1$ we see that

$$|\mathbf{a} \cdot \mathbf{b}| \leqq |\mathbf{a}|\,|\mathbf{b}| \qquad \textit{(Schwarz}^4 \textit{ inequality).} \tag{5}$$

Using this and (2), the reader may prove

$$|\mathbf{a} + \mathbf{b}| \leqq |\mathbf{a}| + |\mathbf{b}| \qquad \textbf{(Triangle inequality).} \tag{6}$$

[4] HERMANN AMANDUS SCHWARZ (1843–1921), German mathematician, who contributed to complex analysis and differential geometry.

A simple direct calculation with inner products shows that

$$|\mathbf{a} + \mathbf{b}|^2 + |\mathbf{a} - \mathbf{b}|^2 = 2(|\mathbf{a}|^2 + |\mathbf{b}|^2) \qquad \textit{(Parallelogram equality).} \tag{7}$$

If vectors **a** and **b** are represented in terms of components, say,

$$\mathbf{a} = a_1\mathbf{i} + a_2\mathbf{j} + a_3\mathbf{k} \qquad \text{and} \qquad \mathbf{b} = b_1\mathbf{i} + b_2\mathbf{j} + b_3\mathbf{k},$$

their inner product is given by the simple basic formula

$$\mathbf{a} \cdot \mathbf{b} = a_1b_1 + a_2b_2 + a_3b_3. \tag{8}$$

We prove (8). Since **i**, **j**, and **k** are unit vectors, we have from (1)

$$\mathbf{i} \cdot \mathbf{i} = 1, \qquad \mathbf{j} \cdot \mathbf{j} = 1, \qquad \mathbf{k} \cdot \mathbf{k} = 1,$$

and since they are orthogonal, it follows from Theorem 1 that

$$\mathbf{i} \cdot \mathbf{j} = 0, \qquad \mathbf{j} \cdot \mathbf{k} = 0, \qquad \mathbf{k} \cdot \mathbf{i} = 0.$$

Hence if we substitute those representations of **a** and **b** into $\mathbf{a} \cdot \mathbf{b}$ and use (4a*) and (4b), we first have a sum of nine inner products,

$$\mathbf{a} \cdot \mathbf{b} = a_1b_1\mathbf{i} \cdot \mathbf{i} + a_1b_2\mathbf{i} \cdot \mathbf{j} + \cdots + a_3b_3\mathbf{k} \cdot \mathbf{k}.$$

Since six of these products are zero, we obtain (8).

Before we consider some applications of inner products, let us introduce one more concept. Suppose that **a** and **b** ($\neq \mathbf{0}$) are any given vectors and let γ denote the angle between them. Then the real number

$$p = |\mathbf{a}| \cos \gamma$$

is called the **component** *of* **a** *in the direction of* **b** or the **projection** *of* **a** *in the direction of* **b**. If $\mathbf{a} = \mathbf{0}$, then γ is undefined, and we set $p = 0$.

It follows that $|p|$ is the length of the orthogonal projection of **a** on a straight line l in the direction of **b**. p may be positive, zero, or negative (Fig. 127).

From this definition we see that in particular the components of a vector **a** in the directions of the unit vectors **i**, **j**, **k** of the fundamental triad associated

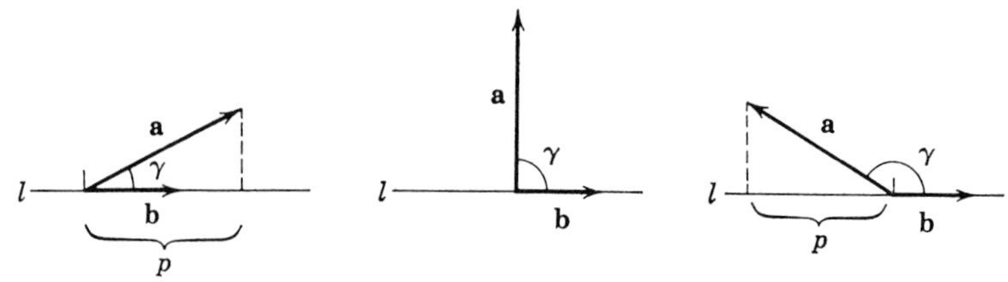

Fig. 127. Component of a vector **a** in the direction of a vector **b**

with a Cartesian coordinate system are the components a_1, a_2, a_3 of **a** as defined in Sec. 6.2. This shows that our present use of the term "component" is merely a slight generalization of the previous one.

From (3) we obtain

$$p = |\mathbf{a}| \cos \gamma = \frac{\mathbf{a} \cdot \mathbf{b}}{|\mathbf{b}|} \qquad (\mathbf{b} \neq \mathbf{0}) \tag{9}$$

and if in particular **b** is a unit vector, then we simply have

$$p = \mathbf{a} \cdot \mathbf{b}. \tag{10}$$

The following examples may illustrate the usefulness of inner products. Various other applications will be considered later.

Example 1. Work done by a force

Consider a particle on which a constant force **a** acts. Let the particle be given a displacement **d.** Then the work W done by **a** in the displacement is defined as the product of $|\mathbf{d}|$ and the component of **a** in the direction of **d,** that is,

$$W = |\mathbf{a}|\,|\mathbf{d}| \cos \alpha = \mathbf{a} \cdot \mathbf{d} \tag{11}$$

where α is the angle between **d** and **a** (Fig. 128).

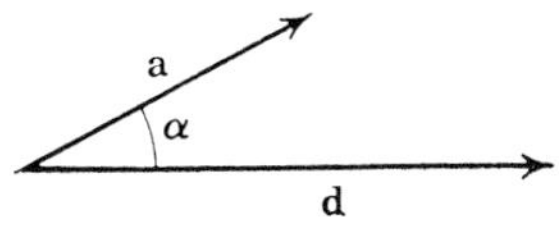

Fig. 128. Work done by a force

Example 2. Orthogonal straight lines in a plane

Find a representation of the straight line L_1 through the point P: (1, 3) in the xy-plane and perpendicular to the line L_2 represented by $x - 2y + 2 = 0$.

Any straight line L_1 in the xy-plane can be represented in the form $a_1x + a_2y = c$. If $c = 0$, then L_1 passes through the origin. If $c \neq 0$, then $a_1x + a_2y = 0$ represents a line L_1^* through the origin and parallel to L_1. The position vector of a point on L_1^* is $\mathbf{r} = x\mathbf{i} + y\mathbf{j}$. Introducing the vector $\mathbf{a} = a_1\mathbf{i} + a_2\mathbf{j}$, we see from (8) that the representation of L_1^* may be written

$$\mathbf{a} \cdot \mathbf{r} = 0.$$

Certainly $\mathbf{a} \neq \mathbf{0}$ and, by Theorem 1, the vector **a** is perpendicular to **r** and, therefore, perpendicular to the line L_1^*. It is called a **normal vector** to L_1^*. Since L_1 and L_1^* are parallel, **a** is also a normal

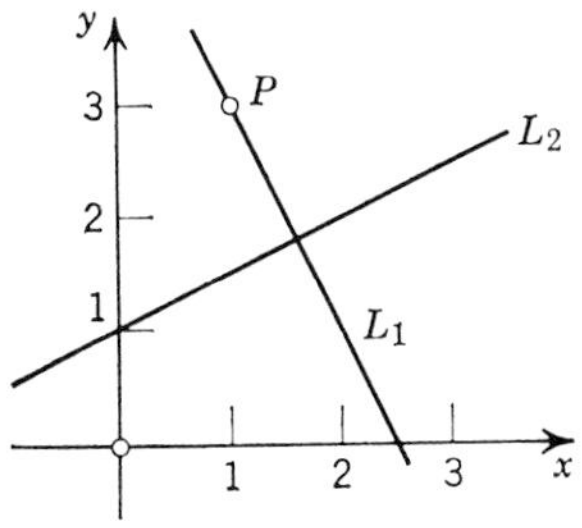

Fig. 129. Example 2

vector to L_1. Hence two lines L_1 and L_2: $b_1x + b_2y = d$ are perpendicular or *orthogonal* if and only if their normal vectors $\mathbf{a}$ and $\mathbf{b} = b_1\mathbf{i} + b_2\mathbf{j}$ are orthogonal, that is, $\mathbf{a} \cdot \mathbf{b} = 0$. Note that this implies that the slopes of the lines are negative reciprocals.

In our case, $\mathbf{b} = \mathbf{i} - 2\mathbf{j}$, and a vector perpendicular to $\mathbf{b}$ is $\mathbf{a} = 2\mathbf{i} + \mathbf{j}$. Hence the representation of L_1 must be of the form $2x + y = c$, and by substituting the coordinates of P we obtain $c = 5$. This yields the solution (cf. Fig. 129)

$$y = -2x + 5.$$

Example 3. Normal vector to a plane

Find a unit vector perpendicular to the plane $4x + 2y + 4z = -7$.

Any plane in space can be represented in the form

$$a_1x + a_2y + a_3z = c. \tag{12}$$

The position vector of a point in the plane is $\mathbf{r} = x\mathbf{i} + y\mathbf{j} + z\mathbf{k}$. Introducing the vector $\mathbf{a} = a_1\mathbf{i} + a_2\mathbf{j} + a_3\mathbf{k}$ and using (8), we may write (12) in the form

$$\mathbf{a} \cdot \mathbf{r} = c. \tag{13}$$

Certainly $\mathbf{a} \neq \mathbf{0}$, and the unit vector in the direction of $\mathbf{a}$ is

$$\mathbf{n} = \frac{\mathbf{a}}{|\mathbf{a}|}.$$

Dividing by $|\mathbf{a}|$, we obtain from (13)

$$\mathbf{n} \cdot \mathbf{r} = p \qquad \text{where} \qquad p = \frac{c}{|\mathbf{a}|}. \tag{14}$$

From (10) we see that p is the projection of $\mathbf{r}$ in the direction of $\mathbf{n}$, and this projection has the same constant value $c/|\mathbf{a}|$ for the position vector $\mathbf{r}$ of any point in the plane. Clearly this holds if and only if $\mathbf{n}$ is perpendicular to the plane. $\mathbf{n}$ is called a **unit normal vector** to the plane (the other being $-\mathbf{n}$). Furthermore, from this and the definition of projection it follows that $|p|$ is the distance of the plane from the origin. Representation (14) is called **Hesse's**[5] **normal form** of a plane. An illustration is given in Fig. 130. In our case,

$$\mathbf{a} = 4\mathbf{i} + 2\mathbf{j} + 4\mathbf{k}, \qquad c = -7, \qquad |\mathbf{a}| = 6, \qquad \mathbf{n} = \tfrac{1}{6}\mathbf{a} = \tfrac{2}{3}\mathbf{i} + \tfrac{1}{3}\mathbf{j} + \tfrac{2}{3}\mathbf{k},$$

and the plane has the distance 7/6 from the origin.

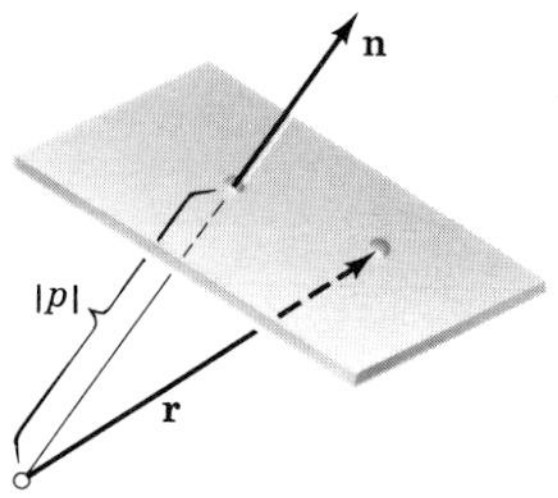

Fig. 130. Normal vector to a plane

[5]LUDWIG OTTO HESSE (1811–1874), German mathematician, who contributed to the theory of curves and surfaces.

Problems for Sec. 6.5

Let $\mathbf{a} = \mathbf{i} + 3\mathbf{j} + 2\mathbf{k}$, $\mathbf{b} = -\mathbf{j} + 4\mathbf{k}$, and $\mathbf{c} = 3\mathbf{i} - 4\mathbf{j} - \mathbf{k}$. Find

1. $\mathbf{a} \cdot \mathbf{b}$, $\mathbf{b} \cdot \mathbf{a}$
2. $|\mathbf{a}|$, $|\mathbf{b}|$, $|\mathbf{c}|$
3. $(\mathbf{a} + \mathbf{b}) \cdot \mathbf{c}$, $\mathbf{a} \cdot \mathbf{c} + \mathbf{b} \cdot \mathbf{c}$
4. $\mathbf{a} \cdot (\mathbf{b} - \mathbf{c})$, $\mathbf{a} \cdot (\mathbf{c} - \mathbf{b})$
5. $|\mathbf{a} + \mathbf{b}|$, $|\mathbf{b} + \mathbf{c}|$
6. $2\mathbf{a} \cdot 3\mathbf{b}$, $6\mathbf{a} \cdot \mathbf{b}$
7. $|\mathbf{a} + \mathbf{c}|$, $|\mathbf{a}| + |\mathbf{c}|$
8. $2\mathbf{a} \cdot (\mathbf{a} - 2\mathbf{b})$
9. $-\mathbf{a} \cdot (2\mathbf{c} - \mathbf{b})$

10. What laws do Probs. 1 and 3 illustrate?

11. Find $\mathbf{a} \cdot \mathbf{b}$ where $\mathbf{a} = 2\mathbf{i}$ and $\mathbf{b} = \mathbf{i} + \mathbf{j}$, $\mathbf{b} = \mathbf{j}$, $\mathbf{b} = -\mathbf{i} + \mathbf{j}$, and sketch a figure similar to Fig. 126.

12. If $\mathbf{a}$ is given and $\mathbf{a} \cdot \mathbf{b} = \mathbf{a} \cdot \mathbf{c}$, can we conclude that $\mathbf{b} = \mathbf{c}$?

Find the work done by a force $\mathbf{p}$ acting on a particle, if the particle is displaced from a point A to a point B along the straight segment AB, where

13. A: $(0, 0, 0)$, B: $(0, 3, 0)$, $\mathbf{p} = \mathbf{i} + \mathbf{j} + \mathbf{k}$
14. A: $(0, -1, 2)$, B: $(0, 0, 0)$, $\mathbf{p} = \mathbf{k}$
15. A: $(8, -2, -3)$, B: $(-2, 0, 6)$, $\mathbf{p} = 3\mathbf{i} - 2\mathbf{j} + 4\mathbf{k}$
16. A: $(2, 2, 2)$, B: $(4, 4, 4)$, $\mathbf{p} = -4\mathbf{j} + 2\mathbf{k}$
17. A: $(4, 1, 3)$, B: $(5, 2, 5)$, $\mathbf{p} = \mathbf{i} + \mathbf{j} - \mathbf{k}$

18. Why is the work in Prob. 17 equal to zero?

19. For what values of a_1 are $\mathbf{a} = a_1\mathbf{i} + 2\mathbf{j}$ and $\mathbf{b} = 3\mathbf{i} + 4\mathbf{j} - \mathbf{k}$ orthogonal vectors?

20. Find a unit vector $a_1\mathbf{i} + a_2\mathbf{j}$ perpendicular to the vector $2\mathbf{i} - 5\mathbf{j}$.

21. Suppose that a particle is displaced from a point A to a point B along the straight segment AB. Show that the work done by two forces acting on the particle is equal to the work done by the resultant of the forces.

22. Using vectors, show that if the diagonals of a rectangle are orthogonal, the rectangle must be a square.

23. Under what conditions will the diagonals of a parallelogram be orthogonal?

24. Are the four diagonals of a cube orthogonal?

25. Show that the planes $x + y + z = 1$ and $3x - 4y + z = 2$ are orthogonal.

26. Find the angle between the planes $x + 2y + z = 2$ and $2x - y + 3z = -4$.

27. Find the angles of the triangle with vertices A: $(1, 1, 0)$, B: $(3, 1, 0)$, C: $(1, 3, 0)$.

Let $\mathbf{a} = 3\mathbf{i} + \mathbf{j} + 4\mathbf{k}$, $\mathbf{b} = \mathbf{i} - \mathbf{j}$, and $\mathbf{c} = \mathbf{i} + 5\mathbf{k}$. Find the cosine of the angle between the following vectors.

28. $\mathbf{a}$, $\mathbf{c}$
29. $\mathbf{a}$, $-\mathbf{c}$
30. $\mathbf{a}$, $\mathbf{b} + \mathbf{c}$
31. $\mathbf{a} + \mathbf{b}$, $\mathbf{a} - \mathbf{b}$

In each case find the component of $\mathbf{a}$ in the direction of $\mathbf{b}$.

32. $\mathbf{a} = \mathbf{i} + \mathbf{j} - \mathbf{k}$, $\mathbf{b} = 3\mathbf{i} + 7\mathbf{k}$
33. $\mathbf{a} = 4\mathbf{j} + 3\mathbf{k}$, $\mathbf{b} = -4\mathbf{j} - 3\mathbf{k}$
34. $\mathbf{a} = 4\mathbf{i} + \mathbf{j}$, $\mathbf{b} = -\mathbf{i} + 4\mathbf{j} + 3\mathbf{k}$
35. $\mathbf{a} = \mathbf{i} + \mathbf{j} + 2\mathbf{k}$, $\mathbf{b} = 6\mathbf{k}$

36. Show that if $\mathbf{a}$, $\mathbf{b}$, $\mathbf{c}$ are orthogonal unit vectors, then $\mathbf{a} + \mathbf{b} + \mathbf{c}$ makes the same angle with each of the vectors.

37. Deduce the law of cosines by using vectors $\mathbf{a}$, $\mathbf{b}$, and $\mathbf{a} - \mathbf{b}$.

38. Let $\mathbf{a} = \cos\alpha\,\mathbf{i} + \sin\alpha\,\mathbf{j}$ and $\mathbf{b} = \cos\beta\,\mathbf{i} + \sin\beta\,\mathbf{j}$, where $0 \leqq \alpha \leqq \beta \leqq 2\pi$. Show that $\mathbf{a}$ and $\mathbf{b}$ are unit vectors. Use (8) to obtain the trigonometric identity for $\cos(\beta - \alpha)$.

39. Prove the triangle inequality (6).

40. Prove the parallelogram equality (7).

6.6 Inner Product Spaces

In Sec. 6.4 we obtained the definition of a vector space by taking some of the basic properties of vector addition and multiplication by scalars in three-dimensional space. It is quite interesting that the dot product (Sec. 6.5) may be used in a similar fashion and motivates the concept of a *real inner product space.* By definition, this is a real vector space on which there is defined an inner product satisfying (4), Sec. 6.5; in detail:

Definition of a real inner product space

A real vector space V is called a *real inner product space* (or *real pre-Hilbert*[6] *space*) if it has the following property. With every pair of vectors $\mathbf{a}$ and $\mathbf{b}$ in V there is associated a real number, which is denoted by $(\mathbf{a}, \mathbf{b})$ and is called the **inner product** (or *scalar product*) of $\mathbf{a}$ and $\mathbf{b}$, such that the following axioms are satisfied.

I. For all scalars q_1 and q_2 and all vectors $\mathbf{a}, \mathbf{b}, \mathbf{c}$ in V,

$$(q_1\mathbf{a} + q_2\mathbf{b}, \mathbf{c}) = q_1(\mathbf{a}, \mathbf{c}) + q_2(\mathbf{b}, \mathbf{c}) \qquad \textit{(Linearity)}.$$

II. For all vectors $\mathbf{a}$ and $\mathbf{b}$ in V,

$$(\mathbf{a}, \mathbf{b}) = (\mathbf{b}, \mathbf{a}) \qquad \textit{(Symmetry)}.$$

III. For every $\mathbf{a}$ in V,

$$\begin{aligned} &(\mathbf{a}, \mathbf{a}) \geqq 0, \text{ and} \\ &(\mathbf{a}, \mathbf{a}) = 0 \quad \text{if and only if} \quad \mathbf{a} = \mathbf{0} \end{aligned} \qquad \textit{(Positive-definiteness)}.$$

Definition of orthogonality

Vectors $\mathbf{a}$ and $\mathbf{b}$ in an inner product space V are called *orthogonal* if

$$(\mathbf{a}, \mathbf{b}) = 0.$$

∎

Note that this definition is suggested and motivated by Theorem 1 in Sec. 6.5.

Using the inner product, with every element $\mathbf{a}$ in V we may associate a number which is denoted by $\|\mathbf{a}\|$, is defined by

$$\|\mathbf{a}\| = \sqrt{(\mathbf{a}, \mathbf{a})} \qquad (\geqq 0),$$

and is called the **norm** of $\mathbf{a}$. From (2), Sec. 6.5, we see that this generalizes the concept of a length. In fact, the dot product is an inner product in the sense of

[6]DAVID HILBERT (1862–1943), great German mathematician, whose work was of basic importance to higher algebra and number theory, integral equations, calculus of variations, functional analysis and mathematical logic. His "Foundations of Geometry" helped the axiomatic method to gain general recognition and acceptance. The famous twenty-three Hilbert's problems (presented in a talk in 1900) had considerable influence on the development of modern mathematics.

our new definition,

$$(\mathbf{a}, \mathbf{b}) = \mathbf{a} \cdot \mathbf{b},$$

and in our new notation, formula (2), Sec. 6.5, can be written

$$\|\mathbf{a}\| = |\mathbf{a}| = \sqrt{(\mathbf{a}, \mathbf{a})} = \sqrt{\mathbf{a} \cdot \mathbf{a}}.$$

From the axioms of an inner product and the definition of a norm, one can derive [cf. (5)–(7) in Sec. 6.5]

$$|(\mathbf{a}, \mathbf{b})| \leqq \|\mathbf{a}\| \, \|\mathbf{b}\| \qquad \textit{(Schwarz inequality)},$$

from this

$$\|\mathbf{a} + \mathbf{b}\| \leqq \|\mathbf{a}\| + \|\mathbf{b}\| \qquad \textit{(Triangle inequality)}$$

and by a simple direct calculation

$$\|\mathbf{a} + \mathbf{b}\|^2 + \|\mathbf{a} - \mathbf{b}\|^2 = 2(\|\mathbf{a}\|^2 + \|\mathbf{b}\|^2) \qquad \textit{(Parallelogram equality)}.$$

We shall not go into details, because these belong to more advanced courses (on functional analysis, in particular Hilbert spaces, approximation theory etc.; cf. [A10] in Appendix 1). However, we want to illustrate the generality of the concept of an inner product space by two important examples. The first is *n-dimensional Euclidean space,* which is the inner product space of all real ordered n-tuples $\mathbf{a} = (a_1, \cdots, a_n)$, $\mathbf{b} = (b_1, \cdots, b_n)$, etc., with the inner product [suggested by (8) in Sec. 6.5]

$$(\mathbf{a}, \mathbf{b}) = a_1 b_1 + a_2 b_2 + \cdots + a_n b_n.$$

The second example is the inner product space of all continuous function $f(x)$, $g(x)$, $\cdots$ defined on an interval $\alpha \leqq x \leqq \beta$ with the inner product defined by

$$(f, g) = \int_\alpha^\beta f(x) g(x) \, dx.$$

Note that in this case orthogonality is identical with orthogonality as defined in Sec. 4.7.

6.7 Vector Product (Cross Product)

Various applications suggest the introduction of another kind of vector multiplication in which the product of two vectors is again a vector. This so-called **vector product** or **cross product** of two vectors **a** and **b** is written

$$\mathbf{a} \times \mathbf{b}$$

and is a vector **v** which is defined as follows.

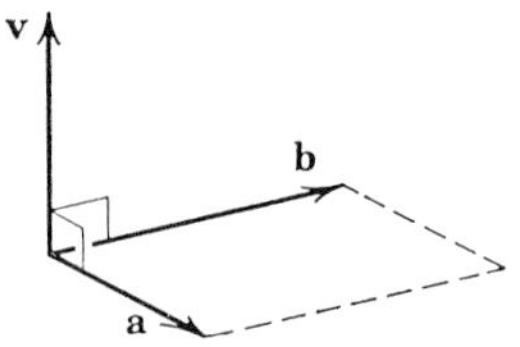

Fig. 131. Vector product

Definition of vector product

If **a** and **b** have the same or opposite direction or one of these vectors is the zero vector, then $\mathbf{v} = \mathbf{a} \times \mathbf{b} = \mathbf{0}$.

In any other case, $\mathbf{v} = \mathbf{a} \times \mathbf{b}$ is the vector whose length is equal to the area of the parallelogram with **a** and **b** as adjacent sides and whose direction is perpendicular to both **a** and **b** and is such that **a, b, v,** in this order, form a right-handed triple or right-handed triad, as shown in Fig. 131. ∎

The term **right-handed** comes from the fact that the vectors **a, b, v,** in this order, assume the same sort of orientation as the thumb, index finger, and middle finger of the right hand when these are held as shown in Fig. 132. We may also say that if **a** is rotated into the direction of **b** through the angle α ($<\pi$), then **v** advances in the same direction as a right-handed screw would if turned in the same way (cf. Fig. 133).

The parallelogram with **a** and **b** as adjacent sides has the area $|\mathbf{a}|\,|\mathbf{b}| \sin \gamma$, where γ is the angle between **a** and **b** (cf. Fig. 131). We thus obtain

$$(1) \qquad |\mathbf{v}| = |\mathbf{a}|\,|\mathbf{b}| \sin \gamma.$$

Let $\mathbf{a} \times \mathbf{b} = \mathbf{v}$ and let $\mathbf{b} \times \mathbf{a} = \mathbf{w}$. Then, by definition, $|\mathbf{v}| = |\mathbf{w}|$, and in order that **b, a, w** form a right-handed triple we must have $\mathbf{w} = -\mathbf{v}$ (Fig. 134). This implies

$$(2) \qquad \mathbf{b} \times \mathbf{a} = -(\mathbf{a} \times \mathbf{b});$$

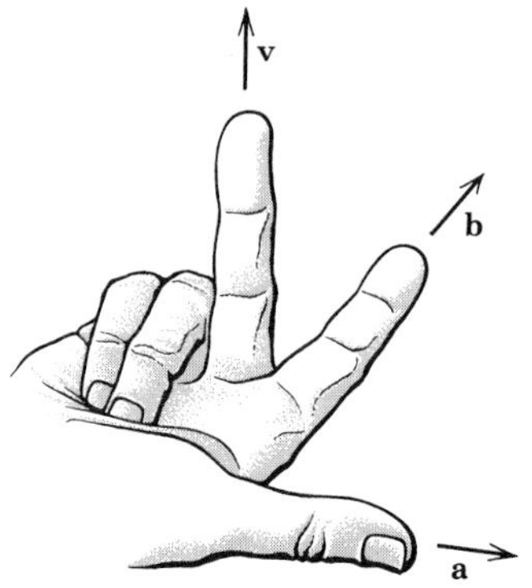

Fig. 132. Right-handed triple of vectors **a, b, v**

Fig. 133. Right-handed screw

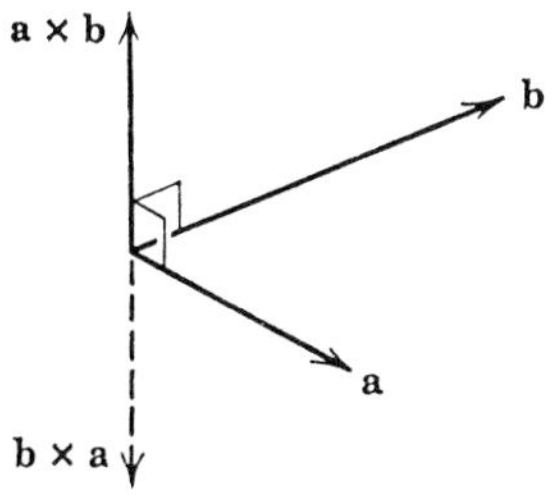

Fig. 134. Anticommutativity of cross multiplication

that is, *cross multiplication of vectors is* **not commutative** *but anticommutative.* Hence the order of the factors in a vector product is of great importance and must be carefully observed.

From the definition it follows that for any constant k,

$$(k\mathbf{a}) \times \mathbf{b} = k(\mathbf{a} \times \mathbf{b}) = \mathbf{a} \times (k\mathbf{b}). \tag{3}$$

Furthermore, cross multiplication is distributive with respect to vector addition, that is,

$$\begin{aligned} \mathbf{a} \times (\mathbf{b} + \mathbf{c}) &= (\mathbf{a} \times \mathbf{b}) + (\mathbf{a} \times \mathbf{c}), \\ (\mathbf{a} + \mathbf{b}) \times \mathbf{c} &= (\mathbf{a} \times \mathbf{c}) + (\mathbf{b} \times \mathbf{c}). \end{aligned} \tag{4}$$

The proof will be given in the next section. More complicated cross products will be considered later, but we want to mention now that cross multiplication is **not associative,** that is, in general

$$\mathbf{a} \times (\mathbf{b} \times \mathbf{c}) \neq (\mathbf{a} \times \mathbf{b}) \times \mathbf{c}.$$

From (1) in this section and (1) in Sec. 6.5 it follows that

$$|\mathbf{v}|^2 = |\mathbf{a}|^2|\mathbf{b}|^2 \sin^2 \gamma = |\mathbf{a}|^2|\mathbf{b}|^2(1 - \cos^2 \gamma) = (\mathbf{a} \cdot \mathbf{a})(\mathbf{b} \cdot \mathbf{b}) - (\mathbf{a} \cdot \mathbf{b})^2.$$

Taking square roots, we thus obtain a useful formula for the length of a vector product:

$$|\mathbf{a} \times \mathbf{b}| = \sqrt{(\mathbf{a} \cdot \mathbf{a})(\mathbf{b} \cdot \mathbf{b}) - (\mathbf{a} \cdot \mathbf{b})^2}. \tag{5}$$

6.8 Vector Products in Terms of Components

We shall now represent a vector product in terms of the components of its factors with respect to a Cartesian coordinate system. In this connection it is important to note that there are two types of such systems, depending on the orientation of the axes, namely, right-handed and left-handed. The definitions are as follows.

A Cartesian coordinate system is called **right-handed,** if the corresponding unit vectors **i, j, k** in the positive directions of the axes form a right-handed triple (Fig. 135*a*); it is called **left-handed,** if these vectors form a *left-handed triple,* that is, assume the same sort of orientation as the thumb, index finger, and middle finger of the left hand (Fig. 135*b*).

In applications we preferably use right-handed systems.

With respect to a given right-handed or left-handed Cartesian coordinate system, let a_1, a_2, a_3 and b_1, b_2, b_3 be the components of two vectors **a** and **b,** respectively. We want to express the components v_1, v_2, v_3 of the product vector

$$\mathbf{v} = \mathbf{a} \times \mathbf{b}$$

in terms of the components of **a** and **b.** We need only consider the case when $\mathbf{v} \neq \mathbf{0}$. Since **v** is perpendicular to both **a** and **b,** it follows from Theorem 1, Sec. 6.5, that $\mathbf{a} \cdot \mathbf{v} = 0$ and $\mathbf{b} \cdot \mathbf{v} = 0$ or, by (8) in Sec. 6.5,

$$\begin{aligned} a_1v_1 + a_2v_2 + a_3v_3 &= 0, \\ b_1v_1 + b_2v_2 + b_3v_3 &= 0. \end{aligned} \tag{1}$$

Multiplying the first equation by b_3, the last by a_3, and subtracting, we obtain

$$(a_3b_1 - a_1b_3)v_1 = (a_2b_3 - a_3b_2)v_2.$$

Multiplying the first equation of (1) by b_1, the last by a_1, and subtracting, we obtain

$$(a_1b_2 - a_2b_1)v_2 = (a_3b_1 - a_1b_3)v_3.$$

We can easily verify that these two equations are satisfied by

$$v_1 = c(a_2b_3 - a_3b_2), \qquad v_2 = c(a_3b_1 - a_1b_3), \qquad v_3 = c(a_1b_2 - a_2b_1), \tag{2}$$

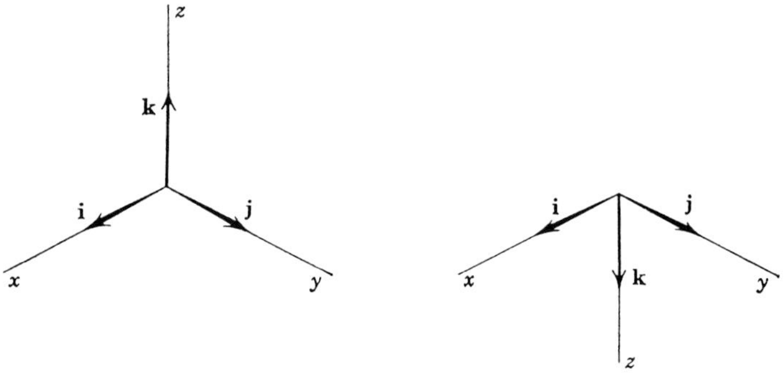

(*a*) Right-handed. (*b*) Left-handed.

Fig. 135. The two types of Cartesian coordinate systems

where c is a constant. The reader may verify by inserting that (2) also satisfies (1). Now each of the equations in (1) represents a plane through the origin in $v_1v_2v_3$-space. The vectors **a** and **b** are normal vectors of these planes (cf. Example 3 in Sec. 6.5). Since $\mathbf{v} \neq \mathbf{0}$, these vectors are not parallel and the two planes do not coincide. Hence their intersection is a straight line L through the origin. Since (2) is a solution of (1) and, for varying c, represents a straight line, we conclude that (2) represents L, and every solution of (1) must be of the form (2). In particular, the components of **v** must be of this form, where c is to be determined. From (2) we obtain

$$|\mathbf{v}|^2 = v_1^2 + v_2^2 + v_3^2 = c^2[(a_2b_3 - a_3b_2)^2 + (a_3b_1 - a_1b_3)^2 + (a_1b_2 - a_2b_1)^2].$$

This can be written

$$|\mathbf{v}|^2 = c^2[(a_1^2 + a_2^2 + a_3^2)(b_1^2 + b_2^2 + b_3^2) - (a_1b_1 + a_2b_2 + a_3b_3)^2],$$

as the reader may readily verify. Using (8) in Sec. 6.5, we thus have

$$|\mathbf{v}|^2 = c^2[(\mathbf{a} \cdot \mathbf{a})(\mathbf{b} \cdot \mathbf{b}) - (\mathbf{a} \cdot \mathbf{b})^2].$$

By comparing this with (5) in the last section we conclude that $c = \pm 1$.

From now on it will be essential whether the coordinate system is right-handed or left-handed. We first consider the case of a right-handed system, and we want to show that then $c = +1$. This can be done as follows.

If we change the lengths and directions of **a** and **b** continuously and so that at the end $\mathbf{a} = \mathbf{i}$ and $\mathbf{b} = \mathbf{j}$ (Fig. 135*a*), then **v** will change its length and direction continuously, and at the end, $\mathbf{v} = \mathbf{i} \times \mathbf{j} = \mathbf{k}$. Obviously we may effect the change so that both **a** and **b** remain different from the zero vector and are not parallel at any instant. Then **v** is never equal to the zero vector, and since the change is continuous and c can only assume the values $+1$ or -1, it follows that at the end c must have the same value as before. Now at the end $\mathbf{a} = \mathbf{i}$, $\mathbf{b} = \mathbf{j}$, $\mathbf{v} = \mathbf{k}$ and, therefore, $a_1 = 1$, $b_2 = 1$, $v_3 = 1$, and the other components in (2) are zero. Hence, from (2), $v_3 = c = +1$. Noting that the expressions in parentheses in (2) may be written in the form of second-order determinants, we may sum up our result as follows.

With respect to a right-handed Cartesian coordinate system,

$$\mathbf{a} \times \mathbf{b} = (a_2b_3 - a_3b_2)\mathbf{i} + (a_3b_1 - a_1b_3)\mathbf{j} + (a_1b_2 - a_2b_1)\mathbf{k}$$

or, written in terms of second-order determinants,[7]

$$\text{(3)} \qquad \mathbf{a} \times \mathbf{b} = \mathbf{i}\begin{vmatrix} a_2 & a_3 \\ b_2 & b_3 \end{vmatrix} + \mathbf{j}\begin{vmatrix} a_3 & a_1 \\ b_3 & b_1 \end{vmatrix} + \mathbf{k}\begin{vmatrix} a_1 & a_2 \\ b_1 & b_2 \end{vmatrix},$$

where a_1, a_2, a_3, and b_1, b_2, b_3 are the components of **a** *and* **b**, *respectively.*

[7] Second- and third-order determinants are usually considered in elementary calculus. Readers unfamiliar with them may consult Sec. 7.9.

For memorizing, it is useful to note that (3) can be interpreted as the expansion of the determinant

$$\mathbf{a} \times \mathbf{b} = \begin{vmatrix} \mathbf{i} & \mathbf{j} & \mathbf{k} \\ a_1 & a_2 & a_3 \\ b_1 & b_2 & b_3 \end{vmatrix} \tag{4}$$

by the first row, but we should keep in mind that this is not an ordinary determinant because the elements of the first row are vectors.

In a *left-handed* Cartesian coordinate system, $\mathbf{i} \times \mathbf{j} = -\mathbf{k}$ (Fig. 135*b*), and the above reasoning leads to $c = -1$ in (2). Hence *with respect to a left-handed Cartesian coordinate system,*

$$\mathbf{a} \times \mathbf{b} = - \begin{vmatrix} \mathbf{i} & \mathbf{j} & \mathbf{k} \\ a_1 & a_2 & a_3 \\ b_1 & b_2 & b_3 \end{vmatrix} . \tag{5}$$

Example 1

With respect to a right-handed Cartesian coordinate system, suppose that $\mathbf{a} = 4\mathbf{i} - \mathbf{k}$ and $\mathbf{b} = -2\mathbf{i} + \mathbf{j} + 3\mathbf{k}$. Then

$$\mathbf{a} \times \mathbf{b} = \begin{vmatrix} \mathbf{i} & \mathbf{j} & \mathbf{k} \\ 4 & 0 & -1 \\ -2 & 1 & 3 \end{vmatrix} = \mathbf{i} - 10\mathbf{j} + 4\mathbf{k}.$$ ▮

We may now prove the **distributive law** (4) in the previous section. From (3) it follows that the first component of $\mathbf{a} \times (\mathbf{b} + \mathbf{c})$ is

$$\begin{vmatrix} a_2 & a_3 \\ b_2 + c_2 & b_3 + c_3 \end{vmatrix} = a_2(b_3 + c_3) - a_3(b_2 + c_2)$$

$$= (a_2b_3 - a_3b_2) + (a_2c_3 - a_3c_2)$$

$$= \begin{vmatrix} a_2 & a_3 \\ b_2 & b_3 \end{vmatrix} + \begin{vmatrix} a_2 & a_3 \\ c_2 & c_3 \end{vmatrix} .$$

The expression on the right is the first component of $\mathbf{a} \times \mathbf{b} + \mathbf{a} \times \mathbf{c}$. For the other two components of that vector the consideration is similar. This proves the first of the relations (4) in the previous section, and the second relation in (4) can be proved by the same argument. ▮

The reader may prove the following criterion for linear dependence and independence of two vectors.

Theorem 1

Two vectors form a linearly dependent set if and only if their vector product is the zero vector.

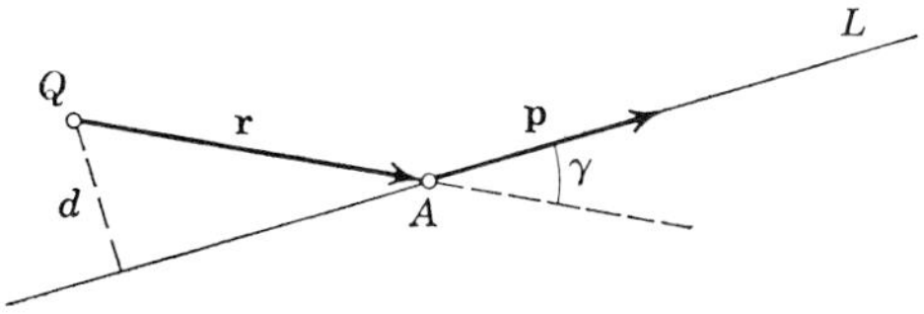

Fig. 136. Moment of a force

The definition of cross multiplication is suggested by many applications, two of which may be illustrated by the following examples. Further physical and geometrical applications will be considered later.

Example 2. Moment of a force

In mechanics the moment m of a force $\mathbf{p}$ about a point Q is defined as the product $m = |\mathbf{p}|d$, where d is the (perpendicular) distance between Q and the line of action L of $\mathbf{p}$ (Fig. 136). If $\mathbf{r}$ is the vector from Q to any point A on L, then $d = |\mathbf{r}| \sin \gamma$ (Fig. 136) and

$$m = |\mathbf{r}|\,|\mathbf{p}| \sin \gamma.$$

Since γ is the angle between $\mathbf{r}$ and $\mathbf{p}$,

$$m = |\mathbf{r} \times \mathbf{p}|,$$

as follows from (1) in the last section. The vector

$$\mathbf{m} = \mathbf{r} \times \mathbf{p} \tag{6}$$

is called the **moment vector** or **vector moment** of $\mathbf{p}$ about Q. Its magnitude is m, and its direction is that of the axis of the rotation about Q which $\mathbf{p}$ has the tendency to produce.

Example 3. Velocity of a rotating body

A rotation of a rigid body B in space can be simply and uniquely described by a vector $\mathbf{w}$ as follows. The direction of $\mathbf{w}$ is that of the axis of rotation and such that the rotation appears clockwise, if one looks from the initial point of $\mathbf{w}$ to its terminal point. The length of $\mathbf{w}$ is equal to the **angular speed** ω (>0) of the rotation, that is, the linear (or tangential) speed of a point of B divided by its distance from the axis of rotation.

Let P be any point of B and d its distance from the axis. Then P has the speed ωd. Let $\mathbf{r}$ be the position vector of P referred to a coordinate system with origin 0 on the axis of rotation. Then $d = |\mathbf{r}| \sin \gamma$ where γ is the angle between $\mathbf{w}$ and $\mathbf{r}$. Therefore,

$$\omega d = |\mathbf{w}|\,|\mathbf{r}| \sin \gamma = |\mathbf{w} \times \mathbf{r}|.$$

From this and the definition of vector product we see that the velocity vector $\mathbf{v}$ of P can be represented in the form (Fig. 137)

$$\mathbf{v} = \mathbf{w} \times \mathbf{r}. \tag{7}$$

This simple formula is useful for determining $\mathbf{v}$ at any point of B.

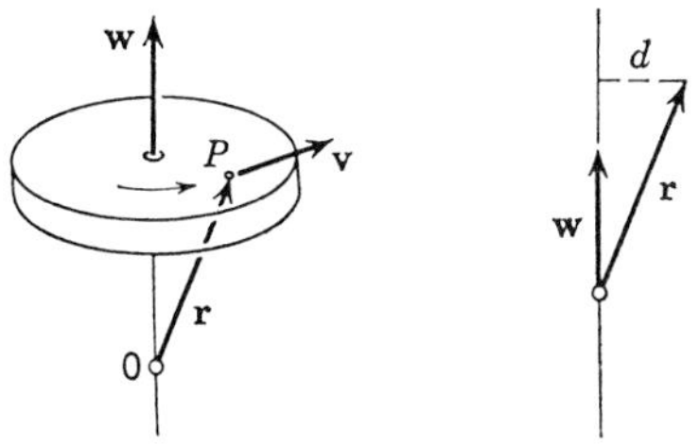

Fig. 137. Rotation of a rigid body

Problems for Sec. 6.8

With respect to a right-handed Cartesian coordinate system, let $\mathbf{a} = \mathbf{i} + 2\mathbf{j} - 3\mathbf{k}$, $\mathbf{b} = \mathbf{i} + 2\mathbf{j}$, and $\mathbf{c} = -\mathbf{i} + \mathbf{j}$. Find

1. $\mathbf{a} \times \mathbf{b}$, $\mathbf{b} \times \mathbf{a}$

2. $\mathbf{b} \times \mathbf{c}$, $|\mathbf{b} \times \mathbf{c}|$, $|\mathbf{c} \times \mathbf{b}|$

3. $\mathbf{a} \times \mathbf{c}$, $|\mathbf{a} \times \mathbf{c}|$, $\mathbf{a} \cdot \mathbf{c}$

4. $(\mathbf{a} + 2\mathbf{b}) \times \mathbf{c}$, $(\frac{1}{2}\mathbf{a} + \mathbf{b}) \times 2\mathbf{c}$

5. $(\mathbf{c} - \mathbf{a}) \times 2\mathbf{b}$, $(\mathbf{a} - \mathbf{c}) \times 2\mathbf{b}$

6. $\mathbf{a} \times (\mathbf{b} + \mathbf{c})$, $\mathbf{a} \times \mathbf{b} + \mathbf{a} \times \mathbf{c}$

7. $\mathbf{a} \times (2\mathbf{b} + 3\mathbf{c})$, $2\mathbf{a} \times \mathbf{b} + 3\mathbf{a} \times \mathbf{c}$

8. $(\mathbf{a} + \mathbf{b}) \times \mathbf{c}$, $\mathbf{a} \times \mathbf{c} - \mathbf{c} \times \mathbf{b}$

9. $\mathbf{a} \times (\mathbf{b} - \mathbf{c})$, $\mathbf{b} \times \mathbf{a} + \mathbf{a} \times \mathbf{c}$

10. $\mathbf{c} \times \mathbf{c}$, $\mathbf{a} \times \mathbf{c} + \mathbf{c} \times \mathbf{a}$

11. $(\mathbf{a} + 2\mathbf{b}) \times \mathbf{b}$, $\mathbf{a} \times (2\mathbf{a} + \mathbf{b})$

12. $(\mathbf{a} \times \mathbf{b}) \times \mathbf{c}$, $\mathbf{a} \times (\mathbf{b} \times \mathbf{c})$

13. What properties of cross multiplication do Probs. 1, 4 and 6 illustrate?

14. Show that $\mathbf{i} \times \mathbf{j} = \mathbf{k}$, $\mathbf{k} \times \mathbf{i} = \mathbf{j}$, $\mathbf{j} \times \mathbf{k} = \mathbf{i}$, assuming right-handed coordinates. Find $(\mathbf{i} \times \mathbf{j}) \times \mathbf{k}$, $\mathbf{i} \times (\mathbf{j} \times \mathbf{k})$.

Find the area of the parallelogram of which the given vectors are adjacent sides.

15. $\mathbf{i} + \mathbf{j}$, $\mathbf{i} - \mathbf{j}$

16. $\mathbf{i} + 2\mathbf{j} - \mathbf{k}$, $\mathbf{j} + \mathbf{k}$

17. $4\mathbf{i} - \mathbf{j} - \mathbf{k}$, $\mathbf{i} + 2\mathbf{j}$

18. $\mathbf{i} - \mathbf{j} + \mathbf{k}$, $\mathbf{i} + \mathbf{j} - \mathbf{k}$

Find the area of the parallelogram that has the following vertices in the xy-plane.

19. $(0, 0)$, $(2, 2)$, $(-1, 1)$, $(1, 3)$

20. $(1, 2)$, $(0, 0)$, $(2, 6)$, $(1, 4)$

21. $(-4, 2)$, $(-6, 5)$, $(-3, 6)$, $(-5, 9)$

22. $(8, -3)$, $(10, 1)$, $(9, 0)$, $(11, 4)$

Find the area of the triangle that has the following vertices.

23. $(0, 0, 0)$, $(0, 1, 0)$, $(1, 1, 0)$

24. $(3, 2, 0)$, $(1, -1, 0)$, $(2, 3, 0)$

25. $(0, 4, 0)$, $(1, 1, 1)$, $(-2, 1, -3)$

26. $(4, -2, 6)$, $(6, -1, 7)$, $(5, 0, 5)$

Using (5), Sec. 6.7, find $|\mathbf{a} \times \mathbf{b}|$, where

27. $\mathbf{a} = \mathbf{i} + \mathbf{j} - \mathbf{k}$, $\mathbf{b} = 3\mathbf{k}$

28. $\mathbf{a} = 2\mathbf{i} + 4\mathbf{k}$, $\mathbf{b} = 5\mathbf{j} - \mathbf{k}$

29. $\mathbf{a} = 3\mathbf{i} - 4\mathbf{j}$, $\mathbf{b} = \mathbf{i} - \mathbf{j} - \mathbf{k}$

30. $\mathbf{a} = 5\mathbf{i} + 2\mathbf{j} + \mathbf{k}$, $\mathbf{b} = 2\mathbf{j} - \mathbf{k}$

Are the following vectors parallel? Orthogonal?

31. $-2\mathbf{i} + 4\mathbf{j}$, $3\mathbf{k}$

32. $3\mathbf{i} + 2\mathbf{j} + \mathbf{k}$, $6\mathbf{i} + 4\mathbf{j} + 2\mathbf{k}$

33. $\mathbf{i} + \mathbf{j}$, $\mathbf{i} - \mathbf{j}$

34. $\mathbf{i} - 2\mathbf{j} + 3\mathbf{k}$, $3\mathbf{i} + 5\mathbf{j}$

35. $-\mathbf{i} + \mathbf{j} - 3\mathbf{k}$, $4\mathbf{i} - 4\mathbf{j} + 12\mathbf{k}$

36. $\mathbf{i} + \mathbf{j} - 2\mathbf{k}$, $-3\mathbf{i} - 3\mathbf{j} + 6\mathbf{k}$

Find two unit vectors perpendicular to both vectors.

37. $\mathbf{i}$, $\mathbf{j}$

38. $\mathbf{i} - \mathbf{j} + 2\mathbf{k}$, $2\mathbf{i} + 3\mathbf{k}$

39. $2\mathbf{j} - 3\mathbf{k}$, $2\mathbf{i}$

40. $3\mathbf{i} + 2\mathbf{j} + 5\mathbf{k}$, $3\mathbf{j} - \mathbf{k}$

Find a unit normal vector to the plane through the following points.

41. $(0, 0, 0)$, $(0, 1, 0)$, $(0, 0, 1)$

42. $(-1, 2, 3)$, $(1, 1, 1)$, $(2, -1, 3)$

43. $(0, -1, 0)$, $(2, 2, 1)$, $(1, 0, 1)$

44. $(3, 2, 8)$, $(1, 3, 3)$, $(0, 2, 1)$

45. Find a vector $\mathbf{v}$ parallel to the intersection of the planes $2x + 3y + 4z = 0$ and $x - y + z = 2$.

46. Find a vector **v** parallel to the plane $x + y + z = 1$ and perpendicular to the line $y = x$, $z = 0$.

A force **p** acts on a line through a point A. Find the moment vector **m** of **p** about a point Q, where the force, the point A and the point Q are:

47. $\mathbf{i} + \mathbf{j}$, $(0, 0, 0)$, $(0, 1, 0)$

48. $2\mathbf{i} + \mathbf{j}$, $(1, 1, 1)$, $(2, 2, 2)$

49. $\mathbf{i}$, $(1, 1, 0)$, $(-5, 1, 0)$

50. $3\mathbf{i} - \mathbf{j} + 2\mathbf{k}$, $(0, -1, 4)$, $(3, 0, 2)$

6.9 Scalar Triple Product. Other Repeated Products

Repeated products of vectors having three or more factors occur frequently in applications. The most important of these products is the **scalar triple product** or *mixed triple product* $\mathbf{a} \cdot (\mathbf{b} \times \mathbf{c})$ of three vectors. With respect to any right-handed Cartesian coordinate system, let

$$\mathbf{a} = a_1\mathbf{i} + a_2\mathbf{j} + a_3\mathbf{k}, \qquad \mathbf{b} = b_1\mathbf{i} + b_2\mathbf{j} + b_3\mathbf{k}, \qquad \mathbf{c} = c_1\mathbf{i} + c_2\mathbf{j} + c_3\mathbf{k}.$$

Then from (4), Sec. 6.8, it follows that

$$\mathbf{a} \cdot (\mathbf{b} \times \mathbf{c}) = (a_1\mathbf{i} + a_2\mathbf{j} + a_3\mathbf{k}) \cdot \begin{vmatrix} \mathbf{i} & \mathbf{j} & \mathbf{k} \\ b_1 & b_2 & b_3 \\ c_1 & c_2 & c_3 \end{vmatrix}.$$

From this and (8), Sec. 6.5, we see that the scalar triple product takes the form

$$\mathbf{a} \cdot (\mathbf{b} \times \mathbf{c}) = \begin{vmatrix} a_1 & a_2 & a_3 \\ b_1 & b_2 & b_3 \\ c_1 & c_2 & c_3 \end{vmatrix}. \tag{1}$$

The scalar triple product $\mathbf{a} \cdot (\mathbf{b} \times \mathbf{c})$ will be denoted by $(\mathbf{a\ b\ c})$.

Since interchanging of two rows reverses the sign of the determinant, we have

$$(\mathbf{a\ b\ c}) = -(\mathbf{b\ a\ c}), \qquad \text{etc.} \tag{2}$$

Interchanging twice, we obtain

$$(\mathbf{a\ b\ c}) = (\mathbf{b\ c\ a}) = (\mathbf{c\ a\ b}). \tag{3}$$

Now, by definition

$$(\mathbf{a\ b\ c}) = \mathbf{a} \cdot (\mathbf{b} \times \mathbf{c}), \qquad (\mathbf{c\ a\ b}) = \mathbf{c} \cdot (\mathbf{a} \times \mathbf{b}),$$

and since dot multiplication is commutative, the last expression is equal to $(\mathbf{a} \times \mathbf{b}) \cdot \mathbf{c}$. Therefore,

$$\mathbf{a} \cdot (\mathbf{b} \times \mathbf{c}) = (\mathbf{a} \times \mathbf{b}) \cdot \mathbf{c}. \tag{4}$$

Furthermore, for any constant k,

$$(k\mathbf{a}\ \mathbf{b}\ \mathbf{c}) = k(\mathbf{a}\ \mathbf{b}\ \mathbf{c}). \tag{5}$$

The absolute value of the scalar triple product $(\mathbf{a}\ \mathbf{b}\ \mathbf{c})$ *has a simple geometrical interpretation. It is equal to the volume of the parallelepiped P with* **a, b, c** *as adjacent edges.*

Indeed, from (1) in Sec. 6.5 we obtain

$$(\mathbf{a}\ \mathbf{b}\ \mathbf{c}) = \mathbf{a} \cdot (\mathbf{b} \times \mathbf{c}) = |\mathbf{a}|\,|\mathbf{b} \times \mathbf{c}| \cos \beta$$

where β is the angle between **a** and the product vector $\mathbf{b} \times \mathbf{c}$. Now $|\mathbf{b} \times \mathbf{c}|$ is the area of the base of P, and the altitude h of P is equal to the absolute value of $|\mathbf{a}| \cos \beta$ (Fig. 138). This proves our statement.

From this geometrical consideration it follows that the value of the scalar triple product is a real number which is independent of the choice of right-handed Cartesian coordinates in space. But we should keep in mind that for left-handed Cartesian coordinate systems we have to use (5) instead of (4), Sec. 6.8, and this leads to a minus sign in front of the determinant in (1). We may also say that the value of the determinant is **invariant** under transformations of right-handed into right-handed, or left-handed into left-handed, Cartesian coordinate systems, but is multiplied by -1 under a transition from a right-handed to a left-handed system (or conversely).

Example 1. Tetrahedron

Find the volume of the tetrahedron with **a, b, c** as adjacent edges where, with respect to right-handed Cartesian coordinates,

$$\mathbf{a} = \mathbf{i} + 2\mathbf{k}, \qquad \mathbf{b} = 4\mathbf{i} + 6\mathbf{j} + 2\mathbf{k}, \qquad \mathbf{c} = 3\mathbf{i} + 3\mathbf{j} - 6\mathbf{k}.$$

The volume V of the parallelepiped having **a, b, c** as adjacent edges is obtained from the scalar triple product

$$(\mathbf{a}\ \mathbf{b}\ \mathbf{c}) = \begin{vmatrix} 1 & 0 & 2 \\ 4 & 6 & 2 \\ 3 & 3 & -6 \end{vmatrix} = -54$$

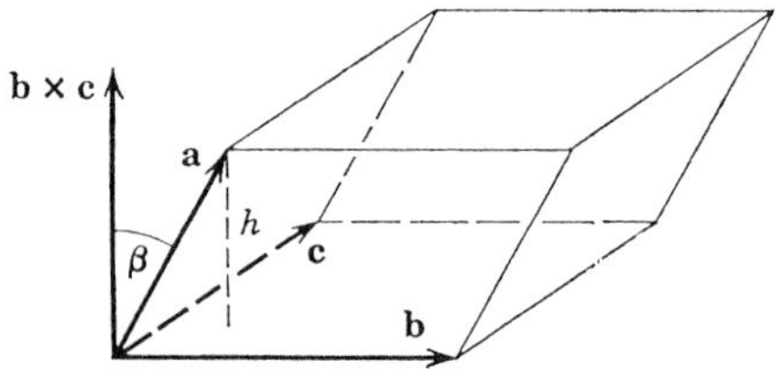

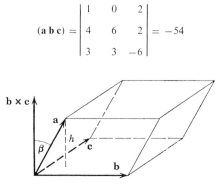

Fig. 138. Geometrical interpretation of a scalar triple product

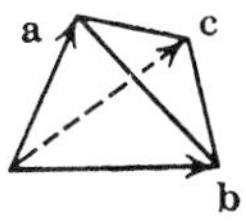

Fig. 139. Tetrahedron

and $V = 54$; the minus sign indicates that the vectors **a, b, c,** in this order, form a left-handed triple. The volume of the tetrahedron is $\frac{1}{6}$ of that of the parallelepiped, namely, 9. ∎

From the geometrical interpretation of a scalar triple product we also obtain a useful criterion for linear dependence and independence of three vectors. These vectors form a linearly dependent set if and only if they are coplanar (including the special case that they are collinear; cf. Sec. 6.4). This implies

Theorem 1 (Linear dependence)
Three vectors form a linearly dependent set if and only if their scalar triple product is zero.

Other repeated products occurring in applications may be expressed in terms of dot, vector, and scalar triple products. An important corresponding formula is

$$\mathbf{b} \times (\mathbf{c} \times \mathbf{d}) = (\mathbf{b} \cdot \mathbf{d})\mathbf{c} - (\mathbf{b} \cdot \mathbf{c})\mathbf{d} \tag{6}$$

(proof below). It implies the **identity of Lagrange**

$$(\mathbf{a} \times \mathbf{b}) \cdot (\mathbf{c} \times \mathbf{d}) = (\mathbf{a} \cdot \mathbf{c})(\mathbf{b} \cdot \mathbf{d}) - (\mathbf{a} \cdot \mathbf{d})(\mathbf{b} \cdot \mathbf{c}) \tag{7}$$

(whose proof is left to the reader; cf. Prob. 39, below), as well as (cf. Prob. 40)

$$(\mathbf{a} \times \mathbf{b}) \times (\mathbf{c} \times \mathbf{d}) = (\mathbf{a}\ \mathbf{b}\ \mathbf{d})\mathbf{c} - (\mathbf{a}\ \mathbf{b}\ \mathbf{c})\mathbf{d}. \tag{8}$$

Before we prove (6), we note that (6) also implies

$$(\mathbf{b} \times \mathbf{c}) \times \mathbf{d} = -\mathbf{d} \times (\mathbf{b} \times \mathbf{c}) = (\mathbf{d} \cdot \mathbf{b})\mathbf{c} - (\mathbf{d} \cdot \mathbf{c})\mathbf{b}.$$

This shows that in general $\mathbf{b} \times (\mathbf{c} \times \mathbf{d})$ and $(\mathbf{b} \times \mathbf{c}) \times \mathbf{d}$ are different vectors, that is, *cross multiplication is* **not associative** and the parentheses in (6) are important and cannot be omitted. For instance, with respect to a right-handed Cartesian coodinate system,

$$(\mathbf{i} \times \mathbf{j}) \times \mathbf{j} = \mathbf{k} \times \mathbf{j} = -\mathbf{i} \qquad \text{but} \qquad \mathbf{i} \times (\mathbf{j} \times \mathbf{j}) = \mathbf{0}.$$

Proof of (6). We choose a right-handed Cartesian coordinate system such that the x-axis has the direction of **d** and the xy-plane contains **c.** Then the vectors in (6) are of the form

$$\mathbf{b} = b_1\mathbf{i} + b_2\mathbf{j} + b_3\mathbf{k}, \qquad \mathbf{c} = c_1\mathbf{i} + c_2\mathbf{j}, \qquad \mathbf{d} = d_1\mathbf{i}.$$

Hence $\mathbf{c} \times \mathbf{d} = -c_2 d_1\mathbf{k}$ and, furthermore,

$$\mathbf{b} \times (\mathbf{c} \times \mathbf{d}) = \begin{vmatrix} \mathbf{i} & \mathbf{j} & \mathbf{k} \\ b_1 & b_2 & b_3 \\ 0 & 0 & -c_2 d_1 \end{vmatrix} = -b_2 c_2 d_1 \mathbf{i} + b_1 c_2 d_1 \mathbf{j}.$$

On the other hand, we obtain

$$(\mathbf{b} \cdot \mathbf{d})\mathbf{c} - (\mathbf{b} \cdot \mathbf{c})\mathbf{d} = b_1 d_1 (c_1 \mathbf{i} + c_2 \mathbf{j}) - (b_1 c_1 + b_2 c_2) d_1 \mathbf{i} = b_1 c_2 d_1 \mathbf{j} - b_2 c_2 d_1 \mathbf{i}.$$

This proves (6) for our special coordinate system. Now the length and direction of a vector and a vector product, and the value of an inner product are independent of the choice of the coordinates. Furthermore, the representation of $\mathbf{b} \times (\mathbf{c} \times \mathbf{d})$ in terms of $\mathbf{i}, \mathbf{j}, \mathbf{k}$ will be the same for right-handed and left-handed systems, because of the double cross multiplication. Hence, (6) holds in any Cartesian coodinate system, and the proof is complete. ∎

Problems for Sec. 6.9

Find the value of the scalar triple product of the given vectors (taken in the given order), whose components are referred to right-handed Cartesian coordinates.

1. $\mathbf{i}, \mathbf{j}, \mathbf{k}$

2. $\mathbf{j}, \mathbf{k}, \mathbf{i}$

3. $\mathbf{i}, \mathbf{k}, \mathbf{j}$

4. $2\mathbf{i}, 3\mathbf{j} + \mathbf{k}, 4\mathbf{i} - \mathbf{k}$

5. $\mathbf{j}, \mathbf{j} + \mathbf{k}, 3\mathbf{i} - 2\mathbf{j}$

6. $3\mathbf{i} - \mathbf{j} + \mathbf{k}, \mathbf{j} + \mathbf{k}, -\mathbf{i} + 2\mathbf{j} - 3\mathbf{k}$

7. $2\mathbf{j} + 3\mathbf{k}, \mathbf{i} + 4\mathbf{j}, 6\mathbf{k}$

8. $4\mathbf{i} + 2\mathbf{j} + \mathbf{k}, \mathbf{i} - \mathbf{j} - \mathbf{k}, 3\mathbf{i} + 2\mathbf{j} + 4\mathbf{k}$

9. $2\mathbf{i} - 3\mathbf{k}, -\mathbf{i}, 3\mathbf{i} - \mathbf{j} + 5\mathbf{k}$

10. $\mathbf{i} + \mathbf{j} + \mathbf{k}, 3\mathbf{i}, 4\mathbf{j} + 5\mathbf{k}$

Are the following vectors linearly dependent or independent?

11. $3\mathbf{i}, -\mathbf{j}$

12. $\mathbf{i} - 6\mathbf{j} + 2\mathbf{k}, 2\mathbf{j} + 7\mathbf{k}, -2\mathbf{i} + 12\mathbf{j} - 4\mathbf{k}$

13. $\mathbf{i} + \mathbf{j}, \mathbf{i} - \mathbf{j}$

14. $\mathbf{i}, \mathbf{0}, \mathbf{j}$

15. $\mathbf{i}, \mathbf{j}, \mathbf{k}, 3\mathbf{i} + 2\mathbf{j} - \mathbf{k}$

16. $4\mathbf{i} + 5\mathbf{j}, \mathbf{i} + 2\mathbf{j}, -\mathbf{i} + 3\mathbf{j}$

17. $\mathbf{i} + \mathbf{j}, \mathbf{j} + \mathbf{k}, \mathbf{k} + \mathbf{i}$

18. $\mathbf{i} + \mathbf{k}, 3\mathbf{i} - 5\mathbf{k}, 8\mathbf{k}$

19. Determine λ such that $\mathbf{a} = \mathbf{i} + \mathbf{j} + \mathbf{k}$, $\mathbf{b} = 2\mathbf{i} - 4\mathbf{k}$, and $\mathbf{c} = \mathbf{i} + \lambda\mathbf{j} + 3\mathbf{k}$ are coplanar.

20. Do the points $(4, -2, 1)$, $(5, 1, 6)$, $(2, 2, -5)$, and $(3, 5, 0)$ lie in a plane?

21. Determine λ and μ such that the points $(-1, 3, 2)$, $(-4, 2, -2)$, and $(5, \lambda, \mu)$ lie on a straight line.

22. A system of three homogeneous linear equations in three unknowns has a nontrivial solution if and only if the determinant of the coefficients of the system is zero. Using this familiar theorem, prove Theorem 1.

Find the volume of the parallelepiped that has the given vectors as adjacent edges.

23. $\mathbf{i}, -\mathbf{j}, 3\mathbf{k}$

24. $2\mathbf{j} + \mathbf{k}, \mathbf{i} - \mathbf{j}, -\mathbf{j} + 4\mathbf{k}$

25. $\mathbf{i} + \mathbf{j}, \mathbf{i} - \mathbf{j}, -\mathbf{k}$

26. $\mathbf{i}, 2\mathbf{i} + 4\mathbf{j} - \mathbf{k}, -\mathbf{i} + \mathbf{j} + 3\mathbf{k}$

27. $3\mathbf{i} + 2\mathbf{k}, -4\mathbf{j}, \mathbf{i} + \mathbf{j} - \mathbf{k}$

28. $\mathbf{j} + \mathbf{k}, \mathbf{i} + \mathbf{j}, \mathbf{i} - 3\mathbf{k}$

Find the volume of the tetrahedron that has the following vertices.

29. $(0, 1, 1)$, $(1, 0, 0)$, $(2, 2, 3)$, $(-1, 0, 4)$

30. $(2, 1, 8)$, $(3, 2, 9)$, $(2, 1, 4)$, $(3, 3, 10)$

31. $(0, 0, 0)$, $(1, 0, 0)$, $(0, 1, 0)$, $(0, 0, 1)$

32. $(-1, 0, 1)$, $(4, 4, 5)$, $(0, 1, 0)$, $(2, 2, 0)$

With respect to a right-handed Cartesian coordinate system, let $\mathbf{a} = \mathbf{i} + 2\mathbf{j} - \mathbf{k}$, $\mathbf{b} = 3\mathbf{i} - 4\mathbf{k}$, $\mathbf{c} = -\mathbf{i} + \mathbf{j}$, and $\mathbf{d} = 2\mathbf{i} - \mathbf{j} + 3\mathbf{k}$. Find

33. $(\mathbf{a} \times \mathbf{b}) \times \mathbf{c}$, $(\mathbf{a} \times \mathbf{b}) \cdot \mathbf{c}$

34. $(\mathbf{b} \times \mathbf{c}) \times \mathbf{d}$, $\mathbf{d} \times (\mathbf{b} \times \mathbf{c})$

35. $(\mathbf{a} \times \mathbf{c}) \times \mathbf{d}$, $(\mathbf{a} \times \mathbf{d}) \times \mathbf{c}$

36. $(\mathbf{b} \times \mathbf{b}) \times \mathbf{c}$, $\mathbf{b} \times (\mathbf{b} \times \mathbf{c})$

37. $(\mathbf{a} \times \mathbf{b}) \times (\mathbf{c} \times \mathbf{d})$

38. $(\mathbf{a} \times \mathbf{c}) \cdot (\mathbf{b} \times \mathbf{d})$, $(\mathbf{c} \times \mathbf{a}) \cdot (\mathbf{d} \times \mathbf{b})$

39. Prove (7). *Hint.* Take the dot product of **a** and (6).

40. Derive (8) from (6).

CHAPTER

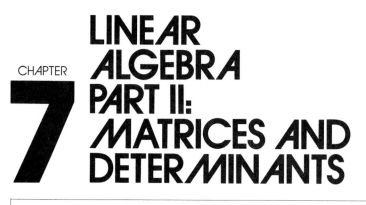

7 LINEAR ALGEBRA PART II: MATRICES AND DETERMINANTS

A matrix is a rectangular array of numbers. Such arrays occur in various branches of applied mathematics. In many cases they form the coefficients of linear transformations (examples in Sec. 7.1) or systems of linear equations (simultaneous linear equations) arising, for instance, from electrical networks, frameworks in mechanics, curve fitting in statistics, and transportation problems. Matrices are useful because they enable us to consider an array of many numbers as a single object, denote it by a single symbol, and perform calculations with these symbols in a very compact form. The "mathematical shorthand" thus obtained is very elegant and powerful and is suitable for various practical problems. It entered engineering mathematics over fifty years ago and is of increasing importance in various branches.

In this chapter we first introduce matrices and related concepts (Sec. 7.1), define the algebraic operations for matrices (Secs. 7.2–7.4), and consider systems of linear equations (solution by elimination in Sec. 7.5, existence in Sec. 7.7, solution by Cramer's rule in Sec. 7.11), determinants (Secs. 7.9 and 7.10), bilinear and other forms (Sec. 7.12), matrix eigenvalue problems (Secs. 7.13 and 7.14) and systems of differential equations (Sec. 7.15). Other important concepts in this chapter are the rank of a matrix (Secs. 7.6 and 7.11) and the inverse (Sec. 7.8).

Numerical methods are presented in Secs. 19.9–19.14 which are independent of the other sections in Chap. 19, so that they can be studied immediately after the corresponding material in the present chapter.

Prerequisite for this chapter: Chap. 6.
Sections which may be omitted in a shorter course: 7.12–7.15.
References: Appendix 1, Part C.
Answers to problems: Appendix 2.

7.1 Basic Concepts

Matrices are rectangular arrays of numbers enclosed in parentheses; for instance,

$$\begin{pmatrix} 2 & 0.4 & 8 \\ 5 & -3 & 1 \end{pmatrix}, \quad \begin{pmatrix} 0 \\ 4 \end{pmatrix}, \quad (a_1 \quad a_2 \quad a_3).$$

The definition of a matrix will be given below, and rules of calculations with matrices will be defined and explained in the next sections.

Matrices occur in connection with linear transformations or systems of linear equations. For example, consider

$$\begin{aligned} 20x_1 - 3x_2 &= y_1 \\ 0.5x_1 + 7x_2 &= y_2 \end{aligned} \tag{1}$$

If y_1 and y_2 are given numbers, this is a system of two linear equations in two unknowns x_1, x_2. The coefficients are 20, -3, 0.5, 7 and, to recognize their position we may arrange them in the way they occur in the equations, that is, in the form of a matrix,

$$\begin{pmatrix} 20 & -3 \\ 0.5 & 7 \end{pmatrix}. \tag{2}$$

If x_1, x_2 and y_1, y_2 are variable quantities (for instance, coordinates corresponding to two different coordinate systems in the plane), then our pair of formulas (1) represents a linear transformation. This transformation is completely determined by its four coefficients, that is, by the matrix (2). We may denote the matrix by a single letter, for instance **A**, and write it in the form

$$\mathbf{A} = \begin{pmatrix} a_{11} & a_{12} \\ a_{21} & a_{22} \end{pmatrix}$$

where $a_{11} = 20$, $a_{12} = -3$, $a_{21} = 0.5$, $a_{22} = 7$. The notation has the advantage that we can immediately see the position of a coefficient a_{jk} because the first subscript is the number of the row (horizontal line) and the second is the number of the column (vertical line) of a_{jk}. Let us now give the general definition of a matrix. (Rules of calculations with matrices will be defined and explained in the next sections.)

A rectangular array of (real or complex) numbers of the form

$$\begin{pmatrix} a_{11} & a_{12} & \cdots & a_{1n} \\ a_{21} & a_{22} & \cdots & a_{2n} \\ \cdot & \cdot & \cdots & \cdot \\ a_{m1} & a_{m2} & \cdots & a_{mn} \end{pmatrix}$$

is called a **matrix.** The numbers $a_{11}, \cdots, a_{mn}$ are called the **elements** of the matrix. The horizontal lines are called **rows** or *row vectors,* and the vertical lines are called **columns** or *column vectors* of the matrix. A matrix with m rows and n columns is called an $(m \times n)$ *matrix* (read "m by n matrix").

Matrices will be denoted by capital (upper case) bold-faced letters **A, B,** etc., or by (a_{jk}), (b_{jk}), etc., that is, by writing the general element of the matrix, enclosed in parentheses.

In the **double-subscript notation** *for the elements, the first subscript always denotes the row and the second subscript the column containing the given element.*

A matrix

$$(a_1 \quad a_2 \quad \cdots \quad a_n)$$

having only one row is called a **row matrix** or *row vector.* A matrix

$$\begin{pmatrix} b_1 \\ b_2 \\ \vdots \\ b_m \end{pmatrix}$$

having only one column is called a **column matrix** or *column vector.* Row and column matrices will be denoted by small (lower case) bold-faced letters. This agrees with our notation for vectors (Chap. 6), whose components may be written in the form of a row matrix or a column matrix.

The elements of a matrix may be real or complex numbers. If all the elements of a matrix are real, the matrix is called a **real matrix.**

A matrix having the same number of rows and columns is called a **square matrix,** and the number of rows is called its **order.** A square matrix of order n is also called an $(n \times n)$ *matrix* (read "n by n matrix"). The diagonal containing the elements $a_{11}, a_{22}, \cdots, a_{nn}$ of such a matrix is called the **principal diagonal.** Square matrices are of particular importance.

A definition of **equality of matrices** is suggested by that for vectors in Chap. 6 (cf. Sec. 6.2, at the end). Two matrices $\mathbf{A} = (a_{jk})$ and $\mathbf{B} = (b_{jk})$ are *equal* if and only if **A** and **B** have the same number of rows and the same number of columns and corresponding elements are equal, that is,

$$a_{jk} = b_{jk} \qquad \text{for all occurring } j \text{ and } k.$$

Then we write

$$\mathbf{A} = \mathbf{B}.$$

Any matrix obtained by omitting some rows and columns from a given $(m \times n)$ matrix **A** is called a **submatrix** of **A.**

As a matter of convention and convenience the notion "submatrix" includes **A** itself (as the matrix obtained from **A** by omitting no rows or columns).

Example 1. Submatrices of a matrix

The matrix

$$\begin{pmatrix} a_{11} & a_{12} & a_{13} \\ a_{21} & a_{22} & a_{23} \end{pmatrix}$$

contains three (2×2) submatrices, namely,

$$\begin{pmatrix} a_{11} & a_{12} \\ a_{21} & a_{22} \end{pmatrix}, \quad \begin{pmatrix} a_{11} & a_{13} \\ a_{21} & a_{23} \end{pmatrix}, \quad \begin{pmatrix} a_{12} & a_{13} \\ a_{22} & a_{23} \end{pmatrix},$$

two (1×3) submatrices (the two row vectors), three (2×1) submatrices (the column vectors), six (1×2) submatrices, namely,

$$(a_{11} \quad a_{12}), \qquad (a_{11} \quad a_{13}), \qquad (a_{12} \quad a_{13}),$$

$$(a_{21} \quad a_{22}), \qquad (a_{21} \quad a_{23}), \qquad (a_{22} \quad a_{23}),$$

and six (1×1) submatrices, $(a_{11}), (a_{12}), \cdots, (a_{23})$.

7.2 Addition of Matrices, Multiplication by Numbers

Since we want to do calculations with matrices, we shall now introduce algebraic operations for matrices. We shall define two such operations, which are called *addition of matrices* and *multiplication of matrices by numbers*. These definitions are motivated and suggested by various applications (to be considered later in this chapter) as well as by the corresponding definitions for vectors in Sec. 6.3, which they are supposed to include as special cases.

Addition of matrices is defined only for matrices having the same number of rows and the same number of columns, and is then defined as follows. The **sum** of two $(m \times n)$ matrices $\mathbf{A} = (a_{jk})$ and $\mathbf{B} = (b_{jk})$ is the $(m \times n)$ matrix $\mathbf{C} = (c_{jk})$ with elements

$$\text{(1)} \qquad c_{jk} = a_{jk} + b_{jk} \qquad \begin{matrix} j = 1, \cdots, m \\ k = 1, \cdots, n \end{matrix}$$

and is written

$$\mathbf{C} = \mathbf{A} + \mathbf{B}.$$

We see that $\mathbf{A} + \mathbf{B}$ is obtained by adding corresponding elements of $\mathbf{A}$ and $\mathbf{B}$.

Example 1. Addition of matrices

If

$$\mathbf{A} = \begin{pmatrix} -4 & 6 & 3 \\ 0 & 1 & 2 \end{pmatrix} \quad \text{and} \quad \mathbf{B} = \begin{pmatrix} 5 & -1 & 0 \\ 3 & 1 & 0 \end{pmatrix}, \quad \text{then} \quad \mathbf{A} + \mathbf{B} = \begin{pmatrix} 1 & 5 & 3 \\ 3 & 2 & 2 \end{pmatrix}. \quad \blacksquare$$

We define the $(m \times n)$ **zero matrix 0** to be the $(m \times n)$ matrix with all elements zeros. Then we see from the definition that matrix addition enjoys properties which are quite similar to those of the addition of real numbers,

$$\text{(2)} \qquad \begin{aligned} &\text{(a)} \quad \mathbf{A} + \mathbf{B} = \mathbf{B} + \mathbf{A} \\ &\text{(b)} \quad (\mathbf{U} + \mathbf{V}) + \mathbf{W} = \mathbf{U} + (\mathbf{V} + \mathbf{W}) \qquad \text{(written } \mathbf{U} + \mathbf{V} + \mathbf{W}\text{)} \\ &\text{(c)} \quad \mathbf{A} + \mathbf{0} = \mathbf{A} \\ &\text{(d)} \quad \mathbf{A} + (-\mathbf{A}) = \mathbf{0} \end{aligned}$$

where $-\mathbf{A} = (-a_{jk})$ is the $(m \times n)$ matrix obtained by multiplying every element of $\mathbf{A}$ by -1 and is called the **negative** of $\mathbf{A}$.

Instead of $\mathbf{A} + (-\mathbf{B})$ we simply write $\mathbf{A} - \mathbf{B}$ and call this matrix the **difference** of $\mathbf{A}$ and $\mathbf{B}$. Obviously, its elements are obtained by subtracting corresponding elements of $\mathbf{A}$ and $\mathbf{B}$.

Multiplication of matrices by numbers is defined as follows. The **product** of an $(m \times n)$ matrix $\mathbf{A}$ by a number q is denoted by $q\mathbf{A}$ or $\mathbf{A}q$ and is the $(m \times n)$ matrix obtained by multiplying every element of $\mathbf{A}$ by q, that is,

$$(3) \qquad q\mathbf{A} = \mathbf{A}q = \begin{pmatrix} qa_{11} & qa_{12} & \cdots & qa_{1n} \\ qa_{21} & qa_{22} & \cdots & qa_{2n} \\ \cdot & \cdot & \cdots & \cdot \\ qa_{m1} & qa_{m2} & \cdots & qa_{mn} \end{pmatrix}$$

Example 2. Multiplication of matrices by numbers

If

$$\mathbf{A} = \begin{pmatrix} 2.7 & -1.8 \\ 0.9 & 3.6 \end{pmatrix}, \quad \text{then} \quad \mathbf{A} + \mathbf{A} = 2\mathbf{A} = \begin{pmatrix} 5.4 & -3.6 \\ 1.8 & 7.2 \end{pmatrix} \quad \text{and} \quad \frac{10}{9}\mathbf{A} = \begin{pmatrix} 3 & -2 \\ 1 & 4 \end{pmatrix}. \quad \blacksquare$$

From the definitions we see that for any $(m \times n)$ matrices (with fixed m and n) and any numbers,

$$(4) \qquad \begin{aligned} &\text{(a)} \quad q(\mathbf{A} + \mathbf{B}) = q\mathbf{A} + q\mathbf{B} \\ &\text{(b)} \quad (c + k)\mathbf{A} = c\mathbf{A} + k\mathbf{A} \\ &\text{(c)} \quad c(k\mathbf{A}) = (ck)\mathbf{A} \qquad \text{(written } ck\mathbf{A}\text{)} \\ &\text{(d)} \quad 1\mathbf{A} = \mathbf{A}. \end{aligned}$$

Note that $(-1)\mathbf{A} = -\mathbf{A}$, the negative of $\mathbf{A}$.

Formulas (2) and (4) express the properties which are characteristic of a vector space (cf. Sec. 6.4). This gives

Theorem 1 (Vector spaces of matrices)

All real (or complex) $(m \times n)$ matrices (with m and n fixed) form a real (or complex) vector space of dimension mn. A basis consists of the mn matrices $\mathbf{E}_{rs}$ $(r = 1, \cdots, m;\ s = 1, \cdots, n)$, where $\mathbf{E}_{rs}$ is the $(m \times n)$ matrix with $e_{rs} = 1$ (that is, the element in the rth row and sth column equals 1) and all the other elements zero.

Problems on addition and multiplication by numbers are included at the end of the next section.

7.3 Transpose of a Matrix. Special Matrices

In this short section we want to define another simple operation on a matrix and introduce certain special types of matrices which are of practical importance.

Transposition of a matrix is defined as follows. The **transpose** $\mathbf{A}^\mathsf{T}$ of an $(m \times n)$ matrix $\mathbf{A} = (a_{jk})$ is the $(n \times m)$ matrix obtained by interchanging the rows and columns in $\mathbf{A}$, that is, the jth row of $\mathbf{A}$ becomes the jth column of $\mathbf{A}^\mathsf{T}$:

$$\text{(1)} \qquad \mathbf{A}^\mathsf{T} = (a_{kj}) = \begin{pmatrix} a_{11} & a_{21} & \cdots & a_{m1} \\ a_{12} & a_{22} & \cdots & a_{m2} \\ \cdot & \cdot & \cdots & \cdot \\ a_{1n} & a_{2n} & \cdots & a_{mn} \end{pmatrix}.$$

The reader may prove that

$$\text{(2)} \qquad (\mathbf{A} + \mathbf{B})^\mathsf{T} = \mathbf{A}^\mathsf{T} + \mathbf{B}^\mathsf{T}.$$

Example 1. Transposition of a matrix

If

$$\mathbf{A} = \begin{pmatrix} 5 & -8 & 1 \\ 4 & 0 & 0 \end{pmatrix}, \qquad \text{then} \qquad \mathbf{A}^\mathsf{T} = \begin{pmatrix} 5 & 4 \\ -8 & 0 \\ 1 & 0 \end{pmatrix}.$$

Example 2. Transposition of a row matrix

The transpose of a row matrix is a column matrix and conversely. For instance, if

$$\mathbf{b} = (7 \quad 5 \quad -2), \qquad \text{then} \qquad \mathbf{b}^\mathsf{T} = \begin{pmatrix} 7 \\ 5 \\ -2 \end{pmatrix}.$$ ∎

A **real** square matrix $\mathbf{A} = (a_{jk})$ is said to be **symmetric** if it is equal to its transpose,

$$\text{(3)} \qquad \mathbf{A}^\mathsf{T} = \mathbf{A}, \qquad \text{that is,} \qquad a_{kj} = a_{jk} \qquad (j, k = 1, \cdots, n).$$

A **real** square matrix $\mathbf{A} = (a_{jk})$ is said to be **skew-symmetric** if

$$\text{(4)} \qquad \mathbf{A}^\mathsf{T} = -\mathbf{A}, \qquad \text{that is,} \qquad a_{kj} = -a_{jk} \qquad (j, k = 1, \cdots, n).$$

(Similar notions for complex matrices will be defined later, in Sec. 7.12). Note that for $k = j$ in (4) we have $a_{jj} = -a_{jj}$, which implies that the elements in the principal diagonal of a skew-symmetric matrix are all zero.

Any real square matrix **A** may be written as the sum of a symmetric matrix **R** and a skew-symmetric matrix **S**, where

$$\mathbf{R} = \tfrac{1}{2}(\mathbf{A} + \mathbf{A}^\mathsf{T}) \qquad \text{and} \qquad \mathbf{S} = \tfrac{1}{2}(\mathbf{A} - \mathbf{A}^\mathsf{T}).$$

Example 3. Symmetric and skew-symmetric matrices

The matrices

$$\mathbf{A} = \begin{pmatrix} -3 & 1 & 5 \\ 1 & 0 & -2 \\ 5 & -2 & 4 \end{pmatrix} \quad \text{and} \quad \mathbf{B} = \begin{pmatrix} 0 & -4 & 1 \\ 4 & 0 & -5 \\ -1 & 5 & 0 \end{pmatrix}$$

are symmetric and skew-symmetric, respectively. The matrix

$$\mathbf{A} = \begin{pmatrix} 2 & 3 \\ 5 & -1 \end{pmatrix}$$

is neither symmetric nor skew-symmetric. **A** may be written in the form $\mathbf{A} = \mathbf{R} + \mathbf{S}$ where

$$\mathbf{R} = \tfrac{1}{2}(\mathbf{A} + \mathbf{A}^\mathsf{T}) = \begin{pmatrix} 2 & 4 \\ 4 & -1 \end{pmatrix} \quad \text{and} \quad \mathbf{S} = \tfrac{1}{2}(\mathbf{A} - \mathbf{A}^\mathsf{T}) = \begin{pmatrix} 0 & -1 \\ 1 & 0 \end{pmatrix}$$

are symmetric and skew-symmetric, respectively. ∎

A square matrix $\mathbf{A} = (a_{jk})$ whose elements above the principal diagonal (or below the principal diagonal) are all zero is called a **triangular matrix.**

For example,

$$\mathbf{T}_1 = \begin{pmatrix} 1 & 0 & 0 \\ -2 & 3 & 0 \\ 5 & 0 & 2 \end{pmatrix} \quad \text{and} \quad \mathbf{T}_2 = \begin{pmatrix} 1 & 6 & -1 \\ 0 & 2 & 3 \\ 0 & 0 & 4 \end{pmatrix}$$

are triangular matrices.

A square matrix $\mathbf{A} = (a_{jk})$ whose elements above and below the principal diagonal are all zero, that is, $a_{jk} = 0$ for all $j \neq k$, is called a **diagonal matrix.**

For example,

$$\begin{pmatrix} 2 & 0 & 0 \\ 0 & 1 & 0 \\ 0 & 0 & -4 \end{pmatrix} \quad \text{and} \quad \begin{pmatrix} 2 & 0 & 0 \\ 0 & 2 & 0 \\ 0 & 0 & 2 \end{pmatrix}$$

are diagonal matrices.

An n-rowed diagonal matrix whose elements in the principal diagonal are all 1 is called a **unit matrix** (for reasons to be explained in Sec. 7.8) and is denoted by $\mathbf{I}_n$ or simply by **I**. For example, the (3×3) unit matrix is

$$\mathbf{I} = \begin{pmatrix} 1 & 0 & 0 \\ 0 & 1 & 0 \\ 0 & 0 & 1 \end{pmatrix}.$$

Problems for Secs. 7.1–7.3

Let $\mathbf{A} = \begin{pmatrix} 2 & 1 \\ 1 & 3 \end{pmatrix}$, $\mathbf{B} = \begin{pmatrix} -2 & 0 \\ 3 & 4 \end{pmatrix}$, $\mathbf{C} = \begin{pmatrix} 3 & 2 & 0 \\ 1 & 0 & 4 \end{pmatrix}$. Find

1. $\mathbf{A} + \mathbf{B}$, $\mathbf{B} + \mathbf{A}$

2. $\mathbf{A} - \mathbf{B}$, $\mathbf{B} - \mathbf{A}$

3. $4\mathbf{A} - 2\mathbf{B}$, $2(2\mathbf{A} - \mathbf{B})$

4. $\mathbf{A}^\mathsf{T}$, $\mathbf{B}^\mathsf{T}$, $\mathbf{C}^\mathsf{T}$, $(\mathbf{B}^\mathsf{T})^\mathsf{T}$

5. $(\mathbf{A} + \mathbf{B})^\mathsf{T}$, $\mathbf{A}^\mathsf{T} + \mathbf{B}^\mathsf{T}$

6. $\mathbf{A} + \mathbf{A}^\mathsf{T}$, $\mathbf{A} - \mathbf{A}^\mathsf{T}$

7. Is $\mathbf{C} + \mathbf{C}^\mathsf{T}$ defined?

8. Is $\mathbf{A} + \mathbf{C}$ defined?

9. Are $\mathbf{A}$ and $\mathbf{B}$ symmetric?

10. Represent $\mathbf{B}$ as the sum of a symmetric and a skew-symmetric matrix.

Let $\mathbf{M} = \begin{pmatrix} 2 & 0 & 0 \\ -1 & 1 & 0 \\ 3 & -4 & 0 \end{pmatrix}$, $\mathbf{N} = \begin{pmatrix} 3 & 8 \\ 4 & -1 \\ 7 & 2 \end{pmatrix}$, $\mathbf{P} = \begin{pmatrix} 2 & 1 & 4 \\ 8 & 0 & 0 \\ 3 & -1 & 1 \end{pmatrix}$. Find

11. $\mathbf{M} + \mathbf{P}$, $\mathbf{P} + \mathbf{M}$

12. $\mathbf{M} - \mathbf{P}$, $\mathbf{P} - \mathbf{M}$

13. $3\mathbf{M} + 2\mathbf{P}$

14. $\mathbf{M} + \mathbf{M}^\mathsf{T}$, $\mathbf{M} - \mathbf{M}^\mathsf{T}$

15. $(\mathbf{M} - \mathbf{P})^\mathsf{T}$, $\mathbf{M}^\mathsf{T} - \mathbf{P}^\mathsf{T}$

16. $(3\mathbf{P})^\mathsf{T}$, $3\mathbf{P}^\mathsf{T}$

17. Are $\mathbf{M} + \mathbf{N}$, $\mathbf{M} + \mathbf{N}^\mathsf{T}$, $\mathbf{M} + \mathbf{P}^\mathsf{T}$, $\mathbf{N} + \mathbf{C}^\mathsf{T}$ (see before) defined?

18. Prove (2) and (4) in Sec. 7.2. Give illustrative examples.

19. Prove $(\mathbf{A} + \mathbf{B})^\mathsf{T} = \mathbf{A}^\mathsf{T} + \mathbf{B}^\mathsf{T}$ and illustrate it with two examples.

20. Prove $(a\mathbf{A})^\mathsf{T} = a\mathbf{A}^\mathsf{T}$, $(\mathbf{A}^\mathsf{T})^\mathsf{T} = \mathbf{A}$ and give an example for each formula.

21. Prove that the symmetric two-rowed square matrices form a vector space. What is the dimension? (Cf. Sec. 6.4.)

22. Do the skew-symmetric two-rowed square matrices form a vector space? The two-rowed square matrices with nonnegative real elements?

23. Find a basis of the vector space of all real two-rowed square matrices. Of the vector space of all real symmetric two-rowed square matrices.

24. Find all n-rowed square matrices which are both symmetric and skew-symmetric.

25. Show that all real symmetric n-rowed square matrices form a vector space, determine the dimension of that space and find a basis.

26. **(Nodal incidence matrix)** Matrices have various engineering applications, as we shall see. For instance, they may be used to characterize connections (in electrical networks, nets of roads connecting cities, etc.). The figure shows an electrical

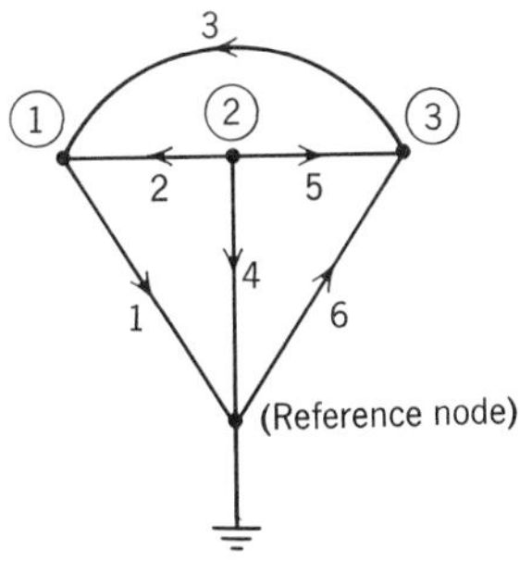

Branch	1	2	3	4	5	6
Node ①	1	−1	−1	0	0	0
Node ②	0	1	0	1	1	0
Node ③	0	0	1	0	−1	−1

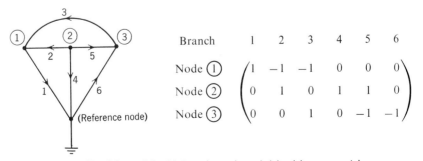

Problem 26. Network and nodal incidence matrix

network having 6 *branches* (connections) and 4 *nodes* (points where two or more branches come together). One node is the *reference node* (node whose voltage is 0 since the node is grounded). We number the other nodes and number and direct the branches. This we do arbitrarily. The network can now be described by a matrix $\mathbf{A} = (a_{jk})$, where

$$a_{jk} = \begin{cases} +1 \text{ if branch } k \text{ leaves node } \textcircled{j} \\ -1 \text{ if branch } k \text{ enters node } \textcircled{j} \\ \;\;0 \text{ if branch } k \text{ does not touch node } \textcircled{j}. \end{cases}$$

A is called the *nodal incidence matrix* of the network. Show that for the network in the figure, **A** has the given form.

27. Find the nodal incidence matrix of the network shown in the figure.

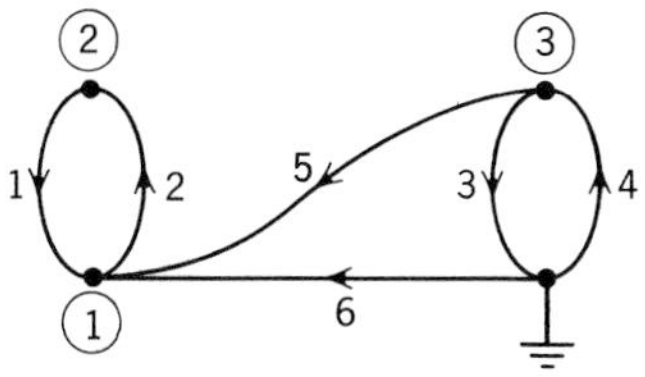

Problem 27. Electrical network

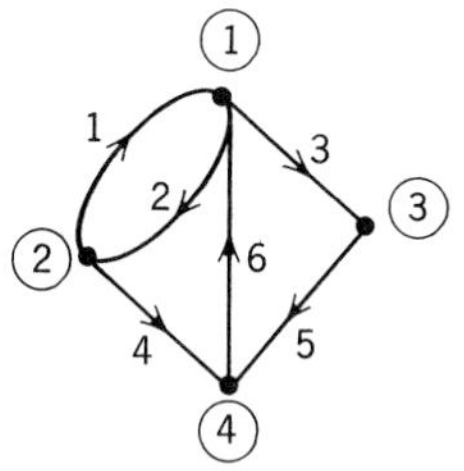

Problem 29. Net of one-way streets

28. Draw a sketch of the network whose nodal incidence matrix is

$$\mathbf{A} = \begin{pmatrix} 1 & 0 & 0 & -1 \\ -1 & 1 & 0 & 0 \\ 0 & -1 & 1 & 0 \end{pmatrix}.$$

29. Methods of electrical network analysis have applications in other fields, too. Determine the analogue of the nodal incidence matrix for the net of one-way streets (directions as indicated by the arrows) shown in the figure.

30. (Mesh incidence matrix) A network can also be characterized by the *mesh incidence matrix* $\mathbf{M} = (m_{jk})$, where

$$m_{jk} = \begin{cases} +1 \text{ if branch } k \text{ is in mesh } \boxed{j} \text{ and has the same orientation} \\ -1 \text{ if branch } k \text{ is in mesh } \boxed{j} \text{ and has the opposite orientation} \\ \;\;0 \text{ if branch } k \text{ is not in mesh } \boxed{j} \end{cases}$$

where a *mesh* is a loop with no branch in its interior (or in its exterior). Here, the meshes are numbered and directed (oriented) in an arbitrary fashion. Show that for the network in the figure on the next page,

$$\mathbf{M} = \begin{pmatrix} 1 & 1 & 0 & -1 & 0 & 0 \\ 0 & 0 & 0 & 1 & -1 & 1 \\ 0 & -1 & 1 & 0 & 1 & 0 \\ 1 & 0 & 1 & 0 & 0 & 1 \end{pmatrix}.$$

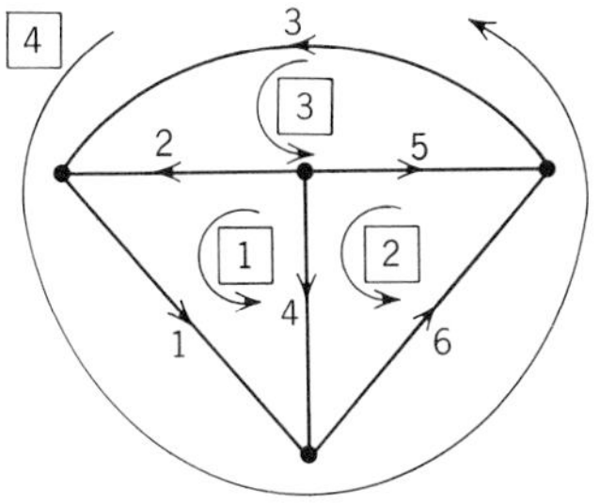

Problem 30. Network

7.4 Matrix Multiplication

We shall now define the multiplication of a matrix by a matrix. The definition will be suggested by the use of matrices in connection with linear transformations. To illustrate this, let us start with a very simple case.

We consider three coordinate systems in the plane which we denote as the w_1w_2-system, the x_1x_2-system, and the y_1y_2-system, and we assume that these systems are related by transformations

$$\text{(1)} \qquad \begin{aligned} y_1 &= a_{11}x_1 + a_{12}x_2 \\ y_2 &= a_{21}x_1 + a_{22}x_2 \end{aligned}$$

and

$$\text{(2)} \qquad \begin{aligned} x_1 &= b_{11}w_1 + b_{12}w_2 \\ x_2 &= b_{21}w_1 + b_{22}w_2. \end{aligned}$$

which are called *linear transformations*. Clearly, the y_1y_2-coordinates can be obtained directly from the w_1w_2-coordinates by a single linear transformation of the form

$$\text{(3)} \qquad \begin{aligned} y_1 &= c_{11}w_1 + c_{12}w_2 \\ y_2 &= c_{21}w_1 + c_{22}w_2. \end{aligned}$$

The coefficients $c_{11}, c_{12}, c_{21}, c_{22}$ of this transformation can be found by inserting (2) into (1),

$$\begin{aligned} y_1 &= a_{11}(b_{11}w_1 + b_{12}w_2) + a_{12}(b_{21}w_1 + b_{22}w_2) \\ y_2 &= a_{21}(b_{11}w_1 + b_{12}w_2) + a_{22}(b_{21}w_1 + b_{22}w_2) \end{aligned}$$

and comparing this and (3). The result is

$$\begin{aligned} c_{11} &= a_{11}b_{11} + a_{12}b_{21} \qquad & c_{12} &= a_{11}b_{12} + a_{12}b_{22} \\ c_{21} &= a_{21}b_{11} + a_{22}b_{21} \qquad & c_{22} &= a_{21}b_{12} + a_{22}b_{22} \end{aligned}$$

or briefly

$$c_{jk} = a_{j1}b_{1k} + a_{j2}b_{2k} = \sum_{i=1}^{2} a_{ji}b_{ik} \qquad j, k = 1, 2. \tag{4}$$

Now the coefficient matrices of the transformations (1) and (2) are

$$\mathbf{A} = \begin{pmatrix} a_{11} & a_{12} \\ a_{21} & a_{22} \end{pmatrix} \quad \text{and} \quad \mathbf{B} = \begin{pmatrix} b_{11} & b_{12} \\ b_{21} & b_{22} \end{pmatrix},$$

and we define the product **AB** (in this order) of **A** and **B** to be the coefficient matrix

$$\mathbf{C} = \begin{pmatrix} c_{11} & c_{12} \\ c_{21} & c_{22} \end{pmatrix}$$

of the "composed transformation" (3), that is,

$$\mathbf{C} = \mathbf{AB}$$

where the elements of **C** are given by (4).

The following is worth noting. If we write **A** in terms of row vectors, say,

$$\mathbf{A} = \begin{pmatrix} \mathbf{a}_1 \\ \mathbf{a}_2 \end{pmatrix} \quad \text{where} \quad \begin{aligned} \mathbf{a}_1 &= (a_{11} \quad a_{12}) \\ \mathbf{a}_2 &= (a_{21} \quad a_{22}) \end{aligned}$$

and **B** in terms of column vectors, say,

$$\mathbf{B} = (\mathbf{b}_1 \quad \mathbf{b}_2) \quad \text{where} \quad \mathbf{b}_1 = \begin{pmatrix} b_{11} \\ b_{21} \end{pmatrix}, \quad \mathbf{b}_2 = \begin{pmatrix} b_{12} \\ b_{22} \end{pmatrix},$$

then the elements of **C** may simply be written as dot products:

$$\begin{aligned} c_{11} &= \mathbf{a}_1 \cdot \mathbf{b}_1, & c_{12} &= \mathbf{a}_1 \cdot \mathbf{b}_2, \\ c_{21} &= \mathbf{a}_2 \cdot \mathbf{b}_1, & c_{22} &= \mathbf{a}_2 \cdot \mathbf{b}_2. \end{aligned}$$

Example 1. Multiplication of matrices

Let

$$\mathbf{A} = \begin{pmatrix} 2 & 1 \\ 3 & 4 \end{pmatrix} \quad \text{and} \quad \mathbf{B} = \begin{pmatrix} 1 & -2 \\ 5 & 3 \end{pmatrix}.$$

Then, by (4),

$$\mathbf{AB} = \begin{pmatrix} 2 & 1 \\ 3 & 4 \end{pmatrix}\begin{pmatrix} 1 & -2 \\ 5 & 3 \end{pmatrix} = \begin{pmatrix} 2\cdot 1 + 1\cdot 5 & 2\cdot(-2) + 1\cdot 3 \\ 3\cdot 1 + 4\cdot 5 & 3\cdot(-2) + 4\cdot 3 \end{pmatrix} = \begin{pmatrix} 7 & -1 \\ 23 & 6 \end{pmatrix}. \qquad \blacksquare$$

To arrive at the definition of the multiplication of two matrices in the general case, we now extend the preceding consideration to two more general linear transformations, say, the transformation

$$\begin{aligned} y_1 &= a_{11}x_1 + \cdots + a_{1n}x_n \\ y_2 &= a_{21}x_1 + \cdots + a_{2n}x_n \\ &\cdots\cdots\cdots\cdots\cdots\cdots \\ y_m &= a_{m1}x_1 + \cdots + a_{mn}x_n \end{aligned} \tag{5}$$

and the transformation

$$\begin{aligned} x_1 &= b_{11}w_1 + \cdots + b_{1p}w_p \\ x_2 &= b_{21}w_1 + \cdots + b_{2p}w_p \\ &\cdots\cdots\cdots\cdots\cdots\cdots \\ x_n &= b_{n1}w_1 + \cdots + b_{np}w_p. \end{aligned} \tag{6}$$

By inserting (6) into (5) we find that $y_1, \cdots, y_m$ can be expressed in terms of $w_1, \cdots, w_p$ in the form

$$\begin{aligned} y_1 &= c_{11}w_1 + \cdots + c_{1p}w_p \\ y_2 &= c_{21}w_1 + \cdots + c_{2p}w_p \\ &\cdots\cdots\cdots\cdots\cdots\cdots \\ y_m &= c_{m1}w_1 + \cdots + c_{mp}w_p \end{aligned} \tag{7}$$

where the coefficients are given by the formula

$$c_{jk} = a_{j1}b_{1k} + a_{j2}b_{2k} + \cdots + a_{jn}b_{nk}.$$

The requirement that the product **AB** (in this order) of the coefficient matrices $\mathbf{A} = (a_{jk})$ and $\mathbf{B} = (b_{jk})$ of the transformations (5) and (6) be the coefficient matrix $\mathbf{C} = (c_{jk})$ of the "composed transformation" (7) leads to the following definition.

Definition. Multiplication of matrices

Let $\mathbf{A} = (a_{jk})$ be an $(m \times n)$ matrix and $\mathbf{B} = (b_{jk})$ an $(r \times p)$ matrix. Then the product **AB** (in this order) is defined only when $r = n$ and is the $(m \times p)$ matrix $\mathbf{C} = (c_{jk})$ whose elements are

$$c_{jk} = a_{j1}b_{1k} + a_{j2}b_{2k} + \cdots + a_{jn}b_{nk} = \sum_{i=1}^{n} a_{ji}b_{ik}. \tag{8}$$

Instead of **AA** we simply write $\mathbf{A}^2$, etc. ∎

We see that c_{jk} is the scalar product (dot product) of the j-th row vector of the first matrix, **A**, *and the k-th column vector of the second matrix,* **B**; *that is, if we write*

$$\mathbf{A} = \begin{pmatrix} \mathbf{a}_1 \\ \mathbf{a}_2 \\ \vdots \\ \mathbf{a}_m \end{pmatrix} \quad \text{and} \quad \mathbf{B} = (\mathbf{b}_1 \quad \mathbf{b}_2 \quad \cdots \quad \mathbf{b}_p)$$

where $\mathbf{a}_1, \cdots, \mathbf{a}_m$ are the row vectors of **A** *and $\mathbf{b}_1, \cdots, \mathbf{b}_p$ are the column vectors of* **B**, *then*

$$\mathbf{C} = \begin{pmatrix} \mathbf{a}_1 \cdot \mathbf{b}_1 & \mathbf{a}_1 \cdot \mathbf{b}_2 & \cdots & \mathbf{a}_1 \cdot \mathbf{b}_p \\ \mathbf{a}_2 \cdot \mathbf{b}_1 & \mathbf{a}_2 \cdot \mathbf{b}_2 & \cdots & \mathbf{a}_2 \cdot \mathbf{b}_p \\ \cdot & \cdot & \cdots & \cdot \\ \mathbf{a}_m \cdot \mathbf{b}_1 & \mathbf{a}_m \cdot \mathbf{b}_2 & \cdots & \mathbf{a}_m \cdot \mathbf{b}_p \end{pmatrix}. \tag{9}$$

The process of matrix multiplication is, therefore, conveniently referred to as the **multiplication of rows into columns.** Cf. Fig. 140.

Example 2. Multiplication of matrices

Let

$$\mathbf{A} = \begin{pmatrix} 3 & 2 & -1 \\ 0 & 4 & 6 \end{pmatrix} \quad \text{and} \quad \mathbf{B} = \begin{pmatrix} 1 & 0 & 2 \\ 5 & 3 & 1 \\ 6 & 4 & 2 \end{pmatrix}.$$

Then, by (8), the product **AB** is

$$\mathbf{AB} = \begin{pmatrix} 3 & 2 & -1 \\ 0 & 4 & 6 \end{pmatrix} \begin{pmatrix} 1 & 0 & 2 \\ 5 & 3 & 1 \\ 6 & 4 & 2 \end{pmatrix} = \begin{pmatrix} 7 & 2 & 6 \\ 56 & 36 & 16 \end{pmatrix},$$

whereas the product **BA** is not defined. ∎

The multiplication of matrices by vectors and the matrix multiplication of vectors may be illustrated by the following examples.

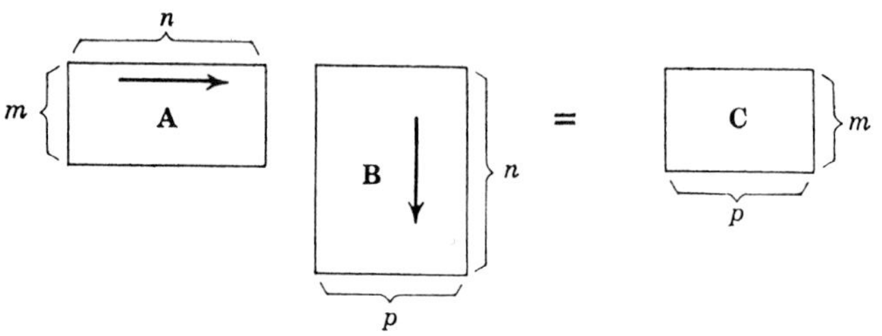

Fig. 140. Matrix multiplication. **AB = C**

Example 3

$$\begin{pmatrix} 3 & 4 & 2 \\ 6 & 0 & -1 \\ -5 & -2 & 1 \end{pmatrix} \begin{pmatrix} 1 \\ 3 \\ 2 \end{pmatrix} = \begin{pmatrix} 19 \\ 4 \\ -9 \end{pmatrix}$$

Example 4

$$(3 \quad 6 \quad 1) \begin{pmatrix} 1 \\ 2 \\ 4 \end{pmatrix} = (19), \qquad \begin{pmatrix} 1 \\ 2 \\ 4 \end{pmatrix} (3 \quad 6 \quad 1) = \begin{pmatrix} 3 & 6 & 1 \\ 6 & 12 & 2 \\ 12 & 24 & 4 \end{pmatrix}$$

Note that the first of these two products corresponds to the dot product as defined in Sec. 6.5. ∎

Desk calculations can be checked by the use of the sums of the elements of the columns. In Example 2,

$$\begin{pmatrix} 3 & 2 & -1 \\ 0 & 4 & 6 \end{pmatrix} \begin{pmatrix} 1 & 0 & 2 \\ 5 & 3 & 1 \\ 6 & 4 & 2 \end{pmatrix} = \begin{pmatrix} 7 & 2 & 6 \\ 56 & 36 & 16 \end{pmatrix}$$

$$\text{Sum} \quad 3 \quad 6 \quad 5 \qquad\qquad\qquad\qquad 63 \quad 38 \quad 22$$

Check:

$$3 \cdot 1 + 6 \cdot 5 + 5 \cdot 6 = 63$$
$$3 \cdot 0 + 6 \cdot 3 + 5 \cdot 4 = 38$$
$$3 \cdot 2 + 6 \cdot 1 + 5 \cdot 2 = 22$$

Figure 141 shows an arrangement of the matrices **A** and **B** in Example 2 and their product **AB**, which is very convenient for desk calculations. The point is that each element c_{jk} of the product matrix occupies the intersection of that row of the first matrix and that column of the last matrix which are used for computing c_{jk}. The extension of the arrangement to computations of products of three and more matrices is obvious (Fig. 141).

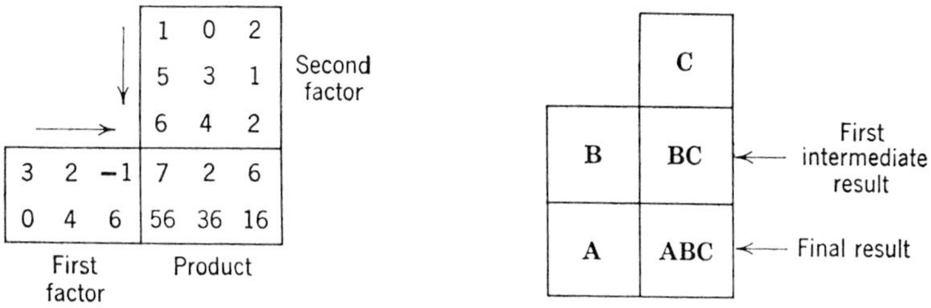

Fig. 141. Matrix multiplication

Theorem 1 (Properties of matrix multiplication)
(a) *Matrix multiplication is associative and is distributive with respect to addition of matrices, that is,*

$$
\begin{aligned}
&(k\mathbf{A})\mathbf{B} = k(\mathbf{AB}) = \mathbf{A}(k\mathbf{B}) && (\textit{written } k\mathbf{AB} \textit{ or } \mathbf{A}k\mathbf{B})\\
&\mathbf{A}(\mathbf{BC}) = (\mathbf{AB})\mathbf{C} && (\textit{written } \mathbf{ABC})\\
&(\mathbf{A} + \mathbf{B})\mathbf{C} = \mathbf{AC} + \mathbf{BC}\\
&\mathbf{C}(\mathbf{A} + \mathbf{B}) = \mathbf{CA} + \mathbf{CB}
\end{aligned}
\tag{10}
$$

provided **A, B,** *and* **C** *are such that the expressions on the left are defined.* (*k is any number.*)
(b) *Matrix multiplication is* ***not commutative;*** *that is, if* **A** *and* **B** *are matrices such that both* **AB** *and* **BA** *are defined, then*

$$\mathbf{AB} \neq \mathbf{BA} \qquad \textit{in general.}$$

(c) *The cancellation law is not true, in general, that is*

$$\mathbf{AB} = \mathbf{0} \qquad \textit{\textbf{does not necessarily imply}} \qquad \mathbf{A} = \mathbf{0} \textit{ or } \mathbf{B} = \mathbf{0}.$$

Proof. The proof of (*a*) is left as an exercise. (*b*) can be seen from Example 4 (above) or from the following example:

$$\begin{pmatrix} 1 & 0 \\ 0 & 0 \end{pmatrix}\begin{pmatrix} 0 & 1 \\ 1 & 0 \end{pmatrix} = \begin{pmatrix} 0 & 1 \\ 0 & 0 \end{pmatrix} \quad \text{but} \quad \begin{pmatrix} 0 & 1 \\ 1 & 0 \end{pmatrix}\begin{pmatrix} 1 & 0 \\ 0 & 0 \end{pmatrix} = \begin{pmatrix} 0 & 0 \\ 1 & 0 \end{pmatrix}.$$

(*c*) can be seen from the following example:

$$\begin{pmatrix} 1 & 1 \\ 2 & 2 \end{pmatrix}\begin{pmatrix} -1 & 1 \\ 1 & -1 \end{pmatrix} = \begin{pmatrix} 0 & 0 \\ 0 & 0 \end{pmatrix}. \qquad \blacksquare$$

Properties (*b*) and (*c*) are quite unusual because they have no counterparts in the usual multiplication of numbers and should, therefore, be carefully observed. Property (*c*) will be considered in more detail in Sec. 7.8. The reader may also use the above example to illustrate that $\mathbf{AB} = \mathbf{0}$ does *not* imply $\mathbf{BA} = \mathbf{0}$.

Property (*b*) implies that *the order of the factors in a matrix product must be carefully observed,* and only in certain cases will the equation $\mathbf{AB} = \mathbf{BA}$ hold. To be precise we say that, in the product **AB,** the matrix **B** is **premultiplied** by the matrix **A,** or, alternatively, that **A** is **postmultiplied** by **B.**

A diagonal matrix whose diagonal elements are all equal is called a **scalar matrix.** Hence a scalar matrix is of the form

$$\mathbf{S} = \begin{pmatrix} k & 0 & \cdots & 0 \\ 0 & k & \cdots & \cdot \\ \cdot & \cdot & \cdots & \cdot \\ 0 & 0 & \cdots & k \end{pmatrix}$$

where k is any number. If $\mathbf{S}$ is n-rowed and $\mathbf{A}$ is any $(n \times n)$ matrix, then

$$\mathbf{AS} = \mathbf{SA} = k\mathbf{A}; \tag{11}$$

that is, the multiplication of an n-rowed square matrix $\mathbf{A}$ by an n-rowed scalar matrix has the same effect as the multiplication of $\mathbf{A}$ by a number, and $\mathbf{S}$ commutes with any $(n \times n)$ matrix. The simple proof is left to the reader. In particular, for the n-rowed unit matrix $\mathbf{I}$ (cf. Sec. 7.3) we have

$$\mathbf{AI} = \mathbf{IA} = \mathbf{A}.$$

Transposition of a product. *The transpose of a product equals the product of the transposed factors, taken in reverse order,*

$$(\mathbf{AB})^{\mathsf{T}} = \mathbf{B}^{\mathsf{T}}\mathbf{A}^{\mathsf{T}}. \tag{12}$$

The proof follows from the definition of matrix multiplication and is left to the reader.

Column vectors of a product. If $\mathbf{b}_1, \cdots, \mathbf{b}_p$ are the column vectors of a matrix $\mathbf{B}$ and the product $\mathbf{AB}$ is defined, then from the definition of matrix multiplication it follows immediately that the product matrix $\mathbf{AB}$ has the column vectors $\mathbf{Ab}_1, \cdots, \mathbf{Ab}_p$, that is,

$$\mathbf{AB} = (\mathbf{Ab}_1 \quad \mathbf{Ab}_2 \quad \cdots \quad \mathbf{Ab}_p). \tag{13}$$

Linear transformations. By using the definition of matrix multiplication we may now write the linear transformation (5) in the simple form

$$\mathbf{y} = \mathbf{Ax} \tag{5'}$$

where $\mathbf{A} = (a_{jk})$ and

$$\mathbf{x} = \begin{pmatrix} x_1 \\ x_2 \\ \vdots \\ x_n \end{pmatrix}, \qquad \mathbf{y} = \begin{pmatrix} y_1 \\ y_2 \\ \vdots \\ y_m \end{pmatrix}.$$

Similarly, (6) may be written

$$\mathbf{x} = \mathbf{Bw} \tag{6'}$$

where $\mathbf{w}$ is the column matrix with elements $w_1, \cdots, w_p$. By inserting (6′) into (5′) and using the associative law of matrix multiplication (cf. Theorem 1) we obtain

$$\mathbf{y} = \mathbf{A}(\mathbf{B}\mathbf{w}) = \mathbf{A}\mathbf{B}\mathbf{w} = \mathbf{C}\mathbf{w}, \tag{7'}$$

in agreement with (7).

We mention that this consideration can be generalized and use the opportunity to introduce a few concepts of general interest. Let X and Y be any vector spaces (cf. Sec. 6.4). To each vector $\mathbf{x}$ in X we assign a unique vector $\mathbf{y}$ in Y. Then we say that a **mapping** (or **transformation** or **operator**) of X into Y is given. Such a mapping is denoted by a capital letter, say F. The vector $\mathbf{y}$ in Y assigned to a vector $\mathbf{x}$ in X is called the **image** of $\mathbf{x}$ and is denoted by $F(\mathbf{x})$ [or $F\mathbf{x}$, without parentheses]. More generally, if S is any subset of X, the set of the images of all the elements of S is called the **image** of S and is denoted by $F(S)$.

F is said to be **injective** or **one-to-one** if different vectors in X have different images in Y. The mapping F is said to be **surjective** or a mapping of X **onto** Y if every vector $\mathbf{y}$ in Y is the image of at least one vector in X. The mapping F is said to be **bijective** if F is both injective and surjective.

F is called a **linear mapping** or **linear transformation** if for all vectors $\mathbf{x}$ and $\mathbf{z}$ in X and for every scalar α,

$$\begin{aligned} F(\mathbf{x} + \mathbf{z}) &= F(\mathbf{x}) + F(\mathbf{z}) \\ F(\alpha\mathbf{x}) &= \alpha F(\mathbf{x}). \end{aligned} \tag{14}$$

For instance, (5) and (6) are linear transformations in the sense of this more general definition.

So far, X and Y were any vector spaces. From now on we assume that $X = R^n$, the vector space of all ordered n-tuples of real numbers, and $Y = R^m$, the vector space of all ordered m-tuples of real numbers.

Let $\mathbf{e}_{(1)}, \cdots, \mathbf{e}_{(n)}$ be a basis of $X = R^n$. Then every vector $\mathbf{x}$ in X has a unique representation

$$\mathbf{x} = x_1\mathbf{e}_{(1)} + \cdots + x_n\mathbf{e}_{(n)},$$

which we may write $\mathbf{x} = (x_1 \ \cdots \ x_n)^{\mathsf{T}}$. If F is any linear transformation of X into Y, then

$$F(\mathbf{x}) = F(x_1\mathbf{e}_{(1)} + \cdots + x_n\mathbf{e}_{(n)}) = x_1F(\mathbf{e}_{(1)}) + \cdots + x_nF(\mathbf{e}_{(n)}).$$

Hence a linear transformation of $X = R^n$ into $Y = R^m$ is uniquely determined by the images of the elements of a basis of X.

(5) and (5′) show that every matrix determines a linear transformation. Conversely, let us choose a basis $\mathbf{e}_{(1)}, \cdots, \mathbf{e}_{(n)}$ of $X = R^n$ where each $\mathbf{e}_{(k)}$ has all components 0 except for the kth component, which is 1, that is,

$$\mathbf{e}_{(1)} = \begin{pmatrix} 1 \\ 0 \\ \vdots \\ 0 \end{pmatrix}, \quad \mathbf{e}_{(2)} = \begin{pmatrix} 0 \\ 1 \\ \vdots \\ 0 \end{pmatrix}, \quad \cdots, \quad \mathbf{e}_{(n)} = \begin{pmatrix} 0 \\ 0 \\ \vdots \\ 1 \end{pmatrix}.$$

Similarly, for $Y = R^m$ we choose a basis $\mathbf{g}_{(1)}, \cdots, \mathbf{g}_{(m)}$, where $\mathbf{g}_{(j)}$ has all components 0 except for the jth component, which is 1, that is,

$$\mathbf{g}_{(1)} = \begin{pmatrix} 1 \\ 0 \\ \vdots \\ 0 \end{pmatrix}, \quad \mathbf{g}_{(2)} = \begin{pmatrix} 0 \\ 1 \\ \vdots \\ 0 \end{pmatrix}, \quad \cdots, \quad \mathbf{g}_{(m)} = \begin{pmatrix} 0 \\ 0 \\ \vdots \\ 1 \end{pmatrix}.$$

Then if F is any given linear transformation of X into Y, there is an $(m \times n)$ matrix $\mathbf{A}$ such that for every $\mathbf{x}$ in X and its image $\mathbf{y} = F(\mathbf{x})$,

$$\mathbf{y} = \mathbf{A}\mathbf{x}.$$

In fact, F is determined by the images $\mathbf{y}_{(1)} = F(\mathbf{e}_{(1)}), \cdots, \mathbf{y}_{(n)} = F(\mathbf{e}_{(n)})$; writing $\mathbf{y}_{(j)} = (\eta_1^{(j)}, \cdots, \eta_m^{(j)})^\mathsf{T}$ and setting $\mathbf{y}_{(1)} = \mathbf{A}\mathbf{e}_{(1)}$, that is,

$$\begin{pmatrix} \eta_1^{(1)} \\ \cdot \\ \cdot \\ \cdot \\ \eta_m^{(1)} \end{pmatrix} = \begin{pmatrix} a_{11} & \cdots & a_{1n} \\ a_{21} & \cdots & a_{2n} \\ \cdot & \cdots & \cdot \\ a_{m1} & \cdots & a_{mn} \end{pmatrix} \begin{pmatrix} 1 \\ 0 \\ \vdots \\ 0 \end{pmatrix}$$

we can determine the elements of the first column of $\mathbf{A}$, namely, $a_{11} = \eta_1^{(1)}$, $a_{21} = \eta_2^{(1)}$, etc. From $\mathbf{y}_{(2)} = \mathbf{A}\mathbf{e}_{(2)}$ we determine the elements of the second column of $\mathbf{A}$, and so on. We say that the matrix $\mathbf{A}$ *represents* the linear transformation F with respect to those bases. The purpose of such a representation is to replace one object of study by another object whose properties are more readily apparent.

Example 5. Linear transformations

Interpreted as transformations of Cartesian coordinates in the plane, the matrices

$$\begin{pmatrix} 0 & 1 \\ 1 & 0 \end{pmatrix}, \quad \begin{pmatrix} 1 & 0 \\ 0 & -1 \end{pmatrix}, \quad \begin{pmatrix} -1 & 0 \\ 0 & -1 \end{pmatrix}, \quad \begin{pmatrix} a & 0 \\ 0 & 1 \end{pmatrix}$$

represent a reflection in the line $x_2 = x_1$, a reflection in the x_1-axis, a reflection in the origin and a stretch (when $a > 1$, or a contraction when $0 < a < 1$) in the x_1-direction, respectively.

Example 6. Linear transformations

Our discussion at the end of this section is simpler than it may look at first sight. To see this, find the matrix $\mathbf{A}$ representing the linear transformation which maps (x_1, x_2) onto $(2x_1 - 5x_2, 3x_1 + 4x_2)$.

Since

$$\begin{pmatrix}1\\0\end{pmatrix} \text{ and } \begin{pmatrix}0\\1\end{pmatrix} \text{ are mapped onto } \begin{pmatrix}2\\3\end{pmatrix} \text{ and } \begin{pmatrix}-5\\4\end{pmatrix},$$

respectively, we obtain, according to our discussion,

$$\mathbf{A} = \begin{pmatrix}2 & -5\\3 & 4\end{pmatrix}.$$

We check this, finding

$$\begin{pmatrix}y_1\\y_2\end{pmatrix} = \begin{pmatrix}2 & -5\\3 & 4\end{pmatrix}\begin{pmatrix}x_1\\x_2\end{pmatrix} = \begin{pmatrix}2x_1 - 5x_2\\3x_1 + 4x_2\end{pmatrix}.$$

The reader may obtain **A** also at once by writing the given transformation in the form

$$y_1 = 2x_1 - 5x_2$$
$$y_2 = 3x_1 + 4x_2.$$

Problems for Sec. 7.4

Let

$$\mathbf{A} = \begin{pmatrix}1\\0\\2\end{pmatrix}, \quad \mathbf{B} = \begin{pmatrix}2 & 0\\0 & 3\\1 & -1\end{pmatrix}, \quad \mathbf{C} = \begin{pmatrix}1 & -1 & 0\\0 & 2 & 3\\4 & 0 & -1\end{pmatrix}.$$

Find those of the following expressions which are defined.

1. $\mathbf{AB}$, $\mathbf{A^TB}$, $\mathbf{BA}$, $\mathbf{B^TA}$ **2.** $\mathbf{BC}$, $\mathbf{B^TC}$, $\mathbf{CB}$, $\mathbf{C^TB}$ **3.** $\mathbf{A}^2$, $\mathbf{B}^2$, $\mathbf{C}^2$

4. $\mathbf{AA^T}$, $\mathbf{A^TA}$ **5.** $\mathbf{BB^T}$, $\mathbf{B^TB}$ **6.** $\mathbf{CC^T}$, $\mathbf{C^TC}$

7. $\mathbf{C}^2$, $\mathbf{C}^4$, $\mathbf{C}^4 - 3\mathbf{C}^2 + 2\mathbf{I}$ **8.** $\mathbf{BCA}$, $\mathbf{B^TCA}$ **9.** $\mathbf{A^TCA}$

10. Find real two-rowed square matrices (as many as you can) whose square is **I**, the unit matrix.

11. (Idempotent matrix) A matrix **A** is said to be *idempotent* if $\mathbf{A}^2 = \mathbf{A}$. Give examples of idempotent matrices, different from the zero or unit matrix.

12. Let $\mathbf{A} = \begin{pmatrix}1 & -2 & 4\\3 & 1 & 5\\2 & 4 & 0\end{pmatrix}$ and $\mathbf{B} = \begin{pmatrix}2 & 4 & -2\\-1 & -2 & 1\\-1 & -2 & 1\end{pmatrix}$. Find **AB** and **BA**.

13. (Rotation) Show that the linear transformation $\mathbf{y} = \mathbf{Ax}$ with matrix

$$\mathbf{A} = \begin{pmatrix}\cos\theta & -\sin\theta\\\sin\theta & \cos\theta\end{pmatrix}$$

is a counterclockwise rotation of the Cartesian x_1x_2-coordinate system in the plane about the origin.

14. Find **A** such that $\mathbf{y} = \mathbf{Ax}$ represents a clockwise rotation in the plane through an angle of 45°.

15. Show that in Prob. 13,

$$\mathbf{A}^n = \begin{pmatrix} \cos n\theta & -\sin n\theta \\ \sin n\theta & \cos n\theta \end{pmatrix}.$$

What does this result mean geometrically?

16. Give a geometrical interpretation (as in Example 5) of the linear transformations with matrices

$$\begin{pmatrix} 1 & 0 \\ 0 & 0 \end{pmatrix}, \quad \begin{pmatrix} 0 & 0 \\ 0 & 1 \end{pmatrix}, \quad \begin{pmatrix} -1 & 0 \\ 0 & 1 \end{pmatrix}, \quad \begin{pmatrix} 1 & 0 \\ 0 & 3 \end{pmatrix}, \quad \begin{pmatrix} a & 0 \\ 0 & a \end{pmatrix} \quad (a > 1).$$

17. Represent each of the transformations

$$\begin{aligned} y_1 &= 2x_1 - x_2 \\ y_2 &= -x_1 + 2x_2 \end{aligned} \qquad \text{and} \qquad \begin{aligned} x_1 &= w_1 + w_2 \\ x_2 &= w_1 - w_2 \end{aligned}$$

by the use of matrices and find the composite transformation which expresses y_1, y_2 in terms of w_1, w_2.

18. Give a geometrical interpretation (as in Example 5) of the linear transformation $\mathbf{x} = \mathbf{Bw}$ with matrix

$$\mathbf{B} = \begin{pmatrix} 1 & 0 \\ 0 & 2 \end{pmatrix}.$$

Find the composite transformation $\mathbf{y} = \mathbf{ABw}$ with $\mathbf{A}$ given in Prob. 13. Show that $\mathbf{AB} \neq \mathbf{BA}$. What does this mean geometrically?

19. **(Associative algebra)** A *real* (or complex) *associative algebra* is defined to be a set A of elements $\mathbf{A}, \mathbf{B}, \mathbf{C}, \cdots$ such that: (*a*) A is a real (or complex) vector space (cf. Sec. 6.4). (*b*) A multiplication is defined on the elements of A such that with any ordered pair of elements $\mathbf{A}, \mathbf{B}$ in A, there is associated a unique element of A, written $\mathbf{AB}$ and called the *product* of $\mathbf{A}$ and $\mathbf{B}$, and this multiplication satisfies (10). Show that the set of all real $(n \times n)$ matrices (n fixed) is an associative algebra.

20. Do the real symmetric (3×3) matrices form a vector space? An associative algebra?

7.5 Systems of Linear Equations. Gauss Elimination

A *system of m linear equations* (or *set of m simultaneous linear equations*) *in n unknowns* $x_1, \cdots, x_n$ is a set of equations of the form

$$\begin{aligned} a_{11}x_1 + \cdots + a_{1n}x_n &= b_1 \\ a_{21}x_1 + \cdots + a_{2n}x_n &= b_2 \\ \cdots\cdots\cdots\cdots&\cdots\cdots \\ a_{m1}x_1 + \cdots + a_{mn}x_n &= b_m \end{aligned} \tag{1}$$

The a_{ik} are given numbers, which are called the **coefficients** of the system. The b_i are also given numbers. If the b_i are all zero, then (1) is called a **homogeneous system.** If at least one b_i is not zero, then (1) is called a **nonhomogeneous system.**

A **solution** of (1) is a set of numbers $x_1, \cdots, x_n$ which satisfy all the m equations. A **solution vector** of (1) is a vector $\mathbf{x}$ whose components constitute a solution of (1). If the system (1) is homogeneous, it has at least the **trivial solution** $x_1 = 0, \cdots, x_n = 0$.

From the definition of matrix multiplication we see that the m equations of (1) may be written as a single vector equation

$$\mathbf{Ax} = \mathbf{b}, \tag{2}$$

where the **coefficient matrix** $\mathbf{A} = (a_{ik})$ is the $(m \times n)$ matrix

$$\mathbf{A} = \begin{pmatrix} a_{11} & a_{12} & \cdots & a_{1n} \\ a_{21} & a_{22} & \cdots & a_{2n} \\ \cdot & \cdot & \cdots & \cdot \\ a_{m1} & a_{m2} & \cdots & a_{mn} \end{pmatrix}, \quad \text{and} \quad \mathbf{x} = \begin{pmatrix} x_1 \\ \cdot \\ \cdot \\ \cdot \\ x_n \end{pmatrix} \quad \text{and} \quad \mathbf{b} = \begin{pmatrix} b_1 \\ \cdot \\ \cdot \\ \cdot \\ b_m \end{pmatrix}$$

are column vectors. We assume that the coefficients a_{ik} are not all zero, so that $\mathbf{A}$ is not a zero matrix. Note that $\mathbf{x}$ has n components, whereas $\mathbf{b}$ has m components. The matrix

$$\mathbf{B} = \begin{pmatrix} a_{11} & \cdots & a_{1n} & b_1 \\ \cdot & \cdots & \cdot & \cdot \\ \cdot & \cdots & \cdot & \cdot \\ a_{m1} & \cdots & a_{mn} & b_m \end{pmatrix}$$

is called the **augmented matrix** of the system (1). We see that $\mathbf{B}$ is obtained by augmenting $\mathbf{A}$ by the column $\mathbf{b}$. The matrix $\mathbf{B}$ determines the system (1) completely, because it contains all the given numbers appearing in (1).

Example 1. Geometric interpretation. Existence of solutions

If $m = n = 2$, we have two equations in two unknowns x_1, x_2

$$a_{11}x_1 + a_{12}x_2 = b_1$$
$$a_{21}x_1 + a_{22}x_2 = b_2$$

If we interpret x_1, x_2 as coordinates in the x_1x_2-plane, then each of the two equations represents a straight line, and (x_1, x_2) is a solution if and only if the point P with coordinates x_1, x_2 lies on both lines. Hence there are three possible cases:

(*a*) No solution if the lines are parallel.
(*b*) Precisely one solution if they intersect.
(*c*) Infinitely many solutions if they coincide.

For instance,

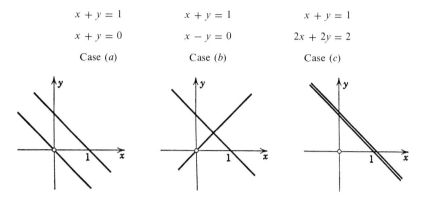

$x + y = 1$	$x + y = 1$	$x + y = 1$
$x + y = 0$	$x - y = 0$	$2x + 2y = 2$
Case (a)	Case (b)	Case (c)

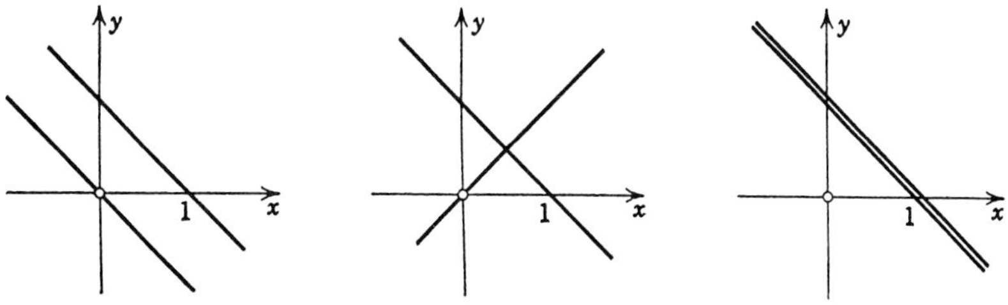

If the system is homogeneous, Case (a) cannot happen, because then those two straight lines pass through the origin, whose coordinates 0, 0 constitute the trivial solution. The reader may consider three equations in three unknowns as representations of three planes in space and discuss the various possible cases in a similar fashion. ∎

Our simple example illustrates that a system (1) may not always have a solution, and relevant problems are as follows. Does a given system (1) have a solution? Under what conditions does it have precisely one solution? If it has more than one solution, how can the set of all solutions be characterized? How can the solutions be obtained?

In the present section we shall discuss a method for obtaining solutions by a systematic process of elimination, which is called the **Gauss elimination** (or **Gauss algorithm**). We first explain the process by some typical examples. Since a system of linear equations is completely determined by its augmented matrix, the process of elimination can be performed by merely considering the matrices. To see this correspondence, we shall write systems of equations and augmented matrices side by side.

Example 2. Gauss elimination

Solve the system of three linear equations in four unknowns

$$(3)\qquad \begin{aligned} 3.0x_1 + 2.0x_2 + 2.0x_3 - 5.0x_4 &= 8.0 \\ 0.6x_1 + 1.5x_2 + 1.5x_3 - 5.4x_4 &= 2.7 \\ 1.2x_1 - 0.3x_2 - 0.3x_3 + 2.4x_4 &= 2.1 \end{aligned} \qquad \begin{pmatrix} 3.0 & 2.0 & 2.0 & -5.0 & 8.0 \\ 0.6 & 1.5 & 1.5 & -5.4 & 2.7 \\ 1.2 & -0.3 & -0.3 & 2.4 & 2.1 \end{pmatrix}$$

First step. *Elimination of x_1 from the second and third equations* by subtracting

$0.6/3.0 = 0.2$ times the first equation from the second equation,

$1.2/3.0 = 0.4$ times the first equation from the third equation.

This gives a new system of equations

$$(4)\qquad \begin{aligned} 3.0x_1 + 2.0x_2 + 2.0x_3 - 5.0x_4 &= 8.0 \\ 1.1x_2 + 1.1x_3 - 4.4x_4 &= 1.1 \\ -1.1x_2 - 1.1x_3 + 4.4x_4 &= -1.1 \end{aligned} \qquad \begin{pmatrix} 3.0 & 2.0 & 2.0 & -5.0 & 8.0 \\ 0 & 1.1 & 1.1 & -4.4 & 1.1 \\ 0 & -1.1 & -1.1 & 4.4 & -1.1 \end{pmatrix}$$

Second step. *Elimination of x_2 from the third equation of* (4) by subtracting

$-1.1/1.1 = -1$ times the second equation from the third equation.

This gives

$$\begin{aligned} 3.0x_1 + 2.0x_2 + 2.0x_3 - 5.0x_4 &= 8.0 \\ 1.1x_2 + 1.1x_3 - 4.4x_4 &= 1.1 \\ 0 &= 0 \end{aligned} \qquad \begin{pmatrix} 3.0 & 2.0 & 2.0 & -5.0 & 8.0 \\ 0 & 1.1 & 1.1 & -4.4 & 1.1 \\ 0 & 0 & 0 & 0 & 0 \end{pmatrix}$$

From the second equation, $x_2 = 1 - x_3 + 4x_4$. From this and the first equation, $x_1 = 2 - x_4$. Since x_3 and x_4 remain arbitrary, we have infinitely many solutions; if we choose a value of x_3 and a value of x_4, then the corresponding values of x_1 and x_2 are uniquely determined.

Example 3. Gauss elimination if no solution exists

What will happen if we apply the Gauss elimination illustrated in Example 2 to a system of equations which has no solution? The answer is that in this case the method will show this fact by producing a contradiction. For instance, consider

$$\begin{aligned} 3x_1 + 2x_2 + x_3 &= 3 \\ 2x_1 + x_2 + x_3 &= 0 \\ 6x_1 + 2x_2 + 4x_3 &= 6 \end{aligned} \qquad \begin{pmatrix} 3 & 2 & 1 & 3 \\ 2 & 1 & 1 & 0 \\ 6 & 2 & 4 & 6 \end{pmatrix}$$

First step. *Elimination of x_1 from the second and third equations* by subtracting

$2/3$ times the first equation from the second equation,

$6/3 = 2$ times the first equation from the third equation.

This gives

$$\begin{aligned} 3x_1 + 2x_2 + x_3 &= 3 \\ -\tfrac{1}{3}x_2 + \tfrac{1}{3}x_3 &= -2 \\ -2x_2 + 2x_3 &= 0 \end{aligned} \qquad \begin{pmatrix} 3 & 2 & 1 & 3 \\ 0 & -\tfrac{1}{3} & \tfrac{1}{3} & -2 \\ 0 & -2 & 2 & 0 \end{pmatrix}$$

Second step. *Elimination of x_2 from the third equation* gives

$$\begin{aligned} 3x_1 + 2x_2 + x_3 &= 3 \\ -\tfrac{1}{3}x_2 + \tfrac{1}{3}x_3 &= -2 \\ 0 &= 12 \end{aligned} \qquad \begin{pmatrix} 3 & 2 & 1 & 3 \\ 0 & -\tfrac{1}{3} & \tfrac{1}{3} & -2 \\ 0 & 0 & 0 & 12 \end{pmatrix}$$

This shows that the system has no solution.

Example 4. Gauss elimination if a unique solution exists

Solve the system

$$\begin{aligned} -x_1 + x_2 + 2x_3 &= 2 \\ 3x_1 - x_2 + x_3 &= 6 \\ -x_1 + 3x_2 + 4x_3 &= 4 \end{aligned} \qquad \begin{pmatrix} -1 & 1 & 2 & 2 \\ 3 & -1 & 1 & 6 \\ -1 & 3 & 4 & 4 \end{pmatrix}$$

First step. *Elimination of x_1 from equations* 2 *and* 3 gives

$$\begin{aligned} -x_1 + x_2 + 2x_3 &= 2 \\ 2x_2 + 7x_3 &= 12 \\ 2x_2 + 2x_3 &= 2 \end{aligned} \qquad \begin{pmatrix} -1 & 1 & 2 & 2 \\ 0 & 2 & 7 & 12 \\ 0 & 2 & 2 & 2 \end{pmatrix}$$

Second step. *Elimination of x_2 from equation* 3 gives

$$\begin{aligned} -x_1 + x_2 + 2x_3 &= 2 \\ 2x_2 + 7x_3 &= 12 \\ -5x_3 &= -10 \end{aligned} \qquad \begin{pmatrix} -1 & 1 & 2 & 2 \\ 0 & 2 & 7 & 12 \\ 0 & 0 & -5 & -10 \end{pmatrix}$$

Beginning with the last equation, we obtain successively $x_3 = 2$, $x_2 = -1$, $x_1 = 1$. We see that the system has a unique solution. ∎

The form of the system and the matrix in the last step of the Gauss elimination is called the **echelon form.** Thus in Example 4 the echelon forms of the coefficient matrix and the augmented matrix are

$$\begin{pmatrix} -1 & 1 & 2 \\ 0 & 2 & 7 \\ 0 & 0 & -5 \end{pmatrix} \quad \text{and} \quad \begin{pmatrix} -1 & 1 & 2 & 2 \\ 0 & 2 & 7 & 12 \\ 0 & 0 & -5 & -10 \end{pmatrix}$$

The method illustrated in these examples is called the **Gauss elimination** or **Gauss algorithm.**[1] In terms of general formulas for the general system (1) it can be formulated as follows.

First step. *Elimination of x_1 from the second, third, $\cdots$ mth equation.* We may assume that the order of the equations and the order of the unknowns in each equation are such that $a_{11} \neq 0$. The variable x_1 can then be eliminated from the second, $\cdots$, mth equation by subtracting

a_{21}/a_{11} times the first equation from the second equation,

a_{31}/a_{11} times the first equation from the third equation,

etc. This gives a new system of equations of the form

$$\begin{aligned} a_{11}x_1 + a_{12}x_2 + \cdots + a_{1n}x_n &= b_1 \\ c_{22}x_2 + \cdots + c_{2n}x_n &= b_2^* \\ &\vdots \\ c_{m2}x_2 + \cdots + c_{mn}x_n &= b_m^*. \end{aligned} \tag{5}$$

Any solution of (1) is a solution of (5) and conversely, because each equation of (5) was derived from two equations of (1) and, by reversing the process, (1) may be obtained from (5).

Second step. *Elimination of x_2 from the third, $\cdots$, mth equation in* (5). If the coefficients $c_{22}, \cdots, c_{mn}$ in (5) are not all zero, we may assume that the order of the equations and the unknowns is such that $c_{22} \neq 0$. Then we may eliminate x_2 from the third, fourth, $\cdots$, mth equation of (5) by subtracting

[1]We present the Gauss elimination in its classical form; further reduction of the amount of computational work was obtained by Banachiewicz and Cholesky. These and other methods are discussed in Ref. [C2]. Cf. also Sec. 19.9.

c_{32}/c_{22} times the second equation from the third equation,

c_{42}/c_{22} times the second equation from the fourth equation, etc.

The further steps are now obvious. In the third step we eliminate x_3, in the fourth step we eliminate x_4, etc.

This process will terminate only when no equations are left or when the coefficients of all the unknowns in the remaining equations are zero. We then have a system of the form

$$
\begin{aligned}
a_{11}x_1 + a_{12}x_2 + \cdots\cdots + a_{1n}x_n &= b_1 \\
c_{22}x_2 + \cdots\cdots + c_{2n}x_n &= b_2^* \\
&\vdots \\
k_{rr}x_r + \cdots + k_{rn}x_n &= \widetilde{b}_r \\
0 &= \widetilde{b}_{r+1} \\
&\vdots \\
0 &= \widetilde{b}_m
\end{aligned}
\tag{6}
$$

where $r \leqq m$ (and $a_{11} \neq 0, c_{22} \neq 0, \cdots, k_{rr} \neq 0$). We see that there are three possible cases:

(a) No solution if $r < m$ and one of the numbers $\widetilde{b}_{r+1}, \cdots, \widetilde{b}_m$ is not zero. This is illustrated by Example 3, where $r = 2 < m = 3$ and $\widetilde{b}_{r+1} = \widetilde{b}_3 = 12$.

(b) Precisely one solution if $r = n$ and $\widetilde{b}_{r+1}, \cdots, \widetilde{b}_m$, if present, are zero. This solution is obtained by solving the nth equation of (6) for x_n, then the $(n-1)$th equation for x_{n-1} and so on up the line. See Example 4, where $r = n = 3$ and $m = 3$, too.

(c) Infinitely many solutions if $r < n$ and $\widetilde{b}_{r+1}, \cdots, \widetilde{b}_m$, if present, are zero. Then any of these solutions is obtained by choosing values at pleasure for the unknowns $x_{r+1}, \cdots, x_n$, solving the rth equation for x_r, then the $(r-1)$th equation for x_{r-1} and so on up the line. Example 2 illustrates this case.

The operations in the Gauss elimination are called *elementary operations.* **Elementary operations for equations** are:

Interchange of two equations.
Multiplication of an equation by a nonzero constant.
Addition of a constant multiple of one equation to another equation.

Instead of equations we may consider augmented matrices (cf. Example 2). Then we have the following corresponding **elementary row operations for matrices:**

Interchange of two rows.
Multiplication of a row by a nonzero constant.
Addition of a constant multiple of one row to another row.

A system S_1 of linear equations is said to be **equivalent** to a system S_2 of linear equations if S_1 can be obtained from S_2 by (finitely many) elementary operations. From our discussion we have the following result.

Theorem 1 (Equivalent systems)
Equivalent systems of linear equations have the same sets of solutions.

A matrix $\mathbf{A}_1$ is said to be **row-equivalent** to a matrix $\mathbf{A}_2$ if $\mathbf{A}_1$ can be obtained from $\mathbf{A}_2$ by finitely many row operations. Thus the augmented matrices of (1) and (6) are row-equivalent.

These discussions and concepts will be helpful in Sec. 7.7, where we shall give a more detailed characterization of solution sets of simultaneous linear equations.

Problems for Sec. 7.5

Solve the following systems of linear equations by the Gauss elimination.

1. $\begin{aligned} 2x + y &= 4 \\ 5x - 2y &= 1 \end{aligned}$

2. $\begin{aligned} 3x + y &= 11 \\ x - y &= 5 \end{aligned}$

3. $\begin{aligned} x + y &= 0 \\ 3x - 4y &= 1 \end{aligned}$

4. $\begin{aligned} 4x - y + z &= 0 \\ x + 2y - z &= 0 \\ 3x + y + 5z &= 0 \end{aligned}$

5. $\begin{aligned} -x + y + 2z &= 0 \\ 3x + 4y + z &= 0 \\ 2x + 5y + 3z &= 0 \end{aligned}$

6. $\begin{aligned} 2x + y + z &= 7 \\ x - y + z &= 0 \\ 4x + 2y - 3z &= 4 \end{aligned}$

7. $\begin{aligned} 4x + 6y + z &= 2 \\ 2x + y - 4z &= 3 \\ 3x - 2y + 5z &= 8 \end{aligned}$

8. $\begin{aligned} 3x - 7y + z &= 4 \\ x + 5y + z &= 0 \\ 2x - y + 3z &= -2 \end{aligned}$

9. $\begin{aligned} 3x + y - 2z &= -3 \\ x - y + 2z &= -1 \\ -4x + 3y - 6z &= 4 \end{aligned}$

10. $\begin{aligned} x - 2y + z &= 3 \\ x \qquad + 3z &= 11 \\ -2y + z &= 1 \end{aligned}$

11. $\begin{aligned} -3x + y - z &= 2 \\ 2x + 3y + z &= 0 \\ x + 5y + 2z &= 6 \end{aligned}$

12. $\begin{aligned} 9x + 4y + 3z &= -1 \\ 5x + y + 2z &= 1 \\ 7x + 3y + 4z &= 1 \end{aligned}$

13. $\begin{aligned} 3x - 3y - 7z &= -4 \\ x - y + 2z &= 3 \end{aligned}$

14. $\begin{aligned} 7x - 4y - 2z &= -6 \\ 16x + 2y + z &= 3 \end{aligned}$

15. $\begin{aligned} x - 3y + 2z &= 2 \\ 5x - 15y + 7z &= 10 \end{aligned}$

16. $\begin{aligned} 7w + 2x - 2y - 6z &= 6 \\ w + 3x - 3y - 9z &= -10 \end{aligned}$

17. $\begin{aligned} 3w - 15x - 3y - 11z &= -27 \\ 5w - 25x - 5y - 7z &= -45 \end{aligned}$

18. $\begin{aligned} 4w + 3x - 9y + z &= 1 \\ -w + 2x - 13y + 3z &= 3 \\ 3w - x + 8y - 2z &= -2 \end{aligned}$

19. $\begin{aligned} 3w - 6x - y - z &= 0 \\ w - 2x + 5y - 3z &= 0 \\ 2w - 4x + 3y - z &= 3 \end{aligned}$

20. $\begin{aligned} w + x + y \qquad &= 3 \\ -3w - 17x + y + 2z &= 1 \\ 4w \qquad + 8y - 5z &= 1 \\ -5x - 2y + z &= 1 \end{aligned}$

21. $\begin{aligned} w - x + 3y - 3z &= 3 \\ -5w + 2x - 5y + 4z &= -5 \\ -3w - 4x + 7y - 2z &= 7 \\ 2w + 3x + y - 11z &= 1 \end{aligned}$

22. Using Kirchhoff's laws, show that the equations for the unknown currents I_1, I_2, I_3 in the figure on the next page are

$$I_1 - I_2 - I_3 = 0, \quad R_2I_2 - R_3I_3 = 0, \quad (R_1 + R_4)I_1 + R_2I_2 = E_0$$

and determine these currents in terms of R_1, R_2, R_3, R_4, and E_0.

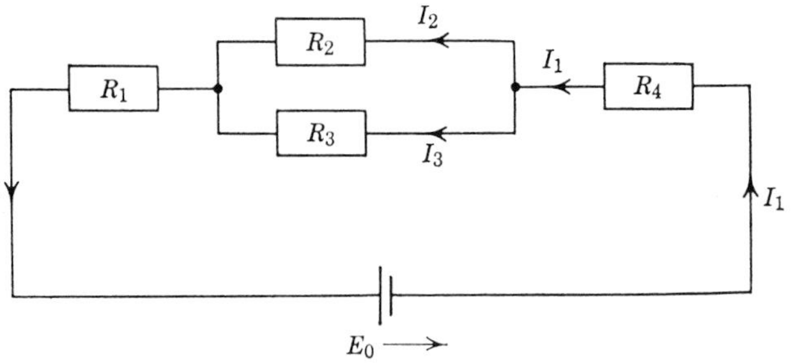

Problem 22. Network

Using Kirchhoff's laws, find the currents in the following networks.

23.

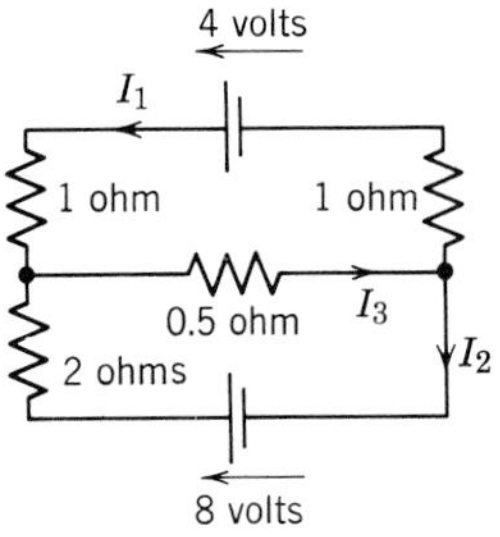

24.

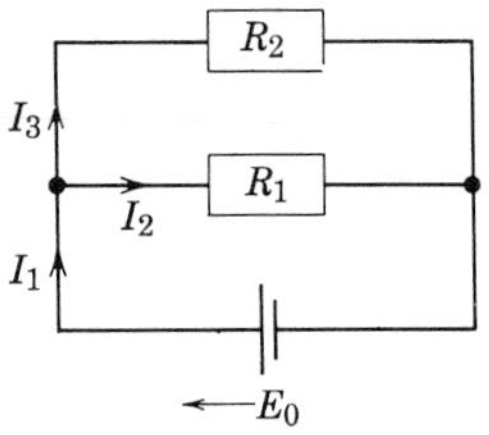

25.

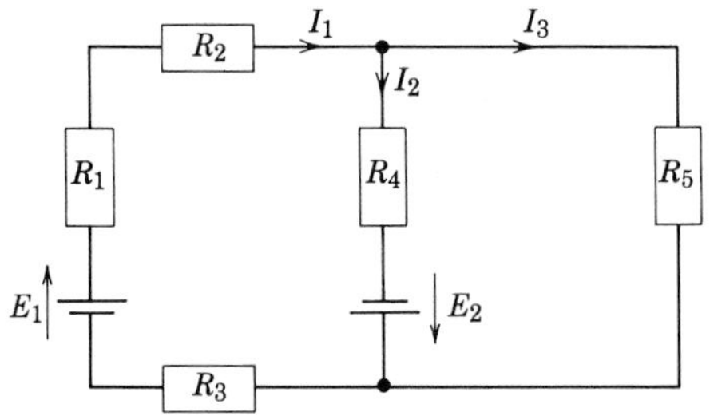

26.

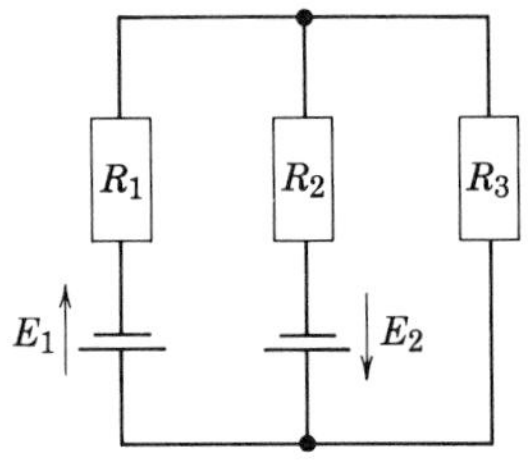

27. **(Wheatstone bridge)** Show that if $R_x/R_3 = R_1/R_2$ in the figure, then $I = 0$. (R_0 is the resistance of the instrument by which I is measured.)

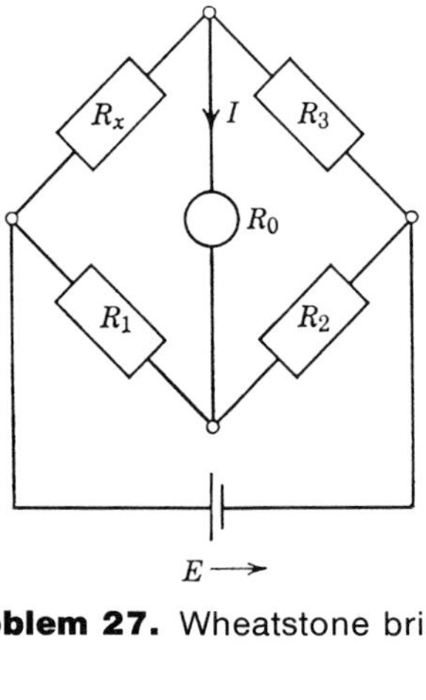

Problem 27. Wheatstone bridge

28. (Traffic flow) Methods of electrical circuit analysis have applications to other fields. For instance, applying the analogue of Kirchhoff's current law, find the traffic flow (cars per hour) in the net of one-way streets (in the directions indicated by the arrows) shown in the figure. Is the solution unique?

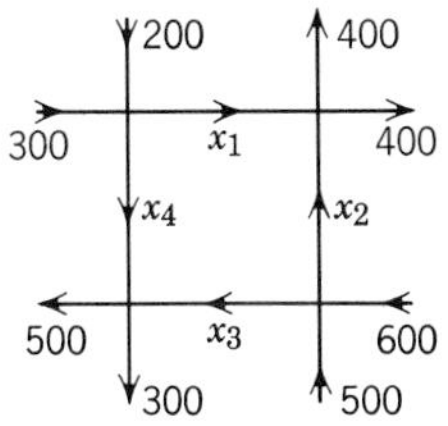

Problem 28. Net of one-way streets

29. (Supply and demand) Determine the equilibrium solution ($D = S$) of the one-commodity market with linear model

$$D = 13 - 2P$$
$$S = 3P - 7$$

where D, S, P mean demand, supply and price of the commodity.

30. Determine the equilibrium solution ($D_1 = S_1, D_2 = S_2$) of the two-commodity market with linear model

$$D_1 = 20 - 2P_1 - P_2$$
$$S_1 = 4P_1 - P_2 + 2$$
$$D_2 = 5P_1 - 2P_2 + 8$$
$$S_2 = 3P_2 - 2$$

where D, S, P mean demand, supply, price, and the subscripts 1 and 2 refer to the first and second commodity, respectively.

7.6 Rank of a Matrix

In the last section we have seen how to solve a system of linear equations, but we also know that there may be no solutions or a single solution or more than just one solution. So we ask whether we can make general statements about these *problems of existence and uniqueness*. The answer is yes, the main tool will be the concept of the rank of a matrix to be introduced now, and those statements will be given in the next section.

The maximum number of linearly independent row vectors of a matrix $\mathbf{A} = (a_{jk})$ is called the **rank** of $\mathbf{A}$ and is denoted by

$$\text{rank } \mathbf{A}.$$

The rank of a zero matrix is 0, by definition. (Linear independence of vectors is defined in Sec. 6.4.)

For example,

$$(1) \qquad \mathbf{A} = \begin{pmatrix} 3 & 0 & 2 & 2 \\ -1 & 7 & 4 & 9 \\ 7 & -7 & 0 & -5 \end{pmatrix}$$

has rank $\mathbf{A} = 2$ since the last row is a linear combination of the two others (twice the first row minus the second), which are linearly independent.

Note further that rank $\mathbf{A} = 0$ if and only if $\mathbf{A} = \mathbf{0}$. This follows directly from the definition.

In our proposed discussion of the existence and uniqueness of solutions of systems of linear equations we shall need

Theorem 1 (Rank in terms of column vectors)
The rank of a matrix $\mathbf{A} = (a_{jk})$ equals the maximum number of linearly independent column vectors of $\mathbf{A}$.

Proof. Let rank $\mathbf{A} = r$. Then, by definition, $\mathbf{A}$ has a linearly independent set of r row vectors, call them $\mathbf{v}_{(1)}, \cdots, \mathbf{v}_{(r)}$, and all row vectors $\mathbf{a}_{(1)}, \cdots, \mathbf{a}_{(m)}$ of $\mathbf{A}$ are linear combinations of those independent ones, say,

$$\begin{aligned} \mathbf{a}_{(1)} &= c_{11}\mathbf{v}_{(1)} + c_{12}\mathbf{v}_{(2)} + \cdots + c_{1r}\mathbf{v}_{(r)} \\ \mathbf{a}_{(2)} &= c_{21}\mathbf{v}_{(1)} + c_{22}\mathbf{v}_{(2)} + \cdots + c_{2r}\mathbf{v}_{(r)} \\ &\vdots \\ \mathbf{a}_{(m)} &= c_{m1}\mathbf{v}_{(1)} + c_{m2}\mathbf{v}_{(2)} + \cdots + c_{mr}\mathbf{v}_{(r)} \end{aligned}$$

These are vector equations. Each of them is equivalent to n equations for corresponding components. Denoting the components of $\mathbf{v}_{(1)}$ by $v_{11}, \cdots, v_{1n}$, the components of $\mathbf{v}_{(2)}$ by $v_{21}, \cdots, v_{2n}$, etc., and similarly for the vectors on the left-hand side, we thus have

$$\begin{aligned} a_{1k} &= c_{11}v_{1k} + c_{12}v_{2k} + \cdots + c_{1r}v_{rk} \\ a_{2k} &= c_{21}v_{1k} + c_{22}v_{2k} + \cdots + c_{2r}v_{rk} \\ &\vdots \\ a_{mk} &= c_{m1}v_{1k} + c_{m2}v_{2k} + \cdots + c_{mr}v_{rk} \end{aligned}$$

where $k = 1, \cdots, n$. This can be written

$$\begin{pmatrix} a_{1k} \\ a_{2k} \\ \vdots \\ a_{mk} \end{pmatrix} = v_{1k}\begin{pmatrix} c_{11} \\ c_{21} \\ \vdots \\ c_{m1} \end{pmatrix} + v_{2k}\begin{pmatrix} c_{12} \\ c_{22} \\ \vdots \\ c_{m2} \end{pmatrix} + \cdots + v_{rk}\begin{pmatrix} c_{1r} \\ c_{2r} \\ \vdots \\ c_{mr} \end{pmatrix}$$

where $k = 1, \cdots, n$. The vector on the left is the kth column vector of $\mathbf{A}$. Hence the equation shows that each column vector of $\mathbf{A}$ is a linear combination of the r vectors on the right. Hence the maximum number of linearly independent column vectors of $\mathbf{A}$ cannot exceed r, which is the maximum number of linearly independent row vectors of $\mathbf{A}$, by the definition of a rank.

Now the same conclusion applies to the transpose $\mathbf{A}^\mathsf{T}$ of $\mathbf{A}$. Since the row vectors of $\mathbf{A}^\mathsf{T}$ are the column vectors of $\mathbf{A}$, and the column vectors of $\mathbf{A}^\mathsf{T}$ are the row vectors of $\mathbf{A}$, that conclusion means that the maximum number of linearly independent row vectors of $\mathbf{A}$ (which is r) cannot exceed the maximum number of linearly independent column vectors of $\mathbf{A}$. Hence that number must equal r, and the proof is complete. ∎

What does this theorem mean with respect to our matrix $\mathbf{A}$ in (1)? Since we have rank $\mathbf{A} = 2$, the column vectors should contain two linearly independent ones, and the other two should be linear combinations of them. Indeed, the first two column vectors are linearly independent, and

$$\begin{pmatrix} 2 \\ 4 \\ 0 \end{pmatrix} = \frac{2}{3}\begin{pmatrix} 3 \\ -1 \\ 7 \end{pmatrix} + \frac{2}{3}\begin{pmatrix} 0 \\ 7 \\ -7 \end{pmatrix} \quad \text{and} \quad \begin{pmatrix} 2 \\ 9 \\ -5 \end{pmatrix} = \frac{2}{3}\begin{pmatrix} 3 \\ -1 \\ 7 \end{pmatrix} + \frac{29}{21}\begin{pmatrix} 0 \\ 7 \\ -7 \end{pmatrix}.$$

This is easy to verify but not so easy to see. Imagining that $\mathbf{A}^\mathsf{T}$ were given, we realize that the determination of a rank by a direct application of the definition is not the proper way, unless a matrix is sufficiently simple. This suggests asking whether we can "simplify" (transform) a matrix without altering its rank.

We claim that *elementary row operations* (Sec. 7.5) *do not alter the rank of a matrix* $\mathbf{A}$.

For the first operation (interchange of two row vectors) this is clear. The second operation (multiplication of a row vector by a nonzero constant) does not alter the rank either, since it does not alter the maximum number of linearly independent row vectors. Finally, the third operation is the addition of c times a row vector $\mathbf{a}_{(j)}$, say, to another row vector, say, $\mathbf{a}_{(i)}$. This produces from $\mathbf{A}$ a matrix $\tilde{\mathbf{A}}$ with the same row vectors, except for the ith, which is of the form $\tilde{\mathbf{a}}_{(i)} = \mathbf{a}_{(i)} + c\mathbf{a}_{(j)}$. From this we see that the spans (Sec. 6.4) of the row vectors of $\mathbf{A}$ and $\tilde{\mathbf{A}}$ are identical. Hence they have the same dimension r, the rank of $\mathbf{A}$. Thus rank $\tilde{\mathbf{A}}$ = rank $\mathbf{A}$, by definition. Remembering from the last section that *row-equivalent matrices* are those which can be obtained from each other by finitely many elementary row operations, our result is

Theorem 2

Row-equivalent matrices have the same rank.

This theorem tells us what we can do in order to determine the rank of a matrix $\mathbf{A}$, namely, we can reduce $\mathbf{A}$ to echelon form (Sec. 7.5), using the technique of the Gauss elimination, because this leaves the rank unchanged, by Theorem 2, and from the echelon form we can recognize the rank directly.

For instance, given (1), we obtain successively

$$\mathbf{A} = \begin{pmatrix} 3 & 0 & 2 & 2 \\ -1 & 7 & 4 & 9 \\ 7 & -7 & 0 & -5 \end{pmatrix} \qquad \text{(given)}$$

$$\begin{pmatrix} 3 & 0 & 2 & 2 \\ 0 & 7 & \frac{14}{3} & \frac{29}{3} \\ 0 & -7 & -\frac{14}{3} & -\frac{29}{3} \end{pmatrix} \qquad \begin{matrix} \\ \text{Row 2} + \frac{1}{3}\text{ Row 1} \\ \text{Row 3} - \frac{7}{3}\text{ Row 1} \end{matrix}$$

$$\begin{pmatrix} 3 & 0 & 2 & 2 \\ 0 & 7 & \frac{14}{3} & \frac{29}{3} \\ 0 & 0 & 0 & 0 \end{pmatrix} \qquad \begin{matrix} \\ \\ \text{Row 3} + \text{Row 2} \end{matrix}$$

The last matrix is in echelon form. From the row vectors and Theorem 2 we see immediately that rank $\mathbf{A} \leqq 2$, and rank $\mathbf{A} = 2$ by Theorem 1 since the first two column vectors are certainly linearly independent.

It is of practical importance that this method of determining the rank has applications in connection with the determination of linear dependence and independence of vectors. The key to this is the following theorem, which results immediately from the definition of a rank.

Theorem 3 (Linear dependence and independence)

p vectors $\mathbf{x}_{(1)}, \cdots, \mathbf{x}_{(p)}$ in the vector space R^n (cf. Sec. 7.4) *are linearly independent if the matrix with row vectors $\mathbf{x}_{(1)}, \cdots, \mathbf{x}_{(p)}$ has rank p; they are linearly dependent if that rank is less than p.*

Since each of those p vectors has n components, that matrix, call it $\mathbf{A}$, has p rows and n columns; and if $n < p$, then by Theorem 1 we have rank $\mathbf{A} \leqq n < p$, so that Theorem 3 yields the remarkable

Theorem 4

p vectors with $n < p$ components are always linearly dependent.

For instance, three or more vectors in the plane are linearly dependent, and so are four or more vectors in space.

Problems for Sec. 7.6

Determine the rank of the following matrices.

1. $\begin{pmatrix} 1 & 3 \\ 3 & 9 \end{pmatrix}$ **2.** $\begin{pmatrix} 3 & 1 & 2 \\ 4 & 0 & 6 \end{pmatrix}$ **3.** $\begin{pmatrix} 7 & 1 & 0 \\ -14 & -2 & 0 \end{pmatrix}$

4. $\begin{pmatrix} -4 & 1 \\ 3 & 2 \end{pmatrix}$ **5.** $\begin{pmatrix} 1 & 7 & -1 \\ 3 & 0 & 4 \end{pmatrix}$ **6.** $\begin{pmatrix} 4 & 2 & 0 & 3 \\ 6 & 1 & 0 & 0 \end{pmatrix}$

7. $\begin{pmatrix} 2 & 2 \\ 0 & 1 \\ 4 & 5 \end{pmatrix}$ **8.** $\begin{pmatrix} 3 & -3 & 0 \\ 1 & 4 & 5 \\ 4 & 4 & 8 \end{pmatrix}$ **9.** $\begin{pmatrix} 6 & 1 & 8 & 3 \\ 2 & 3 & 0 & 2 \\ 4 & -1 & -8 & -3 \end{pmatrix}$

10. Illustrate Theorem 1 with an example.

11. Show that rank $\mathbf{A}^{\mathsf{T}}$ = rank $\mathbf{A}$.

12. Show that in Theorem 3 we may replace "row vectors" by "column vectors."

13. What does it mean geometrically if a (2×2) matrix has rank 1? If a (3×3) matrix has rank 2?

14. Show by an example that rank $\mathbf{A}$ = rank $\mathbf{B}$ does *not* imply rank $\mathbf{A}^2$ = rank $\mathbf{B}^2$.

In each case, find whether the given vectors are linearly dependent or independent.

15. $(2, 1), (1, 2)$

16. $(3, 4), (-9, -12)$

17. $(0, 0), (2, -2)$

18. $(-3, 7, 8), (0, 2, 1)$

19. $(2, 0, 4), (3, 0, 2), (8, 0, -1)$

20. $(0, 1, 0), (1, 0, 0), (1, 1, 0)$

21. $(3, 2, 7), (2, 4, 1), (1, -2, 6)$

22. $(1, 2, 3), (4, -1, 6), (5, 1, 9)$

For what values of k are the following vectors linearly dependent?

23. $(1, 2), (3, k)$

24. $(1, 2, 3), (4, 5, 6), (7, 8, k)$

25. $(1, 2, 3, 4), (5, 6, 7, 8), (9, 10, 11, 12), (13, 14, 15, k)$

7.7 Systems of Linear Equations: Existence and General Properties of Solutions

Using the concept of a rank defined in the previous section, we can now state the

Fundamental Theorem for systems of linear equations

(a) *A system of m linear equations*

$$
\text{(1)}\qquad
\begin{aligned}
a_{11}x_1 + a_{12}x_2 + \cdots + a_{1n}x_n &= b_1 \\
a_{21}x_1 + a_{22}x_2 + \cdots + a_{2n}x_n &= b_2 \\
\cdots\cdots\cdots\cdots\cdots\cdots\cdots\cdots\cdots \\
a_{m1}x_1 + a_{m2}x_2 + \cdots + a_{mn}x_n &= b_m
\end{aligned}
$$

in n unknowns $x_1, \cdots, x_n$ has solutions if and only if the coefficient matrix $\mathbf{A}$ *and the augmented matrix* $\mathbf{B}$, *that is (cf. also Sec. 7.5)*

$$\mathbf{A} = \begin{pmatrix} a_{11} & \cdots & a_{1n} \\ \cdot & \cdots & \cdot \\ \cdot & \cdots & \cdot \\ a_{m1} & \cdots & a_{mn} \end{pmatrix} \quad \textit{and} \quad \mathbf{B} = \begin{pmatrix} a_{11} & \cdots & a_{1n} & b_1 \\ \cdot & \cdots & \cdot & \cdot \\ \cdot & \cdots & \cdot & \cdot \\ a_{m1} & \cdots & a_{mn} & b_m \end{pmatrix},$$

have the same rank.

(b) *If this rank r equals n, there is precisely one solution.*

(c) *If $r < n$, there are infinitely many solutions all of which are obtained by determining r suitable unknowns (whose submatrix of coefficients must have rank r) in terms of the remaining $n - r$ unknowns, to which arbitrary values can be assigned.*

(d) *If solutions exist, they can all be obtained by the Gauss elimination* (cf. Sec. 7.5). (This elimination may be started without first looking at the ranks of **A** and **B**.)

Proof. **(a)** We can write the system (1) in the form $\mathbf{Ax} = \mathbf{b}$ or in terms of the column vectors $\mathbf{c}_{(1)}, \cdots, \mathbf{c}_{(n)}$ of **A**:

$$\text{(2)} \qquad \mathbf{c}_{(1)}x_1 + \mathbf{c}_{(2)}x_2 + \cdots + \mathbf{c}_{(n)}x_n = \mathbf{b}.$$

Since **B** is obtained by attaching to **A** the additional column **b**, Theorem 1 in Sec. 7.6 implies that rank **B** equals rank **A** or rank **A** + 1. Now if (1) has a solution **x**, then (2) shows that **b** must be a linear combination of those column vectors. Hence rank **B** cannot exceed rank **A**, so that rank **B** = rank **A**.

Conversely, if rank **B** = rank **A**, then **b** must be a linear combination of the column vectors of **A**, say,

$$\mathbf{b} = \alpha_1\mathbf{c}_{(1)} + \cdots + \alpha_n\mathbf{c}_{(n)}$$

since otherwise rank **B** = rank **A** + 1. But this means that (1) has a solution, namely, $x_1 = \alpha_1, \cdots, x_n = \alpha_n$.

(b) If rank $\mathbf{A} = r = n$, then the set $C = \{\mathbf{c}_{(1)}, \cdots, \mathbf{c}_{(n)}\}$ is linearly independent, by Theorem 1 in the previous section. It follows that the representation (2) of **b** is unique because

$$\mathbf{c}_{(1)}x_1 + \cdots + \mathbf{c}_{(n)}x_n = \mathbf{c}_{(1)}\tilde{x}_1 + \cdots + \mathbf{c}_{(n)}\tilde{x}_n$$

implies

$$(x_1 - \tilde{x}_1)\mathbf{c}_{(1)} + \cdots + (x_n - \tilde{x}_n)\mathbf{c}_{(n)} = \mathbf{0}$$

and $x_1 - \tilde{x}_1 = 0, \cdots, x_n - \tilde{x}_n = 0$ by the linear independence. Hence the scalars $x_1, \cdots, x_n$ in (2) are uniquely determined, that is, the solution of (1) is unique.

(c) If rank **A** = rank $\mathbf{B} = r < n$, by Theorem 1, Sec. 7.6, there is a linearly independent set K of r column vectors of **A** such that the other $n - r$ column vectors of **A** are linear combinations of those vectors. We renumber the columns and unknowns, denoting the renumbered quantities by $\widehat{\ }$, so that $\{\widehat{\mathbf{c}}_{(1)}, \cdots, \widehat{\mathbf{c}}_{(r)}\}$ is that linearly independent set K. Then (2) becomes

$$\widehat{\mathbf{c}}_{(1)}\widehat{x}_1 + \cdots + \widehat{\mathbf{c}}_{(n)}\widehat{x}_n = \mathbf{b},$$

$\widehat{\mathbf{c}}_{(r+1)}, \cdots, \widehat{\mathbf{c}}_{(n)}$ are linear combinations of the vectors of K, and so are the vectors $\widehat{x}_{r+1}\widehat{\mathbf{c}}_{(r+1)}, \cdots, \widehat{x}_n\widehat{\mathbf{c}}_{(n)}$. Expressing these vectors in terms of the vectors of K and collecting terms, we can write the system in the form

$$\widehat{\mathbf{c}}_{(1)}y_1 + \cdots + \widehat{\mathbf{c}}_{(r)}y_r = \mathbf{b} \tag{3}$$

with $y_j = \widehat{x}_j + \beta_j$, where β_j results from the terms $\widehat{x}_{r+1}\widehat{\mathbf{c}}_{(r+1)}, \cdots, \widehat{x}_n\widehat{\mathbf{c}}_{(n)}$; here, $j = 1, \cdots, r$. Since the system has a solution, there are $y_1, \cdots, y_r$ satisfying (3). These scalars are unique since K is linearly independent. Choosing $\widehat{x}_{r+1}, \cdots, \widehat{x}_n$ fixes the β_j and corresponding $\widehat{x}_j = y_j - \beta_j$, $j = 1, \cdots, r$.

(d) was shown in Sec. 7.6 and is restated here as a reminder. This completes the proof. ∎

The theorem is illustrated by the examples in Sec. 7.5: in Example 2 we have rank $\mathbf{A}$ = rank $\mathbf{B} = 2 < n = 4$ and can choose x_3 and x_4 arbitrarily; in Example 3 there is no solution since rank $\mathbf{A} = 2 <$ rank $\mathbf{B} = 3$, and in Example 4 there is a unique solution since rank $\mathbf{A}$ = rank $\mathbf{B} = n = 3$.

The system in (1) is said to be **homogeneous** if all the b_j's on the right-hand side are zero. Otherwise it is said to be **nonhomogeneous.** (Cf. also Sec. 7.5.) From the Fundamental Theorem we readily obtain the following results.

Theorem 2 (Homogeneous system)

A homogeneous system of linear equations

$$\begin{aligned} a_{11}x_1 + a_{12}x_2 + \cdots + a_{1n}x_n &= 0 \\ a_{21}x_1 + a_{22}x_2 + \cdots + a_{2n}x_n &= 0 \\ \cdots\cdots\cdots\cdots\cdots\cdots\cdots\cdots& \\ a_{m1}x_1 + a_{m2}x_2 + \cdots + a_{mn}x_n &= 0 \end{aligned} \tag{4}$$

always has the **trivial solution** $x_1 = 0, \cdots, x_n = 0$. *Nontrivial solutions exist if and only if* rank $\mathbf{A}$ *is less than n. If rank* $\mathbf{A} = r < n$, *the solutions form a vector space of dimension* $n - r$ *(cf. Sec. 6.4). In particular, if* $\mathbf{x}_{(1)}$ *and* $\mathbf{x}_{(2)}$ *are solution vectors of* (4), *then* $\mathbf{x} = c_1\mathbf{x}_{(1)} + c_2\mathbf{x}_{(2)}$, *where* c_1 *and* c_2 *are any constants, is a solution vector of* (4). (*This* **does not hold** *for nonhomogeneous systems.*)

Proof. The first proposition is obvious and is in agreement with the fact that for a homogeneous system the matrix of the coefficients and the augmented matrix have the same rank. The solution vectors form a vector space because if $\mathbf{x}_{(1)}$ and $\mathbf{x}_{(2)}$ are any of them, then $\mathbf{A}\mathbf{x}_{(1)} = \mathbf{0}$, $\mathbf{A}\mathbf{x}_{(2)} = \mathbf{0}$, and this implies $\mathbf{A}(\mathbf{x}_{(1)} + \mathbf{x}_{(2)}) = \mathbf{A}\mathbf{x}_{(1)} + \mathbf{A}\mathbf{x}_{(2)} = \mathbf{0}$ as well as $\mathbf{A}(c\mathbf{x}_{(1)}) = c\mathbf{A}\mathbf{x}_{(1)} = \mathbf{0}$, where c is arbitrary. If rank $\mathbf{A} = r < n$, the Fundamental Theorem implies that we can choose $n - r$ suitable unknowns, call them $x_{r+1}, \cdots, x_n$, in an arbitrary fashion, and every solution is obtained in this way. It follows that a basis of solutions is $\mathbf{y}_{(1)}, \cdots, \mathbf{y}_{(n-r)}$, where the solution vector $\mathbf{y}_{(j)}, j = 1, \cdots, n - r$, is obtained by choosing $x_{r+j} = 1$ and the other $x_{r+1}, \cdots, x_n$ zero; the corresponding $x_1, \cdots, x_r$ are then determined. This proves that the vector space of all solutions has dimension $n - r$ and completes the proof. ∎

We mention that the vector space of all solutions of (4) is called the **null space** of the coefficient matrix **A**, and its dimension is called the **nullity** of **A**. In terms of these concepts, Theorem 2 states that

$$\text{rank } \mathbf{A} + \text{nullity } \mathbf{A} = n$$

where n is the number of unknowns (number of columns of **A**). If rank $\mathbf{A} = n$, then nullity $\mathbf{A} = 0$, so that the system has only the trivial solution. If rank $\mathbf{A} = r < n$, then nullity $\mathbf{A} = n - r > 0$, so that we have nontrivial solutions which form a vector space of dimension $n - r > 0$.

Note that rank $\mathbf{A} \leqq m$ in (4), by the definition, so that rank $\mathbf{A} < n$ when $m < n$. By Theorem 2 this proves the following theorem, which is of considerable practical importance.

Theorem 3 (System with fewer equations than unknowns)
A homogeneous system of linear equations with fewer equations than unknowns always has nontrivial solutions.

If a nonhomogeneous system of linear equations has solutions, their totality can be characterized as follows.

Theorem 4 (Nonhomogeneous system)
If a nonhomogeneous system of linear equations of the form (1) *has solutions, then all these solutions are of the form*

$$\mathbf{x} = \mathbf{x}_0 + \mathbf{x}_h$$

where $\mathbf{x}_0$ *is any fixed solution of* (1) *and* $\mathbf{x}_h$ *runs through all the solutions of the corresponding homogeneous system* (4).

Proof. Let $\mathbf{x}$ be any given solution of (1) and $\mathbf{x}_0$ an arbitrarily chosen solution of (1). Then $\mathbf{Ax} = \mathbf{b}$, $\mathbf{Ax}_0 = \mathbf{b}$ and, therefore,

$$\mathbf{A}(\mathbf{x} - \mathbf{x}_0) = \mathbf{Ax} - \mathbf{Ax}_0 = \mathbf{0}.$$

This shows that the difference $\mathbf{x} - \mathbf{x}_0$ between any solution $\mathbf{x}$ of (1) and any fixed solution $\mathbf{x}_0$ of (1) is a solution of (4), say, $\mathbf{x}_h$. Hence all solutions of (1) are obtained by letting $\mathbf{x}_h$ run through all the solutions of the homogeneous system (4), and the proof is complete. ∎

7.8 The Inverse of a Matrix

In this section we consider exclusively *square* matrices.

The **inverse** of an $(n \times n)$ matrix $\mathbf{A} = (a_{jk})$ is denoted by $\mathbf{A}^{-1}$ and is an $(n \times n)$ matrix such that

$$\mathbf{AA}^{-1} = \mathbf{A}^{-1}\mathbf{A} = \mathbf{I}, \tag{1}$$

where $\mathbf{I}$ is the n-rowed unit matrix.

If $\mathbf{A}$ has an inverse, then $\mathbf{A}$ is called a **nonsingular matrix.** If $\mathbf{A}$ has no inverse, then $\mathbf{A}$ is called a **singular matrix.**

If $\mathbf{A}$ *has an inverse, the inverse is unique.*

Indeed, if both $\mathbf{B}$ and $\mathbf{C}$ are inverses of $\mathbf{A}$, then

$$\mathbf{AB} = \mathbf{I}, \qquad \mathbf{CA} = \mathbf{I},$$

and we obtain

$$\mathbf{B} = \mathbf{IB} = (\mathbf{CA})\mathbf{B} = \mathbf{C}(\mathbf{AB}) = \mathbf{CI} = \mathbf{C},$$

proving the uniqueness.

Note that, as a consequence of this proof, the proof of either $\mathbf{AB} = \mathbf{I}$ or $\mathbf{BA} = \mathbf{I}$ is enough to assert that $\mathbf{B}$ is the inverse of a given matrix $\mathbf{A}$.

Let us motivate the concept of an inverse and find conditions for the existence. For this purpose we start from $\mathbf{A}$ and consider a corresponding **linear transformation** (Sec. 7.4)

$$\mathbf{y} = \mathbf{Ax}, \tag{2}$$

where the vectors $\mathbf{x}$ and $\mathbf{y}$ are column vectors with n components $x_1, \cdots, x_n$ and $y_1, \cdots, y_n$, respectively. If the inverse matrix $\mathbf{A}^{-1}$ exists, we may premultiply (2) by $\mathbf{A}^{-1}$ and have $\mathbf{A}^{-1}\mathbf{y} = \mathbf{A}^{-1}\mathbf{Ax} = \mathbf{Ix} = \mathbf{x}$, the **inverse transformation**

$$\mathbf{x} = \mathbf{A}^{-1}\mathbf{y}, \tag{3}$$

which expresses $\mathbf{x}$ in terms of $\mathbf{y}$, and involves the inverse $\mathbf{A}^{-1}$.

For a given $\mathbf{y}$, equation (2) may be regarded as a system of n linear equations in n unknowns $x_1, \cdots, x_n$, and we know from the Fundamental Theorem in the previous section that it has a unique solution if and only if $\mathbf{A}$ has rank n, the greatest possible rank for an $(n \times n)$ matrix. This proves

Theorem 1 (Existence of the inverse)
For an $(n \times n)$ *matrix* $\mathbf{A}$, *the inverse* $\mathbf{A}^{-1}$ *exists if and only if* rank $\mathbf{A} = n$. *Hence* $\mathbf{A}$ *is nonsingular if* rank $\mathbf{A} = n$, *and is singular if* rank $\mathbf{A} < n$.

We want to show that the practical determination of the inverse $\mathbf{A}^{-1}$ of $\mathbf{A}$ amounts to solving a linear system $\mathbf{Ax} = \mathbf{y}$. A corresponding method is the Gauss elimination (Sec. 7.5). We know that this method produces from the given square matrix **A** a triangular matrix (the echelon form). From the latter we can easily obtain a diagonal form by further elementary operations, that is, we can reduce

$$\mathbf{Ax} = \mathbf{y} = \mathbf{Iy} \qquad \text{to the form} \qquad \mathbf{Ix} = \mathbf{x} = \mathbf{A}^{-1}\mathbf{y}$$

by starting from **A** and **I** and ending up with **I** and $\mathbf{A}^{-1}$. This slight extension of the Gauss elimination is sometimes called the **Gauss–Jordan elimination.** It suffices to explain a convenient arrangement of the calculations in terms of a typical example.

(A representation of the inverse in terms of determinants will be given in Sec. 7.11.)

Example 1. Inverse of a matrix

Find the inverse $\mathbf{A}^{-1}$ of

$$\mathbf{A} = \begin{pmatrix} -1 & 1 & 2 \\ 3 & -1 & 1 \\ -1 & 3 & 4 \end{pmatrix}.$$

We start from

$$\left(\begin{array}{ccc} -1 & 1 & 2 \\ 3 & -1 & 1 \\ -1 & 3 & 4 \end{array}\right) \left|\ \left(\begin{array}{ccc} 1 & 0 & 0 \\ 0 & 1 & 0 \\ 0 & 0 & 1 \end{array}\right)\right.$$

and obtain the subsequent matrices by performing the indicated row operations on the two immediately preceding matrices.

$$\left(\begin{array}{ccc} -1 & 1 & 2 \\ 0 & 2 & 7 \\ 0 & 2 & 2 \end{array}\right) \left|\ \left(\begin{array}{ccc} 1 & 0 & 0 \\ 3 & 1 & 0 \\ -1 & 0 & 1 \end{array}\right)\right. \qquad \begin{array}{l} \\ \text{Row 2} + 3\ \text{Row 1} \\ \text{Row 3} - \text{Row 1} \end{array}$$

$$\left(\begin{array}{ccc} -1 & 1 & 2 \\ 0 & 2 & 7 \\ 0 & 0 & -5 \end{array}\right) \left|\ \left(\begin{array}{ccc} 1 & 0 & 0 \\ 3 & 1 & 0 \\ -4 & -1 & 1 \end{array}\right)\right. \qquad \begin{array}{l} \\ \\ \text{Row 3} - \text{Row 2} \end{array}$$

This was the Gauss elimination and agrees with Example 4 in Sec. 7.5. The next steps extend the Gauss elimination as indicated above.

$$\left(\begin{array}{ccc} 1 & -1 & 0 \\ 0 & 1 & 0 \\ 0 & 0 & 1 \end{array}\right) \left|\ \left(\begin{array}{ccc} \frac{3}{5} & \frac{2}{5} & -\frac{2}{5} \\ -\frac{13}{10} & -\frac{2}{10} & \frac{7}{10} \\ \frac{4}{5} & \frac{1}{5} & -\frac{1}{5} \end{array}\right)\right. \qquad \begin{array}{l} -\text{Row 1} - \frac{2}{5}\ \text{Row 3} \\ \frac{1}{2}\ \text{Row 2} + \frac{7}{10}\ \text{Row 3} \\ -\frac{1}{5}\ \text{Row 3} \end{array}$$

$$\left(\begin{array}{ccc} 1 & 0 & 0 \\ 0 & 1 & 0 \\ 0 & 0 & 1 \end{array}\right) \left|\ \left(\begin{array}{ccc} -\frac{7}{10} & \frac{2}{10} & \frac{3}{10} \\ -\frac{13}{10} & -\frac{2}{10} & \frac{7}{10} \\ \frac{4}{5} & \frac{1}{5} & -\frac{1}{5} \end{array}\right)\right. \qquad \begin{array}{l} \text{Row 1} + \text{Row 2} \\ \\ \\ \end{array}$$

The matrix on the right is the inverse of **A** because, in terms of systems of linear equations, we started from $\mathbf{Ax} = \mathbf{y} = \mathbf{Iy}$ and reduced this to $\mathbf{x} = \mathbf{Ix} = \mathbf{A}^{-1}\mathbf{y}$. Check:

$$\begin{pmatrix} -1 & 1 & 2 \\ 3 & -1 & 1 \\ -1 & 3 & 4 \end{pmatrix}\begin{pmatrix} -0.7 & 0.2 & 0.3 \\ -1.3 & -0.2 & 0.7 \\ 0.8 & 0.2 & -0.2 \end{pmatrix} = \begin{pmatrix} 1 & 0 & 0 \\ 0 & 1 & 0 \\ 0 & 0 & 1 \end{pmatrix}.$$

Hence $\mathbf{AA}^{1} = \mathbf{I}$. We need not check $\mathbf{A}^{-1}\mathbf{A} = \mathbf{I}$, as was mentioned after the proof of the uniqueness of inverses, earlier in this section. ∎

In conclusion of this section we list a few useful formulas for inverses.

For a nonsingular (2×2) matrix we obtain

$$\textbf{(4)} \qquad \mathbf{A} = \begin{pmatrix} a_{11} & a_{12} \\ a_{21} & a_{22} \end{pmatrix}, \qquad \mathbf{A}^{-1} = \frac{1}{\det \mathbf{A}}\begin{pmatrix} a_{22} & -a_{12} \\ -a_{21} & a_{11} \end{pmatrix},$$

where $\det \mathbf{A} = a_{11}a_{22} - a_{12}a_{21}$ and will be discussed in the next section. Indeed, one can readily verify that (1) holds.

Similarly, for a nonsingular diagonal matrix we simply have

$$(5) \qquad \mathbf{A} = \begin{pmatrix} a_{11} & \cdots & 0 \\ \cdot & \cdots & \cdot \\ \cdot & \cdots & \cdot \\ 0 & \cdots & a_{nn} \end{pmatrix}, \qquad \mathbf{A}^{-1} = \begin{pmatrix} 1/a_{11} & \cdots & 0 \\ \cdot & \cdots & \cdot \\ \cdot & \cdots & \cdot \\ 0 & \cdots & 1/a_{nn} \end{pmatrix};$$

the elements of $\mathbf{A}^{-1}$ in the principal diagonal are the reciprocals of those of **A**.

Example 2. Inverse of a two-rowed square matrix

$$\mathbf{A} = \begin{pmatrix} 3 & 1 \\ 2 & 4 \end{pmatrix}, \qquad \mathbf{A}^{-1} = \frac{1}{10}\begin{pmatrix} 4 & -1 \\ -2 & 3 \end{pmatrix} = \begin{pmatrix} 0.4 & -0.1 \\ -0.2 & 0.3 \end{pmatrix}.$$

Example 3. Inverse of a diagonal matrix

$$\mathbf{A} = \begin{pmatrix} -0.5 & 0 & 0 \\ 0 & 4 & 0 \\ 0 & 0 & 1 \end{pmatrix}, \qquad \mathbf{A}^{-1} = \begin{pmatrix} -2 & 0 & 0 \\ 0 & 0.25 & 0 \\ 0 & 0 & 1 \end{pmatrix}$$ ∎

The inverse of the inverse is the given matrix **A**, that is

$$(6) \qquad (\mathbf{A}^{-1})^{-1} = \mathbf{A}.$$

The simple proof is left to the reader (Prob. 3).

The inverse of a product **AC** can be obtained by inverting each factor and multiplying the results *in reverse order:*

$$(7) \qquad (\mathbf{AC})^{-1} = \mathbf{C}^{-1}\mathbf{A}^{-1}.$$

To prove (7), we start from (1), with $\mathbf{A}$ replaced by $\mathbf{AC}$, that is,

$$\mathbf{AC}(\mathbf{AC})^{-1} = \mathbf{I}.$$

By premultiplying this by $\mathbf{A}^{-1}$ and using $\mathbf{A}^{-1}\mathbf{A} = \mathbf{I}$ we obtain

$$\mathbf{C}(\mathbf{AC})^{-1} = \mathbf{A}^{-1}.$$

If we premultiply this by $\mathbf{C}^{-1}$, the result follows.

Of course, (7) may be generalized to products of more than two matrices; by induction we obtain

$$(\mathbf{AC} \cdots \mathbf{PQ})^{-1} = \mathbf{Q}^{-1}\mathbf{P}^{-1} \cdots \mathbf{C}^{-1}\mathbf{A}^{-1}. \tag{8}$$

Furthermore, we can now obtain more information about the strange fact that for matrix multiplication, the cancellation law is not true, in general; that is,

$$\mathbf{AB} = \mathbf{0} \quad \text{does not necessarily imply} \quad \mathbf{A} = \mathbf{0} \text{ or } \mathbf{B} = \mathbf{0}.$$

This was stated in Sec. 7.4 (Theorem 1) and illustrated by

$$\begin{pmatrix} 1 & 1 \\ 2 & 2 \end{pmatrix}\begin{pmatrix} -1 & 1 \\ 1 & -1 \end{pmatrix} = \begin{pmatrix} 0 & 0 \\ 0 & 0 \end{pmatrix}. \tag{9}$$

Each of the two matrices has rank less than $n = 2$. This is typical since we can now prove the following

Theorem 2

If $\mathbf{A} \neq \mathbf{0}$ *and* $\mathbf{B} \neq \mathbf{0}$ *are* $(n \times n)$ *matrices such that* $\mathbf{AB} = \mathbf{0}$, *then both* $\mathbf{A}$ *and* $\mathbf{B}$ *have rank less than* n.

Proof. Suppose that $\mathbf{AB} = \mathbf{0}$ but rank $\mathbf{A} = n$. Then $\mathbf{A}^{-1}$ exists by Theorem 1, and premultiplication of $\mathbf{AB} = \mathbf{0}$ by $\mathbf{A}^{-1}$ gives $\mathbf{A}^{-1}\mathbf{AB} = \mathbf{B} = \mathbf{0}$, which contradicts $\mathbf{B} \neq \mathbf{0}$. ∎

Note that interchanging the factors in (9) gives

$$\begin{pmatrix} -1 & 1 \\ 1 & -1 \end{pmatrix}\begin{pmatrix} 1 & 1 \\ 2 & 2 \end{pmatrix} = \begin{pmatrix} 1 & 1 \\ -1 & -1 \end{pmatrix}.$$

Hence, if both $\mathbf{A}$ and $\mathbf{B}$ have rank less than n and $\mathbf{AB} = \mathbf{0}$, it does *not* follow that $\mathbf{BA} = \mathbf{0}$.

From the proof of Theorem 2 we also have

Theorem 3

If $\mathbf{A}$ *and* $\mathbf{B}$ *are* $(n \times n)$ *matrices such that* $\mathbf{AB} = \mathbf{0}$, *and if* $\mathbf{A}$ *has rank* n, *then* $\mathbf{B} = \mathbf{0}$.

Furthermore, if $\mathbf{AB} = \mathbf{0}$ and we set $\mathbf{B} = \mathbf{S} - \mathbf{T}$, then

$$\mathbf{AB} = \mathbf{A}(\mathbf{S} - \mathbf{T}) = \mathbf{0} \qquad \text{or} \qquad \mathbf{AS} = \mathbf{AT}.$$

From this and Theorem 3 we obtain

Theorem 4

If **A**, **S** *and* **T** *are* $(n \times n)$ *matrices and* rank $\mathbf{A} = n$, *then*

$$\mathbf{AS} = \mathbf{AT} \qquad \textit{implies} \qquad \mathbf{S} = \mathbf{T}.$$

Problems for Sec. 7.8

1. Verify (4) by showing that $\mathbf{AA}^{-1} = \mathbf{A}^{-1}\mathbf{A} = \mathbf{I}$.
2. Verify (5).
3. Prove (6).

Find the inverse and check the result.

4. $\begin{pmatrix} -3 & 5 \\ 2 & 1 \end{pmatrix}$ 5. $\begin{pmatrix} 3 & -1 \\ -5 & 2 \end{pmatrix}$ 6. $\begin{pmatrix} \cos\theta & -\sin\theta \\ \sin\theta & \cos\theta \end{pmatrix}$

7. $\begin{pmatrix} 4 & -2 & 2 \\ -2 & -4 & 4 \\ -4 & 2 & 8 \end{pmatrix}$ 8. $\begin{pmatrix} 0 & 0 & 1 \\ 0 & 1 & 0 \\ 1 & 0 & 0 \end{pmatrix}$ 9. $\begin{pmatrix} 0 & 1 & 0 \\ 1 & 0 & 0 \\ 0 & 0 & 1 \end{pmatrix}$

10. $\begin{pmatrix} 3 & -1 & 1 \\ -15 & 6 & -5 \\ 5 & -2 & 2 \end{pmatrix}$ 11. $\begin{pmatrix} 5 & -1 & 5 \\ 0 & 2 & 0 \\ -5 & 3 & -15 \end{pmatrix}$ 12. $\begin{pmatrix} 4 & 0 & 0 \\ 0 & 0.2 & 0 \\ 0 & 0 & -3 \end{pmatrix}$

Find the inverse of the given linear transformation.

13. $\begin{aligned} x^* &= 19x + 2y - 9z \\ y^* &= -4x - y + 2z \\ z^* &= -2x + z \end{aligned}$

14. $\begin{aligned} x^* &= -2x - 2y + 7z \\ y^* &= 4x + 3y - 12z \\ z^* &= -x + 2z \end{aligned}$

15. $\begin{aligned} x^* &= x + 2y + 5z \\ y^* &= -y + 2z \\ z^* &= 2x + 4y + 11z \end{aligned}$

16. $\begin{aligned} x^* &= 2x + 4y + z \\ y^* &= x + 2y + z \\ z^* &= 3x + 4y + 2z \end{aligned}$

17. Show that $(\mathbf{A}^2)^{-1} = (\mathbf{A}^{-1})^2$.
18. Find the inverse of the square of the matrices in Probs. 7 and 10.
19. Show that $(\mathbf{A}^{-1})^{\mathsf{T}} = (\mathbf{A}^{\mathsf{T}})^{-1}$.
20. Show that the inverse of a nonsingular symmetric matrix is symmetric.

7.9 Determinants of Second and Third Order

This section is independent of the other sections in this chapter. It is for reference in connection with other chapters, and it also serves to motivate the discussions in the next section where general determinants and their use will be considered.

Determinants of second order can be introduced and used in connection with systems of two linear equations

$$\text{(1)} \qquad \begin{aligned} a_{11}x_1 + a_{12}x_2 &= b_1 \\ a_{21}x_1 + a_{22}x_2 &= b_2 \end{aligned}$$

in two unknowns x_1, x_2, if we proceed as follows. To solve this system, we multiply the first equation by a_{22}, the second by $-a_{12}$, and add, finding

$$(a_{11}a_{22} - a_{21}a_{12})x_1 = b_1a_{22} - b_2a_{12}.$$

Then we multiply the first equation of (1) by $-a_{21}$, the second by a_{11}, and add again, finding

$$(a_{11}a_{22} - a_{21}a_{12})x_2 = a_{11}b_2 - a_{21}b_1.$$

If $a_{11}a_{22} - a_{21}a_{12}$ is not zero, we may divide and obtain the desired result

$$\text{(2)} \qquad x_1 = \frac{b_1a_{22} - b_2a_{12}}{a_{11}a_{22} - a_{21}a_{12}}, \qquad x_2 = \frac{a_{11}b_2 - a_{21}b_1}{a_{11}a_{22} - a_{21}a_{12}}.$$

The expression in the denominators is written in the form

$$\begin{vmatrix} a_{11} & a_{12} \\ a_{21} & a_{22} \end{vmatrix}$$

and is called a **determinant of second order;** thus

$$\text{(3)} \qquad \begin{vmatrix} a_{11} & a_{12} \\ a_{21} & a_{22} \end{vmatrix} = a_{11}a_{22} - a_{21}a_{12}.$$

The four numbers a_{11}, a_{12}, a_{21}, a_{22} are called the **elements** of the determinant. The elements in a horizontal line are said to form a **row** and the elements in a vertical line are said to form a **column** of the determinant.

We may now write the solution (2) of the system (1) in the form

$$\text{(4)} \qquad x_1 = \frac{D_1}{D}, \qquad x_2 = \frac{D_2}{D} \qquad (D \neq 0)$$

where

$$D = \begin{vmatrix} a_{11} & a_{12} \\ a_{21} & a_{22} \end{vmatrix}, \qquad D_1 = \begin{vmatrix} b_1 & a_{12} \\ b_2 & a_{22} \end{vmatrix}, \qquad D_2 = \begin{vmatrix} a_{11} & b_1 \\ a_{21} & b_2 \end{vmatrix}.$$

This formula is called **Cramer's rule.**[2] Note that D_1 is obtained by replacing the first column of D by the column with elements b_1, b_2, and D_2 is obtained by replacing the second column of D by that column.

[2]GABRIEL CRAMER (1704–1752), Swiss mathematician, also known by his book on the theory of curves, which appeared 1750 in Geneva.

If both b_1 and b_2 are zero, the system is said to be **homogeneous.** *In this case it has at least the "trivial solution" $x_1 = 0, x_2 = 0$. It has further solutions if and only if $D = 0$.*

If at least b_1 or b_2 is not zero, the system is said to be **nonhomogeneous.** *Then if $D \neq 0$, it has precisely one solution, which is obtained from* (4).

Determinants of third order occur in connection with systems of three linear equations

$$\begin{aligned} a_{11}x_1 + a_{12}x_2 + a_{13}x_3 &= b_1 \\ a_{21}x_1 + a_{22}x_2 + a_{23}x_3 &= b_2 \\ a_{31}x_1 + a_{32}x_2 + a_{33}x_3 &= b_3 \end{aligned} \tag{5}$$

in three unknowns x_1, x_2, x_3. To obtain an equation involving only x_1, we multiply the equations by

$$a_{22}a_{33} - a_{32}a_{23}, \qquad -(a_{12}a_{33} - a_{32}a_{13}), \qquad a_{12}a_{23} - a_{22}a_{13},$$

respectively. We see that these expressions may be written as second-order determinants:

$$M_{11} = \begin{vmatrix} a_{22} & a_{23} \\ a_{32} & a_{33} \end{vmatrix}, \qquad -M_{21} = -\begin{vmatrix} a_{12} & a_{13} \\ a_{32} & a_{33} \end{vmatrix}, \qquad M_{31} = \begin{vmatrix} a_{12} & a_{13} \\ a_{22} & a_{23} \end{vmatrix}.$$

Adding the resulting equations, we obtain

$$(a_{11}M_{11} - a_{21}M_{21} + a_{31}M_{31})x_1 = b_1M_{11} - b_2M_{21} + b_3M_{31}. \tag{6}$$

Two further equations containing only x_2 and x_3, respectively, may be obtained in a similar manner.

To simplify our notation, we now define a **determinant of third order** by the equation

$$D = \begin{vmatrix} a_{11} & a_{12} & a_{13} \\ a_{21} & a_{22} & a_{23} \\ a_{31} & a_{32} & a_{33} \end{vmatrix} = a_{11}\begin{vmatrix} a_{22} & a_{23} \\ a_{32} & a_{33} \end{vmatrix} - a_{21}\begin{vmatrix} a_{12} & a_{13} \\ a_{32} & a_{33} \end{vmatrix} + a_{31}\begin{vmatrix} a_{12} & a_{13} \\ a_{22} & a_{23} \end{vmatrix}. \tag{7}$$

We see that

$$D = a_{11}M_{11} - a_{21}M_{21} + a_{31}M_{31},$$

the coefficient of x_1 in (6), and if we write out the second-order determinants in (7), we obtain

$$D = a_{11}a_{22}a_{33} - a_{11}a_{32}a_{23} + a_{21}a_{32}a_{13} - a_{21}a_{12}a_{33} + a_{31}a_{12}a_{23} - a_{31}a_{22}a_{13}. \tag{8}$$

Obviously the determinant on the right-hand side of (7) which is multiplied by a_{i1}, $i = 1, 2$, or 3, is obtained from D by omitting the first column and the ith row of D.

We see that (6) may now be written

$$Dx_1 = D_1 \qquad \text{where} \qquad D_1 = \begin{vmatrix} b_1 & a_{12} & a_{13} \\ b_2 & a_{22} & a_{23} \\ b_3 & a_{32} & a_{33} \end{vmatrix}.$$

Similarly, the aforementioned equation containing only x_2 may be written

$$Dx_2 = D_2 \qquad \text{where} \qquad D_2 = \begin{vmatrix} a_{11} & b_1 & a_{13} \\ a_{21} & b_2 & a_{23} \\ a_{31} & b_3 & a_{33} \end{vmatrix},$$

and the equation containing only x_3 may be written

$$Dx_3 = D_3 \qquad \text{where} \qquad D_3 = \begin{vmatrix} a_{11} & a_{12} & b_1 \\ a_{21} & a_{22} & b_2 \\ a_{31} & a_{32} & b_3 \end{vmatrix}.$$

Note that the elements of D are arranged in the same order as they occur as coefficients in the equations of (5), and $D_j, j = 1, 2$, or 3, is obtained from D by replacing the jth column by the column with elements b_1, b_2, b_3, the expressions on the right sides of the equations of (5).

It follows that if $D \neq 0$, then system (5) has the unique solution

$$\text{(9)} \qquad x_1 = \frac{D_1}{D}, \qquad x_2 = \frac{D_2}{D}, \qquad x_3 = \frac{D_3}{D} \qquad \textbf{(Cramer's rule).}$$

If (5) *is homogeneous, that is,* $b_1 = b_2 = b_3 = 0$, *it has at least the trivial solution* $x_1 = x_2 = x_3 = 0$, *and nontrivial solutions exist if and only if* $D = 0$.

If (5) *is nonhomogeneous and* $D \neq 0$, *it has precisely one solution, which is obtained from* (9).

Example 1. Cramer's rule

Solve by Cramer's rule:

$$\begin{aligned} 2x_1 - x_2 + 2x_3 &= 2 \\ x_1 + 10x_2 - 3x_3 &= 5 \\ -x_1 + x_2 + x_3 &= -3. \end{aligned}$$

The determinant of the system is

$$D = \begin{vmatrix} 2 & -1 & 2 \\ 1 & 10 & -3 \\ -1 & 1 & 1 \end{vmatrix} = 46.$$

The determinants in the numerators in (9) are

$$D_1 = \begin{vmatrix} 2 & -1 & 2 \\ 5 & 10 & -3 \\ -3 & 1 & 1 \end{vmatrix} = 92, \quad D_2 = \begin{vmatrix} 2 & 2 & 2 \\ 1 & 5 & -3 \\ -1 & -3 & 1 \end{vmatrix} = 0, \quad D_3 = \begin{vmatrix} 2 & -1 & 2 \\ 1 & 10 & 5 \\ -1 & 1 & -3 \end{vmatrix} = -46.$$

Therefore, $x_1 = 2$, $x_2 = 0$, and $x_3 = -1$. ∎

We shall now list the most important properties of our determinants. The proofs follow from (7) by direct calculation. (In the next section we shall see that determinants of any order n have quite similar properties.)

(A) *The value of a determinant is not altered if its rows are written as columns in the same order.* Example:

$$\begin{vmatrix} 1 & 3 & 0 \\ 2 & 6 & 4 \\ -1 & 0 & 2 \end{vmatrix} = \begin{vmatrix} 1 & 2 & -1 \\ 3 & 6 & 0 \\ 0 & 4 & 2 \end{vmatrix} = -12. \tag{10}$$

(B) *If any two rows (or two columns) of a determinant are interchanged, the value of the determinant is multiplied by* -1. Example:

$$\begin{vmatrix} 2 & 6 & 4 \\ 1 & 3 & 0 \\ -1 & 0 & 2 \end{vmatrix} = - \begin{vmatrix} 1 & 3 & 0 \\ 2 & 6 & 4 \\ -1 & 0 & 2 \end{vmatrix} = 12. \tag{11}$$

The second-order determinant obtained from D [cf. (7)] by deleting one row and one column is called the **minor** of the element which belongs to the deleted row and column. Example: The minors of a_{21} and a_{22} in D are

$$\begin{vmatrix} a_{12} & a_{13} \\ a_{32} & a_{33} \end{vmatrix} \quad \text{and} \quad \begin{vmatrix} a_{11} & a_{13} \\ a_{31} & a_{33} \end{vmatrix},$$

respectively. The **cofactor** of the element of D in the ith row and the kth column is defined as $(-1)^{i+k}$ times the minor of that element. Example: The cofactors of a_{21} and a_{22} are

$$-\begin{vmatrix} a_{12} & a_{13} \\ a_{32} & a_{33} \end{vmatrix} \quad \text{and} \quad \begin{vmatrix} a_{11} & a_{13} \\ a_{31} & a_{33} \end{vmatrix},$$

respectively. The signs $(-1)^{i+k}$ form a checkerboard pattern:

$$\begin{matrix} + & - & + \\ - & + & - \\ + & - & + \end{matrix}$$

Furthermore, we see that we may now write (7) in the form

$$D = a_{11}C_{11} + a_{21}C_{21} + a_{31}C_{31},$$

where C_{i1} is the cofactor of a_{i1} in D. From this and the properties **(A)** and **(B)** we obtain the following property.

(C) *The determinant D may be developed by any row or column, that is, it may be written as the sum of the three elements of any row (or column) each multiplied by its cofactor.* For example, the development of D by its second row is

$$D = -a_{21}\begin{vmatrix} a_{12} & a_{13} \\ a_{32} & a_{33} \end{vmatrix} + a_{22}\begin{vmatrix} a_{11} & a_{13} \\ a_{31} & a_{33} \end{vmatrix} - a_{23}\begin{vmatrix} a_{11} & a_{12} \\ a_{31} & a_{32} \end{vmatrix}.$$

From **(C)** we obtain

(D) *A factor of the elements of any row (or column) can be placed before the determinant.* Example:

$$\begin{vmatrix} 4 & 6 & 1 \\ 3 & -9 & 2 \\ -1 & 12 & 5 \end{vmatrix} = \begin{vmatrix} 4 & 2\cdot 3 & 1 \\ 3 & -3\cdot 3 & 2 \\ -1 & 4\cdot 3 & 5 \end{vmatrix} = 3\begin{vmatrix} 4 & 2 & 1 \\ 3 & -3 & 2 \\ -1 & 4 & 5 \end{vmatrix}.$$

From the properties **(B)** and **(D)** we may draw the following conclusion.

(E) *If corresponding elements of two rows (or columns) of a determinant are proportional, the value of the determinant is zero.*

The following property is basic for simplifying determinants to be evaluated.

(F) *The value of a determinant remains unaltered if the elements of one row (or column) are altered by adding to them any constant multiple of the corresponding elements in any other row (or column).* Example:

$$\begin{vmatrix} -6 & 21 & -30 \\ 1 & -3 & 5 \\ 2 & 7 & -4 \end{vmatrix} = \begin{vmatrix} -6+1\cdot 7 & 21-3\cdot 7 & -30+5\cdot 7 \\ 1 & -3 & 5 \\ 2 & 7 & -4 \end{vmatrix}$$

$$= \begin{vmatrix} 1 & 0 & 5 \\ 1 & -3 & 5 \\ 2 & 7 & -4 \end{vmatrix} = \begin{vmatrix} 1 & 0 & 5 \\ 0 & -3 & 0 \\ 2 & 7 & -4 \end{vmatrix} = -3\begin{vmatrix} 1 & 5 \\ 2 & -4 \end{vmatrix} = 42.$$

(G) *If each element of a row (or column) of a determinant is expressed as a binomial, the determinant can be written as the sum of two determinants.* Example:

$$(12)\qquad \begin{vmatrix} 4x+3 & 2 & 1 \\ x & 3 & 4 \\ 2x-1 & 1 & -1 \end{vmatrix} = \begin{vmatrix} 4x & 2 & 1 \\ x & 3 & 4 \\ 2x & 1 & -1 \end{vmatrix} + \begin{vmatrix} 3 & 2 & 1 \\ 0 & 3 & 4 \\ -1 & 1 & -1 \end{vmatrix}$$

By applying the product rule of differentiation we obtain the following property.

(H) *If the elements of a determinant are differentiable functions of a variable, the derivative of the determinant may be written as a sum of three determinants,*

$$\frac{d}{dx}\begin{vmatrix} f & g & h \\ p & q & r \\ u & v & w \end{vmatrix} = \begin{vmatrix} f' & g' & h' \\ p & q & r \\ u & v & w \end{vmatrix} + \begin{vmatrix} f & g & h \\ p' & q' & r' \\ u & v & w \end{vmatrix} + \begin{vmatrix} f & g & h \\ p & q & r \\ u' & v' & w' \end{vmatrix},$$

where primes denote derivatives with respect to x.

Problems for Sec. 7.9

Evaluate

1. $\begin{vmatrix} 3 & -1 \\ 2 & 4 \end{vmatrix}$ **2.** $\begin{vmatrix} 0 & 3 \\ 5 & 7 \end{vmatrix}$ **3.** $\begin{vmatrix} \cos\theta & \sin\theta \\ -\sin\theta & \cos\theta \end{vmatrix}$

4. $\begin{vmatrix} 1 & 2 & 3 \\ 4 & 5 & 6 \\ 7 & 8 & 9 \end{vmatrix}$ **5.** $\begin{vmatrix} -4 & 18 & 7 \\ 0 & -1 & 4 \\ 0 & 0 & 6 \end{vmatrix}$ **6.** $\begin{vmatrix} 6 & 13 & -2 \\ 5 & 37 & -5 \\ 1 & 13 & -2 \end{vmatrix}$

7. $\begin{vmatrix} 1 & c & -b \\ -c & 1 & a \\ b & -a & 1 \end{vmatrix}$ **8.** $\begin{vmatrix} 1 & a & a^2 \\ 1 & b & b^2 \\ 1 & c & c^2 \end{vmatrix}$ **9.** $\begin{vmatrix} a-b & m-n & x-y \\ b-c & n-p & y-z \\ c-a & p-m & z-x \end{vmatrix}$

10. Show that the straight line through two given points P_1: (x_1, y_1) and P_2: (x_2, y_2) in the xy-plane is

$$\begin{vmatrix} x & y & 1 \\ x_1 & y_1 & 1 \\ x_2 & y_2 & 1 \end{vmatrix} = 0.$$

Derive from this the familiar formula

$$\frac{x - x_1}{x_1 - x_2} = \frac{y - y_1}{y_1 - y_2}.$$

7.10 Determinants of Arbitrary Order

A determinant of arbitrary order n is a scalar associated with an n-rowed square matrix as explained below. It can be defined in several equivalent ways. One possible way is suggested by a process of solving n linear equations in n unknowns. For $n = 2$ and $n = 3$ this was shown in the previous section, and the generalization to arbitrary n will be discussed in the next section.

Determinants have various applications in engineering mathematics, although their importance has decreased since they are often impractical in numerical work. Indeed, Cramer's rule (Secs. 7.9 and 7.11) is certainly not the *practical* answer to the problem of solving systems of linear equations numerically, and there are numerically better methods (cf. Secs. 7.5, 19.9–19.11).

In the present section we define determinants of nth order and consider their most important properties.

A determinant of order n is written as a square array of n^2 quantities enclosed between vertical bars,

$$D = \begin{vmatrix} a_{11} & a_{12} & \cdots & a_{1n} \\ a_{21} & a_{22} & \cdots & a_{2n} \\ \cdot & \cdot & \cdots & \cdot \\ \cdot & \cdot & \cdots & \cdot \\ a_{n1} & a_{n2} & \cdots & a_{nn} \end{vmatrix} \tag{1}$$

and has a certain value defined below. The quantities $a_{11}, \cdots, a_{nn}$ which are numbers (or sometimes functions), are called the **elements** of the determinant. The horizontal lines of elements are called **rows;** the vertical lines are called **columns.** The sloping line of elements extending from a_{11} to a_{nn} is called the **principal diagonal** of the determinant.

By deleting the ith row and the kth column from the determinant D we obtain an $(n - 1)$th order determinant (a square array of $n - 1$ rows and $n - 1$ columns between vertical bars), which is called the **minor** of the element a_{ik} (which belongs to the deleted row and column) and is denoted by M_{ik}.

The minor M_{ik} multiplied by $(-1)^{i+k}$ is called the **cofactor** of a_{ik} in D and will be denoted by C_{ik}; thus

$$C_{ik} = (-1)^{i+k} M_{ik}. \tag{2}$$

For example, in the third-order determinant

$$\begin{vmatrix} a_{11} & a_{12} & a_{13} \\ a_{21} & a_{22} & a_{23} \\ a_{31} & a_{32} & a_{33} \end{vmatrix}$$

we have

$$C_{11} = M_{11} = \begin{vmatrix} a_{22} & a_{23} \\ a_{32} & a_{33} \end{vmatrix}, \qquad C_{32} = -M_{32} = -\begin{vmatrix} a_{11} & a_{13} \\ a_{21} & a_{23} \end{vmatrix}, \text{ etc.}$$

We are now in a position to define an nth-order determinant in a way which is most convenient for practical purposes.

Definition. Determinant of order n

The symbol

$$D = \begin{vmatrix} a_{11} & a_{12} & \cdots & a_{1n} \\ a_{21} & a_{22} & \cdots & a_{2n} \\ \cdot & \cdot & \cdots & \cdot \\ \cdot & \cdot & \cdots & \cdot \\ a_{n1} & a_{n2} & \cdots & a_{nn} \end{vmatrix}$$

is called a **determinant of order** n. For $n = 1$ it means a_{11}. For $n \geqq 2$ it means the sum of the products of the elements of any row or column and their respective cofactors; that is,

(3a) $$D = a_{i1}C_{i1} + a_{i2}C_{i2} + \cdots + a_{in}C_{in} \qquad (i = 1, 2, \cdots, \text{ or } n)$$

or

(3b) $$D = a_{1k}C_{1k} + a_{2k}C_{2k} + \cdots + a_{nk}C_{nk} \qquad (k = 1, 2, \cdots, \text{ or } n).$$ ▮

In this way, D is defined in terms of n determinants of order $n - 1$, each of which is, in turn, defined in terms of $n - 1$ determinants of order $n - 2$, and so on; we finally arrive at second-order determinants, in which the cofactors of the elements are single elements of D. We see that *if the elements of D are numbers, the value of D will be a number.*

Furthermore, it follows from the definition that we may **develop** *D by any row or column,* that is, choose in (3) the elements of any row or column, similarly when developing the cofactors in (3), and so on. Consequently, we have to show that the definition is ***unambiguous,*** that is, yields the same value for D no matter which columns or rows we choose. This proof will be given at the end of the section.

For later use we note that (3) may also be written in terms of minors [cf. (2)]

(4a) $$D = \sum_{k=1}^{n} (-1)^{i+k} a_{ik} M_{ik} \qquad (i = 1, 2, \cdots, \text{ or } n)$$

(4b) $$D = \sum_{i=1}^{n} (-1)^{i+k} a_{ik} M_{ik} \qquad (k = 1, 2, \cdots, \text{ or } n).$$

Example 1

Let

$$D = \begin{vmatrix} 1 & 3 & 0 \\ 2 & 6 & 4 \\ -1 & 0 & 2 \end{vmatrix}.$$

The development by the first row is

$$D = 1\begin{vmatrix} 6 & 4 \\ 0 & 2 \end{vmatrix} - 3\begin{vmatrix} 2 & 4 \\ -1 & 2 \end{vmatrix} = 1(12 - 0) - 3(4 + 4) = -12.$$

The development by the first column is

$$D = 1\begin{vmatrix} 6 & 4 \\ 0 & 2 \end{vmatrix} - 2\begin{vmatrix} 3 & 0 \\ 0 & 2 \end{vmatrix} - 1\begin{vmatrix} 3 & 0 \\ 6 & 4 \end{vmatrix} = 12 - 12 - 12 = -12,$$

etc. ∎

From our definition we may now readily obtain the most important properties of determinants, as follows.

Since the same value is obtained whether we expand a determinant by any row or any column, we have

Theorem 1 (Transposition)
The value of a determinant is not altered if its rows are written as columns, in the same order.

An example is given in formula (10) of the preceding section.

Theorem 2 (Multiplication by a constant)
If all the elements of one row (or one column) of a determinant are multiplied by the same factor k, the value of the new determinant is k times the value of the given determinant.

Proof. Expand the determinant by that row (or column) whose elements are multiplied by k. ∎

Note that kD equals a determinant obtained by multiplying the elements of *just one row* (or column) of D by k, whereas in the case of a matrix, $k\mathbf{A}$ equals the matrix obtained by multiplying *all the elements of* $\mathbf{A}$ by k.

Theorem 2 can be used for simplifying a given determinant; cf. the following example.

Example 2

$$\begin{vmatrix} 1 & 3 & 0 \\ 2 & 6 & 4 \\ -1 & 0 & 2 \end{vmatrix} = 2\begin{vmatrix} 1 & 3 & 0 \\ 1 & 3 & 2 \\ -1 & 0 & 2 \end{vmatrix} = 6\begin{vmatrix} 1 & 1 & 0 \\ 1 & 1 & 2 \\ -1 & 0 & 2 \end{vmatrix} = 12\begin{vmatrix} 1 & 1 & 0 \\ 1 & 1 & 1 \\ -1 & 0 & 1 \end{vmatrix} = -12$$ ∎

From Theorem 2, with $k = 0$, or directly, by expanding, we obtain

Theorem 3
If all the elements of a row (or a column) of a determinant are zero, the value of the determinant is zero.

Theorem 4
If each element of a row (or a column) of a determinant is expressed as a binomial, the determinant can be written as the sum of two determinants, for example,

$$\begin{vmatrix} a_1 + d_1 & b_1 & c_1 \\ a_2 + d_2 & b_2 & c_2 \\ a_3 + d_3 & b_3 & c_3 \end{vmatrix} = \begin{vmatrix} a_1 & b_1 & c_1 \\ a_2 & b_2 & c_2 \\ a_3 & b_3 & c_3 \end{vmatrix} + \begin{vmatrix} d_1 & b_1 & c_1 \\ d_2 & b_2 & c_2 \\ d_3 & b_3 & c_3 \end{vmatrix}.$$

Proof. Expand the determinant by the row (or column) whose terms are binomials. ∎

Another example is given in (12), Sec. 7.9. The generalization of Theorem 4 to the case of elements which are sums of more than two terms is obvious.

Theorem 5 (Interchange of rows or columns)
If any two rows (or two columns) of a determinant are interchanged, the value of the determinant is multiplied by -1.

Proof. The proof is by induction. We see that the theorem holds for determinants of order $n = 2$. Assuming that it holds for determinants of order $n - 1$, we will show that it holds for determinants of order n.

Let D be of order n and E obtained from D by interchanging two rows. Expand D and E by a row that is not one of those interchanged, call it the ith row. Then, by (4a),

$$D = \sum_{k=1}^{n} (-1)^{i+k} a_{ik} M_{ik}, \qquad E = \sum_{k=1}^{n} (-1)^{i+k} a_{ik} N_{ik}$$

where N_{ik} is obtained from the minor M_{ik} of a_{ik} in D by interchanging two rows. Since these minors are of order $n - 1$, the induction hypothesis applies and gives $N_{ik} = -M_{ik}$. From this we obtain the statement for rows because

$$E = \sum_{k=1}^{n} (-1)^{i+k} a_{ik}(-M_{ik}) = -\sum_{k=1}^{n} (-1)^{i+k} a_{ik} M_{ik} = -D.$$

The proof for columns is similar. ∎

An example is given in formula (11) of the preceding section.

Theorem 6 (Proportional rows or columns)
If corresponding elements of two rows (or two columns) of a determinant are proportional, the value of the determinant is zero.

Proof. Let the elements of the ith and jth rows of D be proportional, say, $a_{ik} = ca_{jk}$, $k = 1, \cdots, n$. If $c = 0$, then $D = 0$. Let $c \neq 0$. By Theorem 2,

$$D = cB$$

where the ith and jth rows of B are identical. Interchange these rows. Then, by Theorem 5, B goes over into $-B$. On the other hand, since the rows are identical, the new determinant is still B. Thus $B = -B$, $B = 0$, and $D = 0$. ∎

Example 3

$$\begin{vmatrix} 3 & 6 & -4 \\ 1 & -1 & 3 \\ -6 & -12 & 8 \end{vmatrix} = 0$$

Before evaluating a determinant it is advisable to simplify it. This may be done by Theorem 2 and

Theorem 7 (Addition of a row or column)

The value of a determinant is left unchanged, if the elements of a row (or column) are altered by adding to them any constant multiple of the corresponding elements in any other row (or column, respectively).

Proof. Apply Theorem 4 to the determinant that results from the given addition. This yields a sum of two determinants; one is the original determinant and the other contains two proportional rows. According to Theorem 6, the second determinant is zero, and the proof is complete. ▮

Example 4

Evaluate

$$D = \begin{vmatrix} 1 & 24 & 21 & 93 \\ 2 & -37 & -1 & 194 \\ -2 & 35 & 0 & -171 \\ -3 & 177 & 63 & 234 \end{vmatrix}.$$

Add the second row to the third, add three times the first row to the last, subtract twice the first row from the second, and develop the resulting determinant by the first column:

$$D = \begin{vmatrix} 1 & 24 & 21 & 93 \\ 0 & -85 & -43 & 8 \\ 0 & -2 & -1 & 23 \\ 0 & 249 & 126 & 513 \end{vmatrix} = \begin{vmatrix} -85 & -43 & 8 \\ -2 & -1 & 23 \\ 249 & 126 & 513 \end{vmatrix}.$$

Add three times the first row to the last row:

$$D = \begin{vmatrix} -85 & -43 & 8 \\ -2 & -1 & 23 \\ -6 & -3 & 537 \end{vmatrix}.$$

Subtract twice the second column from the first, and then develop the resulting determinant by the first column:

$$D = \begin{vmatrix} 1 & -43 & 8 \\ 0 & -1 & 23 \\ 0 & -3 & 537 \end{vmatrix} = \begin{vmatrix} -1 & 23 \\ -3 & 537 \end{vmatrix} = -537 + 69 = -468. \quad ▮$$

The determinant of the elements of an n-rowed square matrix $\mathbf{A} = (a_{jk})$ is called the **determinant of the matrix A** and will be denoted by det **A**; thus

$$\det \mathbf{A} = \begin{vmatrix} a_{11} & \cdots & a_{1n} \\ \cdot & \cdots & \cdot \\ \cdot & \cdots & \cdot \\ a_{n1} & \cdots & a_{nn} \end{vmatrix}.$$

We shall now prove the important theorem that the determinant of the product of two n-rowed square matrices equals the product of the determinants of the matrices. As a theorem on determinants this means that the product of two nth-order determinants can be written as an nth-order determinant whose elements are obtained in the same fashion as the elements of a product matrix.

Theorem 8 (Determinant of a product of matrices)
Let $\mathbf{A} = (a_{ik})$ and $\mathbf{B} = (b_{ik})$ be n-rowed square matrices. Then

$$\det(\mathbf{AB}) = \det \mathbf{A} \det \mathbf{B}, \tag{5}$$

which is equivalent to

$$\begin{vmatrix} a_{11} & \cdots & a_{1n} \\ \cdot & \cdots & \cdot \\ \cdot & \cdots & \cdot \\ a_{n1} & \cdots & a_{nn} \end{vmatrix} \begin{vmatrix} b_{11} & \cdots & b_{1n} \\ \cdot & \cdots & \cdot \\ \cdot & \cdots & \cdot \\ b_{n1} & \cdots & b_{nn} \end{vmatrix} = \begin{vmatrix} c_{11} & \cdots & c_{1n} \\ \cdot & \cdots & \cdot \\ \cdot & \cdots & \cdot \\ c_{n1} & \cdots & c_{nn} \end{vmatrix}, \tag{5*}$$

where the element c_{ik} in the ith row and kth column is

$$c_{ik} = a_{i1}b_{1k} + a_{i2}b_{2k} + \cdots + a_{in}b_{nk}, \tag{5'}$$

the scalar product of the ith row vector of the first determinant and the kth column vector of the second determinant.

Proof of Theorem 8 for n = 2. Let

$$D_1 = \begin{vmatrix} a_{11} & a_{12} \\ a_{21} & a_{22} \end{vmatrix}, \qquad D_2 = \begin{vmatrix} b_{11} & b_{12} \\ b_{21} & b_{22} \end{vmatrix}, \qquad \text{and } D = D_1 D_2.$$

We first show that

$$D = D_1 D_2 = \begin{vmatrix} a_{11} & a_{12} & 0 & 0 \\ a_{21} & a_{22} & 0 & 0 \\ -1 & 0 & b_{11} & b_{12} \\ 0 & -1 & b_{21} & b_{22} \end{vmatrix}. \tag{6}$$

To prove (6), develop the determinant on the right by the first row and then the resulting two determinants again by their first rows:

$$D = a_{11}\begin{vmatrix} a_{22} & 0 & 0 \\ 0 & b_{11} & b_{12} \\ -1 & b_{21} & b_{22} \end{vmatrix} - a_{12}\begin{vmatrix} a_{21} & 0 & 0 \\ -1 & b_{11} & b_{12} \\ 0 & b_{21} & b_{22} \end{vmatrix}$$

$$= a_{11}a_{22}\begin{vmatrix} b_{11} & b_{12} \\ b_{21} & b_{22} \end{vmatrix} - a_{12}a_{21}\begin{vmatrix} b_{11} & b_{12} \\ b_{21} & b_{22} \end{vmatrix} = (a_{11}a_{22} - a_{12}a_{21})D_2 = D_1D_2.$$

We now transform the determinant in (6) as follows. We add the third row, multiplied by a_{11}, and the fourth row, multiplied by a_{12}, to the first row; then the first row takes the form

$$0 \quad 0 \quad a_{11}b_{11} + a_{12}b_{21} \quad a_{11}b_{12} + a_{12}b_{22}.$$

We now add the third row, multiplied by a_{21}, and the fourth row, multiplied by a_{22}, to the second row; then the second row becomes

$$0 \quad 0 \quad a_{21}b_{11} + a_{22}b_{21} \quad a_{21}b_{12} + a_{22}b_{22}.$$

Altogether,

$$D = \begin{vmatrix} 0 & 0 & a_{11}b_{11} + a_{12}b_{21} & a_{11}b_{12} + a_{12}b_{22} \\ 0 & 0 & a_{21}b_{11} + a_{22}b_{21} & a_{21}b_{12} + a_{22}b_{22} \\ -1 & 0 & b_{11} & b_{12} \\ 0 & -1 & b_{21} & b_{22} \end{vmatrix}.$$

Developing this determinant by the first column, and then the resulting third-order determinant again by its first column, we obtain

$$D = D_1D_2 = \begin{vmatrix} a_{11}b_{11} + a_{12}b_{21} & a_{11}b_{12} + a_{12}b_{22} \\ a_{21}b_{11} + a_{22}b_{21} & a_{21}b_{12} + a_{22}b_{22} \end{vmatrix},$$

which is (5*) with $n = 2$.

Proof of Theorem 8 for arbitrary n. The steps are quite similar to those in the case $n = 2$. Let $D_1 = \det \mathbf{A}$, $D_2 = \det \mathbf{B}$, and $D = D_1D_2$, as before. We first show that

$$\text{(6*)} \qquad D = \begin{vmatrix} a_{11} & \cdot & \cdots & a_{1n} & 0 & \cdots & 0 \\ \cdot & \cdot & \cdots & \cdot & \cdot & \cdots & \cdot \\ \cdot & \cdot & \cdots & \cdot & \cdot & \cdots & \cdot \\ a_{n1} & \cdot & \cdots & a_{nn} & 0 & \cdots & 0 \\ -1 & 0 & \cdots & 0 & b_{11} & \cdots & b_{1n} \\ 0 & -1 & \cdots & 0 & \cdot & \cdots & \cdot \\ \cdot & \cdot & \cdots & \cdot & \cdot & \cdots & \cdot \\ 0 & 0 & \cdots & -1 & b_{n1} & \cdots & b_{nn} \end{vmatrix}.$$

To prove this, develop the determinant by the first row, the resulting determinants of order $2n - 1$ again by their first rows, etc. After n steps, the result will be of the form

$$(\cdots)\begin{vmatrix} b_{11} & \cdots & b_{1n} \\ \cdot & \cdots & \cdot \\ b_{n1} & \cdots & b_{nn} \end{vmatrix}$$

where $(\cdots)$ is the representation of the first of the given determinants in terms of products of its elements. Then D in (6*) is transformed as follows. Add to the first row

the $(n + 1)$th row multiplied by a_{11},

the $(n + 2)$th row multiplied by a_{12},

. . . .

the $2n$th row multiplied by a_{1n}.

The first row will then become

$$\underbrace{0 \cdots 0}_{\substack{\text{first } n \\ \text{elements}}} \quad \underbrace{a_{11}b_{11} + a_{12}b_{21} + \cdots + a_{1n}b_{n1}}_{(n+1)\text{th element} = c_{11}} \quad \cdots \quad \underbrace{a_{11}b_{1n} + \cdots + a_{1n}b_{nn}}_{\text{last element} = c_{1n}}$$

Transforming the second, third, $\cdots$, nth row in a similar fashion, we altogether obtain

$$D = \begin{vmatrix} 0 & \cdots & 0 & c_{11} & \cdots & c_{1n} \\ \cdot & \cdots & \cdot & \cdot & \cdots & \cdot \\ 0 & \cdots & 0 & c_{n1} & \cdots & c_{nn} \\ -1 & \cdots & 0 & b_{11} & \cdots & b_{1n} \\ \cdot & \cdots & \cdot & \cdot & \cdots & \cdot \\ 0 & \cdots & -1 & b_{n1} & \cdots & b_{nn} \end{vmatrix}$$

where c_{ik} is given by (5′). Developing this determinant by the first column, the resulting $(2n - 1)$th order determinant again by its first column, etc., the result will be the determinant on the right side of (5*) multiplied by $(-1)^n$ and, if n is odd, by another $(-1)^n$ because then in each of those n successive developments the cofactor of the element -1 in the first column of the determinant to be developed is equal to minus one times the minor of this element. Hence if n is odd, that determinant is multiplied by the factor $(-1)^n(-1)^n = +1$, and, if n is even, by $(-1)^n = +1$. This completes the proof. ∎

Example 5.

$$\begin{vmatrix} 2 & 4 & 3 \\ 6 & 10 & 14 \\ 4 & 7 & 9 \end{vmatrix}\begin{vmatrix} 4 & 0 & 5 \\ -2 & 1 & -1 \\ 3 & 0 & 4 \end{vmatrix} = \begin{vmatrix} 9 & 4 & 18 \\ 46 & 10 & 76 \\ 29 & 7 & 49 \end{vmatrix}$$

Proof that the definition of a determinant is unambiguous. We have to show that the definition of a determinant D, given at the beginning of the section, is unambiguous, that is, yields the same value of D no matter which rows or columns we choose.

For a second-order determinant

$$D = \begin{vmatrix} a_{11} & a_{12} \\ a_{21} & a_{22} \end{vmatrix}$$

this is immediately clear, because then we have only four possible forms of (3), namely, the development by

$$\text{the first row: } D = a_{11}a_{22} + a_{12}(-a_{21}),$$

$$\text{the last row: } D = a_{21}(-a_{12}) + a_{22}a_{11},$$

$$\text{the first column: } D = a_{11}a_{22} + a_{21}(-a_{12}),$$

$$\text{the last column: } D = a_{12}(-a_{21}) + a_{22}a_{11}.$$

We see that we always obtain the same value of D, which value agrees with (3) in Sec. 7.9.

We shall now prove that *our definition of a determinant D of arbitrary order n yields the same value of D no matter which row or column is chosen.*

We shall prove first that *the same result is obtained no matter which* ***row*** *is chosen.*

The proof is by induction. The statement is true for a second-order determinant (see before). Assuming that it is true for a determinant of order $n - 1$, we prove that it is true for a determinant D of order n.

For this purpose we expand D, given in the definition, in terms of each of two arbitrary rows, say the ith and the jth, and compare the results. Without loss of generality let us assume $i < j$.

First expansion. We expand D by the ith row. A typical term in this expansion is

$$a_{ik}C_{ik} = a_{ik} \cdot (-1)^{i+k} M_{ik}. \tag{7}$$

The minor M_{ik} of a_{ik} in D is an $(n - 1)$th order determinant. By the induction hypothesis we may expand it by any row. We expand it by the row corresponding to the jth row of D. This row contains the elements a_{jl} $(l \neq k)$. It is the $(j - 1)$th row of M_{ik}, because M_{ik} does not contain elements of the ith row of D, and $i < j$. We have to distinguish between two cases as follows.

Case I. If $l < k$, then the element a_{jl} belongs to the lth column of M_{ik} (cf. Fig. 142). Hence the term involving a_{jl} in this expansion is

$$a_{jl} \cdot (\text{cofactor of } a_{jl} \text{ in } M_{ik}) = a_{jl} \cdot (-1)^{(j-1)+l} M_{ikjl} \tag{8}$$

where M_{ikjl} is the minor of a_{jl} in M_{ik}. Since this minor is obtained from M_{ik} by deleting the row and column of a_{jl}, it is obtained from D by deleting the ith and jth rows and the kth and lth columns of D. We insert the expansions of the M_{ik} into that of D. Then it follows from (7) and (8) that the terms of the resulting representation of D are of the form

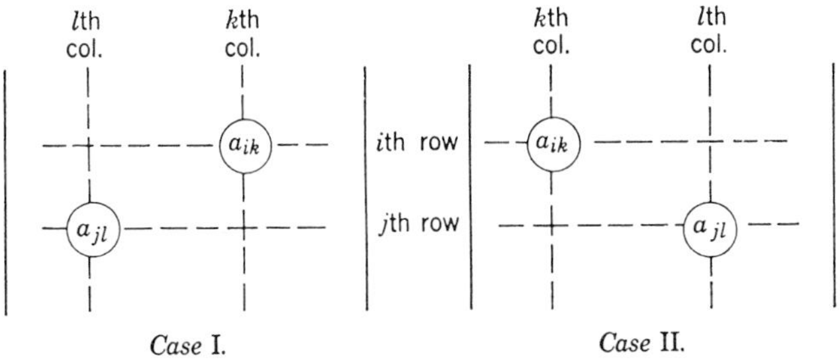

Fig. 142. Cases I and II of the two expansions of D

$$a_{ik}a_{jl} \cdot (-1)^b M_{ikjl} \qquad (l < k) \tag{9a}$$

where

$$b = i + k + j + l - 1.$$

Case II. If $l > k$, the only difference is that then a_{jl} belongs to the $(l - 1)$th column of M_{ik}, because M_{ik} does not contain elements of the kth column of D, and $k < l$. This causes an additional minus sign in (8), and, instead of (9a), we therefore obtain

$$-a_{ik}a_{jl} \cdot (-1)^b M_{ikjl} \qquad (l > k) \tag{9b}$$

where b is the same as before.

Second expansion. We now expand D at first by the jth row. A typical term in this expansion is

$$a_{jl}C_{jl} = a_{jl} \cdot (-1)^{j+l} M_{jl}. \tag{10}$$

By the induction hypothesis we may expand the minor M_{jl} of a_{jl} in D by its ith row, which corresponds to the ith row of D, since $j > i$.

Case I. If $k > l$, the element a_{ik} in that row belongs to the $(k - 1)$th column of M_{jl}, because M_{jl} does not contain elements of the lth column of D, and $l < k$ (cf. Fig. 142). Hence the term involving a_{ik} in this expansion is

$$a_{ik} \cdot (\text{cofactor of } a_{ik} \text{ in } M_{jl}) = a_{ik} \cdot (-1)^{i+(k-1)} M_{ikjl}, \tag{11}$$

where the minor M_{ikjl} of a_{ik} in M_{jl} is obtained by deleting the ith and jth rows and the kth and lth columns of D [and is, therefore, identical with M_{ikjl} in (8), so that our notation is consistent]. We insert the expansions of the M_{jl} into that of D. It follows from (10) and (11) that this yields a representation whose terms are identical with those given by (9a) when $l < k$.

Case II. If $k < l$, then a_{ik} belongs to the kth column of M_{jl}, we obtain an additional minus sign, and the result agrees with that characterized by (9b).

We have shown that the two expansions of D consist of the same terms, and this proves our statement concerning rows.

The proof of the statement concerning *columns* is quite similar; if we expand D in terms of two arbitrary columns, say, the kth and the lth, we find that the general term involving $a_{jl}a_{ik}$ is exactly the same as before. This proves that not only all column expansions of D yield the same value, but also that their common value is equal to the common value of the row expansions of D.

This completes the proof and shows that *our definition of an nth order determinant is unambiguous.* ∎

Problems for Sec. 7.10

Evaluate

1. $\begin{vmatrix} 1 & 3 & 1 & -2 \\ 2 & 0 & -2 & -4 \\ -1 & 1 & 0 & 1 \\ 2 & 5 & 3 & 6 \end{vmatrix}$

2. $\begin{vmatrix} -4 & 1 & 4 & -2 \\ -5 & -8 & 4 & -2 \\ 8 & 16 & -6 & 15 \\ -2 & 8 & -3 & 7 \end{vmatrix}$

3. $\begin{vmatrix} b^2+c^2 & a^2 & a^2 \\ b^2 & c^2+a^2 & b^2 \\ c^2 & c^2 & a^2+b^2 \end{vmatrix}$

4. $\begin{vmatrix} b^2+c^2 & ab & ca \\ ab & c^2+a^2 & bc \\ ca & bc & a^2+b^2 \end{vmatrix}$

5. Prove that the definition of a determinant D of order n yields the same value of D no matter which column is chosen [cf. (3b)]. *Hint.* Model your work somewhat after that for rows given in the text.

6. Carry out the proof of Theorem 2 in detail.

7. Carry out the proof of Theorem 4 in detail.

8. Show that $\det(k\mathbf{A}) = k^n \det \mathbf{A}$, where $\mathbf{A}$ is an n-rowed square matrix.

9. Show that three points (x_1, y_1), (x_2, y_2), and (x_3, y_3) lie on a straight line if and only if

$$\begin{vmatrix} x_1 & y_1 & 1 \\ x_2 & y_2 & 1 \\ x_3 & y_3 & 1 \end{vmatrix} = 0.$$

10. Show that the equation of the circle through three given points (x_1, y_1), (x_2, y_2), and (x_3, y_3) in the xy-plane is

$$\begin{vmatrix} x^2+y^2 & x & y & 1 \\ x_1^2+y_1^2 & x_1 & y_1 & 1 \\ x_2^2+y_2^2 & x_2 & y_2 & 1 \\ x_3^2+y_3^2 & x_3 & y_3 & 1 \end{vmatrix} = 0.$$

7.11 Rank in Terms of Determinants. Cramer's Rule

The rank of a matrix is a very important concept, as we can see from Sec. 7.7. From Sec. 7.6 we know that the rank of a matrix $\mathbf{A}$ is the maximum number of linearly independent row or column vectors of $\mathbf{A}$. It is remarkable, or perhaps even surprising, that we can use determinants for characterizing the rank of a matrix. This characterization is often used for *defining* the rank. We formulate it as follows, assuming rank $\mathbf{A} > 0$ (since rank $\mathbf{A} = 0$ if and only if $\mathbf{A} = \mathbf{0}$; cf. Sec. 7.6).

Theorem 1 (Rank in terms of determinants)

An $(m \times n)$ matrix $\mathbf{A} = (a_{jk})$ has rank $r \geqq 1$ if and only if $\mathbf{A}$ has an $(r \times r)$ submatrix with nonzero determinant whereas the determinant of every square submatrix with $r + 1$ or more rows which $\mathbf{A}$ has (or does not have!) is zero.

In particular, if $\mathbf{A}$ is a square matrix, $\mathbf{A}$ is nonsingular, so that the inverse $\mathbf{A}^{-1}$ of $\mathbf{A}$ exists, if and only if

$$\det \mathbf{A} \neq 0.$$

Proof. The key lies in the fact that elementary row operations (Sec. 7.5), which do not alter the rank (by Theorem 2 in Sec. 7.6), also do not alter the property of a determinant of being zero or not zero, since the determinant is multiplied by

(*i*) -1 if we interchange two rows (Theorem 5, Sec. 7.10),
(*ii*) $c \neq 0$ if we multiply a row by $c \neq 0$ (Theorem 2, Sec. 7.10),
(*iii*) 1 if we add a multiple of a row to another row (Theorem 7, Sec. 7.10).

Let $\tilde{\mathbf{A}}$ denote the echelon form of $\mathbf{A}$ (cf. Sec. 7.5). $\tilde{\mathbf{A}}$ has r nonzero row vectors (which are the first r row vectors) if and only if rank $\mathbf{A} = r$. Let $\tilde{\mathbf{R}}$ be the $(r \times r)$ submatrix of $\tilde{\mathbf{A}}$ consisting of the r^2 elements which are simultaneously in the first r rows and the first r columns of $\tilde{\mathbf{A}}$. Since $\tilde{\mathbf{R}}$ is triangular and has all diagonal elements different from zero, $\det \tilde{\mathbf{R}} \neq 0$. Since $\tilde{\mathbf{R}}$ is obtained from the corresponding $(r \times r)$ submatrix $\mathbf{R}$ of $\mathbf{A}$ by elementary operations, $\det \boldsymbol{R} \neq 0$. Similarly, $\det \mathbf{S} = 0$ for a square submatrix $\mathbf{S}$ of $r + 1$ or more rows possibly contained in $\mathbf{A}$, since the corresponding submatrix $\tilde{\mathbf{S}}$ of $\tilde{\mathbf{A}}$ must contain a row of zeros, so that $\det \tilde{\mathbf{S}} = 0$ by Theorem 3 in Sec. 7.10. This proves the assertion of the theorem for an $(m \times n)$ matrix.

If $\mathbf{A}$ is square, say, an $(n \times n)$ matrix, the statement just proved implies that rank $\mathbf{A} = n$ if and only if $\mathbf{A}$ has an $(n \times n)$ submatrix with a nonzero determinant; but this means that we must have $\det \mathbf{A} \neq 0$. ∎

Using this theorem, we shall now derive Cramer's rule, which represents solutions of systems of linear equations as quotients of determinants. For numerical work, other formulas are preferable (cf. Secs. 7.5, 19.9–19.11), but Cramer's rule is of interest, for instance, if the equations contain parameters.

Cramer's Theorem (Solution of linear equations by determinants)

If the determinant $D = \det \mathbf{A}$ of a system of n linear equations

$$
\begin{aligned}
a_{11}x_1 + a_{12}x_2 + \cdots + a_{1n}x_n &= b_1 \\
a_{21}x_1 + a_{22}x_2 + \cdots + a_{2n}x_n &= b_2 \\
\cdots\cdots\cdots\cdots\cdots\cdots\cdots\cdots\cdots\cdots \\
a_{n1}x_1 + a_{n2}x_2 + \cdots + a_{nn}x_n &= b_n
\end{aligned}
\tag{1}
$$

in the same number of unknowns $x_1, \cdots, x_n$ is not zero, the system has precisely one solution. This solution is given by the formulas

$$x_1 = \frac{D_1}{D}, \quad x_2 = \frac{D_2}{D}, \quad \cdots, \quad x_n = \frac{D_n}{D} \qquad \textbf{(Cramer's rule)} \tag{2}$$

where D_k is the determinant obtained from D by replacing in D the kth column by the column with the elements $b_1, \cdots, b_n$.

Hence if (1) is homogeneous and $D \neq 0$, it has only the trivial solution $x_1 = 0$, $x_2 = 0, \cdots, x_n = 0$. If $D = 0$, the homogeneous system has also nontrivial solutions.

Proof. From Theorem 1 and the Fundamental Theorem in Sec. 7.7 it follows that (1) has a unique solution since

$$D = \det \mathbf{A} = \begin{vmatrix} a_{11} & \cdots & a_{1n} \\ \cdot & \cdots & \cdot \\ \cdot & \cdots & \cdot \\ a_{n1} & \cdots & a_{nn} \end{vmatrix} \neq 0$$

implies rank $\mathbf{A} = n$. We prove (2). Developing D by the kth column, we obtain

$$D = a_{1k}C_{1k} + a_{2k}C_{2k} + \cdots + a_{nk}C_{nk} \tag{3}$$

where C_{ik} is the cofactor of the element a_{ik} in D. If we replace the elements of the kth column of D by any other numbers, we obtain a new determinant, say, $\tilde{D}$. Clearly, its development by the kth column will be of the form (3), with $a_{1k}, \cdots, a_{nk}$ replaced by those new elements and the cofactors C_{ik} as before. In particular, if we choose as new elements the elements $a_{1l}, \cdots, a_{nl}$ of the lth column of D (where $l \neq k$), then the development of the resulting determinant $\tilde{D}$ becomes

$$a_{1l}C_{1k} + a_{2l}C_{2k} + \cdots + a_{nl}C_{nk} = 0 \qquad (l \neq k), \tag{4}$$

because $\tilde{D}$ has two identical columns and is zero (cf. Theorem 6 in Sec. 7.10). If we multiply the first equation in (1) by C_{1k}, the second by $C_{2k}, \cdots$, the last by C_{nk} and add the resulting equations, we first obtain

$$C_{1k}(a_{11}x_1 + \cdots + a_{1n}x_n) + \cdots + C_{nk}(a_{n1}x_1 + \cdots + a_{nn}x_n)$$
$$= b_1C_{1k} + \cdots + b_nC_{nk}.$$

The expression on the left may be written

$$x_1(a_{11}C_{1k} + \cdots + a_{n1}C_{nk}) + \cdots + x_n(a_{1n}C_{1k} + \cdots + a_{nn}C_{nk}).$$

From (3) we conclude that the factor of x_k in this representation is equal to D, and from (4) it follows that the factor of x_l $(l \neq k)$ is zero. Thus

$$x_kD = b_1C_{1k} + b_2C_{2k} + \cdots + b_nC_{nk}.$$

Since $D \neq 0$, we may divide, finding

$$x_k = \frac{1}{D}(b_1C_{1k} + b_2C_{2k} + \cdots + b_nC_{nk}) = \frac{D_k}{D} \tag{5}$$

where $k = 1, \cdots, n$. This is identical with (2).

Furthermore, if $D \neq 0$ as before and (1) is homogeneous, then $D_1 = 0, \cdots, D_n = 0$, so that (2) yields the trivial solution. If $D = 0$ and (1) is homogeneous, then rank $\mathbf{A} < n$ by Theorem 1, so that nontrivial solutions exist by **(c)** in the Fundamental Theorem (Sec. 7.7). ∎

An example is included in Sec. 7.9.

We want to mention that it is possible (although not very practical) to apply Cramer's rule to systems of m linear equations

$$\begin{aligned} a_{11}x_1 + a_{12}x_2 + \cdots + a_{1n}x_n &= b_1 \\ a_{21}x_1 + a_{22}x_2 + \cdots + a_{2n}x_n &= b_2 \\ \cdots\cdots\cdots\cdots\cdots\cdots\cdots\cdots\cdots& \\ a_{m1}x_1 + a_{m2}x_2 + \cdots + a_{mn}x_n &= b_m \end{aligned}$$

in n unknowns. If the matrix of the coefficients $\mathbf{A}$ and the augmented matrix $\mathbf{B}$ both have rank r, we know from the Fundamental Theorem that we may assign arbitrary values to $n - r$ suitable unknowns, call them $x_{r+1}, \cdots, x_n$, such that the submatrix of the coefficients of the other unknowns $x_1, \cdots, x_r$ has rank r. Then, by the definition of a rank, $\mathbf{A}$ and $\mathbf{B}$ have r linearly independent row vectors, say, the first r row vectors (after a rearrangement if necessary), and if $r < m$, then each of the other row vectors is a linear combination of those. It follows that the corresponding $m - r$ equations can be reduced to the form $0 = 0$ by elementary operations. From this and Theorem 1 in Sec. 7.5 we see that we may omit those $m - r$ equations from our system. We can now write the reduced system in the form

$$\begin{aligned} a_{11}x_1 + \cdots + a_{1r}x_r &= b_1 - (a_{1,r+1}x_{r+1} + \cdots + a_{1n}x_n) \\ a_{21}x_1 + \cdots + a_{2r}x_r &= b_2 - (a_{2,r+1}x_{r+1} + \cdots + a_{2n}x_n) \\ \cdots\cdots\cdots\cdots\cdots\cdots\cdots&\cdots\cdots\cdots\cdots\cdots\cdots\cdots \\ a_{r1}x_1 + \cdots + a_{rr}x_r &= b_r - (a_{r,r+1}x_{r+1} + \cdots + a_{rn}x_n) \end{aligned}$$

(where, if $r = n$, the expressions on the right are $b_1, \cdots, b_r$) and solve it for $x_1, \cdots, x_r$ by Cramer's rule.

Example 1. Cramer's rule

Solve

$$\begin{aligned} 3x_1 + 2x_2 + 2x_3 - 5x_4 &= 8 \\ 2x_1 + 5x_2 + 5x_3 - 18x_4 &= 9 \\ 4x_1 - x_2 - x_3 + 8x_4 &= 7. \end{aligned}$$

The matrix of the coefficients and the augmented matrix have rank 2. We may omit the last equation (why?) and write the reduced system in the form

$$\begin{aligned} 3x_1 + 2x_2 &= 8 - 2x_3 + 5x_4 \\ 2x_1 + 5x_2 &= 9 - 5x_3 + 18x_4. \end{aligned}$$

Then Cramer's rule gives $x_1 = 2 - x_4$, $x_2 = 1 - x_3 + 4x_4$, where x_3 and x_4 are arbitrary. ∎

As an important consequence of Cramer's theorem, we may now express the elements of the inverse of a matrix as follows.

Theorem 3 (Inverse of a matrix)

The inverse of a nonsingular $(n \times n)$ matrix $\mathbf{A} = (a_{jk})$ is given by

$$\text{(6)} \qquad \mathbf{A}^{-1} = \frac{1}{\det \mathbf{A}} \begin{pmatrix} A_{11} & A_{21} & \cdots & A_{n1} \\ A_{12} & A_{22} & \cdots & A_{n2} \\ \cdot & \cdot & \cdots & \cdot \\ A_{1n} & A_{2n} & \cdots & A_{nn} \end{pmatrix},$$

where A_{jk} is the cofactor of a_{jk} in $\det \mathbf{A}$. (Note well that in $\mathbf{A}^{-1}$, the cofactor A_{jk} occupies the same place as a_{kj} (not a_{jk}) does in $\mathbf{A}$.)

Proof. We denote the right-hand side of (6) by $\mathbf{B}$ and show that $\mathbf{BA} = \mathbf{I}$. This entails $\mathbf{B} = \mathbf{A}^{-1}$, as was pointed out in Sec. 7.8 in connection with the uniqueness of the inverse. We write

$$\text{(7)} \qquad \mathbf{BA} = \mathbf{G} = (g_{kl}).$$

Here, by the definition of matrix multiplication,

$$\text{(8)} \qquad g_{kl} = \sum_{s=1}^{n} \frac{A_{sk}}{\det \mathbf{A}} a_{sl} = \frac{1}{\det \mathbf{A}} \sum_{s=1}^{n} A_{sk} a_{sl}.$$

For $l = k$ the last sum is the development of $D = \det \mathbf{A}$ by the kth column. Hence

$$g_{kk} = \frac{1}{\det \mathbf{A}} \sum_{s=1}^{n} A_{sk} a_{sk} = \frac{1}{\det \mathbf{A}} \det \mathbf{A} = 1.$$

For $l \neq k$ this is a similar development of the determinant $\tilde{D}$ obtained from D by replacing the kth column of D with the lth column of D, so that $\tilde{D}$ has two identical columns and is zero:

$$g_{kl} = \frac{1}{\det \mathbf{A}} \sum_{s=1}^{n} A_{sk} a_{sl} = \frac{1}{\det \mathbf{A}} \tilde{D} = 0 \qquad (k \neq l).$$

Hence $\mathbf{BA} = \mathbf{G} = \mathbf{I}$ in (7), so that $\mathbf{B} = \mathbf{A}^{-1}$. ∎

Problems for Sec. 7.11

Using Theorem 1, find the rank.

1. $\begin{pmatrix} 4 & 1 \\ 3 & 2 \end{pmatrix}$ **2.** $\begin{pmatrix} 3 & 7 & -1 \\ 1 & 0 & 2 \end{pmatrix}$ **3.** $\begin{pmatrix} 4 & 2 & 0 & -12 \\ -2 & -1 & 0 & 6 \end{pmatrix}$

4. $\begin{pmatrix} 4 & 2 \\ 1 & 0 \\ -3 & 10 \end{pmatrix}$ **5.** $\begin{pmatrix} 3 & 1 & 4 \\ 0 & 5 & 8 \\ -3 & 4 & 4 \end{pmatrix}$ **6.** $\begin{pmatrix} 8 & 1 & 3 & 6 \\ 0 & 3 & 2 & 2 \\ -8 & -1 & -3 & 4 \end{pmatrix}$

Solve by Cramer's rule and by the Gauss elimination (Sec. 7.5).

7. $\begin{aligned} 7x + 3y &= 13 \\ 3x + 7y &= 17 \end{aligned}$ **8.** $\begin{aligned} -x - 2y &= -5 \\ 5x + 4y &= 1 \end{aligned}$ **9.** $\begin{aligned} x - 3y &= -10 \\ 10x - 5y &= 0 \end{aligned}$

10. $\begin{aligned} x + y + z &= 0 \\ 2x + 5y + 3z &= 1 \\ -x + 2y + z &= 2 \end{aligned}$ **11.** $\begin{aligned} 2x + y + 2z &= -1 \\ x + z &= -1 \\ -x + 3y - 2z &= 7 \end{aligned}$ **12.** $\begin{aligned} x - y + 2z &= 2 \\ 3x + y - z &= 3 \\ 2x + 2y - 3z &= 1 \end{aligned}$

13. Write down the general formulas for the solution of (1) with $n = 2$ as obtained by the Gauss elimination (Sec. 7.5) and compare them with those given by Cramer's rule.

14. For what values of λ does the following system have a nontrivial solution? (This is a problem of great interest in applications, as we shall see in Sec. 7.13.)

$$\begin{aligned} (a_{11} - \lambda)x_1 + \quad a_{12}x_2 \quad &= 0 \\ a_{11}x_1 \quad + (a_{22} - \lambda)x_2 &= 0 \end{aligned}$$

Using Theorem 3, find the inverse. Check the answer.

15. $\begin{pmatrix} -1 & 5 \\ 2 & 3 \end{pmatrix}$ **16.** $\begin{pmatrix} \cos\theta & \sin\theta \\ -\sin\theta & \cos\theta \end{pmatrix}$ **17.** $\begin{pmatrix} 1 & 2 \\ 3 & 4 \end{pmatrix}$

18. $\begin{pmatrix} 0.5 & -0.1 & 0.5 \\ 0 & 0.2 & 0 \\ -0.5 & 0.3 & -1.5 \end{pmatrix}$ **19.** $\begin{pmatrix} 0 & 0 & c \\ 0 & b & 0 \\ a & 0 & 0 \end{pmatrix}$ **20.** $\begin{pmatrix} 2 & 0 & -1 \\ 5 & 1 & 0 \\ 0 & 1 & 3 \end{pmatrix}$

21. Can you think of a geometric argument for obtaining the inverse in Prob. 16?

22. Obtain (4), Sec. 7.8, from the present Theorem 3.

23. Obtain (5), Sec. 7.8, from Theorem 3.

24. (*a*) Show that the product of two ($n \times n$) matrices is singular if and only if at least one of the two matrices is singular. (*b*) Show that the sum of two nonsingular ($n \times n$) matrices may be singular, and the sum of two singular matrices may be nonsingular.

25. Using $\mathbf{A}^{-1}$ as given by (6), show that $\mathbf{AA}^{-1} = \mathbf{I}$.

Four-terminal network. Consider a *four-terminal network* (as shown in the figure) in

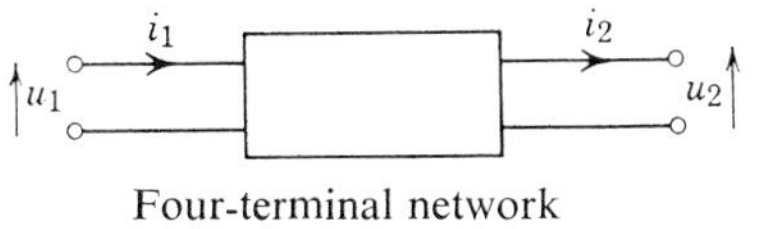

Four-terminal network

which an input signal is applied to one pair of terminals and the output signal is taken from the other pair. Assume that the network is linear, that is, the currents i_1 and i_2 are linear functions of u_1 and u_2, say

$$
\begin{aligned}
i_1 &= a_{11}u_1 + a_{12}u_2 \\
i_2 &= a_{21}u_1 + a_{22}u_2.
\end{aligned}
\tag{11}
$$

26. Show that (11) may be written $\mathbf{i} = \mathbf{Au}$, where

$$
\mathbf{i} = \begin{pmatrix} i_1 \\ i_2 \end{pmatrix}, \qquad \mathbf{A} = \begin{pmatrix} a_{11} & a_{12} \\ a_{21} & a_{22} \end{pmatrix}, \qquad \mathbf{u} = \begin{pmatrix} u_1 \\ u_2 \end{pmatrix}
$$

and $\mathbf{u} = \mathbf{A}^{-1}\mathbf{i}$.

27. Show that the input potential and current may be expressed as linear functions of the output potential and current, say,

$$
\begin{aligned}
u_1 &= t_{11}u_2 + t_{12}i_2, \\
i_1 &= t_{21}u_2 + t_{22}i_2,
\end{aligned}
\quad \text{or} \quad \mathbf{v}_1 = \mathbf{T}\mathbf{v}_2 \quad \text{where} \quad \mathbf{v}_1 = \begin{pmatrix} u_1 \\ i_1 \end{pmatrix}, \ \mathbf{v}_2 = \begin{pmatrix} u_2 \\ i_2 \end{pmatrix},
\tag{12}
$$

and the "transmission matrix" $\mathbf{T}$ is

$$
\mathbf{T} = \begin{pmatrix} t_{11} & t_{12} \\ t_{21} & t_{22} \end{pmatrix} = \frac{1}{a_{21}} \begin{pmatrix} -a_{22} & 1 \\ -\det \mathbf{A} & a_{11} \end{pmatrix}.
$$

Hint. To verify the last expression, start from (11); express u_1 in terms of u_2 and i_2; then express i_1 in terms of u_2 and i_2; finally, compare the resulting representations with (12).

28. Express $\mathbf{A}$ in terms of the elements of $\mathbf{T}$.

In each case show that the given transmission matrix $\mathbf{T}$ corresponds to the indicated four-terminal network.

29.

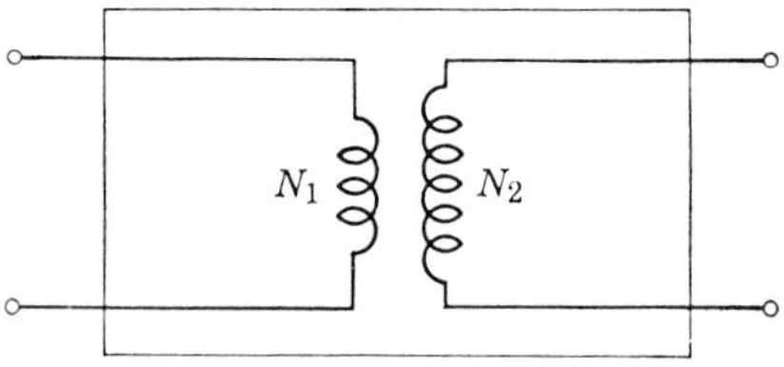

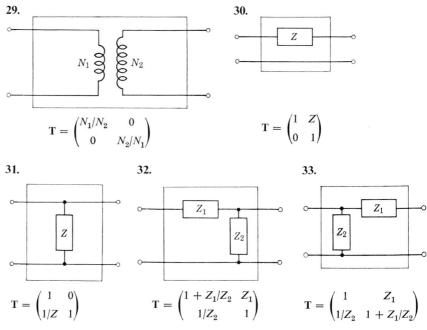

$$\mathbf{T} = \begin{pmatrix} N_1/N_2 & 0 \\ 0 & N_2/N_1 \end{pmatrix}$$

30.

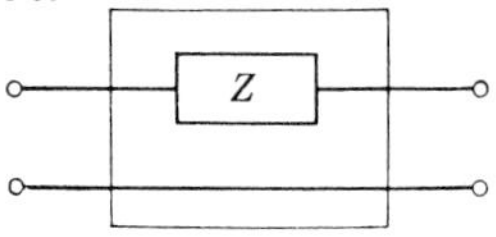

$$\mathbf{T} = \begin{pmatrix} 1 & Z \\ 0 & 1 \end{pmatrix}$$

31.

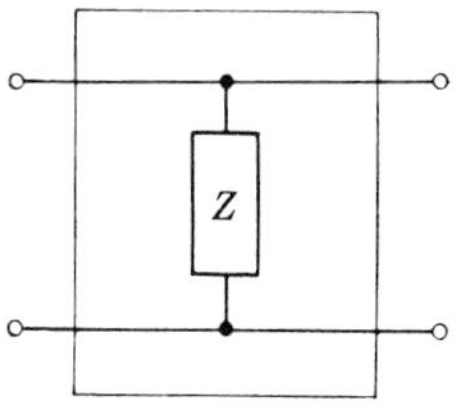

$$\mathbf{T} = \begin{pmatrix} 1 & 0 \\ 1/Z & 1 \end{pmatrix}$$

32.

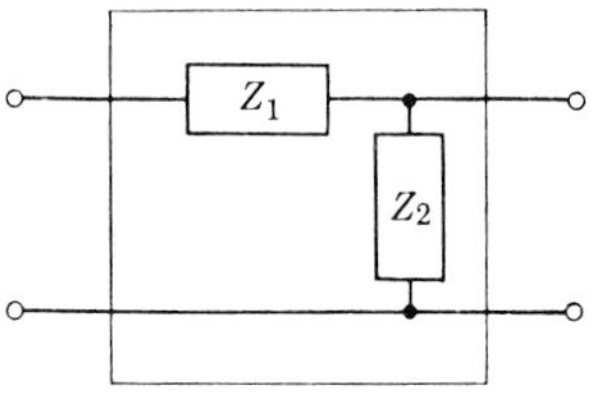

$$\mathbf{T} = \begin{pmatrix} 1 + Z_1/Z_2 & Z_1 \\ 1/Z_2 & 1 \end{pmatrix}$$

33.

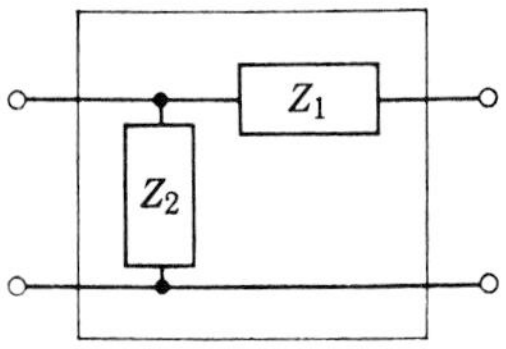

$$\mathbf{T} = \begin{pmatrix} 1 & Z_1 \\ 1/Z_2 & 1 + Z_1/Z_2 \end{pmatrix}$$

34. If two four-terminal networks are connected in cascade as shown in the figure, prove that the resulting network may be regarded as a four-terminal network for which $\mathbf{v}_1 = \mathbf{T}\mathbf{v}_2$ where

$$\mathbf{v}_1 = \begin{pmatrix} u_1 \\ i_1 \end{pmatrix}, \qquad \mathbf{v}_2 = \begin{pmatrix} u_2 \\ i_2 \end{pmatrix}, \qquad \mathbf{T} = \mathbf{T}_1\mathbf{T}_2.$$

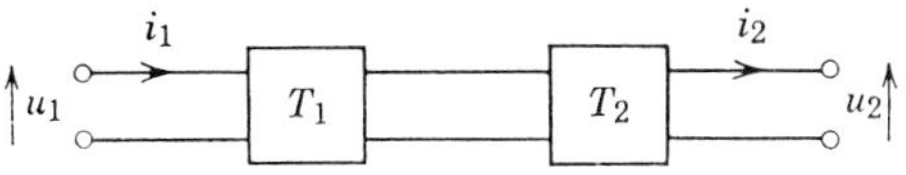

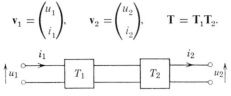

Four-terminal networks in cascade

35. Applying the result of Prob. 34, obtain the matrix in Prob. 32 from those in Prob. 30 (with $Z = Z_1$) and Prob. 31 (with $Z = Z_2$). Obtain the matrix in Prob. 33 from those in Probs. 30 and 31.

7.12 Bilinear, Quadratic, Hermitian and Skew-Hermitian Forms

Bilinear, quadratic, Hermitian and skew-Hermitian forms appear in applications from time to time, so that the engineer should know about them. These forms will also be needed in Sec. 7.14.

An expression of the form

$$B = \sum_{j=1}^{n} \sum_{k=1}^{n} a_{jk} x_j y_k \tag{1}$$

(where the a_{jk} are numbers) is called a **bilinear form** in the $2n$ variables $x_1, \cdots, x_n$ and $y_1, \cdots, y_n$. Writing B out, we have

$$\begin{aligned} B = \quad & a_{11}x_1y_1 + a_{12}x_1y_2 + \cdots + a_{1n}x_1y_n \\ & + a_{21}x_2y_1 + a_{22}x_2y_2 + \cdots + a_{2n}x_2y_n \\ & \cdots\cdots\cdots\cdots\cdots\cdots\cdots\cdots\cdots \\ & + a_{n1}x_ny_1 + a_{n2}x_ny_2 + \cdots + a_{nn}x_ny_n. \end{aligned} \tag{1'}$$

The n-rowed square matrix $\mathbf{A} = (a_{jk})$ is called the *coefficient matrix* of the form. Introducing the vectors

$$\mathbf{x} = \begin{pmatrix} x_1 \\ \vdots \\ x_n \end{pmatrix} \quad \text{and} \quad \mathbf{y} = \begin{pmatrix} y_1 \\ \vdots \\ y_n \end{pmatrix},$$

we see from the definition of matrix multiplication that we may write

$$B = \mathbf{x}^{\mathsf{T}}\mathbf{A}\mathbf{y} \tag{2}$$

where $\mathbf{x}^{\mathsf{T}} = (x_1 \cdots x_n)$ is the transpose of $\mathbf{x}$.

Example 1. Inner product

If $\mathbf{A}$ is the unit matrix $\mathbf{I}$, then

$$B = \mathbf{x}^{\mathsf{T}}\mathbf{I}\mathbf{y} = \mathbf{x}^{\mathsf{T}}\mathbf{y} = x_1y_1 + x_2y_2 + \cdots + x_ny_n;$$

that is, in this case, B is the inner product (scalar product) of the vectors $\mathbf{x}$ and $\mathbf{y}$, which are assumed to be real. ∎

If $\mathbf{y} = \mathbf{x}$, then (1) is called a **quadratic form** in the n variables $x_1, \cdots, x_n$. Denoting this form by Q, we have

$$Q = \sum_{j=1}^{n}\sum_{k=1}^{n} a_{jk}x_jx_k. \tag{3}$$

Writing Q out and taking corresponding terms $a_{jk}x_jx_k$ and $a_{kj}x_kx_j$ together, we obtain

$$\begin{aligned} Q = a_{11}x_1^2 + (a_{12} + a_{21})x_1x_2 + \cdots &+ (a_{1n} + a_{n1})x_1x_n \\ + a_{22}x_2^2 \qquad\qquad &+ \cdots + (a_{2n} + a_{n2})x_2x_n \\ &+ \cdots\cdots\cdots\cdots\cdots \\ &\qquad\qquad + a_{nn}x_n^2. \end{aligned}$$

We may now set

$$\tfrac{1}{2}(a_{jk} + a_{kj}) = c_{jk}.$$

Then $c_{kj} = c_{jk}$ and $c_{jk} + c_{kj} = a_{jk} + a_{kj}$ so that we may write

$$Q = \sum_{j=1}^{n}\sum_{k=1}^{n} c_{jk}x_jx_k.$$

If $\mathbf{A}$ is real, then the coefficient matrix $\mathbf{C} = (c_{jk})$ in this new representation is a real symmetric matrix (cf. Sec. 7.3). This shows that *any real quadratic form Q in n variables $x_1, \cdots, x_n$ may be written*

$$Q = \mathbf{x}^{\mathsf{T}}\mathbf{C}\mathbf{x}$$

where $\mathbf{C}$ is a real symmetric matrix. The reader may prove that the correspondence between these forms Q and symmetric matrices $\mathbf{C}$ is then one-to-one, that is, to each such form Q there corresponds exactly one symmetric matrix $\mathbf{C}$, and conversely.

Example 2. Quadratic form

The quadratic form

$$Q = \mathbf{x}^{\mathsf{T}}\mathbf{A}\mathbf{x} = 2x_1^2 + x_1x_2 - 3x_2^2$$

has the coefficient matrix

$$\mathbf{A} = (a_{jk}) = \begin{pmatrix} 2 & 1 \\ 0 & -3 \end{pmatrix}.$$

Hence the corresponding symmetric coefficient matrix is

$$\mathbf{C} = \tfrac{1}{2}(\mathbf{A} + \mathbf{A}^{\mathsf{T}}) = \begin{pmatrix} 2 & \frac{1}{2} \\ \frac{1}{2} & -3 \end{pmatrix}.$$ ∎

We introduce the following notation. *Let* $\mathbf{A} = (a_{jk})$ *be any matrix. Then* $\overline{\mathbf{A}}$ *denotes the matrix* $(\bar{a}_{jk})$ *which is obtained from* $\mathbf{A}$ *by replacing each element* a_{jk} *by its complex conjugate* $\bar{a}_{jk}$.

A square matrix $\mathbf{A} = (a_{jk})$ for which the transpose equals the complex conjugate, that is,

$$\text{(4)} \qquad \mathbf{A}^{\mathsf{T}} = \overline{\mathbf{A}} \qquad (\text{that is, } a_{kj} = \bar{a}_{jk})$$

is called a **Hermitian**[3] **matrix.**

From (4) we see that the elements in the principal diagonal of a Hermitian matrix are always real. Furthermore, if all the elements of a Hermitian matrix $\mathbf{A}$ are real, then (4) assumes the form $\mathbf{A}^{\mathsf{T}} = \mathbf{A}$, which means that *a real Hermitian matrix is a symmetric matrix, and so Hermitian matrices are a natural generalization of real symmetric matrices.*

A form

$$H = \bar{\mathbf{x}}^{\mathsf{T}}\mathbf{A}\mathbf{x} \qquad (\mathbf{A} \text{ Hermitian})$$

is called a **Hermitian form;** here the n components of the vector $\mathbf{x}$ may be real or complex variables. Obviously, this is a generalization of a real quadratic form.

From the definition of matrix multiplication it follows that

$$H = \sum_{j=1}^{n} \sum_{k=1}^{n} a_{jk}\bar{x}_j x_k .$$

Example 3. Hermitian form

The matrix

$$\mathbf{A} = \begin{pmatrix} 2 & 3+i \\ 3-i & 1 \end{pmatrix}$$

is Hermitian, and the corresponding Hermitian form is

$$\begin{aligned} H = \bar{\mathbf{x}}^{\mathsf{T}}\mathbf{A}\mathbf{x} &= (\bar{x}_1 \quad \bar{x}_2)\begin{pmatrix} 2 & 3+i \\ 3-i & 1 \end{pmatrix}\begin{pmatrix} x_1 \\ x_2 \end{pmatrix} \\ &= 2\bar{x}_1x_1 + (3+i)\bar{x}_1x_2 + (3-i)\bar{x}_2x_1 + \bar{x}_2x_2 \\ &= 2|x_1|^2 + |x_2|^2 + 2\,\mathrm{Re}\,[(3+i)\bar{x}_1x_2] \end{aligned}$$

where Re denotes the real part. The last expression shows that for every choice of the vector $\mathbf{x}$ the value of H is a real number. We shall now prove that any Hermitian form has this remarkable property. ∎

[3]CHARLES HERMITE (1822–1901), French mathematician, is known by his work in algebra and number theory.

Theorem 1 (Hermitian form)
For every choice of the vector $\mathbf{x}$ *the value of a Hermitian form* $H = \overline{\mathbf{x}}^{\mathsf{T}}\mathbf{A}\mathbf{x}$ ($\mathbf{A}$ *Hermitian) is a real number.*

Proof. Using (4), we obtain

$$\overline{H} = \overline{(\overline{\mathbf{x}}^{\mathsf{T}}\mathbf{A}\mathbf{x})} = \mathbf{x}^{\mathsf{T}}\overline{\mathbf{A}}\overline{\mathbf{x}} = \mathbf{x}^{\mathsf{T}}\mathbf{A}^{\mathsf{T}}\overline{\mathbf{x}}.$$

The expression on the right is a scalar. Hence transposition does not change its value. Using (12) in Sec. 7.4, we thus have

$$\mathbf{x}^{\mathsf{T}}\mathbf{A}^{\mathsf{T}}\overline{\mathbf{x}} = (\mathbf{x}^{\mathsf{T}}\mathbf{A}^{\mathsf{T}}\overline{\mathbf{x}})^{\mathsf{T}} = \overline{\mathbf{x}}^{\mathsf{T}}\mathbf{A}\mathbf{x} = H.$$

Hence, $\overline{H} = H$, which means that H is real. This completes the proof. ∎

A square matrix $\mathbf{A} = (a_{jk})$ for which

$$\mathbf{A}^{\mathsf{T}} = -\overline{\mathbf{A}} \qquad (\text{that is, } a_{kj} = -\overline{a}_{jk}) \tag{5}$$

is called a **skew-Hermitian matrix.** Obviously this is a generalization of a real skew-symmetric matrix, because if all the elements of a skew-Hermitian matrix $\mathbf{A}$ are real, then (5) assumes the form $\mathbf{A}^{\mathsf{T}} = -\mathbf{A}$; that is, $\mathbf{A}$ is then a real skew-symmetric matrix (cf. Sec. 7.3).

A form

$$S = \overline{\mathbf{x}}^{\mathsf{T}}\mathbf{A}\mathbf{x} \qquad (\mathbf{A} \text{ skew-Hermitian}) \tag{6}$$

is called a **skew-Hermitian form.** The reader may prove the following theorem.

Theorem 2 (Skew-Hermitian form)
For every choice of $\mathbf{x}$ *the value of a skew-Hermitian form is a pure imaginary number or zero.*

Problems for Sec. 7.12

1. Show that any square matrix may be written as the sum of a Hermitian matrix and a skew-Hermitian matrix.
2. If $\mathbf{A}$ and $\mathbf{B}$ are n-rowed Hermitian matrices and a and b are any *real* numbers, show that $\mathbf{C} = a\mathbf{A} + b\mathbf{B}$ is a Hermitian matrix.
3. Let $\mathbf{A}$ and $\mathbf{B}$ be skew-Hermitian matrices. Under what conditions is $\mathbf{C} = a\mathbf{A} + b\mathbf{B}$ a skew-Hermitian matrix? (a and b are numbers.)
4. Show that the elements of the principle diagonal of a skew-Hermitian matrix are pure imaginary or zero.

Quadratic forms. Find a real symmetric matrix $\mathbf{C}$ such that $Q = \mathbf{x}^{\mathsf{T}}\mathbf{C}\mathbf{x}$, where Q equals

5. $6x_1^2 - 4x_1x_2 + 2x_2^2$
6. $(x_1 - x_2)^2$
7. $5x_1^2 - 2x_1x_2 + x_2^2$
8. $(x_1 + x_2)^2 - x_3^2$
9. $(x_1 - x_2 + x_3)^2$
10. $x_1^2 + 2x_1x_2 + 3x_2^2 + 6x_2x_3 + 2x_3^2$

11. Show that if the variables of a quadratic form $Q = \mathbf{x}^T\mathbf{C}\mathbf{x}$ undergo a linear transformation, say, $\mathbf{x} = \mathbf{P}\mathbf{y}$, then $Q = \mathbf{y}^T\mathbf{A}\mathbf{y}$ where $\mathbf{A} = \mathbf{P}^T\mathbf{C}\mathbf{P}$.

12. Consider the quadratic form $Q = \mathbf{x}^T\mathbf{C}\mathbf{x}$ ($\mathbf{C}$ symmetric) in three variables x_1, x_2, x_3. Find $\partial Q/\partial x_1$, $\partial Q/\partial x_2$, $\partial Q/\partial x_3$ and show that these are the components of the vector $2\mathbf{C}\mathbf{x}$.

13. (Definiteness) A real quadratic form $Q = \mathbf{x}^T\mathbf{C}\mathbf{x}$ and its symmetric matrix $\mathbf{C} = (c_{jk})$ are said to be **positive definite** if $Q > 0$ for all $(x_1, \cdots, x_n) \neq (0, \cdots, 0)$. A necessary and sufficient condition for positive definiteness is that all the determinants

$$C_1 = c_{11}, \quad C_2 = \begin{vmatrix} c_{11} & c_{12} \\ c_{21} & c_{22} \end{vmatrix}, \quad C_3 = \begin{vmatrix} c_{11} & c_{12} & c_{13} \\ c_{21} & c_{22} & c_{23} \\ c_{31} & c_{32} & c_{33} \end{vmatrix}, \cdots, \quad C_n = \det \mathbf{C}$$

are positive (cf. Ref. [C4]). Show that the form in Prob. 5 is positive definite.

14. Test the forms in Probs. 6 and 10 for positive definiteness.

Hermitian forms. Find $H = \bar{\mathbf{x}}^T\mathbf{A}\mathbf{x}$, where

15. $\mathbf{A} = \begin{pmatrix} 1 & 0 \\ 0 & 1 \end{pmatrix}, \quad \mathbf{x} = \begin{pmatrix} 1+i \\ 1-i \end{pmatrix}$

16. $\mathbf{A} = \begin{pmatrix} 0 & i \\ -i & 0 \end{pmatrix}, \quad \mathbf{x} = \begin{pmatrix} 1 \\ i \end{pmatrix}$

17. $\mathbf{A} = \begin{pmatrix} -1 & 5+i \\ 5-i & 2 \end{pmatrix}, \quad \mathbf{x} = \begin{pmatrix} x_1 \\ x_2 \end{pmatrix}$

18. $\mathbf{A} = \begin{pmatrix} a & b \\ \bar{b} & c \end{pmatrix}, \quad \mathbf{x} = \begin{pmatrix} 3i \\ 4 \end{pmatrix}$, a, c real

19. $\mathbf{A} = \begin{pmatrix} 1 & -i & 2i \\ i & 1 & 0 \\ -2i & 0 & 1 \end{pmatrix}, \quad \mathbf{x} = \begin{pmatrix} 0 \\ 1 \\ i \end{pmatrix}$

20. $\mathbf{A} = \begin{pmatrix} 1 & i & 0 \\ -i & 0 & 3 \\ 0 & 3 & 2 \end{pmatrix}, \quad \mathbf{x} = \begin{pmatrix} x_1 \\ x_2 \\ x_3 \end{pmatrix}$

Skew-Hermitian forms. Find $S = \bar{\mathbf{x}}^T\mathbf{A}\mathbf{x}$, where

21. $\mathbf{A} = \begin{pmatrix} i & 0 \\ 0 & -i \end{pmatrix}, \quad \mathbf{x} = \begin{pmatrix} 1 \\ i \end{pmatrix}$

22. $\mathbf{A} = \begin{pmatrix} 5i & 2+i \\ -2+i & i \end{pmatrix}, \quad \mathbf{x} = \begin{pmatrix} 2i \\ 3 \end{pmatrix}$

23. $\mathbf{A} = \begin{pmatrix} 2i & 4 \\ -4 & 0 \end{pmatrix}, \quad \mathbf{x} = \begin{pmatrix} x_1 \\ x_2 \end{pmatrix}$

24. $\mathbf{A} = \begin{pmatrix} i\alpha & \mu + i\nu \\ -\mu + i\nu & i\beta \end{pmatrix}, \mathbf{x} = \begin{pmatrix} -i \\ 4 \end{pmatrix}$

25. Prove Theorem 2.

7.13 Eigenvalues. Eigenvectors

From the standpoint of engineering applications, eigenvalue problems are among the most important problems in connection with matrices, and the number of research papers on corresponding numerical methods for computers is enormous. The basic concepts are as follows.

Let $\mathbf{A} = (a_{jk})$ be a given square n-rowed matrix and consider the vector equation

$$\mathbf{Ax} = \lambda\mathbf{x} \tag{1}$$

where λ is a number.

It is clear that the zero vector $\mathbf{x} = \mathbf{0}$ is a solution of (1) for any value of λ. A value of λ for which (1) has a solution $\mathbf{x} \neq \mathbf{0}$ is called an **eigenvalue**[4] or **characteristic value** (or *latent root*) of the matrix **A**. The corresponding solutions $\mathbf{x} \neq \mathbf{0}$ of (1) are called **eigenvectors** or **characteristic vectors** of **A** corresponding to that eigenvalue λ. The set of the eigenvalues is called the **spectrum** of **A**. The largest of the absolute values of the eigenvalues of **A** is called the **spectral radius** of **A**.

The problem of determining the eigenvalues and eigenvectors of a matrix is called an *eigenvalue problem*.[5] Problems of this type occur in connection with physical and technical applications. Therefore, the student should know the fundamental ideas and concepts which are important in this field of mathematics. During the last two decades various new methods for the approximate determination of eigenvalues have been developed and other methods which have been known for still a longer time have been put into a form which is suitable for electronic computers. Cf. Refs. [C3], [C15] in Appendix 1.

Let us mention two simple problems which lead to an equation of the form (1).

Example 1. Vibrating system

The vertical motion of the mechanical system in Fig. 143 (no damping, masses of springs neglected) is governed by the simultaneous differential equations

$$\begin{aligned} \ddot{y}_1 &= -3y_1 + 2(y_2 - y_1) \\ \ddot{y}_2 &= -2(y_2 - y_1) \end{aligned} \tag{2}$$

where $y_1(t)$ and $y_2(t)$ are the displacements of the two masses such that $y_1 = 0, y_2 = 0$ correspond to the position of static equilibrium. The derivation of these equations is similar to that in Sec. 2.6. We may write the system in the form

$$\begin{aligned} \ddot{y}_1 &= -5y_1 + 2y_2 \\ \ddot{y}_2 &= 2y_1 - 2y_2 \end{aligned}$$

or as a single vector equation

$$\ddot{\mathbf{y}} = \mathbf{Ay},$$

where $\mathbf{y} = \begin{pmatrix} y_1 \\ y_2 \end{pmatrix}$ and $\mathbf{A} = \begin{pmatrix} -5 & 2 \\ 2 & -2 \end{pmatrix}$

To solve this equation we substitute

$$\mathbf{y} = \mathbf{x}e^{\omega t}. \tag{3}$$

This yields

$$\omega^2 \mathbf{x} e^{\omega t} = \mathbf{Ax}e^{\omega t}$$

or

$$\mathbf{Ax} = \lambda\mathbf{x} \qquad \text{where } \lambda = \omega^2.$$

Hence, for (3) to be a solution of (2), not identically zero, $\omega^2 = \lambda$ must be an eigenvalue of **A** and the vector **x** in (3) must be a corresponding eigenvector.

[4] German: Eigenwert

[5] More precisely: an algebraic eigenvalue problem, because there are other eigenvalue problems involving a differential equation (see Secs. 4.8 and 11.3) or an integral equation.

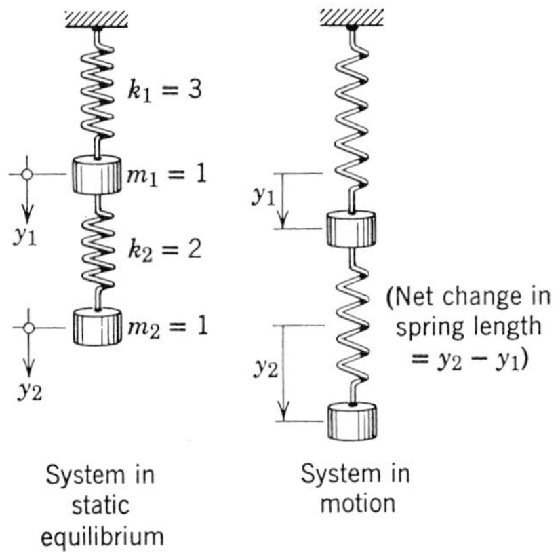

Fig. 143. Example 1

Example 2. Linear transformation
Given a linear transformation

$$\mathbf{y} = \mathbf{A}\mathbf{x} \qquad (\mathbf{A} \text{ real}),$$

does there exist a real vector $\mathbf{x} \neq \mathbf{0}$ for which the corresponding vector $\mathbf{y}$ is real and has the same direction and sense as $\mathbf{x}$? This means we want to find a vector $\mathbf{x}$ such that

$$\mathbf{y} = \mathbf{A}\mathbf{x} = \lambda\mathbf{x} \qquad (\lambda \text{ real and positive}).$$

Clearly, such a vector exists if and only if $\mathbf{A}$ has a real positive eigenvalue. $\mathbf{x}$ is then an eigenvector corresponding to that eigenvalue. ▮

Let us consider (1). If $\mathbf{x}$ is any vector, then the vectors $\mathbf{x}$ and $\mathbf{A}\mathbf{x}$ will, in general, be linearly independent. If $\mathbf{x}$ is an eigenvector, then $\mathbf{x}$ and $\mathbf{A}\mathbf{x}$ are linearly dependent; corresponding components of $\mathbf{x}$ and $\mathbf{A}\mathbf{x}$ are then proportional, the factor of proportionality being the eigenvalue λ.

We shall now demonstrate that *any n-rowed square matrix has at least* 1 *and at most n distinct* (*real or complex*) *eigenvalues.*

For this purpose we write (1) out:

$$\begin{aligned}
a_{11}x_1 + \cdots + a_{1n}x_n &= \lambda x_1 \\
a_{21}x_1 + \cdots + a_{2n}x_n &= \lambda x_2 \\
\cdots\cdots\cdots&\cdots\cdots \\
a_{n1}x_1 + \cdots + a_{nn}x_n &= \lambda x_n.
\end{aligned}$$

By transferring the terms on the right-hand side to the left-hand side we obtain

$$\begin{aligned}
(a_{11} - \lambda)x_1 + a_{12}x_2 + \cdots + a_{1n}x_n &= 0 \\
a_{21}x_1 + (a_{22} - \lambda)x_2 + \cdots + a_{2n}x_n &= 0 \\
\cdots\cdots\cdots\cdots\cdots\cdots&\cdots \\
a_{n1}x_1 + a_{n2}x_2 + \cdots + (a_{nn} - \lambda)x_n &= 0.
\end{aligned} \tag{4}$$

In matrix notation,

$$(\mathbf{A} - \lambda \mathbf{I})\mathbf{x} = \mathbf{0}.$$

By Cramer's theorem in Sec. 7.11, this homogeneous system of linear equations has a nontrivial solution if and only if the corresponding determinant of the coefficients is zero:

$$\textbf{(5)} \qquad D(\lambda) = \det(\mathbf{A} - \lambda \mathbf{I}) = \begin{vmatrix} a_{11} - \lambda & a_{12} & \cdots & a_{1n} \\ a_{21} & a_{22} - \lambda & \cdots & a_{2n} \\ \cdot & \cdot & \cdots & \cdot \\ a_{n1} & a_{n2} & \cdots & a_{nn} - \lambda \end{vmatrix} = 0.$$

$D(\lambda)$ is called the **characteristic determinant,** and (5) is called the **characteristic equation** corresponding to the matrix **A.** By developing $D(\lambda)$ we obtain a polynomial of nth degree in λ. This is called the **characteristic polynomial** corresponding to **A.**

We have thus obtained the following important result.

Theorem 1 (Eigenvalues)

The eigenvalues of a square matrix **A** *are the roots of the corresponding characteristic equation* (5).

An eigenvalue which is a root of mth order of the characteristic polynomial is called an *eigenvalue of mth* **order** of the corresponding matrix.

Once the eigenvalues have been determined, corresponding eigenvectors can be determined from the system (4). Since the system is homogeneous, it is clear that *if* **x** *is an eigenvector of* **A,** *then* $k\mathbf{x}$, *where k is any constant, not zero, is also an eigenvector of* **A** *corresponding to the same eigenvalue.*

Example 3. Eigenvalues and eigenvectors

Determine the eigenvalues and eigenvectors of the matrix

$$\mathbf{A} = \begin{pmatrix} 5 & 4 \\ 1 & 2 \end{pmatrix}.$$

The characteristic equation

$$D(\lambda) = \begin{vmatrix} 5 - \lambda & 4 \\ 1 & 2 - \lambda \end{vmatrix} = \lambda^2 - 7\lambda + 6 = 0$$

has the roots $\lambda_1 = 6$ and $\lambda_2 = 1$. For $\lambda = \lambda_1$ the system (4) assumes the form

$$-x_1 + 4x_2 = 0$$
$$x_1 - 4x_2 = 0.$$

Thus $x_1 = 4x_2$, and

$$\mathbf{x}_1 = \begin{pmatrix} 4 \\ 1 \end{pmatrix}$$

is an eigenvector of $\mathbf{A}$ corresponding to the eigenvalue λ_1. In the same way we find that an eigenvector of $\mathbf{A}$ corresponding to λ_2 is

$$\mathbf{x}_2 = \begin{pmatrix} 1 \\ -1 \end{pmatrix}$$

$\mathbf{x}_1$ and $\mathbf{x}_2$ are linearly independent vectors.

Example 4. Complex eigenvalues

The matrix

$$\mathbf{A} = \begin{pmatrix} a & b \\ -b & a \end{pmatrix} \qquad (a, b \text{ real}, b \neq 0)$$

has the complex conjugate eigenvalues $\lambda_1 = a + ib$ and $\lambda_2 = a - ib$. Corresponding eigenvectors are

$$\mathbf{x}_1 = \begin{pmatrix} 1 \\ i \end{pmatrix} \quad \text{and} \quad \mathbf{x}_2 = \begin{pmatrix} 1 \\ -i \end{pmatrix}$$

If $b = 0$, then $\lambda_1 = \lambda_2 = a$; the matrix $\mathbf{A}$ has just one eigenvalue (which is then of the second order), and every vector $\mathbf{x} \neq \mathbf{0}$ with two components is an eigenvector of $\mathbf{A}$.

Example 5

The matrix

$$\mathbf{A} = \begin{pmatrix} -2 & 2 & -3 \\ 2 & 1 & -6 \\ -1 & -2 & 0 \end{pmatrix}$$

has the eigenvalues $\lambda_1 = 5$ and $\lambda_2 = \lambda_3 = -3$. The vector

$$\mathbf{x}_1 = \begin{pmatrix} 1 \\ 2 \\ -1 \end{pmatrix}$$

is an eigenvector of $\mathbf{A}$ corresponding to the eigenvalue 5, and the vectors

$$\mathbf{x}_2 = \begin{pmatrix} -2 \\ 1 \\ 0 \end{pmatrix} \quad \text{and} \quad \mathbf{x}_3 = \begin{pmatrix} 3 \\ 0 \\ 1 \end{pmatrix}$$

are two linearly independent eigenvectors of $\mathbf{A}$ corresponding to the eigenvalue -3. This agrees with the fact that, for $\lambda = -3$, the matrix $\mathbf{A} - \lambda\mathbf{I}$ has rank 1 and so, by Theorem 2 in Sec. 7.7, a basis of solutions of the corresponding system (4), viz.,

$$\begin{aligned} x_1 + 2x_2 - 3x_3 &= 0 \\ 2x_1 + 4x_2 - 6x_3 &= 0 \\ -x_1 - 2x_2 + 3x_3 &= 0 \end{aligned}$$

consists of two linearly independent vectors. ∎

Example 5 shows that *to an eigenvalue there may correspond several linearly independent eigenvectors*. The maximum number of linearly independent eigenvectors corresponding to an eigenvalue λ is called the (geometric) **multiplicity** of λ.

Problems for Sec. 7.13

1. Do there exist matrices which have no eigenvalues at all?

Find the eigenvalues and eigenvectors of the following matrices.

2. $\begin{pmatrix} 1 & 0 \\ 0 & 2 \end{pmatrix}$ **3.** $\begin{pmatrix} 8 & -4 \\ 2 & 2 \end{pmatrix}$ **4.** $\begin{pmatrix} 0 & 0 \\ 0 & -1 \end{pmatrix}$ **5.** $\begin{pmatrix} 0 & 0 \\ 0 & 0 \end{pmatrix}$

6. $\begin{pmatrix} -4 & 2 \\ -1 & -1 \end{pmatrix}$ **7.** $\begin{pmatrix} 6 & 8 \\ 8 & -6 \end{pmatrix}$ **8.** $\begin{pmatrix} -1 & 0 \\ -2 & 1 \end{pmatrix}$ **9.** $\begin{pmatrix} 1.5 & 2 \\ 2 & -1.5 \end{pmatrix}$

10. $\begin{pmatrix} 2 & 0 & 0 \\ 0 & 4 & 0 \\ 0 & 0 & 3 \end{pmatrix}$ **11.** $\begin{pmatrix} a & 0 & 0 \\ 0 & b & 0 \\ 0 & 0 & c \end{pmatrix}$ **12.** $\begin{pmatrix} 3 & 1 & 4 \\ 0 & 2 & 6 \\ 0 & 0 & 5 \end{pmatrix}$ **13.** $\begin{pmatrix} 0 & 1 & 0 \\ 1 & 0 & 0 \\ 0 & 0 & 1 \end{pmatrix}$

Find the eigenvalues of the following matrices.

14. $\begin{pmatrix} 13 & 0 & -15 \\ -3 & 4 & 9 \\ 5 & 0 & -7 \end{pmatrix}$ **15.** $\begin{pmatrix} 26 & 2 & 4 \\ -2 & 21 & 2 \\ 2 & 4 & 28 \end{pmatrix}$ **16.** $\begin{pmatrix} 1.5 & -1.0 & -2.0 \\ -0.4 & 1.2 & 0.4 \\ -0.3 & -0.6 & -0.2 \end{pmatrix}$

17. Show that the eigenvalues of a triangular matrix **A** are equal to the elements of the principal diagonal of **A**.

18. Show that the transpose $\mathbf{A}^\mathsf{T}$ has the same eigenvalues as **A**.

19. Show that if **A** has real elements, then the eigenvalues of **A** are real or complex conjugates in pairs.

20. Show that if **A** is a real n-rowed square matrix and n is odd, then **A** has at least one real eigenvalue.

21. Show that the constant term of $D(\lambda)$ equals det **A**.

22. Show that $\lambda = 0$ is an eigenvalue of a square matrix **A** if and only if **A** is singular.

23. Show that if **A** has the eigenvalues $\lambda_1, \cdots, \lambda_n$, then $k\mathbf{A}$ has the eigenvalues $k\lambda_1, \cdots, k\lambda_n$. Using this, obtain the eigenvalues in Prob. 3 from those in Prob. 6.

24. Show that if **A** has the eigenvalues $\lambda_1, \cdots, \lambda_n$, then $\mathbf{A}^m$ (m a nonnegative integer) has the eigenvalues $\lambda_1{}^m, \cdots, \lambda_n{}^m$.

25. Show that if **A** has the eigenvalues $\lambda_1, \cdots, \lambda_n$, then

$$k_m\mathbf{A}^m + k_{m-1}\mathbf{A}^{m-1} + \cdots + k_1\mathbf{A} + k_0\mathbf{I}$$

has the eigenvalues

$$k_m\lambda_j{}^m + k_{m-1}\lambda_j{}^{m-1} + \cdots + k_1\lambda_j + k_0 \qquad (j = 1, \cdots, n).$$

7.14 Eigenvalues of Hermitian, Skew-Hermitian, and Unitary Matrices

A square matrix $\mathbf{A} = (a_{jk})$ for which

$$\mathbf{A}^{\mathsf{T}} = \overline{\mathbf{A}}^{-1} \tag{1}$$

is called a **unitary matrix.**

A real unitary matrix $\mathbf{A}$ is called an **orthogonal matrix.** For such a matrix, (1) assumes the form

$$\mathbf{A}^{\mathsf{T}} = \mathbf{A}^{-1}, \tag{2}$$

that is, *an orthogonal matrix is a real matrix for which the inverse equals its transpose.*

A system of vectors $\mathbf{x}_1, \cdots, \mathbf{x}_n$ for which

$$\bar{\mathbf{x}}_j^{\mathsf{T}}\mathbf{x}_k = \delta_{jk} = \begin{cases} 0 & \text{when } j \neq k \\ 1 & \text{when } j = k \end{cases} \qquad (j, k = 1, \cdots, n) \tag{3}$$

is called a **unitary system.**

Clearly, if these vectors are real, then $\bar{\mathbf{x}}_j^{\mathsf{T}}\mathbf{x}_k = \mathbf{x}_j^{\mathsf{T}}\mathbf{x}_k$ is the inner product (scalar product) of $\mathbf{x}_j$ and $\mathbf{x}_k$ in the elementary sense, and the condition (3) means that $\mathbf{x}_1, \cdots, \mathbf{x}_n$ are orthogonal unit vectors. *Hence a unitary system of real vectors is a system of orthogonal (mutually perpendicular) unit vectors.*

Theorem 1 (Unitary matrix)

The column vectors (and also the row vectors) of a unitary matrix form a unitary system.

Proof. Let $\mathbf{A}$ be a unitary matrix with column vectors $\mathbf{a}_1, \cdots, \mathbf{a}_n$. Then from (1) it follows that

$$\mathbf{A}^{-1}\mathbf{A} = \overline{\mathbf{A}}^{\mathsf{T}}\mathbf{A} = \begin{pmatrix} \bar{\mathbf{a}}_1^{\mathsf{T}} \\ \cdot \\ \cdot \\ \cdot \\ \bar{\mathbf{a}}_n^{\mathsf{T}} \end{pmatrix} (\mathbf{a}_1 \cdots \mathbf{a}_n) = \begin{pmatrix} \bar{\mathbf{a}}_1^{\mathsf{T}}\mathbf{a}_1 & \bar{\mathbf{a}}_1^{\mathsf{T}}\mathbf{a}_2 & \cdots & \bar{\mathbf{a}}_1^{\mathsf{T}}\mathbf{a}_n \\ \bar{\mathbf{a}}_2^{\mathsf{T}}\mathbf{a}_1 & \bar{\mathbf{a}}_2^{\mathsf{T}}\mathbf{a}_2 & \cdots & \bar{\mathbf{a}}_2^{\mathsf{T}}\mathbf{a}_n \\ \cdot & \cdot & \cdots & \cdot \\ \bar{\mathbf{a}}_n^{\mathsf{T}}\mathbf{a}_1 & \bar{\mathbf{a}}_n^{\mathsf{T}}\mathbf{a}_2 & \cdots & \bar{\mathbf{a}}_n^{\mathsf{T}}\mathbf{a}_n \end{pmatrix} = \mathbf{I}$$

so that

$$\bar{\mathbf{a}}_j^{\mathsf{T}}\mathbf{a}_k = \begin{cases} 0 & \text{when } j \neq k, \\ 1 & \text{when } j = k. \end{cases}$$

This shows that the column vectors of a unitary matrix form a unitary system. From (1) it follows that the inverse $\mathbf{A}^{-1}$ of a unitary matrix $\mathbf{A}$ is unitary (cf. Prob. 9) and, furthermore, the column vectors of $\mathbf{A}^{-1}$ are the conjugates of the row vectors of $\mathbf{A}$. Hence the row vectors of $\mathbf{A}$ form a unitary system. ∎

With respect to orthogonal matrices, Theorem 1 means that *the row vectors (and also the column vectors) of an orthogonal matrix* $\mathbf{A} = (a_{jk})$ *form a system of orthogonal unit vectors, that is*

$$\sum_{m=1}^{n} a_{jm}a_{km} = \delta_{jk}. \tag{4}$$

Orthogonal matrices are of great practical importance because **orthogonal transformations,** that is, linear transformations $\mathbf{y} = \mathbf{Ax}$ with an orthogonal matrix as the coefficient matrix, have the following property.

Theorem 2 (Orthogonal transformation)

An orthogonal transformation leaves the inner product

$$(\mathbf{x}, \mathbf{y}) = \mathbf{x}^\mathsf{T}\mathbf{y} \qquad \textit{and thus the norm} \qquad \|\mathbf{x}\| = \sqrt{\mathbf{x}^\mathsf{T}\mathbf{x}}$$

invariant. In particular, in three-dimensional Euclidean space an orthogonal transformation preserves the length of any vector and the angle between any two vectors; hence it is a rotation (possibly combined with a reflection), and the coefficient matrix of a rotation of a Cartesian coordinate system is an orthogonal matrix.

Proof. Let $\mathbf{A}$ be orthogonal and $\mathbf{u} = \mathbf{Ax}$ and $\mathbf{v} = \mathbf{Ay}$. Using (2) in this section and (12) in Sec. 7.4, we obtain

$$(\mathbf{u}, \mathbf{v}) = \mathbf{u}^\mathsf{T}\mathbf{v} = (\mathbf{Ax})^\mathsf{T}\mathbf{Ay} = \mathbf{x}^\mathsf{T}\mathbf{A}^\mathsf{T}\mathbf{Ay} = \mathbf{x}^\mathsf{T}\mathbf{A}^{-1}\mathbf{Ay} = \mathbf{x}^\mathsf{T}\mathbf{Iy} = \mathbf{x}^\mathsf{T}\mathbf{y} = (\mathbf{x}, \mathbf{y}). \quad \blacksquare$$

We shall now consider the eigenvalues of Hermitian, skew-Hermitian, and unitary matrices, and prove the general characterization illustrated in Fig. 144.

Theorem 3 (Eigenvalues)

(a) *The eigenvalues of a Hermitian matrix are real.*

(b) *The eigenvalues of a skew-Hermitian matrix are pure imaginary or zero.*

(c) *The eigenvalues of a unitary matrix have the absolute value* 1.

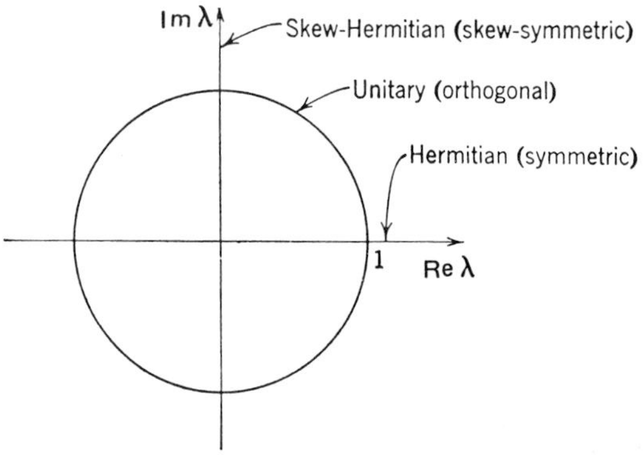

Fig. 144. Location of the eigenvalues of Hermitian, skew-Hermitian, and unitary matrices in the complex λ-plane

Proof. (**a**) and (**b**). Let λ be an eigenvalue of $\mathbf{A}$. Then by definition there is a vector $\mathbf{x} \neq \mathbf{0}$ such that $\mathbf{Ax} = \lambda\mathbf{x}$. From this we obtain

$$\bar{\mathbf{x}}^{\mathsf{T}}\mathbf{Ax} = \bar{\mathbf{x}}^{\mathsf{T}}\lambda\mathbf{x} = \lambda\bar{\mathbf{x}}^{\mathsf{T}}\mathbf{x}.$$

Since $\mathbf{x} \neq \mathbf{0}$, it follows that $\bar{\mathbf{x}}^{\mathsf{T}}\mathbf{x} \neq 0$ and we may divide, finding

$$\lambda = \frac{\bar{\mathbf{x}}^{\mathsf{T}}\mathbf{Ax}}{\bar{\mathbf{x}}^{\mathsf{T}}\mathbf{x}}.$$

The denominator is real. If $\mathbf{A}$ is Hermitian, the numerator is real (Theorem 1 in Sec. 7.12) and so is λ. If $\mathbf{A}$ is skew-Hermitian, the numerator is pure imaginary or zero (Theorem 2 in Sec. 7.12) and so is λ.

(**c**) Let $\mathbf{U}$ be unitary, λ an eigenvalue of $\mathbf{U}$, and $\mathbf{x}$ a corresponding eigenvector. Then

$$\mathbf{Ux} = \lambda\mathbf{x}. \tag{5}$$

Taking the conjugate transpose and using (12) in Sec. 7.4, we obtain

$$(\overline{\mathbf{Ux}})^{\mathsf{T}} = \bar{\mathbf{x}}^{\mathsf{T}}\bar{\mathbf{U}}^{\mathsf{T}} = \bar{\lambda}\bar{\mathbf{x}}^{\mathsf{T}}.$$

Since $\mathbf{U}$ is unitary, $\bar{\mathbf{U}}^{\mathsf{T}} = \mathbf{U}^{-1}$, and our equation can be written

$$\bar{\mathbf{x}}^{\mathsf{T}}\mathbf{U}^{-1} = \bar{\lambda}\bar{\mathbf{x}}^{\mathsf{T}}. \tag{6}$$

By postmultiplying the left-hand side of (6) by the left-hand side of (5) and the right-hand side of (6) by the right-hand side of (5) we obtain

$$\bar{\mathbf{x}}^{\mathsf{T}}\mathbf{U}^{-1}\mathbf{Ux} = \bar{\lambda}\bar{\mathbf{x}}^{\mathsf{T}}\lambda\mathbf{x} = \lambda\bar{\lambda}\bar{\mathbf{x}}^{\mathsf{T}}\mathbf{x}.$$

In this equation we have $\mathbf{U}^{-1}\mathbf{U} = \mathbf{I}$ and $\lambda\bar{\lambda} = |\lambda|^2$. Hence the equation becomes

$$\bar{\mathbf{x}}^{\mathsf{T}}\mathbf{x} = |\lambda|^2\bar{\mathbf{x}}^{\mathsf{T}}\mathbf{x}.$$

Since $\mathbf{x} \neq \mathbf{0}$, it follows that $\bar{\mathbf{x}}^{\mathsf{T}}\mathbf{x} \neq 0$. Hence we must have $|\lambda|^2 = 1$. Thus $|\lambda| = 1$, and the proof is complete. ∎

An important particular case of Theorem 3 is

Theorem 4

The eigenvalues of a symmetric matrix are real. The eigenvalues of a skew-symmetric matrix are pure imaginary or zero. The eigenvalues of an orthogonal matrix have the absolute value 1 *and are real or complex conjugates in pairs.*

Numerical methods for matrix eigenvalue problems are presented in Secs. 19.13 and 19.14, which are independent of the other sections in Chap. 19, so that they can be studied now.

Problems for Sec. 7.14

Show that the following matrices are orthogonal.

1. $\begin{pmatrix} \cos\theta & -\sin\theta \\ \sin\theta & \cos\theta \end{pmatrix}$ **2.** $\begin{pmatrix} \cos\theta & -\sin\theta & 0 \\ \sin\theta & \cos\theta & 0 \\ 0 & 0 & 1 \end{pmatrix}$ **3.** $\begin{pmatrix} 0 & 1 & 0 \\ 1 & 0 & 0 \\ 0 & 0 & 1 \end{pmatrix}$

4. Interpret the transformation $\mathbf{y} = \mathbf{Ax}$ geometrically, where $\mathbf{A}$ is the matrix in Prob. 1 and the components of $\mathbf{x}$ and $\mathbf{y}$ are Cartesian coordinates.

5. Same task as in Prob. 4, where $\mathbf{A}$ is the matrix in Prob. 2.

6. Show that if $\mathbf{A}$ is orthogonal, then $\det \mathbf{A} = 1$ or -1. Give examples.

Show that the following matrices are unitary and determine their eigenvalues and eigenvectors.

7. $\begin{pmatrix} 1/\sqrt{2} & i/\sqrt{2} \\ -i/\sqrt{2} & -1/\sqrt{2} \end{pmatrix}$ **8.** $\begin{pmatrix} 1/\sqrt{3} & i\sqrt{2/3} \\ -i\sqrt{2/3} & -1/\sqrt{3} \end{pmatrix}$

9. Show that the inverse of a unitary matrix is unitary. Verify this for the matrix in Prob. 7.

10. Show that the product of two n-rowed unitary matrices is unitary. Verify this for the matrices in Probs. 7 and 8.

11. Show that the eigenvectors of a real symmetric matrix corresponding to different eigenvalues are orthogonal. Give an example.

12. **(Similarity transformation)** Consider the linear transformation $\mathbf{y} = \mathbf{Ax}$ where $\mathbf{A}$ is square and set $\mathbf{x} = \mathbf{T}\tilde{\mathbf{x}}$, $\mathbf{y} = \mathbf{T}\tilde{\mathbf{y}}$ where $\mathbf{T}$ is nonsingular. Show that then

$$\tilde{\mathbf{y}} = \tilde{\mathbf{A}}\tilde{\mathbf{x}} \qquad \text{where} \qquad \tilde{\mathbf{A}} = \mathbf{T}^{-1}\mathbf{AT}.$$

(This transformation which transforms $\mathbf{A}$ into $\tilde{\mathbf{A}}$ is called a *similarity transformation*, and $\mathbf{A}$ and $\tilde{\mathbf{A}}$ are called *similar matrices*.)

13. Show that if $\tilde{\mathbf{A}} = \mathbf{T}^{-1}\mathbf{AT}$ where $\mathbf{A}$ is any square matrix, then $\det \tilde{\mathbf{A}} = \det \mathbf{A}$. *Hint.* Use (5), Sec. 7.10.

14. Let $\mathbf{A}$ be a given square matrix, and let $\mathbf{X}$ be the matrix whose column vectors $\mathbf{x}_1, \cdots, \mathbf{x}_n$ are eigenvectors of $\mathbf{A}$ corresponding to the eigenvalues $\lambda_1, \cdots, \lambda_n$ of $\mathbf{A}$. Show that

$$\mathbf{AX} = (\lambda_1\mathbf{x}_1 \cdots \lambda_n\mathbf{x}_n) = \mathbf{XD},$$

where $\mathbf{D}$ is the diagonal matrix with elements $\lambda_1, \cdots, \lambda_n$ in the principal diagonal.

15. If the $\mathbf{x}_i$ in Prob. 14 are linearly independent, show that $\mathbf{X}^{-1}\mathbf{AX} = \mathbf{D}$.

16. **(Trace)** The sum of the elements in the principal diagonal of a square matrix $\mathbf{C} = (c_{ik})$ is called the *trace* of $\mathbf{C}$. Thus

$$\text{trace } \mathbf{C} = c_{11} + c_{22} + \cdots + c_{nn}.$$

Let $\mathbf{A} = (a_{ik})$ and $\mathbf{B} = (b_{ik})$ be two n-rowed square matrices. Show that

$$\text{trace } \mathbf{AB} = \sum_{i=1}^{n}\sum_{l=1}^{n} a_{il}b_{li} = \text{trace } \mathbf{BA}.$$

17. If $\tilde{\mathbf{A}} = \mathbf{T}^{-1}\mathbf{AT}$, show that trace $\tilde{\mathbf{A}}$ = trace $\mathbf{A}$.

18. Show that the matrices $\mathbf{A}$ and $\tilde{\mathbf{A}} = \mathbf{T}^{-1}\mathbf{AT}$ have the same eigenvalues.

19. If $\mathbf{x}$ is an eigenvector of $\mathbf{A}$ corresponding to an eigenvalue λ, show that $\mathbf{y} = \mathbf{T}^{-1}\mathbf{x}$ is an eigenvector of $\tilde{\mathbf{A}} = \mathbf{T}^{-1}\mathbf{AT}$ corresponding to the same eigenvalue λ.

20. Verify the statement of Prob. 18 for the matrices

$$\mathbf{A} = \begin{pmatrix} 4 & -2 \\ 1 & 1 \end{pmatrix} \quad \text{and} \quad \tilde{\mathbf{A}} = \mathbf{T}^{-1}\mathbf{AT} \quad \text{where} \quad \mathbf{T} = \begin{pmatrix} 2 & 1 \\ 5 & 3 \end{pmatrix}.$$

21. **(Normal matrix)** By definition, a *normal matrix* is a square matrix which commutes with its complex conjugate transpose, that is,

$$\mathbf{A}\bar{\mathbf{A}}^{\mathsf{T}} = \bar{\mathbf{A}}^{\mathsf{T}}\mathbf{A}.$$

Show that Hermitian, skew-Hermitian and unitary matrices are normal.

Under what conditions are the following matrices normal?

22. $\begin{pmatrix} 0 & b \\ c & 0 \end{pmatrix}$

23. $\begin{pmatrix} 0 & a & 0 \\ 0 & 0 & b \\ c & 0 & 0 \end{pmatrix}$

24. **(Stochastic matrix)** By definition, a *stochastic matrix* is a real square matrix with nonnegative elements such that for each row the sum of the elements is 1. Show that every stochastic matrix has 1 as one of its eigenvalues.

25. The elements of a stochastic matrix may be probabilities, for instance, probabilities for the motion of people from a city to its suburbs and conversely. Consider the matrix

		Final State	
		City	Suburb
Initial State	City	0.95	0.05
	Suburb	0.02	0.98

where *final state* means the state one year later. Let $(C \quad S) = (100\,000 \quad 100\,000)$ be the initial state, where C means *city* and S *suburb*. What can we expect after one year? After two years? (Disregard population increases.)

7.15 Systems of Linear Differential Equations

Systems of differential equations arise in various engineering problems. A first impression of this fact was given in Secs. 3.1 and 7.13. Simpler systems can be treated without the use of matrices (cf. Sec. 3.1). However, for large systems or for discussing general properties of solutions and developing a systematic theory, the use of matrices is of great value and convenience.

In this section we shall illustrate the role of matrices and eigenvalues in connection with systems of linear differential equations.

Example 1. Vibrating system

Determine the motion of the mechanical system in Fig. 145 satisfying the initial conditions

$$y_1(0) = 1, \quad y_2(0) = 2, \quad \dot{y}_1(0) = -2\sqrt{6}, \quad \dot{y}_2(0) = \sqrt{6}.$$

This system was considered in Example 1, Sec. 7.13, and we know that the corresponding system of differential equations can be written

$$\ddot{\mathbf{y}} = \mathbf{A}\mathbf{y}$$

where

$$\mathbf{y} = \begin{pmatrix} y_1 \\ y_2 \end{pmatrix} \quad \text{and} \quad \mathbf{A} = \begin{pmatrix} -5 & 2 \\ 2 & -2 \end{pmatrix}.$$

Substituting $\mathbf{y} = \mathbf{x}e^{\omega t}$ and canceling out $e^{\omega t}$, we obtain

$$\mathbf{A}\mathbf{x} = \lambda\mathbf{x} \qquad (\lambda = \omega^2).$$

The characteristic equation is

$$\det(\mathbf{A} - \lambda\mathbf{I}) = \begin{vmatrix} -5-\lambda & 2 \\ 2 & -2-\lambda \end{vmatrix} = \lambda^2 + 7\lambda + 6 = 0.$$

The roots are $\lambda_1 = -1$ and $\lambda_2 = -6$. These are the eigenvalues of **A**. Corresponding eigenvectors are found to be

$$\mathbf{x}_1 = \begin{pmatrix} 1 \\ 2 \end{pmatrix}, \qquad \mathbf{x}_2 = \begin{pmatrix} 1 \\ -\frac{1}{2} \end{pmatrix}.$$

To $\lambda_1 = -1$ and $\lambda_2 = -6$ there correspond $\omega_1 = \pm i$ and $\omega_2 = \pm i\sqrt{6}$. Hence a solution is

(1) $$\mathbf{y}(t) = a_1\mathbf{x}_1 \cos t + b_1\mathbf{x}_1 \sin t + a_2\mathbf{x}_2 \cos \sqrt{6}t + b_2\mathbf{x}_2 \sin \sqrt{6}t,$$

where a_1, b_1, a_2, b_2 are arbitrary constants. In terms of components this can be written

(2) $$\begin{aligned} y_1(t) &= a_1 \cos t + b_1 \sin t + a_2 \cos \sqrt{6}t + b_2 \sin \sqrt{6}t \\ y_2(t) &= 2a_1 \cos t + 2b_1 \sin t - \tfrac{1}{2}a_2 \cos \sqrt{6}t - \tfrac{1}{2}b_2 \sin \sqrt{6}t, \end{aligned}$$

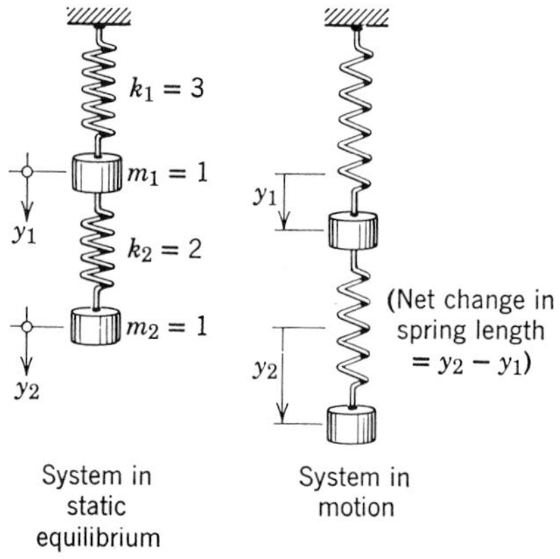

Fig. 145. Mechanical system in Example 1

as can be seen from the form of $\mathbf{x}_1$ and $\mathbf{x}_2$. From (1) and the initial conditions $y_1(0) = 1, y_2(0) = 2$, written in vector form, we have

$$\mathbf{y}(0) = a_1\mathbf{x}_1 + a_2\mathbf{x}_2 = a_1\begin{pmatrix}1\\2\end{pmatrix} + a_2\begin{pmatrix}1\\-\frac{1}{2}\end{pmatrix} = \begin{pmatrix}1\\2\end{pmatrix}.$$

Hence $a_1 = 1$ and $a_2 = 0$. Substitution of this into (1) and differentiation gives

$$\dot{\mathbf{y}}(t) = -\mathbf{x}_1 \sin t + b_1\mathbf{x}_1 \cos t + b_2\sqrt{6}\,\mathbf{x}_2 \cos \sqrt{6}t.$$

From this and the other two initial conditions $\dot{y}_1(0) = -2\sqrt{6}$, $\dot{y}_2(0) = \sqrt{6}$ we obtain

$$\dot{\mathbf{y}}(0) = b_1\mathbf{x}_1 + b_2\sqrt{6}\,\mathbf{x}_2 = b_1\begin{pmatrix}1\\2\end{pmatrix} + b_2\sqrt{6}\begin{pmatrix}1\\-\frac{1}{2}\end{pmatrix} = \begin{pmatrix}-2\sqrt{6}\\\sqrt{6}\end{pmatrix}.$$

Hence $b_1 = 0$ and $b_2 = -2$. The solution of our problem is

$$\mathbf{y}(t) = \mathbf{x}_1 \cos t - 2\mathbf{x}_2 \sin \sqrt{6}t.$$

In terms of components,

$$y_1(t) = \cos t - 2 \sin \sqrt{6}t$$

$$y_2(t) = 2 \cos t + \sin \sqrt{6}t.$$

Another way of treating the present system of equations is the reduction to a single equation. This was shown in Sec. 3.1, and our present solution (2) agrees with the solution (9) given there.

A third approach—not very profitable in the present case but very important in principle—is the transformation of the given system into a larger system of first-order equations. This is quite simple. All we have to do is to set $\dot{y}_1 = y_3$ and $\dot{y}_2 = y_4$. Then $\ddot{y}_1 = \dot{y}_3, \ddot{y}_2 = \dot{y}_4$, and we altogether have the four first-order equations

$$\begin{aligned}\dot{y}_1 &= y_3\\ \dot{y}_2 &= y_4\\ \dot{y}_3 &= -5y_1 + 2y_2\\ \dot{y}_4 &= 2y_1 - 2y_2\end{aligned} \tag{3}$$

This can be written $\dot{\tilde{\mathbf{y}}} = \tilde{\mathbf{A}}\tilde{\mathbf{y}}$ where

$$\tilde{\mathbf{A}} = \begin{pmatrix}0 & 0 & 1 & 0\\0 & 0 & 0 & 1\\-5 & 2 & 0 & 0\\2 & -2 & 0 & 0\end{pmatrix}. \tag{4}$$

Substituting $\tilde{\mathbf{y}} = \tilde{\mathbf{x}}e^{\tilde{\lambda}t}$ and $\dot{\tilde{\mathbf{y}}} = \tilde{\lambda}\tilde{\mathbf{x}}e^{\tilde{\lambda}t}$, we have

$$\tilde{\mathbf{A}}\tilde{\mathbf{x}} = \tilde{\lambda}\tilde{\mathbf{x}}.$$

From this we obtain the characteristic equation

$$\tilde{\lambda}^4 + 7\tilde{\lambda}^2 + 6 = 0.$$

Hence $\tilde{\lambda}^2 = -1$ or -6, so that the four eigenvalues of $\tilde{\mathbf{A}}$ are $i, -i, i\sqrt{6}, -i\sqrt{6}$. Corresponding eigenvectors are found to be

$$\tilde{\mathbf{x}}_1 = \begin{pmatrix}1\\2\\i\\2i\end{pmatrix}, \quad \tilde{\mathbf{x}}_2 = \begin{pmatrix}1\\2\\-i\\-2i\end{pmatrix}, \quad \tilde{\mathbf{x}}_3 = \begin{pmatrix}1\\-1/2\\i\sqrt{6}\\-i\sqrt{3/2}\end{pmatrix}, \quad \tilde{\mathbf{x}}_4 = \begin{pmatrix}1\\-1/2\\-i\sqrt{6}\\i\sqrt{3/2}\end{pmatrix}.$$

The corresponding solution of (3) is

$$\tilde{\mathbf{y}} = c_1\tilde{\mathbf{x}}_1 e^{it} + c_2\tilde{\mathbf{x}}_2 e^{-it} + c_3\tilde{\mathbf{x}}_3 e^{i\sqrt{6}t} + c_4\tilde{\mathbf{x}}_4 e^{-i\sqrt{6}t}.$$

In terms of components,

$$y_1 = c_1 e^{it} + c_2 e^{-it} + c_3 e^{i\sqrt{6}t} + c_4 e^{-i\sqrt{6}t}$$

$$y_2 = 2c_1 e^{it} + 2c_2 e^{-it} - \tfrac{1}{2}c_3 e^{i\sqrt{6}t} - \tfrac{1}{2}c_4 e^{-i\sqrt{6}t}$$

and $y_3 = \dot{y}_1$, $y_4 = \dot{y}_2$. This can be written

$$y_1 = a_1 \cos t + b_1 \sin t + a_2 \cos \sqrt{6}t + b_2 \sin \sqrt{6}t$$

$$y_2 = 2a_1 \cos t + 2b_1 \sin t - \tfrac{1}{2}a_2 \cos \sqrt{6}t - \tfrac{1}{2}b_2 \sin \sqrt{6}t,$$

in agreement with (2). ∎

The first-order system (3) in our example is of the form

$$\dot{y}_j = \sum_{k=1}^{n} a_{jk} y_k + h_j(t) \qquad j = 1, \cdots, n$$

and we want to say a few words about a general theory for such systems. Ordinarily the a_{jk}'s depend on t, but, for simplicity, we assume that they are constant. (In Example 1 we have $h_j = 0$.) In matrix notation, the system is

$$\dot{\mathbf{y}} = \mathbf{A}\mathbf{y} + \mathbf{h}. \tag{5}$$

We assume that $\mathbf{A}$ has n linearly independent eigenvectors $\mathbf{x}_1, \cdots, \mathbf{x}_n$. (For instance, this holds for a symmetric matrix, which often occurs due to symmetry in the problem.) Let $\lambda_1, \cdots, \lambda_n$ denote the corresponding eigenvalues. (Note that some of these may be equal.) Then

$$\mathbf{A}\mathbf{x}_1 = \lambda_1\mathbf{x}_1, \qquad \cdots, \qquad \mathbf{A}\mathbf{x}_n = \lambda_n\mathbf{x}_n$$

or, written as a single matrix equation,

$$(\mathbf{A}\mathbf{x}_1 \quad \mathbf{A}\mathbf{x}_2 \quad \cdots \quad \mathbf{A}\mathbf{x}_n) = (\lambda_1\mathbf{x}_1 \quad \lambda_2\mathbf{x}_2 \quad \cdots \quad \lambda_n\mathbf{x}_n).$$

Introducing

$$\mathbf{X} = (\mathbf{x}_1 \quad \cdots \quad \mathbf{x}_n),$$

the matrix with column vectors $\mathbf{x}_1, \cdots, \mathbf{x}_n$, we may write this

$$\mathbf{A}\mathbf{X} = \mathbf{D}\mathbf{X}, \tag{6}$$

where $\mathbf{D}$ is the diagonal matrix with principal diagonal elements $\lambda_1, \cdots, \lambda_n$. To introduce $\mathbf{X}$ into (5), we set

$$\mathbf{y} = \mathbf{X}\mathbf{z} \tag{7}$$

Then $\dot{\mathbf{y}} = \mathbf{X}\dot{\mathbf{z}}$ and (5) becomes

$$\mathbf{X}\dot{\mathbf{z}} = \mathbf{A}\mathbf{X}\mathbf{z} + \mathbf{h}.$$

Assuming $\mathbf{X}$ to be nonsingular, we obtain

$$\dot{\mathbf{z}} = \mathbf{X}^{-1}\mathbf{A}\mathbf{X}\mathbf{z} + \mathbf{X}^{-1}\mathbf{h}.$$

Since $\mathbf{X}^{-1}\mathbf{A}\mathbf{X} = \mathbf{D}$ by (6), we simply have

$$\dot{\mathbf{z}} = \mathbf{D}\mathbf{z} + \mathbf{X}^{-1}\mathbf{h}. \tag{8}$$

In terms of components, taking $\mathbf{Dz}$ to the left, we can write

$$\dot{z}_j - \lambda_j z_j = r_j(t) \qquad (j = 1, \cdots, n), \tag{8*}$$

where, because of the definition of matrix multiplication, $r_j(t)$ is a linear combination, with constant coefficients, of $h_1(t), \cdots, h_n(t)$. By Sec. 1.7, the solution of (8*) is

$$z_j(t) = e^{\lambda_j t}\left(\int e^{-\lambda_j t} r_j(t)\, dt + c_j\right) \tag{9}$$

where c_j is a constant which is arbitrary. As in Sec. 1.7 we can assign to c_j a unique value by prescribing an initial condition

$$\mathbf{y}(t_0) = \mathbf{y}_0. \tag{10}$$

Indeed, by (7), this determines

$$\mathbf{z}(t_0) = \mathbf{z}_0 = \mathbf{X}^{-1}\mathbf{y}_0$$

or, in terms of components, $z_j(t_0) = z_{0j}$, where z_{0j} is the jth component of the constant vector $\mathbf{z}_0$.

Remembering how we obtained the general solution in the case of a single first-order linear equation in Sec. 1.7, we see that we can formulate our result as follows.

Existence and uniqueness theorem for systems

Let $\mathbf{h}(t)$ *in* (5) *be continuous in an interval* $\alpha < t < \beta$ *and* t_0 *any given point in that interval. Assume that* $\mathbf{A}$ *has a linearly independent set of n eigenvectors. Then the initial value problem* (5), (10) *has a unique solution* $\mathbf{y}(t)$ *on that interval.*

Real symmetric and, more generally, Hermitian matrices satisfy that condition, as can be shown, and since in many applications the occurring matrices are Hermitian, this theorem is of considerable practical importance.

In applications, it is often important to know whether the solutions tend to 0 as $t \to \infty$. This is the problem of the *stability of solutions*. Our discussion shows the following. If $\mathbf{h} = \mathbf{0}$ in (5), then $\mathbf{X}^{-1}\mathbf{h} = \mathbf{0}$ in (8), hence $r_j = 0$ in (8*) and $z_j = c_j e^{\lambda_j t}$ in (9). This proves

Theorem 2 (Stability)

All solutions of the system (5) *with* $\mathbf{h} = \mathbf{0}$ *approach* $\mathbf{0}$ *as* $t \to \infty$ *if and only if all the eigenvalues of* $\mathbf{A}$ *have negative real parts.*

The reader may want to see an illustration of our previous general reasoning by a simple special example:

Example 2

Solve the initial value problem consisting of the equations

$$\dot{y}_1 = 5y_1 + 8y_2 + 1$$
$$\dot{y}_2 = -6y_1 - 9y_2 + t$$

and the initial conditions $y_1(0) = 4$, $y_2(0) = -3$.

In matrix notation we have $\dot{\mathbf{y}} = \mathbf{Ay} + \mathbf{h}$, where

$$\mathbf{A} = \begin{pmatrix} 5 & 8 \\ -6 & -9 \end{pmatrix}, \qquad \mathbf{h} = \begin{pmatrix} 1 \\ t \end{pmatrix}.$$

The characteristic equation yields the eigenvalues $\lambda_1 = -1$ and $\lambda_2 = -3$. We now determine corresponding eigenvectors, finding

$$\mathbf{x}_1 = \begin{pmatrix} 1 \\ -3/4 \end{pmatrix}, \qquad \mathbf{x}_2 = \begin{pmatrix} 1 \\ -1 \end{pmatrix}.$$

Hence

$$\mathbf{X} = (\mathbf{x}_1 \quad \mathbf{x}_2) = \begin{pmatrix} 1 & 1 \\ -3/4 & -1 \end{pmatrix}, \qquad \mathbf{X}^{-1} = \begin{pmatrix} 4 & 4 \\ -3 & -4 \end{pmatrix}$$

and (8) takes the form

$$\dot{\mathbf{z}} = \mathbf{Dz} + \mathbf{X}^{-1}\mathbf{h}$$

where

$$\mathbf{D} = \begin{pmatrix} -1 & 0 \\ 0 & -3 \end{pmatrix}, \qquad \mathbf{X}^{-1}\mathbf{h} = \begin{pmatrix} 4 + 4t \\ -3 - 4t \end{pmatrix}.$$

We take $\mathbf{Dz}$ to the left, write the equation in terms of components and use (9):

$$\dot{z}_1 + z_1 = 4 + 4t, \quad \text{solution} \quad z_1 = c_1 e^{-t} + 4t$$
$$\dot{z}_2 + 3z_2 = -3 - 4t, \quad \text{solution} \quad z_2 = c_2 e^{-3t} - \tfrac{5}{9} - \tfrac{4}{3}t.$$

From this we obtain

$$\mathbf{y} = \mathbf{Xz} = \begin{pmatrix} 1 & 1 \\ -3/4 & -1 \end{pmatrix}\begin{pmatrix} z_1 \\ z_2 \end{pmatrix} = c_1\begin{pmatrix} 1 \\ -3/4 \end{pmatrix} e^{-t} + c_2\begin{pmatrix} 1 \\ -1 \end{pmatrix} e^{-3t} + \frac{1}{9}\begin{pmatrix} -5 + 24t \\ 5 - 15t \end{pmatrix}.$$

By the initial conditions,

$$\mathbf{y}(0) = c_1\begin{pmatrix} 1 \\ -3/4 \end{pmatrix} + c_2\begin{pmatrix} 1 \\ -1 \end{pmatrix} + \begin{pmatrix} -5/9 \\ 5/9 \end{pmatrix} = \begin{pmatrix} 4 \\ -3 \end{pmatrix}.$$

This yields $c_1 = 4$, $c_2 = 5/9$. Hence the solution of our problem is

$$y_1 = 4e^{-t} + \tfrac{5}{9}e^{-3t} - \tfrac{5}{9} + \tfrac{8}{3}t$$
$$y_2 = -3e^{-t} - \tfrac{5}{9}e^{-3t} + \tfrac{5}{9} - \tfrac{5}{3}t.$$

This completes the present example which illustrates our reasoning leading to the existence and uniqueness theorem (see before).

Problems for Sec. 7.15

Solve the following initial value problems.

1. $\dot{y}_1 = 4y_1 - 2y_2$, $\dot{y}_2 = y_1 + y_2$ $\quad y_1(0) = 3,\ y_2(0) = 2$

2. $\dot{y}_1 = 3y_1 + 4y_2$, $\dot{y}_2 = 4y_1 - 3y_2$ $\quad y_1(0) = 1,\ y_2(0) = 3$

3. $\dot{y}_1 = y_1 + 2y_2$, $\dot{y}_2 = -8y_1 + 11y_2$ $\quad y_1(0) = 1,\ y_2(0) = 1$

4. $\dot{y}_1 = y_2$, $\dot{y}_2 = y_1 + 3y_3$, $\dot{y}_3 = y_2$ $\quad y_1(0) = 2,\ y_2(0) = 0,\ y_3(0) = 2$

5. $\dot{y}_1 = 3y_1$, $\dot{y}_2 = 5y_1 + 4y_2$, $\dot{y}_3 = 3y_1 + 6y_2 + y_3$ $\quad y_1(0) = 2,\ y_2(0) = -10,\ y_3(0) = -27$

6. In Example 1, find and graph the solutions satisfying the initial conditions $y_1(0) = 1$, $y_2(0) = 2$, $\dot{y}_1(0) = 0$, $\dot{y}_2(0) = 0$.

7. Show that for general m_1, m_2, k_1, k_2, the mechanical system in Fig. 145 is governed by $\ddot{\mathbf{y}} = \mathbf{A}\mathbf{y}$, where

$$\mathbf{A} = \begin{pmatrix} -\dfrac{k_1 + k_2}{m_1} & \dfrac{k_2}{m_1} \\[2ex] \dfrac{k_2}{m_2} & -\dfrac{k_2}{m_2} \end{pmatrix}$$

8. Show that, for any positive k_1, k_2, m_1, m_2, the eigenvalues of the matrix in Prob. 7 are negative real. What does this imply with respect to motions of the system?

Find the solutions of the differential equations in Prob. 7, assuming that

9. $m_1 = m_2 = 1$, $k_1 = 9$, $k_2 = 6$, $y_1(0) = -2$, $y_2(0) = 1$, $\dot{y}_1(0) = 0$, $\dot{y}_2(0) = 0$.

10. $m_1 = m_2 = 1$, $k_1 = 6$, $k_2 = 4$, $y_1(0) = 0$, $y_2(0) = 0$, $\dot{y}_1(0) = \sqrt{2}$, $\dot{y}_2(0) = 2\sqrt{2}$.

11. In Example 1, give a derivation of the eigenvalues and eigenvectors of the matrix $\tilde{\mathbf{A}}$.

12. **(Vibrations)** Show that the vertical vibrations of the mechanical system in the figure (masses of springs and damping neglected) are governed by the system of differential equations

$$\begin{aligned} \ddot{y}_1 &= -ky_1 + k(y_2 - y_1) \\ \ddot{y}_2 &= -k(y_2 - y_1) - ky_2 \end{aligned} \quad \text{thus} \quad \ddot{\mathbf{y}} = \mathbf{A}\mathbf{y} \quad \text{where} \quad \mathbf{A} = \begin{pmatrix} -2k & k \\ k & -2k \end{pmatrix}.$$

13. To solve the vector equation in Prob. 12, substitute $\mathbf{y} = \mathbf{x}e^{\omega t}$. Show that this leads to the eigenvalue problem $\mathbf{A}\mathbf{x} = \lambda\mathbf{x}$ where $\lambda = \omega^2$. Find the eigenvalues and eigenvectors. Find the solution of the vector equation that satisfies the initial conditions $y_1(0) = 1$, $y_2(0) = 1$, $\dot{y}_1(0) = \sqrt{3k}$, $\dot{y}_2(0) = -\sqrt{3k}$.

14. In Prob. 12, the eigenvalues are negative real. Show that this still holds if we replace the springs by springs with any (positive) spring constants k_1, k_2, k_3.

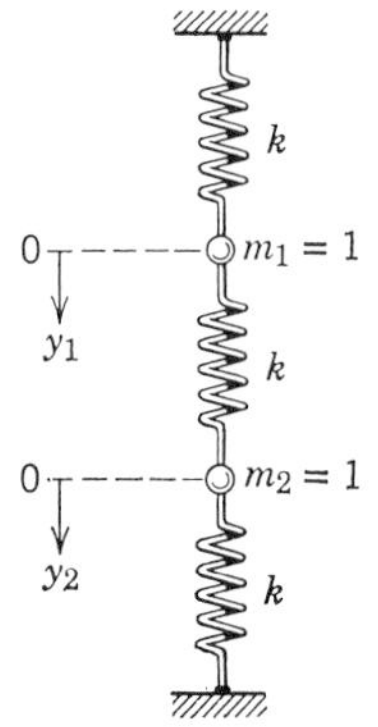

Problem 12 (System in static equilibrium)

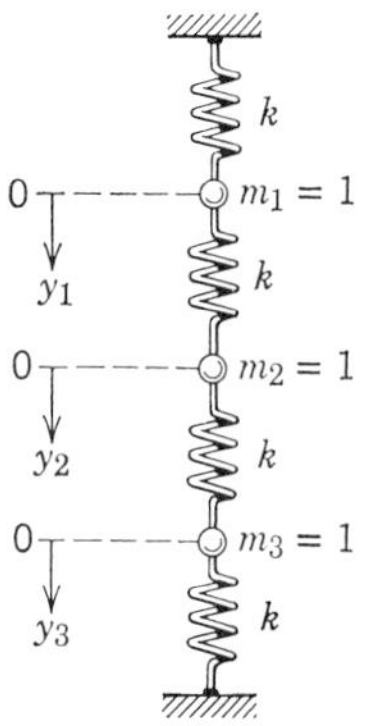

Problem 15 (System in static equilibrium)

15. Show that the vertical vibrations of the mechanical system in the figure are governed by $\ddot{\mathbf{y}} = \mathbf{A}\mathbf{y}$, where

$$\mathbf{y} = \begin{pmatrix} y_1 \\ y_2 \\ y_3 \end{pmatrix}, \qquad \mathbf{A} = \begin{pmatrix} -(k_1 + k_2) & k_2 & 0 \\ k_2 & -(k_2 + k_3) & k_5 \\ 0 & k_3 & -(k_3 + k_4) \end{pmatrix}$$

and substitution of $\mathbf{y} = \mathbf{x}e^{\omega t}$ yields $\mathbf{A}\mathbf{x} = \lambda\mathbf{x}$ where $\lambda = \omega^2$.

16. Verify that the coefficients of the characteristic equation

$$\lambda^3 + a\lambda^2 + b\lambda + c = 0,$$

corresponding to the matrix **A** in Prob. 15, are positive, and conclude that the eigenvalues of **A** must be negative real. What does this result mean with respect to the motion of the mechanical system in Prob. 15?

17. Let **A** be the matrix in Prob. 15, with $k_1 = k_2 = k_3 = k_4 = k$. Show that the corresponding matrix $\mathbf{B} = (1/k)\mathbf{A} + 2\mathbf{I}$ has the eigenvalues 0, $+\sqrt{2}$, and $-\sqrt{2}$. Conclude from this that **A** has the eigenvalues $-2k$, $(-2 + \sqrt{2})k$, and $(-2 - \sqrt{2})k$.

18. Give a detailed proof of Theorem 2.

19. Verify all the steps in Example 2.

20. Solve Example 2 by reducing the system to a single equation.

CHAPTER

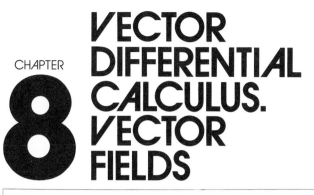

8 VECTOR DIFFERENTIAL CALCULUS. VECTOR FIELDS

Vector fields can be given in terms of vector functions (Sec. 8.1) and have various physical and geometrical applications. The basic concepts of differential calculus can be extended to vector functions in a simple and natural fashion (Sec. 8.2). Vector functions are useful for representing and investigating curves (Secs. 8.3–8.5) and their application in mechanics (Sec. 8.6). Physically and geometrically important concepts in connection with scalar and vector fields are the gradient (Sec. 8.8), divergence (Sec. 8.10), and curl (Sec. 8.11). (Corresponding integral theorems will be considered in Chap. 9.)

Prerequisite for this chapter: Chap. 6.
Sections which may be omitted in a shorter course: 8.4–8.6, 8.9.
References: Appendix 1, Part C.
Answers to problems: Appendix 2.

8.1 Scalar Fields and Vector Fields

A **scalar function** is a function which is defined at each point of a certain set of points in space and whose values are real numbers depending only on the points in space but not on the particular choice of the coordinate system. In most applications the domain of definition D of a scalar function f will be a curve, a surface, or a three-dimensional region in space. The function f associates with each point in D a scalar, a real number, and we say that a **scalar field** is given in D.

If we introduce coordinates x, y, z, then f may be represented in terms of the coordinates, and we write $f(x, y, z)$, but we should keep in mind that the value of f at any point P is independent of the particular choice of coordinates. In order to indicate this fact it is also customary to write $f(P)$ instead of $f(x, y, z)$. The function f may also depend on parameters such as time.

Example 1. Scalar function

The distance $f(P)$ of any point P from a fixed point P_0 in space is a scalar function whose domain of definition D is the whole space. $f(P)$ defines a scalar field in space. If we introduce a Cartesian coordinate system and P_0 has the coordinates x_0, y_0, z_0, then f is given by the well-known formula

$$f(P) = f(x, y, z) = \sqrt{(x - x_0)^2 + (y - y_0)^2 + (z - z_0)^2}$$

where x, y, z are the coordinates of P. If we replace the given Cartesian coordinate system by another such system, then the values of the coordinates of P and P_0 will in general change, but $f(P)$ will have the same value as before. Hence $f(P)$ is a scalar function. The direction cosines of the line through P and P_0 are not scalars because their values will depend on the choice of the coordinate system.

Example 2. Scalar fields

The temperature T within a body B is a scalar function. It defines a scalar field, namely, the temperature field in B. The function T may depend on the time or other parameters. Other examples of scalar fields are the pressure within a region through which a compressible fluid is flowing, and the density of the air of the earth's atmosphere. ∎

If to each point P of a certain set of points in space (for example, the points of a curve, a surface, or a three-dimensional region) a vector $\mathbf{v}(P)$ is assigned, then a **vector field** is said to be given at those points, and $\mathbf{v}(P)$ is called a **vector function.** Some illustrative examples are shown in Figs. 146–149.

If we introduce Cartesian coordinates x, y, z, then we may write

$$\mathbf{v}(x, y, z) = v_1(x, y, z)\mathbf{i} + v_2(x, y, z)\mathbf{j} + v_3(x, y, z)\mathbf{k},$$

but we should keep in mind that $\mathbf{v}$ depends only on the points of its domain of definition, and at any such point defines the same vector for every choice of the coordinate system.

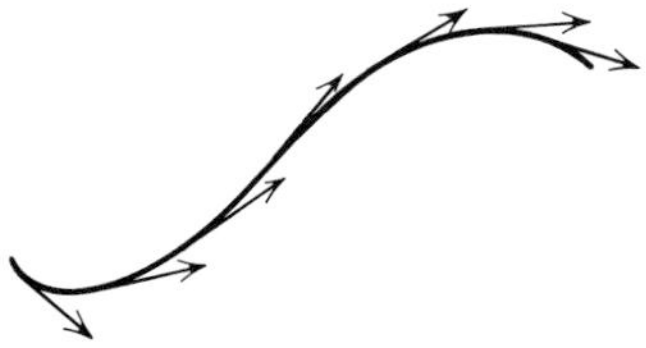

Fig. 146. Field of tangent vectors of a curve

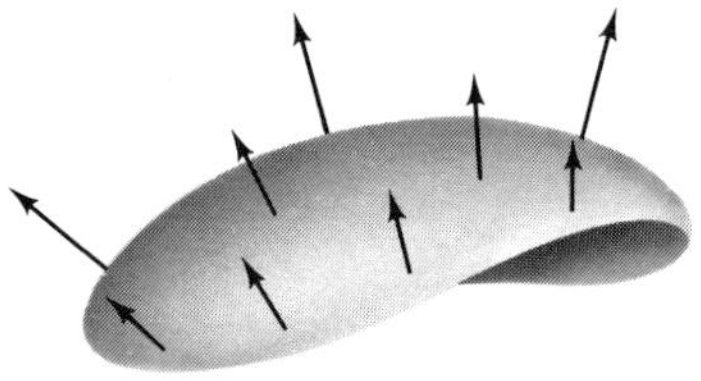

Fig. 147. Field of normal vectors of a surface

Example 3. Vector field (Velocity field)

At any instant the velocity vectors $\mathbf{v}(P)$ of a rotating body B constitute a vector field, the so-called *velocity field* of the rotation. If we introduce a Cartesian coordinate system having the origin on the axis of rotation, then (cf. Example 3 in Sec. 6.8)

$$\mathbf{v}(x, y, z) = \mathbf{w} \times (x\mathbf{i} + y\mathbf{j} + z\mathbf{k}) \tag{1}$$

where x, y, z are the coordinates of any point P of B at the instant under consideration. If the coordinates are such that the z-axis is the axis of rotation and $\mathbf{w}$ points in the positive z-direction, then $\mathbf{w} = \omega\mathbf{k}$ and

$$\mathbf{v} = \begin{vmatrix} \mathbf{i} & \mathbf{j} & \mathbf{k} \\ 0 & 0 & \omega \\ x & y & z \end{vmatrix} = \omega(-y\mathbf{i} + x\mathbf{j}).$$

An example of a rotating body and the corresponding velocity field are shown in Fig. 148.

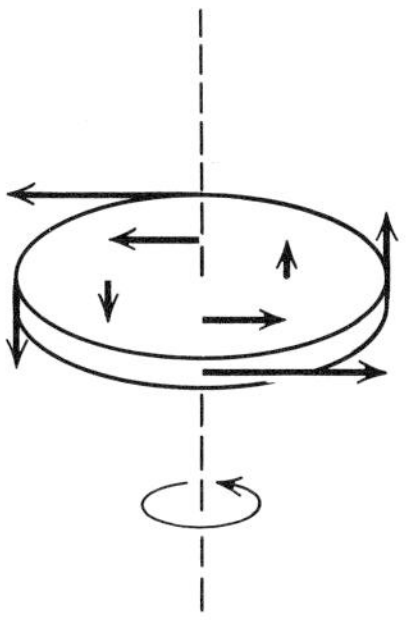

Fig. 148. Velocity field of a rotating body

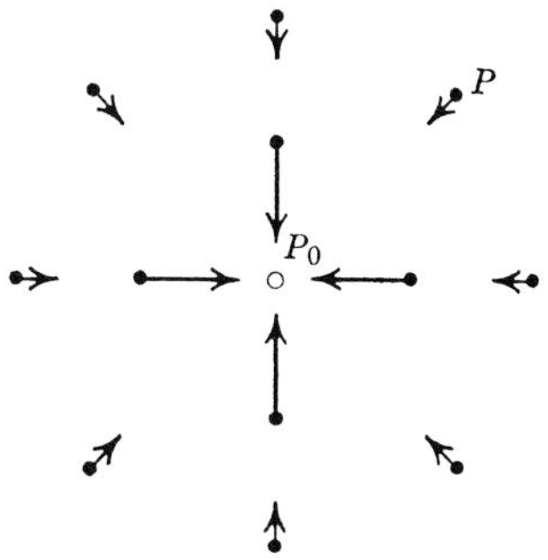

Fig. 149. Gravitational field

Example 4. Vector field (Field of force)

Let a particle A of mass M be fixed at a point P_0 and let a particle B of mass m to be free to take up various positions P in space. Then A attracts B. According to **Newton's law of gravitation** the corresponding gravitational force $\mathbf{p}$ is directed from P to P_0, and its magnitude is proportional to $1/r^2$ where r is the distance between P and P_0, say,

$$|\mathbf{p}| = \frac{c}{r^2}, \qquad c = GMm, \tag{2}$$

where $G (= 6.67 \cdot 10^{-8}\ \mathrm{cm^3/gm \cdot sec^2})$ is the gravitational constant. Hence $\mathbf{p}$ defines a vector field in space. If we introduce Cartesian coordinates such that P_0 has the coordinates x_0, y_0, z_0 and P has the coordinates x, y, z, then by the Pythagorean theorem,

$$r = \sqrt{(x - x_0)^2 + (y - y_0)^2 + (z - z_0)^2} \qquad (\geqq 0).$$

Assuming that $r > 0$ and introducing the vector

$$\mathbf{r} = (x - x_0)\mathbf{i} + (y - y_0)\mathbf{j} + (z - z_0)\mathbf{k},$$

we have $|\mathbf{r}| = r$, and $-\mathbf{r}/r$ is a unit vector in the direction of $\mathbf{p}$; the minus sign indicates that $\mathbf{p}$ is directed from P to P_0 (Fig. 149). From this and (2) we obtain

$$\mathbf{p} = |\mathbf{p}|\left(-\frac{\mathbf{r}}{r}\right) = -c\frac{\mathbf{r}}{r^3} = -c\frac{x - x_0}{r^3}\mathbf{i} - c\frac{y - y_0}{r^3}\mathbf{j} - c\frac{z - z_0}{r^3}\mathbf{k}. \tag{3}$$

This vector function describes the gravitational force acting on B.

Problems for Sec. 8.1

Determine the **level curves** $p(x, y) = const$ (curves of constant pressure or *isobars*) of the pressure fields in the xy-plane given by the following functions. Plot some of the isobars.

1. $p = y$ **2.** $p = xy$ **3.** $p = x^2 - y^2$

4. $p = \ln(x^2 + y^2)$ **5.** $p = \arctan\frac{y}{x}$ **6.** $p = e^x \cos y$

Find the isotherms (curves of constant temperature T) of the following temperature fields in the xy-plane.

7. $T = x + y$ **8.** $T = x - y$ **9.** $T = x^2 + 4y^2$

Consider the scalar field (temperature field) determined by $T(x, y) = 3x^2y - y^3$. Find:

10. The temperature at the points (1, 1), (−2, 2) and (3, 4).

11. The region in which $T > 0$. (Graph the region.)

12. T on the lines $y = x$ and $y = -x$. (Graph these two functions.)

13. A formula for T on the unit circle $x^2 + y^2 = 1$.

14. Sketch a figure of $T(x, y)$ as a surface in space.

Find the **level surfaces** (surfaces $f = const$) of the scalar fields in space given by the following functions.

15. $f = x + y + z$

16. $f = x + y$

17. $f = x^2 + y^2 + z^2$

18. $f = (x^2 + y^2 + z^2)^{-1}$

19. $f = x^2 + y^2 - z$

20. $f = x^2 + 2y^2 + 4z^2$

Consider the pressure field in space given by $p(x, y, z) = 3x^2 + y^2 + z^2$. Find:

21. The level surfaces. (Graph some of them.)

22. A formula for the pressure on the surface $xyz = 1$.

23. The level curves in the plane $z = 1$. (Graph some of them.)

24. The region in which $3 \leqq p(x, y, z) \leqq 12$.

Draw figures (similar to Fig. 149) of the vector fields in the xy-plane given by the following vector functions **v**.

25. $\mathbf{v} = 2\mathbf{i}$

26. $\mathbf{v} = 3\mathbf{i} - 2\mathbf{j}$

27. $\mathbf{v} = x\mathbf{i} + y\mathbf{j}$

28. $\mathbf{v} = y\mathbf{i} + x\mathbf{j}$

29. $\mathbf{v} = 6xy\mathbf{i} + 3(x^2 - y^2)\mathbf{j}$

30. $\mathbf{v} = 2x\mathbf{i} - 2y\mathbf{j}$

Find and graph the curves on which **v** has constant length and the curves on which **v** has constant direction, where

31. $\mathbf{v} = x\mathbf{i} + 2y\mathbf{j}$

32. $\mathbf{v} = (x^2 - y^2)\mathbf{i} + 2xy\mathbf{j}$

33. $\mathbf{v} = y\mathbf{i} + x^2\mathbf{j}$

34. $\mathbf{v} = (x - y)\mathbf{i} + (x + y)\mathbf{j}$

35. Find the surfaces on which $\mathbf{v} = 3x\mathbf{i} + 2z\mathbf{j} + y\mathbf{k}$ has constant length.

8.2 Vector Calculus

The basic concepts of calculus, such as convergence, continuity, and differentiability, can be introduced to vector analysis in a simple and natural way as follows.

An infinite sequence of vectors $\mathbf{a}_{(n)}, n = 1, 2, \dots$, is said to **converge** if there is a vector **a** such that

$$\lim_{n\to\infty} |\mathbf{a}_{(n)} - \mathbf{a}| = 0. \tag{1}$$

a is called the **limit vector** of that sequence, and we write

$$\lim_{n\to\infty} \mathbf{a}_{(n)} = \mathbf{a}. \tag{2}$$

Clearly, if a Cartesian coordinate system has been introduced, then that sequence of vectors converges to **a** if and only if the three sequences of the components of the vectors converge to the corresponding components of **a**. The simple proof is left to the reader.

Similarly, a vector function $\mathbf{u}(t)$ of a real variable t is said to have the **limit l** as t approaches t_0, if $\mathbf{u}(t)$ is defined in some *neighborhood*[1] of t_0 (possibly except at t_0) and

$$\lim_{t\to t_0} |\mathbf{u}(t) - \mathbf{l}| = 0. \tag{3}$$

Then we write

$$\lim_{t\to t_0} \mathbf{u}(t) = \mathbf{l}. \tag{4}$$

A vector function $\mathbf{u}(t)$ is said to be **continuous** at $t = t_0$ if it is defined in some neighborhood of t_0 and

$$\lim_{t\to t_0} \mathbf{u}(t) = \mathbf{u}(t_0). \tag{5}$$

If we introduce a Cartesian coordinate system, we may represent $\mathbf{u}(t)$ in the form

$$\mathbf{u}(t) = u_1(t)\mathbf{i} + u_2(t)\mathbf{j} + u_3(t)\mathbf{k}.$$

Then $\mathbf{u}(t)$ is continuous at t_0 if and only if its three components are continuous at t_0.

A vector function $\mathbf{u}(t)$ is said to be **differentiable** at a point t if the limit

$$\mathbf{u}'(t) = \lim_{\Delta t\to 0} \frac{\mathbf{u}(t + \Delta t) - \mathbf{u}(t)}{\Delta t} \tag{6}$$

exists. The vector $\mathbf{u}'(t)$ is called the **derivative** of $\mathbf{u}(t)$. Cf. Fig. 150. (The curve in this figure is the locus of the heads of the arrows representing $\mathbf{u}$ for values of the independent variable in some interval containing t and $t + \Delta t$.)

Suppose that a Cartesian coordinate system has been introduced. Then $\mathbf{u}(t)$ is differentiable at a point t if and only if its three components are differentiable at that t, that is, if and only if the derivatives

$$u_m'(t) = \lim_{\Delta t\to 0} \frac{u_m(t + \Delta t) - u_m(t)}{\Delta t} \qquad (m = 1, 2, 3)$$

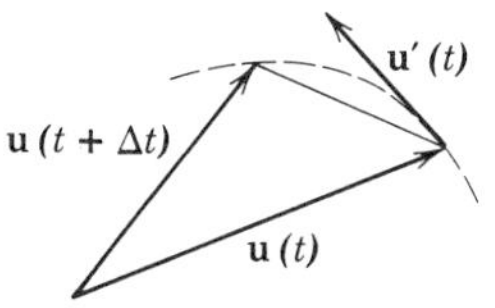

Fig. 150. Derivative of a vector function

[1] That is, in some interval on the t-axis containing t_0 as an interior point.

exist. Then

$$\mathbf{u}'(t) = u_1'(t)\mathbf{i} + u_2'(t)\mathbf{j} + u_3'(t)\mathbf{k}; \tag{7}$$

that is, *to differentiate a vector function one differentiates each component separately.*

The familiar rules of differentiation yield corresponding rules for differentiating vector functions, for example,

$$(c\mathbf{u})' = c\mathbf{u}' \ (c \text{ constant}), \qquad (\mathbf{u} + \mathbf{v})' = \mathbf{u}' + \mathbf{v}'$$

and in particular

$$(\mathbf{u} \cdot \mathbf{v})' = \mathbf{u}' \cdot \mathbf{v} + \mathbf{u} \cdot \mathbf{v}' \tag{8}$$

$$(\mathbf{u} \times \mathbf{v})' = \mathbf{u}' \times \mathbf{v} + \mathbf{u} \times \mathbf{v}' \tag{9}$$

$$(\mathbf{u}\ \mathbf{v}\ \mathbf{w})' = (\mathbf{u}'\ \mathbf{v}\ \mathbf{w}) + (\mathbf{u}\ \mathbf{v}'\ \mathbf{w}) + (\mathbf{u}\ \mathbf{v}\ \mathbf{w}'). \tag{10}$$

The simple proofs are left to the reader. In (9), the order of the vectors must be carefully observed because cross multiplication is not commutative.

Example 1. Derivative of a vector function of constant length

Let $\mathbf{u}(t)$ be a vector function whose length is constant, say, $|\mathbf{u}(t)| = c$. Then $|\mathbf{u}|^2 = \mathbf{u} \cdot \mathbf{u} = c^2$, and $(\mathbf{u} \cdot \mathbf{u})' = 2\mathbf{u} \cdot \mathbf{u}' = 0$, by differentiation [cf. (8)]. This yields the following result. *The derivative of a vector function* $\mathbf{u}(t)$ *of constant length is either the zero vector or is perpendicular to* $\mathbf{u}(t)$. ▮

Important applications of derivatives will be discussed in the following sections.

From our considerations the way of introducing partial differentiation to vector analysis is obvious. Let the components of a vector function

$$\mathbf{u} = u_1\mathbf{i} + u_2\mathbf{j} + u_3\mathbf{k}$$

be differentiable functions of n variables $t_1, \cdots, t_n$. Then the **partial derivative** of $\mathbf{u}$ with respect to t_l is denoted by $\partial\mathbf{u}/\partial t_l$ and is defined as the vector function

$$\frac{\partial \mathbf{u}}{\partial t_l} = \frac{\partial u_1}{\partial t_l}\mathbf{i} + \frac{\partial u_2}{\partial t_l}\mathbf{j} + \frac{\partial u_3}{\partial t_l}\mathbf{k}.$$

Similarly,

$$\frac{\partial^2 \mathbf{u}}{\partial t_l \partial t_m} = \frac{\partial^2 u_1}{\partial t_l \partial t_m}\mathbf{i} + \frac{\partial^2 u_2}{\partial t_l \partial t_m}\mathbf{j} + \frac{\partial^2 u_3}{\partial t_l \partial t_m}\mathbf{k},$$

and so on.

Example 2. Partial derivatives

Let $\mathbf{r}(t_1, t_2) = a \cos t_1 \mathbf{i} + a \sin t_1 \mathbf{j} + t_2\mathbf{k}$. Then

$$\frac{\partial \mathbf{r}}{\partial t_1} = -a \sin t_1 \mathbf{i} + a \cos t_1 \mathbf{j}, \qquad \frac{\partial \mathbf{r}}{\partial t_2} = \mathbf{k}.$$

Note that if $\mathbf{r}(t_1, t_2)$ is interpreted as a position vector, it represents a cylinder of revolution of radius a, having the z-axis as axis of rotation. (Representations of surfaces will be considered in Sec. 9.5.)

Problems for Sec. 8.2

Find $\mathbf{u}'$, $|\mathbf{u}'|$, $\mathbf{u}''$, and $|\mathbf{u}''|$, where $\mathbf{u}$ equals

1. $\mathbf{a} + \mathbf{b}t$
2. $t\mathbf{i} + t^2\mathbf{j}$
3. $\cos t\,\mathbf{i} + \sin t\,\mathbf{j}$
4. $\cos t\,\mathbf{i} + 4\sin t\,\mathbf{j} + \mathbf{k}$
5. $t\mathbf{i} + t^2\mathbf{j} + t^3\mathbf{k}$
6. $2\cos t\,\mathbf{i} + 2\sin t\,\mathbf{j} + t\,\mathbf{k}$
7. $e^t\mathbf{i} + e^{-t}\mathbf{j}$
8. $e^{-t}(\cos t\,\mathbf{i} + \sin t\,\mathbf{j})$
9. $\sin 3t\,(\mathbf{i} + \mathbf{j})$

Let $\mathbf{u} = t\mathbf{i} + 2t^2\mathbf{k}$, $\mathbf{v} = t^3\mathbf{j} + t\mathbf{k}$, and $\mathbf{w} = \mathbf{i} + t\mathbf{j} + t^2\mathbf{k}$. Find

10. $(\mathbf{u} \cdot \mathbf{v})'$
11. $(\mathbf{u} \times \mathbf{v})'$
12. $(\mathbf{u}\ \mathbf{v}\ \mathbf{w})'$
13. $[(\mathbf{u} \times \mathbf{v}) \times \mathbf{w}]'$
14. $[\mathbf{u} \times (\mathbf{v} \times \mathbf{w})]'$
15. $[(\mathbf{u} + \mathbf{v}) \cdot \mathbf{w}]'$

In each case find the first partial derivatives with respect to x, y, z.

16. $x\mathbf{i} + 2y\mathbf{j}$
17. $(x^2 - y^2)\mathbf{i} + 2xy\mathbf{j}$
18. $x^2\mathbf{i} + y^2\mathbf{j} + z^2\mathbf{k}$
19. $yz\mathbf{i} + zx\mathbf{j} + xy\mathbf{k}$
20. $(x + y)\mathbf{j} + (x - y)\mathbf{k}$
21. $x^2y\mathbf{i} + y^2z\mathbf{j} + z^2x\mathbf{k}$

22. Using (8) in Sec. 6.5, prove (8). Prove (9).
23. Derive (10) from (8) and (9).
24. Give a direct proof of (10).
25. Find formulas similar to (8) and (9) for $(\mathbf{u} \cdot \mathbf{v})''$ and $(\mathbf{u} \times \mathbf{v})''$.
26. Show that if $\mathbf{u}(t)$ is a unit vector and $\mathbf{u}'(t) \neq \mathbf{0}$, then $\mathbf{u}$ and $\mathbf{u}'$ are orthogonal.
27. Show that the equation $\mathbf{u}'(t) = \mathbf{c}$ has the solution $\mathbf{u}(t) = \mathbf{c}t + \mathbf{b}$ where $\mathbf{c}$ and $\mathbf{b}$ are constant vectors.
28. Show that $\mathbf{u}(t) = \mathbf{b}e^{\lambda t} + \mathbf{c}e^{-\lambda t}$ satisfies the equation $\mathbf{u}'' - \lambda^2\mathbf{u} = \mathbf{0}$. ($\mathbf{b}$ and $\mathbf{c}$ are constant vectors.)
29. Show that $\left(\dfrac{\mathbf{u}}{|\mathbf{u}|}\right)' = \dfrac{\mathbf{u}'(\mathbf{u} \cdot \mathbf{u}) - \mathbf{u}(\mathbf{u} \cdot \mathbf{u}')}{(\mathbf{u} \cdot \mathbf{u})^{3/2}}$.
30. Differentiate $(t\mathbf{i} + t^2\mathbf{j})/|t\mathbf{i} + t^2\mathbf{j}|$.

8.3 Curves

As an important application of vector calculus, let us now consider some basic facts about curves in space. The student will know that curves occur in many considerations in calculus as well as in physics, for example, as paths of moving particles. The consideration will be a part of an important branch of mathematics, which is called **differential geometry** and which may be defined as the study of curves and surfaces by means of calculus. Cf. Ref. [C8] in Appendix 1.

A Cartesian coordinate system being given, we may represent a curve C by a vector function (Fig. 151 on the next page)

$$\mathbf{r}(t) = x(t)\mathbf{i} + y(t)\mathbf{j} + z(t)\mathbf{k}; \tag{1}$$

to each value t_0 of the real variable t there corresponds a point of C having the position vector $\mathbf{r}(t_0)$, that is, the coordinates $x(t_0)$, $y(t_0)$, $z(t_0)$.

A representation of the form (1) is called a **parametric representation** of the curve C, and t is called the *parameter* of this representation. This type of representation is useful in many applications, for example, in mechanics where the variable t may be time.

Other types of representations of curves in space are

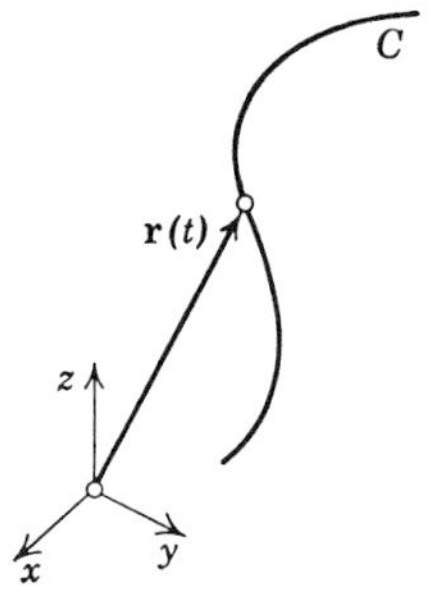

Fig. 151. Parametric representation of a curve

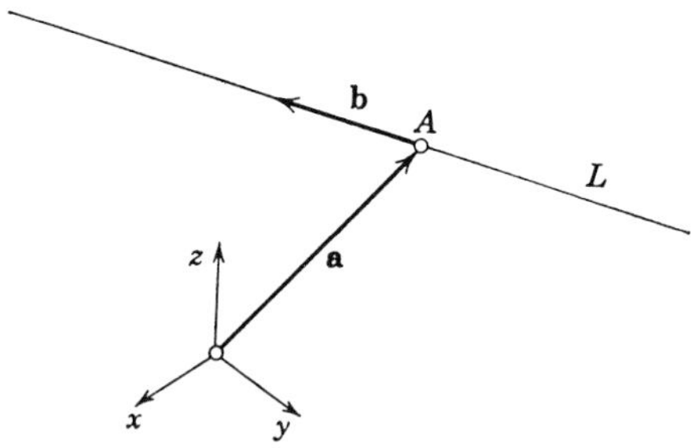

Fig. 152. Parametric representation of a straight line

$$y = f(x), \qquad z = g(x) \tag{2}$$

and

$$F(x, y, z) = 0, \qquad G(x, y, z) = 0. \tag{3}$$

By setting $x = t$ we may write (2) in the form (1), namely

$$\mathbf{r}(t) = t\mathbf{i} + f(t)\mathbf{j} + g(t)\mathbf{k}.$$

In (3), each equation represents a surface, and the curve is the intersection of the two surfaces.

A **plane curve** is a curve which lies in a plane in space. A curve which is not a plane curve is called a **twisted curve.**

Example 1. Straight line

Any straight line L can be represented in the form

$$\mathbf{r}(t) = \mathbf{a} + t\mathbf{b} = (a_1 + tb_1)\mathbf{i} + (a_2 + tb_2)\mathbf{j} + (a_3 + tb_3)\mathbf{k} \tag{4}$$

where **a** and **b** are constant vectors. L passes through the point A with position vector $\mathbf{r} = \mathbf{a}$ and has the direction of **b** (Fig. 152). If **b** is a unit vector, its components are the *direction cosines* of L, and in this case, $|t|$ measures the distance of the points of L from A.

Example 2. Ellipse, circle

The vector function

$$\mathbf{r}(t) = a \cos t\, \mathbf{i} + b \sin t\, \mathbf{j} \tag{5}$$

represents an ellipse in the xy-plane with center at the origin and principal axes in the direction of the x and y axes. In fact, since $\cos^2 t + \sin^2 t = 1$, we obtain from (5)

$$\frac{x^2}{a^2} + \frac{y^2}{b^2} = 1, \qquad z = 0.$$

If $b = a$, then (5) represents a *circle* of radius a.

Example 3. Circular helix

The twisted curve C represented by the vector function

$$\mathbf{r}(t) = a \cos t\, \mathbf{i} + a \sin t\, \mathbf{j} + ct\mathbf{k} \qquad (c \neq 0) \tag{6}$$

is called a *circular helix*. It lies on the cylinder $x^2 + y^2 = a^2$. If $c > 0$, the helix is shaped like a right-handed screw (Fig. 153). If $c < 0$, it looks like a left-handed screw (Fig. 154). ∎

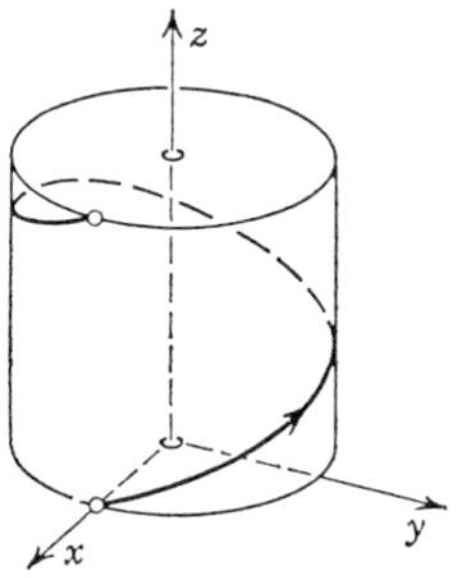

Fig. 153. Right-handed circular helix

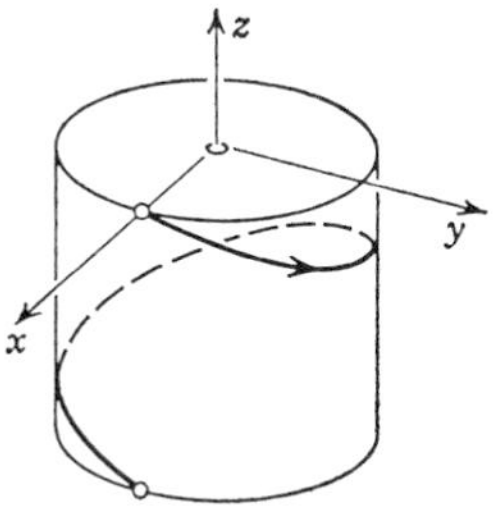

Fig. 154. Left-handed circular helix

The portion between any two points of a curve is often called an **arc** *of a curve*. For the sake of simplicity we shall use the single term "curve" to denote an entire curve as well as an arc of a curve.

A curve may have self-intersections; the points of intersection are called *multiple points* of the curve. Two examples are shown in Fig. 155. A curve having no multiple points is called a **simple curve.**

Example 4. Simple and nonsimple curves

Ellipses and helices are simple curves. The curve represented by

$$\mathbf{r}(t) = (t^2 - 1)\mathbf{i} + (t^3 - t)\mathbf{j}$$

is not simple since it has a double point at the origin; this point corresponds to the two values $t = 1$ and $t = -1$. ∎

We finally mention that a given curve C may be represented by various vector functions. For example, if C is represented by (1) and we set $t = h(t^*)$, then we obtain a new vector function $\tilde{\mathbf{r}}(t^*) = \mathbf{r}(h(t^*))$ representing C, provided $h(t^*)$ takes on all the values of t occurring in (1).

Example 5. Change of parameter

The parabola $y = x^2$ in the xy-plane may be represented by the vector function

$$\mathbf{r}(t) = t\mathbf{i} + t^2\mathbf{j} \qquad (-\infty < t < \infty).$$

If we set $t = -2t^*$, we obtain another representation of the parabola:

$$\tilde{\mathbf{r}}(t^*) = \mathbf{r}(-2t^*) = -2t^*\mathbf{i} + 4t^{*2}\mathbf{j}.$$

If we set $t = t^{*2}$, we obtain

$$\tilde{\mathbf{r}}(t^*) = t^{*2}\mathbf{i} + t^{*4}\mathbf{j},$$

but this function represents only the portion of the parabola in the first quadrant, because $t^{*2} \geqq 0$ for all t^*.

Fig. 155. Curves having double points

Problems for Sec. 8.3

Find a parametric representation of the straight line through a point A in the direction of a vector $\mathbf{b}$, where

1. A: $(0, 0, 0)$, $\mathbf{b} = \mathbf{i} + \mathbf{j}$

2. A: $(1, 3, 2)$, $\mathbf{b} = -\mathbf{i} + \mathbf{k}$

3. A: $(2, 1, 0)$, $\mathbf{b} = 2\mathbf{j} + \mathbf{k}$

4. A: $(0, 4, 1)$, $\mathbf{b} = \mathbf{i} - \mathbf{j} + 2\mathbf{k}$

Find a parametric representation of the straight line through the points A and B, where

5. A: $(0, 0, 0)$, B: $(1, 1, 1)$

6. A: $(-1, 8, 3)$, B: $(1, 0, 0)$

7. A: $(1, 5, 3)$, B: $(0, 2, -1)$

8. A: $(1, 4, 2)$, B: $(1, 4, -2)$

Find a parametric representation of the straight line represented by

9. $y = x, \quad z = 0$

10. $7x - 3y + z = 14, \; 4x - 3y - 2z = -1$

11. $x + y = 0, \quad x - z = 0$

12. $x + y + z = 1, \quad y - z = 0$

Represent the following curves in parametric form and sketch these curves.

13. $x^2 + y^2 = 1, \quad z = 0$

14. $y = x^4, \quad z = 0$

15. $y = x^2, \quad z = x^3$

16. $x^2 + y^2 - 2x - 4y = -1, \; z = 0$

17. $4(x + 1)^2 + y^2 = 4, \quad z = 0$

18. $x^2 + y^2 = 4, \quad z = e^x$

19. Determine the orthogonal projections of the circular helix (6) in the coordinate planes.

Sketch figures of the curves represented by the following functions $\mathbf{r}(t)$.

20. $t\mathbf{i} + 2t\mathbf{j} - t\mathbf{k}$

21. $2 \sin t\,\mathbf{i} + 2 \cos t\,\mathbf{j}$

22. $(1 + \cos t)\mathbf{i} + \sin t\,\mathbf{j}$

23. $\cos t\,\mathbf{i} + 2 \sin t\,\mathbf{j}$

24. $\cos t\,\mathbf{i} + \sin t\,\mathbf{j} + t\,\mathbf{k}$

25. $t\mathbf{i} + t^2\mathbf{j} + t^3\mathbf{k}$

8.4 Arc Length

To define the length of a simple curve C we may proceed as follows. We inscribe in C a broken line of n chords joining the two endpoints of C as shown in Fig. 156. This we do for each positive integer n in an arbitrary way but such that the maximum chord-length approaches zero as n approaches infinity. The lengths of these lines of chords can be obtained from the theorem of Pythagoras. If the sequence of these lengths $l_1, l_2, \cdots$ is convergent with limit l, then C is said to be **rectifiable**, and l is called the **length** of C.

If C is not simple but consists of finitely many rectifiable simple curves, the

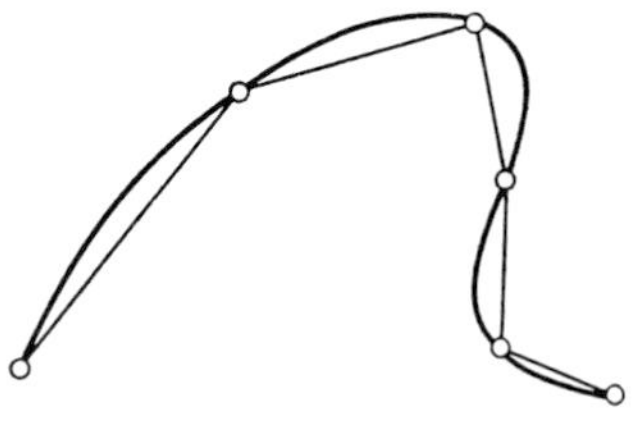

Fig. 156. Length of a curve

length of C is defined to be the sum of the lengths of those curves.

If C can be represented by a continuously differentiable[2] vector function

$$\mathbf{r} = \mathbf{r}(t) \qquad (a \leqq t \leqq b),$$

then it can be shown that C is rectifiable, and its length l is given by the integral

$$\textbf{(1)} \qquad l = \int_a^b \sqrt{\dot{\mathbf{r}} \cdot \dot{\mathbf{r}}}\, dt \qquad \left(\dot{\mathbf{r}} = \frac{d\mathbf{r}}{dt}\right),$$

whose value is independent of the choice of the parametric representation. The proof is quite similar to that for plane curves usually considered in elementary integral calculus (cf. Ref. [A14]) and can be found in Ref. [C8] in Appendix 1.

If we replace the fixed upper limit b in (1) with a variable upper limit t, the integral becomes a function of t, say, $s(t)$; denoting the variable of integration by t^*, we have

$$\textbf{(2)} \qquad s(t) = \int_a^t \sqrt{\dot{\mathbf{r}} \cdot \dot{\mathbf{r}}}\, dt^* \qquad \left(\dot{\mathbf{r}} = \frac{d\mathbf{r}}{dt^*}\right).$$

This function $s(t)$ is called the *arc length function* or, simply, the **arc length** of C.

From our consideration it follows that, geometrically, for a fixed $t = t_0 \geqq a$, the arc length $s(t_0)$ is the length of the portion of C between the points corresponding to $t = a$ and $t = t_0$. For $t = t_0 < a$ we have $s(t_0) < 0$ and that length is $-s(t_0)$.

The arc length s may serve as a parameter in parametric representations of curves. We shall see that this leads to a simplification of various formulas.

The constant a in (2) may be replaced by another constant; that is, the point of the curve corresponding to $s = 0$ may be chosen in an arbitrary manner. The sense corresponding to increasing values of s is called the **positive sense** on C; in this fashion any representation $\mathbf{r}(s)$ or $\mathbf{r}(t)$ of C defines a certain **orientation** of C. Obviously, there are two ways of *orienting* C, and it is not difficult to see that the transition from one orientation to the opposite orientation can be effected by a transformation of the parameter whose derivative is negative.

From (2) we obtain

$$(3) \qquad \left(\frac{ds}{dt}\right)^2 = \frac{d\mathbf{r}}{dt} \cdot \frac{d\mathbf{r}}{dt} = \left(\frac{dx}{dt}\right)^2 + \left(\frac{dy}{dt}\right)^2 + \left(\frac{dz}{dt}\right)^2.$$

It is customary to write

$$d\mathbf{r} = dx\,\mathbf{i} + dy\,\mathbf{j} + dz\,\mathbf{k}$$

and

$$\textbf{(4)} \qquad ds^2 = d\mathbf{r} \cdot d\mathbf{r} = dx^2 + dy^2 + dz^2.$$

ds is called the **linear element** of C.

[2] "*Continuously differentiable*" means that the derivative exists and is continuous; "*twice continuously differentiable*" means that the first and second derivatives exist and are continuous, and so on.

Example 1. Circle. Arc length as parameter
In the case of the circle

$$\mathbf{r}(t) = a \cos t\, \mathbf{i} + a \sin t\, \mathbf{j}$$

we have $\dot{\mathbf{r}} = -a \sin t\, \mathbf{i} + a \cos t\, \mathbf{j}$, $\dot{\mathbf{r}} \cdot \dot{\mathbf{r}} = a^2$, and therefore,

$$s(t) = \int_0^t a\, dt^* = at.$$

Hence $t(s) = s/a$ and a representation of the circle with the arc length s as parameter is

$$\mathbf{r}\left(\frac{s}{a}\right) = a \cos \frac{s}{a}\, \mathbf{i} + a \sin \frac{s}{a}\, \mathbf{j}.$$

The circle is oriented in the counterclockwise sense, which corresponds to increasing values of s. Setting $s = -\tilde{s}$ and remembering that $\cos(-\alpha) = \cos \alpha$ and $\sin(-\alpha) = -\sin \alpha$, we obtain

$$\mathbf{r}\left(-\frac{\tilde{s}}{a}\right) = a \cos \frac{\tilde{s}}{a}\, \mathbf{i} - a \sin \frac{\tilde{s}}{a}\, \mathbf{j};$$

we have $ds/d\tilde{s} = -1 < 0$, and the circle is now oriented in the clockwise sense.

Problems for Sec. 8.4

Graph the following curves and find their lengths.

1. *Catenary* $y = \cosh x$, $z = 0$, from $x = 0$ to $x = 1$
2. *Circular helix* $\mathbf{r}(t) = a \cos t\, \mathbf{i} + a \sin t\, \mathbf{j} + ct\mathbf{k}$ from $(a, 0, 0)$ to $(a, 0, 2\pi c)$
3. *Semicubical parabola* $y = x^{3/2}$, $z = 0$, from $(0, 0, 0)$ to $(4, 8, 0)$
4. Four-cusped *hypocycloid* $\mathbf{r}(t) = a \cos^3 t\, \mathbf{i} + a \sin^3 t\, \mathbf{j}$, total length
5. *Involute of circle* $\mathbf{r}(t) = (\cos t + t \sin t)\mathbf{i} + (\sin t - t \cos t)\mathbf{j}$ from $(1, 0, 0)$ to $(-1, \pi, 0)$
6. $\mathbf{r}(t) = e^t \cos t\, \mathbf{i} + e^t \sin t\, \mathbf{j}$, $0 \leqq t \leqq \pi/2$

7. If a plane curve is represented in the form $y = f(x)$, $z = 0$, using (1) show that its length between $x = a$ and $x = b$ is

$$l = \int_a^b \sqrt{1 + y'^2}\, dx.$$

8. Using the formula in Prob. 7, find the length of a circle of radius a.
9. Show that if a plane curve is represented in polar coordinates $\rho = \sqrt{x^2 + y^2}$ and $\theta = \arctan(y/x)$, then $ds^2 = \rho^2\, d\theta^2 + d\rho^2$.

Using the formula in Prob. 9, find the lengths of the following curves.

10. Circle of radius a, total length
11. $\rho = e^\theta$, $0 \leqq \theta \leqq \pi$
12. $\rho = \theta^2$, $0 \leqq \theta \leqq \pi/2$
13. *Cardioid* $\rho = a(1 - \cos \theta)$. (Graph this curve.)
14. $\rho = 1 + \cos \theta$, $0 \leqq \theta \leqq \pi/2$

15. If a curve is represented by a parametric representation, show that a transformation of the parameter whose derivative is negative reverses the orientation.

8.5 Tangent. Curvature and Torsion

The *tangent* to a curve C at a point P of C is defined as the limiting position of the straight line L through P and another point Q of C as Q approaches P along the curve (Fig. 157).

Suppose that C is represented by a continuously differentiable vector function $\mathbf{r}(t)$ where t is any parameter. Let P and Q correspond to t and $t + \Delta t$, respectively. Then L has the direction of the vector

$$[\mathbf{r}(t + \Delta t) - \mathbf{r}(t)]/\Delta t.$$

Hence, if the vector

$$\dot{\mathbf{r}} = \lim_{\Delta t \to 0} \frac{\mathbf{r}(t + \Delta t) - \mathbf{r}(t)}{\Delta t} \tag{1}$$

is not the zero vector, it has the direction of the tangent to C at P. It points in the direction of increasing values of t, and its sense, therefore, depends on the orientation of the curve. $\dot{\mathbf{r}}$ is called a *tangent vector* of C at P, and the corresponding unit vector

$$\mathbf{u} = \frac{\dot{\mathbf{r}}}{|\dot{\mathbf{r}}|} \tag{2}$$

is called the **unit tangent vector** to C at P.

If in particular C is represented by $\mathbf{r}(s)$, where s is the arc length, it follows from (3), Sec. 8.4, that the derivative $d\mathbf{r}/ds$ is a unit vector, and (2) becomes

$$\mathbf{u} = \mathbf{r}' = \frac{d\mathbf{r}}{ds}. \tag{3}$$

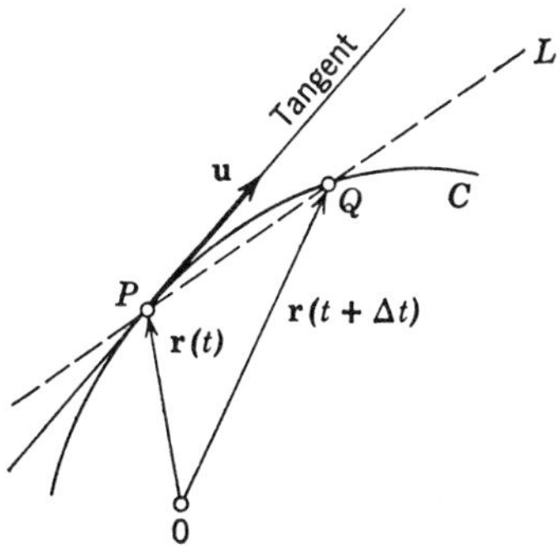

Fig. 157. Tangent to a curve

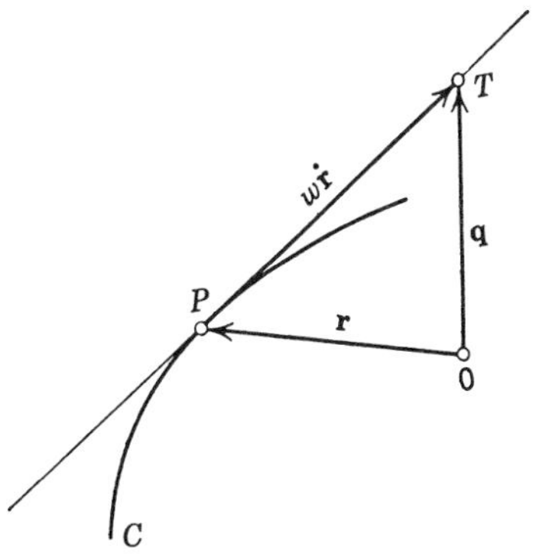

Fig. 158. Representation of the tangent to a curve

Clearly the position vector of a point T on the tangent is the sum of the position vector $\mathbf{r}$ of P and a vector in the direction of the tangent. Hence a parametric representation of the tangent is (Fig. 158)

$$\mathbf{q}(w) = \mathbf{r} + w\dot{\mathbf{r}} \tag{4}$$

where both $\mathbf{r}$ and $\dot{\mathbf{r}}$ depend on P and the parameter w is a real variable.

Consider a given curve C, represented by a three times continuously differentiable vector function $\mathbf{r}(s)$ (cf. footnote 2 in Sec. 8.4) where s is the arc length. Then

$$\kappa(s) = |\mathbf{u}'(s)| = |\mathbf{r}''(s)| \qquad (\kappa \geqq 0) \tag{5}$$

is called the **curvature** of C. If $\kappa \neq 0$, the unit vector $\mathbf{p}$ in the direction of $\mathbf{u}'(s)$ is

$$\mathbf{p} = \frac{\mathbf{u}'}{\kappa} \qquad (\kappa > 0) \tag{6}$$

and is called the **unit principal normal vector** of C. From Example 1 in Sec. 8.2 we see that $\mathbf{p}$ is perpendicular to $\mathbf{u}$. The vector

$$\mathbf{b} = \mathbf{u} \times \mathbf{p} \qquad (\kappa > 0) \tag{7}$$

is called the **unit binormal vector** of C. From the definition of a vector product it follows that $\mathbf{u}, \mathbf{p}, \mathbf{b}$ constitute a right-handed triple of orthogonal unit vectors (Secs. 6.3 and 6.7). This triple is called the **trihedron** of C at the point under consideration (Fig. 159). The three straight lines through that point in the directions of $\mathbf{u}, \mathbf{p}, \mathbf{b}$ are called the **tangent,** the **principal normal,** and the **binormal** of C. Figure 159 also shows the names of the three planes spanned by each pair of those vectors.

If the derivative $\mathbf{b}'$ is not the zero vector, it is perpendicular to $\mathbf{b}$ (cf. Example 1 in Sec. 8.2). It is also perpendicular to $\mathbf{u}$. In fact, by differentiating $\mathbf{b} \cdot \mathbf{u} = 0$ we have $\mathbf{b}' \cdot \mathbf{u} + \mathbf{b} \cdot \mathbf{u}' = 0$; hence $\mathbf{b}' \cdot \mathbf{u} = 0$ because $\mathbf{b} \cdot \mathbf{u}' = 0$. Consequently, $\mathbf{b}'$ is of the form $\mathbf{b}' = \alpha\mathbf{p}$ where α is a scalar. It is customary to set $\alpha = -\tau$. Then

$$\mathbf{b}' = -\tau\mathbf{p} \qquad (\kappa > 0). \tag{8}$$

The scalar function τ is called the **torsion** of C. Scalar multiplication of both sides of (8) by $\mathbf{p}$ yields

$$\tau(s) = -\mathbf{p}(s) \cdot \mathbf{b}'(s). \tag{9}$$

The concepts just introduced are basic in the theory and application of curves. Let us illustrate them by a typical example. Further applications will follow later.

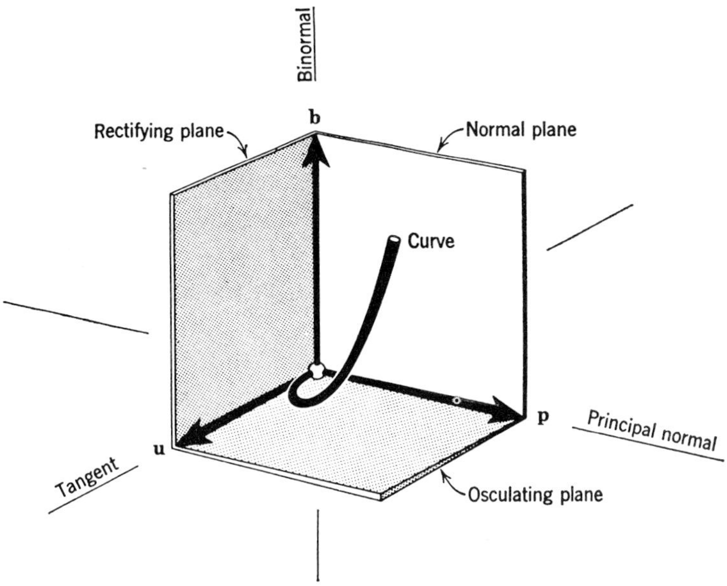

Fig. 159. Trihedron

Example 1. Circular helix

In the case of the circular helix (6) in Sec. 8.3 we obtain the arc length $s = t\sqrt{a^2 + c^2}$. Hence we may represent the helix in the form

$$\mathbf{r}(s) = a\cos\frac{s}{K}\mathbf{i} + a\sin\frac{s}{K}\mathbf{j} + c\frac{s}{K}\mathbf{k} \quad \text{where} \quad K = \sqrt{a^2 + c^2}.$$

It follows that

$$\mathbf{u}(s) = \mathbf{r}'(s) = -\frac{a}{K}\sin\frac{s}{K}\mathbf{i} + \frac{a}{K}\cos\frac{s}{K}\mathbf{j} + \frac{c}{K}\mathbf{k}$$

$$\mathbf{r}''(s) = -\frac{a}{K^2}\cos\frac{s}{K}\mathbf{i} - \frac{a}{K^2}\sin\frac{s}{K}\mathbf{j}$$

$$\kappa = |\mathbf{r}''| = \sqrt{\mathbf{r}''\cdot\mathbf{r}''} = \frac{a}{K^2} = \frac{a}{a^2 + c^2}$$

$$\mathbf{p}(s) = \frac{\mathbf{r}''(s)}{\kappa(s)} = -\cos\frac{s}{K}\mathbf{i} - \sin\frac{s}{K}\mathbf{j}$$

$$\mathbf{b}(s) = \mathbf{u}(s) \times \mathbf{p}(s) = \frac{c}{K}\sin\frac{s}{K}\mathbf{i} - \frac{c}{K}\cos\frac{s}{K}\mathbf{j} + \frac{a}{K}\mathbf{k}$$

$$\mathbf{b}'(s) = \frac{c}{K^2}\cos\frac{s}{K}\mathbf{i} + \frac{c}{K^2}\sin\frac{s}{K}\mathbf{j}$$

$$\tau(s) = -\mathbf{p}(s)\cdot\mathbf{b}'(s) = \frac{c}{K^2} = \frac{c}{a^2 + c^2}.$$

Hence the circular helix has constant curvature and torsion. If $c > 0$ (right-handed helix, cf. Fig. 153), then $\tau > 0$, and if $c < 0$ (left-handed helix, cf. Fig. 154), then $\tau < 0$. ∎

Since **u**, **p**, and **b** are linearly independent vectors, we may represent any vector in space as a linear combination of these vectors. Hence if the derivatives **u**′, **p**′, and **b**′ exist, they may be represented in that fashion. The corresponding formulas are

$$\text{(10)}\qquad \begin{aligned} &\text{(a)} \quad \mathbf{u}' = \qquad \kappa\mathbf{p} \\ &\text{(b)} \quad \mathbf{p}' = -\kappa\mathbf{u} \qquad + \tau\mathbf{b} \\ &\text{(c)} \quad \mathbf{b}' = \qquad -\tau\mathbf{p} \end{aligned}$$

They are called the **formulas of Frenet.** (10a) follows from (6), and (10c) is identical with (8). To derive (10b) we note that, by the definition of a vector product,

$$\mathbf{p} = \mathbf{b} \times \mathbf{u}, \qquad \mathbf{p} \times \mathbf{u} = -\mathbf{b}, \qquad \mathbf{b} \times \mathbf{p} = -\mathbf{u}.$$

Differentiating the first of these formulas and using (10a) and (10c), we obtain

$$\mathbf{p}' = \mathbf{b}' \times \mathbf{u} + \mathbf{b} \times \mathbf{u}' = -\tau\mathbf{p} \times \mathbf{u} + \mathbf{b} \times \kappa\mathbf{p} = -\tau(-\mathbf{b}) + \kappa(-\mathbf{u}),$$

which proves (10b).

Problems for Sec. 8.5

Find a parametric representation of the tangent of the following curves at the given point P.

1. $\mathbf{r}(t) = \cos t\,\mathbf{i} + \sin t\,\mathbf{j}$, $\quad P$: $(-1/\sqrt{2}, 1/\sqrt{2})$
2. $\mathbf{r}(t) = t\,\mathbf{i} + t^2\mathbf{j} + t^3\mathbf{k}$, $\quad P$: $(1, 1, 1)$
3. $\mathbf{r}(t) = \cos t\,\mathbf{i} + \sin t\,\mathbf{j} + 2t\,\mathbf{k}$, $\quad P$: $(1, 0, 4\pi)$
4. $\mathbf{r}(t) = 4\cos t\,\mathbf{i} + 4\sin t\,\mathbf{j}$, $\quad P$: $(2\sqrt{2}, -2\sqrt{2}, 0)$

5. Show that in Example 1, the angle between $\mathbf{u}$ and the z-axis is constant.
6. Show that straight lines are the only curves whose unit tangent vector is constant.
7. Show that the curvature of a straight line is identically zero.
8. Show that if the curve C is represented by $\mathbf{r}(t)$ where t is any parameter, then the curvature is

$$\text{(5')} \qquad \kappa = \frac{\sqrt{(\dot{\mathbf{r}} \cdot \dot{\mathbf{r}})(\ddot{\mathbf{r}} \cdot \ddot{\mathbf{r}}) - (\dot{\mathbf{r}} \cdot \ddot{\mathbf{r}})^2}}{(\dot{\mathbf{r}} \cdot \dot{\mathbf{r}})^{3/2}}.$$

9. Show that the curvature of a circle of radius a equals $1/a$.
10. Find the curvature of the ellipse $\mathbf{r}(t) = a\cos t\,\mathbf{i} + b\sin t\,\mathbf{j}$.
11. Using (5′), show that for a curve $y = y(x)$ in the xy-plane,

$$\kappa = |y''|/(1 + y'^2)^{3/2} \qquad (y' = dy/dx, \text{ etc.}).$$

12. Show that the torsion of a plane curve (with $\kappa > 0$) is identically zero.
13. Using (7) and (9), show that

$$\text{(9')} \qquad \tau = (\mathbf{u}\ \mathbf{p}\ \mathbf{p}') \qquad (\kappa > 0).$$

14. Using (6), show that (9′) may be written

$$\text{(9'')} \qquad \tau = (\mathbf{r}'\ \mathbf{r}''\ \mathbf{r}''')/\kappa^2 \qquad (\kappa > 0).$$

15. Show that if the curve C is represented by $\mathbf{r}(t)$ where t is any parameter, then (9″) becomes

$$\text{(9''')} \qquad \tau = \frac{(\dot{\mathbf{r}}\ \ddot{\mathbf{r}}\ \dddot{\mathbf{r}})}{(\dot{\mathbf{r}} \cdot \dot{\mathbf{r}})(\ddot{\mathbf{r}} \cdot \ddot{\mathbf{r}}) - (\dot{\mathbf{r}} \cdot \ddot{\mathbf{r}})^2} \qquad (\kappa > 0).$$

8.6 Velocity and Acceleration

Let $\mathbf{r}(t)$ be the position vector of a moving particle P in space, where t is time. Then $\mathbf{r}(t)$ represents the path C of P. From the previous section we know that the vector

$$\mathbf{v} = \dot{\mathbf{r}} = \frac{d\mathbf{r}}{dt} \tag{1}$$

is tangent to C and, therefore, points in the instantaneous direction of motion of P. From (3) in Sec. 8.4 we see that

$$|\mathbf{v}| = \sqrt{\dot{\mathbf{r}} \cdot \dot{\mathbf{r}}} = \frac{ds}{dt}$$

where s is the arc length, which measures the distance of P from a fixed point ($s = 0$) on C along the curve. Hence ds/dt is the **speed** of P. The vector $\mathbf{v}$ is, therefore, called the **velocity vector**[3] of the motion.

The derivative of the velocity vector is called the **acceleration vector** and will be denoted by $\mathbf{a}$; thus

$$\mathbf{a}(t) = \dot{\mathbf{v}}(t) = \ddot{\mathbf{r}}(t). \tag{2}$$

Example 1. Centripetal acceleration

The vector function

$$\mathbf{r}(t) = R \cos \omega t\, \mathbf{i} + R \sin \omega t\, \mathbf{j} \qquad (\omega > 0)$$

represents a circle C of radius R with center at the origin of the xy-plane and describes a motion of a particle P in the counterclockwise sense. The velocity vector

$$\mathbf{v} = \dot{\mathbf{r}} = -R\omega \sin \omega t\, \mathbf{i} + R\omega \cos \omega t\, \mathbf{j}$$

(cf. Fig. 160 on the next page) is tangent to C, and its magnitude, the speed

$$|\mathbf{v}| = \sqrt{\dot{\mathbf{r}} \cdot \dot{\mathbf{r}}} = R\omega$$

is constant. The **angular speed** (speed divided by the distance R from the center) is equal to ω. The acceleration vector is

$$\mathbf{a} = \dot{\mathbf{v}} = -R\omega^2 \cos \omega t\, \mathbf{i} - R\omega^2 \sin \omega t\, \mathbf{j} = -\omega^2 \mathbf{r}. \tag{3}$$

We see that there is an acceleration of constant magnitude $|\mathbf{a}| = \omega^2 R$ toward the origin. This is called the **centripetal acceleration.** It results from the fact that the velocity vector is changing direction at a constant rate. The **centripetal force** is $m\mathbf{a}$, where m is the mass of P. The opposite vector $-m\mathbf{a}$ is called the **centrifugal force,** and the two forces are in equilibrium at each instant of the motion. ∎

[3]When no confusion is likely to arise, the word *velocity* is often used to denote the speed, the length of $\mathbf{v}$.

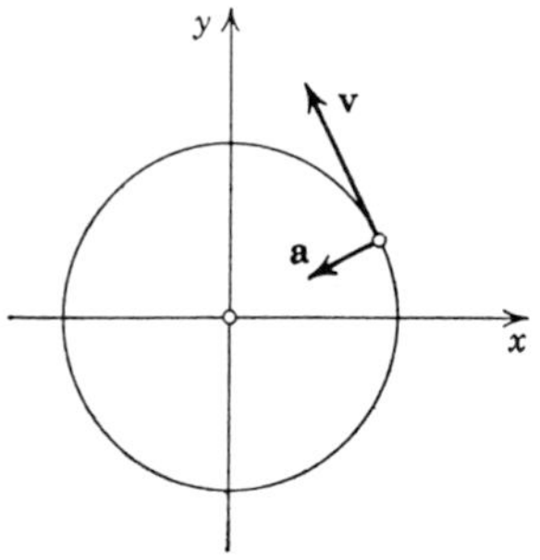

Fig. 160. Centripetal acceleration

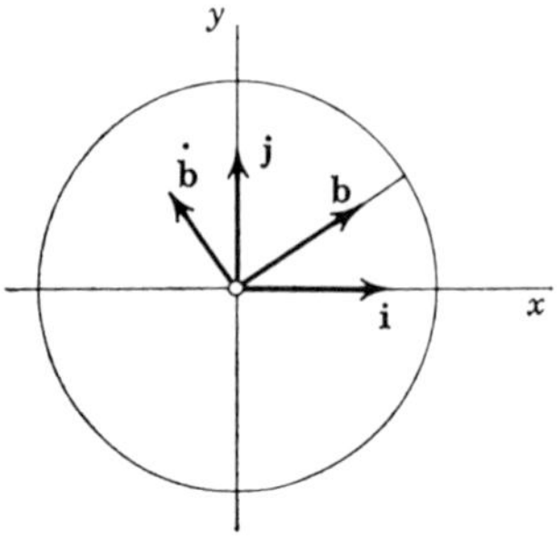

Fig. 161. Motion in Example 2

It is clear that **a** is the time rate of change of **v**. In Example 1 we have $|\mathbf{v}| = const$, but $|\mathbf{a}| \neq 0$. This illustrates that the magnitude of **a** is not in general the rate of change of $|\mathbf{v}|$. The reason is that, in general, **a** is not tangent to the path C. In fact, by applying the chain rule of differentiation to (1), and denoting derivatives with respect to s by primes, we have

$$\mathbf{v} = \frac{d\mathbf{r}}{dt} = \frac{d\mathbf{r}}{ds}\frac{ds}{dt} = \mathbf{r}'\frac{ds}{dt}$$

and, by differentiating this again,

$$\mathbf{a} = \frac{d\mathbf{v}}{dt} = \frac{d}{dt}\left(\mathbf{r}'\frac{ds}{dt}\right) = \mathbf{r}''\left(\frac{ds}{dt}\right)^2 + \mathbf{r}'\frac{d^2s}{dt^2}. \tag{4}$$

Since $\mathbf{r}'$ is the unit tangent vector **u** to C (Sec. 8.5) and its derivative $\mathbf{u}' = \mathbf{r}''$ is perpendicular to **u** (Sec. 8.5), formula (4) is a decomposition of the acceleration vector into its normal component $\mathbf{r}''\dot{s}^2$ and its tangential component $\mathbf{r}'\ddot{s}$. From this we see that if and only if the normal component is zero, $|\mathbf{a}|$ equals the time rate of change of $|\mathbf{v}| = \dot{s}$ (except for the sign), because then $|\mathbf{a}| = |\mathbf{r}'|\,|\ddot{s}| = |\ddot{s}|$.

Example 2. Coriolis[4] acceleration

A particle P moves on a disk towards the edge, the position vector being

$$\mathbf{r}(t) = t\,\mathbf{b} \tag{5}$$

where **b** is a unit vector, rotating together with the disk with constant angular speed ω in the counterclockwise sense (Fig. 161). Find the acceleration **a** of P.

Because of the rotation, **b** is of the form

$$\mathbf{b}(t) = \cos\omega t\,\mathbf{i} + \sin\omega t\,\mathbf{j}. \tag{6}$$

Differentiating (5), we obtain the velocity

$$\mathbf{v} = \dot{\mathbf{r}} = \mathbf{b} + t\,\dot{\mathbf{b}}. \tag{7}$$

Obviously **b** is the velocity of P relative to the disk, and $t\,\dot{\mathbf{b}}$ is the additional velocity due to the rotation. Differentiating once more, we obtain the acceleration

[4]GUSTAVE GASPARD CORIOLIS (1792–1843), French physicist.

$$\mathbf{a} = \dot{\mathbf{v}} = 2\dot{\mathbf{b}} + t\,\ddot{\mathbf{b}}. \tag{8}$$

In the last term of (8) we have $\ddot{\mathbf{b}} = -\omega^2\mathbf{b}$, as follows by differentiating (6). Hence this acceleration $t\,\ddot{\mathbf{b}}$ is directed toward the center of the disk, and from Example 1 we see that this is the centripetal acceleration due to the rotation. In fact, the distance of P from the center is equal to t which, therefore, plays the role of R in Example 1.

The most interesting and probably unexpected term in (8) is $2\dot{\mathbf{b}}$, the so-called **Coriolis acceleration,** which results from the interaction of the rotation of the disk and the motion of P on the disk. It has the direction of $\dot{\mathbf{b}}$, that is, it is tangential to the edge of the disk and, referred to the fixed xy-coordinate system, it points in the direction of the rotation. If P is a person of mass m_0 walking on the disk according to (5), he will feel a force $-2m_0\dot{\mathbf{b}}$ in the opposite direction, that is, against the sense of the rotation.

Example 3. Superposition of two rotations

Find the acceleration of a particle P moving on a "meridian" M of a rotating sphere with constant speed relative to the sphere.

The motion of P on M can be described analytically in the form

$$\mathbf{r}(t) = R\cos\gamma t\,\mathbf{b} + R\sin\gamma t\,\mathbf{k} \tag{9}$$

where R is the radius of the sphere, γ (>0) the angular speed of P on M, $\mathbf{b}$ a horizontal unit vector in the plane of M (Fig. 162), and $\mathbf{k}$ the unit vector in the positive z-direction. Since $\mathbf{b}$ rotates together with the sphere, it is of the form

$$\mathbf{b} = \cos\omega t\,\mathbf{i} + \sin\omega t\,\mathbf{j} \tag{10}$$

where ω (>0) is the angular speed of the sphere and $\mathbf{i}$ and $\mathbf{j}$ are the unit vectors in the positive x and y directions, which are fixed in space. By differentiating (9) we obtain the velocity

$$\mathbf{v} = \dot{\mathbf{r}} = R\cos\gamma t\,\dot{\mathbf{b}} - \gamma R\sin\gamma t\,\mathbf{b} + \gamma R\cos\gamma t\,\mathbf{k}. \tag{11}$$

By differentiating this again we obtain the acceleration

$$\mathbf{a} = \dot{\mathbf{v}} = R\cos\gamma t\,\ddot{\mathbf{b}} - 2\gamma R\sin\gamma t\,\dot{\mathbf{b}} - \gamma^2 R\cos\gamma t\,\mathbf{b} - \gamma^2 R\sin\gamma t\,\mathbf{k}, \tag{12}$$

where, by (10),

$$\dot{\mathbf{b}} = -\omega\sin\omega t\,\mathbf{i} + \omega\cos\omega t\,\mathbf{j},$$

$$\ddot{\mathbf{b}} = -\omega^2\cos\omega t\,\mathbf{i} - \omega^2\sin\omega t\,\mathbf{j} = -\omega^2\mathbf{b}.$$

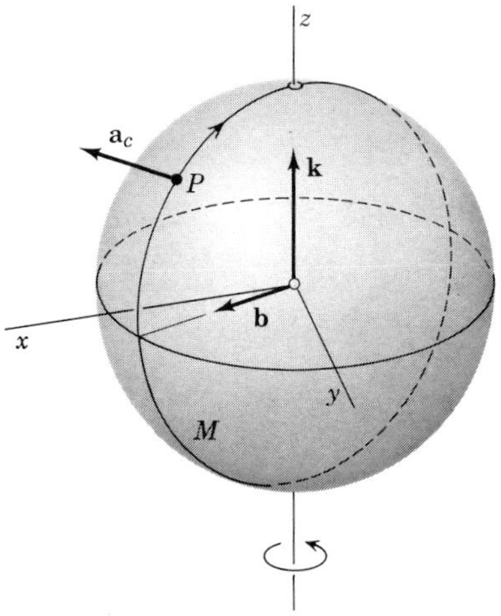

Fig. 162. Superposition of two rotations

From (9) we see that the sum of the last two terms in (12) is equal to $-\gamma^2\mathbf{r}$, and (12) becomes

$$\mathbf{a} = -\omega^2 R \cos \gamma t\, \mathbf{b} - 2\gamma R \sin \gamma t\, \dot{\mathbf{b}} - \gamma^2 \mathbf{r}. \tag{13}$$

The first term on the right is the centripetal acceleration caused by the rotation of the sphere, and the last term is the centripetal acceleration resulting from the rotation of P on M. The second term is the **Coriolis acceleration**

$$\mathbf{a}_c = -2\gamma R \sin \gamma t\, \dot{\mathbf{b}}. \tag{14}$$

On the "Northern hemisphere," $\sin \gamma t > 0$ [cf. (9)] and because of the minus sign, $\mathbf{a}_c$ is directed opposite to $\dot{\mathbf{b}}$, that is, tangential to the surface of the sphere, perpendicular to M, and opposite to the rotation of the sphere. Its magnitude $2\gamma R\,|\sin \gamma t|\,\omega$ is maximum at the "North pole" and zero at the equator. If P is a fly of mass m_0 walking according to (9), it will feel a force $-m_0\mathbf{a}_c$, opposite to $m_0\mathbf{a}_c$; this is quite similar to the force felt in Example 2. This force tends to let the fly deviate from the path M to the right. On the "Southern hemisphere," $\sin \gamma t < 0$ and that force acts in the opposite direction, tending to let the fly deviate from M to the left. This effect can be observed in connection with missiles and other projectiles. The flow of the air towards an area of low pressure also shows these deviations.

Problems for Sec. 8.6

Let $\mathbf{r}(t)$ be the position vector of a moving particle where $t\ (\geqq 0)$ is the time. Describe the geometric shape of the path and find the velocity vector, the speed, and the acceleration vector.

1. $\mathbf{r} = t\,\mathbf{i}$

2. $\mathbf{r} = t^2\mathbf{k}$

3. $\mathbf{r} = (2t - t^2)\mathbf{i}$

4. $\mathbf{r} = t^3\mathbf{i}$

5. $\mathbf{r} = \sin t\,\mathbf{j}$

6. $\mathbf{r} = 3\cos 2t\,\mathbf{i} + 3\sin 2t\,\mathbf{j}$

7. $\mathbf{r} = \cos t^2\,\mathbf{i} + \sin t^2\,\mathbf{j}$

8. $\mathbf{r} = 3\cos 2t\,\mathbf{i} + 2\sin 3t\,\mathbf{j}$

9. $\mathbf{r} = \cos t\,\mathbf{i} + \sin t\,\mathbf{j} + t\mathbf{k}$

10. $\mathbf{r} = (1 + t^3)\mathbf{i} + 2t^3\mathbf{j} + (2 - t^3)\mathbf{k}$

11. Find the centripetal acceleration of the moon towards the earth, assuming that the orbit of the moon is a circle of radius 239,000 miles $= 3.85 \cdot 10^8$ meters and the time for one complete revolution is 27.3 days $= 2.36 \cdot 10^6$ sec.

12. Find the Coriolis acceleration in Example 2 with (5) replaced by $\mathbf{r} = t^2\mathbf{b}$.

13. Find the motion for which the acceleration vector is constant.

14. Obtain (3) by differentiating (7), Sec. 6.8.

15. If $\mathbf{r}(t)$ represents the path of a moving particle and t is the time, what does a parametric transformation $t = \varphi(\tilde{t})$ mean in terms of mechanics?

8.7 Chain Rule and Mean Value Theorem for Functions of Several Variables

We shall now consider some facts about functions of several variables which will be needed in the following sections. For the sake of simplicity we shall formulate everything for functions of two variables; the generalization to functions of three or more variables will be obvious. Familiarity of the student with the concept of a partial derivative will be assumed.

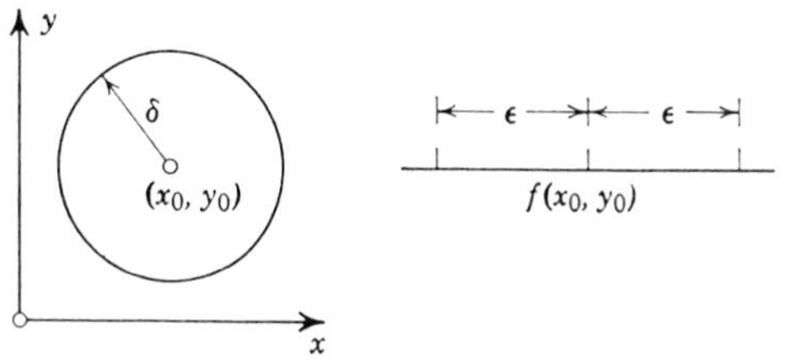

Fig. 163. Continuity of a function of two variables

A function $f(x, y)$ is said to be **continuous** *at a point* (x_0, y_0) if f is defined in a *neighborhood*[5] of that point and if for any positive number ϵ (no matter how small, but not zero) we can find a positive number δ such that

$$|f(x, y) - f(x_0, y_0)| < \epsilon$$

for all (x, y) in that neighborhood for which

$$(x - x_0)^2 + (y - y_0)^2 < \delta^2.$$

Geometrically speaking, continuity of $f(x, y)$ at (x_0, y_0) means that for each interval of length 2ϵ with midpoint $f(x_0, y_0)$ we can find a circular disk with nonzero radius δ and center (x_0, y_0) in that neighborhood such that for every point (x, y) in the disk the corresponding function value $f(x, y)$ lies in that interval (Fig. 163).

From elementary calculus we know that if w is a differentiable function of x, and x is a differentiable function of t, then

$$\frac{dw}{dt} = \frac{dw}{dx}\frac{dx}{dt}.$$

This so-called chain rule of differentiation can be generalized as follows.

Theorem 1 (Chain rule)

Let $w = f(x, y)$ be continuous and have continuous first partial derivatives at every point of a domain[6] D in the xy-plane. Let $x = x(t)$ and $y = y(t)$ be differentiable functions of a variable in some interval T such that, for each t in T, the point $[x(t), y(t)]$ lies in D. Then $w = f[x(t), y(t)]$ is a differentiable function for all t in T, and

$$\frac{dw}{dt} = \frac{\partial w}{\partial x}\frac{dx}{dt} + \frac{\partial w}{\partial y}\frac{dy}{dt}. \tag{1}$$

[5]That is, in some circular disk $(x - x_0)^2 + (y - y_0)^2 < r^2$, $(r > 0)$, in the xy-plane.

[6]A **domain** D is an open connected point set, where "connected" means that any two points of D can be joined by a broken line of finitely many linear segments all of whose points belong to D, and "open" means that every point of D has a neighborhood all of whose points belong to D. For example, the interior of a rectangle or a circle is a domain.

Proof. We choose at t in T and Δt so small that $t + \Delta t$ is also in T, and set

$$\Delta x = x(t + \Delta t) - x(t), \qquad \Delta y = y(t + \Delta t) - y(t), \tag{2}$$

and furthermore

$$\Delta w = f(x + \Delta x, y + \Delta y) - f(x, y).$$

By adding and subtracting a term this may be written

$$\Delta w = [f(x + \Delta x, y + \Delta y) - f(x, y + \Delta y)] + [f(x, y + \Delta y) - f(x, y)].$$

If we apply the mean value theorem for a function of a single variable (cf. Ref. [A14]) to each of the two expressions in the brackets, we obtain

$$\Delta w = \Delta x \frac{\partial f}{\partial x}\bigg|_{x_1,\, y+\Delta y} + \Delta y \frac{\partial f}{\partial y}\bigg|_{x,\, y_1} \tag{3}$$

where x_1 lies between x and $x + \Delta x$, and y_1 lies between y and $y + \Delta y$. Dividing (3) by Δt on both sides, letting Δt approach 0, and noting that $\partial f/\partial x$ and $\partial f/\partial y$ are assumed to be continuous, we obtain (1). ∎

This theorem may now immediately be extended as follows.

Theorem 2

Let $w = f(x, y)$ be continuous and have continuous first partial derivatives in a domain D of the xy-plane. Let $x = x(u, v)$ and $y = y(u, v)$ be functions which have first partial derivatives in a domain B of the uv-plane, which is such that, for any point (u, v) in B, the corresponding point $[x(u, v), y(u, v)]$ lies in D. Then the function $w = f(x(u, v), y(u, v))$ is defined in B and has first partial derivatives with respect to u and v in B, and

$$\begin{aligned} \frac{\partial w}{\partial u} &= \frac{\partial w}{\partial x}\frac{\partial x}{\partial u} + \frac{\partial w}{\partial y}\frac{\partial y}{\partial u}, \\ \frac{\partial w}{\partial v} &= \frac{\partial w}{\partial x}\frac{\partial x}{\partial v} + \frac{\partial w}{\partial y}\frac{\partial y}{\partial v}. \end{aligned} \tag{4}$$

The proof follows immediately from Theorem 1 by keeping one of the two variables u and v constant.

From elementary calculus we know that if a function $f(x)$ is differentiable, then

$$f(x_0 + h) - f(x_0) = h\frac{df}{dx},$$

the derivative being evaluated at a suitable point between x_0 and $x_0 + h$ (cf. Ref. [A14]). This so-called mean-value theorem of differential calculus can be extended to functions of two variables as follows.

Theorem 3 (Mean-value theorem)

Let $f(x, y)$ be continuous and have continuous first partial derivatives in a domain D. Furthermore, let (x_0, y_0) and $(x_0 + h, y_0 + k)$ be points in D such that the line segment joining these points lies in D (Fig. 164). Then

$$f(x_0 + h, y_0 + k) - f(x_0, y_0) = h\frac{\partial f}{\partial x} + k\frac{\partial f}{\partial y}, \tag{5}$$

the partial derivatives being evaluated at a suitable point of that segment.

Proof. Let

$$x = x_0 + th, \qquad y = y_0 + tk \qquad (0 \leqq t \leqq 1)$$

and

$$F(t) = f(x_0 + th, y_0 + tk).$$

Then

$$f(x_0 + h, y_0 + k) = F(1), \qquad f(x_0, y_0) = F(0).$$

By the mean value theorem for a function of a single variable there is a value t_1 between 0 and 1 such that

$$f(x_0 + h, y_0 + k) - f(x_0, y_0) = F(1) - F(0) = F'(t_1). \tag{6}$$

We now apply Theorem 1. Since $dx/dt = h$ and $dy/dt = k$, we obtain

$$F'(t_1) = \frac{\partial f}{\partial x}h + \frac{\partial f}{\partial y}k, \tag{7}$$

the derivatives on the right being evaluated at the point $(x_0 + t_1 h, y_0 + t_1 k)$, which lies on the line segment with endpoints (x_0, y_0) and $(x_0 + h, y_0 + k)$. By inserting (7) into (6) we obtain (5), and the proof is complete. ∎

For a function $f(x, y, z)$ of three variables, satisfying conditions analogous to those in Theorem 3, the consideration is quite similar and leads to the formula

$$f(x_0 + h, y_0 + k, z_0 + l) - f(x_0, y_0, z_0) = h\frac{\partial f}{\partial x} + k\frac{\partial f}{\partial y} + l\frac{\partial f}{\partial z}, \tag{8}$$

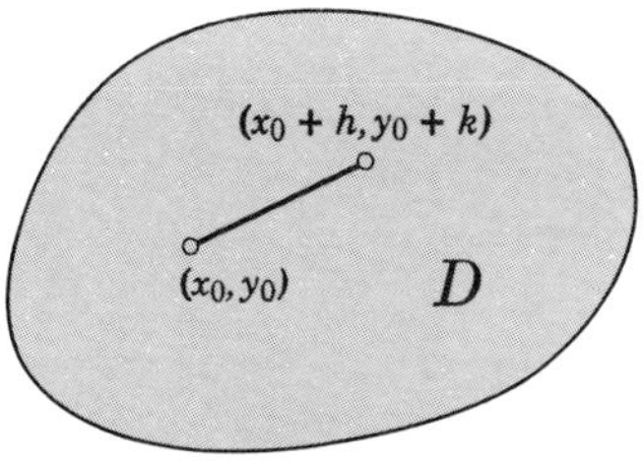

Fig. 164. Mean value theorem

the partial derivatives being evaluated at a suitable point of the segment with endpoints (x_0, y_0, z_0) and $(x_0 + h, y_0 + k, z_0 + l)$.

Problems for Sec. 8.7

Using (1), find dw/dt where

1. $w = x + y, \quad x = t^2, \quad y = \ln t$

2. $w = \sqrt{x^2 + y^2}, \quad x = e^{2t}, \quad y = e^{-2t}$

3. $w = x/y, \quad x = g(t), \quad y = h(t)$

4. $w = x^y, \quad x = \cos t, \quad y = \sin t$

5. Let $w = f(x, y, z)$ where x, y, z are functions of t. Show that under conditions similar to those in Theorem 1,

$$\frac{dw}{dt} = \frac{\partial w}{\partial x}\frac{dx}{dt} + \frac{\partial w}{\partial y}\frac{dy}{dt} + \frac{\partial w}{\partial z}\frac{dz}{dt}. \tag{9}$$

Using (9), find dw/dt where

6. $w = x^2 + y^2 + z^2, \quad x = e^{2t}\cos 2t, \quad y = e^{2t}\sin 2t, \quad z = e^{4t}$

7. $w = (x^2 + y^2 + z^2)^{-1/2}, \quad x = \cos t, \quad y = \sin t, \quad z = t$

8. Prove Theorem 2.

Find $\partial w/\partial u$ and $\partial w/\partial v$ where

9. $w = \ln(x^2 + y^2), \quad x = e^u \cos v, \quad y = e^u \sin v$

10. $w = x^2 - y^2, \quad x = u^2 - v^2, \quad y = 2uv$

11. $w = xy, \quad x = e^u \cos v, \quad y = e^u \sin v$

12. Derive (8).

13. Let $w = f(x, y)$ and $x = r\cos\theta$, $y = r\sin\theta$. Show that

$$\left(\frac{\partial w}{\partial r}\right)^2 + \frac{1}{r^2}\left(\frac{\partial w}{\partial \theta}\right)^2 = \left(\frac{\partial w}{\partial x}\right)^2 + \left(\frac{\partial w}{\partial y}\right)^2.$$

14. Let $w = f(v, z)$ and $v = x + ct$, $z = x - ct$, where c is a constant. Show that, granted sufficient differentiability,

$$c^2 w_{xx} - w_{tt} = 4c^2 w_{vz}.$$

15. Let $w = f(x, y)$ and $x = r\cos\theta$, $y = r\sin\theta$. Show that

$$w_{xx} + w_{yy} = w_{rr} + \frac{1}{r}w_r + \frac{1}{r^2}w_{\theta\theta}.$$

8.8 Directional Derivative. Gradient of a Scalar Field

We consider a scalar field in space given by a scalar function $f(P) = f(x, y, z)$ (cf. Sec. 8.1). We know that the first partial derivatives of f are the rates of change of f in the directions of the coordinate axes. It seems unnatural to restrict attention to these three directions, and we may ask for the rate of change of f in any direction. This simple idea leads to the notion of a directional derivative.

To define that derivative we choose a point P in space and a direction at P, given by a unit vector **b.** Let C be the ray from P in the direction of **b,** and let Q be a point on C, whose distance from P is s (Fig. 165). Then if the limit

$$\text{(1)} \qquad \frac{\partial f}{\partial s} = \lim_{s \to 0} \frac{f(Q) - f(P)}{s} \qquad (s = \text{distance between } P \text{ and } Q)$$

exists, it is called the **directional derivative** of f *at* P *in the direction of* **b.** Obviously, $\partial f/\partial s$ is the rate of change of f at P in the direction of **b.**

Another notation for $\partial f/\partial s$ also used in the literature is

$$D_{\mathbf{b}} f$$

where D suggests *differentiation* and **b** indicates the direction.

In this way there are now infinitely many directional derivatives of f at P, each corresponding to a certain direction. But, a Cartesian coordinate system being given, we may represent any such derivative in terms of the first partial derivatives of f at P as follows. If P has the position vector **a,** then the ray C can be represented in the form

$$\text{(2)} \qquad \mathbf{r}(s) = x(s)\mathbf{i} + y(s)\mathbf{j} + z(s)\mathbf{k} = \mathbf{a} + s\mathbf{b} \qquad (s \geqq 0),$$

and $\partial f/\partial s$ is the derivative of the function $f[x(s), y(s), z(s)]$ with respect to the arc length s of C. Hence, assuming that f has continuous first partial derivatives and applying the chain rule (Theorem 1 in the last section), we obtain

$$\text{(3)} \qquad \frac{\partial f}{\partial s} = \frac{\partial f}{\partial x}x' + \frac{\partial f}{\partial y}y' + \frac{\partial f}{\partial z}z'$$

where primes denote derivatives with respect to s (which are evaluated at $s = 0$). Now from (2),

$$\mathbf{r}' = x'\mathbf{i} + y'\mathbf{j} + z'\mathbf{k} = \mathbf{b}.$$

This suggests that we introduce the vector

$$\text{(4)} \qquad \operatorname{grad} f = \frac{\partial f}{\partial x}\mathbf{i} + \frac{\partial f}{\partial y}\mathbf{j} + \frac{\partial f}{\partial z}\mathbf{k}$$

and write (3) in the form of an inner product (dot product):

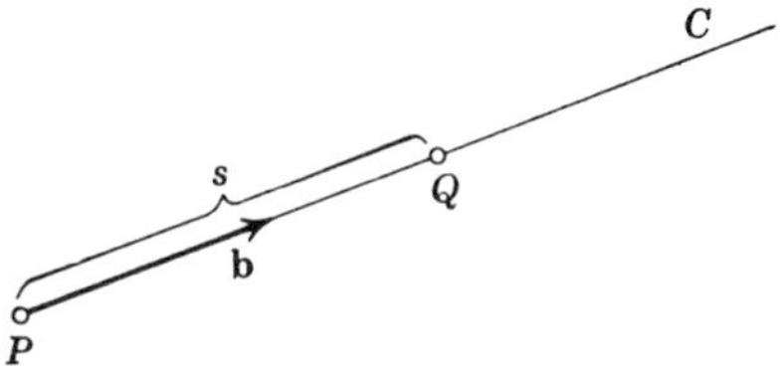

Fig. 165. Directional derivative.

$$\frac{\partial f}{\partial s} = \mathbf{b} \cdot \operatorname{grad} f \qquad (|\mathbf{b}| = 1). \tag{5}$$

The vector grad f is called the **gradient** of the scalar function f.

By introducing the *differential operator*

$$\nabla = \frac{\partial}{\partial x}\mathbf{i} + \frac{\partial}{\partial y}\mathbf{j} + \frac{\partial}{\partial z}\mathbf{k}$$

(read **nabla** or "del") we may write

$$\operatorname{grad} f = \nabla f = \frac{\partial f}{\partial x}\mathbf{i} + \frac{\partial f}{\partial y}\mathbf{j} + \frac{\partial f}{\partial z}\mathbf{k}.$$

The notation ∇f for the gradient is frequently used in the engineering literature.

If in particular $\mathbf{b}$ has the direction of the positive x-axis, then $\mathbf{b} = \mathbf{i}$, and

$$\frac{\partial f}{\partial s} = \mathbf{b} \cdot \operatorname{grad} f = \frac{\partial f}{\partial x}\mathbf{i} \cdot \mathbf{i} = \frac{\partial f}{\partial x}.$$

Similarly, the directional derivative in the positive y-direction is $\partial f/\partial y$, etc.

Example 1. Directional derivative

Find the directional derivative $\partial f/\partial s$ of $f(x, y, z) = 2x^2 + 3y^2 + z^2$ at the point P: (2, 1, 3) in the direction of the vector $\mathbf{a} = \mathbf{i} - 2\mathbf{k}$. We obtain

$$\operatorname{grad} f = 4x\mathbf{i} + 6y\mathbf{j} + 2z\mathbf{k}, \qquad \text{and at } P, \qquad \operatorname{grad} f = 8\mathbf{i} + 6\mathbf{j} + 6\mathbf{k}.$$

Since $|\mathbf{a}| = \sqrt{5}$, the unit vector in the direction of $\mathbf{a}$ is

$$\mathbf{b} = \frac{\mathbf{a}}{|\mathbf{a}|} = \frac{1}{\sqrt{5}}\mathbf{i} - \frac{2}{\sqrt{5}}\mathbf{k}.$$

Therefore,

$$\frac{\partial f}{\partial s} = (8\mathbf{i} + 6\mathbf{j} + 6\mathbf{k}) \cdot \left(\frac{1}{\sqrt{5}}\mathbf{i} - \frac{2}{\sqrt{5}}\mathbf{k}\right) = -\frac{4}{\sqrt{5}}.$$

The minus sign indicates that f decreases in the direction under consideration. ∎

We want to show that *the length and direction of* grad f *are independent of the particular choice of Cartesian coordinates.*

This is, of course, not obvious, because (4) involves partial derivatives, which depend on the choice of the coordinates, so that we do not know yet whether the corresponding expression

$$\frac{\partial f}{\partial x^*}\mathbf{i}^* + \frac{\partial f}{\partial y^*}\mathbf{j}^* + \frac{\partial f}{\partial z^*}\mathbf{k}^*$$

with respect to other Cartesian coordinates x^*, y^*, z^* (and corresponding unit vectors $\mathbf{i}^*, \mathbf{j}^*, \mathbf{k}^*$) will have the same length and direction as (4).

To prove our proposition, we may reason as follows. By the definition of a scalar function, the value of f at a point P depends on P but is independent of the coordinates, and s, the arc length of that ray C, is also independent of the choice of coordinates. Hence $\partial f/\partial s$ is independent of the particular choice of coordinates. From (5) we obtain

$$\frac{\partial f}{\partial s} = |\mathbf{b}|\,|\text{grad}\, f| \cos\gamma = |\text{grad}\, f| \cos\gamma$$

where γ is the angle between $\mathbf{b}$ and grad f. We see that $\partial f/\partial s$ is maximum when $\cos\gamma = 1$, $\gamma = 0$, and then $\partial f/\partial s = |\text{grad}\, f|$. This shows that the length and direction of grad f are independent of the coordinates, and we have the following result.

Theorem 1 (Gradient)

Let $f(P) = f(x, y, z)$ be a scalar function having continuous first partial derivatives. Then grad f *exists and its length and direction are independent of the particular choice of Cartesian coordinates in space. If at a point P the gradient of f is not the zero vector, it has the direction of maximum increase of f at P.*

Another important geometrical characterization of the gradient can be obtained as follows. Consider a differentiable scalar function $f(x, y, z)$ in space. Suppose that for each constant c the equation

$$f(x, y, z) = c = const \tag{6}$$

represents a surface S in space. Then, by letting c assume all values, we obtain a family of surfaces, which are called the **level surfaces** of the function f. Since, by the definition of a function, our function f has a unique value at each point in space, it follows that through each point in space there passes one, and only one, level surface of f. We remember that a curve C in space may be represented in the form (cf. Sec. 8.3)

$$\mathbf{r}(t) = x(t)\mathbf{i} + y(t)\mathbf{j} + z(t)\mathbf{k}. \tag{7}$$

If we now require that C lies on S, then the functions $x(t)$, $y(t)$, and $z(t)$ in (7) must be such that

$$f[x(t), y(t), z(t)] = c;$$

cf. (6). By differentiating this with respect to t and using the chain rule (Sec. 8.7) we obtain

$$\frac{\partial f}{\partial x}\dot{x} + \frac{\partial f}{\partial y}\dot{y} + \frac{\partial f}{\partial z}\dot{z} = (\text{grad}\, f) \cdot \dot{\mathbf{r}} = 0, \tag{8}$$

where the vector

$$\dot{\mathbf{r}} = \dot{x}\mathbf{i} + \dot{y}\mathbf{j} + \dot{z}\mathbf{k}$$

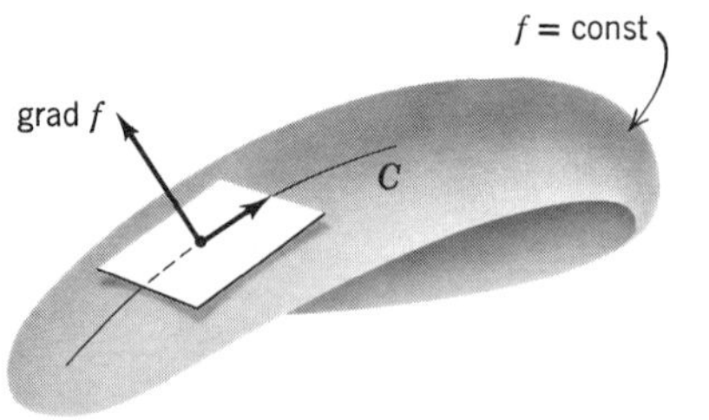

Fig. 166. Level surface and gradient

is tangent to C (cf. Sec. 8.5). If we consider curves on S passing through a point P of S in various directions, their tangents at P will, in general, lie in the same plane which touches S at P. This plane is called the **tangent plane** of S at P. The straight line through P and perpendicular to the tangent plane is called the **normal** of S at P (Fig. 166). From (8) and Theorem 1, Sec. 6.5, we thus obtain the following result.

Theorem 2 (Gradient and surface normal)

Let f be a scalar function which is defined and differentiable in a domain D in space, and let P be any point in D on a level surface S of f. Then if the gradient of f at P is not the zero vector, it is perpendicular to S at P, that is, it has the direction of the normal to S at P.

Example 2. Normal to a plane curve

The level curves $f = const$ of $f(x, y) = \ln(x^2 + y^2)$ are concentric circles about the origin. The gradient

$$\text{grad } f = \frac{\partial f}{\partial x}\mathbf{i} + \frac{\partial f}{\partial y}\mathbf{j} = \frac{2x}{x^2 + y^2}\mathbf{i} + \frac{2y}{x^2 + y^2}\mathbf{j}$$

has the direction of the normals to the circles, and its direction corresponds to that of the maximum increase of f. For example, at the point P: (2, 1), we have (cf. Fig. 167)

$$\text{grad } f = 0.8\mathbf{i} + 0.4\mathbf{j}.$$

Example 3. Normal to a surface

Find a unit normal vector $\mathbf{n}$ of the cone of revolution $z^2 = 4(x^2 + y^2)$ at the point P: (1, 0, 2). We may regard the cone as the level surface $f = 0$ of the function $f(x, y, z) = 4(x^2 + y^2) - z^2$. Then

$$\text{grad } f = 8x\mathbf{i} + 8y\mathbf{j} - 2z\mathbf{k} \qquad \text{and at } P, \qquad \text{grad } f = 8\mathbf{i} - 4\mathbf{k}.$$

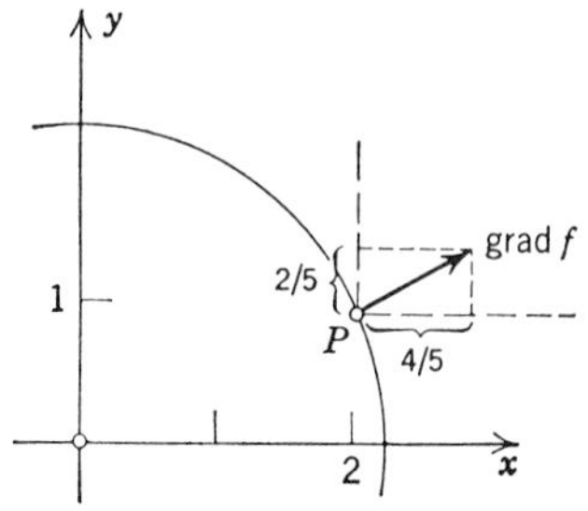

Fig. 167. Normal to a circle

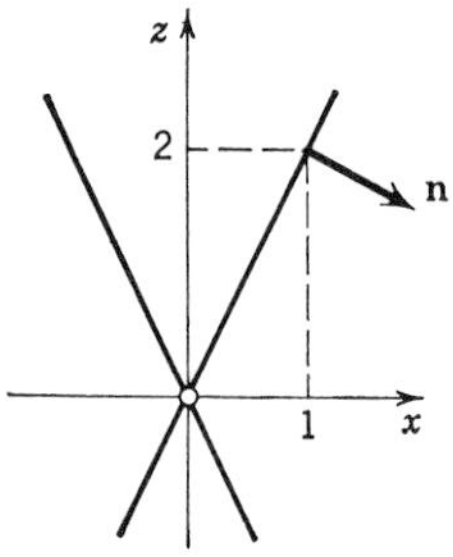

Fig. 168. Intersection of the xz-plane and the cone in Example 3

Hence, by Theorem 2,

$$\mathbf{n} = \frac{\text{grad } f}{|\text{grad } f|} = \frac{2}{\sqrt{5}}\mathbf{i} - \frac{1}{\sqrt{5}}\mathbf{k}$$

is a unit normal vector of the cone at P, and $-\mathbf{n}$ is the other one (cf. Fig. 168). ∎

Some of the vector fields occurring in physics are given by vector functions which can be obtained as the gradients of suitable scalar functions. Such a scalar function is then called a *potential function* or **potential** of the corresponding vector field. The use of potentials simplifies the investigation of those vector fields considerably. To obtain a first impression of this approach to vector fields, let us consider an important example.

Example 4. Gravitational field. Laplace's equation

In Example 4, Sec. 8.1, we have seen that, according to Newton's law of gravitation, the force of attraction between two particles is

$$\mathbf{p} = -c\frac{\mathbf{r}}{r^3} = -c\left(\frac{x - x_0}{r^3}\mathbf{i} + \frac{y - y_0}{r^3}\mathbf{j} + \frac{z - z_0}{r^3}\mathbf{k}\right) \tag{9}$$

where

$$r = \sqrt{(x - x_0)^2 + (y - y_0)^2 + (z - z_0)^2}$$

is the distance between the two particles and c is a constant. Observing that

$$\frac{\partial}{\partial x}\left(\frac{1}{r}\right) = -\frac{2(x - x_0)}{2[(x - x_0)^2 + (y - y_0)^2 + (z - z_0)^2]^{3/2}} = -\frac{x - x_0}{r^3} \tag{10a}$$

and similarly

$$\frac{\partial}{\partial y}\left(\frac{1}{r}\right) = -\frac{y - y_0}{r^3}, \qquad \frac{\partial}{\partial z}\left(\frac{1}{r}\right) = -\frac{z - z_0}{r^3}, \tag{10b}$$

we see that $\mathbf{p}$ is the gradient of the scalar function

$$f(x, y, z) = \frac{c}{r} \qquad (r > 0);$$

that is, f is a potential of that gravitational field.

By differentiating (10) we find

$$\frac{\partial^2}{\partial x^2}\left(\frac{1}{r}\right) = -\frac{1}{r^3} + \frac{3(x - x_0)^2}{r^5}, \qquad \frac{\partial^2}{\partial y^2}\left(\frac{1}{r}\right) = -\frac{1}{r^3} + \frac{3(y - y_0)^2}{r^5},$$

$$\frac{\partial^2}{\partial z^2}\left(\frac{1}{r}\right) = -\frac{1}{r^3} + \frac{3(z - z_0)^2}{r^5}.$$

Since the sum of the three expressions on the right is zero, we see that the potential $f = c/r$ satisfies the equation

$$\frac{\partial^2 f}{\partial x^2} + \frac{\partial^2 f}{\partial y^2} + \frac{\partial^2 f}{\partial z^2} = 0. \tag{11}$$

This important partial differential equation is called **Laplace's equation;** it will be considered in detail in Chaps. 11 and 18. The expression on the left is called the **Laplacian** of f and is denoted by $\nabla^2 f$ or Δf. The differential operator

$$\nabla^2 = \Delta = \frac{\partial^2}{\partial x^2} + \frac{\partial^2}{\partial y^2} + \frac{\partial^2}{\partial z^2}$$

(read "nabla squared" or "delta") is called the **Laplace operator.** Using this operator, we may write (11) in the form

$$\nabla^2 f = 0.$$

It can be shown that the field of force produced by any distribution of masses is given by a vector function which is the gradient of a scalar function f, and f satisfies (11) in any region of space which is free of matter.

There are other laws in physics which are of the same form as Newton's law of gravitation. For example, in electrostatics the force of attraction (or repulsion) between two particles of opposite (or like) charges Q_1 and Q_2 is

$$\mathbf{p} = k\frac{\mathbf{r}}{r^3} \qquad \textbf{(Coulomb's law)}$$

where $k = Q_1Q_2/4\pi\epsilon$, and ϵ is the dielectric constant. Hence $\mathbf{p}$ is the gradient of the potential $f = -k/r$, and f satisfies (11) when $r > 0$. ∎

If the vector function defining a vector field is the gradient of a scalar function, the field is said to be **conservative,** because, as we shall see in Sec. 9.12, in such a field the work done in displacing a particle from a point P_1 to a point P_2 in the field depends only on P_1 and P_2 but not on the path along which the particle is displaced from P_1 to P_2. We shall also see that not every field is conservative.

Problems for Sec. 8.8

Find the gradient ∇f, where

1. $f = 2x - y$
2. $f = e^x \cos y$
3. $f = \sin x \cosh y$
4. $f = x^2 - y^2$
5. $f = \ln(x^2 + y^2)$
6. $f = \arctan(y/x)$
7. $f = xyz$
8. $f = yz + zx + xy$
9. $f = x^2 + y^2 + z^2$
10. $f = (x^2 + y^2 + z^2)^{-1/2}$
11. $f = e^{xyz}$
12. $f = \cos(x^2 + y^2 + z^2)$

Graph some level curves $f = const.$ Find ∇f and indicate ∇f by arrows at some points of the level curves.

13. $f = 2x - 3y$
14. $f = y/x$
15. $f = xy$
16. $f = x^2y^2$
17. $f = 9x^2 + y^2$
18. $f = 4x^2 + 9y^2$

Find and graph a normal vector to the given plane curve at a point P: (x, y).

19. $y = x, \quad P: (3, 3)$
20. $y = x^2, \quad P: (-2, 4)$
21. $y = 3x - 4, \quad P: (1, -1)$
22. $x^2 + y^2 = 25, \quad P: (3, 4)$
23. $y^2 = x^3, \quad P: (4, -8)$
24. $3x^2 - 2y^2 = 1, \quad P: (1, 1)$

Find a normal vector to the given surface at a point P: (x, y, z).

25. $x + y + z = 0, \quad P: (1, 1, -2)$
26. $2x + 3y - z = 1, \quad P: (1, 2, 7)$
27. $z = x^2 + y^2, \quad P: (2, 2, 8)$
28. $x^2 + y^2 + z^2 = 4, \quad P: (2, -2, 0)$
29. $x^2 + 3y^2 + 2z^2 = 6, \quad P: (2, 0, 1)$
30. $z = xy, \quad P: (3, -1, -3)$

Find a scalar function f such that $\mathbf{v} = \nabla f$.

31. $\mathbf{v} = \mathbf{i} - \mathbf{j} + \mathbf{k}$
32. $\mathbf{v} = x\mathbf{i} + y\mathbf{j} + \mathbf{k}$
33. $\mathbf{v} = x\mathbf{i} + 2y\mathbf{j} + z\mathbf{k}$
34. $\mathbf{v} = yz\mathbf{i} + zx\mathbf{j} + xy\mathbf{k}$
35. $\mathbf{v} = (x\mathbf{i} + y\mathbf{j})/(x^2 + y^2)$
36. $\mathbf{v} = e^x \sin y\,\mathbf{i} + e^x \cos y\,\mathbf{j}$

37. Find the directional derivative of $f = x^2 + y^2$ at the point P: (2, 2) in the directions of the vectors $\mathbf{i}$, $\mathbf{i} + \mathbf{j}$, $\mathbf{j}$, $-\mathbf{i} + \mathbf{j}$, $-\mathbf{i}$, $-\mathbf{i} - \mathbf{j}$, $-\mathbf{j}$, and $\mathbf{i} - \mathbf{j}$.

Find the directional derivative of f at P in the direction of $\mathbf{a}$, where

38. $f = x^2 - y^2$, P: (2, 3), $\mathbf{a} = \mathbf{i} + \mathbf{j}$

39. $f = 2x + 3y$, P: (0, 2), $\mathbf{a} = 3\mathbf{j}$

40. $f = x^2 + y^2$, P: (1, 2), $\mathbf{a} = \mathbf{i} - \mathbf{j}$

41. $f = y/x$, P: (3, −1), $\mathbf{a} = 9\mathbf{i} - 3\mathbf{j}$

42. $f = x + 2y - z$, P: (1, 4, 0), $\mathbf{a} = \mathbf{j} - \mathbf{k}$

43. $f = x^2 + y^2 + z^2$, P: (2, 0, 3), $\mathbf{a} = 2\mathbf{i} - \mathbf{j}$

44. What are the directional derivatives of a function $f(x, y, z)$ in the x, y, and z directions? In the negative x, y, and z directions?

45. Let r be the distance from a fixed point P to a point Q: (x, y, z). What are the level surfaces of r? Show that ∇r is a unit vector in the direction from P to Q.

46. Show that the functions in Probs. 1–6 are solutions of Laplace's equation.

Gradient of product etc. Granted sufficient differentiability, show that

47. $\nabla(fg) = f\nabla g + g\nabla f$

48. $\nabla(f^n) = nf^{n-1}\nabla f$

49. $\nabla(f/g) = (g\nabla f - f\nabla g)/g^2$

50. $\nabla^2(fg) = g\nabla^2 f + 2\nabla f \cdot \nabla g + f\nabla^2 g$

8.9 Transformation of Coordinate Systems and Vector Components

We shall now characterize the transformations which carry Cartesian coordinate systems into Cartesian coordinate systems and investigate the change of vector components under such a transformation. This problem is basic for theoretical as well as practical reasons.[7] We shall need the results in the next two sections.

Let x, y, z and x^*, y^*, z^* be any two systems of Cartesian coordinates. Let

$$\text{(1)} \quad \text{(a)}\ \mathbf{v} = v_1\mathbf{i} + v_2\mathbf{j} + v_3\mathbf{k} \qquad \text{and} \qquad \text{(b)}\ \mathbf{v} = v_1^*\mathbf{i}^* + v_2^*\mathbf{j}^* + v_3^*\mathbf{k}^*$$

be the representations of a given vector $\mathbf{v}$ in these two coordinate systems; here $\mathbf{i}, \mathbf{j}, \mathbf{k}$ and $\mathbf{i}^*, \mathbf{j}^*, \mathbf{k}^*$ are unit vectors in the positive x, y, z and x^*, y^*, z^* directions, respectively. We want to express the components v_1^*, v_2^*, v_3^* in terms of the components v_1, v_2, v_3, and conversely.

From (1a) we obtain

$$\text{(2)} \qquad \mathbf{i}^* \cdot \mathbf{v} = v_1\mathbf{i}^* \cdot \mathbf{i} + v_2\mathbf{i}^* \cdot \mathbf{j} + v_3\mathbf{i}^* \cdot \mathbf{k}.$$

[7] We shall keep the present section completely independent of Chap. 7, but readers familiar with matrices should recognize that we are dealing with orthogonal transformations and matrices and Theorems 1 and 2 follow from Theorem 2 in Sec. 7.14.

Similarly, by taking the scalar product of (1b) and $\mathbf{i}^*$, we have

$$\mathbf{i}^* \cdot \mathbf{v} = v_1^* \mathbf{i}^* \cdot \mathbf{i}^* + v_2^* \mathbf{i}^* \cdot \mathbf{j}^* + v_3^* \mathbf{i}^* \cdot \mathbf{k}^*.$$

Since the first scalar product on the right is 1 and the others are zero,

$$\mathbf{i}^* \cdot \mathbf{v} = v_1^*.$$

From this and (2) it follows that

$$v_1^* = \mathbf{i}^* \cdot \mathbf{i}\, v_1 + \mathbf{i}^* \cdot \mathbf{j}\, v_2 + \mathbf{i}^* \cdot \mathbf{k}\, v_3.$$

Similarly,

$$v_2^* = \mathbf{j}^* \cdot \mathbf{i}\, v_1 + \mathbf{j}^* \cdot \mathbf{j}\, v_2 + \mathbf{j}^* \cdot \mathbf{k}\, v_3,$$

$$v_3^* = \mathbf{k}^* \cdot \mathbf{i}\, v_1 + \mathbf{k}^* \cdot \mathbf{j}\, v_2 + \mathbf{k}^* \cdot \mathbf{k}\, v_3.$$

Hence the components of a vector $\mathbf{v}$ in a Cartesian coordinate system can be expressed as linear functions of the components of $\mathbf{v}$ with respect to another Cartesian coordinate system.

To write our transformation formulas in a simpler form we adopt the notation

$$\begin{aligned} &\mathbf{i}^* \cdot \mathbf{i} = c_{11} && \mathbf{i}^* \cdot \mathbf{j} = c_{12} && \mathbf{i}^* \cdot \mathbf{k} = c_{13} \\ &\mathbf{j}^* \cdot \mathbf{i} = c_{21} && \mathbf{j}^* \cdot \mathbf{j} = c_{22} && \mathbf{j}^* \cdot \mathbf{k} = c_{23} \\ &\mathbf{k}^* \cdot \mathbf{i} = c_{31} && \mathbf{k}^* \cdot \mathbf{j} = c_{32} && \mathbf{k}^* \cdot \mathbf{k} = c_{33}. \end{aligned} \tag{3}$$

Then we have

$$\begin{aligned} v_1^* &= c_{11}v_1 + c_{12}v_2 + c_{13}v_3 \\ v_2^* &= c_{21}v_1 + c_{22}v_2 + c_{23}v_3 \\ v_3^* &= c_{31}v_1 + c_{32}v_2 + c_{33}v_3. \end{aligned} \tag{4}$$

Using summation signs, we may write this more briefly

$$v_k^* = \sum_{l=1}^{3} c_{kl} v_l, \qquad k = 1, 2, 3. \tag{4'}$$

A similar consideration leads to the inverse formulas

$$\begin{aligned} v_1 &= c_{11}v_1^* + c_{21}v_2^* + c_{31}v_3^* \\ v_2 &= c_{12}v_1^* + c_{22}v_2^* + c_{32}v_3^* \\ v_3 &= c_{13}v_1^* + c_{23}v_2^* + c_{33}v_3^* \end{aligned} \tag{5}$$

or, more briefly,

$$v_l = \sum_{m=1}^{3} c_{ml} v_m^*, \qquad l = 1, 2, 3. \tag{5'}$$

Note that (4) and (5) contain the same coefficients c_{kl}, but these coefficients (except for c_{11}, c_{22}, c_{33}) occupy different positions in (4) and (5).

The geometrical interpretation of those nine coefficients c_{kl} is very simple. Since **i** and **i*** are unit vectors, it follows from (1) in Sec. 6.5 that $c_{11} = \mathbf{i}^* \cdot \mathbf{i}$ is the cosine of the angle between the positive x^* and x axes. Similarly, $c_{12} = \mathbf{i}^* \cdot \mathbf{j}$ is the cosine of the angle between the positive x^* and y axes, and so on.

The coefficients c_{kl} satisfy certain important relations which we shall now derive. By inserting (5′) into (4′) we find

$$(6) \qquad v_k^* = \sum_{l=1}^{3} c_{kl} v_l = \sum_{l=1}^{3} c_{kl} \sum_{m=1}^{3} c_{ml} v_m^* = \sum_{m=1}^{3} v_m^* \left(\sum_{l=1}^{3} c_{kl} c_{ml} \right),$$

where $k = 1, 2, 3$. For example, if $k = 1$, this becomes

$$v_1^* = v_1^* \left(\sum_{l=1}^{3} c_{1l} c_{1l} \right) + v_2^* \left(\sum_{l=1}^{3} c_{1l} c_{2l} \right) + v_3^* \left(\sum_{l=1}^{3} c_{1l} c_{3l} \right).$$

In order that this relation hold for any vector $\mathbf{v} = v_1^* \mathbf{i}^* + v_2^* \mathbf{j}^* + v_3^* \mathbf{k}^*$, the first sum must be 1 and the other two sums must be zero. For $k = 2$ and $k = 3$ the situation is similar. Consequently, (6) holds for any vector if and only if

$$(7) \qquad \sum_{l=1}^{3} c_{kl} c_{ml} = \begin{cases} 0 & (k \neq m) \\ 1 & (k = m). \end{cases}$$

Using the so-called *Kronecker*[8] *symbol* or **Kronecker delta**

$$\delta_{km} = \begin{cases} 0 & (k \neq m) \\ 1 & (k = m) \end{cases}$$

we may write (7) in the form

$$(7') \qquad \sum_{l=1}^{3} c_{kl} c_{ml} = \delta_{km} \qquad (k, m = 1, 2, 3).$$

Forming three vectors with components

$$c_{11}, c_{12}, c_{13} \qquad c_{21}, c_{22}, c_{23} \qquad c_{31}, c_{32}, c_{33}$$

we see that the left side of (7′) is the scalar product of two of these vectors, and (7′) implies that these vectors are orthogonal unit vectors. Hence their scalar triple product has the value +1 or −1; that is,

[8]LEOPOLD KRONECKER (1823–1891), German mathematician, who made important contributions to algebra and the theory of numbers.

$$\begin{vmatrix} c_{11} & c_{12} & c_{13} \\ c_{21} & c_{22} & c_{23} \\ c_{31} & c_{32} & c_{33} \end{vmatrix} = \pm 1. \tag{8}$$

We mention, without proof, that if both coordinate systems under consideration are right-handed (or both left-handed), then the determinant has the value $+1$, whereas if one system is right-handed and the other left-handed, the determinant has the value -1. We may now sum up our result as follows.

Theorem 1 (Transformation law for vector components)
The components v_1, v_2, v_3 and v_1^*, v_2^*, v_3^* of any vector $\mathbf{v}$ *with respect to any two Cartesian coordinate systems can be obtained from each other by means of* (4) *and* (5), *where the coefficients c_{kl} are given by* (3) *and satisfy* (7) *and* (8).

From our consideration of vector components we may now immediately obtain the formulas for the transformation of a Cartesian coordinate system into any other such system as follows.

If the xyz and $x^*y^*z^*$ coordinate systems have the same origin, then $\mathbf{v}$ may be bound at the origin and regarded as the position vector of its terminal point Q. If (x, y, z) and (x^*, y^*, z^*) are the coordinates of Q with respect to the two coordinate systems, then in (4) and (5)

$$v_1 = x, \quad v_2 = y, \quad v_3 = z \qquad \text{and} \qquad v_1^* = x^*, \quad v_2^* = y^*, \quad v_3^* = z^*.$$

Consequently, (4) and (5), written in terms of x, y, z and x^*, y^*, z^* instead of v_1, v_2, v_3 and v_1^*, v_2^*, v_3^*, represent the transformations between those two coordinate systems with common origin.

The most general transformation of a Cartesian coordinate system into another such system may be decomposed into a transformation of the type just considered and a translation. Under a translation, corresponding coordinates differ merely by a constant. We thus obtain

Theorem 2 (Transformation law for Cartesian coordinates)
The transformation of any Cartesian xyz-coordinate system into any other Cartesian $x^*y^*z^*$-coordinate system is of the form

$$\begin{aligned} x^* &= c_{11}x + c_{12}y + c_{13}z + b_1 \\ y^* &= c_{21}x + c_{22}y + c_{23}z + b_2 \\ z^* &= c_{31}x + c_{32}y + c_{33}z + b_3 \end{aligned} \tag{9}$$

and conversely

$$\begin{aligned} x &= c_{11}x^* + c_{21}y^* + c_{31}z^* + \widetilde{b}_1 \\ y &= c_{12}x^* + c_{22}y^* + c_{32}z^* + \widetilde{b}_2 \\ z &= c_{13}x^* + c_{23}y^* + c_{33}z^* + \widetilde{b}_3 \end{aligned} \tag{10}$$

where the coefficients c_{kl} *are given by* (3) *and satisfy* (7) *and* (8) *and* $b_1, b_2, b_3, \tilde{b}_1, \tilde{b}_2, \tilde{b}_3$ *are constants.*

Important applications of our results will be considered in the following two sections.

Problems for Sec. 8.9

1. Interpret all the coefficients in (4′) geometrically.

Determine the constants c_{kl} and b_k such that (9) represents:

2. A translation which carries the origin into the point $(1, 3, 2)$

3. A translation which carries the point $(2, -1, 0)$ into the point $(1, 4, 2)$

4. A reflection in the xy-plane

5. A reflection in the plane $y = x$

6. A rotation about the x-axis through an angle θ

7. A rotation such that the positive x^*, y^*, z^* axes coincide with the positive y, z, x axes, respectively

8. What are the values of the determinant (8) in Probs. 2–7?

9. Derive (5).

10. From (1), Sec. 6.5, it follows that the value of a scalar product is independent of the particular choice of Cartesian coordinates. Using (4) or (5) in this section, show that this follows also from (8), Sec. 6.5.

8.10 Divergence of a Vector Field

Let $\mathbf{v}(x, y, z)$ be a differentiable vector function, where x, y, z are Cartesian coordinates in space, and let v_1, v_2, v_3 be the components of $\mathbf{v}$. Then the function

$$\text{div}\,\mathbf{v} = \frac{\partial v_1}{\partial x} + \frac{\partial v_2}{\partial y} + \frac{\partial v_3}{\partial z} \tag{1}$$

is called the **divergence** *of* $\mathbf{v}$ or the *divergence of the vector field defined by* $\mathbf{v}$.

Another common notation for the divergence of $\mathbf{v}$ is $\nabla \cdot \mathbf{v}$,

$$\begin{aligned}\text{div}\,\mathbf{v} = \nabla \cdot \mathbf{v} &= \left(\frac{\partial}{\partial x}\mathbf{i} + \frac{\partial}{\partial y}\mathbf{j} + \frac{\partial}{\partial z}\mathbf{k}\right) \cdot (v_1\mathbf{i} + v_2\mathbf{j} + v_3\mathbf{k}) \\ &= \frac{\partial v_1}{\partial x} + \frac{\partial v_2}{\partial y} + \frac{\partial v_3}{\partial z},\end{aligned}$$

with the understanding that the "product" $(\partial/\partial x)v_1$ in the dot product means the partial derivative $\partial v_1/\partial x$, etc. This is a convenient notation, but nothing more. Note that $\nabla \cdot \mathbf{v}$ means the scalar div $\mathbf{v}$ whereas ∇f means the vector grad f defined in Sec. 8.8.

For example, if

$$\mathbf{v} = 3xz\mathbf{i} + 2xy\mathbf{j} - yz^2\mathbf{k}, \qquad \text{then} \qquad \operatorname{div} \mathbf{v} = 3z + 2x - 2yz.$$

We shall see later that the divergence has an important physical meaning. Clearly the values of a function which characterizes a physical or geometrical property must be independent of the particular choice of coordinates; that is, those values must be invariant with respect to coordinate transformations.

Theorem 1 (Invariance of the divergence with respect to coordinate transformations)

The values of div $\mathbf{v}$ *depend only on the points in space (and, of course, on* $\mathbf{v}$*) but not on the particular choice of the coordinates in* (1), *so that with respect to other Cartesian coordinates* x^*, y^*, z^* *and corresponding components* v_1^*, v_2^*, v_3^* *of* $\mathbf{v}$ *the function* div $\mathbf{v}$ *is given by*

$$\operatorname{div} \mathbf{v} = \frac{\partial v_1^*}{\partial x^*} + \frac{\partial v_2^*}{\partial y^*} + \frac{\partial v_3^*}{\partial z^*}. \tag{2}$$

Proof. We shall derive (2) from (1). To simplify our formulas in this proof, we adopt the notation

$$x_1 = x, \quad x_2 = y, \quad x_3 = z \qquad \text{and} \qquad x_1^* = x^*, \quad x_2^* = y^*, \quad x_3^* = z^*.$$

Then we may write formula (9) in Sec. 8.9 in the short form

$$x_k^* = \sum_{l=1}^{3} c_{kl} x_l + b_k \qquad (k = 1, 2, 3). \tag{3}$$

The chain rule for functions of several variables (Sec. 8.7) yields

$$\frac{\partial v_l}{\partial x_l} = \sum_{k=1}^{3} \frac{\partial v_l}{\partial x_k^*} \frac{\partial x_k^*}{\partial x_l}. \tag{4}$$

In this sum, $\partial x_k^*/\partial x_l = c_{kl}$, as follows from (3). By (5′) in Sec. 8.9,

$$v_l = \sum_{m=1}^{3} c_{ml} v_m^*.$$

By differentiating this we have

$$\frac{\partial v_l}{\partial x_k^*} = \sum_{m=1}^{3} c_{ml} \frac{\partial v_m^*}{\partial x_k^*}.$$

If we substitute these expressions into (4), we obtain

$$\frac{\partial v_l}{\partial x_l} = \sum_{k=1}^{3} \sum_{m=1}^{3} c_{ml} \frac{\partial v_m^*}{\partial x_k^*} c_{kl} \qquad (l = 1, 2, 3).$$

By adding these three formulas we find

$$\operatorname{div} \mathbf{v} = \sum_{l=1}^{3} \frac{\partial v_l}{\partial x_l} = \sum_{k=1}^{3} \sum_{m=1}^{3} \sum_{l=1}^{3} c_{kl} c_{ml} \frac{\partial v_m^*}{\partial x_k^*}.$$

Because of (7′) in Sec. 8.9 this reduces to

$$\operatorname{div} \mathbf{v} = \sum_{k=1}^{3} \sum_{m=1}^{3} \delta_{km} \frac{\partial v_m^*}{\partial x_k^*} = \frac{\partial v_1^*}{\partial x_1^*} + \frac{\partial v_2^*}{\partial x_2^*} + \frac{\partial v_3^*}{\partial x_3^*}.$$

We see that the expression on the right is identical with that in (2), and the proof is complete. ∎

If $f(x, y, z)$ is a twice differentiable scalar function, then

$$\operatorname{grad} f = \frac{\partial f}{\partial x}\mathbf{i} + \frac{\partial f}{\partial y}\mathbf{j} + \frac{\partial f}{\partial z}\mathbf{k}$$

and by (1),

$$\operatorname{div}(\operatorname{grad} f) = \frac{\partial^2 f}{\partial x^2} + \frac{\partial^2 f}{\partial y^2} + \frac{\partial^2 f}{\partial z^2}.$$

The expression on the right is the Laplacian of f (cf. Sec. 8.8). Thus

$$\textbf{(5)} \qquad \operatorname{div}(\operatorname{grad} f) = \nabla^2 f.$$

Example 1. Gravitational force

The gravitational force $\mathbf{p}$ in Example 4, Sec. 8.8, is the gradient of the scalar function $f(x, y, z) = c/r$, which satisfies Laplace's equation $\nabla^2 f = 0$. According to (5), this means that $\operatorname{div} \mathbf{p} = 0$ $(r > 0)$. ∎

The following example, taken from hydrodynamics, may serve as an introductory illustration of the physical significance of the divergence of a vector field. A more detailed physical interpretation of the divergence will be given later (in Sec. 9.9).

Example 2. Motion of a compressible fluid

We consider the motion of a fluid in a region R having no *sources* or *sinks* in R, that is, no points at which fluid is produced or disappears. The concept of *fluid state* is meant to cover also gases and vapors. Fluids in the restricted sense, or liquids, have a small compressibility which can be neglected in many problems. Gases and vapors have a large compressibility, that is, their density ρ (= mass per unit volume) depends on the coordinates x, y, z in space (and may depend on time t). We assume that our fluid is compressible.

We consider the flow through a small rectangular parallelepiped W of dimensions[9] Δx, Δy, Δz with edges parallel to the coordinate axes (Fig. 169). W has the volume $\Delta V = \Delta x \, \Delta y \, \Delta z$. Let

$$\mathbf{v} = v_1\mathbf{i} + v_2\mathbf{j} + v_3\mathbf{k}$$

be the velocity vector of the motion. We set

$$\mathbf{u} = \rho\mathbf{v} = u_1\mathbf{i} + u_2\mathbf{j} + u_3\mathbf{k} \tag{6}$$

and assume that **u** and **v** are continuously differentiable vector functions of x, y, z, and t. Let us calculate the change in the mass included in W by considering the **flux** across the boundary, that is, the total loss of mass leaving W per unit time. Consider the flow through the left-hand face of W, whose area is $\Delta x \Delta z$. The components v_1 and v_3 of **v** are parallel to that face and contribute nothing to that flow. Hence the mass of fluid entering through that face during a short time interval Δt is given approximately by

$$(\rho v_2)_y \Delta x \, \Delta z \, \Delta t = (u_2)_y \, \Delta x \, \Delta z \, \Delta t$$

where the subscript y indicates that this expression refers to the left-hand face. The mass of fluid leaving the parallelepiped W through the opposite face during the same time interval is approximately $(u_2)_{y+\Delta y} \, \Delta x \, \Delta z \, \Delta t$, where the subscript $y + \Delta y$ indicates that this expression refers to the right-hand face. The difference

$$\Delta u_2 \, \Delta x \, \Delta z \, \Delta t = \frac{\Delta u_2}{\Delta y} \Delta V \, \Delta t \qquad [\Delta u_2 = (u_2)_{y+\Delta y} - (u_2)_y]$$

is the approximate loss of mass. Two similar expressions are obtained by considering the other two pairs of parallel faces of W. If we add these three expressions, we find that the total loss of mass in W during the time interval Δt is approximately

$$\left(\frac{\Delta u_1}{\Delta x} + \frac{\Delta u_2}{\Delta y} + \frac{\Delta u_3}{\Delta z}\right) \Delta V \, \Delta t$$

where

$$\Delta u_1 = (u_1)_{x+\Delta x} - (u_1)_x \qquad \text{and} \qquad \Delta u_3 = (u_3)_{z+\Delta z} - (u_3)_z.$$

This loss of mass in W is caused by the time rate of change of the density and is thus equal to

$$-\frac{\partial \rho}{\partial t} \Delta V \, \Delta t.$$

If we equate both expressions, divide the resulting equation by $\Delta V \, \Delta t$, and let Δx, Δy, Δz and Δt approach zero, then we obtain

$$\operatorname{div} \mathbf{u} = \operatorname{div} (\rho \mathbf{v}) = -\frac{\partial \rho}{\partial t}$$

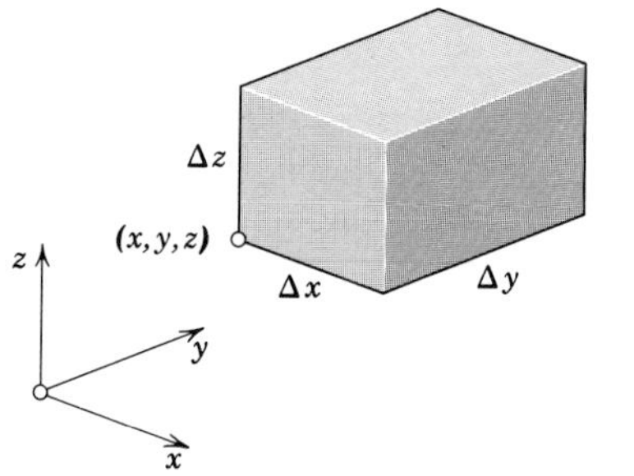

Fig. 169. Physical interpretation of the divergence

[9] It is a standard usage to indicate small quantities by Δ; this has, of course, nothing to do with the Laplacian.

or

$$\frac{\partial \rho}{\partial t} + \operatorname{div}(\rho \mathbf{v}) = 0. \tag{7}$$

This important relation is called the *condition for the conservation of mass* or the **continuity equation** *of a compressible fluid flow.*

If the flow is *steady,* that is, independent of time t, then $\partial \rho/\partial t = 0$, and the continuity equation is

$$\operatorname{div}(\rho \mathbf{v}) = 0. \tag{8}$$

If the fluid is incompressible, then the density ρ is constant, and (8) becomes

$$\operatorname{div} \mathbf{v} = 0. \tag{9}$$

This relation is known as the **condition of incompressibility.** It expresses the fact that the balance of outflow and inflow for a given volume element is zero at any time. Clearly the assumption that the flow has no sources or sinks in R is essential to our argument.

Problems for Sec. 8.10

Find the divergence of the following vector functions.

1. $x\mathbf{i} + y\mathbf{j}$

2. $y^2\mathbf{i} + x^2\mathbf{j}$

3. $(x^2 - 3y^2)\mathbf{i}$

4. $x^2\mathbf{i} + y^2\mathbf{j} - z\mathbf{k}$

5. $xyz(\mathbf{i} + \mathbf{j} + \mathbf{k})$

6. $x^2\mathbf{i} + y^2\mathbf{j} + z^2\mathbf{k}$

7. $yz^2\mathbf{i} - zx^2\mathbf{k}$

8. $(x\mathbf{i} + y\mathbf{j} + z\mathbf{k})/(x^2 + y^2 + z^2)^{3/2}$

9. Is it obvious from (1) that div $\mathbf{v}$ is a scalar?

Formulas for the divergence. Show that

10. $\operatorname{div}(k\mathbf{v}) = k \operatorname{div} \mathbf{v}$ (k constant)

11. $\operatorname{div}(f\mathbf{v}) = f \operatorname{div} \mathbf{v} + \mathbf{v} \cdot \nabla f$

12. $\operatorname{div}(f\nabla g) = f\nabla^2 g + \nabla f \cdot \nabla g$

13. $\operatorname{div}(f\nabla g) - \operatorname{div}(g\nabla f) = f\nabla^2 g - g\nabla^2 f$

14. Show that the continuity equation (7) may be written

$$\frac{\partial \rho}{\partial t} + \rho \operatorname{div} \mathbf{v} + \mathbf{v} \cdot \nabla \rho = 0.$$

Using the formula in Prob. 11, find the divergence of

15. $e^x(\sin y\,\mathbf{i} + \cos y\,\mathbf{j})$

16. $(x\mathbf{i} + y\mathbf{j} + z\mathbf{k})/(x^2 + y^2 + z^2)^{3/2}$

17. Consider a steady flow whose velocity vector is $\mathbf{v} = y\mathbf{i}$. Show that it has the following properties. The flow is incompressible. The particles which at time $t = 0$ are in the cube bounded by the planes $x = 0$, $x = 1$, $y = 0$, $y = 1$, $z = 0$, $z = 1$ occupy at $t = 1$ the volume 1.

18. Consider a steady flow having the velocity $\mathbf{v} = x\mathbf{i}$. Show that the individual particles have the position vectors $\mathbf{r}(t) = c_1e^t\mathbf{i} + c_2\mathbf{j} + c_3\mathbf{k}$, where c_1, c_2, c_3 are constants, the flow is compressible, and the particles which at $t = 0$ fill the cube in Prob. 17 occupy at $t = 1$ the volume e.

Find the directional derivative of div $\mathbf{u}$ at the point P: (4, 4, 2) in the direction of the corresponding outer normal of the sphere $x^2 + y^2 + z^2 = 36$ where

19. $\mathbf{u} = x^4\mathbf{i} + y^4\mathbf{j} + z^4\mathbf{k}$

20. $\mathbf{u} = xz\mathbf{i} + yx\mathbf{j} + zy\mathbf{k}$

8.11 Curl of a Vector Field

Let x, y, z be right-handed Cartesian coordinates in space, and let

$$\mathbf{v}(x, y, z) = v_1\mathbf{i} + v_2\mathbf{j} + v_3\mathbf{k}$$

be a differentiable vector function. Then the function

$$\text{(1)}\qquad \operatorname{curl} \mathbf{v} = \nabla \times \mathbf{v} = \begin{vmatrix} \mathbf{i} & \mathbf{j} & \mathbf{k} \\ \dfrac{\partial}{\partial x} & \dfrac{\partial}{\partial y} & \dfrac{\partial}{\partial z} \\ v_1 & v_2 & v_3 \end{vmatrix}$$

$$= \left(\frac{\partial v_3}{\partial y} - \frac{\partial v_2}{\partial z}\right)\mathbf{i} + \left(\frac{\partial v_1}{\partial z} - \frac{\partial v_3}{\partial x}\right)\mathbf{j} + \left(\frac{\partial v_2}{\partial x} - \frac{\partial v_1}{\partial y}\right)\mathbf{k}$$

is called the **curl** *of the vector function* $\mathbf{v}$ or the *curl of the vector field defined by* $\mathbf{v}$. In the case of a left-handed Cartesian coordinate system, the symbolic determinant in (1) is preceded by a minus sign, in agreement with (5) in Sec. 6.8.

Instead of curl $\mathbf{v}$ the notation rot $\mathbf{v}$ is also used in the literature.

Theorem 1 (Invariance of the curl)
The length and direction of curl $\mathbf{v}$ *are independent of the particular choice of Cartesian coordinate systems in space.*

The proof of this theorem is included in the answer to Prob. 19.

The curl plays an important role in many applications. Its significance will be explained in more detail in Sec. 9.11. At present we confine ourselves to some simple examples and remarks.

Example 1. Rotation of a rigid body

We have seen in Example 3, Sec. 6.8, that a rotation of a rigid body B about a fixed axis in space can be described by a vector $\mathbf{w}$ of magnitude ω in the direction of the axis of rotation, where ω (> 0) is the angular speed of the rotation, and $\mathbf{w}$ is directed so that the rotation appears clockwise if we look in the direction of $\mathbf{w}$. According to (7), Sec. 6.8, the velocity field of the rotation can be represented in the form

$$\mathbf{v} = \mathbf{w} \times \mathbf{r}$$

where $\mathbf{r}$ is the position vector of a moving point with respect to a Cartesian coordinate system having the origin on the axis of rotation. Let us choose right-handed Cartesian coordinates such that $\mathbf{w} = \omega\mathbf{k}$; that is, the axis of rotation is the z-axis. Then (cf. Example 3 in Sec. 8.1)

$$\mathbf{v} = \mathbf{w} \times \mathbf{r} = -\omega y\mathbf{i} + \omega x\mathbf{j}$$

and, therefore,

$$\operatorname{curl} \mathbf{v} = \begin{vmatrix} \mathbf{i} & \mathbf{j} & \mathbf{k} \\ \dfrac{\partial}{\partial x} & \dfrac{\partial}{\partial y} & \dfrac{\partial}{\partial z} \\ -\omega y & \omega x & 0 \end{vmatrix} = 2\omega\mathbf{k},$$

that is,

$$\text{curl } \mathbf{v} = 2\mathbf{w}. \tag{2}$$

Hence, in the case of a rotation of a rigid body, the curl of the velocity field has the direction of the axis of rotation, and its magnitude equals twice the angular speed ω *of the rotation.*

Note that our result does not depend on the particular choice of the Cartesian coordinate system in space. ∎

For any twice continuously differentiable scalar function f,

$$\text{curl } (\text{grad } f) = \mathbf{0}, \tag{3}$$

as can easily be verified by direct calculation. *Hence if a vector function is the gradient of a scalar function, its curl is the zero vector.* Since the curl characterizes the rotation in a field, we also say more briefly that *gradient fields describing a motion are irrotational.* (If such a field occurs in some other connection, not as a velocity field, it is usually called *conservative;* cf. at the end of Sec. 8.8.)

Example 2

The gravitational field in Example 4, Sec. 8.8, has curl $\mathbf{p} = \mathbf{0}$. The field in Example 1 of the present section is not irrotational. A similar velocity field is obtained by stirring coffee in a cup.

Problems for Sec. 8.11

Find curl $\mathbf{v}$ where, with respect to right-handed Cartesian coordinates,

1. $\mathbf{v} = y\mathbf{i} - x\mathbf{j}$

2. $\mathbf{v} = y\mathbf{i} + z\mathbf{j} + x\mathbf{k}$

3. $\mathbf{v} = z^2\mathbf{i} + x^2\mathbf{j} + y^2\mathbf{k}$

4. $\mathbf{v} = x^2\mathbf{i} + y^2\mathbf{j} + z^2\mathbf{k}$

5. $\mathbf{v} = xz\mathbf{i} - yz\mathbf{j}$

6. $\mathbf{v} = (x\mathbf{i} + y\mathbf{j} + z\mathbf{k})/(x^2 + y^2 + z^2)^{3/2}$

In each case the velocity $\mathbf{v}$ of a steady fluid motion is given. Find curl $\mathbf{v}$. Is the motion incompressible? Find the paths of the particles.

7. $\mathbf{v} = x\mathbf{i} + y\mathbf{j}$

8. $\mathbf{v} = y\mathbf{i} - x\mathbf{j}$

9. $\mathbf{v} = y^3\mathbf{i}$

Formulas for curl, div, etc. Show that, granted sufficient differentiability,

10. $\text{curl } (\mathbf{u} + \mathbf{v}) = \text{curl } \mathbf{u} + \text{curl } \mathbf{v}$

11. $\text{div } (\text{curl } \mathbf{v}) = 0$

12. $\text{curl } (f\mathbf{v}) = \text{grad } f \times \mathbf{v} + f \text{ curl } \mathbf{v}$

13. $\text{curl } (\text{grad } f) = \mathbf{0}$

14. $\text{div } (\mathbf{u} \times \mathbf{v}) = \mathbf{v} \cdot \text{curl } \mathbf{u} - \mathbf{u} \cdot \text{curl } \mathbf{v}$

15. $\text{div } (g\nabla f \times f\nabla g) = 0$

With respect to right-handed Cartesian coordinates, let $\mathbf{u} = y\mathbf{i} + z\mathbf{j} + x\mathbf{k}$ and $\mathbf{v} = xy\mathbf{i} + yz\mathbf{j} + zx\mathbf{k}$. Find

16. $\text{curl } (\mathbf{u} \times \mathbf{v})$, $\text{div } (\mathbf{u} \times \mathbf{v})$

17. $\mathbf{u} \times \text{curl } \mathbf{v}$, $\mathbf{v} \times \text{curl } \mathbf{u}$

18. Verify by direct calculation that, in Example 2, curl $\mathbf{p} = \mathbf{0}$.

19. Prove Theorem 1.

20. Using Prob. 12, show that for a vector $\mathbf{w}$ of constant direction, if curl $\mathbf{w} \neq \mathbf{0}$, then curl $\mathbf{w}$ is orthogonal to $\mathbf{w}$.

CHAPTER

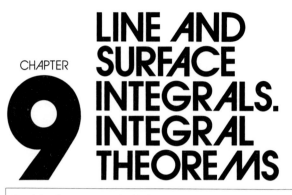

9 LINE AND SURFACE INTEGRALS. INTEGRAL THEOREMS

In this chapter we shall define line integrals and surface integrals and consider some important applications of such integrals, which occur frequently in connection with physical and engineering problems. We shall see that a line integral is a natural generalization of a definite integral, and a surface integral is a generalization of a double integral.

Line integrals can be transformed into double integrals (Sec. 9.4) or into surface integrals (Sec. 9.10), and conversely. Triple integrals can be transformed into surface integrals (Sec. 9.8). These transformations are of great practical importance. The corresponding formulas of Gauss, Green, and Stokes serve as powerful tools in many applications as well as in theoretical problems. We shall see that they also lead to a better understanding of the physical meaning of the divergence and the curl of a vector function.

Prerequisites for this chapter: elementary integral calculus and Chap. 8.
Sections which may be omitted in a shorter course: 9.6, 9.9, 9.11.
References: Appendix 1, Part C.
Answers to problems: Appendix 2.

9.1 Line Integral

The concept of a line integral is a simple and natural generalization of the concept of a definite integral

(1) $$\int_a^b f(x)\,dx.$$

In (1) we integrate along the x-axis from a to b and the integrand f is a function defined at each point between a and b. In the case of a line integral we shall integrate along a curve C in space (or in the plane) and the integrand f will be a function defined at each point of C. (Hence *curve integral* would be a better name, but *line integral* is the standard term.)

The way of defining a line integral is quite similar to the familiar way of defining a definite integral known from calculus. We may proceed as follows.

We consider a curve C in space. We *orient* C by choosing one of the two directions along C as the *positive direction.* The opposite direction along C is then called the *negative direction.* Let A denote the initial point and B the

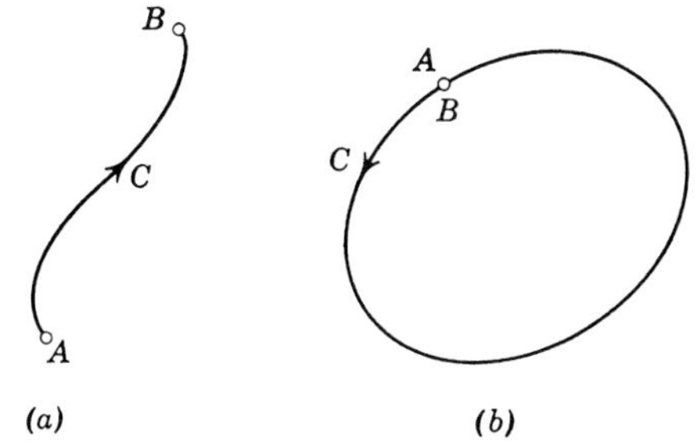

Fig. 170. Oriented curve

terminal point of C under that orientation. (These points may coincide as in Fig. 170b; then C is closed.)

We assume that C is simple (Sec. 8.3) and has a representation

$$\mathbf{r}(s) = x(s)\mathbf{i} + y(s)\mathbf{j} + z(s)\mathbf{k} \qquad (a \leqq s \leqq b) \tag{2}$$

(s the arc length of C; cf. Sec. 8.4) such that $\mathbf{r}(s)$ is continuous and has a continuous first derivative $\mathbf{r}'(s)$ which is different from the zero vector for all s under consideration. Then C is a **smooth curve,** that is, C has a unique tangent at each of its points whose direction varies continuously as we traverse the curve.

Let $f(x, y, z)$ be a given function which is defined (at least) at each point of C, and is a continuous function of s. We subdivide C into n portions in an arbitrary manner (Fig. 171); let P_0 $(= A)$, P_1, P_2, $\cdots$, P_n $(= B)$ be the endpoints of these portions, and let

$$s_0\,(= a) < s_1 < s_2 < \cdots < s_n\,(= b)$$

be the corresponding values of s. Then we choose an arbitrary point on each portion, say, a point Q_1 between P_0 and P_1, a point Q_2 between P_1 and P_2, etc. Taking the values of f at these points Q_1, Q_2, $\cdots$, Q_n, we form the sum

$$J_n = \sum_{m=1}^{n} f(x_m, y_m, z_m)\,\Delta s_m$$

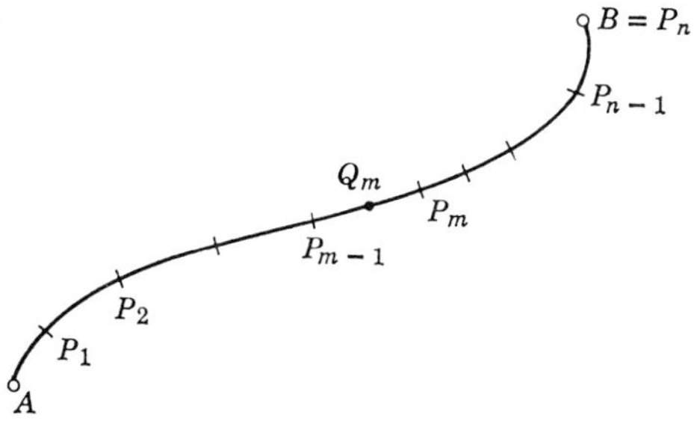

Fig. 171. Subdivision of C

where x_m, y_m, z_m are the coordinates of Q_m and

$$\Delta s_m = s_m - s_{m-1}.$$

This we do for each $n = 2, 3, \cdots$ in a completely independent manner, but so that the greatest Δs_m approaches zero as n approaches infinity. In this way we obtain a sequence of real numbers J_2, J_3, $\cdots$. The limit of this sequence is called the **line integral** *of f along C from A to B,* and is denoted by

$$\int_C f(x, y, z)\, ds.$$

The curve C is called the **path of integration.**

Since, by assumption, f is continuous and C is smooth, that limit exists and is independent of the choice of those subdivisions and points Q_m. In fact, the position of a point P on C is determined by the corresponding value of the arc length s; since A and B correspond to $s = a$ and $s = b$, respectively, we thus have

$$\text{(3)} \qquad \int_C f(x, y, z)\, ds = \int_a^b f[x(s), y(s), z(s)]\, ds.$$

Hence the line integral is equal to the definite integral on the right, for which the statement holds, as is known from calculus.

General assumption

In this book, every path of integration of a line integral is assumed to be **piecewise smooth,** *that is, consists of finitely many smooth curves.*

For a line integral over a *closed* path C, the symbol

$$\oint_C \qquad \left(\text{instead of } \int_C\right)$$

is sometimes used in the literature.

From the definition it follows that familiar properties of ordinary definite integrals are equally valid for line integrals:

$$\text{(a)} \quad \int_C kf\, ds = k \int_C f\, ds \qquad (k \text{ constant})$$

$$\text{(4)} \qquad \text{(b)} \quad \int_C (f + g)\, ds = \int_C f\, ds + \int_C g\, ds$$

$$\text{(c)} \quad \int_C f\, ds = \int_{C_1} f\, ds + \int_{C_2} f\, ds$$

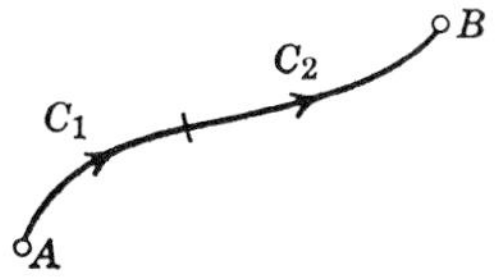

Fig. 172. Formula (4c)

where in (4c) the path C is subdivided into two arcs C_1 and C_2 which have the same orientation as C (Fig. 172). In (4b) the orientation of C is the same in all three integrals. If the sense of integration along C is reversed, the value of the integral is multiplied by -1.

Examples will be considered in the following sections.

9.2 Evaluation of Line Integrals

A line integral is evaluated by reducing it to a definite integral. This reduction is quite simple and is done by means of the representation of the path of integration C, as follows.

If C is represented by

$$\mathbf{r}(s) = x(s)\mathbf{i} + y(s)\mathbf{j} + z(s)\mathbf{k} \qquad a \leqq s \leqq b$$

(s the arc length of C), we use [cf. (3) in Sec. 9.1]

$$\int_C f(x, y, z)\, ds = \int_a^b f[x(s), y(s), z(s)]\, ds. \tag{1}$$

If C is represented by

$$\mathbf{r}(t) = x(t)\mathbf{i} + y(t)\mathbf{j} + z(t)\mathbf{k} \qquad t_0 \leqq t \leqq t_1$$

(t any parameter), we use

$$\int_C f(x, y, z)\, ds = \int_{t_0}^{t_1} f[x(t), y(t), z(t)] \frac{ds}{dt}\, dt, \tag{2}$$

where, by (3) in Sec. 8.4,

$$\frac{ds}{dt} = \sqrt{\dot{\mathbf{r}} \cdot \dot{\mathbf{r}}} = \sqrt{\dot{x}^2 + \dot{y}^2 + \dot{z}^2}.$$

Here we assume $\mathbf{r}(t)$ and $\dot{\mathbf{r}}(t)$ to be continuous and $\dot{\mathbf{r}}(t) \neq \mathbf{0}$, in agreement with our general assumption in the previous section.

We derive (2). Instead of $\mathbf{r}(t)$ we write momentarily

$$\tilde{\mathbf{r}}(t) = \tilde{x}(t)\mathbf{i} + \tilde{y}(t)\mathbf{j} + \tilde{z}(t)\mathbf{k}.$$

From this we obtain the arc length $s(t)$. We now set $\mathbf{r}(s(t)) = \tilde{\mathbf{r}}(t)$, that is, $x(s(t)) = \tilde{x}(t)$, etc. Then the familiar substitution rule for a definite integral yields in (1) on the right

$$\int_a^b f[x(s), y(s), z(s)]\, ds = \int_{t_0}^{t_1} f[\tilde{x}(t), \tilde{y}(t), \tilde{z}(t)] \frac{ds}{dt}\, dt.$$

This is the integral in (2), except for the notation. ∎

Formula (2) takes care of most applications since in most practical cases, a representation $\mathbf{r}(t)$ will be available or can readily be obtained.

Example 1. Illustration of (1)

Integrate $f(x, y) = 2xy^2$ over the circular arc (Fig. 173)

$$\mathbf{r}(s) = \cos s\,\mathbf{i} + \sin s\,\mathbf{j} \qquad 0 \leqq s \leqq \pi/2.$$

Since $x(s) = \cos s$ and $y(s) = \sin s$, formula (1) yields

$$\int_C f(x, y)\, ds = \int_C 2xy^2\, ds = \int_0^{\pi/2} 2 \cos s \sin^2 s\, ds$$

$$= 2\int_0^1 u^2\, du = \frac{2}{3} \qquad (\sin s = u).$$

Example 2. Illustration of (2)

Evaluate $\int_C xy^3\, ds$, where C is the segment of the line $y = 2x$ in the xy-plane from A: $(-1, -2, 0)$ to B: $(1, 2, 0)$.

We can represent C in the form[1]

$$\mathbf{r}(t) = t\mathbf{i} + 2t\mathbf{j} \qquad -1 \leqq t \leqq 1.$$

Then

$$\dot{\mathbf{r}} = \mathbf{i} + 2\mathbf{j} \qquad \text{and} \qquad \frac{ds}{dt} = \sqrt{\dot{\mathbf{r}} \cdot \dot{\mathbf{r}}} = \sqrt{5}.$$

On C we have $xy^3 = t(2t)^3 = 8t^4$. Hence

$$\int_C xy^3\, ds = 8\sqrt{5} \int_{-1}^1 t^4\, dt = \frac{16}{\sqrt{5}} \approx 7.16.$$

Example 3. Illustration of (2) for a space curve

Evaluate $\int_C (x^2 + y^2 + z^2)^2\, ds$ where C is the arc of the circular helix (Sec. 8.3)

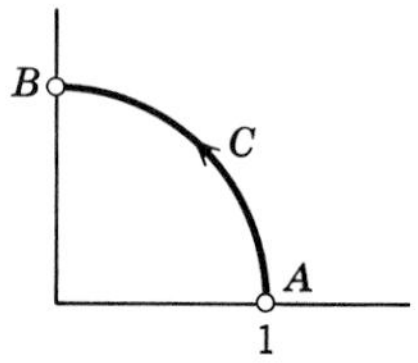

Fig. 173. Example 1

[1]Of course, since $x = t$ we may write x in place of t.

$$\mathbf{r}(t) = \cos t\,\mathbf{i} + \sin t\,\mathbf{j} + 3t\mathbf{k}$$

from A: $(1, 0, 0)$ to B: $(1, 0, 6\pi)$. We find $ds/dt = \sqrt{10}$. On C,

$$(x^2 + y^2 + z^2)^2 = [\cos^2 t + \sin^2 t + (3t)^2]^2 = (1 + 9t^2)^2.$$

Since C corresponds to $0 \leqq t \leqq 2\pi$, we thus have

$$\int_C (x^2 + y^2 + z^2)^2\,ds = \sqrt{10}\int_0^{2\pi} (1 + 9t^2)^2\,dt$$
$$= \sqrt{10}\,[2\pi + 6(2\pi)^3 + \tfrac{81}{5}(2\pi)^5] \approx 506\,400.$$ ∎

Line integrals which involve empirically given functions or lead to complicated definite integrals may be evaluated by using numerical methods of integration.

In many applications the integrands of line integrals are of the form

$$(3) \qquad g(x, y, z)\frac{dx}{ds}, \qquad g(x, y, z)\frac{dy}{ds}, \qquad \text{or} \qquad g(x, y, z)\frac{dz}{ds},$$

where dx/ds, dy/ds, and dz/ds are the derivatives of the functions occurring in the parametric representation of the path of integration. Then we simply write

$$\textbf{(4)} \qquad \int_C g(x, y, z)\frac{dx}{ds}\,ds = \int_C g(x, y, z)\,dx,$$

and similarly in the other two cases. For sums of these types of integrals along the same path C we adopt the simplified notation

$$\textbf{(5)} \qquad \int_C f\,dx + \int_C g\,dy + \int_C h\,dz = \int_C (f\,dx + g\,dy + h\,dz).$$

Using the representation of C, we may eliminate two of the three independent variables in the integrand and then evaluate the resulting definite integral in which the remaining independent variable is the variable of integration.

Example 4. Illustration of (4) and (5)

Evaluate the line integral $I = \int_C [x^2y\,dx + (x - z)\,dy + xyz\,dz]$ where C is the arc of the parabola $y = x^2$ in the plane $z = 2$ from A: $(0, 0, 2)$ to B: $(1, 1, 2)$; cf. Fig. 174.

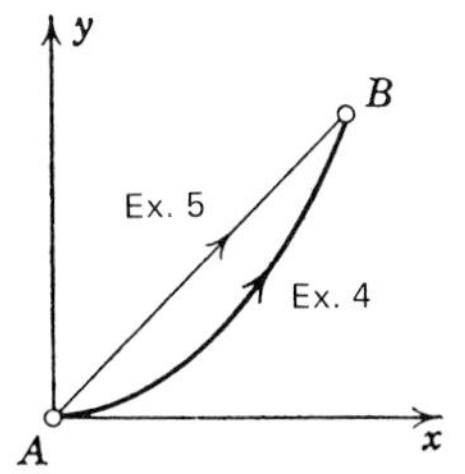

Fig. 174. Paths in Examples 4 and 5

Since $y = x^2$, we have $dy/dx = 2x$ or $dy = 2x\,dx$. Since $z = 2$ is constant, it follows that the integral of the last term in the integrand is zero. Thus

$$I = \int_0^1 [x^2x^2\,dx + (x-2)2x\,dx] = \int_0^1 (x^4 + 2x^2 - 4x)\,dx = -\frac{17}{15}.$$

Example 5. Integration over different paths with the same endpoints
Evaluate the line integral I in Example 4 where C now is the segment of the straight line $y = x$, $z = 2$ from A: $(0, 0, 2)$ to B: $(1, 1, 2)$. We now have $dy = dx$, so that

$$I = \int_0^1 (x^3 + x - 2)\,dx = -\frac{5}{4}.$$ ∎

In Examples 4 and 5 the integrands and the endpoints of the paths are the same, but the values of I are different. This illustrates the important fact that, *in general, the value of a line integral of a given function depends not only on the endpoints but also on the geometric shape of the path of integration.* We shall consider this basic fact in Sec. 9.12.

In many cases, the functions f, g, h in (5) are the components v_1, v_2, v_3 of a vector function

$$\mathbf{v} = v_1\mathbf{i} + v_2\mathbf{j} + v_3\mathbf{k}.$$

Then

$$v_1\,dx + v_2\,dy + v_3\,dz = \left(v_1\frac{dx}{ds} + v_2\frac{dy}{ds} + v_3\frac{dz}{ds}\right)ds,$$

the expression in parentheses being the scalar product of the vector $\mathbf{v}$ and the unit tangent vector

$$\frac{d\mathbf{r}}{ds} = \frac{dx}{ds}\mathbf{i} + \frac{dy}{ds}\mathbf{j} + \frac{dz}{ds}\mathbf{k} \qquad \text{(cf. Sec. 8.5),}$$

where $\mathbf{r}(s)$ represents the path of integration C. Therefore,

$$(6) \qquad \int_C (v_1\,dx + v_2\,dy + v_3\,dz) = \int_C \mathbf{v}\cdot\frac{d\mathbf{r}}{ds}\,ds.$$

This is sometimes written

$$\int_C \mathbf{v}\cdot\frac{d\mathbf{r}}{ds}\,ds = \int_C \mathbf{v}\cdot d\mathbf{r}$$

where

$$d\mathbf{r} = dx\,\mathbf{i} + dy\,\mathbf{j} + dz\,\mathbf{k},$$

but we should note that this is merely a convenient notation.

Example 6. Work done by a force

Consider a particle on which a variable force $\mathbf{p}$ acts. Let the particle be displaced along a given path C in space. Then the work W done by $\mathbf{p}$ in this displacement is given by the line integral

$$W = \int_C \mathbf{p} \cdot d\mathbf{r}, \tag{7}$$

the integration being taken in the sense of the displacement. This definition of work is suggested by that in Example 1, Sec. 6.5, together with the definition of a line integral as the limit of a sum.

We may introduce the time t as the variable of integration. Then

$$d\mathbf{r} = \frac{d\mathbf{r}}{dt}\,dt = \mathbf{v}\,dt$$

where $\mathbf{v}$ is the velocity vector (cf. Sec. 8.6). Hence the line integral (7) becomes

$$W = \int_{t_0}^{t_1} \mathbf{p} \cdot \mathbf{v}\,dt \tag{7'}$$

where t_0 and t_1 are the initial and final values of t. Furthermore, by Newton's second law (Sec. 2.6) we have

$$\mathbf{p} = m\ddot{\mathbf{r}} = m\dot{\mathbf{v}},$$

where m is the mass of the particle. If we insert this into (7′), we obtain

$$W = \int_{t_0}^{t_1} m\dot{\mathbf{v}} \cdot \mathbf{v}\,dt = \int_{t_0}^{t_1} \frac{d}{dt}\left(\frac{m}{2}\mathbf{v} \cdot \mathbf{v}\right) dt = \int_{t_0}^{t_1} \frac{d}{dt}\left(\frac{m}{2}|\mathbf{v}|^2\right) dt = \frac{m}{2}|\mathbf{v}|^2 \Big|_{t_0}^{t_1};$$

that is, *the work done equals the gain in kinetic energy*. This is a basic law in mechanics.

Problems for Sec. 9.2

Orienting C so that the sense of integration becomes the positive sense on C, evaluate $\int_C (x^2 + y^2)\,ds$:

1. Over the path $y = 2x$ from $(0, 0)$ to $(1, 2)$

2. Over the path $y = -x$ from $(1, -1)$ to $(2, -2)$

3. Counterclockwise along the circle $x^2 + y^2 = 4$ from $(2, 0)$ to $(0, 2)$

4. Counterclockwise around the circle $x^2 + y^2 = 1$ from $(1, 0)$ to $(1, 0)$

5. Over the x-axis from $(0, 0)$ to $(1, 0)$ and then parallel to the y-axis from $(1, 0)$ to $(1, 1)$

6. Over the y-axis from $(0, 0)$ to $(0, 1)$ and then parallel to the x-axis from $(0, 1)$ to $(1, 1)$

7. Evaluate $\int_C x^{-1}(y + z)\,ds$ where C is the arc of the circle $x^2 + y^2 = 4$, $z = 0$, from $(2, 0, 0)$ to $(\sqrt{2}, \sqrt{2}, 0)$ (counterclockwise).

Evaluate $\int_C (y^2\,dx - 2x^2\,dy)$:

8. Along the straight line from $(0, 2)$ to $(1, 1)$

9. Along the parabola $y = x^2$ from $(0, 0)$ to $(2, 4)$

10. Counterclockwise along the circle $x^2 + y^2 = 1$ from $(1, 0)$ to $(0, 1)$

Find the work done by the force $\mathbf{p} = x\mathbf{i} - z\mathbf{j} + 2y\mathbf{k}$ in the displacement:

11. Along the y-axis from 0 to 1

12. Along the parabola $y = 2x^2$, $z = 2$ from $(0, 0, 2)$ to $(1, 2, 2)$

13. Along the curve $z = y^4$, $x = 1$ from $(1, 0, 0)$ to $(1, 1, 1)$

14. Along the closed path consisting of the three straight segments from $(0, 0, 0)$ to $(1, 1, 0)$ to $(1, 1, 1)$ and back to $(0, 0, 0)$

15. Along the straight line $y = x$, $z = x$ from $(1, 1, 1)$ to $(3, 3, 3)$

16. Along the curve $y = x$, $z = x^2$ from $(0, 0, 0)$ to $(1, 1, 1)$

17. Along the curve $y = x^3$, $z = 2$ from $(1, 1, 2)$ to $(2, 8, 2)$

18. Let $\mathbf{p}$ be a vector function defined at all points of a curve C, and suppose that $|\mathbf{p}|$ is bounded, say, $|\mathbf{p}| < M$ on C, where M is some positive number. Show that

$$\left| \int_C \mathbf{p} \cdot d\mathbf{r} \right| < Ml \tag{8}$$

where l is the length of C.

19. Using (8), find an upper bound for the absolute value of the work W done by the force $\mathbf{p} = x\mathbf{i} + y^2\mathbf{j}$ in the displacement along the straight line from $(0, 0, 0)$ to $(1, 1, 0)$. Find W by integration and compare the results.

20. Using the trapezoidal rule (Sec. 19.6), evaluate $\int_C f(x, y)\, ds$ along the straight line $y = x$ in the xy-plane from $(0, 0)$ to $(1, 1)$, where $f(x, y)$ is given by the values $f(0, 0) = 1.0$, $f(\frac{1}{4}, \frac{1}{4}) = 1.5$, $f(\frac{1}{2}, \frac{1}{2}) = 1.7$, $f(\frac{3}{4}, \frac{3}{4}) = 1.5$, $f(1, 1) = 1.0$.

9.3 Double Integrals

In our further consideration we shall need the concept of a double integral. Although the reader will be familiar with double integrals from calculus, we shall now present a brief review.

In the case of a definite integral

$$\int_a^b f(x)\, dx$$

the integrand is a function $f(x)$ which exists for all x in an interval $a \leqq x \leqq b$ of the x-axis. In the case of a double integral the integrand is a function $f(x, y)$ which is given for all (x, y) in a closed bounded[2] region R of the xy-plane.

The definition of the double integral is quite similar to that of the definite integral. We subdivide the region R by drawing parallels to the x and y axes (Fig. 175). We number those rectangles which are within R from 1 to n. In each such rectangle we choose a point, say, (x_k, y_k) in the kth rectangle, and then we form the sum

$$J_n = \sum_{k=1}^{n} f(x_k, y_k)\, \Delta A_k$$

[2] "Closed" means that the boundary is part of the region, and "bounded" means that the region can be enclosed in a circle of sufficiently large radius.

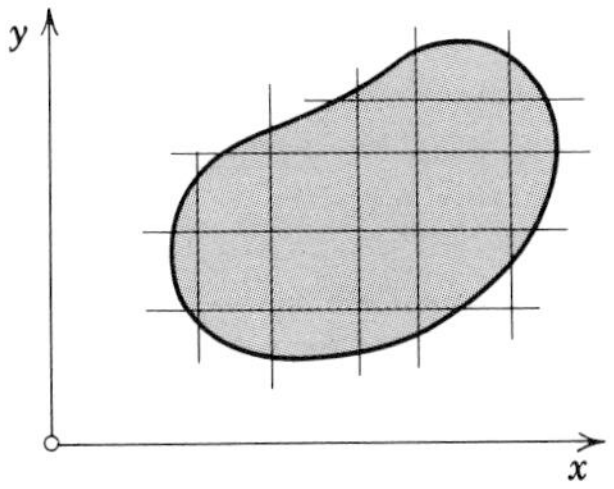

Fig. 175. Subdivision of R

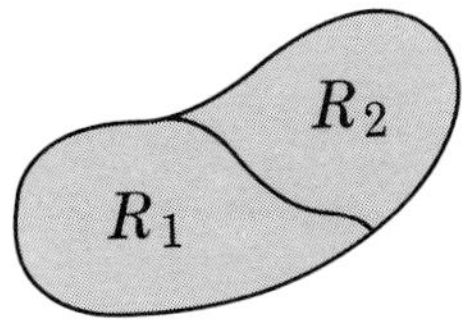

Fig. 176. Formula (1)

where ΔA_k is the area of the kth rectangle. This we do for larger and larger positive integers n in a completely independent manner but so that the length of the maximum diagonal of the rectangles approaches zero as n approaches infinity. In this fashion we obtain a sequence of real numbers $J_{n_1}, J_{n_2}, \cdots$. Assuming that $f(x, y)$ is continuous in R and R is bounded by finitely many smooth curves (cf. Sec. 9.1) it can be shown[3] that the sequence converges and its limit is independent of the choice of subdivisions and corresponding points (x_k, y_k). This limit is called the **double integral** *of $f(x, y)$ over the region R*, and is denoted by the symbol

$$\iint_R f(x, y)\, dx\, dy.$$

From the definition it follows that double integrals enjoy properties which are quite similar to those of definite integrals. Let f and g be functions of x and y, defined and continuous in a region R. Then

$$\iint_R kf\, dx\, dy = k \iint_R f\, dx\, dy \qquad (k \text{ constant})$$

(1)
$$\iint_R (f + g)\, dx\, dy = \iint_R f\, dx\, dy + \iint_R g\, dx\, dy$$

$$\iint_R f\, dx\, dy = \iint_{R_1} f\, dx\, dy + \iint_{R_2} f\, dx\, dy \qquad \text{(cf. Fig. 176).}$$

Furthermore, there exists at least one point (x_0, y_0) in R such that

$$\iint_R f(x, y)\, dx\, dy = f(x_0, y_0)\, A, \tag{2}$$

where A is the area of R; this is called the **mean value theorem** *for double integrals.*

[3]Cf. Ref. [A3] in Appendix 1.

Double integrals over a region R may be evaluated by two successive integrations as follows.

Suppose that R can be described by inequalities of the form

$$a \leqq x \leqq b, \qquad g(x) \leqq y \leqq h(x)$$

(Fig. 177), so that $y = g(x)$ and $y = h(x)$ represent the boundary of R. Then

$$\iint_R f(x, y)\, dx\, dy = \int_a^b \left[\int_{g(x)}^{h(x)} f(x, y)\, dy\right] dx. \tag{3}$$

We first integrate the inner integral

$$\int_{g(x)}^{h(x)} f(x, y)\, dy.$$

In this definite integral, x plays the role of a parameter, and the result of the integration will be a function of x, say, $F(x)$. By integrating $F(x)$ over x from a to b we then obtain the value of the double integral in (3).

Similarly, if R can be described by inequalities of the form

$$c \leqq y \leqq d, \qquad p(y) \leqq x \leqq q(y)$$

(Fig. 178), then we obtain

$$\iint_R f(x, y)\, dx\, dy = \int_c^d \left[\int_{p(y)}^{q(y)} f(x, y)\, dx\right] dy; \tag{4}$$

we now integrate first over x and then the resulting function of y from c to d.

If R cannot be represented by those inequalities, but can be subdivided into finitely many portions which have that property, we may integrate $f(x, y)$ over each portion separately and add the results; this will give us the value of the double integral of $f(x, y)$ over that region R.

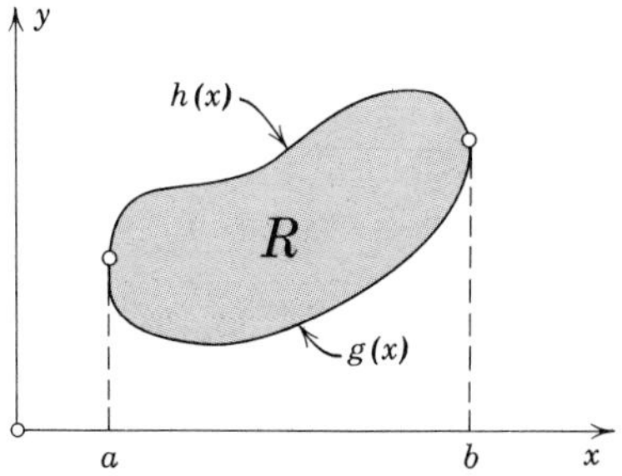

Fig. 177. Evaluation of a double integral

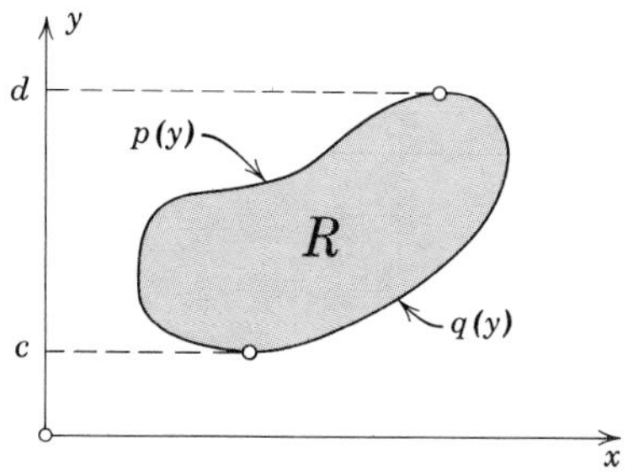

Fig. 178. Evaluation of a double integral

Applications of double integrals

Double integrals have various geometrical and physical applications. For example, the **area** A of R is

$$A = \iint_R dx\,dy.$$

The **volume** V beneath the surface $z = f(x, y)$ (>0) and above the region R in the xy-plane is (Fig. 179)

$$V = \iint_R f(x, y)\,dx\,dy,$$

because the term $f(x_k, y_k)\,\Delta A_k$ in J_n on p. 416 represents the volume of a rectangular parallelepiped with base ΔA_k and altitude $f(x_k, y_k)$.

Let $f(x, y)$ be the density (= mass per unit area) of a distribution of mass in the xy-plane. Then the total mass M in R is

$$M = \iint_R f(x, y)\,dx\,dy,$$

the **center of gravity** of the mass in R has the coordinates

$$\bar{x} = \frac{1}{M}\iint_R xf(x, y)\,dx\,dy \qquad \text{and} \qquad \bar{y} = \frac{1}{M}\iint_R yf(x, y)\,dx\,dy,$$

the **moments of inertia** I_x and I_y of the mass in R about the x and y axes, respectively, are

$$I_x = \iint_R y^2 f(x, y)\,dx\,dy, \qquad I_y = \iint_R x^2 f(x, y)\,dx\,dy,$$

and the *polar moment of inertia* about the origin is

$$I_0 = I_x + I_y = \iint_R (x^2 + y^2) f(x, y)\,dx\,dy.$$

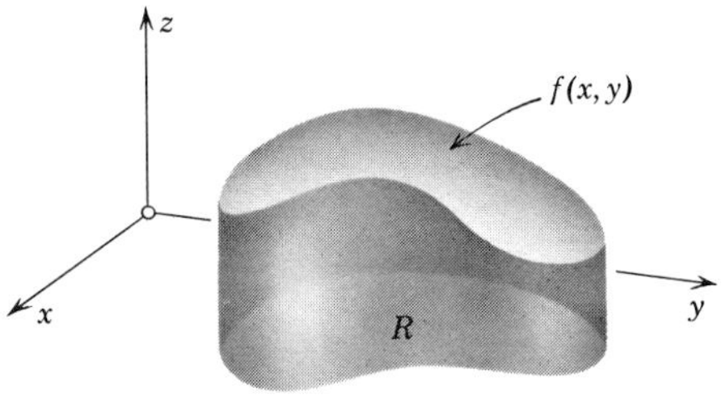

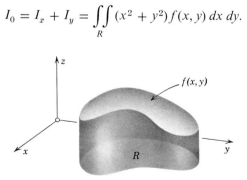

Fig. 179. Double integral as volume

Example 1. Applications of the double integral

Let $f(x, y) = 1$ be the density of mass in the region R: $0 \leqq y \leqq \sqrt{1 - x^2}$, $0 \leqq x \leqq 1$ (Fig. 180). Find the center of gravity and the moments of inertia I_x, I_y, and I_0. The total mass in R is

$$M = \iint_R dx\,dy = \int_0^1 \left[\int_0^{\sqrt{1-x^2}} dy\right] dx = \int_0^1 \sqrt{1 - x^2}\,dx = \int_0^{\pi/2} \cos^2\theta\,d\theta = \frac{\pi}{4}$$

($x = \sin\theta$), which is the area of R. The coordinates of the center of gravity are

$$\bar{x} = \frac{4}{\pi}\iint_R x\,dx\,dy = \frac{4}{\pi}\int_0^1 \left[\int_0^{\sqrt{1-x^2}} x\,dy\right] dx = \frac{4}{\pi}\int_0^1 x\sqrt{1 - x^2}\,dx = -\frac{4}{\pi}\int_1^0 z^2\,dz = \frac{4}{3\pi}$$

($\sqrt{1 - x^2} = z$), and $\bar{y} = \bar{x}$, for reasons of symmetry. Furthermore,

$$I_x = \iint_R y^2\,dx\,dy = \int_0^1 \left[\int_0^{\sqrt{1-x^2}} y^2\,dy\right] dx = \frac{1}{3}\int_0^1 (\sqrt{1 - x^2})^3\,dx$$

$$= \frac{1}{3}\int_0^{\pi/2} \cos^4\theta\,d\theta = \frac{\pi}{16}, \qquad I_y = \frac{\pi}{16}, \qquad I_0 = I_x + I_y = \frac{\pi}{8} \approx 0.3927.$$

A simpler way of calculating I_x will be shown in Example 2. ∎

It will often be necessary to change the variables of integration in double integrals. We know from calculus that in the case of a definite integral

$$\int_a^b f(x)\,dx$$

a new variable of integration u can be introduced by setting

$$x = x(u),$$

where the function $x(u)$ is continuous and has a continuous derivative in some interval $\alpha \leqq u \leqq \beta$ such that $x(\alpha) = a$, $x(\beta) = b$ [or $x(\alpha) = b$, $x(\beta) = a$] and $x(u)$ varies between a and b when u varies between α and β. Then

$$\text{(5)} \qquad \int_a^b f(x)\,dx = \int_\alpha^\beta f(x(u))\frac{dx}{du}\,du.$$

For example, let $f(x) = \sqrt{1 - x^2}$, $a = 0$, $b = 1$, and set $x = \sin u$. Then

$$f[x(u)] = \sqrt{1 - \sin^2 u} = \cos u, \qquad \frac{dx}{du} = \cos u, \qquad \alpha = 0, \qquad \beta = \frac{\pi}{2},$$

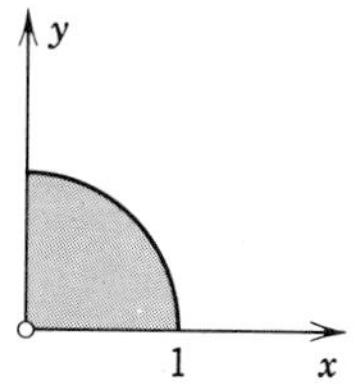

Fig. 180. Example 1

and

$$\int_0^1 \sqrt{1-x^2}\,dx = \int_0^{\pi/2} \cos^2 u\,du = \frac{\pi}{4}.$$

In the case of a double integral

$$\iint_R f(x, y)\,dx\,dy$$

new variables of integration u, v can be introduced by setting

$$x = x(u, v), \qquad y = y(u, v)$$

where the functions $x(u, v)$ and $y(u, v)$ are continuous and have continuous first partial derivatives in some region R^* in the uv-plane so that each point (u_0, v_0) in R^* corresponds to a point $[x(u_0, v_0), y(u_0, v_0)]$ in R and conversely, and furthermore the **Jacobian**[4]

$$J = \frac{\partial(x, y)}{\partial(u, v)} = \begin{vmatrix} \dfrac{\partial x}{\partial u} & \dfrac{\partial x}{\partial v} \\[2ex] \dfrac{\partial y}{\partial u} & \dfrac{\partial y}{\partial v} \end{vmatrix}$$

is either positive throughout R^* or negative throughout R^*. Then

(6) $$\iint_R f(x, y)\,dx\,dy = \iint_{R^*} f[x(u, v), y(u, v)] \left| \frac{\partial(x, y)}{\partial(u, v)} \right| du\,dv;$$

that is, the integrand is expressed in terms of u and v, and $dx\,dy$ is replaced by $du\,dv$ times the absolute value of the Jacobian J. For the proof see Ref. [A2] in Appendix 1.

For example, **polar coordinates** r and θ can be introduced by setting

$$x = r\cos\theta, \qquad y = r\sin\theta.$$

Then

$$J = \frac{\partial(x, y)}{\partial(r, \theta)} = \begin{vmatrix} \cos\theta & -r\sin\theta \\ \sin\theta & r\cos\theta \end{vmatrix} = r,$$

and

(7) $$\iint_R f(x, y)\,dx\,dy = \iint_{R^*} f(r\cos\theta, r\sin\theta) r\,dr\,d\theta,$$

where R^* is the region in the $r\theta$-plane corresponding to R in the xy-plane.

[4] Named after the German mathematician CARL GUSTAV JACOB JACOBI (1804–1851), who made important contributions to elliptic functions, partial differential equations, mechanics, and the calculus of variations.

Example 2. Double integral in polar coordinates
Using (7), we obtain for I_x in Example 1

$$I_x = \iint_R y^2\,dx\,dy = \int_0^{\pi/2}\int_0^1 r^2 \sin^2\theta\, r\,dr\,d\theta = \int_0^{\pi/2} \sin^2\theta\,d\theta \int_0^1 r^3\,dr = \frac{\pi}{4}\cdot\frac{1}{4} = \frac{\pi}{16}.$$

Example 3. Change of variables in a double integral
Evaluate the double integral

$$\iint_R (x^2 + y^2)\,dx\,dy$$

where R is the square in Fig. 181. The shape of R suggests the transformation $x + y = u$, $x - y = v$. Then $x = \frac{1}{2}(u + v)$, $y = \frac{1}{2}(u - v)$, the Jacobian is

$$J = \frac{\partial(x, y)}{\partial(u, v)} = \begin{vmatrix} \frac{1}{2} & \frac{1}{2} \\ \frac{1}{2} & -\frac{1}{2} \end{vmatrix} = -\tfrac{1}{2},$$

R corresponds to the square $0 \leqq u \leqq 2$, $0 \leqq v \leqq 2$, and, therefore,

$$\iint_R (x^2 + y^2)\,dx\,dy = \int_0^2\int_0^2 \tfrac{1}{2}(u^2 + v^2)\tfrac{1}{2}\,du\,dv = \frac{8}{3}.$$

Problems for Sec. 9.3

Evaluate and describe the region of integration.

1. $\int_0^1\int_0^1 (x^2 + y^2)\,dy\,dx$
2. $\int_0^1\int_0^{\sqrt{1-x^2}} (x^2 + y^2)\,dy\,dx$
3. $\int_0^1\int_{x^2}^{x} (1 - xy)\,dy\,dx$
4. $\int_{-\pi}^{\pi}\int_{-1}^{1} xy\,dx\,dy$
5. $\int_0^1\int_y^{y^2+1} x^2y\,dx\,dy$
6. $\int_0^{\pi}\int_0^{1-\cos\theta} r\,dr\,d\theta$

Find the volume of the following regions in space.

7. The tetrahedron cut from the first octant by the plane $x + y + z = 1$
8. The tetrahedron cut from the first octant by the plane $6x + 3y + 2z = 6$
9. The region bounded by the cylinders $x^2 + y^2 = 1$ and $y^2 + z^2 = 1$
10. The region bounded by the surfaces $y = x^2$, $x = y^2$ and the planes $z = 0$, $z = 3$

Find the coordinates $\bar{x}, \bar{y}$ of the center of gravity of a mass of density $f(x, y)$ in a region R, where

11. $f(x, y) = 1$, R: the rectangle $0 \leqq x \leqq 2$, $0 \leqq y \leqq 4$
12. $f(x, y) = 1$, R: the region $x^2 + y^2 \leqq a^2$ in the first quadrant

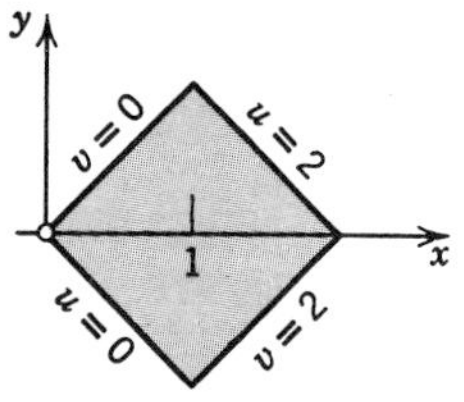

Fig. 181. Region in Example 3

13. $f(x, y) = xy$, R: the rectangle $0 \leqq x \leqq 4$, $0 \leqq y \leqq 2$

14. $f(x, y) = x^2 + y^2$, R: the region $x^2 + y^2 \leqq 1$ in the first quadrant

Find the moments of inertia I_x, I_y, I_0 of a mass of density $f(x, y) = 1$ in a region R shown in the following figures (which the engineer is likely to need in applications).

15.

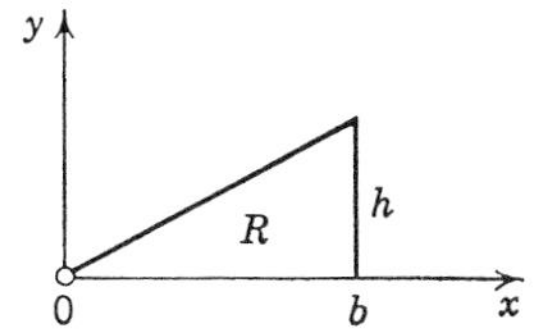

16.

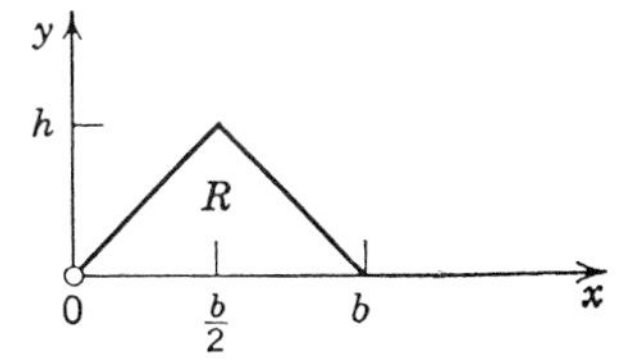

17.

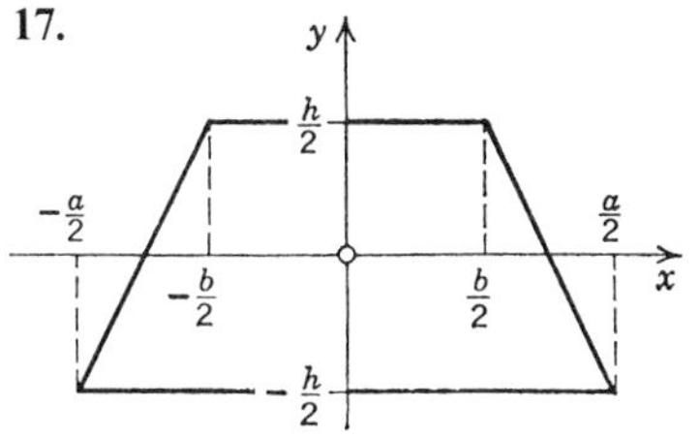

18.

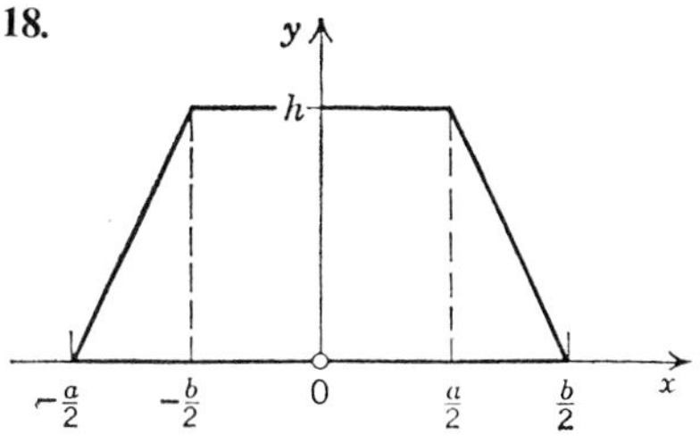

Using polar coordinates, evaluate $\iint_R f(x, y)\, dx\, dy$, where

19. $f = x + y$, R: $x^2 + y^2 \leqq 9$, $x \geqq 0$

20. $f = \cos(x^2 + y^2)$, R: $x^2 + y^2 \leqq \pi/2$, $x \geqq 0$

21. $f = (x^2 + y^2)^2$, R: $x^2 + y^2 \leqq 4$, $y \geqq 0$

22. $f = e^{-x^2-y^2}$, R: the annulus bounded by $x^2 + y^2 = 1$ and $x^2 + y^2 = 4$

In each case, find the Jacobian and give a geometrical reason for the result.

23. Translation $x = u + a$, $y = v + b$

24. Expansion $x = au$, $y = bv$

25. Rotation $x = u \cos \phi - v \sin \phi$, $y = u \sin \phi + v \cos \phi$

9.4 Transformation of Double Integrals into Line Integrals

Double integrals over a plane region may be transformed into line integrals over the boundary of the region and conversely. This transformation is of practical interest because it may help to make the evaluation of an integral easier. It also helps in the theory whenever one wants to switch from one type of integral to the other. The transformation can be done by the following basic theorem.

Green's[5] theorem in the plane (Transformation of double integrals into line integrals and conversely)

Let R be a closed bounded region in the xy-plane whose boundary C consists of finitely many smooth curves. Let $f(x, y)$ and $g(x, y)$ be functions which are

[5] GEORGE GREEN (1793–1841), English mathematician.

continuous and have continuous partial derivatives $\partial f/\partial y$ and $\partial g/\partial x$ everywhere in some domain containing R. Then

$$\text{(1)} \qquad \iint_R \left(\frac{\partial g}{\partial x} - \frac{\partial f}{\partial y}\right) dx\, dy = \int_C (f\, dx + g\, dy),$$

the integration being taken along the entire boundary C of R such that R is on the left as one advances in the direction of integration (cf. Fig. 182).

Proof. We first prove the theorem for a *special region R* which can be represented in both of the forms

$$a \leqq x \leqq b, \qquad u(x) \leqq y \leqq v(x),$$

and

$$c \leqq y \leqq d, \qquad p(y) \leqq x \leqq q(y)$$

(Fig. 183). Using (3) in Sec. 9.3, we obtain

$$\text{(2)} \qquad \iint_R \frac{\partial f}{\partial y}\, dx\, dy = \int_a^b \left[\int_{u(x)}^{v(x)} \frac{\partial f}{\partial y}\, dy\right] dx.$$

We integrate the inner integral:

$$\int_{u(x)}^{v(x)} \frac{\partial f}{\partial y}\, dy = f(x, y)\Big|_{y=u(x)}^{y=v(x)} = f[x, v(x)] - f[x, u(x)].$$

By inserting this into (2) we find

$$\iint_R \frac{\partial f}{\partial y}\, dx\, dy = \int_a^b f[x, v(x)]\, dx - \int_a^b f[x, u(x)]\, dx$$

$$= -\int_a^b f[x, u(x)]\, dx - \int_b^a f[x, v(x)]\, dx.$$

Since $y = u(x)$ represents the oriented curve C^* (Fig. 183*a*) and $y = v(x)$

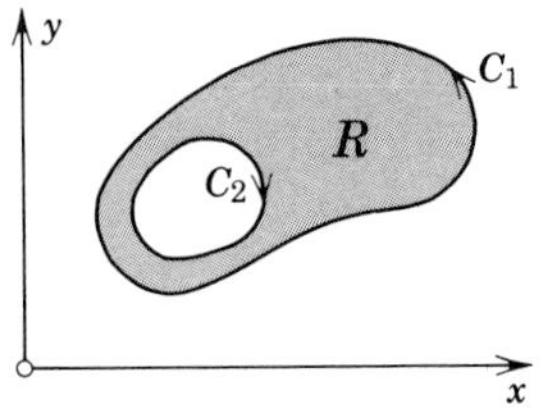

Fig. 182. Region *R* whose boundary *C* consists of two parts; C_1 is traversed in the counterclockwise sense, while C_2 is traversed in the clockwise sense

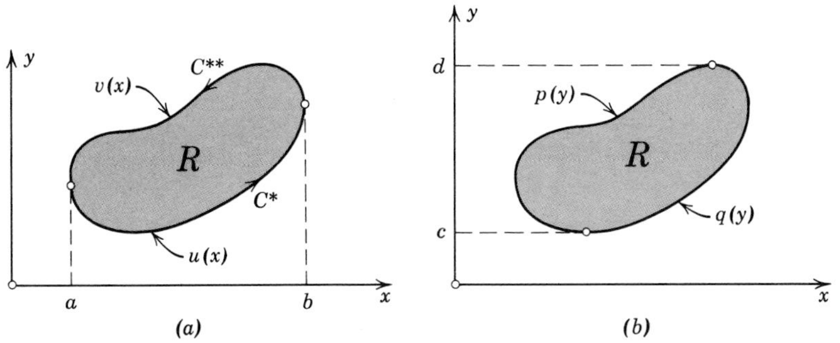

Fig. 183. Example of a special region for Green's theorem

represents C^{**}, the integrals on the right may be written as line integrals over C^* and C^{**}, and, therefore,

$$(3) \qquad \iint_R \frac{\partial f}{\partial y}\,dx\,dy = -\int_{C^*} f(x, y)\,dx - \int_{C^{**}} f(x, y)\,dx = -\oint_C f(x, y)\,dx.$$

If portions of C are segments parallel to the y-axis (such as $\tilde{C}$ and $\tilde{\tilde{C}}$ in Fig. 184), then the result is the same as before, because the integrals over these portions are zero and may be added to the integrals over C^* and C^{**} to obtain the integral over the whole boundary C in (3). Similarly, using (4), Sec. 9.3, we obtain (cf. Fig. 183*b*)

$$\begin{aligned}\iint_R \frac{\partial g}{\partial x}\,dx\,dy &= \int_c^d \left[\int_{p(y)}^{q(y)} \frac{\partial g}{\partial x}\,dx\right] dy \\ &= \int_c^d g[q(y), y]\,dy + \int_d^c g[p(y), y]\,dy \\ &= \oint_C g(x, y)\,dy.\end{aligned}$$

From this and (3), the formula (1) follows, and the theorem is proved for special regions.

We now prove the theorem for a region R which itself is not a special region but can be subdivided into finitely many special regions (Fig. 185). In this case

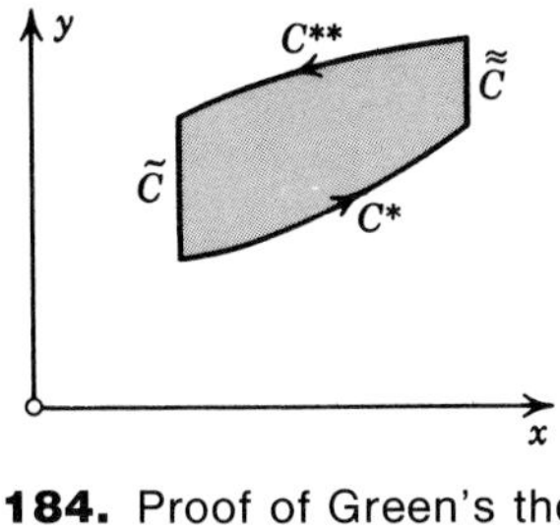

Fig. 184. Proof of Green's theorem

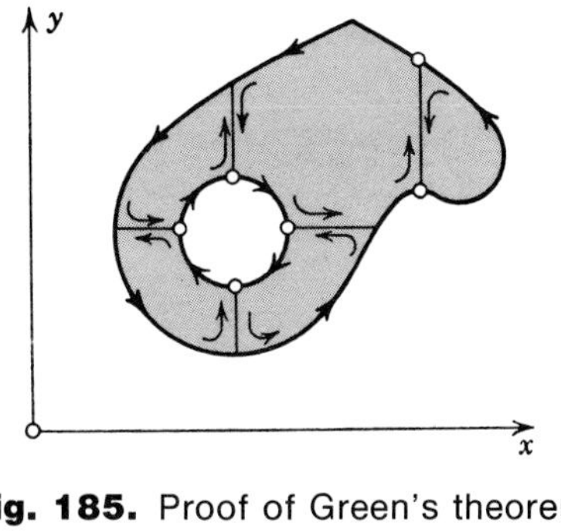

Fig. 185. Proof of Green's theorem

we apply the theorem to each subregion and then add the results; the left-hand members add up to the integral over R while the right-hand members add up to the line integral over C plus integrals over the curves introduced for subdividing R. Each of the latter integrals occurs twice, taken once in each direction. Hence these two integrals cancel each other, and we are left with the line integral over C.

The proof thus far covers all regions which are of interest in engineering problems. To prove the theorem for the most general region R satisfying the conditions in the theorem, we must approximate R by a region of the type just considered and then use a limiting process. For details see Ref. [A2] in Appendix 1. ∎

Green's theorem will be of basic importance in our further consideration. For the time being we consider a few simple illustrative examples.

Example 1. Area of a plane region as a line integral over the boundary
In (1), let $f = 0$ and $g = x$. Then

$$\iint_R dx\,dy = \int_C x\,dy.$$

The integral on the left is the area A of R. Similarly, let $f = -y$ and $g = 0$; then from (1),

$$A = \iint_R dx\,dy = -\int_C y\,dx.$$

By adding both formulas we obtain

$$A = \frac{1}{2}\int_C (x\,dy - y\,dx), \tag{4}$$

the integration being taken as indicated in Green's theorem. This interesting formula expresses the area of R in terms of a line integral over the boundary. It has various applications; for example, the theory of certain planimeters is based upon this formula; cf. Ref. [G20] in Appendix 1.

Example 2. Area of a plane region in polar coordinates
Let r and θ be polar coodinates defined by $x = r\cos\theta$, $y = r\sin\theta$. Then

$$dx = \cos\theta\,dr - r\sin\theta\,d\theta, \qquad dy = \sin\theta\,dr + r\cos\theta\,d\theta,$$

and (4) assumes the form

$$A = \frac{1}{2}\int_C r^2\,d\theta. \tag{5}$$

This formula is well known from calculus.

As an application of (5), consider the cardioid $r = a(1 - \cos\theta)$ where $0 \leqq \theta \leqq 2\pi$ (Fig. 186). We find

$$A = \frac{a^2}{2}\int_0^{2\pi} (1 - \cos\theta)^2\,d\theta = \frac{3\pi}{2}a^2.$$

Example 3. Transformation of a double integral of the Laplacian of a function into a line integral of its normal derivative
Let $w(x, y)$ be a function which is continuous and has continuous first and second partial derivatives in a domain of the xy-plane containing a region R of the type indicated in Green's theorem. We set

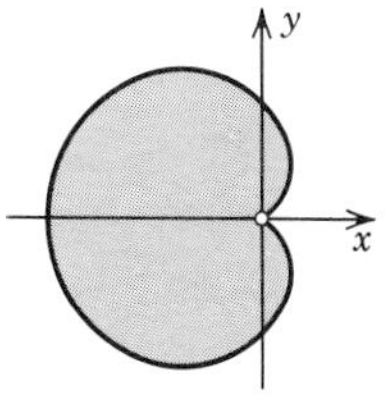

Fig. 186. Cardioid

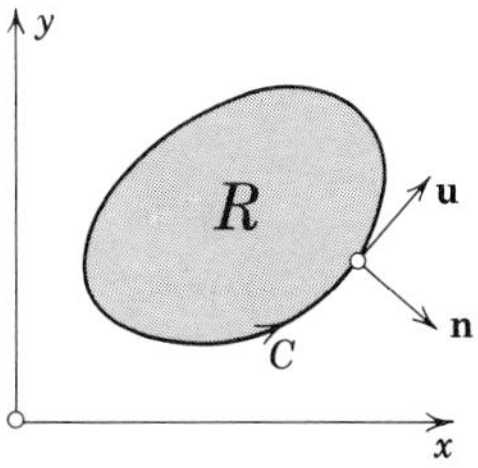

Fig. 187. Example 3

$f = -\partial w/\partial y$ and $g = \partial w/\partial x$. Then $\partial f/\partial y$ and $\partial g/\partial x$ are continuous in R, and

$$\frac{\partial g}{\partial x} - \frac{\partial f}{\partial y} = \frac{\partial^2 w}{\partial x^2} + \frac{\partial^2 w}{\partial y^2} = \nabla^2 w, \tag{6}$$

the Laplacian of w (cf. Sec. 8.8). Furthermore, using those expressions for f and g, we obtain

$$\int_C (f\,dx + g\,dy) = \int_C \left(f\frac{dx}{ds} + g\frac{dy}{ds}\right) ds = \int_C \left(-\frac{\partial w}{\partial y}\frac{dx}{ds} + \frac{\partial w}{\partial x}\frac{dy}{ds}\right) ds, \tag{7}$$

where s is the arc length of C, and C is oriented as shown in Fig. 187. The integrand of the last integral may be written as the dot product (scalar product) of the vectors

$$\text{grad } w = \frac{\partial w}{\partial x}\mathbf{i} + \frac{\partial w}{\partial y}\mathbf{j} \qquad \text{and} \qquad \mathbf{n} = \frac{dy}{ds}\mathbf{i} - \frac{dx}{ds}\mathbf{j};$$

that is,

$$-\frac{\partial w}{\partial y}\frac{dx}{ds} + \frac{\partial w}{\partial x}\frac{dy}{ds} = (\text{grad } w) \cdot \mathbf{n}. \tag{8}$$

The vector $\mathbf{n}$ is a unit normal vector to C, because the vector

$$\mathbf{u} = \frac{d\mathbf{r}}{ds} = \frac{dx}{ds}\mathbf{i} + \frac{dy}{ds}\mathbf{j}$$

(cf. Sec. 8.5) is the unit tangent vector to C, and $\mathbf{u} \cdot \mathbf{n} = 0$. Furthermore, it is not difficult to see that $\mathbf{n}$ is directed to the exterior of C. From this and (5) in Sec. 8.8 it follows that the expression on the right-hand side of (8) is the derivative of w in the direction of the outward normal to C. Denoting this directional derivative by $\partial w/\partial n$ and taking (6), (7), and (8) into account, we obtain from Green's theorem the useful integral formula

$$\iint_R \nabla^2 w\,dx\,dy = \int_C \frac{\partial w}{\partial n}\,ds. \tag{9}$$

Further important applications and consequences of Green's theorem in the plane will be considered in the following sections.

Problems for Sec. 9.4

Using (a) direct calculation, (b) Green's theorem, evaluate

1. $\int_C (y^2\,dx + x^2\,dy)$, C: the boundary of the square $-1 \leqq x \leqq 1$, $-1 \leqq y \leqq 1$ (counterclockwise)

2. $\int_C (y\,dx - x\,dy)$, C: the boundary of the square $0 \leqq x \leqq 1$, $0 \leqq y \leqq 1$ (clockwise)

3. $\int_C [(3x^2 + y)\,dx + 4y^2\,dy]$, C: the boundary of the triangle with vertices (0, 0), (1, 0), (0, 2) (counterclockwise)

4. $\int_C [y^3\,dx + (x^3 + 3y^2x)\,dy]$ C: the boundary of the region between $y = x^2$ and $y = x$, where $0 \leqq x \leqq 1$ (counterclockwise)

5. $\int_C (-xy^2\,dx + x^2y\,dy)$ C: the boundary of the region in the first quadrant bounded by $y = 1 - x^2$ (counterclockwise)

Using Green's theorem, evaluate the line integral $\int_C (f\,dx + g\,dy)$ counterclockwise around the given contour C, where $f\,dx + g\,dy$ equals

6. $(x - y)\,dx - x^2\,dy$, C: the boundary of the square $0 \leqq x \leqq 2$, $0 \leqq y \leqq 2$
7. $(x^3 - 3y)\,dx + (x + \sin y)\,dy$, C as in Prob. 3
8. $(e^x - 3y)\,dx + (e^y + 6x)\,dy$, C: $x^2 + 4y^2 = 4$
9. $(x^2 - \cosh y)\,dx + (y + \sin x)\,dy$, C: the boundary of the rectangle $0 \leqq x \leqq \pi$, $0 \leqq y \leqq 1$
10. $-y^3\,dx + x^3\,dy$, C: $x^2 + y^2 = 1$
11. $(\cos x \sin y - xy)\,dx + \sin x \cos y\,dy$, C as in Prob. 10.
12. $x^{-1}e^y\,dx + (e^y \ln x + 2x)\,dy$, C: the boundary of the region bounded by $y = x^4 + 1$ and $y = 2$

Using one of the formulas in Example 1, find the area of the following plane regions.

13. The interior of the ellipse $x^2/a^2 + y^2/b^2 = 1$
14. The region in the first quadrant bounded by $y = x$, $y = 1/x$, and $y = x/4$
15. The region bounded by $y = x^2$ and $y = x + 2$

Using (9), evaluate $\int_C \frac{\partial w}{\partial n}\,ds$ where

16. $w = x^2 + 3y^2$, C: $x^2 + y^2 = 4$
17. $w = 3x^2y - y^3$, C as in Prob. 2
18. $w = e^x + e^y$, C: the boundary of the rectangle $0 \leqq x \leqq 2$, $0 \leqq y \leqq 1$

19. If $w(x, y)$ satisfies Laplace's equation $\nabla^2 w = 0$ in a region R, show that

$$\iint_R \left[\left(\frac{\partial w}{\partial x}\right)^2 + \left(\frac{\partial w}{\partial y}\right)^2\right] dx\,dy = \int_C w \frac{\partial w}{\partial n}\,ds \tag{10}$$

where $\partial w/\partial n$ is defined as in Example 3. *Hint.* Model your work somewhat after that of Example 3.

Using (10), evaluate $\int_C w \frac{\partial w}{\partial n}\,ds$ where

20. $w = x$, C as in Prob. 2
21. $w = e^x \cos y$, C as in Prob. 2

22. Introducing $\mathbf{v} = g\mathbf{i} - f\mathbf{j}$, show that the formula in Green's theorem may be written

$$\iint_R \operatorname{div} \mathbf{v}\,dx\,dy = \int_C \mathbf{v} \cdot \mathbf{n}\,ds \tag{11}$$

where $\mathbf{n}$ is the outward unit normal vector to the curve C (Fig. 158) and s is the arc length of C.

23. Verify (11) when $\mathbf{v} = x\mathbf{i} + y\mathbf{j}$ and C is the circle $x^2 + y^2 = 1$.

24. Show that the formula in Green's theorem may be written

$$\textbf{(12)} \qquad \iint_R (\operatorname{curl} \mathbf{v}) \cdot \mathbf{k}\, dx\, dy = \int_C \mathbf{v} \cdot \mathbf{u}\, ds$$

where $\mathbf{k}$ is a unit vector perpendicular to the xy-plane, $\mathbf{u}$ is the unit tangent vector to C, and s is the arc length of C.

25. Verify (12) when $\mathbf{v} = -y\mathbf{i} + x\mathbf{j}$ and C is the boundary of the triangle with vertices at $(0, 0)$, $(1, 0)$, $(1, 1)$.

9.5 Surfaces

In Sec. 9.7 we shall consider surface integrals. For this we must know some basic facts about surfaces which we shall now explain and illustrate by simple examples.

A surface S may be represented in the form

$$\textbf{(1)} \qquad f(x, y, z) = 0,$$

where x, y, z are Cartesian coordinates in space. Then the gradient of f is normal to S (cf. Theorem 2 in Sec. 8.8), provided $\operatorname{grad} f \neq \mathbf{0}$. Consequently, for S to have a unique normal at each point whose direction depends continuously on the points of S, we must require that *f has continuous first partial derivatives, and at each point at least one of these three derivatives is not zero.* Then the vector

$$\textbf{(2)} \qquad \mathbf{n} = \frac{\operatorname{grad} f}{|\operatorname{grad} f|}$$

is a unit normal vector of S (and $-\mathbf{n}$ is the other).

Example 1. Unit normal vector

The sphere $x^2 + y^2 + z^2 - a^2 = 0$ has the unit normal vector

$$\mathbf{n}(x, y, z) = \frac{x}{a}\mathbf{i} + \frac{y}{a}\mathbf{j} + \frac{z}{a}\mathbf{k}. \qquad \blacksquare$$

Sometimes it is convenient to use an *explicit representation*

$$(3) \qquad z = g(x, y)$$

of a given surface. Clearly, by writing $z - g(x, y) = 0$ we immediately obtain an *implicit representation* of the form (1).

A surface S can also be represented by a **parametric representation**

$$\textbf{(4)} \qquad \mathbf{r}(u, v) = x(u, v)\mathbf{i} + y(u, v)\mathbf{j} + z(u, v)\mathbf{k}.$$

Here, u and v are two independent real variables, called the *parameters* of the representation, on which the vector function $\mathbf{r}(u, v)$ depends. $\mathbf{r}(u, v)$ is the position vector of the points of S; its tip ranges over S as (u, v) varies in some region R in the uv-plane. To each (u_0, v_0) in R there corresponds a point of S with position vector $\mathbf{r}(u_0, v_0)$. Hence the surface S is the image of the plane region R in the uv-plane (cf. Fig. 188). This is similar to a parametric representation $\mathbf{r}(t)$ of a curve C discussed in Sec. 8.3, where C is the image of a line segment (interval on the t-axis); cf. Fig. 188. The only essential difference is that in the case of a surface we need *two* parameters, whereas for a curve we had only *one* parameter.

In our further work, surfaces must have certain geometric properties. To guarantee this, we make the following

Assumptions

$\mathbf{r}(u, v)$ *in* (4) *is continuous and has continuous first partial derivatives* $\mathbf{r}_u$ *and* $\mathbf{r}_v$ *in a domain of the uv-plane which includes the region R, and R is simply connected*[6] *and bounded. Furthermore,*

$$\mathbf{r}_u \times \mathbf{r}_v \neq \mathbf{0} \tag{5}$$

everywhere in R.

By definition, a **smooth surface** S is a surface which has a unique normal whose direction depends continuously on the points of S.

In the next section we shall see that under our assumptions, S represented by $\mathbf{r}(u, v)$ is smooth.

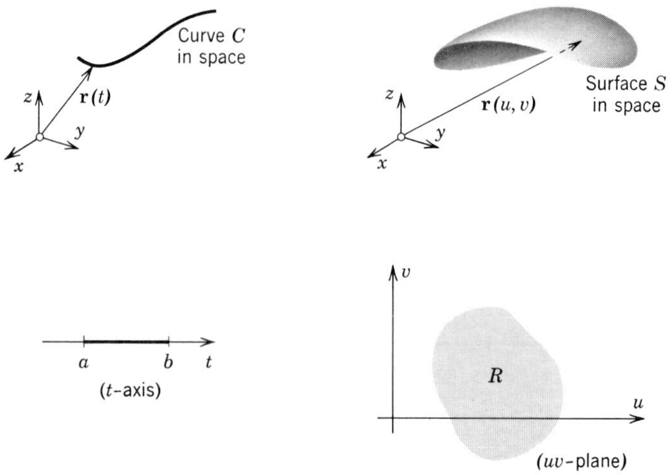

Fig. 188. Parametric representations of a curve and a surface

[6]That is, every closed curve in R can be continuously shrunk to any point in R without leaving R. "Bounded" is explained in Sec. 9.3, footnote 2.

A **piecewise smooth surface** is one which can be subdivided into *finitely many* portions each of which is smooth. For example, the boundary surface of a cube is piecewise smooth. A sphere is a smooth surface.

Example 2. Parametric representation of a sphere
The sphere in Example 1 can be represented in the form

$$\mathbf{r}(u, v) = a \cos v \cos u\, \mathbf{i} + a \cos v \sin u\, \mathbf{j} + a \sin v\, \mathbf{k} \tag{6}$$

where $0 \leqq u \leqq 2\pi$, $-\pi/2 \leqq v \leqq \pi/2$; that is,

$$x = a \cos v \cos u, \qquad y = a \cos v \sin u, \qquad z = a \sin v.$$

The curves $u = const$ and $v = const$ are the "meridians" and "parallels" on S (cf. Fig. 189). The relation (5) is satisfied everywhere except at the "poles" $v = -\pi/2$ and $v = \pi/2$. *The representation* (6) *is used in geography for measuring the latitude and longitude of points on the globe.*

Problems for Sec. 9.5

What surfaces are represented by the following parametric representations? What are the **coordinate curves** (curves $u = const$ and $v = const$) on the surface?

1. $\mathbf{r} = u\mathbf{i} + v\mathbf{j}$

2. $\mathbf{r} = u \cos v\, \mathbf{i} + u \sin v\, \mathbf{j}$

3. $\mathbf{r} = \cos u\, \mathbf{i} + \sin u\, \mathbf{j} + v\mathbf{k}$

4. $\mathbf{r} = u\mathbf{i} + v\mathbf{j} + uv\mathbf{k}$

5. $\mathbf{r} = 4 \cos u\, \mathbf{i} + \sin u\, \mathbf{j} + v\mathbf{k}$

6. $\mathbf{r} = u\mathbf{i} + v\mathbf{j} + (u + v)\mathbf{k}$

7. $\mathbf{r} = u \cos v\, \mathbf{i} + u \sin v\, \mathbf{j} + u\mathbf{k}$

8. $\mathbf{r} = u \cos v\, \mathbf{i} + u \sin v\, \mathbf{j} + u^2\mathbf{k}$

Find a parametric representation of the following surfaces.

9. The xz-plane

10. The plane $z = x$

11. The plane $x + y + z = 1$

12. The cylinder of revolution $y^2 + z^2 = a^2$

13. The parabolic cylinder $z = x^2$

14. The elliptic cylinder $y^2 + 9z^2 = 9$

Represent the following surfaces in the form (1) and sketch figures.

15. The ellipsoid $\mathbf{r} = a \cos v \cos u\, \mathbf{i} + b \cos v \sin u\, \mathbf{j} + c \sin v\, \mathbf{k}$

16. The elliptic paraboloid $\mathbf{r} = au \cos v\, \mathbf{i} + bu \sin v\, \mathbf{j} + u^2\mathbf{k}$

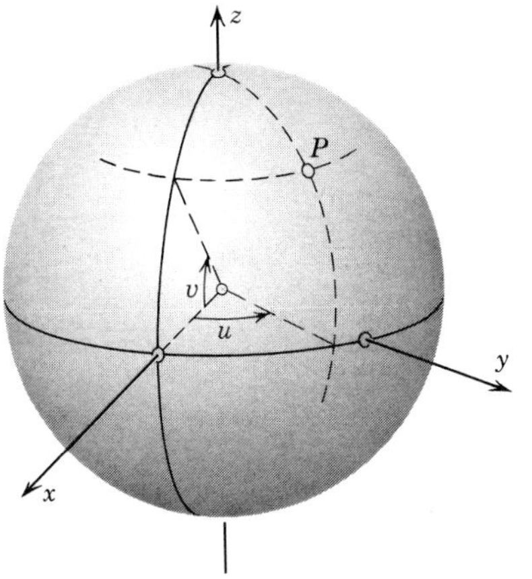

Fig. 189. Parametric representation of a sphere

17. The hyperbolic paraboloid $\mathbf{r} = au \cosh v\, \mathbf{i} + bu \sinh v\, \mathbf{j} + u^2\mathbf{k}$

18. The hyperboloid $\mathbf{r} = a \sinh u \cos v\, \mathbf{i} + b \sinh u \sin v\, \mathbf{j} + c \cosh u\, \mathbf{k}$

19. Find a unit normal vector of the surface in Prob. 4.

20. Find the points in Example 2, Prob. 2, and Prob. 7 at which (5) is not satisfied.

9.6 Tangent Plane. First Fundamental Form. Area

If a surface S is represented in the form $\mathbf{r} = \mathbf{r}(u, v)$, then a curve on S can be represented by a pair of continuous functions

$$u = g(t), \qquad v = h(t) \tag{1}$$

of a real parameter t.

Example 1. Cylinder of revolution and circular helix

The vector function $\mathbf{r}(u, v) = a \cos u\, \mathbf{i} + a \sin u\, \mathbf{j} + v\mathbf{k}$ represents a cylinder of revolution S of radius a. The equations $u = t$, $v = ct$ represent a circular helix on S. In fact, by substituting these equations into the representation of S we obtain (cf. Example 3 in Sec. 8.3)

$$\mathbf{r}[u(t), v(t)] = a \cos t\, \mathbf{i} + a \sin t\, \mathbf{j} + ct\mathbf{k}. \qquad \blacksquare$$

Let S be a smooth surface represented by a vector function $\mathbf{r}(u, v)$ and let C be a curve on S represented in the form (1). Then C, considered as a curve in space, is represented by the vector function

$$\mathbf{r}(t) = \mathbf{r}[u(t), v(t)]. \tag{2}$$

Suppose that both functions in (1) have continuous first derivatives such that for each t at least one of these derivatives is not zero. Then C has a tangent at each of its points whose direction depends continuously on the points, and a tangent vector to C is

$$\dot{\mathbf{r}}(t) = \frac{d\mathbf{r}}{dt} = \mathbf{r}_u \dot{u} + \mathbf{r}_v \dot{v}.$$

From (5) in the last section it follows that the vectors $\mathbf{r}_u = \partial\mathbf{r}/\partial u$ and $\mathbf{r}_v = \partial\mathbf{r}/\partial v$ are linearly independent. Hence they determine a plane. This plane is called the **tangent plane** of S at the corresponding point P of S and will be denoted by $T(P)$. It touches S at P. From (2) it follows that *$T(P)$ contains the tangent to any curve on S through P, at that point P* (see also Sec. 8.8).

The straight line through P and perpendicular to $T(P)$ is called the **normal** to S at P. Since the vectors $\mathbf{r}_u$ and $\mathbf{r}_v$ lie in $T(P)$, the unit vector

$$\mathbf{n} = \frac{\mathbf{r}_u \times \mathbf{r}_v}{|\mathbf{r}_u \times \mathbf{r}_v|} \tag{3}$$

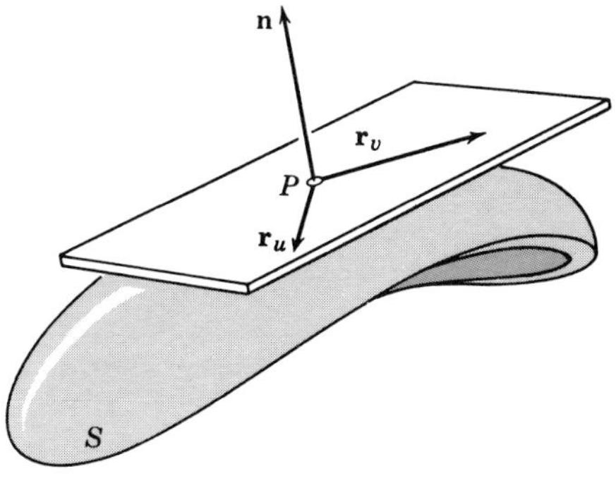

Fig. 190. Tangent plane and normal vector

(Fig. 190) is perpendicular to $T(P)$. This vector is called the **unit normal vector** of S at P. Its sense depends on the choice of the coordinates u and v; the transformation $u = -\bar{u}$, $v = \bar{v}$ or any other transformation whose Jacobian (cf. Sec. 9.3) has a negative value, reverses the sense of **n**.

We shall now determine the *linear element* of a curve C, represented in the form (1), on a surface S, represented by $\mathbf{r}(u, v)$. We have

$$d\mathbf{r} = \mathbf{r}_u\, du + \mathbf{r}_v\, dv,$$

and, by (4) in Sec. 8.4,

$$\begin{aligned} ds^2 &= d\mathbf{r} \cdot d\mathbf{r} = (\mathbf{r}_u\, du + \mathbf{r}_v\, dv) \cdot (\mathbf{r}_u\, du + \mathbf{r}_v\, dv) \\ &= \mathbf{r}_u \cdot \mathbf{r}_u\, du^2 + 2\mathbf{r}_u \cdot \mathbf{r}_v\, du\, dv + \mathbf{r}_v \cdot \mathbf{r}_v\, dv^2. \end{aligned}$$

If we use the standard notation

$$\text{(4)} \qquad E = \mathbf{r}_u \cdot \mathbf{r}_u \qquad F = \mathbf{r}_u \cdot \mathbf{r}_v \qquad G = \mathbf{r}_v \cdot \mathbf{r}_v$$

the last expression assumes the form

$$\text{(5)} \qquad ds^2 = E\, du^2 + 2F\, du\, dv + G\, dv^2.$$

This quadratic differential form is called the **first fundamental form** of S.

Example 2. First fundamental form for polar coordinates

The vector function

$$\mathbf{r}(u, v) = u \cos v\, \mathbf{i} + u \sin v\, \mathbf{j}$$

represents the xy-plane, and u and v are polar coordinates. By differentiation we obtain

$$\mathbf{r}_u = \cos v\, \mathbf{i} + \sin v\, \mathbf{j}, \qquad \mathbf{r}_v = -u \sin v\, \mathbf{i} + u \cos v\, \mathbf{j}.$$

Hence $E = 1$, $F = 0$, $G = u^2$, and the first fundamental form corresponding to polar coordinates $u = \rho$ and $v = \theta$ is

$$\text{(6)} \qquad ds^2 = d\rho^2 + \rho^2\, d\theta^2. \qquad \blacksquare$$

We shall now see that *the first fundamental form is of basic importance because it enables us to measure lengths, angles between curves, and areas on the corresponding surface S.*

Length. From (1) and (4) in Sec. 8.4 and (5) in the present section it follows that a curve

$$C: u(t), v(t), \qquad a \leqq t \leqq b,$$

on a surface S: $\mathbf{r}(u, v)$ has length

$$l = \int_a^b \sqrt{\dot{\mathbf{r}} \cdot \dot{\mathbf{r}}}\, dt = \int_a^b \frac{ds}{dt}\, dt = \int_a^b \sqrt{E\dot{u}^2 + 2F\dot{u}\dot{v} + G\dot{v}^2}\, dt. \tag{7}$$

Angle. Consider two curves

$$C_1: u = g(t),\ v = h(t) \qquad \text{and} \qquad C_2: u = p(t),\ v = q(t)$$

on a surface S: $\mathbf{r}(u, v)$ which intersect at a point P of S. The vectors

$$\mathbf{a} = \frac{d}{dt}\mathbf{r}[g(t), h(t)] = \mathbf{r}_u\dot{g} + \mathbf{r}_v\dot{h}$$

and

$$\mathbf{b} = \frac{d}{dt}\mathbf{r}[p(t), q(t)] = \mathbf{r}_u\dot{p} + \mathbf{r}_v\dot{q}$$

at P are tangent to C_1 and C_2, respectively. The angle of intersection between C_1 and C_2 at P is defined as the angle γ between **a** and **b**, and from (3) in Sec. 6.5 we have

$$\cos\gamma = \frac{\mathbf{a} \cdot \mathbf{b}}{|\mathbf{a}|\,|\mathbf{b}|} \tag{8}$$

where

$$\mathbf{a} \cdot \mathbf{b} = (\mathbf{r}_u\dot{g} + \mathbf{r}_v\dot{h}) \cdot (\mathbf{r}_u\dot{p} + \mathbf{r}_v\dot{q}) = E\dot{g}\dot{p} + F(\dot{g}\dot{q} + \dot{h}\dot{p}) + G\dot{h}\dot{q}$$

and similarly

$$|\mathbf{a}| = \sqrt{\mathbf{a} \cdot \mathbf{a}} = \sqrt{E\dot{g}^2 + 2F\dot{g}\dot{h} + G\dot{h}^2},$$

$$|\mathbf{b}| = \sqrt{\mathbf{b} \cdot \mathbf{b}} = \sqrt{E\dot{p}^2 + 2F\dot{p}\dot{q} + G\dot{q}^2}.$$

This important result shows that the angle between two intersecting curves on a surface can be expressed in terms of E, F, G and the derivatives of the functions representing the curves, evaluated at the point of intersection.

Area. The area A of a surface S: $\mathbf{r}(u, v)$ is defined by the double integral

$$A = \iint_R |\mathbf{r}_u \times \mathbf{r}_v|\, du\, dv \tag{9}$$

over the region R in the uv-plane corresponding to that surface. The expression

$$dA = |\mathbf{r}_u \times \mathbf{r}_v|\, du\, dv \tag{10}$$

is called the *element of area.*

Formula (9) is suggested by the fact that, according to the definition of a vector product, the small parallelogram in Fig. 191 has area

$$\Delta A = |\mathbf{r}_u\, \Delta u \times \mathbf{r}_v\, \Delta v| = |\mathbf{r}_u \times \mathbf{r}_v|\, \Delta u\, \Delta v.$$

The integral (9) is obtained by subdividing S into parts $S_1, \cdots, S_n$, approximating each part S_k by a portion of the tangent plane of S at a point in S_k, and forming the sum of all the approximating areas. This is done for each $n = 1, 2, \cdots$ so that the dimensions of the largest S_k approach zero as n approaches infinity. The limit of those sums is the integral (9). For details see Ref. [C8] in Appendix 1.

It is quite simple to verify our claim that the first fundamental form enables us to measure area. All we have to do is to express (9) in terms of E, F, G. From (5) in Sec. 6.7 and (4) in the current section we obtain

$$|\mathbf{r}_u \times \mathbf{r}_v|^2 = (\mathbf{r}_u \cdot \mathbf{r}_u)(\mathbf{r}_v \cdot \mathbf{r}_v) - (\mathbf{r}_u \cdot \mathbf{r}_v)^2 = EG - F^2. \tag{11}$$

Hence (9) can be written

$$A = \iint_R \sqrt{EG - F^2}\, du\, dv \tag{9*}$$

and (10) can be written

$$dA = \sqrt{EG - F^2}\, du\, dv. \tag{10*}$$

Example 3. Torus

The vector function

$$\mathbf{r}(u, v) = (a + b \cos v) \cos u\, \mathbf{i} + (a + b \cos v) \sin u\, \mathbf{j} + b \sin v\, \mathbf{k} \qquad (a > b > 0)$$

represents a torus (Fig. 192). This surface is obtained by rotating a circle C about a fixed straight line A so that the plane of C always passes through A, and C does not intersect A. Using (4), we find

$$E = (a + b \cos v)^2, \qquad F = 0, \qquad G = b^2.$$

Hence

$$|\mathbf{r}_u \times \mathbf{r}_v|^2 = EG - F^2 = b^2(a + b \cos v)^2,$$

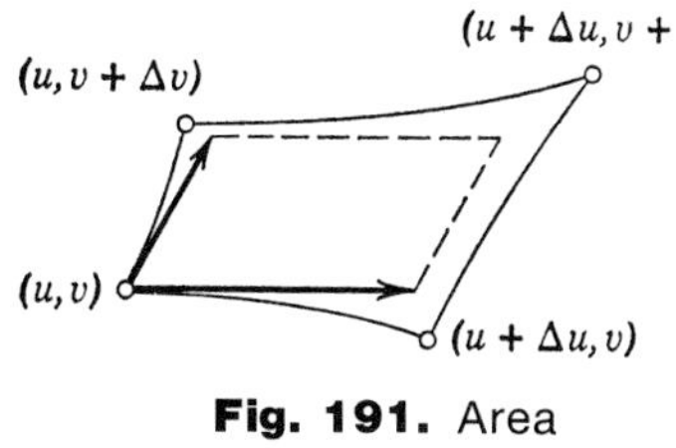

Fig. 191. Area

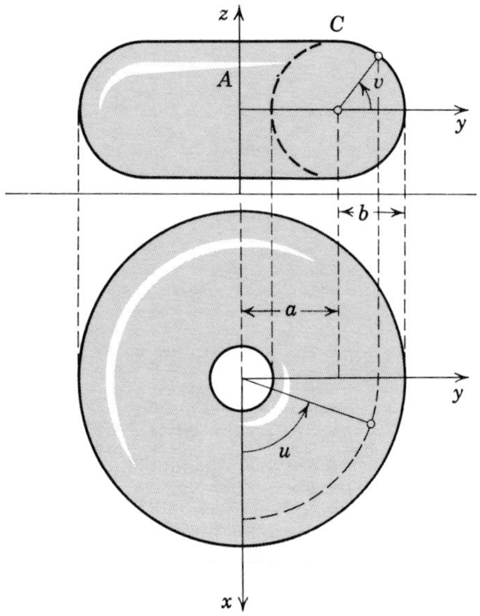

Fig. 192. Torus

and the total area of the torus is

$$A = \int_0^{2\pi} \int_0^{2\pi} b(a + b\cos v)\, du\, dv = 4\pi^2 ab. \qquad ▮$$

Area of a surface $z = g(x, y)$. Suppose that a surface S is given by a representation

$$z = g(x, y).$$

Then we may set $x = u$ and $y = v$ and write the representation in parametric form:

$$\mathbf{r}(u, v) = u\mathbf{i} + v\mathbf{j} + g(u, v)\mathbf{k}.$$

Then the partial derivatives of $\mathbf{r}$ with respect to u and v are

$$\text{(12)} \qquad \mathbf{r}_u = \mathbf{i} + g_u\mathbf{k}, \qquad \mathbf{r}_v = \mathbf{j} + g_v\mathbf{k}.$$

It follows that the coefficients of the first fundamental form are

$$E = 1 + g_u^2, \qquad F = g_u g_v, \qquad G = 1 + g_v^2.$$

Hence

$$|\mathbf{r}_u \times \mathbf{r}_v|^2 = EG - F^2 = 1 + g_u^2 + g_v^2.$$

Since $u = x$ and $v = y$, the integral (9) now assumes the form

$$\text{(9')} \qquad A = \iint_{\bar{S}} \sqrt{1 + \left(\frac{\partial g}{\partial x}\right)^2 + \left(\frac{\partial g}{\partial y}\right)^2}\, dx\, dy$$

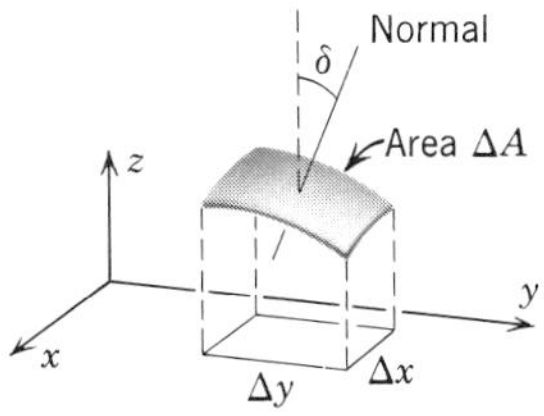

Fig. 193. Derivation of (10″)

where $\bar{S}$ is the orthogonal projection of S into the xy-plane. Clearly,

$$(10') \qquad dA = \sqrt{1 + \left(\frac{\partial g}{\partial x}\right)^2 + \left(\frac{\partial g}{\partial y}\right)^2}\, dx\, dy.$$

For later use we show that this can be written

$$(10'') \qquad dA = \sec \delta \, dx \, dy$$

where δ is the *acute* angle between the z-axis and the undirected normal to S. Figure 193 explains the geometrical reason: a small "parallelogram" of area ΔA on the surface projects onto a small rectangle of area $\overline{\Delta A} = \Delta A \cos \delta$ in the xy-plane, so that, since $\overline{\Delta A} = \Delta x \Delta y$, we obtain

$$\Delta A = \overline{\Delta A} \sec \delta = \sec \delta \, \Delta x \Delta y.$$

A proof of (10″) runs as follows. Let $\mathbf{a} = \mathbf{r}_u \times \mathbf{r}_v$. Using (12) and noting that $u = x$, $v = y$, we calculate from the definition of a vector product

$$\mathbf{a} = \frac{\partial \mathbf{r}}{\partial x} \times \frac{\partial \mathbf{r}}{\partial y} = -\frac{\partial g}{\partial x}\mathbf{i} - \frac{\partial g}{\partial y}\mathbf{j} + \mathbf{k}, \qquad |\mathbf{a}| = \sqrt{1 + \left(\frac{\partial g}{\partial x}\right)^2 + \left(\frac{\partial g}{\partial y}\right)^2}.$$

Hence $\mathbf{a} \cdot \mathbf{k} = 1$. On the other hand, $\mathbf{a} \cdot \mathbf{k} = |\mathbf{a}| \cos \delta^*$, where δ^* is the angle between $\mathbf{a}$ and the positive z-axis. Together $|\mathbf{a}| \cos \delta^* = 1$. Clearly $\cos \delta^* > 0$ and therefore $\delta^* < \pi/2$, that is, δ^* is acute and thus equal to δ. Hence

$$|\mathbf{a}| \cos \delta = 1 \qquad \text{or} \qquad \sec \delta = |\mathbf{a}| \qquad \left(\delta < \frac{\pi}{2}\right)$$

and (10″) is established.

Problems for Sec. 9.6

1. Show that the tangent plane of a surface S: $\mathbf{r}(u, v)$ at a point P can be represented in the form

$$\mathbf{r}^*(p, q) = \mathbf{r} + p\mathbf{r}_u + q\mathbf{r}_v \qquad (\mathbf{r}, \mathbf{r}_u, \mathbf{r}_v \text{ refer to } P)$$

or by a scalar triple product (Sec. 6.9)

$$(\mathbf{r}^* - \mathbf{r} \quad \mathbf{r}_u \quad \mathbf{r}_v) = 0.$$

Find a representation of the tangent plane and a unit normal vector of a surface S at a point P if S is represented in the form

2. $z = g(x, y)$

3. $f(x, y, z) = 0$

4. Let $\mathbf{n}(u, v)$ be the unit normal vector (3) of a surface S: $\mathbf{r}(u, v)$. Set $u = -\tilde{u}, v = \tilde{v}$ and show that the unit normal vector corresponding to the resulting representation is $-\mathbf{n}(u, v)$.

Find a representation of the tangent plane of the following surfaces at P_0: (x_0, y_0, z_0).

5. $z = xy, \quad P_0$: $(1, 1, 1)$

6. $x^2 + y^2 + z^2 = 25, \quad P_0$: $(3, 4, 0)$

7. $z = x^2, \quad P_0$: $(2, 1, 4)$

8. $x^2 + y^2 = 8, \quad P_0$: $(2, 2, 3)$

9. $z = x^2 + y^2, \quad P_0$: $(2, 1, 5)$

10. $3x^2 + 2y^2 + z^2 = 20, \quad P_0$: $(1, 2, 3)$

Find the first fundamental form of the following surfaces.

11. $\mathbf{r} = u\mathbf{i} + v\mathbf{j}$

12. $\mathbf{r} = u\mathbf{i} + v\mathbf{j} + u^2\mathbf{k}$

13. The sphere (6) in Sec. 9.5

14. The torus in Example 3

15. $\mathbf{r} = u\mathbf{i} + v\mathbf{j} + uv\mathbf{k}$

16. $\mathbf{r} = u \cos v\, \mathbf{i} + u \sin v\, \mathbf{j} + u^2\mathbf{k}$

17. **(Orthogonal coordinates)** Show that the coordinate curves $u = const$ and $v = const$ on a surface $\mathbf{r} = \mathbf{r}(u, v)$ intersect at right angles if and only if $F = \mathbf{r}_u \cdot \mathbf{r}_v = 0$.

18. Under what conditions do the coordinate curves $p = const$ and $q = const$ in the solution of Prob. 1 intersect at right angles?

Using (9), find the area of the following surfaces.

19. $x^2 + y^2 = 1, \quad 0 \leqq z \leqq 1$

20. The sphere (6) in Sec. 9.5

21. $z = x^2 + y^2, \quad 0 \leqq z \leqq 1$

22. $z^2 = x^2 + y^2, \quad -1 \leqq z \leqq 1$

23. **(Theorem of Pappus)** Show that the area A in Example 3 can be obtained by the *theorem of Pappus,* which states that the area of a surface of revolution equals the product of the length of a meridian C and the length of the path of the center of gravity of C when C is rotated through the angle 2π.

24. Using the theorem of Pappus, find the center of gravity $(0, \bar{y})$ of a semicircle $x^2 + y^2 = a^2, y \geqq 0$.

25. Using the theorem of Pappus, find the area of the cone $z^2 = x^2 + y^2$ $(0 \leqq z \leqq 1)$.

9.7 Surface Integrals

The concept of a surface integral is a natural generalization of the concept of a double integral considered in Sec. 9.3. Surface integrals occur in many applications, for example, in connection with the center of gravity of a curved lamina, the potential due to charges distributed on surfaces, etc.

The definition of a surface integral parallels that of a double integral. Let S be a portion of a surface of finite area, and let $f(x, y, z)$ be a function which is defined and continuous on S. We subdivide S into n parts $S_1, \cdots, S_n$ of areas $\Delta A_1, \cdots, \Delta A_n$. In each part S_k we choose an arbitrary point P_k with coordinates x_k, y_k, z_k and form the sum

(1) $$J_n = \sum_{k=1}^{n} f(x_k, y_k, z_k)\, \Delta A_k.$$

This we do for each $n = 1, 2, \cdots$ in an arbitrary manner, but so that the largest part S_k shrinks to a point as n approaches infinity. The infinite sequence of numbers $J_1, J_2, \cdots$ has a limit which is independent of the choice of subdivisions and points P_k; the proof is similar to that in the case of a double integral. This limit is called the **surface integral** *of $f(x, y, z)$ over S* and is denoted by

(2) $$\iint_S f(x, y, z)\, dA.$$

To evaluate the surface integral (2), *we may reduce it to a double integral as follows.*

If S is represented in parametric form by a vector function $\mathbf{r}(u, v)$, then $dA = |\mathbf{r}_u \times \mathbf{r}_v|\, du\, dv = \sqrt{EG - F^2}\, du\, dv$, cf. (10) in the previous section. Hence

(3) $$\begin{aligned}\iint_S f(x, y, z)\, dA &= \iint_R f[x(u, v),\ y(u, v),\ z(u, v)]|\mathbf{r}_u \times \mathbf{r}_v|\, du\, dv \\ &= \iint_R f[x(u, v),\ y(u, v),\ z(u, v)]\sqrt{EG - F^2}\, du\, dv\end{aligned}$$

where R is the region corresponding to S in the uv-plane.

Similarly, if S is represented in the form $z = g(x, y)$, then from (10′) in the previous section it follows that

(4) $$\iint_S f(x, y, z)\, dA = \iint_{\bar{S}} f[x, y, g(x, y)]\sqrt{1 + \left(\frac{\partial g}{\partial x}\right)^2 + \left(\frac{\partial g}{\partial y}\right)^2}\, dx\, dy.$$

Example 1. Moment of inertia

Find the moment of inertia I of a homogeneous spherical lamina S: $x^2 + y^2 + z^2 = a^2$ of mass M about the z-axis.

If a mass is distributed over a surface S and $\mu(x, y, z)$ is the density of the mass (= mass per unit area), then the moment of inertia I of the mass with respect to a given axis L is defined by the surface integral

(5) $$I = \iint_S \mu D^2\, dA$$

where $D(x, y, z)$ is the distance of the point (x, y, z) from L.

Since, in the present example, μ is constant and S has the area $A = 4\pi a^2$, we have

$$\mu = \frac{M}{A} = \frac{M}{4\pi a^2}.$$

Representing S by (6), Sec. 9.5, we obtain from (4) in the previous section

$$E = a^2 \cos^2 v, \qquad F = 0, \qquad G = a^2,$$

and from this

$$dA = |\mathbf{r}_u \times \mathbf{r}_v|\, du\, dv = \sqrt{EG - F^2}\, du\, dv = a^2 \cos v\, du\, dv.$$

Furthermore, the square of the distance of a point (x, y, z) from the z-axis becomes

$$D^2 = x^2 + y^2 = a^2 \cos^2 v.$$

Hence we obtain the result

$$I = \iint_S \mu D^2\, dA = \frac{M}{4\pi a^2} \int_{-\pi/2}^{\pi/2} \int_0^{2\pi} a^4 \cos^3 v\, du\, dv = \frac{Ma^2}{2} \int_{-\pi/2}^{\pi/2} \cos^3 v\, dv = \frac{2Ma^2}{3}. \quad \blacksquare$$

In various applications there occur surface integrals for which the concept of orientation of a surface is essential. Therefore we shall now consider this concept, starting with the case of a smooth surface (cf. Sec. 9.5).

Let S be a smooth surface, and let P be any point of S. Then we may choose a unit normal vector $\mathbf{n}$ of S at P. The direction of $\mathbf{n}$ is then called the *positive normal direction* of S at P. Obviously there are two possibilities in choosing $\mathbf{n}$.

A smooth surface S is said to be **orientable** if the positive normal direction, when given at an arbitrary point P_0 of S, can be continued in a unique and continuous way to the entire surface.

Hence the surface S is orientable provided there does not exist a closed curve C on S through P_0 such that the positive normal direction reverses when it is displaced continuously from P_0 along C and back to P_0.

A sufficiently small portion of a smooth surface is always orientable. However, this may not hold in the large. There are nonorientable surfaces. A well-known example of such a surface is the **Möbius**[7] **strip** shown in Fig. 194. When a normal vector, which is given at P_0, is displaced continuously along the curve C in Fig. 194, the resulting normal vector upon returning to P_0 is opposite to the original vector at P_0. A model of a Möbius strip can be made by taking a long rectangular piece of paper, making a half-twist and sticking the shorter sides together so that the two points A and the two points B in Fig. 194 coincide.

If a smooth surface S is orientable, then we may *orient* S by choosing one of the two possible directions of the normal vector $\mathbf{n}$.

If the boundary of S is a simple closed curve C, then we may associate with each of the two possible orientations of S an orientation of C, as shown in Fig. 195*a*. Using this simple idea, we may now readily extend the concept of orientation to piecewise smooth surfaces as follows.

A piecewise smooth surface S is said to be **orientable** if we can orient each

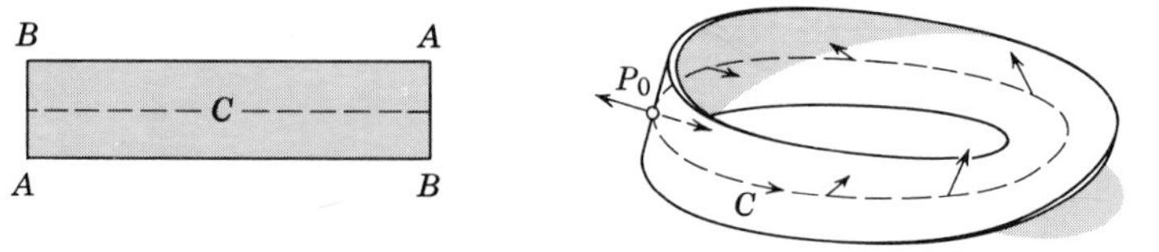

Fig. 194. Möbius strip

[7] AUGUST FERDINAND MÖBIUS (1790–1868), German mathematician, who made important contributions to the theory of surfaces and projective geometry.

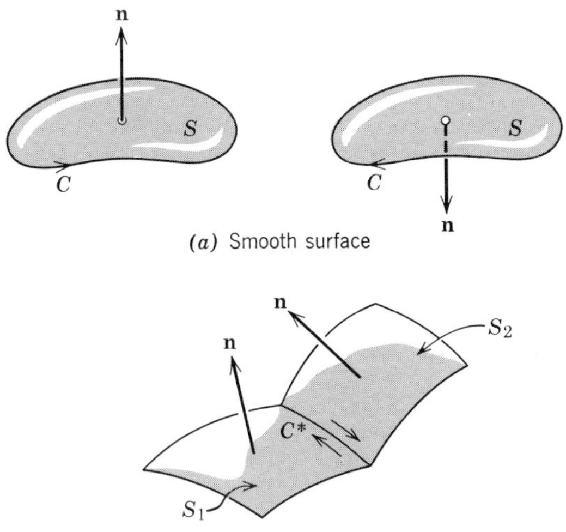

(a) Smooth surface

(b) Piecewise smooth surface

Fig. 195. Orientation of a surface

smooth piece of S in such a manner that along each curve C^* which is a common boundary of two pieces S_1 and S_2 the positive direction of C^* relative to S_1 is opposite to the positive direction of C^* relative to S_2.

Figure 195*b* illustrates the situation for a surface consisting of two smooth pieces.

Let S be a given orientable surface. We orient S by choosing a unit normal vector $\mathbf{n}$. Denoting the angles between $\mathbf{n}$ and the positive x, y, and z axes by α, β, and γ, respectively, we have

$$\mathbf{n} = \cos\alpha\,\mathbf{i} + \cos\beta\,\mathbf{j} + \cos\gamma\,\mathbf{k}. \tag{6}$$

Let $u_1(x, y, z)$, $u_2(x, y, z)$, and $u_3(x, y, z)$ be given functions which are defined and continuous at every point of S. Then the integrals to be considered are usually written in the form

$$\iint_S u_1\,dy\,dz, \qquad \iint_S u_2\,dz\,dx, \qquad \iint_S u_3\,dx\,dy,$$

and *by definition* this means

$$\begin{aligned}
&\iint_S u_1\,dy\,dz = \iint_S u_1 \cos\alpha\,dA,\\
(7)\qquad &\iint_S u_2\,dz\,dx = \iint_S u_2 \cos\beta\,dA,\\
&\iint_S u_3\,dx\,dy = \iint_S u_3 \cos\gamma\,dA.
\end{aligned}$$

It is clear that the value of such an integral depends on the choice of $\mathbf{n}$, that is, on the orientation of S. The transition to the opposite orientation corresponds to the multiplication of the integral by -1, because then the components $\cos\alpha$, $\cos\beta$, and $\cos\gamma$ of $\mathbf{n}$ are multiplied by -1.

Sums of three such integrals may be written in a simple form by using vector notation. In fact, if we introduce the vector

$$\mathbf{u} = u_1\mathbf{i} + u_2\mathbf{j} + u_3\mathbf{k},$$

then we obtain from the definition (7) the formula

$$\textbf{(8)}\qquad \begin{aligned}&\iint_S (u_1\,dy\,dz + u_2\,dz\,dx + u_3\,dx\,dy)\\ &\qquad = \iint_S (u_1\cos\alpha + u_2\cos\beta + u_3\cos\gamma)\,dA = \iint_S \mathbf{u}\cdot\mathbf{n}\,dA.\end{aligned}$$

To evaluate the integrals (7) we may reduce them to double integrals over a plane region, as follows.

If S can be represented in the form $z = h(x, y)$ and is oriented such that $\mathbf{n}$ points upward, then γ is acute. Hence in (10″) of the previous section we have $\delta = \gamma$, so that by the above definition (7) we immediately obtain

$$\text{(9a)}\qquad \iint_S u_3(x, y, z)\,dx\,dy = +\iint_{\bar{R}} u_3[x, y, h(x, y)]\,dx\,dy$$

where $\bar{R}$ is the orthogonal projection of S in the xy-plane. If $\mathbf{n}$ points downward, then γ is obtuse and we obtain

$$\text{(9b)}\qquad \iint_S u_3(x, y, z)\,dx\,dy = -\iint_{\bar{R}} u_3[x, y, h(x, y)]\,dx\,dy.$$

For the other two integrals in (7) the situation is quite similar.

If S is represented in parametric form

$$\mathbf{r}(u, v) = x(u, v)\mathbf{i} + y(u, v)\mathbf{j} + z(u, v)\mathbf{k},$$

then the normal vector is [cf. (3) in the last section]

$$\text{(10)}\qquad \text{(a)}\quad \mathbf{n} = +\frac{\mathbf{r}_u \times \mathbf{r}_v}{|\mathbf{r}_u \times \mathbf{r}_v|} \qquad \text{or} \qquad \text{(b)}\quad \mathbf{n} = -\frac{\mathbf{r}_u \times \mathbf{r}_v}{|\mathbf{r}_u \times \mathbf{r}_v|},$$

depending on the orientation. Now $\cos\gamma = \mathbf{k}\cdot\mathbf{n}$ by (6) and $dA = |\mathbf{r}_u \times \mathbf{r}_v|\,du\,dv$ by (10), Sec. 9.6. Together,

$$\cos\gamma\, dA = \mathbf{k}\cdot\mathbf{n}\, dA = \pm\mathbf{k}\cdot(\mathbf{r}_u \times \mathbf{r}_v)\, du\, dv = \pm\begin{vmatrix} 0 & 0 & 1 \\ x_u & y_u & z_u \\ x_v & y_v & z_v \end{vmatrix} du\, dv$$

$$= \pm\frac{\partial(x, y)}{\partial(u, v)}\, du\, dv,$$

where the last expression involves the Jacobian (cf. Sec. 9.3). Hence in (7),

$$\text{(11)}\qquad \iint_S u_3(x, y, z)\, dx\, dy = \pm\iint_R u_3[x(u, v), y(u, v), z(u, v)]\frac{\partial(x, y)}{\partial(u, v)}\, du\, dv$$

with the plus sign when S is oriented such that (10a) holds and the minus sign in the case of the opposite orientation. Here R is the region corresponding to S in the uv-plane.

Problems for Sec. 9.7

Representing S in parametric form and using (3), evaluate $\iint_S f(x, y, z)\, dA$ where

1. $f = x + 1, \quad S\colon x^2 + y^2 = 1, \quad 0 \leqq z \leqq 3$

2. $f = 8x, \quad S\colon z = x^2, \quad 0 \leqq x \leqq 2, \quad -1 \leqq y \leqq 2$

3. $f = xy, \quad S\colon z = xy, \quad 0 \leqq x \leqq 1, \quad 0 \leqq y \leqq 1$

4. $f = \arctan(y/x), \quad S\colon z = x^2 + y^2, \quad 1 \leqq z \leqq 4, \quad x \geqq 0, y \geqq 0$

5. $f = 3x^3 \sin y, \quad S\colon z = x^3, \quad 0 \leqq x \leqq 1, \quad 0 \leqq y \leqq \pi$

6. $f = x + y + z, \quad S\colon x^2 + y^2 = 1, \quad 0 \leqq z \leqq 2$

7. $f = xy, \quad S\colon x^2 + y^2 = 4, \quad -1 \leqq z \leqq 1$

8. $f = \cos x + \sin y$, S: the portion of $x + y + z = 1$ in the first octant

9. $f = x(z^2 + 12y - y^4), \quad S\colon z = y^2, \quad 0 \leqq x \leqq 1, \quad 0 \leqq y \leqq 1$

10. $f = (x^2 + y^2)^2, \quad S\colon z = (x^2 + y^2)^2, \quad x^2 + y^2 \leqq 1$

11. Justify the following formulas for the mass M and the center of gravity $(\bar{x}, \bar{y}, \bar{z})$ of a lamina S of density (mass per unit area) $\sigma(x, y, z)$ in space:

$$M = \iint_S \sigma\, dA, \quad \bar{x} = \frac{1}{M}\iint_S x\sigma\, dA, \quad \bar{y} = \frac{1}{M}\iint_S y\sigma\, dA, \quad \bar{z} = \frac{1}{M}\iint_S z\sigma\, dA.$$

12. Justify the following formulas for the moments of inertia of the lamina in Prob. 11 about the x-, y-, and z-axes, respectively:

$$I_x = \iint_S (y^2 + z^2)\sigma\, dA, \quad I_y = \iint_S (x^2 + z^2)\sigma\, dA, \quad I_z = \iint_S (x^2 + y^2)\sigma\, dA.$$

Find the moment of inertia of a lamina S of density 1 about an axis A where

13. $S\colon x^2 + y^2 = 1, \quad 0 \leqq z \leqq h$, $\quad A$: the z-axis

14. $S\colon z^2 = x^2 + y^2, \quad 0 \leqq z \leqq h$, $\quad A$: the z-axis

15. S: the torus in Example 3, Sec. 9.6, $\quad A$: the z-axis

16. S as in Prob. 15, A: the line $x = a$ in the xz-plane

17. S as in Prob. 15, A: the line $x = a + b$ in the xz-plane

18. **(Steiner's[8] theorem)** If I_A is the moment of inertia of a mass distribution of total mass M with respect to an axis A through the center of gravity, show that its moment of inertia I_B with respect to an axis B, which is parallel to A and has the distance k from it, is

$$I_B = I_A + k^2 M.$$

19. Using Steiner's theorem and the solution of Prob. 15, solve Probs. 16 and 17.

20. Construct a paper model of a Möbius strip. What happens if you cut it along the curve C in Fig. 194?

9.8 Triple Integrals. Divergence Theorem of Gauss

The triple integral is a generalization of the double integral introduced in Sec. 9.3. For defining this integral we consider a function $f(x, y, z)$ defined in a bounded closed[9] region T of space. We subdivide T by planes parallel to the three coordinate planes. Then we number the parallelepipeds inside T from 1 to n. In each such parallelepiped we choose an arbitrary point, say, (x_k, y_k, z_k) in the kth parallelepiped, and form the sum

$$J_n = \sum_{k=1}^{n} f(x_k, y_k, z_k)\,\Delta V_k$$

where ΔV_k is the volume of the kth parallelepiped. This we do for larger and larger positive integers n in an arbitrary manner, but so that the lengths of the edges of the largest parallelepiped of subdivision approach zero as n approaches infinity. In this way we obtain a sequence of real numbers $J_{n_1}, J_{n_2}, \cdots$. We assume that $f(x, y, z)$ is continuous in a domain containing T, and T is bounded by finitely many *smooth surfaces*.[10] Then it can be shown (cf. Ref. [A3]) that the sequence converges to a limit which is independent of the choice of subdivisions and corresponding points (x_k, y_k, z_k). This limit is called the **triple integral** *of $f(x, y, z)$ over the region T*, and is denoted by

$$\iiint_T f(x, y, z)\,dx\,dy\,dz \qquad \text{or} \qquad \iiint_T f(x, y, z)\,dV.$$

We shall now show that the triple integral of the divergence of a continuously differentiable vector function $\mathbf{u}$ over a region T in space can be transformed into a surface integral of the normal component of $\mathbf{u}$ over the boundary surface S of T. This can be done by means of *the divergence theorem, which is the*

[8] JACOB STEINER (1796–1863), Swiss geometer.

[9] Explained in footnote 2, Sec. 9.3 (with "sphere" instead of "circle").

[10] Cf. Sec. 9.5.

three-dimensional analogue of Green's theorem in the plane considered in Sec. 9.4. The divergence theorem is of basic importance in various theoretical and practical considerations.

Divergence theorem of Gauss (Transformation of volume integrals into surface integrals and conversely)

Let T be a closed bounded region in space whose boundary is a piecewise smooth[11] *orientable surface S. Let* $\mathbf{u}(x, y, z)$ *be a vector function which is continuous and has continuous first partial derivatives in some domain containing T. Then*

$$\iiint_T \operatorname{div} \mathbf{u}\, dV = \iint_S u_n\, dA \tag{1}$$

where

$$u_n = \mathbf{u} \cdot \mathbf{n} \tag{2}$$

is the component of $\mathbf{u}$ *in the direction of the outer normal of S with respect to T, and* $\mathbf{n}$ *is the outer unit normal vector of S.*

Remark. If we write $\mathbf{u}$ and $\mathbf{n}$ in terms of components, say,

$$\mathbf{u} = u_1\mathbf{i} + u_2\mathbf{j} + u_3\mathbf{k} \qquad \text{and} \qquad \mathbf{n} = \cos\alpha\, \mathbf{i} + \cos\beta\, \mathbf{j} + \cos\gamma\, \mathbf{k},$$

so that α, β, and γ are the angles between $\mathbf{n}$ and the positive x, y, and z axes, respectively, formula (1) takes the form

$$\iiint_T \left(\frac{\partial u_1}{\partial x} + \frac{\partial u_2}{\partial y} + \frac{\partial u_3}{\partial z}\right) dx\, dy\, dz = \iint_S (u_1 \cos\alpha + u_2 \cos\beta + u_3 \cos\gamma)\, dA. \tag{3*}$$

Because of (8) in the last section this may be written

$$\iiint_T \left(\frac{\partial u_1}{\partial x} + \frac{\partial u_2}{\partial y} + \frac{\partial u_3}{\partial z}\right) dx\, dy\, dz = \iint_S (u_1\, dy\, dz + u_2\, dz\, dx + u_3\, dx\, dy). \tag{3}$$

Proof of the divergence theorem. Clearly, (3*) is true, if the following three relations hold simultaneously:

$$\iiint_T \frac{\partial u_1}{\partial x} dx\, dy\, dz = \iint_S u_1 \cos\alpha\, dA, \tag{4}$$

$$\iiint_T \frac{\partial u_2}{\partial y} dx\, dy\, dz = \iint_S u_2 \cos\beta\, dA, \tag{5}$$

$$\iiint_T \frac{\partial u_3}{\partial z} dx\, dy\, dz = \iint_S u_3 \cos\gamma\, dA. \tag{6}$$

[11] Cf. Sec. 9.5.

We first prove (6) for a *special region* T which is bounded by a piecewise smooth orientable surface S and has the property that any straight line parallel to any one of the coordinate axes and intersecting T has at most *one* segment (or a single point) in common with T. This implies that T can be represented in the form

$$g(x, y) \leqq z \leqq h(x, y) \tag{7}$$

where (x, y) varies in the orthogonal projection $\bar{R}$ of T in the xy-plane. Clearly, $z = g(x, y)$ represents the "bottom" S_2 of S (Fig. 196) whereas $z = h(x, y)$ represents the "top" S_1 of S, and there may be a remaining vertical portion S_3 of S. (The portion S_3 may degenerate into a curve, as for a sphere.)

To prove (6), we use (7). Since $\mathbf{u}$ is continuously differentiable in some domain containing T, we have

$$\iiint_T \frac{\partial u_3}{\partial z} dx\, dy\, dz = \iint_{\bar{R}} \left[\int_{g(x,y)}^{h(x,y)} \frac{\partial u_3}{\partial z} dz \right] dx\, dy. \tag{8}$$

We integrate the inner integral:

$$\int_g^h \frac{\partial u_3}{\partial z} dz = u_3(x, y, h) - u_3(x, y, g).$$

Hence the left-hand side of (8) equals

$$\iint_{\bar{R}} u_3[x, y, h(x, y)]\, dx\, dy - \iint_{\bar{R}} u_3[x, y, g(x, y)]\, dx\, dy. \tag{9}$$

Let us show that this is equal to the right-hand side of (6). On the lateral portion S_3 of S (Fig. 196) we have $\gamma = \pi/2$ and $\cos \gamma = 0$. Hence this portion does not contribute to the surface integral in (6), and

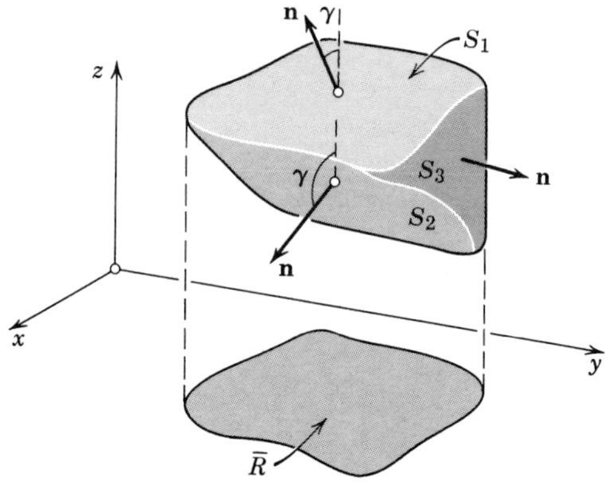

Fig. 196. Example of a special region

$$\iint_S u_3 \cos\gamma \, dA = \iint_{S_1} u_3 \cos\gamma \, dA + \iint_{S_2} u_3 \cos\gamma \, dA$$

(cf. Fig. 196). On S_1 in Fig. 196 the angle γ is acute. Hence by (10″) at the end of Sec. 9.6, with $\delta = \gamma$, we have $dA = \sec\gamma \, dx \, dy$. Since $\cos\gamma \sec\gamma = 1$, we thus obtain

$$\iint_{S_1} u_3 \cos\gamma \, dA = \iint_{\bar{R}} u_3[x, y, h(x, y)] \, dx \, dy,$$

which is identical with the first double integral in (9). On S_2, the angle γ is obtuse (Fig. 196) so that $\pi - \gamma$ corresponds to the acute angle δ in (10″), Sec. 9.6, and, therefore,

$$dA = \sec(\pi - \gamma) \, dx \, dy = -\sec\gamma \, dx \, dy.$$

Hence

$$\iint_{S_2} u_3 \cos\gamma \, dA = -\iint_{\bar{R}} u_3[x, y, g(x, y)] \, dx \, dy,$$

which is identical with the last term in (9). This proves (6).

The relations (4) and (5) then follow by merely relabeling the variables and using the fact that, by assumption, T has representations similar to (7), namely,

$$\tilde{g}(y, z) \leqq x \leqq \tilde{h}(y, z) \qquad \text{and} \qquad g^*(z, x) \leqq y \leqq h^*(z, x).$$

This establishes the divergence theorem for special regions.

For any region T which can be subdivided into *finitely many* special regions by means of auxiliary surfaces, the theorem follows by adding the result for each part separately; this procedure is analogous to that in the proof of Green's theorem in Sec. 9.4. The surface integrals over the auxiliary surfaces cancel in pairs, and the sum of the remaining surface integrals is the surface integral over the whole boundary S of T; the volume integrals over the parts of T add up to the volume integral over T.

The divergence theorem is now proved for any bounded region which is of interest in practical problems. The extension to the most general region T of the type characterized in the theorem would require a certain limit process; this is similar to the situation in the case of Green's theorem in Sec. 9.4. ∎

Green's theorem in the plane (Sec. 9.4) can be used for evaluating line integrals. Similarly, the divergence theorem is suitable for evaluating surface integrals, as may be illustrated by the following example.

Example 1. Evaluation of a surface integral by the divergence theorem

By transforming to a triple integral evaluate

$$I = \iint_S (x^3 \, dy \, dz + x^2 y \, dz \, dx + x^2 z \, dx \, dy)$$

where S is the closed surface consisting of the cylinder $x^2 + y^2 = a^2$ $(0 \leqq z \leqq b)$ and the circular disks $z = 0$ and $z = b$ $(x^2 + y^2 \leqq a^2)$.

In (3) we now have

$$u_1 = x^3, \qquad u_2 = x^2 y, \qquad u_3 = x^2 z.$$

Hence, if we make use of the symmetry of the region T bounded by S, the triple integral in (3) assumes the form

$$\iiint_T (3x^2 + x^2 + x^2)\, dx\, dy\, dz = 4 \cdot 5 \int_0^b \int_0^a \int_0^{\sqrt{a^2-y^2}} x^2\, dx\, dy\, dz.$$

The integral over x equals $\frac{1}{3}(a^2 - y^2)^{3/2}$. Setting $y = a \cos t$, we have

$$dy = -a \sin t\, dt, \qquad (a^2 - y^2)^{3/2} = a^3 \sin^3 t,$$

and the integral over y becomes

$$\frac{1}{3} \int_0^a (a^2 - y^2)^{3/2}\, dy = -\frac{1}{3} a^4 \int_{\pi/2}^0 \sin^4 t\, dt = \frac{\pi}{16} a^4.$$

The integral over z contributes the factor b, and therefore

$$I = 4 \cdot 5 \frac{\pi}{16} b a^4 = \frac{5\pi}{4} a^4 b.$$ ∎

9.9 Consequences and Applications of the Divergence Theorem

The divergence theorem has various applications and important consequences, some of which may be illustrated by the subsequent examples. In these examples, the regions and functions are assumed to satisfy the conditions under which the divergence theorem holds, and in each case, $\mathbf{n}$ is the *outward* unit normal vector of the boundary surface of the region, as before.

Example 1. Representation of the divergence independent of coordinates

Dividing both sides of (1) in Sec. 9.8 by the volume $V(T)$ of the region T, we obtain

$$\text{(1)} \qquad \frac{1}{V(T)} \iiint_T \operatorname{div} \mathbf{u}\, dV = \frac{1}{V(T)} \iint_{S(T)} u_n\, dA$$

where $S(T)$ is the boundary surface of T. The basic properties of the triple integral are essentially the same as those of the double integral considered in Sec. 9.3. In particular the **mean value theorem** *for triple integrals* asserts that for any continuous function $f(x, y, z)$ in the region T under consideration there is a point Q: (x_0, y_0, z_0) in T such that

$$\iiint_T f(x, y, z)\, dV = f(x_0, y_0, z_0) V(T).$$

Setting $f = \operatorname{div} \mathbf{u}$ and using (1), we have

$$\text{(2)} \qquad \frac{1}{V(T)} \iiint_T \operatorname{div} \mathbf{u}\, dV = \operatorname{div} \mathbf{u}(x_0, y_0, z_0).$$

Let P: (x_1, y_1, z_1) be any fixed point in T, and let T shrink down onto P, so that the maximum distance $d(T)$ of the points of T from P approaches zero. Then Q must approach P, and from (1) and (2) it follows that the divergence of $\mathbf{u}$ at P is

$$\text{div } \mathbf{u}(x_1, y_1, z_1) = \lim_{d(T)\to 0} \frac{1}{V(T)} \iint_{S(T)} u_n \, dA. \tag{3}$$

This formula is sometimes used as a definition of the divergence. While the definition of the divergence in Sec. 8.10 involves coordinates, formula (3) is independent of coordinates. Hence *from (3) it follows immediately that the divergence is independent of the particular choice of Cartesian coordinates.*

Example 2. Physical interpretation of the divergence

By means of the divergence theorem we may obtain an intuitive interpretation of the divergence of a vector. For this purpose we consider the flow of an incompressible fluid (cf. p. 405) of constant density $\rho = 1$ which is *stationary* or *steady,* that is, does not vary with time. Such a flow is determined by the field of its velocity vector $\mathbf{v}(P)$ at any point P.

Let S be the boundary surface of a region T in space, and let $\mathbf{n}$ be the outward unit normal vector of S. The mass of fluid which flows through a small portion ΔS of S of area ΔA per unit time from the interior of S to the exterior is equal to $v_n \Delta A$ where[12] $v_n = \mathbf{v} \cdot \mathbf{n}$ is the normal component of $\mathbf{v}$ in the direction of $\mathbf{n}$, taken at a suitable point of ΔS. Consequently, the total mass of fluid which flows across S from T to the outside per unit of time is given by the surface integral

$$\iint_S v_n \, dA.$$

Hence this integral represents the total flow out of T, and the integral

$$\frac{1}{V} \iint_S v_n \, dA, \tag{4}$$

where V is the volume of T, represents the average flow out of T. Since the flow is steady and the fluid is incompressible, the amount of fluid flowing outward must be continuously supplied. Hence, if the value of the integral (4) is different from zero, there must be **sources** (*positive sources and negative sources, called* **sinks**) in T, that is, points where fluid is produced or disappears.

If we let T shrink down to a fixed point P in T, we obtain from (4) the *source intensity* at P represented by the right-hand side of (3) (with u_n replaced by v_n). From this and (3) it follows that *the divergence of the velocity vector* $\mathbf{v}$ *of a steady incompressible flow is the source intensity of the flow at the corresponding point.* There are no sources in T if and only if div $\mathbf{v} \equiv 0$; in this case,

$$\iint_{S^*} v_n \, dA = 0$$

for any closed surface S^* in T.

Example 3. Heat flow. Heat equation

We know that in a body heat will flow in the direction of decreasing temperature. Physical experiments show that the rate of flow is proportional to the gradient of the temperature. This means that the velocity $\mathbf{v}$ of the heat flow in a body is of the form

$$\mathbf{v} = -K \text{ grad } U, \tag{5}$$

where $U(x, y, z, t)$ is the temperature, t is the time, and K is called the *thermal conductivity* of the body; in ordinary physical circumstances K is a constant.

[12] Note that v_n may be negative at a certain point, which means that fluid *enters* the interior of S at such a point.

Let R be a region in the body and let S be its boundary surface. Then the amount of heat leaving R per unit of time is

$$\iint_S v_n \, dA,$$

where $v_n = \mathbf{v} \cdot \mathbf{n}$ is the component of $\mathbf{v}$ in the direction of the outer unit normal vector $\mathbf{n}$ of S. This expression is obtained in a fashion similar to that in the preceding example. From (5) and the divergence theorem we obtain

$$\text{(6)} \qquad \iint_S v_n \, dA = -K \iiint_R \operatorname{div} (\operatorname{grad} U) \, dx \, dy \, dz = -K \iiint_R \nabla^2 U \, dx \, dy \, dz;$$

cf. (5) in Sec. 8.10.

On the other hand, the total amount of heat H in R is

$$H = \iiint_R \sigma \rho U \, dx \, dy \, dz,$$

where the constant σ is the specific heat of the material of the body and ρ is the density (= mass per unit volume) of the material. Hence the time rate of decrease of H is

$$-\frac{\partial H}{\partial t} = -\iiint_R \sigma \rho \frac{\partial U}{\partial t} \, dx \, dy \, dz,$$

and this must be equal to the above amount of heat leaving R; from (6) we thus have

$$-\iiint_R \sigma \rho \frac{\partial U}{\partial t} \, dx \, dy \, dz = -K \iiint_R \nabla^2 U \, dx \, dy \, dz$$

or

$$\iiint_R \left(\sigma \rho \frac{\partial U}{\partial t} - K \nabla^2 U \right) dx \, dy \, dz = 0.$$

Since this holds for any region R in the body, the integrand (if continuous) must be zero everywhere; that is

$$\text{(7)} \qquad \frac{\partial U}{\partial t} = c^2 \nabla^2 U \qquad\qquad c^2 = \frac{K}{\sigma \rho}.$$

This partial differential equation is called the **heat equation;** it is fundamental for heat conduction. Methods for solving problems in heat conduction will be considered in Chap. 11.

Example 4. A basic property of solutions of Laplace's equation

Consider the formula in the divergence theorem:

$$\text{(8*)} \qquad \iiint_T \operatorname{div} \mathbf{u} \, dV = \iint_S u_n \, dA.$$

Assume that $\mathbf{u}$ is the gradient of a scalar function, say, $\mathbf{u} = \operatorname{grad} f$. Then [cf. (5) in Sec. 8.10]

$$\operatorname{div} \mathbf{u} = \operatorname{div} (\operatorname{grad} f) = \nabla^2 f.$$

Furthermore,

$$u_n = \mathbf{u} \cdot \mathbf{n} = \mathbf{n} \cdot \operatorname{grad} f,$$

and from (5) in Sec. 8.8 we see that the right side is the directional derivative of f in the outward normal direction of S. If we denote this derivative by $\partial f/\partial n$, formula (8*) becomes

(8) $$\iiint_T \nabla^2 f\, dV = \iint_S \frac{\partial f}{\partial n} dA.$$

Obviously this is the three-dimensional analogue of the formula (9) in Sec. 9.4.

Taking into account the assumptions under which the divergence theorem holds, we immediately obtain from (8) the following result.

Theorem 1 (A property of solutions of Laplace's equation)

Let $f(x, y, z)$ be a solution of Laplace's equation

$$\nabla^2 f = \frac{\partial^2 f}{\partial x^2} + \frac{\partial^2 f}{\partial y^2} + \frac{\partial^2 f}{\partial z^2} = 0$$

in some domain D, and suppose that the second partial derivatives of f are continuous in D. Then the integral of the normal derivative of f over any piecewise smooth[13] closed orientable surface S in D is zero.

Example 5. Green's theorem

Let f and g be scalar functions such that $\mathbf{u} = f \operatorname{grad} g$ satisfies the assumptions of the divergence theorem in some region T. Then

$$\operatorname{div} \mathbf{u} = \operatorname{div}(f \operatorname{grad} g) = f\nabla^2 g + \operatorname{grad} f \cdot \operatorname{grad} g$$

(cf. Prob. 12 at the end of Sec. 8.10). Furthermore,

$$\mathbf{u} \cdot \mathbf{n} = \mathbf{n} \cdot (f \operatorname{grad} g) = f(\mathbf{n} \cdot \operatorname{grad} g).$$

The expression $\mathbf{n} \cdot \operatorname{grad} g$ is the directional derivative of g in the direction of the outward normal vector $\mathbf{n}$ of the surface S in the divergence theorem. If we denote this derivative by $\partial g/\partial n$, the formula in the divergence theorem becomes

(9) $$\iiint_T (f\nabla^2 g + \operatorname{grad} f \cdot \operatorname{grad} g)\, dV = \iint_S f \frac{\partial g}{\partial n} dA.$$

This formula is called **Green's first formula** or (together with the assumptions) the *first form of Green's theorem*.

By interchanging f and g we obtain a similar formula. On subtracting this formula from (9) we find

(10) $$\iiint_T (f\nabla^2 g - g\nabla^2 f)\, dV = \iint_S \left(f\frac{\partial g}{\partial n} - g\frac{\partial f}{\partial n} \right) dA.$$

This formula is called **Green's second formula** or (together with the assumptions) the *second form of Green's theorem*.

Example 6. Uniqueness of solutions of Laplace's equation

Suppose that f satisfies the assumptions in Theorem 1 and is zero everywhere on a piecewise smooth closed orientable surface S in D. Then, setting $g = f$ in (9) and denoting the interior of S by T, we obtain

$$\iiint_T \operatorname{grad} f \cdot \operatorname{grad} f\, dV = \iiint_T |\operatorname{grad} f|^2\, dV = 0.$$

[13] Cf. Sec. 9.5.

Since by assumption $|\text{grad } f|$ is continuous in T and on S and is non-negative, it must be zero everywhere in T. Hence $f_x = f_y = f_z = 0$, and f is constant in T and, because of continuity, equal to its value 0 on S. This proves

Theorem 2

If a function $f(x, y, z)$ satisfies the assumptions of Theorem 1 and is zero at all points of a piecewise smooth closed orientable surface S in D, then it vanishes identically in the region T bounded by S.

This theorem has an important consequence. Let f_1 and f_2 be functions which satisfy the assumptions of Theorem 1 and take on the same values on S. Then their difference $f_1 - f_2$ satisfies those assumptions and has the value 0 everywhere on S. Consequently, from Theorem 2 it follows that $f_1 - f_2 = 0$ throughout T, and we have the following result.

Theorem 3
(Uniqueness theorem for solutions of Laplace's equation)

Let f be a solution of Laplace's equation which has continuous second partial derivatives in a domain D, and let T be a region in D which satisfies the assumptions of the divergence theorem. Then f is uniquely determined in T by its values on the boundary surface S of T.

Problems for Sec. 9.9

Find the volume of the following regions by triple integration.

1. The tetrahedron with vertices $(0, 0, 0)$, $(2, 0, 0)$, $(0, 1, 0)$, $(0, 0, 3)$

2. The region in the first octant bounded by $y = x$, $y = x^2$, and $z = 1 - x$

3. The region between the paraboloid $z = 1 - x^2 - y^2$ and the xy-plane

4. The region bounded by the cylinders $x^2 + y^2 = 1$ and $y^2 + z^2 = 1$.

Find the total mass of a mass distribution of density σ in a region T where

5. $\sigma = xyz$, T: the cube $0 \leqq x \leqq 1$, $0 \leqq y \leqq 1$, $0 \leqq z \leqq 1$

6. $\sigma = x + y + 2z$, T: the tetrahedron with vertices $(0, 0, 0)$, $(1, 0, 0)$, $(0, 1, 0)$, $(0, 0, 1)$

7. $\sigma = xy$, T: the region in Prob. 6

8. $\sigma = z$, T: the region in the first octant bounded by $y = 1 - x^2$ and $z = x$

Find the moment of inertia

$$I_z = \iiint_T (x^2 + y^2)\, dx\, dy\, dz$$

of a mass of density 1 in a region T about the z-axis, where T is

9. The cube $0 \leqq x \leqq c$, $0 \leqq y \leqq c$, $0 \leqq z \leqq c$

10. The cylinder $x^2 + y^2 \leqq c^2$, $0 \leqq z \leqq h$

11. The cone $x^2 + y^2 \leqq z^2$, $0 \leqq z \leqq h$

12. The interior of the sphere $x^2 + y^2 + z^2 = c^2$

13. Using the divergence theorem, show that the volume V of a region T bounded by a surface S is

$$V = \iint_S x\, dy\, dz = \iint_S y\, dz\, dx = \iint_S z\, dx\, dy = \frac{1}{3} \iint_S (x\, dy\, dz + y\, dz\, dx + z\, dx\, dy).$$

14. Verify the formulas in Prob. 13 for a cube.

15. Same task as in Prob. 14 for the cylinder $x^2 + y^2 \leqq 1$, $0 \leqq z \leqq h$.

16. Using $\mathbf{u} = x\mathbf{i} + y\mathbf{j} + z\mathbf{k}$ in (1), Sec. 9.8, show that a region T with boundary surface S has the volume

$$V = \frac{1}{3}\iint_S r\cos\theta\, dA$$

where r is the distance of a variable point P: (x, y, z) on S from the origin O and θ is the angle between the directed line OP and the outer normal to S at P.

17. Find the volume of a sphere of radius a by means of the formula in Prob. 16.

Evaluate the following surface integrals by the divergence theorem, assuming that S is oriented as in that theorem.

18. $\iint_S [(x + z)\, dy\, dz + (y + z)\, dz\, dx + (x + y)\, dx\, dy]$, S: $x^2 + y^2 + z^2 = 4$

19. $\iint_S (x^2\, dy\, dz + y^2\, dz\, dx + z^2\, dx\, dy)$, S: the surface of the cube in Prob. 5

20. $\iint_S (e^x\, dy\, dz - ye^x\, dz\, dx + 3z\, dx\, dy)$, S: the surface of the cylinder in Prob. 10

21. $\iint_S (yz\, dy\, dz + zx\, dz\, dx + xy\, dx\, dy)$, S as in Prob. 18

22. $\iint_S xyz\, dy\, dz$, S: the parallelepiped $0 \leqq x \leqq 3$, $0 \leqq y \leqq 2$, $0 \leqq z \leqq 1$

23. $\iint_S [\sin x\, dy\, dz + (2 - \cos x) y\, dz\, dx]$, S as in Prob. 22

24. $\iint_S [(y\cos^2 x + y^3)\, dz\, dx + z(\sin^2 x - 3y^2)\, dx\, dy]$, S as in Prob. 18

Let T be a closed bounded region in space and S its boundary surface. Using the divergence theorem, prove the following statements.

25. If g is **harmonic**[14] in a domain containing T, then

$$\iint_S \frac{\partial g}{\partial n}\, dA = 0.$$

26. If g is harmonic in a domain containing T, then

$$\iint_S g\frac{\partial g}{\partial n}\, dA = \iiint_T |\text{grad } g|^2\, dV.$$

27. If g is harmonic in a domain containing T and $\partial g/\partial n = 0$ on S, then g is constant in T.

28. If f and g are harmonic in a domain containing T and $\partial f/\partial n = \partial g/\partial n$ on S, then $f = g + c$ in T, where c is a constant.

[14]This means that g is a solution of Laplace's equation, and its second partial derivatives are continuous in T.

29. If f and g are harmonic in a domain containing T, then

$$\iint_S \left(f\frac{\partial g}{\partial n} - g\frac{\partial f}{\partial n}\right) dA = 0.$$

30. Show that the Laplacian can be represented independently of all coordinate systems in the form

$$\nabla^2 f = \lim_{d(T)\to 0} \frac{1}{V(T)} \iint_{S(T)} \frac{\partial f}{\partial n}\, dA$$

where $d(T)$ is the maximum distance of the points of a region T bounded by $S(T)$ from the point at which the Laplacian is evaluated and $V(T)$ is the volume of T. *Hint.* Put $\mathbf{u} = \operatorname{grad} f$ in (3) and use (5), Sec. 8.8, with $\mathbf{b} = \mathbf{n}$, the outer unit normal vector to S.

9.10 Stokes's Theorem

In Sec. 9.4 it was shown that double integrals over a plane region can be transformed into line integrals over the boundary curve of the region. Generalizing this result, we shall now consider the corresponding problem in the case of a surface integral.

Stokes's theorem[15] (Transformation of surface integrals into line integrals and conversely)

Let S be a piecewise smooth[16] oriented surface in space and let the boundary of S be a piecewise smooth simple closed curve C. Let $\mathbf{v}(x, y, z)$ *be a continuous vector function which has continuous first partial derivatives in a domain in space which contains S. Then*

$$\iint_S (\operatorname{curl} \mathbf{v})_n\, dA = \int_C v_t\, ds; \tag{1}$$

here $(\operatorname{curl} \mathbf{v})_n = (\operatorname{curl} \mathbf{v}) \cdot \mathbf{n}$ *is the component of* curl $\mathbf{v}$ *in the direction of a unit normal vector* $\mathbf{n}$ *of S; the integration around C is taken in the sense shown in Fig. 197, and* v_t *is the component of* $\mathbf{v}$ *in the direction of the tangent vector of C in Fig. 197.*

Proof. We first prove Stokes's theorem for a surface S which can be represented simultaneously in the forms

(2) (a) $z = f(x, y)$, (b) $y = g(x, z)$, (c) $x = h(y, z)$,

[15] GEORGE GABRIEL STOKES (1819–1903), Irish mathematician and physicist, who made important contributions to the theory of infinite series and several branches of theoretical physics.

[16] Cf. Secs. 9.1 and 9.5.

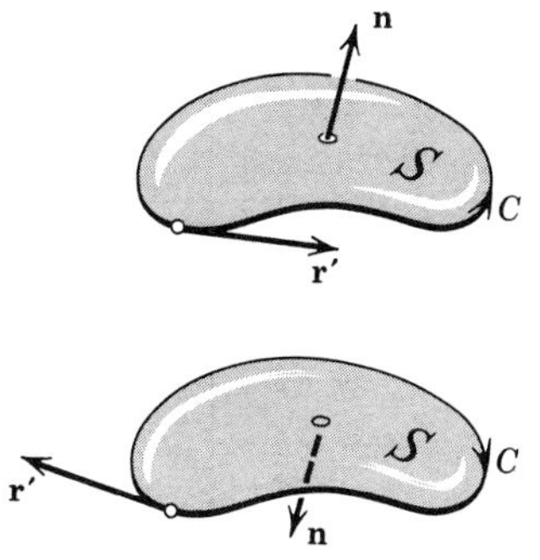

Fig. 197. Stokes's theorem

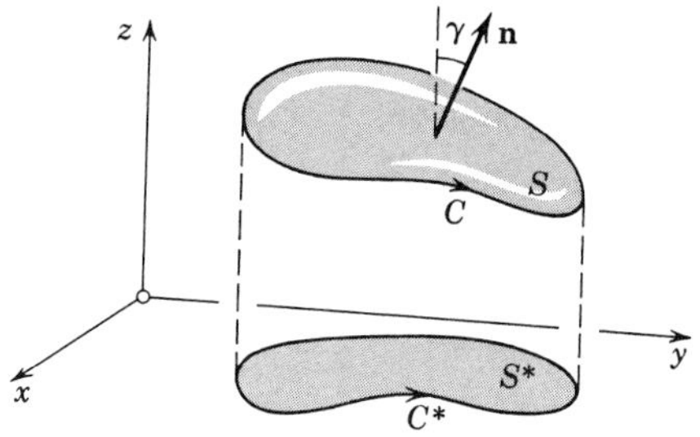

Fig. 198. Proof of Stokes's theorem

where f, g, and h are continuous functions of the respective variables and have continuous first partial derivatives. Let

$$\mathbf{n} = \cos\alpha\,\mathbf{i} + \cos\beta\,\mathbf{j} + \cos\gamma\,\mathbf{k} \tag{3*}$$

be the "upper" unit normal vector of S (Fig. 198), and let $\mathbf{v} = v_1\mathbf{i} + v_2\mathbf{j} + v_3\mathbf{k}$. If we represent C in the form $\mathbf{r} = \mathbf{r}(s)$ where the arc length s increases in the direction of integration, the unit tangent vector is

$$\frac{d\mathbf{r}}{ds} = \frac{dx}{ds}\mathbf{i} + \frac{dy}{ds}\mathbf{j} + \frac{dz}{ds}\mathbf{k},$$

and therefore,

$$v_t = \mathbf{v} \cdot \frac{d\mathbf{r}}{ds} = v_1\frac{dx}{ds} + v_2\frac{dy}{ds} + v_3\frac{dz}{ds}.$$

From this we obtain

$$v_t\,ds = \mathbf{v} \cdot \frac{d\mathbf{r}}{ds}\,ds = v_1\,dx + v_2\,dy + v_3\,dz.$$

Consequently, if we use the representation of the curl in terms of right-handed Cartesian coordinates [cf. (1), Sec. 8.11], the formula in Stokes's theorem may be written

$$\begin{aligned}&\iint_S \left[\left(\frac{\partial v_3}{\partial y} - \frac{\partial v_2}{\partial z}\right)\cos\alpha + \left(\frac{\partial v_1}{\partial z} - \frac{\partial v_3}{\partial x}\right)\cos\beta + \left(\frac{\partial v_2}{\partial x} - \frac{\partial v_1}{\partial y}\right)\cos\gamma\right] dA \\ &\qquad = \int_C (v_1\,dx + v_2\,dy + v_3\,dz),\end{aligned} \tag{3}$$

where α, β, and γ are defined in connection with (3*).

We prove that in (3) the integrals over the terms involving v_1 are equal,

$$\iint_S \left(\frac{\partial v_1}{\partial z}\cos\beta - \frac{\partial v_1}{\partial y}\cos\gamma\right) dA = \int_C v_1\,dx. \tag{4}$$

Let S^* be the orthogonal projection of S in the xy-plane and let C^* be its boundary which is oriented as shown in Fig. 198. Using the representation (2a) of S, we may write the line integral over C as a line integral over C^*:

$$\int_C v_1(x, y, z)\, dx = \int_{C^*} v_1[x, y, f(x, y)]\, dx.$$

We now apply Green's theorem in the plane (Sec. 9.4) to the functions $v_1[x, y, f(x, y)]$ and 0 [instead of f and g in Sec. 9.4]. Then

$$\int_{C^*} v_1[x, y, f(x, y)]\, dx = -\iint_{S^*} \frac{\partial v_1}{\partial y}\, dx\, dy.$$

In the integral on the right,

$$\frac{\partial v_1[x, y, f(x, y)]}{\partial y} = \frac{\partial v_1(x, y, z)}{\partial y} + \frac{\partial v_1(x, y, z)}{\partial z}\frac{\partial f}{\partial y} \qquad [z = f(x, y)]$$

and, thus,

$$(5) \qquad \int_C v_1(x, y, z)\, dx = -\iint_{S^*} \left(\frac{\partial v_1}{\partial y} + \frac{\partial v_1}{\partial z}\frac{\partial f}{\partial y}\right) dx\, dy.$$

We prove that the integral on the left-hand side of (4) equals the integral on the right-hand side of (5). In the former integral we introduce x and y as variables of integration. Writing (2a) in the form

$$F(x, y, z) = z - f(x, y) = 0,$$

we obtain

$$\text{grad } F = -\frac{\partial f}{\partial x}\mathbf{i} - \frac{\partial f}{\partial y}\mathbf{j} + \mathbf{k}.$$

Denoting the length of grad F by a, we have

$$a = |\text{grad } F| = \sqrt{1 + \left(\frac{\partial f}{\partial x}\right)^2 + \left(\frac{\partial f}{\partial y}\right)^2}.$$

Since grad F is normal to S, we obtain

$$\mathbf{n} = \pm\frac{\text{grad } F}{a}$$

where $\mathbf{n}$ is defined as before. But the components of both $\mathbf{n}$ and grad F in the positive z-direction are positive, and, therefore,

$$\mathbf{n} = +\frac{\text{grad } F}{a}.$$

Hence from the preceding representations of $\mathbf{n}$ and grad F in terms of components with respect to the xyz-coordinate system we see that

$$\cos\alpha = -\frac{1}{a}\frac{\partial f}{\partial x}, \qquad \cos\beta = -\frac{1}{a}\frac{\partial f}{\partial y}, \qquad \cos\gamma = \frac{1}{a}.$$

Furthermore, from (10′) at the end of Sec. 9.6 it follows that in (4) we have $dA = a\,dx\,dy$. Hence

$$\iint_S \left(\frac{\partial v_1}{\partial z}\cos\beta - \frac{\partial v_1}{\partial y}\cos\gamma\right) dA = \iint_{S^*} \left(\frac{\partial v_1}{\partial z}\left(-\frac{1}{a}\frac{\partial f}{\partial y}\right) - \frac{\partial v_1}{\partial y}\frac{1}{a}\right) a\,dx\,dy.$$

The integral on the right equals the right-hand side of (5), and (4) is proved.

If $-\mathbf{n}$ were chosen for the positive normal direction, then by assumption the sense of integration along C would be reversed and the result would be the same as before. This shows that (4) holds for both choices of the positive normal of S.

Using the representations (2b) and (2c) of S and reasoning exactly as before, we obtain

$$\iint_S \left(\frac{\partial v_2}{\partial x}\cos\gamma - \frac{\partial v_2}{\partial z}\cos\alpha\right) dA = \int_C v_2\,dy \tag{6}$$

and

$$\iint_S \left(\frac{\partial v_3}{\partial y}\cos\alpha - \frac{\partial v_3}{\partial x}\cos\beta\right) dA = \int_C v_3\,dz. \tag{7}$$

By adding (4), (6), and (7) we obtain (1). This proves Stokes's theorem for a surface S which can be represented simultaneously in the forms (2a), (2b), and (2c).

As in the proof of the divergence theorem, our result may be immediately extended to a surface S which can be decomposed into finitely many pieces, each of which is of the type considered before. This covers most of the cases which are of practical interest. The proof in the case of a most general surface S satisfying the assumptions of the theorem would require a limit process; this is similar to the situation in the case of Green's theorem in Sec. 9.4. ∎

9.11 Consequences and Applications of Stokes's Theorem

Example 1. Green's theorem in the plane as a special case of Stokes's theorem
Let $\mathbf{v} = v_1\mathbf{i} + v_2\mathbf{j}$ be a vector function which is continuously differentiable in a domain in the xy-plane containing a simply-connected bounded closed region S whose boundary C is a piecewise smooth simple closed curve. Then, according to (1) in Sec. 8.11,

$$(\operatorname{curl}\mathbf{v})_n = \frac{\partial v_2}{\partial x} - \frac{\partial v_1}{\partial y}.$$

Furthermore, $v_t\, ds = v_1\, dx + v_2\, dy$, and (1) in the previous section takes the form

$$\iint_S \left(\frac{\partial v_2}{\partial x} - \frac{\partial v_1}{\partial y}\right) dA = \int_C (v_1\, dx + v_2\, dy).$$

This shows that Green's theorem in the plane (Sec. 9.4) is a special case of Stokes's theorem.

Example 2. Physical interpretation of the curl

Let S_r be a circular disk of radius r and center P bounded by the circle C_r (Fig. 199), and let $\mathbf{v}(Q) \equiv \mathbf{v}(x, y, z)$ be a continuously differentiable vector function in a domain containing S_r. Then by Stokes's theorem and the mean value theorem for surface integrals,

$$\int_{C_r} v_t\, ds = \iint_{S_r} (\text{curl } \mathbf{v})_n\, dA = [\text{curl } \mathbf{v}(P^*)]_n A_r$$

where A_r is the area of S_r and P^* is a suitable point of S_r. This may be written

$$[\text{curl } \mathbf{v}(P^*)]_n = \frac{1}{A_r} \int_{C_r} v_t\, ds.$$

In the case of a fluid motion with velocity $\mathbf{v}$, the integral

$$\int_{C_r} v_t\, ds$$

is called the **circulation** of the flow around C_r; it measures the extent to which the corresponding fluid motion is a rotation around the circle C_r. If we now let r approach zero, we find

(1) $$[\text{curl } \mathbf{v}(P)]_n = \lim_{r \to 0} \frac{1}{A_r} \int_{C_r} v_t\, ds;$$

that is, the component of the curl in the positive normal direction can be regarded as the *specific circulation* (circulation per unit area) of the flow in the surface at the corresponding point.

Example 3. Evaluation of a line integral by Stokes's theorem

Evaluate $\int_C v_t\, ds$, where C is the circle $x^2 + y^2 = 4$, $z = -3$, oriented in the counterclockwise sense as viewed from the origin, and, with respect to right-handed Cartesian coordinates,

$$\mathbf{v} = y\mathbf{i} + xz^3\mathbf{j} - zy^3\mathbf{k}.$$

As a surface S bounded by C we can take the plane circular disk $x^2 + y^2 \leqq 4$ in the plane $z = -3$. Then $\mathbf{n}$ in Stokes's theorem points in the positive z-direction; thus $\mathbf{n} = \mathbf{k}$. Hence $(\text{curl } \mathbf{v})_n$ is simply the component of curl $\mathbf{v}$ in the positive z-direction. Since $\mathbf{v}$ with $z = -3$ has the components $v_1 = y$, $v_2 = -27x$, $v_3 = 3y^3$, we thus obtain

$$(\text{curl } \mathbf{v})_n = \frac{\partial v_2}{\partial x} - \frac{\partial v_1}{\partial y} = -27 - 1 = -28.$$

Hence the integral over S in Stokes's theorem equals -28 times the area 4π of the disk S. This yields the answer $-28 \cdot 4\pi = -112\pi \approx -352$.

To appreciate the savings, the reader may evaluate the integral directly, starting from a parametric representation of C, calculating $v_t = \mathbf{v} \cdot \mathbf{u}$ and then integrating with respect to the arc length s; here, $\mathbf{v}$ is taken at the points of C and $\mathbf{u}$ is the unit tangent vector of C in the direction of integration.

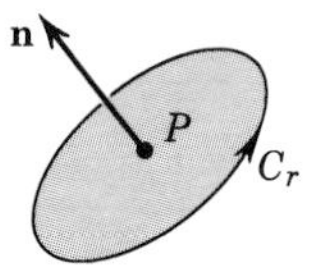

Fig. 199. Example 2

Problems for Sec. 9.11

Evaluate $\iint_S (\text{curl } \mathbf{v})_n \, dA$, where

1. $\mathbf{v} = z\mathbf{i} + x\mathbf{j}$, S: the square $0 \leqq x \leqq 1$, $0 \leqq y \leqq 1$, $z = 1$
2. $\mathbf{v} = z\mathbf{i} + x\mathbf{j} + y\mathbf{k}$, S as in Prob. 1
3. $\mathbf{v} = -y^3\mathbf{i} + x^3\mathbf{j}$, S: the circular disk $x^2 + y^2 \leqq 1$, $z = 0$

4. Verify Stokes's theorem for $\mathbf{v}$ and S in Prob. 1.
5. Verify Stokes's theorem for $\mathbf{v}$ and S in Prob. 3.
6. Show that if $\mathbf{v}$ and S satisfy the assumptions of Stokes's theorem and $\mathbf{v} = \text{grad} f$, then $\int_C v_t \, ds = 0$, where C is the boundary of S.

Evaluate $\int_C v_t \, ds$ by Stokes's theorem, where, with respect to right-handed Cartesian coordinates,

7. $\mathbf{v} = 2y\mathbf{i} + z\mathbf{j} + 3y\mathbf{k}$, C: the intersection of $x^2 + y^2 + z^2 = 6z$ and $z = x + 3$, oriented in the clockwise sense as viewed from the origin
8. $\mathbf{v} = -3y\mathbf{i} + 3x\mathbf{j} + z\mathbf{k}$, C: the circle $x = \cos \alpha$, $y = \sin \alpha$, $z = 1$ $(0 \leqq \alpha \leqq 2\pi)$
9. $\mathbf{v} = y^2\mathbf{i} + x^2\mathbf{j} - (x + z)\mathbf{k}$, C: the boundary of the triangle with vertices at $(0, 0, 0)$, $(1, 0, 0)$, $(1, 1, 0)$, (counterclockwise)
10. $\mathbf{v} = 4z\mathbf{i} - 2x\mathbf{j} + 2x\mathbf{k}$, C: the intersection of $x^2 + y^2 = 1$ and $z = y + 1$, oriented in the clockwise sense as viewed from the origin
11. $\mathbf{v} = x^2\mathbf{i} + y^2\mathbf{j} + z^2\mathbf{k}$, C: the intersection of $x^2 + y^2 + z^2 = a^2$ and $z = y^2$

Using (3) in the last section, evaluate the following line integrals in the clockwise sense (as viewed from the origin) around the given contour C.

12. $\int_C (z \, dx + x \, dy + y \, dz)$, C: the boundary of the triangle with vertices $(1, 0, 0)$, $(0, 1, 0)$, $(0, 0, 1)$
13. $\int_C (yz \, dx + xz \, dy + xy \, dz)$, C: the intersection of $x^2 + y^2 = 1$ and $z = y^2$
14. $\int_C (\sin z \, dx - \cos x \, dy + \sin y \, dz)$, C: the boundary of the rectangle $0 \leqq x \leqq \pi$, $0 \leqq y \leqq 1$, $z = 3$
15. $\int_C (e^x \, dx + 2y \, dy - dz)$, C: $x^2 + y^2 = 4$, $z = 2$

9.12 Line Integrals Independent of Path

In Sec. 9.2 we have seen that the value of a line integral

$$\int_C (f \, dx + g \, dy + h \, dz) \tag{1}$$

will in general depend not only on the endpoints P and Q of the path C but also on C; that is, if we integrate from P to Q along different paths, we shall, in general, obtain different values of the integral. We shall now see under what conditions that value depends only on P and Q but does not depend on the path C from P to Q. This problem is of great importance. We first state the following definition.

Let $f(x, y, z)$, $g(x, y, z)$, and $h(x, y, z)$ be functions which are defined and continuous in a domain D of space. Then a line integral of the form (1) is said to be **independent of path** *in D*, if for every pair of endpoints P and Q in D the value of the integral is the same for all paths C in D starting from P and ending at Q. This value will then, in general, depend on the choice of the points P and Q but not on the choice of the path joining them.

In this connection we recall that, by definition, a function is a **single-valued** relation, that is, to each point in its domain of definition it assigns a *single* value in its range. This will be essential in our present discussion.

For formulating our results, the following two concepts will be convenient. An expression of the form

$$f\,dx + g\,dy + h\,dz, \tag{2}$$

where f, g, h are functions defined in a domain D in space, is called a *first-order* **differential form** in three variables. This form is said to be **exact** or *an exact differential in D*, if it is the differential

$$du = \frac{\partial u}{\partial x}dx + \frac{\partial u}{\partial y}dy + \frac{\partial u}{\partial z}dz \tag{3}$$

of a differentiable function $u(x, y, z)$ everywhere in D; that is,

$$f\,dx + g\,dy + h\,dz = du. \tag{4}$$

By comparing (3) and (4) we see that *the form* (2) *is exact in D if and only if there is a differentiable function $u(x, y, z)$ such that*

$$f = \frac{\partial u}{\partial x}, \qquad g = \frac{\partial u}{\partial y}, \qquad h = \frac{\partial u}{\partial z} \tag{5}$$

everywhere in D.

In vector language this means that *the form* (2) *is exact in D if and only if the vector function*

$$\mathbf{v} = f\mathbf{i} + g\mathbf{j} + h\mathbf{k}$$

is the gradient of a function $u(x, y, z)$ in D:

$$\mathbf{v} = \operatorname{grad} u = \frac{\partial u}{\partial x}\mathbf{i} + \frac{\partial u}{\partial y}\mathbf{j} + \frac{\partial u}{\partial z}\mathbf{k}. \tag{5'}$$

We shall now prove that exactness is a necessary and sufficient condition for independence of path, as follows.

Theorem 1 (Exactness and independence of path)
Let $f(x, y, z)$, $g(x, y, z)$, and $h(x, y, z)$ be continuous in a domain D in space. Then the line integral

$$\int_C (f\,dx + g\,dy + h\,dz) \tag{6}$$

is independent of path in D if and only if the differential form under the integral sign is exact in D.

Remark. Note that (6) can be written

$$\int_C \mathbf{v} \cdot d\mathbf{r} \tag{6'}$$

where $\mathbf{v} = f\mathbf{i} + g\mathbf{j} + h\mathbf{k}$ and $d\mathbf{r} = dx\,\mathbf{i} + dy\,\mathbf{j} + dz\,\mathbf{k}$.

Proof of the theorem. **(a)** Suppose that the integral under consideration is independent of path in D. We choose any fixed point P: (x_0, y_0, z_0) and a point Q: (x, y, z) in D. Then we define a function $u(x, y, z)$ by the formula

$$u(x, y, z) = u_0 + \int_P^Q (f\,dx^* + g\,dy^* + h\,dz^*), \tag{7}$$

where u_0 is a constant and we integrate along an arbitrary path in D from P to Q. Since the integral is independent of path and P is fixed, it does indeed depend only on the coordinates x, y, z of the endpoint Q and defines a function $u(x, y, z)$ in D. It remains to show that from (7) the relations (5) follow, which means that $f\,dx + g\,dy + h\,dz$ is exact in D. We prove the first of these three relations. Since the integral is independent of path, we may integrate from P to a point Q_1: (x_1, y, z) and then parallel to the x-axis along the line segment from Q_1 to Q (Fig. 200); here Q_1 is chosen so that the whole segment Q_1Q lies in D. Then

$$u(x, y, z) = u_0 + \int_P^{Q_1} (f\,dx^* + g\,dy^* + h\,dz^*) + \int_{Q_1}^{Q} (f\,dx^* + g\,dy^* + h\,dz^*).$$

We take the partial derivative with respect to x. Since P and Q_1 do not depend on x, that derivative of the first integral is zero. Since on the segment Q_1Q, both y and z are constant, the last integral may be written as the definite integral

$$\int_{x_1}^{x} f(x^*, y, z)\,dx^*.$$

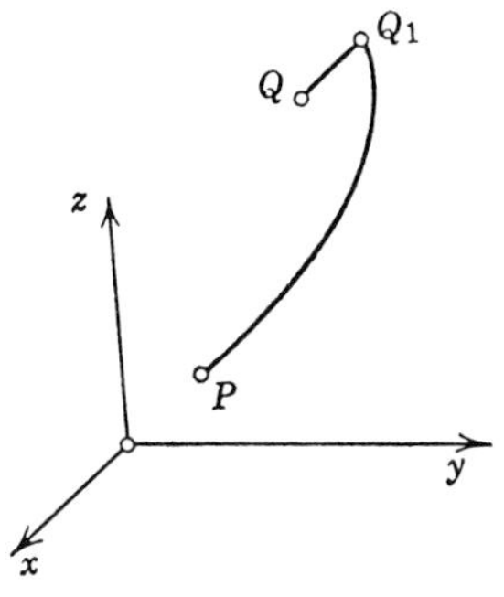

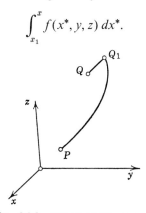

Fig. 200. Proof of Theorem 1

Hence the partial derivative of that integral with respect to x is $f(x, y, z)$, and the first of the relations (5) is proved. The other two relations can be proved by the same argument.

(b) Conversely, suppose that $f\,dx + g\,dy + h\,dz$ is exact in D. Then (5) holds in D for some function u. Let C be any path from P to Q in D, and let

$$\mathbf{r}(t) = x(t)\mathbf{i} + y(t)\mathbf{j} + z(t)\mathbf{k} \qquad (t_0 \leqq t \leqq t_1)$$

be a parametric representation of C such that P corresponds to $t = t_0$ and Q corresponds to $t = t_1$. Then

$$\int_P^Q (f\,dx + g\,dy + h\,dz) = \int_P^Q \left(\frac{\partial u}{\partial x}dx + \frac{\partial u}{\partial y}dy + \frac{\partial u}{\partial z}dz\right)$$

$$= \int_{t_0}^{t_1} \frac{du}{dt}dt = u[x(t), y(t), z(t)]\Bigg|_{t=t_0}^{t=t_1} = u(Q) - u(P).$$

This shows that the value of the integral is simply the difference of the values of u at the two endpoints of C and is, therefore, independent of the path C. Theorem 1 is proved. ∎

The last formula in the proof,

$$\text{(8)} \qquad \int_P^Q (f\,dx + g\,dy + h\,dz) = u(Q) - u(P),$$

is the analogue of the usual formula

$$\int_a^b f(x)\,dx = F(x)\Bigg|_a^b = F(b) - F(a) \qquad [\text{where } F'(x) = f(x)]$$

for evaluating definite integrals which is known from elementary calculus. Formula (8) should be applied whenever a line integral is independent of path. The practical use of this formula will be explained at the end of this section, in Example 3.

We remember that the work W done by a (variable) force $\mathbf{p} = f\mathbf{i} + g\mathbf{j} + h\mathbf{k}$ in the displacement of a particle along a path C in space is given by the line integral (7) in Sec. 9.2,

$$W = \int_C \mathbf{p} \cdot d\mathbf{r} = \int_C (f\,dx + g\,dy + h\,dz),$$

the integration being taken in the sense of the displacement. From Theorem 1 we conclude that W depends only on the endpoints P and Q of the path C if and only if the differential form under the integral sign is exact; and this form is exact if and only if $\mathbf{p}$ is the gradient of a scalar function u. In this case the field of force given by $\mathbf{p}$ is said to be **conservative** (cf. also at the end of Sec. 8.8).

Example 1. Work done by a variable force
Find the work W done by the force $\mathbf{p} = yz\mathbf{i} + xz\mathbf{j} + xy\mathbf{k}$ in the displacement of a particle along the straight segment C from P: (1, 1, 1) to Q: (3, 3, 2). The work is given by the integral

$$W = \int_C (yz\,dx + xz\,dy + xy\,dz).$$

This integral is of the form (1) where

$$f = yz, \qquad g = xz, \qquad h = xy,$$

and (5) holds for $u = xyz$. Hence W is independent of path and we may use (8), finding

$$W = u(Q) - u(P) = u(3, 3, 2) - u(1, 1, 1) = 18 - 1 = 17. \qquad \blacksquare$$

From Theorem 1 we may immediately derive the following important result.

Theorem 2 (Independence of path)
Let f, g, h be continuous in a domain D of space. Then the line integral

$$\int_C (f\,dx + g\,dy + h\,dz)$$

is independent of path in D if and only if it is zero on every simple closed path in D.

Proof. **(a)** Let C be a simple closed path in D and suppose that the integral is independent of path in D. We subdivide C into two arcs C_1 and C_2 (Fig. 201). Then

$$\begin{aligned}(9)\qquad &\oint_C (f\,dx + g\,dy + h\,dz)\\ &= \int_{C_1} (f\,dx + g\,dy + h\,dz) + \int_{C_2} (f\,dx + g\,dy + h\,dz).\end{aligned}$$

Because of independence of path,

$$\begin{aligned}\int_{C_2} (f\,dx + g\,dy + h\,dz) &= \int_{C_1^*} (f\,dx + g\,dy + h\,dz)\\ &= -\int_{C_1} (f\,dx + g\,dy + h\,dz),\end{aligned}$$

where C_1^* denotes C_1 traced in the reverse direction. From this it follows that the integral on the left-hand side of (9) is zero.

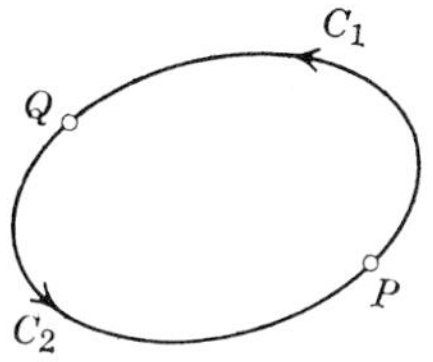

Fig. 201. Proof of Theorem 2

(b) Conversely, suppose that the integral under consideration is zero on every simple closed path in D. Let P and Q be any two points of D, and let C_1 and C_2 be two paths in D which join P and Q and do not cross (Fig. 201). C_1 and C_2 together form a simple closed path C. Therefore,

$$\int_{C_1} (f\,dx + g\,dy + h\,dz) + \int_{C_2} (f\,dx + g\,dy + h\,dz) = \oint_C (f\,dx + g\,dy + h\,dz) = 0.$$

From this we obtain

$$\int_{C_2} (f\,dx + g\,dy + h\,dz) = -\int_{C_1} (f\,dx + g\,dy + h\,dz) = \int_{C_1^*} (f\,dx + g\,dy + h\,dz).$$

This completes the proof of Theorem 2. ∎

The extension of this theorem to the case of closed paths crossing a finite number of times is immediate.

To make effective practical use of the simple Theorem 1 we need a criterion by which we can decide whether the differential form under the integral sign is exact. Such a criterion will be given in Theorem 3, below. For formulating this theorem we shall need the following concept.

A domain D is called **simply connected** if every closed curve in D can be continuously shrunk to any point in D without leaving D.

For example, the interior of a sphere or a cube, the interior of a sphere with finitely many points removed, and the domain between two concentric spheres are simply connected, while the interior of a torus (cf. Sec. 9.6) and the interior of a cube with one space diagonal removed are not simply connected.

Theorem 3 (Criterion for exactness and independence of path) *Let $f(x, y, z)$, $g(x, y, z)$, and $h(x, y, z)$ be continuous functions having continuous first partial derivatives in a domain D in space. If the line integral*

$$\int_C (f\,dx + g\,dy + h\,dz) \tag{10}$$

is independent of path in D (and, consequently, $f\,dx + g\,dy + h\,dz$ is exact in D), then

$$\frac{\partial h}{\partial y} = \frac{\partial g}{\partial z}, \qquad \frac{\partial f}{\partial z} = \frac{\partial h}{\partial x}, \qquad \frac{\partial g}{\partial x} = \frac{\partial f}{\partial y} \tag{11}$$

or, in vector language,

$$\operatorname{curl} \mathbf{v} = \mathbf{0} \qquad (\mathbf{v} = f\mathbf{i} + g\mathbf{j} + h\mathbf{k}) \tag{11'}$$

everywhere in D. Conversely, if D is simply connected and (11) *holds everywhere in D, then the integral* (10) *is independent of path in D* (*and, consequently, the differential form* $f\,dx + g\,dy + h\,dz$ *is exact in D*).

Proof. **(a)** Suppose that (10) is independent of path in D. Then, by Theorem 1, the form $f\,dx + g\,dy + h\,dz$ is exact in D and there is a function u such that, by (5′),

$$\mathbf{v} = f\mathbf{i} + g\mathbf{j} + h\mathbf{k} = \text{grad } u$$

in D. From this and (3) in Sec. 8.11 we obtain

$$\text{curl } \mathbf{v} = \text{curl (grad } u) = \mathbf{0}.$$

(b) Conversely, suppose that D is simply connected and (11′) holds everywhere in D. Let C be any simple closed path in D. Since D is simply connected, we can find a surface S in D bounded by C, Stokes's theorem (Sec. 9.10) is applicable, and we obtain

$$\int_C (f\,dx + g\,dy + h\,dz) = \int_C v_t\,ds = \iint_S (\text{curl } \mathbf{v})_n\,dA = 0$$

for proper direction on C and normal $\mathbf{n}$ on S. From this and Theorem 2 it follows that the integral (10) is independent of path in D. This completes the proof. ∎

We see that in the case of a line integral

$$\int_C (f\,dx + g\,dy)$$

in the xy-plane, (11) reduces to the single relation

$$\frac{\partial g}{\partial x} = \frac{\partial f}{\partial y}. \tag{11*}$$

The assumption that D be simply connected is essential and cannot be omitted. This can be seen from the following example.

Example 2. On the assumption of simply connectedness in Theorem 3

Let

$$f = -\frac{y}{x^2 + y^2}, \qquad g = \frac{x}{x^2 + y^2}, \qquad h = 0.$$

Differentiation shows that (11*) is satisfied in any domain of the xy-plane not containing the origin, for example, in the domain D: $\frac{1}{2} < \sqrt{x^2 + y^2} < \frac{3}{2}$ shown in Fig. 202. Clearly, D is not simply connected. If the integral

$$I = \int_C (f\,dx + g\,dy) = \int_C \frac{-y\,dx + x\,dy}{x^2 + y^2}$$

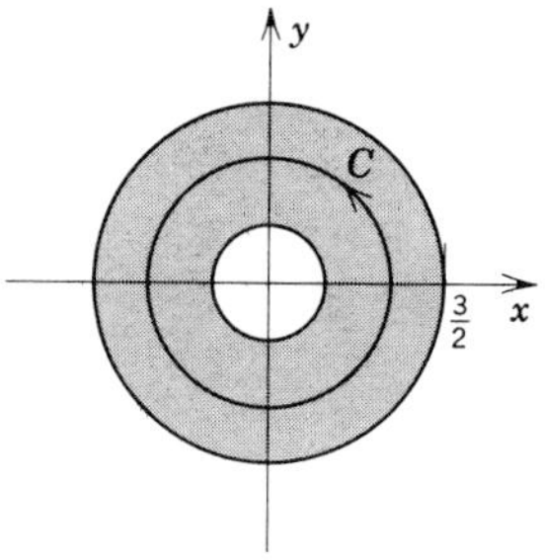

Fig. 202. Example 2

were independent of path in D, then $I = 0$ on any closed curve in D, for example, on the circle $x^2 + y^2 = 1$. But setting $x = r\cos\theta$, $y = r\sin\theta$, and noting that the circle is represented by $r = 1$, we easily obtain

$$I = \int_0^{2\pi} d\theta = 2\pi,$$

the integration being taken once around the circle in the counterclockwise sense. In fact, since D is not simply connected, we cannot apply Theorem 3 and conclude that I is independent of path in D. Furthermore, we see that

$$f\,dx + g\,dy = du \qquad \text{where} \qquad u = \arctan\frac{y}{x} = \theta,$$

but u is not single-valued in D. On the other hand, if we take the so-called "principal value" of u, defined by $-\pi < u \leqq \pi$, then u is not differentiable (not even continuous) at the points of the negative x-axis in D; this was required in connection with the exactness of a differential form (see above). ∎

If a line integral is independent of path, it can be evaluated by means of the formula (8):

Example 3. Evaluation of a line integral in the case of independence of path
Evaluate

$$I = \int_C [2xyz^2\,dx + (x^2z^2 + z\cos yz)\,dy + (2x^2yz + y\cos yz)\,dz]$$

on any path C from P: $(0, 0, 1)$ to Q: $(1, \pi/4, 2)$.

From Theorem 3 it follows that the integral is independent of path in space. Since $f = 2xyz^2$, we obtain from the first relation (5)

$$u = \int f\,dx = x^2yz^2 + a(y, z). \tag{12}$$

From this and the second relation (5) it follows that

$$\frac{\partial u}{\partial y} = x^2z^2 + \frac{\partial a}{\partial y} = g = x^2z^2 + z\cos yz.$$

Hence we must have

$$\frac{\partial a}{\partial y} = z\cos yz, \qquad a = \sin yz + c(z).$$

From this, the third relation (5), and (12) we obtain

$$\frac{\partial u}{\partial z} = 2x^2yz + y\cos yz + \frac{dc}{dz} = h = 2x^2yz + y\cos yz.$$

Hence $dc/dz = 0$, $c = const$ and, therefore,

$$u(x, y, z) = x^2yz^2 + a = x^2yz^2 + \sin yz + c.$$

The reader should take another look at the integral and try to find this u by inspection. Formula (8) now yields the result

$$I = [x^2yz^2 + \sin yz + c]\Big|_P^Q = \pi + \sin\frac{\pi}{2} = \pi + 1.$$

Problems for Sec. 9.12

Are the following differential forms exact?

1. $x\,dx - y\,dy + z\,dz$

2. $x^2\,dx + 2xy\,dy + y^2\,dz$

3. $e^y\,dx + e^x\,dy + e^z\,dz$

4. $(y + z\cos x)\,dx + x\,dy + \sin x\,dz$

5. $z\cos yz\,dy + y\cos yz\,dz$

6. $y^3\,dx + z^3\,dy + x^3\,dz$

7. $yz\,dx + xz\,dy + xy\,dz$

8. $y^2z^3\,dx + 2xyz^3\,dy + 3xy^2z^2\,dz$

In each case show that the given differential form is exact and find a function u such that the form equals du.

9. $x\,dx - y\,dy - z\,dz$

10. $dx + dy + dz$

11. $dx + z\,dy + y\,dz$

12. $(z^2 - 2xy)\,dx - x^2\,dy + 2xz\,dz$

13. $\cos x\,dx - 2yz\,dy - y^2\,dz$

14. $6xy\,dx + 3x^2\,dy - 4z\,dz$

Show that the form under the integral sign is exact and evaluate:

15. $\int_{(0,0,0)}^{(1,1,1)} (x\,dx + y\,dy - z\,dz)$

16. $\int_{(1,0,0)}^{(2,1,4)} [yz\,dx + (xz + 1)\,dy + xy\,dz]$

17. $\int_{(0,-1,-1)}^{(\pi,3,2)} (\cos x\,dx + z\,dy + y\,dz)$

18. $\int_{(1,3,2)}^{(4,0,1)} [z\,dx - z\,dy + (x - y)\,dz]$

19. $\int_{(0,2,1)}^{(2,0,1)} [ze^x\,dx + 2yz\,dy + (e^x + y^2)\,dz]$

20. $\int_{(0,0,0)}^{(1,4,2)} (3y^2e^z\,dy + y^3e^z\,dz)$

CHAPTER

10 FOURIER SERIES AND INTEGRALS

Periodic functions occur frequently in engineering problems. Their representation in terms of simple periodic functions, such as sine and cosine, is a matter of great practical importance, which leads to Fourier series. These series, named after the French physicist JOSEPH FOURIER (1768–1830), are a very powerful tool in connection with various problems involving ordinary and partial differential equations.

In the present chapter we shall discuss basic concepts, facts and techniques in connection with Fourier series. Illustrative examples and some important engineering applications will be included. Further applications will be presented in the next chapter on partial differential equations and boundary value problems.

The *theory* of Fourier series is rather complicated, but the *application* of these series is simple. Fourier series are, in a certain sense, more universal than Taylor series, because many *discontinuous* periodic functions of practical interest can be developed in Fourier series, but, of course, do not have Taylor series representations.

The last section of the present chapter is devoted to Fourier integrals. Applications to partial differential equations will be considered in the next chapter (in Sec. 11.6).

Prerequisite for this chapter: elementary integral calculus.
Sections which may be omitted in a shorter course: 10.6–10.8.
References: Appendix 1, Part D.
Answers to problems: Appendix 2.

10.1 Periodic Functions. Trigonometric Series

A function $f(x)$ is said to be **periodic** if it is defined for all real x and if there is some positive number T such that

(1) $$f(x + T) = f(x) \qquad \text{for all } x.$$

The number T is then called a **period**[1] of $f(x)$. The graph of such a function

[1] If a periodic function $f(x)$ has a smallest period $T\,(>0)$, this is often called the *primitive period* of $f(x)$. For example, the primitive periods of $\sin x$ and $\sin 2x$ are 2π and π, respectively. Examples of periodic functions without primitive period are $f = const$ and $f(x) = 0$ (x rational), $f(x) = 1$ otherwise.

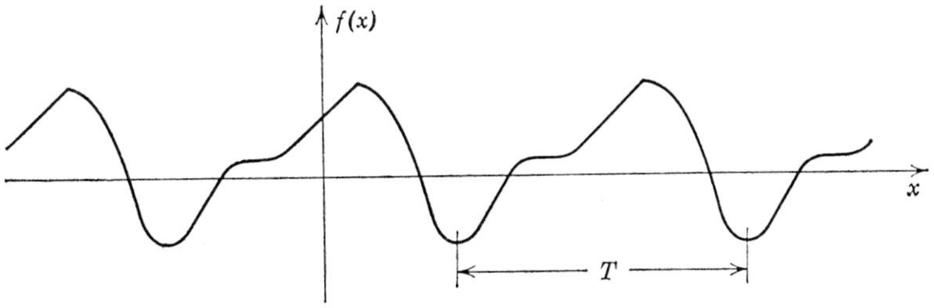

Fig. 203. Periodic function

is obtained by periodic repetition of its graph in any interval of length T (Fig. 203).

From (1) it follows that, if n is any integer,

$$f(x + nT) = f(x) \qquad \text{for all } x.$$

Hence $2T, 3T, 4T, \cdots$ are also periods of $f(x)$. Furthermore, if $f(x)$ and $g(x)$ have period T, then the function

$$h(x) = af(x) + bg(x) \qquad (a, b \text{ constant})$$

has the period T.

Familiar examples of periodic functions are the sine and cosine functions, and we note that the function $f = c = const$ is also a periodic function in the sense of the definition, because it satisfies (1) for every positive T.

Our problem in the first few sections of this chapter will be the representation of various functions of period 2π in terms of the simple functions

$$1, \quad \cos x, \sin x, \quad \cos 2x, \sin 2x, \cdots, \quad \cos nx, \sin nx, \cdots$$

which have the period 2π (Fig. 204). The series which will arise in this connection will be of the form

$$a_0 + a_1 \cos x + b_1 \sin x + a_2 \cos 2x + b_2 \sin 2x + \cdots, \tag{2}$$

where $a_0, a_1, a_2, \cdots, b_1, b_2, \cdots$ are real constants. Such a series is called a **trigonometric series,** and the a_n and b_n are called the **coefficients** of the series.

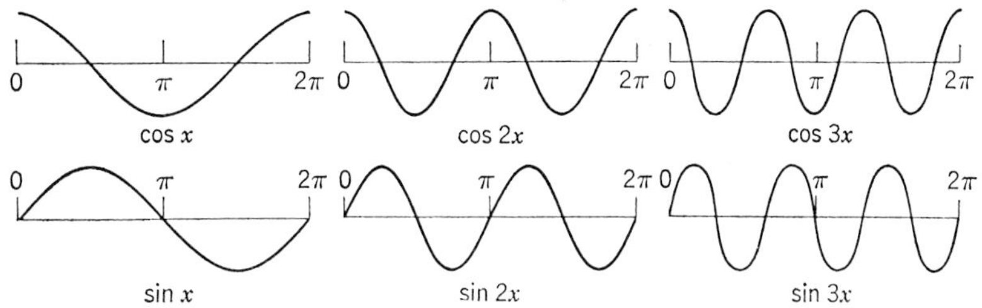

Fig. 204. Cosine and sine functions having the period 2π

We see that each term of the series has the period 2π. Hence *if the series converges, its sum will be a function of period* 2π.

Periodic functions that occur in engineering problems are often rather complicated, and it is desirable to represent these functions in terms of simple periodic functions. We shall see that almost any periodic function $f(x)$ of period 2π that appears in applications, for example, in connection with vibrations, can be represented by a trigonometric series, and we shall derive formulas for the coefficients in (2) in terms of $f(x)$ such that (2) converges and has the sum $f(x)$. Later we shall extend our results to functions of arbitrary period; this extension will turn out to be quite simple.

Problems for Sec. 10.1

Find the smallest positive period T of the following functions.

1. $\cos x,\ \sin x,\ \cos 2x,\ \sin 2x,\ \cos \pi x,\ \sin \pi x,\ \cos 2\pi x,\ \sin 2\pi x$

2. $\cos nx,\ \sin nx,\ \cos\dfrac{2\pi x}{k},\ \sin\dfrac{2\pi x}{k},\ \cos\dfrac{2\pi n x}{k},\ \sin\dfrac{2\pi n x}{k}$

Plot accurate graphs of the following functions.

3. $\sin x,\quad \sin x + \frac{1}{3}\sin 3x,\quad \sin x + \frac{1}{3}\sin 3x + \frac{1}{5}\sin 5x,$

$$f(x) = \begin{cases} -\pi/4 & \text{when } -\pi < x < 0 \\ \pi/4 & \text{when } 0 < x < \pi \end{cases} \quad\text{and}\quad f(x + 2\pi) = f(x)$$

4. $\sin 2\pi x,\quad \sin 2\pi x + \frac{1}{3}\sin 6\pi x,\quad \sin 2\pi x + \frac{1}{3}\sin 6\pi x + \frac{1}{5}\sin 10\pi x$

5. $\sin x,\quad \sin x - \frac{1}{2}\sin 2x,\quad \sin x - \frac{1}{2}\sin 2x + \frac{1}{3}\sin 3x,$
$f(x) = x/2 \quad \text{when} \quad -\pi < x < \pi \quad \text{and} \quad f(x + 2\pi) = f(x)$

6. $-\cos x,\quad -\cos x + \frac{1}{4}\cos 2x,\quad -\cos x + \frac{1}{4}\cos 2x - \frac{1}{9}\cos 3x,$
$f(x) = x^2/4 - \pi^2/12 \quad \text{when} \quad -\pi < x < \pi \quad \text{and} \quad f(x + 2\pi) = f(x)$

7. $f(x) = x^3 \quad \text{when} \quad -\pi < x < \pi \quad \text{and} \quad f(x + 2\pi) = f(x)$

8. $f(x) = e^x \quad \text{when} \quad -\pi < x < \pi \quad \text{and} \quad f(x + 2\pi) = f(x)$

9. $f(x) = \cosh 2x \quad \text{when} \quad -\pi < x < \pi \quad \text{and} \quad f(x + 2\pi) = f(x)$

10. $f(x) = \begin{cases} 0 & \text{when } -\pi < x < 0 \\ x & \text{when } 0 < x < \pi \end{cases} \quad \text{and } f(x + 2\pi) = f(x)$

11. $f(x) = \begin{cases} -x^2 & \text{when } -\pi < x < 0 \\ x^2 & \text{when } 0 < x < \pi \end{cases} \quad \text{and } f(x + 2\pi) = f(x)$

12. $f(x) = \begin{cases} 1 & \text{when } -\pi < x < 0 \\ \cos^2 2x & \text{when } 0 < x < \pi \end{cases} \quad \text{and } f(x + 2\pi) = f(x)$

13. If T is a period of $f(x)$, show that nT, $n = 2, 3, \cdots$ is a period of that function.

14. If $f(x)$ and $g(x)$ have period T, show that $h = af + bg$ (a, b constant) has the period T. Thus all functions of period T form a vector space.

15. Show that the function $f(x) = const$ is a periodic function of period T for every positive T.

16. If $f(x)$ is a periodic function of x of period T, show that $f(ax)$, $a > 0$, is a periodic function of x of period T/a, and $f(x/b)$, $b > 0$, is a periodic function of x of period bT. Verify these results for $f(x) = \sin x$, $a = b = 2$.

Evaluate the following integrals where $n = 0, 1, 2, \cdots$. (These are typical examples of integrals which will be needed in our further work.)

17. $\int_0^{\pi/2} \cos nx\, dx$ **18.** $\int_0^{\pi} \sin nx\, dx$ **19.** $\int_{-\pi/2}^{\pi/2} x \cos nx\, dx$

20. $\int_{-\pi/2}^{\pi/2} x \sin nx\, dx$ **21.** $\int_{-\pi}^{\pi} x \sin nx\, dx$ **22.** $\int_0^{\pi} x \sin nx\, dx$

23. $\int_0^{\pi} x^2 \cos nx\, dx$ **24.** $\int_0^{\pi} e^x \cos nx\, dx$ **25.** $\int_{-\pi}^{0} e^x \sin nx\, dx$

10.2 Fourier Series. Euler Formulas

Let us assume that $f(x)$ is a periodic function of period 2π which can be represented by a trigonometric series

$$f(x) = a_0 + \sum_{n=1}^{\infty} (a_n \cos nx + b_n \sin nx). \tag{1}$$

Given such a function $f(x)$, we want to determine the coefficients a_n and b_n in the corresponding series (1).

We first determine a_0. Integrating on both sides of (1) from $-\pi$ to π, we have

$$\int_{-\pi}^{\pi} f(x)\, dx = \int_{-\pi}^{\pi} \left[a_0 + \sum_{n=1}^{\infty} (a_n \cos nx + b_n \sin nx)\right] dx.$$

If term-by-term integration of the series is allowed,[2] then we obtain

$$\int_{-\pi}^{\pi} f(x)\, dx = a_0 \int_{-\pi}^{\pi} dx + \sum_{n=1}^{\infty} \left(a_n \int_{-\pi}^{\pi} \cos nx\, dx + b_n \int_{-\pi}^{\pi} \sin nx\, dx\right).$$

The first term on the right equals $2\pi a_0$. All the other integrals on the right are zero, as can be readily seen by performing the integrations. Hence our first result is

$$a_0 = \frac{1}{2\pi} \int_{-\pi}^{\pi} f(x)\, dx. \tag{2}$$

We now determine $a_1, a_2, \cdots$ by a similar procedure. We multiply (1) by

[2] This is justified, for instance, in the case of uniform convergence (cf. Theorem 3 in Sec. 16.6).

$\cos mx$, where m is any fixed positive integer, and then integrate from $-\pi$ to π, finding

$$(3)\quad \int_{-\pi}^{\pi} f(x) \cos mx\, dx = \int_{-\pi}^{\pi} \left[a_0 + \sum_{n=1}^{\infty} (a_n \cos nx + b_n \sin nx)\right] \cos mx\, dx.$$

By term-by-term integration the right-hand side becomes

$$a_0 \int_{-\pi}^{\pi} \cos mx\, dx + \sum_{n=1}^{\infty} \left[a_n \int_{-\pi}^{\pi} \cos nx \cos mx\, dx + b_n \int_{-\pi}^{\pi} \sin nx \cos mx\, dx\right].$$

The first integral is zero. By applying (11) in Appendix 3 we obtain

$$\int_{-\pi}^{\pi} \cos nx \cos mx\, dx = \frac{1}{2}\int_{-\pi}^{\pi} \cos (n+m)x\, dx + \frac{1}{2}\int_{-\pi}^{\pi} \cos (n-m)x\, dx,$$

$$\int_{-\pi}^{\pi} \sin nx \cos mx\, dx = \frac{1}{2}\int_{-\pi}^{\pi} \sin (n+m)x\, dx + \frac{1}{2}\int_{-\pi}^{\pi} \sin (n-m)x\, dx.$$

Integration shows that the four terms on the right are zero, except for the last term in the first line which equals π when $n = m$. Since in (3) this term is multiplied by a_m, the right-hand side in (3) is equal to $a_m\pi$, and our second result is

$$(4)\qquad a_m = \frac{1}{\pi}\int_{-\pi}^{\pi} f(x) \cos mx\, dx, \qquad m = 1, 2, \cdots.$$

We finally determine $b_1, b_2, \cdots$ in (1). If we multiply (1) by $\sin mx$, where m is any fixed positive integer, and then integrate from $-\pi$ to π, we have

$$(5)\quad \int_{-\pi}^{\pi} f(x) \sin mx\, dx = \int_{-\pi}^{\pi} \left[a_0 + \sum_{n=1}^{\infty} (a_n \cos nx + b_n \sin nx)\right] \sin mx\, dx.$$

Integrating term-by-term, we see that the right-hand side becomes

$$a_0 \int_{-\pi}^{\pi} \sin mx\, dx + \sum_{n=1}^{\infty} \left[a_n \int_{-\pi}^{\pi} \cos nx \sin mx\, dx + b_n \int_{-\pi}^{\pi} \sin nx \sin mx\, dx\right].$$

The first integral is zero. The next integral is of the type considered before, and we know that it is zero for all $n = 1, 2, \cdots$. For the last integral we obtain

$$\int_{-\pi}^{\pi} \sin nx \sin mx\, dx = \frac{1}{2}\int_{-\pi}^{\pi} \cos (n-m)x\, dx - \frac{1}{2}\int_{-\pi}^{\pi} \cos (n+m)x\, dx.$$

The last term is zero. The first term on the right is zero when $n \neq m$ and is π

when $n = m$. Since in (5) this term is multiplied by b_m, the right-hand side in (5) is equal to $b_m \pi$, and our last result is

$$b_m = \frac{1}{\pi}\int_{-\pi}^{\pi} f(x) \sin mx \, dx, \qquad m = 1, 2, \cdots.$$

Writing n in place of m we altogether have the so-called **Euler formulas:**

$$\text{(6)} \qquad \begin{aligned} &\text{(a)} \quad a_0 = \frac{1}{2\pi}\int_{-\pi}^{\pi} f(x)\, dx \\ &\text{(b)} \quad a_n = \frac{1}{\pi}\int_{-\pi}^{\pi} f(x) \cos nx \, dx \\ &\text{(c)} \quad b_n = \frac{1}{\pi}\int_{-\pi}^{\pi} f(x) \sin nx \, dx \end{aligned} \qquad n = 1, 2, \cdots,$$

Note that because of the periodicity of the integrands the interval of integration in (6) may be replaced by any other interval of length 2π, for instance, by the interval $0 \leqq x \leqq 2\pi$.

A periodic function $f(x)$ with period 2π being given, we may compute the a_n and b_n by (6) and form the trigonometric series

$$\text{(7)} \qquad a_0 + a_1 \cos x + b_1 \sin x + \cdots + a_n \cos nx + b_n \sin nx + \cdots.$$

This series is then called the **Fourier series** corresponding to $f(x)$, and its coefficients obtained from (6) are called the **Fourier coefficients** of $f(x)$.

From the definition of a definite integral it follows that, if $f(x)$ is continuous or merely piecewise continuous (continuous except for finitely many finite jumps in the interval of integration), the integrals in (6) exist and we may compute the Fourier coefficients of $f(x)$ by (6). The remaining question, whether the Fourier series thus obtained converges and has the sum $f(x)$, will be considered later in this section.

Let us illustrate the practical use of (6) by a simple example. Numerous other examples will occur in the following sections.

Example 1. Square wave

Find the Fourier coefficients of the periodic function $f(x)$ in Fig. 205a. The analytic representation is

$$f(x) = \begin{cases} -k & \text{when} \quad -\pi < x < 0 \\ k & \text{when} \quad 0 < x < \pi \end{cases} \qquad \text{and} \qquad f(x + 2\pi) = f(x).$$

Functions of this type may occur as external forces acting on mechanical systems, electromotive forces in electric circuits, etc.

From (6a) we obtain $a_0 = 0$. This can also be seen without integration since the area under the curve of $f(x)$ between $-\pi$ and π is zero. From (6b),

$$a_n = \frac{1}{\pi}\int_{-\pi}^{\pi} f(x) \cos nx \, dx = \frac{1}{\pi}\left[\int_{-\pi}^{0} (-k) \cos nx \, dx + \int_{0}^{\pi} k \cos nx \, dx\right]$$

$$= \frac{1}{\pi}\left[-k\frac{\sin nx}{n}\bigg|_{-\pi}^{0} + k\frac{\sin nx}{n}\bigg|_{0}^{\pi}\right] = 0$$

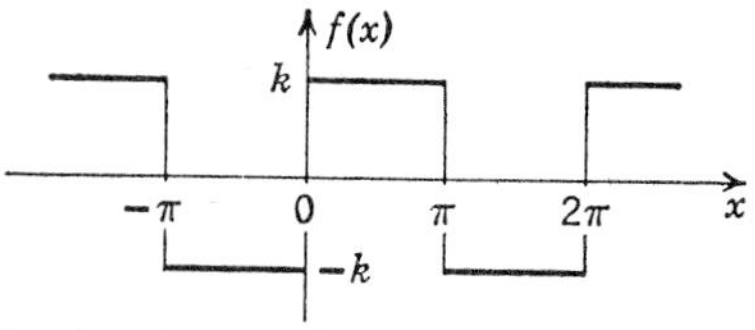

(a) The given function $f(x)$ (Periodic square wave)

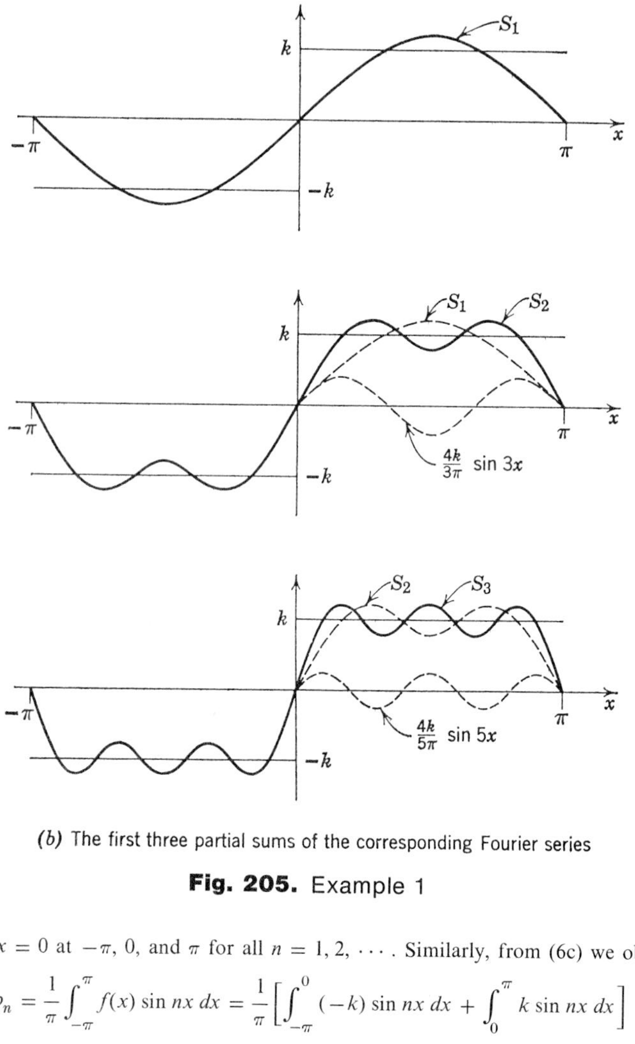

(b) The first three partial sums of the corresponding Fourier series

Fig. 205. Example 1

because $\sin nx = 0$ at $-\pi$, 0, and π for all $n = 1, 2, \cdots$. Similarly, from (6c) we obtain

$$b_n = \frac{1}{\pi}\int_{-\pi}^{\pi} f(x)\sin nx\,dx = \frac{1}{\pi}\left[\int_{-\pi}^{0}(-k)\sin nx\,dx + \int_{0}^{\pi} k\sin nx\,dx\right]$$

$$= \frac{1}{\pi}\left[k\frac{\cos nx}{n}\bigg|_{-\pi}^{0} - k\frac{\cos nx}{n}\bigg|_{0}^{\pi}\right].$$

Since $\cos(-\alpha) = \cos\alpha$ and $\cos 0 = 1$, this yields

$$b_n = \frac{k}{n\pi}[\cos 0 - \cos(-n\pi) - \cos n\pi + \cos 0] = \frac{2k}{n\pi}(1 - \cos n\pi).$$

Now, $\cos\pi = -1$, $\cos 2\pi = 1$, $\cos 3\pi = -1$ etc., in general,

$$\cos n\pi = \begin{cases} -1 & \text{for odd } n, \\ 1 & \text{for even } n, \end{cases} \qquad \text{and thus} \qquad 1 - \cos n\pi = \begin{cases} 2 & \text{for odd } n, \\ 0 & \text{for even } n. \end{cases}$$

Hence the Fourier coefficients b_n of our function are

$$b_1 = \frac{4k}{\pi}, \qquad b_2 = 0, \qquad b_3 = \frac{4k}{3\pi}, \qquad b_4 = 0, \qquad b_5 = \frac{4k}{5\pi}, \cdots,$$

and since the a_n are zero, the corresponding Fourier series is

$$\frac{4k}{\pi}\left(\sin x + \frac{1}{3}\sin 3x + \frac{1}{5}\sin 5x + \cdots\right).$$

The partial sums are

$$S_1 = \frac{4k}{\pi}\sin x, \qquad S_2 = \frac{4k}{\pi}\left(\sin x + \frac{1}{3}\sin 3x\right), \qquad \text{etc.,}$$

and their graphs in Fig. 205 seem to indicate that the series is convergent and has the sum $f(x)$, the given function. We notice that at $x = 0$ and $x = \pi$, the points of discontinuity of $f(x)$, all partial sums have the value zero, the arithmetic mean of the values $-k$ and k of our function.

Furthermore, assuming that $f(x)$ is the sum of the series and setting $x = \pi/2$, we have

$$f\left(\frac{\pi}{2}\right) = k = \frac{4k}{\pi}\left(1 - \frac{1}{3} + \frac{1}{5} - + \cdots\right)$$

or

$$1 - \frac{1}{3} + \frac{1}{5} - \frac{1}{7} + - \cdots = \frac{\pi}{4}.$$

This is a famous result by Leibniz (obtained about 1673 from geometrical considerations). It illustrates that the values of various series with constant terms can be obtained by evaluating Fourier series at specific points. ∎

The class of functions which can be represented by Fourier series is surprisingly large and general. Corresponding sufficient conditions covering almost any conceivable engineering application are as follows.

Theorem 1 (Representation by a Fourier series)

If a periodic function $f(x)$ with period 2π is piecewise continuous[3] in the interval $-\pi \leqq x \leqq \pi$ and has a left- and right-hand derivative[4] at each point of that interval, then the corresponding Fourier series (7) [with coefficients (6)] is convergent. Its sum is $f(x)$, except at a point x_0 at which $f(x)$ is discontinuous and the sum of the series is the average of the left- and right-hand limits[4] of $f(x)$ at x_0.

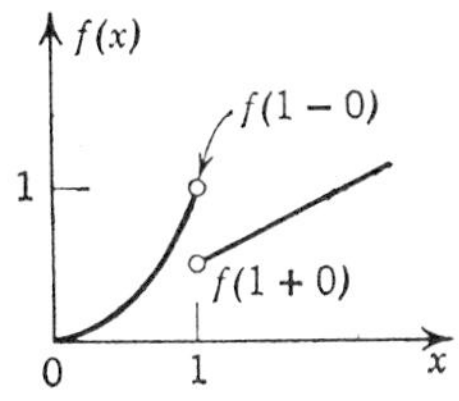

Fig. 206. Left- and right-hand limits

$$f(1 - 0) = 1,$$

$$f(1 + 0) = \tfrac{1}{2}$$

of the function

$$f(x) = \begin{cases} x^2 & \text{when } x < 1 \\ x/2 & \text{when } x > 1 \end{cases}$$

[3] Definition in Sec. 5.1.

[4] The **left-hand limit** of $f(x)$ at x_0 is defined as the limit of $f(x)$ as x approaches x_0 from the left and is frequently denoted by $f(x_0 - 0)$. Thus

$$f(x_0 - 0) = \lim_{h\to 0} f(x_0 - h) \text{ as } h \to 0 \text{ through positive values.}$$

The **right-hand limit** is denoted by $f(x_0 + 0)$ and

$$f(x_0 + 0) = \lim_{h\to 0} f(x_0 + h) \text{ as } h \to 0 \text{ through positive values.}$$

The **left-** and **right-hand derivatives** of $f(x)$ at x_0 are defined as the limits of

$$\frac{f(x_0 - h) - f(x_0 - 0)}{-h} \quad \text{and} \quad \frac{f(x_0 + h) - f(x_0 + 0)}{h},$$

respectively, as $h \to 0$ through positive values. Of course if $f(x)$ is continuous at x_0, the last term in both numerators is simply $f(x_0)$.

Remark. If the Fourier series *corresponding to* a function $f(x)$ converges with the sum $f(x)$ as characterized in Theorem 1, the series will be called the Fourier series ***of*** $f(x)$, we write

$$f(x) = a_0 + a_1 \cos x + b_1 \sin x + \cdots + a_n \cos nx + b_n \sin nx + \cdots,$$

and we say that $f(x)$ is *represented* by this Fourier series. Since the insertion of parentheses in a convergent series yields a new convergent series having the same sum as the original series (proof in Sec. 15.6), we may write more briefly

$$f(x) = a_0 + \sum_{n=1}^{\infty} (a_n \cos nx + b_n \sin nx).$$

Proof of convergence in Theorem 1 for a continuous function f(x) having continuous first and second derivatives. Integrating (6b) by parts, we obtain

$$a_n = \frac{1}{\pi}\int_{-\pi}^{\pi} f(x) \cos nx \, dx = \frac{f(x) \sin nx}{n\pi}\bigg|_{-\pi}^{\pi} - \frac{1}{n\pi}\int_{-\pi}^{\pi} f'(x) \sin nx \, dx.$$

The first term on the right is zero. A second integration by parts gives

$$a_n = \frac{f'(x) \cos nx}{n^2\pi}\bigg|_{-\pi}^{\pi} - \frac{1}{n^2\pi}\int_{-\pi}^{\pi} f''(x) \cos nx \, dx.$$

The first term on the right is zero because of the periodicity and continuity of $f'(x)$. Since f'' is continuous in the interval of integration, we have

$$|f''(x)| < M$$

for an appropriate constant M. Furthermore, $|\cos nx| \leqq 1$. It follows that

$$|a_n| = \frac{1}{n^2\pi}\left|\int_{-\pi}^{\pi} f''(x) \cos nx \, dx\right| < \frac{1}{n^2\pi}\int_{-\pi}^{\pi} M \, dx = \frac{2M}{n^2}.$$

Similarly, $|b_n| < 2M/n^2$ for all n. Hence the absolute value of each term of the Fourier series corresponding to $f(x)$ is at most equal to the corresponding term of the series

$$|a_0| + 2M\left(1 + 1 + \frac{1}{2^2} + \frac{1}{2^2} + \frac{1}{3^2} + \frac{1}{3^2} + \cdots\right)$$

which is convergent. Hence that Fourier series converges and the proof is complete. (Readers already familiar with uniform convergence will see that, by the Weierstrass test in Sec. 16.6, under our present assumptions the Fourier series converges uniformly, and our derivation of (6) by integrating term-by-term is then justified by Theorem 3 of that section.)

The proofs of convergence in the case of a piecewise continuous function $f(x)$ and of the last statement of Theorem 1 can be found in more advanced texts, for example in Ref. [D4]. ∎

Problems for Sec. 10.2

Find the Fourier series of the function $f(x)$ which is assumed to have the period 2π and plot accurate graphs of the first three partial sums,[5] where $f(x)$ equals

1.

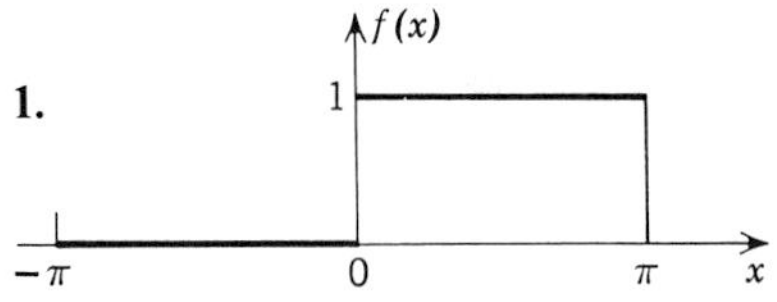

2.

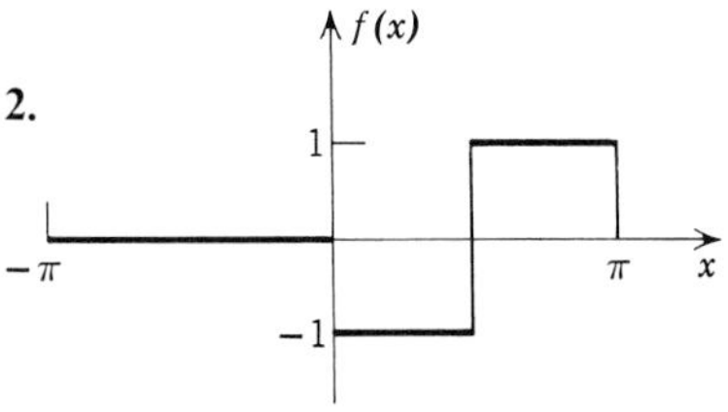

3.

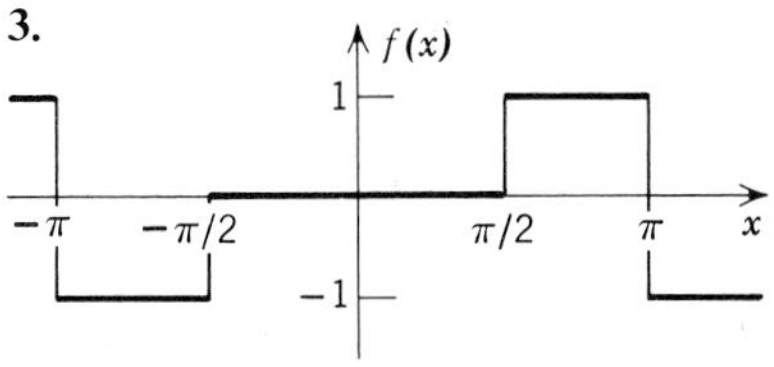

4.

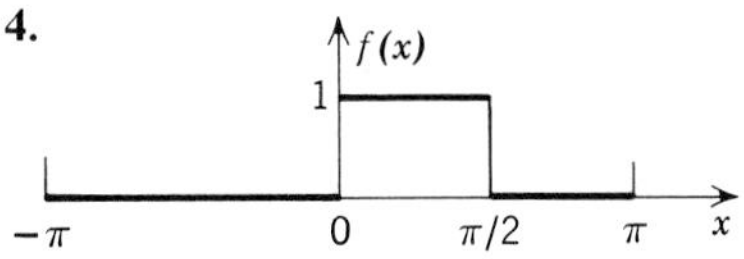

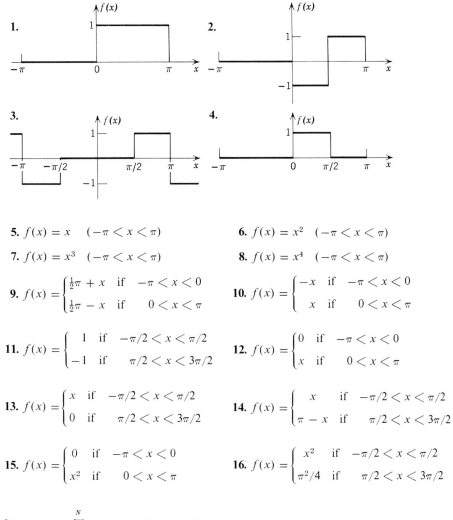

5. $f(x) = x \quad (-\pi < x < \pi)$

6. $f(x) = x^2 \quad (-\pi < x < \pi)$

7. $f(x) = x^3 \quad (-\pi < x < \pi)$

8. $f(x) = x^4 \quad (-\pi < x < \pi)$

9. $f(x) = \begin{cases} \frac{1}{2}\pi + x & \text{if} \quad -\pi < x < 0 \\ \frac{1}{2}\pi - x & \text{if} \quad 0 < x < \pi \end{cases}$

10. $f(x) = \begin{cases} -x & \text{if} \quad -\pi < x < 0 \\ x & \text{if} \quad 0 < x < \pi \end{cases}$

11. $f(x) = \begin{cases} 1 & \text{if} \quad -\pi/2 < x < \pi/2 \\ -1 & \text{if} \quad \pi/2 < x < 3\pi/2 \end{cases}$

12. $f(x) = \begin{cases} 0 & \text{if} \quad -\pi < x < 0 \\ x & \text{if} \quad 0 < x < \pi \end{cases}$

13. $f(x) = \begin{cases} x & \text{if} \quad -\pi/2 < x < \pi/2 \\ 0 & \text{if} \quad \pi/2 < x < 3\pi/2 \end{cases}$

14. $f(x) = \begin{cases} x & \text{if} \quad -\pi/2 < x < \pi/2 \\ \pi - x & \text{if} \quad \pi/2 < x < 3\pi/2 \end{cases}$

15. $f(x) = \begin{cases} 0 & \text{if} \quad -\pi < x < 0 \\ x^2 & \text{if} \quad 0 < x < \pi \end{cases}$

16. $f(x) = \begin{cases} x^2 & \text{if} \quad -\pi/2 < x < \pi/2 \\ \pi^2/4 & \text{if} \quad \pi/2 < x < 3\pi/2 \end{cases}$

[5] That is, $a_0 + \sum_{n=1}^{N} (a_n \cos nx + b_n \sin nx)$ for $N = 1, 2, 3$.

17. If $f(x)$ has the Fourier coefficients a_n, b_n, show that $kf(x)$, where k is a constant, has the Fourier coefficients ka_n, kb_n.

18. Show that if $f(x)$ has the Fourier coefficients a_n, b_n and $g(x)$ has the Fourier coefficients a_n^*, b_n^*, then $f(x) + g(x)$ has the Fourier coefficients $a_n + a_n^*$, $b_n + b_n^*$.

19. Using Probs. 17 and 18 and the results of Probs. 1 and 2, find the Fourier series in Prob. 4.

20. Verify the last statement in Theorem 1 concerning the discontinuities for the function in Prob. 11.

10.3 Functions Having Arbitrary Period

The transition from functions having period 2π to functions having any period T is quite simple, because it can be effected by a change of scale.

In fact, suppose that $f(t)$ has period T. Then we can introduce a new variable x such that $f(t)$, as a function of x, has period 2π. If we set

$$\text{(1)} \qquad \text{(a)} \quad t = \frac{T}{2\pi}x \qquad \text{so that} \qquad \text{(b)} \quad x = \frac{2\pi}{T}t,$$

then $x = \pm\pi$ corresponds to $t = \pm T/2$. This means that f, as a function of x, has period 2π. Hence if f has a Fourier series, this series must be of the form

$$\text{(2)} \qquad f(t) = f\left(\frac{T}{2\pi}x\right) = a_0 + \sum_{n=1}^{\infty} (a_n \cos nx + b_n \sin nx),$$

with coefficients obtained from the Euler formulas (6) in the last section:

$$a_0 = \frac{1}{2\pi}\int_{-\pi}^{\pi} f\left(\frac{T}{2\pi}x\right) dx, \qquad a_n = \frac{1}{\pi}\int_{-\pi}^{\pi} f\left(\frac{T}{2\pi}x\right) \cos nx \, dx,$$

$$b_n = \frac{1}{\pi}\int_{-\pi}^{\pi} f\left(\frac{T}{2\pi}x\right) \sin nx \, dx.$$

We could use these formulas directly, but the change to t simplifies calculation. Since

$$x = \frac{2\pi}{T}t, \qquad \text{we have} \qquad dx = \frac{2\pi}{T}dt,$$

and the interval of integration on the x-axis corresponds to the interval

$$-\frac{T}{2} \leqq t \leqq \frac{T}{2}.$$

Consequently, we see that the Fourier coefficients of the periodic function $f(t)$ of period T are given by the **Euler formulas**

$$\text{(3)} \qquad \begin{aligned} &\textbf{(a)} \quad a_0 = \frac{1}{T}\int_{-T/2}^{T/2} f(t)\,dt \\ &\textbf{(b)} \quad a_n = \frac{2}{T}\int_{-T/2}^{T/2} f(t)\cos\frac{2n\pi t}{T}\,dt \\ &\textbf{(c)} \quad b_n = \frac{2}{T}\int_{-T/2}^{T/2} f(t)\sin\frac{2n\pi t}{T}\,dt \end{aligned} \qquad n = 1, 2, \cdots .$$

The Fourier series (2) with x expressed in terms of t becomes

$$\text{(4)} \qquad f(t) = a_0 + \sum_{n=1}^{\infty}\left(a_n \cos\frac{2n\pi}{T}t + b_n \sin\frac{2n\pi}{T}t\right).$$

The interval of integration in (3) may be replaced by any interval of length T, for example, by the interval $0 \leqq t \leqq T$.

Example 1. Periodic square wave

Find the Fourier series of the function (cf. Fig. 207)

$$f(t) = \begin{cases} 0 & \text{when} \quad -2 < t < -1 \\ k & \text{when} \quad -1 < t < 1 \\ 0 & \text{when} \quad 1 < t < 2 \end{cases} \qquad T = 4.$$

From (3a) and (3b) we obtain

$$a_0 = \frac{1}{4}\int_{-2}^{2} f(t)\,dt = \frac{1}{4}\int_{-1}^{1} k\,dt = \frac{k}{2},$$

$$a_n = \frac{1}{2}\int_{-2}^{2} f(t)\cos\frac{n\pi}{2}t\,dt = \frac{1}{2}\int_{-1}^{1} k\cos\frac{n\pi}{2}t\,dt = \frac{2k}{n\pi}\sin\frac{n\pi}{2}.$$

Thus $a_n = 0$ when n is even, $a_n = 2k/n\pi$ when $n = 1, 5, 9, \cdots$, and $a_n = -2k/n\pi$ when $n = 3, 7, 11, \cdots$. From (3c) we find that $b_n = 0$ for $n = 1, 2, \cdots$. Hence the result is

$$f(t) = \frac{k}{2} + \frac{2k}{\pi}\left(\cos\frac{\pi}{2}t - \frac{1}{3}\cos\frac{3\pi}{2}t + \frac{1}{5}\cos\frac{5\pi}{2}t - + \cdots\right).$$

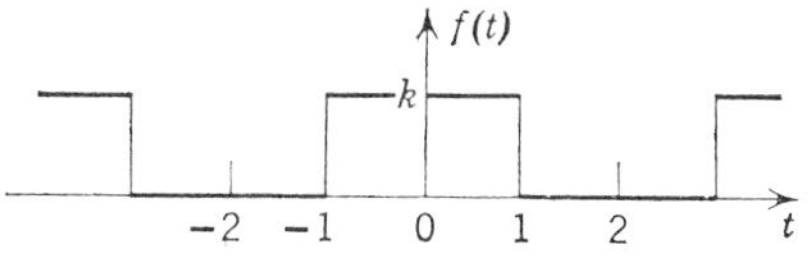

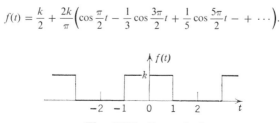

Fig. 207. Example 1

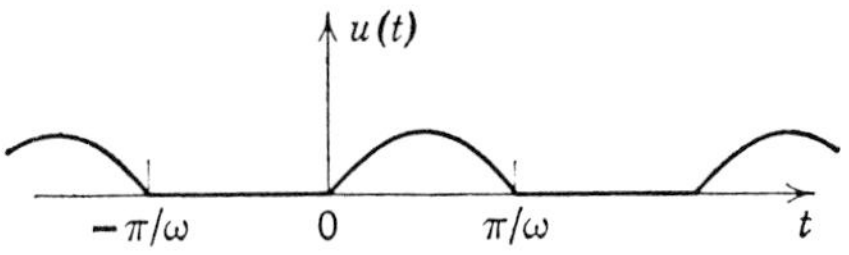

Fig. 208. Half-wave rectifier

Example 2. Half-wave rectifier

A sinusoidal voltage $E \sin \omega t$ is passed through a half-wave rectifier which clips the negative portion of the wave (Fig. 208). Find the Fourier series of the resulting periodic function

$$u(t) = \begin{cases} 0 & \text{when} \quad -T/2 < t < 0, \\ E \sin \omega t & \text{when} \quad 0 < t < T/2, \end{cases} \qquad T = \frac{2\pi}{\omega}.$$

Since $u = 0$ when $-T/2 < t < 0$, we obtain from (3a)

$$a_0 = \frac{\omega}{2\pi} \int_0^{\pi/\omega} E \sin \omega t \, dt = \frac{E}{\pi}$$

and from (3b), by using (11) in Appendix 3, with $x = \omega t$ and $y = n\omega t$,

$$a_n = \frac{\omega}{\pi} \int_0^{\pi/\omega} E \sin \omega t \cos n\omega t \, dt = \frac{\omega E}{2\pi} \int_0^{\pi/\omega} [\sin (1 + n)\omega t + \sin (1 - n)\omega t] \, dt.$$

When $n = 1$, the integral on the right is zero, and when $n = 2, 3, \cdots$, we readily obtain

$$a_n = \frac{\omega E}{2\pi} \left[-\frac{\cos (1 + n)\omega t}{(1 + n)\omega} - \frac{\cos (1 - n)\omega t}{(1 - n)\omega} \right]_0^{\pi/\omega}$$

$$= \frac{E}{2\pi} \left(\frac{-\cos (1 + n)\pi + 1}{1 + n} + \frac{-\cos (1 - n)\pi + 1}{1 - n} \right).$$

When n is odd, this is equal to zero, and for even n we have

$$a_n = \frac{E}{2\pi} \left(\frac{2}{1 + n} + \frac{2}{1 - n} \right) = -\frac{2E}{(n - 1)(n + 1)\pi} \qquad (n = 2, 4, \cdots).$$

In a similar fashion we find from (3c) that $b_1 = E/2$ and $b_n = 0$ for $n = 2, 3, \cdots$. Consequently,

$$u(t) = \frac{E}{\pi} + \frac{E}{2} \sin \omega t - \frac{2E}{\pi} \left(\frac{1}{1 \cdot 3} \cos 2\omega t + \frac{1}{3 \cdot 5} \cos 4\omega t + \cdots \right).$$

Problems for Sec. 10.3

1. Show that each term in (4) has the period T.
2. Show that in (3) the interval of integration may be replaced by any interval of length T.
3. Find the Fourier series of the periodic function which is obtained by passing the voltage $v(t) = 2 \cos 100\pi t$ through a half-wave rectifier.

Find the Fourier series of the periodic function $f(t)$, of period T, and plot accurate graphs of $f(t)$ and the first three partial sums.

4. $f(t) = 0 \quad (-2 < t < 0), \quad f(t) = 1 \quad (0 < t < 2), \quad T = 4$
5. $f(t) = 1 \quad (-1 < t < 1), \quad f(t) = 0 \quad (1 < t < 3), \quad T = 4$
6. $f(t) = 1 \quad (-1 < t < 0), \quad f(t) = -1 \quad (0 < t < 1), \quad T = 2$

7. $f(t) = t \quad (-1 < t < 1), \quad T = 2$

8. $f(t) = |t| \quad (-2 < t < 2), \quad T = 4$

9. $f(t) = t \quad (0 < t < 1), \quad f(t) = 1 - t \quad (1 < t < 2), \quad T = 2$

10. $f(t) = 1/2 \quad (-1 < t < 0), \quad f(t) = -t \quad (0 < t < 1), \quad T = 2$

11. $f(t) = t \quad (-\pi/8 < t < \pi/8), \quad f(t) = (\pi/4) - t \quad (\pi/8 < t < 3\pi/8), \quad T = \pi/2$

12. $f(t) = (1/2) + t \quad (-1/2 < t < 0), \quad f(t) = (1/2) - t \quad (0 < t < 1/2), \quad T = 1$

13. $f(t) = t^2 \quad (-1 < t < 1), \quad T = 2$

14. $f(t) = \sin \pi t \quad (0 < t < 1), \quad T = 1$

15. $f(t) = 1 \quad (-2 < t < 0), \quad f(t) = e^{-t} \quad (0 < t < 2), \quad T = 4$

16. Obtain the Fourier series in Prob. 4 directly from that in Example 2.

17. Obtain the Fourier series in Prob. 6 directly from that in Example 1, Sec. 10.2.

18. Obtain the Fourier series in Prob. 7 directly from that in Prob. 5, Sec. 10.2.

19. Obtain the Fourier series in Prob. 8 directly from that in Prob. 10, Sec. 10.2.

20. Obtain the Fourier series in Prob. 13 directly from that in Prob. 6, Sec. 10.2.

10.4 Even and Odd Functions

Unnecessary work (and corresponding sources of errors) in determining Fourier coefficients of a function can be avoided if the function is odd or even.

We first remember that a function $y = g(x)$ is said to be **even** if

$$g(-x) = g(x) \qquad \text{for all } x.$$

The graph of such a function is symmetric with respect to the y-axis (Fig. 209). A function $h(x)$ is said to be **odd** if

$$h(-x) = -h(x) \qquad \text{for all } x.$$

(Cf. Fig. 210.) The function $\cos nx$ is even, while $\sin nx$ is odd.

If $g(x)$ is an even function, then

$$(1) \qquad \int_{-T/2}^{T/2} g(x)\,dx = 2\int_{0}^{T/2} g(x)\,dx \qquad (g \text{ even}).$$

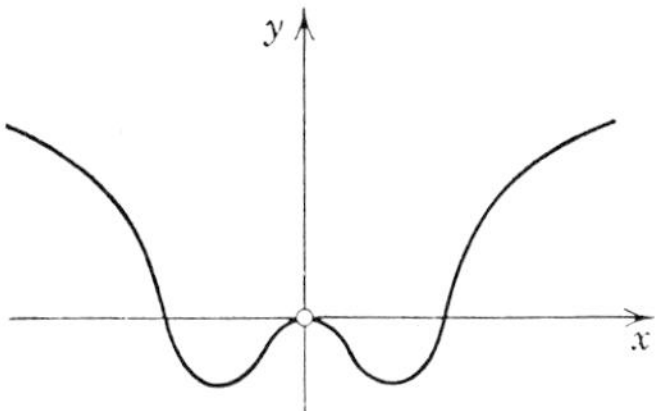

Fig. 209. Even function

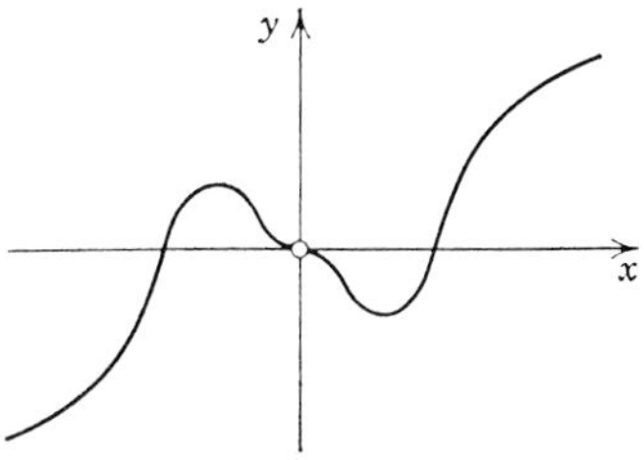

Fig. 210. Odd function

If $h(x)$ is an odd function, then

$$\int_{-T/2}^{T/2} h(x)\,dx = 0 \qquad (h \text{ odd}). \tag{2}$$

The formulas (1) and (2) are obvious from the graphs of g and h, and we leave the formal proofs to the student.

The product $q = gh$ of an even function g and an odd function h is odd, because

$$q(-x) = g(-x)h(-x) = g(x)[-h(x)] = -q(x).$$

Hence if $f(t)$ is even, then the integrand $f \sin(2n\pi t/T)$ in (3c) of the last section is odd, and $b_n = 0$. Similarly, if $f(t)$ is odd, then $f \cos(2n\pi t/T)$ in (3b), Sec. 10.3, is odd, and $a_n = 0$. From this and (1) we obtain

Theorem 1. Fourier series of even and odd functions

The Fourier series of an even function $f(t)$ having period T is a **"Fourier cosine series"**

$$f(t) = a_0 + \sum_{n=1}^{\infty} a_n \cos \frac{2n\pi}{T} t \qquad (f \text{ even}) \tag{3}$$

with coefficients

$$a_0 = \frac{2}{T}\int_0^{T/2} f(t)\,dt, \qquad a_n = \frac{4}{T}\int_0^{T/2} f(t)\cos\frac{2n\pi}{T}t\,dt, \qquad n = 1, 2, \cdots. \tag{4}$$

The Fourier series of an odd function $f(t)$ having period T is a **"Fourier sine series"**

$$f(t) = \sum_{n=1}^{\infty} b_n \sin \frac{2n\pi}{T} t \qquad (f \text{ odd}) \tag{5}$$

with coefficients

$$b_n = \frac{4}{T}\int_0^{T/2} f(t)\sin\frac{2n\pi}{T}t\,dt. \tag{6}$$

In particular, the Fourier series of an even function $f(x)$ of period 2π is a Fourier cosine series

$$f(x) = a_0 + a_1 \cos x + a_2 \cos 2x + a_3 \cos 3x + \cdots \tag{3*}$$

with coefficients

$$a_0 = \frac{1}{\pi}\int_0^{\pi} f(x)\,dx, \qquad a_n = \frac{2}{\pi}\int_0^{\pi} f(x)\cos nx\,dx, \qquad n = 1, 2, \cdots. \tag{4*}$$

Similarly, the Fourier series of an odd function $f(x)$ of period 2π is a Fourier sine series

$$(5^*) \qquad f(x) = b_1 \sin x + b_2 \sin 2x + b_3 \sin 3x + \cdots$$

with coefficients

$$(6^*) \qquad b_n = \frac{2}{\pi}\int_0^{\pi} f(x) \sin nx\, dx, \qquad n = 1, 2, \cdots .$$

For instance, $f(x)$ in Example 1, Sec. 10.2, is odd and, therefore, is represented by a Fourier sine series.

Further simplifications result from

Theorem 2 (Sum of functions)

The Fourier coefficients of a sum $f_1 + f_2$ are the sums of corresponding Fourier coefficients of f_1 and f_2.

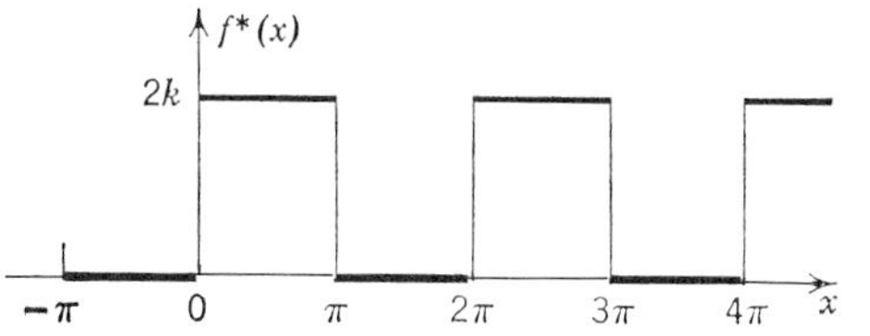

Fig. 211. Example 1

Example 1. Rectangular pulse

The function $f^*(x)$ in Fig. 211 is the sum of the function $f(x)$ in Example 1 of Sec. 10.2 and the constant k. Hence, from that example and Theorem 2 we conclude that

$$f^*(x) = k + \frac{4k}{\pi}\left(\sin x + \frac{1}{3}\sin 3x + \frac{1}{5}\sin 5x + \cdots\right).$$

Example 2. Saw-toothed wave

Find the Fourier series of the function (Fig. 212)

$$f(x) = x + \pi \qquad \text{when} \qquad -\pi < x < \pi \qquad \text{and} \qquad f(x + 2\pi) = f(x).$$

We may write

$$f = f_1 + f_2 \qquad \text{where} \qquad f_1 = x \qquad \text{and} \qquad f_2 = \pi.$$

The Fourier coefficients of f_2 are zero, except for the first one (the constant term), which is π. Hence, by Theorem 2, the Fourier coefficients a_n, b_n are those of f_1, except for a_0 which is π. Since f_1 is odd, $a_n = 0$ for $n = 1, 2, \cdots$, and

$$b_n = \frac{2}{\pi}\int_0^{\pi} f_1(x) \sin nx\, dx = \frac{2}{\pi}\int_0^{\pi} x \sin nx\, dx.$$

Integrating by parts we obtain

$$b_n = \frac{2}{\pi}\left[\frac{-x \cos nx}{n}\Bigg|_0^{\pi} + \frac{1}{n}\int_0^{\pi} \cos nx\, dx\right] = -\frac{2}{n}\cos n\pi.$$

Hence $b_1 = 2$, $b_2 = -2/2$, $b_3 = 2/3$, $b_4 = -2/4$, $\cdots$, and the Fourier series representation of $f(x)$ is

$$f(x) = \pi + 2\left(\sin x - \frac{1}{2}\sin 2x + \frac{1}{3}\sin 3x - + \cdots\right).$$

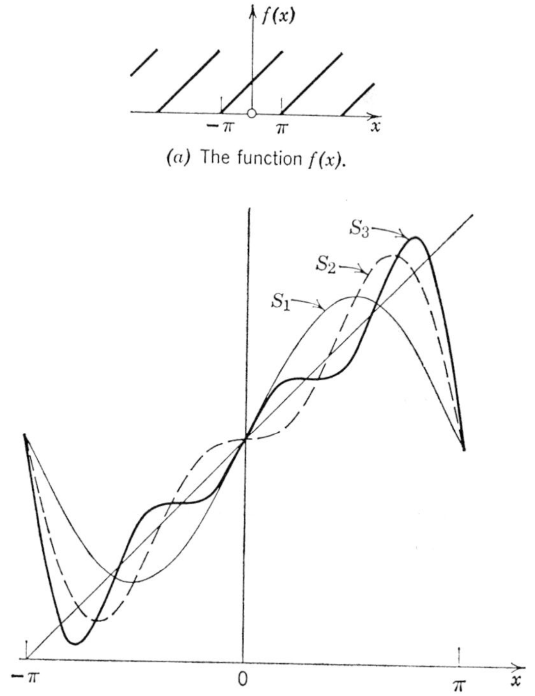

(*a*) The function $f(x)$.

(*b*) Partial sums $S_n(x)$.

Fig. 212. Example 2

Problems for Sec. 10.4

1. Prove Theorem 2.

Are the following functions even, odd, or neither even nor odd?

2. $x + x^2$, $|x|$, e^x, e^{x^2}, $\sin^2 x$, $x \sin x$, $\ln x$, $x \cos x$

3. $|x^3|$, $x \cos nx$, $x^2 \cos nx$, $\cosh x$, $\sinh x$, $\sin x + \cos x$, $x|x|$

Are the following functions $f(x)$ which are assumed to be periodic, of period 2π, even, odd, or neither even nor odd?

4. $f(x) = \begin{cases} 0 & \text{if} \quad -\pi < x < 0 \\ x & \text{if} \quad 0 < x < \pi \end{cases}$

5. $f(x) = \begin{cases} -x^2 & \text{if} \quad -\pi < x < 0 \\ x^2 & \text{if} \quad 0 < x < \pi \end{cases}$

6. $f(x) = \begin{cases} x & \text{if} \quad -\pi/2 < x < \pi/2 \\ 0 & \text{if} \quad \pi/2 < x < 3\pi/2 \end{cases}$

7. $f(x) = \begin{cases} 0 & \text{if} \quad -\pi < x < -\pi/2 \\ x & \text{if} \quad -\pi/2 < x < \pi \end{cases}$

8. $f(x) = |\sin x| \quad (-\pi < x < \pi)$

9. $f(x) = e^{|x|} \quad (-\pi < x < \pi)$

10. $f(x) = e^{-|x|} \quad (-\pi < x < \pi)$

11. $f(x) = x|x| \quad (-\pi < x < \pi)$

12. $f(x) = x \quad (0 < x < 2\pi)$

13. $f(x) = x^2 \quad (0 < x < 2\pi)$

14. Find all functions which are both even and odd.

Prove:

15. If $f(x)$ is odd, then $|f(x)|$ and $f^2(x)$ are even.

16. If $f(x)$ is even, then $|f(x)|$, $f^2(x)$, and $f^3(x)$ are even.

17. The sum of odd functions is odd.

18. The product of two odd functions is even.

19. The sum and the product of even functions are even.

20. If $g(x)$ is any function, defined for all x, then $p(x) = [g(x) + g(-x)]/2$ is even and $q(x) = [g(x) - g(-x)]/2$ is odd, and $g(x) = p(x) + q(x)$.

Represent the following functions as the sum of an even and an odd function.

21. e^x **22.** $1/(1 - x)$ **23.** $x/(1 - x)$ **24.** $(1 + x)/(1 - x)$

Find the Fourier series of the following functions which are assumed to have the period 2π.

25. $f(x) = x \quad (-\pi < x < \pi)$

26. $f(x) = |x| \quad (-\pi < x < \pi)$

27. $f(x) = x^3 \quad (-\pi < x < \pi)$

28. $f(x) = |\sin x| \quad (-\pi < x < \pi)$

29. $f(x) = \begin{cases} x & \text{if } -\pi/2 < x < \pi/2 \\ \pi - x & \text{if } \pi/2 < x < 3\pi/2 \end{cases}$

30. $f(x) = \begin{cases} x & \text{if } 0 < x < \pi \\ \pi - x & \text{if } \pi < x < 2\pi \end{cases}$

31. $f(x) = \begin{cases} x - \pi & \text{if } 0 < x < \pi \\ -x & \text{if } \pi < x < 2\pi \end{cases}$

32. $f(x) = \begin{cases} -x^2 & \text{if } -\pi < x < 0 \\ x^2 & \text{if } 0 < x < \pi \end{cases}$

33. Show that $f(x) = x^2 \; (-\pi < x \leqq \pi)$, $f(x + 2\pi) = f(x)$ has the Fourier series

$$f(x) = \frac{\pi^2}{3} - 4\left(\cos x - \frac{1}{4}\cos 2x + \frac{1}{9}\cos 3x - + \cdots\right).$$

34. Setting $x = \pi$ in Prob. 33, obtain the following famous result by Euler:

$$\sum_{n=1}^{\infty} \frac{1}{n^2} = 1 + \frac{1}{4} + \frac{1}{9} + \frac{1}{16} + \cdots = \frac{\pi^2}{6}. \tag{7}$$

35. Using Prob. 33, show that

$$\sum_{n=1}^{\infty} \frac{(-1)^{n+1}}{n^2} = 1 - \frac{1}{4} + \frac{1}{9} - \frac{1}{16} + - \cdots = \frac{\pi^2}{12}. \tag{8}$$

10.5 Half-Range Expansions

In various physical and engineering problems there is a practical need for applying Fourier series to functions $f(t)$ which are defined merely on some finite interval. Typical examples will arise in the next chapter (Secs. 11.3 and 11.5) in connection with partial differential equations. Then $f(t)$ will be defined on an interval $0 \leqq t \leqq l$, and on this interval we want to represent $f(t)$ by a Fourier series. For this purpose we may use Theorem 1 in the last section. We may let the interval $0 \leqq t \leqq l$ correspond to the interval of integration $0 \leqq t \leqq T/2$, that is, we may set $T/2 = l$ or $T = 2l$. Using (4) in Sec. 10.4, we then obtain a Fourier cosine series which represents an even periodic function $f_1(t)$ of period $T = 2l$. By construction, $f_1(t) = f(t)$ on $0 \leqq t \leqq l$. For this reason $f_1(t)$ is called

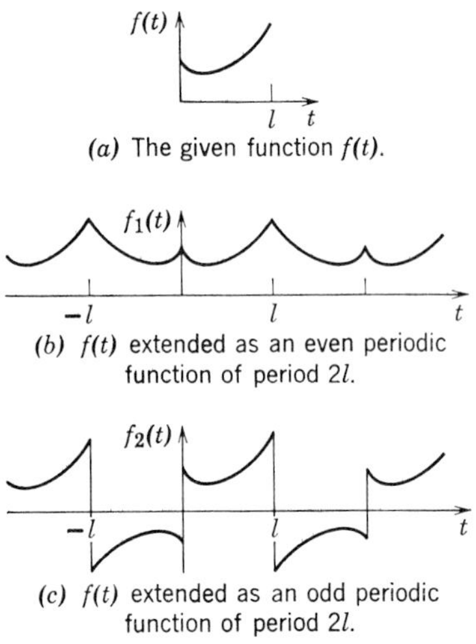

f(t)
l t
(a) The given function f(t).
f1(t)
-l
l
t
(b) f(t) extended as an even periodic function of period 2l.
f2(t)
-l
l
t
(c) f(t) extended as an odd periodic function of period 2l.

Fig. 213. Periodic extensions

the *even periodic extension of* $f(t)$ of period $T = 2l$. An illustration is given in Fig. 213*b*. From (3) and (4) in Sec. 10.4 with $T = 2l$ we see that

$$f(t) = a_0 + \sum_{n=1}^{\infty} a_n \cos \frac{n\pi}{l} t \qquad (0 \leqq t \leqq l) \tag{1}$$

and the coefficients are

$$a_0 = \frac{1}{l} \int_0^l f(t)\, dt, \qquad a_n = \frac{2}{l} \int_0^l f(t) \cos \frac{n\pi}{l} t\, dt, \qquad n = 1, 2, \cdots. \tag{2}$$

Instead of (4) in Sec. 10.4 we may equally well use (6) in Sec. 10.4 (with $T = 2l$, as before). Then we obtain a Fourier sine series, which represents an odd periodic function, say, $f_2(t)$, of period $T = 2l$. By construction, $f_2(t) = f(t)$ on $0 \leqq t \leqq l$, and $f_2(t)$ is called the *odd periodic extension of* $f(t)$ *of period* $T = 2l$. An illustration is given in Fig. 213*c*. From (5) and (6) in Sec. 10.4 we see that

$$f(t) = \sum_{n=1}^{\infty} b_n \sin \frac{n\pi}{l} t \qquad (0 \leqq t \leqq l) \tag{3}$$

and the coefficients are

$$b_n = \frac{2}{l} \int_0^l f(t) \sin \frac{n\pi}{l} t\, dt, \qquad n = 1, 2, \cdots. \tag{4}$$

The series (1) and (3) with coefficients (2) and (4) are called **half-range expansions** of the given function $f(t)$.

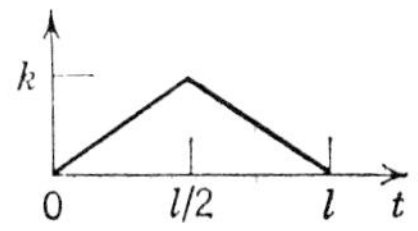

Fig. 214. The given function in Example 1

Example 1. Triangular pulse

Find the half-range expansions of the function (Fig. 214)

$$f(t) = \begin{cases} \dfrac{2k}{l}t & \text{when} \quad 0 < t < \dfrac{l}{2} \\[2ex] \dfrac{2k}{l}(l - t) & \text{when} \quad \dfrac{l}{2} < t < l \end{cases}$$

From (2) we obtain

$$a_0 = \frac{1}{l}\left[\frac{2k}{l}\int_0^{l/2} t\,dt + \frac{2k}{l}\int_{l/2}^{l} (l - t)\,dt\right] = \frac{k}{2},$$

$$a_n = \frac{2}{l}\left[\frac{2k}{l}\int_0^{l/2} t\cos\frac{n\pi}{l}t\,dt + \frac{2k}{l}\int_{l/2}^{l} (l - t)\cos\frac{n\pi}{l}t\,dt\right].$$

Now by integration by parts,

$$\int_0^{l/2} t\cos\frac{n\pi}{l}t\,dt = \frac{lt}{n\pi}\sin\frac{n\pi}{l}t\,\bigg|_0^{l/2} - \frac{l}{n\pi}\int_0^{l/2}\sin\frac{n\pi}{l}t\,dt$$

$$= \frac{l^2}{2n\pi}\sin\frac{n\pi}{2} + \frac{l^2}{n^2\pi^2}\left(\cos\frac{n\pi}{2} - 1\right).$$

Similarly,

$$\int_{l/2}^{l} (l - t)\cos\frac{n\pi}{l}t\,dt = -\frac{l^2}{2n\pi}\sin\frac{n\pi}{2} - \frac{l^2}{n^2\pi^2}\left(\cos n\pi - \cos\frac{n\pi}{2}\right).$$

By inserting these two results we obtain

$$a_n = \frac{4k}{n^2\pi^2}\left(2\cos\frac{n\pi}{2} - \cos n\pi - 1\right).$$

Thus,

$$a_2 = -16k/2^2\pi^2, \qquad a_6 = -16k/6^2\pi^2, \qquad a_{10} = -16k/10^2\pi^2, \cdots$$

and $a_n = 0$ when $n \neq 2, 6, 10, 14, \cdots$. Hence the first half-range expansion of $f(t)$ is

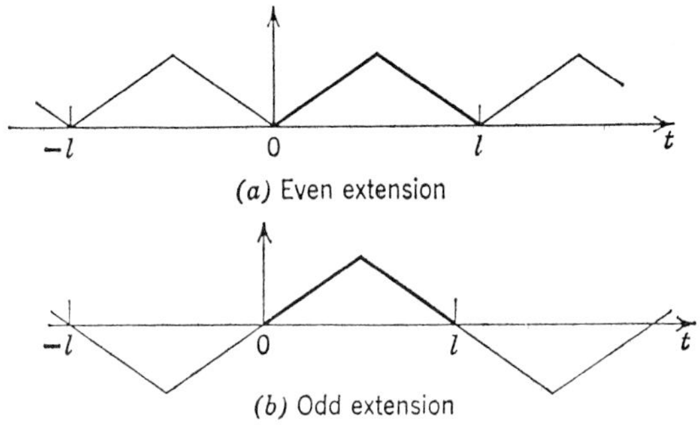

Fig. 215. Periodic extensions of $f(t)$ in Example 1

$$f(t) = \frac{k}{2} - \frac{16k}{\pi^2}\left(\frac{1}{2^2}\cos\frac{2\pi}{l}t + \frac{1}{6^2}\cos\frac{6\pi}{l}t + \cdots\right).$$

This series represents the even periodic extension of $f(t)$ shown in Fig. 215*a*.

Similarly, from (4),

$$b_n = \frac{8k}{n^2\pi^2}\sin\frac{n\pi}{2} \tag{5}$$

and the other half-range expansion of $f(t)$ is

$$f(t) = \frac{8k}{\pi^2}\left(\frac{1}{1^2}\sin\frac{\pi}{l}t - \frac{1}{3^2}\sin\frac{3\pi}{l}t + \frac{1}{5^2}\sin\frac{5\pi}{l}t - + \cdots\right).$$

This series represents the odd periodic extension of $f(t)$ shown in Fig. 215*b*.

Problems for Sec. 10.5

Represent the following functions $f(t)$ by a Fourier cosine series and graph the corresponding periodic extension of $f(t)$.

1. $f(t) = 1 \quad (0 < t < l)$

2. $f(t) = t \quad (0 < t < l)$

3. $f(t) = \begin{cases} 1 & \text{if} \quad 0 < t < l/2 \\ 0 & \text{if} \quad l/2 < t < l \end{cases}$

4. $f(t) = \begin{cases} 0 & \text{if} \quad 0 < t < l/2 \\ 1 & \text{if} \quad l/2 < t < l \end{cases}$

5. $f(t) = 1 - \frac{t}{l} \quad (0 < t < l)$

6. $f(t) = t^2 \quad (0 < t < l)$

7. $f(t) = t^3 \quad (0 < t < l)$

8. $f(t) = e^t \quad (0 < t < l)$

9. $f(t) = \sin\frac{\pi}{l}t \quad (0 < t < l)$

10. $f(t) = \sin\frac{\pi t}{2l} \quad (0 < t < l)$

Represent the following functions $f(t)$ by a Fourier sine series and graph the corresponding periodic extension of $f(t)$.

11. $f(t) = 1 \quad (0 < t < l)$

12. $f(t) = t \quad (0 < t < l)$

13. $f(t) = \begin{cases} 1 & \text{if} \quad 0 < t < l/2 \\ 0 & \text{if} \quad l/2 < t < l \end{cases}$

14. $f(t) = \begin{cases} 1/2 & \text{if} \quad 0 < t < l/2 \\ 3/2 & \text{if} \quad l/2 < t < l \end{cases}$

15. $f(t) = t^2 \quad (0 < t < l)$

16. $f(t) = t^3 \quad (0 < t < l)$

17. (Complex form of the Fourier series, complex Fourier coefficients) Using the formula $e^{i\theta} = \cos\theta + i\sin\theta$ (cf. Sec. 2.4), show that

$$\cos nx = \frac{1}{2}(e^{inx} + e^{-inx}), \qquad \sin nx = \frac{1}{2i}(e^{inx} - e^{-inx})$$

and the Fourier series

$$f(x) = a_0 + \sum_{n=1}^{\infty}(a_n \cos nx + b_n \sin nx)$$

may be written in the form

$$f(x) = c_0 + \sum_{n=1}^{\infty}(c_n e^{inx} + k_n e^{-inx}) \tag{6}$$

where $c_0 = a_0$, $c_n = (a_n - ib_n)/2$, $k_n = (a_n + ib_n)/2$, $n = 1, 2, \cdots$. Using (6), Sec. 10.2, show that

$$c_n = \frac{1}{2\pi}\int_{-\pi}^{\pi} f(x)e^{-inx}\,dx, \qquad k_n = \frac{1}{2\pi}\int_{-\pi}^{\pi} f(x)e^{inx}\,dx, \qquad n = 1, 2, \cdots.$$

Introducing the notation $k_n = c_{-n}$, show that (6) may be written

$$\text{(7)} \quad f(x) = \sum_{n=-\infty}^{\infty} c_n e^{inx}, \qquad c_n = \frac{1}{2\pi}\int_{-\pi}^{\pi} f(x)e^{-inx}\,dx, \qquad n = 0, \pm 1, \pm 2, \cdots.$$

This is the so-called *complex form* of the Fourier series, and the c_n are called the *complex Fourier coefficients* of $f(x)$.

18. Using (7), show that the complex form of the Fourier series of the function

$$f(x) = e^x \qquad \text{when} \qquad -\pi < x < \pi \qquad \text{and} \qquad f(x + 2\pi) = f(x)$$

is

$$f(x) = \frac{\sinh \pi}{\pi} \sum_{n=-\infty}^{\infty} (-1)^n \frac{1 + in}{1 + n^2} e^{inx}.$$

19. Obtain from the series in Prob. 18 the real Fourier series

$$f(x) = K\left[\frac{1}{2} - \frac{1}{1 + 1^2}(\cos x - \sin x) + \frac{1}{1 + 2^2}(\cos 2x - 2 \sin 2x) - + \cdots\right]$$

where $K = (2 \sinh \pi)/\pi$.

20. Show that the complex Fourier coefficients of an odd function are pure imaginary and those of an even function are real.

10.6 Determination of Fourier Coefficients without Integration

In several of our previous examples, relatively complicated and lengthy integrations led to relatively simple expressions for the Fourier coefficients a_n and b_n. This raises the question whether there might be a simpler way of obtaining Fourier coefficients. There is, and we want to show that the Fourier coefficients of a periodic function which is represented by polynomials can be obtained in terms of the jumps of the function and its derivatives. Of course, this is a big advantage, and the corresponding formulas are of great practical importance, because by applying them integrations are avoided (except for a_0, which must be determined as before).

By a *jump j* of a function $g(x)$ at a point x_0 we mean the difference between the right-hand and left-hand limits (Sec. 10.2) of $g(x)$ at x_0; that is

$$\text{(1)} \qquad j = g(x_0 + 0) - g(x_0 - 0).$$

It follows that an upward jump is positive and a downward jump is negative;

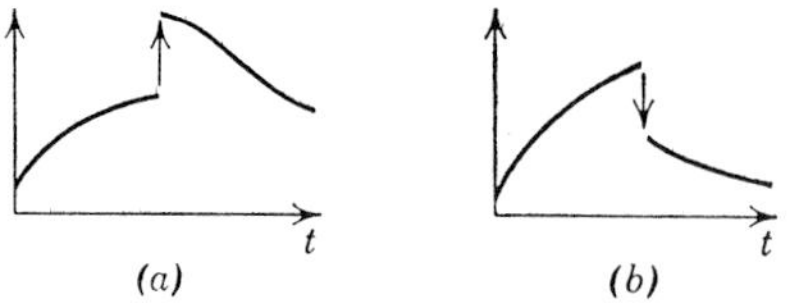

Fig. 216. Jump of a function. (*a*) Positive jump. (*b*) Negative jump

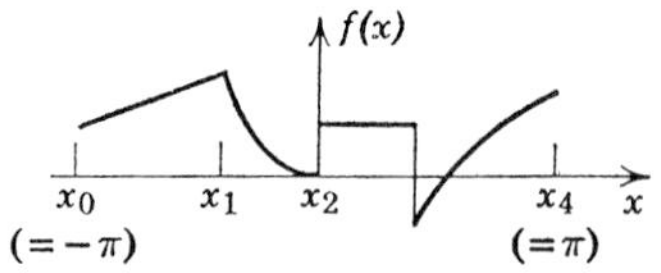

Fig. 217. Example of a representation of the form (2) (with $m = 4$)

see Fig. 216.

Let $f(x)$ be a function which has period 2π and is represented by polynomials $p_1, \cdots, p_m$ in the interval $-\pi < x < \pi$, say (Fig. 217)

$$
(2) \qquad f(x) = \begin{cases} p_1(x) \text{ when } x_0 < x < x_1, & (x_0 = -\pi) \\ p_2(x) \text{ when } x_1 < x < x_2, & \\ \vdots & \\ p_m(x) \text{ when } x_{m-1} < x < x_m \ (= \pi). & \end{cases}
$$

Then f may have jumps at $x_0, x_1, \cdots, x_m$, and the same is true for the derivatives $f', f'', \cdots$. We choose the following notation.

$$
(3) \qquad \begin{aligned} j_s &= \text{jump of } f \text{ at } x_s, \\ j_s' &= \text{jump of } f' \text{ at } x_s, \\ j_s'' &= \text{jump of } f'' \text{ at } x_s, \quad \text{etc.} \end{aligned} \qquad (s = 1, 2, \cdots, m)
$$

Of course, if f is continuous at x_s, then $j_s = 0$, and for the derivatives the situation is similar, so that some of the numbers $j_s, j_s', \cdots$ in (3) may be zero.

Example 1. Jumps of a function and its derivatives

Let

$$
f(x) = \begin{cases} 0 & \text{when} \quad -\pi < x < 0, \\ x^2 & \text{when} \quad 0 < x < \pi. \end{cases}
$$

We plot graphs of f and its derivatives (Fig. 218),

$$
f' = \begin{cases} 0 \\ 2x \end{cases} \qquad f'' = \begin{cases} 0 \\ 2 \end{cases} \qquad f''' = 0.
$$

We see that the jumps are

	Jump at $x_1 = 0$	Jump at $x_2 = \pi$
f	$j_1 = 0$	$j_2 = -\pi^2$
f'	$j_1' = 0$	$j_2' = -2\pi$
f''	$j_1'' = 2$	$j_2'' = -2$

Note that the jumps at $x = -\pi$ are not listed, because they are taken into account at $x = \pi$, the other end of the interval of periodicity. ∎

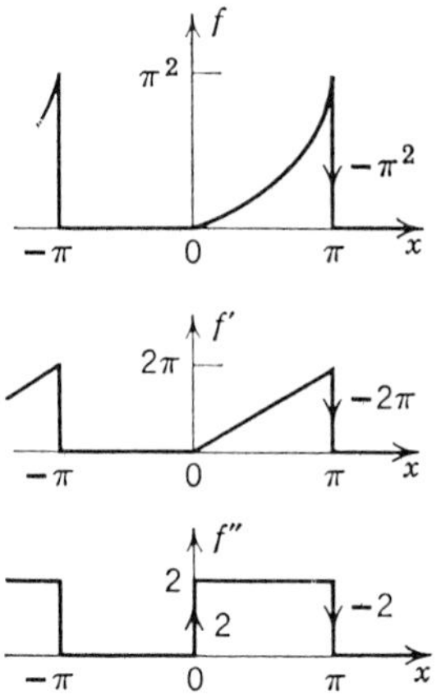

Fig. 218. $f(x)$ and derivatives in Example 1

To derive the desired formula for the Fourier coefficients $a_1, a_2, \cdots$ of f, given by (2), we start from the Euler formula (6b) in Sec. 10.2:

$$\pi a_n = \int_{-\pi}^{\pi} f \cos nx \, dx. \tag{4}$$

Since f is represented by (2), we write the integral as the sum of m integrals:

$$\pi a_n = \int_{x_0}^{x_1} + \int_{x_1}^{x_2} + \cdots + \int_{x_{m-1}}^{x_m} = \sum_{s=1}^{m} \int_{x_{s-1}}^{x_s} f \cos nx \, dx, \tag{5}$$

where $x_0 = -\pi$ and $x_m = \pi$. Integration by parts yields

$$\int_{x_{s-1}}^{x_s} f \cos nx \, dx = \frac{f}{n} \sin nx \Big|_{x_{s-1}}^{x_s} - \frac{1}{n} \int_{x_{s-1}}^{x_s} f' \sin nx \, dx. \tag{6}$$

Now comes an important point: the evaluation of the first expression on the right. $f(x)$ may be discontinuous at x_s (Fig. 219), and we have to take the left-hand limit $f(x_s - 0)$ of f at x_s. Similarly, at x_{s-1} we have to take the right-hand limit $f(x_{s-1} + 0)$. Hence the first expression on the right-hand side of (6) equals

$$\frac{1}{n}[f(x_s - 0) \sin nx_s - f(x_{s-1} + 0) \sin nx_{s-1}].$$

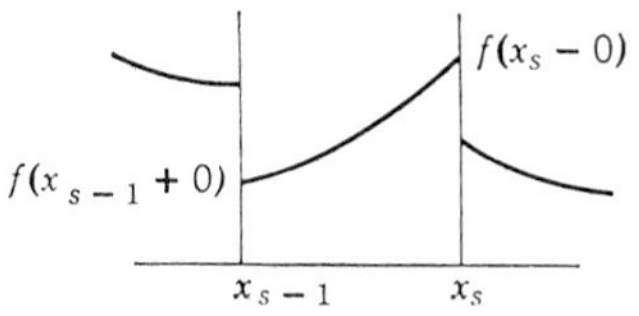

Fig. 219. The formula (6)

Consequently, by inserting (6) into (5) and using the short notations $S_0 = \sin nx_0$, $S_1 = \sin nx_1$, etc., we obtain

$$\begin{aligned}\pi a_n = \frac{1}{n}[f(x_1 - 0)S_1 - f(x_0 + 0)S_0 + f(x_2 - 0)S_2 - f(x_1 + 0)S_1 \\ + \cdots + f(x_m - 0)S_m - f(x_{m-1} + 0)S_{m-1}] \\ - \frac{1}{n}\sum_{s=1}^{m}\int_{x_{s-1}}^{x_s} f' \sin nx \, dx.\end{aligned} \tag{7}$$

If we collect terms with the same S, the expression in the brackets becomes

$$\begin{aligned}-f(x_0 + 0)S_0 + [f(x_1 - 0) - f(x_1 + 0)]S_1 \\ + [f(x_2 - 0) - f(x_2 + 0)]S_2 + \cdots + f(x_m - 0)S_m.\end{aligned} \tag{8}$$

The expressions in the brackets in (8) are the jumps of f, multiplied by -1. Furthermore, because of periodicity, $S_0 = S_m$ and $f(x_0) = f(x_m)$, so that we may combine the first and the last term in (8). Thus (8) is equal to

$$-j_1 S_1 - j_2 S_2 - \cdots - j_m S_m,$$

and from (7) we therefore have the intermediate result

$$\pi a_n = -\frac{1}{n}\sum_{s=1}^{m} j_s \sin nx_s - \frac{1}{n}\sum_{s=1}^{m}\int_{x_{s-1}}^{x_s} f' \sin nx \, dx. \tag{9}$$

By applying the same procedure to the integrals on the right-hand side of (9) we find

$$\sum_{s=1}^{m}\int_{x_{s-1}}^{x_s} f' \sin nx \, dx = \frac{1}{n}\sum_{s=1}^{m} j_s' \cos nx_s + \frac{1}{n}\sum_{s=1}^{m}\int_{x_{s-1}}^{x_s} f'' \cos nx \, dx. \tag{10}$$

Continuing in this fashion, we obtain integrals involving higher and higher derivatives of f. Since f is represented by polynomials and the $(r + 1)$th derivative of a polynomial of degree r is identically zero, we shall reach the point where no integrals are left, and this will happen after finitely many steps. By inserting (10) and the analogous formulas obtained in the further steps into (9) we obtain the desired formula

$$\begin{aligned}a_n = \frac{1}{n\pi}\Bigg[-\sum_{s=1}^{m} j_s \sin nx_s - \frac{1}{n}\sum_{s=1}^{m} j_s' \cos nx_s \\ + \frac{1}{n^2}\sum_{s=1}^{m} j_s'' \sin nx_s + \frac{1}{n^3}\sum_{s=1}^{m} j_s''' \cos nx_s - - + + \cdots\Bigg],\end{aligned} \tag{11a}$$

where $n = 1, 2, \cdots$ (and a_0 must be obtained by integration as before). In precisely the same fashion we obtain from (6c) in Sec. 10.2

$$
\textbf{(11b)}\quad
\begin{aligned}
b_n = \frac{1}{n\pi}\Bigg[\sum_{s=1}^{m} j_s \cos nx_s - \frac{1}{n}\sum_{s=1}^{m} j_s{}' \sin nx_s \\
- \frac{1}{n^2}\sum_{s=1}^{m} j_s{}'' \cos nx_s + \frac{1}{n^3}\sum_{s=1}^{m} j_s{}''' \sin nx_s + - - + + \cdots\Bigg].
\end{aligned}
$$

To avoid errors it is practical to graph $f(x)$ and its derivatives and list the jumps in a table as shown in Example 1.

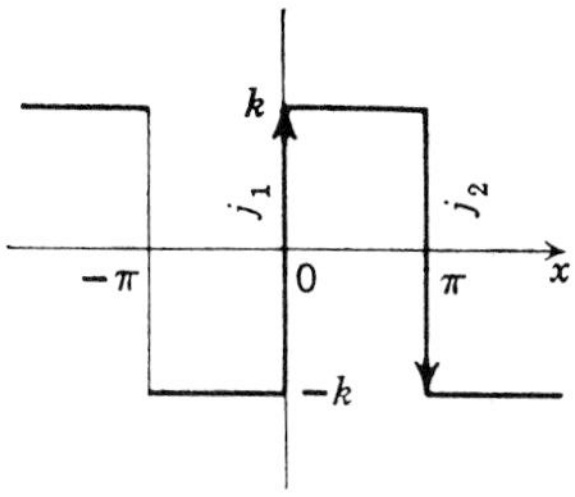

Fig. 220. Example 2

Example 2. Periodic square wave
Find the Fourier coefficients of the function (Fig. 220)

$$f(x) = \begin{cases} -k & \text{when} \quad -\pi < x < 0, \\ k & \text{when} \quad 0 < x < \pi. \end{cases}$$

We see that $f' \equiv 0$ and the jumps of f are

	Jump at $x_1 = 0$	Jump at $x_2 = \pi$
f	$j_1 = 2k$	$j_2 = -2k$

f is odd. Hence $a_n = 0$, and from (11b),

$$
\begin{aligned}
b_n &= \frac{1}{n\pi}[j_1 \cos nx_1 + j_2 \cos nx_2] = \frac{1}{n\pi}[2k \cos 0 - 2k \cos n\pi] \\
&= \frac{2k}{n\pi}(1 - \cos n\pi) = \begin{cases} 4k/n\pi & \text{for odd } n \\ 0 & \text{for even } n \end{cases} \qquad \text{(cf. Example 1, Sec. 10.2).}
\end{aligned}
$$

Example 3
Find the Fourier series of the function in Example 1 of the current section. By integration,

$$a_0 = \frac{1}{2\pi}\int_0^{\pi} x^2\, dx = \frac{\pi^2}{6}.$$

From (11a),

$$a_n = \frac{1}{n\pi}\left[\pi^2 \sin n\pi + \frac{2\pi}{n}\cos n\pi + \frac{1}{n^2}(2 \sin 0 - 2 \sin n\pi)\right] = \frac{2}{n^2}\cos n\pi.$$

Hence $a_1 = -2/1^2$, $a_2 = 2/2^2$, $a_3 = -2/3^2$, $\cdots$. From (11b) it follows that

$$b_n = \frac{1}{n\pi}\left[-\pi^2 \cos n\pi + \frac{2\pi}{n}\sin n\pi - \frac{1}{n^2}(2\cos 0 - 2\cos n\pi)\right]$$

$$= -\frac{\pi}{n}\cos n\pi + \frac{2}{n^3\pi}(\cos n\pi - 1).$$

Hence,

$$b_1 = \pi - \frac{4}{\pi}, \qquad b_2 = -\frac{\pi}{2}, \qquad b_3 = \frac{\pi}{3} - \frac{4}{3^3\pi}, \qquad b_4 = -\frac{\pi}{4}, \cdots,$$

and the Fourier series is

$$f(x) = \frac{\pi^2}{6} - 2\cos x + \left(\pi - \frac{4}{\pi}\right)\sin x + \frac{1}{2}\cos 2x - \frac{\pi}{2}\sin 2x + \cdots.$$

Problems for Sec. 10.6

1. Derive (11b) from (6c), Sec. 10.2.

Using (11), find the Fourier series of the functions $f(x)$ in

2. Probs. 1–4, Sec. 10.2

3. Probs. 9–12, Sec. 10.2

4. Probs. 5–8, Sec. 10.2

5. Probs. 13–16, Sec. 10.2

6. Show that in the case of a function $f(t)$ having period T the formulas corresponding to (11) are

(12a)
$$a_n = \frac{1}{n\pi}\left[-\sum_{s=1}^{m} j_s \sin\frac{2n\pi}{T}t_s - \frac{T}{2n\pi}\sum_{s=1}^{m} j_s' \cos\frac{2n\pi}{T}t_s + \left(\frac{T}{2n\pi}\right)^2 \sum_{s=1}^{m} j_s'' \sin\frac{2n\pi}{T}t_s + \left(\frac{T}{2n\pi}\right)^3 \sum_{s=1}^{m} j_s''' \cos\frac{2n\pi}{T}t_s - - + + \cdots\right],$$

(12b)
$$b_n = \frac{1}{n\pi}\left[\sum_{s=1}^{m} j_s \cos\frac{2n\pi}{T}t_s - \frac{T}{2n\pi}\sum_{s=1}^{m} j_s' \sin\frac{2n\pi}{T}t_s - \left(\frac{T}{2n\pi}\right)^2 \sum_{s=1}^{m} j_s'' \cos\frac{2n\pi}{T}t_s + \left(\frac{T}{2n\pi}\right)^3 \sum_{s=1}^{m} j_s''' \sin\frac{2n\pi}{T}t_s + - - + + \cdots\right].$$

Using (12), find the Fourier series of the functions $f(t)$ in:

7. Probs. 4–6, Sec. 10.3

8. Probs. 7, 8, Sec. 10.3

9. Probs. 9–11, Sec. 10.3

10. Probs. 12, 13, Sec. 10.3

Using (12), find the Fourier cosine series of the functions $f(t)$ in

11. Probs. 1–4, Sec. 10.5

12. Probs. 5–7, Sec. 10.5

Using (12), find the Fourier sine series of the functions $f(t)$ in

13. Probs. 11–13, Sec. 10.5

14. Probs. 14–16, Sec. 10.5

15. Can (11) be applied to find the Fourier coefficients of the function $f(x) = e^x$ $(0 < x < 2\pi)$, $f(x + 2\pi) = f(x)$?

10.7 Forced Oscillations

Fourier series have important applications in connection with differential equations. Let us consider an important practical problem involving an ordinary differential equation. (Partial differential equations will be considered in the next chapter.)

From Sec. 2.13 we know that forced oscillations of a body of mass m on a spring (cf. Fig. 221) are governed by the equation

$$m\ddot{y} + c\dot{y} + ky = r(t), \tag{1}$$

where k is the spring modulus and c is the damping constant.

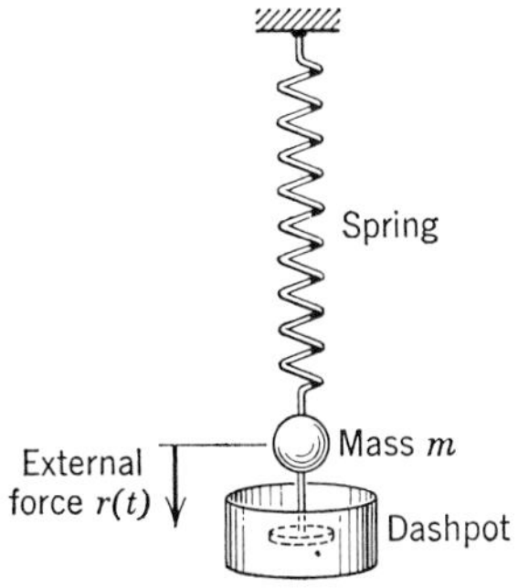

Fig. 221. Vibrating system under consideration

If the external force $r(t)$ is a sine or cosine function and the damping constant c is not zero, the steady-state solution represents a harmonic oscillation having the frequency of the external force.

We shall now see that if $r(t)$ is not a pure sine or cosine function but any other periodic function, then the steady-state solution will represent a superposition of harmonic oscillations having the frequency of $r(t)$ and multiples of this frequency. If the frequency of one of these oscillations is close to the resonant frequency of the vibrating system (cf. Sec. 2.13), then that oscillation may be the dominant part of the response of the system to the external force. Of course, this is quite surprising to an observer not familiar with the corresponding mathematical theory which is highly important in the study of vibrating systems and resonance. Let us illustrate the situation by an example.

Example 1. Forced oscillations under a nonsinusoidal periodic driving force
In (1), let $m = 1$ (gm), $c = 0.02$ (gm/sec), and $k = 25$ (gm/sec^2), so that (1) becomes

$$\ddot{y} + 0.02\dot{y} + 25y = r(t) \tag{2}$$

where $r(t)$ is measured in gm · cm/sec^2. Let (Fig. 222)

$$r(t) = \begin{cases} t + \dfrac{\pi}{2} & \text{when } -\pi < t < 0, \\ -t + \dfrac{\pi}{2} & \text{when } 0 < t < \pi, \end{cases} \qquad r(t + 2\pi) = r(t).$$

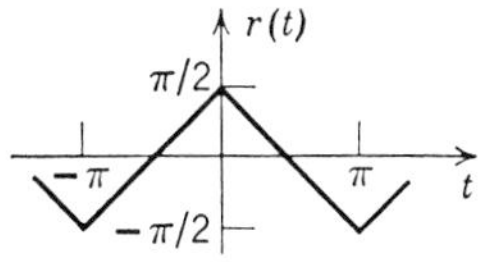

Find the steady-state solution $y(t)$.

Fig. 222. Force in Example 1

We represent $r(t)$ by a Fourier series, finding

$$r(t) = \frac{4}{\pi}\left(\cos t + \frac{1}{3^2}\cos 3t + \frac{1}{5^2}\cos 5t + \cdots\right). \tag{3}$$

Then we consider the differential equation

$$\ddot{y} + 0.02\dot{y} + 25y = \frac{4}{n^2\pi}\cos nt \qquad (n = 1, 3, \cdots) \tag{4}$$

whose right-hand side is a single term of the series (3). From Sec. 2.13 we know that the steady-state solution $y_n(t)$ of (4) is of the form

$$y_n = A_n \cos nt + B_n \sin nt, \tag{5}$$

and by substituting this into (4) we find that

$$A_n = \frac{4(25 - n^2)}{n^2\pi D}, \qquad B_n = \frac{0.08}{n\pi D}, \qquad \text{where} \qquad D = (25 - n^2)^2 + (0.02n)^2. \tag{6}$$

Since the differential equation (2) is linear, we may expect that the steady-state solution is

$$y = y_1 + y_3 + y_5 + \cdots \tag{7}$$

where y_n is given by (5) and (6). In fact, this follows readily by substituting (7) into (2) and using the Fourier series of $r(t)$, provided that termwise differentiation of (7) is permissible. (Readers already familiar with the notion of uniform convergence [Sec. 16.6] may prove that (7) may be differentiated term by term.)

From (6) we find that the amplitude of (5) is

$$C_n = \sqrt{A_n^{\,2} + B_n^{\,2}} = \frac{4}{n^2\pi\sqrt{D}}.$$

Numerical values are

$$C_1 = 0.0530$$
$$C_3 = 0.0088$$
$$C_5 = 0.5100$$
$$C_7 = 0.0011$$
$$C_9 = 0.0003$$

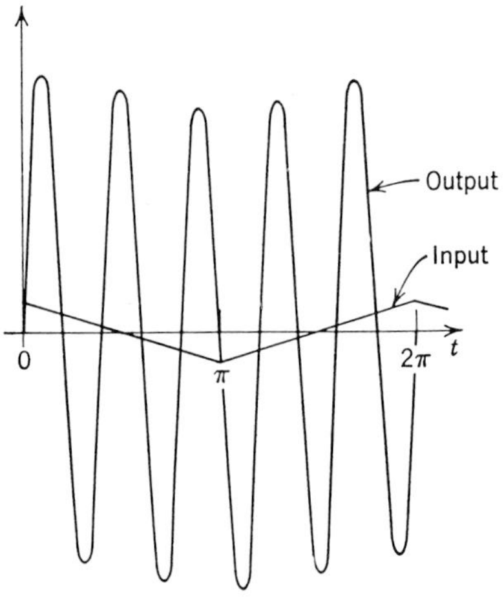

Fig. 223. Input and steady-state output in Example 1

For $n = 5$ the quantity D is very small, the denominator of C_5 is small, and C_5 is so large that y_5 is the dominating term in (7). This implies that the steady-state motion is almost a harmonic oscillation whose frequency equals five times that of the exciting force (Fig. 223).

Problems for Sec. 10.7

Find the general solution of the differential equation $\ddot{y} + \omega^2 y = r(t)$ where:

1. $r(t) = \sin t$, $\omega = 0.5, 0.7, 0.9, 1.1, 1.5, 2.0, 10.0$

2. $r(t) = \cos \alpha t + \cos \beta t \; (\omega^2 \neq \alpha^2, \beta^2)$

3. $r(t) = \sin t + \frac{1}{9}\sin 3t + \frac{1}{25}\sin 5t$, $\omega = 0.5, 0.9, 1.1, 2, 2.9, 3.1, 4, 4.9, 5.1, 6, 8$

4. $r(t) = \begin{cases} t & \text{if } -\pi/2 < t < \pi/2 \\ \pi - t & \text{if } \pi/2 < t < 3\pi/2 \end{cases}$ and $r(t + 2\pi) = r(t)$, $|\omega| \neq 1, 3, 5, \cdots$

5. $r(t) = \dfrac{t^2}{4}$ when $-\pi < t < \pi$ and $r(t + 2\pi) = r(t)$, $|\omega| \neq 0, 1, 2, \cdots$

6. $r(t) = \sum_{n=1}^{N} a_n \cos nt$, $|\omega| \neq 1, 2, \cdots, N$

7. $r(t) = \dfrac{\pi}{4}|\sin t|$ when $-\pi < t < \pi$ and $r(t + 2\pi) = r(t)$, $|\omega| \neq 0, 2, 4, \cdots$

8. $r(t) = \begin{cases} t + \pi & \text{if } -\pi < t < 0 \\ -t + \pi & \text{if } 0 < t < \pi \end{cases}$ and $r(t + 2\pi) = r(t)$, $|\omega| \neq 0, 1, 3, \cdots$

Find the steady-state oscillation corresponding to $\ddot{y} + cy + \dot{y} = r(t)$ where $c > 0$ and

9. $r(t) = K \sin t$

10. $r(t) = \sin 3t$

11. $r(t) = a_n \cos nt$

12. $r(t) = \sum_{n=1}^{N} b_n \sin nt$

13. $r(t) = \dfrac{t}{12}(\pi^2 - t^2)$ when $-\pi < t < \pi$ and $r(t + 2\pi) = r(t)$

14. Find the steady-state current $I(t)$ in the RLC-circuit in the figure where $R =$ 100 ohms, $L = 10$ henrys, $C = 10^{-2}$ farads, $E(t) = 100t(\pi^2 - t^2)$ volts when $-\pi < t < \pi$ and $E(t + 2\pi) = E(t)$. Proceed as follows. Develop $E(t)$ in a Fourier series. $I(t)$ will appear in the form of a trigonometric series. Find the general formulas for the coefficients of this series. Compute numerical values of the first few coefficients. Graph the sum of the first few terms of that series.

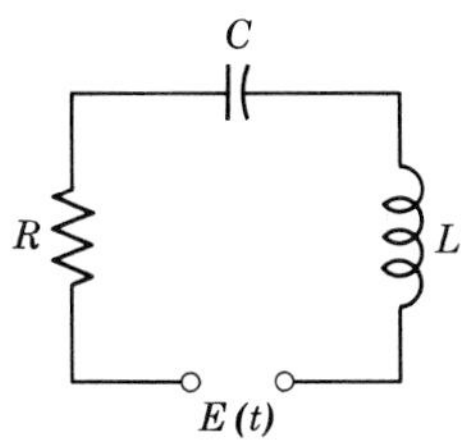

Problem 14. *RLC*-circuit

15. Same task as in Prob. 14 when

$$E(t) = \begin{cases} 100(\pi t + t^2) & \text{if } -\pi < t < 0 \\ 100(\pi t - t^2) & \text{if } 0 < t < \pi \end{cases} \quad \text{and} \quad E(t + 2\pi) = E(t),$$

the other data being as before.

10.8 Approximation by Trigonometric Polynomials. Square Error

Let $f(x)$ be a given function that has period 2π and can be represented by a Fourier series. Then the Nth partial sum of the Fourier series is an approximation to $f(x)$:

$$f(x) \approx a_0 + \sum_{n=1}^{N} (a_n \cos nx + b_n \sin nx). \tag{1}$$

It is natural to ask whether (1) is the "best" approximation to f by a trigonometric polynomial

$$F(x) = \alpha_0 + \sum_{n=1}^{N} (\alpha_n \cos nx + \beta_n \sin nx) \tag{2}$$

with the same fixed N, that is, an approximation for which the "error" is minimum.

Of course, we must first define what we mean by the error E of such an approximation. We want to choose a definition which measures the goodness of agreement between f and F *on the whole interval* $-\pi \leqq x \leqq \pi$. Obviously, the maximum of $|f - F|$ is not suitable for that purpose: in Fig. 224, the function F is a good approximation to f, but $|f - F|$ is large near x_0. We choose for our consideration

$$E = \int_{-\pi}^{\pi} (f - F)^2\, dx. \tag{3}$$

This is the so-called **total square error** of F relative to the function f over the interval $-\pi \leqq x \leqq \pi$. Clearly, $E \geqq 0$.

N being fixed, we want to determine the coefficients in (2) such that E is minimum. We can write (3) in the form

$$E = \int_{-\pi}^{\pi} f^2\, dx - 2\int_{-\pi}^{\pi} fF\, dx + \int_{-\pi}^{\pi} F^2\, dx. \tag{4}$$

By inserting (2) into the last integral and evaluating the occurring integrals as in Sec. 10.2 we readily obtain

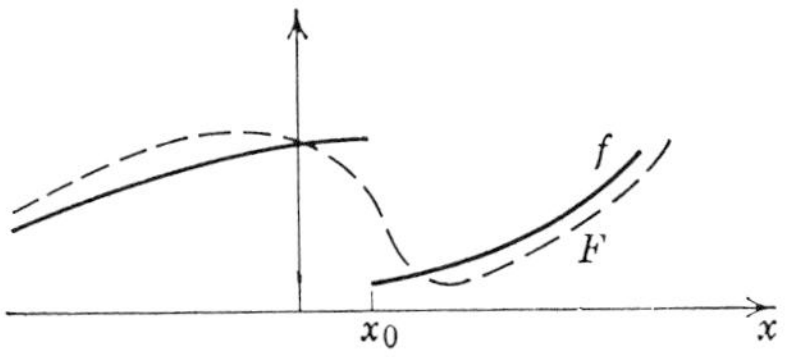

Fig. 224. Error of approximation

$$\int_{-\pi}^{\pi} F^2\,dx = \pi(2\alpha_0^{\ 2} + \alpha_1^{\ 2} + \cdots + \alpha_N^{\ 2} + \beta_1^{\ 2} + \cdots + \beta_N^{\ 2}).$$

By inserting (2) into the second integral in (4) we see that the occurring integrals are those in the Euler formulas (6), Sec. 10.2, and

$$\int_{-\pi}^{\pi} fF\,dx = \pi(2\alpha_0 a_0 + \alpha_1 a_1 + \cdots + \alpha_N a_N + \beta_1 b_1 + \cdots + \beta_N b_N).$$

With these expressions (4) becomes

$$\begin{aligned} E = \int_{-\pi}^{\pi} f^2\,dx &- 2\pi\left[2\alpha_0 a_0 + \sum_{n=1}^{N}(\alpha_n a_n + \beta_n b_n)\right] \\ &+ \pi\left[2\alpha_0^{\ 2} + \sum_{n=1}^{N}(\alpha_n^{\ 2} + \beta_n^{\ 2})\right]. \end{aligned} \tag{5}$$

If we take $\alpha_n = a_n$ and $\beta_n = b_n$ in (2), then from (5) we see that the square error corresponding to this particular choice of the coefficients of F is

$$E^* = \int_{-\pi}^{\pi} f^2\,dx - \pi\left[2a_0^{\ 2} + \sum_{n=1}^{N}(a_n^{\ 2} + b_n^{\ 2})\right]. \tag{6}$$

By subtracting (6) from (5) we obtain

$$E - E^* = \pi\left\{2(\alpha_0 - a_0)^2 + \sum_{n=1}^{N}[(\alpha_n - a_n)^2 + (\beta_n - b_n)^2]\right\}.$$

The terms on the right are squares of real numbers and therefore nonnegative. Hence

$$E - E^* \geqq 0 \qquad \text{or} \qquad E \geqq E^*,$$

and $E = E^*$ if, and only if, $\alpha_0 = a_0, \cdots, \beta_N = b_N$. This proves

Theorem 1 (Minimum square error)
The total square error of F [cf. (2), N fixed] relative to f on the interval $-\pi \leqq x \leqq \pi$ is minimum if and only if the coefficients of F in (2) are the corresponding Fourier coefficients of f. This minimum value is given by (6).

From (6) we see that E^* cannot increase as N increases, but may decrease. Hence, *with increasing N, the partial sums of the Fourier series of f yield better and better approximations to $f(x)$*, considered from the viewpoint of the square error.

Since $E^* \geqq 0$ and (6) holds for any N, we obtain from (6) the important

Bessel inequality[6]

$$2a_0^2 + \sum_{n=1}^{\infty} (a_n^2 + b_n^2) \leqq \frac{1}{\pi} \int_{-\pi}^{\pi} f(x)^2 \, dx \tag{7}$$

for the Fourier coefficients of any function f for which the integral on the right exists.

Problems for Sec. 10.8

1. Let $f(x) = -1$ when $-\pi < x < 0$, $f(x) = 1$ when $0 < x < \pi$, and $f(x + 2\pi) = f(x)$. Find the function $F(x)$ of the form (2) for which the total square error (3) is minimum.
2. Compute the minimum square error in Prob. 1 for $N = 1, 3, 5, 7$. What is the smallest N such that $E^* \leqq 0.2$?
3. Show that the minimum square error (6) is a monotone decreasing function of N.

In each case, find the function $F(x)$ of the form (2) for which the total square error E is minimum and compute this minimum value for $N = 1, 2, \cdots, 5$.

4. $f(x) = x \quad (-\pi < x < \pi)$
5. $f(x) = x^2 \quad (-\pi < x < \pi)$

10.9 The Fourier Integral

Fourier series are powerful tools in treating various problems involving periodic functions. A first illustration of this fact was given in Sec. 10.7, and further important problems in connection with partial differential equations will be considered in Chap. 11. Since, of course, many practical problems do not involve periodic functions, it is desirable to generalize the method of Fourier series to include nonperiodic functions.

Roughly speaking, if we start with a periodic function $f_T(x)$ of period T and let T approach infinity, then the resulting function $f(x)$ is no longer periodic. Let us illustrate this with two examples.

Example 1

Consider the function (Fig. 225)

$$f_T(x) = \begin{cases} 0 & \text{when} \quad -T/2 < x < -1 \\ 1 & \text{when} \quad -1 < x < 1 \\ 0 & \text{when} \quad 1 < x < T/2 \end{cases}$$

having period $T > 2$. For $T \to \infty$ we obtain the function

$$f(x) = \lim_{T \to \infty} f_T(x) = \begin{cases} 1 & \text{when} \quad -1 < x < 1, \\ 0 & \text{otherwise.} \end{cases}$$

[6] It can be shown that for such a function f even the equality sign in (7) holds. Proof in Ref. [D4].

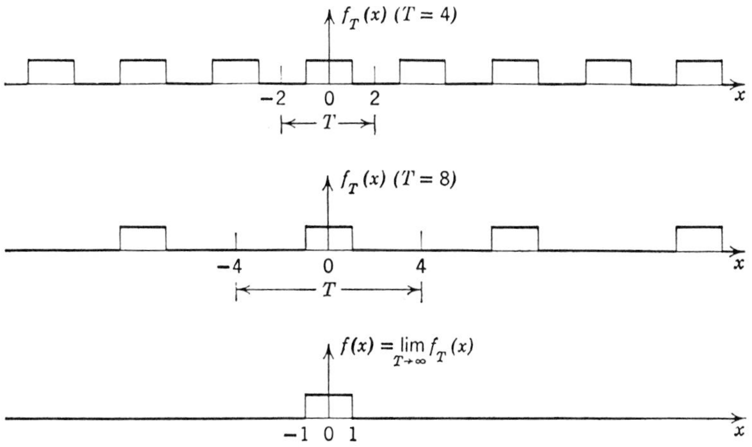

Fig. 225. Example 1

Example 2

Let

$$f_T(x) = e^{-|x|} \quad \text{when} \quad -T/2 < x < T/2 \quad \text{and} \quad f_T(x + T) = f_T(x).$$

Then (cf. Fig. 226)

$$f(x) = \lim_{T\to\infty} f_T(x) = e^{-|x|}.$$ ▮

Now let us start from a periodic function $f_T(x)$ that has period T and can be represented by a Fourier series:

$$f_T(x) = a_0 + \sum_{n=1}^{\infty} \left(a_n \cos \frac{2n\pi}{T} x + b_n \sin \frac{2n\pi}{T} x \right).$$

We transform this representation as follows. We use the short notation

$$w_n = \frac{2n\pi}{T}$$

and insert a_n and b_n according to the Euler formulas (3) in Sec. 10.3, denoting

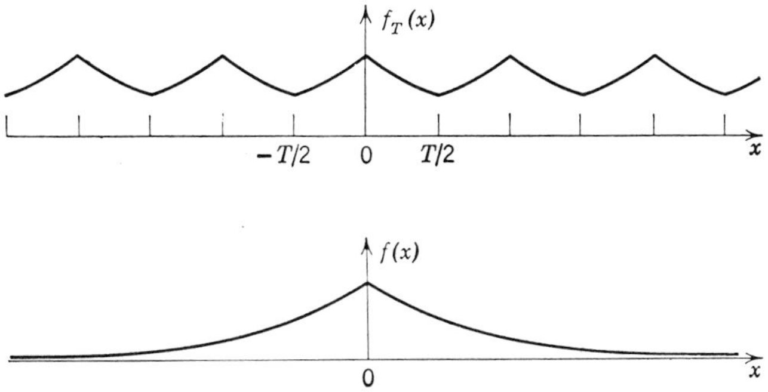

Fig. 226. Example 2

the variable of integration by v. Then we obtain

$$f_T(x) = \frac{1}{T}\int_{-T/2}^{T/2} f_T(v)\,dv + \frac{2}{T}\sum_{n=1}^{\infty}\left[\cos w_n x \int_{-T/2}^{T/2} f_T(v)\cos w_n v\,dv + \sin w_n x \int_{-T/2}^{T/2} f_T(v)\sin w_n v\,dv\right].$$

Now,

$$w_{n+1} - w_n = \frac{2(n+1)\pi}{T} - \frac{2n\pi}{T} = \frac{2\pi}{T},$$

and we set

$$\Delta w = w_{n+1} - w_n = \frac{2\pi}{T}.$$

Then $2/T = \Delta w/\pi$, and we may write that Fourier series in the form

(1) $$f_T(x) = \frac{1}{T}\int_{-T/2}^{T/2} f_T(v)\,dv + \frac{1}{\pi}\sum_{n=1}^{\infty}\left[\cos(w_n x)\,\Delta w \int_{-T/2}^{T/2} f_T(v)\cos w_n v\,dv + \sin(w_n x)\,\Delta w \int_{-T/2}^{T/2} f_T(v)\sin w_n v\,dv\right].$$

This representation is valid for any fixed T, arbitrarily large, but finite.

We now let T approach infinity and assume that the resulting nonperiodic function

$$f(x) = \lim_{T\to\infty} f_T(x)$$

is absolutely integrable on the x-axis, that is, the integral

(2) $$\int_{-\infty}^{\infty} |f(x)|\,dx$$

exists. Then $1/T \to 0$ and the value of the first term on the right side of (1) approaches zero. Furthermore, $\Delta w = 2\pi/T \to 0$ and it seems plausible that the infinite series in (1) becomes an integral from 0 to ∞, which represents $f(x)$, namely,

(3) $$f(x) = \frac{1}{\pi}\int_0^{\infty}\left[\cos wx \int_{-\infty}^{\infty} f(v)\cos wv\,dv + \sin wx \int_{-\infty}^{\infty} f(v)\sin wv\,dv\right]dw.$$

If we introduce the short notations

(4) $$A(w) = \int_{-\infty}^{\infty} f(v)\cos wv\,dv, \qquad B(w) = \int_{-\infty}^{\infty} f(v)\sin wv\,dv,$$

this may be written in the form

(5) $$f(x) = \frac{1}{\pi}\int_0^{\infty} [A(w)\cos wx + B(w)\sin wx]\,dw.$$

This is a representation of $f(x)$ by a so-called **Fourier integral.**

It is clear that our naive approach merely *suggests* the representation (5), but by no means establishes it; in fact, the limit of the series in (1) as Δw approaches zero is not the definition of the integral (3). Sufficient conditions for the validity of (5) are as follows.

Theorem 1 (Fourier integral)
If $f(x)$ is piecewise continuous (cf. Sec. 5.1) in every finite interval and has a right-hand derivative and a left-hand derivative at every point (cf. Sec. 10.2) and the integral (2) exists, then $f(x)$ can be represented by a Fourier integral. At a point where $f(x)$ is discontinuous the value of the Fourier integral equals the average of the left- and right-hand limits of $f(x)$ at that point (cf. Sec. 10.2). (Proof in Ref. [D2]; cf. Appendix 1.)

Example 3. Single pulse, sine integral
Find the Fourier integral representation of the function (Fig. 227)

$$f(x) = \begin{cases} 1 & \text{when} \quad |x| < 1, \\ 0 & \text{when} \quad |x| > 1. \end{cases}$$

From (4) we obtain

$$A(w) = \int_{-\infty}^{\infty} f(v) \cos wv \, dv = \int_{-1}^{1} \cos wv \, dv = \frac{\sin wv}{w}\Big|_{-1}^{1} = \frac{2 \sin w}{w},$$

$$B(w) = \int_{-1}^{1} \sin wv \, dv = 0,$$

and (5) becomes

$$f(x) = \frac{2}{\pi} \int_0^{\infty} \frac{\cos wx \sin w}{w} \, dw. \tag{6}$$

The average of the left- and right-hand limits of $f(x)$ at $x = 1$ is equal to $(1 + 0)/2$, that is, $1/2$. Hence, from (6) and Theorem 1 we obtain the desired answer

$$\int_0^{\infty} \frac{\cos wx \sin w}{w} \, dw = \begin{cases} \pi/2 & \text{when} \quad 0 \leqq x < 1, \\ \pi/4 & \text{when} \quad x = 1, \\ 0 & \text{when} \quad x > 1. \end{cases}$$

We mention that this integral is called **Dirichlet's[7] discontinuous factor.** Let us consider the case $x = 0$, which is of particular interest. When $x = 0$, then

$$\int_0^{\infty} \frac{\sin w}{w} \, dw = \frac{\pi}{2}. \tag{7}$$

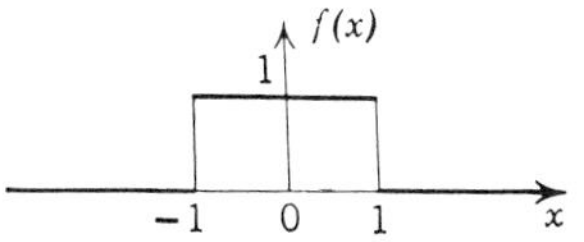

Fig. 227. Example 3

[7] PETER GUSTAV LEJEUNE DIRICHLET (1805–1859), German mathematician, known by his important research work on Fourier series and in number theory.

We see that this integral is the limit of the so-called **sine integral**

$$\text{Si}(z) = \int_0^z \frac{\sin w}{w}\, dw \tag{8}$$

as $z \to \infty$ (z real). The graph of Si(z) is shown in Fig. 228.

In the case of a Fourier series the graphs of the partial sums are approximation curves of the curve of the periodic function represented by the series. Similarly, in the case of the Fourier integral (5), approximations are obtained by replacing ∞ by numbers a. Hence the integral

$$\int_0^a \frac{\cos wx \sin w}{w}\, dw \tag{9}$$

approximates the integral in (6) and therefore $f(x)$. Figure 229 shows oscillations near the points of discontinuity of $f(x)$.

We might expect that these oscillations disappear as a approaches infinity, but this is not true; with increasing a, they are shifted closer to the points $x = \pm 1$. This unexpected behavior, which also occurs in connection with Fourier series, is known as the **Gibbs**[8] **phenomenon.** It can be explained by representing (9) in terms of the sine integral as follows. Using (11) in Appendix 3, we first have

$$\frac{2}{\pi}\int_0^a \frac{\cos wx \sin w}{w}\, dw = \frac{1}{\pi}\int_0^a \frac{\sin (w + wx)}{w}\, dw + \frac{1}{\pi}\int_0^a \frac{\sin (w - wx)}{w}\, dw.$$

In the first integral on the right we set $w + wx = t$. Then $dw/w = dt/t$, and $0 \leqq w \leqq a$ corresponds to $0 \leqq t \leqq (x + 1)a$. In the last integral we set $w - wx = -t$. Then $dw/w = dt/t$, and the interval $0 \leqq w \leqq a$ corresponds to $0 \leqq t \leqq (x - 1)a$. Since $\sin(-t) = -\sin t$, we thus obtain

$$\frac{2}{\pi}\int_0^a \frac{\cos wx \sin w}{w}\, dw = \frac{1}{\pi}\int_0^{(x+1)a} \frac{\sin t}{t}\, dt - \frac{1}{\pi}\int_0^{(x-1)a} \frac{\sin t}{t}\, dt.$$

From this and (8) we see that our integral equals

$$\frac{1}{\pi}\,\text{Si}(a[x + 1]) - \frac{1}{\pi}\,\text{Si}(a[x - 1]),$$

and the oscillations in Fig. 229 result from those in Fig. 228. The increase of a amounts to a transformation of the scale on the axis and causes the shift of the oscillations. For further details see Ref. [D1] in Appendix 1. ∎

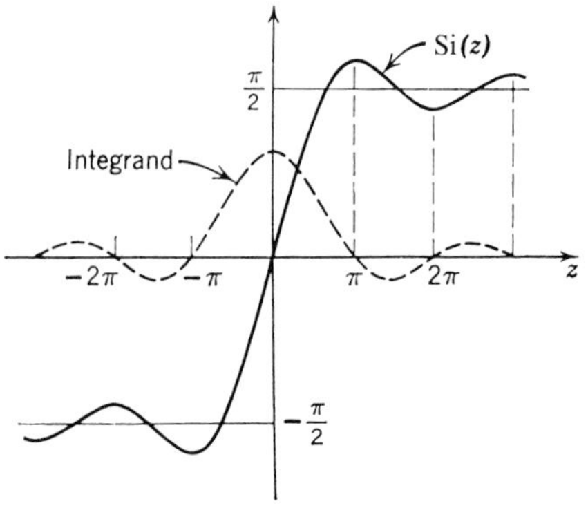

Fig. 228. Sine integral

[8] JOSIAH WILLARD GIBBS (1839–1903), American mathematician, whose work was of great importance to the development of vector analysis and mathematical physics.

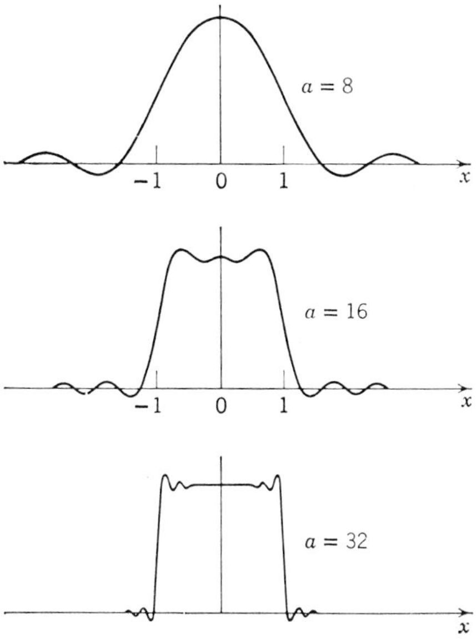

Fig. 229. The integral (9) for $a = 8$, 16 and 32

If $f(x)$ is an even function, then $B(w) = 0$ in (4),

$$A(w) = 2 \int_0^\infty f(v) \cos wv \, dv, \tag{10}$$

and (5) reduces to the simpler form

$$f(x) = \frac{1}{\pi} \int_0^\infty A(w) \cos wx \, dw \qquad (f \text{ even}). \tag{11}$$

If $f(x)$ is odd, then $A(w) = 0$,

$$B(w) = 2 \int_0^\infty f(v) \sin wv \, dv, \tag{12}$$

and (5) becomes

$$f(x) = \frac{1}{\pi} \int_0^\infty B(w) \sin wx \, dw \qquad (f \text{ odd}). \tag{13}$$

These simplifications are quite similar to those in the case of a Fourier series discussed in Sec. 10.4.

Example 4. Laplace integrals

Find the Fourier integral of

$$f(x) = e^{-kx} \quad \text{when} \quad x > 0 \quad \text{and} \quad f(-x) = f(x) \qquad (k > 0)$$

(cf. Fig. 226 in this section, where $k = 1$). Since f is even, we have from (10)

$$A(w) = 2\int_0^\infty e^{-kv} \cos wv \, dv.$$

Now, by integration by parts,

$$\int e^{-kv} \cos wv \, dv = -\frac{k}{k^2 + w^2} e^{-kv} \left(-\frac{w}{k} \sin wv + \cos wv\right).$$

When $v = 0$, the expression on the right equals $-k/(k^2 + w^2)$; when v approaches infinity, it approaches zero because of the exponential factor. Thus

$$A(w) = \frac{2k}{k^2 + w^2},$$

and by substituting this in (11) we obtain the representation

$$f(x) = e^{-kx} = \frac{2k}{\pi}\int_0^\infty \frac{\cos wx}{k^2 + w^2}\, dw \qquad (x > 0, k > 0).$$

From this representation we see that

$$\text{(14)} \qquad \int_0^\infty \frac{\cos wx}{k^2 + w^2}\, dw = \frac{\pi}{2k} e^{-kx} \qquad (x > 0, k > 0).$$

Similarly, from the Fourier integral (13) of the odd function

$$f(x) = e^{-kx} \quad \text{when} \quad x > 0 \quad \text{and} \quad f(-x) = -f(x), \qquad (k > 0)$$

we obtain the result

$$\text{(15)} \qquad \int_0^\infty \frac{w \sin wx}{k^2 + w^2}\, dw = \frac{\pi}{2} e^{-kx} \qquad (x > 0, k > 0).$$

The integrals (14) and (15) are the so-called *Laplace integrals*.

This example illustrates that the Fourier integral representation may be used for evaluating integrals.

Problems for Sec. 10.9

Using the Fourier integral representation, show that

1. $$\int_0^\infty \frac{\cos xw + w \sin xw}{1 + w^2}\, dw = \begin{cases} 0 & \text{when } x < 0 \\ \pi/2 & \text{when } x = 0 \\ \pi e^{-x} & \text{when } x > 0 \end{cases} \qquad \text{[Use (5).]}$$

2. $$\int_0^\infty \frac{1 - \cos \pi w}{w} \sin xw \, dw = \begin{cases} \dfrac{\pi}{2} & \text{when } 0 < x < \pi \\ 0 & \text{when } x > \pi \end{cases} \qquad \text{[Use (13).]}$$

3. $$\int_0^\infty \frac{\cos xw}{1 + w^2}\, dw = \frac{\pi}{2} e^{-x} \quad (x > 0) \qquad \text{[Use (11).]}$$

4. $$\int_0^\infty \frac{\sin w \cos xw}{w}\, dw = \begin{cases} \pi/2 & \text{when } 0 \leqq x < 1 \\ \pi/4 & \text{when } x = 1 \\ 0 & \text{when } x > 1 \end{cases} \qquad \text{[Use (11).]}$$

5. $\displaystyle\int_0^\infty \frac{\cos(\pi w/2)\cos xw}{1-w^2}\,dw = \begin{cases} \frac{\pi}{2}\cos x & \text{when } |x| < \frac{\pi}{2} \\ 0 & \text{when } |x| > \frac{\pi}{2} \end{cases}$ [Use (11).]

6. $\displaystyle\int_0^\infty \frac{w^3 \sin xw}{w^4+4}\,dw = \frac{\pi}{2}e^{-x}\cos x \quad \text{when } x > 0$ [Use (13).]

7. $\displaystyle\int_0^\infty \frac{\sin \pi w \sin xw}{1-w^2}\,dw = \begin{cases} \frac{\pi}{2}\sin x & \text{when } 0 \leqq x \leqq \pi \\ 0 & \text{when } x > \pi \end{cases}$ [Use (13).]

Represent the following functions $f(x)$ in the form (11).

8. $f(x) = \begin{cases} 1 & \text{when } 0 < x < a \\ 0 & \text{when } x > a \end{cases}$

9. $f(x) = \begin{cases} x & \text{when } 0 < x < a \\ 0 & \text{when } x > a \end{cases}$

10. $f(x) = \dfrac{1}{1+x^2}$ [cf. (14)]

11. $f(x) = e^{-x} + e^{-2x}$ $(x > 0)$

12. $f(x) = \begin{cases} x & \text{when } 0 < x < 1 \\ 2-x & \text{when } 1 < x < 2 \\ 0 & \text{when } x > 2 \end{cases}$

13. $f(x) = \begin{cases} x^2 & \text{when } 0 < x < a \\ 0 & \text{when } x > a \end{cases}$

14. Show that $f(x) = 1$ $(0 < x < \infty)$ cannot be represented by a Fourier integral.

If $f(x)$ has the representation (11), show that

15. $f(ax) = \dfrac{1}{\pi a}\displaystyle\int_0^\infty A\left(\frac{w}{a}\right)\cos xw\,dw \quad (a > 0)$

16. $x^2 f(x) = \dfrac{1}{\pi}\displaystyle\int_0^\infty A^*(w)\cos xw\,dw, \qquad A^* = -\frac{d^2A}{dw^2}$

17. Solve Prob. 13 by applying the formula in Prob. 16 to the result of Prob. 8.

18. Show that $xf(x) = \dfrac{1}{\pi}\displaystyle\int_0^\infty B^*(w)\sin xw\,dw$ where $B^* = -\dfrac{dA}{dw}$ and A is given by formula (10).

19. Verify the formula in Prob. 18 for $f(x) = 1$ when $0 < x < a$ and $f(x) = 0$ when $x > a$.

20. **(Complex form of the Fourier integral, Fourier transform)** Using (6) in Appendix 3, show that (5) can be written

$$f(x) = \frac{1}{\pi}\int_0^\infty \left[\int_{-\infty}^\infty f(v)\cos(wx - wv)\,dv\right]dw. \tag{16}$$

Show that the integral from $-\infty$ to ∞ in (16) is an even function of w and (16) may be written

$$f(x) = \frac{1}{2\pi}\int_{-\infty}^\infty \left[\int_{-\infty}^\infty f(v)\cos(wx - wv)\,dv\right]dw.$$

Show that

$$\frac{i}{2\pi}\int_{-\infty}^\infty \left[\int_{-\infty}^\infty f(v)\sin(wx - wv)\,dv\right]dw = 0, \tag{17}$$

so that by addition,

$$f(x) = \frac{1}{2\pi}\int_{-\infty}^{\infty}\left[\int_{-\infty}^{\infty} f(v)\, e^{iw(x-v)}\, dv\right] dw. \tag{18}$$

This is the so-called *complex form* of the Fourier integral. Show that (18) can be written

$$f(x) = \frac{1}{\sqrt{2\pi}}\int_{-\infty}^{\infty} C(w)e^{iwx}\, dw \quad \text{where} \quad C(w) = \frac{1}{\sqrt{2\pi}}\int_{-\infty}^{\infty} f(v)e^{-iwv}\, dv. \tag{19}$$

$f(x)$ is called the *Fourier transform* of $C(w)$, and $C(w)$ is called the *inverse Fourier transform* of $f(x)$. (Elaborate tables of Fourier transforms are contained in [B7]; cf. Appendix 1.)

CHAPTER

11 PARTIAL DIFFERENTIAL EQUATIONS

Partial differential equations arise in connection with various physical and geometrical problems when the functions involved depend on two or more independent variables. These variables may be the time and one or several coordinates in space. The present chapter will be devoted to some of the most important partial differential equations occurring in engineering applications. We shall derive these equations from physical principles and consider methods for solving initial and boundary value problems, that is, methods for obtaining solutions of those equations corresponding to the given physical situations.

In Sec. 11.1 we shall define the notion of a solution of a partial differential equation. Sections 11.2–11.4 will be devoted to the one-dimensional wave equation, governing the motion of a vibrating string. The heat equation will be considered in Secs. 11.5 and 11.6, the two-dimensional wave equation (vibrating membranes) in Secs. 11.7–11.10, and Laplace's equation in Secs. 11.11 and 11.12.

In Sec. 11.13 we shall see that partial differential equations can also be solved by the Laplace transformation introduced in Chap. 5.

Prerequisites for this chapter: ordinary linear differential equations (Chap. 2) and Fourier series (Chap. 10).
Sections which may be omitted in a shorter course: 11.6, 11.9, 11.10.
References: Appendix 1, Part E.
Answers to problems: Appendix 2.

11.1 Basic Concepts

An equation involving one or more partial derivatives of an (unknown) function of two or more independent variables is called a **partial differential equation.** The order of the highest derivative is called the **order** of the equation.

Just as in the case of an ordinary differential equation, we say that a partial differential equation is **linear** if it is of the first degree in the dependent variable and its partial derivatives. If each term of such an equation contains either the dependent variable or one of its derivatives, the equation is said to be **homogeneous;** otherwise it is said to be **nonhomogeneous.**

Example 1. Important linear partial differential equations of the second order

(1) $$\frac{\partial^2 u}{\partial t^2} = c^2 \frac{\partial^2 u}{\partial x^2}$$ *One-dimensional wave equation*

(2) $$\frac{\partial u}{\partial t} = c^2 \frac{\partial^2 u}{\partial x^2}$$ *One-dimensional heat equation*

(3) $$\frac{\partial^2 u}{\partial x^2} + \frac{\partial^2 u}{\partial y^2} = 0$$ *Two-dimensional Laplace equation*

(4) $$\frac{\partial^2 u}{\partial x^2} + \frac{\partial^2 u}{\partial y^2} = f(x, y)$$ *Two-dimensional Poisson equation*

(5) $$\frac{\partial^2 u}{\partial x^2} + \frac{\partial^2 u}{\partial y^2} + \frac{\partial^2 u}{\partial z^2} = 0$$ *Three-dimensional Laplace equation*

Here c is a constant, t is the time, and x, y, z are Cartesian coordinates. Equation (4) (with $f \not\equiv 0$) is nonhomogeneous, while the other equations are homogeneous. ▮

A **solution** *of a partial differential equation in some region R of the space of the independent variables* is a function which has all the partial derivatives appearing in the equation in some domain containing R, and satisfies the equation everywhere in R. (Often one merely requires that that function is continuous on the boundary of R, has those derivatives in the interior of R, and satisfies the equation in the interior of R.)

In general, the totality of solutions of a partial differential equation is very large. For example, the functions

(6) $$u = x^2 - y^2, \qquad u = e^x \cos y, \qquad u = \ln(x^2 + y^2),$$

which are entirely different from each other, are solutions of (3), as the student may verify. We shall see later that the unique solution of a partial differential equation corresponding to a given physical problem will be obtained by the use of additional information arising from the physical situation. For example, in some cases the values of the required solution of the problem on the boundary of some domain will be given ("**boundary conditions**"); in other cases when the time t is one of the variables, the values of the solution at $t = 0$ will be prescribed ("**initial conditions**").

We know that if an *ordinary* differential equation is linear and homogeneous, then from known solutions we can obtain further solutions by superposition. For a homogeneous linear *partial* differential equation the situation is quite similar. In fact, the following theorem holds.

Fundamental Theorem 1
(Linear homogeneous partial differential equations)
If u_1 and u_2 are any solutions of a linear homogeneous partial differential equation in some region, then

$$u = c_1 u_1 + c_2 u_2,$$

where c_1 and c_2 are any constants, is also a solution of that equation in that region.

The proof of this important theorem is quite similar to that of Theorem 1 in Sec. 2.1 and is left to the student.

Problems for Sec. 11.1

Verify that the following functions are solutions of the wave equation (1) for a suitable value of the constant c in (1). Sketch figures of these functions as surfaces in space.

1. $u = x^2 + t^2$ **2.** $u = x^2 + 9t^2$ **3.** $u = \cos t \sin x$

4. $u = \sin t \sin x$ **5.** $u = \cos ct \sin x$ **6.** $u = \sin \omega ct \sin \omega x$

Verify that the following functions (with arbitrary twice differentiable v and w) are solutions of (1), where $c = 1$ in Prob. 7.

7. $u(x, t) = v(x + t) + w(x - t)$ **8.** $u(x, t) = v(x + ct) + w(x - ct)$

Verify that the following functions are solutions of the heat equation (2).

9. $u = e^{-t} \cos x$ **10.** $u = e^{-t} \sin x$ **11.** $u = e^{-\omega^2 c^2 t} \sin \omega x$

Verify that the following functions are solutions of Laplace's equation (3). Sketch figures of the solutions as surfaces in space.

12. $u = x^2 - y^2$ **13.** $u = x^3 - 3xy^2$ **14.** $u = 3x^2y - y^3$

15. $u = e^x \cos y$ **16.** $u = e^x \sin y$ **17.** $u = \ln (x^2 + y^2)$

18. $u = \arctan (y/x)$ **19.** $u = \sin x \sinh y$ **20.** $u = \sin x \cosh y$

21. Verify that $u(x, y) = a \ln (x^2 + y^2) + b$ satisfies Laplace's equation (3) and determine a and b so that u satisfies the boundary conditions $u = 0$ on the circle $x^2 + y^2 = 1$ and $u = 3$ on the circle $x^2 + y^2 = 4$. Sketch a figure of the surface represented by this function.

22. Show that $u = 1/\sqrt{x^2 + y^2 + z^2}$ is a solution of (5).

Relations to ordinary differential equations

If a partial differential equation involves derivatives with respect to one of the independent variables only, we may solve it like an ordinary differential equation, treating the other independent variables as parameters. Solve the following equations where $u = u(x, y)$.

23. $u_x = 0$ **24.** $u_y = 0$ **25.** $u_{xx} = 0$ **26.** $u_{xx} + u = 0$

Solve the following systems of partial differential equations.

27. (*a*) $u_{xx} = 0$, (*b*) $u_{yy} = 0$ **28.** $u_x = 0, \quad u_y = 0$

29. $u_{xx} = 0, \quad u_{xy} = 0, \quad u_{yy} = 0$ **30.** $u_{xx} = 0, \quad u_{xy} = 0$

Setting $u_x = p$, solve

31. $u_{xy} = u_x$ **32.** $u_{xy} + u_x = 0$ **33.** $u_{xy} + u_x + x + y + 1 = 0$

34. Show that if the level curves $z = const$ of a surface $z = z(x, y)$ are straight lines parallel to the x-axis, then z is a solution of the differential equation $z_x = 0$. Give examples.

35. Show that the solutions $z = z(x, y)$ of $yz_x - xz_y = 0$ represent surfaces of revolution. Give examples. *Hint.* Set $x = r \cos \theta$, $y = r \sin \theta$ and show that the equation becomes $z_\theta = 0$.

11.2 Modeling: Vibrating String. One-Dimensional Wave Equation

As a first important partial differential equation, let us derive the equation governing small transverse vibrations of an elastic string, which is stretched to length l and then fixed at the endpoints. Suppose that the string is distorted and then at a certain instant, say, $t = 0$, it is released and allowed to vibrate. The problem is to determine the vibrations of the string, that is, to find its deflection $u(x, t)$ at any point x and at any time $t > 0$; cf. Fig. 230.

When deriving a differential equation corresponding to a given physical problem, we usually have to make simplifying assumptions in order that the resulting equation does not become too complicated. We know this important fact from our considerations of ordinary differential equations, and for partial differential equations the situation is similar.

In our present case we make the following assumptions.

1. *The mass of the string per unit length is constant ("homogeneous string"). The string is perfectly elastic and does not offer any resistance to bending.*

2. *The tension caused by stretching the string before fixing it at the endpoints is so large that the action of the gravitational force on the string can be neglected.*

3. *The motion of the string is a small transverse vibration in a vertical plane, that is, each particle of the string moves strictly vertically, and the deflection and the slope at any point of the string are small in absolute value.*

These assumptions are such that we may expect that the solution $u(x, t)$ of the differential equation to be obtained will reasonably well describe small vibrations of the physical "nonidealized" string of small homogeneous mass under large tension.

To obtain the differential equation we consider the forces acting on a small portion of the string (Fig. 230). Since the string does not offer resistance to bending, the tension is tangential to the curve of the string at each point. Let T_1 and T_2 be the tensions at the endpoints P and Q of that portion. Since there is no motion in horizontal direction, the horizontal components of the tension must be constant. Using the notation shown in Fig. 230 we thus obtain

$$T_1 \cos \alpha = T_2 \cos \beta = T = const. \tag{1}$$

In vertical direction we have two forces, namely the vertical components $-T_1 \sin \alpha$ and $T_2 \sin \beta$ of T_1 and T_2; here the minus sign appears because that component at P is directed downward. By Newton's second law the resultant of those two forces is equal to the mass $\rho \, \Delta x$ of the portion times the acceleration $\partial^2 u/\partial t^2$, evaluated at some point between x and $x + \Delta x$; here ρ is the mass of the undeflected string per unit length, and Δx is the length of the portion of the undeflected string. Hence

$$T_2 \sin \beta - T_1 \sin \alpha = \rho \, \Delta x \frac{\partial^2 u}{\partial t^2}.$$

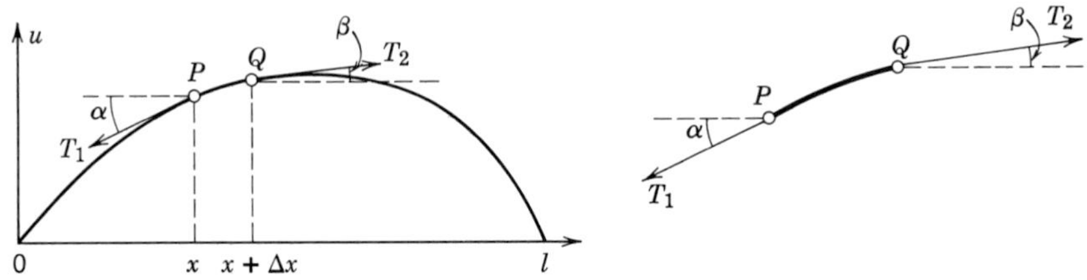

Fig. 230. Vibrating string

By using (1) we obtain

$$\frac{T_2 \sin \beta}{T_2 \cos \beta} - \frac{T_1 \sin \alpha}{T_1 \cos \alpha} = \tan \beta - \tan \alpha = \frac{\rho\, \Delta x}{T} \frac{\partial^2 u}{\partial t^2}. \tag{2}$$

Now $\tan \alpha$ and $\tan \beta$ are the slopes of the curve of the string at x and $x + \Delta x$, that is,

$$\tan \alpha = \left(\frac{\partial u}{\partial x}\right)_x \qquad \text{and} \qquad \tan \beta = \left(\frac{\partial u}{\partial x}\right)_{x+\Delta x}.$$

Here we have to write *partial* derivatives because u depends also on t. Dividing (2) by Δx, we thus have

$$\frac{1}{\Delta x}\left[\left(\frac{\partial u}{\partial x}\right)_{x+\Delta x} - \left(\frac{\partial u}{\partial x}\right)_x\right] = \frac{\rho}{T} \frac{\partial^2 u}{\partial t^2}.$$

If we let Δx approach zero, we obtain the linear homogeneous partial differential equation

$$\frac{\partial^2 u}{\partial t^2} = c^2 \frac{\partial^2 u}{\partial x^2} \qquad c^2 = \frac{T}{\rho}. \tag{3}$$

This is the so-called **one-dimensional wave equation,** which governs our problem. The notation c^2 (instead of c) for the physical constant T/ρ has been chosen to indicate that this constant is positive.

Solutions of the equation will be obtained in the following section.

11.3 Separation of Variables (Product Method)

We have seen that the vibrations of an elastic string are governed by the one-dimensional wave equation

$$\frac{\partial^2 u}{\partial t^2} = c^2 \frac{\partial^2 u}{\partial x^2}, \tag{1}$$

where $u(x, t)$ is the deflection of the string. Since the string is fixed at the ends $x = 0$ and $x = l$, we have the two **boundary conditions**

$$u(0, t) = 0, \qquad u(l, t) = 0 \qquad \text{for all } t. \tag{2}$$

The form of the motion of the string will depend on the initial deflection (deflection at $t = 0$) and on the initial velocity (velocity at $t = 0$). Denoting the initial deflection by $f(x)$ and the initial velocity by $g(x)$, we thus obtain the two **initial conditions**

$$u(x, 0) = f(x) \tag{3}$$

and

$$\left.\frac{\partial u}{\partial t}\right|_{t=0} = g(x). \tag{4}$$

Our problem is now to find a solution of (1) satisfying the conditions (2)–(4). We shall proceed step by step, as follows.

First step. By applying the so-called *product method,* or *method of separating variables,* we shall obtain two ordinary differential equations.

Second step. We shall determine solutions of those equations that satisfy the boundary conditions.

Third step. Those solutions will be composed so that the result will be a solution of the wave equation (1), satisfying also the given initial conditions.

The details are as follows.

First step. The product method yields solutions of (1) of the form

$$u(x, t) = F(x)G(t) \tag{5}$$

which are a product of two functions, each depending only on one of the variables x and t. We shall see later that this method has various applications in engineering mathematics. By differentiating (5) we obtain

$$\frac{\partial^2 u}{\partial t^2} = F\ddot{G} \qquad \text{and} \qquad \frac{\partial^2 u}{\partial x^2} = F''G,$$

where dots denote derivatives with respect to t and primes derivatives with respect to x. By inserting this into (1) we have

$$F\ddot{G} = c^2 F''G.$$

Dividing by c^2FG, we find

$$\frac{\ddot{G}}{c^2 G} = \frac{F''}{F}.$$

The expression on the left involves functions depending only on t while the expression on the right involves functions depending only on x. Hence both expressions must be equal to a constant, say, k, because if the expression on the left is not constant, then changing t will presumably change the value of this expression but certainly not that on the right, since the latter does not depend on t. Similarly, if the expression on the right is not constant, changing x will presumably change the value of this expression but certainly not that on the left. Thus

$$\frac{\ddot{G}}{c^2 G} = \frac{F''}{F} = k.$$

This yields immediately the two ordinary linear differential equations

$$F'' - kF = 0 \tag{6}$$

and

$$\ddot{G} - c^2kG = 0. \tag{7}$$

In these equations, k is still arbitrary.

Second step. We shall now determine solutions F and G of (6) and (7) so that $u = FG$ satisfies (2), that is,

$$u(0, t) = F(0)G(t) = 0, \qquad u(l, t) = F(l)G(t) = 0 \qquad \text{for all } t.$$

Clearly, if $G \equiv 0$, then $u \equiv 0$, which is of no interest. Thus $G \not\equiv 0$ and then

$$\text{(a) } F(0) = 0 \qquad \text{(b) } F(l) = 0. \tag{8}$$

For $k = 0$ the general solution of (6) is $F = ax + b$, and from (8) we obtain $a = b = 0$. Hence $F \equiv 0$, which is of no interest because then $u \equiv 0$. For positive $k = \mu^2$ the general solution of (6) is

$$F = Ae^{\mu x} + Be^{-\mu x},$$

and from (8) we obtain $F \equiv 0$, as before. Hence we are left with the possibility of choosing k negative, say, $k = -p^2$. Then (6) takes the form

$$F'' + p^2F = 0,$$

and the general solution is

$$F(x) = A \cos px + B \sin px.$$

From this and (8) we have

$$F(0) = A = 0 \qquad \text{and then} \qquad F(l) = B \sin pl = 0.$$

We must take $B \neq 0$ since otherwise $F \equiv 0$. Hence $\sin pl = 0$, that is,

$$pl = n\pi \qquad \text{or} \qquad p = \frac{n\pi}{l} \qquad (n \text{ integral}). \tag{9}$$

Setting $B = 1$, we thus obtain infinitely many solutions $F(x) = F_n(x)$,

$$F_n(x) = \sin \frac{n\pi}{l} x \qquad n = 1, 2, \cdots, \tag{10}$$

which satisfy (8). [For negative integral n we obtain essentially the same solutions, except for a minus sign, because $\sin(-\alpha) = -\sin \alpha$.]

k is now restricted to the values $k = -p^2 = -(n\pi/l)^2$, resulting from (9). For these k the equation (7) takes the form

$$\ddot{G} + \lambda_n{}^2 G = 0 \qquad \text{where} \qquad \lambda_n = \frac{cn\pi}{l}.$$

The general solution is

$$G_n(t) = B_n \cos \lambda_n t + B_n{}^* \sin \lambda_n t.$$

Hence the functions $u_n(x, t) = F_n(x)G_n(t)$, written out

$$\textbf{(11)} \qquad u_n(x, t) = (B_n \cos \lambda_n t + B_n{}^* \sin \lambda_n t) \sin \frac{n\pi}{l} x \qquad (n = 1, 2, \cdots),$$

are solutions of (1), satisfying the boundary conditions (2). These functions are called the **eigenfunctions,** or *characteristic functions,* and the values $\lambda_n = cn\pi/l$ are called the **eigenvalues,** or *characteristic values,* of the vibrating string. The set $\{\lambda_1, \lambda_2, \cdots\}$ is called the *spectrum.*

We see that each u_n represents a harmonic motion having the frequency $\lambda_n/2\pi = cn/2l$ cycles per unit time. The motion is called the nth **normal mode** of the string. The first normal mode is known as the *fundamental mode* ($n = 1$), and the others are known as *overtones;* musically they give the octave, octave plus fifth, etc. Since in (11)

$$\sin \frac{n\pi x}{l} = 0 \qquad \text{at} \qquad x = \frac{l}{n}, \frac{2l}{n}, \cdots, \frac{n-1}{n} l,$$

the nth normal mode has $n - 1$ so-called **nodes,** that is, points of the string which do not move (Fig. 231).

Figure 232 shows the second normal mode for various values of t. At any instant the string has the form of a sine wave. When the left part of the string is moving downward the other half is moving upward, and conversely. For the other modes the situation is similar.

Third step. Clearly, a single solution $u_n(x, t)$ will, in general, not satisfy the initial conditions (3) and (4). Now, since the equation (1) is linear and homogeneous, it follows from Fundamental Theorem 1 in Sec. 11.1 that the sum of finitely many solutions u_n is a solution of (1). To obtain a solution that satisfies (3) and (4), we consider the infinite series

$$\textbf{(12)} \qquad u(x, t) = \sum_{n=1}^{\infty} u_n(x, t) = \sum_{n=1}^{\infty} (B_n \cos \lambda_n t + B_n{}^* \sin \lambda_n t) \sin \frac{n\pi}{l} x.$$

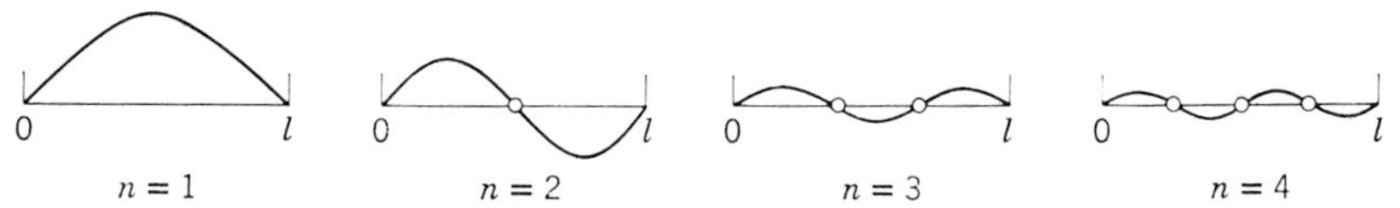

Fig. 231. Normal modes of the vibrating string

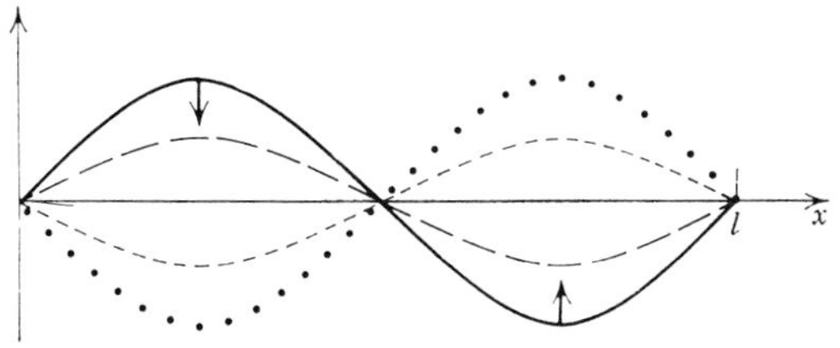

Fig. 232. Second normal mode for various values of t

From this and (3) it follows that

$$u(x, 0) = \sum_{n=1}^{\infty} B_n \sin \frac{n\pi}{l} x = f(x). \tag{13}$$

Hence, in order that (12) satisfy (3), the coefficients B_n must be chosen so that $u(x, 0)$ becomes a half-range expansion of $f(x)$, namely, the Fourier sine series of $f(x)$; that is [cf. (4) in Sec. 10.5]

$$B_n = \frac{2}{l} \int_0^l f(x) \sin \frac{n\pi x}{l} dx, \qquad n = 1, 2, \cdots. \tag{14}$$

Similarly, by differentiating (12) with respect to t and using (4) we find

$$\begin{aligned} \left.\frac{\partial u}{\partial t}\right|_{t=0} &= \left[\sum_{n=1}^{\infty} (-B_n\lambda_n \sin \lambda_n t + B_n{}^*\lambda_n \cos \lambda_n t) \sin \frac{n\pi x}{l}\right]_{t=0} \\ &= \sum_{n=1}^{\infty} B_n{}^*\lambda_n \sin \frac{n\pi x}{l} = g(x). \end{aligned}$$

Hence, in order that (12) satisfy (4), the coefficients $B_n{}^*$ must be chosen so that, for $t = 0$, $\partial u/\partial t$ becomes the Fourier sine series of $g(x)$; thus, by (4) in Sec. 10.5,

$$B_n{}^*\lambda_n = \frac{2}{l} \int_0^l g(x) \sin \frac{n\pi x}{l} dx$$

or, since $\lambda_n = cn\pi/l$,

$$B_n{}^* = \frac{2}{cn\pi} \int_0^l g(x) \sin \frac{n\pi x}{l} dx, \qquad n = 1, 2, \cdots. \tag{15}$$

It follows that $u(x, t)$, given by (12) with coefficients (14) and (15), is a solution of (1) that satisfies the conditions (2)–(4), provided that the series (12) converges and also that the series obtained by differentiating (12) twice (termwise) with respect to x and t, converge and have the sums $\partial^2 u/\partial x^2$ and $\partial^2 u/\partial t^2$, respectively, which are continuous.

Fig. 233. Odd periodic extension of $f(x)$

Hence the solution (12) is at first a purely formal expression, and we shall now establish it. For the sake of simplicity we consider only the case when the initial velocity $g(x)$ is identically zero. Then the B_n^* are zero, and (12) reduces to the form

$$u(x, t) = \sum_{n=1}^{\infty} B_n \cos \lambda_n t \sin \frac{n\pi x}{l}, \qquad \lambda_n = \frac{cn\pi}{l}. \tag{16}$$

It is possible to *sum* this series, that is, to write the result in a closed or finite form. We have [cf. (11) in Appendix 3]

$$\cos \frac{cn\pi}{l} t \sin \frac{n\pi}{l} x = \frac{1}{2}\left[\sin\left\{\frac{n\pi}{l}(x - ct)\right\} + \sin\left\{\frac{n\pi}{l}(x + ct)\right\}\right].$$

Consequently, we may write (16) in the form

$$u(x, t) = \frac{1}{2}\sum_{n=1}^{\infty} B_n \sin\left\{\frac{n\pi}{l}(x - ct)\right\} + \frac{1}{2}\sum_{n=1}^{\infty} B_n \sin\left\{\frac{n\pi}{l}(x + ct)\right\}.$$

These two series are those obtained by substituting $x - ct$ and $x + ct$, respectively, for the variable x in the Fourier sine series (13) for $f(x)$. Therefore,

$$u(x, t) = \tfrac{1}{2}[f^*(x - ct) + f^*(x + ct)], \tag{17}$$

where f^* is the odd periodic extension of f with the period $2l$ (Fig. 233). Since the initial deflection $f(x)$ is continuous on the interval $0 \leqq x \leqq l$ and zero at the endpoints, it follows from (17) that $u(x, t)$ is a continuous function of both variables x and t for all values of the variables. By differentiating (17) we see that $u(x, t)$ is a solution of (1), provided $f(x)$ is twice differentiable on the interval $0 < x < l$, and has one-sided second derivatives at $x = 0$ and $x = l$, which are zero. Under these conditions $u(x, t)$ is established as a solution of (1), satisfying (2)–(4).

If $f'(x)$ and $f''(x)$ are merely piecewise continuous (cf. Sec. 5.1), or if those one-sided derivatives are not zero, then for each t there will be finitely many values of x at which the second derivatives of u appearing in (1) do not exist.

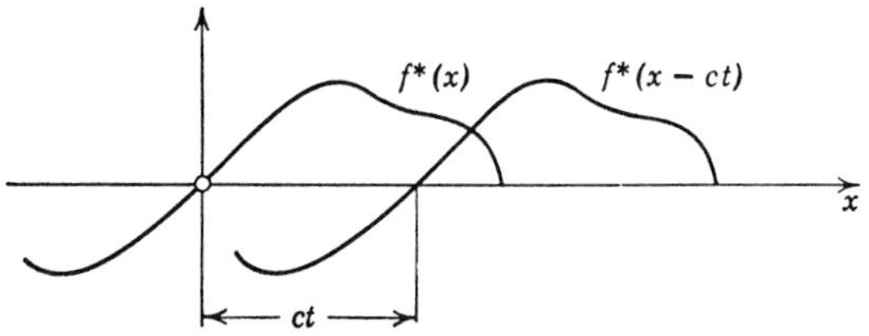

Fig. 234. Interpretation of (17)

Except at these points the wave equation will still be satisfied, and we may then regard $u(x, t)$ as a solution of our problem in a broader sense. For example, the case of a triangular initial deflection (Example 1, below) leads to a solution of this type.

Let us mention a very interesting physical interpretation of (17). The graph of $f^*(x - ct)$ is obtained from the graph of $f^*(x)$ by shifting the latter ct units to the right (Fig. 234). This means that $f^*(x - ct)$ $(c > 0)$ represents a wave which is traveling to the right as t increases. Similarly, $f^*(x + ct)$ represents a wave which is traveling to the left, and $u(x, t)$ is the superposition of these two waves.

Example 1. Vibrating string if the initial deflection is triangular

Find the solution of the wave equation (1) corresponding to the triangular initial deflection

$$f(x) = \begin{cases} \dfrac{2k}{l}x & \text{when} \quad 0 < x < \dfrac{l}{2} \\ \dfrac{2k}{l}(l - x) & \text{when} \quad \dfrac{l}{2} < x < l \end{cases}$$

and initial velocity zero. Since $g(x) \equiv 0$, we have $B_n^* = 0$ in (12), and from Example 1 in Sec. 10.5

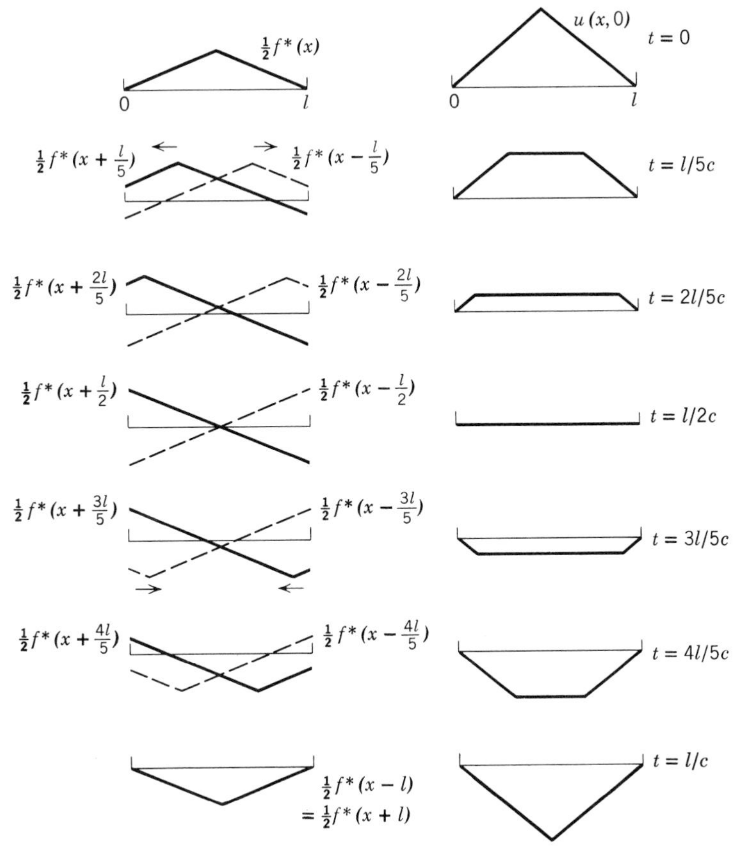

Fig. 235. Solution $u(x, t)$ in Example 1 for various values of t (right part of the figure) as obtained as the superposition of a wave traveling to the right (dashed) and a wave traveling to the left (left part of the figure)

we see that the B_n are given by (5), Sec. 10.5. Thus (12) takes the form

$$u(x, t) = \frac{8k}{\pi^2}\left[\frac{1}{1^2}\sin\frac{\pi}{l}x\cos\frac{\pi c}{l}t - \frac{1}{3^2}\sin\frac{3\pi}{l}x\cos\frac{3\pi c}{l}t + - \cdots\right].$$

For plotting the graph of the solution we may use $u(x, 0) = f(x)$ and the above interpretation of the two functions in the representation (17). This leads to the graph shown in Fig. 235.

Problems for Sec. 11.3

1. In what manner does the frequency of the fundamental mode of the vibrating string depend on the length of the string, the tension, and the mass of the string per unit length?

Find the deflection $u(x, t)$ of the vibrating string (length $l = \pi$, ends fixed, and $c^2 = T/\rho = 1$) corresponding to zero initial velocity and initial deflection:

2. $0.01 \sin x$ **3.** $k \sin 2x$ **4.** $k(\sin x + \sin 3x)$

5.

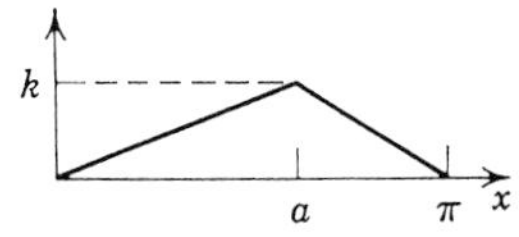

6.

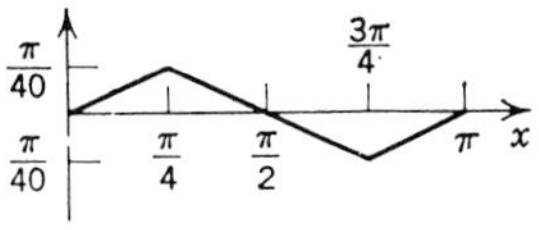

7.

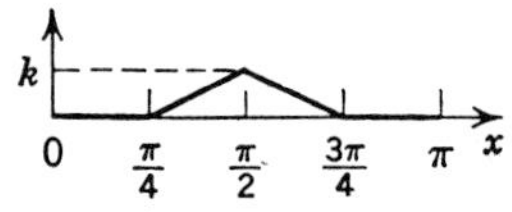

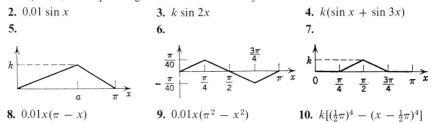

8. $0.01x(\pi - x)$ **9.** $0.01x(\pi^2 - x^2)$ **10.** $k[(\tfrac{1}{2}\pi)^4 - (x - \tfrac{1}{2}\pi)^4]$

11. What is the ratio of the amplitudes of the fundamental mode and the second overtone in Prob. 8? The ratio $a_1^2/(a_1^2 + a_2^2 + \cdots)$? *Hint.* Use (7) in Sec. 10.8, with the equality sign.

Separating variables, find solutions $u(x, y)$ of the following equations.

12. $u_x + u_y = 0$ **13.** $u_x = u_y$ **14.** $u_x = yu_y$

15. $xu_x = yu_y$ **16.** $ayu_x = bxu_y$ **17.** $u_x + u_y = 2(x + y)u$

18. $u_{xy} = u$ **19.** $u_{xx} + u_{yy} = 0$ **20.** $x^2u_{xy} + 3y^2u = 0$

21. Show that substitution of

$$(18) \qquad u(x, t) = \sum_{n=0}^{\infty} G_n(t) \sin\frac{n\pi x}{l}$$

into the wave equation (1) leads to the equation

$$\ddot{G}_n + \lambda_n^2 G = 0, \qquad \lambda_n = \frac{cn\pi}{l}.$$

22. (Forced vibrations of an elastic string) Show that the forced vibrations of an elastic string under an external force $P(x, t)$ per unit length acting normal to the string are governed by the equation

$$u_{tt} = c^2 u_{xx} + \frac{P}{\rho}.$$

Consider and solve the equation in Prob. 22 for the sinusoidal force $P = A\rho \sin \omega t$, as follows.

23. Show that

$$P/\rho = A \sin \omega t = \sum_{n=1}^{\infty} k_n(t) \sin \frac{n\pi x}{l}$$

where $k_n(t) = (2A/n\pi)(1 - \cos n\pi) \sin \omega t$; consequently $k_n = 0$ (n even), and $k_n = (4A/n\pi) \sin \omega t$ (n odd).

24. Show that by substituting the expressions for u in Prob. 21 and P/ρ in Prob. 23 into the equation under consideration we obtain

$$\ddot{G}_n + \lambda_n^2 G_n = \frac{2A}{n\pi}(1 - \cos n\pi) \sin \omega t, \qquad \lambda_n = \frac{cn\pi}{l}.$$

Show that if $\lambda_n^2 \neq \omega^2$, the solution is

$$G_n(t) = B_n \cos \lambda_n t + B_n^* \sin \lambda_n t + \frac{2A(1 - \cos n\pi)}{n\pi(\lambda_n^2 - \omega^2)} \sin \omega t.$$

25. Determine B_n and B_n^* in Prob. 24 so that u satisfies the initial conditions $u(x, 0) = f(x)$, $u_t(x, 0) = 0$.

11.4 D'Alembert's Solution of the Wave Equation

It is interesting to note that the solution (17), Sec. 11.3, of the wave equation

$$\frac{\partial^2 u}{\partial t^2} = c^2 \frac{\partial^2 u}{\partial x^2} \qquad c^2 = \frac{T}{\rho} \tag{1}$$

can be immediately obtained by transforming (1) in a suitable way, namely, by introducing the new independent variables[1]

$$v = x + ct, \qquad z = x - ct. \tag{2}$$

u then becomes a function of v and z, and the derivatives in (1) can be expressed in terms of derivatives with respect to v and z by the use of the chain rule in Sec. 8.7. Denoting partial derivatives by subscripts, we see from (2) that $v_x = 1$ and $z_x = 1$. For simplicity let us denote $u(x, t)$, as a function of v and z, by the same letter u. Then

$$u_x = u_v v_x + u_z z_x = u_v + u_z.$$

By applying the chain rule to the right side and using $v_x = 1$ and $z_x = 1$ we find

$$u_{xx} = (u_v + u_z)_x = (u_v + u_z)_v v_x + (u_v + u_z)_z z_x = u_{vv} + 2u_{vz} + u_{zz}.$$

The other derivative in (1) is transformed by the same procedure, and the result is

$$u_{tt} = c^2(u_{vv} - 2u_{vz} + u_{zz}).$$

[1] We mention that the general theory of partial differential equations provides a systematic way for finding this transformation which will simplify the equation. Cf. Ref. [E14] in Appendix 1.

By inserting these two results in (1) we obtain

$$u_{vz} \equiv \frac{\partial^2 u}{\partial v\, \partial z} = 0. \tag{3}$$

Obviously, the point of the present approach is that the resulting equation (3) can be readily solved by two successive integrations. In fact, integrating with respect to z, we find

$$\frac{\partial u}{\partial v} = h(v)$$

where $h(v)$ is an arbitrary function of v. Integrating this with respect to v, we have

$$u = \int h(v)\, dv + \psi(z)$$

where $\psi(z)$ is an arbitrary function of z. Since the integral is a function of v, say, $\phi(v)$, the solution u is of the form $u = \phi(v) + \psi(z)$. Because of (2) we may write

$$u(x, t) = \phi(x + ct) + \psi(x - ct). \tag{4}$$

This is known as **d'Alembert's solution**[2] of the wave equation (1).

The functions ϕ and ψ can be determined from the initial conditions. Let us illustrate this in the case of zero initial velocity and given initial deflection $u(x, 0) = f(x)$.

By differentiating (4) we have

$$\frac{\partial u}{\partial t} = c\phi'(x + ct) - c\psi'(x - ct) \tag{5}$$

where primes denote derivatives with respect to the *entire* arguments $x + ct$ and $x - ct$, respectively. From (4), (5), and the initial conditions we have

$$u(x, 0) = \phi(x) + \psi(x) = f(x)$$

$$u_t(x, 0) = c\phi'(x) - c\psi'(x) = 0.$$

From the last equation, $\psi' = \phi'$. Hence $\psi = \phi + k$, and from this and the first equation, $2\phi + k = f$ or $\phi = (f - k)/2$. With these functions ϕ and ψ the solution (4) becomes

$$u(x, t) = \tfrac{1}{2}[f(x + ct) + f(x - ct)], \tag{6}$$

in agreement with (17) in Sec. 11.3. The student may show that because of the

[2] JEAN-LE-ROND D'ALEMBERT (1717–1783), French mathematician, who is known for his important work in mechanics.

boundary conditions (2) in that section the function f must be odd and have period $2l$.

Our result shows that the two initial conditions and the boundary conditions determine the solution uniquely.

Problems for Sec. 11.4

1. Express x and t in terms of v and z, cf. (2), and use the result for transforming (3) into (1).

Using (6), sketch a figure (of the type of Fig. 235 in Sec. 11.3) of the deflection $u(x, t)$ of a vibrating string (length $l = 1$, ends fixed) starting with initial velocity zero and the following initial deflection $f(x)$, where k is small, say, $k = 0.01$.

2. $f(x) = k \sin 2\pi x$
3. $f(x) = kx(1 - x)$
4. $f(x) = k(x - x^3)$
5. $f(x) = k(x^2 - x^4)$
6. $f(x) = k(x^3 - x^5)$
7. $f(x) = k \sin^2 \pi x$

Using the indicated transformations, solve the following equations.

8. $xu_{xy} = yu_{yy} + u_y \quad (v = x, z = xy)$
9. $u_{xy} - u_{yy} = 0 \quad (v = x, z = x + y)$
10. $u_{xx} + 2u_{xy} + u_{yy} = 0 \quad (v = x, z = x - y)$
11. $u_{xx} - 2u_{xy} + u_{yy} = 0 \quad (v = x, z = x + y)$
12. $u_{xx} + u_{xy} - 2u_{yy} = 0 \quad (v = x + y, z = 2x - y)$

13. **(Types of linear partial differential equations)** An equation of the form

$$Au_{xx} + 2Bu_{xy} + Cu_{yy} = F(x, y, u, u_x, u_y) \tag{7}$$

is said to be **elliptic** if $AC - B^2 > 0$, **parabolic** if $AC - B^2 = 0$, and **hyperbolic** if $AC - B^2 < 0$. (Here A, B, C may be functions of x and y, and the type of (7) may be different in different parts of the xy-plane.) Show that

Laplace's equation $u_{xx} + u_{yy} = 0$ is elliptic,

the **heat equation** $u_t = c^2u_{xx}$ is parabolic,

the **wave equation** $u_{tt} = c^2u_{xx}$ is hyperbolic,

the **Tricomi equation** $yu_{xx} + u_{yy} = 0$ is of mixed type (elliptic in the upper half plane, parabolic on the x-axis, and hyperbolic in the lower half plane).

14. If (7) is *hyperbolic,* it can be transformed to the *normal form* $u_{vz} = F^*(v, z, u, u_v, u_z)$ by setting $v = \Phi(x, y)$, $z = \Psi(x, y)$ where $\Phi = const$ and $\Psi = const$ are the solutions $y = y(x)$ of $Ay'^2 - 2By' + C = 0$ (cf. Ref. [E14]). Show that in the case of the wave equation (1),

$$\Phi = x + ct, \qquad \Psi = x - ct.$$

15. If (7) is *parabolic,* the substitution $v = x$, $z = \Psi(x, y)$, with Ψ defined as in Prob. 14, reduces it to the *normal form* $u_{vv} = F^*(v, z, u, u_v, u_z)$. Verify this result for the equation in Prob. 10.

16. **(Vibrations of a beam)** It can be shown that the small free vertical vibrations of a uniform beam are governed by the fourth-order equation

$$\frac{\partial^2 u}{\partial t^2} + c^2 \frac{\partial^4 u}{\partial x^4} = 0 \tag{8}$$ (Ref. [E15]),

where $c^2 = EI/\rho A$ (E = Young's modulus of elasticity, I = moment of inertia of the cross section with respect to the y-axis in the figure, ρ = density, A = cross-sectional area). Substituting $u = F(x)G(t)$ into (8) and separating variables, show that

$$F^{(4)}/F = -\ddot{G}/c^2G = \beta^4 = const,$$

$$F(x) = A\cos\beta x + B\sin\beta x + C\cosh\beta x + D\sinh\beta x,$$

$$G(t) = a\cos c\beta^2 t + b\sin c\beta^2 t.$$

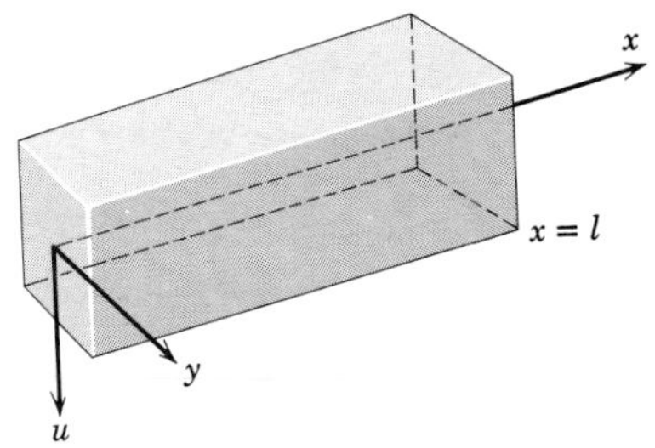

Problem 16. Undeformed beam

17. Find solutions $u_n = F_n(x)G_n(t)$ of (8) corresponding to zero initial velocity and satisfying the boundary conditions

$u(0, t) = 0$, $u(l, t) = 0$ (ends simply supported for all times t),
$u_{xx}(0, t) = 0$, $u_{xx}(l, t) = 0$ (zero moments, hence zero curvature at the ends).

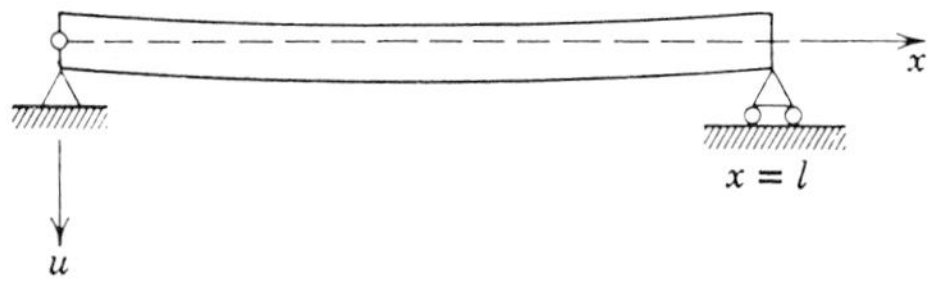

Problem 17

18. Find the solution of (8) that satisfies the conditions in Prob. 17 and the initial condition $u(x, 0) = f(x) = x(l - x)$.

19. What are the boundary conditions if the beam is clamped at both ends?

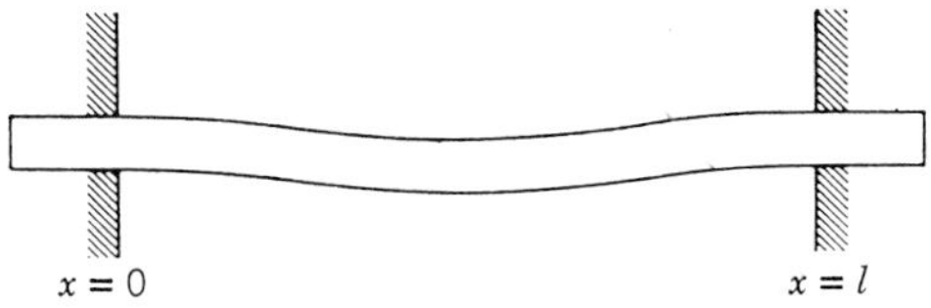

Problem 19

20. Show that $F(x)$ in Prob. 16 satisfies the conditions in Prob. 19, if βl is a root of the equation

$$\cosh\beta l\cos\beta l = 1. \tag{9}$$

Determine approximate solutions of (9).

11.5 One-Dimensional Heat Flow

The heat flow in a body of homogeneous material is governed by the heat equation (cf. Sec. 9.9)

$$\frac{\partial u}{\partial t} = c^2 \nabla^2 u \qquad\qquad c^2 = \frac{K}{\sigma \rho}$$

where $u(x, y, z, t)$ is the temperature in the body, K is the thermal conductivity, σ is the specific heat, and ρ is the density of the material of the body. $\nabla^2 u$ is the Laplacian of u, and with respect to Cartesian coordinates x, y, z,

$$\nabla^2 u = \frac{\partial^2 u}{\partial x^2} + \frac{\partial^2 u}{\partial y^2} + \frac{\partial^2 u}{\partial z^2}.$$

As an important application, let us consider the temperature in a long thin bar or wire of constant cross section and homogeneous material which is oriented along the x-axis (Fig. 236) and is perfectly insulated laterally, so that heat flows in the x-direction only. Then u depends only on x and time t, and the heat equation becomes the so-called **one-dimensional heat equation**

$$\frac{\partial u}{\partial t} = c^2 \frac{\partial^2 u}{\partial x^2}. \tag{1}$$

We shall solve (1) for some important types of boundary and initial conditions. The procedure of solving (1) will be similar to that in the case of the wave equation. The behavior of the solutions will be entirely different from that of the solutions of the wave equation, because (1) involves u_t whereas the wave equation involves u_{tt}. (Hence the classification in Prob. 13 of the last section is not merely a formal matter but has far-reaching consequences with respect to the general behavior of the solutions.)

Let us start with the case when the ends $x = 0$ and $x = l$ of the bar are kept at temperature zero. Then the *boundary conditions* are

$$u(0, t) = 0, \qquad u(l, t) = 0 \qquad \text{for all } t. \tag{2}$$

Note that this has the same form as (2) in Sec. 11.3. Let $f(x)$ be the initial temperature in the bar. Then the *initial condition* is

$$u(x, 0) = f(x), \tag{3}$$

where $f(x)$ is a given function. We shall determine a solution $u(x, t)$ of (1) satisfying (2) and (3).

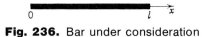

Fig. 236. Bar under consideration

First step. By applying the method of separation of variables we first determine solutions of (1) that satisfy the boundary conditions (2). We start from

$$u(x, t) = F(x)G(t). \tag{4}$$

Substituting this expression and its derivatives into (1), we obtain

$$F\dot{G} = c^2F''G$$

where dots denote derivatives with respect to t and primes denote derivatives with respect to x. We divide this equation by c^2FG, finding

$$\frac{\dot{G}}{c^2G} = \frac{F''}{F}. \tag{5}$$

The expression on the left depends only on t, while the right side depends only on x. As in Sec. 11.3, we conclude that both expressions must be equal to a constant, say, k. The student may show that for $k \geqq 0$ the only solution $u = FG$ that satisfies (2) is $u \equiv 0$. For negative $k = -p^2$ we have from (5)

$$\frac{\dot{G}}{c^2G} = \frac{F''}{F} = -p^2.$$

We see that this yields the two ordinary linear differential equations

$$F'' + p^2F = 0 \tag{6}$$

and

$$\dot{G} + c^2p^2G = 0. \tag{7}$$

Second step. We consider (6). The general solution is

$$F(x) = A\cos px + B\sin px. \tag{8}$$

From (2) it follows that

$$u(0, t) = F(0)G(t) = 0 \qquad \text{and} \qquad u(l, t) = F(l)G(t) = 0.$$

Since $G \equiv 0$ implies $u \equiv 0$, we require that $F(0) = 0$ and $F(l) = 0$. By (8), $F(0) = A$. Thus $A = 0$, and, therefore,

$$F(l) = B\sin pl.$$

We must have $B \neq 0$, since otherwise $F \equiv 0$. Hence the condition $F(l) = 0$ leads to

$$\sin pl = 0 \qquad \text{or} \qquad p = \frac{n\pi}{l}, \qquad n = 1, 2, \cdots.$$

Setting $B = 1$, we thus obtain the following solutions of (6) satisfying (2):

$$F_n(x) = \sin \frac{n\pi x}{l}, \qquad n = 1, 2, \cdots.$$

(As in Sec. 11.3, we need not consider negative integral values of n.)

We now consider the differential equation (7). For the values $p = n\pi/l$ it takes the form

$$\dot{G} + \lambda_n^2 G = 0 \qquad \text{where} \qquad \lambda_n = \frac{cn\pi}{l}.$$

The general solution is

$$G_n(t) = B_n e^{-\lambda_n^2 t}, \qquad n = 1, 2, \cdots,$$

where B_n is a constant. Hence the functions

(9) $$u_n(x, t) = F_n(x)G_n(t) = B_n \sin \frac{n\pi x}{l}\, e^{-\lambda_n^2 t} \qquad n = 1, 2, \cdots.$$

are solutions of the heat equation (1), satisfying (2).

Third step. To obtain a solution also satisfying (3), we consider the series

(10) $$u(x, t) = \sum_{n=1}^{\infty} u_n(x, t) = \sum_{n=1}^{\infty} B_n \sin \frac{n\pi x}{l}\, e^{-\lambda_n^2 t} \qquad \left(\lambda_n = \frac{cn\pi}{l}\right).$$

From this and (3),

$$u(x, 0) = \sum_{n=1}^{\infty} B_n \sin \frac{n\pi x}{l} = f(x).$$

Hence, for (10) to satisfy (3), the coefficients B_n must be chosen such that $u(x, 0)$ becomes a half-range expansion of $f(x)$, namely, the Fourier sine series of $f(x)$; that is [cf. (4) in Sec. 10.5],

(11) $$B_n = \frac{2}{l}\int_0^l f(x) \sin \frac{n\pi x}{l}\, dx \qquad n = 1, 2, \cdots.$$

The solution of our problem can be established, assuming that $f(x)$ is piecewise continuous on the interval $0 \leqq x \leqq l$ (cf. Sec. 5.1), and has one-sided derivatives[3] at all interior points of that interval; that is, under these assumptions the series (10) with coefficients (11) is the solution of our physical problem. The proof, which requires the knowledge of uniform convergence of series, will be given at a later occasion (Probs. 19, 20 at the end of Sec. 16.6).

[3]Cf. footnote 4 in Sec. 10.2.

Because of the exponential factor all the terms in (11) approach zero as t approaches infinity. The rate of decay varies with n.

Example 1

If the initial temperature is

$$f(x) = \begin{cases} x & \text{when} \quad 0 < x < l/2, \\ l - x & \text{when} \quad l/2 < x < l \end{cases}$$

(cf. Fig. 237 where $l = \pi$ and $c = 1$), then we obtain from (11)

$$B_n = \frac{2}{l}\left(\int_0^{l/2} x \sin\frac{n\pi x}{l}\,dx + \int_{l/2}^{l} (l - x)\sin\frac{n\pi x}{l}\,dx\right). \tag{12}$$

Integration yields $B_n = 0$ when n is even and

$$B_n = \frac{4l}{n^2\pi^2} \qquad (n = 1, 5, 9, \cdots),$$

$$B_n = -\frac{4l}{n^2\pi^2} \qquad (n = 3, 7, 11, \cdots).$$

Hence the solution is

$$u(x, t) = \frac{4l}{\pi^2}\left[\sin\frac{\pi x}{l}\, e^{-(c\pi/l)^2 t} - \frac{1}{9}\sin\frac{3\pi x}{l}\, e^{-(3c\pi/l)^2 t} + \cdots\right].$$

The reader may compare Fig. 237 and Fig. 235 in Sec. 11.3.

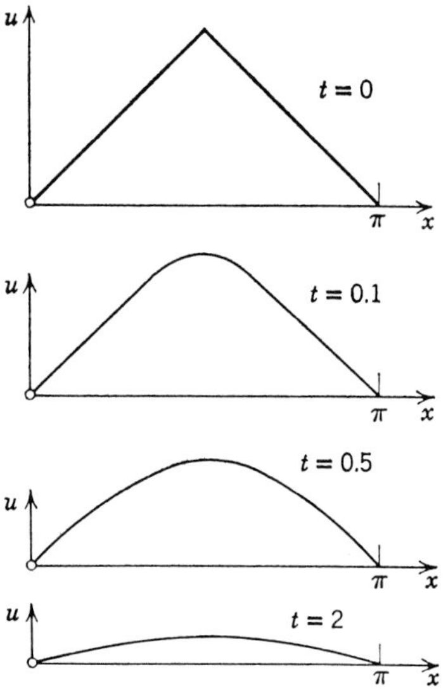

Fig. 237. Solution of Example 1 for various values of t

Problems for Sec. 11.5

1. Compare Figs. 235 and 237 and describe the different behavior of the solutions of the two equations.
2. How does the rate of decay of (9) for fixed n depend on the specific heat, the density, and the thermal conductivity of the material?

3. Graph u_1, u_2, u_3 [cf. (9), with $B_n = 1, c = 1, l = \pi$] as functions of x for the values $t = 0, 1, 2, 3$. Compare the behavior of these functions.

4. Represent the solutions in Prob. 3 graphically as surfaces over the xt-plane.

5. The ends of a rod that satisfies the assumptions in the text are kept at different constant temperatures $u(0, t) = U_1$ and $u(l, t) = U_2$. Find the temperature $u_I(x)$ in the rod after a long time (theoretically: as $t \to \infty$).

Find the temperature $u(x, t)$ in a bar of silver (length 10 cm, constant cross section of area 1 cm^2, density 10.6 gm/cm^3, thermal conductivity 1.04 cal/cm deg sec, specific heat 0.056 cal/gm deg) which is perfectly insulated laterally, whose ends are kept at temperature 0°C and whose initial temperature (in °C) is $f(x)$, where

6. $f(x) = \sin 0.2\pi x$

7. $f(x) = \sin 0.1\pi x$

8. $f(x) = \begin{cases} x & \text{if } 0 < x < 5 \\ 0 & \text{if } 5 < x < 10 \end{cases}$

9. $f(x) = \begin{cases} x & \text{if } 0 < x < 5 \\ 10 - x & \text{if } 5 < x < 10 \end{cases}$

10. $f(x) = x(100 - x^2)$

11. $f(x) = x(10 - x)$

12. Find the temperature $u(x, t)$ in a bar of length l which is perfectly insulated, also at the ends at $x = 0$ and $x = l$, assuming that $u(x, 0) = f(x)$. Physical information: the flux of heat through the faces at the ends is proportional to the values of $\partial u/\partial x$ there. Show that this situation corresponds to the conditions

$$u_x(0, t) = 0, \qquad u_x(l, t) = 0, \qquad u(x, 0) = f(x).$$

Show that the method of separating variables yields the solution

$$u(x, t) = A_0 + \sum_{n=1}^{\infty} A_n \cos \frac{n\pi x}{l} \; e^{-(cn\pi/l)^2 t}$$

where, by (2) in Sec. 10.5,

$$A_0 = \frac{1}{l}\int_0^l f(x)\, dx, \qquad A_n = \frac{2}{l}\int_0^l f(x) \cos \frac{n\pi x}{l}\, dx, \qquad n = 1, 2, \cdots.$$

13. In Prob. 12, $u \to A_0$ as $t \to \infty$. Does this agree with your physical intuition?

Find the temperature in the bar in Prob. 12, if $l = \pi$, $c = 1$, and

14. $f(x) = 1$

15. $f(x) = x^2$

16. $f(x) = \sin x$

17. $f(x) = \begin{cases} x & \text{when } 0 < x < \pi/2 \\ 0 & \text{when } \pi/2 < x < \pi \end{cases}$

18. $f(x) = \begin{cases} 1 & \text{when } 0 < x < \pi/2 \\ 0 & \text{when } \pi/2 < x < \pi \end{cases}$

19. In Prob. 5, let the initial temperature be $u(x, 0) = f(x)$. Show that the temperature for any time $t > 0$ is $u(x, t) = u_I(x) + u_{II}(x, t)$ with u_I as before and

$$u_{II} = \sum_{n=1}^{\infty} B_n \sin \frac{n\pi x}{l} \; e^{-(cn\pi/l)^2 t},$$

where

$$B_n = \frac{2}{l}\int_0^l [f(x) - u_I(x)] \sin \frac{n\pi x}{l}\, dx$$

$$= \frac{2}{l}\int_0^l f(x) \sin \frac{n\pi x}{l}\, dx + \frac{2}{n\pi}[(-1)^n U_2 - U_1].$$

20. Consider the bar in Probs. 6–11. Assume that the ends are kept at 100°C for a long time. Then at some instant, say, at $t = 0$, the temperature at $x = l$ is suddenly changed to 0°C and kept at this value, while the temperature at $x = 0$ is kept at 100° C. What are the temperatures in the middle of the bar at $t = 1, 2, 3, 10, 50$ sec?

11.6 Heat Flow in an Infinite Bar

We shall now consider solutions of the heat equation

$$\frac{\partial u}{\partial t} = c^2 \frac{\partial^2 u}{\partial x^2} \tag{1}$$

in the case of a bar which extends to infinity on both sides (and is laterally insulated, as before). In this case we do not have boundary conditions but only the initial condition

$$u(x, 0) = f(x) \qquad (-\infty < x < \infty) \tag{2}$$

where $f(x)$ is the given initial temperature of the bar.

To solve our present problem we start as in the last section, that is, we substitute $u(x, t) = F(x)G(t)$ into (1). This yields the two ordinary differential equations

$$F'' + p^2 F = 0 \qquad \text{[cf. (6), Sec. 11.5]} \tag{3}$$

and

$$\dot{G} + c^2 p^2 G = 0 \qquad \text{[cf. (7), Sec. 11.5].} \tag{4}$$

Solutions are

$$F(x) = A \cos px + B \sin px \qquad \text{and} \qquad G(t) = e^{-c^2 p^2 t}$$

respectively; here A and B are any constants. Hence a solution of (1) is

$$u(x, t; p) = FG = (A \cos px + B \sin px) e^{-c^2 p^2 t}. \tag{5}$$

[As in the last section, we had to choose the constant of separation k negative, $k = -p^2$, because positive values of k lead to an increasing exponential function in (5), which has no physical meaning.]

Any series of functions (5), found in the usual manner by taking p as multiples of a fixed number, would lead to a function which is periodic in x when $t = 0$. However, since $f(x)$ in (2) is not assumed periodic, it is natural to use Fourier integrals in the present case instead of Fourier series.

Since A and B in (5) are arbitrary, we may consider these quantities as functions of p and write $A = A(p)$ and $B = B(p)$. Since the heat equation is linear and homogeneous, the function

$$(6) \quad u(x, t) = \int_0^\infty u(x, t; p)\, dp = \int_0^\infty [A(p) \cos px + B(p) \sin px]\, e^{-c^2p^2t}\, dp$$

is then a solution of (1), provided this integral exists and can be differentiated twice with respect to x and once with respect to t.

From the representation (6) and the initial condition (2) it follows that

$$(7) \qquad u(x, 0) = \int_0^\infty [A(p) \cos px + B(p) \sin px]\, dp = f(x).$$

Using (4) and (5) in Sec. 10.9, we thus obtain

$$(8) \qquad A(p) = \frac{1}{\pi}\int_{-\infty}^{\infty} f(v) \cos pv\, dv, \qquad B(p) = \frac{1}{\pi}\int_{-\infty}^{\infty} f(v) \sin pv\, dv.$$

According to (16) in Prob. 20, Sec. 10.9, this Fourier integral may be written

$$u(x, 0) = \frac{1}{\pi}\int_0^\infty \left[\int_{-\infty}^{\infty} f(v) \cos (px - pv)\, dv\right] dp,$$

and (6), this section, thus becomes

$$u(x, t) = \frac{1}{\pi}\int_0^\infty \left[\int_{-\infty}^{\infty} f(v) \cos (px - pv)\, e^{-c^2p^2t}\, dv\right] dp.$$

Assuming that we may invert the order of integration, we obtain

$$(9) \qquad u(x, t) = \frac{1}{\pi}\int_{-\infty}^{\infty} f(v) \left[\int_0^\infty e^{-c^2p^2t} \cos (px - pv)\, dp\right] dv.$$

The inner integral can be evaluated by the use of the formula

$$(10) \qquad \int_0^\infty e^{-s^2} \cos 2bs\, ds = \frac{\sqrt{\pi}}{2} e^{-b^2}$$

which will be derived in Sec. 17.4 (Prob. 9). Introducing in (10) a new variable of integration p by setting $s = cp\sqrt{t}$ and choosing

$$b = \frac{x - v}{2c\sqrt{t}}$$

we find that (10) becomes

$$\int_0^\infty e^{-c^2p^2t} \cos (px - pv)\, dp = \frac{\sqrt{\pi}}{2c\sqrt{t}} e^{-(x-v)^2/4c^2t}.$$

By inserting this result into (9) we obtain the representation

$$u(x, t) = \frac{1}{2c\sqrt{\pi t}} \int_{-\infty}^{\infty} f(v) \exp\left\{-\frac{(x-v)^2}{4c^2 t}\right\} dv. \tag{11}$$

We finally introduce the variable of integration $z = (v - x)/2c\sqrt{t}$. Then

$$u(x, t) = \frac{1}{\sqrt{\pi}} \int_{-\infty}^{\infty} f(x + 2cz\sqrt{t})\, e^{-z^2}\, dz. \tag{12}$$

If $f(x)$ is bounded for all values of x and integrable in every finite interval, it can be shown (cf. Ref. [D1]) that the function (11) or (12) satisfies (1) and (2). Hence this function is the required solution in the present case.

Example 1. Temperature in an infinite bar

Find the temperature in the infinite bar if the initial temperature is (Fig. 238)

$$f(x) = \begin{cases} U_0 = const & \text{when } |x| < 1, \\ 0 & \text{when } |x| > 1. \end{cases}$$

From (11) we have

$$u(x, t) = \frac{U_0}{2c\sqrt{\pi t}} \int_{-1}^{1} \exp\left\{-\frac{(x-v)^2}{4c^2 t}\right\} dv.$$

If we introduce the above variable of integration z, then the integration over v from -1 to 1 corresponds to the integration over z from $(-1 - x)/2c\sqrt{t}$ to $(1 - x)/2c\sqrt{t}$, and

$$u(x, t) = \frac{U_0}{\sqrt{\pi}} \int_{-(1+x)/2c\sqrt{t}}^{(1-x)/2c\sqrt{t}} e^{-z^2}\, dz \qquad (t > 0). \tag{13}$$

We mention that this integral is not an elementary function, but can easily be expressed in terms of the error function, whose values have been tabulated. (Table A5 in Appendix 4 contains a few values; larger tables are listed in Ref. [J11] in Appendix 1. Cf. also Probs. 2–7, below.) Figure 239 shows $u(x, t)$ for $U_0 = 100°$ C, $c^2 = 1$ cm^2/sec, and several values of t.

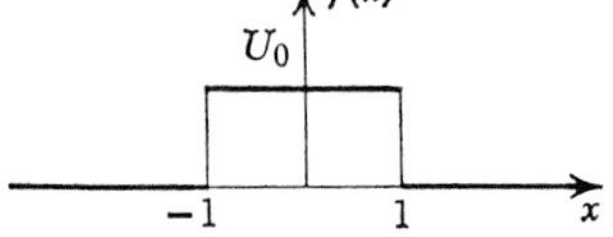

Fig. 238. Initial temperature in Example 1

Problems for Sec. 11.6

1. Graph the temperatures in Example 1 (with $U_0 = 100°$ C and $c^2 = 1$ cm^2/sec) at the points $x = 0.5$, 1, and 1.5 as functions of t. Do the results agree with your physical intuition?

Error function. The error function is defined by the integral

$$\operatorname{erf} x = \frac{2}{\sqrt{\pi}} \int_0^x e^{-w^2}\, dw.$$

It is important in engineering mathematics. To become familiar with the function, the

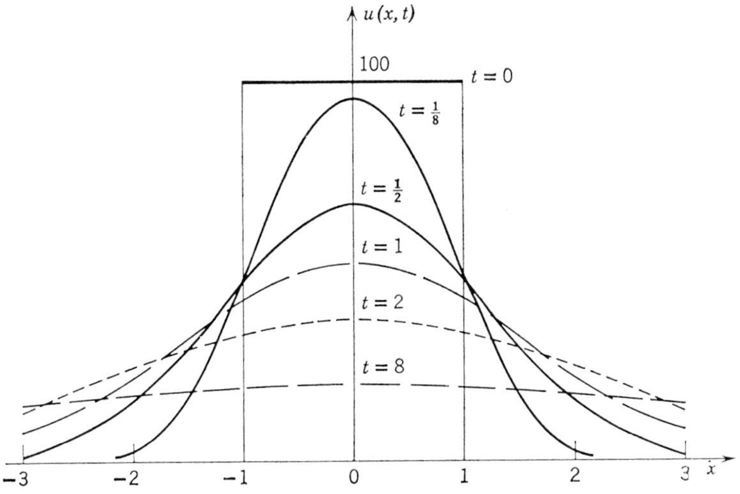

Fig. 239. Solution $u(x, t)$ of Example 1 for $U_0 = 100°C$, $c^2 = 1\ \text{cm}^2/\text{sec}$, and several values of t

student may solve the following problems. (A few formulas are included in Appendix 3, see (35)–(37). Cf. also Example 1 in Sec. 19.15.)

2. Show that $\operatorname{erf} x$ is odd.

3. Show that $\displaystyle\int_a^b e^{-w^2}\,dw = \frac{\sqrt{\pi}}{2}(\operatorname{erf} b - \operatorname{erf} a), \qquad \int_{-b}^{b} e^{-w^2}\,dw = \sqrt{\pi}\operatorname{erf} b.$

4. Using a table of the exponential function, graph the integrand of $\operatorname{erf} x$ (the so-called *bell-shaped curve*).

5. Obtain a small table of $\operatorname{erf} x$ for $x = 0, 0.2, 0.4, \cdots, 1.0, 1.5, 2.0$ by counting squares under the curve in Prob. 4 or some other rough method of approximate integration, and compare the result with the actual values whose first two decimal places are 0.00, 0.22, 0.43, 0.60, 0.74, 0.84, 0.97, 1.00.

6. Obtain the Maclaurin series of $\operatorname{erf} x$ by term-by-term integration of that of the integrand.

7. Show that (13) may be written

$$u(x, t) = \frac{U_0}{2}\left[\operatorname{erf}\frac{1-x}{2c\sqrt{t}} + \operatorname{erf}\frac{1+x}{2c\sqrt{t}}\right] \qquad (t > 0).$$

8. If $f(x) = 1$ when $x > 0$ and $f(x) = 0$ when $x < 0$, show that (12) becomes

$$u(x, t) = \frac{1}{\sqrt{\pi}}\int_{-x/2c\sqrt{t}}^{\infty} e^{-z^2}\,dz \qquad (t > 0).$$

9. It can be shown that $\operatorname{erf}(\infty) = 1$. Using this, show that in Prob. 8,

$$u(x, t) = \tfrac{1}{2} + \tfrac{1}{2}\operatorname{erf}\,(x/2c\sqrt{t}).$$

10. If the bar is semi-infinite, extending from 0 to ∞, the end at $x = 0$ is held at temperature 0 and the initial temperature is $f(x)$, show that the temperature in the bar is

(14) $$u(x, t) = \frac{1}{\sqrt{\pi}}\left[\int_{-x/\tau}^{\infty} f(x + \tau w)e^{-w^2}\, dw - \int_{x/\tau}^{\infty} f(-x + \tau w)e^{-w^2}\, dw\right],$$

where $\tau = 2c\sqrt{t}$.

11. Obtain (14) from (11) by assuming that $f(v)$ in (11) is odd.

12. If $f(x) = 1$ in Prob. 10, show that

$$u(x, t) = \frac{2}{\sqrt{\pi}}\int_{0}^{x/\tau} e^{-w^2}\, dw = \operatorname{erf}\left(\frac{x}{2c\sqrt{t}}\right) \qquad (t > 0).$$

13. What form does (14) take if $f(x) = 1$ when $a < x < b$ (where $a > 0$) and $f(x) = 0$ otherwise?

14. Show that the result of Prob. 12 can be obtained from (11) or (12) by using $f(x) = 1$, when $x > 0$, and $f(x) = -1$ when $x < 0$. What is the reason?

15. Show that in Prob. 12 the times required for any two points to reach the same temperature are proportional to the squares of their distances from the boundary at $x = 0$.

11.7 Modeling: Vibrating Membrane. Two-Dimensional Wave Equation

As another important problem from the field of vibrations, let us consider the motion of a stretched membrane, such as a drumhead. The reader will notice that our present considerations will be similar to those in the case of the vibrating string in Sec. 11.2.

We make the following assumptions:

1. *The mass of the membrane per unit area is constant ("homogeneous membrane"). The membrane is perfectly flexible and is so thin that it does not offer any resistance to bending.*
2. *The membrane is stretched and then fixed along its entire boundary in the xy-plane. The tension per unit length T caused by stretching the membrane is the same at all points and in all directions and does not change during the motion.*
3. *The deflection $u(x, y, t)$ of the membrane during the motion is small compared with the size of the membrane, and all angles of inclination are small.*

Although these assumptions cannot be realized in practice, small transverse vibrations of a thin physical membrane will satisfy these assumptions relatively accurately.

To derive the differential equation which governs the motion of the membrane, we consider the forces acting on a small portion of the membrane as shown in Fig. 240. Since the deflections of the membrane and the angles of inclination are small, the sides of the portion are approximately equal to Δx and Δy. The tension T is the force per unit length. Hence the forces acting on the edges of the portion are approximately $T\,\Delta x$ and $T\,\Delta y$. Since the membrane is perfectly flexible, these forces are tangent to the membrane.

We first consider the horizontal components of the forces. These components

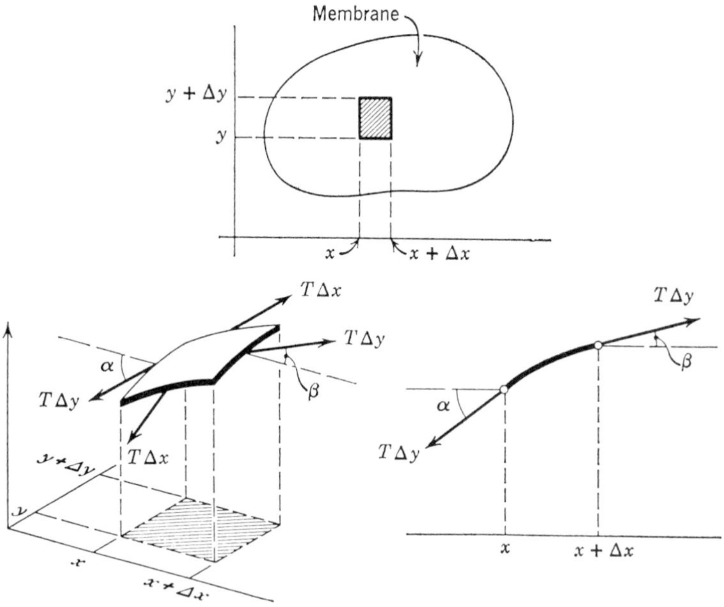

Fig. 240. Vibrating membrane

are obtained by multiplying the forces by the cosines of the angles of inclination. Since these angles are small, their cosines are close to 1. Hence the horizontal components of the forces at opposite edges are approximately equal. Therefore, the motion of the particles of the membrane in horizontal direction will be negligibly small. From this we conclude that we may regard the motion of the membrane as transversal, that is, each particle moves vertically.

The vertical components of the forces along the edges parallel to the yu-plane are[4] (Fig. 240)

$$T\,\Delta y \sin \beta \qquad \text{and} \qquad -T\,\Delta y \sin \alpha;$$

here the minus sign appears because the force on the left edge is directed downward. Since the angles are small, we may replace their sines by their tangents. Hence the resultant of those two vertical components is

$$\begin{aligned} T\,\Delta y(\sin \beta - \sin \alpha) &\approx T\,\Delta y(\tan \beta - \tan \alpha) \\ &= T\,\Delta y[u_x(x + \Delta x, y_1) - u_x(x, y_2)] \end{aligned} \tag{1}$$

where subscripts x denote partial derivatives and y_1 and y_2 are values between y and $y + \Delta y$. Similarly, the resultant of the vertical components of the forces acting on the other two edges of the portion is

$$T\,\Delta x[u_y(x_1, y + \Delta y) - u_y(x_2, y)] \tag{2}$$

where x_1 and x_2 are values between x and $x + \Delta x$.

[4] Note that the angle of inclination varies along the edges, and α and β represent values of that angle at a suitable point of the edges under consideration.

By Newton's second law (cf. Sec. 2.6), the sum of the forces given by (1) and (2) is equal to the mass $\rho\,\Delta A$ of the portion times the acceleration $\partial^2 u/\partial t^2$; here ρ is the mass of the undeflected membrane per unit area and $\Delta A = \Delta x\,\Delta y$ is the area of the portion when it is undeflected. Thus

$$\rho\,\Delta x\,\Delta y \frac{\partial^2 u}{\partial t^2} = T\,\Delta y[u_x(x + \Delta x, y_1) - u_x(x, y_2)]$$
$$+ T\,\Delta x[u_y(x_1, y + \Delta y) - u_y(x_2, y)]$$

where the derivative on the left is evaluated at some suitable point $(\tilde{x}, \tilde{y})$ corresponding to the portion. Division by $\rho\,\Delta x\,\Delta y$ yields

$$\frac{\partial^2 u}{\partial t^2} = \frac{T}{\rho}\left[\frac{u_x(x + \Delta x, y_1) - u_x(x, y_2)}{\Delta x} + \frac{u_y(x_1, y + \Delta y) - u_y(x_2, y)}{\Delta y}\right].$$

If we let Δx and Δy approach zero, then we obtain the partial differential equation

(3) $$\frac{\partial^2 u}{\partial t^2} = c^2\left(\frac{\partial^2 u}{\partial x^2} + \frac{\partial^2 u}{\partial y^2}\right) \qquad c^2 = \frac{T}{\rho}.$$

This equation is called the **two-dimensional wave equation.** The expression in parentheses is the Laplacian $\nabla^2 u$ of u (cf. Sec. 8.8), and (3) can be written

(3′) $$\frac{\partial^2 u}{\partial t^2} = c^2 \nabla^2 u.$$

11.8 Rectangular Membrane

To solve the problem of a vibrating membrane, we have to determine a solution $u(x, y, t)$ of the two-dimensional wave equation

(1) $$\frac{\partial^2 u}{\partial t^2} = c^2\left(\frac{\partial^2 u}{\partial x^2} + \frac{\partial^2 u}{\partial y^2}\right)$$

that satisfies the boundary condition

(2) $u = 0$ on the boundary of the membrane for all $t \geqq 0$

and the two initial conditions

(3) $u(x, y, 0) = f(x, y)$ [given initial displacement $f(x, y)$]

and

(4) $\left.\dfrac{\partial u}{\partial t}\right|_{t=0} = g(x, y)$ [given initial velocity $g(x, y)$].

These conditions are quite similar to those in the case of the vibrating string.

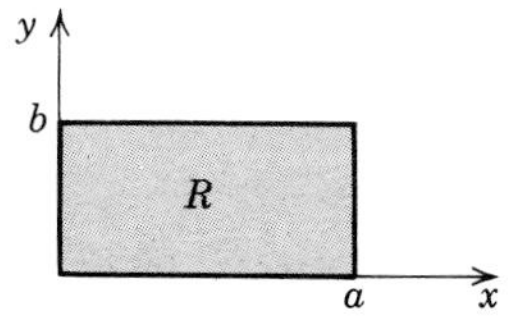

Fig. 241. Rectangular membrane

As a first important case, let us consider the rectangular membrane R shown in Fig. 241.

First step. By applying the method of separation of variables we first determine solutions of (1) that satisfy the condition (2). For this purpose we start from

$$u(x, y, t) = F(x, y)G(t). \tag{5}$$

By substituting this into the wave equation (1) we have

$$F\ddot{G} = c^2(F_{xx}G + F_{yy}G)$$

where subscripts denote partial derivatives and dots denote derivatives with respect to t. Dividing both sides by c^2FG, we find

$$\frac{\ddot{G}}{c^2G} = \frac{1}{F}(F_{xx} + F_{yy}).$$

Since the functions on the left depend only on t while the functions on the right do not depend on t, the expressions on both sides must be equal to a constant. A little investigation shows that only negative values of that constant will lead to solutions which satisfy (2) without being identically zero; this is similar to the procedure in Sec. 11.3. Denoting this negative constant by $-\nu^2$, we thus have

$$\frac{\ddot{G}}{c^2G} = \frac{1}{F}(F_{xx} + F_{yy}) = -\nu^2.$$

From this we obtain the two linear differential equations

$$\ddot{G} + \lambda^2 G = 0 \qquad \text{where} \qquad \lambda = c\nu, \tag{6}$$

and

$$F_{xx} + F_{yy} + \nu^2 F = 0. \tag{7}$$

Now we consider (7) and apply the method of separating variables once more, that is, we determine solutions of (7) of the form

$$F(x, y) = H(x)Q(y) \tag{8}$$

which are zero on the boundary of the membrane. Substitution of (8) into (7) yields

$$\frac{d^2H}{dx^2}Q = -\left(H\frac{d^2Q}{dy^2} + \nu^2 HQ\right).$$

By dividing both sides by HQ we find

$$\frac{1}{H}\frac{d^2H}{dx^2} = -\frac{1}{Q}\left(\frac{d^2Q}{dy^2} + \nu^2 Q\right).$$

The functions on the left depend only on x whereas the functions on the right depend only on y. Hence the expressions on both sides must be equal to a constant. This constant must be negative, say, $-k^2$, because only negative values will lead to solutions that satisfy (2) without being identically zero. Thus

$$\frac{1}{H}\frac{d^2H}{dx^2} = -\frac{1}{Q}\left(\frac{d^2Q}{dy^2} + \nu^2 Q\right) = -k^2.$$

This yields the ordinary differential equations

$$\frac{d^2H}{dx^2} + k^2 H = 0 \tag{9}$$

and

$$\frac{d^2Q}{dy^2} + p^2 Q = 0 \qquad \text{where } p^2 = \nu^2 - k^2. \tag{10}$$

Second step. The general solutions of (9) and (10) are

$$H(x) = A\cos kx + B\sin kx \qquad \text{and} \qquad Q(y) = C\cos py + D\sin py$$

where A, B, C, and D are constants. From (5) and (2) it follows that $F = HQ$ must be zero on the boundary, which corresponds to $x = 0$, $x = a$, $y = 0$, and $y = b$; cf. Fig. 241. This yields the conditions

$$H(0) = 0, \qquad H(a) = 0, \qquad Q(0) = 0, \qquad Q(b) = 0.$$

Therefore, $H(0) = A = 0$, and then

$$H(a) = B\sin ka = 0.$$

We must take $B \neq 0$ since otherwise $H \equiv 0$ and $F \equiv 0$. Hence $\sin ka = 0$ or $ka = m\pi$, that is,

$$k = \frac{m\pi}{a} \qquad (m \text{ integer}).$$

In precisely the same fashion we conclude that $C = 0$ and p must be restricted to the values $p = n\pi/b$ where n is an integer. We thus obtain the solutions

$$H_m(x) = \sin \frac{m\pi x}{a} \qquad \text{and} \qquad Q_n(y) = \sin \frac{n\pi y}{b} \qquad \begin{matrix} m = 1, 2, \cdots, \\ n = 1, 2, \cdots. \end{matrix}$$

(As in the case of the vibrating string, it is not necessary to consider $m, n = -1, -2, \cdots$ since the corresponding solutions are essentially the same as for positive m and n, except for a factor -1.) It follows that the functions

$$\textbf{(11)} \qquad F_{mn}(x, y) = H_m(x)Q_n(y) = \sin \frac{m\pi x}{a} \sin \frac{n\pi y}{b} \qquad \begin{matrix} m = 1, 2, \cdots, \\ n = 1, 2, \cdots, \end{matrix}$$

are solutions of the equation (7) which are zero on the boundary of the rectangular membrane.

Since $p^2 = \nu^2 - k^2$ in (10) and $\lambda = c\nu$ in (6), we have

$$\lambda = c\sqrt{k^2 + p^2}.$$

Hence to $k = m\pi/a$ and $p = n\pi/b$ there corresponds the value

$$(12) \qquad \lambda = \lambda_{mn} = c\pi\sqrt{\frac{m^2}{a^2} + \frac{n^2}{b^2}} \qquad \begin{matrix} m = 1, 2, \cdots, \\ n = 1, 2, \cdots, \end{matrix}$$

in (6), and the corresponding general solution of (6) is

$$G_{mn}(t) = B_{mn} \cos \lambda_{mn} t + B^*_{mn} \sin \lambda_{mn} t.$$

It follows that the functions $u_{mn}(x, y, t) = F_{mn}(x, y)G_{mn}(t)$, written out

$$\textbf{(13)} \qquad u_{mn}(x, y, t) = (B_{mn} \cos \lambda_{mn} t + B^*_{mn} \sin \lambda_{mn} t) \sin \frac{m\pi x}{a} \sin \frac{n\pi y}{b}$$

with λ_{mn} according to (12) are solutions of the wave equation (1), which are zero on the boundary of the rectangular membrane in Fig. 241. These functions are called the **eigenfunctions** or *characteristic functions,* and the numbers λ_{mn} are called the **eigenvalues** or *characteristic values* of the vibrating membrane. The frequency of u_{mn} is $\lambda_{mn}/2\pi$.

It is interesting to note that, depending on a and b, several functions F_{mn} may correspond to the same eigenvalue. Physically this means that there may exist vibrations having the same frequency but entirely different **nodal lines** (curves of points on the membrane which do not move). Let us illustrate this by the following example.

Example 1. Square membrane

Consider the square membrane for which $a = b = 1$. From (12) we see that the eigenvalues are

$$(14) \qquad \lambda_{mn} = c\pi\sqrt{m^2 + n^2}.$$

Hence

$$\lambda_{mn} = \lambda_{nm}$$

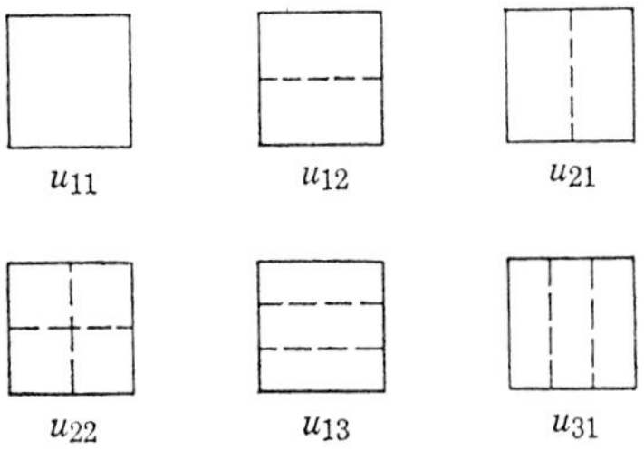

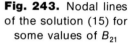

Fig. 242. Nodal lines of the solutions u_{11}, u_{12}, u_{21}, u_{22}, u_{13}, and u_{31} in the case of the square membrane

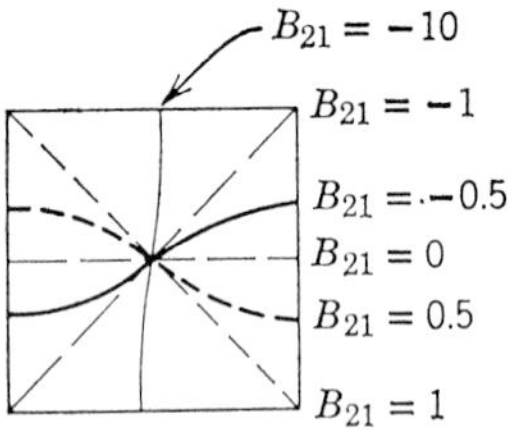

Fig. 243. Nodal lines of the solution (15) for some values of B_{21}

but for $m \neq n$ the corresponding functions

$$F_{mn} = \sin m\pi x \sin n\pi y \qquad \text{and} \qquad F_{nm} = \sin n\pi x \sin m\pi y$$

are certainly different. For example, to $\lambda_{12} = \lambda_{21} = c\pi\sqrt{5}$ there correspond the two functions

$$F_{12} = \sin \pi x \sin 2\pi y \qquad \text{and} \qquad F_{21} = \sin 2\pi x \sin \pi y.$$

Hence the corresponding solutions

$$u_{12} = (B_{12} \cos c\pi\sqrt{5}t + B_{12}^* \sin c\pi\sqrt{5}t)F_{12}$$

and

$$u_{21} = (B_{21} \cos c\pi\sqrt{5}t + B_{21}^* \sin c\pi\sqrt{5}t)F_{21}$$

have the nodal lines $y = \frac{1}{2}$ and $x = \frac{1}{2}$, respectively (cf. Fig. 242). Taking $B_{12} = 1$ and $B_{12}^* = B_{21}^* = 0$, we obtain

$$u_{12} + u_{21} = \cos c\pi\sqrt{5}t\ (F_{12} + B_{21}F_{21}), \tag{15}$$

which represents another vibration corresponding to the eigenvalue $c\pi\sqrt{5}$. The nodal line of this function is the solution of the equation

$$F_{12} + B_{21}F_{21} = \sin \pi x \sin 2\pi y + B_{21} \sin 2\pi x \sin \pi y = 0$$

or, since $\sin 2\alpha = 2 \sin \alpha \cos \alpha$,

$$\sin \pi x \sin \pi y\,(\cos \pi y + B_{21} \cos \pi x) = 0. \tag{16}$$

This solution depends on the value of B_{21} (Fig. 243).

From (14) we see that even more than two functions may correspond to the same numerical value of λ_{mn}. For example, the four functions F_{18}, F_{81}, F_{47}, and F_{74} correspond to the value $\lambda_{18} = \lambda_{81} = \lambda_{47} = \lambda_{74} = c\pi\sqrt{65}$, because

$$1^1 + 8^2 = 4^2 + 7^2 = 65.$$

This happens because 65 can be expressed as the sum of two squares of natural numbers in several ways. According to a theorem by Gauss, this is the case for every sum of two squares among whose prime factors there are at least two different ones of the form $4n + 1$ where n is a positive integer. In our case,

$$65 = 5 \cdot 13 = (4 + 1)(12 + 1). \qquad \blacksquare$$

Third step. To obtain the solution that also satisfies the initial conditions (3) and (4), we proceed in a similar fashion as in Sec. 11.3. We consider the double series[5]

[5] We shall not consider the problems of convergence and uniqueness.

$$
\begin{aligned}
u(x, y, t) &= \sum_{m=1}^{\infty} \sum_{n=1}^{\infty} u_{mn}(x, y, t) \\
&= \sum_{m=1}^{\infty} \sum_{n=1}^{\infty} (B_{mn} \cos \lambda_{mn} t + B^*_{mn} \sin \lambda_{mn} t) \sin \frac{m\pi x}{a} \sin \frac{n\pi y}{b}.
\end{aligned} \tag{17}
$$

From this and (3) we obtain

$$u(x, y, 0) = \sum_{m=1}^{\infty} \sum_{n=1}^{\infty} B_{mn} \sin \frac{m\pi x}{a} \sin \frac{n\pi y}{b} = f(x, y). \tag{18}$$

This series is called a **double Fourier series.** Suppose that $f(x, y)$ can be developed in such a series.[6] Then the Fourier coefficients B_{mn} of $f(x, y)$ in (18) can be determined as follows. Setting

$$K_m(y) = \sum_{n=1}^{\infty} B_{mn} \sin \frac{n\pi y}{b} \tag{19}$$

we may write (18) in the form

$$f(x, y) = \sum_{m=1}^{\infty} K_m(y) \sin \frac{m\pi x}{a}.$$

For fixed y this is the Fourier sine series of $f(x, y)$, considered as a function of x, and from (4) in Sec. 10.5 it follows that the coefficients of this expansion are

$$K_m(y) = \frac{2}{a} \int_0^a f(x, y) \sin \frac{m\pi x}{a} \, dx. \tag{20}$$

Furthermore, (19) is the Fourier sine series of $K_m(y)$, and from (4) in Sec. 10.5 it follows that the coefficients are

$$B_{mn} = \frac{2}{b} \int_0^b K_m(y) \sin \frac{n\pi y}{b} \, dy.$$

From this and (20) we obtain the **generalized Euler formula**

$$B_{mn} = \frac{4}{ab} \int_0^b \int_0^a f(x, y) \sin \frac{m\pi x}{a} \sin \frac{n\pi y}{b} \, dx \, dy \qquad \begin{aligned} m &= 1, 2, \cdots, \\ n &= 1, 2, \cdots, \end{aligned} \tag{21}$$

for the Fourier coefficients of $f(x, y)$ in the double Fourier series (18).

[6] Sufficient conditions: f, $\partial f/\partial x$, $\partial f/\partial y$, $\partial^2 f/\partial x\, \partial y$ continuous in the rectangle R under consideration.

The B_{mn} in (17) are now determined in terms of $f(x, y)$. To determine the B^*_{mn}, we differentiate (17) termwise with respect to t; using (4), we obtain

$$\left.\frac{\partial u}{\partial t}\right|_{t=0} = \sum_{m=1}^{\infty}\sum_{n=1}^{\infty} B^*_{mn}\lambda_{mn} \sin\frac{m\pi x}{a} \sin\frac{n\pi y}{b} = g(x, y).$$

Suppose that $g(x, y)$ can be developed in this double Fourier series. Then, proceeding as before, we find

$$(22) \qquad B^*_{mn} = \frac{4}{ab\lambda_{mn}} \int_0^b \int_0^a g(x, y) \sin\frac{m\pi x}{a} \sin\frac{n\pi y}{b}\, dx\, dy \qquad \begin{array}{l} m = 1, 2, \cdots, \\ n = 1, 2, \cdots. \end{array}$$

The result is that, for (17) to satisfy the initial conditions, the coefficients B_{mn} and B^*_{mn} must be chosen according to (21) and (22).

Problems for Sec. 11.8

1. How does the frequency of a solution (13) change, if the tension of the membrane is increased?
2. Determine and graph the nodal lines of the solutions (13) with $m = 1, 2, 3, 4$ and $n = 1, 2, 3, 4$ in the case $a = b = 1$.
3. Same task as in Prob. 2, when $a = 2$ and $b = 1$.
4. Find further eigenvalues of the square membrane with side 1 such that four different eigenfunctions correspond to each such eigenvalue.
5. Find eigenvalues of the rectangular membrane of sides $a = 2$, $b = 1$ such that two or more different eigenfunctions correspond to each such eigenvalue.
6. Show that, among all rectangular membranes of the same area $A = ab$ and the same c, the square membrane is that for which u_{11} [cf. (13)] has the lowest frequency.
7. Find a similar result as in Prob. 6 for the frequency of a solution (13) with arbitrary fixed m and n.

Represent the following functions $f(x, y)$ $(0 < x < a, 0 < y < b)$ by a double Fourier series of the form (18).

8. $f = 1$
9. $f = x + y$
10. $f = xy$
11. $f = xy(a - x)(b - y)$
12. $f = xy(a^2 - x^2)(b^2 - y^2)$

13. Show that

$$u_{tt} = c^2\nabla^2 u + P/\rho$$

governs the forced vibrations of the membrane, where $P(x, y, t)$ is the external force per unit area acting normal to the xy-plane.

14. Find the eigenfunctions of the rectangular membrane in the figure which is fixed on the boundary.

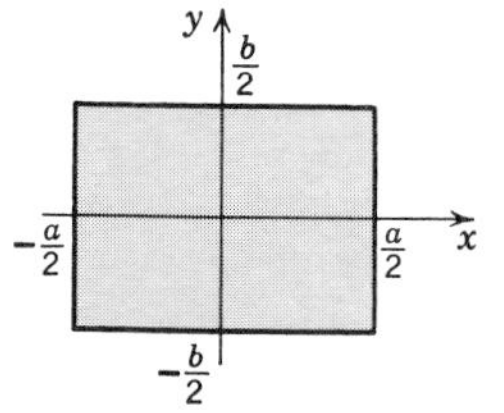

Problem 14. Rectangular membrane

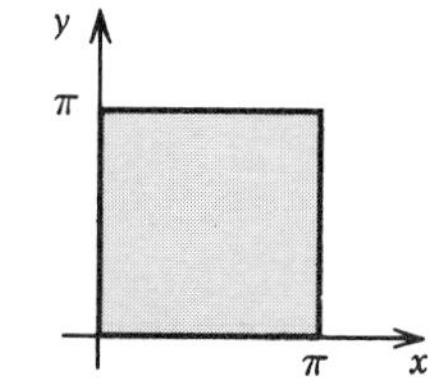

Problem 19. Square plate

Find the deflection $u(x, y, t)$ of the square membrane with $a = b = 1$ and $c = 1$, if the initial velocity is zero and the initial deflection is $f(x, y)$, where

15. $f = 0.01xy(1 - x)(1 - y)$

16. $f = kx(1 - x^2)y(1 - y^2)$

17. $f = k \sin \pi x \sin 2\pi y$

18. $f = k \sin^2 \pi x \sin^2 \pi y$

19. The edges of a thin square plate (see figure) are kept at temperature zero and the faces are perfectly insulated. The initial temperature is assumed to be $u(x, y, 0) = f(x, y)$. By applying the method of separating variables to the two-dimensional heat equation $u_t = c^2\nabla^2 u$ show that the temperature in the plate is

$$u(x, y, t) = \sum_{m=1}^{\infty} \sum_{n=1}^{\infty} B_{mn} \sin mx \sin ny \, e^{-c^2(m^2+n^2)t}$$

where

$$B_{mn} = \frac{4}{\pi^2} \int_0^{\pi} \int_0^{\pi} f(x, y) \sin mx \sin ny \, dx \, dy.$$

20. Find the temperature in the plate of Prob. 19, if $f(x, y) = x(\pi - x)y(\pi - y)$.

11.9 Laplacian in Polar Coordinates

In connection with boundary value problems for partial differential equations, it is a general principle to use coordinates with respect to which the boundary of the region under consideration has a simple representation. In the next section we shall consider circular membranes. Then polar coordinates r and θ, defined by

$$x = r \cos \theta, \qquad y = r \sin \theta,$$

will be appropriate, because the boundary of the membrane can then be represented by the simple equation $r = const.$

When using r and θ we have to transform the Laplacian

$$\nabla^2 u = \frac{\partial^2 u}{\partial x^2} + \frac{\partial^2 u}{\partial y^2}$$

in the wave equation into these new coordinates.

Transformations of differential expressions from one coordinate system into another are frequently required in applications. Therefore, the student should follow our present consideration with great attention.

As in Sec. 11.4, we shall use the chain rule. For the sake of simplicity we shall denote partial derivatives by subscripts and $u(x, y, t)$, as a function of r, θ, t, by the same letter u.

By applying the chain rule (4), Sec. 8.7, we obtain

$$u_x = u_r r_x + u_\theta \theta_x .$$

Differentiating this again with respect to x we first have

$$\text{(1)}\qquad \begin{aligned} u_{xx} &= (u_r r_x)_x + (u_\theta \theta_x)_x \\ &= (u_r)_x r_x + u_r r_{xx} + (u_\theta)_x \theta_x + u_\theta \theta_{xx}. \end{aligned}$$

Now, by applying the chain rule again, we find

$$(u_r)_x = u_{rr} r_x + u_{r\theta}\theta_x \qquad \text{and} \qquad (u_\theta)_x = u_{\theta r} r_x + u_{\theta\theta}\theta_x.$$

To determine the partial derivatives r_x and θ_x, we have to differentiate

$$r = \sqrt{x^2 + y^2} \qquad \text{and} \qquad \theta = \arc\tan\frac{y}{x},$$

finding

$$r_x = \frac{x}{\sqrt{x^2+y^2}} = \frac{x}{r}, \qquad \theta_x = \frac{1}{1+(y/x)^2}\left(-\frac{y}{x^2}\right) = -\frac{y}{r^2}.$$

Differentiating these two formulas again, we obtain

$$r_{xx} = \frac{r - x r_x}{r^2} = \frac{1}{r} - \frac{x^2}{r^3} = \frac{y^2}{r^3}, \qquad \theta_{xx} = -y\left(-\frac{2}{r^3}\right) r_x = \frac{2xy}{r^4}.$$

We substitute all these expressions into (1). Assuming continuity of the first and second partial derivatives, we have $u_{r\theta} = u_{\theta r}$ and by simplifying we obtain

$$\text{(2)}\qquad u_{xx} = \frac{x^2}{r^2}u_{rr} - 2\frac{xy}{r^3}u_{r\theta} + \frac{y^2}{r^4}u_{\theta\theta} + \frac{y^2}{r^3}u_r + 2\frac{xy}{r^4}u_\theta.$$

In a similar fashion it follows that

$$\text{(3)}\qquad u_{yy} = \frac{y^2}{r^2}u_{rr} + 2\frac{xy}{r^3}u_{r\theta} + \frac{x^2}{r^4}u_{\theta\theta} + \frac{x^2}{r^3}u_r - 2\frac{xy}{r^4}u_\theta.$$

By adding (2) and (3) we see that the Laplacian in polar coordinates is

$$\text{(4)}\qquad \nabla^2 u = \frac{\partial^2 u}{\partial r^2} + \frac{1}{r}\frac{\partial u}{\partial r} + \frac{1}{r^2}\frac{\partial^2 u}{\partial \theta^2}.$$

Problems for Sec. 11.9

1. Show that (4) may be written

$$\nabla^2 u = \frac{1}{r}\frac{\partial}{\partial r}\left(r\frac{\partial u}{\partial r}\right) + \frac{1}{r^2}\frac{\partial^2 u}{\partial \theta^2}.$$

2. If u is independent of θ, then (4) reduces to $\nabla^2 u = u_{rr} + u_r/r$. Derive this result directly from the Laplacian in Cartesian coordinates by assuming that u is independent of θ.

3. Transform (4) back into Cartesian coordinates.

4. If x, y are Cartesian coordinates, show that $x^* = x \cos \alpha - y \sin \alpha$, $y^* = x \sin \alpha + y \cos \alpha$ are Cartesian coordinates, and verify by calculation that $\nabla^2 u = u_{x^*x^*} + u_{y^*y^*}$.

5. Express $\nabla^2 u$ in terms of the coordinates $x^* = ax + b, y^* = cy + d$, where x, y are Cartesian coordinates.

6. (Laplacian in cylindrical coordinates) Show that the Laplacian in cylindrical coordinates r, θ, z defined by $x = r \cos \theta$, $y = r \sin \theta$, $z = z$ is

$$\nabla^2 u = u_{rr} + \frac{1}{r} u_r + \frac{1}{r^2} u_{\theta\theta} + u_{zz}.$$

7. Let r, θ, ϕ be *spherical coordinates,* defined by

$$x = r \cos \theta \sin \phi, \qquad y = r \sin \theta \sin \phi, \qquad z = r \cos \phi.$$

If $u(x, y, z)$ is a function of $r = \sqrt{x^2 + y^2 + z^2}$ only, show that

$$\nabla^2 u = u_{rr} + \frac{2}{r} u_r.$$

8. (Laplacian in spherical coordinates) Show that the Laplacian in the spherical coordinates r, θ, ϕ defined in Prob. 7 is

$$\nabla^2 u = u_{rr} + \frac{2}{r} u_r + \frac{1}{r^2} u_{\phi\phi} + \frac{\cot \phi}{r^2} u_\phi + \frac{1}{r^2 \sin^2 \phi} u_{\theta\theta}.$$

9. If the surface of the homogeneous solid sphere $x^2 + y^2 + z^2 \leqq R^2$ is kept at temperature zero and the initial temperature in the sphere is $f(r)$ where $r = \sqrt{x^2 + y^2 + z^2}$, show that the temperature $u(r, t)$ in the sphere is the solution of $u_t = c^2 \left(u_{rr} + \frac{2}{r} u_r \right)$, satisfying the conditions $u(R, t) = 0$, $u(r, 0) = f(r)$.

10. Show that by setting $v = ru$ the formulas in Prob. 9 take the form $v_t = c^2 v_{rr}$, $v(R, t) = 0$, $v(r, 0) = rf(r)$. Include the condition $v(0, t) = 0$ (which holds because u must be bounded at $r = 0$), and solve the resulting problem by separating variables.

11.10 Circular Membrane. Bessel's Equation

We shall now consider vibrations of the circular membrane of radius R (Fig. 244). Using polar coordinates defined by $x = r \cos \theta, y = r \sin \theta$, we see that the

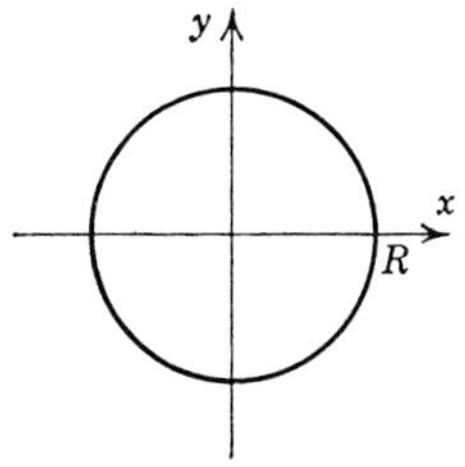

Fig. 244. Circular membrane

wave equation (3′) in Sec. 11.7 takes the form

$$\frac{\partial^2 u}{\partial t^2} = c^2\left(\frac{\partial^2 u}{\partial r^2} + \frac{1}{r}\frac{\partial u}{\partial r} + \frac{1}{r^2}\frac{\partial^2 u}{\partial \theta^2}\right),$$

cf. (4) in Sec. 11.9. In the present section we shall consider solutions $u(r, t)$ of this equation which are radially symmetric, that is, do not depend on θ. Then the wave equation reduces to the simpler form

$$\text{(1)} \qquad \frac{\partial^2 u}{\partial t^2} = c^2\left(\frac{\partial^2 u}{\partial r^2} + \frac{1}{r}\frac{\partial u}{\partial r}\right).$$

Since the membrane is fixed along the boundary $r = R$, we have the boundary condition

$$\text{(2)} \qquad u(R, t) = 0 \qquad \text{for all } t \geqq 0.$$

Solutions not depending on θ will occur if the initial conditions do not depend on θ, that is, if they are of the form

$$\text{(3)} \qquad u(r, 0) = f(r) \qquad \text{[initial deflection } f(r)\text{]}$$

and

$$\text{(4)} \qquad \left.\frac{\partial u}{\partial t}\right|_{t=0} = g(r) \qquad \text{[initial velocity } g(r)\text{]}.$$

First step. Using the method of separating variables, we first determine solutions of (1) that satisfy the boundary condition (2). We start from

$$\text{(5)} \qquad u(r, t) = W(r)G(t).$$

By differentiating and inserting (5) into (1) and dividing the resulting equation by c^2WG we obtain

$$\frac{\ddot{G}}{c^2G} = \frac{1}{W}\left(W'' + \frac{1}{r}W'\right)$$

where dots denote derivatives with respect to t and primes denote derivatives with respect to r. The expressions on both sides must be equal to a constant, and this constant must be negative, say, $-k^2$, in order to obtain solutions that satisfy the boundary condition without being identically zero. Thus,

$$\frac{\ddot{G}}{c^2G} = \frac{1}{W}\left(W'' + \frac{1}{r}W'\right) = -k^2.$$

This yields the two ordinary linear differential equations

$$\text{(6)} \qquad \ddot{G} + \lambda^2 G = 0 \qquad \text{where} \qquad \lambda = ck$$

and

$$W'' + \frac{1}{r}W' + k^2W = 0. \tag{7}$$

Second step. We first consider (7). Introducing the new independent variable $s = kr$ we have $1/r = k/s$,

$$W' = \frac{dW}{dr} = \frac{dW}{ds}\frac{ds}{dr} = \frac{dW}{ds}k \qquad \text{and} \qquad W'' = \frac{d^2W}{ds^2}k^2.$$

By substituting this into (7) and omitting the common factor k^2 we obtain

$$\frac{d^2W}{ds^2} + \frac{1}{s}\frac{dW}{ds} + W = 0.$$

This is **Bessel's equation** (1), Sec. 4.5, with $\nu = 0$. A general solution is (cf. Sec. 4.6)

$$W = C_1J_0(s) + C_2Y_0(s)$$

where J_0 and Y_0 are the Bessel functions of the first and second kind of order zero. Since the deflection of the membrane is always finite while Y_0 becomes infinite as s approaches zero, we cannot use Y_0 and must choose $C_2 = 0$. Clearly $C_1 \neq 0$ since otherwise $W \equiv 0$. We may set $C_1 = 1$, and then

$$W(r) = J_0(s) = J_0(kr). \tag{8}$$

On the boundary $r = R$ we must have $u(R, t) = W(R)G(t) = 0$. Since $G \equiv 0$ would imply $u \equiv 0$, we require that

$$W(R) = J_0(kR) = 0.$$

The Bessel function J_0 has (infinitely many) real zeros. Let us denote the positive zeros of $J_0(s)$ by $s = \alpha_1, \alpha_2, \cdots$ (cf. Fig. 245). We mention that

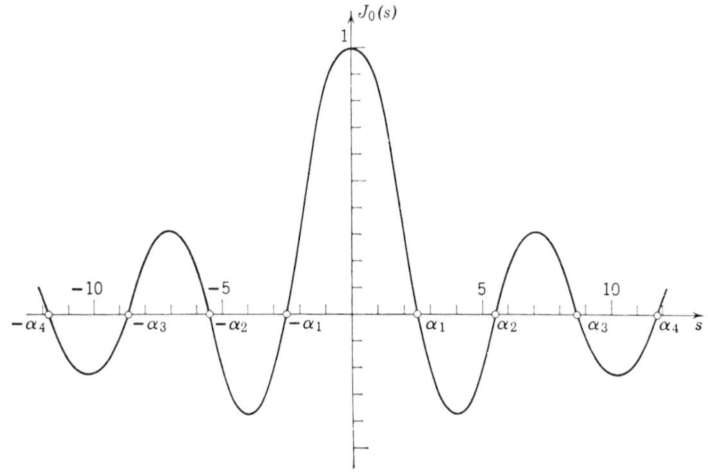

Fig. 245. Bessel function $J_0(s)$

numerical values (exact to 4 decimal places) are

$$\alpha_1 = 2.4048, \quad \alpha_2 = 5.5201, \quad \alpha_3 = 8.6537, \quad \alpha_4 = 11.7915, \quad \alpha_5 = 14.9309.$$

We see that these zeros are irregularly spaced, and the same is true for the further zeros $\alpha_6, \alpha_7, \cdots$. (More extended tables are included in Ref. [A8] in Appendix 1.) Equation (8) now implies

$$kR = \alpha_m \qquad \text{or} \qquad k = k_m = \frac{\alpha_m}{R}, \qquad m = 1, 2, \cdots. \tag{9}$$

Hence the functions

$$W_m(r) = J_0(k_m r) = J_0\left(\frac{\alpha_m}{R} r\right) \qquad m = 1, 2, \cdots \tag{10}$$

are solutions of (7) which vanish at $r = R$.

The corresponding general solutions of (6) with $\lambda = \lambda_m = ck_m$ are

$$G_m(t) = a_m \cos \lambda_m t + b_m \sin \lambda_m t.$$

Hence the functions

$$u_m(r, t) = W_m(r)G_m(t) = (a_m \cos \lambda_m t + b_m \sin \lambda_m t)J_0(k_m r) \tag{11}$$

where $m = 1, 2, \cdots$, are solutions of the wave equation (1), satisfying the boundary condition (2). These are the *eigenfunctions* of our problem, and the corresponding *eigenvalues* are λ_m.

The vibration of the membrane corresponding to u_m is called the mth **normal mode;** it has the frequency $\lambda_m/2\pi$ cycles per unit time. Since the zeros of J_0 are not regularly spaced on the axis (in contrast to the zeros of the sine functions appearing in the case of the vibrating string), the sound of a drum is entirely different from that of a violin. The forms of the normal modes can easily be obtained from Fig. 245 and are shown in Fig. 246. For $m = 1$, all the points of

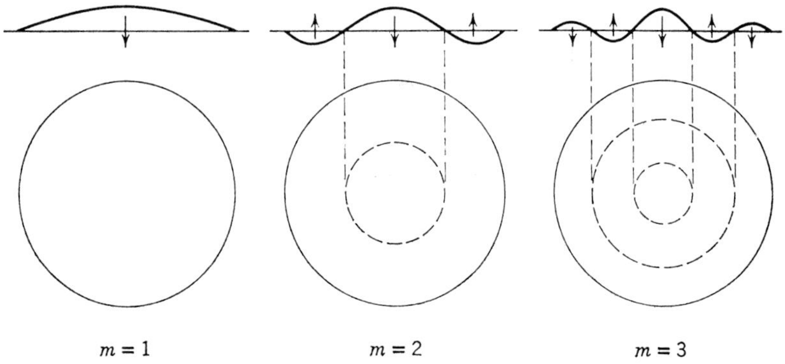

Fig. 246. Normal modes of the circular membrane in the case of vibrations independent of the angle

the membrane move upward (or downward) at the same time. For $m = 2$, the situation is as follows. The function

$$W_2(r) = J_0\left(\frac{\alpha_2}{R}r\right)$$

is zero for $\alpha_2 r/R = \alpha_1$ or $r = \alpha_1 R/\alpha_2$. The circle $r = \alpha_1 R/\alpha_2$ is, therefore, a nodal line, and when at some instant the central part of the membrane moves upward, the outer part ($r > \alpha_1 R/\alpha_2$) moves downward, and conversely. The solution $u_m(r, t)$ has $m - 1$ nodal lines which are concentric circles (Fig. 246).

Third step. To obtain a solution that also satisfies the initial conditions (3) and (4), we may proceed as in the case of the vibrating string, that is, we consider the series[7]

$$(12)\quad u(r, t) = \sum_{m=1}^{\infty} W_m(r)G_m(t) = \sum_{m=1}^{\infty} (a_m \cos \lambda_m t + b_m \sin \lambda_m t)J_0\left(\frac{\alpha_m}{R}r\right).$$

Setting $t = 0$ and using (3), we obtain

$$(13)\qquad u(r, 0) = \sum_{m=1}^{\infty} a_m J_0\left(\frac{\alpha_m}{R}r\right) = f(r).$$

Hence, for (12) to satisfy (3), the a_m must be the coefficients of the Fourier-Bessel series which represents $f(r)$ in terms of $J_0(\alpha_m r/R)$, that is [cf. (9) in Sec. 4.9],

$$a_m = \frac{2}{R^2 J_1^{\,2}(\alpha_m)} \int_0^R r f(r) J_0\left(\frac{\alpha_m}{R}r\right) dr \qquad m = 1, 2, \cdots .$$

Differentiability of $f(r)$ in the interval $0 \leqq r \leqq R$ is sufficient for the existence of the development (13), cf. Ref. [B18]. The coefficients b_m in (12) can be determined from (4) in a similar fashion. To obtain numerical values of a_m and b_m, we may apply one of the usual methods of approximate integration, using tables of J_0 and J_1.

Problems for Sec. 11.10

1. Determine numerical values of the radii of the nodal lines of u_2 and u_3 [cf. (11)] when $R = 1$.
2. Same question as in Prob. 1, for u_4, u_5, and u_6.
3. Sketch a figure similar to Fig. 246, for u_4, u_5, u_6.
4. If the tension of the membrane is increased how does the frequency of each normal mode (11) change?
5. Show that for (12) to satisfy (4),

$$b_m = \frac{2}{c\alpha_m R J_1^{\,2}(\alpha_m)} \int_0^R r g(r) J_0(\alpha_m r/R)\, dr, \qquad m = 1, 2, \cdots .$$

[7] We shall not consider the problem of convergence and uniqueness.

6. Is it possible that, for fixed c and R, two or more functions u_m [cf. (11)] that have different nodal lines correspond to the same eigenvalue?

7. (Vibrations depending on r and θ) Show that substitution of $u = F(r, \theta)G(t)$ into the wave equation

$$u_{tt} = c^2\left(u_{rr} + \frac{1}{r}u_r + \frac{1}{r^2}u_{\theta\theta}\right) \tag{14}$$

leads to

$$\ddot{G} + \lambda^2 G = 0, \qquad \text{where } \lambda = ck, \tag{15}$$

$$F_{rr} + \frac{1}{r}F_r + \frac{1}{r^2}F_{\theta\theta} + k^2 F = 0. \tag{16}$$

8. Show that substitution of $F = W(r)Q(\theta)$ in (16) yields

$$Q'' + n^2 Q = 0, \tag{17}$$

$$r^2 W'' + rW' + (k^2 r^2 - n^2)W = 0. \tag{18}$$

9. Show that $Q(\theta)$ must be periodic with period 2π, and, therefore, $n = 0, 1, \cdots$ in (17) and (18). Show that this yields the solutions $Q_n = \cos n\theta$, $Q_n^* = \sin n\theta$, $W_n = J_n(kr)$, $n = 0, 1, \cdots$.

10. Show that the boundary condition

$$u(R, \theta, t) = 0 \tag{19}$$

leads to the values $k = k_{mn} = \alpha_{mn}/R$ where $s = \alpha_{mn}$ is the mth positive zero of $J_n(s)$.

11. Show that solutions of (14) which satisfy (19) are

$$\begin{aligned} u_{mn} &= (A_{mn} \cos ck_{mn}t + B_{mn} \sin ck_{mn}t)J_n(k_{mn}r) \cos n\theta, \\ u_{mn}^* &= (A_{mn}^* \cos ck_{mn}t + B_{mn}^* \sin ck_{mn}t)J_n(k_{mn}r) \sin n\theta. \end{aligned} \tag{20}$$

12. Show that $u_{m0}^* \equiv 0$ and u_{m0} is identical with (11) in the current section.

13. Show that u_{mn} has $m + n - 1$ nodal lines.

Graph the nodal lines of the following solutions.

14. u_{3n}, $n = 1, 2, 3$ **15.** u_{4n}, $n = 1, 2, 3$ **16.** u_{mn}^*, $m, n = 1, 2, 3$

17. Show that the initial condition $u_t(r, \theta, 0) = 0$ leads to $B_{mn} = 0$, $B_{mn}^* = 0$ in (20).

Find the deflection $u(r, t)$ of the circular membrane of radius $R = 1$, if $c = 1$, the initial velocity is zero, and the initial deflection is $f(r)$ where

18. $f = 0.1 J_0(\alpha_2 r)$ **19.** $f = k(1 - r^2)$ **20.** $f = k(1 - r^4)$

Hint. Remember Probs. 25–32, Sec. 4.9.

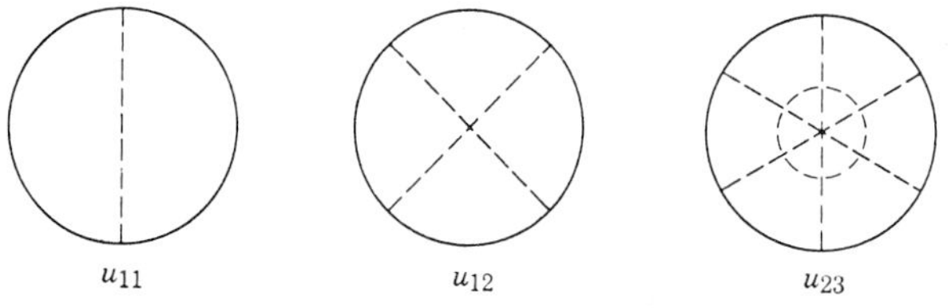

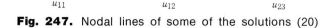

Fig. 247. Nodal lines of some of the solutions (20)

11.11 Laplace's Equation. Potential

One of the most important partial differential equations in physics is Laplace's equation

$$\nabla^2 u = 0. \tag{1}$$

Here $\nabla^2 u$ is the Laplacian of u. With respect to Cartesian coordinates x, y, z in space,

$$\nabla^2 u = \frac{\partial^2 u}{\partial x^2} + \frac{\partial^2 u}{\partial y^2} + \frac{\partial^2 u}{\partial z^2}. \tag{2}$$

The theory of the solutions of Laplace's equation is called **potential theory.** Solutions of (1) that have *continuous* second-order partial derivatives are called **harmonic functions.**

The two-dimensional case, when u depends on two variables only, can most conveniently be treated by methods of complex analysis and will be considered in Sec. 12.5 and Chap 18.

To illustrate the importance of Laplace's equation in engineering mathematics, let us mention some basic applications.

Laplace's equation occurs in connection with gravitational forces. In Example 4, Sec. 8.8, we have seen that if a particle A of mass M is fixed at a point (ξ, η, ζ) and another particle B of mass m is at a point (x, y, z), then A attracts B, the gravitational force being the gradient of the scalar function

$$u(x, y, z) = \frac{c}{r}, \qquad c = GMm = const,$$

$$r = \sqrt{(x-\xi)^2 + (y-\eta)^2 + (z-\zeta)^2} \qquad (> 0).$$

This function is called the **potential** of the gravitational field, and it satisfies Laplace's equation.

The extension to the potential and force due to a continuous distribution of mass is quite direct. If a mass of density $\rho(\xi, \eta, \zeta)$ is distributed throughout a region R in space, then the corresponding potential u at a point (x, y, z) not occupied by mass is defined to be

$$u(x, y, z) = k \iiint_R \frac{\rho}{r}\, d\xi\, d\eta\, d\zeta \qquad (k > 0). \tag{3}$$

Since $1/r$ $(r > 0)$ is a solution of (1), that is, $\nabla^2(1/r) = 0$, and ρ does not depend on x, y, z, we obtain

$$\nabla^2 u = k \iiint_R \rho \nabla^2 \left(\frac{1}{r}\right) d\xi\, d\eta\, d\zeta = 0,$$

that is, the gravitational potential defined by (3) satisfies Laplace's equation at any point which is not occupied by matter.

In *electrostatics* the electrical force of attraction or repulsion between charged particles is governed by *Coulomb's law* (cf. Sec. 8.8), which is of the same mathematical form as Newton's law of gravitation. From this it follows that the field created by a distribution of electrical charges can be described mathematically by a potential function which satisfies Laplace's equation at any point not occupied by charges.

In Chap. 18 we shall see that Laplace's equation also appears in the theory of incompressible fluid flow.

Furthermore, the basic equation in problems of heat conduction is the heat equation (cf. Secs. 9.9 and 11.5)

$$u_t = c^2 \nabla^2 u,$$

and if the temperature u is independent of time t ("*steady state*"), this equation reduces to Laplace's equation.

In most applications leading to Laplace's equation, it is required to solve a **boundary value problem,** that is, to determine the solution of (1) satisfying given boundary conditions on certain surfaces. It is then necessary to introduce coordinates in space such that those surfaces can be represented in a simple manner. This requires the transformation of the Laplacian (2) into other coordinate systems. Of course, such a transformation is quite similar to that in the case of the Laplacian of a function of two variables (cf. Sec. 11.9).

From (4) in Sec. 11.9 it follows immediately that the Laplacian of a function u in **cylindrical coordinates**[8] (cf. Fig. 248)

$$\textbf{(4)} \qquad r = \sqrt{x^2 + y^2} \qquad \theta = \arctan \frac{y}{x}, \qquad z = z$$

is

$$\textbf{(5)} \qquad \nabla^2 u = \frac{\partial^2 u}{\partial r^2} + \frac{1}{r}\frac{\partial u}{\partial r} + \frac{1}{r^2}\frac{\partial^2 u}{\partial \theta^2} + \frac{\partial^2 u}{\partial z^2}.$$

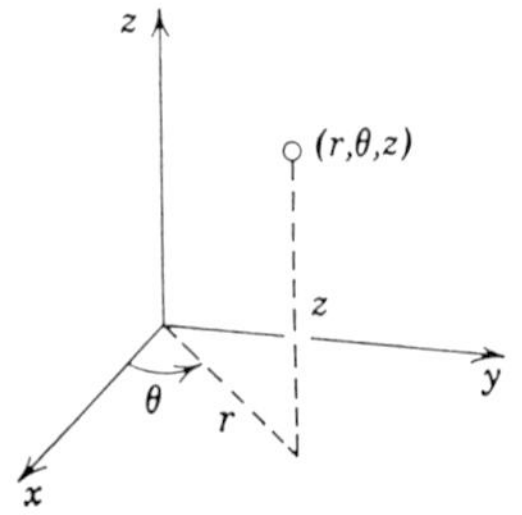

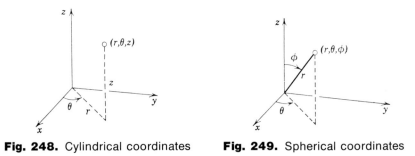

Fig. 248. Cylindrical coordinates

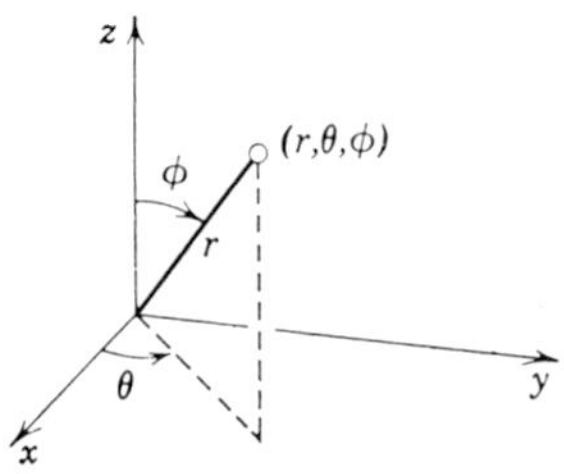

Fig. 249. Spherical coordinates

[8]Observe that θ is not completely determined by the ratio y/x but we must also take the signs of x and y into account.

Other important coordinates are **spherical coordinates**[9] r, θ, ϕ which are related to Cartesian coordinates as follows (cf. Fig. 249):

$$x = r\cos\theta\sin\phi, \qquad y = r\sin\theta\sin\phi, \qquad z = r\cos\phi. \tag{6}$$

The Laplacian of a function u in spherical coordinates is

$$\nabla^2 u = \frac{\partial^2 u}{\partial r^2} + \frac{2}{r}\frac{\partial u}{\partial r} + \frac{1}{r^2}\frac{\partial^2 u}{\partial \phi^2} + \frac{\cot\phi}{r^2}\frac{\partial u}{\partial \phi} + \frac{1}{r^2\sin^2\phi}\frac{\partial^2 u}{\partial \theta^2}. \tag{7}$$

This may also be written

$$\nabla^2 u = \frac{1}{r^2}\left[\frac{\partial}{\partial r}\left(r^2\frac{\partial u}{\partial r}\right) + \frac{1}{\sin\phi}\frac{\partial}{\partial\phi}\left(\sin\phi\frac{\partial u}{\partial\phi}\right) + \frac{1}{\sin^2\phi}\frac{\partial^2 u}{\partial\theta^2}\right]. \tag{7'}$$

This formula can be derived in a manner similar to that in Sec. 11.9; the details are left as an exercise for the reader.

Problems for Sec. 11.11

1. Find the Laplacian in rectangular coordinates $x^* = ax, y^* = by, z^* = cz$, where x, y, z are Cartesian coordinates.
2. Starting from (2), verify by calculation that with respect to the Cartesian coordinates defined by (9), (10), Sec. 8.9,

$$\nabla^2 u = u_{x^*x^*} + u_{y^*y^*} + u_{z^*z^*}.$$

Show that the following functions $u = f(x, y)$ satisfy the Laplace equation and plot some of the equipotential lines $u = const.$

3. $x^2 - y^2$
4. $x^3 - 3xy^2$
5. $x/(x^2 + y^2)$
6. $y/(x^2 + y^2)$
7. $(x^2 - y^2)/(x^2 + y^2)^2$

8. Express the spherical coordinates defined by (6) in terms of Cartesian coordinates.
9. Verify (5) by transforming $\nabla^2 u$ back into Cartesian coordinates.
10. Verify that $u = c/r$ satisfies Laplace's equation in spherical coordinates.
11. Show that the only solution of $\nabla^2 u = 0$ depending only on $r = \sqrt{x^2 + y^2 + z^2}$ is $u = c/r + k$; here c and k are constants.
12. Determine c and k in Prob. 11 such that u represents the electrostatic potential between two concentric spheres of radii $r_1 = 10$ cm and $r_2 = 2$ cm kept at the potentials $U_1 = 110$ volts and $U_2 = 10$ volts, respectively.
13. Show that the only solution of the two-dimensional Laplace equation depending only on $r = \sqrt{x^2 + y^2}$ is $u = c\ln r + k$.
14. Find the electrostatic potential between two coaxial cylinders of radii $r_1 = 10$ cm and $r_2 = 2$ cm kept at the potentials $U_1 = 110$ volts and $U_2 = 10$ volts, respectively. Graph and compare the solutions of Probs. 12 and 14.
15. If $u(r, \theta)$ satisfies $\nabla^2 u = 0$, show that $v(r, \theta) = u(r^{-1}, \theta)$ satisfies $\nabla^2 v = 0$. (r and θ are polar coordinates.)

[9] Sometimes also called *polar coordinates*.

16. Let r, θ, ϕ be spherical coordinates. If $u(r, \theta, \phi)$ satisfies $\nabla^2 u = 0$, show that $v(r, \theta, \phi) = r^{-1}u(r^{-1}, \theta, \phi)$ satisfies $\nabla^2 v = 0$.

17. Show that substitution of $u = U(x, y, z)e^{-i\omega t}$ $(i = \sqrt{-1})$ into the three-dimensional wave equation $u_{tt} = c^2\nabla^2 u$ yields the so-called **Helmholtz equation**[10]

$$\nabla^2 U + k^2 U = 0 \qquad k = \omega/c.$$

Find the steady-state (time-independent) temperature distribution:

18. Between two parallel plates $x = x_0$ and $x = x_1$ kept at the temperatures u_0 and u_1, respectively.

19. Between two coaxial circular cylinders of radii r_0 and r_1 kept at the temperatures u_0 and u_1, respectively.

20. Between two concentric spheres of radii r_0 and r_1 kept at the temperatures u_0 and u_1, respectively.

11.12 Laplace's Equation in Spherical Coordinates. Legendre's Equation

Let us consider a typical boundary value problem that involves Laplace's equation in spherical coordinates. Suppose that a sphere S of radius R is kept at a fixed distribution of electric potential

$$u(R, \theta, \phi) = f(\phi) \tag{1}$$

where r, θ, ϕ are the spherical coordinates defined in the last section, with the origin at the center of S, and $f(\phi)$ is a given function. We wish to find the potential u at all points in space which is assumed to be free of further charges. Since the potential on S is independent of θ, so is the potential in space. Thus $\partial^2 u/\partial\theta^2 = 0$, and from (7′) in Sec. 11.11 we see that Laplace's equation reduces to

$$\frac{\partial}{\partial r}\left(r^2\frac{\partial u}{\partial r}\right) + \frac{1}{\sin\phi}\frac{\partial}{\partial\phi}\left(\sin\phi\,\frac{\partial u}{\partial\phi}\right) = 0. \tag{2}$$

Furthermore, at infinity the potential will be zero:

$$\lim_{r\to\infty} u(r, \phi) = 0. \tag{3}$$

We shall solve the boundary value problem consisting of the equation (2) and the boundary conditions (1) and (3) by the method of separating variables. Substituting a solution of the form

$$u(r, \phi) = G(r)H(\phi)$$

[10] HERMANN VON HELMHOLTZ (1821–1894), German physicist, known by his important work in thermodynamics, hydrodynamics and acoustics.

into (2), and dividing the resulting equation by GH, we obtain

$$\frac{1}{G}\frac{d}{dr}\left(r^2\frac{dG}{dr}\right) = -\frac{1}{H\sin\phi}\frac{d}{d\phi}\left(\sin\phi\,\frac{dH}{d\phi}\right).$$

By the usual argument, the two sides of this equation must be equal to a constant, say, k, so that

$$\frac{1}{\sin\phi}\frac{d}{d\phi}\left(\sin\phi\,\frac{dH}{d\phi}\right) + kH = 0 \tag{4}$$

and

$$\frac{1}{G}\frac{d}{dr}\left(r^2\frac{dG}{dr}\right) = k.$$

The last equation may be written

$$r^2G'' + 2rG' - kG = 0.$$

This is Cauchy's equation, and from Sec. 2.7 we know that it has solutions of the form $G = r^\alpha$. These solutions will have a particularly simple form if we change our notation and write $n(n+1)$ for k. In fact, then the equation becomes

$$r^2G'' + 2rG' - n(n+1)G = 0 \tag{5}$$

where n is still arbitrary. By substituting $G = r^\alpha$ into (5) we have

$$[\alpha(\alpha - 1) + 2\alpha - n(n+1)]\,r^\alpha = 0.$$

The zeros of the expression in brackets are $\alpha = n$ and $\alpha = -n - 1$. Hence we obtain the solutions

$$G_n(r) = r^n \qquad \text{and} \qquad G_n{}^*(r) = \frac{1}{r^{n+1}}. \tag{6}$$

Introducing $k = n(n+1)$ in (4) and setting

$$\cos\phi = w,$$

we have $\sin^2\phi = 1 - w^2$ and

$$\frac{d}{d\phi} = \frac{d}{dw}\frac{dw}{d\phi} = -\sin\phi\,\frac{d}{dw}.$$

Consequently, (4) takes the form

$$\frac{d}{dw}\left[(1 - w^2)\frac{dH}{dw}\right] + n(n+1)H = 0 \tag{7}$$

or

$$(1 - w^2)\frac{d^2H}{dw^2} - 2w\frac{dH}{dw} + n(n+1)H = 0. \tag{7'}$$

This is *Legendre's equation* (cf. Sec. 4.3). For integral[11] $n = 0, 1, \cdots$, the Legendre polynomials

$$H = P_n(w) = P_n(\cos\phi), \qquad n = 0, 1, \cdots,$$

are solutions of (7). We thus obtain the following two sequences of solutions $u = GH$ of Laplace's equation (2):

$$\text{(8*)} \qquad u_n(r, \phi) = A_n r^n P_n(\cos\phi), \qquad u_n^*(r, \phi) = \frac{B_n}{r^{n+1}} P_n(\cos\phi)$$

where $n = 0, 1, \cdots$, and A_n and B_n are constants.

To find a solution of (2), valid at points *inside* the sphere and satisfying (1), we consider the series[12]

$$\textbf{(8)} \qquad u(r, \phi) = \sum_{n=0}^{\infty} A_n r^n P_n(\cos\phi).$$

For (8) to satisfy (1) we must have

$$\text{(9)} \qquad u(R, \phi) = \sum_{n=0}^{\infty} A_n R^n P_n(\cos\phi) = f(\phi);$$

that is, (9) must be the generalized Fourier series of $f(\phi)$ in terms of Legendre polynomials. From (6) and (7) in Sec. 4.7 and (2) in Sec. 4.9 it follows that

$$A_n R^n = \frac{2n+1}{2} \int_{-1}^{1} \tilde{f}(w) P_n(w)\, dw$$

where $\tilde{f}(w)$ denotes $f(\phi)$ as a function of $w = \cos\phi$. Since $dw = -\sin\phi\, d\phi$, and the limits of integration -1 and 1 correspond to $\phi = \pi$ and $\phi = 0$, respectively, we also have

$$\textbf{(10)} \qquad A_n = \frac{2n+1}{2R^n} \int_0^{\pi} f(\phi) P_n(\cos\phi) \sin\phi\, d\phi, \qquad n = 0, 1, \cdots.$$

Thus the series (8) with coefficients (10) is the solution of our problem for points inside the sphere.

To find the solution *exterior* to the sphere, we cannot use the functions

[11] So far, n was arbitrary since k was arbitrary. It can be shown that the restriction of n to real integers is necessary to make the solution of (7) continuous, together with its derivative of the first order, in the interval $-1 \leqq w \leqq 1$ or $0 \leqq \phi \leqq \pi$.

[12] Convergence will not be considered. It can be shown that if $f(\phi)$ and $f'(\phi)$ are piecewise continuous in the interval $0 \leqq \phi \leqq \pi$, the series (8) with coefficients (10) can be differentiated termwise twice with respect to r and with respect to ϕ and the resulting series converge and represent $\partial^2 u/\partial r^2$ and $\partial^2 u/\partial \phi^2$, respectively. Hence the series (8) with coefficients (10) is then the solution of our problem inside the sphere.

$u_n(r, \phi)$ because these functions do not satisfy (3), but we may use the functions $u_n^*(r, \phi)$, which satisfy (3), and proceed as before. This leads to the solution

$$u(r, \phi) = \sum_{n=0}^{\infty} \frac{B_n}{r^{n+1}} P_n(\cos \phi) \qquad (r \geqq R) \tag{11}$$

with coefficients

$$B_n = \frac{2n + 1}{2} R^{n+1} \int_0^{\pi} f(\phi) P_n(\cos \phi) \sin \phi \, d\phi. \tag{12}$$

Problems for Sec. 11.12

1. Verify by substitution that $u_n(r, \phi)$ and $u_n^*(r, \phi)$, $n = 0, 1, 2$, in (8*) are solutions of (2).
2. Find the surfaces on which the functions u_1, u_2, u_3 are zero.
3. Graph the functions $P_n(\cos \phi)$ for $n = 0, 1, 2$, [cf. (11′), Sec. 4.3].
4. Graph the functions $P_3(\cos \phi)$ and $P_4(\cos \phi)$.

Let r, θ, ϕ be the spherical coordinates used in the text. Find the potential in the interior of the sphere $R = 1$, assuming that there are no charges in the interior and the potential on the surface is $f(\phi)$ where

5. $f(\phi) = 1$
6. $f(\phi) = \cos \phi$
7. $f(\phi) = \cos^2 \phi$
8. $f(\phi) = \cos^3 \phi$
9. $f(\phi) = \cos 2\phi$
10. $f(\phi) = \cos 3\phi$
11. $f(\phi) = 10 \cos^3 \phi - 3 \cos^2 \phi - 5 \cos \phi - 1$

12. Show that in Prob. 5, the potential exterior to the sphere is the same as that of a point charge at the origin.
13. Graph the intersections of the equipotential surfaces in Prob. 6 with the xz-plane.
14. Find the potential exterior to the sphere in Probs. 5–11.
15. Verify by substitution that the solution of Prob. 7 satisfies (2).
16. **(Transmission line equations)** Consider a long cable or telephone wire (see figure) which is imperfectly insulated so that leaks occur along the entire length of the cable. The source S of the current $i(x, t)$ in the cable is at $x = 0$, the receiving end T at $x = l$. The current flows from S to T, through the load, and returns to the

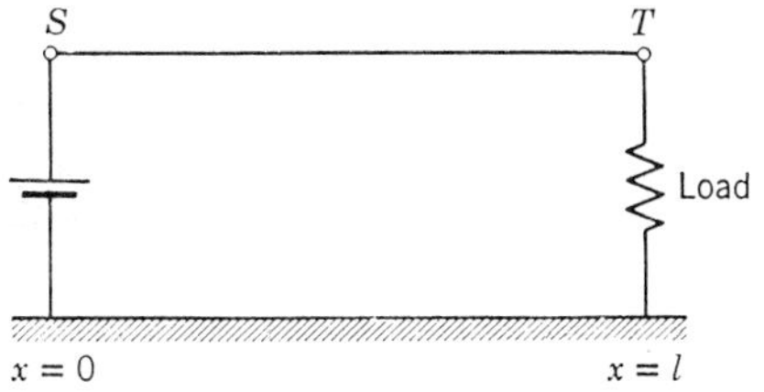

Problem 16. Transmission line

ground. Let the constants R, L, C, and G denote the resistance, inductance, capacitance to ground and conductance to ground, respectively, of the cable per unit length. Show that

$$-\frac{\partial u}{\partial x} = Ri + L\frac{\partial i}{\partial t} \qquad \textbf{(First transmission line equation)}$$

where $u(x, t)$ is the potential in the cable. *Hint.* Apply Kirchhoff's voltage law to a small portion of the cable between x and $x + \Delta x$ (difference of the potentials at x and $x + \Delta x =$ resistive drop + inductive drop).

17. Show that for the cable in Prob. 16,

$$-\frac{\partial i}{\partial x} = Gu + C\frac{\partial u}{\partial t} \qquad \textbf{(Second transmission line equation).}$$

Hint. Use Kirchhoff's current law (difference of the currents at x and $x + \Delta x =$ loss due to leakage to ground + capacitive loss).

18. Show that elimination of i from the transmission line equations leads to

$$u_{xx} = LCu_{tt} + (RC + GL)u_t + RGu.$$

19. **(Telegraph equations)** For a submarine cable G is negligible and the frequencies are low. Show that this leads to the so-called *submarine cable equations* or **telegraph equations**

$$u_{xx} = RCu_t, \qquad i_{xx} = RCi_t.$$

20. **(High frequency line equations)** Show that in the case of alternating currents of high frequencies the equation in Prob. 18 and the analogous equation for the current i can be approximated by the so-called **high frequency line equations**

$$u_{xx} = LCu_{tt}, \qquad i_{xx} = LCi_{tt}.$$

11.13 Laplace Transformation Applied to Partial Differential Equations

Readers familiar with the Laplace transformation (Chap. 5) may wonder whether that method can also be used for solving *partial* differential equations. The answer is yes, and the basic idea may be explained in terms of two typical examples (below). (More complicated problems become accessible by the use of methods of complex analysis; cf. Ref. [B3] listed in Appendix 1. Sometimes the Fourier transformation (cf. Sec. 10.9) is preferable; for a discussion of this point, see Ref. [B16].)

We remember from Chap. 5 that in the case of an *ordinary* differential equation, the Laplace transformation yields an algebraic equation for the transform. We shall see that in the case of a *partial* differential equation in two independent variables, the method yields an ordinary differential equation for the transform. This is so because we transform the given equation with respect to *one* of the independent variables, usually t, so that derivatives with respect to the other independent variable slip into the transformed equation. If the coefficients of the given equation do not depend on t, the transformation will simplify the problem.

Example 1. A first-order equation

Solve the problem

$$\frac{\partial w}{\partial x} + x\frac{\partial w}{\partial t} = 0, \qquad w(x, 0) = 0, \quad w(0, t) = t \qquad (t \geqq 0). \tag{1}$$

We write w since we need u to denote the unit step function (Sec. 5.3).

Solution. We take the Laplace transform of (1) with respect to t. By (1) in Sec. 5.2,

$$\mathscr{L}\left\{\frac{\partial w}{\partial x}\right\} + x[s\mathscr{L}\{w\} - w(x, 0)] = 0. \tag{2}$$

Here $w(x, 0) = 0$. In the first term we assume that we may interchange integration and differentiation:

$$\mathscr{L}\left\{\frac{\partial w}{\partial x}\right\} = \int_0^\infty e^{-st}\frac{\partial w}{\partial x}\,dt = \frac{\partial}{\partial x}\int_0^\infty e^{-st}w(x, t)\,dt = \frac{\partial}{\partial x}\mathscr{L}\{w(x, t)\}. \tag{3}$$

Writing $W(x, s) = \mathscr{L}\{w(x, t)\}$, we thus obtain from (2)

$$\frac{\partial W}{\partial x} + xsW = 0.$$

This may be regarded as an *ordinary* differential equation with x as the independent variable since derivatives with respect to s do not occur in the equation. The general solution is (Sec. 1.7)

$$W(x, s) = c(s)e^{-sx^2/2}.$$

Since $\mathscr{L}\{t\} = 1/s^2$, the condition $w(0, t) = t$ yields $W(0, s) = 1/s^2$, that is,

$$W(0, s) = c(s) = 1/s^2.$$

Hence

$$W(x, s) = \frac{1}{s^2}e^{-sx^2/2}.$$

Now $\mathscr{L}^{-1}\{1/s^2\} = t$, so that the second shifting theorem (Sec. 5.3) with $a = x^2/2$ gives

$$w(x, t) = \left(t - \frac{x^2}{2}\right)u_{x^2/2}(t) = \begin{cases} 0 & \text{if } t < x^2/2 \\ t - \frac{1}{2}x^2 & \text{if } t > x^2/2 \end{cases} \tag{4}$$

Since we proceeded formally, we have to verify that (4) satisfies (1). We leave this to the reader.

Example 2. Semi-infinite string

Find the displacement[13] $w(x, t)$ of an elastic string subject to the following conditions.

(*i*) The string is initially at rest on the x-axis from $x = 0$ to ∞ (*"semi-infinite string"*).

(*ii*) For time $t > 0$ the left end of the string is moved in a given fashion (Fig. 250):

$$w(0, t) = f(t) = \begin{cases} \sin t & \text{if } 0 \leqq t \leqq 2\pi \\ 0 & \text{otherwise} \end{cases}$$

(*iii*) Furthermore,

$$\lim_{x\to\infty} w(x, t) = 0 \qquad \text{for } t \geqq 0.$$

Of course there is no infinite string, but our model describes a long string or rope (of negligible weight) with its right end fixed far out on the x-axis.

[13] We write w since we need u to denote the unit step function.

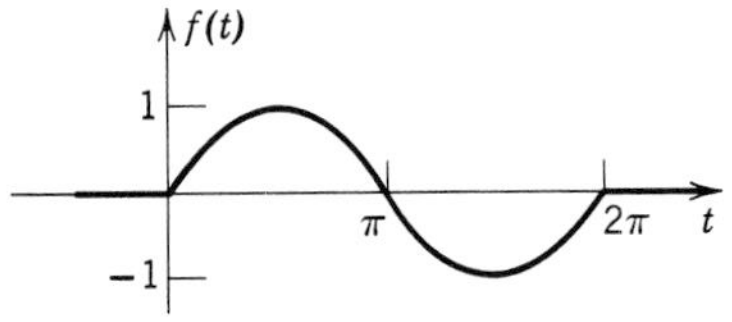

Fig. 250. Motion of the left end of the string in Example 2 as a function of time t

Solution. We have to solve the wave equation (Sec. 11.2)

$$\frac{\partial^2 w}{\partial t^2} = c^2 \frac{\partial^2 w}{\partial x^2} \qquad c^2 = \frac{T}{\rho} \tag{5}$$

under the "boundary conditions"

$$w(0, t) = f(t), \qquad \lim_{x \to \infty} w(x, t) = 0 \qquad (t \geqq 0) \tag{6}$$

with f as given above, and the initial conditions

$$w(x, 0) = 0 \tag{7}$$

$$\left. \frac{\partial w}{\partial t} \right|_{t=0} = 0. \tag{8}$$

We take the Laplace transform *with respect to t*. By (2) in Sec. 5.2,

$$\mathcal{L}\left\{\frac{\partial^2 w}{\partial t^2}\right\} = s^2 \mathcal{L}\{w\} - sw(x, 0) - \left.\frac{\partial w}{\partial t}\right|_{t=0} = c^2 \mathcal{L}\left\{\frac{\partial^2 w}{\partial x^2}\right\}.$$

Two terms drop out, by (7) and (8). On the right we assume that we may interchange integration and differentiation:

$$\mathcal{L}\left\{\frac{\partial^2 w}{\partial x^2}\right\} = \int_0^\infty e^{-st} \frac{\partial^2 w}{\partial x^2}\, dt = \frac{\partial^2}{\partial x^2} \int_0^\infty e^{-st} w(x, t)\, dt = \frac{\partial^2}{\partial x^2} \mathcal{L}\{w(x, t)\}.$$

Writing $W(x, s) = \mathcal{L}\{w(x, t)\}$, we thus obtain

$$s^2 W = c^2 \frac{\partial^2 W}{\partial x^2}$$

or

$$\frac{\partial^2 W}{\partial x^2} - \frac{s^2}{c^2} W = 0.$$

Since this equation contains only a derivative with respect to x, it may be regarded as an ordinary differential equation for $W(x, s)$ considered as a function of x. A general solution is

$$W(x, s) = A(s) e^{sx/c} + B(s) e^{-sx/c}. \tag{9}$$

From (6) we obtain, writing $F(s) = \mathcal{L}\{f(t)\}$,

$$W(0, s) = \mathcal{L}\{w(0, t)\} = \mathcal{L}\{f(t)\} = F(s)$$

and, assuming that the order of integrating with respect to t and taking the limit as $x \to \infty$ can be interchanged,

$$\lim_{x \to \infty} W(x, s) = \lim_{x \to \infty} \int_0^\infty e^{-st} w(x, t)\, dt$$

$$= \int_0^\infty e^{-st} \lim_{x \to \infty} w(x, t)\, dt = 0.$$

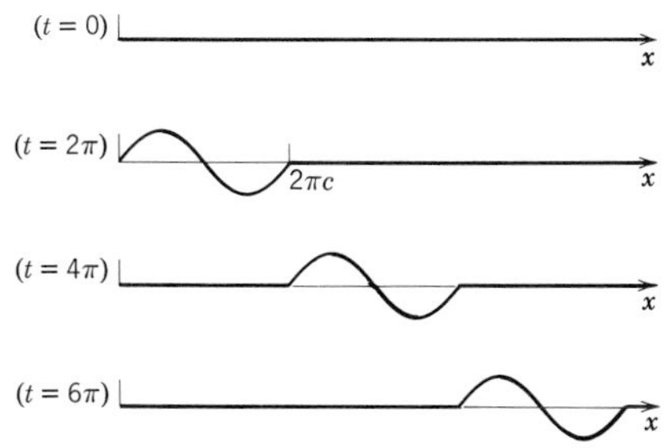

Fig. 251. Traveling wave in Example 2

This implies $A(s) = 0$ in (9) because $c > 0$, so that for every fixed positive s the function $e^{sx/c}$ increases as x increases. Note that we may assume $s > 0$ since a Laplace transform generally exists for *all* s greater than some fixed γ (Sec. 5.2). Hence we have

$$W(0, s) = B(s) = F(s),$$

so that (9) becomes

$$W(x, s) = F(s)e^{-sx/c}.$$

From the second shifting theorem (Sec. 5.3) with $a = x/c$ we obtain the inverse transform (Fig. 251)

$$w(x, t) = f\left(t - \frac{x}{c}\right)u_{x/c}(t), \tag{10}$$

that is,

$$w(x, t) = \sin\left(t - \frac{x}{c}\right) \quad \text{if} \quad \frac{x}{c} < t < \frac{x}{c} + 2\pi \quad \text{or} \quad ct > x > (t - 2\pi)c$$

and zero otherwise. This is a single sine wave traveling to the right with speed c. Note that a point x remains at rest until $t = x/c$, the time needed to reach that x if one starts at $t = 0$ (start of the motion of the left end) and travels with speed c. The result agrees with our physical intuition. Since we proceeded formally, we must verify that (10) satisfies the given conditions. We leave this step to the reader.

Problems for Sec. 11.13

1. Sketch a figure similar to Fig. 251 if $c = 1$ and f is "triangular" as in Example 1, Sec. 11.3, with $k = l/2 = 1$.
2. How does the speed of the wave in Example 2 depend on the tension and the mass of the string?
3. Verify the solution in Example 2. What traveling wave do we obtain in Example 2 if we impose a (nonterminating) sinusoidal motion of the left end starting at $t = 0$?

Solve by the Laplace transformation:

4. $\dfrac{\partial u}{\partial x} + 2x\dfrac{\partial u}{\partial t} = 2x,\ u(x, 0) = 1,\ u(0, t) = 1$

5. $x\dfrac{\partial u}{\partial x} + \dfrac{\partial u}{\partial t} = xt,\ u(x, 0) = 0$ if $x \geqq 0$, $u(0, t) = 0$ if $t \geqq 0$.

6. Solve Prob. 5 by another method.

Find the temperature $w(x, t)$ in a semi-infinite laterally insulated bar extending from $x = 0$ along the x-axis to ∞, assuming that the initial temperature is zero, $w(x, t) \to 0$ as $x \to \infty$ for every fixed $t \geqq 0$, and $w(0, t) = f(t)$. Proceed as follows.

7. Set up the model and show that the Laplace transformation leads to

$$sW(x, s) = c^2 \frac{\partial^2 W}{\partial x^2} \qquad W = \mathscr{L}\{w\}$$

and

$$W(x, s) = F(s)e^{-\sqrt{s}\,x/c} \qquad F = \mathscr{L}\{f\}.$$

8. Applying the convolution theorem in Prob. 7, show that

$$w(x, t) = \frac{x}{2c\sqrt{\pi}} \int_0^t f(t - \tau)\tau^{-3/2} e^{-x^2/4c^2\tau}\, d\tau.$$

9. Let $w(0, t) = f(t) = u_0(t)$ (Sec. 5.3). Denote the corresponding w, W and F by w_0, W_0 and F_0. Show that then in Prob. 8,

$$w_0(x, t) = \frac{x}{2c\sqrt{\pi}} \int_0^t \tau^{-3/2} e^{-x^2/4c^2\tau}\, d\tau = 1 - \operatorname{erf}\left(\frac{x}{2c\sqrt{t}}\right)$$

with the error function erf as defined in the problem set for Sec. 11.6.

10. **(Duhamel's formula)** Show that in Prob. 9,

$$W_0(x, s) = \frac{1}{s} e^{-\sqrt{s}\,x/c}$$

and by the convolution theorem

$$w(x, t) = \int_0^t f(t - \tau) \frac{\partial w_0}{\partial \tau}\, d\tau.$$

[This is one form of what is known as *Duhamel's formula*. It expresses the solution w corresponding to any given f in terms of the solution w_0 corresponding to the simple $f(t) = u_0(t)$.]

CHAPTER

12 COMPLEX NUMBERS. COMPLEX ANALYTIC FUNCTIONS

Many engineering problems may be treated and solved by methods of complex analysis. Roughly speaking, these problems can be subdivided into two large classes. The first class consists of "elementary problems" for which the knowledge of complex numbers gained in college algebra and calculus is sufficient. For example, many applications in connection with electric circuits and mechanical vibrating systems are of this type. The second class of problems requires a detailed knowledge of the theory of complex analytic functions ("theory of functions," for short) and the powerful and elegant methods used in this branch of mathematics. Interesting problems in the theory of heat, in fluid dynamics, and in electrostatics belong to this category.

The following chapters will be devoted to the major parts of the theory of complex analytic functions and their applications. We shall see that the importance of these functions in engineering mathematics has the following three main roots.

1. *The real and imaginary parts of an analytic function are solutions of Laplace's equation in two independent variables. Consequently, two-dimensional potential problems can be treated by methods developed in connection with analytic functions.*
2. *Many complicated real and complex integrals that occur in applications can be evaluated by methods of complex integration.*
3. *The majority of nonelementary functions appearing in engineering mathematics are analytic functions, and the consideration of these functions for complex values of the independent variable leads to a much deeper and more detailed knowledge of their properties.*

In the present chapter we shall consider complex numbers and analytic functions and their general properties. The second half of the chapter will be devoted to the most important elementary complex functions.

Prerequisites for this chapter: elementary calculus.
References: Appendix 1, Part F.
Answers to problems: Appendix 2.

12.1 Complex Numbers

It was observed early in history that there are equations which are not satisfied by any real number, for instance,

$$x^2 + 3 = 0, \qquad x^2 - 10x + 40 = 0.$$

This led to the introduction of complex numbers.[1]

Definition

A **complex number** z is an ordered pair (x, y) of real numbers x, y and we write

$$z = (x, y).$$

We call x the **real part** of z and y the **imaginary part** of z and write

$$\operatorname{Re} z = x, \qquad \operatorname{Im} z = y.$$

For example, Re $(4, -3) = 4$ and Im $(4, -3) = -3$. Furthermore, we define two complex numbers $z_1 = (x_1, y_1)$ and $z_2 = (x_2, y_2)$ to be **equal** if and only if their real parts are equal and their imaginary parts are equal.

Addition of complex numbers $z_1 = (x_1, y_1)$ and $z_2 = (x_2, y_2)$ is defined by the rule

$$z_1 + z_2 = (x_1, y_1) + (x_2, y_2) = (x_1 + x_2, y_1 + y_2). \tag{1}$$

Multiplication is defined by the rule

$$z_1 z_2 = (x_1, y_1)(x_2, y_2) = (x_1 x_2 - y_1 y_2, x_1 y_2 + x_2 y_1). \tag{2}$$

We shall say more about these arithmetical operations below.

Representation in the form $z = x + iy$

A complex number whose imaginary part is zero is of the form $(x, 0)$. For such numbers we simply have from (1) and (2)

$$(x_1, 0) + (x_2, 0) = (x_1 + x_2, 0)$$

$$(x_1, 0)(x_2, 0) = (x_1 x_2, 0),$$

as for real numbers. This suggests identifying $(x, 0)$ with the real number x.

[1]First to use complex numbers for this purpose was the Italian mathematician GIROLAMO CARDANO (1501–1576), who found the formula for solving cubic equations. The term "complex number" was introduced by the great German mathematician CARL FRIEDRICH GAUSS (cf. the footnote in Sec. 4.4), who also paved the way for a general and systematic use of complex numbers.

Hence the complex number system is an *extension* of the real number system. Furthermore, the complex number

$$i = (0, 1)$$

is called the **imaginary unit.** By (2), for every real y,

$$iy = (0, 1)(y, 0) = (0, y).$$

Noting that, by (1), for $z = (x, y)$ we have

$$(x, y) = (x, 0) + (0, y),$$

writing x for $(x, 0)$ and using $(0, y) = iy$, we thus obtain the convenient form

$$z = x + iy$$

which is used in practice almost exclusively. An important property of i is

$$i^2 = -1. \tag{3}$$

This follows from (2). Indeed, $i^2 = (0, 1)(0, 1) = (-1, 0) = -1$.

Complex plane

This is a geometrical representation of complex numbers as points in the plane. It is of great importance in applications. The idea is quite simple and natural. We choose two perpendicular coordinate axes, the horizontal x-axis, called the **real axis,** and the vertical y-axis, called the **imaginary axis.** On both axes we choose the same unit of length (Fig. 252). This is called a **Cartesian coordinate system.**[2] We now plot $z = (x, y) = x + iy$ as the point P with coordinates x, y. The xy-plane in which the complex numbers are represented in this way is called the **complex plane** or *Argand diagram.*[3]

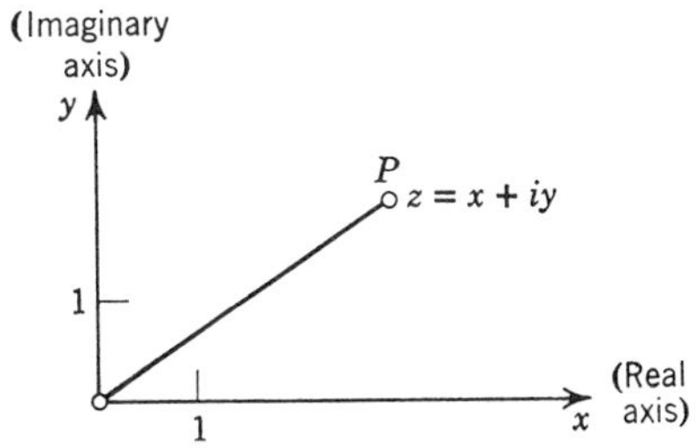

Fig. 252. The complex plane

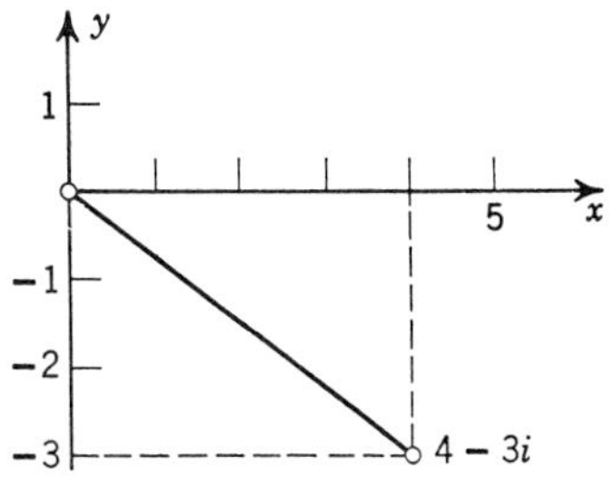

Fig. 253. The number $4 - 3i$ in the complex plane

[2]Named after the French philosopher and mathematician RENATUS CARTESIUS (latinized for RENÉ DESCARTES (1596–1650)), who invented analytic geometry.

[3]JEAN ROBERT ARGAND (1768–1822), French mathematician. His paper on the complex plane appeared in 1806, nine years after a similar memoir by the Norwegian mathematician CASPAR WESSEL (1745–1818).

Instead of saying "the point represented by z in the complex plane" we say briefly and simply **"the point z in the complex plane."** This will cause no misunderstandings.

Arithmetical operations

We can now make use of the notation $z = x + iy$ and of the complex plane.

Addition. The sum $z_1 + z_2$ in (1) can now be written

$$z_1 + z_2 = (x_1 + x_2) + i(y_1 + y_2). \tag{4}$$

We see that addition of complex numbers is in accordance with the "*parallelogram law*" by which forces are added in mechanics (Fig. 254).

Subtraction. This operation is defined as the inverse operation of addition; that is, the difference $z_1 - z_2$ is the complex number z for which $z_1 = z + z_2$. Obviously (cf. Fig. 255)

$$z_1 - z_2 = (x_1 - x_2) + i(y_1 - y_2). \tag{5}$$

Multiplication. The product $z_1 z_2$ in (2) can now be written

$$z_1 z_2 = (x_1 + iy_1)(x_2 + iy_2) = (x_1 x_2 - y_1 y_2) + i(x_1 y_2 + x_2 y_1). \tag{6}$$

This is easy to remember since it is obtained formally by the rules of arithmetic for real numbers and using (3), that is, $i^2 = ii = -1$.

Division. This operation is defined as the inverse operation of multiplication; that is, the quotient $z = z_1/z_2$ is the complex number $z = x + iy$ for which

$$z_1 = zz_2 = (x + iy)(x_2 + iy_2) \qquad (z_2 \neq 0). \tag{7}$$

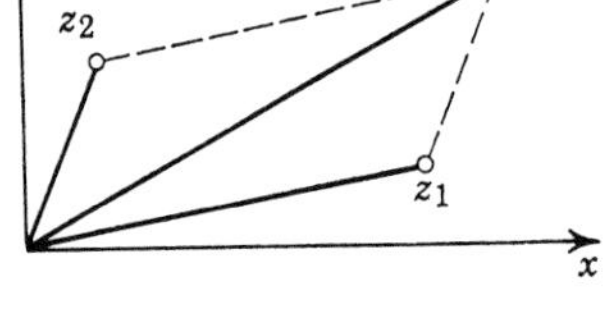

Fig. 254. Addition of complex numbers

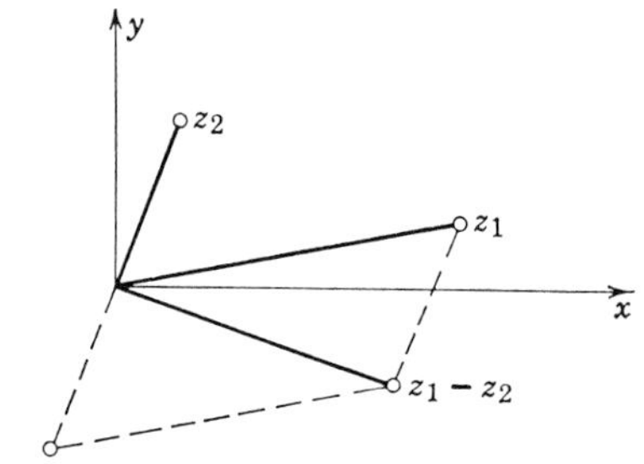

Fig. 255. Subtraction of complex numbers

We show that for $z_2 \neq 0$ the quotient $z = x + iy = z_1/z_2$ is given by

$$(8^*) \qquad x = \frac{x_1x_2 + y_1y_2}{x_2^2 + y_2^2}, \qquad y = \frac{x_2y_1 - x_1y_2}{x_2^2 + y_2^2} \qquad (z_2 \neq 0).$$

The practical rule for getting (8*) is the multiplication of both the numerator and the denominator of the quotient z_1/z_2 by $x_2 - iy_2$ and simplification:

$$\textbf{(8)} \quad z = \frac{x_1 + iy_1}{x_2 + iy_2} = \frac{(x_1 + iy_1)(x_2 - iy_2)}{(x_2 + iy_2)(x_2 - iy_2)} = \frac{x_1x_2 + y_1y_2}{x_2^2 + y_2^2} + i\frac{x_2y_1 - x_1y_2}{x_2^2 + y_2^2}.$$

For example, if $z_1 = 2 - 3i$ and $z_2 = -5 + i$, then

$$\frac{2 - 3i}{-5 + i} = \frac{(2 - 3i)(-5 - i)}{(-5 + i)(-5 - i)} = \frac{-10 - 2i + 15i - 3}{25 + 1} = -\frac{1}{2} + \frac{i}{2},$$

a result which the reader may check by showing that

$$zz_2 = \left(-\frac{1}{2} + \frac{i}{2}\right)(-5 + i) = 2 - 3i = z_1.$$

A proof of (8*) runs as follows. From (6) we see that (7) can be written

$$x_1 + iy_1 = (x_2x - y_2y) + i(y_2x + x_2y).$$

By the definition of equality the real parts and the imaginary parts on both sides must be equal:

$$x_1 = x_2x - y_2y$$

$$y_1 = y_2x + x_2y.$$

This is a system of two linear equations in the unknowns x and y. Assuming that x_2 and y_2 are not both zero (briefly written $z_2 \neq 0$), we obtain the unique solution (8*). ∎

Example 1. Addition, subtraction, multiplication, division

Let $z_1 = 2 - 3i$ and $z_2 = -5 + i$. Then

$$z_1 + z_2 = -3 - 2i, \quad z_1 - z_2 = 7 - 4i, \quad z_1z_2 = -7 + 17i$$

and (see before) $z_1/z_2 = (-1 + i)/2$.

Properties of the arithmetic operations

From the familiar laws for *real* numbers we obtain for any complex numbers z_1, z_2, z_3 the laws

$$\left.\begin{aligned} z_1 + z_2 &= z_2 + z_1 \\ z_1 z_2 &= z_2 z_1 \end{aligned}\right\} \quad (\textit{Commutative laws})$$

$$\left.\begin{aligned} (z_1 + z_2) + z_3 &= z_1 + (z_2 + z_3) \\ (z_1 z_2) z_3 &= z_1 (z_2 z_3) \end{aligned}\right\} \quad (\textit{Associative laws})$$

(9)

$$z_1(z_2 + z_3) = z_1 z_2 + z_1 z_3 \qquad (\textit{Distributive law})$$

$$0 + z = z + 0 = z$$

$$z + (-z) = (-z) + z = 0$$

$$z \cdot 1 = z$$

where $0 = (0, 0)$ and $-z = -x - iy$.

Complex conjugate numbers

Let $z = x + iy$ be any complex number. Then $x - iy$ is called the *conjugate* of z and is denoted by $\bar{z}$. Thus,

$$z = x + iy, \qquad \bar{z} = x - iy.$$

For example, the conjugate of $z = 5 + 2i$ is $\bar{z} = 5 - 2i$ (Fig. 256). Furthermore, by addition and subtraction,

$$z + \bar{z} = 2x, \qquad z - \bar{z} = 2iy.$$

This yields the important formulas

$$\text{(10)} \qquad \operatorname{Re} z = x = \frac{1}{2}(z + \bar{z}), \qquad \operatorname{Im} z = y = \frac{1}{2i}(z - \bar{z}).$$

In particular, for a real number $z = x$ we have $z = \bar{z}$. For $z = 0 + iy = iy$ we have $\bar{z} = -z$. Such a number whose real part is zero is called a **pure imaginary number.** It corresponds to a point on the imaginary axis.

We also have

$$\text{(11)} \qquad \begin{aligned} \overline{(z_1 + z_2)} &= \bar{z}_1 + \bar{z}_2, & \overline{(z_1 - z_2)} &= \bar{z}_1 - \bar{z}_2, \\ \overline{(z_1 z_2)} &= \bar{z}_1 \bar{z}_2, & \overline{\left(\frac{z_1}{z_2}\right)} &= \frac{\bar{z}_1}{\bar{z}_2}. \end{aligned}$$

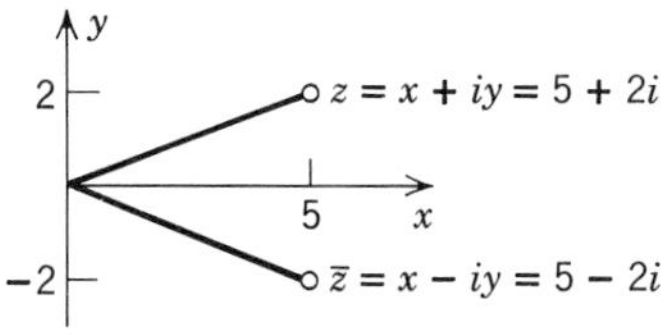

Fig. 256. Complex conjugate numbers

Problems for Sec. 12.1

1. **(Powers of the imaginary unit)** Show that

$$i^2 = -1, \quad i^3 = -i, \quad i^4 = 1, \quad i^5 = i, \quad \cdots$$

(12)

$$\frac{1}{i} = -i, \quad \frac{1}{i^2} = -1, \quad \frac{1}{i^3} = i, \quad \cdots .$$

Let $z_1 = 3 + 4i$ and $z_2 = 5 - 2i$. Find (in the form $x + iy$)

2. $(z_1 - z_2)^2$ **3.** z_1/z_2 **4.** $1/z_1^{\,2}$ **5.** $z_2/2z_1$

Find

6. $\operatorname{Re}\dfrac{1}{2+i}$ **7.** $\operatorname{Im}\dfrac{2+i}{3+4i}$ **8.** $\operatorname{Re} z^3$ **9.** $\operatorname{Im} z^4$

10. $\operatorname{Re}\dfrac{(1+i)^2}{3+2i}$ **11.** $\operatorname{Im}\dfrac{2-i}{4-3i}$ **12.** $\operatorname{Im}\dfrac{1}{z^2}$ **13.** $\operatorname{Re}\dfrac{z}{\bar{z}}$

14. Prove the commutative laws in (9).
15. Prove the associative laws in (9).
16. Prove the distributive law in (9).
17. If a product of two complex numbers is zero, show that at least one factor is zero.

Prove:

18. Any number is equal to the conjugate of its conjugate.
19. $\overline{iz} = -i\bar{z}$.
20. z is real if and only if $\bar{z} = z$.
21. z is pure imaginary if and only if $\bar{z} = -z$.
22. z is real or pure imaginary if and only if $(\bar{z})^2 = z^2$.
23. The formulas (11).
24. $\operatorname{Re}(iz) = -\operatorname{Im} z$, $\operatorname{Im}(iz) = \operatorname{Re} z$.
25. Find formulas for $\operatorname{Re}(\overline{iz})$ and $\operatorname{Im}(\overline{iz})$ similar to those in Prob. 24.

12.2 Polar Form of Complex Numbers. Triangle Inequality

We introduce polar coordinates r, θ in the complex plane by setting

$$x = r\cos\theta, \qquad y = r\sin\theta. \tag{1}$$

Then any complex number $z = x + iy \neq 0$ may be written

$$z = r\cos\theta + ir\sin\theta = r(\cos\theta + i\sin\theta). \tag{2}$$

This is known as the **polar form** or *trigonometric form* of a complex number.

r is called the **absolute value** or **modulus** of z and is denoted by $|z|$. Thus (Fig. 257)

$$(3) \qquad |z| = r = \sqrt{x^2 + y^2} = \sqrt{z\bar{z}} \qquad (\geqq 0).$$

The directed angle measured from the positive x-axis to OP is called the **argument** of z and is denoted by arg z; *angles will be measured in radians, and positive in the counterclockwise sense.* As in trigonometry we have

$$(4) \qquad \arg z = \theta = \arcsin \frac{y}{r} = \arccos \frac{x}{r} = \arctan \frac{y}{x}.$$

Note that for $z = 0$ the angle $\theta = \arg z$ is not defined. This is the reason for the above condition $z \neq 0$.

Geometrically, $|z|$ is the distance of the point z from the origin (Fig. 257). Hence the inequality

$$|z_1| > |z_2|$$

means that the point z_1 is farther from the origin than the point z_2, and $|z_1 - z_2|$ is the distance between the points z_1 and z_2 (Fig. 258).

We note in passing that for complex numbers (not on the real axis), inequalities such as $z_1 < z_2$ or $z_1 \geqq z_2$ make no sense.

For given $z\ (\neq 0)$, the argument θ is determined only up to integer multiples of 2π. The value of θ which lies in the interval

$$-\pi < \theta \leqq \pi$$

is called the **principal value** of the argument of z.

Example 1. Polar form of complex numbers. Principal value

Let $z = 1 + i$. Then

$$z = \sqrt{2}\left(\cos\frac{\pi}{4} + i \sin\frac{\pi}{4}\right), \qquad |z| = \sqrt{2}, \qquad \arg z = \frac{\pi}{4} \pm 2n\pi \qquad (n = 0, 1, \cdots).$$

The principal value of arg z is $\pi/4$ (Fig. 259).

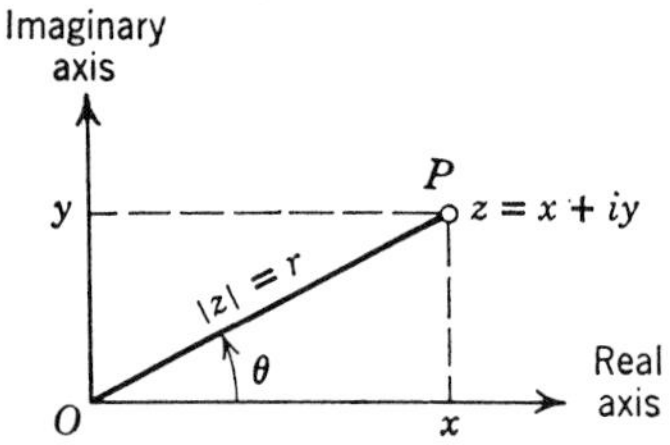

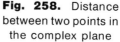

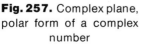

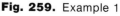

Fig. 257. Complex plane, polar form of a complex number

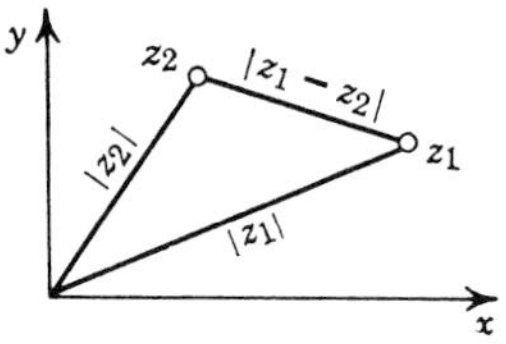

Fig. 258. Distance between two points in the complex plane

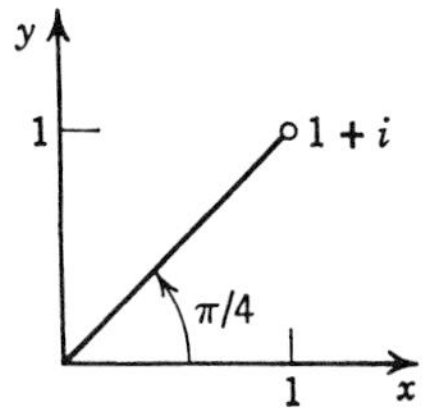

Fig. 259. Example 1

Example 2. Polar form of complex numbers. Principal value

Let $z = 3 + 3\sqrt{3}\,i$. Then $z = 6\left(\cos\frac{\pi}{3} + i\sin\frac{\pi}{3}\right)$, the absolute value of z is $|z| = 6$, and the principal value of $\arg z$ is $\pi/3$. ∎

The polar form of complex numbers is particularly useful in analyzing multiplication and division. Let

$$z_1 = r_1(\cos\theta_1 + i\sin\theta_1) \qquad \text{and} \qquad z_2 = r_2(\cos\theta_2 + i\sin\theta_2).$$

By (6), Sec. 12.1, the product is

$$z_1z_2 = r_1r_2[(\cos\theta_1\cos\theta_2 - \sin\theta_1\sin\theta_2) + i(\sin\theta_1\cos\theta_2 + \cos\theta_1\sin\theta_2)].$$

Because of the familiar addition theorems for the sine and cosine this is simply

$$z_1z_2 = r_1r_2[\cos(\theta_1 + \theta_2) + i\sin(\theta_1 + \theta_2)]. \tag{5}$$

We thus obtain the important rules

$$|z_1z_2| = |z_1|\,|z_2| \tag{6}$$

and

$$\arg(z_1z_2) = \arg z_1 + \arg z_2 \qquad \text{(up to multiples of } 2\pi). \tag{7}$$

Similarly, from the definition of division it follows that

$$\left|\frac{z_1}{z_2}\right| = \frac{|z_1|}{|z_2|} \tag{8}$$

and

$$\arg\frac{z_1}{z_2} = \arg z_1 - \arg z_2 \qquad \text{(up to multiples of } 2\pi). \tag{9}$$

Example 3. Formula of De Moivre

From (6) and (7) we obtain

$$z^n = r^n(\cos\theta + i\sin\theta)^n = r^n(\cos n\theta + i\sin n\theta)$$

and from this the so-called **formula of De Moivre**[4]

$$(\cos\theta + i\sin\theta)^n = \cos n\theta + i\sin n\theta.$$ ∎

Inequalities

For any complex numbers we have the important **triangle inequality** (Fig. 260)

$$|z_1 + z_2| \leqq |z_1| + |z_2|, \tag{10}$$

which we shall use quite frequently. This inequality follows by noting that the

[4]ABRAHAM DE MOIVRE (1667–1754), French mathematician, who introduced imaginary quantities in trigonometry and contributed to the theory of mathematical probability.

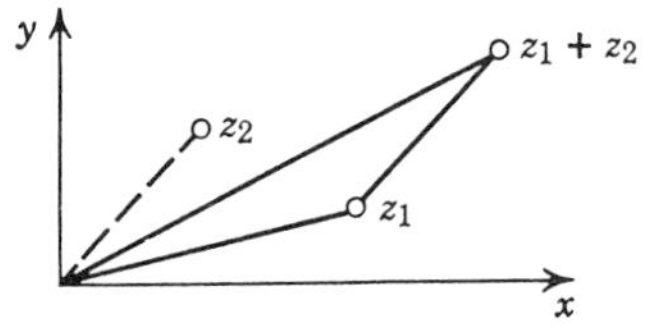

Fig. 260. Triangle inequality

points 0, z_1 and $z_1 + z_2$ are the vertices of a triangle[5] (Fig. 260) with sides $|z_1|$, $|z_2|$ and $|z_1 + z_2|$, and one side cannot exceed the sum of the other two sides. A formal proof is left to the reader (Prob. 19).

By induction the triangle inequality can be extended to arbitrary sums:

$$|z_1 + z_2 + \cdots + z_n| \leqq |z_1| + |z_2| + \cdots + |z_n|; \tag{11}$$

that is, *the absolute value of a sum is at most equal to the sum of the absolute values of the terms.*

Other useful inequalities are

$$|\mathrm{Re}\, z| \leqq |z|, \qquad |\mathrm{Im}\, z| \leqq |z|. \tag{12}$$

They follow by noting that for any $z = x + iy$ we have

$$|z| = \sqrt{x^2 + y^2} \geqq |x| \qquad \text{and similarly} \qquad |z| \geqq |y|.$$

Problems for Sec. 12.2

Find

1. $|1 + i|^2$ **2.** $|-3i|$ **3.** $|\cos\theta + i\sin\theta|$

4. $\left|\dfrac{1 + 4i}{4 + i}\right|$ **5.** $\left|\dfrac{z - 1}{z + 1}\right|$ **6.** $\left|\dfrac{(3 + 4i)^4}{(3 - 4i)^3}\right|$ **7.** $\left|\dfrac{z}{\bar{z}}\right|$

Determine the principal value of the arguments of

8. -3 **9.** $3 + 3i$ **10.** $1 - i\sqrt{3}$ **11.** $-4i$

Represent in polar form:

12. $-3 - 3i$ **13.** -5 **14.** $-4i$ **15.** $1/(4 + 3i)$

16. Verify the triangle inequality for $z_1 = -1 - i$, $z_2 = -2 + 3i$.

17. Verify the triangle inequality for $z_1 = 5 + 2i$, $z_2 = 1 - 4i$.

18. Under what conditions does the equality sign in the triangle inequality hold?

19. Prove the triangle inequality.

20. Using the triangle inequality, prove that $|z_1 + z_2| \geqq |z_1| - |z_2|$.

21. Show that $(|x| + |y|)/\sqrt{2} \leqq |z| \leqq |x| + |y|$. Give examples.

22. Show that $|z_1 + z_2|^2 + |z_1 - z_2|^2 = 2(|z_1|^2 + |z_2|^2)$.

23. Verify (12) for $z = (4 + 5i)^2$.

[5] Which may degenerate if z_1 and z_2 lie on the same straight line through the origin.

24. **(Multiplication by *i*)** Show that multiplication of a complex number by i corresponds to a counterclockwise rotation of the corresponding vector through the angle $\pi/2$.
25. Using the polar form of complex numbers, devise a geometrical construction for the product of two complex numbers. Give an illustrative example, say, $(1 + i)(3 + 2i)$.

12.3 Curves and Regions in the Complex Plane

Curves and regions in the complex plane will be needed quite frequently in our further work. To prepare for this, let us consider some important types of curves and regions and their representations by equations and inequalities. This will also help us to become more familiar with the complex plane.

Since the distance between two points z and a is $|z - a|$, it follows that a circle C of radius ρ with center at a point a (Fig. 261) can be represented in the form

$$|z - a| = \rho. \tag{1}$$

Consequently, the inequality

$$|z - a| < \rho \tag{2}$$

holds for any point z inside C; that is, (2) represents the interior of C. Such a region is called a **circular disk,** or, more precisely, an *open* circular disk, in contrast to the *closed* circular disk

$$|z - a| \leqq \rho,$$

which consists of the interior of C and C itself. The open circular disk (2) is also called a **neighborhood** of the point a. Obviously, a has infinitely many such neighborhoods, each of which corresponds to a certain value of ρ (> 0).

Similarly, the inequality

$$|z - a| > \rho$$

represents the exterior of the circle C. Furthermore, the region between two concentric circles of radii ρ_1 and ρ_2 ($> \rho_1$) can be represented in the form

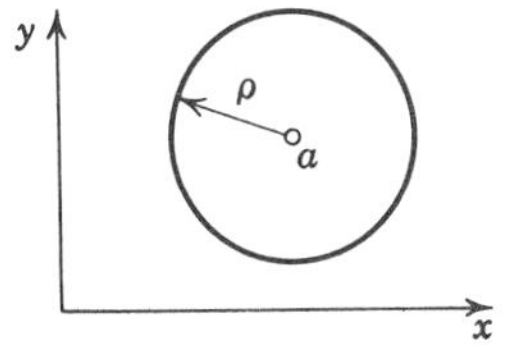

Fig. 261. Circle in the complex plane

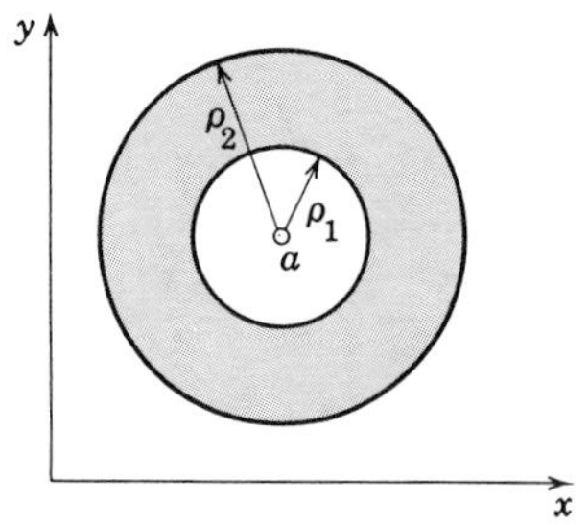

Fig. 262. Annulus in the complex plane

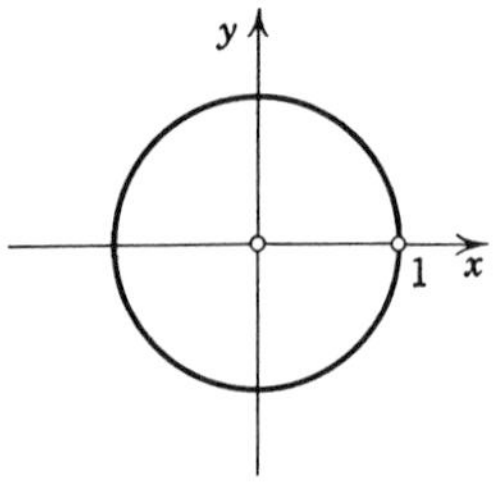

Fig. 263. Unit circle

$$\rho_1 < |z - a| < \rho_2,$$

where a is the center of the circles. Such a region is called an *open circular ring* or *open* **annulus** (Fig. 262).

The equation

$$|z| = 1$$

represents the so-called **unit circle,** that is, the circle of radius 1 with center at the origin (Fig. 263). This circle will play an important role in various considerations.

Example 1. Circular disk

Determine the region in the complex plane given by $|z - 3 + i| \leqq 4$.

Solution. The inequality is valid precisely for all z whose distance from $a = 3 - i$ does not exceed 4. Hence this is a closed circular disk of radius 4 with center at $3 - i$.

Example 2. Unit circle and unit disk

Determine each of the regions

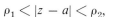

$$\text{(a) } |z| < 1 \qquad \text{(b) } |z| \leqq 1 \qquad \text{(c) } |z| > 1.$$

Solution. (a) The interior of the unit circle. This is called the *open unit disk.*

(b) The unit circle and its interior. This is called the *closed unit disk.*

(c) The exterior of the unit circle. ▮

It is important that the student becomes completely familiar with representations of curves and regions in the complex plane. Therefore, he should pay particular attention to the problems included in the present section. This will make his future work less difficult.

We finally define some notions which are of general importance and will be used in our further work.

The term **set of points** in the complex plane means any sort of collection of finitely or infinitely many points. For example, the solutions of a quadratic equation, the points on a line, and the points in the interior of a circle are sets.

A set S is called **open** if every point of S has a neighborhood every point of which belongs to S. For example, the points in the interior of a circle or a square form an open set, and so do the points of the "right half-plane" $\text{Re } z = x > 0$. The set consisting of the points in the interior of a circle C and on C ("closed circular disk") is not open, because no neighborhood of a point of C lies entirely in the set.

A set S in the complex plane is called **closed** if its complement is open. The **complement** of a set S in the complex plane is the set of all points in the complex plane which do not belong to S. For example, the points on and inside the unit circle form a closed set. (Cf. also Example 2.)

A set is called **bounded** if all of its points lie within a circle of sufficiently large radius. For example, the points in a rectangle form a bounded set, while the points on a straight line do not form a bounded set.

An open set S is said to be **connected** if any two of its points can be joined by a broken line of finitely many line segments all of whose points belong to S. An open connected set is called a **domain.** Thus the interior of a circle is a domain.

A **boundary point** of a set S is a point every neighborhood of which contains both points which belong to S and points which do not belong to S. For example, the boundary points of an annulus are the points on the two bounding circles. Clearly, if a set S is open, then no boundary point belongs to S; if S is closed, then every boundary point belongs to S.

A **region** is a set consisting of a domain plus, perhaps, some or all of its boundary points. (The reader is warned that some authors use the term "region" for what we call a domain [following the modern standard terminology], and others make no distinction between the two terms.)

Problems for Sec. 12.3

Determine and graph the loci represented by

1. $\operatorname{Im} z \geqq -1$ **2.** $\operatorname{Im}(z^2) \leqq 2$ **3.** $\operatorname{Re}(z^2) \geqq 1$

4. $|\arg z| < \pi/4$ **5.** $-\pi < \operatorname{Re} z < \pi$ **6.** $|1/z| > 1$

7. $\left|\dfrac{z+1}{z-1}\right| = 2$ **8.** $\left|\dfrac{z+i}{z-i}\right| = 1$ **9.** $\left|\dfrac{z+1}{z-1}\right| = 1$

10. $\operatorname{Im} \dfrac{z+1}{z-1} \leqq 2$ **11.** $\operatorname{Re} \dfrac{1}{z} > 1$ **12.** $\operatorname{Im} \dfrac{2z+1}{4z-4} \leqq 1$

13. Let z_1 and z_2 be any complex numbers. What is the locus of the points $\alpha z_1 + \beta z_2$, where α and β are real and nonnegative and $\alpha + \beta = 1$?

14. Show that the hyperbola $x^2 - y^2 = 1$ can be written $z^2 + \bar{z}^2 = 2$.

15. The equation $|z - 1| + |z + 1| = 2\sqrt{2}$ represents an ellipse. (*a*) Show this by a geometrical argument. (*b*) Show this algebraically.

12.4 Complex Function. Limit. Derivative. Analytic Function

We now introduce some basic concepts of complex analysis, namely, functions of a complex variable and limits and derivatives for such functions. The reader will notice that these concepts are similar to those in calculus. We then define (complex) analytic functions. These functions play a central role in complex analysis.

We begin by defining a function of a complex variable.

Let S be a set of complex numbers. By a **function** *defined on S* we mean a rule which assigns to each z in S a *unique* complex number w. We then write

$$w = f(z)$$

or $w = g(z)$, etc., or sometimes simply $w(z)$. Here z varies in S and is called a **complex variable.** The set S is called the *domain of definition*[6] of $f(z)$. The set of complex numbers which $w = f(z)$ assumes as z varies on S is called the *range of values* of the function $w = f(z)$.

Let u and v be the real and imaginary parts of w. Then, since w depends on $z = x + iy$, it is clear that in general u depends on x and y, and so does v. We may, therefore, write

$$w = f(z) = u(x, y) + iv(x, y),$$

and this shows that a complex function $f(z)$ is equivalent to two real functions $u(x, y)$ and $v(x, y)$, each depending on the two real variables x and y.

Example 1. Function of a complex variable

Let

$$w = f(z) = z^2 + 3z.$$

Then

$$u(x, y) = \text{Re}\, f(z) = x^2 - y^2 + 3x \qquad \text{and} \qquad v(x, y) = \text{Im}\, f(z) = 2xy + 3y.$$

At $z = x + iy = 1 + 3i$ the function has the value

$$(1 + 3i)^2 + 3(1 + 3i) = -5 + 15i,$$

and we may write

$$f(1 + 3i) = -5 + 15i, \qquad u(1, 3) = -5, \qquad v(1, 3) = 15.$$

Similarly, $f(1 + i) = 3 + 5i$, etc. Obviously, our function is defined for all z.

Example 2. Function of a complex variable

Determine the value of $f(z) = 3\bar{z} = 3x - 3iy$ at $z = 2 + 4i$.

Solution. Since $x = 2$ and $y = 4$, we obtain $f(2 + 4i) = 6 - 12i$. ■

A function $f(z)$ is said to have the **limit** l as z approaches z_0 if $f(z)$ is defined in a neighborhood of z_0 (except perhaps at z_0 itself) and if for every positive real number ϵ (no matter how small but not zero) we can find a positive real number δ such that for all values $z \neq z_0$ in the disk $|z - z_0| < \delta$,

$$|f(z) - l| < \epsilon. \tag{1}$$

[6]This is a standard term, but it is somewhat poor since S need not be a domain as defined in Sec. 12.3 (that is, an open and connected set), although it is a domain in many cases.

In the literature on complex analysis, one sometimes uses relations such that to a value of z there may correspond more than one value of w, and it is customary to call such a relation a function (a "multivalued function"). We shall not adopt this convention but assume that all occurring functions are ***single-valued*** relations, that is, functions in the usual sense.

$f(z)$ denotes the value of f at z, but it is a convenient abuse of language to talk about *the function* $f(z)$ (instead of *the function* f), thereby exhibiting the notation for the independent variable.

This means that the values of $f(z)$ are as close as desired to l for all z which are sufficiently close to z_0 (Fig. 264), and we write

$$\lim_{z \to z_0} f(z) = l. \tag{2}$$

Note well that this definition of a limit implies that z may approach z_0 from *any* direction in the complex plane. Hence this is more restrictive than the definition of a limit in real calculus. (Explain!)

If a limit exists, it is unique. (Cf. Prob. 12.)

A function $f(z)$ is said to be **continuous** at $z = z_0$ if $f(z_0)$ is defined and

$$\lim_{z \to z_0} f(z) = f(z_0). \tag{3}$$

Note that by the definition of a limit this implies that $f(z)$ is defined in some neighborhood of z_0.

$f(z)$ is said to be *continuous in a domain* if it is continuous at each point of that domain.

A function $f(z)$ is said to be **differentiable** at a point $z = z_0$ if the limit

$$f'(z_0) = \lim_{\Delta z \to 0} \frac{f(z_0 + \Delta z) - f(z_0)}{\Delta z} \tag{4}$$

exists. This limit is then called the **derivative** of $f(z)$ at the point $z = z_0$.

Setting $z_0 + \Delta z = z$, we have $\Delta z = z - z_0$ and may also write

$$f'(z_0) = \lim_{z \to z_0} \frac{f(z) - f(z_0)}{z - z_0}. \tag{4'}$$

Remember that the definition of a limit implies that $f(z)$ is defined (at least) in a neighborhood of z_0. Also, by that definition, z may approach z_0 from any direction. Hence differentiability at z_0 means that, along whatever path z approaches z_0, the quotient in (4′) always approaches a certain value and all these values are equal. This fact will be quite important in our later considerations.

By the definition of a limit, (4′) means that there is a complex number $f'(z_0)$ for which, an $\epsilon > 0$ being given, we can find a $\delta > 0$ such that

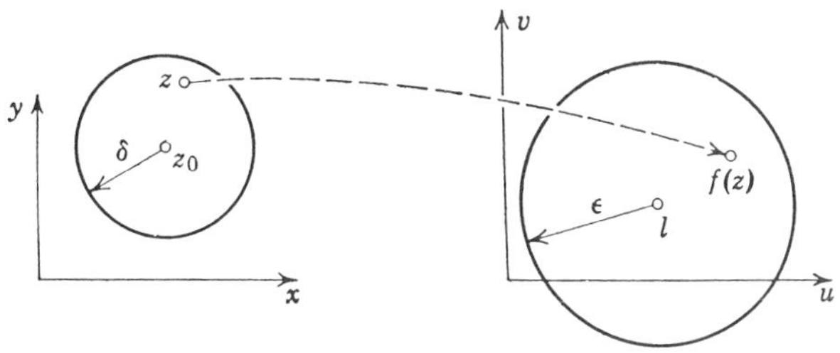

Fig. 264. Limit

(5) $$\left|\frac{f(z) - f(z_0)}{z - z_0} - f'(z_0)\right| < \epsilon \quad \text{when} \quad |z - z_0| < \delta.$$

If $f(z)$ is differentiable at z_0, it is continuous at z_0. (Cf. Prob. 28.)

Example 3. Differentiability. Derivative

The function $f(z) = z^2$ is differentiable for all z and has the derivative $f'(z) = 2z$ because

$$f'(z) = \lim_{\Delta z \to 0} \frac{(z + \Delta z)^2 - z^2}{\Delta z} = \lim_{\Delta z \to 0} (2z + \Delta z) = 2z.$$ ∎

All the familiar rules of real differential calculus, such as the rules for differentiating a constant, integer powers of z, sums, products, and quotients of differentiable functions, and the chain rule for differentiating a function of a function, continue to hold in complex.

In fact, the corresponding proofs are literally the same.

Example 4. $\bar{z}$ not differentiable

It is important to note that there are many simple functions which do not have a derivative at any point. For instance, $f(z) = \bar{z} = x - iy$ is such a function. Indeed, setting $\Delta z = \Delta x + i\Delta y$, we have

(6) $$\frac{f(z + \Delta z) - f(z)}{\Delta z} = \frac{[(x + \Delta x) - i(y + \Delta y)] - (x - iy)}{\Delta x + i\,\Delta y} = \frac{\Delta x - i\,\Delta y}{\Delta x + i\,\Delta y}.$$

If $\Delta y = 0$, this is $+1$. If $\Delta x = 0$, this is -1. Hence (6) approaches $+1$ along path I in Fig. 265 but -1 along path II in Fig. 265. By definition, the limit of (6) as $\Delta z \to 0$ does not exist at any z.

This example may be surprising, but it merely illustrates that differentiability of a complex function is a rather severe requirement. ∎

We now reach the main goal of this section, which is the introduction of the following concept.

Definition (Analyticity)

A function $f(z)$ is said to be *analytic in a domain D* if $f(z)$ is defined and differentiable at all points of D. The function $f(z)$ is said to be *analytic at a point* $z = z_0$ in D if $f(z)$ is analytic in a neighborhood (cf. Sec. 12.3) of z_0.

Hence analyticity of $f(z)$ at z_0 means that $f(z)$ has a derivative at every point in some neighborhood of z_0 (including z_0 itself since, by definition, z_0 is a point of all its neighborhoods). This concept is motivated by the fact that it is of no practical interest when a function is differentiable merely at a single point z_0 but not in some neighborhood of z_0.

An older term for *analytic in D* is *regular in D*, and a more modern term is *holomorphic in D*.

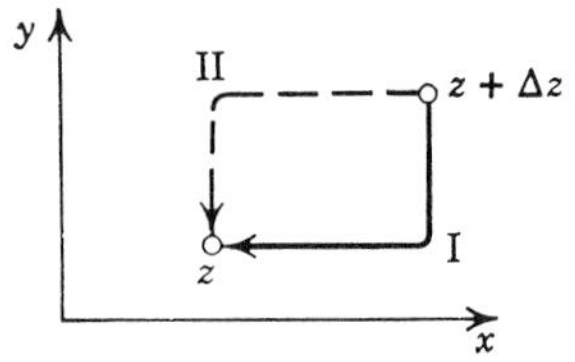

Fig. 265. Paths in (6)

We shall also use the term **analytic function,** without reference to a specific domain; by this term we shall mean a function which is analytic in *some* domain D.

Example 5. Polynomials
The integral powers $1, z, z^2, \cdots$ and, more generally, **polynomials,** that is, functions of the form

$$f(z) = c_0 + c_1 z + c_2 z^2 + \cdots + c_n z^n$$

where $c_0, \cdots, c_n$ are complex constants, are analytic in the entire complex plane. The function $f(z) = 1/(1 - z)$ is analytic everywhere except at $z = 1$. ∎

Complex analysis is concerned exclusively with analytic functions, and although many simple functions are not analytic, the large variety of remaining functions will yield a branch of mathematics which is most beautiful from the theoretical point of view and most useful for practical purposes.

Problems for Sec. 12.4

Find $f(2 + i)$, $f(3i)$, $f(-4 + i)$ where $f(z)$ equals

1. $3z^2 + z$ **2.** $1/z^2$ **3.** $(z + 1)/(z - 1)$

Find the real and imaginary parts of the following functions.

4. $f(z) = 2z^2 - 3z$ **5.** $f(z) = 1/(1 - z)$ **6.** $f(z) = z^3 - z^2 + z$

Suppose that z varies in a region R in the z-plane. Find the (precise) region in the w-plane in which the corresponding values of $w = f(z)$ lie, and show the two regions graphically.

7. $f(z) = 3z$, $|\arg z| < \pi/2$ **8.** $f(z) = 1/z$, $\operatorname{Re} z > 0$

9. $f(z) = z^2$, $|z| > 3$ **10.** $f(z) = z^3$, $|\arg z| \leqq \pi/4$

11. $f(z) = 1/z^2$, $|\arg z| \leqq \pi/4$

12. If $\lim_{z \to z_0} f(z)$ exists, show that this limit is unique.

13. Prove that (2) is equivalent to the pair of relations

$$\lim_{z \to z_0} \operatorname{Re} f(z) = \operatorname{Re} l, \qquad \lim_{z \to z_0} \operatorname{Im} f(z) = \operatorname{Im} l.$$

14. If $\lim_{z \to z_0} f(z) = l$ and $\lim_{z \to z_0} g(z) = p$, show that

$$\lim_{z \to z_0} [f(z) + g(z)] = \lim_{z \to z_0} f(z) + \lim_{z \to z_0} g(z) = l + p,$$

$$\lim_{z \to z_0} [f(z)g(z)] = \lim_{z \to z_0} f(z) \lim_{z \to z_0} g(z) = lp.$$

Are the following functions continuous at the origin?

15. $f(z) = \operatorname{Re} z/|z|$ when $z \neq 0$, $f(0) = 0$

16. $f(z) = \operatorname{Im} z/(1 + |z|)$ when $z \neq 0$, $f(0) = 0$

17. $f(z) = (\operatorname{Re} z)^2/|z|$ when $z \neq 0$, $f(0) = 0$

18. Using the definition of limit, show that $f(z) = z^2$ is continuous.

Show that

19. $[af(z) + bg(z)]' = af'(z) + bg'(z)$ **20.** $[f(z)g(z)]' = f'(z)g(z) + f(z)g'(z)$

Differentiate the following functions.

21. $(z^2 + 1)^2$ **22.** $1/(1 - z)$ **23.** $(z + 1)^2/(z^2 + 1)$

Find the value of the derivative at z_0:

24. $f(z) = z^3 - 2z,\ z_0 = 1 - i$ **25.** $f(z) = (z + 2i)/(z - 2i),\ z_0 = 3 + i$

26. $f(z) = (z^2 - 1)^2,\ z_0 = i$ **27.** $f(z) = iz^2 + (1 + i)z,\ z_0 = -2 + i$

28. If $f(z)$ is differentiable at z_0, show that $f(z)$ is continuous at z_0.

29. Show that $f(z) = \operatorname{Re} z = x$ is not differentiable at any z.

30. Show that $f(z) = |z|^2$ is differentiable only at $z = 0$. *Hint.* Use the relation $|z + \Delta z|^2 = (z + \Delta z)(\overline{z} + \overline{\Delta z})$.

12.5 Cauchy–Riemann Equations. Laplace's Equation

We shall now derive a basic criterion for analyticity of a complex function

$$w = f(z) = u(x, y) + iv(x, y). \tag{1}$$

We show that if $f(z)$ is analytic in a domain D, then u and v satisfy the so-called Cauchy–Riemann equations (5) (see below) everywhere in D; conversely,[7] if u and v are continuous and have first partial derivatives which satisfy (5) everywhere in D, then $f(z)$ is analytic in D. This is our program. The details are as follows.

Suppose $f(z)$ to be defined and continuous in some neighborhood of an arbitrary fixed point z and differentiable at that z. Then, by definition, the derivative

$$f'(z) = \lim_{\Delta z \to 0} \frac{f(z + \Delta z) - f(z)}{\Delta z} \tag{2}$$

at that point z exists. Here, Δz may approach 0 along any path in a neighborhood of z. We may set $\Delta z = \Delta x + i\,\Delta y$. Choosing path I in Fig. 266, we let $\Delta y \to 0$ first and then $\Delta x \to 0$. After Δy becomes zero, $\Delta z = \Delta x$, and by (1),

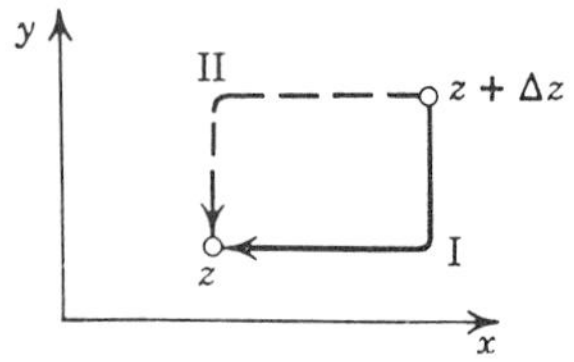

Fig. 266. Paths in (2)

[7]This assumption can be weakened; see D. Menchoff, *Les conditions de monogénéité*. Paris: Hermann, 1936. We shall give a proof under the stronger assumption that the first partial derivatives of u and v are continuous.

$$f'(z) = \lim_{\Delta x \to 0} \frac{u(x + \Delta x, y) + iv(x + \Delta x, y) - [u(x, y) + iv(x, y)]}{\Delta x}$$

$$= \lim_{\Delta x \to 0} \frac{u(x + \Delta x, y) - u(x, y)}{\Delta x} + i \lim_{\Delta x \to 0} \frac{v(x + \Delta x, y) - v(x, y)}{\Delta x}.$$

Since $f'(z)$ exists, the last two real limits exist. They are the partial derivatives of u and v with respect to x. Hence $f'(z)$ can be written

$$f'(z) = \frac{\partial u}{\partial x} + i \frac{\partial v}{\partial x}. \tag{3}$$

Similarly, if we choose path II in Fig. 266, we let $\Delta x \to 0$ first and then $\Delta y \to 0$. After Δx becomes zero, $\Delta z = i\,\Delta y$, and

$$f'(z) = \lim_{\Delta y \to 0} \frac{u(x, y + \Delta y) - u(x, y)}{i\,\Delta y} + i \lim_{\Delta y \to 0} \frac{v(x, y + \Delta y) - v(x, y)}{i\,\Delta y};$$

that is,

$$f'(z) = -i \frac{\partial u}{\partial y} + \frac{\partial v}{\partial y} \tag{4}$$

because $1/i = -i$. The existence of $f'(z)$ thus implies the existence of the four partial derivatives in (3) and (4).

Of course, if $f'(z)$ exists, as we assumed, we can now calculate it by (3) or by (4). More important, by equating the real and the imaginary parts of the right-hand sides of (3) and (4) we obtain

$$\frac{\partial u}{\partial x} = \frac{\partial v}{\partial y} \quad \text{and} \quad \frac{\partial u}{\partial y} = -\frac{\partial v}{\partial x}. \tag{5}$$

These basic relations are called the **Cauchy-Riemann differential equations.**[8]

We may sum up our result as follows.

Theorem 1 (Cauchy–Riemann equations)

Let $f(z) = u(x, y) + iv(x, y)$ be defined and continuous in some neighborhood of a point $z = x + iy$ and differentiable at z itself. Then at that point, the first order partial derivatives of u and v exist and satisfy the Cauchy–Riemann equations (5).

Hence if $f(z)$ is analytic in a domain D, those partial derivatives exist and satisfy (5) *at all points of D.*

[8]Cf. the footnote in Sec. 2.7. BERNHARD RIEMANN (1826–1866), German mathematician, who developed what may be called the "geometrical approach" to complex analysis, based upon the Cauchy–Riemann equations and conformal mapping, in contrast to the German mathematician KARL WEIERSTRASS (1815–1897), who based complex analysis on power series (cf. Secs. 16.1, 16.2). Riemann developed also the so-called Riemannian geometry, which is the mathematical base of Einstein's theory of relativity. His important work gave the impetus to many ideas in modern mathematics (in particular, in topology and functional analysis). Cf. N. Bourbaki, *Elements of Mathematics, General Topology,* Part 1, pp. 162–166. Paris; Hermann, 1966.

Example 1. Cauchy–Riemann equations

The function $f(z) = z^2 = x^2 - y^2 + 2ixy$ is analytic for all z, and $f'(z) = 2z$. We have $u = x^2 - y^2$, $v = 2xy$,

$$\frac{\partial u}{\partial x} = 2x, \qquad \frac{\partial u}{\partial y} = -2y, \qquad \frac{\partial v}{\partial x} = 2y, \qquad \frac{\partial v}{\partial y} = 2x,$$

and the Cauchy–Riemann equations are satisfied for all x, y. ∎

The Cauchy–Riemann equations are fundamental because they are not only necessary but also sufficient for a function to be analytic. More precisely, the following theorem holds. (The conditions in this theorem are sufficient for analyticity, but not necessary. Less restrictive conditions are given in the book quoted in footnote 7 on p. 580.)

Theorem 2 (Cauchy–Riemann equations)

If two real-valued continuous functions $u(x, y)$ and $v(x, y)$ of two real variables x and y have continuous first partial derivatives that satisfy the Cauchy–Riemann equations in some domain D, then the complex function $f(z) = u(x, y) + iv(x, y)$ is analytic in D.

Proof. Let P: (x, y) be any fixed point in D. Since D is a domain, it contains a neighborhood of P. In this neighborhood we may now choose a point Q: $(x + \Delta x, y + \Delta y)$ such that the straight-line segment PQ is in D. Because of our continuity assumptions we may apply the mean-value theorem in Sec. 8.7. This yields

$$\begin{aligned} u(x + \Delta x, y + \Delta y) - u(x, y) &= \Delta x \left(\frac{\partial u}{\partial x}\right)_{M_1} + \Delta y \left(\frac{\partial u}{\partial y}\right)_{M_1}, \\ v(x + \Delta x, y + \Delta y) - v(x, y) &= \Delta x \left(\frac{\partial v}{\partial x}\right)_{M_2} + \Delta y \left(\frac{\partial v}{\partial y}\right)_{M_2}, \end{aligned} \tag{6}$$

the derivatives being evaluated at suitable points M_1 and M_2 of that segment. We set

$$f(z) = u(x, y) + iv(x, y), \qquad \Delta z = \Delta x + i\,\Delta y, \qquad \Delta f = f(z + \Delta z) - f(z).$$

Then we obtain from (6)

$$\Delta f = \Delta x \left(\frac{\partial u}{\partial x}\right)_{M_1} + \Delta y \left(\frac{\partial u}{\partial y}\right)_{M_1} + i \left[\Delta x \left(\frac{\partial v}{\partial x}\right)_{M_2} + \Delta y \left(\frac{\partial v}{\partial y}\right)_{M_2}\right].$$

Using the Cauchy–Riemann equations, we may replace $\partial u/\partial y$ by $-\partial v/\partial x$ and $\partial v/\partial y$ by $\partial u/\partial x$, finding

$$\Delta f = \Delta x \left(\frac{\partial u}{\partial x}\right)_{M_1} + i\,\Delta y \left(\frac{\partial u}{\partial x}\right)_{M_2} + i \left[\Delta x \left(\frac{\partial v}{\partial x}\right)_{M_2} + i\,\Delta y \left(\frac{\partial v}{\partial x}\right)_{M_1}\right].$$

Using $\Delta z = \Delta x + i\Delta y$, we see that this can be written

$$\Delta f = \Delta z\left(\frac{\partial u}{\partial x}\right)_{M_1} + i\Delta y\left\{\left(\frac{\partial u}{\partial x}\right)_{M_2} - \left(\frac{\partial u}{\partial x}\right)_{M_1}\right\}$$
$$+ i\left[\Delta z\left(\frac{\partial v}{\partial x}\right)_{M_1} + \Delta x\left\{\left(\frac{\partial v}{\partial x}\right)_{M_2} - \left(\frac{\partial v}{\partial x}\right)_{M_1}\right\}\right].$$

We divide by Δz on both sides and let Δz approach zero. Then, since the partial derivatives on the right are continuous, they approach $\partial u/\partial x$ and $\partial v/\partial x$, evaluated at (x, y). Furthermore, since $|\Delta x/\Delta z| \leqq 1$, $|\Delta y/\Delta z| \leqq 1$, the limit of the expression on the right exists and is independent of the path along which Δz approaches zero. We see that it equals the right-hand side of (3). This means that $f(z)$ is analytic in D, and the proof is complete. ▮

These theorems are of great practical importance since we can now easily decide whether or not a given complex function is analytic.

Example 2. Cauchy–Riemann equations

Let $f(z) = \operatorname{Re} z = x$. Then

$$u = x \qquad \text{and} \qquad v = 0.$$

The equations (5) are not satisfied, that is, $\operatorname{Re} z$ is not analytic. Similarly, $\operatorname{Im} z$ is not analytic. Further simple functions which are not analytic are included in the problems. ▮

We mention that if we use the polar form $z = r(\cos\theta + i\sin\theta)$ and set $f(z) = u(r, \theta) + iv(r, \theta)$, then the Cauchy–Riemann equations are

(7)
$$\frac{\partial u}{\partial r} = \frac{1}{r}\frac{\partial v}{\partial \theta}, \qquad \frac{\partial v}{\partial r} = -\frac{1}{r}\frac{\partial u}{\partial \theta}.$$

Example 3. Cauchy–Riemann equations in polar form

Let $f(z) = z^3 = r^3(\cos 3\theta + i\sin 3\theta)$. Then

$$u = r^3\cos 3\theta, \qquad v = r^3\sin 3\theta.$$

Hence

$$\frac{\partial u}{\partial r} = 3r^2\cos 3\theta = \frac{1}{r}\frac{\partial v}{\partial \theta}$$

$$\frac{\partial v}{\partial r} = 3r^2\sin 3\theta = -\frac{1}{r}\frac{\partial u}{\partial \theta}.$$

By (7) this confirms that $f(z) = z^3$ is analytic for all $z \neq 0$. (The function z^3 is also analytic at $z = 0$, as we know.) ▮

We shall now discover a practically important relation between complex analysis and the Laplace equation in two variables. We shall prove later (in Sec. 14.6) that the derivative of an analytic function $f(z) = u(x, y) + iv(x, y)$ is itself analytic. By this important fact $u(x, y)$ and $v(x, y)$ will have continuous partial derivatives of all orders. In particular, the mixed second derivatives of these functions will be equal:

$$\frac{\partial^2 u}{\partial x\,\partial y} = \frac{\partial^2 u}{\partial y\,\partial x}, \qquad \frac{\partial^2 v}{\partial x\,\partial y} = \frac{\partial^2 v}{\partial y\,\partial x}.$$

Differentiating the Cauchy–Riemann equations, we thus obtain

$$\frac{\partial^2 u}{\partial x^2} = \frac{\partial^2 v}{\partial x\, \partial y}, \qquad \frac{\partial^2 u}{\partial y^2} = -\frac{\partial^2 v}{\partial x\, \partial y},$$

$$\frac{\partial^2 u}{\partial x\, \partial y} = \frac{\partial^2 v}{\partial y^2}, \qquad \frac{\partial^2 u}{\partial x\, \partial y} = -\frac{\partial^2 v}{\partial x^2}.$$

This yields the following important result.

Theorem 3 (Laplace's equation)
The real part and the imaginary part of a complex function

$$f(z) = u(x, y) + iv(x, y)$$

that is analytic in a domain D are solutions of Laplace's equation,

$$\nabla^2 u = \frac{\partial^2 u}{\partial x^2} + \frac{\partial^2 u}{\partial y^2} = 0, \qquad \nabla^2 v = \frac{\partial^2 v}{\partial x^2} + \frac{\partial^2 v}{\partial y^2} = 0,$$

in D and have continuous second partial derivatives in D.

This is one of the main reasons for the great practical importance of complex analysis in engineering mathematics, as we shall see in Chaps. 13 and 18.

A solution of Laplace's equation having *continuous* second-order partial derivatives is called a **harmonic function** (cf. also Sec. 11.11). Hence the real and imaginary parts of an analytic function are harmonic functions.

If two harmonic functions $u(x, y)$ and $v(x, y)$ satisfy the Cauchy–Riemann equations in a domain D, that is, if u and v are the real and imaginary parts of an analytic function $f(z)$ in D, then $v(x, y)$ is said to be a **conjugate harmonic function** of $u(x, y)$ in D. (Of course this use of the word "conjugate" is different from that employed in defining $\bar{z}$, the conjugate of a complex number z.)

A conjugate of a given harmonic function can be obtained from the Cauchy–Riemann equations, as may be illustrated by the following example.

Example 4. Conjugate harmonic function
The function $u = x^2 - y^2$ is harmonic, and we have $\partial u/\partial x = 2x$, $\partial u/\partial y = -2y$. Hence a conjugate of u must satisfy

$$\frac{\partial v}{\partial y} = 2x, \qquad \frac{\partial v}{\partial x} = 2y.$$

By integrating the first equation with respect to y we obtain

$$v = 2xy + h(x)$$

where $h(x)$ depends only on x. Substituting this in the last equation, we have $h'(x) = 0$ and, therefore, $h = c = const.$ Hence, the most general conjugate function of $x^2 - y^2$ is $2xy + c$, where c is a real constant, and the most general analytic function with the real part $x^2 - y^2$ is $x^2 - y^2 + i(2xy + c) = z^2 + ic$.

Problems for Sec. 12.5

Using (3) or (4), find the derivative of the following functions.

1. $f(z) = az + b$ **2.** $f(z) = z^2$ **3.** $f(z) = 1/z$

4. $f(z) = \dfrac{1}{1-z}$ **5.** $f(z) = z + \dfrac{1}{z}$ **6.** $f(z) = \dfrac{z+1}{z-2}$

7. Show that, in addition to (3) and (4),

$$f'(z) = \frac{\partial u}{\partial x} - i\frac{\partial u}{\partial y}, \qquad f'(z) = \frac{\partial v}{\partial y} + i\frac{\partial v}{\partial x}. \tag{8}$$

Verify that the following pairs of functions satisfy the Cauchy–Riemann equations.

8. $u = x$, $v = y$ **9.** $u = e^x \cos y$, $v = e^x \sin y$

10. $u = x^3 - 3xy^2$, $v = 3x^2y - y^3$

Are the following functions analytic?

11. $f(z) = z^3 + z$ **12.** $f(z) = \operatorname{Im} z$ **13.** $f(z) = \bar{z}$

14. $f(z) = |z|^2$ **15.** $f(z) = 1/(1 - z)$ **16.** $f(z) = e^x(\cos y + i \sin y)$

17. $f(z) = e^x \cos y$ **18.** $f(z) = 1/z^2$ **19.** $f(z) = \arg z$

20. Derive (7) from (5).

21. Using (7), show that $f(z) = z^4$ is analytic.

22. Using (7), show that $f(z) = 1/z^2$ $(z \neq 0)$ is analytic.

Show that the following functions are harmonic and find a corresponding analytic function $f(z) = u(x, y) + iv(x, y)$.

23. $u = x$ **24.** $v = xy$ **25.** $u = xy$

26. $u = \sin x \cosh y$ **27.** $v = -\sin x \sinh y$ **28.** $u = x/(x^2 + y^2)$

29. Under what conditions is $u = ax^3 + bx^2y + cxy^2 + ky^3$ harmonic?

30. Under what condition is $e^{\alpha x} \cos \beta y$ harmonic?

31. If v is a conjugate harmonic function of u, show that u is a conjugate harmonic function of $-v$.

32. Under what condition is $\cos \alpha x \cosh \beta y$ harmonic?

33. If $f(z)$ is analytic in a domain D and $|f(z)| = const$ in D, show that $f(z) = const$.

34. If $f(z)$ is analytic in a domain D and $\operatorname{Re} f(z) = const$ in D, show that $f = const$.

35. If $f(z)$ is analytic in a domain D and $f'(z) = 0$ everywhere in D, show that $f(z) = const$.

12.6 Rational Functions. Root

The remaining sections of this chapter will be devoted to the most important elementary complex functions, such as powers, exponential function, logarithm, trigonometric functions, etc. We shall see that these functions can easily be defined in such a way that, for real values of the independent variable, the functions become identical with the familiar real functions. Some of the complex functions have interesting properties, which do not show when the independent variable is restricted to real values. The student should follow the consideration with great care, because these elementary functions will frequently be needed in applications. Furthermore, a detailed knowledge of these special functions will be helpful later for a better understanding of our more general considerations.

Some of these functions will be analytic in the entire complex plane. Such a function is called an *entire function.*

The powers

$$w = z^n \qquad n = 0, 1, \cdots \tag{1}$$

are analytic in the entire plane. Hence they are entire functions. The same is true for a function of the form

$$w = c_0 + c_1 z + c_2 z^2 + \cdots + c_n z^n \qquad (c_n \neq 0) \tag{2}$$

where $c_0, \cdots, c_n$ are (complex or real) constants. Such a function is called a **polynomial** or an *entire rational function.* The exponent n is called the *degree* of the polynomial. The study of polynomials is the principal subject of classical algebra.

A quotient of two polynomials $p(z)$ and $q(z)$ is called a (*fractional*) **rational function.** Such a function

$$w = \frac{p(z)}{q(z)} \tag{3}$$

is analytic for every z for which $q(z)$ is not zero; here we assume that common factors of $p(z)$ and $q(z)$ have been canceled. A rational function of the particularly simple form

$$\frac{c}{(z - z_0)^m} \qquad (c \neq 0)$$

where both c and z_0 are complex numbers and m is a positive integer, is called a *partial fraction.* It is proved in algebra that every rational function can be represented as the sum of a polynomial and finitely many partial fractions.

If $z = w^n$ $(n = 1, 2, \cdots)$, then to each value of w there corresponds one value of z. We shall immediately see that to a given $z \neq 0$ there correspond precisely n distinct values of w. Each of these values is called an ***n*th root** of z, and we write

$$w = \sqrt[n]{z}. \tag{4}$$

Hence this symbol is *multivalued,* namely, *n-valued,* in contrast to the usual conventions made in *real* calculus. The n values of $\sqrt[n]{z}$ can easily be determined as follows. We set

$$w = R(\cos\phi + i\sin\phi) \qquad \text{and} \qquad z = r(\cos\theta + i\sin\theta).$$

Then from de Moivre's formula (cf. Sec. 12.2) it follows that

$$z = w^n = R^n(\cos n\phi + i \sin n\phi) = r(\cos\theta + i \sin\theta).$$

By equating the absolute values on both sides we have

$$R^n = r \qquad \text{or} \qquad R = \sqrt[n]{r}$$

where the root is real positive and thus uniquely determined. By equating the arguments we obtain

$$n\phi = \theta + 2k\pi \qquad \text{or} \qquad \phi = \frac{\theta}{n} + \frac{2k\pi}{n}$$

where k is an integer. Consequently, $\sqrt[n]{z}$, for $z \neq 0$, has the n distinct values

(5) $$\sqrt[n]{z} = \sqrt[n]{r}\left(\cos\frac{\theta + 2k\pi}{n} + i \sin\frac{\theta + 2k\pi}{n}\right), \qquad k = 0, 1, \cdots, n-1.$$

These n values lie on a circle of radius $\sqrt[n]{r}$ with center at the origin and constitute the vertices of a regular polygon of n sides.

The value of $\sqrt[n]{z}$ obtained by taking the principal value of arg z (cf. Sec. 12.2) and $k = 0$ in (5) is called the **principal value** *of* $w = \sqrt[n]{z}$.

Example 1. Square root

$w = \sqrt{z}$ has the two values

$$z_1 = \sqrt{r}\left(\cos\frac{\theta}{2} + i \sin\frac{\theta}{2}\right)$$

and

$$z_2 = \sqrt{r}\left[\cos\left(\frac{\theta}{2} + \pi\right) + i \sin\left(\frac{\theta}{2} + \pi\right)\right] = -z_1$$

which lie symmetric with respect to the origin. For instance,

$$\sqrt{4i} = \pm 2\left(\cos\frac{\pi}{4} + i \sin\frac{\pi}{4}\right) = \pm(\sqrt{2} + i\sqrt{2}).$$

Example 2. Cube root of a positive real number

If z is *positive real,* then $w = \sqrt[3]{z}$ has the real value $\sqrt[3]{r}$ and the conjugate complex values

$$\sqrt[3]{r}\left(\cos\frac{2\pi}{3} + i \sin\frac{2\pi}{3}\right) = \sqrt[3]{r}\left(-\frac{1}{2} + \frac{\sqrt{3}}{2}i\right)$$

and

$$\sqrt[3]{r}\left(\cos\frac{4\pi}{3} + i \sin\frac{4\pi}{3}\right) = \sqrt[3]{r}\left(-\frac{1}{2} - \frac{\sqrt{3}}{2}i\right).$$

For instance, $\sqrt[3]{1} = 1$, $-\frac{1}{2} \pm \frac{1}{2}\sqrt{3}\,i$ (Fig. 267). Clearly these are the roots of the equation $w^3 - 1 = 0$.

Example 3. *n*th root of unity

From (5) we obtain

(6) $$\sqrt[n]{1} = \cos\frac{2k\pi}{n} + i \sin\frac{2k\pi}{n}, \qquad k = 0, 1, \cdots n-1.$$

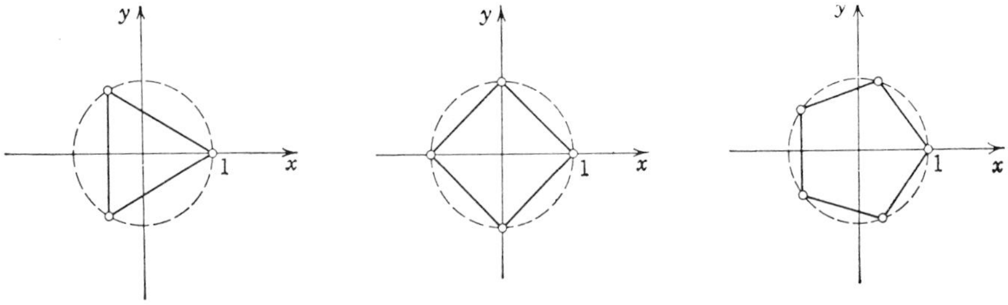

Fig. 267. $\sqrt[3]{1}$ **Fig. 268.** $\sqrt[4]{1}$ **Fig. 269.** $\sqrt[5]{1}$

If ω denotes the value corresponding to $k = 1$, then the n values of $\sqrt[n]{1}$ can be written as $1, \omega, \omega^2, \cdots, \omega^{n-1}$. These values are the vertices of a regular polygon of n sides inscribed in the unit circle, with one vertex at the point 1. Each of these n values is called an *nth root of unity*. For instance, $\sqrt[4]{1}$ has the values $1, i, -1$, and $-i$ (Fig. 268). Figure 269 shows $\sqrt[5]{1}$.

If w_1 is any nth root of an arbitrary complex number z, then

$$w_1, \quad w_1\omega, \quad w_1\omega^2, \quad \cdots, \quad w_1\omega^{n-1}$$

are the n values of $\sqrt[n]{z}$, because the multiplication of a number w_1 by ω^k corresponds to an increase of the argument of w_1 by $2k\pi/n$.

Problems for Sec. 12.6

Determine all values of the following roots and plot the corresponding points in the complex plane.

1. $\sqrt{i}$ **2.** $\sqrt{-i}$ **3.** $\sqrt{-4}$ **4.** $\sqrt{1 - i\sqrt{3}}$
5. $\sqrt[3]{-1}$ **6.** $\sqrt[3]{i}$ **7.** $\sqrt[3]{-i}$ **8.** $\sqrt[3]{1 + i}$
9. $\sqrt[4]{-1}$ **10.** $\sqrt[5]{-1}$ **11.** $\sqrt[6]{-1}$ **12.** $\sqrt[8]{1}$

Find and plot all solutions of the following equations.

13. $z^3 = 27$ **14.** $z^4 + 5z^2 = 36$ **15.** $z^4 + 81 = 0$ **16.** $z^6 + 7z^3 = 8$

17. Find the sum of the n nth roots of unity (a) for $n = 2$ and $n = 4$, (b) for general n.

18. (Square root) Prove that

$$\sqrt{z} = \pm[\sqrt{(|z| + x)/2} + (\operatorname{sign} y)i\sqrt{(|z| - x)/2}] \qquad (z = x + iy)$$

where $\operatorname{sign} y = 1$ if $y \geqq 0$, $\operatorname{sign} y = -1$ if $y < 0$, and all square roots of positive numbers are taken with the positive sign. *Hint.* Set $\sqrt{z} = w = u + iv$, square, separate into two real equations, express u^2 and v^2 in terms of x and y.

Using the result of Prob. 18, find

19. $\sqrt{4i}$ **20.** $\sqrt{3 + 4i}$ **21.** $\sqrt{-5 + 12i}$ **22.** $\sqrt{-8 - 6i}$

Using the result of Prob. 18, solve the following equations.

23. $z^2 + z + 1 = i$ **24.** $z^2 - 3z + 3 = i$
25. $z^2 - (5 + i)z + 8 + i = 0$ **26.** $z^4 - 3(1 + 2i)z^2 = 8 - 6i$

27. Transform to a single equation for $x + iy$ and solve

$$x^4 - 6x^2y^2 + y^4 = 4, \qquad xy(x^2 - y^2) = 1 \qquad (x, y \text{ real}).$$

28. Represent $z^4 + 4$ as a product of quadratic factors with real coefficients.

29. Represent $z^4 + 1$ as a product of quadratic factors with real coefficients.

30. Let P be a regular polygon of n sides with vertices on the unit circle. Find the product of the lengths of the $n - 1$ straight-line segments which join a fixed vertex of P with the $n - 1$ other vertices.

12.7 Exponential Function

The real exponential function e^x (also written $\exp x$) has the properties

$$(1) \qquad (e^x)' = e^x$$

$$(2) \qquad e^{x_1+x_2} = e^{x_1}e^{x_2}$$

and the Maclaurin[9] series

$$(3) \qquad e^x = 1 + x + \frac{x^2}{2!} + \frac{x^3}{3!} + \cdots .$$

The exponential function for complex $z = x + iy$ is denoted by e^z and is defined in terms of the real functions e^x, $\cos y$ and $\sin y$, as follows:

$$\textbf{(4)} \qquad e^z = e^x(\cos y + i \sin y) \qquad \text{(also written } \exp z\text{)}.$$

This definition is suggested by the following facts. When $z = x$ is real, then $e^z = e^x$. From the Cauchy–Riemann equations it follows that e^z is analytic for all z. From (3) in Sec. 12.5 we see that

$$(e^z)' = \frac{\partial}{\partial x}(e^x \cos y) + i\frac{\partial}{\partial x}(e^x \sin y) = e^x \cos y + ie^x \sin y,$$

that is,

$$\textbf{(5)} \qquad (e^z)' = e^z.$$

Furthermore, setting $z_1 = x_1 + iy_1$ and $z_2 = x_2 + iy_2$ and applying the addition formulas for the sine and cosine [cf. (6) in Appendix 3], we readily find

$$\textbf{(6)} \qquad e^{z_1+z_2} = e^{z_1}e^{z_2},$$

which is the analogue of (2). In particular, when $z_1 = x$ and $z_2 = iy$, we have

$$(7) \qquad e^z = e^{x+iy} = e^x e^{iy}.$$

We shall see later (in Chap. 16) that complex analytic functions have Taylor series developments which are similar to the familiar Taylor series of real

[9] COLIN MACLAURIN (1698–1746), Scots mathematician.

functions. It will then easily follow that if we replace x in (3) by z, we obtain the Maclaurin[10] series of e^z.

From (4) we obtain the **Euler formula**

$$e^{iy} = \cos y + i \sin y. \tag{8}$$

This shows that the polar form of a complex number $z = x + iy$ (cf. Sec. 12.2) may now be written

$$z = r(\cos \theta + i \sin \theta) = re^{i\theta}. \tag{9}$$

Furthermore, from (8) we see that

$$|e^{iy}| = \sqrt{\cos^2 y + \sin^2 y} = 1, \tag{10}$$

that is, for pure imaginary exponents the exponential function has absolute value one, a result which the student should remember. Consequently, by (4),

$$|e^z| = e^x, \qquad \arg e^z = y. \tag{11}$$

Since $\cos 2\pi = 1$ and $\sin 2\pi = 0$, we obtain from (8)

$$e^{2\pi i} = 1. \tag{12}$$

Similarly,

$$e^{\pi i} = e^{-\pi i} = -1, \qquad e^{\pi i/2} = i, \qquad e^{-\pi i/2} = -i. \tag{13}$$

From (12) and (6) we find

$$e^{z+2\pi i} = e^z e^{2\pi i} = e^z,$$

which shows that e^z *is periodic with the imaginary period* $2\pi i$. Thus

$$e^{z \pm 2n\pi i} = e^z \qquad (n = 0, 1, \cdots). \tag{14}$$

Because of the periodicity all the values which $w = e^z$ can assume are already assumed in the strip (Fig. 270)

$$-\pi < y \leqq \pi. \tag{15}$$

This infinite strip is called a **fundamental region** of e^z.

From (6) we obtain $e^z e^{-z} = e^0 = 1$, which implies

$$e^z \neq 0 \qquad \text{for all } z. \tag{16}$$

[10]This series may be used for *defining* e^z. For this purpose one would first have to consider Secs. 16.1 and 16.2. The properties of e^z follow from the series, as shown in Example 2, Sec. 16.4. Our present approach has the advantage that integrals (Chap. 14) will appear earlier and we shall have a closer connection between the general consideration of series (Chap. 15) and the Taylor and Laurent series (Chap. 16), for which integrals will be needed. We also avoid irritating poorer students by series at such an early stage.

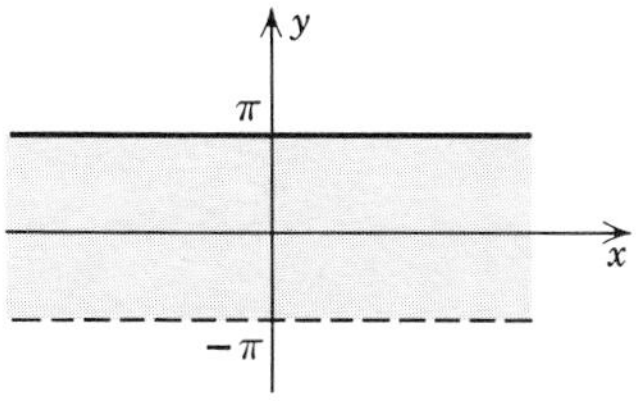

Fig. 270. Fundamental region of the exponential function e^z

Problems for Sec. 12.7

1. Using the Cauchy-Riemann equations, show that e^z is analytic for all z.
2. Derive (6) as explained in the text.

Find the value of e^z when z equals

3. $\pi i/4$ **4.** $-\pi i/4$ **5.** $1 + i$ **6.** $3 + \pi i$

Find the real and imaginary parts of

7. e^{2z} **8.** e^{-4z} **9.** e^{z^2} **10.** e^{z^3}

Write the following expressions in the polar form (9).

11. $\sqrt{i}, \sqrt{-i}$ **12.** $3 - 4i$ **13.** $2 - 2i$ **14.** $\sqrt{z}, \sqrt[n]{z}$

Find all solutions of the following equations and plot some of the solutions in the complex plane.

15. $e^z = 1$ **16.** $e^z = 3$ **17.** $e^z = -2$ **18.** $e^{z^2} = 1$

Find all values of z such that

19. $e^{\bar{z}} = \overline{e^z}$ **20.** $e^{i\bar{z}} = \overline{e^{iz}}$ **21.** $|e^{-2z}| < 1$ **22.** e^z is real

23. Show that $u = e^{xy} \cos(x^2/2 - y^2/2)$ is harmonic and find a conjugate.
24. It is interesting that the conditions $f'(z) = f(z), f(x + i0) = e^x$ determine $f(z)$ (assumed analytic for all z) uniquely, namely, $f(z) = e^z$ as defined by (4). Prove this. (Use the Cauchy-Riemann equations.)
25. Consider the behavior of e^z as $|z| \to \infty$ along different rays, say, for $\arg z = 0$, $\pi/2$, and π.

12.8 Trigonometric and Hyperbolic Functions

From Euler's formula (8) in the last section we obtain

$$\cos x = \frac{1}{2}(e^{ix} + e^{-ix}), \qquad \sin x = \frac{1}{2i}(e^{ix} - e^{-ix}) \qquad (x \text{ real}).$$

This suggests the following definitions for complex $z = x + iy$:

(1) $$\cos z = \frac{1}{2}(e^{iz} + e^{-iz}), \qquad \sin z = \frac{1}{2i}(e^{iz} - e^{-iz}).$$

Furthermore, in agreement with the familiar definitions from real calculus we define

(2) $$\tan z = \frac{\sin z}{\cos z}, \qquad \cot z = \frac{\cos z}{\sin z}$$

and

(3) $$\sec z = \frac{1}{\cos z}, \qquad \csc z = \frac{1}{\sin z}.$$

Since e^z is analytic for all z, the same is true for the functions $\cos z$ and $\sin z$. The functions $\tan z$ and $\sec z$ are analytic except at the points where $\cos z$ is zero, and $\cot z$ and $\csc z$ are analytic except where $\sin z$ is zero.

The functions $\cos z$ and $\sec z$ are even, and those other functions are odd:

(4) $$\begin{aligned} \cos(-z) &= \cos z & \sin(-z) &= -\sin z \\ \cot(-z) &= -\cot z & \tan(-z) &= -\tan z \end{aligned}$$

etc. Since the exponential function is periodic, the trigonometric functions are also periodic, and we have

(5) $$\begin{aligned} \cos(z \pm 2n\pi) &= \cos z & \sin(z \pm 2n\pi) &= \sin z \\ \tan(z \pm n\pi) &= \tan z & \cot(z \pm n\pi) &= \cot z \end{aligned}$$

where $n = 0, 1, \cdots$.

From the definitions it follows immediately that *all the familiar formulas for the real trigonometric functions continue to hold for complex values.* For example,

(6) $$\frac{d}{dz}\cos z = -\sin z, \qquad \frac{d}{dz}\sin z = \cos z, \qquad \frac{d}{dz}\tan z = \sec^2 z,$$

and

(7) $$\begin{aligned} \cos(z_1 \pm z_2) &= \cos z_1 \cos z_2 \mp \sin z_1 \sin z_2 \\ \sin(z_1 \pm z_2) &= \sin z_1 \cos z_2 \pm \sin z_2 \cos z_1 \\ \sin^2 z + \cos^2 z &= 1. \end{aligned}$$

From (1) we see that the **Euler formula** *is valid for complex values,*

(8) $$e^{iz} = \cos z + i \sin z.$$

Using (7), we may readily represent $\cos z$ and $\sin z$ in terms of real functions. We first have

(9) $$\begin{aligned} \cos(x + iy) &= \cos x \cos iy - \sin x \sin iy \\ \sin(x + iy) &= \sin x \cos iy + \cos x \sin iy. \end{aligned}$$

Now from (1) and the definition of the hyperbolic cosine and sine it follows that

$$\cos iy = \frac{1}{2}(e^{-y} + e^{y}) = \cosh y, \qquad \sin iy = \frac{1}{2i}(e^{-y} - e^{y}) = i \sinh y.$$

This yields the desired representations

$$\textbf{(10)} \qquad \begin{aligned} \cos(x + iy) &= \cos x \cosh y - i \sin x \sinh y \\ \sin(x + iy) &= \sin x \cosh y + i \cos x \sinh y \end{aligned}$$

which are useful for numerical computation of $\cos z$ and $\sin z$.

The **hyperbolic cosine** and **sine** of a complex variable z are defined by the formulas

$$\textbf{(11)} \qquad \cosh z = \tfrac{1}{2}(e^{z} + e^{-z}), \qquad \sinh z = \tfrac{1}{2}(e^{z} - e^{-z}),$$

in agreement with the familiar definitions for a real variable [cf. (17) in Appendix 3]. These functions are analytic in the entire plane.

From (11) and (1) it follows that

$$\textbf{(12)} \qquad \cosh z = \cos(iz), \qquad \sinh z = -i \sin(iz).$$

As in real calculus we define

$$(13) \qquad \tanh z = \frac{\sinh z}{\cosh z}, \qquad \coth z = \frac{\cosh z}{\sinh z},$$

$$(14) \qquad \operatorname{sech} z = \frac{1}{\cosh z}, \qquad \operatorname{csch} z = \frac{1}{\sinh z}.$$

The absolute value $|f(z)|$ of an analytic function $f(z)$, $z = x + iy$, is a real function of the two real variables x and y and can, therefore, be represented by a surface in three-dimensional space; to each point (x, y) in the xy-plane there

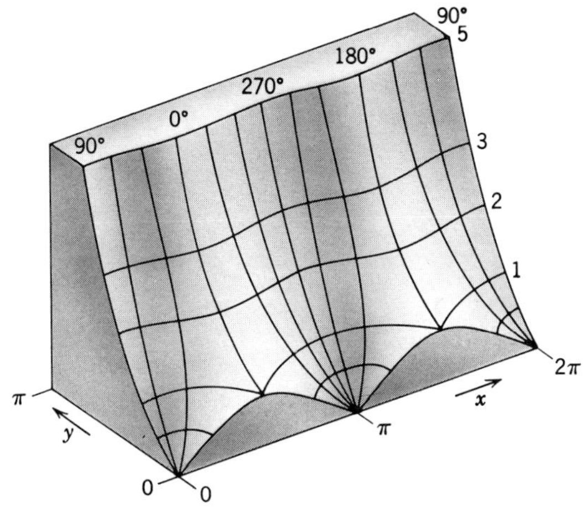

Fig. 271. Modular surface of sin z

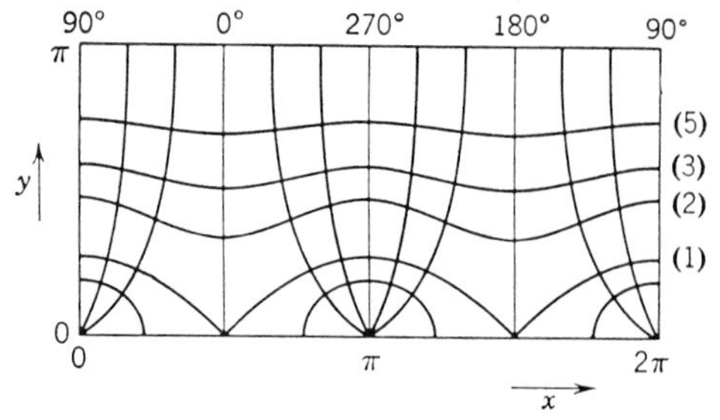

Fig. 272. Curves $|\sin z| = const$ and $\arg(\sin z) = const$ in the z-plane

corresponds a point with Cartesian coordinates $(x, y, |f|)$ in space, and these points form that surface which is called the **modular surface** of $f(z)$. On this surface we may draw some of the curves $|f| = const$ and $\arg f = const$; this yields a useful graphical representation of an analytic function and a geometrical illustration of its behavior.

Figure 271 shows a portion of the modular surface of $f(z) = \sin z$ and some curves $|\sin z| = const$ and $\arg(\sin z) = const$ on the surface. Figure 272 shows the orthogonal projection of these curves into the xy-plane; this graphical representation of the complex sine function is well known and used in electrical engineering.

Modular surfaces have interesting geometrical properties which are considered in Ref. [C8] in Appendix 1.

Problems for Sec. 12.8

1. Show that $\cos z$, $\sin z$, $\cosh z$, and $\sinh z$ are analytic for all z.

2. Prove (4). **3.** Obtain (5) from (1). **4.** Prove (6).

5. Prove (7). **6.** Prove (12).

Find

7. $|\cos z|$ **8.** $|\sin z|$ **9.** $|\tan z|$

10. $\text{Re} \tan z$ **11.** $\text{Re} \cot z$ **12.** $\text{Re} \sec z$

Using Table A1 in Appendix 4, calculate

13. $\sin i$ **14.** $\cosh i$ **15.** $\cos(1 + 2i)$

16. $\sin(1.7 + 1.5i)$ **17.** $\sinh(2 + i)$ **18.** $\cos(1.7 + 1.5i)$

Find all solutions of the following equations.

19. $\cos z = 5$ **20.** $\sin z = 1000$ **21.** $\cosh z = 0$

22. $\sinh z = 0$ **23.** $\cosh z = 0.5$ **24.** $\sin z = i \sinh 1$

Show that:

25. $\cos z = \cosh iz$ and $\sin z = -i \sinh iz$

26. $(\cosh z)' = \sinh z$ and $(\sinh z)' = \cosh z$

27. $\cosh z = \cosh x \cos y + i \sinh x \sin y$ and $\sinh z = \sinh x \cos y + i \cosh x \sin y$

28. $|\sinh y| \leqq |\sin z| \leqq \cosh y$ and $|\sinh y| \leqq |\cos z| \leqq \cosh y$

29. $\cosh^2 z - \sinh^2 z = 1$

30. $\cot z \neq \pm i$ for all z

31. $\tanh z$ has period πi.

32. All z for which $\cos z = 0$ are real.

33. All z for which $\sin z = 0$ are real.

34. Using (1), show that $\cos^4 \theta = \frac{1}{8}(\cos 4\theta + 4\cos 2\theta + 3)$.

35. Using (8), Sec. 12.7, and $1 + z + z^2 + \cdots = 1/(1 - z)$, $z = e^{i\theta}/2$, show that

$$1 + \frac{1}{2}\cos\theta + \frac{1}{4}\cos 2\theta + \frac{1}{8}\cos 3\theta + \cdots = \frac{4 - 2\cos\theta}{5 - 4\cos\theta}.$$

12.9 Logarithm. General Power

The *natural logarithm* of $z = x + iy$ is denoted by $\ln z$ (sometimes also by $\log z$) and is defined as the inverse of the exponential function; that is, $w = \ln z$ is defined by the relation

$$e^w = z \tag{1}$$

for each $z \neq 0$.

Setting $w = u + iv$ and $z = |z|\, e^{i\theta} = re^{i\theta}$ in (1), we have

$$e^w = e^{u+iv} = e^u e^{iv} = re^{i\theta}.$$

The absolute values on both sides must be equal. Now $e^u e^{iv}$ has the absolute value e^u since v is real, so that $|e^{iv}| = 1$; cf. (10) in Sec. 12.7. Since $re^{i\theta}$ has the absolute value r, we thus obtain

$$e^u = |z| = r \qquad \text{or} \qquad u = \ln |z|$$

where $\ln |z|$ is the elementary real natural logarithm of the positive number $|z|$. Similarly, the arguments on both sides must be equal:

$$v = \theta = \arg z.$$

Therefore

$$\ln z = \ln |z| + i \arg z = \ln \sqrt{x^2 + y^2} + i \arg(x + iy). \tag{2}$$

Since the argument of z is determined only up to multiples of 2π, *the complex natural logarithm is infinitely many-valued.*

The value of $\ln z$ corresponding to the principal value of $\arg z$, that is

$$-\pi < \arg z \leqq \pi \qquad \text{(Sec. 12.2)}$$

is called the **principal value** of $\ln z$ and is often denoted by $\operatorname{Ln} z$.

Obviously the other values of $\ln z$ are then of the form

$$\ln z = \operatorname{Ln} z \pm 2n\pi i \qquad (n = 1, 2, \cdots); \tag{3}$$

they have the same real part, and their imaginary parts differ by multiples of 2π, in agreement with the fact that e^z is periodic with the imaginary period $2\pi i$.

Also, if z is real positive, the principal value of $\arg z$ is zero, and the principal value $\operatorname{Ln} z$ is identical with the real natural logarithm known from elementary calculus. If z is real negative, the principal value of $\arg z$ is π, and then

$$\operatorname{Ln} z = \ln |z| + \pi i.$$

Example 1. Natural logarithm. Principal value

$$\ln(-1) = \pm\pi i, \ \pm 3\pi i, \ \pm 5\pi i, \ \cdots, \qquad \operatorname{Ln}(-1) = \pi i$$

$$\ln i = \frac{\pi}{2}i, \ -\frac{3\pi}{2}i, \ \frac{5\pi}{2}i, \ -\frac{7\pi}{2}i, \ \frac{9\pi}{2}i, \ \cdots, \qquad \operatorname{Ln} i = \frac{\pi}{2}i$$

$$\operatorname{Ln}(-i) = -\frac{\pi}{2}i, \qquad \operatorname{Ln}(-2-2i) = \ln\sqrt{8} - \frac{3}{4}\pi i.$$

∎

The familiar relations for the natural logarithm continue to hold for complex values, that is,

$$\text{(4)} \qquad \text{(a)}\ \ln(z_1 z_2) = \ln z_1 + \ln z_2, \qquad \text{(b)}\ \ln(z_1/z_2) = \ln z_1 - \ln z_2$$

but these relations are to be understood in the sense that each value of one side is also contained among the values of the other side.

Example 2. Illustration of the functional relations (4) in complex

Let

$$z_1 = z_2 = e^{\pi i} = -1.$$

If we take

$$\ln z_1 = \ln z_2 = \pi i,$$

then (4a) holds provided we write $\ln(z_1 z_2) = \ln 1 = 2\pi i$; it is not true for the principal value, $\operatorname{Ln}(z_1 z_2) = \operatorname{Ln} 1 = 0$. ∎

By applying (3), Sec. 12.5, to (2), this section, we find

$$\frac{d}{dz}\ln z = \frac{\partial}{\partial x}\ln\sqrt{x^2+y^2} + i\frac{\partial}{\partial x}(\arg z)$$

$$= \frac{x}{x^2+y^2} + i\frac{1}{1+(y/x)^2}\left(-\frac{y}{x^2}\right) = \frac{x-iy}{x^2+y^2} = \frac{1}{z};$$

that is, the derivative of the natural logarithm is

$$\text{(5)} \qquad \frac{d}{dz}\ln z = \frac{1}{z} \qquad (z \neq 0).$$

Hence the principal value $\operatorname{Ln} z$ $(z \neq 0)$, which is single-valued and thus a function in the usual sense, is analytic in the domain $-\pi < \arg z < \pi$ of the z-plane, that is, everywhere except at points of the negative real axis (where its imaginary part is not even continuous but has a jump of magnitude 2π).

General powers of a complex number $z = x + iy$ $(\neq 0)$ are defined by the formula

$$z^c = e^{c \ln z} \qquad (c \text{ complex}, z \neq 0). \tag{6}$$

Since $\ln z$ is infinitely many-valued, z^c will, in general, be multivalued. The particular value

$$z^c = e^{c \operatorname{Ln} z}$$

is called the *principal value* of z^c.

If $c = n = 1, 2, \cdots$, then z^n is single-valued and identical with the usual nth power of z. If $c = -1, -2, \cdots$, the situation is similar.

If $c = 1/n$ where $n = 2, 3, \cdots$, then

$$z^c = \sqrt[n]{z} = e^{(1/n) \ln z} \qquad (z \neq 0),$$

the exponent is determined up to multiples of $2\pi i/n$, and we obtain the n distinct values of the nth root, in agreement with the result in Sec. 12.6. If $c = p/q$, the quotient of two positive integers, the situation is similar, and z^c has only finitely many distinct values. However, if c is real irrational or genuinely complex, then z^c is infinitely many-valued.

Example 3. General power

$$i^i = e^{i \ln i} = e^{i[(\pi/2)i \pm 2n\pi i]} = e^{-(\pi/2) \mp 2n\pi}$$

All these values are real, and the principal value $(n = 0)$ is $e^{-\pi/2}$. ∎

It is a *convention* that for real positive $z = x$ the expression z^c means $e^{c \ln x}$ where $\ln x$ is the elementary real natural logarithm (that is, the principal value of $\ln z$ $(z = x > 0)$ in the sense of our definition). Also, if $z = e$, the base of the natural logarithm, $z^c = e^c$ is *conventionally* regarded as the unique value obtained from (4) in Sec. 12.7.

From (6) we see that for any complex number a,

$$a^z = e^{z \ln a}. \tag{7}$$

Problems for Sec. 12.9

1. Verify (4) for $z_1 = i$, $z_2 = -1$.
2. Using (7), Sec. 12.5, show that $\operatorname{Ln} z$ $(z \neq 0)$ is analytic in the region $-\pi < \theta < \pi$ where θ is the principal value of $\arg z$.
3. Show that $\operatorname{Ln} z$ is not continuous on the negative real axis.
4. Show that $e^{\ln z} = z$, $\ln(e^z) = z \pm 2n\pi i$, $n = 0, 1, \cdots$.

Determine all values of the given expressions and plot some of them in the complex plane.

5. $\ln 1$
6. $\ln 2$
7. $\ln i$
8. $\ln e$
9. $\ln(ie)$
10. $\ln(-ie)$
11. $\ln(e^i)$
12. $\ln(e^{-2})$

Solve the following equations for z.

13. $\ln z = -\pi i/2$ **14.** $\ln z = \pi i/2$ **15.** $\ln z = 1 + \pi i$ **16.** $\ln z = (1 + i)\pi$

Using Table A1 in Appendix 4, calculate the principal value $\operatorname{Ln} z$ when z equals

17. $(1 - i)^2$ **18.** $\sqrt{2} + i\sqrt{2}$ **19.** -7 **20.** $3 + i\sqrt{27}$

Find the principal value of

21. $(2i)^{1/2}$ **22.** $(1 + i)^i$ **23.** $(1 + i)^{1-i}$ **24.** $(1 - i)^{1+i}$

25. 3^{3-i} **26.** 2^{2i} **27.** $(2 - i)^{1+i}$ **28.** 2^{3+2i}

The **inverse sine** $w = \sin^{-1} z$ is defined as the function which satisfies the relation $\sin w = z$. The **inverse cosine** $w = \cos^{-1} z$ is defined as the function which satisfies the relation $\cos w = z$. The other inverse trigonometric and hyperbolic functions are defined and denoted in a similar fashion. Using representations in terms of exponential functions ($\sin w = (e^{iw} - e^{-iw})/2i$, etc.), show that

29. $\sin^{-1} z = -i \ln (iz + \sqrt{1 - z^2})$ **30.** $\cos^{-1} z = -i \ln (z + \sqrt{z^2 - 1})$

31. $\cosh^{-1} z = \ln (z + \sqrt{z^2 - 1})$ **32.** $\sinh^{-1} z = \ln (z + \sqrt{z^2 + 1})$

33. $\tan^{-1} z = \dfrac{i}{2} \ln \dfrac{i + z}{i - z}$ **34.** $\tanh^{-1} z = \dfrac{1}{2} \ln \dfrac{1 + z}{1 - z}$

35. Show that $w = \sin^{-1} z$ is infinitely many-valued, and if w_1 is one of these values, the others are of the form $w_1 \pm 2n\pi$ and $\pi - w_1 \pm 2n\pi$, $n = 0, 1, \cdots$. (The *principal value* of $w = u + iv = \sin^{-1} z$ is defined to be the value for which $-\pi/2 \leqq u \leqq \pi/2$ when $v \geqq 0$ and $-\pi/2 < u < \pi/2$ when $v < 0$.)

CHAPTER

13 CONFORMAL MAPPING

If a complex function $w = f(z)$ is defined in a domain D of the z-plane, then to each point in D there corresponds a point in the w-plane. In this way we have a *mapping* of D onto the range of values of $f(z)$ in the w-plane. This "geometric approach" to complex analysis helps us to "visualize" the nature of a complex function by considering the manner in which the function maps certain curves and regions.

We shall see that if $f(z)$ is an *analytic* function, the mapping given by $f(z)$ is *conformal* (angle-preserving), except at points where the derivative $f'(z)$ is zero.

Conformal mapping is important in engineering mathematics, because it is a standard method for solving boundary value problems in two-dimensional potential theory by transforming a given complicated region into a simpler one.

We shall first define and explain the concept of mapping and then consider the mappings corresponding to elementary analytic functions. Applications will be included in this chapter as well as in Chap. 18.

Prerequisites for this chapter: Chap. 12.
Sections which may be omitted in a shorter course: Secs. 13.4, 13.6.
References: Appendix 1, Part F.
Answers to problems: Appendix 2.

13.1 Mapping

A continuous *real* function $y = f(x)$ of a real variable x can be exhibited graphically by plotting a curve in the Cartesian xy-plane; this curve is called the *graph* of the function. In the case of a *complex* function

$$w = f(z) = u(x, y) + iv(x, y) \qquad (z = x + iy) \tag{1}$$

the situation is more complicated, because each of the complex variables w and z is represented geometrically by the points in the complex plane. This suggests the use of two separate complex planes for the two variables: one the z-plane, in which the point $z = x + iy$ is to be plotted, and the other the w-plane, in which the corresponding point $w = u + iv$ is to be plotted. In this way the function f assigns to each z in its domain a value $w = f(z)$ in the w-plane. The relation thus defined is called a **mapping** (or *transformation*) of the domain of f **into** the w-plane or, more precisely, a mapping of the domain of f **onto**[1] the range of f in the w-plane.

[1] The general terminology is as follows. A mapping of a set A into a set B is called **surjective** or a mapping of A **onto** B if every element of B is the image of at least one element of A. The mapping is said to be **injective** or **one-to-one** if different elements of A have different images in B. The mapping is said to be **bijective** if it is both surjective and injective.

The point $w_0 = f(z_0)$ corresponding to a point z_0 is called the *image point* or **image** of the point z_0 with respect to the mapping defined by $f(z)$. If z moves along some curve C and $f(z)$ is continuous (not a constant), the corresponding point $w = f(z)$ will in general travel along a curve C^* in the w-plane. This curve is then called the *image* of the curve C, and the word "image" applies also to regions or other point sets.

We shall see that the properties of such mappings can be investigated by considering curves (and regions) in the z-plane and their images in the w-plane, and conversely. This will give more information about the functions than the consideration of individual points and their images.

Although two separate planes are used to represent w and z it is sometimes convenient to think of the mapping as effected in one plane and use such familiar terms as translation and rotation. For example, the mapping $w = z + 3$ may be interpreted as a translation which moves each point and configuration in the z-plane three units to the right.

To investigate the specific properties of a mapping defined by a given analytic function $w = u + iv = f(z)$, we may consider the images of the straight lines $x = const$ and $y = const$ in the w-plane. Another possibility is the study of the images of the circles $|z| = const$ and the straight lines through the origin. As a third possibility, we may consider the curves defined by $u(x, y) = const$ and $v(x, y) = const$ in the z-plane. These curves are called the **level curves** of u and v. Or we may consider simple figures (squares, rectangles, etc.) and their images.

To obtain a feeling for what is going on, let us consider some typical examples, starting with very simple mappings.

Example 1. Linear transformation $w = az + b$

The mapping

$$w = z + b \tag{1}$$

is a **translation.** Figure 273 illustrates (1) with $b = 2 + i$. It shows a rectangle and its image, which are congruent (why?); A has the image A^*, etc. Denoting points in this way is helpful, particularly in more complicated mappings. (1) with $b = 0$ is the **identity transformation**

$$w = z.$$

It maps every point onto itself.

The mapping

$$w = az \qquad (|a| = 1)$$

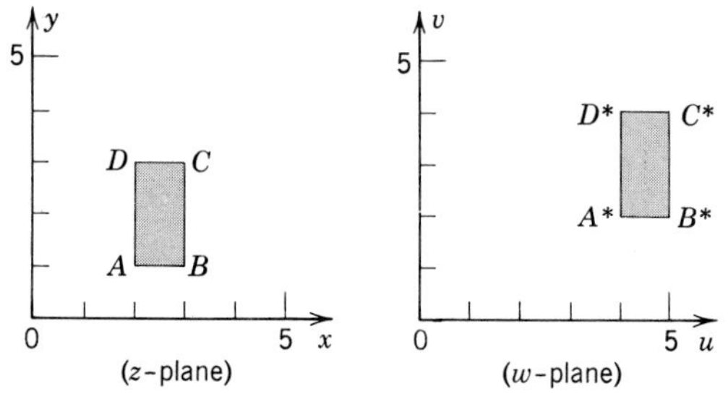

Fig. 273. Translation $w = z + 2 + i$

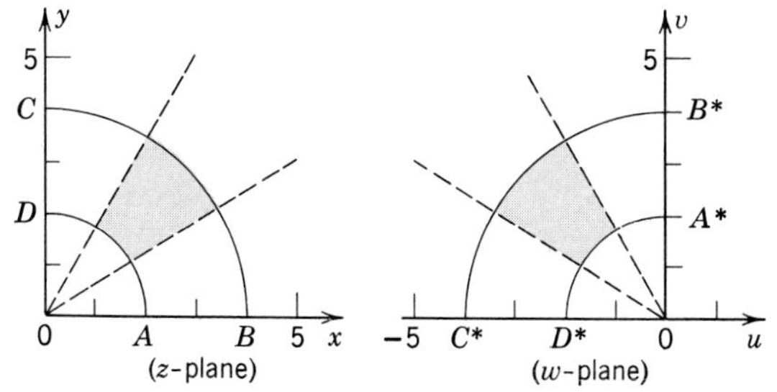

Fig. 274. Rotation $w = iz$. (Angle of rotation $\pi/2$, counterclockwise)

is a **rotation** through a fixed angle $\arg a$. Figure 274 shows $w = iz$. The mapping

$$w = az \qquad (a \text{ positive real})$$

is a uniform **expansion** if $a > 1$ or a uniform **contraction** if $0 < a < 1$. The mapping

$$w = az \qquad (a \text{ arbitrary}) \tag{2}$$

is a *rotation* (through the angle $\arg a$) *combined with a uniform expansion or contraction.* The mapping

$$w = az + b \tag{3}$$

is called a **linear transformation.** This is a rotation and an expansion (or contraction) $w_1 = az$ followed by a translation $w = w_1 + b$. Figure 275 shows $w = (1 + i)z + 2i$, which is a rotation through the angle $\pi/4$ (counterclockwise) and an expansion by the factor $|1 + i| = \sqrt{2}$ followed by a translation vertically upward.

Example 2. Mapping $w = z^2$

We want to discuss properties of the mapping

$$w = z^2. \tag{4}$$

In this case the simplest procedure is the use of polar representations. Indeed, if we set $z = re^{i\theta}$ and $w = Re^{i\phi}$, then (4) becomes $Re^{i\phi} = r^2 e^{2i\theta}$. It follows that

$$R = r^2 \qquad \text{and} \qquad \phi = 2\theta.$$

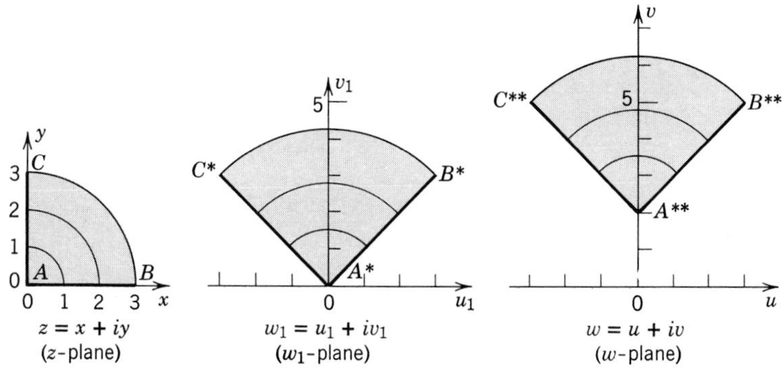

Fig. 275. Linear transformation $w = (1 + i)z + 2i$, consisting of the rotation and expansion $w_1 = (1 + i)z$ followed by the translation $w = w_1 + 2i$

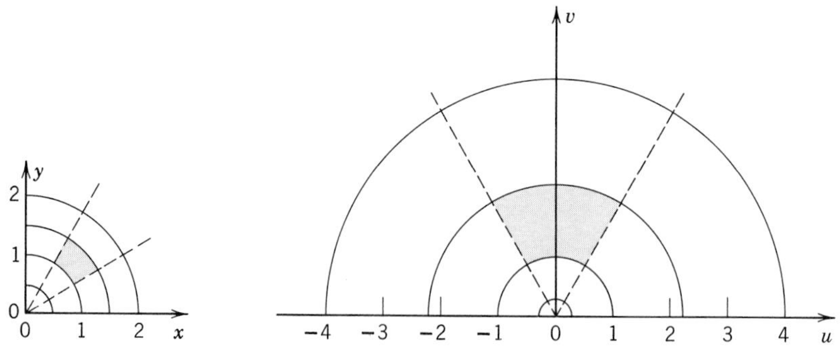

Fig. 276. Mapping $w = z^2$

Hence circles $r = r_0 = const$ are mapped onto circles $R = r_0{}^2 = const$, and rays $\theta = \theta_0 = const$ onto rays $\phi = 2\theta_0 = const$. In particular, the positive real axis ($\theta = 0$) is mapped onto the positive real axis in the w-plane, and the positive imaginary axis ($\theta = \pi/2$) in the z-plane is mapped onto the negative real axis in the w-plane. The angles at the origin are doubled under the mapping. The first quadrant $0 \leqq \theta \leqq \pi/2$ is mapped onto the entire upper half of the w-plane (Fig. 276).

In rectangular coordinates the transformation $w = z^2$ becomes

$$u + iv = x^2 - y^2 + 2xyi.$$

By separating the real and the imaginary parts we obtain

$$u = x^2 - y^2, \qquad v = 2xy. \tag{5}$$

We see that the **level curves** of u and v are equilateral hyperbolas with the lines $y = \pm x$ and the coordinate axes for asymptotes. We observe that these curves are the orthogonal trajectories of each other (cf. Sec. 1.10). In Fig. 277, the two shaded domains in the z-plane are both mapped onto the shaded rectangle in the w-plane. Clearly, every point $w \neq 0$ is the image of precisely two points in the z-plane.

Finally, we may use (5) for determining the images of the straight lines $x = const$ and $y = const$. The line $x = c = const$ has the image

$$u = c^2 - y^2, \qquad v = 2cy.$$

We can eliminate y from these equations, finding

$$v^2 = 4c^2(c^2 - u).$$

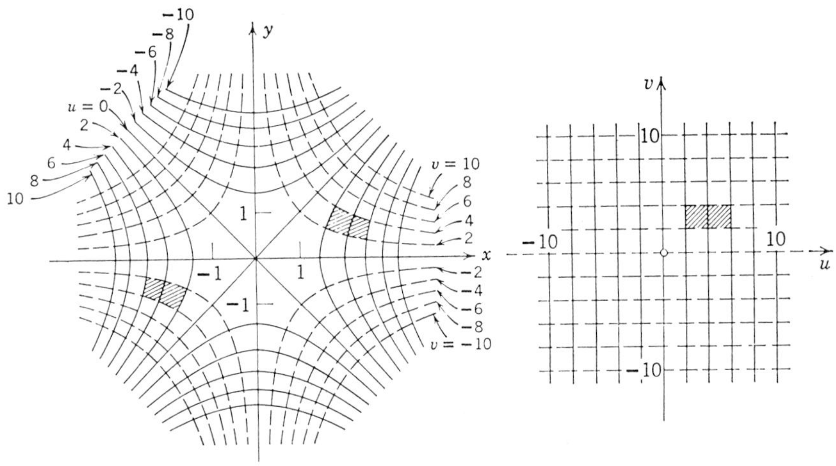

Fig. 277. Level curves of u and v in the case of the mapping $w = z^2$

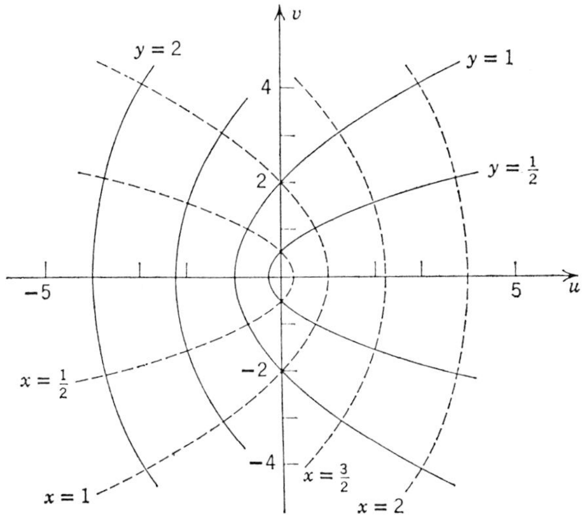

Fig. 278. Mapping $w = z^2$. Images of the straight lines $x = const$ (dashed) and $y = const$

This is a parabola with focus at the origin and opening to the left. Similarly, the image of a line $y = k = const$ can be represented in the form

$$v^2 = 4k^2(k^2 + u).$$

This is a parabola with focus at the origin and opening to the right (Fig. 278). ▮

The other powers

$$w = z^n, \qquad n = 3, 4, \cdots \tag{6}$$

may be considered in a similar fashion. Of course, the level curves, etc., are then represented by more complicated equations. The angular region $0 \leqq \arg z \leqq \pi/n$ is mapped onto the upper half of the w-plane (Fig. 279).

The mappings given by negative integer powers $1/z$, $1/z^2$, $\cdots$ can also be discussed by the use of polar coordinates. In practice, the most important case is given in the following example.

Example 3. Mapping $w = 1/z$. Inversion

We consider

$$w = \frac{1}{z} \qquad z \neq 0. \tag{7}$$

Using again $z = re^{i\theta}$ and $w = Re^{i\phi}$, we have from (7)

$$R = \frac{1}{r}, \qquad \phi = -\theta \qquad (r \neq 0). \tag{7'}$$

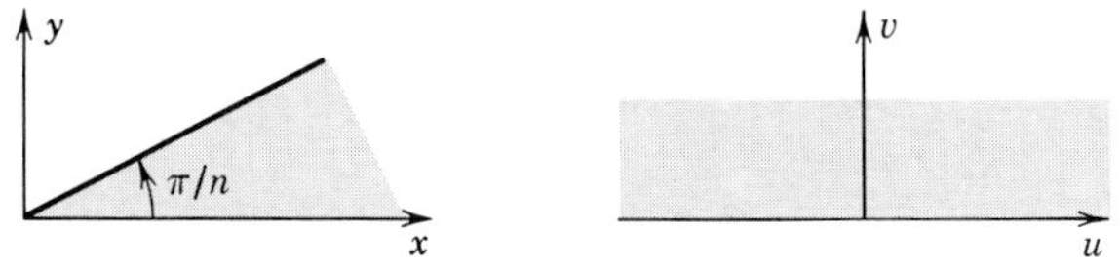

Fig. 279. Mapping defined by $w = z^n$

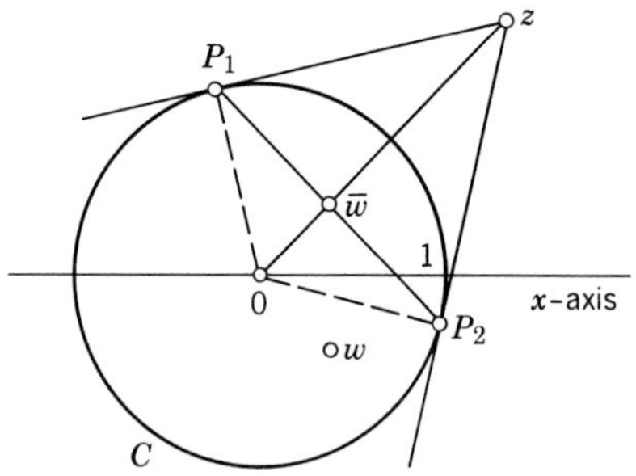

Fig. 280. Geometrical construction of $w = 1/z$. Here, $\bar{w}$ is the intersection of Oz and P_1P_2, where P_1 and P_2 are the points of contact of the tangents to the unit circle C passing through z.

From this we see that a point $w = 1/z$ $(z \neq 0)$ lies on the ray from the origin through $\bar{z}$, at the distance $1/|z|$ from the origin.

We mention that $w = 1/z$ can be obtained geometrically from z by an *inversion in the unit circle* (Fig. 280) followed by a reflection in the x-axis. The reader may prove this, using similar triangles.

Fig. 281 shows that $w = 1/z$ maps horizontal and vertical straight lines onto circles or straight lines. Even the following is true.

$w = 1/z$ maps every straight line or circle onto a circle or straight line.

Proof. Every straight line or circle in the z-plane can be written

$$A(x^2 + y^2) + Bx + Cy + D = 0 \qquad (A, B, C, D \text{ real}).$$

$A = 0$ gives a straight line and $A \neq 0$ a circle. In terms of z and $\bar{z}$, the equation becomes

$$Az\bar{z} + B\frac{z + \bar{z}}{2} + C\frac{z - \bar{z}}{2i} + D = 0.$$

Now $w = 1/z$. Substitution of $z = 1/w$ and multiplication by $w\bar{w}$ gives

$$A + B\frac{\bar{w} + w}{2} + C\frac{\bar{w} - w}{2i} + Dw\bar{w} = 0$$

or, in terms of u and v,

$$A + Bu - Cv + D(u^2 + v^2) = 0.$$

This represents a circle (if $D \neq 0$) or a straight line (if $D = 0$) in the w-plane. ∎

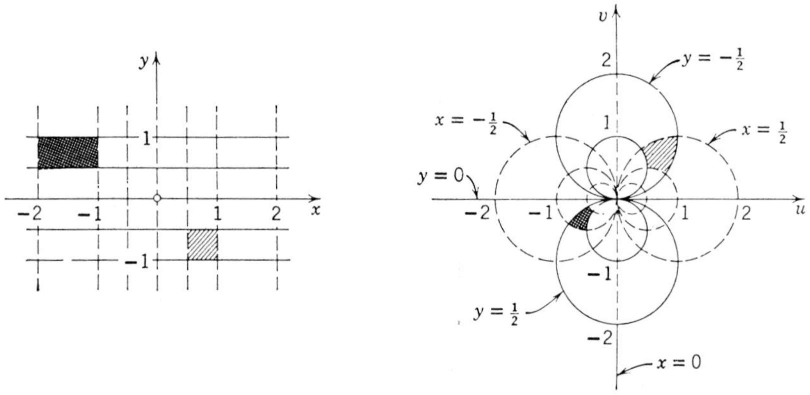

Fig. 281. Mapping $w = 1/z$

Problems for Sec. 13.1

Consider the mapping $w = (1 + i)z - 2$. Find and plot the images of the given curves or regions.

1. $x = 0, 1, 2, 3$ **2.** $y = -1, 0, 1, 2$ **3.** $|z - 2| \leqq 2$

Consider the mapping $w = u + iv = z^2$. Find and plot the images of the following curves.

4. $y = x, y = -x$ **5.** $x = 0, 1, 2, 3$ **6.** $y = 0, 1, 2, 3$
7. $y = 1 + x$ **8.** $y = 1 - x$ **9.** $y^2 = x^2 + 1$

Plot the images of the following regions under the mapping $w = z^2$.

10. $|z| > 2$ **11.** $|z| \leqq 4$ **12.** $|\arg z| < \pi/4$
13. $1 < x < 2$ **14.** $0 \leqq y \leqq 1$ **15.** $-\pi/4 < \arg z < \pi/2$

Find the images of the following circles and straight lines under the mapping $w = 1/z$.

16. $|z| = 1$ **17.** $|z + 1| = 1$ **18.** $|z - 1| = 1$
19. $|z - 2i| = 2$ **20.** $y = x - 1$ **21.** $x = 1$

22. Find the image of the region $-2 < x < -1$, $-1 < y < 1$ under the mapping $w = 1/z$.

23. Find the image of the region $1 < x < 2$ under the mapping $w = 1/z$.

24. Consider $w = 1/z$. What straight lines are mapped onto (*a*) straight lines, (*b*) circles? What circles are mapped onto (*c*) straight lines, (*d*) circles?

25. Show that under the mapping $w = 1/z$, the center of a given circle is not mapped onto the center of the image circle.

26. Find $1/(3 + 4i)$ from $3 + 4i$ by means of the geometrical construction explained in connection with the transformation $w = 1/z$.

27. Find and plot the images of the angular region $0 \leqq \arg z \leqq \pi/4$ in the case of the transformations $w = z$, $w = iz$, $w = -iz$, $w = z^2$, $w = iz^2$, $w = -z^2$, $w = -iz^2$, and $w = z^3$.

28. Find and plot the image of the region in Prob. 27 under the transformations $w = 1/z$, $w = i/z$, $w = 1/z^2$, and $w = i/z^2$.

29. Find an analytic function $w = u + iv = f(z)$ which maps the half plane $x \geqq 0$ onto the region $u \geqq 2$ such that $z = 0$ corresponds to $w = 2 + i$.

30. Find an analytic function $w = u + iv = f(z)$ which maps the angular region $0 < \arg z < \pi/3$ onto the region $u < 1$.

13.2 Conformal Mapping

We shall now consider the most important geometrical property of the mappings defined by analytic functions, namely, their conformality.

A mapping in the plane is said to be *angle-preserving*, or **conformal**, if it preserves angles between oriented curves in magnitude as well as in sense, that is, the images of any two intersecting oriented curves, taken with their corresponding orientation, make the same angle of intersection as the curves, both in magnitude and direction. Here the angle between two oriented curves is defined to be the angle α $(0 \leqq \alpha \leqq \pi)$ between their oriented tangents at the point of intersection (Fig. 282).

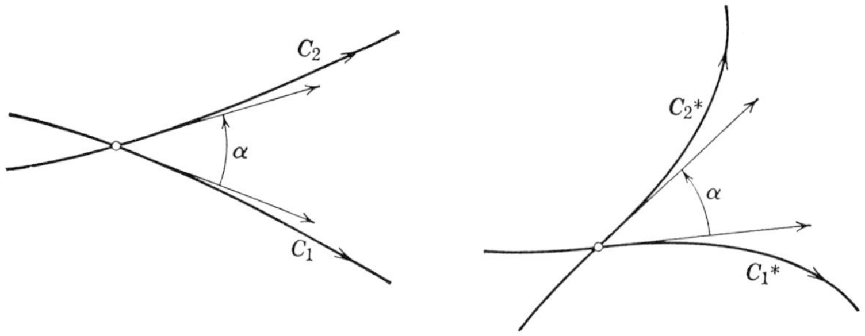

Fig. 282. Curves C_1 and C_2 and their respective images C_1^* and C_2^* under a conformal mapping

We want to show that a mapping $w = f(z)$ is conformal at every point where $f(z)$ is analytic, except at points where the derivative $f'(z)$ is zero. Such a point is called a **critical point.** For instance, in the case of $f(z) = z^2$ we have $f'(z) = 2z = 0$ at $z = 0$, where the mapping is not conformal since angles are doubled there (cf. Example 2 in Sec. 13.1).

Of course, for our purpose we have to consider curves and their images. A curve C in the complex z-plane can be represented in the form

$$z(t) = x(t) + iy(t) \tag{1}$$

where t is a real parameter. For example, the function

$$z(t) = r\cos t + ir\sin t$$

represents the circle $|z| = r$, the function

$$z(t) = t + it^2$$

represents the parabola $y = x^2$, etc. The direction of increasing values of t in (1) is called the *positive direction* or *positive sense* on C. In this way (1) defines an *orientation* on C. We assume that $z(t)$ in (1) is differentiable and the derivative $\dot{z}(t)$ is continuous and nowhere zero. Then C has a unique tangent at each of its points and is called a **smooth curve.** The sense on each tangent corresponding to the positive sense on C is called the *positive sense* on that tangent, which is then oriented.

In fact, the tangent to C at a point $z_0 = z(t_0)$ is defined as the limiting position of the straight line through z_0 and another point $z_1 = z(t_0 + \Delta t)$ as z_1 approaches z_0 along C, that is, as $\Delta t \to 0$. (Cf. also Sec. 8.5.) Now the number $z_1 - z_0$ can be represented by the vector from z_0 to z_1 (Fig. 283), and the vector corresponding to $(z_1 - z_0)/\Delta t$, where $\Delta t > 0$, has the same direction as that vector. It follows that the vector corresponding to

$$\dot{z}(t_0) = \left.\frac{dz}{dt}\right|_{t_0} = \lim_{\Delta t \to 0} \frac{z_1 - z_0}{\Delta t} = \lim_{\Delta t \to 0} \frac{z(t_0 + \Delta t) - z(t_0)}{\Delta t} \tag{2}$$

is tangent to C at z_0, and the angle between this vector and the positive x-axis is $\arg \dot{z}(t_0)$.

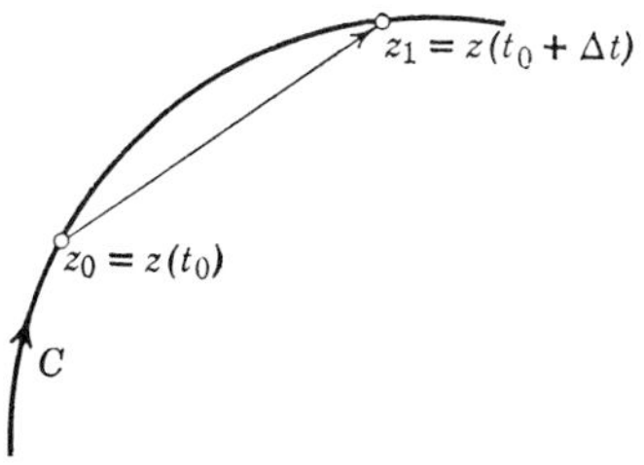

Fig. 283. Derivation of formula (2)

Consider now the mapping given by a nonconstant analytic function $w = f(z) = u(x, y) + iv(x, y)$ defined in a domain containing C. Then the image of C under this mapping is a curve C^* in the w-plane represented by

$$w(t) = f[z(t)].$$

The point $z_0 = z(t_0)$ corresponds to the point $w(t_0)$ of C^*, and $\dot{w}(t_0)$ represents a tangent vector to C^* at this point. Now by the chain rule,

$$\frac{dw}{dt} = \frac{df}{dz}\frac{dz}{dt}. \tag{3}$$

Hence, if $f'(z_0) \neq 0$, we see that $\dot{w}(t_0) \neq 0$ and C^* has a unique tangent at $w(t_0)$, the angle between the tangent vector $\dot{w}(t_0)$ and the positive u-axis being $\arg \dot{w}(t_0)$. Since the argument of a product equals the sum of the arguments of the factors, we have from (3)

$$\arg \dot{w}(t_0) = \arg f'(z_0) + \arg \dot{z}(t_0).$$

Thus under the mapping the directed tangent to C at z_0 is rotated through the angle

$$\arg \dot{w}(t_0) - \arg \dot{z}(t_0) = \arg f'(z_0), \tag{4}$$

the angle between those two tangent vectors to C and C^*. Since the expression on the right is independent of the choice of C, we see that this angle is independent of C; that is, the transformation $w = f(z)$ rotates the tangents of all the curves through z_0 through the *same* angle $\arg f'(z_0)$. Hence, two curves through z_0 which form a certain angle at z_0 are mapped upon curves forming the same angle, in sense as well as in magnitude, at the image point w_0 of z_0. This proves the following basic result.

Theorem 1 (Conformal mapping)

The mapping defined by an analytic function $f(z)$ is conformal, except at points where the derivative $f'(z)$ is zero.

Example 1. Conformality of $w = z^2$

The mapping $w = z^2$ is conformal except at $z = 0$ where $w' = 2z = 0$. This is illustrated by Figs. 276 and 278 in which the image curves intersect at right angles, except at $z = 0$ where the angles are doubled under the mapping, since every ray $\arg z = c = const$ transforms into a ray $\arg w = 2c$ (cf. Fig. 276 in the preceding section). ∎

We further note that, by the definition of a derivative, we have

$$\lim_{z \to z_0} \left| \frac{f(z) - f(z_0)}{z - z_0} \right| = |f'(z_0)|.$$

Therefore, the mapping $w = f(z)$ magnifies the lengths of short lines by approximately the factor $|f'(z_0)|$. The image of a small figure *conforms* to the original figure in the sense that it has approximately the same shape. However, since $f'(z)$ varies from point to point, a *large* figure may have an image whose shape is quite different from that of the original figure.

We want to mention that, by (3) in Sec. 12.5 and the Cauchy–Riemann equations,

$$(5') \qquad |f'(z)|^2 = \left| \frac{\partial u}{\partial x} + i \frac{\partial v}{\partial x} \right|^2 = \left(\frac{\partial u}{\partial x} \right)^2 + \left(\frac{\partial v}{\partial x} \right)^2 = \frac{\partial u}{\partial x} \frac{\partial v}{\partial y} - \frac{\partial u}{\partial y} \frac{\partial v}{\partial x},$$

that is,

$$(5) \qquad |f'(z)|^2 = \begin{vmatrix} \dfrac{\partial u}{\partial x} & \dfrac{\partial u}{\partial y} \\ \dfrac{\partial v}{\partial x} & \dfrac{\partial v}{\partial y} \end{vmatrix} = \frac{\partial(u, v)}{\partial(x, y)}$$

where the determinant is the so-called *Jacobian* (cf. Sec. 9.3) of the transformation $w = f(z)$, written in real form

$$u = u(x, y), \qquad v = v(x, y).$$

Hence, the condition $f'(z_0) \neq 0$ implies that the Jacobian is not zero at z_0. This condition suffices that the restriction of the mapping $w = f(z)$ to a sufficiently small neighborhood N_0 of z_0 is **one-to-one** or injective, that is, different points in N_0 have different images; a proof is included in Ref. [A2] listed in Appendix 1. (Cf. also footnote 1 in Sec. 13.1.)

Example 2

The mapping $w = z^2$ is one-to-one in a sufficiently small neighborhood of any point $z \neq 0$. In a neighborhood of $z = 0$ it is not one-to-one. The full z-plane is mapped onto the w-plane so that each point $w \neq 0$ is the image of two points in the z-plane. For instance, the points $z = 1$ and $z = -1$ are both mapped onto $w = 1$, and, more generally, z_1 and $-z_1$ have the same image point $w = z_1^2$. ∎

The practical importance of conformal mapping results from the fact that harmonic functions of two real variables (cf. Sec. 12.5) remain harmonic under a change of variables arising from a conformal transformation (Theorem 2, below). This has important consequences. Suppose that it is required to solve a **boundary value problem** in connection with a two-dimensional potential, that is, to determine a solution of Laplace's equation (in two independent variables) in a given region D which assumes given values on the boundary of D. It may be possible to find a conformal mapping which transforms D into some simpler region D^*, such as a circular disk or a half-plane. Then we may solve Laplace's equation subject to the transformed boundary conditions in D^*. The resulting solution when carried back to D by the use of that mapping will be the solution of the original problem. This powerful method is justified by the following theorem.

Theorem 2 (Harmonic functions and conformal mapping)
A harmonic function $h(x, y)$ remains harmonic under a change of the variables arising from a one-to-one conformal transformation given by an analytic function $w = f(z)$.

First proof, assuming the existence of a conjugate harmonic. Let $h(x, y)$ be harmonic in a domain D and $g(x, y)$ a conjugate[2] of $h(x, y)$ in D, so that $h + ig$ is an analytic function $H(z)$ of $z = x + iy$ in D. By assumption, the mapping $w = f(z) = u(x, y) + iv(x, y)$ is one-to-one and conformal. Hence the image D^* of D is a domain; also $f'(z) \neq 0$ in D, and the inverse function $z = F(w)$ which maps D^* onto D exists. $F(w)$ is analytic in D^*; indeed, the derivative is

$$\frac{dF}{dw} = \frac{1}{df/dz}.$$

The proof of this formula is similar to that in real calculus. Hence $H(F(w))$ is an analytic function of w in D^*. Its real part is $h(x(u, v), y(u, v))$ and is a harmonic function of u, v in D^*.

Second proof, without that assumption. Let $h(x, y)$ be harmonic in D. As before we use $z = x + iy = F(w)$ to obtain $h(x(u, v), y(u, v))$. We denote this function, considered as a function of u, v, again by h, for simplicity, and show that it is harmonic in D^*, the image of D under $w = f(z)$. We apply the chain rule, denoting partial derivatives by subscripts:

$$h_x = h_u u_x + h_v v_x.$$

We apply the chain rule once more and underscore the terms which will drop out when we form $h_{xx} + h_{yy}$:

$$\begin{aligned} h_{xx} &= \underline{h_u u_{xx}} + (h_{uu} u_x + \underline{h_{uv} v_x}) u_x \\ &\quad + \underline{h_v v_{xx}} + (\underline{h_{vu} u_x} + h_{vv} v_x) v_x. \end{aligned}$$

h_{yy} is the same expression with each x replaced by a y. We now form the sum $h_{xx} + h_{yy}$, noting that

$$u_{xx} + u_{yy} = 0, \qquad v_{xx} + v_{yy} = 0$$

[2] Cf. Sec. 12.5. We mention without proof that a conjugate harmonic exists if D is simply connected (definition in Sec. 9.12).

(since $w = u + iv$ is analytic) and in the sum, $h_{vu} = h_{uv}$ is multiplied by

$$u_x v_x + u_y v_y$$

which is zero by the Cauchy–Riemann equations. There remains

$$h_{xx} + h_{yy} = h_{uu}(u_x{}^2 + u_y{}^2) + h_{vv}(v_x{}^2 + v_y{}^2).$$

By the Cauchy–Riemann equations this equals

$$(h_{uu} + h_{vv})(u_x{}^2 + v_x{}^2).$$

Hence by (5′),

$$\frac{\partial^2 h}{\partial x^2} + \frac{\partial^2 h}{\partial y^2} = |f'(z)|^2 \left(\frac{\partial^2 h}{\partial u^2} + \frac{\partial^2 h}{\partial v^2}\right). \tag{6}$$

Since $f'(z) \neq 0$ by the conformality and the left side is 0 in D by assumption, the expression in the parentheses must be 0 in D^*. ∎

When we use the method of conformal mapping in potential theory, the difficulty is to find an analytic function which maps a given region onto a simpler one. For this purpose we need some experience and a detailed knowledge of the mapping properties of the elementary analytic functions. Therefore, we shall consider the most important elementary functions from this point of view.

Problems for Sec. 13.2

1. Why do the images of the curves $|z| = const$ and $\arg z = const$ under a mapping by an analytic function intersect at right angles?

2. What is the reason that the level curves $u = const$ and $v = const$ of an analytic function $w = u + iv = f(z)$ intersect at right angles at each point at which $f'(z) \neq 0$?

3. Does the mapping $w = \bar{z} = x - iy$ preserve angles in size as well as in sense?

Represent the following curves in the z-plane ($z = x + iy$) in the form $z = z(t)$.

4. $x^2 + y^2 = 4$ **5.** $y = 1/x$ **6.** $y = 4x^2$

7. $x^2 - y^2 = 1$ **8.** $y = ax + b$ **9.** $(x - 1)^2 + (y + 2)^2 = 9$

Determine the points in the z-plane at which the mapping $w = f(z)$ fails to be conformal, where $f(z)$ equals

10. z^4 **11.** $\sin z$ **12.** $z + z^{-1}$ $(z \neq 0)$

13. e^{z^2} **14.** $z^4 - z^2$ **15.** $z^2 + az + b$

Verify (5) for the following functions $f(z) = u(x, y) + iv(x, y)$.

16. e^z **17.** $\cos z$ **18.** $z^2 - 3z$

19. Write down all the formulas in the second proof of Theorem 2 and carry out all the steps in detail.

20. Verify Theorem 2 for $h(x, y) = x/(x^2 + y^2)$, $f(z) = 2z + 1$.

13.3 Linear Fractional Transformations

By a **linear fractional transformation** (or **Möbius transformation**) we mean a transformation of the form

$$w = \frac{az + b}{cz + d} \qquad (ad - bc \neq 0), \tag{1}$$

where the constants a, b, c, d are real or complex numbers. The condition $ad - bc \neq 0$ becomes understandable if we consider the derivative of (1):

$$w' = \frac{a(cz + d) - c(az + b)}{(cz + d)^2} = \frac{ad - bc}{(cz + d)^2}.$$

This shows that $ad - bc \neq 0$ implies $w' \neq 0$, which entails conformality, whereas $ad - bc = 0$ leads to the uninteresting case $w' \equiv 0$ or $w = const$, which will be excluded in our further consideration.

Special cases of (1) already discussed in Sec. 13.1 are the *translation*

$$w = z + b, \tag{2}$$

the *rotation and expansion or contraction*

$$w = az, \tag{3}$$

the *inversion and reflection in the x-axis*

$$w = \frac{1}{z} \tag{4}$$

and the *linear transformation*

$$w = az + b. \tag{5}$$

Extended complex plane. This is an important matter which can be motivated by (1), as follows. From (1) we see that to each z for which $cz + d \neq 0$ there corresponds precisely one complex number w. Suppose that $c \neq 0$. Then to the value $z = -d/c$, for which $cz + d = 0$, there does not correspond a number w. This situation suggests that we attach an "improper point" to the w-plane. This point is called the **point at infinity** and is denoted by the symbol ∞ (*infinity*). The complex plane together with the point ∞ is called the **extended complex plane.** The complex plane without that improper point is called the **finite complex plane.** We now let $w = \infty$ be the image point of $z = -d/c$ $(c \neq 0)$ under the mapping defined by (1). When $c = 0$ in (1), then $a \neq 0$ and $d \neq 0$ (why?), and we let $w = \infty$ be the image point of $z = \infty$, the improper point of the extended z-plane.

The inverse mapping of (1) is obtained by solving (1) for z; we find

$$z = \frac{-dw + b}{cw - a}. \tag{6}$$

When $c \neq 0$, then $cw - a = 0$ for $w = a/c$, and we let $w = a/c$ correspond to $z = \infty$. When $c = 0$, then $a \neq 0$ and $d \neq 0$, and we let $w = \infty$ correspond to $z = \infty$, as before. It follows that every mapping (1) is a one-to-one mapping of the extended z-plane onto the extended w-plane; we say, that *every linear fractional transformation* (1) *maps "the extended plane in a one-to-one manner onto itself."*

Our present discussion suggests the following:

General remark. If $z = \infty$, then the right-hand side of (1) becomes the meaningless expression $(a \cdot \infty + b)/(c \cdot \infty + d)$. We assign to it the value $w = a/c$ when $c \neq 0$ and $w = \infty$ when $c = 0$.

Fixed points. A *fixed point* of a mapping $w = f(z)$ is a point whose image is the same complex number; that is, the fixed points are obtained from

$$w = f(z) = z.$$

Hence the fixed points of (1) are obtained from the equation

$$z = \frac{az + b}{cz + d}$$

or

$$cz^2 - (a - d)z - b = 0. \tag{7}$$

This is a quadratic equation in z whose coefficients all vanish if and only if the mapping is the identity (in this case, $a = d \neq 0$, $b = c = 0$). Consequently, we have

Theorem 1 (Fixed points)

A linear fractional transformation, not the identity, has at most two fixed points. If a linear fractional transformation is known to have three or more fixed points, it must be the identity.

Special linear fractional transformations of practical importance and further general properties of linear fractional transformations will be discussed in the following section.

Problems for Sec. 13.3

Find the fixed points of the following mappings.

1. $w = iz$ **2.** $w = -iz + 4$ **3.** $w = z^2$

4. $w = (z - i)^2$ **5.** $w = z^3$ **6.** $w = iz^2$

7. $w = \dfrac{3z - 1}{z + 3}$ **8.** $w = \dfrac{5z + 4}{z + 5}$ **9.** $w = \dfrac{2iz - 1}{z + 2i}$

Find a linear fractional transformation whose fixed points are

10. $1, -1$ **11.** $i, -i$ **12.** 1

13. Find all linear fractional transformations whose fixed points are $-i$ and i.

14. Find all linear fractional transformations whose fixed points are -1 and 1.

15. Find all linear fractional transformations without fixed points in the finite plane. [Use (7).]

13.4 Special Linear Fractional Transformations

In this section we illustrate how we can determine linear fractional transformations

$$w = \frac{az + b}{cz + d} \qquad (ad - bc \neq 0) \tag{1}$$

for mapping certain simple domains onto others, and how we can discuss properties of (1). Of help in such a discussion is

Theorem 1 (Circles and straight lines)

Every linear fractional transformation (4) *maps the totality of circles and straight lines in the z-plane onto the totality of circles and straight lines in the w-plane.*

Proof. This is trivial for a translation or rotation, fairly obvious for a uniform expansion or contraction and, by Example 3 of Sec. 13.1, true for $w = 1/z$. It also holds for composites of such mappings. From this it follows for (1) with $c \neq 0$, because (1) with $c \neq 0$ can be written

$$w = K\frac{1}{cz + d} + \frac{a}{c} \qquad \text{where} \qquad K = -\frac{ad - bc}{c}, \tag{2}$$

and setting

$$w_1 = cz, \qquad w_2 = w_1 + d, \qquad w_3 = \frac{1}{w_2}, \qquad w_4 = Kw_3,$$

we have $w = w_4 + a/c$. This shows that (1) is a composite of the special cases (2)–(4), Sec. 13.3. ∎

The linear fractional transformation (1) depends on three essential constants, namely, the ratios of any three of the constants a, b, c, d to the fourth. The requirement that three distinct points in the z-plane have specified images in the w-plane leads to a unique linear fractional transformation, as follows.

Theorem 2 (Three points and their images given)
Three given distinct points z_1, z_2, z_3 can always be mapped onto three prescribed distinct points w_1, w_2, w_3 by one, and only one, linear fractional transformation $w = f(z)$. This mapping is given implicitly by the equation

$$\text{(3)} \qquad \frac{w - w_1}{w - w_3} \cdot \frac{w_2 - w_3}{w_2 - w_1} = \frac{z - z_1}{z - z_3} \cdot \frac{z_2 - z_3}{z_2 - z_1}.$$

(If one of these points is the point ∞, the quotient of those two differences which contain this point must be replaced by 1.*)*

Proof. Equation (3) is of the form $F(w) = G(z)$ where F and G denote fractional linear functions of the respective variables. From this we readily obtain $w = f(z) = F^{-1}[G(z)]$ where F^{-1} denotes the inverse function of F. Since the inverse of a linear fractional transformation and the composite of linear fractional transformations are linear fractional transformations (cf. Prob. 3 at the end of the section), $w = f(z)$ is a linear fractional transformation. Furthermore, from (3) we see that

$$F(w_1) = 0, \qquad F(w_2) = 1, \qquad F(w_3) = \infty,$$

$$G(z_1) = 0, \qquad G(z_2) = 1, \qquad G(z_3) = \infty.$$

Hence, $w_1 = f(z_1)$, $w_2 = f(z_2)$, $w_3 = f(z_3)$. This proves the existence of a linear fractional transformation $w = f(z)$ which maps z_1, z_2, z_3 onto w_1, w_2, w_3, respectively.

We prove that $w = f(z)$ is uniquely determined. Suppose that $w = g(z)$ is another linear fractional transformation which maps z_1, z_2, z_3 onto w_1, w_2, w_3, respectively. Then its inverse $g^{-1}(w)$ maps w_1 onto z_1, w_2 onto z_2, and w_3 onto z_3. Consequently, the composite mapping $H = g^{-1}[f(z)]$ maps each of the points z_1, z_2, and z_3 onto itself; that is, it has three distinct fixed points z_1, z_2, z_3. From Theorem 1 in the preceding section it follows that H is the identity mapping, and, therefore, $g(z) \equiv f(z)$.

The last statement of the theorem follows from the general remark in the preceding section. This completes the proof. ∎

Mapping of half-planes onto disks. This is a task of practical interest, for instance in potential problems. Without loss of generality, let us map the upper half-plane $y \geqq 0$ onto the unit disk $|w| \leqq 1$. The boundary of that half-plane is the x-axis; clearly, it must be mapped onto the unit circle $|w| = 1$. This gives the idea: to find a mapping, choose three points on the x-axis, prescribe their images on that circle and apply Theorem 2. Make sure that the half-plane $y \geqq 0$ is mapped onto the interior but not onto the exterior of that circle.

Example 1. Mapping of a half-plane onto a disk

Find the linear fractional transformation (1) which maps $z_1 = -1$, $z_2 = 0$, $z_3 = 1$ onto $w_1 = -1$, $w_2 = -i$, $w_3 = 1$, respectively. From (3) we obtain

$$\frac{w - (-1)}{w - 1} \cdot \frac{-i - 1}{-i - (-1)} = \frac{z - (-1)}{z - 1} \cdot \frac{0 - 1}{0 - (-1)}$$

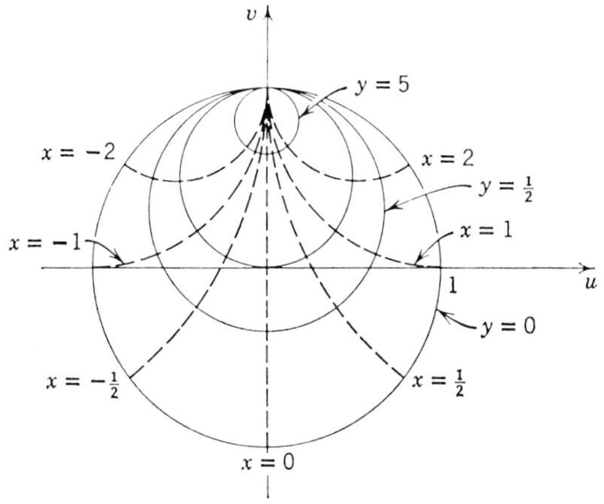

Fig. 284. Linear fractional transformation in Example 1

or

$$w = \frac{z - i}{-iz + 1}. \tag{4}$$

Let us show that we can determine the specific properties of such a mapping without difficult calculations. The images of the lines $x = const$ and $y = const$ are obtained as follows. The point $z = i$ corresponds to $w = 0$, and $z = \infty$ corresponds to $w = i$. If $z = iy$, then we have $w = i(y - 1)/(y + 1)$; that is, the positive imaginary axis is mapped onto the segment $u = 0$, $-1 \leqq v \leqq 1$. Since the mapping is conformal and straight lines are mapped onto circles or straight lines, the lines $y = const$ are mapped onto circles through the image of $z = \infty$, that is, onto circles through $w = i$ and with center on the v-axis. It follows that, for the same reasons, the lines $x = const$ are mapped onto circles which are orthogonal to the image circles of the lines $y = const$ (Fig. 284). The lower half-plane corresponds to the exterior of the unit circle $|w| = 1$.

Example 2. Occurrence of ∞

Determine the linear fractional transformation which maps $z_1 = 0, z_2 = 1, z_3 = \infty$ onto $w_1 = -1$, $w_2 = -i$, $w_3 = 1$, respectively. From (3) we obtain the desired mapping

$$w = \frac{z - i}{z + i}. \tag{5}$$

This is sometimes called the *Cayley transformation*. In this case, (3) gave at first the quotient $(1 - \infty)/(z - \infty)$, which we had to replace by 1. ∎

Mappings of half-planes onto half-planes. This is another task of practical interest. We may map the upper half-plane $y \geqq 0$ onto the upper half-plane $v \geqq 0$, as a typical case. Then the x-axis must be mapped onto the u-axis.

Example 3. Mapping of a half-plane onto a half-plane

Find the linear fractional transformation which maps the points $z_1 = -2, z_2 = 0, z_3 = 2$ onto the points $w_1 = \infty$, $w_2 = \frac{1}{4}$, $w_3 = \frac{3}{8}$, respectively. From (3) we obtain

$$w = \frac{z + 1}{2z + 4}, \tag{6}$$

as the reader may verify. What is the image of the x-axis? ∎

Mapping of disks onto disks. This is a third class of practical problems. We may map the unit disk in the z-plane onto the unit disk in the w-plane. It can be readily verified that the function

$$w = \frac{z - z_0}{cz - 1}, \qquad c = \bar{z}_0, \qquad |z_0| < 1 \tag{7}$$

is of the desired type and maps the point z_0 onto the center $w = 0$ (cf. Prob. 6).

Example 4. Mapping of the unit disk onto the unit disk

Let $z_0 = \frac{1}{2}$. Then by (7),

$$w(z) = \frac{2z - 1}{z - 2}.$$

The real axes correspond to each other; in particular,

$$w(-1) = 1, \qquad w(0) = \tfrac{1}{2}, \qquad w(1) = -1.$$

Since the mapping is conformal and straight lines are mapped onto circles or straight lines and $w(\infty) = 2$, the images of the lines $x = const$ are circles through $w = 2$ with centers on the u-axis; the lines $y = const$ are mapped onto circles which are orthogonal to the aforementioned circles (Fig. 285). ∎

Mappings of angular regions onto the unit disk may be obtained by combining linear fractional transformations and transformations of the form $w = z^n$, where n is an integer greater than 1.

Example 5. Mapping of an angular region onto the unit disk

Map the angular region D: $-\pi/6 \leqq \arg z \leqq \pi/6$ onto the unit disk $|w| \leqq 1$. We may proceed as follows. The mapping $t = z^3$ maps D onto the right half of the t-plane. Then we may apply a linear fractional transformation which maps this half-plane onto the unit disk, for example, the transformation

$$w = i\frac{t - 1}{t + 1}.$$

By inserting $t = z^3$ into this mapping we find

$$w = i\frac{z^3 - 1}{z^3 + 1};$$

this mapping has the required properties (Fig. 286).

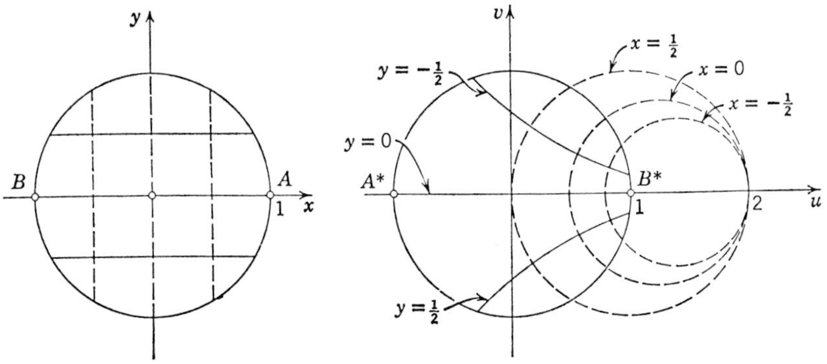

Fig. 285. Mapping in Example 4

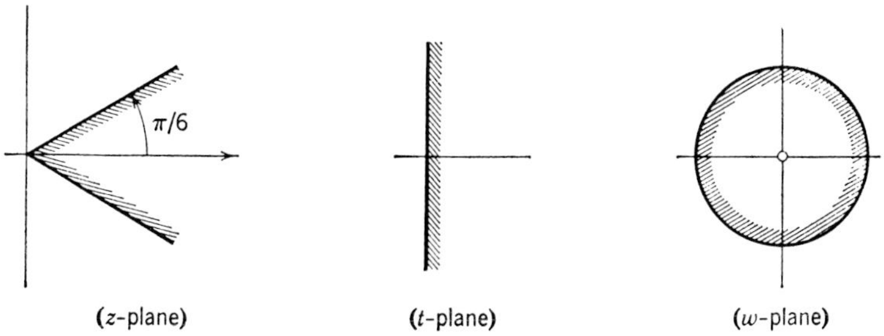

Fig. 286. Mapping in Example 5

Problems for Sec. 13.4

1. Represent $w = (z + i)/(iz + 4)$ as a composite of mappings of the type (2), (3), (4), Sec. 13.3.
2. Prove Theorem 1 for $w = az$, where $a \neq 0$.
3. Show that the composite of two linear fractional transformations is a linear fractional transformation.
4. Derive (5) from (3).
5. Represent (5) as a composite of mappings of the type (2), (3), (4), Sec. 13.3.
6. Prove the statement involving (7).
7. Find a linear fractional transformation which maps $|z| \leqq 1$ onto $|w| \leqq 1$ such that $z = i/4$ is mapped onto $w = 0$ and graph the images of the lines $x = const$ and $y = const$.
8. Find the inverse of (4). Show that (4) maps the lines $x = c = const$ onto circles with centers on the line $v = 1$.

Find the linear fractional transformation which maps

9. $0, 1, \infty$ onto $\infty, 1, 0$, respectively.
10. $0, 1, i$ onto $2, 3, 2 + i$, respectively.
11. $0, 1, 2$ onto $1, 1/2, 1/3$, respectively.
12. $-1, 0, 1$ onto $0, -1, \infty$, respectively.
13. $-i, 0, i$ onto $\infty, -1, 0$, respectively.
14. $0, i, 2i$ onto $\infty, 2i, 5i/2$, respectively.
15. $-1, i, 1 + i$ onto $0, \infty, 2 + i$, respectively.
16. $i, 0, 1$ onto $i/2, 0, (1 + i)/2$, respectively.
17. $1/2, 1, 3$ onto $\infty, 4, 6/5$, respectively.
18. $0, 1, \infty$ onto $\infty, (1 - i)/2, 1/2$, respectively.

Find all linear fractional transformations $w(z)$ with the following property.

19. $z_1 = 0$ is a fixed point.
20. $z_1 = 0$ and $z_2 = \infty$ are fixed points.
21. The x-axis is mapped onto the u-axis.

22. Determine the linear fractional transformation with fixed points -1 and 1 which maps 0 onto ip, where p is real. Discuss the cases $p = 0$ and $p = 1$.
23. Find an analytic function which maps the second quadrant of the z-plane onto the interior of the unit circle in the w-plane.

24. Find an analytic function $w = f(z)$ which maps the region $2 \leqq y \leqq x + 1$ onto the unit disk $|w| \leqq 1$.

25. Find an analytic function $w = f(z)$ which maps the region $0 \leqq \arg z \leqq \pi/4$ onto the unit disk $|w| \leqq 1$.

13.5 Mapping by Other Elementary Functions

We shall now consider the mapping properties of some further important special functions.

The **exponential function** (Sec. 12.7)

$$w = e^z \tag{1}$$

defines a mapping which is conformal everywhere, because its derivative is different from zero at any point. If we set $w = Re^{i\phi}$, then

$$Re^{i\phi} = e^{x+iy} = e^x e^{iy}$$

and (1) can be written in the form

$$R = e^x, \qquad \phi = y. \tag{2}$$

From this we see that the lines $x = a = const$ are mapped onto the circles $R = e^a$, and the lines $y = c$ are mapped onto the rays $\phi = c$. Since $e^z \neq 0$ for all z, the point $w = 0$ is not an image of any point z. A rectangular region, say, $a \leqq x \leqq b$, $c \leqq y \leqq d$, is mapped onto the region

$$e^a \leqq R \leqq e^b, \qquad c \leqq \phi \leqq d$$

bounded by portions of rays and circles (Fig. 287).

The fundamental strip $-\pi < y \leqq \pi$ is mapped onto the full w-plane (cut along the negative real axis). More generally, every horizontal strip bounded by two lines $y = c$ and $y = c + 2\pi$ is mapped onto the full w-plane. This illustrates the fact that e^z is periodic with the period $2\pi i$.

The horizontal strip $0 \leqq y \leqq \pi$ is mapped onto the upper half of the w-plane. The boundary $y = 0$ is mapped onto the positive half of the u-axis, and the line

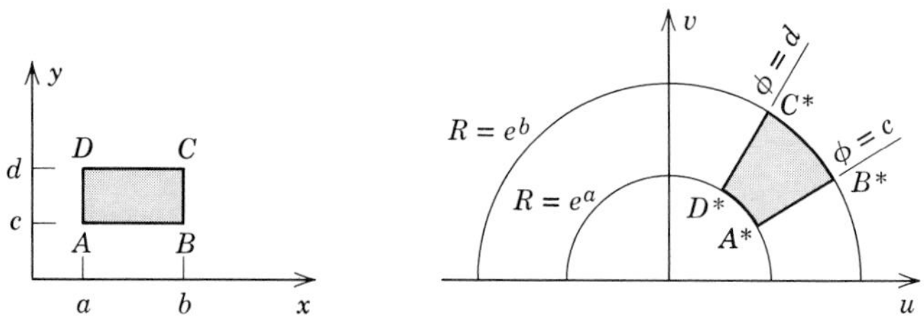

Fig. 287. Mapping by $w = e^z$

$y = \pi$ onto the negative half of the u-axis, as follows from (2). The segment from 0 to πi is mapped onto the semicircle $|w| = 1$, $v \geqq 0$. The left half ($x \leqq 0$) of our strip is mapped onto the region $|w| \leqq 1$, $v \geqq 0$, and the right half ($x \geqq 0$) of the strip is mapped onto the exterior of that semicircle $|w| = 1$ in the upper half of the w-plane (Fig. 288).

Since the **natural logarithm** $w = u + iv = \ln z$ is the inverse relation of the exponential function, the properties of the corresponding conformal mapping can be easily obtained from those of the exponential function by interchanging the roles of the z and the w planes in the preceding considerations. The principal value $w = \operatorname{Ln} z$ thus maps the z-plane (cut along the negative real axis) onto the horizontal strip $-\pi < v \leqq \pi$ of the w-plane. Further details of the mapping will be discussed in Example 2 of the next section.

The **sine function** (Sec. 12.8)

$$w = u + iv = \sin z = \sin x \cosh y + i \cos x \sinh y \tag{3}$$

where

$$u = \sin x \cosh y, \qquad v = \cos x \sinh y, \tag{4}$$

is periodic. Hence, the mapping (4) is certainly not one-to-one if we consider it in the full xy-plane. We restrict z to the infinite strip S defined by $-\pi/2 \leqq x \leqq \pi/2$. Since $f'(z) = \cos z$ is zero at $z = \pm\pi/2$, the mapping is not conformal at these two points. From (4) we see that the boundary of S is mapped into the u-axis. The segment $-\pi/2 \leqq x \leqq \pi/2$ of the x-axis maps onto the segment $-1 \leqq u \leqq 1$ of the u-axis, the line $x = -\pi/2$ maps onto $u \leqq -1$, $v = 0$, and the line $x = \pi/2$ maps onto $u \geqq 1$, $v = 0$. The line segment $y = c > 0$, $-\pi/2 \leqq x \leqq \pi/2$ maps onto the portion of the ellipse

$$u = \cosh c \sin x, \qquad v = \sinh c \cos x$$

or

$$\frac{u^2}{\cosh^2 c} + \frac{v^2}{\sinh^2 c} = 1 \tag{5}$$

in the upper half of the w-plane. The line segment $y = -c$, $-\pi/2 \leqq x \leqq \pi/2$ ($c > 0$) maps onto the lower half of the ellipse (5). The foci of the ellipse are at $w = \pm 1$, and we see that they are independent of c. Consequently, if we let c vary, we obtain a family of confocal ellipses. The rectangular region defined by $-\pi/2 < x < \pi/2$, $-c < y < c$ is thus mapped onto the interior of the

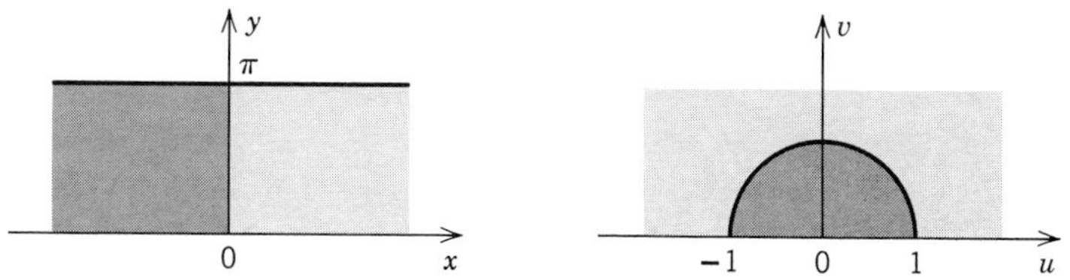

Fig. 288. Mapping by $w = e^z$

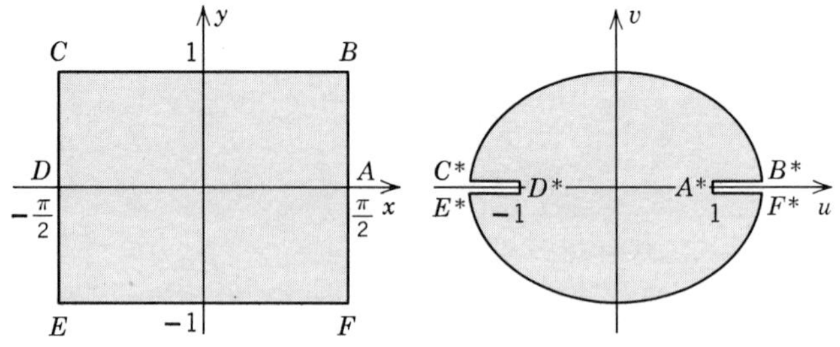

Fig. 289. Mapping by $w = \sin z$

ellipse (5); but note that the image of the boundary consists of the ellipse and the two segments of the x-axis, as shown in Fig. 289 (where $c = 1$). The image points of points on the vertical parts of the boundary coincide in pairs. In particular, $B^* = F^*$ and $C^* = E^*$.

The rectangle $-\pi < x < \pi$, $c < y < d$ $(c > 0)$ maps onto an elliptic ring cut along the negative v-axis (Fig. 290). The straight lines $x = const$ $(-\pi/2 < x < \pi/2)$ map onto confocal hyperbolas which intersect those ellipses at right angles, and the y-axis maps onto the v-axis.

The **cosine function**

$$w = \cos z = \sin\left(z + \frac{\pi}{2}\right) \tag{6}$$

defines the same mapping as $\sin z$, preceded by a translation to the right through $\pi/2$ units.

The **hyperbolic function**

$$w = \sinh z = -i \sin (iz) \tag{7}$$

defines a transformation which is a rotation $t = iz$ followed by the mapping $p = \sin t$ and another rotation $w = -ip$.

Similarly, the transformation

$$w = \cosh z = \cos (iz) \tag{8}$$

is a rotation $t = iz$ followed by the mapping $w = \cos t$.

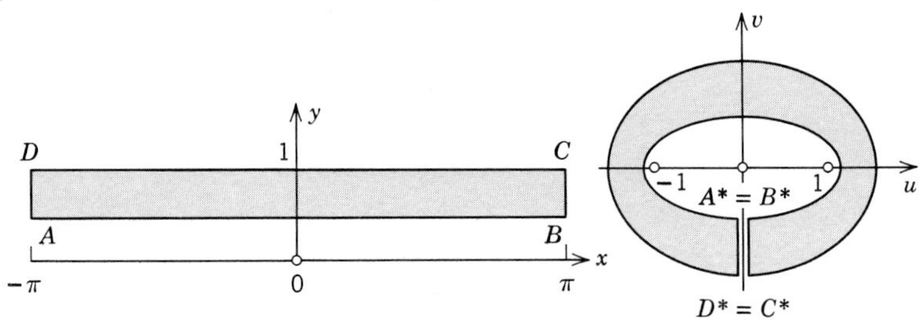

Fig. 290. Mapping by $w = \sin z$

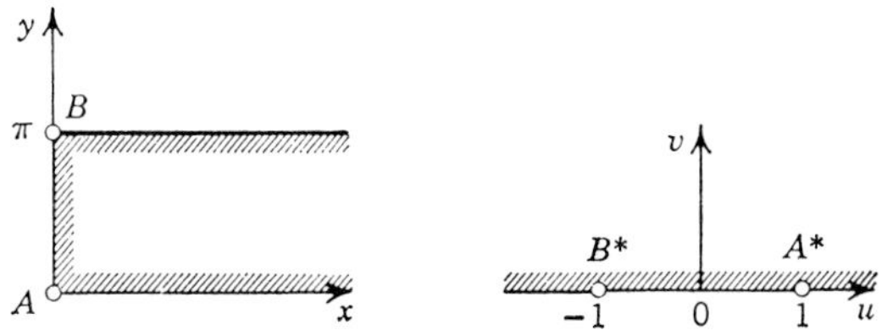

Fig. 291. Mapping in Example 1

Example 1. Mapping of a semi-infinite strip onto a half-plane

Find the image of the semi-infinite strip $x \geqq 0, 0 \leqq y \leqq \pi$ (Fig. 291) under the mapping (8). We set $w = u + iv$. Since $\cosh 0 = 1$, the point $z = 0$ is mapped onto $w = 1$. For real $z = x \geqq 0$, $\cosh z$ is real and increases monotone from 1 as x increases. Hence the positive x-axis is mapped onto the portion $u \geqq 1$ of the u-axis. For pure imaginary $z = iy$ we have $\cosh iy = \cos y$. It follows that the left boundary of the strip is mapped onto the segment $1 \geqq u \geqq -1$ of the u-axis, the point $z = \pi i$ corresponding to

$$w = \cosh i\pi = \cos \pi = -1.$$

On the upper boundary of the strip, $y = \pi$, and since $\sin \pi = 0$, $\cos \pi = -1$, it follows that this part of the boundary is mapped onto the portion $u \leqq -1$ of the u-axis. Hence the boundary of the strip is mapped onto the u-axis. It is not difficult to see that the interior of the strip is mapped onto the upper half of the w-plane, and the mapping is one-to-one.

Example 2. A boundary value problem

Find the temperature $T(x, y)$ in the strip considered in Example 1, if the temperature on the boundary is

$T = T_0$ on the segment from 0 to πi,

$T = 0$ on the upper and lower boundaries.

Since T does not depend on time ("steady-state temperature distribution"), the heat equation reduces to Laplace's equation (cf. Sec. 11.11)

$$\nabla^2 T = \frac{\partial^2 T}{\partial x^2} + \frac{\partial^2 T}{\partial y^2} = 0,$$

and we have to find a solution of this equation satisfying those boundary conditions.

To solve this problem we map the strip by means of (8) onto the upper half of the w-plane. Since the segment $0 \leqq y \leqq \pi$ of the y-axis is mapped onto the segment $-1 \leqq u \leqq 1$ of the u-axis, the boundary conditions in the w-plane are (Fig. 292):

$T = T_0$ on the segment from -1 to 1,

$T = 0$ on the other parts of the u-axis.

The real and imaginary parts of analytic functions are solutions of Laplace's equation, and we have to find such a solution $T(u, v)$ in the upper half of the w-plane which satisfies those boundary conditions. For this purpose we consider the functions

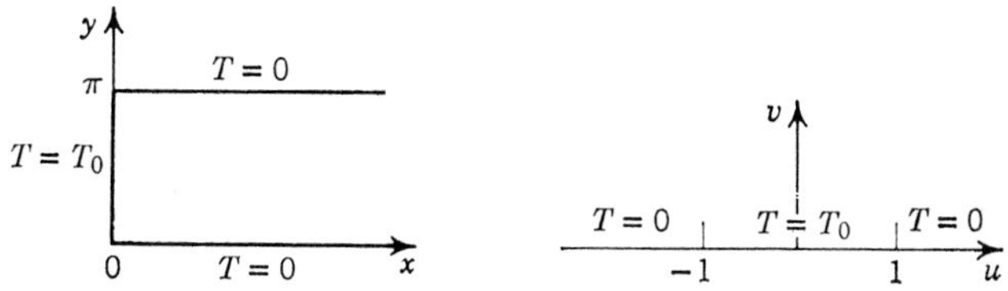

Fig. 292. Boundary conditions in Example 2

(9*)
$$\mathrm{Ln}\,(w+1) = \ln|w+1| + i\phi_1, \qquad \phi_1 = \arg(w+1) = \arctan\frac{v}{u+1},$$
$$\mathrm{Ln}\,(w-1) = \ln|w-1| + i\phi_2, \qquad \phi_2 = \arg(w-1) = \arctan\frac{v}{u-1}.$$

Since $\phi_1(u, v)$ and $\phi_2(u, v)$ are harmonic functions, $\phi_2 - \phi_1$ is harmonic. If $w = u$ is real and smaller than -1, then $\phi_2 - \phi_1 = \pi - \pi = 0$; for real $w = u$ in the interval $-1 < u < 1$ we have $\phi_2 - \phi_1 = \pi - 0 = \pi$, and for real $w = u > 1$ we have $\phi_2 - \phi_1 = 0 - 0 = 0$. Hence the function

$$T(u, v) = \frac{T_0}{\pi}(\phi_2 - \phi_1) \tag{9}$$

is harmonic in the half-plane $v > 0$ and satisfies those boundary conditions in the w-plane. Since $\tan\phi_1 = v/(u+1)$ and $\tan\phi_2 = v/(u-1)$, it follows that

$$\tan(\phi_2 - \phi_1) = \frac{\tan\phi_2 - \tan\phi_1}{1 + \tan\phi_1 \tan\phi_2} = \frac{2v}{u^2 + v^2 - 1},$$

and (9) takes the form

$$T(u, v) = \frac{T_0}{\pi}\arctan\frac{2v}{u^2 + v^2 - 1}. \tag{10}$$

The function $w = \cosh z$ maps the strip under consideration onto the half-plane $v \geqq 0$, and we have

$$w = u + iv = \cosh(x + iy) = \cosh x \cos y + i \sinh x \sin y.$$

Separating the real and imaginary parts on both sides, we may write

$$u = \cosh x \cos y, \qquad v = \sinh x \sin y.$$

From these expressions for u and v it follows that in (10),

$$u^2 + v^2 - 1 = \cosh^2 x \cos^2 y + \sinh^2 x \sin^2 y - 1 = \sinh^2 x - \sin^2 y.$$

By inserting this and the expression for v into (10) and denoting $T(u(x, y), v(x, y))$ by $T^*(x, y)$ we have

$$T^*(x, y) = \frac{T_0}{\pi}\arctan\frac{2\sinh x \sin y}{\sinh^2 x - \sin^2 y}.$$

Noting that the numerator and the denominator are the imaginary part and the real part of the function $(\sinh x + i \sin y)^2$, we may write

$$T^*(x, y) = \frac{T_0}{\pi}\arg[(\sinh x + i\sin y)^2] = \frac{2T_0}{\pi}\arg(\sinh x + i \sin y).$$

Hence the solution of our problem is

$$T^*(x, y) = \frac{2T_0}{\pi}\arctan\frac{\sin y}{\sinh x}. \tag{11}$$

This function is harmonic in the interior of our strip (cf. Theorem 2 in Sec. 13.2) and it satisfies the boundary conditions. Indeed, $T^* = 0$ when $y = 0$ or $y = \pi$, and $T^* = T_0$ when $x = 0$. The isotherms (= curves of constant temperature) are the curves

$$\frac{\sin y}{\sinh x} = const.$$ ∎

In Example 2 we transformed the real potential $T(u, v)$ into the real potential $T^*(x, y)$ by using the function $w = u(x, y) + iv(x, y)$ which maps the half-plane onto the given region. In many cases problems of this type become simpler by working with a **complex potential,** that is, by taking a complex analytic function $F(w)$ such that the real potential $T(u, v)$ is the real or imaginary part of $F(w)$, and transforming F instead of T. It is clear that such a complex potential F may be readily obtained from T by determining a conjugate harmonic function of T (cf. at the end of Sec. 12.5). Let us illustrate this "method of complex potentials" in the case of our previous example. A detailed discussion of complex potentials will be presented later (in Chap. 18).

Example 3. Complex potential

From (9*) we see that the real potential

$$T(u, v) = \frac{T_0}{\pi}(\phi_2 - \phi_1)$$

in Example 2 [cf. (9)] is the imaginary part of the complex potential

$$F(w) = \frac{T_0}{\pi}[\operatorname{Ln}(w-1) - \operatorname{Ln}(w+1)] = \frac{T_0}{\pi}\operatorname{Ln}\frac{w-1}{w+1}. \tag{12}$$

The mapping function in Example 2 is

$$w = \cosh z = \tfrac{1}{2}(e^z + e^{-z}).$$

From this we see that in (12),

$$\frac{w-1}{w+1} = \frac{\cosh z - 1}{\cosh z + 1} = \frac{e^z + e^{-z} - 2}{e^z + e^{-z} + 2} = \frac{(e^{z/2} - e^{-z/2})^2}{(e^{z/2} + e^{-z/2})^2} = \tanh^2 \frac{z}{2}.$$

By inserting this into (12) and denoting $F(w(z))$ by $F^*(z)$ and $\tanh(z/2)$ by H, we have

$$F^*(z) = \frac{T_0}{\pi}\operatorname{Ln}\tanh^2\frac{z}{2} = \frac{2T_0}{\pi}\operatorname{Ln}\tanh\frac{z}{2} = \frac{2T_0}{\pi}\operatorname{Ln} H. \tag{13}$$

Because of (2) in Sec. 12.9 this can be written

$$F^*(z) = \frac{2T_0}{\pi}\operatorname{Ln} H = \frac{2T_0}{\pi}\left(\ln|H| + i \arctan\frac{\operatorname{Im} H}{\operatorname{Re} H}\right). \tag{14}$$

$F^*(z)$ is the complex potential in the strip in Example 2, and its imaginary part is the solution of our problem. By using the definition of H in terms of exponential functions we find that

$$H = \frac{\sinh x + i \sin y}{\cosh x + \cos y}. \tag{15}$$

From this and (14) we have

$$T^*(x, y) = \operatorname{Im} F^*(z) = \frac{2T_0}{\pi}\arctan\frac{\sin y}{\sinh x},$$

in agreement with our result in Example 2.

Furthermore, from (15) it follows that

$$|H|^2 = \frac{\sinh^2 x + \sin^2 y}{(\cosh x + \cos y)^2},$$

and from (14) we see that the real part of $F^*(z)$ becomes

$$S^*(x, y) = \operatorname{Re} F^*(z) = \frac{T_0}{\pi}\ln\frac{\sinh^2 x + \sin^2 y}{(\cosh x + \cos y)^2}.$$

The curves $S^* = const$ intersect the isotherms $T^* = const$ at right angles and, therefore, are the curves along which the heat flows. It is a great advantage that the *complex* potential yields *both* families of curves.

Problems for Sec. 13.5

Find and graph the images of the following regions under the mapping $w = e^z$.

1. $-1 < x < 1, \quad -\pi/2 < y < \pi/2$

2. $0 < x < 2, \quad 0 < y < \pi$

3. $0 < x < 1, \quad 0 < y < 1$

4. $-3 < x < -2, \quad 0 < y < \pi/4$

5. $-2 \leqq x \leqq 2, \quad -\pi \leqq y \leqq -\pi/2$

6. $-1 \leqq x \leqq 3, \quad -\pi \leqq y \leqq \pi$

7. Find an analytic function which maps the region bounded by the positive x and y axes and the hyperbola $xy = \pi/2$ in the first quadrant onto the upper half-plane. *Hint.* First map that region onto a horizontal strip.

Find and graph the images of the following regions under the mapping $w = \sin z$.

8. $0 < x < \pi/2, \quad 0 < y < 2$

9. $-\pi/2 < x < \pi/2, \quad 1 < y < 2$

10. $-\pi/2 < x < \pi/2, \quad 0 < y < 1$

11. $0 < x < 2\pi, \quad 1 < y < 2$

12. Find and plot the images of the lines $x = 0, \pm\pi/6, \pm\pi/3, \pm\pi/2$ under the mapping $w = \sin z$.

13. Describe the transformation $w = \cosh z$ in terms of the transformation $w = \sin z$ and rotations and translations.

14. Show that $w = \text{Ln}\,\dfrac{z-1}{z+1}$ maps the upper half plane onto the horizontal strip $0 \leqq \text{Im}\, w \leqq \pi$ as shown in the figure.

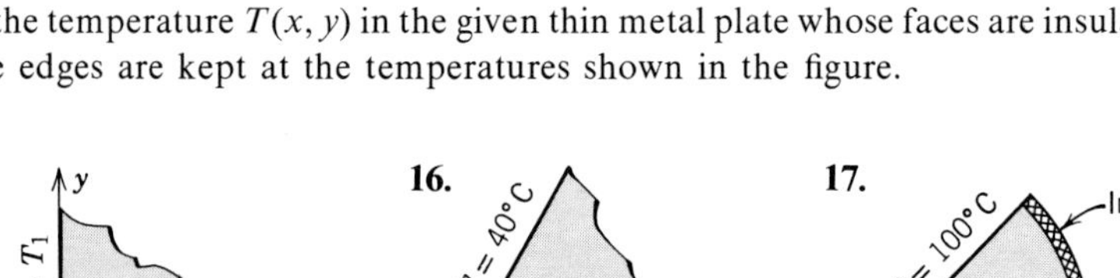

Problem 14

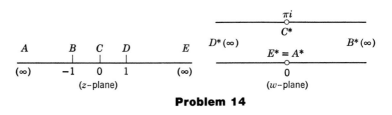

Find the temperature $T(x, y)$ in the given thin metal plate whose faces are insulated and whose edges are kept at the temperatures shown in the figure.

15.

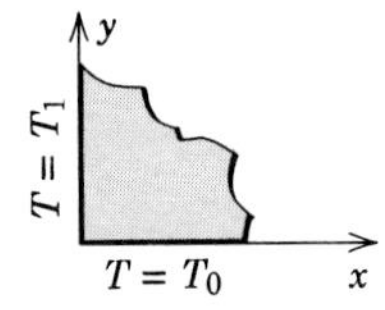

16.

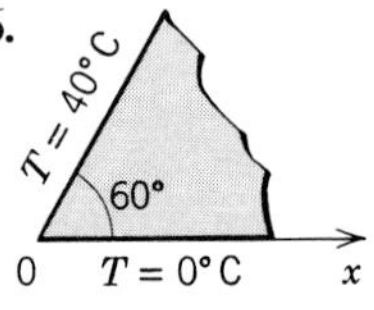

17.

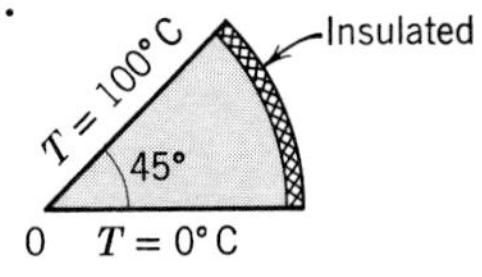

18.

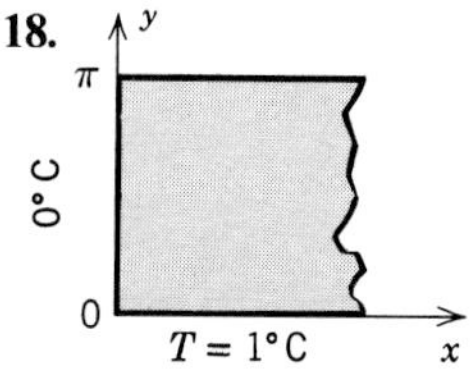

19.

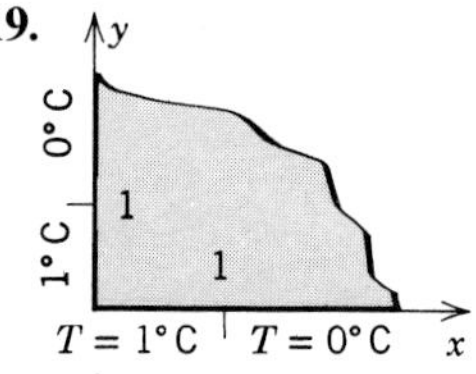

20.

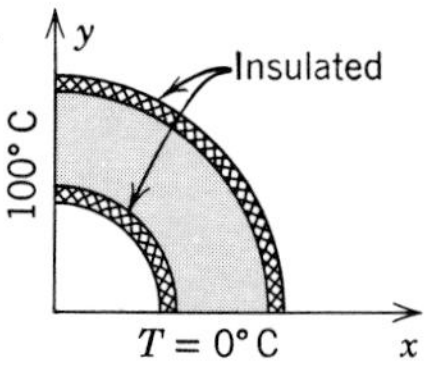

13.6 Riemann Surfaces

We consider the mapping defined by

$$w = u + iv = z^2 \tag{1}$$

(cf. Sec. 13.1). This mapping is conformal, except at $z = 0$ where $w' = 2z$ is zero. The angles at $z = 0$ are doubled under the mapping. The right half of the z-plane (including the positive y-axis) maps onto the full w-plane cut along the negative half of the u-axis; the mapping is one-to-one. Similarly, the left half of the z-plane (including the negative y-axis) is mapped onto the cut w-plane in a one-to-one manner.

Obviously, the mapping of the full z-plane is not one-to-one, because every point $w \neq 0$ corresponds to precisely two points z. In fact, if z_1 is one of these points, then the other is $-z_1$. For example, $z = i$ and $z = -i$ have the same image, namely, $w = -1$, etc. Hence, the w-plane is "covered twice" by the image of the z-plane. We say that the full z-plane is mapped onto the *doubly covered* w-plane. We can still give our imagination the necessary support as follows.

We imagine one of the two previously obtained copies of the cut w-plane to be placed upon the other so that the upper sheet is the image of the right half of the z-plane, and the lower sheet is the image of the left half of the z-plane; we denote these half-planes by R and L, respectively. When we pass from R to L, the corresponding image point should pass from the upper to the lower sheet. For this reason we join the two sheets crosswise along the cut, that is, along the negative real axis. (This construction can be carried out only in imagination, since the penetration of the two sheets of a material model can only be imperfectly realized.) The two origins are fastened together. The configuration thus obtained is called a **Riemann surface.** On it every point $w \neq 0$ appears twice, at superposed positions, and the origin appears precisely once. The function $w = z^2$ now maps the full z-plane onto this Riemann surface in a one-to-one manner, and the mapping is conformal, except for the "winding point" or **branch point** at $w = 0$ (Fig. 293). Such a branch point connecting two sheets is said to be of the *first order.* (More generally, a branch point connecting n sheets is said to be of **order** $n - 1$.)

We now consider the double-valued relation

$$w = \sqrt{z}. \tag{2}$$

To each $z \neq 0$ there correspond two values w, one of which is the principal

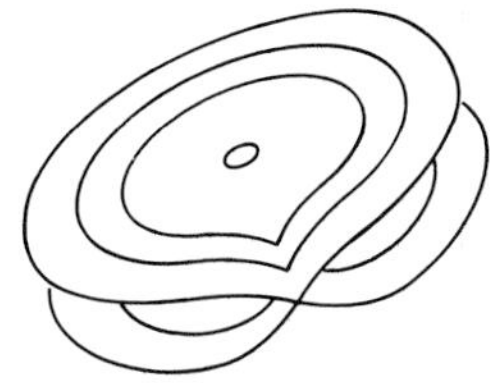

Fig. 293. Example of a Riemann surface

value. If we replace the z-plane by the two-sheeted Riemann surface just considered, then each complex number $z \neq 0$ is represented by two points of the surface at superposed positions. We let one of these points correspond to the principal value—for example, the point in the upper sheet—and the other to the other value. Then (2) becomes single-valued, that is, (2) is a function of the points of the Riemann surface, and to any continuous motion of z on the surface there corresponds a continuous motion of the corresponding point in the w-plane. The function maps the sheet corresponding to the principal value onto the right half of the w-plane and the other sheet onto the left half of the w-plane.

Let us consider some further important examples.

Example 1. Riemann surface of $\sqrt[n]{z}$

In the case of the relation

$$w = \sqrt[n]{z} \qquad n = 3, 4, \cdots \tag{3}$$

we need a Riemann surface consisting of n sheets and having a branch point of order $n - 1$ at $z = 0$. One of the sheets corresponds to the principal value and the other $n - 1$ sheets to the other $n - 1$ values.

Example 2. Riemann surface of the natural logarithm

For every $z \neq 0$ the relation

$$w = \ln z = \operatorname{Ln} z + 2n\pi i \qquad (n = 0, \pm 1, \pm 2, \cdots, z \neq 0) \tag{4}$$

is infinitely many-valued. Hence (4) defines a function on a Riemann surface consisting of infinitely many sheets. The function $w = \operatorname{Ln} z$ corresponds to one of these sheets. On this sheet the argument θ of z ranges in the interval $-\pi < \theta \leqq \pi$ (cf. Sec. 12.9). The sheet is cut along the negative ray of the real axis, and the upper edge of the slit is joined to the lower edge of the next sheet, which corresponds to the interval $\pi < \theta \leqq 3\pi$, that is, to the function $w = \operatorname{Ln} z + 2\pi i$. In this way each value of n in (4) corresponds to precisely one of these infinitely many sheets. The function $w = \operatorname{Ln} z$ maps the corresponding sheet onto the horizontal strip $-\pi < v \leqq \pi$ in the w-plane. The next sheet is mapped onto the neighboring strip $\pi < v \leqq 3\pi$, etc. The function $w = \ln z$ thus maps all the sheets of the corresponding Riemann surface onto the entire w-plane, the correspondence between the points $z \neq 0$ of the Riemann surface and those of the w-plane being one-to-one.

Example 3. Mapping $w = z + z^{-1}$. Airfoils

Let us consider the mapping defined by

$$w = z + \frac{1}{z} \qquad (z \neq 0) \tag{5}$$

which is important in aerodynamics (see below). Since

$$w' = 1 - \frac{1}{z^2} = \frac{(z + 1)(z - 1)}{z^2}$$

the mapping is conformal except at the points $z = 1$ and $z = -1$; these points correspond to $w = 2$ and $w = -2$, respectively. From (5) we find

$$z = \frac{w}{2} \pm \sqrt{\frac{w^2}{4} - 1} = \frac{w}{2} \pm \frac{1}{2}\sqrt{(w + 2)(w - 2)}. \tag{6}$$

Hence, the points $w = 2$ and $w = -2$ are branch points of the first order of $z = z(w)$. To any value w ($\neq 2$, $\neq -2$) there correspond two values of z. Consequently, (5) maps the z-plane onto a two-sheeted Riemann surface, the two sheets being connected crosswise from $w = -2$ to $w = 2$ (Fig. 294), and this mapping is one-to-one. We set $z = re^{i\theta}$ and determine the images of the curves $r = const$ and $\theta = const$. From (5) we obtain

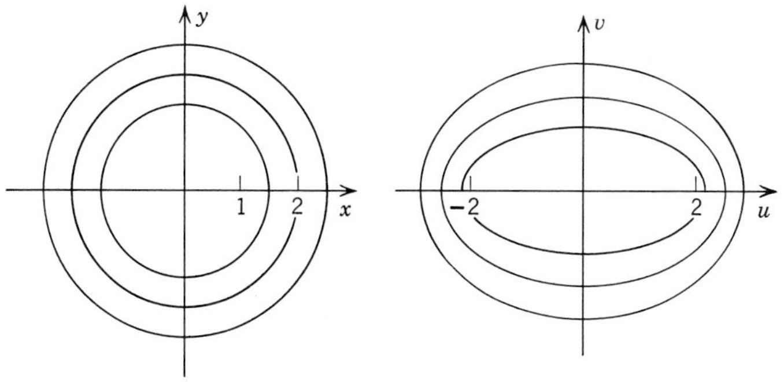

Fig. 294. Example 3

$$w = u + iv = re^{i\theta} + \frac{1}{r}e^{-i\theta} = \left(r + \frac{1}{r}\right)\cos\theta + i\left(r - \frac{1}{r}\right)\sin\theta.$$

By equating the real and imaginary parts on both sides we have

$$u = \left(r + \frac{1}{r}\right)\cos\theta, \qquad v = \left(r - \frac{1}{r}\right)\sin\theta. \tag{7}$$

From this we find

$$\frac{u^2}{a^2} + \frac{v^2}{b^2} = 1, \qquad \text{where} \qquad a = r + \frac{1}{r}, \qquad b = \left|r - \frac{1}{r}\right|.$$

The circles $r = const$ are thus mapped onto ellipses whose principal axes lie in the u and v axes and have the lengths $2a$ and $2b$, respectively. Since $a^2 - b^2 = 4$, independent of r, these ellipses are confocal, with foci at $w = -2$ and $w = 2$. The unit circle $r = 1$ maps onto the line segment from $w = -2$ to $w = 2$. For every $r \neq 1$ the two circles with radii r and $1/r$ map onto the same ellipse in the w-plane, corresponding to the two sheets of the Riemann surface. Hence, the interior of the unit circle $|z| = 1$ corresponds to one sheet, and the exterior to the other.

Furthermore, from (7) we obtain

$$\frac{u^2}{\cos^2\theta} - \frac{v^2}{\sin^2\theta} = -4. \tag{8}$$

The lines $\theta = const$ are thus mapped onto the hyperbolas which are the orthogonal trajectories of those ellipses. The real axis, that is, the rays $\theta = 0$ and $\theta = \pi$, are mapped onto the part of the real axis from $w = 2$ via ∞ to $w = -2$. The y-axis is mapped onto the v-axis. Any other pair of rays $\theta = \theta_0$ and $\theta = \theta_0 + \pi$ is mapped onto the two branches of the same hyperbola.

An exterior region of one of the above ellipses is free of branch points and corresponds to either the interior or the exterior of the corresponding circle in the z-plane, depending on the sheet of the Riemann surface to which the region belongs. In particular the full w-plane corresponds to the interior or the exterior of the unit circle $|z| = 1$, as was mentioned before.

The mapping (5) transforms suitable circles into airfoils with a sharp trailing edge whose interior angle is zero; these airfoils are known as **Joukowski[3] airfoils.** Since the airfoil to be obtained has a sharp edge, it is clear that the circle to be mapped must pass through one of the points $z = \pm 1$ where the mapping is not conformal. Let us choose a circle C through $z = -1$ and such that $z = 1$ lies inside C. The simplest way of determining the image is the graphical vector addition of the vectors corresponding to z and $1/z$ (where the latter can be obtained from the former as shown in Sec. 13.1). This leads to the result shown in Fig. 295. Further details are included in Ref. [F9] in Appendix 1. ∎

A more elaborate treatment of Riemann surfaces can be found in Ref. [F10].

[3] NIKOLAI JEGOROVICH JOUKOWSKI (1847–1921), Russian mathematician.

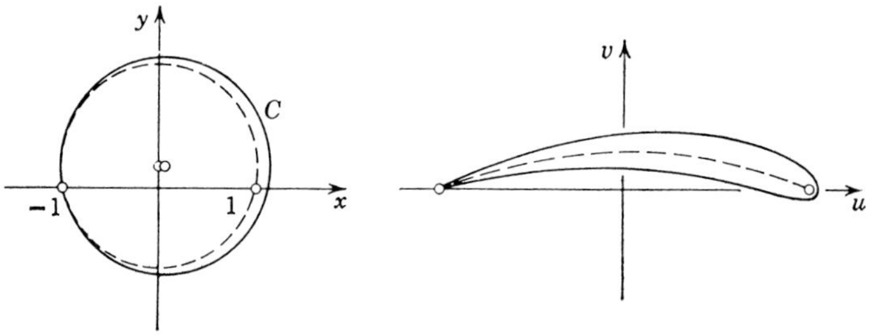

Fig. 295. Joukowski airfoil

Problems for Sec. 13.6

1. Consider $w = \sqrt{z}$. Find the path of the image point w of a point z which moves twice around the unit circle, starting from the initial position $z = 1$.
2. Show that the Riemann surface of $w = \sqrt[3]{z}$ consists of three sheets and has a branch point of second order at $z = 0$. Find the path of the image point w of a point z which moves three times around the unit circle starting from the initial position $z = 1$.
3. Consider the Riemann surfaces of $w = \sqrt[4]{z}$ and $w = \sqrt[5]{z}$ in a fashion similar to that in Prob. 2.
4. Make a sketch, similar to Fig. 293, of the Riemann surfaces of $\sqrt[3]{z}$ and $\sqrt[4]{z}$.
5. Find the images of the annuli $\frac{1}{2} < |z| < 1$, $1 < |z| < 2$, and $2 < |z| < 3$ under the mapping $w = z + 1/z$.
6. Determine the path of the image of a point z under the mapping $w = \ln z$ as z moves several times around the unit circle.
7. Show that the Riemann surface of $w = \sqrt{(z-1)(z-4)}$ has branch points at $z = 1$ and $z = 4$ and consists of two sheets which may be cut along the line segment from 1 to 4 and joined crosswise. *Hint.* Introduce polar coordinates $z - 1 = r_1 e^{i\theta_1}$, $z - 4 = r_2 e^{i\theta_2}$.
8. Show that the Riemann surface of $w = \sqrt{(1 - z^2)(4 - z^2)}$ has four branch points and two sheets which may be joined crosswise along the segments $-2 \leqq x \leqq -1$ and $1 \leqq x \leqq 2$ of the x-axis.

Determine the location of the branch points and the number of sheets of the Riemann surfaces of the following functions.

9. $w = i\sqrt{z}$
10. $w = \sqrt{z - i}$
11. $w = \sqrt[3]{z - i}$
12. $\sqrt[3]{2z + 3i}$
13. $\sqrt{z^2 + 1}$
14. $\sqrt{z(z-1)(z+1)}$
15. $\sqrt{(z-a)(z-b)}$, $a \neq b$
16. $1 + z + \sqrt{z}$
17. $\ln(z - a)$
18. $e^{\sqrt{z}}$
19. $\sqrt{e^z}$
20. $\sqrt{\sqrt[3]{z} - 1}$

CHAPTER

14 COMPLEX INTEGRALS

Integrals in the complex plane are important for two reasons. The practical reason is that in applications there occur real integrals which can be evaluated by complex integration, while the usual methods of real integral calculus are not successful. The other reason is of a theoretical nature and results from the fact that the method of complex integration yields proofs of some basic properties of analytic functions (in particular the existence of higher derivatives), which would be very difficult to prove without using integrals. This situation indicates a basic difference between real and complex calculus.

In this chapter we shall first define complex integrals. The most fundamental result will be Cauchy's integral theorem (Sec. 14.3) from which the important Cauchy integral formula (Sec. 14.5) will follow. In Sec. 14.6 we shall prove that if a function is analytic, it has derivatives of any order. This means that in this respect complex analytic functions behave much more simply than real functions of a real variable.

(Integration by means of residues and applications to real integrals will be considered in Chap. 17.)

Prerequisite for this chapter: Chap. 12.
References: Appendix 1, Part F.
Answers to problems: Appendix 2.

14.1 Line Integral in the Complex Plane

As in real calculus, we distinguish between definite integrals and indefinite integrals or antiderivatives. An **indefinite integral** is a function whose derivative equals a given analytic function in a region. By inverting known differentiation formulas we may find many types of indefinite integrals.

Let us now define *definite integrals,* or *line integrals,* of a complex function $f(z)$ where $z = x + iy$. We shall see that this definition will be a natural generalization of the familiar definition of a real definite integral, and the consideration will be similar to that in Sec. 9.1. In the case of a definite integral the path of integration is an interval on the real axis. In the case of a complex definite integral we shall integrate along a curve[1] in the complex plane.

Let C be a smooth curve (cf. Sec. 13.2) in the complex z-plane. Then we may represent C in the form

$$z(t) = x(t) + iy(t) \qquad (a \leqq t \leqq b) \tag{1}$$

[1] Actually along a portion, or arc, of a curve. For the sake of simplicity we shall use the single term "curve" to denote an entire curve as well as a portion of it.

where $z(t)$ has a continuous derivative $\dot{z}(t) \neq 0$ for all t, so that C is rectifiable (cf. Sec. 8.4) and has a unique tangent at each point. We remember that the positive direction along C corresponds to the sense of increasing values of the parameter t.

Let $f(z)$ be a continuous function which is defined (at least) at each point of C. We subdivide ("partition") the interval $a \leqq t \leqq b$ in (1) by points

$$t_0(=a),\ t_1,\ \cdots,\ t_{n-1},\ t_n\ (=b)$$

where $t_0 < t_1 < \cdots < t_n$. To this there corresponds a subdivision of C by points (cf. Fig. 296)

$$z_0,\ z_1,\ \cdots,\ z_{n-1},\ z_n\ (=Z),$$

where $z_j = z(t_j)$. On each portion of subdivision of C we choose an arbitrary point, say, a point ζ_1 between z_0 and z_1 (that is, $\zeta_1 = z(t)$ where $t_0 \leqq t \leqq t_1$), a point ζ_2 between z_1 and z_2, etc. Then we form the sum

$$S_n = \sum_{m=1}^{n} f(\zeta_m)\,\Delta z_m \tag{2}$$

where

$$\Delta z_m = z_m - z_{m-1}.$$

This we do for each $n = 2, 3, \cdots$ in a completely independent manner, but such that the greatest $|\Delta z_m|$ approaches zero as n approaches infinity. In this way we obtain a sequence of complex numbers $S_2, S_3, \cdots$. The limit of this sequence is called the **line integral** (or simply the *integral*) of $f(z)$ along the oriented curve C and is denoted by

$$\int_C f(z)\,dz. \tag{3}$$

The curve C is called the **path of integration.**

Throughout the following considerations we shall assume that all paths of integration for complex line integrals are ***piecewise smooth****, that is, consist of finitely many smooth curves.*

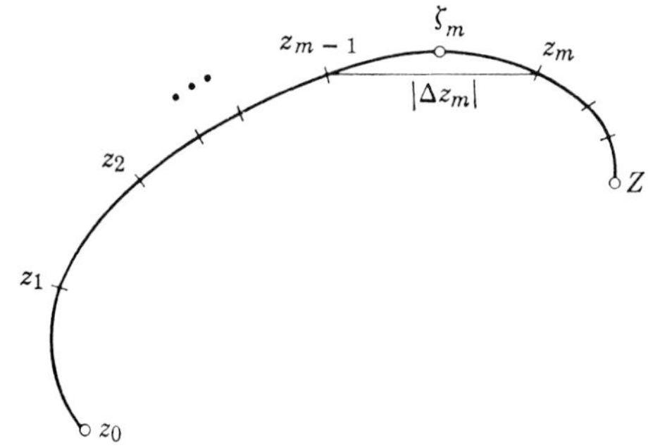

Fig. 296. Complex line integral

From our assumptions the existence of the line integral (3) follows. In fact, let $f(z) = u(x, y) + iv(x, y)$ and set

$$\zeta_m = \xi_m + i\eta_m \qquad \text{and} \qquad \Delta z_m = \Delta x_m + i\,\Delta y_m.$$

Then (2) may be written

$$S_n = \sum (u + iv)(\Delta x_m + i\,\Delta y_m) \tag{4}$$

where $u = u(\xi_m, \eta_m)$, $v = v(\xi_m, \eta_m)$ and we sum over m from 1 to n. We may now split up S_n into four sums:

$$S_n = \sum u\,\Delta x_m - \sum v\,\Delta y_m + i[\sum u\,\Delta y_m + \sum v\,\Delta x_m].$$

These sums are real. Since f is continuous, u and v are continuous. Hence, if we let n approach infinity in the aforementioned way, then the greatest Δx_m and Δy_m will approach zero and each sum on the right becomes a real line integral:

$$\lim_{n\to\infty} S_n = \int_C f(z)\,dz = \int_C u\,dx - \int_C v\,dy + i\left[\int_C u\,dy + \int_C v\,dx\right]. \tag{5}$$

This shows that the line integral (3) exists and its value is independent of the choice of subdivisions and intermediate points ζ_m.

Furthermore, as in Sec. 9.2 we may convert each of those real line integrals to a definite integral by the use of the representation (1) of the curve C:

$$\int_C f(z)\,dz = \int_a^b u\dot{x}\,dt - \int_a^b v\dot{y}\,dt + i\left[\int_a^b u\dot{y}\,dt + \int_a^b v\dot{x}\,dt\right] \tag{6}$$

where $u = u[x(t), y(t)]$, $v = v[x(t), y(t)]$, and dots denote derivatives with respect to t.

Without causing misinterpretation, we may also write

$$\int_C f(z)\,dz = \int_a^b (u + iv)(\dot{x} + i\dot{y})\,dt$$

or more briefly

$$\int_C f(z)\,dz = \int_a^b f[z(t)]\,\dot{z}(t)\,dt. \tag{6*}$$

Let us consider some basic examples.

Example 1. Integral of 1/z around the unit circle

Integrate $f(z) = 1/z$ once around the unit circle C in the counterclockwise sense, starting from $z = 1$. We may represent C in the form

$$z(t) = \cos t + i \sin t \qquad (0 \leqq t \leqq 2\pi). \tag{7}$$

Then we have

$$\dot{z}(t) = -\sin t + i \cos t,$$

and by (6*) the integral under consideration becomes

$$\int_C \frac{dz}{z} = \int_0^{2\pi} \frac{1}{\cos t + i \sin t}(-\sin t + i \cos t)\, dt = i \int_0^{2\pi} dt = 2\pi i,$$

a fundamental result which we shall use in our further consideration.

Clearly, instead of (7) we may write more simply

$$(7') \qquad z(t) = e^{it} \qquad (0 \leqq t \leqq 2\pi).$$

Then we obtain by differentiation

$$\dot{z}(t) = ie^{it}, \qquad dz = ie^{it}\, dt,$$

and from this the same result as before:

$$(8) \qquad \int_C \frac{dz}{z} = \int_0^{2\pi} \frac{1}{e^{it}} ie^{it}\, dt = i \int_0^{2\pi} dt = 2\pi i.$$

Example 2. Integral of a nonanalytic function

Integrate $f(z) = \text{Re}\, z = x$ along the segment from $z_0 = 0$ to $Z = 1 + i$ (path C_1 in Fig. 297).

The segment may be represented in the form

$$z(t) = x(t) + iy(t) = (1 + i)t \qquad (0 \leqq t \leqq 1).$$

Then

$$f[z(t)] = \text{Re}\, z(t) = x(t) = t, \qquad dz = (1 + i)\, dt.$$

Therefore, we obtain the result

$$\int_{C_1} \text{Re}\, z\, dz = \int_0^1 t(1 + i)\, dt = (1 + i) \int_0^1 t\, dt = \tfrac{1}{2}(1 + i).$$

Let us now integrate $f(z) = \text{Re}\, z = x$ along the real axis from 0 to 1, and then vertically to $1 + i$ (path C_2 in Fig. 297). We may represent the first part of C_2 by

$$z = z(t) = t \qquad (0 \leqq t \leqq 1)$$

and the last part in the form

$$z(t) = 1 + i(t - 1) \qquad (1 \leqq t \leqq 2).$$

Then the entire path corresponds to the interval $0 \leqq t \leqq 2$. On the first part, $\text{Re}\, z = t$, $dz = dt$, and on the last part, $\text{Re}\, z = 1$, $dz = i\, dt$. Therefore,

$$\int_{C_2} \text{Re}\, z\, dz = \int_0^1 t\, dt + \int_1^2 i\, dt = \tfrac{1}{2} + i.$$

We note that the last part of C_2 could be represented equally well in the form

$$z(t) = 1 + it \qquad (0 \leqq t \leqq 1);$$

then the limits of the last integral are 0 and 1, and its value is the same as before.

By comparing our two results we see that in the case of the function $\text{Re}\, z$ (which is not analytic) the value of the integral depends not only on the endpoints of the path, but also on its geometric shape.

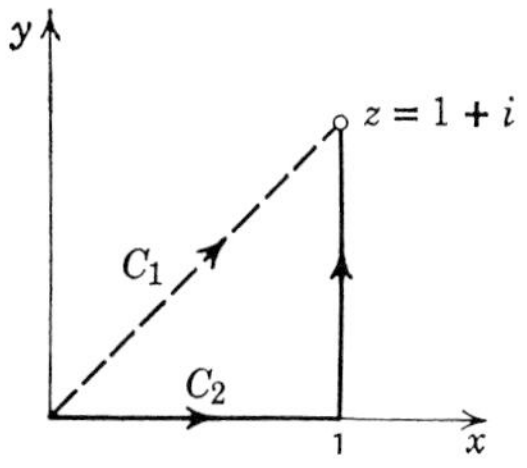

Fig. 297. Paths in Example 2

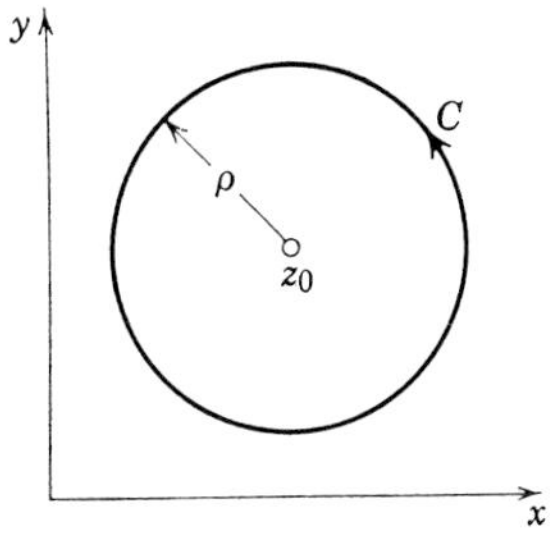

Fig. 298. Path in Example 3

Example 3. Integral of integer powers

Let $f(z) = (z - z_0)^m$ where m is an integer and z_0 is a constant. Integrate in the counterclockwise sense around the circle C of radius ρ with center at z_0 (cf. Fig. 298).

We may represent C in the form

$$z(t) = z_0 + \rho(\cos t + i \sin t) = z_0 + \rho e^{it} \qquad (0 \leqq t \leqq 2\pi).$$

Then we have

$$(z - z_0)^m = \rho^m e^{imt}, \qquad dz = i\rho e^{it}\,dt,$$

and we obtain

$$\int_C (z - z_0)^m\,dz = \int_0^{2\pi} \rho^m e^{imt} i\rho e^{it}\,dt = i\rho^{m+1} \int_0^{2\pi} e^{i(m+1)t}\,dt.$$

The case $m = -1$ was considered in Example 1, and when $m \neq -1$, we obtain [cf. (14) in Sec. 12.7]

$$\int_0^{2\pi} e^{i(m+1)t}\,dt = \left[\frac{e^{i(m+1)t}}{i(m+1)}\right]_0^{2\pi} = 0 \qquad (m \neq -1, \text{ integer}).$$

The result is

$$(9) \qquad \int_C (z - z_0)^m\,dz = \begin{cases} 2\pi i & (m = -1), \\ 0 & (m \neq -1 \text{ and integer}). \end{cases}$$

Example 4. A direct application of the definition

Let $f(z) = k = const$, and C be any curve joining two points z_0 and Z. In this case we may use the definition of the line integral as the limit of the sum S_n given by (2). We have

$$S_n = \sum_{m=1}^{n} k\,\Delta z_m = k[(z_1 - z_0) + (z_2 - z_1) + \cdots + (Z - z_{n-1})] = k(Z - z_0).$$

From this we immediately obtain the result

$$\int_C k\,dz = \lim_{n\to\infty} S_n = k(Z - z_0).$$

We note that the value of this integral depends only on z_0 and Z, not on the choice of the path joining these points. If in particular C is closed, then $z_0 = Z$ and the integral is zero.

Example 5. Another direct application of the definition
Let $f(z) = z$ and C be any curve joining two points z_0 and Z. We again use (2). Taking $\zeta_m = z_m$ we obtain

$$S_n = \sum_{m=1}^{n} z_m \Delta z_m = z_1(z_1 - z_0) + z_2(z_2 - z_1) + \cdots + Z(Z - z_{n-1}).$$

Taking $\zeta_m = z_{m-1}$ we find

$$S_n^* = \sum_{m=1}^{n} z_{m-1} \Delta z_m = z_0(z_1 - z_0) + z_1(z_2 - z_1) + \cdots + z_{n-1}(Z - z_{n-1}).$$

Addition of these two sums yields $S_n + S_n^* = Z^2 - z_0^2$, as can easily be verified. Hence

$$\lim_{n\to\infty} (S_n + S_n^*) = 2\int_{z_0}^{Z} z\, dz = Z^2 - z_0^2.$$

It follows that

$$\int_{z_0}^{Z} z\, dz = \tfrac{1}{2}(Z^2 - z_0^2)$$

along every path joining z_0 and Z. If, in particular, C is a closed path, then $z_0 = Z$ and

$$\int_C z\, dz = 0. \tag{10}$$

Note that this result follows also from Green's theorem (cf. Sec. 9.4) by the use of formula (6) in this section.

Problems for Sec. 14.1

Represent the line segments from A to B in the form $z = z(t)$ where

1. A: $z = 0$, B: $z = 1 + 2i$
2. A: $z = 0$, B: $z = 4 - 8i$
3. A: $z = 1 + i$, B: $z = 3 - 4i$
4. A: $z = 3i$, B: $z = 4 - i$
5. A: $z = -2 + i$, B: $z = -2 + 4i$
6. A: $z = 3 - i$, B: $z = 5 - 7i$

Represent the following curves in the form $z = z(t)$.

7. $|z - 1 + 2i| = 3$
8. $y = x^2$ from $(0, 0)$ to $(3, 9)$
9. $x^2 + 9y^2 = 9$
10. $4(x - 1)^2 + 9(y + 2)^2 = 36$
11. $y = 1/x$ from $(1, 1)$ to $(4, 1/4)$
12. $y = -1 + 3/x$ from $(1, 2)$ to $(3, 0)$

What curves are represented by the following functions?

13. $1 + (2 - i)t, \quad 0 \leqq t \leqq 1$
14. $i - t + 3it, \quad -1 \leqq t \leqq 2$
15. $3i + 3e^{it}, \quad 0 \leqq t \leqq \pi$
16. $4 + 2e^{it}, \quad -\pi \leqq t \leqq 0$
17. $t + 3t^2 i, \quad -1 \leqq t \leqq 2$
18. $i + t - t^3 i, \quad 0 \leqq t \leqq 3$

Integrate $3z^2$ along the line segment:

19. From 0 to $2i$
20. From 0 to $1 + i$
21. From $1 + i$ to $3 - 4i$
22. From $3i$ to $4 - i$

Integrate

23. $3z^2$ around the boundary of the triangle with vertices 0, 1, i (counterclockwise).

24. $z + 1/z$ around the unit circle (clockwise).

25. z from 1 vertically to $1 + i$ and then horizontally to $2 + i$.

26. $az + b$ from 0 along the line segment to $-4 - 4i$.

27. Evaluate $\int_C (z - 3)^{-1}\,dz$, C: $|z - 3| = 2$, (*a*) counterclockwise, (*b*) clockwise.

28. Evaluate $\int_C \operatorname{Re} z\,dz$ around the circle $|z| = r$ (counterclockwise).

29. Evaluate $\int_C |z|\,dz$ from A: $z = -i$ to B: $z = i$ along (*a*) the line segment AB, (*b*) the unit circle in the left half plane, (*c*) the unit circle in the right half plane.

30. Evaluate $\int_C (1/\sqrt{z})\,dz$ from 1 to -1 (*a*) along the upper semicircle $|z| = 1$, (*b*) along the lower semicircle $|z| = 1$, where $\sqrt{z}$ is the principal value of the square root.

14.2 Basic Properties of the Complex Line Integral

From the definition of a complex line integral as the limit of a sum we may immediately obtain the following properties.

If we decompose the path C into two portions C_1 and C_2 (Fig. 299), *then*

$$\int_C f(z)\,dz = \int_{C_1} f(z)\,dz + \int_{C_2} f(z)\,dz. \tag{1}$$

If we reverse the sense of integration, the sign of the value of the integral changes:

$$\int_{z_0}^{Z} f(z)\,dz = -\int_{Z}^{z_0} f(z)\,dz; \tag{2}$$

here the path C with endpoints z_0 and Z is the same; on the left we integrate from z_0 to Z, and on the right from Z to z_0.

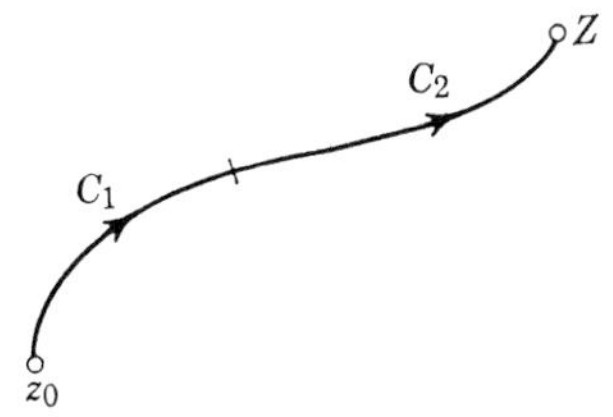

Fig. 299. Formula (1)

A sum of two (or more) functions may be integrated term by term, and constant factors may be taken out from under the integral sign, that is,

$$\int_C [k_1 f_1(z) + k_2 f_2(z)]\, dz = k_1 \int_C f_1(z)\, dz + k_2 \int_C f_2(z)\, dz. \tag{3}$$

There will be a frequent necessity for estimating the absolute value of complex line integrals. The basic formula is

$$\left| \int_C f(z)\, dz \right| \leqq Ml \tag{4}$$

where l is the length of the path C and M is a real constant such that $|f(z)| \leqq M$ everywhere on C.

To prove (4), we apply (11), Sec. 12.2, to the sum S_n defined by (2), Sec. 14.1, finding

$$|S_n| = \left| \sum_{m=1}^{n} f(\zeta_m)\, \Delta z_m \right| \leqq \sum_{m=1}^{n} |f(\zeta_m)|\, |\Delta z_m| \leqq M \sum_{m=1}^{n} |\Delta z_m|.$$

Now $|\Delta z_m|$ is the length of the chord whose endpoints are z_{m-1} and z_m, cf. Fig. 296 in Sec. 14.1. The sum on the right thus represents the length L of the broken line of chords whose endpoints are $z_0, z_1, \cdots, z_n (=Z)$. If n approaches infinity in such a way that the greatest $|\Delta z_m|$ approaches zero, then L approaches the length l of the curve C, by the definition of the length of a curve (cf. Sec. 8.4). From this, formula (4) follows.

Example 1. Application of formula (4)

Let $f(z) = 1/z$ and integrate once around the circle $|z| = \rho$. Then $l = 2\pi\rho$ and $|f(z)| = 1/\rho$ on the circle. Hence, by (4),

$$\left| \int_C \frac{dz}{z} \right| \leqq \frac{1}{\rho} 2\pi\rho = 2\pi \qquad \text{(cf. Example 1, Sec. 14.1).}$$

Example 2. Estimation of another integral

The path C_2 in Example 2, Sec. 14.1, has the length $l = 2$, and $|\text{Re}\, z| \leqq 1$ on C_2. Therefore, by (4),

$$\left| \int_{C_2} \text{Re}\, z\, dz \right| \leqq 2.$$

Problems for Sec. 14.2

1. Verify (1) for $f(z) = 1/z$ where C is the unit circle, C_1 its upper half, and C_2 its lower half.
2. Verify (2) for $f(z) = z^2$ where C is the line segment from $-1 - i$ to $1 + i$.
3. Verify (3) for $k_1 f_1 + k_2 f_2 = 3z - z^2$ where C is the upper half of the unit circle from 1 to -1.
4. Verify (4) for $f(z) = 1/z$ and C as in Prob. 3.

Evaluate $\int_C f(z)\,dz$ where

5. $f(z) = az + b$, C as in Prob. 2

6. $f(z) = z^3 + 2z^{-1}$, C the unit circle (counterclockwise)

7. $f(z) = z^2 + 3z^{-4}$, C as in Prob. 3

8. $f(z) = 2z - z^{-1} + 2z^{-2}$, C the unit circle (counterclockwise)

9. $f(z) = e^z$, C the line segment from 0 to $1 + \pi i/2$

10. $f(z) = (z-1)^{-1} + 2(z-1)^{-2}$, C the circle $|z - 1| = 4$ (clockwise)

11. $f(z) = \cos z$, C the line segment from πi to $2\pi i$

12. $f(z) = \sin z$, C as in Prob. 11

13. $f(z) = \sin z$, C the line segment from 0 to i

14. $f(z) = \sinh z$, C as in Prob. 13

15. $f(z) = \cosh z$, C as in Prob. 13

Using (4), find upper bounds for the following integrals where C is the line segment from 0 to $3 + 4i$.

16. $\int_C z\,dz$ **17.** $\int_C e^z\,dz$ **18.** $\int_C \operatorname{Ln}(z+1)\,dz$ **19.** $\int_C (z+1)^{-1}\,dz$

20. Find a better bound in Prob. 16 by decomposing C into two arcs.

14.3 Cauchy's Integral Theorem

Cauchy's integral theorem is very important in complex analysis and has various theoretical and practical consequences. To state this theorem, we shall need the following concepts.

A domain D in the complex plane is called a **simply connected domain** if every simple closed curve in D (that is, a closed curve in D without self-intersections) encloses only points of D. A domain which is not simply connected is said to be *multiply connected.*

For example, the interior of a circle ("circular disk"), ellipse or square is simply connected. More generally, the interior of a simple closed curve is simply connected. A circular ring (cf. Sec. 12.3) is multiply connected (more precisely: doubly connected[2]).

Furthermore, a domain D is said to be *bounded* if D lies entirely in some circle about the origin. Otherwise, D is said to be *unbounded.*

Cauchy's integral theorem

If $f(z)$ is analytic[3] in a simply connected bounded domain D, then for every simple closed path C in D,

$$\int_C f(z)\,dz = 0. \tag{1}$$

[2] A bounded domain is said to be *p-fold connected* if its boundary consists of p closed connected sets without common points. For the annulus, $p = 2$, because the boundary consists of two circles having no points in common.

[3] Remember that, by definition, a function is a single-valued relation. Cf. Sec. 12.4.

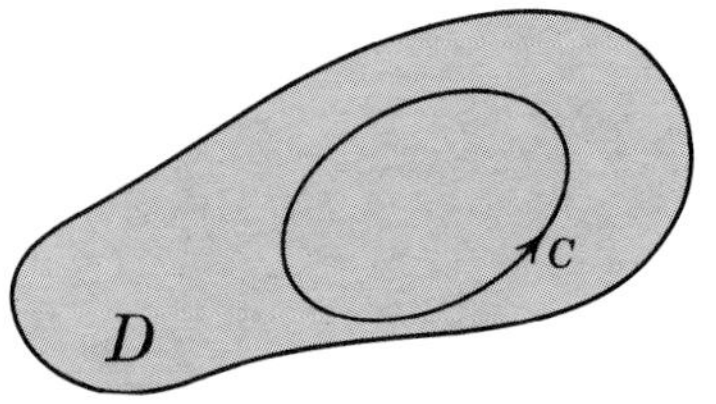

Fig. 300. Cauchy's integral theorem

Cauchy's proof. From (5), Sec. 14.1, we have

$$\int_C f(z)\,dz = \int_C (u\,dx - v\,dy) + i\int_C (u\,dy + v\,dx).$$

$f(z)$ is analytic and, therefore $f'(z)$ exists. Cauchy made the *additional assumption* that $f'(z)$ is continuous. Then u and v have continuous first partial derivatives in D, as follows from (3) and (4), Sec. 12.5. Green's theorem (cf. Sec. 9.4) (with u and $-v$ instead of f and g) is applicable, and

$$\int_C (u\,dx - v\,dy) = \iint_R \left(-\frac{\partial v}{\partial x} - \frac{\partial u}{\partial y}\right) dx\,dy$$

where R is the region bounded by C. The second Cauchy–Riemann equation (Sec. 12.5) shows that the integrand on the right is identically zero. Hence the integral on the left is zero. In the same fashion it follows by the use of the first Cauchy–Riemann equation that the last integral in the above formula is zero. This completes Cauchy's proof. ∎

Goursat's[4] proof. Goursat proved Cauchy's theorem without assuming that $f'(z)$ is continuous. This progress is quite important. We start with the case when C is the boundary of a triangle. We orient it in the counterclockwise sense. By joining the midpoints of the sides of the triangle we subdivide it into four congruent triangles (Fig. 301). Then

$$\int_C f\,dz = \int_{C_{\mathrm{I}}} f\,dz + \int_{C_{\mathrm{II}}} f\,dz + \int_{C_{\mathrm{III}}} f\,dz + \int_{C_{\mathrm{IV}}} f\,dz$$

where $C_{\mathrm{I}}, \cdots, C_{\mathrm{IV}}$ are the boundaries of the triangles; in fact, on the right side

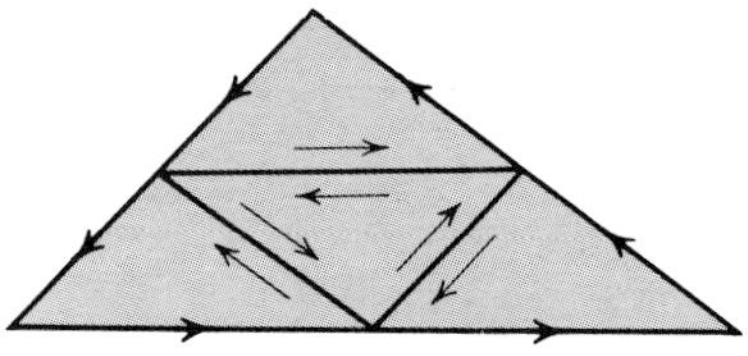

Fig. 301. Proof of Cauchy's integral theorem

[4]EDOUARD GOURSAT (1858–1936), French mathematician.

we integrate along each of the three segments of subdivision in both possible directions, the corresponding integrals cancel out in pairs, and the sum of the integrals on the right equals the integral on the left. Among the four integrals on the right there must be one, call its boundary C_1, for which

$$\left|\int_C f\,dz\right| \leqq 4\left|\int_{C_1} f\,dz\right|$$

because not each of the absolute values of the four integrals can be smaller than 1/4 of the absolute value of the sum of those integrals. This follows readily from (11) in Sec. 12.2.

We subdivide the triangle bounded by C_1 as before and select a triangle of subdivision with boundary C_2 for which

$$\left|\int_{C_1} f\,dz\right| \leqq 4\left|\int_{C_2} f\,dz\right|, \qquad \text{and then} \qquad \left|\int_C f\,dz\right| \leqq 4^2\left|\int_{C_2} f\,dz\right|.$$

Continuing in this fashion, we obtain a sequence of triangles $T_1, T_2, \cdots$ with boundaries $C_1, C_2, \cdots$ which are similar and such that T_n lies in T_m when $n > m$, and

$$\left|\int_C f\,dz\right| \leqq 4^n\left|\int_{C_n} f\,dz\right|, \qquad n = 1, 2, \cdots. \tag{2}$$

Let z_0 be the point which belongs to all these triangles. Since f is differentiable at $z = z_0$, the derivative $f'(z_0)$ exists, and we may write

$$f(z) = f(z_0) + (z - z_0)f'(z_0) + h(z)(z - z_0). \tag{3}$$

Consequently, by integrating over the boundary C_n of the triangle T_n we have

$$\int_{C_n} f(z)\,dz = \int_{C_n} f(z_0)\,dz + \int_{C_n} (z - z_0)f'(z_0)\,dz + \int_{C_n} h(z)(z - z_0)\,dz.$$

Since $f(z_0)$ and $f'(z_0)$ are constants, it follows from the results in Examples 4 and 5, Sec. 14.1, that the first two integrals on the right are zero. Hence,

$$\int_{C_n} f(z)\,dz = \int_{C_n} h(z)(z - z_0)\,dz.$$

By dividing (3) by $z - z_0$, transposing two terms to the left, and taking absolute values we obtain

$$\left|\frac{f(z) - f(z_0)}{z - z_0} - f'(z_0)\right| = |h(z)|.$$

From this and (5) in Sec. 12.4 we see that for a given positive number ϵ we can find a positive number δ such that

$$|h(z)| < \epsilon \qquad \text{when} \qquad |z - z_0| < \delta.$$

We may now take n so large that the triangle T_n lies in the disk $|z - z_0| < \delta$. Let l_n be the length of C_n. Then $|z - z_0| \leqq l_n/2$ for all z on C_n and z_0 in T_n. By applying (4) in Sec. 14.2 we thus obtain

$$\left|\int_{C_n} f(z)\,dz\right| = \left|\int_{C_n} h(z)(z - z_0)\,dz\right| < \epsilon \frac{l_n}{2} l_n = \frac{\epsilon}{2} l_n^{\,2}. \tag{4}$$

Let l be the length of C. Then the path C_1 has the length $l_1 = l/2$, the path C_2 has the length $l_2 = l_1/2 = l/4$, etc., and C_n has the length

$$l_n = \frac{l}{2^n}.$$

From (2) and (4) we thus obtain

$$\left|\int_C f\,dz\right| \leqq 4^n \left|\int_{C_n} f\,dz\right| < 4^n \frac{\epsilon}{2} l_n^{\,2} = 4^n \frac{\epsilon}{2} \frac{l^2}{4^n} = \frac{\epsilon}{2} l^2.$$

By choosing ϵ (> 0) sufficiently small we can make the expression on the right as small as we please, while the expression on the left is the definite value of an integral. From this we conclude that this value must be zero, and the proof is complete.

The proof for *the case in which C is the boundary of a polygon* follows from the previous proof by subdividing the polygon into triangles (Fig. 302). The integral corresponding to each such triangle is zero. The sum of these integrals is equal to the integral over C, because we integrate along each segment of subdivision in both directions, the corresponding integrals cancel out in pairs, and we are left with the integral over C.

The case of a general simple closed path C can be reduced to the preceding one by inscribing in C a closed polygon P of chords, which approximates C "sufficiently accurately," and it can be shown that there is a polygon P such that the integral over P differs from that over C by less than any preassigned positive real number ϵ, no matter how small. The details of this proof are somewhat involved and can be found in Ref. [F6] in Appendix 1. ∎

Example 1

$$\int_C e^z\,dz = 0$$

for any closed path, because e^z is analytic for all z.

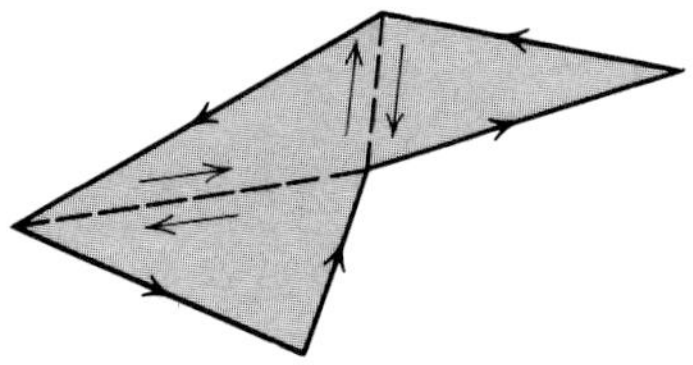

Fig. 302. Proof of Cauchy's integral theorem for a polygon

Example 2

$$\int_C \frac{dz}{z^2} = 0,$$

where C is the unit circle (cf. Sec. 14.1). This result does not follow from Cauchy's theorem, because $f(z) = 1/z^2$ is not analytic at $z = 0$. Hence *the condition that f be analytic in D is sufficient rather than necessary for* (1) *to be true.*

Example 3

$$\int_C \frac{dz}{z} = 2\pi i,$$

the integration being taken around the unit circle in the counterclockwise sense (cf. Sec. 14.1). C lies in the annulus $\frac{1}{2} < |z| < \frac{3}{2}$ where $1/z$ is analytic, but this domain is not simply connected, so that Cauchy's theorem cannot be applied. Hence *the condition that the domain D be simply connected is quite essential.* ∎

If we subdivide the path C in Cauchy's theorem into two arcs C_1 and $C_2{}^*$ (Fig. 303), then (1) takes the form

$$\int_C f\,dz = \int_{C_1} f\,dz + \int_{C_2*} f\,dz = 0.$$

Consequently,

$$-\int_{C_2*} f\,dz = \int_{C_1} f\,dz. \tag{5'}$$

If we reverse the sense of integration along $C_2{}^*$, then the integral over $C_2{}^*$ is multiplied by -1, and we obtain (cf. Fig. 304)

$$\int_{C_2} f(z)\,dz = \int_{C_1} f(z)\,dz. \tag{5}$$

Hence, if f is analytic in D, and C_1 and C_2 are any paths in D joining two points in D and having no further points in common, then (5) holds.

If those paths C_1 and C_2 have finitely many points in common (Fig. 305), then (5) continues to hold; this follows by applying the previous result to the portions of C_1 and C_2 between each pair of consecutive points of intersection.

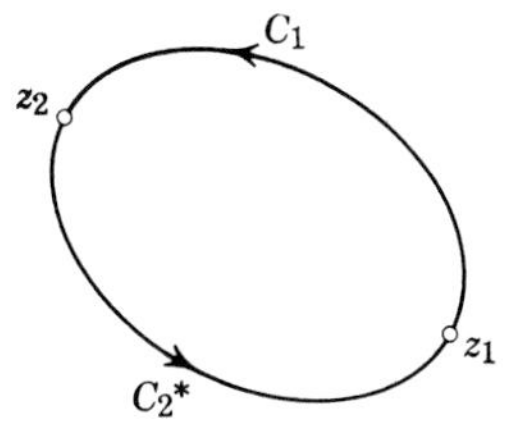

Fig. 303. Formula (5′)

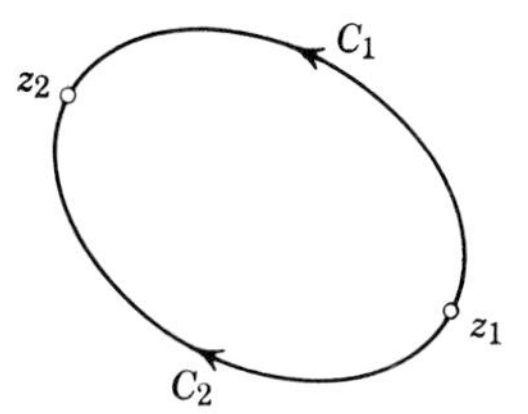

Fig. 304. Formula (5)

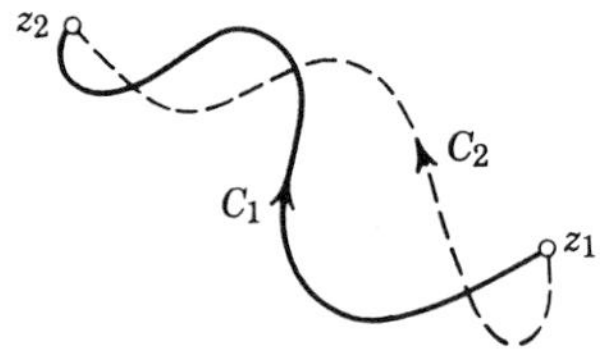

Fig. 305. Paths having finitely many intersections

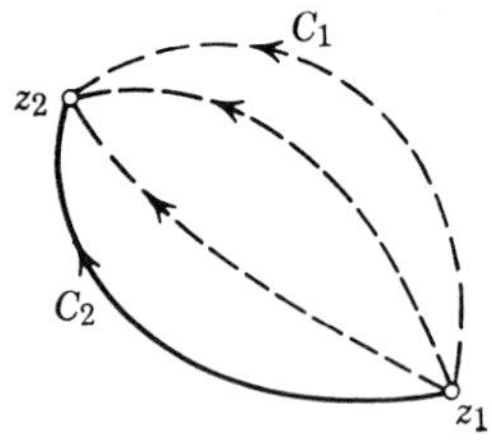

Fig. 306. Continuous deformation of path

It is even true that (5) *holds for any paths entirely in the simply connected domain D where $f(z)$ is analytic and joining any points z_1 and z_2 in D. Then we say that the integral of $f(z)$ from z_1 to z_2 is* **independent of path in *D*.** (Of course, the value of the integral depends on the choice of z_1 and z_2.)

The *proof* would require an additional consideration of the case in which C_1 and C_2 have infinitely many points of intersection, and will not be presented here.

We may imagine that the path C_2 in (5) was obtained from C_1 by a continuous deformation (Fig. 306). It follows that in a given integral we may impose a continuous deformation on the path of integration (keeping the endpoints fixed); as long as we do not pass through a point where $f(z)$ is not analytic, the value of the line integral will not change under such a deformation. This is often called the **principle of deformation of path.**

A multiply connected domain D^* can be cut so that the resulting domain (D^* without the points of the cut or cuts) becomes simply connected. For a doubly connected domain D^* we need one cut $\tilde{C}$ (Fig. 307). If $f(z)$ is analytic in D^* and at each point of C_1 and C_2 then, since C_1, C_2, and $\tilde{C}$ bound a simply connected domain, it follows from Cauchy's theorem that the integral of f taken over C_1, $\tilde{C}$, C_2 in the sense indicated by the arrows in Fig. 307 has the value zero. Since we integrate along $\tilde{C}$ in both directions, the corresponding integrals cancel out, and we obtain

$$\int_{C_1} f(z)\,dz + \int_{C_2} f(z)\,dz = 0 \tag{6}$$

where one of the curves is traversed in the counterclockwise sense and the other in the opposite sense.

Equation (6) may also be written

$$\int_{C_1} f(z)\,dz = \int_{C_2} f(z)\,dz, \tag{7}$$

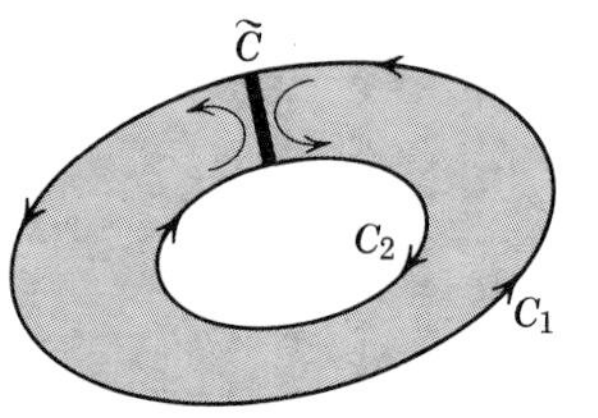

Fig. 307. Doubly connected domain

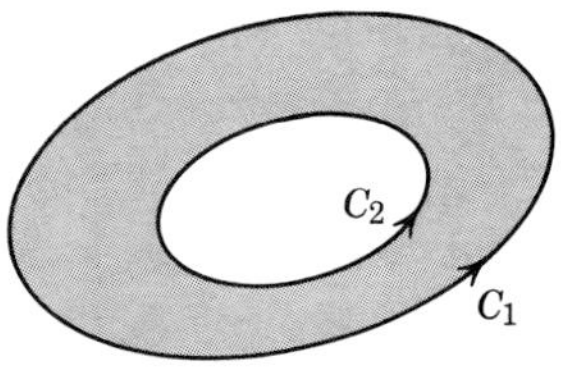

Fig. 308. Paths in (7)

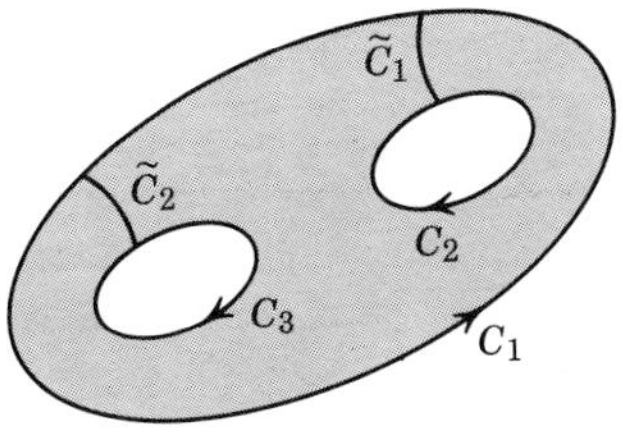

Fig. 309. Triply connected domain

where C_1 and C_2 are now traversed in the same sense (Fig. 308). We remember that (7) holds under the assumption that $f(z)$ is analytic in the domain bounded by C_1 and C_2 and at each point of C_1 and C_2.

For more complicated domains we may need more than one cut, but the basic idea remains the same as before. For instance, for the triply connected domain in Fig. 309,

$$\int_{C_1} f(z)\,dz + \int_{C_2} f(z)\,dz + \int_{C_3} f(z)\,dz = 0$$

where C_2 and C_3 are traversed in the same sense and C_1 is traversed in the opposite sense.

We mention that a simple closed path is sometimes called a *contour,* and integrals along such a path are known as **contour integrals.**

Example 4

Let C_1 be the unit circle $|z| = 1$ and C_2 the circle $|z| = 1/2$. Then by (7),

$$\int_{C_2} \frac{dz}{z} = \int_{C_1} \frac{dz}{z} = 2\pi i \qquad \text{(cf. Example 1 in Sec. 14.1),}$$

where both paths are traversed in the counterclockwise sense.

Example 5

From Example 3 in Sec. 14.1 and the principle of deformation of path it follows that

$$\int_C (z - z_0)^m\,dz = \begin{cases} 2\pi i & (m = -1) \\ 0 & (m \neq -1 \text{ and integer}) \end{cases}$$

where C is any contour containing the point z_0 in its interior and the integration is taken around C in the counterclockwise sense.

Problems for Sec. 14.3

1. Verify Cauchy's theorem for $\int_C z^2\,dz$ where C is the boundary of the triangle with vertices 0, 2, and $2i$.
2. Show that the integral of $1/z^3$ around the unit circle is zero. Does this follow from Cauchy's theorem?
3. For what simple closed paths is the integral of $1/z$ equal to zero?

In each case find the value of the integral of the given function around the unit circle in the counterclockwise sense and indicate whether Cauchy's theorem may be applied.

4. $f(z) = 1/z^4$ **5.** $f(z) = e^{-z}$ **6.** $f(z) = |z|$

7. $f(z) = \operatorname{Im} z$ **8.** $f(z) = \operatorname{Re} z$ **9.** $f(z) = \tanh z$

10. $f(z) = 1/(z^2 + 2)$ **11.** $f(z) = 1/\bar{z}$ **12.** $f(z) = z^2 \sec z$

13. Obtain the answer to Prob. 8 from that to Prob. 7 and Cauchy's theorem applied to $f(z) = z$.

14. Using the principle of deformation of path and

$$\frac{2z-1}{z^2-z} = \frac{1}{z} + \frac{1}{z-1}, \quad \text{show that} \quad \int_C \frac{2z-1}{z^2-z}\,dz = \int_C \frac{dz}{z} + \int_C \frac{dz}{z-1} = 4\pi i$$

where C is the path shown in the figure.

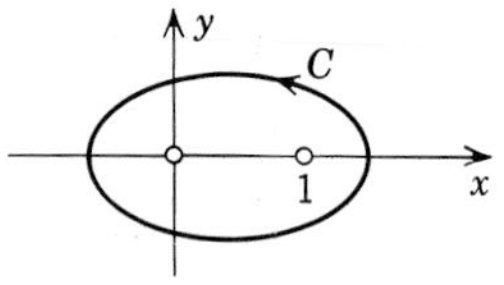

Problem 14

15. Integrate $f(z) = \bar{z}/|z|$ around the circle (*a*) $|z| = 2$, (*b*) $|z| = 4$, in the counterclockwise sense. Can the result in (*b*) be obtained from that in (*a*) by the principle of deformation of path?

Evaluate the following integrals. (*Hint.* If necessary, represent the integrand in terms of partial fractions.)

16. $\int_C \frac{dz}{z}$, C: $|z - 2| = 1$ (clockwise)

17. $\int_C \frac{z^2 - z + 1}{z^3 - z^2}\,dz$, C: (*a*) $|z| = 2$, (*b*) $|z| = 1/2$ (counterclockwise)

18. $\int_C \frac{dz}{z^2 - 1}$, C: (*a*) $|z| = 2$, (*b*) $|z - 1| = 1$ (clockwise)

19. $\int_C \frac{z}{z^2 + 1}\,dz$, C: (*a*) $|z| = 2$, (*b*) $|z + i| = 1$ (counterclockwise)

20. $\int_C \frac{dz}{z^2 + 1}$, C: (*a*) $|z + i| = 1$, (*b*) $|z - i| = 1$ (counterclockwise)

21. $\int_C \frac{e^z}{z}\,dz$, C consists of $|z| = 2$ (counterclockwise) and $|z| = 1$ (clockwise)

22. $\int_C \frac{\cos z}{z^2}\,dz$, C: $|z - 2i| = 1$ (counterclockwise)

23. $\int_C \frac{3z + 1}{z^3 - z}\,dz$, C: (*a*) $|z| = 1/2$, (*b*) $|z| = 2$ (counterclockwise)

24. $\int_C \frac{2z^3 + z^2 + 4}{z^4 + 4z^2}\,dz$, C: $|z - 2| = 4$ (clockwise)

25. $\int_C \frac{dz}{z^4 + 4z^2}$, C consists of $|z| = \frac{3}{2}$ (clockwise) and $|z| = 1$ (counterclockwise)

14.4 Evaluation of Line Integrals by Indefinite Integration

Using Cauchy's integral theorem, we want to show that in many cases complex line integrals can be evaluated by a very simple method, namely, by indefinite integration.

Suppose that $f(z)$ is analytic in a simply connected domain D, and let z_0 be any point in D which we keep fixed. Then the integral

$$\int_{z_0}^{z} f(z^*)\, dz^*$$

is a function of z for all paths which lie in D and join z_0 and z, and we may write

$$F(z) = \int_{z_0}^{z} f(z^*)\, dz^*. \tag{1}$$

Let us prove that $F(z)$ *is an analytic function of* z *in* D, *and* $F'(z) = f(z)$.

We keep z fixed. Since D is a domain, a neighborhood N of z belongs to D. In N we choose a point $z + \Delta z$ such that the line segment with endpoints z and $z + \Delta z$ is in N, hence in D. From (1) we then obtain

$$F(z + \Delta z) - F(z) = \int_{z_0}^{z+\Delta z} f(z^*)\, dz^* - \int_{z_0}^{z} f(z^*)\, dz^* = \int_{z}^{z+\Delta z} f(z^*)\, dz^*,$$

where we may integrate from z to $z + \Delta z$ along that segment (Fig. 310). Hence

$$\frac{F(z + \Delta z) - F(z)}{\Delta z} - f(z) = \frac{1}{\Delta z}\int_{z}^{z+\Delta z} [f(z^*) - f(z)]\, dz^*$$

because z is kept fixed and, thus,

$$-\frac{1}{\Delta z}\int_{z}^{z+\Delta z} f(z)\, dz^* = -\frac{f(z)}{\Delta z}\int_{z}^{z+\Delta z} dz^* = -f(z).$$

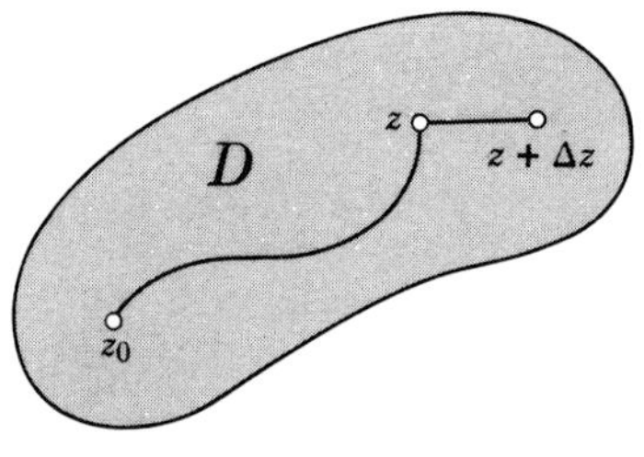

Fig. 310. Path of integration

Now $f(z)$ is continuous. Hence, an $\epsilon > 0$ being given, we can find a $\delta > 0$ such that

$$|f(z^*) - f(z)| < \epsilon \qquad \text{when} \qquad |z^* - z| < \delta.$$

Consequently, if $|\Delta z| < \delta$, then

$$\left|\frac{F(z+\Delta z) - F(z)}{\Delta z} - f(z)\right| = \frac{1}{|\Delta z|}\left|\int_z^{z+\Delta z} [f(z^*) - f(z)]\, dz^*\right|$$
$$< \frac{\epsilon}{|\Delta z|}\left|\int_z^{z+\Delta z} dz^*\right| = \epsilon;$$

that is,

$$F'(z) = \lim_{\Delta z \to 0} \frac{F(z+\Delta z) - F(z)}{\Delta z} = f(z). \tag{2}$$

From (1) it follows that if z_0 is replaced by another fixed point in D, the function $F(z)$ is changed by an additive constant. From (2) we see that $F(z)$ is an indefinite integral or antiderivative of $f(z)$, written

$$F(z) = \int f(z)\, dz;$$

that is, $F(z)$ is an analytic function in D whose derivative is $f(z)$.

If $F'(z) = f(z)$ and $G'(z) = f(z)$, then $F'(z) - G'(z) \equiv 0$ in D. Hence. the function $F(z) - G(z)$ is constant (cf. Prob. 35 at the end of Sec. 12.5). That is, the two indefinite integrals $F(z)$ and $G(z)$ differ only by a constant. In view of (1), we have for any points a and b in D and any path in D from a to b,

$$\int_a^b f(z)\, dz = \int_{z_0}^b f(z)\, dz - \int_{z_0}^a f(z)\, dz = F(b) - F(a),$$

as in the case of real definite integrals, but *it is important that the paths of integration lie in a simply connected domain D where $f(z)$ is analytic.*

We may sum up our result as follows.

Theorem 1 (Evaluation of line integrals by indefinite integration) *If $f(z)$ is analytic in a simply connected domain D, then there exists an indefinite integral of $f(z)$ in the domain D, that is, an analytic function $F(z)$ such that $F'(z) = f(z)$ in D, and for all paths in D joining two points a and b in D,*

$$\int_a^b f(z)\, dz = F(b) - F(a). \tag{3}$$

This theorem enables us to evaluate complex line integrals by indefinite integration. Note that in (3) we can take *any* indefinite integral $F(z)$ of $f(z)$ in D, since $F(z)$ is unique up to an additive constant.

Example 1

$$\int_{i}^{1+4i} z^2\,dz = \left[\frac{z^3}{3}\right]_{i}^{1+4i} = \frac{1}{3}[(1+4i)^3 - i^3] = -\frac{47}{3} - 17i.$$

Example 2

$$\int_{i}^{\pi/2} \cos z\,dz = \sin z \,\bigg|_{i}^{\pi/2} = \sin\frac{\pi}{2} - \sin i = 1 - i\sinh 1.$$

Problems for Sec. 14.4

Evaluate:

1. $\int_{i}^{2+i} z\,dz$
2. $\int_{1}^{3i} z^2\,dz$
3. $\int_{i}^{1} (z+1)^2\,dz$
4. $\int_{1-i}^{1+i} z^3\,dz$
5. $\int_{1}^{1+\pi i} e^z\,dz$
6. $\int_{\pi i}^{2\pi i} e^{2z}\,dz$
7. $\int_{0}^{i} ze^{z^2}\,dz$
8. $\int_{1-\pi i}^{1+\pi i} e^{z/2}\,dz$
9. $\int_{0}^{\pi i} \cos z\,dz$
10. $\int_{0}^{2\pi i} \sin 2z\,dz$
11. $\int_{0}^{\pi i} z\cos z^2\,dz$
12. $\int_{0}^{\pi i} z\cos z\,dz$
13. $\int_{-\pi i}^{\pi i} \sin^2 z\,dz$
14. $\int_{1-i}^{1+i} \cos z\,dz$
15. $\int_{0}^{3i} \cosh z\,dz$
16. $\int_{i}^{1+i} \sinh z\,dz$
17. $\int_{0}^{2i} \sinh z\,dz$
18. $\int_{i}^{3i} z\sinh z^2\,dz$
19. $\int_{-i}^{i} z\cosh z^2\,dz$
20. $\int_{-\pi i}^{\pi i} z\cosh z\,dz$

14.5 Cauchy's Integral Formula

The most important consequence of Cauchy's integral theorem is Cauchy's integral formula. This formula and its conditions of validity may be stated as follows.

Theorem 1 (Cauchy's integral formula)
Let $f(z)$ be analytic in a simply connected domain D. Then for any point z_0 in D and any simple closed path C in D which encloses z_0 (Fig. 311),

(1) $$\oint_C \frac{f(z)}{z - z_0}\,dz = 2\pi i f(z_0) \qquad \textbf{(Cauchy's integral formula)},$$

the integration being taken in the counterclockwise sense.

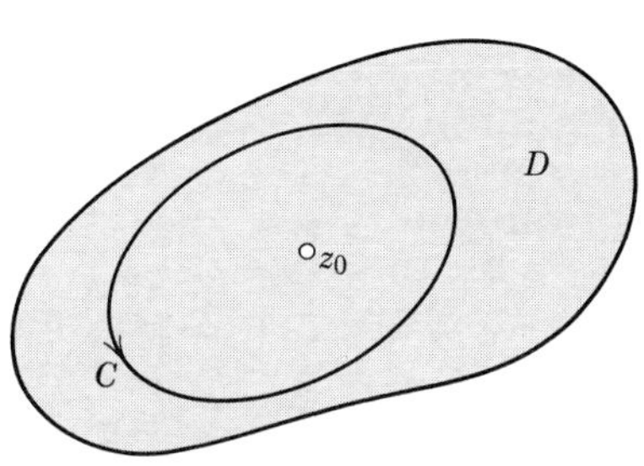

Fig. 311. Cauchy's integral formula

Proof. Writing $f(z) = f(z_0) + [f(z) - f(z_0)]$ and remembering that a constant factor may be taken out from under the integral sign, we have

$$(2) \qquad \int_C \frac{f(z)}{z - z_0} dz = f(z_0) \int_C \frac{dz}{z - z_0} + \int_C \frac{f(z) - f(z_0)}{z - z_0} dz.$$

From Example 5 in Sec. 14.3 it follows that the first term on the right is equal to $2\pi i f(z_0)$. Hence (1) holds, provided the last integral in (2) is zero. Now the integrand of this integral is analytic in D except at the point z_0. We may thus replace C by a small circle K with center at z_0, without changing the value of the integral (Fig. 312). Since $f(z)$ is analytic, it is continuous. Hence, an $\epsilon > 0$ being given, we can find a $\delta > 0$ such that

$$|f(z) - f(z_0)| < \epsilon \qquad \text{for all } z \text{ in the disk } |z - z_0| < \delta.$$

Choosing the radius ρ of K smaller than δ, we thus have the inequality

$$\left| \frac{f(z) - f(z_0)}{z - z_0} \right| < \frac{\epsilon}{\rho}$$

at each point of K. The length of K is $2\pi\rho$. Hence, by (4) in Sec. 14.2,

$$\left| \int_K \frac{f(z) - f(z_0)}{z - z_0} dz \right| < \frac{\epsilon}{\rho} 2\pi\rho = 2\pi\epsilon.$$

Since ϵ can be made arbitrarily small, it follows that the last integral on the right side of (2) has the value zero, and the theorem is proved. ∎

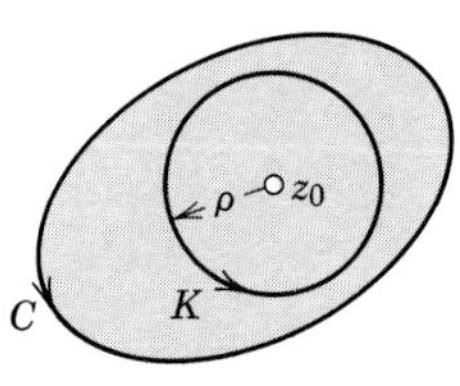

Fig. 312. Proof of Cauchy's integral formula

Example 1. Integration around different closed paths

Integrate

$$\frac{z^2 + 1}{z^2 - 1}$$

in the counterclockwise sense around a circle of radius 1 with center at the point

(a) $z = 1$ (b) $z = \frac{1}{2}$ (c) $z = -1$ (d) $z = i$.

(a) The integral under consideration may be written

$$\int_C \frac{z^2+1}{z^2-1}\,dz = \int_C \frac{z^2+1}{z+1}\,\frac{dz}{z-1}.$$

The expression on the right is of the form (1) where $z_0 = 1$ and

$$f(z) = \frac{z^2+1}{z+1}.$$

The point $z_0 = 1$ lies inside the circle C under consideration, and $f(z)$ is analytic inside and on C. [The point $z = -1$, where $f(z)$ is not analytic, lies outside C.] Hence, by Cauchy's integral formula,

$$\int_C \frac{z^2+1}{z^2-1}\,dz = \int_C \frac{z^2+1}{z+1}\,\frac{dz}{z-1} = 2\pi i\left[\frac{z^2+1}{z+1}\right]_{z=1} = 2\pi i.$$

(b) We obtain the same result as before because the given function is analytic except at the points $z = 1$ and $z = -1$, and we may obtain the circle (b) from that in the case (a) by a continuous deformation (even a translation) without passing over a point where the given function is not analytic.

(c) We may write

$$\int_C \frac{z^2+1}{z^2-1}\,dz = \int_C \frac{z^2+1}{z-1}\,\frac{dz}{z+1}.$$

The integral on the right is of the form (1) where $z_0 = -1$ and

$$f(z) = \frac{z^2+1}{z-1}.$$

The point $z_0 = -1$ lies inside the circle C now under consideration, and $f(z)$ is analytic inside and on C. Therefore, by (1),

$$\int_C \frac{z^2+1}{z^2-1}\,dz = \int_C \frac{z^2+1}{z-1}\,\frac{dz}{z+1} = 2\pi i\left[\frac{z^2+1}{z-1}\right]_{z=-1} = -2\pi i.$$

(d) The given function is analytic everywhere inside the circle now under consideration and on that circle. Hence, by Cauchy's integral theorem, the integral has the value zero. ∎

In the case of a *multiply connected domain* we may proceed in a manner similar to that in Sec. 14.3. For instance, if $f(z)$ is analytic on C_1 and C_2 and in the ring-shaped domain bounded by C_1 and C_2 (Fig. 313) and z_0 is any point in that domain, then

$$f(z_0) = \frac{1}{2\pi i}\int_{C_1} \frac{f(z)}{z-z_0}\,dz - \frac{1}{2\pi i}\int_{C_2} \frac{f(z)}{z-z_0}\,dz, \tag{3}$$

where both integrals are taken in the counterclockwise sense.

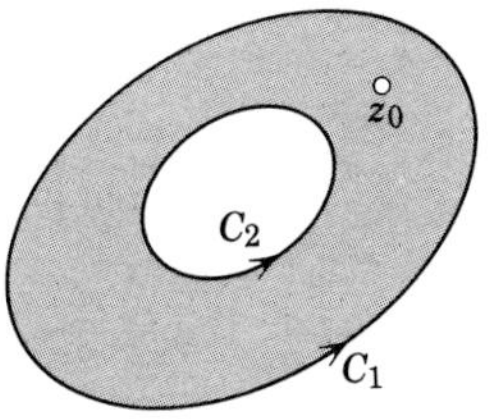

Fig. 313. Formula (3)

Problems for Sec. 14.5

Integrate $z^2/(z^2+1)$ in the counterclockwise sense around the circle

1. $|z+i|=1$ **2.** $|z-i|=1/2$ **3.** $|z|=2$ **4.** $|z|=1/2$

Integrate $z^2/(z^4-1)$ in the counterclockwise sense around the circle

5. $|z-1|=1$ **6.** $|z+i|=1$ **7.** $|z-i|=1/2$ **8.** $|z|=2$

Integrate the following functions in the counterclockwise sense around the unit circle.

9. $1/z$ **10.** $1/(z^2+4)$ **11.** $1/(4z+i)$ **12.** e^z/z
13. $e^{2z}/(z+2i)$ **14.** $e^{z^2}/(2z-i)$ **15.** $(\cos z)/z$ **16.** $(\sin z)/z$
17. $(e^z-1)/z$ **18.** $(\sinh z)/z$ **19.** $(\cosh 3z)/z$ **20.** $(z-2)^{-1}\sin z$

14.6 The Derivatives of an Analytic Function

From the assumption that a *real* function of a real variable is once differentiable nothing follows about the existence of derivatives of higher order. We shall now see that from the assumption that a *complex* function has a first derivative in a domain D there follows the existence of derivatives of all orders in D. This means that in this respect complex analytic functions behave much more simply than real functions which are once differentiable.

Theorem 1 (Derivatives of an analytic function)
If $f(z)$ is analytic in a domain D, then it has derivatives of all orders in D which are then also analytic functions in D. The values of these derivatives at a point z_0 in D are given by the formulas

$$f'(z_0)=\frac{1}{2\pi i}\int_C \frac{f(z)}{(z-z_0)^2}\,dz, \tag{1'}$$

$$f''(z_0)=\frac{2!}{2\pi i}\int_C \frac{f(z)}{(z-z_0)^3}\,dz, \tag{1''}$$

and in general

$$f^{(n)}(z_0)=\frac{n!}{2\pi i}\int_C \frac{f(z)}{(z-z_0)^{n+1}}\,dz \qquad (n=1,2,\cdots); \tag{1}$$

here C is any simple closed path in D which encloses z_0 and whose full interior belongs to D; the curve C is traversed in the counterclockwise sense (*Fig.* 314).

Remark. For memorizing (1), it is useful to observe that these formulas are obtained formally by differentiating the Cauchy formula (1), Sec. 14.5, under the integral sign *with respect to z_0*.

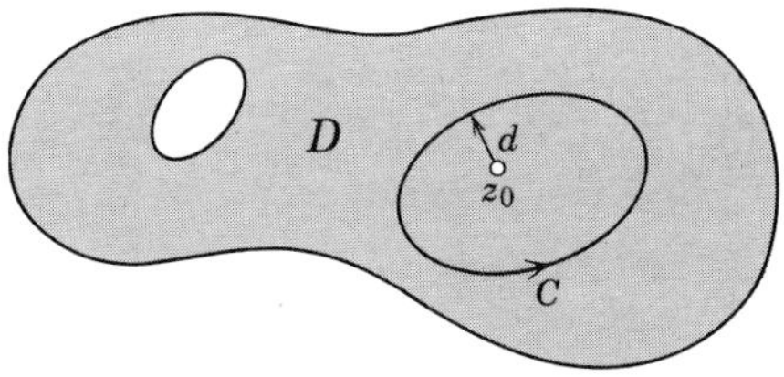

Fig. 314. Theorem 1

Proof of Theorem 1. We prove (1′). By definition,

$$f'(z_0) = \lim_{\Delta z \to 0} \frac{f(z_0 + \Delta z) - f(z_0)}{\Delta z}.$$

From this and Cauchy's integral formula (1), Sec. 14.5, it follows that

$$(2) \qquad f'(z_0) = \lim_{\Delta z \to 0} \frac{1}{2\pi i\,\Delta z}\left[\int_C \frac{f(z)}{z - (z_0 + \Delta z)}\,dz - \int_C \frac{f(z)}{z - z_0}\,dz\right].$$

Straightforward calculation shows that

$$\frac{1}{\Delta z}\left[\frac{1}{z - (z_0 + \Delta z)} - \frac{1}{z - z_0}\right] = \frac{1}{(z - z_0)^2} + \frac{\Delta z}{(z - z_0 - \Delta z)(z - z_0)^2}.$$

Hence (2) can be written

$$f'(z_0) = \frac{1}{2\pi i}\int_C \frac{f(z)}{(z - z_0)^2}\,dz + \lim_{\Delta z \to 0} \frac{\Delta z}{2\pi i}\int_C \frac{f(z)}{(z - z_0 - \Delta z)(z - z_0)^2}\,dz.$$

Consequently, (1′) will be established if we show that the last expression on the right has the value zero. This is what we shall now do.

On C the function $f(z)$ is continuous. Hence $f(z)$ is bounded in absolute value on C, say, $|f(z)| < M$. Let d be the distance of the point or points of C which are closest to z_0. Then for all z on C,

$$|z - z_0| \geqq d \qquad \text{and} \qquad \frac{1}{|z - z_0|} \leqq \frac{1}{d}.$$

Furthermore, if $|\Delta z| \leqq d/2$, then for all z on C we have the inequality

$$|z - z_0 - \Delta z| \geqq \frac{d}{2} \qquad \text{and} \qquad \frac{1}{|z - z_0 - \Delta z|} \leqq \frac{2}{d}.$$

Denoting the length of C by L and using (4) in Sec. 14.2, we thus obtain

$$\left|\frac{\Delta z}{2\pi i}\int_C \frac{f(z)}{(z - z_0 - \Delta z)(z - z_0)^2}\,dz\right| < \frac{|\Delta z|}{2\pi}\frac{M}{\frac{1}{2}dd^2}L \qquad \left(|\Delta z| \leqq \frac{d}{2}\right).$$

If Δz approaches zero, the expression on the right approaches zero. This proves (1′). Note that we used Cauchy's integral formula (1), Sec. 14.5, but if all we had known about $f(z_0)$ is the fact that it can be represented by (1), Sec. 14.5, our argument would have established the existence of the derivative $f'(z_0)$ of $f(z)$. This implies that formula (1″) can be proved by a similar argument, and the general formula (1) follows by induction. ∎

Using Theorem 1, let us prove the converse of Cauchy's theorem.

Morera's[5] theorem

If $f(z)$ is continuous in a simply connected domain D and if

$$\int_C f(z)\,dz = 0 \tag{3}$$

for every closed path in D, then $f(z)$ is analytic in D.

Proof. In Sec. 14.4 it was shown that if $f(z)$ is analytic in D, then

$$F(z) = \int_{z_0}^{z} f(z^*)\,dz^*$$

is analytic in D and $F'(z) = f(z)$. In the proof we used only the continuity of $f(z)$ and the property that its integral around every closed path in D is zero; from these assumptions we concluded that $F(z)$ is analytic. By Theorem 1, the derivative of $F(z)$ is analytic, that is, $f(z)$ is analytic in D, and Morera's theorem is proved. ∎

We shall now derive an important inequality. In (1), let C be a circle of radius r with center at z_0, and let M be the maximum of $|f(z)|$ on C. Then, by applying (4), Sec. 14.2, to (1) we find

$$|f^{(n)}(z_0)| = \frac{n!}{2\pi}\left|\int_C \frac{f(z)}{(z-z_0)^{n+1}}\,dz\right| \leqq \frac{n!}{2\pi} M \frac{1}{r^{n+1}} 2\pi r.$$

This yields **Cauchy's inequality**

$$|f^{(n)}(z_0)| \leqq \frac{n!M}{r^n}. \tag{4}$$

Using (4), let us prove the following interesting and basic result.

Liouville's theorem

If $f(z)$ is analytic and bounded in absolute value for all z in the finite complex plane (cf. Sec. 13.3), *then $f(z)$ is a constant.*

[5] GIACINTO MORERA (1856–1909), Italian mathematician.

Proof. By assumption, $|f(z)|$ is bounded, say, $|f(z)| < K$ for all finite z. Using (4), we see that $|f'(z_0)| < K/r$. Since this is true for every r, we can take r as large as we please and conclude that $f'(z_0) = 0$. Since z_0 is arbitrary, $f'(z) = 0$ for all finite z, and $f(z)$ is a constant (cf. Prob. 35 at the end of Sec. 12.5). This completes the proof. ∎

Problems for Sec. 14.6

Using Theorem 1, integrate the following functions in the counterclockwise sense around the unit circle. (n in Probs. 8, 12, 13 is a positive integer.)

1. $z^2/(2z-1)^2$ **2.** $z^2/(2z-1)^4$ **3.** $z^3/(2z+i)^3$ **4.** $z^4/(z-3i)^2$

5. $z/(4z+i)^2$ **6.** e^z/z^2 **7.** e^z/z^3 **8.** e^z/z^n

9. $ze^z/(4z+\pi i)^2$ **10.** $z^{-2}\cos z$ **11.** $z^{-2}\sin z$ **12.** $z^{-2n}\cos z$

13. $z^{-2n-1}\cos z$ **14.** e^{z^2}/z^3 **15.** $z^{-2}e^{-z}\sin z$ **16.** e^{z^3}/z^3

17. If $f(z)$ is not a constant and is analytic for all (finite) z, and R and M are any positive real numbers (no matter how large), show that there exist values of z for which $|z| > R$ and $|f(z)| > M$. *Hint.* Use Liouville's theorem.

18. If $f(z)$ is a polynomial of degree $n > 0$ and M an arbitrary positive real number (no matter how large), show that there exists a positive real number R such that $|f(z)| > M$ for *all* $|z| > R$.

19. Show that $f(z) = e^z$ has the property characterized in Prob. 17, but does not have that characterized in Prob. 18.

20. Prove the **Fundamental theorem of algebra:** If $f(z)$ is a polynomial in z, not a constant, then $f(z) = 0$ for at least one value of z. *Hint.* Assume $f(z) \neq 0$ for all z and apply the result of Prob. 17 to $g = 1/f$.

CHAPTER

15 SEQUENCES AND SERIES

This chapter is a self-contained presentation of the fundamental concepts and facts on complex and real sequences and series.

Secs. 15.1 and 15.2 contain the basic definitions.

Secs. 15.3–15.5 concern convergence tests and Sec. 15.6 is devoted to operations on series, such as termwise addition and rearrangements.

Secs. 15.3–15.6 are optional for a reader who is primarily interested in complex analytic functions and has some experience with real series, so that he can speed up his study by going from Sec. 15.2 directly to Chap. 16 (on power series, Taylor series and Laurent series, which are basic in complex analysis). That is, the reader may skip Secs. 15.3–15.6 and use them only for reference when necessary.

Special classes of series are considered in

Chap. 10 **(Fourier series)**

Chap. 16 **(Power series)**

Sec. 19.15 **(Asymptotic expansions).**

Prerequisites for this chapter: Secs. 12.1–12.5.

Sections which may be omitted in a shorter course: Secs. 15.3–15.6.

References: [A2], [A3], [A7], [A9], [A15] in Appendix 1.

Answers to Problems: Appendix 2.

15.1 Sequences

Series, in particular power series, play a basic role in complex analysis. To introduce them, we begin with the definition of a sequence and related concepts. We shall see that most of the definitions and theorems for *complex* sequences and series are similar to those for *real* sequences and series used in calculus.

If to each positive integer n there is assigned a number z_n, then these numbers

$$z_1, z_2, \cdots, z_n, \cdots$$

are said to form an *infinite sequence* or, briefly, a **sequence,** and the numbers z_n are called the **terms** of the sequence.

A sequence whose terms are real is called a *real sequence.*

At times it is convenient to number the terms of a sequence starting with 0, with 2, or with some other integer.

A sequence $z_1, z_2, \cdots$ is said to **converge** or to **be convergent** if there is a number c with the following property. For every positive real ϵ (no matter how small but not zero) we can find an integer N such that

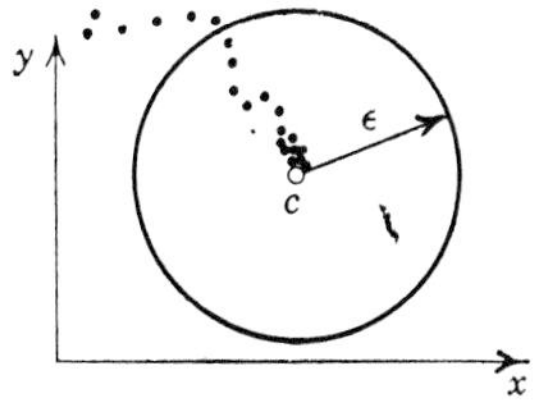

Fig. 315. Convergent complex sequence

(1) $$|z_n - c| < \epsilon \qquad \text{for all } n > N.$$

c is called the **limit** of the sequence. Then we write

$$\lim_{n\to\infty} z_n = c$$

or simply

$$z_n \to c \qquad (n \to \infty)$$

and we say that the sequence *converges to c* or *has the limit c*.

A sequence which is not convergent is said to *be* **divergent** or to **diverge.**

The condition (1) has a simple geometrical interpretation. It means that each term z_n with $n > N$ lies in the open circular disk of radius ϵ with center at c (Fig. 315), and at most finitely many terms z_n do not lie in that disk, no matter how small the radius ϵ of the disk is chosen. Of course, N will depend on the choice of ϵ, in general.

In the case of a *real* sequence, the condition (1) means geometrically that each term z_n with $n > N$ lies between $c - \epsilon$ and $c + \epsilon$ (Fig. 316), and at most finitely many terms of the sequence do not lie in that interval.

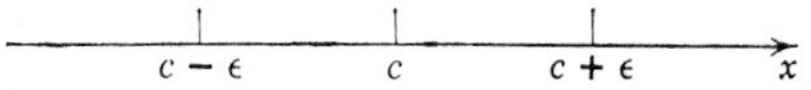

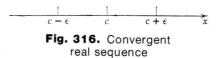

Fig. 316. Convergent real sequence

Example 1. Convergent and divergent sequences

The sequence whose terms are $z_n = 1 + \frac{2}{n}$ is $3, 2, \frac{5}{3}, \frac{6}{4}, \frac{7}{5}, \cdots$. It is convergent and has the limit $c = 1$. In fact, in (1)

$$z_n - c = 1 + \frac{2}{n} - 1 = \frac{2}{n} \quad \text{and} \quad \frac{2}{n} < \epsilon \quad \text{when} \quad \frac{n}{2} > \frac{1}{\epsilon} \quad \text{or} \quad n > \frac{2}{\epsilon}.$$

For example, choosing $\epsilon = 0.01$ we have $2/n < 0.01$ when $n > 200$.

The sequences $1, 2, 3, \cdots$ and $\frac{1}{4}, \frac{3}{4}, \frac{1}{5}, \frac{4}{5}, \frac{1}{6}, \frac{5}{6}, \cdots$ are divergent.

The sequence whose terms are

$$z_n = 2 - \frac{1}{n} + i\left(1 + \frac{2}{n}\right)$$

is

$$1 + 3i, \qquad \tfrac{3}{2} + 2i, \qquad \tfrac{5}{3} + \tfrac{5}{3}i, \qquad \tfrac{7}{4} + \tfrac{3}{2}i, \cdots,$$

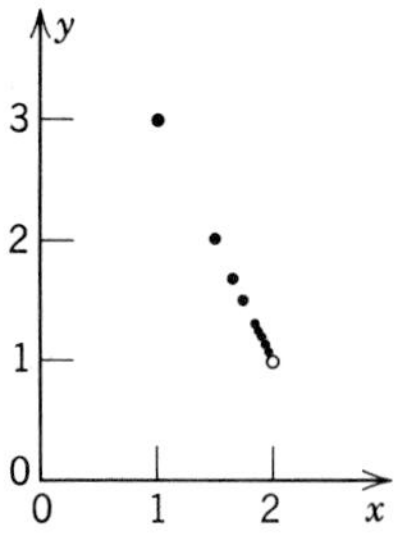

Fig. 317. Last sequence in Example 1

cf. Fig. 317. This sequence is convergent, the limit being $c = 2 + i$. Indeed, in (1) we have

$$|z_n - c| = \left|\frac{2n-1}{n} + i\frac{n+2}{n} - (2+i)\right| = \left|-\frac{1}{n} + \frac{2i}{n}\right| = \frac{\sqrt{5}}{n}$$

$$\text{and} \quad \frac{\sqrt{5}}{n} < \epsilon \quad \text{when} \quad \frac{n}{\sqrt{5}} > \frac{1}{\epsilon} \quad \text{or} \quad n > \frac{\sqrt{5}}{\epsilon} = \frac{2.236 \cdots}{\epsilon}.$$

For instance, choosing $\epsilon = \frac{1}{100}$, we have $|z_n - c| < \epsilon$ when $n = 224, 225, \cdots$. ∎

A complex sequence $z_1, z_2, z_3, \cdots$ being given, we may set $z_n = x_n + iy_n$ and consider the sequence of the real parts and the sequence of the imaginary parts, that is,

$$x_1, x_2, x_3 \cdots \quad \text{and} \quad y_1, y_2, y_3, \cdots.$$

For instance, in the case of the last sequence in Example 1 the two sequences are

$$1, \tfrac{3}{2}, \tfrac{5}{3}, \tfrac{7}{4}, \cdots \quad \text{and} \quad 3, 2, \tfrac{5}{3}, \tfrac{3}{2}, \cdots.$$

We observe that they converge to 2 and 1, respectively; these are the real and imaginary parts of the limit of the given complex sequence. This is typical. Indeed, it illustrates

Theorem 1 (Sequences of the real and imaginary parts)
A sequence $z_1, z_2, \cdots, z_n, \cdots$ of complex numbers $z_n = x_n + iy_n$ $(n = 1, 2, \cdots)$ converges to $c = a + ib$ if and only if the sequence of the real parts $x_1, x_2, \cdots$ converges to a and the sequence of the imaginary parts $y_1, y_2, \cdots$ converges to b.

Proof. If $|z_n - c| < \epsilon$, then $z_n = x_n + iy_n$ is within the circle of radius ϵ about $c = a + ib$ so that necessarily (Fig. 318*a*)

$$|x_n - a| < \epsilon, \qquad |y_n - b| < \epsilon.$$

Thus convergence $z_n \to c$ as $n \to \infty$ implies convergence $x_n \to a$ and $y_n \to b$.
Conversely, if $x_n \to a$ and $y_n \to b$ as $n \to \infty$, then for a given $\epsilon > 0$ we can choose N so large that, for every $n > N$,

$$|x_n - a| < \frac{\epsilon}{2}, \qquad |y_n - b| < \frac{\epsilon}{2}.$$

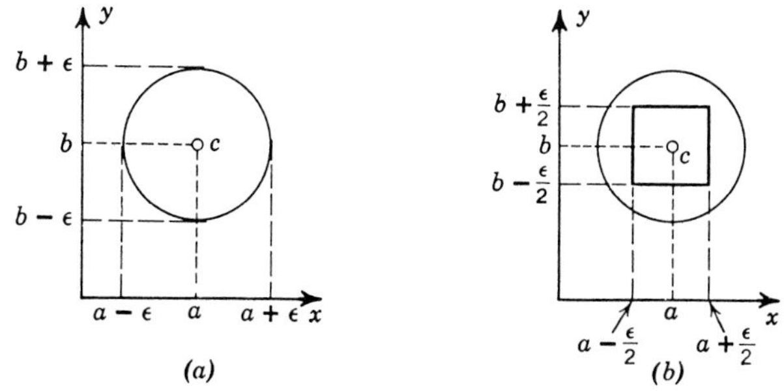

Fig. 318. Proof of Theorem 1

These two inequalities imply that $z_n = x_n + iy_n$ lies in a square with center c and side ϵ. Hence, z_n must lie within a circle of radius ϵ about c (Fig. 318*b*). This completes the proof. ∎

This theorem shows that, by studying the real and imaginary parts, the convergence of complex sequences can be referred back to that of real sequences.

A sequence $z_1, z_2, \cdots$ is said to be **bounded,** if there is a positive number K such that all the terms of the sequence lie in a disk of radius K about the origin, that is

$$|z_n| < K \qquad \text{for all } n.$$

A sequence which is not bounded is said to be **unbounded.**

Using this concept, divergence can often be seen from the following simple

Theorem 2

Every convergent sequence is bounded. Hence if a sequence is unbounded, it diverges.

Proof. Let the sequence $z_1, z_2, \cdots$ be convergent with the limit c. Then we may choose an $\epsilon > 0$ and find a corresponding N such that every z_n with $n > N$ lies in the disk of radius ϵ with center c, and at most finitely many terms may not lie in that disk. It is clear that we can find a circle of radius K about the origin so large that the disk as well as those finitely many terms lie within that circle. This proves that the sequence is bounded. ∎

Note well that boundedness is not sufficient for convergence. For instance, the sequence $1, 0, 1, 0, \cdots$ is bounded but diverges. (Why?) Unbounded sequences are

$$1, 2, 3, 4, \cdots \qquad \text{and} \qquad \tfrac{1}{2}, 2, \tfrac{1}{3}, 3, \tfrac{1}{4}, 4, \cdots.$$

Hence they diverge, by Theorem 2.

Problems for Sec. 15.1

Find and plot the first few terms of the sequence $z_1, z_2, \cdots, z_n, \cdots$ where z_n equals

1. $n/(n + 3)$ **2.** $2n/(n^2 + 1)$ **3.** i^n/n^3

4. $in/(n + 1)$ **5.** $i^n n^2/(n + i)$ **6.** $(-1)^n + 2n\pi i$

7. Find the first few terms and the limit of the sequence $z_1 = 1$, $z_2 = i/2$, $z_n = iz_{n-2}z_{n-1}$ $(n = 3, 4, \cdots)$.

Are the following sequences $z_1, z_2, \cdots, z_n, \cdots$ bounded? Convergent? In the case of convergence find the limit.

8. $z_n = i^n$ **9.** $z_n = i^n/n$ **10.** $z_n = in/(n+1)$

11. $z_n = n^2/(n+i)$ **12.** $z_n = (-1)^n/n^3$ **13.** $z_n = e^{in\pi/4}$

14. **(Uniqueness of limit)** Show that if a sequence converges, its limit is unique.

15. Prove convergence in Prob. 3, proceeding as in Example 1.

16. Illustrate Theorem 1 by the sequence whose terms are given by the formula $z_n = (n^2 - 1)/(2n^2 + 1) + in/(n+2)$.

17. Show that a complex sequence $z_1, z_2, \cdots$ is bounded if and only if the two corresponding sequences of the real parts and the imaginary parts are bounded.

18. If the sequence $z_1, z_2, \cdots$ is convergent with the limit 0, and if the sequence $b_1, b_2, \cdots$ is such that $|b_n| \leqq K|z_n|$ for some fixed $K > 0$ and all n, show that the sequence $b_1, b_2, \cdots$ is convergent with the limit 0.

19. If $z_1, z_2, \cdots$ converges with the limit l and $z_1^*, z_2^*, \cdots$ converges with the limit l^*, show that $z_1 + z_1^*, z_2 + z_2^*, \cdots$ converges with the limit $l + l^*$.

20. Show that under the assumptions of Prob. 19 the sequence $z_1z_1^*, z_2z_2^*, \cdots$ converges with the limit ll^*.

15.2 Series

Let $w_1, w_2, \cdots, w_m, \cdots$ be a sequence of numbers, complex or real. Then we may consider the *infinite series* or, briefly, **series**

$$\sum_{m=1}^{\infty} w_m = w_1 + w_2 + w_3 + \cdots. \tag{1}$$

The w_m are called the **terms** of the series. The sum of the first n terms is

$$s_n = w_1 + w_2 + \cdots + w_n. \tag{2}$$

This expression is called the nth **partial sum** of the series (1). Clearly if we omit the terms of s_n from (1), the remaining expression is

$$R_n = w_{n+1} + w_{n+2} + w_{n+3} + \cdots. \tag{3}$$

This is called the **remainder** *of the series* (1) *after the n-th term.*

In this way we have now associated with the series (1) the sequence of its partial sums $s_1, s_2, s_3, \cdots$. If this sequence is convergent, say,

$$\lim_{n\to\infty} s_n = s,$$

then the series (1) is said to **converge** or to *be convergent,* the number s is called its **value** or *sum,* and we write

$$s = \sum_{m=1}^{\infty} w_m = w_1 + w_2 + \cdots.$$

If the sequence of the partial sums diverges, the series (1) is said to **diverge** or to *be divergent.*

If the series (1) converges and has the value s, then

$$s = s_n + R_n \qquad \text{or} \qquad R_n = s - s_n. \tag{4}$$

And from the definition of the convergence of a sequence it follows that $|R_n|$ can be made as small as we please, by taking n large enough. In many cases it will be impossible to find the sum s of a convergent series. Then for computational purposes we must use a partial sum s_n as an approximation of s, and by estimating the remainder R_n we may obtain information about the degree of accuracy of the approximation.

Example 1. Convergent and divergent series

The series

$$\sum_{m=1}^{\infty} \frac{1}{2^m} = \frac{1}{2} + \frac{1}{4} + \frac{1}{8} + \cdots$$

converges and has the value 1, because

$$s_n = \frac{1}{2} + \frac{1}{4} + \cdots + \frac{1}{2^n} = 1 - \frac{1}{2^n} \qquad \text{and} \qquad \lim_{n\to\infty} s_n = 1.$$

The series

$$\sum_{m=1}^{\infty} m = 1 + 2 + 3 + \cdots$$

diverges. The series

$$\sum_{m=0}^{\infty} (-1)^m = 1 - 1 + 1 - + \cdots$$

diverges, because

$$s_0 = 1, \qquad s_1 = 1 - 1 = 0, \qquad s_2 = 1 - 1 + 1 = 1, \qquad \text{etc.,}$$

and the sequence 1, 0, 1, 0, $\cdots$ is divergent.

The **harmonic series**

$$\sum_{m=1}^{\infty} \frac{1}{m} = 1 + \frac{1}{2} + \frac{1}{3} + \cdots$$

diverges. In fact,

$$s_n = 1 + \frac{1}{2} + \cdots + \frac{1}{n},$$

and s_n equals the sum of the areas of the n rectangles in Fig. 319 on the next page. This area is greater than the area A_n under the corresponding portion of the curve $y = 1/x$, and

$$A_n = \int_1^{n+1} \frac{dx}{x} = \ln(n+1) \to \infty \qquad \text{as} \qquad n \to \infty.$$

Since $s_n > A_n$, it follows that $s_n \to \infty$ as $n \to \infty$, which means divergence. ∎

From Theorem 1 in the preceding section we immediately obtain the corresponding theorem for series:

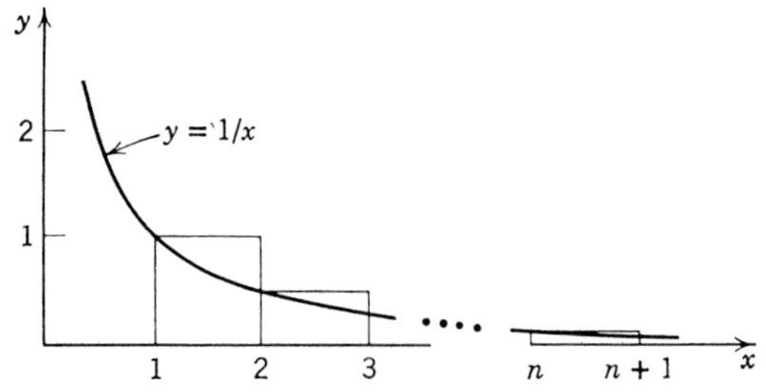

Fig. 319. Example 1

Theorem 1 (Series of the real and imaginary parts)
Let $w_m = u_m + iv_m$. *Then the series*

$$\sum_{m=1}^{\infty} w_m = w_1 + w_2 + w_3 + \cdots$$

converges and has the value $s = a + ib$ *if and only if the series of the real parts and the series of the imaginary parts, that is*

$$\sum_{m=1}^{\infty} u_m = u_1 + u_2 + \cdots \qquad \text{and} \qquad \sum_{m=1}^{\infty} v_m = v_1 + v_2 + \cdots,$$

converge and have the values a and b, respectively.

We see that this theorem gives a relation between complex and real series. A more important relation is based upon the following concept.

A series $w_1 + w_2 + \cdots$ is said to be **absolutely convergent** if the corresponding series

$$\sum_{m=1}^{\infty} |w_m| = |w_1| + |w_2| + \cdots \tag{5}$$

(whose terms are real and nonnegative) converges.

If the series $w_1 + w_2 + \cdots$ converges but (5) diverges, then the series is called, more precisely, *conditionally convergent.*

Example 2. Absolutely and conditionally convergent series

The series

$$\tfrac{1}{2} - \tfrac{1}{4} + \tfrac{1}{8} - \tfrac{1}{16} + - \cdots$$

is absolutely convergent since the corresponding series (5) is convergent (cf. Example 1). The series

$$1 - \tfrac{1}{2} + \tfrac{1}{3} - \tfrac{1}{4} + - \cdots$$

is only conditionally convergent, because the series itself converges (by the familiar Leibniz test for real series, to be reviewed in Sec. 15.4), but the corresponding series (5) is the harmonic series, which diverges (cf. Example 1). ∎

The following property of an absolutely convergent series is almost obvious.

Theorem 2
If a series $w_1 + w_2 + \cdots$ is absolutely convergent, it is convergent.

We shall obtain a simple proof of this theorem from the so-called Cauchy convergence principle at the end of the next section.

We finally state another simple theorem which is often quite useful.

Theorem 3
If a series $w_1 + w_2 + \cdots$ converges, then

$$\lim_{m\to\infty} w_m = 0. \tag{6}$$

Hence a series which does not satisfy (6) *diverges.*

Proof. Let $w_1 + w_2 + \cdots$ be convergent with the sum s. Then

$$w_{n+1} = s_{n+1} - s_n,$$

and

$$\lim_{n\to\infty} (s_{n+1} - s_n) = \lim_{n\to\infty} s_{n+1} - \lim_{n\to\infty} s_n = s - s = 0. \quad \blacksquare$$

Note well that (6) is only necessary for convergence but not sufficient. In fact, the harmonic series in Example 1 satisfies (6) but diverges. The second and third series in Example 1 do not satisfy (6); hence they diverge.

15.3 Cauchy Convergence Principle for Sequences and Series

Before we use a sequence or a series, we must find out whether it converges.[1] A direct application of the definition may be difficult since in most cases we do not know the limit in advance. The Cauchy convergence principle overcomes this difficulty, since it is a criterion for convergence which can be applied without knowing that limit.

In the proof of the Cauchy convergence principle we shall use the famous Bolzano–Weierstrass theorem (Theorem 1, below), and for stating the latter we shall need the following concept, which is of general interest.

A point a is called a **limit point** of a sequence $z_1, z_2, \cdots$ if, given an $\epsilon > 0$ (no matter how small),

$$|z_n - a| < \epsilon \qquad \text{for infinitely many } n. \tag{1}$$

[1] A reader with primary interest in complex analytic functions and some experience in real sequences and series is advised to read again the introduction to this chapter (Chap. 15) about the arrangement of the material. The reader may then perhaps continue immediately with Chap. 16 on power series, Taylor series and Laurent series, using Secs. 15.3–15.6 only for reference when needed.

Geometrically speaking, this means that infinitely many terms of the sequence lie within the circle of radius ϵ about a, no matter how small the radius ϵ is chosen.

Note that if (1) holds, there may still be infinitely many terms which do not lie within that circle, and the sequence may not be convergent. In fact, if a sequence converges, its limit is a limit point (why?) and is the only limit point of the sequence. And if a sequence has more than one limit point, it diverges.

Note further that a number which appears infinitely often in a sequence is a limit point of that sequence, by the definition.

Let us illustrate the situation. Recall that boundedness was defined near the end of Sec. 15.1.

Example 1. Limit points, convergence, boundedness

Sequence	Limit points at:	Convergent or divergent	Bounded or unbounded
$1, 2, 3, \cdots$	(none)	divergent	unbounded
$\frac{1}{2}, \frac{2}{3}, \frac{3}{4}, \frac{4}{5}, \cdots$	1	convergent	bounded
$\frac{1}{2}, 2, \frac{1}{3}, 3, \frac{1}{4}, 4, \cdots$	0	divergent	unbounded
$\frac{1}{4}, \frac{3}{4}, \frac{1}{5}, \frac{4}{5}, \frac{1}{6}, \frac{5}{6}, \cdots$	0 and 1	divergent	bounded

We observe that the two bounded sequences in this example have limit points. This illustrates the important

Theorem 1 (Bolzano[2] and Weierstrass[3])

A bounded infinite sequence $z_1, z_2, z_3, \cdots$ in the complex plane has at least one limit point.

Proof. It is obvious that we need both conditions: a finite sequence cannot have a limit point, and the sequence $1, 2, 3, \cdots$, which is infinite but not bounded, has no limit point. To prove the theorem, consider a bounded infinite sequence $z_1, z_2, \cdots$ and let K be such that $|z_n| < K$ for all n. If only finitely many values of the z_n are different, then, since the sequence is infinite, some number z must occur infinitely many times in the sequence, and, by definition, this number is a limit point of the sequence.

We may now turn to the case when the sequence contains infinitely many different terms. We draw the large square Q_0 in Fig. 320 which contains all z_n. We subdivide Q_0 into four congruent squares. Clearly, at least one of these squares (each taken with its complete boundary) must contain infinitely many terms of the sequence. The square of this type with the lowest number (1, 2, 3, or 4) will be denoted by Q_1. This is the first step. In the next step we subdivide Q_1 into four congruent squares and select a square Q_2 according to the same rule, and so on. This yields an infinite sequence of squares $Q_0, Q_1, Q_2, \cdots, Q_n, \cdots$ with the property that the side of Q_n approaches zero as n approaches infinity, and Q_m contains all Q_n with $n > m$. It is not difficult to see that the number

[2] BERNHARD BOLZANO (1781–1848), German mathematician, a pioneer in the study of point sets.

[3] Cf. footnote 8 in Sec. 12.5.

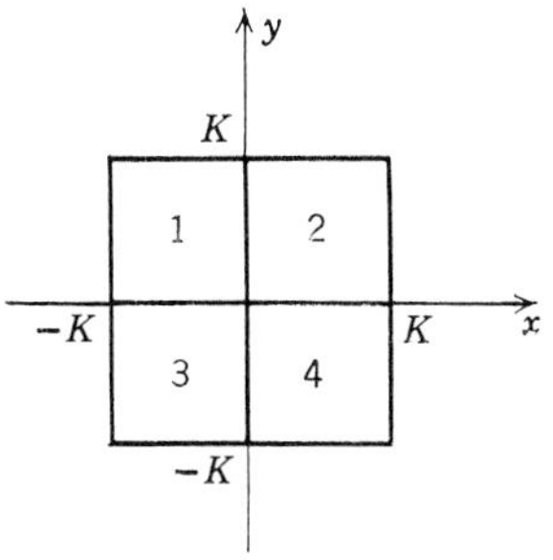

Fig. 320. Proof of Theorem 1

which belongs to all these squares,[4] call it $z = a$, is a limit point of the sequence. In fact, given an $\epsilon > 0$, we can choose an N so large that the side of the square Q_N is less than ϵ and, since Q_N contains infinitely many z_n we have $|z_n - a| < \epsilon$ for infinitely many n. This completes the proof. ∎

We are now ready to state the main theorem of this section:

Theorem 2 (Cauchy convergence principle for sequences)
A sequence $z_1, z_2, z_3, \cdots$ is convergent if and only if for every positive number ϵ we can find a number N (which may depend on ϵ) such that

$$|z_m - z_n| < \epsilon \qquad \text{when } m > N,\ n > N; \tag{2}$$

(that is, any two terms z_m, z_n with $m > N$, $n > N$ must have a distance of less than ϵ from each other).

Proof. **(a)** Let the sequence $z_1, z_2, \cdots$ be convergent with the limit c. Then, for a given $\epsilon > 0$, we can find an N such that

$$|z_n - c| < \frac{\epsilon}{2} \qquad \text{for every } n > N.$$

Hence, when $m > N$, $n > N$, then by the triangle inequality

$$|z_m - z_n| = |(z_m - c) - (z_n - c)| \leqq |z_m - c| + |z_n - c| < \frac{\epsilon}{2} + \frac{\epsilon}{2} = \epsilon,$$

which shows that, if the sequence converges, then (2) holds.

(b) Conversely, consider a sequence $z_1, z_2, \cdots$ for which (2) holds. We first show that the sequence is bounded. In fact, choose a fixed ϵ and a fixed $n = n_0 > N$ in (2). Then (2) implies that every z_m with $m > N$ lies in a disk of radius ϵ about z_{n_0}, and only finitely many terms of the sequence do not lie in that disk. It is clear that we can find a circle about the origin so large that the disk and those finitely many z_n lie within the circle. This shows that the

[4]The fact that such a unique number $z = a$ exists seems to be obvious, but it actually follows from an axiom of the real number system, the so-called *Cantor–Dedekind axiom*. Cf. footnote 5 in the next section.

sequence is bounded, and from the Bolzano-Weierstrass theorem we conclude that it has at least one limit point, call it L.

We show that the sequence is convergent with the limit L. The definition of a limit point implies that, an $\epsilon > 0$ being given, $|z_n - L| < \epsilon/2$ for infinitely many n. Since (2) holds for any $\epsilon > 0$, when an $\epsilon > 0$ is given we can find an N^* such that $|z_m - z_n| < \epsilon/2$ for $m > N^*$, $n > N^*$. Choose a fixed $n > N^*$ such that $|z_n - L| < \epsilon/2$, and let m be any integer greater than N^*. Then, by the triangle inequality,

$$|z_m - L| = |(z_m - z_n) + (z_n - L)| \leqq |z_m - z_n| + |z_n - L| < \frac{\epsilon}{2} + \frac{\epsilon}{2} = \epsilon,$$

that is, $|z_m - L| < \epsilon$ for all $m > N^*$; by definition, this means that the sequence is convergent with the limit L. This completes the proof. ∎

Given a series $w_1 + w_2 + \cdots$, we may apply the present theorem to the sequence of the partial sums s_n. Then the inequality (2) becomes

$$|s_m - s_n| < \epsilon \qquad (m > N, n > N)$$

or, if we set $m = n + p$,

$$|s_{n+p} - s_n| < \epsilon \qquad (n > N, p = 1, 2, \cdots).$$

Now by the definition of a partial sum,

$$s_{n+p} - s_n = w_{n+1} + w_{n+2} + \cdots + w_{n+p}.$$

This proves the basic

Theorem 3 (Cauchy convergence principle for series)
A series $w_1 + w_2 + \cdots$ is convergent if and only if for every given $\epsilon > 0$ (no matter how small) we can find an N (which will depend on ϵ, in general) such that

$$|w_{n+1} + w_{n+2} + \cdots + w_{n+p}| < \epsilon \qquad \textit{for every } n > N \textit{ and } p = 1, 2, \cdots.$$

As a first application of this important theorem let us prove a result already stated above (Theorem 2 in the previous section):

Theorem 4
If a series $w_1 + w_2 + \cdots$ is absolutely convergent, it is convergent.

Proof. By the generalized triangle inequality (11) in Sec. 12.2 we have

$$(3) \qquad |w_{n+1} + \cdots + w_{n+p}| \leqq |w_{n+1}| + |w_{n+2}| + \cdots + |w_{n+p}|.$$

Since, by assumption, the series $|w_1| + |w_2| + \cdots$ converges, it follows from Theorem 3 that the right-hand side of (3) becomes less than any given $\epsilon > 0$ for each $n > N$ (N sufficiently large) and $p = 1, 2, \cdots$. Hence the same is true for the left-hand side of (3) and, by the same theorem, this proves convergence of the series $w_1 + w_2 + \cdots$. ∎

Problems for Sec. 15.3

Are the following sequences $z_1, z_2, \cdots, z_n, \cdots$ bounded? Convergent? Find their limit points.

1. $z_n = (2i)^n$ **2.** $z_n = 1 + i^n$ **3.** $z_n = (-1)^n + i/n$
4. $z_n = e^{in\pi/2}$ **5.** $z_n = i^n \cos n\pi$ **6.** $z_n = i^n \cosh n\pi$
7. $z_n = (1 - i)^n$ **8.** $z_n = (1 + i)^{2n}/n$ **9.** $z_n = (3 + 4i)^n/n!$
10. $z_n = \pi i + \sin n\pi$ **11.** $z_n = (\cos n\pi)/\sqrt{n}$ **12.** $z_n = i^n n^2$
13. $z_1 = 1,\ z_2 = 2,\ z_3 = 3,\ z_n = z_{n-3} - z_{n-2} + z_{n-1}\ (n = 4, 5, \cdots)$
14. $z_1 = 1/3,\ z_2 = 1/4,\ z_n = z_{n-2}/z_{n-1}\ (n = 3, 4, \cdots)$
15. $z_1 = 1,\ z_2 = i,\ z_n = z_{n-2}z_{n-1}\ (n = 3, 4, \cdots)$

15.4 Monotone Real Sequences. Leibniz Test for Real Series

This section contains two theorems for real sequences and series which do not have analogues for complex sequences and series. Both theorems are of considerable practical interest.

A **real** sequence $x_1, x_2, \cdots, x_n, \cdots$ is said to be **monotone increasing** if

$$x_1 \leqq x_2 \leqq x_3 \leqq \cdots .$$

Similarly, the sequence is said to be **monotone decreasing** if

$$x_1 \geqq x_2 \geqq x_3 \geqq \cdots .$$

A sequence which is either monotone increasing or monotone decreasing is called a **monotone sequence.**

For instance, the divergent sequence $1, 2, 3, \cdots$ is monotone and unbounded. The convergent sequence $\frac{1}{2}, \frac{2}{3}, \frac{3}{4}, \cdots$ is monotone and bounded, and we shall prove that these two properties are sufficient for convergence:

Theorem 1 (Convergence of a real sequence)
If a real sequence is bounded and monotone, it converges.

Proof. Let $x_1, x_2, \cdots$ be a bounded monotone increasing sequence. Then its terms are smaller than some number B and, since $x_1 \leqq x_n$ for all n, they lie in the interval $x_1 \leqq x_n \leqq B$, which will be denoted by I_0. We bisect I_0, that is, we subdivide it into two parts of equal length. If the right half (together with its endpoints) contains terms of the sequence, we denote it by I_1. If it does not contain terms of the sequence, then the left half of I_0 (together with its endpoints) is called I_1. This is the first step.

In the second step we bisect I_1, select one half by the same rule, and call it I_2, and so on (see Fig. 321 on the next page).

In this way we obtain shorter and shorter intervals $I_0, I_1, I_2, \cdots$ with the following properties. Each I_m contains all I_n for $n > m$. No term of the sequence

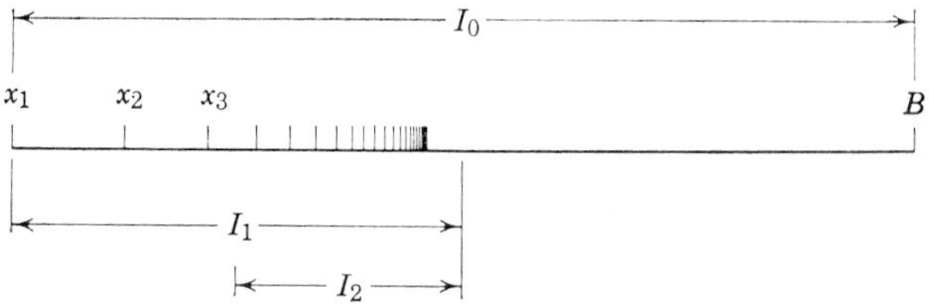

Fig. 321. Proof of Theorem 1

lies to the right of I_m, and, since the sequence is monotone increasing, all x_n with n greater than some number N lie in I_m; of course, N will depend on m, in general. The lengths of the I_m approach zero as m approaches infinity. Hence there is precisely one number, call it L, which lies in all those intervals,[5] and we may now easily prove that the sequence is convergent with the limit L.

In fact, given an $\epsilon > 0$, we choose an m such that the length of I_m is less than ϵ. Then L and all the x_n with $n > N(m)$ lie in I_m, and, therefore, $|x_n - L| < \epsilon$ for all those n. This completes the proof for an increasing sequence. For a decreasing sequence the proof is the same, except for a suitable interchange of "left" and "right" in the construction of those intervals. ∎

We shall now prove another practically useful theorem, which concerns real series whose terms have alternating sign and decrease in absolute value. The theorem states conditions sufficient for convergence and gives estimates for the remainders.

Theorem 2 (Leibniz test for real series)

Let $u_1, u_2, \cdots$ be real and

$$(1) \qquad \text{(a)} \quad u_1 \geqq u_2 \geqq u_3 \geqq \cdots, \qquad \text{(b)} \quad \lim_{m \to \infty} u_m = 0.$$

Then the series

$$u_1 - u_2 + u_3 - u_4 + - \cdots$$

converges, and for the remainder R_n after the nth term we have the estimate

$$(2) \qquad |R_n| \leqq u_{n+1}.$$

Proof. Let s_n be the nth partial sum of the series. Then, because of (1a),

$$s_1 = u_1, \qquad s_2 = u_1 - u_2 \leqq s_1,$$

$$s_3 = s_2 + u_3 \geqq s_2, \qquad s_3 = s_1 - (u_2 - u_3) \leqq s_1,$$

[5]This statement seems to be obvious, but actually it is not; it may be regarded as an axiom of the real number system in the following form. Let $J_1, J_2, \cdots$ be closed intervals such that each J_m contains all J_n with $n > m$, and the lengths of the J_m approach zero as m approaches infinity. Then there is precisely one real number which is contained in all those intervals. This is the so-called **Cantor-Dedekind Axiom,** named after the German mathematicians, GEORG CANTOR (1845–1918), the creator of set theory, and RICHARD DEDEKIND (1831–1916). For further details see Ref. [A2] in Appendix 1. (An interval I is said to be **closed** if its two endpoints are regarded as points belonging to I. It is said to be **open** if the endpoints are not regarded as points of I.)

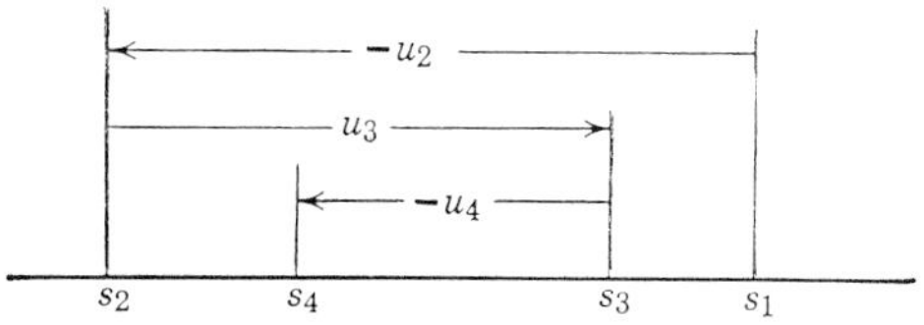

Fig. 322. Proof of Theorem 2 (Leibniz test)

so that $s_2 \leqq s_3 \leqq s_1$. Proceeding in this fashion, we conclude that (Fig. 322)

$$s_1 \geqq s_3 \geqq s_5 \geqq \cdots \geqq s_6 \geqq s_4 \geqq s_2 \tag{3}$$

which shows that the odd partial sums form a bounded monotone sequence, and so do the even partial sums. Hence, by Theorem 1, both sequences converge, say,

$$\lim_{n\to\infty} s_{2n+1} = s, \qquad \lim_{n\to\infty} s_{2n} = s^*.$$

Now, since $s_{2n+1} - s_{2n} = u_{2n+1}$, we readily see that (1b) implies

$$s - s^* = \lim_{n\to\infty} s_{2n+1} - \lim_{n\to\infty} s_{2n} = \lim_{n\to\infty} (s_{2n+1} - s_{2n}) = \lim_{n\to\infty} u_{2n+1} = 0.$$

Hence $s^* = s$, and the series converges with the sum s.

We shall now prove the estimate (2) for the remainder. Since $s_n \to s$, it follows from (3) that

$$s_{2n+1} \geqq s \geqq s_{2n} \qquad \text{and also} \qquad s_{2n-1} \geqq s \geqq s_{2n}.$$

By subtracting s_{2n} and s_{2n-1}, respectively, we obtain

$$s_{2n+1} - s_{2n} \geqq s - s_{2n} \geqq 0, \qquad 0 \geqq s - s_{2n-1} \geqq s_{2n} - s_{2n-1}.$$

In these inequalities, the first expression is equal to u_{2n+1}, the last is equal to $-u_{2n}$, and the expressions between the inequality signs are the remainders R_{2n} and R_{2n-1}. Thus the inequalities may be written

$$u_{2n+1} \geqq R_{2n} \geqq 0, \qquad 0 \geqq R_{2n-1} \geqq -u_{2n}$$

and we see that they imply (2). This completes the proof. ∎

Problems for Sec. 15.4

Are the following real sequences bounded or not, convergent or not, monotone or not? Determine their limit points.

1. $1, \frac{1}{3}, \frac{1}{9}, \frac{1}{27}, \cdots$

2. $2, -\frac{1}{2}, 3, -\frac{1}{3}, 4, -\frac{1}{4}, \cdots$

3. $2, 2^2, 2^3, 2^4, \cdots$

4. $1, \sqrt{2}, \sqrt[3]{3}, \sqrt[4]{4}, \cdots$

5. $\frac{7}{6}, \frac{11}{6}, \frac{8}{7}, \frac{13}{7}, \frac{9}{8}, \frac{15}{8}, \cdots$

6. $\ln 1, \ln 2, \ln 3, \cdots$

7. $4/1!, 4^2/2!, 4^3/3!, \cdots$

8. $a, a^2, a^3, \cdots$

9. $c, 2c^2, 3c^3, \cdots \quad (|c| < 1)$

10. $c, 2^2c^2, 3^2c^3, 4^2c^4, \cdots \quad (|c| < 1)$

Are the following series convergent or divergent?

11. $1 - \frac{1}{3} + \frac{1}{9} - \frac{1}{27} + - \cdots$

12. $1 - \frac{1}{\sqrt{2}} + \frac{1}{\sqrt{3}} - \frac{1}{\sqrt{4}} + - \cdots$

13. $\frac{3}{2} - \frac{4}{3} + \frac{5}{4} - \frac{6}{5} + - \cdots$

14. $\frac{1}{\ln 2} + \frac{1}{\ln 3} + \frac{1}{\ln 4} + \cdots$

Show that the following series converge. How many terms are needed for computing the sum s with an error less than 0.01?

15. $s = \frac{1}{2} - \frac{1}{4} + \frac{1}{8} - + \cdots$

16. $s = 1 - \frac{1}{3} + \frac{1}{6} - \frac{1}{12} + \frac{1}{24} - + \cdots$

17. $s = 1 - \frac{1}{3} + \frac{1}{9} - \frac{1}{27} + - \cdots$

18. $s = 1 - 1 + \frac{1}{2!} - \frac{1}{3!} + \frac{1}{4!} - + \cdots$

19. $s = 1 - \frac{1}{3!} + \frac{1}{5!} - \frac{1}{7!} + - \cdots$

20. $s = 1 - \frac{1}{2!} + \frac{1}{4!} - \frac{1}{6!} + - \cdots$

15.5 Tests for Convergence and Divergence of Series

Before we use an infinite series for computational or other purposes we must know whether it converges. In most cases that arise in engineering mathematics, this question may be answered by applying one of the various tests for convergence and divergence. These tests, therefore, are of great practical interest.

A simple test for divergence (Theorem 3 in Sec. 15.2) and the so-called Leibniz test for real series were already considered. The following theorem is a source of various criteria for convergence.

Theorem 1 (Comparison test)

If a series $w_1 + w_2 + \cdots$ is given and we can find a converging series $b_1 + b_2 + \cdots$ with nonnegative real terms such that

$$|w_n| \leqq b_n \qquad \text{for } n = 1, 2, \cdots, \tag{1}$$

then the given series is absolutely convergent.

Proof. Since the series $b_1 + b_2 + \cdots$ converges, it follows from Theorem 3 in Sec. 15.3 that for any given $\epsilon > 0$ we can find an N such that

$$b_{n+1} + \cdots + b_{n+p} < \epsilon \qquad \text{for every } n > N \text{ and } p = 1, 2, \cdots.$$

From this and (1) we conclude that

$$|w_{n+1}| + \cdots + |w_{n+p}| \leqq b_{n+1} + \cdots + b_{n+p} < \epsilon$$

for those n and p. Consequently, from Theorem 3 in Sec. 15.3 it follows that the series $|w_1| + |w_2| + \cdots$ is convergent, and the given series converges absolutely. This completes the proof. ∎

For deriving two important tests from Theorem 1 let us prove

Theorem 2 (Geometric series)

The geometric series

$$\sum_{m=0}^{\infty} q^m = 1 + q + q^2 + \cdots$$

converges with the sum $1/(1 - q)$ *when* $|q| < 1$ *and diverges when* $|q| \geqq 1$.

Proof. When $|q| \geqq 1$, then $|q^n| \geqq 1$ and divergence follows from Theorem 3 in Sec. 15.2. Now let $|q| < 1$. The nth partial sum is

$$s_n = 1 + q + \cdots + q^n.$$

From this,

$$qs_n = \quad q + \cdots + q^n + q^{n+1}.$$

By subtraction,

$$s_n - qs_n = (1 - q)s_n = 1 - q^{n+1}$$

since the other terms cancel in pairs. Now $1 - q \neq 0$ because $q \neq 1$, and we may solve for s_n, finding

$$s_n = \frac{1 - q^{n+1}}{1 - q} = \frac{1}{1 - q} - \frac{q^{n+1}}{1 - q}. \tag{2}$$

Since $|q| < 1$, the last term approaches zero as $n \to \infty$. Hence the series is convergent and has the value $1/(1 - q)$. This completes the proof. ∎

From Theorems 1 and 2 we shall now derive two important tests, the ratio test and the root test.

Theorem 3 (Ratio test)

Consider the series

$$w_1 + w_2 + w_3 + \cdots .$$

Suppose that $w_n \neq 0$ *for* $n = 1, 2, \cdots$ *and the sequence of the ratios*

$$\left|\frac{w_{n+1}}{w_n}\right| \qquad n = 1, 2, \cdots$$

converges with the limit L. Then the given series is

absolutely convergent, if $L < 1$,

divergent, if $L > 1$.

(*If* $L = 1$, *the test fails*).

Proof. By assumption

$$\lim_{n\to\infty} k_n = L \qquad \text{where} \qquad k_n = \left|\frac{w_{n+1}}{w_n}\right|.$$

Clearly, k_n and L are real. By the definition of a limit, for a given $\epsilon > 0$ we can find an N such that every k_n with $n > N$ lies between $L - \epsilon$ and $L + \epsilon$, that is,

$$\text{(3)} \qquad \text{(a)} \quad k_n < L + \epsilon \qquad \text{(b)} \quad k_n > L - \epsilon \qquad (n > N).$$

We first consider the case $L < 1$. We set $L + \epsilon = q$ and choose $\epsilon = (1 - L)/2$. Then $\epsilon > 0$, and (3a) becomes

$$k_n < q = L + \frac{1 - L}{2} = \frac{1 + L}{2}.$$

Since $L < 1$, we have $q < 1$. Now we can write

$$\text{(4)} \qquad \begin{aligned} &|w_{N+1}| + |w_{N+2}| + |w_{N+3}| + \cdots \\ &\quad = |w_{N+1}|\left(1 + \left|\frac{w_{N+2}}{w_{N+1}}\right| + \left|\frac{w_{N+3}}{w_{N+2}}\right|\left|\frac{w_{N+2}}{w_{N+1}}\right| + \cdots\right) \\ &\quad = |w_{N+1}|(1 + k_{N+1} + k_{N+2}k_{N+1} + k_{N+3}k_{N+2}k_{N+1} + \cdots). \end{aligned}$$

Since $k_n < q < 1$, each term of this series is less than the corresponding term of the geometric series

$$|w_{N+1}|(1 + q + q^2 + q^3 + \cdots).$$

By Theorem 2 this series converges because $q < 1$. From Theorem 1 it follows that the series in (4) is convergent. Hence the series $|w_1| + |w_2| + \cdots$ is convergent. This implies that the series $w_1 + w_2 + \cdots$ is absolutely convergent.

We now consider the case $L > 1$. We choose $\epsilon = (L - 1)/2$. Clearly, $\epsilon > 0$, and (3b) becomes

$$k_n > L - \epsilon = \frac{1 + L}{2} > 1 \qquad (n > N),$$

that is,

$$k_n = \left|\frac{w_{n+1}}{w_n}\right| > 1 \qquad \text{or} \qquad |w_{n+1}| > |w_n| \qquad (n > N).$$

The last inequality shows that the terms are increasing in absolute value. From this and Theorem 3, Sec. 15.2, it follows that the series is divergent.

For $L = 1$ the series may converge or diverge so that the test fails. This is illustrated by the harmonic series (Example 1 in Sec. 15.2)

$$\sum_{n=1}^{\infty} \frac{1}{n} = 1 + \frac{1}{2} + \frac{1}{3} + \cdots$$

which diverges and for which

$$\frac{w_{n+1}}{w_n} = \frac{n}{n+1} \quad \to \quad L = 1 \qquad \text{as } n \to \infty,$$

and by the series

$$\sum_{n=1}^{\infty} \frac{1}{n^2} = 1 + \frac{1}{4} + \frac{1}{9} + \cdots \tag{5}$$

which converges and for which

$$\frac{w_{n+1}}{w_n} = \frac{n^2}{(n+1)^2} \quad \to \quad L = 1 \qquad \text{as } n \to \infty.$$

Convergence of (5) may be shown as follows. The nth partial sum is

$$s_n = 1 + \frac{1}{4} + \frac{1}{9} + \cdots + \frac{1}{n^2}.$$

Clearly, $s_n > 0$ and (Fig. 323)

$$s_n \leqq 1 + \int_1^n \frac{dx}{x^2} = 2 - \frac{1}{n},$$

which shows that the sequence of the partial sums is bounded. Since the terms of the series are positive, the sequence is monotone increasing, and convergence follows from Theorem 1 in the preceding section. ∎

The following test is more general than the ratio test, but its application is often more difficult.

Theorem 4 (Root test)
Consider the series

$$w_1 + w_2 + w_3 + \cdots.$$

Suppose that the sequence of the roots

$$\sqrt[n]{|w_n|} \qquad n = 1, 2, \cdots$$

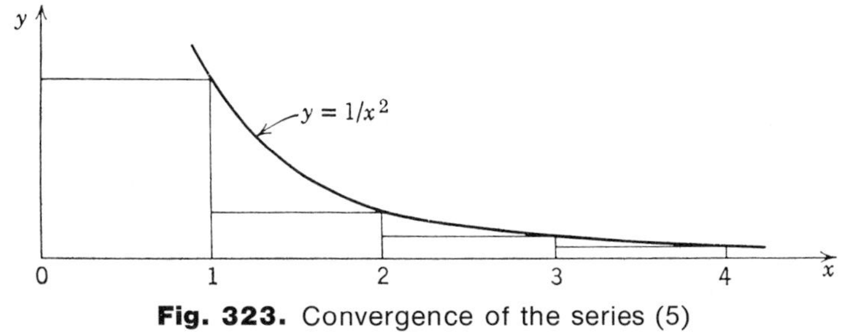

Fig. 323. Convergence of the series (5)

is convergent with the limit L. Then the given series is

$$\textit{absolutely convergent, if } L < 1,$$

$$\textit{divergent, if } L > 1.$$

(*If* $L = 1$, *the test fails*).

Proof. If $L < 1$, then, as in the proof for the ratio test, we can choose $q < 1$ and find a corresponding N such that

$$k_n^* \equiv \sqrt[n]{|w_n|} < q < 1 \qquad \text{for every } n > N.$$

From this we obtain

$$|w_n| < q^n < 1 \qquad (n > N),$$

and the series $|w_N| + |w_{N+1}| + \cdots$ converges by comparison with the geometric series. Hence, the series $w_1 + w_2 + \cdots$ converges absolutely.

If $L > 1$, then $\sqrt[n]{|w_n|} > 1$ for all sufficiently large n. Hence $|w_n| > 1$ for those n, and divergence follows from Theorem 3 in Sec. 15.2.

If $L = 1$, the test fails. This is illustrated by the harmonic series for which

$$\sqrt[n]{\frac{1}{n}} = \frac{1}{n^{1/n}} = \frac{1}{e^{(1/n)\ln n}} \to \frac{1}{e^0} = 1 \qquad (n \to \infty)$$

because $(1/n) \ln n \to 0$, and the series (5), for which

$$\sqrt[n]{\frac{1}{n^2}} = \frac{1}{n^{2/n}} = \frac{1}{e^{(2/n)\ln n}} \to \frac{1}{e^0} = 1 \qquad (n \to \infty)$$

because $(2/n) \ln n \to 0$ as $n \to \infty$. ∎

Example 1. Application of the ratio and root tests

Test the series

$$\sum_{n=1}^{\infty} \frac{n^2}{2^n} = \frac{1}{2} + 1 + \frac{9}{8} + 1 + \frac{25}{32} + \cdots.$$

We have

$$w_n = \frac{n^2}{2^n}, \qquad w_{n+1} = \frac{(n+1)^2}{2^{n+1}}, \qquad \frac{w_{n+1}}{w_n} = \frac{(n+1)^2}{2n^2} \to \frac{1}{2} \qquad (n \to \infty),$$

and convergence follows from the ratio test. We may also apply the root test:

$$\sqrt[n]{\frac{n^2}{2^n}} = \frac{n^{2/n}}{2} = \frac{e^{(2/n)\ln n}}{2} \to \frac{e^0}{2} = \frac{1}{2} \qquad (n \to \infty).$$

This confirms our result.

Example 2. Application of the ratio test
Is the series

$$\sum_{n=0}^{\infty} \frac{(3-4i)^n}{n!} = 1 + (3-4i) + \frac{1}{2!}(3-4i)^2 + \cdots$$

convergent or divergent? We have

$$|w_n| = \frac{|3-4i|^n}{n!} = \frac{5^n}{n!}, \qquad |w_{n+1}| = \frac{5^{n+1}}{(n+1)!}, \qquad \left|\frac{w_{n+1}}{w_n}\right| = \frac{5}{n+1} \to 0$$

as $n \to \infty$, and convergence follows from the ratio test.

We finally mention that both the ratio test and the root test can be generalized to the case when the sequences with the terms $|w_{n+1}/w_n|$ and $\sqrt[n]{|w_n|}$, respectively, are not convergent.

Theorem 5 (Ratio test)

Consider the series

$$w_1 + w_2 + w_3 + \cdots,$$

assuming $w_n \neq 0$ *for* $n = 1, 2, \cdots$. *If for every n greater than some number N*

$$\left|\frac{w_{n+1}}{w_n}\right| \leqq q, \tag{6}$$

where q is some fixed number less than 1, *then the given series converges absolutely. If*

$$\left|\frac{w_{n+1}}{w_n}\right| \geqq 1 \qquad \textit{for every } n > N, \tag{7}$$

the series diverges.

Proof. In the first part of the proof of Theorem 3, convergence followed from the existence of a number $q < 1$ such that (6) holds for all $n > N$. Hence, for the present theorem, convergence follows by the same argument. The last statement of the theorem follows from the fact that, by (7), $|w_{n+1}| \geqq |w_n|$ and Theorem 3 in Sec. 15.2. ∎

Example 3. Application of Theorem 5, failure of Theorem 3
In the series

$$1 + \frac{1}{2} + \frac{1}{8} + \frac{1}{16} + \frac{1}{64} + \frac{1}{128} + \frac{1}{512} + \cdots$$

the odd terms and the even terms each form a geometric series with ratio $\frac{1}{8}$. Convergence follows from Theorem 5, because the ratios of successive terms are

$$\frac{1}{2}, \frac{1}{4}, \frac{1}{2}, \frac{1}{4}, \cdots.$$

We see that the sequence of these ratios does not converge. Hence Theorem 3 cannot be applied, and our example illustrates that Theorem 5 is more general than Theorem 3. ∎

Theorem 6 (Root test)

Consider the series

$$w_1 + w_2 + w_3 + \cdots .$$

If for every n greater than some number N we have

$$\sqrt[n]{|w_n|} \leqq q, \tag{8}$$

where q is some fixed number less than 1, *then the given series converges absolutely. If for infinitely many n we have*

$$\sqrt[n]{|w_n|} \geqq 1, \tag{9}$$

then that series diverges.

Proof. If (8) holds, then

$$|w_n| \leqq q^n < 1 \qquad (n > N)$$

and the series $|w_N| + |w_{N+1}| + \cdots$ converges by comparison with the geometric series. Hence, the series $w_1 + w_2 + \cdots$ converges absolutely. If (9) holds, then $|w_n| \geqq 1$ for infinitely many n, and divergence follows from Theorem 3 in Sec. 15.2. ∎

In the last two theorems it is essential for convergence that $|w_{n+1}/w_n|$ *and* $\sqrt[n]{|w_n|}$, *respectively, should be ultimately less than or equal to a definite number* $q < 1$. *It does not at all suffice for convergence that we should have*

$$\left|\frac{w_{n+1}}{w_n}\right| < 1 \qquad \text{or} \qquad \sqrt[n]{|w_n|} < 1$$

for sufficiently large n. For example, in the case of the harmonic series in Sec. 15.2 we have

$$\frac{w_{n+1}}{w_n} = \frac{1/(n+1)}{1/n} = \frac{n}{n+1} < 1 \qquad \text{and} \qquad \sqrt[n]{\frac{1}{n}} < 1,$$

but the series diverges.

Problems for Sec. 15.5

Are the following series convergent or divergent?

1. $1 + \dfrac{1}{\sqrt{2}} + \dfrac{1}{\sqrt{3}} + \dfrac{1}{\sqrt{4}} + \cdots$

2. $\dfrac{1}{1\cdot 2} + \dfrac{1}{2\cdot 3} + \dfrac{1}{3\cdot 4} + \cdots$

3. $\frac{1}{\sqrt{1 \cdot 2}} + \frac{1}{\sqrt{2 \cdot 3}} + \frac{1}{\sqrt{3 \cdot 4}} + \cdots$ **4.** $1 + 10i + \frac{(10i)^2}{2!} + \frac{(10i)^3}{3!} + \cdots$

5. $1 + \frac{i}{2} + \frac{i^2}{2^2} + \frac{i^3}{2^3} + \cdots$ **6.** $1 + i + i^2 + i^3 + \cdots$

Are the following series convergent or divergent?

7. $\sum_{n=1}^{\infty} \frac{n^n}{n!}$ **8.** $\sum_{n=0}^{\infty} \frac{(1+i)^n}{n!}$ **9.** $\sum_{n=0}^{\infty} \frac{(10+5i)^n}{n!}$ **10.** $\sum_{n=0}^{\infty} \frac{(3+i)^{2n}}{(2n)!}$

11. $\sum_{n=0}^{\infty} n\left(\frac{i}{2}\right)^n$ **12.** $\sum_{n=0}^{\infty} \left(\frac{4+3i}{6}\right)^n$ **13.** $\sum_{n=0}^{\infty} \left(\frac{3-4i}{4}\right)^n$ **14.** $\sum_{n=0}^{\infty} \frac{(-1)^n(10i)^{2n}}{(2n)!}$

15. $\sum_{n=1}^{\infty} \frac{i^n}{2^n n}$ **16.** $\sum_{n=1}^{\infty} \frac{n+1}{2^n n}$ **17.** $\sum_{n=1}^{\infty} \frac{(n!)^2}{(2n)!}$ **18.** $\sum_{n=1}^{\infty} \frac{2^n n!}{n^n}$

19. Determine how many terms are needed to compute the sum s of the geometric series $1 + q + q^2 + \cdots$ with an error less than 0.01, when $q = 0.25$, $q = 0.5$, $q = 0.9$.

20. If $|w_{n+1}/w_n| \leqq q < 1$, so that the series $w_1 + w_2 + \cdots$ converges by the ratio test (Theorem 5), show that the remainder $R_n = w_{n+1} + w_{n+2} + \cdots$ satisfies $|R_n| \leqq |w_{n+1}|/(1 - q)$. (*Hint.* Use the fact that the ratio test is a comparison of the series $w_1 + w_2 + \cdots$ with the geometric series.) Using the result, find how many terms suffice for computing the sum s of the series in Prob. 16 with an error not exceeding 0.05 and compute s to this accuracy.

15.6 Operations on Series

We shall now consider some simple operations which are frequently used in working with series.

Let us start with the addition of series and show that two convergent series may be added term by term.

Theorem 1 (Termwise addition and subtraction)

If two series $w_1 + w_2 + \cdots$ and $z_1 + z_2 + \cdots$ are convergent with the respective sums s and s^*, then the series

$$(1) \qquad \sum_{n=1}^{\infty} (w_n + z_n), \qquad \sum_{n=1}^{\infty} (w_n - z_n), \qquad \text{and} \qquad \sum_{n=1}^{\infty} kw_n$$

where k is any constant, are convergent and have the sums $s + s^*$, $s - s^*$, and ks, respectively.

Proof. The partial sums of the two given series are

$$s_n = w_1 + \cdots + w_n, \qquad s_n^* = z_1 + \cdots + z_n,$$

and by the definition of convergence of a series we have

$$\lim_{n\to\infty} s_n = s, \qquad \lim_{n\to\infty} s_n^* = s^*.$$

Now the nth partial sum of the first series in (1) is

$$S_n = s_n + s_n^* = (w_1 + z_1) + \cdots + (w_n + z_n).$$

From this we obtain

$$\lim_{n\to\infty} S_n = \lim_{n\to\infty} (s_n + s_n^*) = \lim_{n\to\infty} s_n + \lim_{n\to\infty} s_n^* = s + s^*.$$

Hence this series converges and has the sum $s + s^*$. The other statements can be proved by a similar argument. ∎

The next operation to be considered is that of inserting parentheses in a given series. This operation is called **grouping.**

For example, by grouping the terms of the series $w_1 + w_2 + w_3 + \cdots$ we may obtain the series $(w_1 + w_2) + (w_3 + w_4) + \cdots$ whose terms are $W_n = w_{2n-1} + w_{2n}$ where $n = 1, 2, \cdots$.

Of course, in the case of a *finite* series (that is, a series with only finitely many terms) we may insert parentheses without changing the sum of the series. We shall prove that the same holds for a convergent infinite series. It is interesting to note that for a divergent series, grouping may produce convergence. For instance, by grouping the divergent series

$$1 - 1 + 1 - 1 + - \cdots$$

(cf. Example 1 in Sec. 15.2) we obtain the convergent series

$$(1 - 1) + (1 - 1) + \cdots = 0 + 0 + \cdots$$

whose sum is zero.

Theorem 2 (Grouping)

If a series converges, then insertion of parentheses yields a new convergent series having the same sum as the original series.

Proof. Obviously, the partial sums of the new series are obtained by skipping certain partial sums of the given series. For example, the partial sums of the series

$$(w_1 + w_2 + w_3) + (w_4 + w_5 + w_6) + \cdots$$

are the partial sums $s_3, s_6, s_9, \cdots$ of the series $w_1 + w_2 + \cdots$. Now if the partial sums s_n of the given series form a convergent sequence whose limit is s, it follows immediately from the definition of convergence of a sequence that the new sequence obtained by skipping must also converge to s. This proves the theorem. ∎

Example 1. Grouping

By the Leibniz test (cf. Sec. 15.4) the series

$$\sum_{n=1}^{\infty} \frac{(-1)^{n+1}}{n} = 1 - \frac{1}{2} + \frac{1}{3} - \frac{1}{4} + - \cdots$$

converges. Let s be its sum. Then by Theorem 2,

$$\frac{1}{1\cdot 2}+\frac{1}{3\cdot 4}+\frac{1}{5\cdot 6}+\cdots=\left(1-\frac{1}{2}\right)+\left(\frac{1}{3}-\frac{1}{4}\right)+\cdots$$

(2)

$$=\sum_{m=1}^{\infty}\left(\frac{1}{2m-1}-\frac{1}{2m}\right)=s$$

and, similarly,

$$\frac{1\cdot 2+3\cdot 4}{1\cdot 2\cdot 3\cdot 4}+\frac{5\cdot 6+7\cdot 8}{5\cdot 6\cdot 7\cdot 8}+\cdots=\left(1-\frac{1}{2}+\frac{1}{3}-\frac{1}{4}\right)+\left(\frac{1}{5}-\frac{1}{6}+\frac{1}{7}-\frac{1}{8}\right)+\cdots$$

(3)

$$=\sum_{m=1}^{\infty}\left(\frac{1}{4m-3}-\frac{1}{4m-2}+\frac{1}{4m-1}-\frac{1}{4m}\right)=s.$$

▮

The last operation to be considered in this section is that of changing the order of the terms in a series.

Clearly, if a series is finite, the order of its terms may be changed without altering the sum. Also, we may change the order of finitely many terms of a given infinite series: if the given series is divergent, so will be the new series, and if the given series converges, the new series will be convergent with the same sum. This follows immediately from the definitions of convergence and divergence.

We may now ask what happens if we "change the order of infinitely many terms" of a series. Of course we must first define what we mean by such an operation. This can be done as follows.

A series

$$\sum_{n=1}^{\infty} w_n^* = w_1^* + w_2^* + \cdots$$

is said to be a **rearrangement** of a series

$$\sum_{m=1}^{\infty} w_m = w_1 + w_2 + \cdots$$

if there is a one-to-one correspondence between the indices n and m such that $w_n^* = w_m$ for corresponding indices.

For example, the series

$$\frac{1}{2}+1+\frac{1}{4}+\frac{1}{3}+\frac{1}{6}+\frac{1}{5}+\cdots$$

is a rearrangement of the harmonic series

$$1+\frac{1}{2}+\frac{1}{3}+\frac{1}{4}+\frac{1}{5}+\frac{1}{6}+\cdots.$$

The following example illustrates that by rearranging a convergent series we may obtain a convergent series whose sum is different from that of the original series.

Example 2. A rearrangement which changes the sum
We rearrange the convergent series

$$s = 1 - \frac{1}{2} + \frac{1}{3} - \frac{1}{4} + \frac{1}{5} - \frac{1}{6} + \frac{1}{7} - \frac{1}{8} + \frac{1}{9} - + \cdots$$

by writing first two positive terms, then one negative, then again two positive, etc., in the order of their occurrence. This yields the rearrangement

$$1 + \frac{1}{3} - \frac{1}{2} + \frac{1}{5} + \frac{1}{7} - \frac{1}{4} + \frac{1}{9} + \frac{1}{11} - \frac{1}{6} + \frac{1}{13} + \cdots.$$

It can be shown that this new series converges (proof in [A9], p. 76; cf. Appendix 1); we call its sum s^*. We want to show that the sums s and s^* are different. By inserting parentheses in the rearrangement we have

$$s^* = \left(1 + \frac{1}{3} - \frac{1}{2}\right) + \left(\frac{1}{5} + \frac{1}{7} - \frac{1}{4}\right) + \cdots = \sum_{m=1}^{\infty} \left(\frac{1}{4m-3} + \frac{1}{4m-1} - \frac{1}{2m}\right).$$

On the other hand, by multiplying the series (2) by $\frac{1}{2}$ and adding the resulting series and the series (3) term by term (cf. Theorem 1), we get

$$\frac{3s}{2} = \sum_{m=1}^{\infty} \left(\frac{1}{4m-3} - \frac{1}{4m-2} + \frac{1}{4m-1} - \frac{1}{4m} + \frac{1/2}{2m-1} - \frac{1/2}{2m}\right)$$

$$= \sum_{m=1}^{\infty} \left(\frac{1}{4m-3} + \frac{1}{4m-1} - \frac{1}{2m}\right) = s^*.$$

Hence $s^* = 3s/2$. ∎

We note that the series in this example is not absolutely convergent. Let us prove that an absolutely convergent series can always be rearranged without altering the sum.

Theorem 3 (Rearrangement of absolutely convergent series)

If a series converges absolutely, then each of its rearrangements converges absolutely and has the same sum as the original series.

Proof. Let $w_1^* + w_2^* + w_3^* + \cdots$ be any rearrangement of an absolutely convergent series $w_1 + w_2 + w_3 + \cdots$. Then, since every w_m^* equals a w_n for appropriate n and no two m's correspond to the same n, we clearly have

$$\sum_{m=1}^{n} |w_m^*| \leqq \sum_{k=1}^{\infty} |w_k| \qquad \text{for every } n.$$

The sum on the left is the nth partial sum of the series $|w_1^*| + |w_2^*| + \cdots$. Since these partial sums are nonnegative, the inequality shows that they form a bounded sequence. Since $|w_m^*| \geqq 0$, the sequence is monotone increasing and, therefore, convergent (cf. Theorem 1 in Sec. 15.4). Hence the rearrangement $w_1^* + w_2^* + \cdots$ converges absolutely. Let s^* be its sum, and let s be the sum of the original series. We show that $s^* = s$.

From the definition of convergence and from Theorem 3 in Sec. 15.3 when applied to the series $|w_1| + |w_2| + \cdots$ it follows that, when an $\epsilon > 0$ is given, we can find an N such that

$$(4) \qquad \text{(a)} \quad |s_n - s| < \frac{\epsilon}{2} \qquad \text{(b)} \quad |w_{n+1}| + \cdots + |w_{n+p}| < \frac{\epsilon}{2}$$

for every $n > N$ and $p = 1, 2, \cdots$; here s_n is the nth partial sum of the original series. Now for sufficiently large m, the mth partial sum $s_m{}^*$ of the rearrangement will contain all the terms $w_1, \cdots, w_n$ ($n > N$ and fixed) and perhaps some more terms w_r ($r > n$) of the original series. Hence $s_m{}^*$ will be of the form

$$(5) \qquad s_m{}^* = s_n + A_{mn}$$

where A_{mn} is the sum of those additional terms. Let $n + p$ be the greatest subscript of the terms in A_{mn}. Then by (4b), since $n > N$, we have

$$|A_{mn}| \leqq |w_{n+1}| + \cdots + |w_{n+p}| < \frac{\epsilon}{2}.$$

From this and (5) it follows that

$$|s_m{}^* - s_n| = |A_{mn}| < \frac{\epsilon}{2}.$$

Using (4a) and the triangle inequality, we thus obtain

$$|s_m{}^* - s| = |(s_m{}^* - s_n) + (s_n - s)| \leqq |s_m{}^* - s_n| + |s_n - s| < \frac{\epsilon}{2} + \frac{\epsilon}{2} = \epsilon$$

for every sufficiently large m. Hence, the sequence $s_1{}^*, s_2{}^*, \cdots$ converges to s and, therefore, $s^* = s$. This completes the proof. ∎

CHAPTER

16 POWER SERIES, TAYLOR SERIES, LAURENT SERIES

Power series (Sec. 16.1) are the most important type of series in complex analysis. The reason is that they represent analytic functions (Theorem 5, Sec. 16.2) and, conversely, every analytic function has power series representations, called *Taylor series* (Secs. 16.3–16.6). These Taylor series are the complex analogues of the Taylor series in real calculus. Indeed, if we replace the real variable in the latter series by a complex variable, we may *"extend"* or *"continue"* real functions to the complex domain.

The last part of the chapter is devoted to the representation of analytic functions by Laurent series. These are series involving positive and negative integral powers of the independent variable. They are also useful for evaluating complex and real integrals, as we shall see in the next chapter.

Prerequisites for this chapter: Chaps. 12, 14, 15.
Sections which may be omitted in a shorter course: 16.6, 16.8.
References: Appendix 1, Part F.
Answers to problems: Appendix 2.

16.1 Power Series

We have made the definitions in Sec. 15.2 for series of *constant* terms. If the terms of a series are *variable,* say, functions of a variable z, they assume definite values when z is given a fixed value and then all those definitions apply. It is obvious that in the case of such a series of functions of z the partial sums, the remainders, and the sum will be functions of z. Usually such a series will be convergent for some values of z, for example, for all z in some region, and divergent for the other values of z.

In complex analysis, the most important series with variable terms are power series. A **power series**[1] *in powers of* $z - a$ is an infinite series of the form

$$\textbf{(1)}\qquad \sum_{m=0}^{\infty} c_m(z-a)^m = c_0 + c_1(z-a) + c_2(z-a)^2 + \cdots$$

where z is a variable, $c_0, c_1, \cdots$ are constants, called the **coefficients,** and a is a constant, called the **center** of the series.

[1]It should be emphasized that the term "power series" alone usually refers to series of the form (1), including the particular case (2), but *does not include* series of negative powers of z such as $c_1z^{-1} + c_2z^{-2} + \cdots$ or series involving fractional powers of z.

If $a = 0$, we obtain as a particular case a *power series in powers of* z:

$$\sum_{m=0}^{\infty} c_m z^m = c_0 + c_1 z + c_2 z^2 + \cdots . \tag{2}$$

The convergence behavior of a power series can be characterized in a very simple way. Let us start with three examples, which will turn out to be typical.

Example 1. Convergence in a disk. Geometric series

The geometric series

$$\sum_{m=0}^{\infty} z^m = 1 + z + z^2 + \cdots$$

converges absolutely when $|z| < 1$ and diverges when $|z| \geqq 1$ (cf. Theorem 2 in Sec. 15.5).

Example 2. Convergence in the entire finite plane

The power series

$$\sum_{n=0}^{\infty} \frac{z^n}{n!} = 1 + z + \frac{z^2}{2!} + \frac{z^3}{3!} + \cdots$$

is absolutely convergent for every (finite) z, as follows from the ratio test. In fact, for any fixed z,

$$\left| \frac{z^{n+1}/(n+1)!}{z^n/n!} \right| = \frac{|z|}{n+1} \to 0 \qquad \text{as} \qquad n \to \infty.$$

Example 3. Convergence only at the center

The power series

$$\sum_{n=0}^{\infty} n!\, z^n = 1 + z + 2z^2 + 6z^3 + \cdots$$

converges only at the center $z = 0$, but diverges for every $z \neq 0$, In fact, this follows from the ratio test because

$$\left| \frac{(n+1)! z^{n+1}}{n!\, z^n} \right| = (n+1)\,|z| \to \infty \qquad \text{as} \qquad n \to \infty \qquad (z \text{ fixed and } \neq 0). \quad ▮$$

The power series (1) converges when $z = a$, because then $z - a = 0$ and the series reduces to the single term c_0. Example 3 illustrates that in some cases this may be the only value of z for which the series converges. However, if the series (1) converges for some $z_0 \neq a$, then it converges for every z whose distance from the center is less than that of z_0. In fact, the following theorem holds.

Theorem 1. Convergence of a power series

If the power series (1) *converges at a point* $z = z_0$, *it converges absolutely for every* z *for which* $|z - a| < |z_0 - a|$, *that is, for each* z *within the circle through* z_0 *about* a.

Proof. Since the series (1) converges for z_0, Theorem 3 in Sec. 15.2 shows that

$$c_n(z_0 - a)^n \to 0 \qquad \text{as} \qquad n \to \infty.$$

This implies that for $z = z_0$ the terms of the series (1) are bounded, say,

$$|c_n(z_0 - a)^n| < M \qquad \text{for every } n = 0, 1, \cdots .$$

From this we obtain

$$|c_n(z - a)^n| = \left| c_n(z_0 - a)^n \left(\frac{z - a}{z_0 - a}\right)^n \right| < M \left| \frac{z - a}{z_0 - a} \right|^n$$

and, therefore,

$$\sum_{n=0}^{\infty} |c_n(z - a)^n| < \sum_{n=0}^{\infty} M \left| \frac{z - a}{z_0 - a} \right|^n = M \sum_{n=0}^{\infty} \left| \frac{z - a}{z_0 - a} \right|^n . \tag{3}$$

From the assumption $|z - a| < |z_0 - a|$ we have the inequality

$$\left| \frac{z - a}{z_0 - a} \right| < 1,$$

and the series on the right-hand side of (3) is a geometric series which converges. Hence the series on the left-hand side of (3) converges, and the series (1) converges absolutely when $|z - a| < |z_0 - a|$. This completes the proof. ∎

Examples 2 and 3 illustrate that a power series may converge for all z or only for $z = a$. Let us exclude these two cases for a moment. Then, if a power series (1) is given, we may consider all the points z in the complex plane for which the series converges. Let R be the smallest real number such that the distance of each of those points from the center a is at most equal to R. (For instance, $R = 1$ in Example 1.) Then from Theorem 1 it follows that the series converges for all z within the circle of radius R about the center a, that is, for all z for which

$$|z - a| < R, \tag{4}$$

and, by the definition of R, the series diverges for all z for which

$$|z - a| > R.$$

The circle

$$|z - a| = R$$

is called the **circle of convergence,** and its radius R is called the **radius of convergence** of (1). Cf. Fig. 324a.

At the points on the circle of convergence the series may converge or diverge. For instance, in Example 1 we have $R = 1$, and the series diverges at each point of the circle of convergence $|z| = 1$. The power series

$$\sum_{n=1}^{\infty} \frac{z^n}{n} = z + \frac{z^2}{2} + \frac{z^3}{3} + \cdots$$

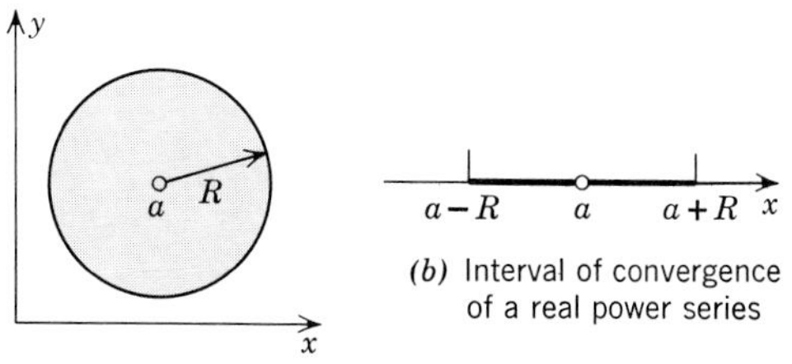

(*a*) Circle of convergence

(*b*) Interval of convergence of a real power series

Fig. 324. Circle of convergence and interval of convergence

converges for $|z| < 1$ and diverges for $|z| > 1$, as follows by the ratio test. Thus $R = 1$. At $z = 1$ it becomes the harmonic series and diverges. At $z = -1$ it becomes $-1 + \frac{1}{2} - \frac{1}{3} + \frac{1}{4} - + \cdots$ and converges (cf. Example 2, Sec. 15.2). This illustrates that a series may converge at some points on the circle of convergence, while at others it may diverge.

Clearly, if we consider a *real* power series (1), that is, if the variable $z = x$, the center, and the coefficients are real, then (4) represents an interval of length $2R$ with midpoint at a on the x-axis, the so-called **interval of convergence** (Fig. 324*b*).

If the series (1) converges for all z (as in Example 2), then we set

$$R = \infty \qquad \text{(and } 1/R = 0\text{);}$$

if it converges only at the center $z = a$ (as in Example 3), then we set

$$R = 0 \qquad \text{(and } 1/R = \infty\text{).}$$

Using these conventions, the radius of convergence R of the power series (1) may be determined from the coefficients of the series as follows.

Theorem 2 (Radius of convergence)

If the sequence $\sqrt[n]{|c_n|}$, $n = 1, 2, \cdots$, *converges with the limit* L, *then the radius of convergence* R *of the power series* (1) *is*

(5a) $$R = \frac{1}{L} \qquad \textbf{(Cauchy–Hadamard formula[2]),}$$

including the case $L = 0$ *where* $R = \infty$ *and the series* (1) *converges for all* z.

If that sequence does not converge but is bounded, then

(5b) $$R = \frac{1}{l}$$

where l *is the greatest of the limit points of the sequence.*

If that sequence is not bounded, then $R = 0$, *and the series converges only for* $z = a$.

[2]Named after the French mathematicians, A. L. CAUCHY (cf. the footnote in Sec. 2.7) and JACQUES HADAMARD (1865–1964).

Proof. If

$$\lim_{n\to\infty} \sqrt[n]{|c_n|} = L \neq 0,$$

then

$$\lim_{n\to\infty} \sqrt[n]{|c_n(z-a)^n|} = |z-a| \lim_{n\to\infty} \sqrt[n]{|c_n|} = |z-a|\, L.$$

Since the terms of the series (1) are $w_n = c_n(z-a)^n$, the root test (Sec. 15.5) shows that the series converges absolutely when

$$|z-a|\, L < 1 \qquad \text{or} \qquad |z-a| < \frac{1}{L} = R$$

but diverges when

$$|z-a|\, L > 1 \qquad \text{or} \qquad |z-a| > \frac{1}{L} = R.$$

If

$$\lim_{n\to\infty} \sqrt[n]{|c_n|} = L = 0,$$

it follows from the definition of a limit that for any given $\epsilon > 0$, for example, for $\epsilon = 1/(2|z_1 - a|)$ with arbitrary fixed z_1, we can find an N such that

$$\sqrt[n]{|c_n|} < \frac{1}{2|z_1-a|} \qquad \text{for every } n > N.$$

From this we have

$$|c_n| < \frac{1}{(2|z_1-a|)^n} \qquad \text{or} \qquad |c_n(z_1-a)^n| < \frac{1}{2^n}.$$

Now, since $\Sigma 2^{-n}$ converges, the comparison test (Sec. 15.5) shows that the series (1) converges absolutely for $z = z_1$. Since z_1 is arbitrary, this means absolute convergence for every finite z, and the proof of the statement involving (5a) is complete.

We next prove the statement involving (5b). The existence of l follows from the Bolzano–Weierstrass theorem (Sec. 15.3), and since $\sqrt[n]{|c_n|} \geqq 0$, we clearly have $l > 0$. From the definition of a limit point it follows that, an $\epsilon > 0$ being given,

$$l - \epsilon < \sqrt[n]{|c_n|} < l + \epsilon \qquad \text{for infinitely many } n.$$

Multiplying this by the positive quantity $|z-a|$, we obtain the inequalities

$$(6) \qquad |z-a|\,(l-\epsilon) < \sqrt[n]{|c_n(z-a)^n|}$$

and

$$(7) \qquad \sqrt[n]{|c_n(z-a)^n|} < |z-a|\,(l+\epsilon).$$

The inequality (7) holds even for all sufficiently large n, say, for $n > N$, because l is the greatest limit point and, therefore, at most finitely many terms can be greater than the expression on the right. We prove that for

$$|z - a| < \frac{1}{l} \tag{8}$$

convergence of the power series (1) follows from (7). In fact, if we choose

$$\epsilon = \frac{1 - l|z - a|}{2|z - a|},$$

then because of (8) we have $\epsilon > 0$, and the inequality (7) takes the form

$$\sqrt[n]{|c_n(z - a)^n|} < \frac{1 + l|z - a|}{2} \qquad (n > N).$$

From (8) we see that the expression on the right is less than 1, and convergence follows by means of the root test (Sec. 15.5, Theorem 6). On the other hand, when

$$|z - a| > \frac{1}{l},$$

then, choosing

$$\epsilon = \frac{l|z - a| - 1}{2|z - a|},$$

we have $\epsilon > 0$, and (6) becomes

$$\sqrt[n]{|c_n(z - a)^n|} > \frac{|z - a|l + 1}{2} > 1.$$

Hence, by the root test, the series diverges for those z. This proves the statement involving (5b).

Finally if the sequence $\sqrt[n]{|c_n|}$ is not bounded, then, by definition, any K being given,

$$\sqrt[n]{|c_n|} > K \qquad \text{for infinitely many } n.$$

Choosing $K = 1/|z - a|$ where $z \neq a$, we see that this inequality becomes

$$\sqrt[n]{|c_n|} > \frac{1}{|z - a|} \qquad \text{or} \qquad \sqrt[n]{|c_n(z - a)^n|} > 1,$$

and divergence follows from Theorem 6 in Sec. 15.5. This completes the proof of Theorem 2. ∎

We shall now consider the operations of addition and multiplication of power series.

Two power series may be added term by term for every z for which both series are convergent. This follows immediately from Theorem 1 in Sec. 15.6.

Let us consider the term-by-term multiplication of two power series

$$(9)\quad \sum_{k=0}^{\infty} a_k z^k = a_0 + a_1 z + \cdots \qquad \text{and} \qquad \sum_{m=0}^{\infty} c_m z^m = c_0 + c_1 z + \cdots .$$

If we multiply each term of the first series by each term of the second series and collect products of like powers of z, we obtain

$$(10)\quad \begin{aligned} &a_0c_0 + (a_0c_1 + a_1c_0)z + (a_0c_2 + a_1c_1 + a_2c_0)z^2 + \cdots \\ &\qquad = \sum_{n=0}^{\infty} (a_0c_n + a_1c_{n-1} + \cdots + a_nc_0)z^n. \end{aligned}$$

This series is called the **Cauchy product** of the series (9).

Theorem 3 (Cauchy product of power series)

The Cauchy product of two power series (9) *is absolutely convergent for each z within the circle of convergence of each of the series* (9). *If the series have the sums $g(z)$ and $h(z)$, respectively, the Cauchy product has the sum*

$$(11)\qquad s(z) = g(z)h(z).$$

Proof. The general term of the product series (10) is

$$p_n = (a_0c_n + a_1c_{n-1} + \cdots + a_nc_0)z^n.$$

Now, by the generalized triangle inequality (11) in Sec. 12.2, we obtain

$$|p_0| + |p_1| = |a_0c_0| + |(a_0c_1 + a_1c_0)z| \leqq (|a_0| + |a_1z|)(|c_0| + |c_1z|),$$

$$|p_0| + |p_1| + |p_2| \leqq (|a_0| + |a_1z| + |a_2z^2|)(|c_0| + |c_1z| + |c_2z^2|),$$

as can be verified by performing the multiplication on the right; in general

$$\begin{aligned} &|p_0| + |p_1| + \cdots + |p_n| \\ &\qquad \leqq (|a_0| + |a_1z| + \cdots + |a_nz^n|)(|c_0| + |c_1z| + \cdots + |c_nz^n|). \end{aligned}$$

If z lies within the circle of convergence of each of the series (9), the sequence of the expressions on the right is bounded, and so is the sequence of the partial sums of the series $|p_0| + |p_1| + \cdots$. Since $|p_n| \geqq 0$, that sequence is also monotone increasing, so that it converges. (Cf. Sec. 15.4, Theorem 1.) Hence that series is convergent, and the product series (10) converges absolutely.

We prove (11). We use the fact that every rearrangement of (10) is absolutely convergent for precisely those z and has the same sum as (10). (Cf. Sec.

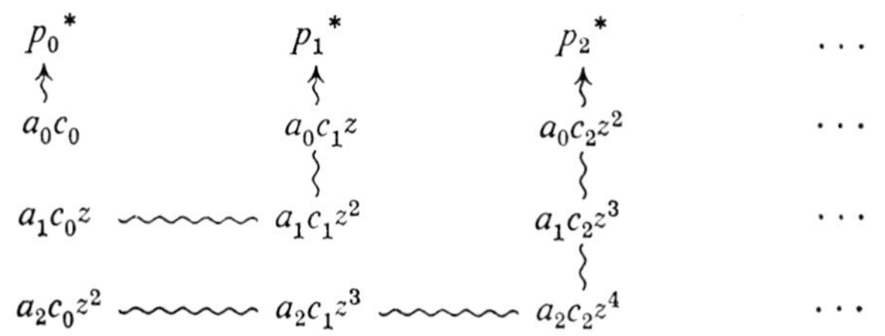

Fig. 325. Proof of Theorem 3

15.6, Theorem 3.) We consider the particular rearrangement $p_0^* + p_1^* + \cdots$ where p_n^* equals (Fig. 325)

$$(a_nc_0 + a_0c_n)z^n + (a_nc_1 + a_1c_n)z^{n+1} + \cdots + (a_nc_{n-1} + a_{n-1}c_n)z^{2n-1} + a_nc_nz^{2n}.$$

Obviously,

$$a_0c_0 = p_0^*, \qquad (a_0 + a_1z)(c_0 + c_1z) = p_0^* + p_1^*$$

and, in general,

$$(a_0 + a_1z + \cdots + a_nz^n)(c_0 + c_1z + \cdots + c_nz^n) = p_0^* + p_1^* + \cdots + p_n^*.$$

By letting n approach infinity we obtain (11), and Theorem 3 is proved. ∎

Example 4. Cauchy product

The geometric series $1 + z + z^2 + \cdots$ has the sum $1/(1 - z)$ when $|z| < 1$. (Cf. Sec. 15.5.) From Theorem 3 we thus obtain

$$\left(\frac{1}{1-z}\right)^2 = \sum_{k=0}^{\infty} z^k \sum_{m=0}^{\infty} z^m = (1 + z + z^2 + \cdots)(1 + z + z^2 + \cdots)$$

$$= 1 + 2z + 3z^2 + \cdots = \sum_{n=0}^{\infty} (n+1)z^n \qquad (|z| < 1).$$

Problems for Sec. 16.1

1. Show that if the sequence $|c_{n+1}/c_n|$, $n = 1, 2, \cdots$, is convergent with the limit L, then the radius of convergence R of the power series (1) is $R = 1/L$ when $L > 0$ and is $R = \infty$ when $L = 0$.
2. Show that if (2) has radius of convergence R (assumed finite), then the radius of convergence of $\Sigma c_m z^{2m}$ is $\sqrt{R}$.

Find the radius of convergence of the following series.

3. $\sum_{n=0}^{\infty} (z - 2i)^n$
4. $\sum_{n=0}^{\infty} \frac{(z-2)^n}{2^n}$
5. $\sum_{n=0}^{\infty} n\left(\frac{z}{3}\right)^n$
6. $\sum_{n=0}^{\infty} \frac{(2z)^n}{n!}$
7. $\sum_{n=0}^{\infty} \frac{z^{2n}}{n!}$
8. $\sum_{n=1}^{\infty} \frac{z^n}{n}$
9. $\sum_{n=0}^{\infty} \frac{(-1)^n}{n!} z^n$
10. $\sum_{n=0}^{\infty} \frac{n^2}{2^n} z^n$

11. $\sum_{n=0}^{\infty} \frac{(2n)!}{(n!)^2} z^n$ **12.** $\sum_{n=0}^{\infty} \frac{z^n}{n^n}$ **13.** $\sum_{n=0}^{\infty} \frac{(-1)^n}{(2n)!} z^{2n}$ **14.** $\sum_{n=1}^{\infty} \frac{z^n}{n^2}$

15. $\sum_{n=0}^{\infty} 6^n(z-i)^n$ **16.** $\sum_{n=0}^{\infty} (n!)^2 z^n$ **17.** $\sum_{n=0}^{\infty} 3^{2n} z^n$ **18.** $\sum_{n=0}^{\infty} \left(\frac{\pi}{2}\right)^n z^n$

19. If $\Sigma c_n z^n$ converges for all finite z, show that $\sqrt[n]{|c_n|} \to 0$ as $n \to \infty$. Give examples.

20. At points on the circle of convergence a series may converge or diverge. Illustrate this for the geometric series and the series in Probs. 8 and 14 at $z = 1$ and $z = -1$.

16.2 Functions Represented by Power Series

The main goal of this section is to show that power series represent analytic functions (Theorem 5, below). The fact that, conversely, *every* analytic function can be represented by power series (called *Taylor series*) will be proved in the next section. Both facts together account for the great importance of power series in complex analysis.

Let $\Sigma_{n=0}^{\infty} c_n z^n$ be an arbitrary power series with nonzero radius of convergence R. Then the sum of this series is a function of z, say, $f(z)$, and we write

$$f(z) = \sum_{n=0}^{\infty} c_n z^n = c_0 + c_1 z + c_2 z^2 + \cdots \qquad (|z| < R). \tag{1}$$

We say that $f(z)$ *is represented by the power series* or that *it is developed in the power series*. For instance, the geometric series represents $f(z) = 1/(1-z)$ in the interior of the unit circle $|z| = 1$. (Cf. Sec. 15.5, Theorem 2.)

Theorem 1 (Continuity)
The function $f(z)$ in (1) *with $R > 0$ is continuous at $z = 0$.*

Proof. We have to show that

$$\lim_{z \to 0} f(z) = f(0) = c_0. \tag{2}$$

We choose an arbitrary positive number $r < R$. Since the series in (1) is absolutely convergent in the disk $|z| < R$, it follows that the series

$$\sum_{n=1}^{\infty} |c_n| \, r^{n-1} = \frac{1}{r} \sum_{n=1}^{\infty} |c_n| \, r^n \qquad (0 < r < R)$$

is convergent. Let K denote its sum. Then we readily obtain

$$|f(z) - c_0| = \left| z \sum_{n=1}^{\infty} c_n z^{n-1} \right| \leqq |z| \sum_{n=1}^{\infty} |c_n| \, |z|^{n-1} \leqq |z| \, K \qquad (0 < |z| \leqq r).$$

An $\epsilon > 0$ being given, $|f(z) - c_0| < \epsilon$ for all z such that $|z| < \delta$ where δ is a positive real number less than both r and ϵ/K. By the definition of a limit, this means that (2) holds, and the proof is complete. ∎

Next we consider the question of **uniqueness** and show that *the same function $f(z)$ cannot be represented by two different power series with the same center.* If $f(z)$ can be developed in a power series with center a, the development is unique. This important fact is frequently used in real and complex analysis. We may formulate it as follows (assuming that $a = 0$, without loss of generality).

Theorem 2 (Identity theorem for power series)
Suppose that

$$\sum_{n=0}^{\infty} a_n z^n \qquad \textit{and} \qquad \sum_{n=0}^{\infty} b_n z^n$$

are power series which are convergent for $|z| < R$ where R is positive, and have the same sum for all these z. Then the series are identical, that is

$$a_n = b_n \qquad \text{for all } n = 0, 1, \cdots. \tag{3}$$

Proof. We proceed by induction. By assumption,

$$a_0 + a_1 z + a_2 z^2 + \cdots = b_0 + b_1 z + b_2 z^2 + \cdots \qquad (|z| < R). \tag{4}$$

We let z approach zero. Then, by Theorem 1, we have $a_0 = b_0$. We assume that $a_n = b_n$ for $n = 0, 1, \cdots, m$. Then, by omitting the first $m + 1$ terms on both sides of (4) and dividing by z^{m+1} ($\neq 0$), we obtain

$$a_{m+1} + a_{m+2} z + a_{m+3} z^2 + \cdots = b_{m+1} + b_{m+2} z + b_{m+3} z^2 + \cdots.$$

By Theorem 1, each of these power series represents a function which is continuous at $z = 0$. Hence $a_{m+1} = b_{m+1}$. This completes the proof. ∎

Let us now consider termwise differentiation and integration of power series. By differentiating the series $c_0 + c_1 z + c_2 z^2 + \cdots$ we obtain the series

$$\sum_{n=1}^{\infty} n c_n z^{n-1} = c_1 + 2c_2 z + 3c_3 z^2 + \cdots. \tag{5}$$

This series is called the *derived series* of the given power series.

Theorem 3 (Termwise differentiation)
The derived series of a power series has the same radius of convergence as the original series.

Proof. Let $nc_n = c_n^*$. Then $\sqrt[n]{|c_n^*|} = \sqrt[n]{n}\sqrt[n]{|c_n|}$. Since $\sqrt[n]{n} \to 1$ as $n \to \infty$, it follows that the sequences $\sqrt[n]{|c_n^*|}$ and $\sqrt[n]{|c_n|}$ either both converge with the same limit or both diverge. If they diverge, they are both unbounded or bounded, and in the latter case their greatest limit points are the same. From this and Theorem 2 in Sec. 16.1 the statement of our present theorem follows. ∎

Example 1

The power series

$$\sum_{n=1}^{\infty} (n+1)z^n$$

has the radius of convergence $R = 1$. This follows by differentiating the geometric series and applying Theorem 3.

Theorem 4 (Termwise integration)

The power series

$$\sum_{n=0}^{\infty} \frac{c_n}{n+1} z^{n+1} = c_0 z + \frac{c_1}{2} z^2 + \frac{c_2}{3} z^3 + \cdots$$

obtained by integrating the series $c_0 + c_1 z + c_2 z^2 + \cdots$ *term by term has the same radius of convergence as the original series.*

The proof is similar to that of Theorem 3.

Power series represent analytic functions, and the derived series (obtained by termwise differentiation) represent the derivatives of those functions. More precisely:

Theorem 5 (Analytic functions. Their derivatives)

A power series with a nonzero radius of convergence R represents an analytic function at every point interior to its circle of convergence. The derivatives of this function are obtained by differentiating the original series term by term; all the series thus obtained have the same radius of convergence as the original series.

Proof. We first prove that for any integer $n \geqq 2$,

$$\text{(6a)} \qquad \frac{b^n - a^n}{b - a} - na^{n-1} = (b - a)A_n$$

where

$$\text{(6b)} \qquad A_n = b^{n-2} + 2ab^{n-3} + 3a^2b^{n-4} + \cdots + (n-1)a^{n-2}.$$

We proceed by induction. A simple calculation shows that (6) is true for $n = 2$. Assuming that (6) holds for $n = k$, we show that it holds for $n = k + 1$. We have

$$\frac{b^{k+1} - a^{k+1}}{b - a} = \frac{b^{k+1} - ba^k + ba^k - a^{k+1}}{b - a} = b\frac{b^k - a^k}{b - a} + a^k.$$

By the induction hypothesis, the right-hand side equals

$$b[(b - a)A_k + ka^{k-1}] + a^k.$$

Direct calculation shows that this is equal to

$$(b - a)\{bA_k + ka^{k-1}\} + ka^k + a^k.$$

From (6b) with $n = k$ we see that the expression in the braces equals

$$b^{k-1} + 2ab^{k-2} + \cdots + (k - 1)ba^{k-2} + ka^{k-1} = A_{k+1}.$$

Hence our result is

$$\frac{b^{k+1} - a^{k+1}}{b - a} = (b - a)A_{k+1} + (k + 1)a^k,$$

which is (6) with $n = k + 1$. This proves (6) for any integer $n \geqq 2$.

To prove the statements in Theorem 5, we consider a representation

$$f(z) = \sum_{n=0}^{\infty} c_n z^n,$$

assuming that the radius of convergence R is not zero, and prove that, for any fixed z with $|z| < R$ and $\Delta z \to 0$ the difference quotient $[f(z + \Delta z) - f(z)]/\Delta z$ approaches the function represented by the derived series (5), which we denote by $f_1(z)$. By termwise addition we first have

$$\frac{f(z + \Delta z) - f(z)}{\Delta z} - f_1(z) = \sum_{n=2}^{\infty} c_n \left[\frac{(z + \Delta z)^n - z^n}{\Delta z} - nz^{n-1}\right].$$

From (6) with $b = z + \Delta z$, $a = z$, and $b - a = \Delta z$ we see that the series on the right can be written

$$\Delta z \sum_{n=2}^{\infty} c_n[(z + \Delta z)^{n-2} + 2z(z + \Delta z)^{n-3} + \cdots + (n - 1)z^{n-2}],$$

and for $|z| \leqq R_0$ and $|z + \Delta z| \leqq R_0$, $R_0 < R$, the absolute value of this cannot exceed

$$(7) \qquad |\Delta z| \sum_{n=2}^{\infty} |c_n|\, n(n - 1)R_0^{n-2},$$

where $n - 1$ is the largest of the coefficients $1, 2, \cdots, n - 1$, and n is the number of terms. The series in (7) is closely related to the second derived series of the series under consideration, evaluated at R_0. Indeed that derived series has the coefficients c_n [instead of the $|c_n|$ in (7)] and converges absolutely at R_0 ($< R$), by Theorem 3 in this section and Theorem 1 in the preceding one. This implies that the series in (7) converges at R_0; let $K(R_0)$ be its value. Then our result may be written

$$\left|\frac{f(z+\Delta z)-f(z)}{\Delta z}-f_1(z)\right| \leqq |\Delta z|\, K(R_0).$$

Letting $\Delta z \to 0$ and noting that R_0 ($< R$) is arbitrary, we conclude that $f(z)$ is analytic at any point interior to the circle of convergence and its derivative is represented by the derived series. From this the statements about the higher derivatives follow by induction, and Theorem 5 is proved. ∎

From Theorem 5 we see that the mth derivative $f^{(m)}(z)$ of the function $f(z)$ represented by (1) is

(8) $$f^{(m)}(z) = \sum_{n=m}^{\infty} n(n-1)\cdots(n-m+1)c_n z^{n-m} \qquad (|z| < R).$$

In the next section we shall see that *every analytic function can be represented by power series.*

Problems for Sec. 16.2

1. If $f(z)$ in (1) is even, show that $c_n = 0$ for odd n. (Use Theorem 2.)
2. If $f(z)$ in (1) is odd, show that $c_n = 0$ for even n. Give examples.
3. Applying Theorem 2 to $(1+z)^p(1+z)^q = (1+z)^{p+q}$ (p and q positive integers), show that

$$\sum_{n=0}^{r}\binom{p}{n}\binom{q}{r-n} = \binom{p+q}{r}.$$

4. Verify Theorems 3 and 4 for the geometric series and for the series in Example 2, Sec. 16.1.

Applying Theorems 3 and 4 to the geometric series, find the radius of convergence of the following series.

5. $\displaystyle\sum_{n=1}^{\infty}\frac{n}{5^n}(z-i)^n$

6. $\displaystyle\sum_{n=1}^{\infty}\frac{(z+2)^n}{n}$

7. $\displaystyle\sum_{n=1}^{\infty}\frac{4^n}{n}(z+2i)^n$

8. $\displaystyle\sum_{n=k}^{\infty}\binom{n}{k}\left(\frac{z}{2}\right)^n$

9. $\displaystyle\sum_{n=0}^{\infty}\left[\binom{n+k}{n}\right]^{-1} z^{n+k}$

10. $\displaystyle\sum_{n=0}^{\infty}\left[\binom{n+k}{n}\right]^{-1}\left(\frac{z}{3}\right)^n$

16.3 Taylor Series

The familiar Taylor series is an effective tool in real calculus and its applications. We shall now see that in complex analysis there is a Taylor expansion which is a generalization of that series and is even more important.

Let us consider a function $f(z)$ which is analytic in a neighborhood of a point $z = a$. Let C be a circle which lies in this neighborhood and has the center a. Then we may apply Cauchy's integral formula (1), Sec. 14.5; writing z and z^* in place of z_0 and z, we have

$$f(z) = \frac{1}{2\pi i}\int_C \frac{f(z^*)}{z^* - z}\,dz^* \tag{1}$$

where z is an arbitrary fixed point inside C and z^* is the complex variable of integration (cf. Fig. 326). We shall now develop $1/(z^* - z)$ in (1) in powers of $z - a$. We first have

$$\frac{1}{z^* - z} = \frac{1}{z^* - a - (z - a)} = \frac{1}{(z^* - a)\left(1 - \dfrac{z - a}{z^* - a}\right)}. \tag{2}$$

For later use we note that since z^* is on C while z is inside C,

$$\left|\frac{z - a}{z^* - a}\right| < 1. \tag{3}$$

From the geometric progression

$$1 + q + q^2 + \cdots + q^n = \frac{1 - q^{n+1}}{1 - q} \qquad (q \neq 1)$$

we obtain the relation

$$\frac{1}{1 - q} = 1 + q + \cdots + q^n + \frac{q^{n+1}}{1 - q}.$$

By setting $q = (z - a)/(z^* - a)$ we find

$$\frac{1}{1 - [(z - a)/(z^* - a)]} = 1 + \frac{z - a}{z^* - a} + \left(\frac{z - a}{z^* - a}\right)^2 + \cdots + \left(\frac{z - a}{z^* - a}\right)^n + \frac{[(z - a)/(z^* - a)]^{n+1}}{(z^* - z)/(z^* - a)}.$$

We insert this into (2), and then (2) into (1). Since z and a are constant, we may

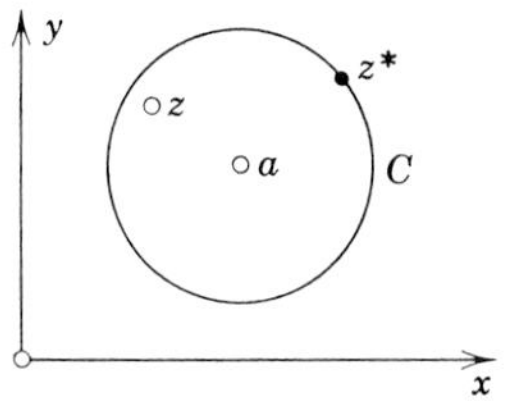

Fig. 326. Representation (1)

take the powers of $z - a$ out from under the integral sign, and (1) takes the form

$$
\begin{aligned}
f(z) = \frac{1}{2\pi i}\int_C \frac{f(z^*)}{z^* - a}\,dz^* + \frac{z-a}{2\pi i}\int_C \frac{f(z^*)}{(z^*-a)^2}\,dz^* + \cdots \\
+ \frac{(z-a)^n}{2\pi i}\int_C \frac{f(z^*)}{(z^*-a)^{n+1}}\,dz^* + R_n(z)
\end{aligned}
\tag{4}
$$

where the last term is given by the formula

$$R_n(z) = \frac{(z-a)^{n+1}}{2\pi i}\int_C \frac{f(z^*)}{(z^*-a)^{n+1}(z^*-z)}\,dz^*. \tag{5}$$

Using (1), Sec. 14.6, we may write this expansion in the form

$$
\begin{aligned}
f(z) = f(a) + \frac{z-a}{1!}f'(a) + \frac{(z-a)^2}{2!}f''(a) + \cdots + \frac{(z-a)^n}{n!}f^{(n)}(a) \\
+ R_n(z).
\end{aligned}
$$

This representation is called **Taylor's[3] formula.** $R_n(z)$ is called the *remainder.* Since the analytic function $f(z)$ has derivatives of all orders, we may take n in (6) as large as we please. If we let n approach infinity, we obtain from (6) the power series

$$f(z) = \sum_{m=0}^{\infty} \frac{f^{(m)}(a)}{m!}(z-a)^m. \tag{7}$$

This series is called the **Taylor series** of $f(z)$ *with center at a.* The particular case where $a = 0$ is called the **Maclaurin[4] series** of $f(z)$.

Clearly, the series (7) will converge and represent $f(z)$ if and only if

$$\lim_{n\to\infty} R_n(z) = 0. \tag{8}$$

To prove (8), we consider (5). Since z^* is on C while z is inside C, we have $|z^* - z| > 0$. Since $f(z)$ is analytic inside C and on C, it follows that the absolute value of $f(z^*)/(z^* - z)$ is bounded, say,

$$\left|\frac{f(z^*)}{z^* - z}\right| < \widetilde{M}$$

for all z^* on C. Let r be the radius of C. Then $|z^* - a| = r$ for all z^* on C, and C has the length $2\pi r$. Hence, by applying (4) in Sec. 14.2 to (5) in the current section we obtain

[3] BROOK TAYLOR (1685–1731), English mathematician, who introduced this formula for functions of a real variable.

[4] COLIN MACLAURIN (1698–1746), Scots mathematician.

$$|R_n| = \frac{|z-a|^{n+1}}{2\pi}\left|\int_C \frac{f(z^*)}{(z^*-a)^{n+1}(z^*-z)}\,dz^*\right|$$

$$< \frac{|z-a|^{n+1}}{2\pi}\widetilde{M}\frac{1}{r^{n+1}}2\pi r = \widetilde{M}r\left|\frac{z-a}{r}\right|^{n+1}.$$

If we let n approach infinity, it follows from (3) that the expression on the right approaches zero. This proves (8) for all z inside C. Since, by Theorem 2 in the preceding section, the representation of $f(z)$ in the form (7) is unique in the sense that (7) is the only power series with center at a which represents the given function $f(z)$, we may sum up our result as follows.

Taylor's theorem

Let $f(z)$ be analytic in a domain D and let $z = a$ be any point in D. Then there exists precisely one power series with center at a which represents $f(z)$; this series is of the form

(9) $$f(z) = \sum_{n=0}^{\infty} b_n(z-a)^n$$

where

$$b_n = \frac{1}{n!}f^{(n)}(a) \qquad n = 0, 1, \cdots;$$

this representation is valid in the largest open disk with center a contained in D. The remainders $R_n(z)$ of (9) can be represented in the form (5). The coefficients satisfy the inequality

(10) $$|b_n| \leqq \frac{M}{r^n}$$

where M is the maximum of $|f(z)|$ on the circle $|z-a| = r$.

Relation (10) follows from Cauchy's inequality (4) in Sec. 14.6.

Practically speaking, (8) means that for all z for which (9) converges, the nth partial sum of (9) will approximate $f(z)$ to any assigned degree of accuracy; we just have to choose n large enough.

From Taylor's theorem we see that the radius of convergence of (9) is at least equal to the shortest distance from a to the boundary of D. It may be larger, but then the series may no longer represent $f(z)$ at all points of D which lie in the interior of the circle of convergence.

One surprising property of complex analytic functions is that they have derivatives of all orders, and now we have discovered the other surprising property that they can always be represented by power series of the form (9). This is not true in general for real functions; there are real functions which have derivatives of all orders but cannot be represented by a power series. (Example: $f(x) = \exp(-1/x^2)$ when $x \neq 0$ and $f(0) = 0$. For details, see Ref. [A2, p. 128] in Appendix 1.)

The relation between our present consideration and that on power series in the preceding section may be established by the following theorem.

Theorem 2

Every power series with a nonzero radius of convergence is the Taylor series of the function represented by that power series.

Proof. Let the power series

$$\sum_{n=0}^{\infty} b_n(z-a)^n$$

have a nonzero radius of convergence R. Then it represents some analytic function $f(z)$ in the disk $|z-a| < R$, that is,

$$f(z) = b_0 + b_1(z-a) + b_2(z-a)^2 + \cdots .$$

From Theorem 5 in Sec. 16.2 it follows that

$$f'(z) = b_1 + 2b_2(z-a) + \cdots$$

and more generally

$$f^{(n)}(z) = n!\, b_n + (n+1)n \cdots 3 \cdot 2b_{n+1}(z-a) + \cdots ;$$

all these series converge in the disk $|z-a| < R$ and represent analytic functions. Hence these functions are continuous at $z = a$. By setting $z = a$ we thus obtain

$$f(a) = b_0, \qquad f'(a) = b_1, \qquad \cdots, \qquad f^{(n)}(a) = n!\, b_n, \cdots .$$

Since these formulas are identical with those in Taylor's theorem, the proof is complete. ∎

A point at which a function $f(z)$ ceases to be analytic is called a **singular point** of $f(z)$; we also say that $f(z)$ has a **singularity** at such a point. More precisely: a point $z = z_0$ is called a *singular point* of $f(z)$, if $f(z)$ is not differentiable at z_0 but if every neighborhood of z_0 contains points at which $f(z)$ is differentiable.

Using this concept, we may say that there is at least one singular point of $f(z)$ on the circle of convergence[5] of the development (9).

Before we discuss examples and practical aspects of Taylor series, let us mention the concepts of **developments around different centers** and **analytic continuation.** Suppose that a power series of powers of $z - a$ with a nonzero radius of convergence R is given and denote its sum by $f(z)$; thus

[5]The radius of convergence of (9) will in general be equal to the distance from a to the nearest singular point of $f(z)$, but it may be larger; for example, Ln z is singular along the negative real axis, and the distance from $a = -1 + i$ to that axis is 1, but the Taylor series of Ln z with center $a = -1 + i$ has radius of convergence $\sqrt{2}$.

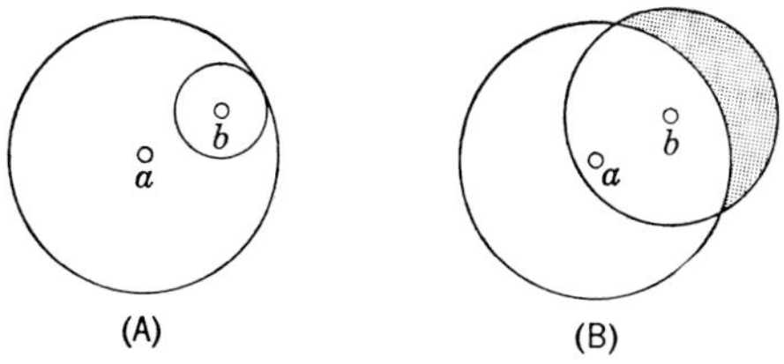

Fig. 327. Development around different centers and analytic continuation

$$f(z) = \sum_{n=0}^{\infty} a_n(z-a)^n. \tag{11}$$

From Theorem 5 in Sec. 16.2 we know that $f(z)$ is analytic in the disk $|z - a| < R$. From Theorem 2 in this section we conclude that the series in (11) is the Taylor series of $f(z)$ with center a. We may now choose any point b in that disk and apply Taylor's theorem to obtain the development

$$f(z) = \sum_{n=0}^{\infty} b_n(z-b)^n \tag{12}$$

with coefficients resulting from (11) by differentiating and setting $z = b$:

$$b_n = \frac{1}{n!} f^{(n)}(b) = \sum_{k=n}^{\infty} \binom{n}{k} a_k (b-a)^k.$$

This new development is valid at least in the disk $|z - b| < R - |b|$ shown in Fig. 327A. However, in many cases the radius of convergence of (12) will be larger than $R - |b|$, as is illustrated in Fig. 327B, so that (12) yields an *"extension"* or *"continuation"* of $f(z)$ to points outside the disk $|z - a| < R$ (shaded in Fig. 327B). This process of extending an analytic function, at first given by a power series in a region of convergence, beyond that region is called **analytic continuation.**

16.4 Taylor Series of Elementary Functions

Example 1. Geometric series

Let $f(z) = 1/(1 - z)$. Then we have $f^{(n)}(z) = n!/(1 - z)^{n+1}$, $f^{(n)}(0) = n!$. Hence the Maclaurin expansion of $1/(1 - z)$ is the geometric series

$$\frac{1}{1-z} = \sum_{n=0}^{\infty} z^n = 1 + z + z^2 + \cdots \qquad (|z| < 1). \tag{1}$$

$f(z)$ is singular at $z = 1$; this point lies on the circle of convergence.

Example 2. Exponential function

We know that the exponential function e^z (Sec. 12.7) is analytic for all z, and $(e^z)' = e^z$. Hence from (9), Sec. 16.3, with $a = 0$, we obtain the Maclaurin series

$$e^z = \sum_{n=0}^{\infty} \frac{z^n}{n!} = 1 + z + \frac{z^2}{2!} + \cdots . \tag{2}$$

This series is also obtained if we replace x in the Maclaurin series of e^x by z.

Let us show how we can prove the multiplication formula

$$e^{z_1}e^{z_2} = e^{z_1+z_2} \tag{3}$$

by using (2). We have

$$e^{z_1}e^{z_2} = \sum_{k=0}^{\infty} \frac{z_1{}^k}{k!} \sum_{m=0}^{\infty} \frac{z_2{}^m}{m!}.$$

Since both series converge absolutely, we may multiply them term by term; the sum of the products for which $k + m = n$ is

$$\frac{z_1{}^n}{n!} + \frac{z_1^{n-1}}{(n-1)!}\frac{z_2}{1!} + \cdots + \frac{z_1}{1!}\frac{z_2^{n-1}}{(n-1)!} + \frac{z_2{}^n}{n!}$$

$$= \frac{1}{n!}\left[z_1{}^n + \binom{n}{1}z_1^{n-1}z_2 + \binom{n}{2}z_1^{n-2}z_2{}^2 + \cdots + z_2{}^n\right] = \frac{(z_1+z_2)^n}{n!}.$$

Hence the product of the two series may be written

$$\sum_{n=0}^{\infty} \frac{(z_1+z_2)^n}{n!} = e^{z_1+z_2}$$

and (3) is proved.

Furthermore, by setting $z = iy$ in (2) and applying Theorem 1 in Sec. 15.6 we obtain

$$e^{iy} = \sum_{n=0}^{\infty} \frac{(iy)^n}{n!} = \sum_{k=0}^{\infty} (-1)^k \frac{y^{2k}}{(2k)!} + i \sum_{k=0}^{\infty} (-1)^k \frac{y^{2k+1}}{(2k+1)!}.$$

Since the series on the right are the familiar Maclaurin expansions of the real functions $\cos y$ and $\sin y$, this represents the Euler formula

$$e^{iy} = \cos y + i \sin y; \tag{4}$$

cf. (8) in Sec. 12.7. Multiplying by e^x and using (3), we obtain the formula (4) in Sec. 12.7 which was used to define e^z. Our present consideration shows that ***one may use (2) for defining e^z*** and derive from (2) all the formulas in Sec. 12.7.

Example 3. Trigonometric and hyperbolic functions

By substituting (2) into (1) of Sec. 12.8 we obtain

$$\begin{aligned} \cos z &= \sum_{n=0}^{\infty} (-1)^n \frac{z^{2n}}{(2n)!} = 1 - \frac{z^2}{2!} + \frac{z^4}{4!} - + \cdots \\ \sin z &= \sum_{n=0}^{\infty} (-1)^n \frac{z^{2n+1}}{(2n+1)!} = z - \frac{z^3}{3!} + \frac{z^5}{5!} - + \cdots . \end{aligned} \tag{5}$$

When $z = x$ these are the familiar Maclaurin series of the real functions $\cos x$ and $\sin x$. Similarly, by substituting (2) into (11), Sec. 12.8, we obtain

$$\text{(6)}\qquad \begin{aligned} \cosh z &= \sum_{n=0}^{\infty} \frac{z^{2n}}{(2n)!} = 1 + \frac{z^2}{2!} + \frac{z^4}{4!} + \cdots \\ \sinh z &= \sum_{n=0}^{\infty} \frac{z^{2n+1}}{(2n+1)!} = z + \frac{z^3}{3!} + \frac{z^5}{5!} + \cdots . \end{aligned}$$

Example 4. Logarithm

From (9) in Sec. 16.3 it follows that

$$\text{(7)}\qquad \operatorname{Ln}(1 + z) = z - \frac{z^2}{2} + \frac{z^3}{3} - + \cdots \qquad (|z| < 1).$$

Replacing z by $-z$ and multiplying both sides by -1, we get

$$\text{(8)}\qquad -\operatorname{Ln}(1 - z) = \operatorname{Ln}\frac{1}{1 - z} = z + \frac{z^2}{2} + \frac{z^3}{3} + \cdots \qquad (|z| < 1).$$

By adding both series we obtain

$$\text{(9)}\qquad \operatorname{Ln}\frac{1 + z}{1 - z} = 2\left(z + \frac{z^3}{3} + \frac{z^5}{5} + \cdots\right) \qquad (|z| < 1).$$

Problems for Sec. 16.4

1. Using (2), prove $(e^z)' = e^z$.
2. Derive (5) and (6) from (2). Obtain (7) from Taylor's theorem.
3. Using (5), show that $\cos z$ is even and $\sin z$ is odd.
4. Using (6), show that $\cosh z \neq 0$ for all real $z = x$.

Find the Taylor series of the following functions about the point $z = a$ and determine the radius of convergence.

5. $\cos 2z$, $a = 0$
6. $\sin z^2$, $a = 0$
7. e^{-z}, $a = 0$
8. e^z, $a = 1$
9. e^z, $a = \pi i$
10. $\sin z$, $a = \pi/2$
11. $\cos z$, $a = -\pi/4$
12. $1/(1 - z)$, $a = -1$
13. $1/z$, $a = -1$
14. $1/(1 - z)$, $a = i$
15. $\cos^2 z$, $a = 0$
16. $\sin^2 z$, $a = 0$

Find the first three terms of the Maclaurin series of the following functions.

17. $\tan z$
18. $e^z \sin z$
19. $z \cot z$

Find the Maclaurin series by integrating that of the integrand term by term. (erf z is called the **error function,** Si(z) the **sine integral,** and S(z) and C(z) the **Fresnel integrals.** Cf. also Sec. 19.15 and Appendix 3.)

20. $\displaystyle\int_0^z \frac{e^t - 1}{t}\,dt$
21. $\displaystyle\int_0^z \frac{1 - \cos t}{t^2}\,dt$
22. $\displaystyle\operatorname{erf} z = \frac{2}{\sqrt{\pi}}\int_0^z e^{-t^2}\,dt$
23. $\displaystyle\operatorname{Si}(z) = \int_0^z \frac{\sin t}{t}\,dt$
24. $\displaystyle\operatorname{S}(z) = \int_0^z \sin t^2\,dt$
25. $\displaystyle\operatorname{C}(z) = \int_0^z \cos t^2\,dt$

16.5 Practical Methods for Obtaining Power Series

In most practical cases the determination of the coefficients of a Taylor series by means of the formula in Taylor's theorem will be complicated or time-consuming. There are a number of simpler practical procedures for that purpose which may be illustrated by the following examples. The uniqueness of the representations thus obtained follows from Theorem 2 in Sec. 16.2.

Example 1. Substitution

Find the Maclaurin series of $f(z) = 1/(1 + z^2)$. By substituting $-z^2$ for z in (1), Sec. 16.4, we obtain

$$\frac{1}{1+z^2} = \frac{1}{1-(-z^2)} = \sum_{n=0}^{\infty} (-z^2)^n = \sum_{n=0}^{\infty} (-1)^n z^{2n} \tag{1}$$

$$= 1 - z^2 + z^4 - z^6 + \cdots \qquad (|z| < 1).$$

Example 2. Integration

Let $f(z) = \tan^{-1} z$. We have $f'(z) = 1/(1 + z^2)$. By integrating (1) term by term and noting that $f(0) = 0$ we find

$$\tan^{-1} z = \sum_{n=0}^{\infty} \frac{(-1)^n}{2n+1} z^{2n+1} = z - \frac{z^3}{3} + \frac{z^5}{5} - + \cdots \qquad (|z| < 1);$$

this series represents the principal value of $w = u + iv = \tan^{-1} z$, defined as that value for which $|u| < \pi/2$.

Example 3. Development by using the geometric series

Develop $1/(c - bz)$ in powers of $z - a$ where $c - ab \neq 0$ and $b \neq 0$. Obviously,

$$\frac{1}{c-bz} = \frac{1}{c-ab-b(z-a)} = \frac{1}{(c-ab)\left[1 - \dfrac{b(z-a)}{c-ab}\right]}.$$

To the last expression we apply (1) in Sec. 16.4 with z replaced by $b(z - a)/(c - ab)$, finding

$$\frac{1}{c-bz} = \frac{1}{c-ab} \sum_{n=0}^{\infty} \left[\frac{b(z-a)}{c-ab}\right]^n = \sum_{n=0}^{\infty} \frac{b^n}{(c-ab)^{n+1}} (z-a)^n$$

$$= \frac{1}{c-ab} + \frac{b}{(c-ab)^2}(z-a) + \frac{b^2}{(c-ab)^3}(z-a)^2 + \cdots.$$

This series converges for

$$\left|\frac{b(z-a)}{c-ab}\right| < 1 \qquad \text{that is,} \qquad |z-a| < \left|\frac{c-ab}{b}\right| = \left|\frac{c}{b} - a\right|.$$

Example 4. Binomial series, reduction by partial fractions

Find the Taylor series of the function

$$f(z) = \frac{2z^2 + 9z + 5}{z^3 + z^2 - 8z - 12}$$

with center at $z = 1$.

Given a rational function, we may first represent it as a sum of partial fractions and then apply the *binomial series*

(2)
$$\frac{1}{(1+z)^m} = (1+z)^{-m} = \sum_{n=0}^{\infty} \binom{-m}{n} z^n$$
$$= 1 - mz + \frac{m(m+1)}{2!}z^2 - \frac{m(m+1)(m+2)}{3!}z^3 + \cdots .$$

Since the function on the left is singular at $z = -1$, the series converges in the disk $|z| < 1$. In our case we obtain

$$f(z) = \frac{1}{(z+2)^2} + \frac{2}{z-3} = \frac{1}{[3+(z-1)]^2} - \frac{2}{2-(z-1)}.$$

This may be written in the form

$$f(z) = \frac{1}{9}\left(\frac{1}{[1+\frac{1}{3}(z-1)]^2}\right) - \frac{1}{1-\frac{1}{2}(z-1)}.$$

By using the binomial series we obtain

$$f(z) = \frac{1}{9}\sum_{n=0}^{\infty}\binom{-2}{n}\left(\frac{z-1}{3}\right)^n - \sum_{n=0}^{\infty}\left(\frac{z-1}{2}\right)^n.$$

We may add the two series on the right term by term. Since the binomial coefficient in the first series equals $(-2)(-3)\cdots(-[n+1])/n! = (-1)^n(n+1)$, we find

$$f(z) = \sum_{n=0}^{\infty}\left[\frac{(-1)^n(n+1)}{3^{n+2}} - \frac{1}{2^n}\right](z-1)^n = -\tfrac{8}{9} - \tfrac{31}{54}(z-1) - \tfrac{23}{108}(z-1)^2 - \cdots .$$

Since $z = 3$ is the singular point of $f(z)$ which is nearest to the center $z = 1$, the series converges in the disk $|z-1| < 2$.

Example 5. Use of differential equations

Find the Maclaurin series of $f(z) = \tan z$. We have $f'(z) = \sec^2 z$ and, therefore,

$$f'(z) = 1 + f^2(z), \qquad f'(0) = 1.$$

Observing that $f(0) = 0$, we obtain by successive differentiation

$$\begin{aligned}
&f'' = 2ff', && f''(0) = 0, && \\
&f''' = 2f'^2 + 2ff'', && f'''(0) = 2, && f'''(0)/3! = 1/3, \\
&f^{(4)} = 6f'f'' + 2ff''', && f^{(4)}(0) = 0, && \\
&f^{(5)} = 6f''^2 + 8f'f''' + 2ff^{(4)}, && f^{(5)}(0) = 16, && f^{(5)}(0)/5! = 2/15, \quad \text{etc.}
\end{aligned}$$

Hence the result is

(3)
$$\tan z = z + \tfrac{1}{3}z^3 + \tfrac{2}{15}z^5 + \tfrac{17}{315}z^7 + \cdots \qquad \left(|z| < \frac{\pi}{2}\right).$$

Example 6. Undetermined coefficients

Find the Maclaurin series of $\tan z$ by using those of $\cos z$ and $\sin z$ (Sec. 16.4). Since $\tan z$ is odd, the desired expansion will be of the form

$$\tan z = b_1 z + b_3 z^3 + b_5 z^5 + \cdots .$$

Using $\sin z = \tan z \cos z$ and inserting those developments, we obtain

$$z - \frac{z^3}{3!} + \frac{z^5}{5!} - + \cdots = (b_1 z + b_3 z^3 + b_5 z^5 + \cdots)\left(1 - \frac{z^2}{2!} + \frac{z^4}{4!} - + \cdots\right).$$

Since $\tan z$ is analytic except at $z = \pm\pi/2, \pm 3\pi/2, \cdots$, its Maclaurin series converges in the disk $|z| < \pi/2$, and for these z we may multiply the two series on the right term by term and arrange the resulting series in powers of z (cf. Theorem 3 in Sec. 16.1). By Theorem 2 in Sec. 16.2 the coefficient of each power of z is the same on both sides. This yields

$$1 = b_1, \qquad -\frac{1}{3!} = -\frac{b_1}{2!} + b_3, \qquad \frac{1}{5!} = \frac{b_1}{4!} - \frac{b_3}{2!} + b_5, \qquad \text{etc.}$$

Therefore $b_1 = 1$, $b_3 = \frac{1}{3}$, $b_5 = \frac{2}{15}$, etc., as before.

Problems for Sec. 16.5

Find the Maclaurin series of the following functions.

1. $\dfrac{1}{1 + z^4}$ **2.** $\dfrac{1}{1 - z^3}$ **3.** $\dfrac{1}{1 + z^3}$

4. $\dfrac{1}{1 - z^6}$ **5.** $\dfrac{1}{(1 + z^2)^2}$ **6.** $\dfrac{4z^2 + 30z + 68}{(z + 4)^2(z - 2)}$

7. $\cos z^2$ **8.** $e^{z^2 - z}$ **9.** e^{z^4}

Find the first few terms of the Maclaurin series of the following functions.

10. $\dfrac{\cos z}{1 - z^2}$ **11.** $e^{z^2} \sin z^2$ **12.** $e^{z^2}/\cos z$

13. $e^{1/(1-z)}$ **14.** $\cos\left(\dfrac{z}{1 - z}\right)$ **15.** $e^{(e^z)}$

Find the Taylor series of the given function about $z = a$.

16. $\dfrac{1}{2z - i}$, $a = -1$ **17.** $\dfrac{1}{4 - 3z}$, $a = 1 + i$ **18.** $\dfrac{2 - 3z}{2z^2 - 3z + 1}$, $a = -1$

19. $\dfrac{1}{(1 + z)^2}$, $a = -i$ **20.** $\dfrac{1}{(2 + 3z^3)^2}$, $a = 0$ **21.** $\tan z$, $a = \dfrac{\pi}{4}$

22. Developing $1/\sqrt{1 - z^2}$ and integrating, show that

$$\sin^{-1} z = z + \left(\frac{1}{2}\right)\frac{z^3}{3} + \left(\frac{1 \cdot 3}{2 \cdot 4}\right)\frac{z^5}{5} + \left(\frac{1 \cdot 3 \cdot 5}{2 \cdot 4 \cdot 6}\right)\frac{z^7}{7} + \cdots \qquad (|z| < 1).$$

Show that this series represents the principal value of $\sin^{-1} z$ (definition in Prob. 35, Sec. 12.9).

23. (Euler numbers) The Maclaurin series

$$\sec z = E_0 - \frac{E_2}{2!}z^2 + \frac{E_4}{4!}z^4 - + \cdots \tag{4}$$

defines the *Euler numbers* E_{2n}. Show that $E_0 = 1$, $E_2 = -1$, $E_4 = 5$, $E_6 = -61$.

24. (Bernoulli numbers) The Maclaurin series

$$\frac{z}{e^z - 1} = 1 + B_1 z + \frac{B_2}{2!}z^2 + \frac{B_3}{3!}z^3 + \cdots \tag{5}$$

defines the *Bernoulli numbers* B_n. Using the method of undetermined coefficients, show that[6]

$$\text{(6)}\quad B_1 = -\frac{1}{2}, \quad B_2 = \frac{1}{6}, \quad B_3 = 0, \quad B_4 = -\frac{1}{30}, \quad B_5 = 0, \quad B_6 = \frac{1}{42}, \cdots.$$

25. Using (1), (2), Sec. 12.8, and (5), show that

$$\text{(7)}\quad \tan z = \frac{2i}{e^{2iz} - 1} - \frac{4i}{e^{4iz} - 1} - i = \sum_{n=1}^{\infty} (-1)^{n-1} \frac{2^{2n}(2^{2n} - 1)}{(2n)!} B_{2n} z^{2n-1}.$$

16.6 Uniform Convergence

Suppose we know that a given series converges in a certain region R. Then the remaining question is whether the convergence is sufficiently rapid throughout the whole region or whether there are points near which the convergence becomes poor. The practical importance of this question in connection with computations is obvious, but we shall see that the theoretical aspect of the question is even more important. To illustrate the situation, let us start with some examples.

Example 1

Suppose we want to compute a table of e^x for real x in the interval $0 \leqq x \leqq 1$, for example, for $x = 0, 0.1, 0.2, \cdots$ and the absolute value of the error of each value should be less than a given number ϵ, say, less than $\frac{1}{2}$ unit of the sixth decimal place. We may use a suitable partial sum

$$s_n = 1 + x + \cdots + \frac{x^n}{n!}$$

of the Maclaurin series. Then the absolute value of the error is equal to $|R_n| = |s - s_n|$ where $s = e^x$, the sum of the series, and we have to choose n such that

$$|s(x) - s_n(x)| < \epsilon \; (= 5 \cdot 10^{-7}).$$

From Prob. 20 in Sec. 15.5 it follows that when $x = 1$, then for $n = 10$, and thus for any $n > N = 9$, we obtain the required accuracy. Now the remainder decreases in absolute value as x $(\geqq 0)$ decreases, and, therefore,

$$|s(x) - s_n(x)| < \epsilon \qquad \text{for} \qquad n > N(\epsilon) \; (= 9) \qquad \text{and all } x$$

under consideration. Note that, of course, N depends on ϵ, and if we want more accurate values so that ϵ is smaller, then N will be larger.

Example 2

In the case of the geometric series $1 + z + z^2 + \cdots$ the remainder is

$$R_n(z) = s(z) - s_n(z) = \sum_{m=n+1}^{\infty} z^m = \frac{z^{n+1}}{1 - z}$$

and becomes arbitrarily large for real $z = x$ (<1) and sufficiently close to 1. Hence, a maximum error ϵ being prescribed, we cannot find a number N *depending only on* ϵ such that we have

[6]For tables, see J. W. L. Glaisher, *Quarterly Journal Math.* **29**, 1898, 1–168.

$|R_n(x)| = |s(x) - s_n(x)| < \epsilon$ for $n > N$ and *all* x in the interval $0 \leqq x < 1$. (Cf. also Prob. 19 at the end of Sec. 15.5.) This result is not quite unexpected because the series diverges at $z = 1$. A really surprising situation will be discussed in the next example.

Example 3

Consider the series

$$x^2 + \frac{x^2}{1+x^2} + \frac{x^2}{(1+x^2)^2} + \frac{x^2}{(1+x^2)^3} + \cdots .$$

Using the formula for the sum of a geometric progression, the student may readily verify that the nth partial sum is

$$s_n(x) = 1 + x^2 - \frac{1}{(1+x^2)^n}.$$

(Some of these sums are shown in Fig. 328.) Hence, if $x \neq 0$, the series has the sum

$$s(x) = \lim_{n\to\infty} s_n(x) = 1 + x^2.$$

If $x = 0$, then $s_n = 0$ for all n and, therefore,

$$s(0) = \lim_{n\to\infty} s_n(0) = 0.$$

This shows that the series converges for all x (even absolutely), but we have the surprising result that the sum is discontinuous (at $x = 0$), although all the terms of the series are continuous functions. Furthermore, when $x \neq 0$ the absolute value of the remainder is

$$|R_n(x)| = |s(x) - s_n(x)| = \frac{1}{(1+x^2)^n}$$

and we see that for a given ϵ (<1) we cannot find an N depending only on ϵ such that $|R_n| < \epsilon$ for all $n > N(\epsilon)$ and all x in the interval $0 \leqq x \leqq 1$. ∎

The series in the examples are of the form

$$(1) \qquad \sum_{n=0}^{\infty} f_n(z) = f_0(z) + f_1(z) + f_2(z) + \cdots .$$

We assume that (1) converges for all z in a region G. Let $s(z)$ be the sum and

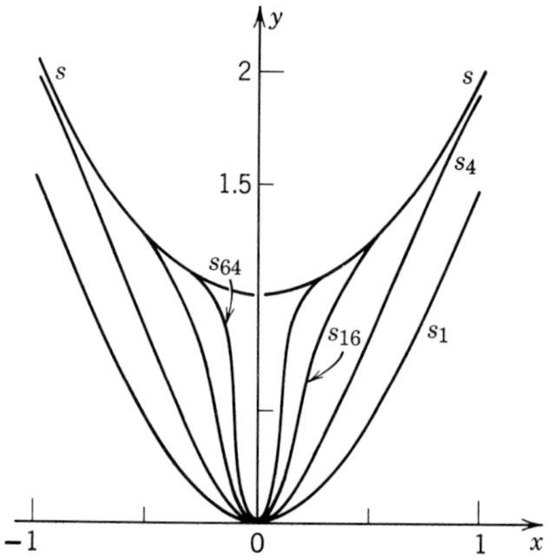

Fig. 328. Partial sums in Example 3

$s_n(z)$ the nth partial sum of (1). We know that convergence of (1) at a point z means that, an $\epsilon > 0$ being given, we can find an $N = N(\epsilon, z)$ such that

$$|s(z) - s_n(z)| < \epsilon \qquad \text{for all } n > N(\epsilon, z).$$

N depends on ϵ and will, in general, also depend on the point z which has been selected for consideration. Now, any $\epsilon > 0$ being given, it may happen that we can find a number $N(\epsilon)$, *independent of* z, such that

$$|s(z) - s_n(z)| < \epsilon \qquad \text{for all } n > N(\epsilon) \textbf{ and all } z \textbf{ in } G.$$

Then the series is said to be **uniformly convergent** *in* G.

Uniformity of convergence is thus a property that refers to an infinite set of values of z, whereas the convergence of a series may be considered for various particular values of z, without reference to other values.

The series in Example 1 is uniformly convergent in the interval $0 \leqq x \leqq 1$ (and, in fact, in any bounded interval). The series in Example 3 is not uniformly convergent in any interval containing the point 0. This shows that *an absolutely convergent series may not be uniformly convergent.* Conversely, *a uniformly convergent series may not be absolutely convergent.* This is illustrated by

Example 4. A uniformly but not absolutely convergent series

The series

$$\sum_{n=1}^{\infty} \frac{(-1)^{n-1}}{x^2 + n} = \frac{1}{x^2 + 1} - \frac{1}{x^2 + 2} + \frac{1}{x^2 + 3} - + \cdots \qquad (x \text{ real})$$

is uniformly convergent for all real x, but not absolutely convergent (cf. Prob. 18). ∎

Example 2 is typical of power series, because for such a series the situation is very simple, as follows.

Theorem 1 (Power series)

A power series

$$\sum_{n=0}^{\infty} c_n(z - a)^n \tag{2}$$

with a nonzero radius of convergence R *is uniformly convergent in every circular disk* $|z - a| \leqq r$ *of radius* $r < R$.

Proof. For $|z - a| \leqq r$ we have

$$|c_{n+1}(z - a)^{n+1} + \cdots + c_{n+p}(z - a)^{n+p}| \leqq |c_{n+1}| \, r^{n+1} + \cdots + |c_{n+p}| \, r^{n+p}. \tag{3}$$

Since (2) converges absolutely when $|z - a| = r < R$, it follows from the Cauchy convergence principle (Sec. 15.3) that, an $\epsilon > 0$ being given, we can find an $N(\epsilon)$ such that

$$|c_{n+1}| \, r^{n+1} + \cdots + |c_{n+p}| \, r^{n+p} < \epsilon \quad \text{for} \quad n > N(\epsilon) \quad \text{and} \quad p = 1, 2, \cdots.$$

From this and (3) we obtain

$$|c_{n+1}(z-a)^{n+1} + \cdots + c_{n+p}(z-a)^{n+p}| < \epsilon$$

for all z in the disk $|z-a| \leqq r$, every $n > N(\epsilon)$, and every $p = 1, 2, \cdots$. Since $N(\epsilon)$ is independent of z, this shows uniform convergence, and the theorem is proved. ∎

While, of course, the sum of finitely many continuous functions is continuous, Example 3 illustrates that the sum of an infinite series of continuous functions may be discontinuous, even if it converges absolutely. But if a series converges *uniformly,* this cannot happen. In fact the following important theorem holds.

Theorem 2 (Continuity)

Let the series

$$\sum_{m=0}^{\infty} f_m(z) = f_0(z) + f_1(z) + \cdots$$

be uniformly convergent in a region G and let $F(z)$ be its sum. Then, if each term $f_m(z)$ is continuous at a point z_0 in G, the function $F(z)$ is continuous at z_0.

Proof. Let $s_n(z)$ be the nth partial sum of the series and $R_n(z)$ the corresponding remainder:

$$s_n = f_0 + f_1 + \cdots + f_n, \qquad R_n = f_{n+1} + f_{n+2} + \cdots .$$

Since the series converges uniformly, for a given $\epsilon > 0$ we can find an $n = N(\epsilon)$ such that

$$|R_N(z)| < \frac{\epsilon}{3} \qquad \text{for all } z \text{ in } G.$$

Since $s_N(z)$ is a sum of finitely many functions which are continuous at z_0, this sum is continuous at z_0. Therefore we can find a $\delta > 0$ such that

$$|s_N(z) - s_N(z_0)| < \frac{\epsilon}{3} \qquad \text{for all } z \text{ in } G \text{ for which } |z - z_0| < \delta.$$

By the triangle inequality (Sec. 12.2) for these z we thus obtain

$$\begin{aligned}|F(z) - F(z_0)| &= |s_N(z) + R_N(z) - [s_N(z_0) + R_N(z_0)]| \\ &\leqq |s_N(z) - s_N(z_0)| + |R_N(z)| + |R_N(z_0)| < \frac{\epsilon}{3} + \frac{\epsilon}{3} + \frac{\epsilon}{3} = \epsilon.\end{aligned}$$

This implies that $F(z)$ is continuous at z_0, and the theorem is proved. ∎

We want to mention that in this theorem uniformity of convergence is a sufficient rather than a necessary condition. This may be illustrated by the following example.

Example 5
Let

$$u_m(x) = \frac{mx}{1 + m^2x^2}$$

and consider the series

$$\sum_{m=1}^{\infty} f_m(x) \qquad \text{where} \qquad f_m(x) = u_m(x) - u_{m-1}(x).$$

The nth partial sum is

$$s_n = u_1 - u_0 + u_2 - u_1 + \cdots + u_n - u_{n-1} = u_n - u_0 = u_n.$$

Hence the series has the sum

$$F(x) = \lim_{n\to\infty} s_n(x) = \lim_{n\to\infty} u_n(x) = 0,$$

which is a continuous function. However, the series is not uniformly convergent in an interval $0 \leqq x \leqq a$, where $a > 0$. In fact, from

$$|F(x) - s_n(x)| = \frac{nx}{1 + n^2x^2} < \epsilon$$

we obtain

$$\frac{nx}{\epsilon} < 1 + n^2x^2 \qquad \text{or} \qquad n^2x^2 - \frac{nx}{\epsilon} + 1 > 0$$

and from this

$$n > \frac{1}{2x\epsilon}(1 + \sqrt{1 - 4\epsilon^2}).$$

For fixed ϵ the right side approaches infinity as x approaches zero, which shows that the series is not uniformly convergent in that interval. ∎

Under what conditions may we integrate a series term by term?

Let us start our discussion with an example which illustrates the important fact that term by term integration of series is not always permissible.

Example 6. A series for which termwise integration is not permissible
Let

$$u_m(x) = mxe^{-mx^2}$$

and consider the series

$$\sum_{m=1}^{\infty} f_m(x) \qquad \text{where} \qquad f_m(x) = u_m(x) - u_{m-1}(x)$$

in the interval $0 \leqq x \leqq 1$. The nth partial sum is

$$s_n = u_1 - u_0 + u_2 - u_1 + \cdots + u_n - u_{n-1} = u_n - u_0 = u_n.$$

Hence the series has the sum

$$F(x) = \lim_{n\to\infty} s_n(x) = \lim_{n\to\infty} u_n(x) = 0 \qquad (0 \leqq x \leqq 1).$$

From this we obtain

$$\int_0^1 F(x)\,dx = 0.$$

On the other hand, by integrating term by term,

$$\sum_{m=1}^{\infty} \int_0^1 f_m(x)\,dx = \lim_{n\to\infty} \sum_{m=1}^{n} \int_0^1 f_m(x)\,dx = \lim_{n\to\infty} \int_0^1 s_n(x)\,dx.$$

Now $s_n = u_n$ and the expression on the right becomes

$$\lim_{n\to\infty} \int_0^1 u_n(x)\,dx = \lim_{n\to\infty} \int_0^1 nxe^{-nx^2}\,dx = \lim_{n\to\infty} \tfrac{1}{2}(1 - e^{-n}) = \tfrac{1}{2},$$

but not 0. This shows that the series under consideration cannot be integrated term by term from $x = 0$ to $x = 1$. ∎

The series in Example 6 is not uniformly convergent in that interval, and we shall now prove that in the case of a uniformly convergent series of continuous functions we may integrate term by term.

Theorem 3 (Termwise integration)

Let

$$F(z) = \sum_{n=0}^{\infty} f_n(z) = f_0(z) + f_1(z) + \cdots$$

be a uniformly convergent series of continuous functions within a region G. Let C be any path in G. Then the series

$$(4) \qquad \sum_{n=0}^{\infty} \int_C f_n(z)\,dz = \int_C f_0(z)\,dz + \int_C f_1(z)\,dz + \cdots$$

is convergent and has the sum $\int_C F(z)\,dz$.

Proof. From Theorem 2 it follows that $F(z)$ is continuous. Let $s_n(z)$ be the nth partial sum of the given series and $R_n(z)$ the corresponding remainder. Then $F = s_n + R_n$ and

$$\int_C F(z)\,dz = \int_C s_n(z)\,dz + \int_C R_n(z)\,dz.$$

Let l be the length of C. Since the given series converges uniformly, for every given $\epsilon > 0$ we can find a number N such that

$$|R_n(z)| < \frac{\epsilon}{l} \qquad \text{for all } n > N \text{ and all } z \text{ in } G.$$

By applying (4), Sec. 14.2, we thus obtain

$$\left|\int_C R_n(z)\,dz\right| < \frac{\epsilon}{l}\,l = \epsilon \qquad \text{for all } n > N.$$

Since $R_n = F - s_n$, this means that

$$\left| \int_C F(z)\,dz - \int_C s_n(z)\,dz \right| < \epsilon \qquad \text{for all } n > N.$$

Hence, the series (4) converges and has the sum indicated in the theorem. This completes the proof. ▮

Theorems 2 and 3 characterize the two most important properties of uniformly convergent series.

Of course, since differentiation and integration are inverse processes, we readily conclude from Theorem 3 that a convergent series may be differentiated term by term, provided the terms of the given series have continuous derivatives and the resulting series is uniformly convergent; more precisely, the following theorem holds.

Theorem 4 (Termwise differentiation)

Let the series $f_0(z) + f_1(z) + f_2(z) + \cdots$ be convergent in a region G and let $F(z)$ be its sum. Suppose that the series $f_0'(z) + f_1'(z) + f_2'(z) + \cdots$ converges uniformly in G and its terms $f_0'(z), f_1'(z), \cdots$ are continuous in G. Then

$$F'(z) = f_0'(z) + f_1'(z) + f_2'(z) + \cdots \qquad \textit{for all } z \textit{ in } G.$$

The simple proof is left to the student (Prob. 10).

Usually uniform convergence is established by a comparison test, which is called the Weierstrass M-test.

Theorem 5 (Weierstrass *M*-test)

If, for all values of z within a region G, the absolute values of the terms of a given series of the form (1) *are, respectively, less than or equal to the corresponding terms in a convergent series of constant terms,*

$$M_0 + M_1 + M_2 + \cdots, \tag{5}$$

then the series (1) *converges uniformly in G.*

The simple proof is left to the student (Prob. 11).

Example 7. Weierstrass *M*-test

The series

$$\sum_{m=1}^{\infty} \frac{\sin mx}{m^2} \qquad (x \text{ real})$$

converges uniformly in any interval. This follows from the Weierstrass test, because for real x,

$$\left| \frac{\sin mx}{m^2} \right| \leqq \frac{1}{m^2}$$

and $\Sigma\, m^{-2}$ is convergent [cf. (5) in Sec. 15.5].

Problems for Sec. 16.6

Prove that the following series converge uniformly in the given regions.

1. $\sum_{n=0}^{\infty} z^n, \quad |z| \leqq 0.99$

2. $\sum_{n=0}^{\infty} \frac{z^n}{n!}, \quad |z| \leqq 10^{30}$

3. $\sum_{n=1}^{\infty} \frac{(n!)^2}{(2n)!} z^n, \quad |z| \leqq 3.9$

4. $\sum_{n=1}^{\infty} \frac{z^n}{n^2}, \quad |z| \leqq 1$

5. $\sum_{n=1}^{\infty} \frac{\sin n|z|}{2^n}, \quad$ all z

6. $\sum_{n=1}^{\infty} \frac{\tanh^n x}{n(n+1)}, \quad$ all real x

7. $\sum_{n=1}^{\infty} \frac{\cos^n |z|}{n^2}, \quad$ all z

8. $\sum_{n=1}^{\infty} \frac{1}{|z| + n^2}, \quad$ all z

9. If the series (1) converges uniformly in a region G, show that it converges uniformly in any portion of G.

10. Derive Theorem 4 from Theorem 3.

11. Give a proof of the Weierstrass M-test (Theorem 5).

12. Determine the smallest integer n such that $|R_n| < 0.01$ in Example 2, when $z = x = 0.5, 0.6, 0.7, 0.8, 0.9$. What does the result mean from the viewpoint of computing $1/(1 - x)$ with an absolute error less than 0.01 by means of the geometric series?

13. Find the series whose nth partial sum is $s_n(x) = nx/(nx + 1)$ and graph s_1, s_2, s_3, s_4, and $s = \lim_{n \to \infty} s_n$ for $x \geqq 0$.

14. Prove that the series in Example 3 is not uniformly convergent in any interval containing the point $x = 0$.

15. Show that $x^2 \sum_{n=1}^{\infty} (1 + x^2)^{-n} = 1$ $(x \neq 0)$ and 0 when $x = 0$ and sketch a figure of the partial sums similar to Fig. 328.

16. Find the precise region of convergence of the series in Example 3 with x replaced by a complex variable z.

17. Show that $1 + \sum_{n=1}^{\infty} (x^n - x^{n-1})$ is not uniformly convergent in the interval $0 \leqq x \leqq 1$. Graph the partial sums s_1, s_2, s_3, s_4.

18. Prove the statement in Example 4.

Heat equation. Show that (10), Sec. 11.5, with coefficients (11) is a solution of the heat equation for $t > 0$, assuming that $f(x)$ is continuous on the interval $0 \leqq x \leqq l$ and has one-sided derivatives at all interior points of that interval. Proceed as follows.

19. Show that $|B_n|$ is bounded, say, $|B_n| < K$ for all n. Conclude that

$$|u_n| < Ke^{-\lambda_n^2 t_0} \qquad \text{when} \qquad t \geqq t_0 > 0$$

and, by the Weierstrass test, the series (10) converges uniformly with respect to x and t when $t \geqq t_0$, $0 \leqq x \leqq l$. Using Theorem 2, show that $u(x, t)$ is continuous when $t \geqq t_0$ and thus satisfies the boundary conditions (2) when $t \geqq t_0$.

20. Show that $|\partial u_n/\partial t| < \lambda_n^2 K e^{-\lambda_n^2 t_0}$ when $t \geqq t_0$ and the series of the expressions on the right converges, by the ratio test. Conclude from this, the Weierstrass test and Theorem 4 that the series (10) can be differentiated term by term with respect to t and the resulting series has the sum $\partial u/\partial t$. Show that (10) can be differentiated twice with respect to x and the resulting series has the sum $\partial^2 u/\partial x^2$. Conclude from this and the result of Prob. 19 that (10) is a solution of the heat equation for all $t \geqq t_0$. (The proof that (10) satisfies the given initial condition can be found in Ref. [D2].)

16.7 Laurent Series

In various applications it is necessary to expand a function $f(z)$ around points where $f(z)$ is singular. Taylor's theorem cannot be applied in such cases. A new type of series, known as *Laurent series*, is necessary. This will be a representation which is valid in an annulus bounded by two concentric circles C_1 and C_2 such that $f(z)$ is analytic in the annulus and at each point of C_1 and C_2 (Fig. 329). As in the case of the Taylor series, $f(z)$ may be singular at some points outside C_1 and, as the essentially new feature, it may also be singular at some points inside C_2.

Laurent's theorem[7]

If $f(z)$ is analytic on two concentric circles C_1 and C_2 with center a and in the annulus between them, then $f(z)$ can be represented by the ***Laurent series***

$$
\begin{aligned}
f(z) &= \sum_{n=0}^{\infty} b_n(z-a)^n + \sum_{n=1}^{\infty} \frac{c_n}{(z-a)^n} \\
&= b_0 + b_1(z-a) + b_2(z-a)^2 + \cdots + \frac{c_1}{z-a} + \frac{c_2}{(z-a)^2} + \cdots
\end{aligned} \tag{1}
$$

where[8]

$$
b_n = \frac{1}{2\pi i}\int_C \frac{f(z^*)}{(z^*-a)^{n+1}}\,dz^*, \qquad c_n = \frac{1}{2\pi i}\int_C (z^*-a)^{n-1} f(z^*)\,dz^*, \tag{2}
$$

each integral being taken in the counterclockwise sense around any simple closed path C which lies in the annulus and encircles the inner circle (Fig. 329).

This series converges and represents $f(z)$ in the open annulus obtained from the given annulus by continuously increasing the circle C_1 and decreasing C_2 until each of the two circles reaches a point where $f(z)$ is singular.

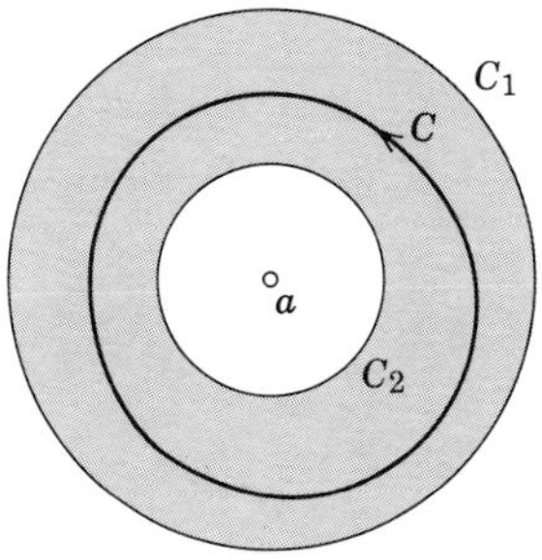

Fig. 329. Laurent's theorem

[7] PIERRE ALPHONSE LAURENT (1813–1854), French mathematician.
[8] We denote the variable of integration by z^* because z is used in $f(z)$.

Remark. Obviously, instead of (1) and (2) we may write simply

$$f(z) = \sum_{n=-\infty}^{\infty} A_n(z-a)^n \tag{1'}$$

where

$$A_n = \frac{1}{2\pi i}\int_C \frac{f(z^*)}{(z^*-a)^{n+1}}\,dz^*. \tag{2'}$$

Proof of Laurent's theorem. Let z be any point in the given annulus. Then from Cauchy's integral formula [cf. (3) in Sec. 14.5] it follows that

$$f(z) = \frac{1}{2\pi i}\int_{C_1} \frac{f(z^*)}{z^*-z}\,dz^* - \frac{1}{2\pi i}\int_{C_2} \frac{f(z^*)}{z^*-z}\,dz^*, \tag{3}$$

where we integrate in the counterclockwise sense. We shall now transform these integrals in a fashion similar to that in Sec. 16.3. Since z lies inside C_1, the first of these integrals is precisely of the same type as the integral (1), Sec. 16.3. By expanding it and estimating the remainder as in Sec. 16.3 we obtain

$$\frac{1}{2\pi i}\int_{C_1} \frac{f(z^*)}{z^*-z}\,dz^* = \sum_{n=0}^{\infty} b_n(z-a)^n \tag{4}$$

where the coefficients are given by the formula

$$b_n = \frac{1}{2\pi i}\int_{C_1} \frac{f(z^*)}{(z^*-a)^{n+1}}\,dz^* \tag{5}$$

and we integrate in the counterclockwise sense. Since a is not a point of the annulus, the functions $f(z^*)/(z^*-a)^{n+1}$ are analytic in the annulus. Hence we may integrate along the path C (cf. in the theorem) instead of C_1, without altering the value of b_n. This proves (2) for all $n \geqq 0$.

In the case of the last integral in (3), the situation is different, since z lies outside C_2. Instead of (3), Sec. 16.3, we now have

$$\left|\frac{z^*-a}{z-a}\right| < 1, \tag{6}$$

that is, we now have to develop $1/(z^*-z)$ in powers of $(z^*-a)/(z-a)$ for the resulting series to be convergent. We find

$$\frac{1}{z^*-z} = \frac{1}{z^*-a-(z-a)} = \frac{-1}{(z-a)\left(1-\dfrac{z^*-a}{z-a}\right)}.$$

By applying the formula for the sum of a finite geometric progression to the last expression we obtain

$$\frac{1}{z^*-z} = -\frac{1}{z-a}\left\{1 + \frac{z^*-a}{z-a} + \left(\frac{z^*-a}{z-a}\right)^2 + \cdots + \left(\frac{z^*-a}{z-a}\right)^n\right\} - \frac{1}{z-z^*}\left(\frac{z^*-a}{z-a}\right)^{n+1}.$$

To get the last integral in (3), we multiply this development by $(-1/2\pi i)f(z^*)$ and integrate over C_2. We readily obtain

$$-\frac{1}{2\pi i}\int_{C_2}\frac{f(z^*)}{z^*-z}\,dz^*$$
$$= \frac{1}{2\pi i}\left\{\frac{1}{z-a}\int_{C_2} f(z^*)\,dz^* + \frac{1}{(z-a)^2}\int_{C_2}(z^*-a)f(z^*)\,dz^* + \cdots + \frac{1}{(z-a)^{n+1}}\int_{C_2}(z^*-a)^n f(z^*)\,dz^*\right\} + R_n^*(z);$$

in this representation the last term is of the form

$$(7) \qquad R_n^*(z) = \frac{1}{2\pi i(z-a)^{n+1}}\int_{C_2}\frac{(z^*-a)^{n+1}}{z-z^*}f(z^*)\,dz^*.$$

In the integrals in the braces we may replace the circle C_2 by the aforementioned path C, without altering their values. This establishes Laurent's theorem provided that

$$(8) \qquad \lim_{n\to\infty} R_n^*(z) = 0.$$

We prove (8). Since $z - z^* \neq 0$ and $f(z)$ is analytic in the annulus and on C_2, the absolute value of the expression $f(z^*)/(z - z^*)$ in (7) is bounded, say,

$$\left|\frac{f(z^*)}{z-z^*}\right| < \widetilde{M} \qquad \text{for all } z^* \text{ on } C_2.$$

By applying (4) in Sec. 14.2 to (7) and denoting the length of C_2 by l we thus obtain

$$|R_n^*(z)| < \frac{1}{2\pi|z-a|^{n+1}}|z^*-a|^{n+1}\,\widetilde{M}l = \frac{\widetilde{M}l}{2\pi}\left|\frac{z^*-a}{z-a}\right|^{n+1}.$$

From (6) we see that the expression on the right approaches zero as n approaches infinity. This proves (8). The representation (1) with coefficients (2) is now established in the given annulus.

Finally let us prove convergence of (1) in the open annulus characterized at the end of the theorem.

We denote the sums of the two series in (1) by $g(z)$ and $h(z)$, and the radii of C_1 and C_2 by r_1 and r_2, respectively. Then $f = g + h$. The first series is a power series. Since it converges in the annulus, it must converge in the entire disk bounded by C_1, and g is analytic in that disk.

Setting $Z = 1/(z - a)$, the last series becomes a power series in Z. The annulus $r_2 < |z - a| < r_1$ then corresponds to the annulus $1/r_1 < |Z| < 1/r_2$, the new series converges in this annulus and, therefore, in the entire disk $|Z| < 1/r_2$. Since this disk corresponds to $|z - a| > r_2$, the exterior of C_2, the given series converges for all z outside C_2, and h is analytic for all these z.

Since $f = g + h$, it follows that g must be singular at all those points outside C_1 where f is singular, and h must be singular at all those points inside C_2 where f is singular. Consequently, the first series converges for all z inside the circle about a whose radius is equal to the distance of that singularity of f outside C_1 which is closest to a. Similarly, the second series converges for all z outside the circle about a whose radius is equal to the maximum distance of the singularities of f inside C_2. The domain common to both of those domains of convergence is the open annulus characterized at the end of the theorem, and the proof is complete. ∎

It follows that if $f(z)$ is analytic inside C_2, the Laurent series reduces to the Taylor series of $f(z)$ with center a. In fact, by applying Cauchy's integral theorem to (2) we see that in this case all the coefficients of the negative powers in (1) are zero.

Furthermore, if $z = a$ is the only singular point of $f(z)$ in C_2, then the Laurent expansion (1) converges for all z in C_1 except at $z = a$. This case occurs frequently and, therefore, is of particular importance. We discuss it later in detail.

The Laurent series of a given analytic function $f(z)$ in its annulus of convergence is unique (cf. Prob. 10, this section). *However, $f(z)$ may have different Laurent series in two annuli with the same center* (cf. Example 2, below).

The uniqueness is important, because Laurent series usually are not obtained by using (2) for determining the coefficients, but by various other methods. Some of these methods are illustrated by the following examples. If a Laurent series is found by any such process, it must be *the* Laurent series of the given function in the given annulus.

Example 1

The Laurent series of $z^2e^{1/z}$ with center 0 can be obtained from (2) in Sec. 16.4. Replacing z by $1/z$ in that series, we find

$$z^2e^{1/z} = z^2\left(1 + \frac{1}{1!\,z} + \frac{1}{2!\,z^2} + \cdots\right) = z^2 + z + \frac{1}{2} + \frac{1}{3!\,z} + \frac{1}{4!\,z^2} + \cdots \qquad (|z| > 0).$$

Example 2. Laurent expansions in different annuli

Find all Laurent series of the function $f(z) = 1/(1 - z^2)$ with center at $z = 1$. We have $1 - z^2 = -(z - 1)(z + 1)$. Using the geometric series

$$\frac{1}{1-q} = \sum_{n=0}^{\infty} q^n \qquad (|q| < 1),$$

we find

$$\text{(a)}\qquad \begin{aligned}\frac{1}{z+1} &= \frac{1}{2 + (z-1)} = \frac{1}{2}\,\frac{1}{\left[1 - \left(-\dfrac{z-1}{2}\right)\right]}\\ &= \frac{1}{2}\sum_{n=0}^{\infty}\left(-\frac{z-1}{2}\right)^n = \sum_{n=0}^{\infty}\frac{(-1)^n}{2^{n+1}}(z-1)^n;\end{aligned}$$

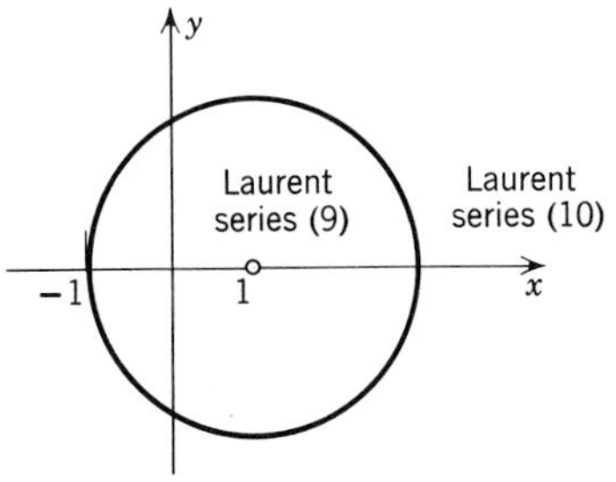

Fig. 330. Example 2

this series converges in the disk $|(z-1)/2| < 1$, that is, $|z-1| < 2$. Cf. Fig. 330. Similarly,

$$\text{(b)}\qquad \begin{aligned}\frac{1}{z+1} &= \frac{1}{(z-1)+2} = \frac{1}{(z-1)}\frac{1}{\left(1+\dfrac{2}{z-1}\right)}\\ &= \frac{1}{z-1}\sum_{n=0}^{\infty}\left(-\frac{2}{z-1}\right)^n = \sum_{n=0}^{\infty}\frac{(-2)^n}{(z-1)^{n+1}};\end{aligned}$$

this series converges for $|2/(z-1)| < 1$, that is, $|z-1| > 2$. Cf. Fig. 330. Hence from (a) we obtain

$$\text{(9)}\qquad \begin{aligned} f(z) &= \frac{-1}{(z-1)(z+1)} = \sum_{n=0}^{\infty}\frac{(-1)^{n+1}}{2^{n+1}}(z-1)^{n-1}\\ &= \frac{-1/2}{z-1} + \frac{1}{4} - \frac{1}{8}(z-1) + \frac{1}{16}(z-1)^2 - + \cdots;\end{aligned}$$

this series converges in the domain $0 < |z-1| < 2$. Similarly, from (b) we obtain

$$\text{(10)}\qquad f(z) = -\sum_{n=0}^{\infty}\frac{(-2)^n}{(z-1)^{n+2}} = -\frac{1}{(z-1)^2} + \frac{2}{(z-1)^3} - \frac{4}{(z-1)^4} + - \cdots + - \cdots.$$

This series converges for $|z-1| > 2$.

Example 3. cot *z*

From Prob. 19 at the end of Sec. 16.4 we immediately have the Laurent series

$$\cot z = \frac{1}{z} - \frac{1}{3}z - \frac{1}{45}z^3 - \frac{2}{945}z^5 - \cdots \qquad (0 < |z| < \pi). \quad ∎$$

If $z = a$ is the only singular point of $f(z)$ in C_2 (cf. Fig. 329), the Laurent series (1) converges in a region of the form

$$\text{(11)}\qquad 0 < |z-a| < R.$$

The singularity of $f(z)$ at $z = a$ is called a **pole** or an **essential singularity** depending on whether this Laurent series (the one that converges in a neighborhood of $z = a$, except at $z = a$) has finitely or infinitely many negative powers. An analytic function whose only singularities in the finite plane are poles is called a **meromorphic function.**

For instance, for determining the kind of singularity of $1/(1-z^2)$ at $z = 1$

(cf. Example 2) we must use (9) but not (10), since it is (9) that converges in a region of the form (11), with $a = 1$. Since (9) has one negative power, that singularity is a pole but not an essential singularity [as we would erroneously conclude from (10)].

A detailed discussion and further examples will follow in the next section.

Problems for Sec. 16.7

Expand the following functions in Laurent series which converge for $0 < |z| < R$ and determine the precise region of convergence.

1. $\dfrac{e^{-z}}{z^3}$ **2.** $\dfrac{e^{1/z^2}}{z^6}$ **3.** $\dfrac{\cos 2z}{z^2}$

4. $\dfrac{1}{z^4(1 + z)}$ **5.** $\dfrac{1}{z^2(1 - z^2)}$ **6.** $\dfrac{1}{z^2(z - 3)}$

7. $\dfrac{\sinh 3z}{z^3}$ **8.** $\dfrac{1}{z^8 + z^4}$ **9.** $\dfrac{1}{z^2(1 + z)^2}$

10. Prove that the Laurent expansion of a given analytic function in a given annulus is unique.

11. Does $\tan(1/z)$ have a Laurent series convergent in a region $0 < |z| < R$?

Find all Taylor series and Laurent series with center $z = a$ and determine the precise regions of convergence.

12. $\dfrac{1}{z^2 + 1}$, $a = -i$ **13.** $\dfrac{1}{z^4}$, $a = 1$ **14.** $\dfrac{1}{z^3}$, $a = i$

15. $\dfrac{1}{z^2 + 1}$, $a = i$ **16.** $\dfrac{1}{1 - z^4}$, $a = -1$ **17.** $\dfrac{4z - 1}{z^4 - 1}$, $a = 0$

18. $\dfrac{\sin z}{(z - \frac{1}{4}\pi)^3}$, $a = \dfrac{\pi}{4}$ **19.** $\dfrac{e^z}{(z - 1)^2}$, $a = 1$ **20.** $\dfrac{4z^2 + 2z - 4}{z^3 - 4z}$, $a = 2$

16.8 Analyticity at Infinity. Zeros and Singularities

In this section we consider zeros and singularities of analytic functions. We shall see that there are different types of singularities, which can be characterized by means of the Laurent series.

Our consideration will take place in the extended complex plane since we also want to investigate the behavior of functions $f(z)$ as $|z| \to \infty$. So let us first recall from Sec. 13.3 that the **extended complex plane** is obtained by attaching an improper point ∞ ("*point at infinity*") to the complex plane. The latter is then called the *finite complex plane,* for distinction. This process was motivated in Sec. 13.3 by the transformation $w = 1/z$; then $z = \infty$ has the image $w = 0$ (and $w = \infty$ has the inverse image $z = 0$).

If we want to investigate a given function $f(z)$ for large $|z|$, we may now set $z = 1/w$ and investigate $f(z) = f(1/w) \equiv g(w)$ in a neighborhood of $w = 0$. We define

$$g(0) = \lim_{w \to 0} g(w). \tag{1}$$

We also define that $f(z)$ is **analytic or singular at infinity** according as $g(w)$ is analytic or singular, respectively, at $w = 0$. (For the concept of a singularity, see Sec. 16.3.)

Example 1. Functions analytic or singular at infinity

The function $f(z) = 1/z^2$ is analytic at infinity since $g(w) = f(1/w) = w^2$ is analytic at $w = 0$. The function $f(z) = z^3$ is singular at infinity because $g(w) = f(1/w) = 1/w^3$ is singular at $w = 0$. The exponential function e^z is singular at infinity since $e^{1/w}$ is singular at $w = 0$. Similarly, the trigonometric functions $\sin z$ and $\cos z$ are singular at infinity. ▮

If a function $f(z)$ is analytic at infinity, we may obtain a corresponding Laurent series in a simple fashion as follows. Suppose that $f(z)$ is analytic in a domain $|z - a| > R$ (the exterior of a circle of radius R about a), also at infinity. We set

$$z = \frac{1}{w} + a. \qquad \text{Then} \qquad z - a = \frac{1}{w}$$

and the function $h(w)$ given by

$$h(w) = f\left(\frac{1}{w} + a\right) = f(z)$$

is analytic in the disk $|z - a| = |1/w| > R$, that is, $|w| < 1/R$. Hence $h(w)$ has a Maclaurin series, say,

$$h(w) = \sum_{n=0}^{\infty} c_n w^n = c_0 + c_1 w + c_2 w^2 + \cdots \qquad \left(|w| < \frac{1}{R}\right).$$

Then by inserting $w = 1/(z - a)$ we readily obtain the Laurent series

$$(2) \qquad f(z) = \sum_{n=0}^{\infty} \frac{c_n}{(z-a)^n} = c_0 + \frac{c_1}{z-a} + \frac{c_2}{(z-a)^2} + \cdots \qquad (|z - a| > R).$$

Riemann Number Sphere

The usual representation of complex numbers z in the complex plane is convenient as long as the absolute values of the numbers are not too large. For large $|z|$ the situation becomes inconvenient, and in this case we may prefer a representation of the complex numbers on a sphere, which was suggested by Riemann and is obtained as follows. (See Fig. 331 on the next page.)

Let S be a sphere of diameter 1 which touches the complex z-plane at the origin (Fig. 331). Let N be the "North pole" of S (the point diametrically opposite to the point of contact between the sphere and the plane). Let P be any point in the finite complex plane. Then the straight segment with endpoints P and N intersects S at a point P^*. We let P and P^* correspond to each other. In this way we obtain a correspondence between the points in the finite complex plane and the points on S, and P^* is the image point of P with respect to this mapping. The complex numbers, first represented in the plane, are now represented by points on S. To each z there corresponds a point on S. Conversely, each point on S represents a complex number z, except for the point N which

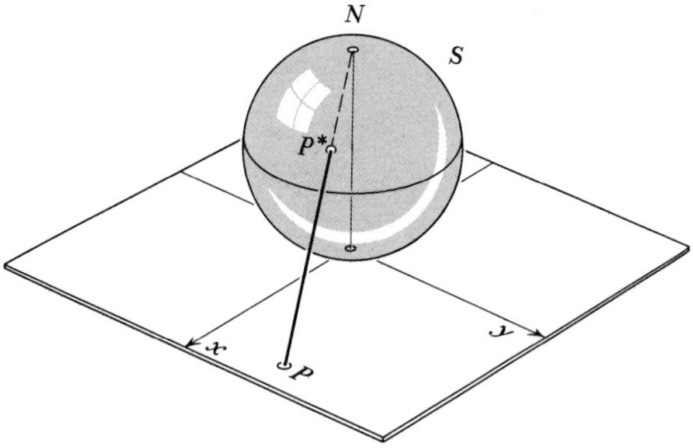

Fig. 331. Riemann number sphere

does not correspond to any point in the complex plane. But if we introduce the improper point $z = \infty$ and let this point correspond to N, the mapping becomes a one-to-one mapping of the extended complex plane onto S. The sphere S is called the **Riemann number sphere.** The particular mapping which we have used is called a **stereographic projection.**

Obviously, the unit circle is mapped onto the "equator" of S. The interior of the unit circle corresponds to the "Southern hemisphere" and the exterior to the "Northern hemisphere." Numbers z whose absolute values are large lie close to the North pole N. The x and y axes (and, more generally, all the straight lines through the origin) are mapped onto "meridians," while circles with center at the origin are mapped onto "parallels." It can be shown that any circle or straight line in the z-plane is mapped onto a circle on S, and, furthermore, that stereographic projection is conformal.

Zeros

If a function $f(z)$ is analytic in a domain D and is zero at a point $z = a$ in D, then $f(z)$ is said to have a **zero** at that point $z = a$. If not only f but also the derivatives $f', \cdots, f^{(n-1)}$ are all zero at $z = a$ and $f^{(n)}(a) \neq 0$, then $f(z)$ is said to have a *zero of* **order** n at the point $z = a$.

$f(z)$ is said to have a *zero of n-th order at infinity* if $f(1/z)$ has such a zero at $z = 0$.

For instance, if $f(a) = 0, f'(a) \neq 0$, then f has a zero of first order or *simple zero* at $z = a$. If $f(a) = f'(a) = 0, f''(a) \neq 0$, then the zero of f at $z = a$ is of second order, etc.

Example 2. Zeros

The function $\sin z$ has simple zeros at $z = 0, \pm\pi, \pm 2\pi, \cdots$. The function $(z - a)^3$ has a zero of third order at $z = a$. The function $1 - \cos z$ has second-order zeros at $z = 0, \pm 2\pi, \pm 4\pi, \cdots$. The function $1/(1 - z)$ has a simple zero at infinity. ∎

Clearly, if $f(z)$ is analytic in some neighborhood of a point $z = a$ and has a zero of nth order at a, it follows from Taylor's theorem in Sec. 16.3 that the coefficients $b_0, \cdots, b_{n-1}$ of its Taylor series with center $z = a$ are zero, and the series is of the form

$$f(z) = b_n(z-a)^n + b_{n+1}(z-a)^{n+1} + \cdots$$
$$(3) \qquad = (z-a)^n[b_n + b_{n+1}(z-a) + b_{n+2}(z-a)^2 + \cdots] \qquad (b_n \neq 0).$$

A point of a point set S is called an **isolated point** of S, if it has a neighborhood which does not contain further points of S. A point b is called an **accumulation point** of S (or a *limit point* of S) if every neighborhood of b (no matter how small) contains at least one point ($\neq b$) of S (and hence infinitely many points of S). Note that b may be, but need not be, a point of S.

Example 3. Isolated and nonisolated points. Accumulation points

The set of the points $z = n$ ($n = 1, 2, \cdots$) consists wholly of isolated points and has no accumulation point in the finite plane.

The set of the points $z = i/n$ ($n = 1, 2, \cdots$) on the imaginary axis consists wholly of isolated points and has one accumulation point, namely, $z = 0$; this point does not belong to the set.

The set of all complex numbers z for which $|z| < 1$ has no isolated points. All points of the set and also the points on the unit circle $|z| = 1$ (which do not belong to the set) are accumulation points of the set.

Theorem 1 (Zeros)

The zeros of an analytic function $f(z)$ ($\not\equiv 0$) are isolated.

Proof. We consider (3). Let $g(z)$ be the analytic function represented by the series in the brackets $[\cdots]$. Since $b_n \neq 0$, we have $g(a) \neq 0$. Consequently, since $g(z)$ is continuous, it is not zero in some neighborhood of $z = a$. It follows that $f(z)$ is not zero in that neighborhood (except at $z = a$), so that $z = a$ is the only zero of $f(z)$ in that neighborhood and is, therefore, isolated. ∎

Singularities

Analytic functions may have different types of singularities.[9] We first remember that a *singular point* of an analytic function $f(z)$ is a point where $f(z)$ ceases to be analytic (cf. Sec. 16.3). We also say that $f(z)$ *is singular*, or *has a singularity*, at that point. The function $f(z)$ is said to be *singular at infinity*, if $f(1/z)$ is singular at $z = 0$.

If $f(z)$ has an isolated singularity at a point $z = a$, then we can represent it by its **Laurent series** (Sec. 16.7)

$$(4) \qquad f(z) = \sum_{n=0}^{\infty} b_n(z-a)^n + \sum_{n=1}^{\infty} \frac{c_n}{(z-a)^n}$$

valid throughout some neighborhood of $z = a$ (except at $z = a$ itself). The last series in (4) is called the **principal part** of $f(z)$ near $z = a$.

It may happen that from some n on, all the coefficients c_n are zero, say, $c_m \neq 0$ and $c_n = 0$ for all $n > m$. Then (4) reduces to the form

$$(5) \qquad f(z) = \sum_{n=0}^{\infty} b_n(z-a)^n + \frac{c_1}{z-a} + \cdots + \frac{c_m}{(z-a)^m} \qquad (c_m \neq 0).$$

[9] We recall that, by definition, a function is a single-valued relation. (Cf. Sec. 12.4.)

In this case, where the principal part consists of finitely many terms, the singularity of f at $z = a$ is called a **pole,** and m is called the **order** of the pole. Poles of the first order are also known as *simple poles.*

If an analytic function f (single-valued in the complex plane) has a singularity other than a pole, this singularity is called an **essential singularity** of f.

Poles are, by definition, isolated singularities. All singularities which are not isolated (for instance, the singularity of $\tan(1/z)$ at $z = 0$) are thus essential singularities. An essential singularity may be isolated or not. If in (4) infinitely many c_n are different from zero, then the singularity of $f(z)$ at $z = a$ is not a pole but an isolated essential singularity.

Example 4. Poles. Essential singularities

The function

$$f(z) = \frac{1}{z(z-2)^5} + \frac{3}{(z-2)^2}$$

has a simple pole at $z = 0$ and a pole of fifth order at $z = 2$. The functions

$$e^{1/z} = \sum_{n=0}^{\infty} \frac{1}{n!\, z^n} = 1 + \frac{1}{z} + \frac{1}{2!\, z^2} + \cdots \tag{6}$$

and

$$\sin\frac{1}{z} = \sum_{n=0}^{\infty} \frac{(-1)^n}{(2n+1)!\, z^{2n+1}} = \frac{1}{z} - \frac{1}{3!\, z^3} + \frac{1}{5!\, z^5} - + \cdots \tag{7}$$

have an isolated essential singularity at $z = 0$.

The function $\tan(1/z)$ has poles at

$$\frac{1}{z} = \pm\frac{\pi}{2}, \pm\frac{3\pi}{2}, \cdots \qquad \text{that is,} \qquad z = \pm\frac{2}{\pi}, \pm\frac{2}{3\pi}, \cdots.$$

The limit point $z = 0$ of these points is thus a nonisolated essential singularity of $\tan(1/z)$.

Example 5. Singularities at infinity

The polynomial $f(z) = 2z + 6z^3$ has a pole of third order at infinity, because

$$f\left(\frac{1}{z}\right) = \frac{2}{z} + \frac{6}{z^3}$$

has such a pole at $z = 0$. More generally, a polynomial of nth degree has a pole of nth order at infinity.

The functions e^z, $\sin z$, $\cos z$ have an isolated essential singularity at infinity, since $e^{1/z}$, $\sin(1/z)$, and $\cos(1/z)$ have an isolated essential singularity at $z = 0$. ∎

A function $f(z)$ which is not analytic at a point $z = a$ but can be made analytic there by assigning some value to $f(z)$ at $z = a$, is said to have a *removable singularity* at $z = a$. Such singularities are not of interest, because they can be removed.

A function which is analytic everywhere in the finite plane is called an **entire function.**

If such a function is also analytic at infinity, it is bounded for all z, and from Liouville's theorem (cf. Sec. 14.6) it follows that it must be a constant. Hence, any entire function which is not a constant must be singular at infinity. For instance, polynomials (of at least first degree), e^z, $\sin z$, and $\cos z$ are entire functions, and they are singular at infinity.

An analytic function whose only singularities in the finite plane are poles is called a **meromorphic function.**

Example 6. Meromorphic functions

Rational functions with nonconstant denominator, $\tan z$, $\cot z$, $\sec z$, and $\csc z$ are meromorphic functions. ▮

The classification of singularities into poles and essential singularities is not merely a formal matter, because the behavior of an analytic function in a neighborhood of an essential singularity is entirely different from that in the neighborhood of a pole.

Example 7. Behavior near a pole

The function $f(z) = 1/z^2$ has a pole at $z = 0$, and $|f(z)| \to \infty$ as $z \to 0$ in any manner. ▮

This example illustrates

Theorem 2 (Poles)

If $f(z)$ is analytic and has a pole at $z = a$, then $|f(z)| \to \infty$ as $z \to a$ in any manner. (Cf. Prob. 30.)

Example 8. Behavior near an essential singularity

The function $f(z) = e^{1/z}$ has an essential singularity at $z = 0$. It has no limit for approach along the imaginary axis; it becomes infinite if $z \to 0$ through positive real values, but it approaches zero if $z \to 0$ through negative real values. It takes on any given value $c = c_0 e^{i\alpha} \neq 0$ in an arbitrarily small neighborhood of $z = 0$. In fact, setting $z = re^{i\theta}$, we must solve the equation

$$e^{1/z} = e^{(\cos\theta - i\sin\theta)/r} = c_0 e^{i\alpha}$$

for r and θ. Equating the absolute values and the arguments, we have $e^{(\cos\theta)/r} = c_0$ or

$$\cos\theta = r \ln c_0 \qquad \text{and} \qquad \sin\theta = -\alpha r.$$

From these two equations and $\cos^2\theta + \sin^2\theta = 1$ we obtain

$$r^2 = \frac{1}{(\ln c_0)^2 + \alpha^2} \qquad \text{and} \qquad \tan\theta = -\frac{\alpha}{\ln c_0}.$$

Hence r can be made arbitrarily small by adding multiples of 2π to α, leaving c unaltered. ▮

This example illustrates the famous

Picard's theorem

If $f(z)$ is analytic and has an isolated essential singularity at a point a, it takes on every value, with at most one exceptional value, in an arbitrarily small neighborhood of a.

In Example 8, the exceptional value is $z = 0$. The proof of Picard's theorem is rather complicated; it can be found in Ref. [F11].

Problems for Sec. 16.8

Are the following functions analytic at infinity?

1. $z^2 + z^{-2}$	**2.** $z^{-3} + z^{-1}$	**3.** e^z, e^{z^2}, e^{-z}
4. $e^{1/z}$, e^{1/z^2}	**5.** $\cos z$, $\sin z$	**6.** $\cosh z$, $\sinh z$
7. $\tan z$, $\cot z$	**8.** $(z^2 + 1)/(z^2 - 1)$	**9.** $z^{-1} \sin z$, $\csc z$

Describe and sketch the images of the following regions on the Riemann number sphere.

10. $\text{Im}\, z \geqq 0$
11. $|z| \geqq 5$
12. $|z| \leqq 2$
13. $1/2 \leqq |z| \leqq 2$
14. $|z| < 3$, $|\arg z| < \pi/4$
15. $-\pi \leqq \text{Im}\, z \leqq \pi$
16. $|\arg z| \leqq 3\pi/4$
17. $|z| > 1$, $|\arg z| < \pi/2$
18. $-\pi/4 < \arg z < 3\pi/4$

19. Describe the relative positions of z, $-z$, $\bar{z}$, $-\bar{z}$ in the complex plane and on the Riemann sphere.

20. Using the method described in connection with (2), develop $1/z^4$ in a Laurent series of negative powers of $z - 1$.

Determine the location and order of the zeros of the following functions.

21. $z^2 - 1$
22. z^3
23. $(z + i)^3$
24. $\cos^2 z$
25. $\sin^3 \pi z$
26. $(\sin z - 1)^2$
27. $\cosh^2 z$
28. $z^2 e^z$
29. $e^z - e^{2z}$

30. Verify Theorem 2 for $f(z) = z^{-2} + z^{-1}$. Prove Theorem 2.

31. Let $f(z)$ have a zero of order n at $z = a$. Show that then $f^2(z)$ has a zero of order $2n$, the derivative $f'(z)$ has a zero of order $n - 1$ (provided $n > 1$) and $1/f(z)$ has a pole of order n at that point. Give examples.

Determine the location and type of singularities of the following functions.

32. $z + z^{-1}$
33. $1/(z + a)^3$
34. $\cosh z$
35. $e^z + e^{1/z}$
36. $\tan \pi z$
37. e^{z^2}/z^5
38. $(\cos z - \sin z)^{-1}$
39. $\sin (1/z^2)$
40. $e^{1/z}/(z - 1)$

CHAPTER

17 INTEGRATION BY THE METHOD OF RESIDUES

Since there are various methods for determining the coefficients of a Laurent series (1), Sec. 16.7, without using the integral formulas (2), Sec. 16.7, we may use the formula for c_1 for evaluating complex integrals in a very elegant and simple fashion. c_1 will be called the *residue* of $f(z)$ at $z = a$. This powerful method may also be applied for evaluating certain real integrals, as we shall see in Secs. 17.3 and 17.4.

Prerequisites for this chapter: Chaps. 12, 14, 16.
References: Appendix 1, Part F.
Answers to problems: Appendix 2.

17.1 Residues

If $f(z)$ is analytic in a neighborhood of a point $z = a$, then, by Cauchy's integral theorem,

$$\int_C f(z)\,dz = 0 \tag{1}$$

for any contour in that neighborhood. If, however, $f(z)$ has an isolated singularity at $z = a$ and a lies in the interior of C, then the integral in (1) will, in general, be different from zero. In this case we may represent $f(z)$ by a Laurent series

$$f(z) = \sum_{n=0}^{\infty} b_n(z-a)^n + \frac{c_{-1}}{z-a} + \frac{c_{-2}}{(z-a)^2} + \cdots \tag{2}$$

which converges in the domain $0 < |z - a| < R$, where R is the distance from a to the nearest singular point of $f(z)$. From (2) in Sec. 16.7 we see that the coefficient c_1 of the power $1/(z - a)$ in this development is

$$c_1 = \frac{1}{2\pi i}\int_C f(z)\,dz.$$

Consequently,

$$\int_C f(z)\,dz = 2\pi i c_1, \tag{3}$$

the integration being taken in the counterclockwise sense around a simple closed path C which lies in the domain $0 < |z - a| < R$ and contains the point $z = a$ in its interior. The coefficient c_1 in the development (2) of $f(z)$ is called the **residue** of $f(z)$ at $z = a$, and we shall use the notation

$$c_1 = \operatorname*{Res}_{z=a} f(z). \tag{4}$$

We have seen that Laurent expansions can be obtained by various methods, without using the integral formulas for the coefficients. *Hence we may determine the residue by one of those methods and then use the formula* (3) *for evaluating contour integrals.*

Example 1. Evaluation of an integral by means of a residue

Integrate the function $f(z) = z^{-4} \sin z$ around the unit circle C in the counterclockwise sense.

From (5) in Sec. 16.4 we obtain the Laurent series

$$f(z) = \frac{\sin z}{z^4} = \frac{1}{z^3} - \frac{1}{3!z} + \frac{z}{5!} - \frac{z^3}{7!} + - \cdots .$$

We see that $f(z)$ has a pole of third order at $z = 0$, the corresponding residue is $c_1 = -1/3!$, and from (3) it follows that

$$\int_C \frac{\sin z}{z^4}\, dz = 2\pi i c_1 = -\frac{\pi i}{3}.$$ ∎

Before we proceed in evaluating integrals we shall develop a simple standard method for determining the residue in the case of a pole.

If $f(z)$ has a **simple pole** at a point $z = a$, the corresponding Laurent series is of the form [cf. (3) in the last section]

$$f(z) = \frac{c_1}{z - a} + b_0 + b_1(z - a) + b_2(z - a)^2 + \cdots \qquad (0 < |z - a| < R)$$

where $c_1 \neq 0$. Multiplying both sides by $z - a$, we have

$$(z - a)f(z) = c_1 + (z - a)[b_0 + b_1(z - a) + \cdots]. \tag{5}$$

If we let z approach a, the right-hand side approaches c_1 and we obtain

$$\operatorname*{Res}_{z=a} f(z) = c_1 = \lim_{z\to a} (z - a)f(z). \tag{6}$$

This is our first result, namely, a formula for the residue in the case of a simple pole.

In the case of a simple pole, another useful formula is obtained as follows. If $f(z)$ has a simple pole at $z = a$, we may set

$$f(z) = \frac{p(z)}{q(z)}$$

where $p(z)$ and $q(z)$ are analytic at $z = a$, $p(a) \neq 0$, and $q(z)$ has a simple zero at $z = a$. Consequently, $q(z)$ can be expanded in a Taylor series of the form

$$q(z) = (z - a)q'(a) + \frac{(z-a)^2}{2!}q''(a) + \cdots .$$

Hence, by (6),

$$\operatorname*{Res}_{z=a} f(z) = \lim_{z \to a} (z - a)\frac{p(z)}{q(z)} = \lim_{z \to a} \frac{(z-a)p(z)}{(z-a)[q'(a) + (z-a)q''(a)/2 + \cdots]};$$

that is, in the case of a simple pole we also have

(7)
$$\operatorname*{Res}_{z=a} f(z) = \operatorname*{Res}_{z=a} \frac{p(z)}{q(z)} = \frac{p(a)}{q'(a)}.$$

Example 2. Residues at simple poles

The function $f(z) = (4 - 3z)/(z^2 - z)$ has simple poles at $z = 0$ and $z = 1$. From (7) we obtain

$$\operatorname*{Res}_{z=0} f(z) = \left[\frac{4-3z}{2z-1}\right]_{z=0} = -4, \qquad \operatorname*{Res}_{z=1} f(z) = \left[\frac{4-3z}{2z-1}\right]_{z=1} = 1.$$ ∎

We shall now consider **poles of higher order.** If $f(z)$ has a pole of order $m > 1$ at a point $z = a$, the corresponding Laurent expansion is of the form

$$f(z) = \frac{c_m}{(z-a)^m} + \frac{c_{m-1}}{(z-a)^{m-1}} + \cdots + \frac{c_2}{(z-a)^2} + \frac{c_1}{z-a}$$
$$+ b_0 + b_1(z-a) + \cdots$$

where $c_m \neq 0$ and the series converges in some neighborhood of $z = a$, except at the point a itself. By multiplying both sides by $(z-a)^m$ we obtain

$$(z-a)^m f(z) = c_m + c_{m-1}(z-a) + \cdots + c_2(z-a)^{m-2} + c_1(z-a)^{m-1}$$
$$+ b_0(z-a)^m + b_1(z-a)^{m+1} + \cdots .$$

This shows that the residue c_1 of $f(z)$ at $z = a$ is now the coefficient of the power $(z-a)^{m-1}$ in the Taylor expansion of the function $g(z) = (z-a)^m f(z)$ with center at $z = a$. Therefore, from Taylor's theorem (Sec. 16.3),

$$c_1 = \frac{g^{(m-1)}(a)}{(m-1)!}.$$

Hence if $f(z)$ has a pole of mth order at $z = a$, the residue is

(8)
$$\operatorname*{Res}_{z=a} f(z) = \frac{1}{(m-1)!} \lim_{z \to a} \left\{ \frac{d^{m-1}}{dz^{m-1}} [(z-a)^m f(z)] \right\}.$$

Example 3. Residue at a pole of higher order

The function

$$f(z) = \frac{2z}{(z+4)(z-1)^2}$$

has a pole of second order at $z = 1$, and from (8) we obtain the corresponding residue

$$\operatorname*{Res}_{z=1} f(z) = \lim_{z\to 1} \frac{d}{dz}[(z-1)^2 f(z)] = \lim_{z\to 1} \frac{d}{dz}\left(\frac{2z}{z+4}\right) = \frac{8}{25}.$$

Of course, in the case of a rational function $f(z)$ the residues can also be determined from the representation of $f(z)$ in terms of partial fractions.

Example 4. Partial fractions

From

$$f(z) = \frac{7z^4 - 13z^3 + z^2 + 4z - 1}{(z^3 + z^2)(z-1)^2} = \frac{3}{z} - \frac{1}{z^2} + \frac{4}{z+1} - \frac{1}{(z-1)^2}$$

we see that

$$\operatorname*{Res}_{z=0} f(z) = 3, \qquad \operatorname*{Res}_{z=-1} f(z) = 4, \qquad \operatorname*{Res}_{z=1} f(z) = 0.$$

Problems for Sec. 17.1

Find the residues at the singular points of the following functions.

1. $\dfrac{1}{1-z}$ **2.** $\dfrac{z+3}{z+1}$ **3.** $\dfrac{1}{z^2}$

4. $\dfrac{z}{1+z^2}$ **5.** $\dfrac{1}{z^2-1}$ **6.** $\dfrac{1}{(z^2-1)^2}$

7. $\dfrac{z}{z^4-1}$ **8.** $\dfrac{1}{z^4-1}$ **9.** $\dfrac{1}{1-e^z}$

10. $\sec z$ **11.** $\cot z$ **12.** $\tan z$

In each case, find the residues at those singular points which lie inside the circle $|z| = 1.5$.

13. $\dfrac{3z^2}{1-z^4}$ **14.** $\dfrac{z-\frac{1}{4}}{z^2+3z+2}$ **15.** $\dfrac{6-4z}{z^3+3z^2}$

16. $\dfrac{1}{(z^4-1)^2}$ **17.** $\dfrac{z+2}{(z+1)(z^2+16)}$ **18.** $\dfrac{4-3z}{z^3-3z^2+2z}$

Evaluate the following integrals where C is the unit circle (counterclockwise).

19. $\displaystyle\int_C e^{1/z}\,dz$ **20.** $\displaystyle\int_C ze^{1/z}\,dz$ **21.** $\displaystyle\int_C \cot z\,dz$

22. $\displaystyle\int_C \tan z\,dz$ **23.** $\displaystyle\int_C \frac{dz}{\sin z}$ **24.** $\displaystyle\int_C \frac{z}{2z+i}\,dz$

25. $\displaystyle\int_C \frac{dz}{\cosh z}$ **26.** $\displaystyle\int_C \frac{z^2-4}{(z-2)^4}\,dz$ **27.** $\displaystyle\int_C \frac{z^2+1}{z^2-2z}\,dz$

28. $\displaystyle\int_C \frac{\sin \pi z}{z^4}\,dz$ **29.** $\displaystyle\int_C \frac{dz}{1-e^z}$ **30.** $\displaystyle\int_C \frac{(z^2+1)}{e^z \sin z}\,dz$

17.2 The Residue Theorem

So far we are in a position to evaluate contour integrals whose integrands have only a single isolated singularity inside the contour of integration. We shall now see that our simple method may easily be extended to the case when the integrand has several isolated singularities inside the contour.

Residue theorem

Let $f(z)$ be a function which is analytic inside a simple closed path C and on C, except for finitely many singular points $a_1, a_2, \cdots, a_m$ inside C. Then

$$\int_C f(z)\,dz = 2\pi i \sum_{j=1}^{m} \operatorname*{Res}_{z=a_j} f(z), \tag{1}$$

the integral being taken in the counterclockwise sense around C.

Proof. We enclose each of the singular points a_j in a circle C_j with radius small enough that those m circles and C are all separated (Fig. 332). Then $f(z)$ is analytic in the multiply connected domain D bounded by C and $C_1, \cdots, C_m$ and on the entire boundary of D. From Cauchy's integral theorem we have

$$\int_C f(z)\,dz + \int_{C_1} f(z)\,dz + \int_{C_2} f(z)\,dz + \cdots + \int_{C_m} f(z)\,dz = 0, \tag{2}$$

the integral along C being taken in the counterclockwise sense and the other integrals in the clockwise sense (cf. Sec. 14.3). We now reverse the sense of integration along $C_1, \cdots, C_m$. Then the signs of the values of these integrals change, and we obtain from (2)

$$\int_C f(z)\,dz = \int_{C_1} f(z)\,dz + \int_{C_2} f(z)\,dz + \cdots + \int_{C_m} f(z)\,dz, \tag{3}$$

all integrals now being taken in the counterclockwise sense. Since, by (3) in the last section,

$$\int_{C_j} f(z)\,dz = 2\pi i \operatorname*{Res}_{z=a_j} f(z),$$

formula (3) yields (1), and the theorem is proved. ∎

This important theorem has various applications in connection with complex and real integrals. We shall first consider some complex integrals.

Example 1. Integration by the residue theorem

The function $(4 - 3z)/(z^2 - z)$ is analytic except at the points 0 and 1 where it has simple poles; the residues are -4 and 1, respectively (cf. Example 2 in the last section). Therefore,

$$\int_C \frac{4 - 3z}{z^2 - z}\,dz = 2\pi i(-4 + 1) = -6\pi i$$

for every simple closed path C which encloses the points 0 and 1, and

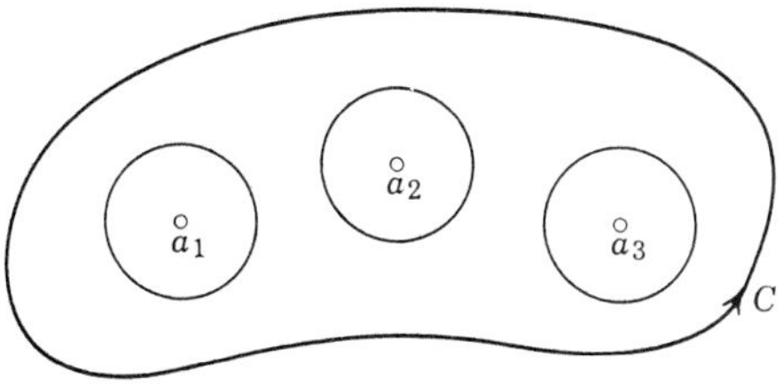

Fig. 332. Residue theorem

$$\int_C \frac{4-3z}{z^2-z}\,dz = 2\pi i(-4) = -8\pi i$$

for any simple closed path C for which $z = 0$ lies inside C and $z = 1$ lies outside C, the integrations being taken in the counterclockwise sense.

Example 2. Integrand having poles of higher order

Integrate $1/(z^3 - 1)^2$ in the counterclockwise sense around the circle $|z - 1| = 1$. The function has poles of second order at $z = 1$, $z = e^{2\pi i/3}$, and $z = e^{-2\pi i/3}$. Only the pole at $z = 1$ lies inside C. Using (8), Sec. 17.1, we thus obtain

$$\int_C \frac{dz}{(z^3-1)^2} = 2\pi i \operatorname*{Res}_{z=1} \frac{1}{(z^3-1)^2} = 2\pi i\left(-\frac{2}{9}\right) = -\frac{4\pi i}{9}.$$

Example 3. Confirmation of an earlier result

Integrate $1/(z - a)^m$ (m a positive integer) in the counterclockwise sense around any simple closed path C enclosing the point $z = a$. We find

$$\operatorname*{Res}_{z=a} \frac{1}{z-a} = 1 \qquad \text{and} \qquad \operatorname*{Res}_{z=a} \frac{1}{(z-a)^m} = 0 \qquad (m = 2, 3, \cdots).$$

Therefore, the result is

$$\int_C \frac{dz}{(z-a)^m} = \begin{cases} 2\pi i & (m = 1) \\ 0 & (m = 2, 3, \cdots), \end{cases}$$

in agreement with Example 3 in Sec. 14.1.

Problems for Sec. 17.2

Integrate $\dfrac{3z^2 + 2z - 4}{z^3 - 4z}$ around the following paths C in the counterclockwise sense.

1. $|z| = 1$ **2.** $|z| = 3$ **3.** $|z - 4| = 1$

Integrate $\dfrac{z+1}{z(z-1)(z-2)}$ in the clockwise sense around the path

4. $|z - 2| = \frac{1}{2}$ **5.** $|z| = \frac{3}{2}$ **6.** $|z - \frac{1}{2}| = \frac{1}{4}$

Evaluate the following integrals where C is the unit circle (counterclockwise).

7. $\displaystyle\int_C \frac{3z}{3z-1}\,dz$ **8.** $\displaystyle\int_C \frac{z}{4z^2-1}\,dz$ **9.** $\displaystyle\int_C \frac{dz}{z^2-2z}$

10. $\displaystyle\int_C \frac{dz}{z^2+4}$ **11.** $\displaystyle\int_C \frac{z+1}{4z^3-z}\,dz$ **12.** $\displaystyle\int_C \frac{z^5-3z^3+1}{(2z+1)(z^2+4)}\,dz$

13. $\displaystyle\int_C \frac{z}{1+9z^2}\,dz$ **14.** $\displaystyle\int_C \frac{z+1}{z^4-2z^3}\,dz$ **15.** $\displaystyle\int_C \frac{(z+4)^3}{z^4+5z^3+6z^2}\,dz$

16. $\displaystyle\int_C \tan z\,dz$ **17.** $\displaystyle\int_C \tan \pi z\,dz$ **18.** $\displaystyle\int_C \frac{6z^2-4z+1}{(z-2)(1+4z^2)}\,dz$

19. $\displaystyle\int_C \tan 2\pi z\,dz$ **20.** $\displaystyle\int_C \frac{\tan \pi z}{z^3}\,dz$ **21.** $\displaystyle\int_C \frac{e^z}{z^2-5z}\,dz$

22. $\displaystyle\int_C \frac{e^z}{\sin z}\,dz$ **23.** $\displaystyle\int_C \frac{e^z}{\cos z}\,dz$ **24.** $\displaystyle\int_C \frac{e^z}{\cos \pi z}\,dz$

25. $\displaystyle\int_C \frac{\cosh z}{z^2-3iz}\,dz$ **26.** $\displaystyle\int_C \coth z\,dz$ **27.** $\displaystyle\int_C \frac{\sinh z}{2z-i}\,dz$

28. $\displaystyle\int_C \cot z\,dz$ **29.** $\displaystyle\int_C \frac{\cot z}{z}\,dz$ **30.** $\displaystyle\int_C \frac{e^{z^2}}{\cos \pi z}\,dz$

17.3 Evaluation of Real Integrals

The residue theorem yields a very elegant and simple method for evaluating certain classes of complicated real integrals, and we shall see that in many cases this method of integration can be made a routine matter.

Integrals of rational functions of cos θ and sin θ

We first consider integrals of the type

(1) $$I = \int_0^{2\pi} R(\cos\theta, \sin\theta)\, d\theta$$

where $R(\cos\theta, \sin\theta)$ is a real rational function of $\cos\theta$ and $\sin\theta$ finite on the interval $0 \leqq \theta \leqq 2\pi$. Setting $e^{i\theta} = z$, we obtain

$$\cos\theta = \tfrac{1}{2}(e^{i\theta} + e^{-i\theta}) = \frac{1}{2}\left(z + \frac{1}{z}\right),$$

$$\sin\theta = \frac{1}{2i}(e^{i\theta} - e^{-i\theta}) = \frac{1}{2i}\left(z - \frac{1}{z}\right),$$

and we see that the integrand becomes a rational function of z, say, $f(z)$. As θ ranges from 0 to 2π, the variable z ranges once around the unit circle $|z| = 1$ in the counterclockwise sense. Since $dz/d\theta = ie^{i\theta}$, we have $d\theta = dz/iz$, and the given integral takes the form

(2) $$I = \int_C f(z)\frac{dz}{iz},$$

the integration being taken in the counterclockwise sense around the unit circle.

Example 1. An integral of the type (1)

Let p be a fixed real number in the interval $0 < p < 1$. We consider

$$\int_0^{2\pi} \frac{d\theta}{1 - 2p\cos\theta + p^2} = \int_C \frac{dz/iz}{1 - 2p\dfrac{1}{2}\left(z + \dfrac{1}{z}\right) + p^2} = \int_C \frac{dz}{i(1 - pz)(z - p)}.$$

The integrand has simple poles at $z = 1/p > 1$ and $z = p < 1$. Only the last pole lies inside the unit circle C, and the residue is

$$\operatorname*{Res}_{z=p} \frac{1}{i(1 - pz)(z - p)} = \left[\frac{1}{i(1 - pz)}\right]_{z=p} = \frac{1}{i(1 - p^2)}.$$

The residue theorem yields

$$\int_0^{2\pi} \frac{d\theta}{1 - 2p\cos\theta + p^2} = 2\pi i \frac{1}{i(1 - p^2)} = \frac{2\pi}{1 - p^2} \qquad (0 < p < 1).$$ ∎

Improper integrals of rational functions

We now consider real integrals of the type

$$\int_{-\infty}^{\infty} f(x)\,dx. \tag{3}$$

Such an integral, for which the interval of integration is not finite, is called an **improper integral,** and it has the meaning

$$\int_{-\infty}^{\infty} f(x)\,dx = \lim_{a\to-\infty}\int_{a}^{0} f(x)\,dx + \lim_{b\to\infty}\int_{0}^{b} f(x)\,dx. \tag{4'}$$

If both limits exist, we may couple the two independent passages to $-\infty$ and ∞, and write[1]

$$\int_{-\infty}^{\infty} f(x)\,dx = \lim_{r\to\infty}\int_{-r}^{r} f(x)\,dx. \tag{4}$$

We assume that the function $f(x)$ in (3) is a real rational function whose denominator is different from zero for all real x and is of degree at least two units higher than the degree of the numerator. Then the limits in (4′) exist, and we may start from (4). We consider the corresponding contour integral

$$\int_C f(z)\,dz \tag{4*}$$

around a path C, as shown in Fig. 333. Since $f(x)$ is rational, $f(z)$ has finitely many poles in the upper half-plane, and if we choose r large enough, C encloses all these poles. By the residue theorem we then obtain

$$\int_C f(z)\,dz = \int_S f(z)\,dz + \int_{-r}^{r} f(x)\,dx = 2\pi i \sum \operatorname{Res} f(z)$$

where the sum consists of all the residues of $f(z)$ at the points in the upper half-plane where $f(z)$ has a pole. From this we have

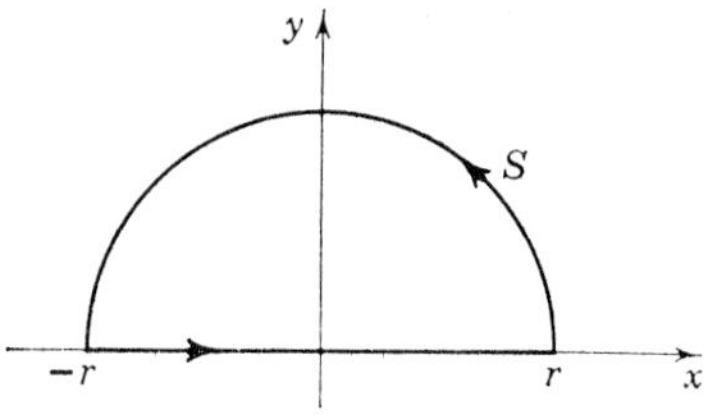

Fig. 333. Path of the contour integral in (4*)

[1]The expression on the right side of (4) is called the **Cauchy principal value** of the integral; it may exist even if the limits in (4′) do not exist. For instance,

$$\lim_{r\to\infty}\int_{-r}^{r} x\,dx = \lim_{r\to\infty}\left(\frac{r^2}{2}-\frac{r^2}{2}\right) = 0, \quad \text{but} \quad \lim_{b\to\infty}\int_{0}^{b} x\,dx = \infty.$$

(5)
$$\int_{-r}^{r} f(x)\,dx = 2\pi i \sum \operatorname{Res} f(z) - \int_S f(z)\,dz.$$

We prove that, if $r \to \infty$, the value of the integral over the semicircle S approaches zero. If we set $z = re^{i\theta}$, then S is represented by $r = const$, and as z ranges along S the variable θ ranges from 0 to π. Since the degree of the denominator of $f(z)$ is at least two units higher than the degree of the numerator, we have

$$|f(z)| < \frac{k}{|z|^2} \qquad (|z| = r > r_0)$$

for sufficiently large constants k and r_0. By applying (4) in Sec. 14.2 we thus obtain

$$\left|\int_S f(z)\,dz\right| < \frac{k}{r^2}\pi r = \frac{k\pi}{r} \qquad (r > r_0).$$

Hence, as r approaches infinity, the value of the integral over S approaches zero, and (4) and (5) yield the result

(6)
$$\int_{-\infty}^{\infty} f(x)\,dx = 2\pi i \sum \operatorname{Res} f(z),$$

the sum being extended over all the residues of $f(z)$ corresponding to the poles of $f(z)$ in the upper half-plane.

Example 2. An improper integral from 0 to ∞

Using (6), let us show that

$$\int_0^{\infty} \frac{dx}{1 + x^4} = \frac{\pi}{2\sqrt{2}}.$$

Indeed, $f(z) = 1/(1 + z^4)$ has four simple poles at the points

$$z_1 = e^{\pi i/4}, \qquad z_2 = e^{3\pi i/4}, \qquad z_3 = e^{-3\pi i/4}, \qquad z_4 = e^{-\pi i/4}.$$

The first two of these poles lie in the upper half-plane (Fig. 334). From (7) in Sec. 17.1 we find

$$\operatorname*{Res}_{z=z_1} f(z) = \left[\frac{1}{(1 + z^4)'}\right]_{z=z_1} = \left[\frac{1}{4z^3}\right]_{z=z_1} = \frac{1}{4}e^{-3\pi i/4} = -\frac{1}{4}e^{\pi i/4},$$

$$\operatorname*{Res}_{z=z_2} f(z) = \left[\frac{1}{(1 + z^4)'}\right]_{z=z_2} = \left[\frac{1}{4z^3}\right]_{z=z_2} = \frac{1}{4}e^{-9\pi i/4} = \frac{1}{4}e^{-\pi i/4}.$$

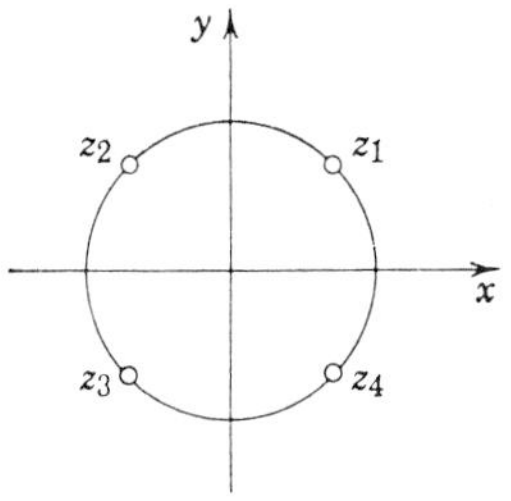

Fig. 334. Example 2

By (1) in Sec. 12.8 and (6) in the current section,

$$\int_{-\infty}^{\infty} \frac{dx}{1 + x^4} = \frac{2\pi i}{4}(-e^{\pi i/4} + e^{-\pi i/4}) = \pi \sin\frac{\pi}{4} = \frac{\pi}{\sqrt{2}}. \tag{7}$$

Since $1/(1 + x^4)$ is an even function,

$$\int_{0}^{\infty} \frac{dx}{1 + x^4} = \frac{1}{2}\int_{-\infty}^{\infty} \frac{dx}{1 + x^4}.$$

From this and (7) the result follows.

Problems for Sec. 17.3

Evaluate the following integrals involving cosine and sine.

1. $\displaystyle\int_0^{2\pi} \frac{d\theta}{2 + \cos\theta}$ **2.** $\displaystyle\int_0^{\pi} \frac{d\theta}{1 + \frac{1}{3}\cos\theta}$ **3.** $\displaystyle\int_0^{\pi} \frac{d\theta}{k + \cos\theta} \quad (k > 1)$

4. $\displaystyle\int_0^{2\pi} \frac{d\theta}{25 - 24\cos\theta}$ **5.** $\displaystyle\int_0^{2\pi} \frac{d\theta}{5 - 3\cos\theta}$ **6.** $\displaystyle\int_0^{2\pi} \frac{\cos\theta}{17 - 8\cos\theta}\,d\theta$

7. $\displaystyle\int_0^{2\pi} \frac{\cos\theta}{3 + \sin\theta}\,d\theta$ **8.** $\displaystyle\int_0^{2\pi} \frac{\cos\theta}{13 - 12\cos 2\theta}\,d\theta$ **9.** $\displaystyle\int_0^{2\pi} \frac{d\theta}{5/4 - \sin\theta}$

10. $\displaystyle\int_0^{2\pi} \frac{\sin^2\theta}{5 - 4\cos\theta}\,d\theta$ **11.** $\displaystyle\int_0^{2\pi} \frac{\cos^2\theta}{26 - 10\cos 2\theta}\,d\theta$ **12.** $\displaystyle\int_0^{2\pi} \frac{\cos^2 3\theta}{5 - 4\cos 2\theta}\,d\theta$

Evaluate the following improper integrals.

13. $\displaystyle\int_{-\infty}^{\infty} \frac{dx}{1 + x^2}$ **14.** $\displaystyle\int_{-\infty}^{\infty} \frac{dx}{(1 + x^2)^2}$ **15.** $\displaystyle\int_{-\infty}^{\infty} \frac{dx}{1 + x^6}$

16. $\displaystyle\int_{-\infty}^{\infty} \frac{dx}{x^4 + 16}$ **17.** $\displaystyle\int_{-\infty}^{\infty} \frac{dx}{(1 + x^2)^3}$ **18.** $\displaystyle\int_{-\infty}^{\infty} \frac{x}{(4 + x^2)^2}\,dx$

19. $\displaystyle\int_{-\infty}^{\infty} \frac{x^3}{1 + x^8}\,dx$ **20.** $\displaystyle\int_{0}^{\infty} \frac{1 + x^2}{1 + x^4}\,dx$ **21.** $\displaystyle\int_{-\infty}^{\infty} \frac{dx}{(x^2 + 1)(x^2 + 9)}$

22. $\displaystyle\int_{-\infty}^{\infty} \frac{dx}{(x^2 + 1)(x^2 + 4)^2}$ **23.** $\displaystyle\int_{-\infty}^{\infty} \frac{x}{(x^2 - 2x + 2)^2}\,dx$ **24.** $\displaystyle\int_{-\infty}^{\infty} \frac{x^2}{(x^2 + 1)(x^2 + 4)}\,dx$

25. Solve Prob. 13 by elementary methods. Obtain the answers to Probs. 18 and 19 without calculation.

17.4 Further Types of Real Integrals

There are further classes of real integrals which can be evaluated by applying the residue theorem to suitable complex integrals. In applications such integrals may arise in connection with integral transformations or representations of special functions. In the present section we shall consider two such classes of integrals. One of them is important in problems involving the Fourier integral representation (Sec. 10.9). The other class consists of real integrals whose integrand is infinite at some point in the interval of integration.

Fourier integrals

Real integrals of the form

(1) $$\int_{-\infty}^{\infty} f(x)\cos sx\,dx \quad \text{and} \quad \int_{-\infty}^{\infty} f(x)\sin sx\,dx \qquad (s \text{ real})$$

occur in connection with the Fourier integral (cf. Sec. 10.9).

If $f(x)$ is a rational function satisfying the assumptions stated in connection with (3), Sec. 17.3, then the integrals (1) may be evaluated in a similar way as the integrals in (3) of the previous section. In fact, we may then consider the corresponding integral

$$\int_C f(z)\,e^{isz}\,dz \qquad (s \text{ real and positive})$$

over the contour C in Fig. 333, and instead of (6), Sec. 17.3, we now obtain

(2) $$\int_{-\infty}^{\infty} f(x)\,e^{isx}\,dx = 2\pi i \sum \operatorname{Res}\,[f(z)\,e^{isz}] \qquad (s > 0)$$

where the sum consists of the residues of $f(z)e^{isz}$ at its poles in the upper half-plane. Equating the real and the imaginary parts on both sides of (2), we have

(3) $$\begin{aligned} \int_{-\infty}^{\infty} f(x)\cos sx\,dx &= -2\pi \sum \operatorname{Im}\operatorname{Res}\,[f(z)\,e^{isz}], \\ \int_{-\infty}^{\infty} f(x)\sin sx\,dx &= 2\pi \sum \operatorname{Re}\operatorname{Res}\,[f(z)\,e^{isz}]. \end{aligned} \qquad (s > 0)$$

We remember that (6), Sec. 17.3, was established by proving that the value of the integral over the semicircle S in Fig. 333 approaches zero as $r \to \infty$. To establish (2) we should now prove the same fact for our present contour integral. This can be done as follows. Since S lies in the upper half-plane $y \geqq 0$ and $s > 0$, it follows that

$$|e^{isz}| = |e^{isx}|\,|e^{-sy}| = e^{-sy} \leqq 1 \qquad (s > 0,\ y \geqq 0).$$

From this we obtain the inequality

$$|f(z)\,e^{isz}| = |f(z)|\,|e^{isz}| \leqq |f(z)| \qquad (s > 0, y \geqq 0)$$

which reduces our present problem to that in the previous section. Continuing as before, we see that the value of the integral under consideration approaches zero as r approaches infinity. This establishes (2).

Example 1. An application of (3)

Let us show that

$$\int_{-\infty}^{\infty} \frac{\cos sx}{k^2 + x^2}\, dx = \frac{\pi}{k} e^{-ks}, \qquad \int_{-\infty}^{\infty} \frac{\sin sx}{k^2 + x^2}\, dx = 0 \qquad (s > 0,\ k > 0).$$

In fact, $e^{isz}/(k^2 + z^2)$ has only one pole in the upper half-plane, namely, a simple pole at $z = ik$, and from (7) in Sec. 17.1 we obtain

$$\operatorname*{Res}_{z=ik} \frac{e^{isz}}{k^2 + z^2} = \left[\frac{e^{isz}}{2z}\right]_{z=ik} = \frac{e^{-ks}}{2ik}.$$

Therefore,

$$\int_{-\infty}^{\infty} \frac{e^{isx}}{k^2 + x^2}\, dx = 2\pi i \frac{e^{-ks}}{2ik} = \frac{\pi}{k} e^{-ks}$$

and this yields the above results [cf. also (14) in Sec. 10.9]. ∎

Other types of real improper integrals

Another kind of improper integral is a definite integral

$$\int_A^B f(x)\, dx \tag{4}$$

whose integrand becomes infinite at a point a in the interval of integration, that is,

$$\lim_{x\to a} |f(x)| = \infty.$$

Then the integral (4) means

$$\int_A^B f(x)\, dx = \lim_{\epsilon\to 0} \int_A^{a-\epsilon} f(x)\, dx + \lim_{\eta\to 0} \int_{a+\eta}^{B} f(x)\, dx \tag{5}$$

where both ϵ and η approach zero independently and through positive values. It may happen that neither of these limits exists when $\epsilon, \eta \to 0$ independently, but

$$\lim_{\epsilon\to 0} \left[\int_A^{a-\epsilon} f(x)\, dx + \int_{a+\epsilon}^{B} f(x)\, dx\right] \tag{6}$$

exists; this is called the **Cauchy principal value** of the integral and is often written

$$\text{pr. v.} \int_A^B f(x)\, dx.$$

For example,

$$\text{pr. v.} \int_{-1}^{1} \frac{dx}{x^3} = \lim_{\epsilon\to 0} \left[\int_{-1}^{-\epsilon} \frac{dx}{x^3} + \int_{\epsilon}^{1} \frac{dx}{x^3}\right] = 0;$$

the principal value exists although the integral itself has no meaning. The whole situation is quite similar to that discussed in the second part of the previous section.

To evaluate improper integrals whose integrands have poles on the real axis, we use a path which avoids these singularities by following small semicircles with centers at the singular points; the procedure may be illustrated by the following example.

Example 2. Integrand having a pole on the real axis. Sine integral
Show that

$$\int_0^\infty \frac{\sin x}{x}\,dx = \frac{\pi}{2}.$$

(This is the limit of the sine integral Si(x) as $x \to \infty$; cf. Sec. 10.9.) We do not consider $(\sin z)/z$ because this function does not behave suitably at infinity. We consider e^{iz}/z, which has a simple pole at $z = 0$, and integrate around the contour shown in Fig. 335. Since e^{iz}/z is analytic inside and on C, it follows from Cauchy's integral theorem that

$$\int_C \frac{e^{iz}}{z}\,dz = 0. \tag{7}$$

We show that the value of the integral over the large semicircle C_1 approaches zero as R approaches infinity. Setting $z = Re^{i\theta}$ we have $dz = iRe^{i\theta}\,d\theta$, $dz/z = i\,d\theta$ and therefore

$$\left|\int_{C_1} \frac{e^{iz}}{z}dz\right| = \left|\int_0^\pi e^{iz}\,i\,d\theta\right| \leqq \int_0^\pi |e^{iz}|\,d\theta \qquad (z = Re^{i\theta}).$$

In the integrand on the right,

$$|e^{iz}| = |e^{iR(\cos\theta + i\sin\theta)}| = |e^{iR\cos\theta}||e^{-R\sin\theta}| = e^{-R\sin\theta}.$$

By inserting this and using $\sin(\pi - \theta) = \sin\theta$ we obtain

$$\int_0^\pi |e^{iz}|\,d\theta = \int_0^\pi e^{-R\sin\theta}\,d\theta = 2\int_0^{\pi/2} e^{-R\sin\theta}\,d\theta$$

$$= 2\left[\int_0^\epsilon e^{-R\sin\theta}\,d\theta + \int_\epsilon^{\pi/2} e^{-R\sin\theta}\,d\theta\right]$$

where ϵ is any value between 0 and $\pi/2$. The absolute value of the integrand in the first and the last integral on the right is at most equal to 1 and $e^{-R\sin\epsilon}$, respectively, because the integrand is a monotone decreasing function of θ in the interval of integration. Consequently, the whole expression on the right is smaller than

$$2\left[\int_0^\epsilon d\theta + e^{-R\sin\epsilon}\int_\epsilon^{\pi/2} d\theta\right] = 2\left[\epsilon + e^{-R\sin\epsilon}\left(\frac{\pi}{2} - \epsilon\right)\right] < 2\epsilon + \pi e^{-R\sin\epsilon}.$$

Altogether,

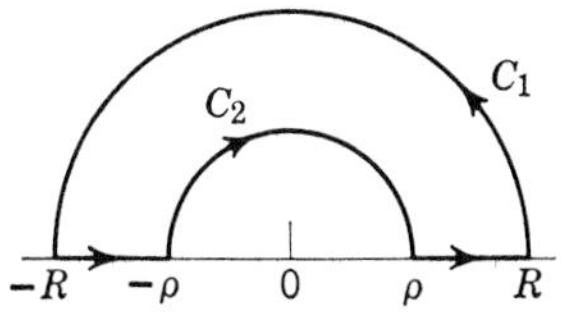

Fig. 335. Example 2

$$\left|\int_{C_1} \frac{e^{iz}}{z}\, dz\right| < 2\epsilon + \pi e^{-R\sin\epsilon}.$$

We first take ϵ arbitrarily small. Then, having fixed ϵ, the last term can be made as small as we please by choosing R sufficiently large. Hence, the value of the integral along C_1 approaches zero as R approaches infinity.

For the integral over the small semicircle C_2 in Fig. 335 we have

$$\int_{C_2} \frac{e^{iz}}{z}\, dz = \int_{C_2} \frac{dz}{z} + \int_{C_2} \frac{e^{iz}-1}{z}\, dz.$$

The first integral on the right has the value $-\pi i$. The integrand of the last integral is analytic at $z = 0$ and, therefore, bounded in absolute value as $\rho \to 0$. From this and (4) in Sec. 14.2 we conclude that the value of that integral approaches zero as $\rho \to 0$. From (7) we thus obtain

$$\text{pr. v.} \int_{-\infty}^{\infty} \frac{e^{ix}}{x}\, dx = -\lim_{\rho\to 0} \int_{C_2} \frac{e^{iz}}{z}\, dz = +\pi i,$$

and by taking the imaginary parts on both sides

$$\text{pr. v.} \int_{-\infty}^{\infty} \frac{\sin x}{x}\, dx = \pi. \tag{8}$$

Now the integrand in (8) is not singular at $x = 0$. Furthermore, since for positive x the function $1/x$ decreases, the areas under the curve of the integrand between two consecutive positive zeros decrease in a monotone fashion, that is, the absolute values of the integrals

$$I_n = \int_{n\pi}^{n\pi+\pi} \frac{\sin x}{x}\, dx \qquad n = 0, 1, \cdots$$

form a monotone decreasing sequence $|I_1|, |I_2|, \cdots$, and $I_n \to 0$ as $n \to \infty$. Since these integrals have alternating sign, it follows from the Leibniz criterion in Sec. 15.4 that the infinite series $I_0 + I_1 + I_2 + \cdots$ converges. Clearly, the sum of the series is the integral

$$\int_0^{\infty} \frac{\sin x}{x}\, dx = \lim_{b\to\infty} \int_0^{b} \frac{\sin x}{x}\, dx,$$

which therefore exists. Similarly, the integral from 0 to $-\infty$ exists. Hence we need not take the principal value in (8), and

$$\int_{-\infty}^{\infty} \frac{\sin x}{x}\, dx = \pi.$$

Since the integrand is an even function, the desired result follows. ∎

Problems for Sec. 17.4

1. Derive (3) from (2).

Evaluate the following real integrals.

2. $\displaystyle\int_{-\infty}^{\infty} \frac{\cos 2x}{(x^2+4)^2}\, dx$ **3.** $\displaystyle\int_{-\infty}^{\infty} \frac{\cos x}{(x^2+1)^2}\, dx$ **4.** $\displaystyle\int_{-\infty}^{\infty} \frac{\sin 3x}{1+x^4}\, dx$

5. $\displaystyle\int_{0}^{\infty} \frac{\cos sx}{x^2+1}\, dx$ **6.** $\displaystyle\int_{-\infty}^{\infty} \frac{\sin 2x}{x^2+x+1}\, dx$ **7.** $\displaystyle\int_{-\infty}^{\infty} \frac{\cos x}{x^4+1}\, dx$

8. $\displaystyle\int_{-\infty}^{\infty} \frac{\cos 4x}{(x^2+1)(x^2+4)}\, dx$

9. Integrating e^{-z^2} around the boundary of the rectangle with vertices $-a, a, a + ib$, $-a + ib$, letting $a \to \infty$, and using

$$\int_0^\infty e^{-x^2}\,dx = \frac{\sqrt{\pi}}{2}, \qquad \text{show that} \qquad \int_0^\infty e^{-x^2}\cos 2bx\,dx = \frac{\sqrt{\pi}}{2}e^{-b^2}.$$

10. Find pr. v. $\displaystyle\int_{-\infty}^{\infty} \frac{dx}{(x+1)(x^2+2)}$ by integrating along the contour shown in the figure.

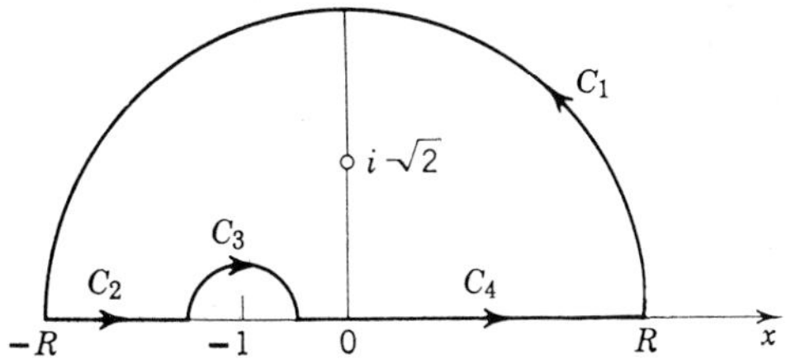

Problem 10

CHAPTER

18 COMPLEX ANALYTIC FUNCTIONS AND POTENTIAL THEORY

Laplace's equation $\nabla^2 u = 0$ is one of the most important partial differential equations in engineering mathematics, because it occurs in connection with gravitational fields (Sec. 8.8), electrostatic fields (Sec. 11.11), steady-state heat conduction (Sec. 13.5), incompressible fluid flow, etc. The theory of the solutions of this equation is called *potential theory*.

In the "two-dimensional case" when u depends only on two Cartesian coordinates x and y, Laplace's equation becomes

$$\nabla^2 u = u_{xx} + u_{yy} = 0.$$

We know that then its solutions are closely related to complex analytic functions (cf. Sec. 12.5).[1] We shall now consider this connection and its consequences in more detail and illustrate it by practical examples taken from hydrodynamics and electrostatics. In Sec. 18.3 we shall see that results about analytic functions can be used for characterizing various general properties of harmonic functions. Finally, we shall derive an important general formula (Poisson's integral formula) for the solution of boundary value problems involving Laplace's equation in a circular disk.

Prerequisites for this chapter: Chaps. 12–15.
References: Appendix 1, Part F.
Answers to problems: Appendix 2.

18.1 Electrostatic Fields

The electrical force of attraction or repulsion between charged particles is governed by Coulomb's law. This force is the gradient of a function u, called the *electrostatic potential,* and at any points free of charges u is a solution of Laplace's equation

$$\nabla^2 u = 0.$$

Cf. Sec. 11.11. The surfaces $u = const$ are called **equipotential surfaces.** At each point P the gradient of u is perpendicular to the surface $u = const$ through P, that is, the electrical force has the direction perpendicular to the equipotential surface.

[1] No such close relation exists in the three-dimensional case.

Example 1. Potential between parallel plates

Find the potential of the field between two parallel conducting plates extending to infinity (Fig. 336), which are kept at potentials U_1 and U_2, respectively. From the shape of the plates it follows that u depends only on x, and Laplace's equation becomes $u'' = 0$. By integrating twice we obtain $u = ax + b$ where the constants a and b are determined by the given boundary values of u on the plates. For example, if the plates correspond to $x = -1$ and $x = 1$, the solution is

$$u(x) = \tfrac{1}{2}(U_2 - U_1)x + \tfrac{1}{2}(U_2 + U_1).$$

The equipotential surfaces are parallel planes.

Example 2. Potential between coaxial cylinders

Find the potential between two coaxial conducting cylinders which extend to infinity on both sides (Fig. 337) and are kept at potentials U_1 and U_2, respectively. Here u depends only on $r = \sqrt{x^2 + y^2}$, for reasons of symmetry, and Laplace's equation becomes

$$ru'' + u' = 0 \qquad \text{[cf. (4) in Sec. 11.9].}$$

By separating variables and integrating we obtain

$$\frac{u''}{u'} = -\frac{1}{r}, \qquad \ln u' = -\ln r + \tilde{a}, \qquad u' = \frac{a}{r}, \qquad u = a \ln r + b,$$

and a and b are determined by the given values of u on the cylinders. Although no infinitely extended conductors exist, the field in our idealized conductor will approximate the field in a long finite conductor in that part which is far away from the ends of the two cylinders. ▮

If the potential u depends only on two Cartesian coordinates x and y, then Laplace's equation becomes

$$\nabla^2 u = \frac{\partial^2 u}{\partial x^2} + \frac{\partial^2 u}{\partial y^2} = 0. \tag{1}$$

The equipotential surfaces $u = const$ appear as *equipotential lines* in the xy-plane.

We make the general assumption that $u(x, y)$ is *harmonic,* that is, its second partial derivatives are continuous. Then, if $v(x, y)$ is a conjugate harmonic function of $u(x, y)$ (cf. Sec. 12.5), the function

$$F(z) = u(x, y) + iv(x, y)$$

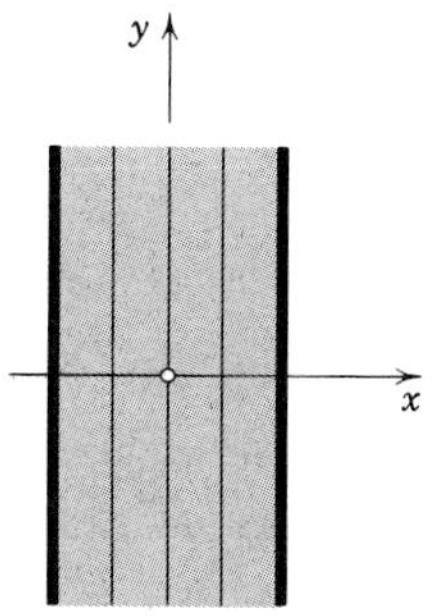

Fig. 336. Potential in Example 1

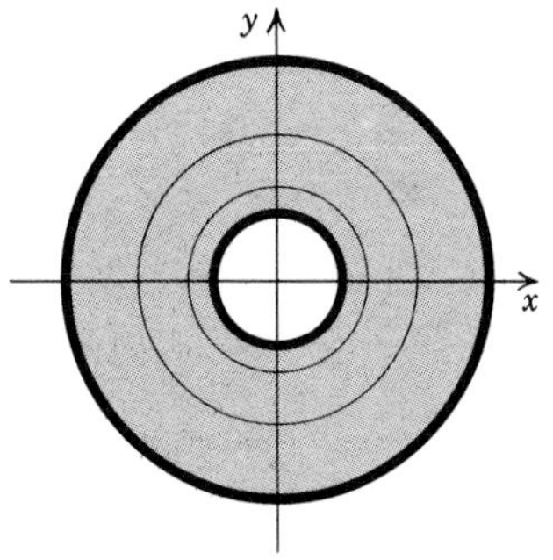

Fig. 337. Potential in Example 2

is an analytic function of $z = x + iy$. This function is called the **complex potential** corresponding to the real potential u. Remember that for a given u the conjugate is uniquely determined, except for an additive real constant.

Since the lines $v = const$ intersect the equipotential lines $u = const$ at right angles [except at points where $F'(z) = 0$], they have the direction of the electrical force and, therefore, are called **lines of force.**

Example 3. Complex potential

In Example 1, a conjugate is $v = ay$. It follows that the complex potential is

$$F(z) = az + b = ax + b + iay,$$

and the lines of force are straight lines parallel to the x-axis.

Example 4. Complex potential

In Example 2 we have

$$u = a \ln r + b = a \ln |z| + b.$$

A conjugate is $v = a \arg z$. The complex potential is $F(z) = a \ln z + b$, and the lines of force are straight lines through the origin. $F(z)$ may also be interpreted as the complex potential of a source line whose trace in the xy-plane is the origin. ∎

Often more complicated potentials can be obtained by superposition. This may be illustrated by

Example 5. Complex potential of a pair of source lines

Determine the complex potential of a pair of oppositely charged source lines of the same strength at the points $z = x_1$ and $z = x_2$. From Examples 2 and 4 it follows that the potential of each of the source lines is

$$u_1 = -c \ln |z - x_1| \qquad \text{and} \qquad u_2 = c \ln |z - x_2|,$$

respectively. These are the real parts of the complex potentials

$$F_1(z) = -c \ln (z - x_1) \qquad \text{and} \qquad F_2(z) = c \ln (z - x_2).$$

Hence, the complex potential of the combination of the two source lines is

$$F(z) = F_1(z) + F_2(z) = c \ln \frac{z - x_2}{z - x_1}. \tag{2}$$

The equipotential lines are the curves

$$u = \operatorname{Re} F(z) = c \ln \frac{|z - x_2|}{|z - x_1|} = const,$$

which are circles, as the reader may show by a straightforward calculation. The lines of force are the curves

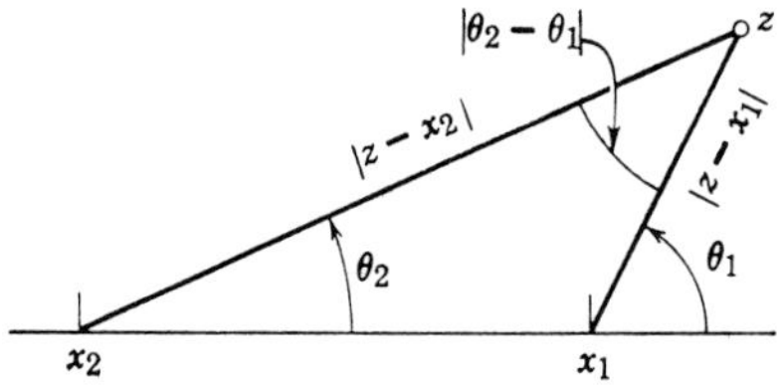

Fig. 338. Example 5

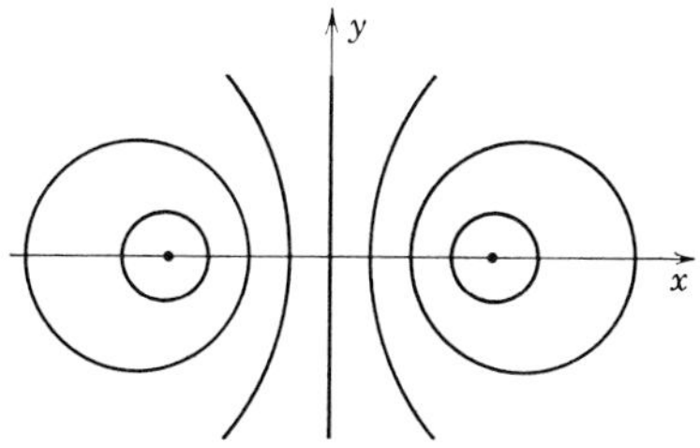

Fig. 339. Potential field of two oppositely charged source lines of the same strength

$$v = \operatorname{Im} F(z) = c \arg \frac{z - x_2}{z - x_1} = c[\arg(z - x_2) - \arg(z - x_1)] = const,$$

that is,

$$v = c(\theta_2 - \theta_1) = const$$

(cf. Fig. 338). Now, $|\theta_2 - \theta_1|$ is the angle between the line segments from z to x_1 and x_2. The lines of force, therefore, are curves along each of which the line segment x_1x_2 appears under a constant angle; these curves are the totality of circular arcs over x_1x_2, as is well known from elementary geometry. The function (2) may also be interpreted as the complex potential of a capacitor consisting of two circular cylinders whose axes are parallel but do not coincide (Fig. 339).

Problems for Sec. 18.1

Find the potential u between two infinite coaxial cylinders of radii r_1 and $r_2 (>r_1)$ which are kept on the potentials U_1 and U_2, respectively, where

1. $r_1 = 1$, $r_2 = 5$, $U_1 = 0$, $U_2 = 100$ V

2. $r_1 = 0.5$, $r_2 = 2$, $U_1 = -110$ V, $U_2 = 110$ V

3. $r_1 = 2$, $r_2 = 20$, $U_1 = 100$ V, $U_2 = 200$ V

4. $r_1 = 3$, $r_2 = 6$, $U_1 = 100$ V, $U_2 = 50$ V

5. Find the equipotential lines of the complex potential $F(z) = 1/z$ and show these lines graphically.

6. Find the potential of two oppositely charged source lines at the points $z = a$ and $z = -a$. Show the equipotential lines graphically.

7. Find the potential of two source lines at $z = a$ and $z = -a$ having the same charge.

8. Show that $F(z) = \cos^{-1} z$ may be interpreted as the complex potential of each of the three configurations in the figure.

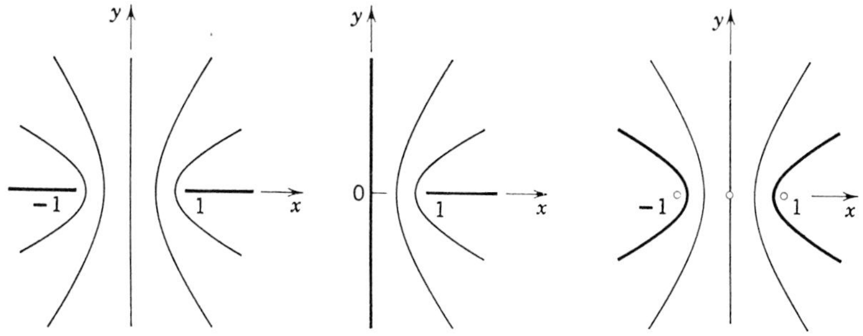

Problem 8

9. Show that $F(z) = \cosh^{-1} z$ may be interpreted as the complex potential between two confocal elliptic cylinders.

10. Find the potential u between the infinite cylinders in the figure, if on the left cylinder, $u = -1$ and on the right, $u = 1$. *Hint.* Use the potential in Prob. 6.

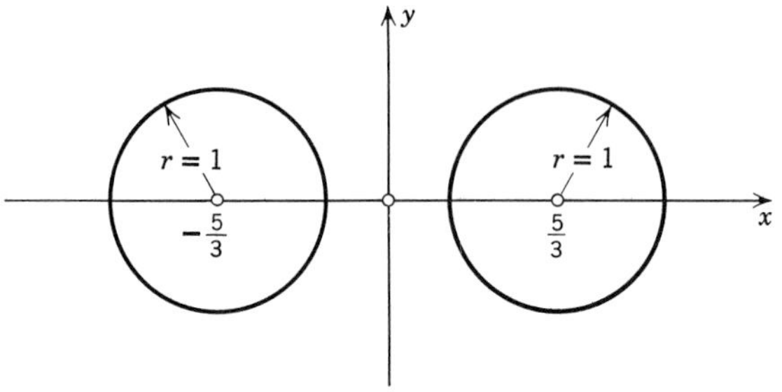

Problem 10

18.2 Two-Dimensional Fluid Flow

Harmonic functions play an important role in hydrodynamics. To illustrate this, let us consider the two-dimensional steady motion of a nonviscous fluid. Here "two-dimensional" means that the motion of the fluid is the same in all planes parallel to the xy-plane, the velocity being parallel to that plane. It then suffices to consider the motion of the fluid in the xy-plane. "Steady" means that the velocity is independent of the time.

At any point (x, y) the flow has a certain velocity which is determined by its magnitude and its direction and is thus a vector. Since in the complex plane any number a represents a vector (the vector from the origin to the point corresponding to a), we may represent the velocity of the flow by a complex variable, say,

$$V = V_1 + iV_2. \tag{1}$$

Then V_1 and V_2 are the components of the velocity in the x and y directions, and V is tangential to the paths of the moving particles of the fluid. Such a path is called a **streamline** of the motion. Cf. Fig. 340.

Consider now an arbitrary given smooth curve C. Let s denote the arc length of C, and let the real variable V_t be the component of the velocity V tangent to C (Fig. 341). Then the value of the line integral

$$\int_C V_t \, ds \tag{2}$$

taken along C in the sense of increasing values of s is called the **circulation** of the fluid along C. Dividing the circulation by the length of C, we obtain the

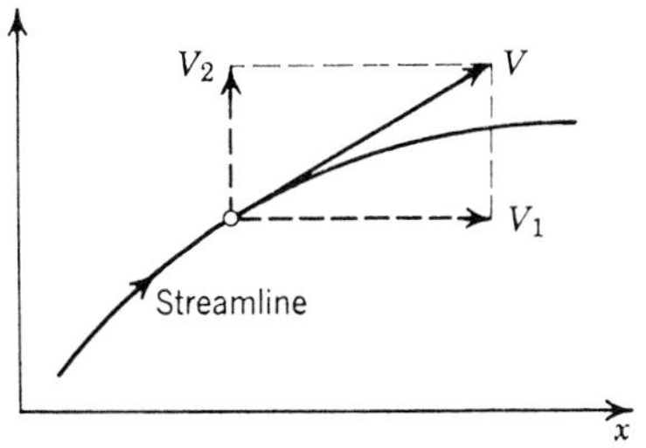

Fig. 340. Velocity

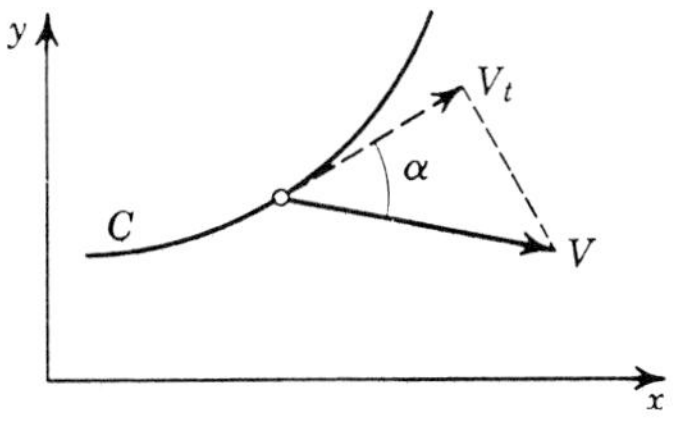

Fig. 341. Tangential component of the velocity with respect to a curve C

mean velocity[2] of the flow along the curve C. Now

$$V_t = |V| \cos \alpha \qquad \text{(cf. Fig. 341).}$$

Consequently, V_t is the scalar product (Sec. 6.5) of V and the unit tangent vector (cf. Sec. 13.2)

$$\frac{dz}{ds} = \frac{dx}{ds} + i\frac{dy}{ds}$$

to C, where $z(s) = x(s) + iy(s)$ is a representation of C. Therefore, the product $V_t\, ds$ may be written

$$V_t\, ds = V \cdot dz = V_1\, dx + V_2\, dy \qquad (dz = dx + i\, dy).$$

(Note well that this is dot multiplication applied to two vectors, not complex multiplication.)

Now let C be a *closed* curve, namely, the boundary curve of a simply connected domain D. Then, if V has continuous partial derivatives in a domain containing D and C it follows from Green's theorem (Sec. 9.4) that the circulation around C can be represented by a double integral,

$$\text{(3)} \qquad \int_C (V_1\, dx + V_2\, dy) = \iint_D \left(\frac{\partial V_2}{\partial x} - \frac{\partial V_1}{\partial y}\right) dx\, dy.$$

The function in the integral on the right permits a simple physical interpretation as follows. Let C be a circle of radius r. The circulation divided by $2\pi r$ then represents the mean velocity of the fluid along C, and the mean angular velocity

[2] *Definitions:* $\dfrac{1}{b-a}\displaystyle\int_a^b f(x)\, dx$ = mean value of f on the interval $a \leqq x \leqq b$.

$\dfrac{1}{l}\displaystyle\int_C f(s)\, ds$ = mean value of f on C. (l = length of C)

$\dfrac{1}{A}\displaystyle\iint_D f(x, y)\, dx\, dy$ = mean value of f on D. (A = area of D)

ω_0 of the fluid about the axis of the circle is obtained by dividing the mean velocity by the radius r; thus

$$\omega_0 = \frac{1}{\pi r^2} \iint_D \frac{1}{2}\left(\frac{\partial V_2}{\partial x} - \frac{\partial V_1}{\partial y}\right) dx\, dy.$$

The expression on the right is the mean value[3] of the function

(4)
$$\omega = \frac{1}{2}\left(\frac{\partial V_2}{\partial x} - \frac{\partial V_1}{\partial y}\right)$$

on the disk D bounded by the circle C. The function ω is called the **rotation,** and 2ω is called the **vorticity** of the motion. If $r \to 0$, the limit of that expression is the value of ω at the center of C. Hence, $\omega(x, y)$ is the limiting angular velocity of a circular element of the fluid as the circle shrinks to the point (x, y). Roughly speaking, if a spherical element of the fluid were suddenly solidified and the surrounding fluid simultaneously annihilated, the element would rotate with the angular velocity ω (cf. also Sec. 8.11).

We consider only *irrotational flows,* that is, flows for which ω is zero everywhere in the region D of the flow, that is

(5)
$$\frac{\partial V_2}{\partial x} - \frac{\partial V_1}{\partial y} = 0,$$

the existence and continuity of the derivatives being assumed.

We further assume that the fluid is *incompressible.* Then

(6)
$$\frac{\partial V_1}{\partial x} + \frac{\partial V_2}{\partial y} = 0 \qquad \text{[cf. (9), Sec. 8.10]}$$

in every region which is free of **sources** or **sinks,** that is, points at which fluid is produced or disappears.

If D is a simply connected domain and the flow is irrotational, it follows from Theorem 3 in Sec. 9.12 that the line integral

(7)
$$\int_C (V_1\, dx + V_2\, dy)$$

is independent of path in D. If we integrate from a fixed point (a, b) in D to a variable point (x, y) in D, the integral thus becomes a function of the point (x, y), say, $\Phi(x, y)$:

(8)
$$\Phi(x, y) = \int_{(a,b)}^{(x,y)} (V_1\, dx + V_2\, dy).$$

The function $\Phi(x, y)$ is called the **velocity potential**[4] of the motion. Since the

[3] Cf. the previous footnote.

[4] Some authors use $-\Phi$ (instead of Φ) as the velocity potential.

integral is independent of path, $V_1\,dx + V_2\,dy$ is an exact differential (Sec. 9.12), namely, the differential of the function $\Phi(x, y)$, that is,

$$V_1\,dx + V_2\,dy = \frac{\partial\Phi}{\partial x}\,dx + \frac{\partial\Phi}{\partial y}\,dy. \tag{9}$$

It follows that

$$V_1 = \frac{\partial\Phi}{\partial x}, \qquad V_2 = \frac{\partial\Phi}{\partial y}, \tag{10}$$

and we see that the velocity vector is the gradient of $\Phi(x, y)$:

$$V = V_1 + iV_2 = \frac{\partial\Phi}{\partial x} + i\frac{\partial\Phi}{\partial y} \qquad \text{(cf. Sec. 8.8).} \tag{11}$$

The curves $\Phi(x, y) = const$ are called **equipotential lines.** Since V is the gradient of Φ, at each point V is perpendicular to the equipotential line through that point (provided $V \neq 0$).

Furthermore, by inserting (10) into (6) we see that Φ satisfies Laplace's equation

$$\nabla^2\Phi = \frac{\partial^2\Phi}{\partial x^2} + \frac{\partial^2\Phi}{\partial y^2} = 0.$$

Let $\Psi(x, y)$ be a conjugate harmonic function of $\Phi(x, y)$. Then at each point the curves

$$\Psi(x, y) = const$$

are perpendicular[5] to the equipotential lines $\Phi(x, y) = const$. Hence, their tangents have the direction of the velocity of the fluid. Consequently, *the curves* $\Psi(x, y) = const$ *are the streamlines of the flow.* The function $\Psi(x, y)$ is called the **stream function** of this flow.

We assume that both Φ and Ψ have continuous second partial derivatives. Then the complex function

$$F(z) = \Phi(x, y) + i\Psi(x, y) \tag{12}$$

is analytic in the region of the flow. This function is called the **complex potential** of the flow. Working with the complex potential is simpler than working with Φ and Ψ separately.

The velocity of the flow can be obtained by differentiating (12) and using the Cauchy-Riemann equations; we find

$$F'(z) = \frac{\partial\Phi}{\partial x} + i\frac{\partial\Psi}{\partial x} = \frac{\partial\Phi}{\partial x} - i\frac{\partial\Phi}{\partial y} = V_1 - iV_2.$$

From this it follows that

$$V = V_1 + iV_2 = \overline{F'(z)}. \tag{13}$$

[5] Except for points where $F'(z) = 0$ [cf. (12)].

In this way *two-dimensional irrotational steady flows of incompressible fluids may be described in terms of analytic functions,* and the methods of complex analysis, such as conformal mapping, may be employed.

In connection with boundary value problems the stream function Ψ is important, because a boundary across which fluid cannot flow is a streamline. Under conformal mapping, a streamline transforms into a streamline in the image plane. Another possibility for obtaining and investigating complicated flows is the *superposition of simple flows.* The sum $F = F_1 + F_2$ of the complex potentials F_1, F_2 of two flows is the complex potential of the flow which is obtained by vector addition of the velocity vectors of the two flows. Clearly, since Laplace's equation is linear and homogeneous, the sum of two harmonic functions is a harmonic function.

Note that while in electrostatics the given boundaries (conducting plates) are equipotential lines, in hydrodynamics the boundaries are streamlines and thus are orthogonal to the equipotential lines.

Let us discuss a typical example. Further applications are included in the problem set.

Example 1. Flow around a corner

The complex potential

$$F(z) = z^2 = x^2 - y^2 + 2ixy \tag{14}$$

describes a flow whose equipotential lines are the hyperbolas

$$\Phi = x^2 - y^2 = const$$

and whose streamlines are the hyperbolas

$$\Psi = 2xy = const.$$

From (13) we obtain the velocity vector

$$V = 2\bar{z} = 2(x - iy), \qquad \text{that is,} \qquad V_1 = 2x, \qquad V_2 = -2y.$$

The speed (magnitude of the velocity) is

$$|V| = \sqrt{V_1^2 + V_2^2} = 2\sqrt{x^2 + y^2}.$$

The flow may be interpreted as the flow in a channel bounded by the positive coordinate axes and a hyperbola, say, $xy = 1$ (Fig. 342). We note that the speed along a streamline S has a minimum at the point P where the cross section of the channel is large.

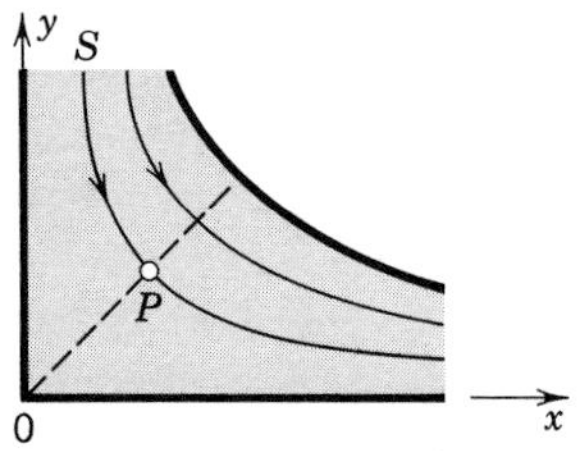

Fig. 342. Flow around a corner

Problems for Sec. 18.2

1. **(Parallel flow)** Show that $F(z) = Kz$ (K positive real) describes a uniform flow to the right, which can be interpreted as a uniform flow between two parallel lines (between two parallel planes in three-dimensional space). Cf. the figure. Find the velocity vector, the streamlines, and the equipotential lines.

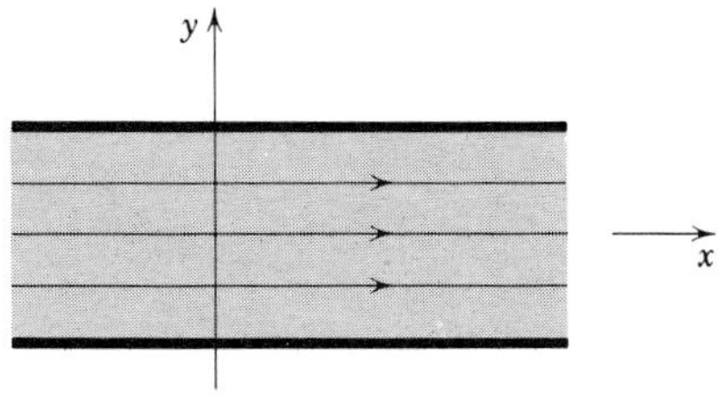

Problem 1. Parallel flow

2. Show that $F(z) = iz^2$ describes a flow around a corner. Find and graph the streamlines and the equipotential lines. Find the velocity vector V.
3. Obtain the flow in Example 1 from that in Prob. 1 by a conformal mapping of the first quadrant onto the upper half-plane.
4. Sketch the streamlines and equipotential lines of the flow with complex potential $F(z) = z^3$. Find the velocity vector V and determine all points at which that vector is parallel to the x-axis.

Consider the flow corresponding to the given complex potential $F(z)$. Show the streamlines and equipotential lines graphically. Find the velocity vector, and determine all points at which that vector is parallel to the x-axis. (Assume k in Prob. 6 to be real.)

5. iz 6. $-ikz$ 7. $(1 + i)z$ 8. $z^2 + z$ 9. iz^3

10. **(Source and sink)** Consider the complex potential $F(z) = (c/2\pi) \ln z$ where c is positive real. Show that $V = (c/2\pi r^2)(x + iy)$ where $r = \sqrt{x^2 + y^2}$, and this implies that the flow is directed radially outward (cf. the figure), so that this potential corresponds to a **point source** at $z = 0$ (that is, a **source line** $x = 0$, $y = 0$ in space). (The constant c is called the **strength** or **discharge** of the source. If c is negative real, the flow is said to have a **sink** at $z = 0$, it is directed radially inward and fluid disappears at the singular point $z = 0$ of the complex potential.)

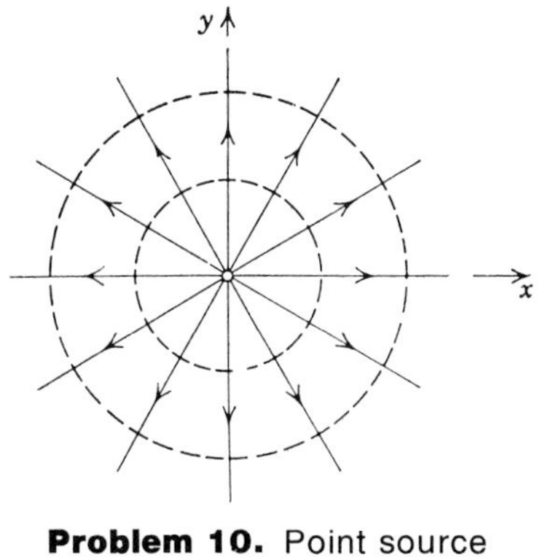

Problem 10. Point source

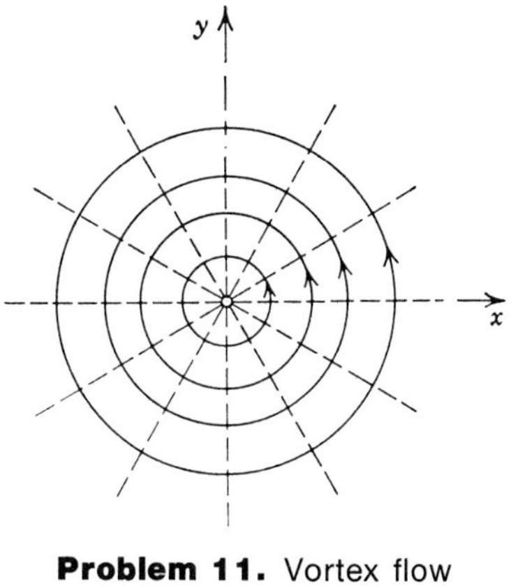

Problem 11. Vortex flow

11. **(Vortex line)** Show that $F(z) = -(iK/2\pi) \ln z$ with real positive K describes a flow which circles around the origin in the counterclockwise sense; cf. the figure on the previous page. (The point $z = 0$ is a **vortex;** the potential increases by K each time we travel around the vortex.)

12. Find the complex potential of a flow which has a point source of strength 1 at $z = -a$.

13. Find the complex potential of a flow which has a point sink of strength 1 at $z = a$.

14. Show that vector addition of the velocity vectors of two flows leads to a flow whose complex potential is obtained by adding those of the two flows.

15. Add the potentials in Probs. 12 and 13 and show the streamlines graphically.

16. Find the streamlines of the flow corresponding to $F(z) = 1/z$. Show that for small $|a|$ the streamlines in Prob. 15 look similar to those in the present problem.

17. Show that $F(z) = \cosh^{-1} z$ corresponds to a flow whose streamlines are confocal hyperbolas with foci at $z = \pm 1$, and the flow may be interpreted as a flow through an aperture. (Cf. the figure.)

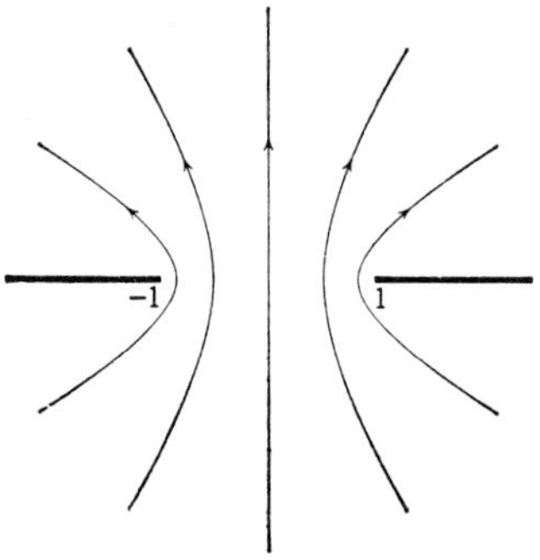

Problem 17. Flow through an aperture

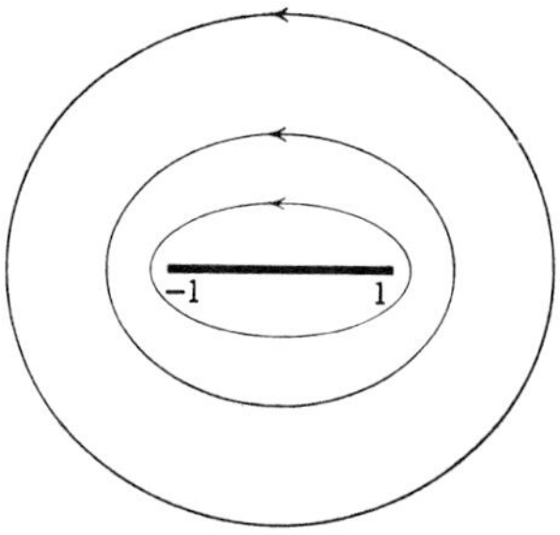

Problem 18. Flow around a plate

18. Show that $F(z) = \cos^{-1} z$ can be interpreted as the complex potential of a flow circulating around an elliptic cylinder or a plate (the straight segment from $z = -1$ to $z = 1$). Show that the streamlines are confocal ellipses with foci at $z = \pm 1$. (Cf. the figure.)

19. **(Flow around a cylinder)** Consider $F(z) = z + z^{-1}$. Setting $z = re^{i\theta}$, show that the streamlines are $(r - r^{-1}) \sin \theta = const$, the streamline $(r - r^{-1}) \sin \theta = 0$ consists of the x-axis and the unit circle, and for large $|z|$ the flow is nearly uniform and parallel, so that it may be interpreted as the flow around a long circular cylinder of unit radius (see the left upper part of the figure). Find the **stagnation points** of the flow (points at which the velocity is zero).

20. **(Flow with circulation around a cylinder)** Show that addition of the potentials in Probs. 11 and 19 gives a flow such that the cylinder wall $|z| = 1$ is a streamline. Find the speed and show that the stagnation points are

$$z = \frac{iK}{4\pi} \pm \sqrt{\frac{-K^2}{16\pi^2} + 1};$$

when $K = 0$ they are at ± 1, as K increases they move up on the unit circle until they unite at $z = i$ ($K = 4\pi$, cf. the figure), and when $K > 4\pi$ they lie on the imaginary axis (one lies in the field of flow and the other one lies inside the cylinder and has no physical meaning).

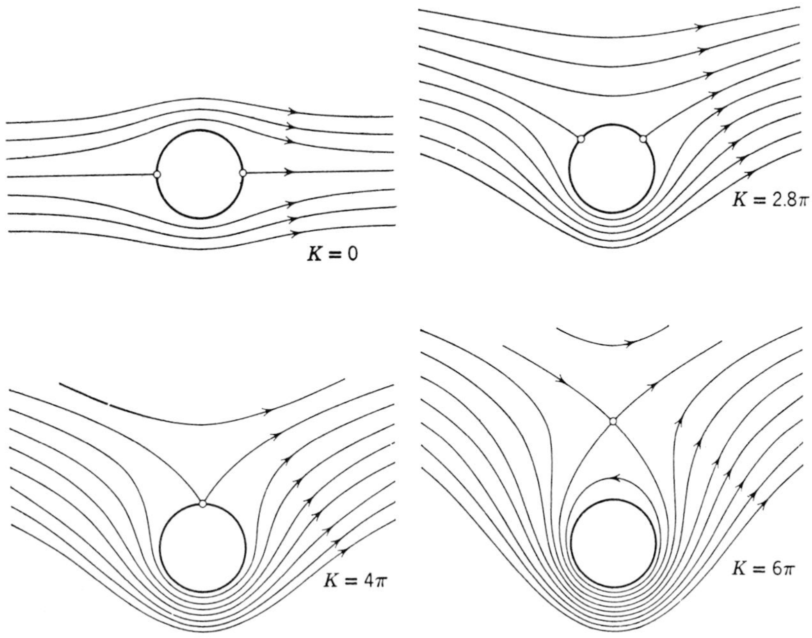

Problems 19 and 20. Flow around a cylinder without circulation ($K = 0$) and with circulation

18.3 General Properties of Harmonic Functions

In this section we want to show how results about complex analytic functions can be used for deriving general properties of harmonic functions.

Let $u(x, y)$ be a given function which is harmonic in a simply connected domain D. Then we may determine a conjugate harmonic function $v(x, y)$ by the Cauchy-Riemann equations, and $f(z) = u(x, y) + iv(x, y)$ is then an analytic function in D. (Cf. Sec. 12.5; cf. also footnote 2 in Sec. 13.2.) This is the connection which we can use for deriving properties of harmonic functions from properties of analytic functions. In the first place, since analytic functions have derivatives of all orders, we immediately obtain

Theorem 1 (Partial derivatives)

A function $u(x, y)$ which is harmonic in a simply connected domain D has partial derivatives of all orders in D.

Furthermore, if $f(z)$ is analytic in a simply connected domain D, then, by Cauchy's integral formula (Sec. 14.5),

$$f(z_0) = \frac{1}{2\pi i}\int_C \frac{f(z)}{z - z_0}\,dz \tag{1*}$$

where C is a simple closed path in D and z_0 lies inside C. Choosing for C a circle

$$z = z_0 + re^{i\phi}$$

in D, we have

$$z - z_0 = re^{i\phi}, \qquad dz = ire^{i\phi}\,d\phi$$

and the integral formula becomes

$$f(z_0) = \frac{1}{2\pi}\int_0^{2\pi} f(z_0 + re^{i\phi})\,d\phi. \tag{1}$$

The right side is the mean value of f on the circle (= value of the integral divided by the length of the interval of integration). This proves

Theorem 2 (Mean value property of analytic functions)
Let $f(z)$ be analytic in a simply connected domain D. Then the value of $f(z)$ at a point z_0 in D is equal to the mean value of $f(z)$ on any circle in D with center at z_0.

Another important property of analytic functions is the following one.

Theorem 3 (Maximum modulus theorem for analytic functions)
Let D be a bounded region and suppose that $f(z)$ is analytic and nonconstant in D and on the boundary of D. Then the absolute value $|f(z)|$ cannot have a maximum at an interior point of D. Consequently, the maximum of $|f(z)|$ is taken on the boundary of D. If $f(z) \neq 0$ in D, the same is true with respect to the minimum of $|f(z)|$.

Proof. We assume that $|f(z)|$ has a maximum at an interior point z_0 of D and show that this leads to a contradiction. Let $|f(z_0)| = M$ be that maximum. Since $f(z)$ is not constant, $|f(z)|$ is not constant. Consequently, we can find a circle C of radius r with center at z_0 such that the interior of C is in D and $|f(z)|$ is smaller than M at some point P of C. Since $|f(z)|$ is continuous, it will be smaller than M on an arc C_1 of C which contains P, say, $|f(z)| \leqq M - \epsilon$ ($\epsilon > 0$) for all z on C_1 (Fig. 343). If C_1 has the length l_1, the complementary arc C_2 of C has the length $2\pi r - l_1$. By applying (4), Sec. 14.2, to (1*), this section, and noting that $|z - z_0| = r$, we thus obtain

$$M = |f(z_0)| \leqq \frac{1}{2\pi}\left|\int_{C_1} \frac{f(z)}{z - z_0}\,dz\right| + \frac{1}{2\pi}\left|\int_{C_2} \frac{f(z)}{z - z_0}\,dz\right|$$

$$\leqq \frac{1}{2\pi}(M - \epsilon)\frac{1}{r}l_1 + \frac{1}{2\pi}M\frac{1}{r}(2\pi r - l_1) = M - \frac{\epsilon l_1}{2\pi r} < M,$$

that is, $M < M$, which is impossible. Hence, our assumption is false and the first statement of the theorem is proved.

We prove the last statement. If $f(z) \neq 0$ in D, then $1/f(z)$ is analytic in D. From the statement already proved it follows that the maximum of $1/|f(z)|$ lies on the boundary of D. But this maximum corresponds to a minimum of $|f(z)|$. This completes the proof. ∎

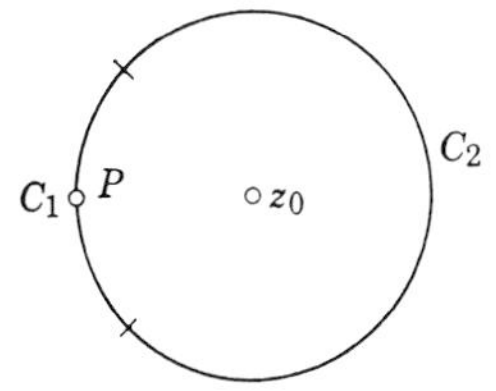

Fig. 343. Proof of Theorem 3

From these theorems we shall now derive corresponding results about harmonic functions.

Theorem 4 (Harmonic functions)

Let D be a simply connected bounded domain and let C be its boundary curve. Then if $u(x, y)$ is harmonic in a domain containing D and C, it has the following properties.

I. The value of $u(x, y)$ at a point (x_0, y_0) of D is equal to the mean value of $u(x, y)$ on any circle in D with center at (x_0, y_0).

II. The value of $u(x, y)$ at the point (x_0, y_0) is equal to the mean value of $u(x, y)$ in any circular disk in D with center at (x_0, y_0). [Cf. footnote 2 in Sec. 18.2.]

III. **(Maximum principle)** *If $u(x, y)$ is nonconstant, it has neither a maximum nor a minimum in D. Consequently, the maximum and the minimum are taken on the boundary of D.*

IV. If $u(x, y)$ is constant on C, then $u(x, y)$ is a constant.

V. If $h(x, y)$ is harmonic in D and on C and if $h(x, y) = u(x, y)$ on C, then $h(x, y) = u(x, y)$ everywhere in D.

Proof. Statement I follows from (1) by taking the real parts on both sides, that is,

$$u(x_0, y_0) = \operatorname{Re} f(x_0 + iy_0) = \frac{1}{2\pi}\int_0^{2\pi} u(x_0 + r\cos\phi, y_0 + r\sin\phi)\, d\phi.$$

If we multiply both sides by r and integrate over r from 0 to r_0 where r_0 is the radius of a circular disk in D with center at (x_0, y_0), then we obtain on the left-hand side $\frac{1}{2}r_0^2 u(x_0, y_0)$ and therefore

$$u(x_0, y_0) = \frac{1}{\pi r_0^2}\int_0^{r_0}\int_0^{2\pi} u(x_0 + r\cos\phi, y_0 + r\sin\phi) r\, d\phi\, dr.$$

This proves the second statement.

We prove statement III. Let $v(x, y)$ be a conjugate harmonic function of $u(x, y)$ in D. Then $f(z) = u(x, y) + iv(x, y)$ is analytic in D, and so is

$$F(z) = e^{f(z)}.$$

The absolute value is

$$|F(z)| = e^{\operatorname{Re} f(z)} = e^{u(x,y)}.$$

From Theorem 3 it follows that $|F(z)|$ cannot have a maximum at an interior point of D. Since e^u is a monotone increasing function of the real variable u, statement III about the maximum of u follows. From this, the statement about the minimum follows by replacing u by $-u$.

If u is constant on C, say, $u = k$, then by III the maximum and the minimum of u are equal. From this, statement IV follows.

If h and u are harmonic in D and on C, then $h - u$ is also harmonic in D and on C, and by assumption, $h - u = 0$ everywhere on C. Hence, by IV, we have $h - u = 0$ everywhere in D, and statement V is proved. This completes the proof of Theorem 4. ∎

The last statement of Theorem 4 is very important. It means that a *harmonic function is uniquely determined in D by its values on the boundary of D*. Usually, $u(x, y)$ is required to be harmonic in D and continuous on the boundary[6] of D. Under these circumstances the maximum principle (Theorem 4, III) is still applicable. The problem of determining $u(x, y)$ when the boundary values are given is known as the **Dirichlet problem** for the Laplace equation in two variables. From Theorem 4, V we have

Theorem 5 (Dirichlet problem)

If for a given region and given boundary values the Dirichlet problem for the Laplace equation in two variables has a solution, the solution is unique.

Problems for Sec. 18.3

1. Verify Theorem 2 for $f(z) = (z + 2)^2$, $z_0 = 1$ and a circle of radius 1 with center at z_0.
2. Verify Theorem 3 for $f(z) = z^2$ and the rectangle $-2 < x < 2$, $-1 < y < 1$.
3. Verify Theorem 3 for $f(z) = e^z$ and any bounded domain.
4. The function $f(x) = \cos x$ has a maximum at $x = 0$. Using Theorem 3, conclude that the modular surface of $f(z) = \cos z$ (cf. Sec. 12.8) cannot have a summit at $z = 0$.
5. If $f(z)$ is analytic (not a constant) in a simply connected domain D, and the curve given by $|f(z)| = c$ (c any fixed constant) lies in D and is closed, show that $f(z) = 0$ at a point in the interior of that curve. Give examples.

18.4 Poisson's Integral Formula

The Dirichlet problem for a circular disk can be solved by the use of the so-called Poisson[7] formula which represents a harmonic function in terms of its values given on the boundary circle of the disk. We shall derive this formula, starting from Cauchy's integral formula

[6] That is, $\lim_{\substack{x \to x_0 \\ y \to y_0}} u(x, y) = u(x_0, y_0)$ where (x_0, y_0) is on the boundary and (x, y) is in D.

[7] SIMÉON DENIS POISSON (1781–1840), French mathematician and physicist.

$$f(z) = \frac{1}{2\pi i}\int_C \frac{f(z^*)}{z^* - z}\,dz^*; \tag{1}$$

here C is the circle represented by

$$z^* = Re^{i\phi} \qquad (0 \leqq \phi \leqq 2\pi)$$

and the function

$$f(z) = u(r, \theta) + iv(r, \theta) \qquad (z = re^{i\theta})$$

is assumed to be analytic in a simply connected region containing C in its interior.

Since $dz^* = iRe^{i\phi}\,d\phi = iz^*\,d\phi$, we obtain from (1)

$$f(z) = \frac{1}{2\pi}\int_0^{2\pi} f(z^*)\frac{z^*}{z^* - z}\,d\phi \qquad (z^* = Re^{i\phi}, z = re^{i\theta}). \tag{2}$$

On the other hand, if we consider a point Z outside C, say, the point $Z = z^*\bar{z}^*/\bar{z}$ (whose absolute value is $R^2/r > R$), then the integrand in (1) is analytic in the disk $|z| \leqq R$ and the integral is zero:

$$0 = \frac{1}{2\pi i}\int_C \frac{f(z^*)}{z^* - Z}\,dz^* = \frac{1}{2\pi}\int_0^{2\pi} f(z^*)\frac{z^*}{z^* - Z}\,d\phi.$$

By inserting $Z = z^*\bar{z}^* / \bar{z}$ and simplifying the fraction we find

$$0 = \frac{1}{2\pi}\int_0^{2\pi} f(z^*)\frac{\bar{z}}{\bar{z} - \bar{z}^*}\,d\phi.$$

Subtracting this from (2) and using

$$\frac{z^*}{z^* - z} - \frac{\bar{z}}{\bar{z} - \bar{z}^*} = \frac{z^*\bar{z}^* - z\bar{z}}{(z^* - z)(\bar{z}^* - \bar{z})}, \tag{3}$$

we obtain

$$f(z) = \frac{1}{2\pi}\int_0^{2\pi} f(z^*)\frac{z^*\bar{z}^* - z\bar{z}}{(z^* - z)(\bar{z}^* - \bar{z})}\,d\phi. \tag{4}$$

From the polar representations of z and z^* we see that the quotient in the integrand is equal to

$$\frac{R^2 - r^2}{(Re^{i\phi} - re^{i\theta})(Re^{-i\phi} - re^{-i\theta})} = \frac{R^2 - r^2}{R^2 - 2Rr\cos(\theta - \phi) + r^2}.$$

Consequently, by taking the real parts on both sides of formula (4) we obtain

Poisson's integral formula

$$\textbf{(5)} \qquad u(r,\theta) = \frac{1}{2\pi}\int_0^{2\pi} u(R,\phi)\frac{R^2 - r^2}{R^2 - 2Rr\cos(\theta - \phi) + r^2}\,d\phi,$$

which represents the harmonic function u in the disk $|z| \leqq R$ in terms of its values $u(R, \phi)$ on the circle which bounds the disk.

It is of practical interest to note that in (5) we may use for $u(R, \phi)$ any function which is merely piecewise continuous on the interval of integration. Formula (5) then *defines* a function $u(r, \theta)$ which is harmonic in the open disk $|z| < R$ and continuous on the circle $|z| = R$. On this circle this function is equal to $u(R, \phi)$, except at points where $u(R, \phi)$ is discontinuous. The proof can be found in Ref. [F1].

From (5) we may obtain an important series development of u in terms of simple harmonic functions. We remember that the quotient in the integrand of (5) was derived from (3), and it is not difficult to see that the right side of (3) is the real part of $(z^* + z)/(z^* - z)$. By the use of the geometric series we obtain

$$(6) \qquad \frac{z^* + z}{z^* - z} = \frac{1 + (z/z^*)}{1 - (z/z^*)} = \left(1 + \frac{z}{z^*}\right)\sum_{n=0}^{\infty}\left(\frac{z}{z^*}\right)^n = 1 + 2\sum_{n=1}^{\infty}\left(\frac{z}{z^*}\right)^n.$$

Since $z = re^{i\theta}$ and $z^* = Re^{i\phi}$, we have

$$(7) \qquad \begin{aligned} \operatorname{Re}\left(\frac{z}{z^*}\right)^n &= \operatorname{Re}\left[\frac{r^n}{R^n}e^{in\theta}e^{-in\phi}\right] = \left(\frac{r}{R}\right)^n \cos(n\theta - n\phi) \\ &= \left(\frac{r}{R}\right)^n (\cos n\theta \cos n\phi + \sin n\theta \sin n\phi). \end{aligned}$$

From (6) and (7) we obtain

$$\operatorname{Re}\frac{z^* + z}{z^* - z} = 1 + 2\sum_{n=1}^{\infty}\frac{r^n}{R^n}(\cos n\theta \cos n\phi + \sin n\theta \sin n\phi).$$

This expression is equal to the quotient in (5), as we have mentioned before, and by inserting the series into (5) and integrating term by term we find

$$\textbf{(8)} \qquad u(r,\theta) = a_0 + \sum_{n=1}^{\infty}\left(\frac{r}{R}\right)^n (a_n \cos n\theta + b_n \sin n\theta)$$

where the coefficients are

$$\textbf{(9)} \qquad \begin{aligned} a_0 &= \frac{1}{2\pi}\int_0^{2\pi} u(R,\phi)\,d\phi, \qquad a_n = \frac{1}{\pi}\int_0^{2\pi} u(R,\phi)\cos n\phi\,d\phi, \\ b_n &= \frac{1}{\pi}\int_0^{2\pi} u(R,\phi)\sin n\phi\,d\phi, \qquad n = 1, 2, \cdots, \end{aligned}$$

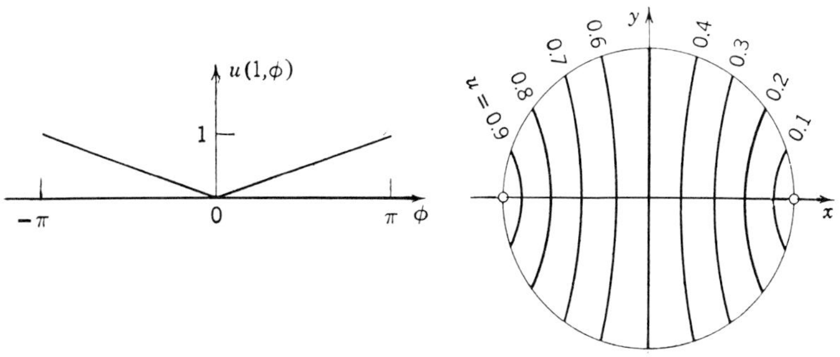

Fig. 344. Potential in Example 1

the Fourier coefficients of $u(R, \phi)$. Note that for $r = R$ the series (8) becomes the Fourier series of $u(R, \phi)$, and therefore the representation (8) will be valid whenever $u(R, \phi)$ can be represented by a Fourier series.

Example 1. Dirichlet problem for the unit disk

Find the potential $u(r, \theta)$ in the unit disk $r < 1$ having the boundary values (Fig. 344)

$$u(1, \phi) = \begin{cases} -\phi/\pi & \text{when } -\pi < \phi < 0 \\ \phi/\pi & \text{when } \quad 0 < \phi < \pi. \end{cases}$$

Since $u(1, \phi)$ is even, $b_n = 0$, and from (9) we obtain $a_0 = \frac{1}{2}$ and

$$a_n = \frac{1}{\pi}\left[-\int_{-\pi}^{0} \frac{\phi}{\pi} \cos n\phi \, d\phi + \int_{0}^{\pi} \frac{\phi}{\pi} \cos n\phi \, d\phi\right] = \frac{2}{n^2\pi^2}(\cos n\pi - 1).$$

Hence, $a_n = -4/n^2\pi^2$ when n is odd, $a_n = 0$ when $n = 2, 4, \cdots$, and the potential is

$$u(r, \theta) = \frac{1}{2} - \frac{4}{\pi^2}\left[r \cos \theta + \frac{r^3}{3^2} \cos 3\theta + \frac{r^5}{5^2} \cos 5\theta + \cdots\right].$$

Problems for Sec. 18.4

1. Verify (3).
2. Show that each term in (8) is a harmonic function in the disk $r^2 < R^2$.

Using (8), find the potential $u(r, \theta)$ in the unit disk $r < 1$ having the given boundary values $u(1, \theta)$. Using the first few terms of the series, compute some values of u and sketch a figure of the equipotential lines.

3. $u(1, \theta) = \sin \theta$
4. $u(1, \theta) = 1 - \cos \theta$
5. $u(1, \theta) = \sin 3\theta$
6. $u(1, \theta) = \cos 2\theta - \cos 4\theta$
7. $u(1, \theta) = 4 \sin^3 \theta$
8. $u(1, \theta) = \theta$
9. $u(1, \theta) = 1$ if $0 < \theta < \pi$ and 0 otherwise
10. $u(1, \theta) = \theta$ if $-\pi/2 < \theta < \pi/2$, $u(1, \theta) = \pi - \theta$ if $\pi/2 < \theta < 3\pi/2$
11. $u(1, \theta) = -\pi/2$ if $-\pi < \theta < -\pi/2$, $u(1, \theta) = \theta$ if $-\pi/2 < \theta < \pi/2$, $u(1, \theta) = \pi/2$ if $\pi/2 < \theta < \pi$
12. $u(1, \theta) = 1$ if $0 < \theta < \pi/2$, $u(1, \theta) = -1$ if $\pi/2 < \theta < \pi$ and 0 otherwise

13. Using (9) in Sec. 16.4, show that the result of Prob. 12 may be written

$$u(r, \theta) = \frac{1}{\pi} \operatorname{Im} \operatorname{Ln} \frac{(1 + iz)(1 + z^2)}{(1 - iz)(1 - z^2)}.$$

14. Show that the potential in Prob. 8 may be written $u(r, \theta) = 2 \operatorname{Im} \operatorname{Ln}(1 + z)$.

15. Using (8), show that the potential $u(r, \theta)$ in the unit disk $r < 1$ having the boundary values

$$u(1, \theta) = \begin{cases} -1 & \text{when } -\pi < \theta < 0 \\ 1 & \text{when } \quad 0 < \theta < \pi \end{cases}$$

is given by the series

$$u(r, \theta) = \frac{4}{\pi}\left(r \sin \theta + \frac{r^3}{3} \sin 3\theta + \frac{r^5}{5} \sin 5\theta + \cdots\right).$$

Compute some values of u by the use of the first few terms of this series and draw some of the equipotential lines. Compare the result with the figure. Graph the lines of force (orthogonal trajectories).

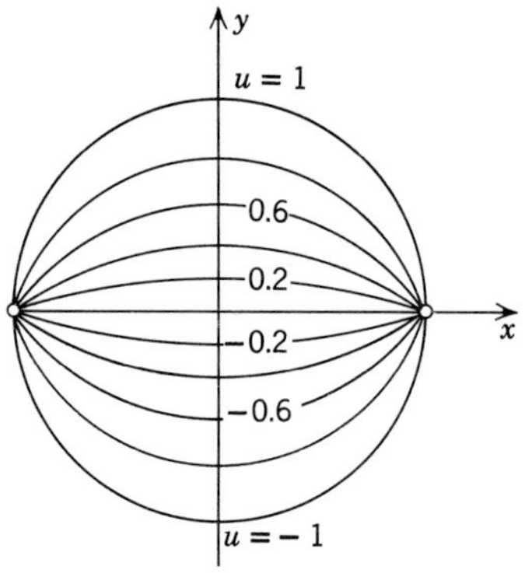

Problem 15

16. Using (9), Sec. 16.4, show that in Prob. 15

$$u(r, \theta) = \frac{2}{\pi} \operatorname{Im} \operatorname{Ln} \frac{1 + z}{1 - z} = \frac{2}{\pi}[\arg(1 + z) - \arg(1 - z)].$$

17. Applying a familiar theorem of elementary geometry to the result of Prob. 16, show that the curves $u = const$ in Prob. 15 are circular arcs.

18. Show that

$$H = 1 + \frac{2}{\pi} \operatorname{Im} \operatorname{Ln} \frac{w + 1}{w - 1} \qquad (w = u + iv)$$

is harmonic in the upper half-plane $v > 0$ and has the values -1 for $-1 < u < 1$ and $+1$ on the remaining part of the u-axis.

19. Show that the linear fractional transformation which maps $w_1 = -1$, $w_2 = 0$, $w_3 = 1$ onto $z_1 = -1$, $z_2 = -i$, $z_3 = 1$, respectively, is

$$z = \frac{w - i}{-iw + 1}.$$

Find the inverse $w = w(z)$, insert it into H in Prob. 18, and show that the resulting harmonic function is that in Prob. 16.

20. Derive Theorem 4 I, Sec. 18.3, from (5).

CHAPTER

19 NUMERICAL ANALYSIS

Engineering mathematics ultimately comes down to numerical results, and the engineering student should, therefore, include in his mathematical equipment a definite knowledge of some fundamental **numerical methods,** that is, methods of obtaining numerical results from given data. The branch of mathematics concerned with the development of such methods is called **numerical analysis.**

The need for such processes for obtaining practical answers to given problems is obvious, because in many cases the answers obtained from the theory may be almost useless for numerical purposes. Typical examples are the integral formula for the solution of a linear first-order differential equation (Sec. 1.7) and Cramer's rule for solving systems of linear algebraic equations in terms of determinants (Secs. 7.9, 7.11). In other cases the theory may merely guarantee the existence of a solution but give no indication of how to obtain it.

The development of numerical analysis is strongly influenced and determined by the advent and use of the automatic computer, which is now indispensable in the mathematical work of every engineer. This development includes the creation of new methods, modification of existing methods to make them more effective in automatic calculation, a theoretical and practical analyzation of algorithms for the standard computational processes and pointing out those algorithms which are satisfactory in various situations. Error analysis is part of that task for the numerical analyst, and he must try to characterize and eliminate arithmetic traps of all kinds which are ready for the unwary.

Prerequisite for this chapter: elementary calculus; for Secs. 19.7 and 19.8 also the elements of differential equations, and for some of the topics in Secs. 19.9–19.14 the elements of matrix algebra.

References: Appendix 1, Part G.

Answers to problems: Appendix 2.

19.1 Errors and Mistakes. Automatic Computers

Since in computations we work with a finite number of digits and carry out finitely many steps, the methods of numerical analysis are **finite processes,** and a numerical result is an **approximate value** of the (unknown) exact result, except for the rare cases where the exact answer is a sufficiently simple rational number and we can use a numerical method that gives the exact answer.

If a^* is an approximate value of a quantity whose exact value is a, then the difference $\epsilon = a^* - a$ is called the *absolute error* of a^* or, briefly, the **error** of a^*.

Hence

$$a^* = a + \epsilon \qquad \text{Approximation} = \text{True value} + \text{error}.$$

The **relative error** ϵ_r of a^* is defined by

$$\epsilon_r = \frac{\epsilon}{a} = \frac{a^* - a}{a} = \frac{\text{Error}}{\text{True value}} \qquad (a \neq 0).$$

Clearly, $\epsilon_r \approx \epsilon/a^*$ if $|\epsilon|$ is much less than $|a^*|$. We may also introduce the quantity $\gamma = a - a^* = -\epsilon$ and call it the *correction,* thus[1]

$$a = a^* + \gamma \qquad \text{True value} = \text{Approximation} + \text{correction}.$$

Finally, an **error bound** for a^* is a number β such that

$$|a^* - a| \leqq \beta, \qquad \text{that is,} \qquad |\epsilon| \leqq \beta.$$

Depending on the source of errors, we may distinguish between experimental errors, truncation errors, and roundoff errors. **Experimental errors** are errors of given data (probably arising from measurements). **Truncation errors** are errors corresponding to the fact that a (finite or infinite) sequence of computational steps necessary to produce an exact result is "truncated" prematurely after a certain number of steps. These errors depend on the computational method and have to be discussed individually with each method. **Rounding errors** are errors arising from the process of rounding off during computation, and we shall now discuss this type of error.

In the decimal notation, every real number is represented by a finite or infinite sequence of decimal digits. For machine computation the number must be replaced by a number of finitely many digits. In automatic digital computers there are two ways in which numbers may be stored, as follows. In a **fixed point** system the numbers are represented with a *fixed number of decimal places,* e.g., 62.358, 0.013, 1.000. In a **floating point** system the numbers are represented with a *fixed number of significant digits,* e.g., 0.6236×10^2, -0.1714×10^{-3}, -0.2000×10^1. **Significant digit** of a number c is any given digit of c, except possibly for zeros to the left of the first nonzero digit that serve only to fix the position of the decimal point. (Thus, any other zero is a significant digit of c.) For instance, each of the numbers 1360, 1.360, 0.001 360 has 4 significant digits.[2]

We shall now state the **rule for rounding off** a number to k decimals. (The rule for rounding off to k significant figures is the same, with "decimal" replaced by "significant figure.")

[1] $\gamma = -\epsilon$ is sometimes called the *error;* the distinction is rather irrelevant, but one should be consistent.

[2] In tables of functions showing k significant digits, it is conventionally assumed that any given value a^* deviates from the corresponding exact value a by at most ± 0.5 unit of the last given digit, unless otherwise stated; e.g. if $a = 1.1996$, then a table with 4 significant digits should show $a^* = 1.200$. Correspondingly, if 12 000 is correct to three digits only, we should better write 120×10^2, etc. "Decimal" is abbreviated by D and "significant digit" by S. For example, 5D means 5 decimals, and 8S means 8 significant digits.

Discard the $(k + 1)$th and all subsequent decimals. If the number thus discarded is less than half a unit in the kth place, leave the kth decimal unchanged ("*rounding down*"). If it is greater than half a unit in the kth place, add one to the kth decimal ("*rounding up*"). If it is exactly half a unit, round off to the nearest *even* decimal. (Example: Rounding off 3.45 and 3.55 to 1 decimal gives 3.4 and 3.6, respectively.)

The last part of the rule is supposed to ensure that in discarding exactly half a decimal, rounding up and rounding down happens about equally often, on the average.

If we round off 1.253 5 to 3, 2, 1 decimals, we get 1.254, 1.25, 1.3, but if 1.25 is rounded off to one decimal, without further information we get 1.2.

Rounding errors may ruin a computation completely. In general, they are the more dangerous the more steps of computations we have to perform. It is therefore important to analyze computational programs for rounding errors to be expected and to find an arrangement of the computations such that the effect of rounding errors is as small as possible. **Guarding figures** are extra figures (one, two, or even more) carried in intermediate stages of a desk calculation with the intention to eliminate rounding errors to some extent. **Scaling** of variables may reduce the number of decimal places carried in a calculation. For instance, $0.002:900 = 0.000\,002\,222$ and we need 9 decimal places to have a result of 4 significant digits, but if we can scale the variables so that the calculation becomes $2:9 = 0.222\,2$, we need only 4 decimals.

We want to emphasize that a careful selection of most effective methods is as important in automatic computation as it is in using a desk calculator, not merely in order to minimize time and cost but also errors. Even for simple problems there may be several ways which differ by their efficiency. Let us illustrate this with an example.

Example 1. Quadratic equation

Find the roots of each of the equations

$$\text{(1)} \qquad \text{(a) } x^2 - 4x + 2 = 0 \qquad \text{and} \qquad \text{(b) } x^2 - 40x + 2 = 0,$$

using 4 significant figures in the calculation.

Formulas for the roots x_1, x_2 of a quadratic equation $ax^2 + bx + c = 0$ are

$$\text{(2)} \qquad x_1 = \frac{1}{2a}(-b + \sqrt{b^2 - 4ac}), \qquad x_2 = \frac{1}{2a}(-b - \sqrt{b^2 - 4ac})$$

and, since $x_1 x_2 = c/a$,

$$\text{(3)} \qquad x_1 \text{ as before,} \qquad x_2 = \frac{c}{ax_1}.$$

For (1a), formula (2) gives $x = 2 \pm \sqrt{2} = 2.000 \pm 1.414$, $x_1 = 3.414$, $x_2 = 0.586$, and (3) gives $x_1 = 3.414$, $x_2 = 2.000/3.414 = 0.5858$, in error by less than one unit in the last digit (cf. Prob. 6). For (1b), formula (2) gives $x = 20 \pm \sqrt{398} = 20.00 \pm 19.95$, $x_1 = 39.95$, $x_2 = 0.05$, which is poor, whereas (3) gives $x_1 = 39.95$, $x_2 = 2.000/39.95 = 0.05006$, in error by less than one unit in the last digit (cf. Prob. 7). ∎

In hand computing we may vary the number of decimal places being carried in the light of intermediate results in order to obtain an answer of the required accuracy with a minimum of work. In automatic computing we can perform a calculation using the maximum possible number of digits (which depends on the design of the machine) and then determine the accuracy of the result obtained, rather than try to produce a result of prescribed accuracy.

In addition to errors a computation may contain **mistakes** because of the fallibility of the programmer, machine operator or user of the desk calculator, or the mechanical or electrical equipment used in the computation. Whereas errors cannot be avoided, mistakes are avoidable in principle. **Checking** of numerical results is quite important, and any numerical method should include checking procedures for final (and intermediate) results.

To safeguard against mistakes in **desk calculation,** any computation should be arranged in a systematic tabular form. We should not use scraps of paper but record all the intermediate results directly on the working sheet (squared paper of full notebook page size or paper ruled in rectangular boxes), and write the numbers in a neat and legible manner. We should not try to do too many steps at once. To correct a mistake we should draw a line through the wrong number and write the corrected value above the wrong number, thus keeping a record of the correction. Each sheet should have on it some explanations to keep the calculation understandable for months and years to come.

In **automatic computation** the situation can be roughly described as follows. An automatic computer can add, subtract, multiply, and divide. It has to be told in detail the steps necessary to solve a given problem. This we do by feeding into the computer a **program** punched on **punch cards** and consisting of all relevant given data and a sequence of **instructions** which the computer executes in a certain order, thereby performing the computational steps that lead to an answer which the computer prints on a sheet of paper. Hence the essential components of a computer are:

(I) an **input** for feeding punch cards and transferring the information on the cards into the memory;

(II) a **memory** or **store** for storing instructions and data, each of which, when stored, is allocated an *address* (position in the memory) and can be produced from the store at will;

(III) a **control unit** for organizing the calculations, that is, arranging for the input of information, the execution of arithmetic operations in the requested order specified by the program, and the output of information;

(IV) an **arithmetic unit** for performing arithmetical operations (addition, subtraction, multiplication, division); and

(V) an **output** for transferring information from the memory onto the output sheet.

Fig. 345 represents the main machine functions in the form of a block diagram.

We first plan a computation in all details. Then we describe it formally in a **"problem-oriented language"** such as FORTRAN (FORmula TRANslation) or ALGOL (ALGOrithmic Language) and transcribe this description onto a deck of punch cards, which we feed into the computer. Using a **compiler,** the computer performs an **automatic translation** of our program into a **machine**

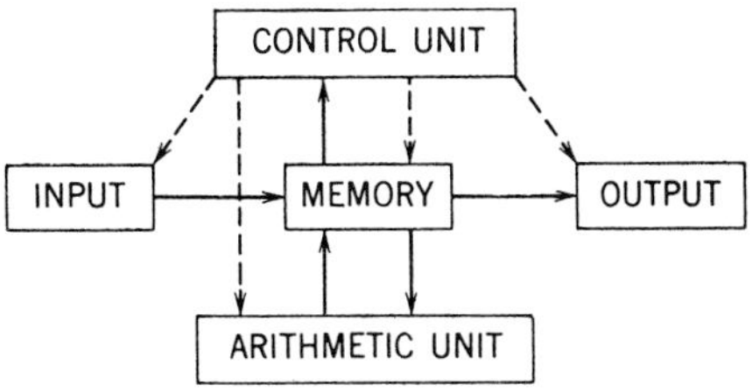

Fig. 345. Components of a computer and main machine functions. Solid arrows represent flow of data or instructions and dashed arrows represent flow of control signals.

language program and executes the latter, thereby carrying out the planned computation. (The machine language program may not appear outside the machine, but if we want to use it again, the computer can punch it on a deck of cards which we can keep.)

Experience shows that the occurrence of mistakes in programming cannot be discounted. The translation process usually includes searches for more or less clerical mistakes which are listed and corrected by the computer, and the corrected program is resubmitted to the computer. However, programs should be prepared with greatest care and checked as far as possible before they reach the stage of machine operation. Detection of mistakes by means of such early checking is more efficient than reliance on programmed checks. Furthermore, it must not be assumed that a program is correct merely because it can be translated and executed, but one should always use a new program to repeat the computations of **pilot programs** that have been thoroughly checked previously.

Problems for Sec. 19.1

1. **(Floating point)** Floating point representations are of the form $x \times 10^y$. Actually, to avoid negative exponents, one adds 50 to the exponent, that is, 0.13×10^{-8} is written 0.13×10^{42}. Write 381.79206 in this form, with 4 significant digits.
2. If β_1 and β_2 are error bounds for a_1^* and a_2^*, show that $\beta = \beta_1 + \beta_2$ is an error bound for the sum $s^* = a_1^* + a_2^*$.
3. Show that β in Prob. 2 is an error bound for the difference $d^* = a_1^* - a_2^*$, and illustrate with an example that in general β cannot be replaced by a smaller number.
4. Show that for small relative errors the relative error of a product is approximately equal to the sum of the relative errors of the factors. Give a simple numerical example.
5. Show that for small relative errors the relative error of a quotient is approximately equal to the difference of the relative errors of the factors.
6. Using Prob. 5, show that in Example 1 the absolute value of the error of $x_2 = 2.000/3.414 = 0.5858$ is less than 0.0001.
7. Using Prob. 5, show that in Example 1 the absolute value of the error of $x_2 = 2.000/39.95 = 0.050\,06$ is less than 0.000 01.
8. Give a simple example illustrating the use of guarding figures.
9. Compute a two-decimal table[3] of $f(x) = x/16$, $x = 0(1)20$ and find out how the rounding error is distributed.

[3] $x = a(h)b$ means that function values are given for $x = a, a + h, a + 2h, \cdots, b$.

10. Compute a three-decimal table of $f(x) = x/3$, $x = 0(1)100$ and find out how the rounding error is distributed.
11. Explain the difficulties in obtaining a four-place table by rounding off the values (*a*) in a five-place table, (*b*) in a six-place table.
12. Illustrate with an example that in computations with a fixed number of significant digits the result of adding numbers depends on the order in which they are added.
13. Let 2.3182 and 0.443 be correctly rounded to the number of digits shown. Determine the smallest interval in which the sum $s = 2.3182 + 0.443$, using true instead of rounded values of the quantities, must lie.
14. Given n numbers $a_1, \cdots, a_n$, where a_j is correctly rounded to D_j decimals. In calculating the sum $a_1 + \cdots + a_n$, retaining $D = \min_j D_j$ decimals, is it essential whether we add first and then round the result or whether we first round each number to D decimals and then add?
15. Let a and b be given and correctly rounded to $2D$ and D decimals, respectively, and let $|a| < |b| < 1$. If we want to compute a/b as a D-decimal quotient, show that dividing and then rounding is preferable to first rounding and then dividing. Give an example.

19.2 Solution of Equations by Iteration

In engineering mathematics we frequently have to find **solutions** of equations of the form

$$f(x) = 0, \tag{1}$$

that is, numbers X_0 such that $f(X_0)$ equals zero; here f is a given function. Examples are $x^2 - 3x + 2 = 0$, $x^3 + x = 1$, $\sin x = 0.5x$, $\tan x = x$, $\cosh x = \sec x$, $\cosh x \cos x = -1$, which can all be written in the form (1). The first two are **algebraic equations** because the corresponding f is a polynomial, and in this case the solutions are also called **roots** of the equations. The other equations are **transcendental equations** because they involve transcendental functions. Formulas that give exact numerical values of the solutions will exist only in very simple cases. In most cases we have to use approximation methods, in particular, iteration methods.

A numerical **iteration method** is a method such that we choose an arbitrary x_0 and compute a sequence $x_0, x_1, x_2, \cdots$ recursively from a relation of the form

$$x_{n+1} = g(x_n) \qquad (n = 0, 1, \cdots) \tag{2}$$

where g is defined in some interval containing x_0 and the range of g lies in that interval. Hence we compute successively $x_1 = g(x_0)$, $x_2 = g(x_1)$, $x_3 = g(x_2)$, $\cdots$.

In this section, both the domain and the range of $g(x)$ will be on the real line. In more general problems x or g or both may be vector variables.

We mention that iteration methods are very important for various types of problems in numerical analysis.

There are several ways of obtaining iteration methods for solving (1), and we shall discuss three particularly important ones.

Algebraic transformations. We may try to transform (1) *algebraically* into the form

$$x = g(x). \tag{3}$$

Then the corresponding iteration procedure is given by (2) with the function g obtained in (3). A solution of (3) is called a **fixed point** of g. To a given equation (1) there may correspond several equations (3) and the behavior of iterative sequences $x_0, x_1, x_2, \cdots$ may differ accordingly (and depend on x_0). Let us illustrate this with a simple example.

Example 1. An iteration process

Set up an iteration process for the equation $f(x) = x^2 - 3x + 1 = 0$. Since we know the solutions

$$x = 1.5 \pm \sqrt{1.25}, \qquad \text{thus} \qquad 2.618\,034 \qquad \text{and} \qquad 0.381\,966,$$

we can watch the behavior of the error as the iteration proceeds. The equation may be written

$$x = g_1(x) = \tfrac{1}{3}(x^2 + 1), \qquad \text{thus} \qquad x_{n+1} = \tfrac{1}{3}(x_n{}^2 + 1), \tag{4a}$$

and if we choose $x_0 = 1$, we obtain the sequence (cf. Fig. 346*a*)

$$x_0 = 1.000, \quad x_1 = 0.667, \quad x_2 = 0.481, \quad x_3 = 0.411, \quad x_4 = 0.390, \cdots$$

which seems to approach the smaller root. If we choose $x_0 = 2$ the situation is similar. If we choose $x_0 = 3$, we obtain the sequence (cf. Fig. 346*a*)

$$x_0 = 3.000, \quad x_1 = 3.333, \quad x_2 = 4.037, \quad x_3 = 5.766, \quad x_4 = 11.414, \cdots$$

which seems to diverge. Our equation may also be written

$$x = g_2(x) = 3 - \frac{1}{x}, \qquad \text{thus} \qquad x_{n+1} = 3 - \frac{1}{x_n}, \tag{4b}$$

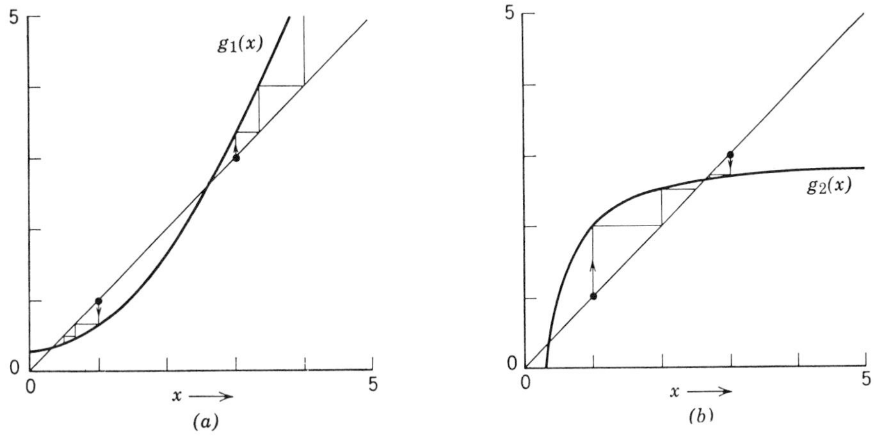

Fig. 346. Example 1, iterations (4a) and (4b)

and if we choose $x_0 = 1$, we obtain the sequence (cf. Fig. 346*b*)

$$x_0 = 1.000, \quad x_1 = 2.000, \quad x_2 = 2.500, \quad x_3 = 2.600, \quad x_4 = 2.615, \cdots$$

which seems to approach the larger solution. Similarly, if we choose $x_0 = 3$, we obtain the sequence (cf. Fig. 346*b*)

$$x_0 = 3.000, \quad x_1 = 2.667, \quad x_2 = 2.625, \quad x_3 = 2.619, \quad x_4 = 2.618, \cdots$$

Our figures indicate that convergence seems to depend on the fact that in a neighborhood of a solution the curve of $g(x)$ is less steep than the straight line $y = x$, and we shall now see that this condition $|g'(x)| < 1$ (slope of $y = x$) is sufficient for convergence. ∎

An iteration process defined by (2) is said to be **convergent** for an x_0 if the corresponding sequence $x_0, x_1, \cdots$ is convergent.

A sufficient condition for convergence is given in the following theorem, which has various practical applications.

Theorem 1 (Convergence)

Let $x = s$ be a solution of $x = g(x)$ and suppose that g has a continuous derivative in some interval J containing s. Then if $|g'(x)| \leqq \alpha < 1$ in J, the iteration process defined by (2) converges for any x_0 in J.

Proof. By the mean value theorem of differential calculus there is a ξ between x and s such that

$$g(x) - g(s) = g'(\xi)(x - s) \qquad (x \text{ in } J).$$

Since $g(s) = s$ and $x_1 = g(x_0)$, $x_2 = g(x_1)$, $\cdots$, we obtain

$$|x_n - s| = |g(x_{n-1}) - g(s)| = |g'(\xi)||x_{n-1} - s| \leqq \alpha\, |x_{n-1} - s|$$
$$\leqq \alpha^2\, |x_{n-2} - s| \leqq \cdots \leqq \alpha^n\, |x_0 - s|$$

and, since $\alpha < 1$, we have $\alpha^n \to 0$ and $|x_n - s| \to 0$ as $n \to \infty$. This completes the proof. ∎

Example 2. An iteration process. Illustration of Theorem 1

Find a solution of $f(x) = x^3 + x - 1 = 0$ by iteration. A rough sketch shows that a real solution lies near $x = 1$. We may write the equation in the form

$$x = g_1(x) = \frac{1}{1 + x^2}, \qquad \text{thus} \qquad x_{n+1} = \frac{1}{1 + x_n^{\,2}}.$$

Then $|g_1'(x)| = 2\,|x|/(1 + x^2)^2 < 1$ for any x, so that we have convergence for any x_0. Choosing $x_0 = 1$, we obtain (cf. Fig. 347)

$$x_1 = 0.500, \quad x_2 = 0.800, \quad x_3 = 0.610, \quad x_4 = 0.729, \quad x_5 = 0.653, \quad x_6 = 0.701, \cdots.$$

The root exact to 6D is $s = 0.682\,328$. The equation may also be written

$$x = g_2(x) = 1 - x^3. \qquad \text{Then} \qquad |g_2'(x)| = 3x^2$$

and this is greater than 1 near the root, so that we cannot expect convergence. The reader may try $x_0 = 1$, $x_0 = 0.5$, $x_0 = 2$ and see what happens. ∎

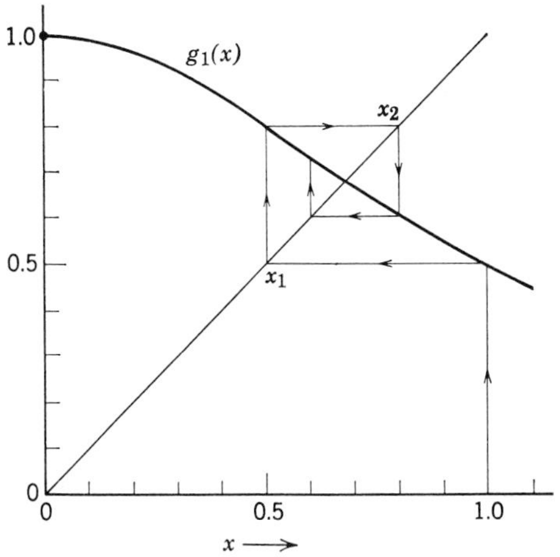

Fig. 347. Iteration in Example 2

Newton's method (also called *Newton-Raphson method*) is another iteration method for solving equations $f(x) = 0$, where f is differentiable. The idea is that we approximate the graph of f by suitable tangents. Using a value x_0 obtained from the graph of f we let x_1 be the point of intersection of the x-axis and the tangent to the curve of f at x_0 (cf. Fig. 348). Then

$$\tan \beta = f'(x_0) = \frac{f(x_0)}{x_0 - x_1}, \qquad \text{hence} \qquad x_1 = x_0 - \frac{f(x_0)}{f'(x_0)}.$$

In the next step we compute (cf. Fig. 348)

$$x_2 = x_1 - \frac{f(x_1)}{f'(x_1)}$$

etc., and the general formula is

$$x_{n+1} = x_n - \frac{f(x_n)}{f'(x_n)} \qquad (n = 0, 1, \cdots). \tag{5}$$

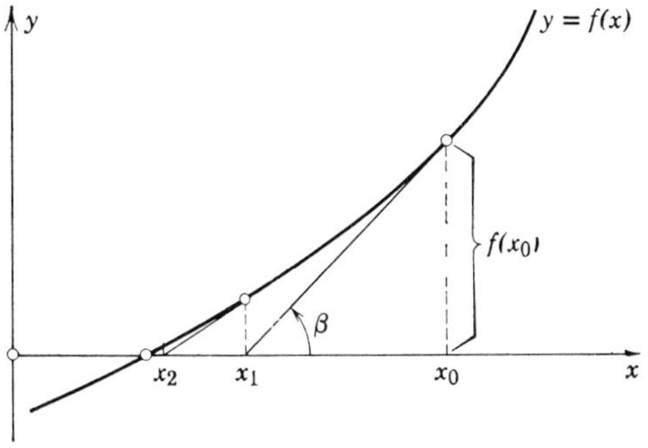

Fig. 348. Newton's method

Example 3. Square root

Set up a Newton iteration for computing the square root x of a given positive number c and apply it to $c = 2$. We have $x = \sqrt{c}$, hence $f(x) = x^2 - c = 0$, $f'(x) = 2x$, and (5) takes the form

$$x_{n+1} = x_n - \frac{x_n^{\,2} - c}{2x_n} = \frac{1}{2}\left(x_n + \frac{c}{x_n}\right).$$

For $c = 2$, choosing $x_0 = 1$, we obtain

$$x_1 = 1.500\,000, \qquad x_2 = 1.416\,667, \qquad x_3 = 1.414\,216, \qquad x_4 = 1.414\,214, \cdots .$$

x_4 is exact to 6D.

Example 4. Iteration for a transcendental equation

Find the positive solution of $2 \sin x = x$. Setting $f(x) = x - 2 \sin x$, we have $f'(x) = 1 - 2 \cos x$, and (5) gives

$$x_{n+1} = x_n - \frac{x_n - 2 \sin x_n}{1 - 2 \cos x_n} = \frac{2(\sin x_n - x_n \cos x_n)}{1 - 2 \cos x_n} = \frac{N_n}{D_n}.$$

From the graph of f we conclude that the solution is near $x_0 = 2$. Using tables of the sine and cosine functions, we compute:

n	x_n	N_n	D_n	x_{n+1}
0	2.000	3.483	1.832	1.901
1	1.901	3.125	1.648	1.896
2	1.896	3.107	1.639	1.896

(The solution to 4D is 1.8955.)

Example 5. Newton's method applied to an algebraic equation

Apply Newton's method to the equation $f(x) = x^3 + x - 1 = 0$. From (5) we have

$$x_{n+1} = x_n - \frac{x_n^{\,3} + x_n - 1}{3x_n^{\,2} + 1} = \frac{2x_n^{\,3} + 1}{3x_n^{\,2} + 1}.$$

Starting from $x_0 = 1$, we obtain

$$x_1 = 0.750\,000, \qquad x_2 = 0.686\,047, \qquad x_3 = 0.682\,340, \qquad x_4 = 0.682\,328, \cdots$$

where x_4 is exact to 6D. A comparison with Example 2 shows that the present convergence is much more rapid. This may motivate the concept of the *order of an iteration process,* which we shall now discuss. ∎

Let $x_{n+1} = g(x_n)$ define an iteration method, and let x_n approximate a solution s of $x = g(x)$. Then $x_n = s + \epsilon_n$, where ϵ_n is the error of x_n. Suppose that g is differentiable a number of times, so that the Taylor formula gives

$$\begin{aligned} x_{n+1} = g(x_n) &= g(s) + g'(s)(x_n - s) + \tfrac{1}{2}g''(s)(x_n - s)^2 + \cdots \\ &= g(s) + g'(s)\epsilon_n + \tfrac{1}{2}g''(s)\epsilon_n^{\,2} + \cdots . \end{aligned}$$

The exponent of ϵ_n in the first nonvanishing term after $g(s)$ is called the **order** of the iteration process defined by g. Since $x_{n+1} - g(s) = x_{n+1} - s = \epsilon_{n+1}$, the error of x_{n+1}, and in the case of convergence ϵ_n will be small for large n, the order is a measure for the speed of convergence.

Newton's iteration is of second order.

In fact, we have

$$g(x) = x - \frac{f(x)}{f'(x)}, \qquad g'(x) = \frac{f(x)f''(x)}{f'(x)^2}$$

and $g'(s) = 0$ because $f(s) = 0$; hence the process is at least of second order. Another differentiation shows that $g''(s) = f''(s)/f'(s)$, which is not zero in general. In Example 2, the process given by $g_1(x)$ is only of first order because $g_1(x) = 1/(1 + x^2)$ and $g_1'(x) = -2x/(1 + x^2)^2$.

If $f'(x) = 0$ near a solution of $f(x) = 0$, then Newton's method may cause difficulties, but these can often be clarified by plotting the curve of $f(x)$ for values of x near the required solution and using the geometrical interpretation of Newton's method. If $f'(x)$ is small near a required solution of $f(x) = 0$, it may be necessary to compute $f(x_n)$ and $f'(x_n)$ to a high degree of accuracy in order to obtain an accurate approximation of x_{n+1}. This indicates that the equation is ill-conditioned. This concept is defined as follows.

An equation $f(x) = 0$ is said to be **ill-conditioned** if a small value of

$$R(s^*) = f(s^*)$$

can be produced by values s^* appreciably different from the exact solution s, or if small variations in the constants appearing in f can produce large variations in the solutions. $R(s^*) = f(s^*)$ is called the **residual** of the equation at s^*. Its smallness gives some indication of the accuracy of s^*, but it is important to realize that small $R(s^*)$ does not necessarily imply that the error of s^* is small.

A third iteration method for solving an equation $f(x) = 0$ is the **method of false position** (*regula falsi*). In this method we approximate the curve of $f(x)$ by a chord, as shown in Fig. 349. The chord intersects the x-axis at

$$\text{(6a)} \qquad x_1 = \frac{x_0 f(b) - bf(x_0)}{f(b) - f(x_0)}$$

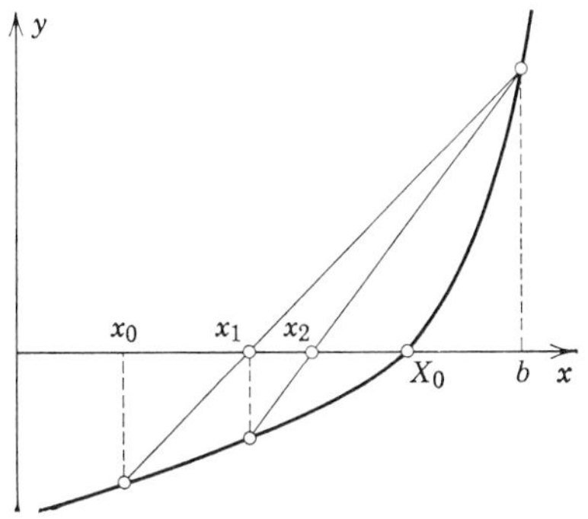

Fig. 349. Method of false position (Regula falsi)

and this is an approximation of the solution X_0 of $f(x) = 0$. In the next step we compute the approximation

$$x_2 = \frac{x_1 f(b) - bf(x_1)}{f(b) - f(x_1)} \tag{6b}$$

etc. Convergence to X_0 can be accelerated if improvements in b can be made in the direction of bringing b closer to X_0. This can usually be done by guesswork.

Example 6. Method of false position

Determine an approximate value of the root of the equation $f(x) = x^3 + x - 1 = 0$ near $x = 1$ (cf. Example 2). Since $f(0.5) = -0.375$ and $f(1) = 1$, we may choose $x_0 = 0.5$ and $b = 1$. Formula (6a) then yields

$$x_1 = \frac{0.5 \cdot 1 - 1 \cdot (-0.375)}{1 - (-0.375)} = 0.64.$$

From (6b) we obtain the more accurate approximation $x_2 = 0.672$, and so on.

Problems for Sec. 19.2

Find a root of the following equations; carry out three steps of Newton's method, starting from the given x_0.

1. $x^3 - 3.9x^2 + 4.79x - 1.881 = 0,\ x_0 = 1$

2. $x^3 - 1.2x^2 + 2x - 2.4 = 0,\ x_0 = 2$

3. The roots of the equation in Prob. 1 are 0.9, 1.1, and 1.9. Although $x_0 = 1$ lies close to 0.9 and 1.1, Newton's method does not yield one of these roots. Why? Choose another x_0 such that the method yields approximations for the root 1.1.

Find all real roots of the following equations by Newton's method.

4. $\cos x = x$ **5.** $x + \ln x = 2$ **6.** $2x + \ln x = 1$

7. $x^4 - 0.1x^3 - 0.82x^2 - 0.1x - 1.82 = 0$

8. Show that in Example 2, $|g_1'(x)|$ is maximum at $\tilde{x} = \pm 1/\sqrt{3}$, the maximum value being $|g'(\tilde{x})| = 3\sqrt{3}/8 = 0.65$.

9. Why do we obtain a monotone sequence in Example 1, but not in Example 2?

10. Perform the iterations indicated at the end of Example 2. Sketch a figure similar to Fig. 347.

11. Find the root of $x^5 = x + 0.2$ near $x = 0$ by transforming the equation algebraically into the form (2) and starting from $x_0 = 0$.

12. The equation in Prob. 11 has a root near $x = 1$. Find this root by writing the equation in the form $x = \sqrt[5]{x + 0.2}$ and iterating, starting with $x_0 = 1$.

13. What happens in Prob. 12 if you write the equation in the form $x = x^5 - 0.2$ and start from $x_0 = 1$?

14. Using iteration, show that the smallest positive root of the equation $x = \tan x$ is 4.49, approximately. *Hint.* Conclude from the graphs of x and $\tan x$ that a root lies close to $x_0 = 3\pi/2$; write the equation in the form $x = \pi + \arctan x$. (Why?)

15. Calculate $\sqrt{5}$ by the iteration in Example 3, starting from $x_0 = 2$ and calculating $x_1, \cdots, x_3$. Calculate the error, using $\sqrt{5} = 2.236\,068$.

16. Show that in Example 3 we have

$$x_{n+1}{}^2 - c = \frac{1}{4}\left(x_n - \frac{c}{x_n}\right)^2,$$

which is a measure of the accuracy. Show that, approximately,

$$\left|x_n - \sqrt{c}\right| \approx \frac{1}{2}\left|x_n - \frac{c}{x_n}\right|$$

and apply this to Prob. 15.

17. Find an interval on the positive x-axis in which the iteration process in Example 3 with $c = 2$ satisfies the condition of Theorem 1.

18. Design a Newton iteration for the cube root. Calculate $\sqrt[3]{7}$, starting from $x_0 = 2$ and performing 3 steps.

19. Design a Newton iteration for computing the kth root of a positive number c.

Find the real roots of the following equations by the method of false position.

20. $x^4 = 2$ **21.** $x^4 = 2x$ **22.** $3 \sin x = 2x$

23. In Prob. 20 the approximate values of the positive root will always be somewhat smaller than the exact value of the root. Why?

24. Newton's method requires the calculation of $f'(x)$. In applications, this may sometimes be quite involved. One way of avoiding $f'(x_n)$ is to replace it by the difference quotient $[f(x_n) - f(x_{n-1})]/(x_n - x_{n-1})$. Explain the relation between the resulting formula and the regula falsi.

25. Show that if g is continuous in a closed interval I and its range lies in I, then the equation $x = g(x)$ has at least one solution in that interval. Illustrate that it may have more than one solution.

19.3 Finite Differences

Finite differences are basic in various branches of numerical analysis, such as interpolation, checking tables, approximation, differentiation, and solution of differential equations. We assume that we are given a **table** of numerical values $f_j = f(x_j)$ of a function f at equally spaced points

$$x_0, \quad x_1 = x_0 + h, \quad x_2 = x_0 + 2h, \quad x_3 = x_0 + 3h, \quad \cdots \qquad (h > 0, \text{ fixed})$$

where the $f(x_j)$ may result from a formula or may be obtained **empirically** by experiments. We may now form the **first differences** by subtracting each function value of $f(x)$ from that for the next greater value of x in the table. An example is shown in Table 19.1, where[3] $f(x) = x^3$, $x = -3(1)3$. Applying the same process of subtraction to the first differences, we obtain the **second differences** of f, and so on. In this **difference table,** each difference is entered, in the appropriate column, midway between the elements in the preceding column from which it is constructed. Decimal points and leading zeros of differences may be omitted (cf. Table 19.2).

[3] $x = a(h)b$ means that function values are given for $x = a, a + h, a + 2h, \cdots, b$.

Table 19.1
Difference Table of $f(x) = x^3$, $x = -3(1)3$

x	$f(x) = x^3$	First Diff.	Second Diff.	Third Diff.	Fourth Diff.
−3	−27				
		19			
−2	−8		−12		
		7		6	
−1	−1		−6		0
		1		6	
0	0		0		0
		1		6	
1	1		6		0
		7		6	
2	8		12		
		19			
3	27				

There are three different **notations for the differences** occurring in a difference table. To avoid confusion, we should say at the beginning that in any concrete numerical case the same numbers occur in the same positions no matter which of the three notations we use. The first (and probably most important) notation is that for **central differences:**

x_{-2}	f_{-2}			
		$\delta f_{-3/2}$		
x_{-1}	f_{-1}		$\delta^2 f_{-1}$	
		$\delta f_{-1/2}$		$\delta^3 f_{-1/2}$
x_0	f_0		$\delta^2 f_0$	
		$\delta f_{1/2}$		$\delta^3 f_{1/2}$
x_1	f_1		$\delta^2 f_1$	
		$\delta f_{3/2}$		
x_2	f_2			

Table 19.2
Values and Differences of $f(x) = 1/x$,
$x = 1(0.2)2$, 4D

x	$f(x) = 1/x$	First Diff.	Second Diff.	Third Diff.
1.0	1.0000			
		−1667		
1.2	0.8333		477	
		−1190		−180
1.4	0.7143		297	
		−893		−98
1.6	0.6250		199	
		−694		−61
1.8	0.5556		138	
		−556		
2.0	0.5000			

Thus $\delta f_{-3/2} = f_{-1} - f_{-2}$, $\delta f_{-1/2} = f_0 - f_{-1}$ and in general

(1) $$\delta f_{m+1/2} = f_{m+1} - f_m,$$

the subscript on the left is the mean of the subscripts on the right. Similarly,

$$\delta^2 f_m = \delta f_{m+1/2} - \delta f_{m-1/2},$$

and so on. Differences with the same subscript appear in a horizontal line. (Note that x_0 need not be the smallest x in the table. For instance, in Table 19.2 we may let $x_0 = 1.6$, then $f_0 = 0.6250$, $\delta f_{1/2} = -0.0694$, $\delta^2 f_0 = 0.0199$, etc.)

The second notation is that for **forward differences:**

$$\begin{array}{ccccc}
x_{-2} & f_{-2} & & & \\
 & & \Delta f_{-2} & & \\
x_{-1} & f_{-1} & & \Delta^2 f_{-2} & \\
 & & \Delta f_{-1} & & \Delta^3 f_{-2} \\
x_0 & f_0 & & \Delta^2 f_{-1} & \\
 & & \Delta f_0 & & \Delta^3 f_{-1} \\
x_1 & f_1 & & \Delta^2 f_0 & \\
 & & \Delta f_1 & & \\
x_2 & f_2 & & &
\end{array}$$

Thus $\Delta f_{-2} = f_{-1} - f_{-2}$, $\Delta f_{-1} = f_0 - f_{-1}$, $\Delta f_0 = f_1 - f_0$ and in general

(2) $$\Delta f_m = f_{m+1} - f_m.$$

Similarly,

$$\Delta^2 f_m = \Delta f_{m+1} - \Delta f_m,$$

and so on. For instance, in Table 19.2, if we let $x_0 = 1.6$, then $f_0 = 0.6250$, $\Delta f_0 = -0.0694$, $\Delta^2 f_0 = 0.0138$. Differences with the same subscript appear on lines sloping downward or *forward* into the table.

The third notation is that for **backward differences:**

$$\begin{array}{ccccc}
x_{-2} & f_{-2} & & & \\
 & & \nabla f_{-1} & & \\
x_{-1} & f_{-1} & & \nabla^2 f_0 & \\
 & & \nabla f_0 & & \nabla^3 f_1 \\
x_0 & f_0 & & \nabla^2 f_1 & \\
 & & \nabla f_1 & & \nabla^3 f_2 \\
x_1 & f_1 & & \nabla^2 f_2 & \\
 & & \nabla f_2 & & \\
x_2 & f_2 & & &
\end{array}$$

Thus $\nabla f_{-1} = f_{-1} - f_{-2}$, $\nabla f_0 = f_0 - f_{-1}$, $\nabla f_1 = f_1 - f_0$ and in general

(3) $$\nabla f_m = f_m - f_{m-1}.$$

Similarly,

$$\nabla^2 f_m = \nabla f_m - \nabla f_{m-1},$$

Table 19.3
Spread of an Error of a Function Value in the Difference Table
$f(x) = \sqrt{x}$, $x = 2.0(0.1)2.6$, 4D [Error in $f(2.3)$.]

x	$\sqrt{x}$	Differences			$\sqrt{x}$	Differences			Spread of error ϵ			
2.0	1.4142				1.4142							
		349				349						
2.1	1.4491		-8		1.4491		-8					
		341		1		341		11				ϵ
2.2	1.4832		-7		1.4832		3				ϵ	
		334		-1		344		-31		ϵ		-3ϵ
2.3	1.5166		-8		1.5176		-28		ϵ		-2ϵ	
		326		1		316		31		$-\epsilon$		3ϵ
2.4	1.5492		-7		1.5492		3				ϵ	
		319		2		319		-8				$-\epsilon$
2.5	1.5811		-5		1.5811		-5					
		314				314						
2.6	1.6125				1.6125							

and so on. Differences with the same subscript lie on lines sloping upward or *backward* into the table. Backward difference notation is often convenient for calculations near the *end* of a tabulated range.

Any specified difference in a table can now be represented by three different symbols. For example, in Table 19.2, if we let $x_0 = 1.6$, then we obtain $-0.0893 = \delta f_{-1/2} = \Delta f_{-1} = \nabla f_0$. In general we have

$$\delta^n f_m = \Delta^n f_{m-n/2} = \nabla^n f_{m+n/2}.$$

Differences may be used for **detecting errors** in tables. An error ϵ in a function value spreads into the differences as shown in Table 19.3. It follows that large fluctuations in the differences indicate possible errors in the function values. Of course, small fluctuation may result from rounding.

Differences are also useful in the **approximation** *of functions by polynomials.* For a polynomial $p_n(x) = a_0x^n + a_1x^{n-1} + \cdots + a_n$ of degree n the nth differences in a table with step h are constant (equal to $n!h^na_0$) and all higher differences are zero, because

$$p_n(x + h) - p_n(x) = a_0[(x + h)^n - x^n] + \cdots = a_0nhx^{n-1} + \cdots$$

is a polynomial of degree $n - 1$; similarly, the second differences are represented by a polynomial of degree $n - 2$ with leading coefficient $a_0n(n - 1)h^2$, etc. It follows that if the nth differences in a given table of a function f are approximately constant over a given range, then the tabular values can be represented approximately in this range by an nth-degree polynomial p_n. Let us illustrate a method for finding p_n when f is given.

Example 1. Approximation

In Table 19.3 the second differences are almost constant (equal to -7). Hence, we may try to find a quadratic approximating polynomial p_2. We first construct a difference table, assuming that the second differences are *exactly* equal to -7 and choosing a function value and a first difference in

Table 19.4
Approximation of $f(x) = \sqrt{x}$ by a Quadratic Polynomial p_2

x	$p_2(x)$	Differences	
2.0	1.4143		
		348	
2.1	1.4491		−7
		341	
2.2	1.4832		−7
		334	
2.3	1.5166		−7
		327	
2.4	1.5493		−7
		320	
2.5	1.5813		−7
		313	
2.6	1.6126		

the middle of the range, say 1.5166 and 334. This gives Table 19.4. The leading coefficient a_0 of p_2 is obtained from $a_0 2! h^2 = a_0 \cdot 2 \cdot 0.1^2 = -0.0007$ (= the second difference); thus $a_0 = -0.0007/0.02 = -0.035$. Hence $p_1(x) = p_2(x) + 0.035x^2$ is of first degree, and from Table 19.4 we calculate that its first differences are constant (0.04915), and we know that they must equal $a_1 h$. Thus $a_1 = 0.04915/0.1 = 0.4915$. Finally, $p_1(x) - 0.4915x = a_2 = 0.5713$, so that we have

$$p_2(x) = -0.0350x^2 + 0.4915x + 0.5713.$$

The example illustrates how differences can be used for judging the quality of an approximation before actually determining a polynomial. For the latter task, there are various other methods, some of which will be discussed in the next section.

Problems for Sec. 19.3

1. In Table 19.2, choose $x_0 = 1.2$ and write all the occurring numbers in terms of the notation (*a*) for central differences, (*b*) for forward differences, (*c*) for backward differences.
2. Calculate a difference table of $f(x) = x^3$ for $x = 0(1)5$. Choose $x_0 = 2$ and write all occurring numbers in terms of the notations (*a*) for central differences, (*b*) for forward differences, (*c*) for backward differences.
3. Show that $\delta^2 f_m = f_{m+1} - 2f_m + f_{m-1}$, $\delta^3 f_{m+1/2} = f_{m+2} - 3f_{m+1} + 3f_m - f_{m-1}$.
4. Calculate $f(x) = 1/(x + 1)$, $x = 0(0.2)1$ to (*a*) 2D, (*b*) 3D, (*c*) 4D, and compare the effect of the rounding errors in the corresponding difference tables.
5. Set up a difference table of $f(x) = x^2$ for $x = 0(1)10$. Do the same, with $f(5) = 25$ replaced by 26, showing first, second, third and fourth differences. Observe the spread of the error.
6. Using differences, check the table:

x	4.0	4.1	4.2	4.3	4.4	4.5
$f(x)$	0.250	0.244	0.242	0.233	0.227	0.222

7. Perform the computations indicated in Example 1.

19.4 Interpolation

If a table of values of a function $f(x)$ is given, it is often necessary to obtain values of $f(x)$ for values of x between the x-values appearing in the table. This problem of obtaining such a value of f from the given tabular values is called **interpolation.** The tabular values of $f(x)$ used in the process are called **pivotal values.** The usual methods of interpolation are based on the assumption that in a neighborhood of the x-value in question, f can be approximated by a polynomial p, whose value at that x is then taken as an approximation of the value of f at that x.

The simplest method is **linear interpolation,** in which we approximate the curve of f by a chord at two adjacent tabular x-values x_0 and x_1 (cf. Fig. 350). Hence as the approximate value of f at $x = x_0 + rh$ we take

$$\text{(1)}\quad f(x) \approx p_1(x) = f_0 + r(f_1 - f_0) = f_0 + r\,\Delta f_0 \qquad \left(r = \frac{x - x_0}{h},\ 0 \leqq r \leqq 1\right).$$

This is known to us from elementary calculus in connection with tables of logarithms or trigonometric functions. For example, to obtain $\ln 9.2$ from a table showing $\ln 9.0 = 2.197$ and $\ln 9.5 = 2.251$, we may first compute $r = 0.2/0.5 = 0.4$ and then

$$\ln 9.2 = \ln 9.0 + 0.4(\ln 9.5 - \ln 9.0) = 2.219.$$

Linear interpolation will be satisfactory as long as the x-values in a table are so close that chords deviate but little from the curve of $f(x)$, say, by less than $\frac{1}{2}$ unit of the last digit in the table for each x between x_0 and x_1.

In **quadratic interpolation** we approximate the curve of the function f between x_0 and $x_2 = x_0 + 2h$ by the quadratic parabola which passes through the points (x_0, f_0), (x_1, f_1), (x_2, f_2) and obtain the more accurate formula

$$\text{(2)}\quad f(x) \approx p_2(x) = f_0 + r\,\Delta f_0 + \frac{r(r-1)}{2}\Delta^2 f_0 \qquad \left(r = \frac{x - x_0}{h},\ 0 \leqq r \leqq 2\right)$$

where $x = x_0 + rh$. For $x = x_0$ $(r = 0)$ the right-hand side equals f_0; for $x = x_1$ $(r = 1)$ it equals $f_0 + \Delta f_0 = f_1$, and for $x = x_2$ $(r = 2)$ it equals

$$f_0 + 2(f_1 - f_0) + [(f_2 - f_1) - (f_1 - f_0)] = f_2.$$

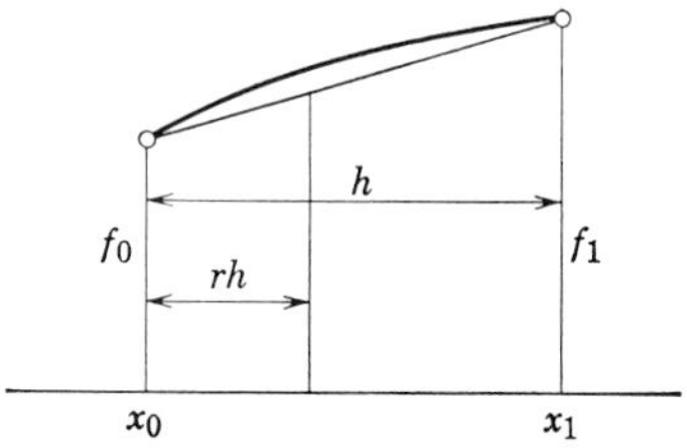

Fig. 350. Linear interpolation

Example 1. Linear and quadratic interpolation

If $\ln 9.0 = 2.1972$ and $\ln 9.5 = 2.2513$ are given, then (1) yields $\ln 9.2 = 2.2188$, exact to 3D only, whereas (2), with $\ln 10.0 = 2.3026$, yields

$$\ln 9.2 = 2.1972 + 0.4 \cdot 0.0541 + \frac{0.4 \cdot (-0.6)}{2}(-0.0028) = 2.2192,$$

which is exact to 4D. ∎

Still more accurate approximations are obtained if we use polynomials of higher degree. A polynomial $p_n(x)$ of degree n is uniquely determined by its values at $n + 1$ distinct values of x. In the present case we need the polynomial p_n such that $p_n(x_0) = f_0, \cdots, p_n(x_n) = f_n$, where $f_0 = f(x_0), \cdots, f_n = f(x_n)$ are the values of f in the table. This polynomial is given in **Newton's** (or *Gregory-Newton's*) **forward-difference interpolation formula**

$$\begin{aligned} f(x) \approx p_n(x) &= f_0 + r\,\Delta f_0 + \frac{r(r-1)}{2!}\Delta^2 f_0 + \frac{r(r-1)(r-2)}{3!}\Delta^3 f_0 + \cdots \\ &\qquad + \frac{r(r-1)\cdots(r-n+1)}{n!}\Delta^n f_0 \end{aligned} \tag{3}$$

$$\left(x = x_0 + rh,\ r = \frac{x - x_0}{h},\ 0 \leqq r \leqq n\right),$$

which includes (1) and (2) as special cases ($n = 1, n = 2$). We must prove that $p_n(x_k) = f_k$ ($k = 0, 1, \cdots, n$). The right-hand side of (3) shows that this follows from

$$f_k = \binom{k}{0} f_0 + \binom{k}{1}\Delta f_0 + \binom{k}{2}\Delta^2 f_0 + \cdots + \binom{k}{k}\Delta^k f_0 \tag{4}$$

where the **binomial coefficients** are defined by

$$\binom{k}{0} = 1, \quad \binom{k}{s} = \frac{k(k-1)(k-2)\cdots(k-s+1)}{s!} \qquad (s \geqq 0,\ \text{integral}) \tag{5}$$

and $s! = 1 \cdot 2 \cdot 3 \cdots s$. In fact, if we set $r = k$ in (3), then the right-hand side of (3) becomes identical with that of (4). Formula (4) can be proved by induction. It is true for $k = 0$. Suppose that it holds for a $k = q$. Then, using (4) with $k = q$ and the formula obtained from it by applying Δ, we have

$$\begin{aligned} f_{q+1} &= f_q + \Delta f_q \\ &= \binom{q}{0} f_0 + \binom{q}{1}\Delta f_0 + \binom{q}{2}\Delta^2 f_0 + \cdots + \binom{q}{q}\Delta^q f_0 \\ &\qquad + \binom{q}{0}\Delta f_0 + \binom{q}{1}\Delta^2 f_0 + \binom{q}{2}\Delta^3 f_0 + \cdots + \binom{q}{q}\Delta^{q+1} f_0. \end{aligned}$$

In this formula, $\Delta^s f_0$ has the coefficient [cf. (5)]

$$\binom{q}{s}+\binom{q}{s-1}=\binom{q+1}{s},$$

which gives (4) with $k = q + 1$ and completes the proof by induction. ∎

A formula similar to (3) but involving backward differences is **Newton's** (or *Gregory-Newton's*) **backward-difference interpolation formula**

$$\text{(6)}\qquad \begin{aligned} f(x) \approx p_n(x) &= f_0 + r\,\nabla f_0 + \frac{r(r+1)}{2!}\nabla^2 f_0 + \cdots \\ &\quad + \frac{r(r+1)\cdots(r+n-1)}{n!}\nabla^n f_0 \end{aligned}$$

with $x = x_0 + rh$, $r = (x - x_0)/h$, $0 \leqq r \leqq n$, as in (3).

The literature on methods and formulas of interpolation using finite differences is enormous. For instance, there are formulas involving only even-order differences. A particularly useful formula of this type is the simplest **Everett formula**

$$\text{(7)}\quad f(x) \approx (1-r)f_0 + rf_1 + \frac{(2-r)(1-r)(-r)}{3!}\delta^2 f_0 + \frac{(r+1)r(r-1)}{3!}\delta^2 f_1$$

where $r = (x - x_0)/h$, $0 \leqq r \leqq 1$. To make its application easy, function tables often include second differences needed in (7).

Example 2. An application of the Everett formula (7)

Find $e^{1.24}$ from (7) and the table

x	e^x	δ^2
1.2	3.3201	333
1.3	3.6693	367

We have $r = 0.04/0.1 = 0.4$, and (7) gives

$$\begin{aligned} e^{1.24} &= 0.6\cdot 3.3201 + 0.4\cdot 3.6693 + \frac{1.6\cdot 0.6\cdot(-0.4)}{6}\cdot 0.0333 + \frac{1.4\cdot 0.4\cdot(-0.6)}{6}\cdot 0.0367 \\ &= 3.4598 - 0.0021 - 0.0021 = 3.4556 \end{aligned}$$

exact to 4D. Note that linear interpolation gives 3.4598, which is exact to only 2D. (The reader may use $e^{1.1} = 3.0042$ and $e^{1.4} = 4.0552$ for checking the given second differences.) ∎

We mention that the general **Everett formula** is

$$\text{(8)}\quad f(x) = qf_0 + rf_1 + \binom{q+1}{3}\delta^2 f_0 + \binom{r+1}{3}\delta^2 f_1 + \binom{q+2}{5}\delta^4 f_0 + \binom{r+2}{5}\delta^4 f_1 +$$

where $r = (x - x_0)/h$, $0 \leqq r \leqq 1$, as before, and $q = 1 - r$. In this formula the ratio of the coefficients of $\delta^4 f_0$ and $\delta^2 f_0$ is

$$\binom{q+2}{5} \Big/ \binom{q+1}{3} = \frac{q^2-4}{20}$$

and, similarly, the ratio of the coefficients of $\delta^4 f_1$ and $\delta^2 f_1$ is $(r^2 - 4)/20$. Both ratios vary but little in the interval from 0 to 1. Hence if we choose a reasonable mean value μ of them, we can include the fourth-difference effect by using (7) with **modified second differences**

$$\delta_m{}^2 f = \delta^2 f + \mu \delta^4 f, \qquad \mu = -0.18393, \tag{9}$$

where the given μ is the preferred value among various possible values discussed in the literature. This method of using modified differences is called **throwback.** The actual error incurred is less than half a unit if the fourth differences are less than 1000 units (of the last digit) and the fifth differences less than 70.

We mention without proof that if $x_0, x_1, \cdots, x_n$ are arbitrarily spaced, the polynomial of degree n through $(x_0, f_0), \cdots, (x_n, f_n)$, where $f_j = f(x_j)$, is given by the right-hand side of **Newton's divided difference interpolation formula**

$$\begin{aligned} f(x) \approx f_0 &+ (x - x_0) f[x_0, x_1] + (x - x_0)(x - x_1) f[x_0, x_1, x_2] + \cdots \\ &+ (x - x_0) \cdots (x - x_{n-1}) f[x_0, \cdots, x_n] \end{aligned} \tag{10a}$$

involving **divided differences,** which are defined iteratively by the relations

$$\begin{aligned} f[x_0, x_1] &= \frac{f(x_1) - f(x_0)}{x_1 - x_0}, \qquad f[x_0, x_1, x_2] = \frac{f[x_1, x_2] - f[x_0, x_1]}{x_2 - x_0}, \cdots \\ f[x_0, \cdots, x_k] &= \frac{f[x_1, \cdots, x_k] - f[x_0, \cdots, x_{k-1}]}{x_k - x_0}. \end{aligned} \tag{10b}$$

If $x_k = x_0 + kh$ (equal spacing), then $f[x_0, \cdots, x_k] = \Delta^k f_0 / k! h^k$, and (10a) takes the form (3).

In interpolation by difference methods we have to calculate differences and may use them for checks if a table is new or doubtful. However, the question of what order of interpolation to use is left unanswered in most tables. **Lagrangian interpolation** is based on **Lagrange's interpolation formula**

$$f(x) \approx L_n(x) = \sum_{k=0}^{n} \frac{l_k(x)}{l_k(x_k)} f_k \tag{11a}$$

where $x_0, \cdots, x_n$ are not necessarily equally spaced and

$$l_0(x) = (x - x_1)(x - x_2) \cdots (x - x_n)$$

(11b) $$l_k(x) = (x - x_0) \cdots (x - x_{k-1})(x - x_{k+1}) \cdots (x - x_n), \qquad 0 < k < n$$

$$l_n(x) = (x - x_0)(x - x_1) \cdots (x - x_{n-1}).$$

Formula (11a) is known as **Lagrange's $n + 1$ point formula.** It can easily be seen that $L_n(x_k) = f_k$ because inspection of (11b) shows that $l_k(x_j) = 0$ when $j \neq k$ and $(l_k(x)/l_k(x_k))f_k = f_k$ when $x = x_k$. The advantage is that we need not compute differences and we see the effect of the various f_k's more directly, but the calculations may be more laborious and the use of the method should be restricted to guaranteed tables. Tables of Lagrangian coefficients $l_k(x)/l_k(x_k)$ for equal spacing have been published; cf. Ref. [J12] in Appendix 1.

Example 3. An application of Lagrange's interpolation formula
Find ln 9.2, using (11) with $n = 3$ and the values

x	9.0	9.5	10.0	11.0
$\ln x$	2.19722	2.25129	2.30259	2.39790

We have $l_0(x) = (x - 9.5)(x - 10)(x - 11)$, $l_1(x) = (x - 9)(x - 10)(x - 11)$, etc., and compute

$$\ln 9.2 = \frac{-0.43200}{-1.00000} \cdot 2.19722 + \frac{0.28800}{0.37500} \cdot 2.25129 + \frac{0.10800}{-0.50000} \cdot 2.30259 + \frac{0.04800}{3.00000} \cdot 2.39790$$

$$= 2.21920 \text{ (exact to 5D)}.$$ ∎

The problem of finding x for given $f(x)$ is known as **inverse interpolation.** If f is differentiable and df/dx is not zero near the point where the inverse interpolation is to be effected, the inverse $x = F(y)$ of $y = f(x)$ exists locally near the given value of f and it may happen that F can be approximated in that neighborhood by a polynomial of moderately low degree. Then we may effect the inverse interpolation by tabulating F as a function of y and apply methods of direct interpolation to F. If $df/dx = 0$ near or at the desired point, it may be useful to solve $p(x) = \tilde{f}$ by iteration; here $p(x)$ is a polynomial that approximates $f(x)$ and $\tilde{f}$ is the given value.

Problems for Sec. 19.4

1. Verify that the parabola (2) passes through (x_0, f_0), (x_1, f_1), (x_2, f_2).
2. Using the values given in the table, find sin 0.26 by (1), that is, by linear interpolation. Show that the first two decimals are exact. (sin 0.26 = 0.257 08.)

x	$\sin x$	1st Diff.	2nd Diff.
0.0	0.000 00		
		198 67	
0.2	0.198 67		−792
		190 75	
0.4	0.389 42		−1553
		175 22	
0.6	0.564 64		−2250
		152 72	
0.8	0.717 36		−2861
		124 11	
1.0	0.841 47		

Problems 2–7

3. Find sin 0.26 by (2), that is, by quadratic interpolation. Show that the first three decimals are exact.
4. Extend the given table to include third and fourth differences. Then calculate sin 0.26 by (3) (*a*) with $n = 3$, (*b*) with $n = 4$. Comparing with the 5D-value $\sin 0.26 = 0.257\,08$, observe that (*a*) the first three decimals, (*b*) all five decimals are exact.
5. Apply the backward-difference interpolation formula (6) (*a*) with $n = 1$, (*b*) with $n = 2$ to obtain sin 0.26. Observe that in both cases, the first two decimals will be exact. Hence the result in (*b*) is poorer than that in Prob. 3. Why?
6. Extend the given table so that you can apply (6) with (*a*) $n = 3$, (*b*) $n = 4$, (*c*) $n = 5$, for calculating sin 0.26; show that you need $\sin x$ for the values $x = -0.6, -0.4, -0.2, 0, 0.2$, and the differences. What property of the sine function makes this extension very easy? Then apply (6), finding (*a*) 0.257 09, (*b*) 0.257 05, (*c*) 0.257 08 (the exact 5D-value). Why are the results poorer than in Prob. 4?
7. Show that the Everett formula (7) gives $\sin 0.26 = 0.257\,07$ with relatively little work.
8. Verify the calculations in Example 2.
9. Interpolate $f(x) = \sqrt{x}$ quadratically by (2), using $f(2.0) = 1.414\,214$, $f(2.3) = 1.516\,575$, $f(2.6) = 1.612\,452$. Compare the result with (4) in Sec. 19.3.
10. Show that

$$\Delta^k f_n = \binom{k}{0} f_{n+k} - \binom{k}{1} f_{n+k-1} + - \cdots + (-1)^k \binom{k}{k} f_n.$$

11. Find $f(3)$, using Lagrange's formula (11) and $f(1) = 2$, $f(2) = 11$, $f(4) = 77$.
12. Using $\ln 8.5 = 2.14007$ and ln 9.0, ln 9.5, ln 10 as in Example 3, compute ln 9.2 (*a*) by means of (3) with $n = 3$ (and $x_0 = 8.5$), (*b*) by means of (6) with $n = 3$ (and $x_0 = 10$).
13. Using $\ln 8.5 = 2.14007$ and ln 9, ln 10, ln 11 as in Example 3 and applying (11) with $n = 3$, compute ln 9.2 and compare the result with that in Example 3.
14. Using the same data as in Prob. 12, compute ln 9.2 (*a*) by means of (7), (*b*) by means of (11) with $n = 3$.
15. Assuming that $x_1 = x_0 + h$, $x_2 = x_0 + 2h$, $x_3 = x_0 + 3h$ and using $r = (x - x_0)/h$, show that (11) with $n = 3$ may be written

$$f(x) \approx -\binom{r-1}{3} f_0 + \frac{r(r-2)(r-3)}{2} f_1 - \frac{r(r-1)(r-3)}{2} f_2 + \binom{r}{3} f_3.$$

16. Applying the formula in Prob. 15, check the result in Prob. 14*b*.
17. **(Checking differences)** Show that the sum of the differences in a column is equal to the difference between the last and the first entries of the preceding column. Apply this partial check to Table 19.2 in Sec. 19.3.

19.5 Splines

Spline approximation is piecewise polynomial approximation. This means that a function $f(x)$ is given on an interval $a \leqq x \leqq b$, and we want to approximate $f(x)$ on that interval by a function $g(x)$ which is obtained as follows. We *partition* $a \leqq x \leqq b$, that is, we subdivide it into subintervals with common endpoints (called *nodes*)

$$a = x_0 < x_1 < \cdots < x_n = b \tag{1}$$

and we require that $g(x)$ in these subintervals is given by polynomials, one polynomial per subinterval, such that at those endpoints, $g(x)$ is several times differentiable. Hence, instead of approximating $f(x)$ by a single polynomial on the entire interval $a \leqq x \leqq b$, we now approximate $f(x)$ by n polynomials. In this way we may obtain approximating functions $g(x)$ which are more suitable in many problems of approximation and interpolation. For instance, they may not be as oscillatory between nodes as a single polynomial on $a \leqq x \leqq b$ often is. Functions $g(x)$ thus obtained are called **splines.** This name is derived from thin rods, called *splines,* which engineers have used for a long time to fit curves through given points. Since splines are of increasing practical importance, we give an introduction to this new field.

The simplest continuous piecewise polynomial approximation would be by piecewise *linear* functions. But such a function is not differentiable at those endpoints, and it is preferable to use functions which have a certain number of derivatives *everywhere* on the interval $a \leqq x \leqq b$.

We shall consider cubic splines, which are perhaps the most important ones from a practical point of view. By definition, a **cubic spline** $g(x)$ on $a \leqq x \leqq b$ corresponding to the partition (1) is a continuous function $g(x)$ which has continuous first and second derivatives everywhere in that interval and, in each subinterval of that partition, is represented by a polynomial of degree not exceeding three. Hence $g(x)$ consists of cubic polynomials, one in each subinterval.

If $f(x)$ is given on $a \leqq x \leqq b$ and a partition (1) has been chosen, we obtain a cubic spline $g(x)$ which approximates $f(x)$ by requiring that

$$g(x_0) = f(x_0), \qquad g(x_1) = f(x_1), \quad \cdots, \qquad g(x_n) = f(x_n), \tag{2}$$

as in the classical interpolation problem of the previous section. We claim that there is a cubic spline $g(x)$ satisfying these conditions (2). And if we also require

$$g'(x_0) = k_0, \qquad g'(x_n) = k_n \tag{3}$$

(k_0 and k_n given numbers), then we have a uniquely determined cubic spline. This is the content of the following existence and uniqueness theorem.

Theorem 1 (Cubic splines)

Let $f(x)$ be defined on the interval $a \leqq x \leqq b$, let a partition (1) *of the interval be given, and let k_0 and k_n be any two given numbers. Then there exists one and only one cubic spline $g(x)$ corresponding to* (1) *and satisfying* (2) *and* (3).

Proof. By definition, on every subinterval I_j given by $x_j \leqq x \leqq x_{j+1}$, the spline $g(x)$ must agree with a cubic polynomial $p_j(x)$ such that

$$p_j(x_j) = f(x_j), \qquad p_j(x_{j+1}) = f(x_{j+1}). \tag{4}$$

We write $1/(x_{j+1} - x_j) = c_j$ and

$$p_j'(x_j) = k_j, \qquad p_j'(x_{j+1}) = k_{j+1}, \tag{5}$$

where k_0 and k_n are given, and $k_1, \cdots, k_{n-1}$ will be determined later. (4) and (5) are four conditions for $p_j(x)$. By direct calculation we can verify that the unique cubic polynomial $p_j(x)$ satisfying (4) and (5) is

$$\begin{aligned} p_j(x) &= f(x_j)c_j^2(x - x_{j+1})^2[1 + 2c_j(x - x_j)] \\ &\quad + f(x_{j+1})c_j^2(x - x_j)^2[1 - 2c_j(x - x_{j+1})] \\ &\quad + k_jc_j^2(x - x_j)(x - x_{j+1})^2 \\ &\quad + k_{j+1}c_j^2(x - x_j)^2(x - x_{j+1}) \end{aligned} \tag{6}$$

Differentiating twice, we obtain

$$p_j''(x_j) = -6c_j^2f(x_j) + 6c_j^2f(x_{j+1}) - 4c_jk_j - 2c_jk_{j+1} \tag{7}$$

$$p_j''(x_{j+1}) = 6c_j^2f(x_j) - 6c_j^2f(x_{j+1}) + 2c_jk_j + 4c_jk_{j+1}. \tag{8}$$

By definition, $g(x)$ has continuous second derivatives. This gives the conditions

$$p_{j-1}''(x_j) = p_j''(x_j) \qquad j = 1, \cdots, n - 1.$$

Using (8) with j replaced by $j - 1$, and (7), we see that these $n - 1$ equations take the form

$$c_{j-1}k_{j-1} + 2(c_{j-1} + c_j)k_j + c_jk_{j+1} = 3[c_{j-1}^2 \nabla f_j + c_j^2 \nabla f_{j+1}], \tag{9}$$

where $\nabla f_j = f(x_j) - f(x_{j-1})$ and $\nabla f_{j+1} = f(x_{j+1}) - f(x_j)$ and $j = 1, \cdots, n - 1$, as before. This system of $n - 1$ linear equations has a unique solution $k_1, \cdots, k_{n-1}$ because all the coefficients of the system are nonnegative, and each element in the principal diagonal is greater than the sum of the other elements in the corresponding row, so that the coefficient determinant cannot be zero. Hence we are able to determine unique values $k_1, \cdots, k_{n-1}$ of the first derivative of $g(x)$ at the nodes. This completes the proof. ∎

Let us illustrate the theorem by a simple example.

Example 1. Spline approximation

Approximate $f(x) = x^4$ on the interval $-1 \leqq x \leqq 1$ by a cubic spline $g(x)$ corresponding to the partition $x_0 = -1$, $x_1 = 0$, $x_2 = 1$ and satisfying (2) and $g'(-1) = f'(-1)$, $g'(1) = f'(1)$.

Solution. We have to determine the coefficients of

$$p_0(x) = a_3x^3 + a_2x^2 + a_1x + a_0 \qquad (-1 \leqq x \leqq 0)$$

$$p_1(x) = b_3x^3 + b_2x^2 + b_1x + b_0 \qquad (0 \leqq x \leqq 1).$$

From $p_0(0) = p_1(0) = f(0) = 0$ we have $a_0 = b_0 = 0$. From $p_0'(0) = p_1'(0)$ we obtain $a_1 = b_1$. From $p_0''(0) = p_1''(0)$ we obtain $a_2 = b_2$. Together,

$$p_0(x) = a_3x^3 + a_2x^2 + a_1x$$

$$p_1(x) = b_3x^3 + a_2x^2 + a_1x$$

The remaining four coefficients can now be determined from the remaining four conditions

$$\begin{aligned} p_0(-1) &= -a_3 + a_2 - a_1 = f(-1) = 1 \\ p_1(1) &= b_3 + a_2 + a_1 = f(1) = 1 \\ p_0'(-1) &= 3a_3 - 2a_2 + a_1 = f'(-1) = -4 \\ p_1'(1) &= 3b_3 + 2a_2 + a_1 = f'(1) = 4. \end{aligned} \tag{10}$$

The solution is $a_1 = 0$, $a_2 = -1$, $a_3 = -2$, $b_3 = 2$. Hence we obtain the spline (Fig. 351)

$$g(x) = \begin{cases} -2x^3 - x^2 & \text{if} \quad -1 \leqq x \leqq 0 \\ 2x^3 - x^2 & \text{if} \quad 0 \leqq x \leqq 1. \end{cases} \tag{11}$$ ∎

Splines have an interesting minimum property, which we shall now derive. Suppose that $f(x)$ in Theorem 1 is continuous and has continuous first and second derivatives on $a \leqq x \leqq b$. Assume that (3) is of the form

$$g'(a) = f'(a), \qquad g'(b) = f'(b) \tag{12}$$

(as in Example 1). Then $f' - g'$ is zero at a and b. Integrating by parts, we thus obtain

$$\int_a^b g''(x)[f''(x) - g''(x)]\, dx = -\int_a^b g'''(x)[f'(x) - g'(x)]\, dx.$$

Since $g'''(x)$ is constant on each subinterval of the partition, evaluating the integral on the right over a subinterval, we obtain $const \cdot [f(x) - g(x)]$, taken at the endpoints of the subinterval, which is zero by (2). Hence that integral on the right is zero. This proves

$$\int_a^b f''(x)g''(x)\, dx = \int_a^b g''(x)^2\, dx.$$

Consequently,

$$\begin{aligned} \int_a^b [f''(x) - g''(x)]^2\, dx &= \int_a^b f''(x)^2\, dx - 2\int_a^b f''(x)g''(x)\, dx + \int_a^b g''(x)^2\, dx \\ &= \int_a^b f''(x)^2\, dx - \int_a^b g''(x)^2\, dx. \end{aligned}$$

The integrand on the left is nonnegative and so is the integral. This yields the inequality

$$\int_a^b f''(x)^2\, dx \geqq \int_a^b g''(x)^2\, dx. \tag{13}$$

Our result can be stated as follows.

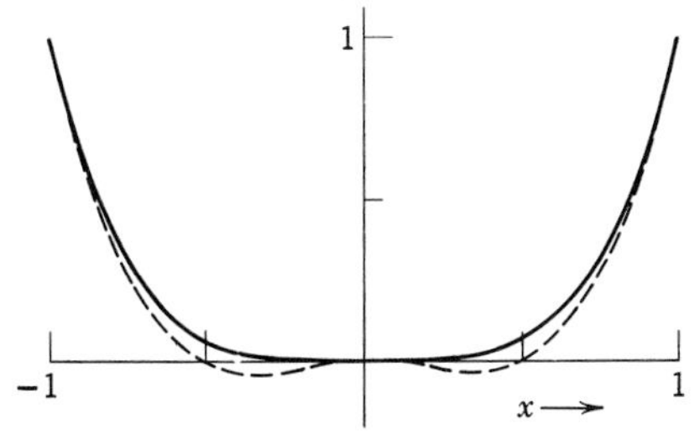

Fig. 351. Function $f(x) = x^4$ and cubic spline $g(x)$ (dashed) in Example 1

Theorem 2 (Minimum property of cubic splines)
Let $f(x)$ be continuous and have continuous first and second derivatives on some interval $a \leqq x \leqq b$. Let $g(x)$ be the cubic spline corresponding to some partition (1) of that interval and satisfying (2) and (12). Then $f(x)$ and $g(x)$ satisfy (13), and equality holds if and only if $f(x)$ is the cubic spline $g(x)$.

Engineers use thin rods called *splines* to fit curves through given points, as was mentioned before, and the strain energy minimized by such splines is proportional, approximately, to the integral of the square of the second derivative of the spline. Hence inequality (13) explains the use of the term *splines* for the functions $g(x)$ considered in this section.

Problems for Sec. 19.5

1. Verify that $p_j(x)$ in (6) satisfies (4) and (5).

2. Derive (7) and (8) from (6).

3. Consider Example 1. Show that, under the conditions stated in the example, formula (6) gives

$$p_0(x) = -2x^3 - x^2 + k_1 x(x+1)^2$$

$$p_1(x) = 2x^3 - x^2 + k_1 x(x-1)^2$$

and (9) gives $k_1 = 0$, so that we obtain (11).

4. Compare the spline $g(x)$ in Example 1 with the quadratic interpolation polynomial $p(x)$ over the whole interval. What are the maximum deviations of $g(x)$ and $p(x)$ from $f(x)$? Comment.

5. Carry out the details of solving (10).

6. Show that the cubic splines corresponding to a given partition of a given interval form a vector space (cf. Sec. 6.4).

7. Show that for a given partition of the form (1) there exist $n + 1$ unique cubic splines $g_0(x), \cdots, g_n(x)$ such that $g_j'(a) = g_j'(b) = 0$ and

$$g_j(x_k) = \delta_{jk} = \begin{cases} 0 & \text{if } j \neq k \\ 1 & \text{if } j = k. \end{cases}$$

8. If a cubic spline is three times continuously differentiable (that is, has continuous first, second and third derivatives), show that it must be a polynomial.

9. It may sometimes happen that a spline is represented by the same polynomial in adjacent subintervals of $a \leqq x \leqq b$. To illustrate this, find the cubic spline $g(x)$ for $f(x) = \sin x$ corresponding to the partition $x_0 = -\pi/2$, $x_1 = 0$, $x_2 = \pi/2$ of the interval $-\pi/2 \leqq x \leqq \pi/2$ and satisfying $g'(-\pi/2) = f'(-\pi/2)$, $g'(\pi/2) = f'(\pi/2)$.

10. A possible geometric interpretation of (13) is that a cubic spline function minimizes the integral of the square of the curvature, at least approximately. Explain.

19.6 Numerical Integration and Differentiation

The problem of **numerical integration** is the numerical evaluation of a definite integral

$$J = \int_a^b f(x)\,dx$$

where a and b are given and f is a function given analytically by a formula or empirically by a table of values.

We know that if f is such that we can find a differentiable function F whose derivative is f, then we can evaluate J by applying the familiar formula

$$J = \int_a^b f(x)\,dx = F(b) - F(a) \qquad [F'(x) = f(x)],$$

and tables of integrals, such as Ref. [J3] listed in Appendix 1, may be helpful for that purpose.

However, in engineering applications there frequently occur integrals whose integrand is an empirical function given by a table, or is such that the integral cannot be represented in terms of finitely many elementary functions (examples in Appendix 3), or the explicit form of F is complicated and not very helpful. Then we may use a numerical method of approximate integration.

Since J equals the area of the region R under the curve of $f(x)$ between a and b (cf. Fig. 352), we may obtain rough approximate values of J by cutting R from cardboard, determine its weight, and divide the result by the weight of a square

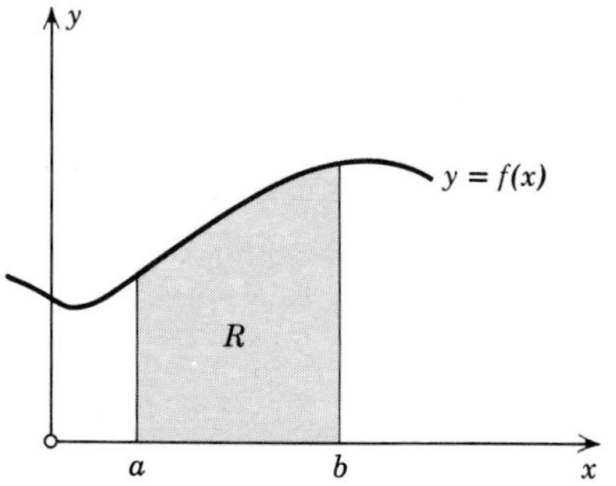

Fig. 352. Geometrical interpretation of a definite integral

of this cardboard of area 1. Another simple method is to draw R on graph paper and count squares. More accurate results can be obtained by using a **planimeter,** whose theory and application are discussed in Ref. [G20] listed in Appendix 1.

Numerical integration methods may be obtained by approximating the integrand f by polynomials. To derive the simplest formula, we subdivide the interval of integration into n equal subintervals of length $h = (b - a)/n$ and approximate f in each such interval by a constant function $f(x_j^*)$, where x_j^* is the midpoint of the interval (cf. Fig. 353). Then f is approximated by a **step function** (piecewise constant function), the n rectangles in Fig. 353 have the areas $f(x_1^*)h, \cdots, f(x_n^*)h$, and we obtain the **rectangular rule**

$$\textbf{(1)} \qquad J = \int_a^b f(x)\,dx \approx h[f(x_1^*) + f(x_2^*) + \cdots + f(x_n^*)] \qquad \left(h = \frac{b-a}{n}\right).$$

If we approximate f by a piecewise linear function (polygon of chords of the curve of f; cf. Fig. 354), we obtain the **trapezoidal rule**

$$\textbf{(2)} \qquad J = \int_a^b f(x)\,dx \approx h[\tfrac{1}{2}f(a) + f(x_1) + f(x_2) + \cdots + f(x_{n-1}) + \tfrac{1}{2}f(b)],$$

where $h = (b - a)/n$, and $x_0\ (= a), x_1, x_2, \cdots, x_{n-1}, x_n\ (= b)$ are the endpoints of the intervals already used in (1); thus $x_j = x_0 + jh$. In fact, the n trapezoids in Fig. 354 have the areas

$$\tfrac{1}{2}[f(a) + f(x_1)]h, \quad \tfrac{1}{2}[f(x_1) + f(x_2)]h, \quad \cdots, \quad \tfrac{1}{2}[f(x_{n-1}) + f(b)]h,$$

and their sum J^* equals the right-hand side of (2).

The error ϵ of J^* is (cf. Sec. 19.1)

$$\epsilon = J^* - J.$$

If $f(x)$ is a linear function, then $\epsilon = 0$. In this case f' is constant, f'' is zero for all x, and it seems plausible that for a general function f (having a continuous second derivative) we may obtain **error bounds** (bounds for ϵ) involving f''. For this purpose we replace b by a variable t and apply (2) with $n = 1$. Then the corresponding error is

$$\epsilon(t) = \frac{t-a}{2}[f(a) + f(t)] - \int_a^t f(x)\,dx.$$

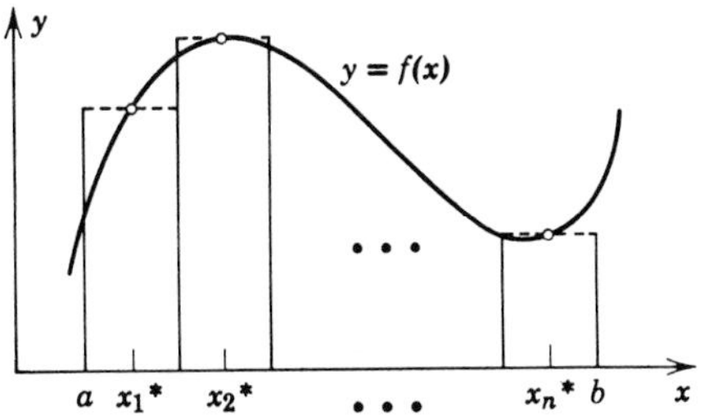

Fig. 353. Rectangular rule

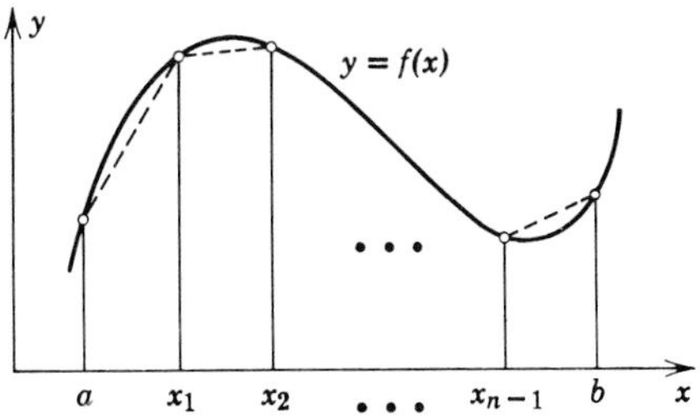

Fig. 354. Trapezoidal rule

We note that $\epsilon(a) = 0$, which is trivial. Differentiation gives

$$\epsilon'(t) = \tfrac{1}{2}[f(a) + f(t)] + \frac{t-a}{2}f'(t) - f(t).$$

We see that $\epsilon'(a) = 0$. Differentiating once more, we have

$$\epsilon''(t) = \tfrac{1}{2}(t-a)f''(t)$$

and can obtain bounds for ϵ'' by replacing f'' by its smallest and largest value in the interval $a \leqq t \leqq b$. Denoting these values by M_2^* and M_2, respectively, and noting that $t - a \geqq 0$, we obtain

$$\tfrac{1}{2}(t-a)M_2^* \leqq \epsilon''(t) \leqq \tfrac{1}{2}(t-a)M_2$$

for all t in that interval. Integrating from a to t, we have

$$\tfrac{1}{4}(t-a)^2M_2^* \leqq \epsilon'(t) - \epsilon'(a) \leqq \tfrac{1}{4}(t-a)^2M_2.$$

Using $\epsilon'(a) = 0$, $\epsilon(a) = 0$, integrating again, and setting $t = a + h$, we see that

$$\tfrac{1}{12}h^3M_2^* \leqq \epsilon(a+h) \leqq \tfrac{1}{12}h^3M_2.$$

For the errors corresponding to the other $n - 1$ subintervals we obtain $n - 1$ similar inequalities, and addition of all the n inequalities gives an inequality for the error ϵ corresponding to the integration from a to b; since $h = (b-a)/n$, we obtain

(3) $$KM_2^* \leqq \epsilon \leqq KM_2 \qquad \text{where} \qquad K = \frac{(b-a)^3}{12n^2}$$

and M_2^* and M_2 are the smallest and largest values of the second derivative of f in the interval of integration.

Example 1. Trapezoidal rule. Error estimate

Evaluate $J = \int_0^1 e^{-x^2}\,dx$ by means of (2) with $n = 10$ and estimate the error. From Table 19.5 we see that

$$J \approx 0.1(0.5 \cdot 1.367\,879 + 6.778\,167) = 0.746\,211.$$

Since $f''(x) = 2(2x^2 - 1)e^{-x^2}$, we have $M_2^* = f''(0) = -2$, $M_2 = f''(1) = 0.735\,759$. Furthermore $K = 1/1200$, and (3) gives

$$-0.001\,667 \leqq \epsilon \leqq 0.000\,614.$$

Hence the exact value of J must lie between

$$0.746\,211 - 0.000\,614 = 0.745\,597 \quad \text{and} \quad 0.746\,211 + 0.001\,667 = 0.747\,878.$$

(Actually, $J = 0.746\,824$, exact to 6D.) ∎

Piecewise constant approximation of f led to the rectangular rule (1), piecewise linear approximation to the trapezoidal rule (2), and piecewise quadratic approximation will give Simpson's rule, which is of great practical importance

Table 19.5
Computation in Example 1

j	x_j	x_j^2	$e^{-x_j^2}$	
0	0	0	1.000 000	
1	0.1	0.01		0.990 050
2	0.2	0.04		0.960 789
3	0.3	0.09		0.913 931
4	0.4	0.16		0.852 144
5	0.5	0.25		0.778 801
6	0.6	0.36		0.697 676
7	0.7	0.49		0.612 626
8	0.8	0.64		0.527 292
9	0.9	0.81		0.444 858
10	1.0	1.00	0.367 879	
		Sums	1.367 879	6.778 167

because it is sufficiently accurate for most problems, but still sufficiently simple. To derive this formula, we subdivide the interval of integration $a \leqq x \leqq b$ into an **even** number of equal subintervals, say, into $2n$ subintervals of length $h = (b - a)/2n$, with endpoints x_0 $(= a)$, $x_1, \cdots, x_{2n-1}, x_{2n}$ $(= b)$. We consider the first two subintervals and approximate $f(x)$ in the interval $x_0 \leqq x \leqq x_2 = x_0 + 2h$ by the Lagrange polynomial $p_2(x)$ through (x_0, f_0), (x_1, f_1), (x_2, f_2), where $f_j = f(x_j)$. From (11a) in Sec. 19.4 we obtain

$$(4^*)\qquad \begin{aligned} p_2(x) = \frac{(x - x_1)(x - x_2)}{(x_0 - x_1)(x_0 - x_2)} f_0 &+ \frac{(x - x_0)(x - x_2)}{(x_1 - x_0)(x_1 - x_2)} f_1 \\ &+ \frac{(x - x_0)(x - x_1)}{(x_2 - x_0)(x_2 - x_1)} f_2. \end{aligned}$$

The denominators are $2h^2$, $-h^2$, and $2h^2$, respectively. Setting $s = (x - x_1)/h$, we have $x - x_0 = (s + 1)h$, $x - x_1 = sh$, $x - x_2 = (s - 1)h$, and

$$p_2(x) = \tfrac{1}{2}s(s - 1)f_0 - (s + 1)(s - 1)f_1 + \tfrac{1}{2}(s + 1)sf_2.$$

We now integrate with respect to x from x_0 to x_2. This corresponds to integrating with respect to s from -1 to 1. Since $dx = h\,ds$, the result of this simple integration is

$$\int_{x_0}^{x_2} f(x)\,dx \approx \int_{x_0}^{x_2} p_2(x)\,dx = h\left(\frac{1}{3}f_0 + \frac{4}{3}f_1 + \frac{1}{3}f_2\right).$$

A similar formula holds for the next two subintervals from x_2 to x_4, and so on. By summing all these n formulas we obtain **Simpson's rule**

$$\textbf{(4)}\qquad \int_a^b f(x)\,dx \approx \frac{h}{3}(f_0 + 4f_1 + 2f_2 + 4f_3 + \cdots + 2f_{2n-2} + 4f_{2n-1} + f_{2n}),$$

where $h = (b - a)/2n$ and $f_j = f(x_j)$. Bounds for the error ϵ_S in (4) can be obtained by a method similar to that in the case of the trapezoidal rule (2), assuming that the fourth derivative of f exists and is continuous in the interval of integration. The result is

$$(5) \qquad CM_4^* \leqq \epsilon_S \leqq CM_4, \qquad \text{where} \qquad C = \frac{(b-a)^5}{180(2n)^4}$$

and M_4^* and M_4 are the smallest and the largest value of the fourth derivative of f in the interval of integration. It is trivial that $\epsilon_S = 0$ for a quadratic polynomial f, but (5) shows that $\epsilon_S = 0$ even for a cubic polynomial.

Example 2. Simpson's rule. Error estimate

Evaluate $J = \int_0^1 e^{-x^2}\,dx$ by Simpson's rule with $2n = 10$ and estimate the error. Since $h = 0.1$, Table 19.6 gives

$$J \approx \frac{0.1}{3}(1.367\,879 + 4 \cdot 3.740\,266 + 2 \cdot 3.037\,901) = 0.746\,825.$$

Estimate of error. Differentiation gives $f^{\text{IV}}(x) = 4(4x^4 - 12x^2 + 3)e^{-x^2}$. By considering the derivative of f^{IV} we find that the smallest value of f^{IV} in the interval of integration occurs at $x = x^* = 2.5 + 0.5\sqrt{10}$ and the largest value at $x = 0$. Computation gives $M_4^* = f^{\text{IV}}(x^*) = -7.359$ and $M_4 = f^{\text{IV}}(0) = 12$. Since $2n = 10$ and $b - a = 1$, we obtain the value $C = 1/1\,800\,000 = 0.000\,000\,56$. Therefore

$$-0.000\,004 \leqq \epsilon_S \leqq 0.000\,006, \qquad \text{and} \qquad 0.746\,818 \leqq J \leqq 0.746\,830,$$

which shows exactness of our approximation to at least 4D. Actually 0.746 825 is exact to 5D, because $J = 0.746\,824$ (exact to 6D).

Note that our present result is much better than that in Example 1 obtained by the trapezoidal rule, whereas the work is nearly the same in both cases. ▮

Simpson's rule is sufficiently accurate for most engineering purposes, and in automatic computation it is preferred to more complicated formulas of higher precision, which are obtained by the use of approximating polynomials of higher degree. We want to mention two such formulas of higher precision,

Table 19.6
Computations in Example 2

j	x_j	x_j^2	$e^{-x_j^2}$		
0	0	0	1.000 000		
1	0.1	0.01		0.990 050	
2	0.2	0.04			0.960 789
3	0.3	0.09		0.913 931	
4	0.4	0.16			0.852 144
5	0.5	0.25		0.778 801	
6	0.6	0.36			0.697 676
7	0.7	0.49		0.612 626	
8	0.8	0.64			0.527 292
9	0.9	0.81		0.444 858	
10	1.0	1.00	0.367 879		
Sums			1.367 879	3.740 266	3.037 901

which are sometimes useful. A cubic through $(x_0, f_0), \cdots, (x_3, f_3)$ yields the **three-eights rule**

$$\int_{x_0}^{x_3} f(x)\,dx \approx \frac{3h}{8}(f_0 + 3f_1 + 3f_2 + f_3), \tag{6}$$

which is exact for cubical polynomials (as is Simpson's rule) and has the advantage that it can be applied to an *odd* number of subintervals (divisible by 3). The polynomial of sixth degree through $(x_0, f_0), \cdots, (x_6, f_6)$ yields a formula with rather complicated coefficients, but a much simpler formula of the same kind, and only slightly inferior in accuracy, is **Weddle's rule**

$$\int_{x_0}^{x_6} f(x)\,dx \approx \frac{3h}{10}(f_0 + 5f_1 + f_2 + 6f_3 + f_4 + 5f_5 + f_6). \tag{7}$$

Our integration formulas discussed involve function values for equidistant x-values and give exact results for polynomials not exceeding a certain degree. More generally, we may set

$$\int_{-1}^{1} f(x)\,dx \approx \sum_{j=1}^{n} A_j f_j \qquad [f_j = f(x_j)] \tag{8}$$

(choosing $a = -1$ and $b = 1$, which may be accomplished by a linear transformation of scale) and determine the $2n$ constants $A_1, \cdots, A_n, x_1, \cdots, x_n$ so that (8) gives exact results for polynomials of degree m as high as possible. Since $2n$ is the number of coefficients of a polynomial of degree $2n - 1$, it follows that $m \leqq 2n - 1$. Gauss has shown that exactness for polynomials of degree $2n - 1$ can be attained if and only if $x_1, \cdots, x_n$ are the n zeros of the **Legendre polynomial** *of degree n* [cf. also Sec. 4.3]

$$P_n(x) = \frac{(2n)!}{2^n(n!)^2}x^n - \frac{(2n-2)!}{2^n 1!(n-1)!(n-2)!}x^{n-2} + \frac{(2n-4)!}{2^n 2!(n-2)!(n-4)!}x^{n-4} - + \cdots,$$

that is

$$P_0 = 1, \qquad P_1(x) = x, \qquad P_2(x) = \tfrac{1}{2}(3x^2 - 1), \qquad P_3(x) = \tfrac{1}{2}(5x^3 - 3x), \quad \cdots,$$

and the coefficients A_j are suitably chosen. Then (8) is called a **Gaussian integration formula.** Its advantage is the high accuracy, its disadvantage the irregular spacing of $x_1, \cdots, x_n$. Hence such a formula is practical when automatic computers are used and the computation of $f(x_j)$ does not depend on the number of decimals in the argument, or when $f(x_j)$ is obtained from an experiment in which the x_j can be set once and for all. The formula is not likely to be practical when the integrand is tabulated at equal intervals, because then preliminary interpolations are necessary to evaluate the $f(x_j)$, and these may outweigh any gain due to the smaller error estimates.

Since the endpoints -1 and 1 of the interval of integration in (8) are not zeros of P_n, they do not occur among $x_1, \cdots, x_n$, and the Gauss formula (8) is called, therefore, an **open formula,** in contrast to a **closed formula** in which the endpoints of the interval of integration do occur. [For example, (2), (4), (6), and (7) are closed.]

We mention that, just as in the case of interpolation, there are numerical integration methods based on differences. A very efficient method uses the **central-difference formula by Gauss**

$$\textbf{(9)} \qquad \int_{x_0}^{x_1} f(x)\,dx \approx \frac{h}{2}\left(f_0 + f_1 - \frac{\delta^2 f_0 + \delta^2 f_1}{12} + \frac{11(\delta^4 f_0 + \delta^4 f_1)}{720}\right).$$

For more details see Refs. [G13], [G16] and [G17] in Appendix 1.

Numerical differentiation

The problem of **numerical differentiation** is the determination of approximate values of the derivative of a function f which is given by a table. Numerical differentiation should be avoided whenever possible because approximate values of derivatives will in general be less accurate than the function values from which they are derived. In fact, the derivative is the limit of the difference quotient, and in the latter we normally subtract two large quantities and divide by a small one; furthermore if a given function f is approximated by a polynomial p, the difference in function values may be small but the derivatives may differ considerably. Hence it is plausible that numerical differentiation is delicate, in contrast to numerical integration, which is not much affected by inaccuracies of function values, because integration is essentially a smoothing process.

We use the notations $f_j' = f'(x_j)$, $f_j'' = f''(x_j)$, etc., and may obtain rough approximation formulas for derivatives by noting that

$$f'(x) = \lim_{h \to 0} \frac{f(x+h) - f(x)}{h}.$$

This suggests

$$\textbf{(10)} \qquad f'_{1/2} \approx \frac{\delta f_{1/2}}{h} = \frac{f_1 - f_0}{h}.$$

Similarly, for the second derivative we obtain

$$(11) \qquad f_1'' \approx \frac{\delta^2 f_1}{h^2} = \frac{f_2 - 2f_1 + f_0}{h^2}$$

and so on.

More accurate approximations are obtained by differentiating suitable Lagrange polynomials. Differentiating (4*) and remembering that the denominators in (4*) are $2h^2$, $-h^2$, $2h^2$, we have

$$f'(x) \approx p_2'(x) = \frac{2x - x_1 - x_2}{2h^2} f_0 - \frac{2x - x_0 - x_2}{h^2} f_1 + \frac{2x - x_0 - x_1}{2h^2} f_2.$$

Evaluating this at x_0, x_1, x_2, we obtain the "three-point formulas"

(12)

$$\text{(a)} \quad f_0' \approx \frac{1}{2h}(-3f_0 + 4f_1 - f_2)$$

$$\text{(b)} \quad f_1' \approx \frac{1}{2h}(-f_0 + f_2)$$

$$\text{(c)} \quad f_2' \approx \frac{1}{2h}(f_0 - 4f_1 + 3f_2).$$

Applying the same idea to Lagrange's five-point formula, we obtain similar formulas, in particular

$$f_2' \approx \frac{1}{12h}(f_0 - 8f_1 + 8f_3 - f_4). \tag{13}$$

Further details and formulas are included in Ref. [G6] listed in Appendix 1.

Problems for Sec. 19.6

Review some integration formulas and methods by integrating

1. $\int \sin^2 x \, dx$ **2.** $\int \cos^2 \omega x \, dx$ **3.** $\int e^{ax} \sin bx \, dx$

4. $\int e^{ax} \cos bx \, dx$ **5.** $\int \tan kx \, dx$ **6.** $\int \ln x \, dx$

7. $\int \frac{dx}{k^2 + x^2}$ **8.** $\int \frac{dx}{\sqrt{a^2 - x^2}}$ **9.** $\int \frac{dx}{x^2(x^2 + 1)^2}$

10. Compute the integral in Example 1 by using (1) with $n = 5$.

11. Compute $\int_0^1 x^3 \, dx$ by the rectangular rule (1) with $n = 5$. What is the error?

12. Compute the integral in Prob. 11 by the trapezoidal rule (2) with $n = 5$. What error bounds are obtained from (3)? What is the actual error of the result? Why is this result larger than the exact value?

13. Find an approximate value of $\ln 2 = \int_1^2 \frac{dx}{x}$ by Simpson's rule with $2n = 4$. Estimate the error by (5).

14. Evaluate $\int_0^1 x^5 \, dx$ by Simpson's rule with $2n = 10$. What error bounds are obtained from (5)? What is the actual error of the result?

Using the values of $\sin x$ in Table A1, Appendix 4, evaluate $\int_0^1 \frac{\sin x}{x} \, dx$:

15. By the rectangular rule (1) with $n = 5$.

16. By the trapezoidal rule (2) with $n = 5$.

17. By (2) with $n = 10$.

18. By Simpson's rule with $2n = 2$ and with $2n = 10$.

19. Determine α and β so that

$$\int_{x_n}^{x_{n+1}} f(x)\,dx \approx h[\alpha f(x_n) + \beta f(x_{n+1})] \qquad h = x_{n+1} - x_n$$

is exact for polynomials of first degree. What formula do you obtain?

20. If $f(x)$ is a polynomial of second order, show that (10) is exact. What does this mean geometrically?

21. Carry out the details of the derivation of (12).

22. Derive (13) as explained in the text.

23. Consider $f(x) = x^4$ for $x_0 = 0, x_1 = 0.2, x_2 = 0.4, x_3 = 0.6, x_4 = 0.8$. Calculate f_2' from (12a), (12b), (12c), (13). Determine the errors. Compare and comment.

24. A "four-point formula" for the derivative is

$$f_2' \approx \frac{1}{6h}(-2f_1 - 3f_2 + 6f_3 - f_4).$$

Apply it to $f(x) = x^4$ with $x_1, \cdots, x_4$ as in Prob. 23, determine the error and compare it with that in the case of (13).

25. The derivative $f'(x)$ can also be approximated in terms of first and higher differences:

$$f'(x_0) \approx \frac{1}{h}\left(\Delta f_0 - \frac{1}{2}\Delta^2 f_0 + \frac{1}{3}\Delta^3 f_0 - \frac{1}{4}\Delta^4 f_0 + - \cdots\right).$$

Obtain this formula by differentiating (3), Sec. 19.4, with respect to r, finding

$$hf'(x) \approx \Delta f_0 + \frac{2r-1}{2!}\Delta^2 f_0 + \frac{3r^2 - 6r + 2}{3!}\Delta^3 f_0 + \cdots,$$

where $x = x_0 + rh$, and setting $r = 0$. Apply the formula for obtaining approximations to $f'(0.4)$ in Prob. 23, using differences up to and including (*a*) first order, (*b*) second order, (*c*) third order, (*d*) fourth order.

19.7 Numerical Methods for First-Order Differential Equations

This section is independent of the other sections in this chapter and may also be discussed immediately after Chap. 1.

From Chap. 1 we know that a *differential equation of the first order* is of the form $F(x, y, y') = 0$, and often it will be possible to write the equation in the *explicit form* $y' = f(x, y)$. An **initial value problem** consists of a differential equation and a condition which the solution must satisfy (or several conditions referring to the same value of x if the equation is of higher order). In this section we shall consider initial value problems of the form

(1) $$y' = f(x, y), \qquad y(x_0) = y_0,$$

assuming f to be such that the problem has a unique solution in some interval containing x_0, and we shall discuss methods for computing numerical values of the solution.

If we can obtain a formula for the solution, we may evaluate it numerically, either directly or by the use of tables. In this approach the book by Kamke (Ref. [B12] listed in Appendix 1) and an index of tables (cf. Ref. [A6] in Appendix 1) will be helpful. If that formula is too complicated or if no formula for the solution is available, we may apply one of the methods discussed in this section.

These methods are **step-by-step methods,** that is, we start from $y_0 = y(x_0)$ and proceed stepwise. In the first step we compute an approximate value y_1 of the solution y of (1) at $x = x_1 = x_0 + h$. In the second step we compute an approximate value y_2 of that solution at $x = x_2 = x_0 + 2h$, etc. Here h is a fixed number, for example, 0.2 or 0.1 or 0.01; principles for the choice of h will be discussed later in this section.

In each step the computations are done by the same formula. Such formulas are suggested by the Taylor series

$$y(x + h) = y(x) + hy'(x) + \frac{h^2}{2}y''(x) + \cdots .$$

From (1) we have $y' = f$. By differentiation, $y'' = f' = \partial f/\partial x + (\partial f/\partial y)y'$, etc., and the Taylor series becomes

$$y(x + h) = y(x) + hf + \frac{h^2}{2}f' + \frac{h^3}{6}f'' + \cdots , \tag{2}$$

where $f, f', f'', \cdots$ are evaluated at $(x, y(x))$.

For small values of h, the higher powers $h^2, h^3, \cdots$ in (2) will be very small. This suggests the crude approximation

$$y(x + h) \approx y(x) + hf$$

and the following iteration process. In the first step we compute

$$y_1 = y_0 + hf(x_0, y_0),$$

which approximates $y(x_1) = y(x_0 + h)$. In the second step we compute

$$y_2 = y_1 + hf(x_1, y_1),$$

which approximates $y(x_2) = y(x_0 + 2h)$, etc., and in general

$$y_{n+1} = y_n + hf(x_n, y_n) \qquad (n = 0, 1, \cdots). \tag{3}$$

This is called the **Euler method** or **Euler-Cauchy method.** Geometrically it is an approximation of the curve of $y(x)$ by a polygon whose first side is tangent to the curve at x_0 (cf. Fig. 355 on the next page).

The method is called a **first-order method,** because in (2) we take only the constant term and the term containing the first power of h. The omission of the

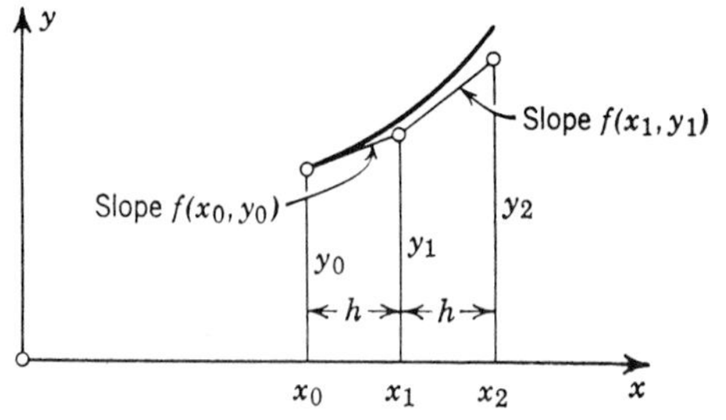

Fig. 355. Euler method

further terms in (2) causes an error, which is called the **truncation error** of the method. For small h, the third and higher powers of h will be small compared with h^2 in the first neglected term in (2), and we therefore say that the **truncation error per step** is of **order** h^2. In addition there are **rounding errors** in this and other methods, which may affect the accuracy of the values $y_1, y_2, \cdots$ more and more as n increases; we shall return to this point in the next section.

The practical value of the Euler method is limited, but since it is simple, it may be helpful for understanding the basic idea of the methods in this section.

Example 1. Euler method

Apply the Euler method to the initial value problem

$$y' = x + y, \qquad y(0) = 0, \tag{4}$$

choosing $h = 0.2$ and computing $y_1, \cdots, y_5$. Here $f(x, y) = x + y$, and (3) becomes

$$y_{n+1} = y_n + 0.2(x_n + y_n).$$

Table 19.7 shows the computations, the values of the exact solution

$$y(x) = e^x - x - 1$$

obtained from (4) in Sec. 1.7, and the error. In practice the exact solution is unknown, but an indication of the accuracy of the values can be obtained by applying the Euler method once more with step $2h = 0.4$ and comparing corresponding approximations. They differ by 0.040 ($x = 0.4$) and 0.110 ($x = 0.8$), as a simple calculation shows. Since the error is of order h^2, a switch from h to $2h$ corresponds to a multiplication by $2^2 = 4$, but since we need only half as many steps as before, the error will only be multiplied by $4/2 = 2$; hence those differences indicate the size of the error (cf. Table 19.7). ∎

Table 19.7
Euler Method Applied to (4) and Error

n	x_n	y_n	$0.2(x_n + y_n)$	Exact Values	Absolute Value of Error
0	0.0	0.000	0.000	0.000	0.000
1	0.2	0.000	0.040	0.021	0.021
2	0.4	0.040	0.088	0.092	0.052
3	0.6	0.128	0.146	0.222	0.094
4	0.8	0.274	0.215	0.426	0.152
5	1.0	0.489		0.718	0.229

By taking more terms in (2) into account we obtain numerical methods of higher order and precision. The corresponding formulas can be represented in such a form that the complicated computation of derivatives of $f(x, y)$ is avoided and is replaced by computing f for one or several suitably chosen auxiliary values of (x, y). Let us discuss two such methods which are of practical importance.

The first method is the so-called **improved Euler method** or **improved Euler-Cauchy method** (sometimes also called **Heun's method**). In each step of this method we first compute the auxiliary value

(5a) $$y^*_{n+1} = y_n + hf(x_n, y_n)$$

and then the new value

(5b) $$y_{n+1} = y_n + \tfrac{1}{2}h[f(x_n, y_n) + f(x_{n+1}, y^*_{n+1})].$$

This method has a simple geometric interpretation. In fact, we may say that in the interval from x_n to $x_n + \frac{1}{2}h$ we approximate the solution y by the straight line through (x_n, y_n) with slope $f(x_n, y_n)$, and then we continue along the straight line with slope $f(x_{n+1}, y^*_{n+1})$ until x reaches x_{n+1} (cf. Fig. 356, where $n = 0$).

The improved Euler–Cauchy method is a **predictor-corrector method,** because in each step we first predict a value by (5a) and then correct it by (5b).

Example 2. Improved Euler method

Apply the improved Euler method to the initial value problem (4) in Example 1, choosing $h = 0.2$, as before. Here (5) becomes

$$y^*_{n+1} = y_n + 0.2(x_n + y_n)$$

$$y_{n+1} = y_n + 0.1[(x_n + y_n) + (x_{n+1} + y^*_{n+1})]$$

and is so simple that it pays to insert the first formula into the second. Simplifying the result, we readily obtain

$$y_{n+1} = 0.12x_n + 0.1x_{n+1} + 1.22y_n.$$

From Table 19.8 on the next page we see that the present results are more accurate than those in Example 1; cf. also Table 19.10 on p. 798. ∎

The improved Euler method is a **second-order method,** *because the truncation error per step is of order* h^3.

In fact, setting $\tilde{f}_n = f(x_n, y(x_n))$ and using (2), we have

(6a) $$y(x_n + h) - y(x_n) = h\tilde{f}_n + \tfrac{1}{2}h^2\tilde{f}_n' + \tfrac{1}{6}h^3\tilde{f}_n'' + \cdots .$$

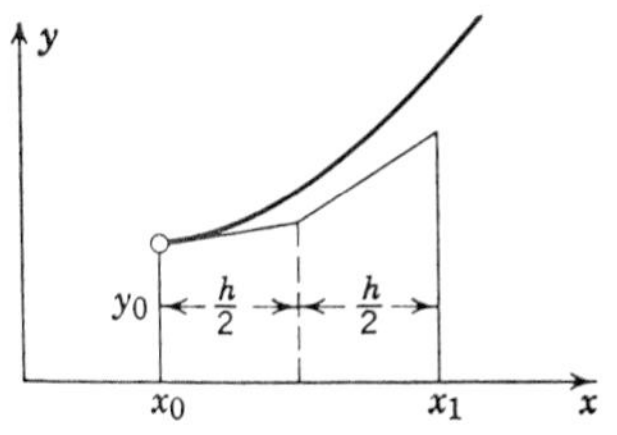

Fig. 356. Improved Euler method

Table 19.8
Improved Euler Method Applied to (4)

n	x_n	y_n	$0.12x_n$	$0.1x_{n+1}$	$1.22y_n$	y_{n+1}
0	0.0	0.0000	0.0000	0.0200	0.0000	0.0200
1	0.2	0.0200	0.0240	0.0400	0.0244	0.0884
2	0.4	0.0884	0.0480	0.0600	0.1078	0.2158
3	0.6	0.2158	0.0720	0.0800	0.2633	0.4153
4	0.8	0.4153	0.0960	0.1000	0.5067	0.7027
5	1.0	0.7027				

Approximating the expression in the brackets in (5b) by $\tilde{f}_n + \tilde{f}_{n+1}$ and using again the Taylor expansion, we obtain from (5b)

$$\text{(6b)} \qquad y_{n+1} - y_n \approx \tfrac{1}{2}h(\tilde{f}_n + \tilde{f}_n + h\tilde{f}_n{}' + \tfrac{1}{2}h^2\tilde{f}_n{}'' + \cdots).$$

Subtraction of (6a) from (6b) gives the truncation error per step

$$\frac{h^3}{4}\tilde{f}_n{}'' - \frac{h^3}{6}\tilde{f}_n{}'' + \cdots = \frac{h^3}{12}\tilde{f}_n{}'' + \cdots .$$

We shall now discuss the **choice of the step** h, which is an important problem in applying a step-by-step method. h should not be too small, because otherwise the number of steps and the rounding errors become large. On the other hand, h should not be too large, because a large h implies a large truncation error per step and, in addition to it, an error caused by the fact that f is evaluated at (x_n, y_n) instead of $(x_n, y(x_n))$. The latter would be zero if f were independent of y, and it will matter the more the faster f varies as y varies, that is, the larger the absolute value of the partial derivative $f_y = \partial f/\partial y$ is. More precisely, denoting that error by φ_n and applying the mean value theorem, we have

$$\varphi_n = f(x_n, y_n) - f(x_n, y(x_n)) = f_y(x_n, \tilde{y})\eta_n$$

where $\eta_n = y_n - y(x_n)$ is the error of y_n and $\tilde{y}$ lies between y_n and $y(x_n)$. Hence the contribution of φ_n to the error of y_{n+1} is approximately $h\varphi_n = hf_y(x_n, \tilde{y})\eta_n$. This suggests to take a close upper bound K of $|f_y|$ in the region of interest and to choose h such that

$$\kappa = hK$$

is not too large. We see that if $|f_y|$ is large (strong dependence of f on y), then K is large and h must be small, which is understandable. (In Examples 1 and 2, $f_y = 1$, $K = 1$, $hK = 0.2$.) If f_y varies very much, we may choose a close upper bound K_n of $|f_y(x_n, \tilde{y})|$ and choose two or even three different values of h in different regions, to keep

$$\kappa_n = hK_n$$

within a certain interval (e.g., $0.1 \leqq \kappa_n \leqq 0.2$), which depends on the desired accuracy; of course, because of the truncation error per step, we cannot let h increase beyond a certain value.

A still more accurate method is the **Runge-Kutta method,**[4] which is of great practical importance. In each step of this method, we first compute four auxiliary quantities

$$\text{(7a)}\qquad \begin{aligned} A_n &= hf(x_n, y_n), & B_n &= hf(x_n + \tfrac{1}{2}h, y_n + \tfrac{1}{2}A_n), \\ C_n &= hf(x_n + \tfrac{1}{2}h, y_n + \tfrac{1}{2}B_n), & D_n &= hf(x_{n+1}, y_n + C_n), \end{aligned}$$

and then the new value

$$\text{(7b)}\qquad y_{n+1} = y_n + \tfrac{1}{6}(A_n + 2B_n + 2C_n + D_n).$$

It can be shown that the truncation error per step is of the order h^5 (cf. Ref. [G20], Appendix 1) and the method is, therefore, a fourth-order method. Note that if f depends only on x, this method reduces to Simpson's rule of integration (Sec. 19.6).

In desk calculations, the frequent calculation of $f(x, y)$ is laborious. In automatic computation this does not matter too much, and the method is well suited because it needs no special starting procedure, makes light demands on the store, requires no estimation, and uses the same straightforward computational procedure several times.

Example 3. Runge–Kutta method

Apply the Runge–Kutta method to the initial value problem (4) in Example 1, choosing $h = 0.2$, as before and computing five steps. Here $f(x, y) = x + y$, and (7a) becomes

$$\begin{aligned} A_n &= 0.2(x_n + y_n), & B_n &= 0.2(x_n + 0.1 + y_n + 0.5A_n), \\ C_n &= 0.2(x_n + 0.1 + y_n + 0.5B_n), & D_n &= 0.2(x_n + 0.2 + y_n + C_n). \end{aligned}$$

Since these expressions are so simple, we find it convenient to insert A_n into B_n, obtaining $B_n = 0.22\,(x_n + y_n) + 0.02$, insert this into C_n, finding $C_n = 0.222\,(x_n + y_n) + 0.022$, and finally insert this into D_n, finding $D_n = 0.2444(x_n + y_n) + 0.0444$. If we use these expressions, (7b) becomes

$$\text{(8)}\qquad y_{n+1} = y_n + 0.2214(x_n + y_n) + 0.0214.$$

Table 19.9 shows the corresponding computations, and from Table 19.10 we see that the values are much more accurate than those in Examples 1 and 2. ∎

Table 19.9
Runge-Kutta Method Applied to (4); Computations by the Use of (8)

n	x_n	y_n	$x_n + y_n$	$0.2214(x_n + y_n)$	$0.2214(x_n + y_n) + 0.0214$
0	0.0	0	0	0	0.021 400
1	0.2	0.021 400	0.221 400	0.049 018	0.070 418
2	0.4	0.091 818	0.491 818	0.108 889	0.130 289
3	0.6	0.222 107	0.822 107	0.182 014	0.203 414
4	0.8	0.425 521	1.225 521	0.271 330	0.292 730
5	1.0	0.718 251			

[4] CARL RUNGE (1856–1927) and WILHELM KUTTA (1867–1944), German mathematicians.

Table 19.10
Comparison of the Accuracy of the Three Methods under Consideration in the Case of the Initial Value Problem (4), with $h = 0.2$

x	$y = e^x - x - 1$	Absolute Value of Error		
		Euler Method (Table 19.7)	Improved Euler Method (Table 19.8)	Runge-Kutta Method (Table 19.9)
0.2	0.021 403	0.021	0.0014	0.000 003
0.4	0.091 825	0.052	0.0034	0.000 007
0.6	0.222 119	0.094	0.0063	0.000 011
0.8	0.425 541	0.152	0.0102	0.000 020
1.0	0.718 282	0.229	0.0156	0.000 031

The **step length** h should not be greater than a certain value H which depends on the accuracy and should otherwise be such that

$$\kappa = hK \qquad (K \text{ a small upper bound for } |\partial f/\partial y|)$$

lies about between 0.1 and 0.2; this is similar to the case of the improved Euler method discussed before. It is an advantage of the Runge-Kutta method that we may control h by means of A_n, B_n, C_n, because from the definition of f_y we have

$$\kappa = hK \approx h|f_y| \approx h \left| \frac{f(x, y^*) - f(x, y^{**})}{y^* - y^{**}} \right|,$$

and if we choose $x = x_n + \frac{1}{2}h$, $y^* = y_n + \frac{1}{2}B_n$, $y^{**} = y_n + \frac{1}{2}A_n$, then we have $y^* - y^{**} = (B_n - A_n)/2$ and

$$\text{(9)} \qquad \kappa \approx \kappa_n = 2\left| \frac{C_n - B_n}{B_n - A_n} \right|.$$

We may now make provision to leave h unchanged as long as, say, $0.05 \leqq \kappa_n \leqq 0.2$, to decrease h by 50% if $\kappa_n > 0.2$, and to double h if $\kappa_n < 0.05$ (if doubling is possible without increasing h beyond a suitably chosen number H, which depends on the desired accuracy).

Another control of h results from performing the computation simultaneously with step $2h$, which corresponds to increasing the truncation error per step by a factor $2^5 = 32$, but since the number of steps decreases, the actual increase is by a factor $2^5/2 = 16$. Hence the error of the first computation (with step h) equals about $\frac{1}{15}$ times the difference δ of corresponding values of y. We may now choose a number ϵ (for example, 1 unit of the last digit that is supposed to be significant) and leave h unchanged if $0.2\epsilon \leqq |\delta| \leqq 10\epsilon$, decrease h by 50% if $|\delta| > 10\epsilon$, and double h if $|\delta| < 0.2\epsilon$; of course, in doubling we must care that the step does not become larger than a suitable number H; this is as before.

Problems for Sec. 19.7

Apply the Euler method to the following differential equations. (Carry out ten steps.)

1. $y' = y$, $y(0) = 1$, $h = 0.1$ **2.** $y' = y$, $y(0) = 1$, $h = 0.01$

3. $y' = 2xy + 1$, $y(0) = 0$, $h = 0.1$ **4.** $y' = 2xy$, $y(0) = 1$, $h = 0.1$

5. Apply the improved Euler method to the initial value problem $y' = 2x, y(0) = 0$, choosing $h = 0.1$. Why are the errors zero?

6. Give some examples such that the improved Euler–Cauchy method yields the exact solution.

7. Repeat the computations in Example 1, choosing $h = 0.1$ (instead of $h = 0.2$), and note that the errors of the values thus obtained are about 50% of those in Example 1 (which are listed in Table 19.10).

8. Same task as in Prob. 7, but with $h = 0.01$ (twenty steps). Compare the error of the value corresponding to $x = 0.2$ with that in Example 1.

9. Repeat the computations in Example 2, choosing $h = 0.1$, and note that the errors of the values thus obtained are about 25% of those in Example 2.

10. Same task as in Prob. 9, but with $h = 0.05$ (eight steps). Compare the error of the value corresponding to $x = 0.4$ with those in Example 2 and Prob. 9.

11. Apply the Euler method to the initial value problem $y' = 1 + y^2$, $y(0) = 0$, choosing $h = 0.1$. Compute five steps and compare the results with the exact values (cf. Table A1 in Appendix 4).

12. Apply the improved Euler method to the initial value problem in Prob. 11, choosing $h = 0.1$. Compute two steps and compare the errors with those of the corresponding values in Prob. 11.

13. Apply the Runge–Kutta method to the initial value problem in Prob. 11, choosing $h = 0.1$. Compute two steps. Compare the results with the exact values 0.100 334 672 and 0.202 710 036.

14. Compute $y = e^x$ for $x = 0, 0.1, 0.2, \cdots, 1.0$ by applying the Runge–Kutta method to the initial value problem in Prob. 1. Show that the first five decimal places of the result are correct (cf. Table A1 in Appendix 4).

15. In Prob. 14 insert A_n into B_n, then B_n into C_n, etc., and show that the resulting formula becomes $y_{n+1} = 1.105\,170\,833 y_n$. Repeat the computations in Prob. 14, using this formula.

16. Apply the Euler method with $h = 0.1$ to $y' = -100y, y(0) = 1$. Show that it gives $y = (-9)^{10x}$, compare this and the exact solution, and comment.

17. Show that the Euler method is also obtained by integrating (1) from x_n to x_{n+1}, approximating the integrand $f[t, y(t)]$ by its value at x_n.

18. In Prob. 17, apply the trapezoidal rule to the integral, to get

$$y(x_n + h) \approx y(x_n) + \frac{h}{2}\{f[x_n, y(x_n)] + f[x_n + h, y(x_n + h)]\}.$$

Then approximate $y(x_n + h)$ by the Euler–Cauchy method. Show that the result is the improved Euler–Cauchy method.

19. Another Euler–Cauchy type method is given by

$$y_{n+1} = y_n + hf(x_n + \tfrac{1}{2}h, y_{n+1}^*),$$

where $y_{n+1}^* = y_n + \frac{1}{2}hf(x_n, y_n)$. Give a geometric motivation of the method. Apply it to (4), choosing $h = 0.2$ and calculating five steps.

20. Kutta's third-order method is defined by

$$y_{n+1} = y_n + \frac{1}{6}(A_n + 4B_n + M_n)$$

where

$$A_n = hf(x_n, y_n), \qquad B_n = hf\left(x_n + \frac{1}{2}h, y_n + \frac{1}{2}A_n\right)$$

$$M_n = hf(x_{n+1}, y_n - A_n + 2B_n).$$

Apply the method to (4) in Example 1. Choose $h = 0.2$ and perform five steps. Compare with Table 19.10.

19.8 Numerical Methods for Second-Order Differential Equations

This section includes numerical methods similar to those in the preceding section, and is independent of the other sections in this chapter. Hence it may also be discussed immediately after Chap. 2.

An **initial value problem** for a second-order differential equation consists of that equation and two conditions (*initial conditions*) referring to the same point. In this section we shall consider two numerical methods for solving initial value problems of the form

$$\text{(1)} \qquad y'' = f(x, y, y'), \qquad y(x_0) = y_0, \qquad y'(x_0) = y_0'$$

assuming f to be such that the problem has a unique solution on some interval containing x_0. The first method is simple (but inaccurate) and serves to illustrate the principle, whereas the second method is of great precision and practical importance.

In both methods we shall obtain approximate values of the solution $y(x)$ of (1) at equidistant points $x_1 = x_0 + h$, $x_2 = x_0 + 2h$, $\cdots$; these values will be denoted by $y_1, y_2, \cdots$, respectively. Similarly, approximate values of the derivative $y'(x)$ at those points will be denoted by $y_1', y_2', \cdots$, respectively.

The methods in the previous section were suggested by the Taylor expansion

$$\text{(2)} \qquad y(x + h) = y(x) + hy'(x) + \frac{h^2}{2}y''(x) + \frac{h^3}{3!}y'''(x) + \cdots,$$

which we shall now use for the same purpose, together with the expansion for the derivative

$$\text{(3)} \qquad y'(x + h) = y'(x) + hy''(x) + \frac{h^2}{2}y'''(x) + \cdots.$$

The roughest numerical method is obtained by neglecting the terms containing y''' and the further terms in (2) and (3); this yields the approximations

$$y(x+h) \approx y(x) + hy'(x) + \frac{h^2}{2}y''(x),$$

$$y'(x+h) \approx y'(x) + hy''(x).$$

In the first step of the method we compute

$$y_0'' = f(x_0, y_0, y_0')$$

from (1), then

$$y_1 = y_0 + hy_0' + \frac{h^2}{2}y_0''$$

which approximates $y(x_1) = y(x_0 + h)$, and furthermore

$$y_1' = y_0' + hy_0''$$

which will be needed in the next step. In the second step we compute

$$y_1'' = f(x_1, y_1, y_1')$$

from (1), then

$$y_2 = y_1 + hy_1' + \frac{h^2}{2}y_1''$$

which approximates $y(x_2) = y(x_0 + 2h)$, and furthermore

$$y_2' = y_1' + hy_1''.$$

In the $(n + 1)$th step we compute

$$y_n'' = f(x_n, y_n, y_n')$$

from (1), then the new value

(4a) $$y_{n+1} = y_n + hy_n' + \frac{h^2}{2}y_n''$$

which is an approximation for $y(x_{n+1})$, and furthermore

(4b) $$y_{n+1}' = y_n' + hy_n''$$

which is an approximation for $y'(x_{n+1})$ needed in the next step.

Note that, geometrically speaking, this method is an approximation of the curve of $y(x)$ by portions of parabolas.

Example 1. An application of the method defined by (4)

Apply (4) to the initial value problem

(5) $$y'' = \tfrac{1}{2}(x + y + y' + 2), \qquad y(0) = 0, \qquad y'(0) = 0,$$

choosing $h = 0.2$. Here (4) becomes

$$y_{n+1} = y_n + 0.2y_n' + 0.02y_n'' \qquad y'_{n+1} = y_n' + 0.2y_n'' \qquad \text{where} \qquad y_n'' = \tfrac{1}{2}(x_n + y_n + y_n' + 2).$$

The computations are shown in Table 19.11. The student may verify that the exact solution is

$$y = e^x - x - 1.$$

Table 19.13 shows that the errors of our approximate values are large. This is typical, because in most practical cases our present method will be too inaccurate. ∎

A much more accurate method is the **Runge-Kutta-Nyström method**[5] which is a generalization of the Runge-Kutta method in Sec. 19.7. We mention without proof that this is a *fourth-order method,* which means that in the Taylor formulas for y and y' the first terms up to and including the term that contains h^4 are given exactly.

In the general step [the $(n + 1)$th step] of the method, we first compute the auxiliary quantities

$$\textbf{(6a)} \quad \begin{aligned} A_n &= \tfrac{1}{2}hf(x_n, y_n, y_n') \\ B_n &= \tfrac{1}{2}hf(x_n + \tfrac{1}{2}h, y_n + \beta_n, y_n' + A_n) \\ C_n &= \tfrac{1}{2}hf(x_n + \tfrac{1}{2}h, y_n + \beta_n, y_n' + B_n) \\ D_n &= \tfrac{1}{2}hf(x_n + h, y_n + \delta_n, y_n' + 2C_n) \end{aligned} \qquad \begin{aligned} &\text{where} \quad \beta_n = \tfrac{1}{2}h(y_n' + \tfrac{1}{2}A_n) \\ \\ &\text{where} \quad \delta_n = h(y_n' + C_n) \end{aligned}$$

then the new value

$$\textbf{(6b)} \quad y_{n+1} = y_n + h(y_n' + K_n) \qquad \text{where} \qquad K_n = \tfrac{1}{3}(A_n + B_n + C_n)$$

which is an approximation for $y(x_{n+1})$, and furthermore,

$$\textbf{(6c)} \quad y'_{n+1} = y_n' + K_n^* \quad \text{where} \quad K_n^* = \tfrac{1}{3}(A_n + 2B_n + 2C_n + D_n)$$

which is an approximation for $y'(x_{n+1})$ needed in the next step.

Table 19.11
Computations in Example 1

n	x_n	y_n	y_n'	$0.2y_n'$	$x_n + y_n + y_n' + 2$	$0.2y_n''$	$0.02y_n''$	$0.2y_n' + 0.02y_n''$
0	0	0.0000	0.0000	0.0000	2.0000	0.2000	0.0200	0.0200
1	0.2	0.0200	0.2000	0.0400	2.4200	0.2420	0.0242	0.0642
2	0.4	0.0842	0.4420	0.0884	2.9262	0.2926	0.0293	0.1177
3	0.6	0.2019	0.7346	0.1469	3.5365	0.3537	0.0354	0.1823
4	0.8	0.3842	1.0883	0.2177	4.2725	0.4273	0.0427	0.2604
5	1.0	0.6446						

[5] E. J. NYSTRÖM, Finnish mathematician. His paper on the method appeared in *Acta Soc. Sci. fennicae,* vol. 50 (1925).

h can be controlled in a fashion similar to that described at the end of the last section, where we now take for δ the larger of δ^* and δ^{**}; here δ^* is $\frac{1}{15}$ times the difference of corresponding values of y, and δ^{**} is $\frac{1}{15}$ times the difference of corresponding values of y'.

Example 2. Runge–Kutta–Nyström method

Apply the Runge–Kutta–Nyström method to the initial value problem (5), choosing $h = 0.2$. Here $f = 0.5(x + y + y' + 2)$, and (6a) becomes

$$A_n = 0.05(x_n + y_n + y_n' + 2),$$

$$B_n = 0.05(x_n + 0.1 + y_n + \beta_n + y_n' + A_n + 2), \qquad \beta_n = 0.1(y_n' + \tfrac{1}{2}A_n),$$

$$C_n = 0.05(x_n + 0.1 + y_n + \beta_n + y_n' + B_n + 2),$$

$$D_n = 0.05(x_n + 0.2 + y_n + \delta_n + y_n' + 2C_n + 2), \qquad \delta_n = 0.2(y_n' + C_n).$$

In the present case the differential equation is simple, and so are the expressions for A_n, B_n, C_n, and D_n. Hence we may insert A_n into B_n, then B_n into C_n, and, finally, C_n into D_n. The result of this elementary calculation is

$$B_n = 0.05[1.0525(x_n + y_n) + 1.152\,5y_n' + 2.205],$$

$$C_n = 0.05[1.055\,125(x_n + y_n) + 1.160\,125y_n' + 2.215\,25],$$

$$D_n = 0.05[1.116\,063\,75(x_n + y_n) + 1.327\,613\,75y_n' + 2.443\,677\,5].$$

From this we may now determine K_n and K_n^* and insert the resulting expressions into (6b) and (6c), finding

$$\text{(7)} \qquad \begin{aligned} y_{n+1} &= y_n + a(x_n + y_n) + by_n' + c \\ y'_{n+1} &= y_n' + a^*(x_n + y_n) + b^*y_n' + c^* \end{aligned}$$

where

$$a = 0.010\,3588 \qquad b = 0.211\,0421 \qquad c = 0.021\,4008$$

$$a^* = 0.105\,5219 \qquad b^* = 0.115\,8811 \qquad c^* = 0.221\,4030$$

Table 19.12 shows the corresponding computations. The errors of the approximate values for $y(x)$ are much smaller than those in Example 1 (cf. Table 19.13). ▮

In addition to the truncation error of a method, there are **rounding errors,** and we want to warn the reader that rounding errors may affect results to a large extent. For instance, the solution of the problem $y'' = y$, $y(0) = 1$, $y'(0) = -1$ is

Table 19.12
Runge–Kutta–Nyström Method (with $h = 0.2$) Applied to the Initial Value Problem (5); Five Steps Computed by the Use of (7)

n	x_n	y_n	y_n'	$a(x_n + y_n) + by_n' + c$	$a^*(x_n + y_n) + b^*y_n' + c^*$
0	0.0	0.000 0000	0.000 0000	0.021 4008	0.221 4030
1	0.2	0.021 4008	0.221 4030	0.070 4196	0.270 4220
2	0.4	0.091 8204	0.491 8250	0.130 2913	0.330 2940
3	0.6	0.222 1117	0.822 1190	0.203 4186	0.403 4219
4	0.8	0.425 5303	1.225 5409	0.292 7365	0.492 7403
5	1.0	0.718 2668	1.718 2812		

Table 19.13
Comparison of Accuracy of the Two Methods Under Consideration in the Case of the Initial Value Problem (5), with $h = 0.2$

x	$y = e^x - x - 1$	Absolute Value of Error	
		Example 1	Table 19.12
0.2	0.021 4028	0.0014	0.000 0020
0.4	0.091 8247	0.0076	0.000 0043
0.6	0.222 1188	0.0202	0.000 0071
0.8	0.425 5409	0.0413	0.000 0106
1.0	0.718 2818	0.0737	0.000 0150

$y = e^{-x}$, but the rounding error will introduce a small multiple of the unwanted solution e^x, which may eventually (after sufficiently many steps) swamp the required solution. This is known as **building-up error.** In our simple example we may avoid it by starting from known values of e^{-x} and its derivative for a large x and compute in the reverse direction, but in more complicated cases considerable experience is needed for avoiding that phenomenon.

Problems for Sec. 19.8

Apply (4) to the following initial value problems (Carry out five steps.)

1. $y'' = y, \quad y(0) = 1, \quad y'(0) = 1, \quad h = 0.1$

2. $y'' = y, \quad y(0) = 1, \quad y'(0) = -1, \quad h = 0.1$

3. $y'' = -y, \quad y(0) = 1, \quad y'(0) = 0, \quad h = 0.1$

4. $y'' = -y, \quad y(0) = 1, \quad y'(0) = 0, \quad h = 0.05$

5. $y'' = -y, \quad y(0) = 0, \quad y'(0) = 1, \quad h = 0.1$

6. Repeat the computation in Example 1, choosing $h = 0.1$, and compare the errors of the values thus obtained with those in Example 1 (listed in Table 19.13).

7. Perform the same task as in Prob. 6, choosing $h = 0.05$.

8. Apply the Runge–Kutta–Nyström method to the initial value problem in Prob. 5. Choose $h = 0.2$, carry out five steps, and compare the results with the exact values (cf. Table A1 in Appendix 4).

9. In Prob. 8 replace $h = 0.2$ by $h = 0.1$, carry out four steps, and compare the results with the corresponding values in Prob. 8 and with the values (exact to 9D)

0.099 833 417, 0.198 669 331, 0.295 520 207, 0.389 418 342.

10. Consider the initial value problem $(1 - x^2)y'' - 2xy' + 2y = 0$, $y(0) = 0$, $y'(0) = 1$. Show that (4) with $h = 0.1$ takes the form

$$y_{n+1} = y_n + 0.1y_n' + 0.01\frac{x_n y_n' - y_n}{1 - x_n^2}, \qquad y'_{n+1} = y_n' + 0.2\frac{x_n y_n' - y_n}{1 - x_n^2}.$$

Carry out five steps. Verify by substitution that the exact solution is $y = x$.

Apply (4) with $h = 0.1$ to the given initial value problem. Carry out five steps. Verify the given exact solution.

11. $y'' = xy' - 3y$, $y(0) = 0$, $y'(0) = -3$. Exact solution $y = x^3 - 3x$.

12. $y'' = xy' - 4y$, $y(0) = 3$, $y'(0) = 0$. Exact solution $y = x^4 - 6x^2 + 3$.

13. $(1 - x^2)y'' - 2xy' + 6y = 0$, $y(0) = -\frac{1}{2}$, $y'(0) = 0$. Exact: $y = \frac{1}{2}(3x^2 - 1)$.

14. Numerical methods for boundary value problems will not be considered in any generality, but the reader may show the following. To solve

$$y'' + f(x)y' + g(x)y = r(x), \qquad y(0) = 0, \quad y(b) = k$$

numerically, apply one of the previous methods to get a solution $Y(x)$ of the equation satisfying the initial conditions $Y(0) = 0$, $Y'(0) = 1$. Then determine by that method a solution $z(x)$ of the homogeneous equation satisfying the initial conditions $z(0) = 0$, $z'(0) = 1$. Show that

$$y(x) = Y(x) + cz(x)$$

with a suitable c satisfies the given problem. What is the condition for c?

15. To solve the boundary value problem

$$y'' + f(x)y' + g(x)y = r(x), \qquad y(a) = k_1, \quad y(b) = k_2$$

numerically, show that one may determine a solution $y_1(x)$ satisfying $y_1(a) = k_1$, $y_1'(a) = c_1$ and a solution $y_2(x)$ satisfying $y_2(a) = k_1$, $y_2'(a) = c_2$ and such that $y_1(b) \neq y_2(b)$ and then obtain the solution of the given problem in the form

$$y(x) = \frac{1}{y_1(b) - y_2(b)}[(k_2 - y_2(b))y_1(x) + (y_1(b) - k_2)y_2(x)].$$

19.9 Systems of Linear Equations. Gauss Elimination

A **system of *m* linear equations** (or *set of m simultaneous linear equations*) *in n unknowns* $x_1, \cdots, x_n$ is a set of equations of the form

$$\begin{aligned} a_{11}x_1 + \cdots + a_{1n}x_n &= b_1 \\ a_{21}x_1 + \cdots + a_{2n}x_n &= b_2 \\ \cdots\cdots\cdots\cdots&\cdots\cdots \\ a_{m1}x_1 + \cdots + a_{mn}x_n &= b_m \end{aligned} \tag{1}$$

where the **coefficients** a_{jk} and the b_j are given numbers. The system is said to be **homogeneous** if all the b_j are zero; otherwise it is said to be **nonhomogeneous.** Readers familiar with matrix multiplication (Sec. 7.4) will see that (1) may be written as a single vector equation

$$\mathbf{Ax} = \mathbf{b} \tag{2}$$

where the **coefficient matrix** $\mathbf{A} = (a_{ik})$ is the $(m \times n)$ matrix

$$\mathbf{A} = \begin{pmatrix} a_{11} & a_{12} & \cdots & a_{1n} \\ a_{21} & a_{22} & \cdots & a_{2n} \\ \cdot & \cdot & \cdots & \cdot \\ a_{m1} & a_{m2} & \cdots & a_{mn} \end{pmatrix}, \quad \text{and} \quad \mathbf{x} = \begin{pmatrix} x_1 \\ \cdot \\ \cdot \\ \cdot \\ x_n \end{pmatrix} \quad \text{and} \quad \mathbf{b} = \begin{pmatrix} b_1 \\ \cdot \\ \cdot \\ \cdot \\ b_m \end{pmatrix}$$

are column vectors. A **solution** of (1) is a set of numbers $x_1, \cdots, x_n$ which satisfy all the m equations, and a **solution vector** of (1) is a vector $\mathbf{x}$ whose components constitute a solution of (1).

The method of solving such a system by determinants (Cramer's rule in Sec. 7.11) is impracticable for large systems, even with efficient methods for evaluating the determinants.

A practical method for the solution of a system of linear equations is the **Gauss elimination.** It suffices to explain it in terms of an example.

Example 1. Gauss elimination

Solve the system

	Check sum
$2w + x + 2y + z = 6$	12
$6w - 6x + 6y + 12z = 36$	54
$4w + 3x + 3y - 3z = -1$	6
$2w + 2x - y + z = 10$	14

First step. We subtract multiples of the first equation from the other three so as to eliminate w in each of these, giving

	Check sum
$-9x + 9z = 18$	18
$x - y - 5z = -13$	-18
$x - 3y = 4$	2

Second step. We subtract multiples of the first of these equations from the other two so as to eliminate x; we obtain

	Check sum
$-y - 4z = -11$	-16
$-3y + z = 6$	4

Third step. We subtract a multiple of the first of these equations from the other so as to eliminate y:

	Check sum
$13z = 39$	52

Final step. By back-substituting we now obtain

$13z = 39$	$z = 3$
$-y - 4\cdot 3 = -11$	$y = -1$
$-9x + 9\cdot 3 = 18$	$x = 1$
$2w + 1 + 2\cdot(-1) + 3 = 6$	$w = 2$

An example in which the number of equations differs from the number of unknowns is included in Sec. 7.5.

A useful check is provided by carrying along, as indicated in the extra column on the right, the sum of the coefficients and terms on the right, operating on these numbers as on the equations, and then checking that the derived equations have the correct sum.

In practice, the full equations at each step are not recorded, but only the matrix of coefficients and the column of terms on the right. ∎

In each step the coefficient of the first unknown in the first equation is called the **pivotal coefficient.** If a pivotal coefficient is small, we have to subtract large multiples of the corresponding equation from the others, and any uncertainty in coefficients will be amplified. This may affect the accuracy of the result. Consequently, in practice it is important that we choose pivotal coefficients which are not too small in absolute value. We can accomplish this by reordering the equations and the variables if necessary. This process is called **pivoting** or **positioning for size.**

Very small pivots can often be avoided by **scaling,** that is, by multiplying equations by suitable factors or changing some of the variables by setting $x_j^* = c_j x_j$ where c_j is a suitable number.

The **inverse** of a nonsingular square matrix **A** may now be determined by solving the n systems

$$\mathbf{A}\mathbf{x} = \mathbf{b}_j \qquad (j = 1, \cdots, n), \tag{3}$$

where $\mathbf{b}_j$ is the jth column of the n-rowed unit matrix. (An iterative method for determining the inverse is included in the next section.)

There are various modifications of the Gauss method. We mention a method based on a result by **Cholesky,** who has shown that if a square matrix **A** and all its leading square submatrices are nonsingular, then **A** can be represented in the form

$$\mathbf{A} = \mathbf{L}\mathbf{U} \tag{4}$$

where **L** and **U** are lower and upper triangular matrices. **L** and **U** are essentially unique, and if we specify the diagonal elements of **L** (or of **U**), they are unique. The point is that **L** and **U** can be obtained without solving simultaneous equations (example below). To solve a system $\mathbf{A}\mathbf{x} = \mathbf{b}$ of n equations in n unknowns, we may now insert (4), having

$$\mathbf{L}\mathbf{U}\mathbf{x} = \mathbf{b},$$

and premultiply by $\mathbf{L}^{-1}$. This gives

$$\mathbf{U}\mathbf{x} = \mathbf{z} \qquad \text{where} \qquad \mathbf{z} = \mathbf{L}^{-1}\mathbf{b} \tag{5}$$

and we see that this is the triangular form of our system. We first determine **z** from [cf. (5)]

$$\mathbf{L}\mathbf{z} = \mathbf{b} \tag{6}$$

and then we solve

$$\mathbf{Ux} = \mathbf{z} \tag{7}$$

for $\mathbf{x}$.

Example 2. Cholesky's method

The reader may verify that the system

$$\begin{aligned} x + 2y + 3z &= 14 \\ 2x + 3y + 4z &= 20 \\ 3x + 4y + z &= 14 \end{aligned}$$

has the solution $x = 1, y = 2, z = 3$, and we shall see how to get this result by Cholesky's method. The coefficient matrix is symmetric. This implies $\mathbf{L} = \mathbf{U}^{\mathsf{T}}$ and we may determine the elements of $\mathbf{U}$ from

$$\begin{pmatrix} 1 & 2 & 3 \\ 2 & 3 & 4 \\ 3 & 4 & 1 \end{pmatrix} = \begin{pmatrix} a_{11} & 0 & 0 \\ a_{12} & a_{22} & 0 \\ a_{13} & a_{23} & a_{33} \end{pmatrix} \begin{pmatrix} a_{11} & a_{12} & a_{13} \\ 0 & a_{22} & a_{23} \\ 0 & 0 & a_{33} \end{pmatrix}$$

by using the definition of matrix multiplication and equating corresponding elements. We obtain successively $a_{11}{}^2 = 1$, say, $a_{11} = 1$, from this $a_{11}a_{12} = a_{12} = 2$, $a_{11}a_{13} = a_{13} = 3$, $a_{12}{}^2 + a_{22}{}^2 = 4 + a_{22}{}^2 = 3$, say, $a_{22} = i\ (= \sqrt{-1})$, from this

$$a_{12}a_{13} + a_{22}a_{23} = 6 + ia_{23} = 4, \qquad a_{23} = 2i$$

and finally

$$a_{13}{}^2 + a_{23}{}^2 + a_{33}{}^2 = 9 - 4 + a_{33}{}^2 = 1, \qquad \text{say, } a_{33} = 2i.$$

Hence (6) has the form

$$\begin{pmatrix} 1 & 0 & 0 \\ 2 & i & 0 \\ 3 & 2i & 2i \end{pmatrix} \begin{pmatrix} z_1 \\ z_2 \\ z_3 \end{pmatrix} = \begin{pmatrix} 14 \\ 20 \\ 14 \end{pmatrix}, \qquad \text{and} \qquad \begin{pmatrix} z_1 \\ z_2 \\ z_3 \end{pmatrix} = \begin{pmatrix} 14 \\ 8i \\ 6i \end{pmatrix}.$$

We now solve

$$\begin{pmatrix} 1 & 2 & 3 \\ 0 & i & 2i \\ 0 & 0 & 2i \end{pmatrix} \begin{pmatrix} x_1 \\ x_2 \\ x_3 \end{pmatrix} = \begin{pmatrix} 14 \\ 8i \\ 6i \end{pmatrix}, \qquad \text{finding} \qquad \begin{pmatrix} x_1 \\ x_2 \\ x_3 \end{pmatrix} = \begin{pmatrix} 1 \\ 2 \\ 3 \end{pmatrix}.$$ ∎

Another variant of the Gauss elimination is the *Gauss–Jordan elimination,* introduced by W. Jordan in 1920, in which resubstitution is avoided by additional calculations which reduce the matrix to "diagonal form," instead of the "triangular form" in the Gauss elimination. Cf. Prob. 14 (below). Because of those additional calculations, the method seems to have no advantage in connection with automatic computing equipment.

Error estimates for the Gauss elimination are discussed in Ref. [G9] listed in Appendix 1.

Problems for Sec. 19.9

Using the Gauss elimination, solve the following systems of linear equations.

1. $5x - 2y = 1$, $6x + 8y = 22$

2. $3x + y = -5$, $2x + 3y = 6$

3. $x - 2y = -8$, $5x + 3y = -1$

4. $x + 2y - 8z = 0$, $2x - 3y + 5z = 0$, $3x + 2y - 12z = 0$

5. $7x - y - 2z = 0$, $9x - y - 3z = 0$, $2x + 4y - 7z = 0$

6. $3x - y + z = -2$, $x + 5y + 2z = 6$, $2x + 3y + z = 0$

7. $3x - 3y - 7z = -4$, $x - y + 2z = 3$

8. $7x - 4y - 2z = -6$, $16x + 2y + z = 3$

9. $x - 3y + 2z = 2$, $5x - 15y + 7z = 10$

10. $3w - 6x - y - z = 0$, $w - 2x + 5y - 3z = 0$, $2w - 4x + 3y - z = 3$

11. $4w + 3x - 9y + z = 1$, $-w + 2x - 13y + 3z = 3$, $3w - x + 8y - 2z = -2$

12. Show that in the case of n equations in n unknowns the number of multiplications and divisions in the Gauss elimination is of the order of $n^3/3$. (To characterize the efficiency of a method, one conventionally counts only multiplications and divisions because the first modern digital computers did additions and subtractions much faster than multiplications and divisions. This is changing at present, but if additions and subtractions are about as numerous as multiplications and divisions, the convention of counting only the latter still makes sense.)

13. In desk calculation, one may record only coefficients and check sums and the operation performed. Show that for the first step in Example 1 we may thus write

2	1	2	1	6	12	Row 1
0	−9	0	9	18	18	Row 2—3 Row 1
0	1	−1	−5	−13	−18	Row 3—2 Row 1
0	1	−3	0	4	2	Row 4—Row 1

Do this for all steps in Prob. 5.

14. (Gauss–Jordan elimination) In Example 1, the Gauss elimination gave

$$\begin{aligned} &\text{(a)} \quad 2w + x + 2y + z = 6 \\ &\text{(b)} \quad -9x + 9z = 18 \\ &\text{(c)} \quad -y - 4z = -11 \\ &\text{(d)} \quad 13z = 39 \end{aligned}$$

In the *Gauss–Jordan elimination,* go on as follows. Use (b) to eliminate x from (a). Then use (c) to eliminate y from (b) (which is not necessary in the present case) and (a). Then use (d) to eliminate z from (a), (b), (c). Show that this yields

$$\begin{aligned} 2w &= 4 \\ -9x &= -9 \\ -y &= 1 \\ 13z &= 39. \end{aligned}$$

Finally, solve each equation for the unknown, to get $w = 2$, $x = 1$, $y = -1$, $z = 3$.

15. Apply the Gauss–Jordan elimination to Prob. 6.

19.10 Systems of Linear Equations. Solution by Iteration

The Gauss elimination in the previous section belongs to the **direct methods** for solving systems of linear equations; these are methods that yield solutions after an amount of computation that can be specified in advance. In contrast, an **indirect** or **iterative method** is one in which we start from an approximation to the true solution and, if successful, obtain better and better approximations from a computational cycle repeated as often as may be necessary for achieving a required accuracy, so that the amount of arithmetic depends upon the accuracy required.

Iterative methods are used mainly in those problems for which convergence is known to be rapid, so that the solution is obtained with much less work than that of a direct method, and for systems of large order but with many zero coefficients, the so-called *"sparse" systems,* for which elimination methods would be relatively laborious and need much storage. Such systems occur in connection with vibrational problems, partial differential equations, and other applications and have given rise to a rapidly growing literature.

An iteration process of practical importance is the **Gauss-Seidel iteration,** which we shall explain in terms of an example. Let us consider the system

$$
\begin{aligned}
w - 0.25x - 0.25y \qquad\qquad &= 0.50 \\
-0.25w + x \qquad\qquad - 0.25z &= 0.50 \\
-0.25w \qquad\quad + y - 0.25z &= 0.25 \\
- 0.25x - 0.25y + z &= 0.25
\end{aligned} \tag{1}
$$

(Equations of this form arise in the numerical solution of partial differential equations.) We write the system in the form

$$
\begin{aligned}
w &= \qquad\quad 0.25x + 0.25y \qquad\quad + 0.50 \\
x &= 0.25w \qquad\qquad\qquad + 0.25z + 0.50 \\
y &= 0.25w \qquad\qquad\qquad + 0.25z + 0.25 \\
z &= \qquad\quad 0.25x + 0.25y \qquad\quad + 0.25
\end{aligned} \tag{2}
$$

and use these equations for the iteration; that is, we start from an approximation to the solution, say, $w_0 = 1$, $x_0 = 1$, $y_0 = 1$, $z_0 = 1$, and compute a presumably better approximation

$$
\begin{aligned}
w_1 &= \qquad\quad 0.25x_0 + 0.25y_0 \qquad\quad + 0.50 = 1.0000 \\
x_1 &= 0.25w_1 \qquad\qquad\qquad + 0.25z_0 + 0.50 = 1.0000 \\
y_1 &= 0.25w_1 \qquad\qquad\qquad + 0.25z_0 + 0.25 = 0.7500 \\
z_1 &= \qquad\quad 0.25x_1 + 0.25y_1 \qquad\quad + 0.25 = 0.6875
\end{aligned} \tag{3}
$$

We see that these equations are obtained from (2) by substituting on the right the most recent approximations. In fact, corresponding elements replace previous ones as soon as they have been computed, so that in the second and third equations we use w_1 (not w_0), and in the last equation of (3) we use x_1 and y_1 (not x_0 and y_0). The next step yields

$$\begin{aligned}
w_2 &= \qquad\qquad 0.25x_1 + 0.25y_1 \qquad\qquad + 0.50 = 0.9375\\
x_2 &= 0.25w_2 \qquad\qquad\qquad + 0.25z_1 + 0.50 = 0.9062\\
y_2 &= 0.25w_2 \qquad\qquad\qquad + 0.25z_1 + 0.25 = 0.6562\\
z_2 &= \qquad\qquad 0.25x_2 + 0.25y_2 \qquad\qquad + 0.25 = 0.6406
\end{aligned}$$

The reader may show that the exact solution is $w = x = 0.875$, $y = z = 0.625$.

We mention without proof that the Gauss–Seidel iteration converges for any choice of the initial approximation if and only if all the eigenvalues (Sec. 7.13) of the "iteration matrix" $\mathbf{C}$ [below in (5)] have absolute value less than 1, and the rate of convergence depends on the **spectral radius** (= maximum of those absolute values). The matrix $\mathbf{C}$ is obtained as follows. Let

$$\mathbf{Ax} = \mathbf{b}$$

be the given system of n equations, where $\mathbf{x}$ is a column vector with components $x_1, \cdots, x_n$, the n unknowns. Let $\mathbf{x}_{(0)}, \mathbf{x}_{(1)}, \cdots$ be the iterative sequence of the successive Gauss–Seidel approximations corresponding to an initial approximation $\mathbf{x}_{(0)}$. The method is said to *converge* for an $\mathbf{x}_{(0)}$ if the corresponding iterative sequence converges to a solution of the given system.

We assume that $a_{jj} = 1$ for $j = 1, \cdots, n$. (Note that this can be achieved if we can rearrange the equations so that no diagonal coefficient is zero; then we may divide each equation by the corresponding diagonal coefficient.) Now we may write $\mathbf{A} = \mathbf{I} + \tilde{\mathbf{L}} + \tilde{\mathbf{U}}$, where $\mathbf{I}$ is the n-rowed unit matrix and $\tilde{\mathbf{L}}$ and $\tilde{\mathbf{U}}$ are respectively lower and upper triangular matrices with null principal diagonals. Substituting this form of $\mathbf{A}$ into $\mathbf{Ax} = \mathbf{b}$, we have $(\mathbf{I} + \tilde{\mathbf{L}} + \tilde{\mathbf{U}})\mathbf{x} = \mathbf{b}$. It is customary to set $\tilde{\mathbf{L}} = -\mathbf{L}$ and $\tilde{\mathbf{U}} = -\mathbf{U}$; then

$$(\mathbf{I} - \mathbf{L} - \mathbf{U})\mathbf{x} = \mathbf{b}, \qquad \text{thus} \qquad (\mathbf{I} - \mathbf{L})\mathbf{x} = \mathbf{b} + \mathbf{Ux}.$$

From this we obtain the Gauss–Seidel formula

$$(4) \qquad (\mathbf{I} - \mathbf{L})\mathbf{x}_{(m+1)} = \mathbf{b} + \mathbf{Ux}_{(m)} \qquad (m = 0, 1, \cdots).$$

In fact, $\mathbf{U}$ is upper triangular and its nonzero elements correspond to those positions at which we have still to use "old" approximations because corresponding "new" ones are not yet available. In contrast, $\mathbf{L}$ is lower triangular and its nonzero elements correspond to those positions at which elements of the "new" approximation $\mathbf{x}_{(m+1)}$ are already available. Solving (4) for $\mathbf{x}_{(m+1)}$, we have

$$(5) \qquad \mathbf{x}_{(m+1)} = (\mathbf{I} - \mathbf{L})^{-1}\mathbf{b} + \mathbf{Cx}_{(m)} \qquad \text{where} \qquad \mathbf{C} = (\mathbf{I} - \mathbf{L})^{-1}\mathbf{U},$$

the aforementioned matrix **C** whose eigenvalues are essential in connection with convergence.

The Gauss–Seidel iteration is a method of **successive corrections** because we replace approximations by corresponding new ones as soon as the latter have been computed. A method is called a method of **simultaneous corrections** if no element of an approximation $\mathbf{x}_{(m+1)}$ is used until all the elements of $\mathbf{x}_{(m+1)}$ have been calculated. A method of this type is the **Jacobi iteration,** which is similar to the Gauss–Seidel iteration, but consists in *not* using improved values until a step has been completed and then replacing $\mathbf{x}_{(m)}$ by $\mathbf{x}_{(m+1)}$ entirely for the next cycle. Hence, if we write $\mathbf{Ax} = \mathbf{b}$ in the form $\mathbf{x} = \mathbf{b} + (\mathbf{I} - \mathbf{A})\mathbf{x}$, the Jacobi iteration in matrix notation is

$$\mathbf{x}_{(m+1)} = \mathbf{b} + (\mathbf{I} - \mathbf{A})\mathbf{x}_{(m)}. \tag{6}$$

This method is largely of theoretical interest. It converges for every choice of $\mathbf{x}_{(0)}$ if and only if the spectral radius of $\mathbf{I} - \mathbf{A}$ is less than 1; here we again assume $a_{jj} = 1$ for $j = 1, \cdots, n$.

Given a system $\mathbf{Ax} = \mathbf{b}$, we may set

$$\mathbf{r} = \mathbf{Ax} - \mathbf{b}$$

and call **r** the **residual.** Clearly, $\mathbf{r} = \mathbf{0}$ if and only if **x** is a solution. Hence $\mathbf{r} \neq \mathbf{0}$ for an approximate solution. In the Gauss–Seidel iteration, at each stage we modify or *relax* a component of an approximate solution in order to reduce a component of **r** to zero. Hence the Gauss–Seidel iteration belongs to a class of methods which are often called **relaxation methods.**

The **inverse** of a nonsingular square matrix **A** may also be determined by an iteration method suggested by the following idea. The reciprocal x of a given number a satisfies $xa = 1$, and if we want to determine x without division, we may apply Newton's method to the function $f(x) = x^{-1} - a$. Since we have $f'(x) = -1/x^2$, the Newton iteration is

$$x_{m+1} = x_m - ({x_m}^{-1} - a)(-{x_m}^2) = x_m(2 - ax_m).$$

This suggests an analogous iteration formula for determining the inverse $\mathbf{X} = \mathbf{A}^{-1}$ of **A**, namely,

$$\mathbf{X}_{(m+1)} = \mathbf{X}_{(m)}(2\mathbf{I} - \mathbf{AX}_{(m)}). \tag{7}$$

The process converges (produces $\mathbf{A}^{-1}$ as $m \to \infty$) if and only if an $\mathbf{X}_{(0)}$ is chosen so that each eigenvalue of $\mathbf{I} - \mathbf{AX}_0$ is of absolute value less than 1 (cf. Ref. [G17] listed in Appendix 1). The method is suitable if the occurring multiplications are simple (e.g., many zeros in **A**). In practice a suitable choice of $\mathbf{A}_{(0)}$ is difficult, if not impossible, and the method is mostly used for improving an inaccurate inverse obtained by another method.

Problems for Sec. 19.10

1. Verify the solution of (1) given in the text. Solve (1) by the Gauss elimination.
2. Compute two more steps of the Gauss–Seidel iteration in the text.
3. Find **C** [cf. (5)] for the system (1).
4. Apply the Jacobi iteration to the system (1); start from $w_0 = 1$, $x_0 = 1$, $y_0 = 1$, $z_0 = 1$, perform two steps, and compare the accuracy of the result with that in the text.

Apply the Gauss–Seidel iteration to the following systems. Perform three steps, starting from 1, 1, 1.

5. $\begin{aligned} 10x - y - z &= 13 \\ x + 10y + z &= 36 \\ -x - y + 10z &= 35 \end{aligned}$

6. $\begin{aligned} 4x + y \quad &= -8 \\ 4y + z &= 2 \\ x \quad + 2z &= 2 \end{aligned}$

7. $\begin{aligned} 4x + 2y + z &= 14 \\ x + 5y - z &= 10 \\ x + y + 8z &= 20 \end{aligned}$

8. Starting from 0, 0, 0, show that for the following system the Gauss–Seidel iteration converges, whereas the Jacobi iteration diverges.

$$\begin{aligned} 2x + y + z &= 4 \\ x + 2y + z &= 4 \\ x + y + 2z &= 4 \end{aligned}$$

9. It is plausible to think that the Gauss–Seidel iteration is better than the Jacobi iteration. Actually the methods are not comparable. Illustrate this surprising fact by showing that for the following system the Jacobi iteration converges whereas the Gauss–Seidel iteration diverges. (*Hint.* Use eigenvalues).

$$\begin{aligned} x \quad + z &= 2 \\ -x + y \quad &= 0 \\ x + 2y - 3z &= 0 \end{aligned}$$

10. Show that for $\mathbf{X}_{(m)} = \mathbf{A}^{-1}$, formula (7) yields $\mathbf{X}_{(m+1)} = \mathbf{A}^{-1}$.
11. Consider an approximation $\mathbf{X}_{(0)}$ to the inverse of a matrix **A**, where

$$\mathbf{X}_{(0)} = \begin{pmatrix} 0.5 & -0.1 & 0.4 \\ 0 & 0.2 & 0 \\ -0.4 & 0.3 & -1.5 \end{pmatrix} \quad \text{and} \quad \mathbf{A} = \begin{pmatrix} 3 & 0 & 1 \\ 0 & 5 & 0 \\ -1 & 1 & -1 \end{pmatrix}$$

Calculate $\mathbf{X}_{(1)}$ by (7). Determine $\mathbf{A}^{-1}$ and show that each element of $\mathbf{X}_{(0)}$ deviates from the corresponding element of $\mathbf{A}^{-1}$ by at most 0.1, whereas for $\mathbf{X}_{(1)}$ that maximum deviation is 0.03.

12. Apply (7) with

$$\mathbf{X}_{(0)} = \begin{pmatrix} 2.9 & -0.9 \\ -4.9 & 1.9 \end{pmatrix} \quad \text{to the matrix} \quad \mathbf{A} = \begin{pmatrix} 2 & 1 \\ 5 & 3 \end{pmatrix},$$

verify that the condition for convergence is satisfied, perform two steps, and compare the results with the exact inverse.

19.11 Systems of Linear Equations. Ill-conditioning

A system of linear equations is said to be **well-conditioned** if small errors in the coefficients or in the solving process have only a small effect on the solution. In this case the solution is relatively strongly indicated by the equations.

A system of linear equations is said to be **ill-conditioned** if small errors in the coefficients or in the solving process have a large effect on the solution. In this case the solution is relatively weakly indicated by the equations.

For instance, two linear equations in two unknowns represent two straight lines. Such a system is ill-conditioned if and only if the angle γ between the lines is small, that is, if and only if the lines are almost parallel. In fact, then a small change in a coefficient may cause a large displacement of the point of intersection of the lines (cf. Fig. 357). For larger systems of linear equations the situation is similar in principle, but no such simple geometrical interpretation is possible, and we would not be able to follow each detail of the situation.

Example 1. An ill-conditioned system

The reader may verify that the system

$$\begin{aligned} 0.9999x - 1.0001y &= 1 \\ x - y &= 1 \end{aligned}$$

has the solution $x = 0.5$, $y = -0.5$, whereas the system

$$\begin{aligned} 0.9999x - 1.0001y &= 1 \\ x - y &= 1 + \epsilon \end{aligned}$$

has the solution $x = 0.5 + 5000.5\epsilon$, $y = -0.5 + 4999.5\epsilon$. This shows that the system is ill-conditioned because a change on the right of magnitude ϵ produces a change in the solution of magnitude 5000ϵ, approximately. ∎

Ill-conditioning may be regarded as an approach toward singularity. It manifests itself by a loss of significant figures during computation, so that it is more difficult to get an accurate inverse or solution of the associated equations. Several measures for ill-conditioning have been proposed, but the common usage of the term is still qualitative.

In the case of ill-conditioning we must use a much larger number of decimal

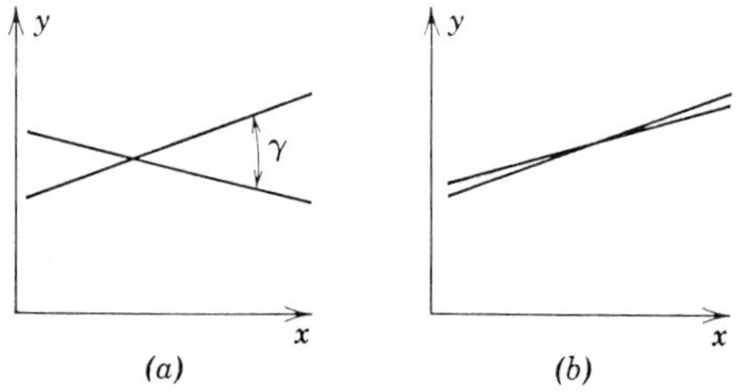

Fig. 357. (*a*) Well-conditioned and (*b*) ill-conditioned system of two linear equations in two unknowns

places in intermediate calculations necessary to determine the solution to a specific number of decimal places (when round-off error is present). If the coefficients of an ill-conditioned system as well as the numbers on the right-hand side can be computed from a formula, we can calculate them, at some small effort, to an arbitrary degree of accuracy, using single precision, double precision, triple precision, etc., so that the situation is not too serious. A much greater handicap arises in cases in which those coefficients and numbers are obtained experimentally with a necessarily limited accuracy. Then we have to accept the fact that inaccuracies in the given data permit the solution to be determined only within relatively wide error limits, and we should perhaps try to reformulate the original problem in terms of another system of equations which is reasonably well-conditioned.

Some symptoms for ill-conditioning are as follows. $|\det \mathbf{A}|$ is small compared with the size of the maximum of the $|a_{jk}|$'s and the terms on the right-hand side of the system. Poor approximations of the solution may produce merely very small residuals (see below). The elements of $\mathbf{A}^{-1}$ are large in absolute value compared with the absolute values of the elements of the solution.

If the principal diagonal elements are large in absolute value compared with the absolute values of the other elements, the system is well-conditioned. If a square matrix with largest elements between 0.1 and 10 has an inverse with largest elements also of order unity, then the associated system of linear equations is well-conditioned.

In the case of ill-conditioning we may want to obtain an improved solution of

$$\mathbf{A}\mathbf{x} = \mathbf{b} \tag{1}$$

from an approximate solution $\mathbf{x}_{(1)}$. To $\mathbf{x}_{(1)}$ there corresponds the **residual**

$$\mathbf{r}_{(1)} = \mathbf{b} - \mathbf{A}\mathbf{x}_{(1)}.$$

We have

$$\mathbf{A}\mathbf{x}_{(1)} = \mathbf{b} - \mathbf{r}_{(1)}$$

and thus

$$\mathbf{A}(\mathbf{x} - \mathbf{x}_{(1)}) = \mathbf{r}_{(1)}. \tag{2}$$

This shows that the correction to be applied to $\mathbf{x}_{(1)}$ to give the solution of (1) is the solution of (2). Unless the system is very ill-conditioned, the components of $\mathbf{r}_{(1)}$ will be smaller than those of $\mathbf{b}$. For details of corresponding automatic computation, see Ref. [G12] in Appendix 1.

Problems for Sec. 19.11

1. In Example 1, compute the determinant of the system, divided by the coefficient which has the largest absolute value. Comment. Can the determinant of an ill-conditioned system have a large value?
2. Setting $\xi = x + y + 1$, $\eta = x - y - 1$, obtain another ill-conditioned system from that in Example 1.
3. Show that the angle γ between two lines represented by a system

$$a_{11}x_1 + a_{12}x_2 = b_1, \qquad a_{21}x_1 + a_{22}x_2 = b_2$$

is given by

$$\tan \gamma = (a_{11}a_{22} - a_{12}a_{21})/(a_{11}a_{21} + a_{12}a_{22})$$

and explain consequences of this formula in connection with ill-conditioning.

4. Show that the solution of the system

$$\begin{aligned} 6x_1 + 7x_2 + 8x_3 &= 21 \\ 7x_1 + 8x_2 + 9x_3 &= 24 \\ 8x_1 + 9x_2 + 9x_3 &= 26 \end{aligned}$$

is $x_1 = 1, x_2 = 1, x_3 = 1$. Compute the determinant of the system and the residual corresponding to $x_1 = -0.8$, $x_2 = 2.9$, $x_3 = 0.7$. Comment.

5. Let

$$\mathbf{A} = \begin{pmatrix} 1.00 & 1.00 \\ 1.00 & 1.01 \end{pmatrix} \quad \text{and} \quad \mathbf{B} = \begin{pmatrix} 111 & -100 \\ -110 & 100 \end{pmatrix}.$$

Show that **AB** is almost equal to the unit matrix, whereas **BA** is not. Comment.

6. Using the Gauss elimination, show that the system

$$\begin{aligned} x + \tfrac{1}{2}y + \tfrac{1}{3}z &= 1 \\ \tfrac{1}{2}x + \tfrac{1}{3}y + \tfrac{1}{4}z &= 0 \\ \tfrac{1}{3}x + \tfrac{1}{4}y + \tfrac{1}{5}z &= 0 \end{aligned}$$

has the solution $x = 9$, $y = -36$, $z = 30$. Rework the problem, assuming that a computer capable of carrying only two significant digits at a time is to do the computations. Compare the results and comment. (The coefficient matrix is called the three-rowed *Hilbert matrix*.)

7. For the $(n \times n)$ Hilbert matrix, the magnitudes of the elements of the inverse grow very rapidly. To get an impression, verify that the (4×4) Hilbert matrix

$$\begin{pmatrix} 1 & \frac{1}{2} & \frac{1}{3} & \frac{1}{4} \\ \frac{1}{2} & \frac{1}{3} & \frac{1}{4} & \frac{1}{5} \\ \frac{1}{3} & \frac{1}{4} & \frac{1}{5} & \frac{1}{6} \\ \frac{1}{4} & \frac{1}{5} & \frac{1}{6} & \frac{1}{7} \end{pmatrix} \text{ has the inverse } \begin{pmatrix} 16 & -120 & 240 & -140 \\ -120 & 1200 & -2700 & 1680 \\ 240 & -2700 & 6480 & -4200 \\ -140 & 1680 & -4200 & 2800 \end{pmatrix}$$

The Hilbert matrix is not a purely academic curiosity, but comparable matrices occur in connection with curve fitting by least squares, as we shall see in the next section.

8. Verify that the inverse of the coefficient matrix **A** of the system

$$\begin{pmatrix} 10 & 7 & 8 & 7 \\ 7 & 5 & 6 & 5 \\ 8 & 6 & 10 & 9 \\ 7 & 5 & 9 & 10 \end{pmatrix} \begin{pmatrix} w \\ x \\ y \\ z \end{pmatrix} = \begin{pmatrix} 32 \\ 23 \\ 33 \\ 31 \end{pmatrix} \quad \text{is} \quad \begin{pmatrix} 25 & -41 & 10 & -6 \\ -41 & 68 & -17 & 10 \\ 10 & -17 & 5 & -3 \\ -6 & 10 & -3 & 2 \end{pmatrix}$$

so that the solution is $w = x = y = z = 1$. Show that $w = 9.2$, $x = -12.6$,

$y = 4.5$, $z = -1.1$ satisfies the system with an error of ± 0.1 in each equation. Conclude from this that **A** is ill-conditioned. Apply the Gauss–Seidel iteration with $\mathbf{x}_0 = \mathbf{0}$; perform two or three steps to see that the convergence is slow. (More about this system can be found in Ref. [G17], p. 242.)

19.12 Method of Least Squares

In **curve fitting** we are given n points (pairs of numbers)

$$(x_1, y_1), \quad \cdots, \quad (x_n, y_n)$$

and we want to determine a function $f(x)$ such that $f(x_j) \approx y_j, j = 1, \cdots, n$. The type of function (for example, polynomials, exponential functions, sine and cosine functions) may be suggested by the nature of the problem (the underlying physical law, for instance), and in many cases a polynomial of a certain degree will be appropriate.

If we require strict equality $f(x_1) = y_1, \cdots, f(x_n) = y_n$ and use polynomials of sufficiently high degree, we may apply one of the methods discussed in Sec. 19.4 in connection with interpolation. However, there are situations where this would not be the appropriate solution of the actual problem. For instance, to the four points

$$(1) \qquad (-1.0, 1.000), \qquad (-0.1, 1.099), \qquad (0.2, 0.808), \qquad (1.0, 1.000)$$

there corresponds the Lagrange polynomial $f(x) = x^3 - x + 1$ (Fig. 358), but if we graph the points, we see that they lie nearly on a straight line. Hence if these values are obtained in an experiment and thus involve an experimental error, and if the nature of the experiment suggests a linear relation, we better fit a straight line through the points (Fig. 358). Such a line may be useful for predicting values to be expected for other values of x. In simple cases a straight line may be fitted by eye, but if the points are scattered, this becomes unreliable and we better use a mathematical principle. A widely used procedure of this type is the **method of least squares** by Gauss. In the present situation it may be formulated as follows.

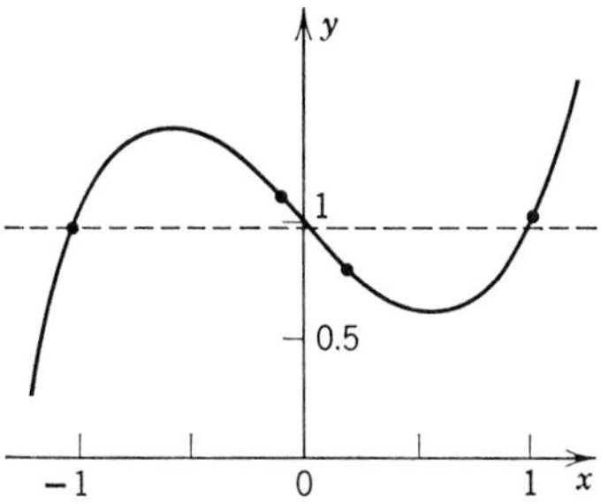

Fig. 358. The approximate fitting of a straight line

Method of least squares. *The straight line*

$$y = a + bx$$

should be fitted through the given points $(x_1, y_1), \cdots, (x_n, y_n)$ *so that the sum of the squares of the distances of those points from the straight line is minimum, where the distance is measured in the vertical direction (the y-direction).*

The point on the line with abscissa x_j has the ordinate $a + bx_j$. Hence its distance from (x_j, y_j) is $|y_j - a - bx_j|$ (cf. Fig. 359) and that sum of squares is

$$q = \sum_{j=1}^{n} (y_j - a - bx_j)^2.$$

q depends on a and b. A necessary condition for q to be minimum is

$$\frac{\partial q}{\partial a} = -2 \sum (y_j - a - bx_j) = 0$$

(2)

$$\frac{\partial q}{\partial b} = -2 \sum x_j(y_j - a - bx_j) = 0$$

(where we sum over j from 1 to n). Thus

$$an + b \sum x_j = \sum y_j$$

(3)

$$a \sum x_j + b \sum x_j^2 = \sum x_j y_j.$$

These equations are called the **normal equations** of our problem.

Example 1. Straight line

Using the method of least squares, fit a straight line to the four points in (1). We calculate

$$n = 4, \quad \sum x_j = 0.1, \quad \sum x_j^2 = 2.05, \quad \sum y_j = 3.907, \quad \sum x_j y_j = 0.0517$$

Hence the normal equations are

$$4a + 0.10b = 3.9070$$

$$0.1a + 2.05b = 0.0517$$

The solution is $a = 0.9773$, $b = -0.0224$, and we obtain the line (Fig. 358)

$$y = 0.9773 - 0.0224x.$$ ∎

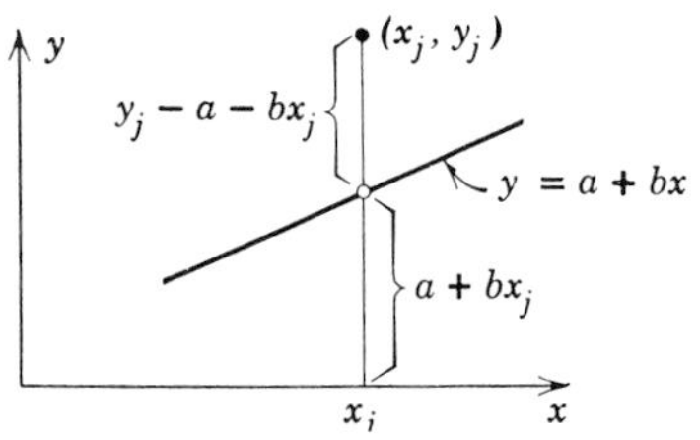

Fig. 359. Vertical distance of a point (x_j, y_j) from a straight line $y = a + bx$

Our approach of curve fitting can be generalized from a polynomial $y = a + bx$ to a polynomial degree m

$$p(x) = b_0 + b_1x + \cdots + b_mx^m.$$

Then q takes the form

$$q = \sum_{j=1}^{n} (y_j - p(x_j))^2$$

and depends on $m + 1$ parameters $b_0, \cdots, b_m$. Instead of (2) we then have $m + 1$ conditions

$$\frac{\partial q}{\partial b_0} = 0, \cdots, \frac{\partial q}{\partial b_m} = 0,$$

which give a system of $m + 1$ normal equations. The reader may show that in the case of a quadratic polynomial

$$p(x) = b_0 + b_1x + b_2x^2 \tag{4}$$

the normal equations are (summation from 1 to n)

$$\begin{aligned} b_0n \quad &+ b_1\sum x_j + b_2\sum x_j^2 = \sum y_j \\ b_0\sum x_j &+ b_1\sum x_j^2 + b_2\sum x_j^3 = \sum x_j y_j \\ b_0\sum x_j^2 &+ b_1\sum x_j^3 + b_2\sum x_j^4 = \sum x_j^2y_j \end{aligned} \tag{5}$$

Problems for Sec. 19.12

In each case plot the given points in the xy-plane and fit a straight line (a) by eye, (b) by the method of least squares.

1. Number of errors per hour of radio telegraphists as a function of the temperature x[°C]

x	26	28	31	33
y	12.0	11.5	15.3	17.3

2. (1, 1), (2, 2 + k), (3, 3), where k is a constant.

3. Iron content y[%] of a certain type of iron ore as a function of the density x[grams/cm^3]

x	2.8	2.9	3.0	3.1	3.2	3.2	3.2	3.3	3.4
y	27	23	30	28	30	32	34	33	30

4. (2, 5), (3, 9), (4, 15), (5, 21)

5. (0, 2.3), (2, 4.1), (4, 5.7), (6, 6.9)

6. (4, 3), (15, 16), (30, 13), (100, 70), (200, 90)

7. If a car is traveling along a straight road with constant speed $v = b_1$ [m/sec], its position y[m] at time t[sec] is $y = b_0 + b_1 t$. Suppose that measurements are

t	0	3	5	8	10
y	0	30	40	70	90

Graph these as points in the ty-plane. Fit a straight line (*a*) by eye, (*b*) by the least-squares method. What estimate of the speed can you obtain from it?

In each case determine the parabola (4) by the method of least squares.

8. $(-1, 2)$, $(0, 0)$, $(0, 1)$, $(1, 2)$

9. $(0, 3)$, $(1, 1)$, $(2, 0)$, $(4, 1)$, $(6, 4)$

10. $(1.4, 7400)$, $(1.8, 7500)$, $(2.3, 7600)$, $(3.0, 7500)$, $(4.0, 7200)$

11. Solve Prob. 8 by the Cholesky method (Sec. 19.9).

12. Determine the normal equations in the case of a polynomial of the third degree.

13. In the least-squares principle involving a polynomial we seek to satisfy as well as possible

$$b_0 + b_1 x_j + b_2 x_j^2 + \cdots + b_m x_j^m = y_j \qquad (j = 1, \cdots, n).$$

Introduce a matrix $\mathbf{C}$ such that this can be written $\mathbf{Cb} = \mathbf{y}$ and show that then the normal equations can be written $\mathbf{C^T C b} = \mathbf{C^T y}$.

14. The coefficient matrix in (5), call it $\mathbf{C}$, is symmetric and positive definite (definition in Sec. 7.12), so that the Cholesky method (Sec. 19.9) is very satisfactory for solving (5) and, similarly, normal equations in the case of polynomials of higher degree. But $\mathbf{C}$ may be ill-conditioned. To understand this, reason as follows. Show that

$$\mathbf{C} = \begin{pmatrix} c_0 & c_1 & c_2 \\ c_1 & c_2 & c_3 \\ c_2 & c_3 & c_4 \end{pmatrix} \qquad \text{where} \quad c_k = \sum_{j=1}^{n} x_j^k.$$

Assume that n is large and the x_j lie in the interval $0 \leqq x \leqq 1$ and are fairly uniformly distributed. Conclude from the definition of an integral that

$$\int_0^1 x^k\,dx \approx \frac{1}{n}\sum_{j=1}^{n} x_j^k = \frac{1}{n} c_k.$$

Integrate to obtain $\mathbf{C} = n\mathbf{H}$, where $\mathbf{H}$ is the (3×3) Hilbert matrix (Sec. 19.11, Prob. 6). Conclude that in the case of a polynomial of degree m one arrives at the $(m + 1)$-rowed Hilbert matrix, which is ill-conditioned for larger m.

15. In Problems of growth it is often required to fit an exponential function $y = b_0 e^{bx}$ by the method of least squares. Show that by taking logarithms this task can be reduced to that of determining a straight line.

19.13 Inclusion of Matrix Eigenvalues

An **eigenvalue** or **characteristic value** (or *latent root*) of an n-rowed (real or complex) square matrix $\mathbf{A} = (a_{jk})$ is defined to be a number λ such that the vector equation

$$\mathbf{Ax} = \lambda\mathbf{x} \tag{1}$$

has a nontrivial solution, that is, a solution $\mathbf{x} \neq \mathbf{0}$, which is then called an **eigenvector** or **characteristic vector** of $\mathbf{A}$ corresponding to that eigenvalue λ. The set of all eigenvalues of $\mathbf{A}$ is called the **spectrum** of $\mathbf{A}$. The eigenvalues of $\mathbf{A}$ are the roots of the characteristic equation

$$D(\lambda) = \det(\mathbf{A} - \lambda\mathbf{I}) = 0, \tag{2}$$

where $\mathbf{I}$ is the n-rowed unit matrix. $D(\lambda)$ is called the **characteristic determinant** and can be represented as a polynomial in λ of degree n, which is called the **characteristic polynomial** corresponding to $\mathbf{A}$. It follows that $\mathbf{A}$ has at least one and at most n numerically different eigenvalues. For more details, see Sec. 7.13.

Given $\mathbf{A}$, we may determine the coefficients of the characteristic polynomial and then determine the roots, for instance, by Newton's method. However, if n is large, both steps involve time-consuming computations, and it is desirable to proceed in a more efficient way. There are essentially two types of corresponding methods:

1. *Methods for obtaining bounds for eigenvalues,*
2. *Methods for computing approximate values for eigenvalues.*

We shall illustrate both types by some standard examples. For further methods see Refs. [C2], [C3], [C16] in Appendix 1.

The following interesting theorem by Gershgorin[6] yields a region consisting of closed circular disks and containing all the eigenvalues of a given matrix.

Theorem 1
Let λ be an eigenvalue of an arbitrary n-rowed square matrix $\mathbf{A} = (a_{jk})$. Then for some integer k $(1 \leqq k \leqq n)$,

$$|a_{kk} - \lambda| \leqq |a_{k1}| + |a_{k2}| + \cdots + |a_{k,k-1}| + |a_{k,k+1}| + \cdots + |a_{kn}|. \tag{3}$$

Proof. Let $\mathbf{x}$ be an eigenvector corresponding to that eigenvalue λ of $\mathbf{A}$. Then

$$\mathbf{Ax} = \lambda\mathbf{x} \qquad \text{or} \qquad (\mathbf{A} - \lambda\mathbf{I})\mathbf{x} = \mathbf{0}. \tag{4}$$

Let x_k be a component of $\mathbf{x}$ which is the greatest in absolute value. Then we

[6] *Bull. Acad. Sciences de l'URSS,* Classe mathém., 7-e série, Leningrad, 1931, p. 749.

have $|x_m/x_k| \leqq 1$ $(m = 1, \cdots, n)$. The vector equation (2) is equivalent to a system of n equations for the n components of the vectors on both sides, and the kth of these n equations is

$$a_{k1}x_1 + \cdots + a_{k,k-1}x_{k-1} + (a_{kk} - \lambda)x_k + a_{k,k+1}x_{k+1} + \cdots + a_{kn}x_n = 0.$$

From this we have

$$a_{kk} - \lambda = -a_{k1}\frac{x_1}{x_k} - \cdots - a_{k,k-1}\frac{x_{k-1}}{x_k} - a_{k,k+1}\frac{x_{k+1}}{x_k} - \cdots - a_{kn}\frac{x_n}{x_k}.$$

By taking absolute values on both sides of this equation, applying the triangle inequality

$$|a + b| \leqq |a| + |b|$$

(where a and b are any complex numbers) and observing that

$$\left|\frac{x_1}{x_k}\right| \leqq 1, \cdots, \quad \left|\frac{x_n}{x_k}\right| \leqq 1$$

we obtain (3), and the theorem is proved. ∎

For each $k = 1, \cdots, n$ the inequality (3) determines a closed circular disk in the complex λ-plane whose center is at a_{kk} and whose radius is given by the expression on the right-hand side of (3). Theorem 1 states that each of the eigenvalues of $\mathbf{A}$ lies in one of these n disks.

Example 1. Application of Gershgorin's theorem

From Theorem 1 it follows that the eigenvalues of the matrix

$$\mathbf{A} = \begin{pmatrix} 26 & -2 & 2 \\ 2 & 21 & 4 \\ 4 & 2 & 28 \end{pmatrix}$$

lie in the three disks (Fig. 360)

$$\begin{aligned} &D_1: \text{ center at } 26, \quad \text{radius } |-2| + 2 = 4, \\ &D_2: \text{ center at } 21, \quad \text{radius } 2 + 4 = 6, \\ &D_3: \text{ center at } 28, \quad \text{radius } 4 + 2 = 6. \end{aligned}$$

(The reader may verify that the eigenvalues of $\mathbf{A}$ are 30, 25, and 20.) ∎

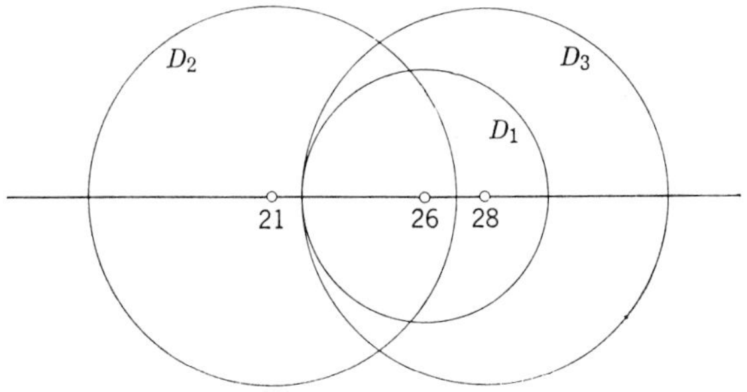

Fig. 360. Example 1

Bounds for the absolute value of the eigenvalues result from the following theorem by Schur[7] which we shall state without proof.

Theorem 2

Let $\mathbf{A} = (a_{jk})$ *be an* $(n \times n)$ *matrix, and let* $\lambda_1, \cdots, \lambda_n$ *be its eigenvalues. Then*

$$\sum_{i=1}^{n} |\lambda_i|^2 \leqq \sum_{j=1}^{n} \sum_{k=1}^{n} |a_{jk}|^2 \qquad \textbf{(Schur's inequality).} \tag{5}$$

In (5) *the equality sign holds if and only if* $\mathbf{A}$ *is such that*

$$\bar{\mathbf{A}}^{\mathsf{T}}\mathbf{A} = \mathbf{A}\bar{\mathbf{A}}^{\mathsf{T}}. \tag{6}$$

Matrices which satisfy (6) are called **normal matrices.** It is not difficult to see that Hermitian, skew-Hermitian, and unitary matrices are normal, and so are real symmetric, skew-symmetric, and orthogonal matrices.

Let λ_m be any eigenvalue of the matrix $\mathbf{A}$ in Theorem 2. Then $|\lambda_m|^2$ is less than or equal to the sum on the left side of (5), and by taking square roots we obtain from (5)

$$|\lambda_m| \leqq \sqrt{\sum_{j=1}^{n} \sum_{k=1}^{n} |a_{jk}|^2}. \tag{7}$$

Example 2. Bounds for eigenvalues obtained from Schur's inequality

For the matrix $\mathbf{A}$ in Example 1 we obtain from (7)

$$|\lambda| \leqq \sqrt{1949} < 44.2.$$

(The eigenvalues of $\mathbf{A}$ are 30, 25, and 20; thus $30^2 + 25^2 + 20^2 = 1925 < 1949$; in fact, $\mathbf{A}$ is not normal.) ∎

The preceding theorems are valid for every real or complex square matrix. There are other theorems which hold for special classes of matrices only. The following theorem by Frobenius,[8] which we state without proof, is of this type.

Theorem 3

Let $\mathbf{A}$ *be a real square matrix whose elements are all positive. Then* $\mathbf{A}$ *has at least one real positive eigenvalue* λ, *and the corresponding eigenvector can be chosen real and such that all its components are positive.*

From this theorem we may derive the following useful result by Collatz.[9]

Theorem 4

Let $\mathbf{A} = (a_{jk})$ *be a real n-rowed square matrix whose elements are all positive. Let* $\mathbf{x}$ *be any real vector whose components* $x_1, \cdots, x_n$ *are positive, and let* $y_1, \cdots, y_n$ *be the components of the vector* $\mathbf{y} = \mathbf{Ax}$. *Then the closed interval on the real axis*

[7] *Mathematische Annalen,* Vol. 66, 1909, p. 488.

[8] *Sitzungsberichte Preuss. Akad. Wiss. Math.-phys. Klasse,* Berlin, 1908, p. 471.

[9] L. Collatz, *Eigenwertaufgaben mit technischen Anwendungen.* Leipzig: Akademische Verlagsgesellschaft, 1949, p. 291.

bounded by the smallest and the largest of the n quotients $q_j = y_j/x_j$ contains at least one eigenvalue of **A**.

Proof. We have $\mathbf{Ax} = \mathbf{y}$ or

$$\mathbf{y} - \mathbf{Ax} = \mathbf{0}. \tag{8}$$

The transpose $\mathbf{A}^\mathsf{T}$ satisfies the conditions of Theorem 3. Hence $\mathbf{A}^\mathsf{T}$ has a positive eigenvalue λ and, corresponding to this eigenvalue, an eigenvector **u** whose components u_j are all positive. Thus $\mathbf{A}^\mathsf{T}\mathbf{u} = \lambda\mathbf{u}$, and by taking the transpose we obtain $\mathbf{u}^\mathsf{T}\mathbf{A} = \lambda\mathbf{u}^\mathsf{T}$. From this and (8),

$$\mathbf{u}^\mathsf{T}(\mathbf{y} - \mathbf{Ax}) = \mathbf{u}^\mathsf{T}\mathbf{y} - \mathbf{u}^\mathsf{T}\mathbf{Ax} = \mathbf{u}^\mathsf{T}(\mathbf{y} - \lambda\mathbf{x}) = 0$$

or written out

$$\sum_{j=1}^{n} u_j(y_j - \lambda x_j) = 0.$$

Since all the components u_j are positive, it follows that

$$\begin{aligned} &y_j - \lambda x_j \geqq 0, \quad \text{that is,} \quad q_j \geqq \lambda \quad \text{for at least one } j, \quad \text{and} \\ &y_j - \lambda x_j \leqq 0, \quad \text{that is,} \quad q_j \leqq \lambda \quad \text{for at least one } j. \end{aligned} \tag{9}$$

Since **A** and $\mathbf{A}^\mathsf{T}$ have the same eigenvalues, λ is an eigenvalue of **A**, and from (9) the statement of the theorem follows. ∎

Example 3. Bounds for eigenvalues from Collatz's theorem

Let

$$\mathbf{A} = \begin{pmatrix} 8 & 1 & 1 \\ 1 & 5 & 2 \\ 1 & 2 & 5 \end{pmatrix}. \quad \text{Choose} \quad \mathbf{x} = \begin{pmatrix} 1 \\ 1 \\ 1 \end{pmatrix}. \quad \text{Then} \quad \mathbf{y} = \begin{pmatrix} 10 \\ 8 \\ 8 \end{pmatrix}.$$

Hence $q_1 = 10$, $q_2 = 8$, $q_3 = 8$, and Theorem 4 implies that one of the eigenvalues of **A** must lie in the interval $8 \leqq \lambda \leqq 10$. Of course, the length of such an interval depends on the choice of **x**. The student may show that $\lambda = 9$ is an eigenvalue of **A**.

Problems for Sec. 19.13

Using Theorem 1, determine and sketch disks which contain the eigenvalues of the following matrices.

1. $\begin{pmatrix} 3 & 4 \\ 4 & -3 \end{pmatrix}$ **2.** $\begin{pmatrix} 5 & 1 \\ 1 & 5 \end{pmatrix}$ **3.** $\begin{pmatrix} 1/\sqrt{2} & i/\sqrt{2} \\ -i/\sqrt{2} & -1/\sqrt{2} \end{pmatrix}$

4. $\begin{pmatrix} 6 & 0 & -3 \\ 0 & 6 & 3 \\ -3 & 3 & 2 \end{pmatrix}$ **5.** $\begin{pmatrix} -9 & 1 & 0 \\ 1 & -9 & 1 \\ 0 & 1 & -9 \end{pmatrix}$ **6.** $\begin{pmatrix} 0 & 0 & 3i \\ 0 & 2-i & 1+i \\ 1+2i & 0 & 0 \end{pmatrix}$

Using (7), obtain an upper bound for the absolute value of the eigenvalues of the matrix:

7. In Prob. 1 **8.** In Prob. 2 **9.** In Prob. 4

10. In Example 3 **11.** In Prob. 5 **12.** In Prob. 6

13. Verify Theorem 3 for the matrix in Prob. 2.

Apply Theorem 4 to the following matrices, choosing the given vectors as vectors **x**.

14. $\begin{pmatrix} 17 & 8 & 1 \\ 8 & 18 & 8 \\ 1 & 8 & 17 \end{pmatrix}, \begin{pmatrix} 1 \\ 1 \\ 1 \end{pmatrix}, \begin{pmatrix} 1 \\ 2 \\ 1 \end{pmatrix}, \begin{pmatrix} 2 \\ 3 \\ 2 \end{pmatrix}$ **15.** $\begin{pmatrix} 3 & 1 & 1 \\ 1 & 3 & 1 \\ 1 & 1 & 3 \end{pmatrix}, \begin{pmatrix} 1 \\ 2 \\ 1 \end{pmatrix}, \begin{pmatrix} 1 \\ 1 \\ 1 \end{pmatrix}$

16. **(Nonzero determinant)** If in each row of a determinant the absolute value of the element in the principal diagonal is greater than the sum of the absolute values of the remaining elements in that row, show that the value of the determinant is different from zero. What does this imply with respect to the solvability of systems of linear equations?

17. **(Inclusion set)** An *inclusion set* for a matrix **A** is a set in the complex plane containing at least one eigenvalue of **A**. What type of inclusion sets do we obtain for unitary matrices by combining Theorem 1 and Theorem 3c, Sec. 7.14?

18. Show that the matrix in Example 1 is not normal.

19. Show that the matrix in Example 3 has the eigenvalues 9, 6, 3, and (5) holds with the equality sign.

20. Using Theorems 1 and 3 in Sec. 7.14, show that for a unitary matrix, Schur's inequality with the equality sign holds.

19.14 Determination of Eigenvalues by Iteration

A standard procedure for computing approximate values of the eigenvalues of an n-rowed square matrix $\mathbf{A} = (a_{jk})$ is the **iteration method for eigenvalues.** In this method we start from any vector $\mathbf{x}_0$ ($\neq \mathbf{0}$) with n components and compute successively

$$\mathbf{x}_1 = \mathbf{A}\mathbf{x}_0, \qquad \mathbf{x}_2 = \mathbf{A}\mathbf{x}_1, \quad \cdots, \qquad \mathbf{x}_s = \mathbf{A}\mathbf{x}_{s-1}.$$

For simplifying notation, we denote $\mathbf{x}_{s-1}$ by $\mathbf{x}$ and $\mathbf{x}_s$ by $\mathbf{y}$, so that $\mathbf{y} = \mathbf{A}\mathbf{x}$. If $\mathbf{A}$ is real symmetric, the following theorem gives an approximation and error bounds.

Theorem 1

Let $\mathbf{A}$ *be an n-rowed real symmetric matrix. Let* $\mathbf{x}$ ($\neq \mathbf{0}$) *be any real vector with n components. Furthermore, let*

$$\mathbf{y} = \mathbf{A}\mathbf{x}, \qquad m_0 = \mathbf{x}^\mathsf{T}\mathbf{x}, \qquad m_1 = \mathbf{x}^\mathsf{T}\mathbf{y}, \qquad m_2 = \mathbf{y}^\mathsf{T}\mathbf{y}.$$

Then the quotient

$$q = \frac{m_1}{m_0} \qquad \textbf{(Rayleigh quotient}^{10}\textbf{)}$$

is an approximation for an eigenvalue[11] λ *of* **A**, *and if we set* $q = \lambda + \epsilon$, *so that* ϵ *is the error of* q, *then*

$$\text{(1)} \qquad |\epsilon| \leqq \sqrt{\frac{m_2}{m_0} - q^2}.$$

Proof. Let δ^2 denote the radicand in (1). Then, since $m_1 = qm_0$, we have

$$\text{(2)} \qquad (\mathbf{y} - q\mathbf{x})^{\mathsf{T}}(\mathbf{y} - q\mathbf{x}) = m_2 - 2qm_1 + q^2 m_0 = m_2 - q^2 m_0 = \delta^2 m_0.$$

Since **A** is real symmetric, it has an orthogonal set of n real unit eigenvectors $\mathbf{z}_1, \cdots, \mathbf{z}_n$ corresponding to the eigenvalues $\lambda_1, \cdots, \lambda_n$, respectively (some of which may be equal). (Proof in Ref. [C14].) Then **x** has a representation of the form

$$\mathbf{x} = a_1 \mathbf{z}_1 + \cdots + a_n \mathbf{z}_n.$$

Now $\mathbf{A}\mathbf{z}_1 = \lambda_1 \mathbf{z}_1$, etc., and we obtain

$$\mathbf{y} = \mathbf{A}\mathbf{x} = a_1 \lambda_1 \mathbf{z}_1 + \cdots + a_n \lambda_n \mathbf{z}_n$$

and, since the $\mathbf{z}_j$ are orthogonal unit vectors,

$$\text{(3)} \qquad m_0 = \mathbf{x}^{\mathsf{T}}\mathbf{x} = a_1^2 + \cdots + a_n^2.$$

It follows that in (2),

$$\mathbf{y} - q\mathbf{x} = a_1(\lambda_1 - q)\mathbf{z}_1 + \cdots + a_n(\lambda_n - q)\mathbf{z}_n.$$

Since the $\mathbf{z}_j$ are orthogonal unit vectors, we thus obtain from (2)

$$\delta^2 m_0 = a_1^2(\lambda_1 - q)^2 + \cdots + a_n^2(\lambda_n - q)^2.$$

Replacing each $(\lambda_j - q)^2$ by the smallest of these terms and using (3), we have

$$\delta^2 m_0 \geqq (\lambda_c - q)^2(a_1^2 + \cdots + a_n^2) = (\lambda_c - q)^2 m_0$$

where λ_c is an eigenvalue to which q is closest. From this, (1) follows, and the theorem is proved. ∎

[10] LORD RAYLEIGH (JOHN WILLIAM STRUTT) (1842–1919), English physicist and mathematician, who made important contributions to various branches of applied mathematics and theoretical physics, in particular the theory of waves, elasticity, and hydrodynamics.

[11] Ordinarily that λ which is greatest in absolute value, but no general statements are possible.

Example 1. An application of Theorem 1

As in Example 3 of the preceding section, let us consider the real symmetric matrix

$$\mathbf{A} = \begin{pmatrix} 8 & 1 & 1 \\ 1 & 5 & 2 \\ 1 & 2 & 5 \end{pmatrix} \quad \text{and choose} \quad \mathbf{x}_0 = \begin{pmatrix} 1 \\ 1 \\ 1 \end{pmatrix}.$$

Then we obtain successively

$$\mathbf{x}_1 = \begin{pmatrix} 10 \\ 8 \\ 8 \end{pmatrix}, \quad \mathbf{x}_2 = \begin{pmatrix} 96 \\ 66 \\ 66 \end{pmatrix}, \quad \mathbf{x}_3 = \begin{pmatrix} 900 \\ 558 \\ 558 \end{pmatrix}, \quad \mathbf{x}_4 = \begin{pmatrix} 8316 \\ 4806 \\ 4806 \end{pmatrix}.$$

Taking $\mathbf{x} = \mathbf{x}_3$ and $\mathbf{y} = \mathbf{x}_4$, we have

$$m_0 = \mathbf{x}^\mathsf{T}\mathbf{x} = 1\,432\,728, \qquad m_1 = \mathbf{x}^\mathsf{T}\mathbf{y} = 12\,847\,896, \qquad m_2 = \mathbf{y}^\mathsf{T}\mathbf{y} = 115\,351\,128.$$

From this we calculate

$$q = \frac{m_1}{m_0} = 8.967, \qquad |\epsilon| \leqq \sqrt{\frac{m_2}{m_0} - q^2} = 0.311.$$

This shows that $q = 8.967$ is an approximation for an eigenvalue which must lie between 8.656 and 9.278. The student may show that an eigenvalue is $\lambda = 9$.

Problems for Sec. 19.14

1. Let

$$\mathbf{A} = \begin{pmatrix} 2 & -1 & 1 \\ -1 & 3 & 2 \\ 1 & 2 & 3 \end{pmatrix} \quad \text{and choose} \quad \mathbf{x}_0 = \begin{pmatrix} 1 \\ 1 \\ 1 \end{pmatrix}.$$

Compute $\mathbf{x}_1$, $\mathbf{x}_2$, $\mathbf{x}_3$. Take $\mathbf{x}_2$ for $\mathbf{x}$ and $\mathbf{x}_3$ for $\mathbf{y}$. Show that q deviates from 5 (the exact value of the largest eigenvalue of $\mathbf{A}$) by 1.7%, approximately. Find a bound for the error of q from (1).

Let $\mathbf{A}$ be as in Prob. 1. Compute $\mathbf{x}_1$, the corresponding q, and a bound for the error, choosing

2. $\mathbf{x}_0 = \begin{pmatrix} 1 \\ 0 \\ 0 \end{pmatrix}$ **3.** $\mathbf{x}_0 = \begin{pmatrix} 0 \\ 1 \\ 0 \end{pmatrix}$ **4.** $\mathbf{x}_0 = \begin{pmatrix} 0 \\ 1 \\ 1 \end{pmatrix}$

5. Show that the eigenvalues of the matrix $\mathbf{A}$ in Prob. 1 are 0, 3, 5, and indicate which eigenvalue is approximated by the results of Probs. 2–4.

Choosing for $\mathbf{x}_0$ the column vector with components 1, 1, 1, 1, compute $\mathbf{x}_1$, $\mathbf{x}_2$ and approximations $q = \mathbf{x}_1{}^\mathsf{T}\mathbf{x}_0/\mathbf{x}_0{}^\mathsf{T}\mathbf{x}_0$, $q = \mathbf{x}_2{}^\mathsf{T}\mathbf{x}_1/\mathbf{x}_1{}^\mathsf{T}\mathbf{x}_1$ and corresponding error bounds for an eigenvalue of each of the following symmetric matrices on the next page.

6. $\begin{pmatrix} 3 & 2 & 0 & 1 \\ 2 & 0 & 5 & -1 \\ 0 & 5 & 2 & 1 \\ 1 & -1 & 1 & 4 \end{pmatrix}$ **7.** $\begin{pmatrix} 1 & 0 & 0 & 1 \\ 0 & 2 & -1 & 0 \\ 0 & -1 & 3 & 0 \\ 1 & 0 & 0 & -1 \end{pmatrix}$ **8.** $\begin{pmatrix} 2 & 0 & 1 & 0 \\ 0 & 0 & 3 & 1 \\ 1 & 3 & 4 & -2 \\ 0 & 1 & -2 & 0 \end{pmatrix}$

9. Using $\mathbf{x}_0^\mathsf{T} = (3 \quad 1 \quad 2)$ and applying iteration to

$$\mathbf{A} = \begin{pmatrix} -2 & -1 & 4 \\ 2 & 1 & -2 \\ -1 & -1 & 3 \end{pmatrix},$$

show that $\mathbf{A}^2\mathbf{x}_0 = \mathbf{x}_0$ and draw conclusions from this result.

10. Can (1) be used in the case of the matrix in Prob. 9?

11. Show that if $\mathbf{x}$ is an eigenvector, then $\epsilon = 0$ in (1).

12. To understand the importance of the error bound (1), consider the matrix

$$\mathbf{A} = \begin{pmatrix} 3 & 4 \\ 4 & -3 \end{pmatrix}, \qquad \text{choose} \qquad \mathbf{x}_0 = \begin{pmatrix} 3 \\ -1 \end{pmatrix}$$

and show that $q = 0$ for all s. Find the eigenvalues and explain what happened. Start again, choosing another $\mathbf{x}_0$.

13. To understand why the Rayleigh quotient q is generally an approximation of that eigenvalue, λ_1, which is largest in absolute value, let

$$\mathbf{x}_0 = c_1\mathbf{z}_1 + \cdots + c_n\mathbf{z}_n$$

(with $\mathbf{z}_1, \cdots, \mathbf{z}_n$ as in the proof of Theorem 1) and show that

$$\mathbf{x} = \mathbf{x}_{s-1} = c_1\lambda_1^{s-1}\mathbf{z}_1 + \cdots + c_n\lambda_n^{s-1}\mathbf{z}_n$$

$$\mathbf{y} = \mathbf{x}_s = c_1\lambda_1{}^s\mathbf{z}_1 + \cdots + c_n\lambda_n{}^s\mathbf{z}_n$$

and, consequently,

$$q = \frac{m_1}{m_0} = \frac{c_1{}^2\lambda_1^{2s-1} + \cdots}{c_1{}^2\lambda_1^{2s-2} + \cdots} \approx \lambda_1.$$

Under what conditions will this be a good approximation?

14. Show that the matrix $\mathbf{A}$ in Example 1 has the eigenvalues $\lambda_1 = 9, \lambda_2 = 6, \lambda_3 = 3$. Apply iteration to $\mathbf{A} - 4.5\mathbf{I}$, starting from $\mathbf{x}_0^\mathsf{T} = (1 \quad 1 \quad 1)$ and taking $\mathbf{x} = \mathbf{x}_3$ and $\mathbf{y} = \mathbf{x}_4$ in (1). Show that $q = 4.4995$, so that 8.9995 approximates λ_1, with an error $|\epsilon| \leqq 0.0393$. Can you explain the reason for the great improvement over the result in Example 1? (For more details about improving the iteration method, see [G8] in Appendix 1.)

15. Let $\mathbf{A}$ be symmetric, with eigenvalues $\lambda_1 > \lambda_2 \geqq \cdots \geqq \lambda_{n-1} > \lambda_n$, assuming $|\lambda_1| > |\lambda_n|$, and let α be a good estimate of λ_1. Then the iteration applied to $\mathbf{B} = \mathbf{A} - \alpha\mathbf{I}$ will in general yield approximations to λ_n. Why is this so, and what does "in general" mean? Apply the method to $\mathbf{A}$ in Prob. 1, using $\alpha = 4.9$ (as suggested by the answer to Prob. 1) and $\mathbf{x}_0^\mathsf{T} = (1 \quad 1 \quad 1)$ and taking $\mathbf{x}_1$ for $\mathbf{x}$ and $\mathbf{x}_2$ for $\mathbf{y}$ in (1).

19.15 Asymptotic Expansions

Asymptotic expansions are (in general divergent) series which are of great practical importance for computing values of a function $f(x)$ for large x. It is clear that the Maclaurin series of $f(x)$, if it exists and converges for large x, is not suitable for that purpose, because the number of terms needed to obtain a certain number of significant digits increases rapidly as x increases. For a Taylor series with center a and large $|x - a|$, the situation is similar. We shall see that the larger x is, the fewer terms of an asymptotic expansion we need for obtaining a required accuracy. On the other hand, the accuracy is limited, and it decreases as x decreases so that asymptotic expansions can be used for large x only.

The variables and functions to be considered in this section are assumed to be real.

A series of the form

$$c_0 + \frac{c_1}{x} + \frac{c_2}{x^2} + \cdots \qquad (c_0, c_1, \cdots \text{ constant})$$

(which need not converge for any value of x) *is called an* **asymptotic expansion,** *or* **asymptotic series,** *of a function* $f(x)$ *which is defined for every sufficiently large value of* x *if, for every fixed* $n = 0, 1, 2, \cdots,$

$$\textbf{(1)} \qquad \left[f(x) - \left(c_0 + \frac{c_1}{x} + \frac{c_2}{x^2} + \cdots + \frac{c_n}{x^n}\right)\right]x^n \to 0 \qquad \text{as } x \to \infty,$$

and we shall then write

$$f(x) \sim c_0 + \frac{c_1}{x} + \frac{c_2}{x^2} + \cdots .$$

If a function $f(x)$ has an asymptotic expansion, then this expansion is unique, because its coefficients $c_0, c_1, \cdots$ are uniquely determined by (1). In fact, from (1) we obtain

$$(1^*) \qquad \begin{aligned} &f(x) - c_0 \to 0 &&\text{or} &&c_0 = \lim_{x\to\infty} f(x), \\ &\left[f(x) - c_0 - \frac{c_1}{x}\right]x \to 0 &&\text{or} &&c_1 = \lim_{x\to\infty} [f(x) - c_0]x, \quad \text{etc.} \end{aligned}$$

On the other hand, different functions may have the same asymptotic expansion. In fact, let $f(x) = e^{-x}$. Then, since $e^{-x} \to 0$, $xe^{-x} \to 0$, etc., we see from (1*) that $c_0 = 0$, $c_1 = 0$, etc. Hence,

$$e^{-x} \sim 0 + \frac{0}{x} + \cdots .$$

Thus, if $g(x)$ has an asymptotic expansion, then $g(x) + e^{-x}$ has certainly the same asymptotic expansion.

For applications it is advantageous to *extend the definition* by writing

$$f(x) \sim g(x) + h(x)\left(c_0 + \frac{c_1}{x} + \frac{c_2}{x^2} + \cdots\right)$$

whenever

$$\frac{f(x) - g(x)}{h(x)} \sim c_0 + \frac{c_1}{x} + \frac{c_2}{x^2} + \cdots$$

in the sense of the preceding definition.

Only in rare cases may we determine the coefficients of an asymptotic expansion directly from (1*). In general, other methods will be more suitable, for example, successive integration by parts.

Example 1. Asymptotic series of the error function

The error function erf x is defined by the integral (cf. Fig. 406 in Appendix 3)

$$\text{(2a)} \qquad \operatorname{erf} x = \frac{2}{\sqrt{\pi}} \int_0^x e^{-t^2}\, dt$$

and its complementary function erfc x by

$$\text{(2b)} \qquad \operatorname{erfc} x = 1 - \operatorname{erf} x = \frac{2}{\sqrt{\pi}} \int_x^\infty e^{-t^2}\, dt = \frac{1}{\sqrt{\pi}} \int_{x^2}^\infty e^{-\tau}\tau^{-1/2}\, d\tau,$$

where $t^2 = \tau$, and erf $\infty = 1$. Repeated integration by parts will lead to integrals of the form

$$\text{(3)} \qquad F_n(x) = \int_{x^2}^\infty e^{-\tau}\tau^{-(2n+1)/2}\, d\tau, \qquad n = 0, 1, \cdots.$$

Note that erfc $x = F_0(x)/\sqrt{\pi}$. By integration by parts we obtain

$$F_n(x) = -e^{-\tau}\tau^{-(2n+1)/2}\Big|_{x^2}^\infty - \frac{2n+1}{2}\int_{x^2}^\infty e^{-\tau}\tau^{-(2n+3)/2}\, d\tau.$$

The integral on the right is $F_{n+1}(x)$ and, therefore,

$$e^{x^2}F_n(x) = \frac{1}{x^{2n+1}} - \frac{2n+1}{2}e^{x^2}F_{n+1}(x), \qquad n = 0, 1, \cdots.$$

Repeated application of this formula yields

$$e^{x^2}F_0(x) = \frac{1}{x} - \frac{1}{2}e^{x^2}F_1(x)$$

$$\cdots\cdots\cdots\cdots\cdots$$

$$\text{(4)} \qquad = \left[\frac{1}{x} - \frac{1}{2x^3} + \frac{1\cdot 3}{2^2x^5} - + \cdots + (-1)^{n-1}\frac{1\cdot 3\cdots(2n-3)}{2^{n-1}x^{2n-1}}\right]$$

$$+ (-1)^n\frac{1\cdot 3\cdots(2n-1)}{2^n}e^{x^2}F_n(x).$$

We show that the series obtained in this way is an asymptotic expansion,

$$e^{x^2}F_0(x) \sim \frac{1}{x} - \frac{1}{2x^3} + \frac{1 \cdot 3}{2^2 x^5} - + \cdots . \tag{5}$$

Let S_{2n-1} denote the expression in the brackets in (4). Then from (4) we obtain

$$[e^{x^2}F_0(x) - S_{2n-1}]x^{2n-1} = K_n e^{x^2} x^{2n-1} F_n(x), \tag{6}$$

where $K_n = (-2)^{-n} 1 \cdot 3 \cdots (2n-1)$. We have to show that for each fixed $n = 1, 2, \cdots$ the expression on the right approaches zero as x approaches infinity. In (3) we have

$$\frac{1}{\tau^{(2n+1)/2}} \leqq \frac{1}{x^{2n+1}} \qquad \text{for all } \tau \geqq x^2.$$

Hence we obtain the inequality

$$F_n(x) = \int_{x^2}^{\infty} \frac{e^{-\tau}}{\tau^{(2n+1)/2}}\, d\tau < \frac{1}{x^{2n+1}} \int_{x^2}^{\infty} e^{-\tau}\, d\tau = \frac{e^{-x^2}}{x^{2n+1}}. \tag{7}$$

From this we immediately see that

$$|K_n| e^{x^2} x^{2n-1} F_n(x) < \frac{|K_n|}{x^2} \to 0 \qquad (x \to \infty).$$

This proves that the series in (5) is an asymptotic expansion of the function on the left-hand side of (5). Since $\operatorname{erf} x = 1 - \operatorname{erfc} x = 1 - F_0(x)/\sqrt{\pi}$, it follows that the desired asymptotic series of the error function is

$$\operatorname{erf} x \sim 1 - \frac{1}{\sqrt{\pi}} e^{-x^2} \left(\frac{1}{x} - \frac{1}{2x^3} + \frac{1 \cdot 3}{2^2 x^5} - \frac{1 \cdot 3 \cdot 5}{2^3 x^7} + - \cdots \right). \tag{8}$$

For large x we thus have the simple approximation

$$\operatorname{erf} x \approx 1 - \frac{1}{\sqrt{\pi}\, x} e^{-x^2}. \tag{8*}$$

From (6) and (7) we see that

$$|e^{x^2}F_0(x) - S_{2n-1}| = \frac{1 \cdot 3 \cdots (2n-1)}{2^n} e^{x^2} F_n(x) < \frac{1 \cdot 3 \cdots (2n-1)}{2^n} \frac{1}{x^{2n+1}},$$

and for sufficiently large x the expression on the right is very small. This shows that for such x, S_{2n-1} is a very good approximation to $e^{x^2}F_0(x)$, and so, for large x, the error function can be computed very accurately by means of (7). It turns out that even for relatively small $|x|$ the results are surprisingly accurate. For instance, (8*) gives

$$\operatorname{erf} 5 \sim 1 - \frac{1}{\sqrt{\pi}} \frac{e^{-25}}{5} = 0.99999\ 99999\ 98433;$$

the first thirteen decimals are correct, the error is three units of the fourteenth decimal. However, we warn the reader that the present case may be misleading; for other asymptotic expansions a similarly favorable situation may arise not at $x = 5$, but only for much larger x, perhaps for $x = 20$ or $x = 100$. ∎

For practical applications it is useful to know that asymptotic series may be added, multiplied, and under certain restrictions also integrated and differentiated term by term. Let us formulate these properties in a precise manner.

Theorem 1 (Addition and multiplication)
If

$$f(x) \sim a_0 + \frac{a_1}{x} + \frac{a_2}{x^2} + \cdots \qquad \text{and} \qquad g(x) \sim b_0 + \frac{b_1}{x} + \frac{b_2}{x^2} + \cdots$$

then the function $Af + Bg$, where A and B are constants, has the asymptotic expansion

$$(9) \qquad Af(x) + Bg(x) \sim Aa_0 + Bb_0 + \frac{Aa_1 + Bb_1}{x} + \frac{Aa_2 + Bb_2}{x^2} + \cdots$$

and the function fg has the asymptotic expansion

$$(10) \qquad f(x)g(x) \sim c_0 + \frac{c_1}{x} + \frac{c_2}{x^2} + \cdots$$

whose coefficients are given by the formula

$$c_n = a_0 b_n + a_1 b_{n-1} + \cdots + a_n b_0.$$

Proof. The simple proof of (9) is left to the reader. We prove the last statement. We must show that for any fixed nonnegative integer n,

$$(11) \qquad (fg - S_n)x^n \to 0 \qquad \text{as } x \to \infty,$$

where

$$S_n(x) = c_0 + \frac{c_1}{x} + \cdots + \frac{c_n}{x^n}$$

with the coefficients $c_0, \cdots, c_n$ given in the theorem.

We choose an arbitrary fixed n and write

$$f(x) = s_n(x) + \frac{h(x)}{x^n} \qquad \text{where} \qquad s_n(x) = a_0 + \frac{a_1}{x} + \cdots + \frac{a_n}{x^n}.$$

Then

$$[f(x) - s_n(x)]x^n = h(x).$$

From this and the definition of an asymptotic expansion it follows that $h(x)$ must approach zero as $x \to \infty$. Similarly, if we write

$$g(x) = s_n^*(x) + \frac{l(x)}{x^n} \qquad \text{where} \qquad s_n^*(x) = b_0 + \frac{b_1}{x} + \cdots + \frac{b_n}{x^n},$$

then $l(x) \to 0$ as $x \to \infty$. From these representations we obtain

$$fg = s_n s_n^* + \frac{h + l}{x^n} + \frac{hl}{x^{2n}}.$$

By multiplying term by term and collecting like powers of x we can readily verify that

$$s_n s_n^* = S_n + T_n$$

where T_n is the sum of terms involving the powers $1/x^{n+1}, \cdots, 1/x^{2n}$. Clearly, $x^n T_n \to 0$ as $x \to \infty$. Now, in (11),

$$fg - S_n = T_n + \frac{h+l}{x^n} + \frac{hl}{x^{2n}}.$$

We multiply by x^n on both sides. Then the resulting expression on the right approaches zero as $x \to \infty$, because $x^n T_n \to 0$, $l \to 0$, and $h \to 0$. Hence the expression on the left-hand side of (11) approaches zero, and the proof is complete. ▮

Theorem 2 (Integration)
Let $f(x)$ be continuous for all sufficiently large x, and let

$$f(x) \sim \frac{c_2}{x^2} + \frac{c_3}{x^3} + \cdots.$$

Then for those x,

$$\int_x^\infty f(t)\,dt \sim \frac{c_2}{x} + \frac{c_3}{2x^2} + \frac{c_4}{3x^3} + \cdots. \tag{12}$$

Proof. Let $F(x)$ denote the integral in (12) and let $S_{n-1}(x)$ denote the integral of

$$s_n(x) = \frac{c_2}{x^2} + \cdots + \frac{c_n}{x^n},$$

that is,

$$S_{n-1}(x) = \int_x^\infty s_n(t)\,dt = \frac{c_2}{x} + \frac{c_3}{2x^2} + \cdots + \frac{c_n}{(n-1)x^{n-1}}.$$

By the definition of an asymptotic expansion, for every $n = 0, 1, \cdots,$

$$|f(x) - s_n(x)|x^n \to 0 \qquad \text{as } x \to \infty.$$

Because of the continuity of f this implies that, given any $\epsilon > 0$ we can find an x_0 such that for all $x > x_0$,

$$|f(x) - s_n(x)|x^n < \epsilon \qquad \text{or} \qquad |f(x) - s_n(x)| < \frac{\epsilon}{x^n}.$$

From this we obtain for all $x > x_0$

$$|F(x) - S_{n-1}(x)| = \left|\int_x^\infty f(t)\,dt - \int_x^\infty s_n(t)\,dt\right| = \left|\int_x^\infty [f(t) - s_n(t)]\,dt\right|$$

$$\leqq \int_x^\infty |f(t) - s_n(t)|\,dt < \epsilon \int_x^\infty \frac{dt}{t^n} = \frac{\epsilon}{(n-1)x^{n-1}}.$$

By multiplying on both sides by the positive quantity x^{n-1} we have

$$|F(x) - S_{n-1}(x)|x^{n-1} < \frac{\epsilon}{n-1} \qquad \text{when} \quad x > x_0(\epsilon).$$

Since ϵ (> 0) may be chosen as small as we please, it follows that

$$|F(x) - S_{n-1}(x)|x^{n-1} \to 0 \qquad \text{as } x \to \infty,$$

and the proof is complete. ∎

If $f(x)$ is continuous for all sufficiently large x and

$$f(x) \sim c_0 + \frac{c_1}{x} + \frac{c_2}{x^2} + \cdots,$$

then we conclude from Theorem 2 that

$$\text{(12*)} \qquad \int_x^\infty \left[f(t) - c_0 - \frac{c_1}{t}\right] dt \sim \frac{c_2}{x} + \frac{c_3}{2x^2} + \frac{c_4}{3x^3} + \cdots.$$

If $f(x)$ has an asymptotic expansion, it does not follow that the derivative $f'(x)$ has an asymptotic expansion. For example, from (1*) we obtain

$$f(x) = e^{-x} \sin(e^x) \sim 0 + \frac{0}{x} + \frac{0}{x^2} + \cdots$$

but the derivative of $f(x)$, which is given by the formula

$$f'(x) = -e^{-x} \sin(e^x) + e^{-x} \cos(e^x)\, e^x = -f(x) + \cos(e^x)$$

does not have an asymptotic expansion. (Why?) However, if the derivative $f'(x)$ of a function $f(x)$ has an asymptotic expansion, it can be obtained by termwise differentiation of that of $f(x)$. In fact, the following theorem holds true.

Theorem 3 (Differentiation)

If

$$\text{(13)} \qquad f(x) \sim c_0 + \frac{c_1}{x} + \frac{c_2}{x^2} + \cdots$$

and $f(x)$ has a continuous derivative $f'(x)$ which has an asymptotic expansion, then this expansion is

(14) $$f'(x) \sim -\frac{c_1}{x^2} - \frac{2c_2}{x^3} - \frac{3c_3}{x^4} - \cdots .$$

Proof. By assumption,

(15) $$f'(x) \sim a_0 + \frac{a_1}{x} + \frac{a_2}{x^2} + \cdots$$

and we have to show that the coefficients a_n are such that (14) and (15) are identical. We first show that $a_0 = 0$ and $a_1 = 0$. From (13) and the definition of an asymptotic expansion we have

(16) $$\text{(a)} \lim_{x\to\infty} f(x) = c_0, \qquad \text{(b)} \lim_{x\to\infty} [f(x) - c_0]x = c_1.$$

The corresponding relations for the asymptotic series (15) are

(17) $$\text{(a)} \lim_{x\to\infty} f'(x) = a_0, \qquad \text{(b)} \lim_{x\to\infty} [f'(x) - a_0]x = a_1.$$

We may relate f and f' by the formula

(18) $$f(x) = \int_{x_0}^{x} f'(t)\, dt + k \qquad [x_0\ (> 0) \text{ and } k \text{ constant}].$$

From this and (16a) it follows that

$$\lim_{x\to\infty} \int_{x_0}^{x} f'(t)\, dt + k = c_0.$$

Now (17a) shows that if a_0 were not zero, the limit of the integral would not exist. Hence $a_0 = 0$. Then (17b) becomes

(17b*) $$\lim_{x\to\infty} xf'(x) = a_1.$$

By definition this means that for any preassigned $\epsilon > 0$ and all sufficiently large x,

(19) $$a_1 - \epsilon < xf'(x) < a_1 + \epsilon \qquad \text{or} \qquad \frac{a_1 - \epsilon}{x} < f'(x) < \frac{a_1 + \epsilon}{x}.$$

From (16b) and (18) we obtain

$$\lim_{x\to\infty} \left(\int_{x_0}^{x} f'(t)\, dt + k - c_0\right)x = c_1.$$

From (19) we see that for $a_1 \neq 0$ this limit does not exist. Hence $a_1 = 0$, and (15) becomes

$$f'(x) \sim \frac{a_2}{x^2} + \frac{a_3}{x^3} + \cdots .$$

Consequently, from (18) and Theorem 2 we now obtain

$$(20)\qquad \begin{aligned} f(x) &= \int_{x_0}^{\infty} f'(t)\,dt - \int_{x}^{\infty} f'(t)\,dt + k \\ &\sim \int_{x_0}^{\infty} f'(t)\,dt + k - \frac{a_2}{x} - \frac{a_3}{2x^3} - \cdots , \end{aligned}$$

the first integral on the right being a constant. If a given function has an asymptotic expansion, it is unique. We may thus compare corresponding terms in (13) and (20), finding $a_2 = -c_1$, $a_3 = -2c_2$, etc. With these coefficients the series (14) and (15) become identical, and the theorem is proved. ▮

If we know that a function $f(x)$ satisfies the assumptions of Theorem 3 and is a solution of a first-order differential equation, then we may determine the coefficients of its asymptotic expansion by substituting (13) and (14) in the differential equation.

Example 2. Exponential integral

The exponential integral $\mathrm{Ei}(x)$ is defined by the formula

$$\mathrm{Ei}(x) = \int_x^{\infty} \frac{e^{-t}}{t}\,dt \qquad (x > 0).$$

To obtain the asymptotic expansion of $\mathrm{Ei}(x)$, we consider the function

$$y = f(x) = e^{x}\,\mathrm{Ei}(x) = e^{x} \int_x^{\infty} \frac{e^{-t}}{t}\,dt \qquad (x > 0).$$

By differentiation we see that $f(x)$ satisfies the linear differential equation

$$(21)\qquad y' - y + \frac{1}{x} = 0.$$

It may be proved directly that this equation has only one solution y such that y and y' exist for positive x and have an asymptotic expansion. By substituting (13) and (14) into (21) and equating the coefficient of each power of x to zero we obtain

$$-c_0 = 0, \quad -c_1 + 1 = 0, \quad -c_1 - c_2 = 0, \quad \cdots, \quad -nc_n - c_{n+1} = 0, \quad \cdots,$$

that is,

$$c_0 = 0, \quad c_1 = 1, \quad c_2 = -1, \quad \cdots, \quad c_{n+1} = (-1)^n n!, \quad \cdots$$

and, therefore,

$$(22)\qquad \mathrm{Ei}(x) = e^{-x} f(x) \sim e^{-x}\left(\frac{1}{x} - \frac{1}{x^2} + \frac{2!}{x^3} - \frac{3!}{x^4} + - \cdots\right).$$

Problems for Sec. 19.15

1. Show that the series $1 + \frac{1}{x} + \frac{1}{2!\,x^2} + \cdots$ [which converges for $|x| > 0$ and represents $e^{1/x}$] is an asymptotic expansion of $e^{1/x}$.

2. Show that $\sin\frac{1}{x} \sim \frac{1}{x} - \frac{1}{3!\,x^3} + \frac{1}{5!\,x^5} - + \cdots$.

Integrating by parts, find the asymptotic expansion of[12]

3. $\mathrm{ci}(x) = \int_x^\infty \frac{\cos t}{t}\,dt$ **(Cosine integral).**

4. $\mathrm{si}(x) = \int_x^\infty \frac{\sin t}{t}\,dt$ **(Complementary sine integral).**

5. $\mathrm{c}(x) = \int_x^\infty \cos t^2\,dt$ **(Complementary Fresnel integral).**

6. $\mathrm{s}(x) = \int_x^\infty \sin t^2\,dt$ **(Complementary Fresnel integral).**

7. $Q(\alpha, x) = \int_x^\infty e^{-t}t^{\alpha-1}\,dt$ **(Incomplete gamma function).**

Using the results of Probs. 4, 5, and 7, find the asymptotic expansion of

8. $\mathrm{Si}(x) = \int_0^x \frac{\sin t}{t}\,dt$ **(Sine integral).** Use $\mathrm{si}(0) = \frac{\pi}{2}$.

9. $\mathrm{C}(x) = \int_0^x \cos t^2\,dt$ **(Fresnel integral).** Use $\mathrm{c}(0) = \frac{1}{2}\sqrt{\frac{\pi}{2}}$.

10. $P(\alpha, x) = \int_0^x e^{-t}t^{\alpha-1}\,dt$ **(Incomplete gamma function).**

11. Obtain the asymptotic expansion of $\operatorname{erf} x$ by showing that $y = \frac{1}{2}\sqrt{\pi}\,e^{x^2}\operatorname{erfc} x$ satisfies $y' - 2xy + 1 = 0$.

12. Obtain the asymptotic expansion of $Q(\frac{1}{2}, x)$ by showing that $y = e^x\sqrt{x}\,Q(\frac{1}{2}, x)$ satisfies $y' = (1/2x + 1)y - 1$.

13. Obtain the asymptotic expansion of $Q(\alpha, x)$ from a differential equation for $y = e^x x^{1-\alpha}Q(\alpha, x)$.

14. Show that

$$\mathrm{Ei}(x) = e^{-x}\int_0^\infty \frac{e^{-u}}{u + x}\,du$$

and obtain from this the series in (22) by developing $1/(u + x)$ in powers of $1/x$ and termwise integration.

15. Show that $\mathrm{Ei}(ix) = \mathrm{ci}(x) - i\,\mathrm{si}(x)$. Then replace x by ix in (22) and separate the real and imaginary parts on both sides; show that this leads to the asymptotic expansions of $\mathrm{ci}(x)$ and $\mathrm{si}(x)$ as obtained in Probs. 3 and 4.

[12] Some formulas for these and related functions are included in Appendix 3. Further formulas can be found in Ref. [A5] (cf. Appendix 1) and references to tables of numerical values in Ref. [A6]. In Prob. 3, *cosine integral* is the usual name since the integral from 0 to x does not exist. (Why?)

CHAPTER

20 PROBABILITY AND STATISTICS

The importance of mathematical statistics in engineering is increasing, in particular in mass production and in the analysis of experimental data. In the present chapter we shall give an introduction to this field. We shall start with the tabular and graphical representation of data (Secs. 20.2, 20.3). Then we shall discuss some fundamental concepts and relations from the theory of mathematical probability (Secs. 20.4–20.11), because probability theory forms the basis of mathematical statistics. The remaining sections (20.12–20.20) will be devoted to some of the most important statistical methods.

Prerequisite for this chapter: elementary differential and integral calculus.
Sections which may be omitted in a shorter course: 20.17–20.20.
References: Appendix 1, Part H.
Answers to problems: Appendix 2.

20.1 Nature and Purpose of Mathematical Statistics

In engineering statistics we are concerned with methods for designing and evaluating experiments to obtain information about practical problems, for example, the inspection of quality of raw material or manufactured products, the comparison of machines and tools or methods used in production, the output of workers, the reaction of consumers to new products, the yield of a chemical process under various conditions, the relation between the iron content of iron ore and the density of the ore, the effectivity of air conditioning systems under various temperatures, the relation between the Rockwell hardness and the carbon content of steel, and so on.

For instance, in a process of mass production (of screws, bolts, light bulbs, typewriter keys, etc.) we ordinarily have *nondefective items,* that is, articles which satisfy the quality requirements, and *defective items,* those which do not satisfy the requirements. Such requirements include maximum and minimum diameters of axle shafts, minimum lifetimes of light bulbs, limiting values of the resistance of resistors in radio and TV production, maximum weights of airmail envelopes, minimum contents of automatically filled bottles, maximum reaction times of switches, and minimum values of strength of yarn.

The reason for the differences in the quality of products is *variation* due to numerous factors (in the raw material, function of automatic machinery, workmanship, etc.) whose influence cannot be predicted, so that the variation must be regarded as a *random variation.* In the case of evaluating the efficiency of production methods and the other preceding examples the situation is similar.

In most cases the inspection of each item of a production is prohibitively expensive and time-consuming. It may even be impossible if it leads to the destruction of the item. Hence instead of inspecting all the manufactured items just a few of them (a "*sample*") are inspected and from this inspection conclusions can be drawn about the totality (the "*population*"). If we draw 100 screws from a lot of 10,000 screws and find that 5 of the 100 screws are defective, we are inclined to conclude that *about* 5% of the lot are defective, provided the screws were drawn "*at random*", that is, so that each screw of the lot had the same "chance" of being drawn. It is clear that such a conclusion is not absolutely certain, that is, we cannot say that the lot contains *precisely* 5% defectives, but in most cases such a precise statement would not be of particular practical interest anyway. We also feel that the conclusion is the more dependable the more randomly selected screws we inspect. Using the theory of mathematical probability we shall see that those somewhat vague intuitive notions can be made precise. Furthermore, this theory will also yield measures for the reliability of conclusions about populations obtained from samples by statistical methods. Hence probability theory forms the basis of statistical methods.

Similarly, to obtain information about the iron content μ of iron ore, we may pick a certain number n of specimens at random and measure the iron content. This yields a sample of n numbers $x_1, \cdots, x_n$ (the results of the n measurements) whose average $\bar{x} = (x_1 + x_2 + \cdots + x_n)/n$ is an approximate value for μ.

Problems of differing natures may require different methods and techniques, but the steps leading from the formulation to the solution of the problem are the same in almost all cases. They are as follows.

1. *Formulation of the problem.* It is important to formulate the problem in a precise fashion and to limit the investigation, so that we can expect a useful answer within a prescribed interval of time, taking into account the cost of the statistical investigation, the skill of the investigators, and the facilities available. This step must also include the creation of a mathematical model based on clear concepts. (For example, we must define what we mean by a defective item in a specific situation.)

2. *Design of experiment.* This step includes the choice of the statistical method to be used in the last step, the sample size n (number of items to be drawn and inspected or number of experiments to be made, etc.), and the physical methods and techniques to be used in the experiment. The goal is to obtain a maximum of information using a minimum of cost and time.

3. *Experimentation or collection of data.* This step should adhere to strict rules.

4. *Tabulation.* In this step we arrange the experimental data in a clear and simple tabular form, and we may represent them graphically by diagrams, bar charts, etc. We also compute numbers that characterize the average size and the spread of the sample values.

5. *Statistical inference.* In this step we use the sample and apply a suitable statistical method for drawing conclusions about the unknown properties of the population so that we obtain the answer to our problem.

20.2 Tabular and Graphical Representation of Samples

In the course of a statistical experiment we usually obtain a sequence of observations (numbers in most cases) which we write down in the order they occur. A typical example is shown in Table 20.1. These data were obtained by making standard test cylinders (diameter 6 in., length 12 in.) of concrete and splitting them 28 days later. We thus have a **sample** consisting of 100 **sample values,** so that the **size** of the sample is $n = 100$.

In the present section we want to learn how to represent samples in a suitable tabular and graphical form. It suffices to discuss the corresponding methods in terms of our sample in Table 20.1.

To see what information is contained in Table 20.1, we shall now order the data. We write 300 (the smallest value), 310, 320, $\cdots$, 440 (the largest value) in a column. Then we run through Table 20.1 line by line and make a tally mark for each occurring number. In this way we obtain a **tally chart** of our sample, consisting of the first two columns of Table 20.2. The number of tallies is then listed in the third column of Table 20.2. It indicates how often the corresponding value x occurs in the sample and is called the *absolute frequency* or, more briefly, the **frequency** of that value x in the sample. Dividing it by the size n of the sample, we obtain the **relative frequency** in column 4; in our case, $n = 100$. Thus $x = 330$ has the frequency 6 and the relative frequency 0.06 or 6%.

If for a certain x we sum all the frequencies corresponding to the sample values which are smaller than or equal to that x, we obtain the **cumulative frequency** corresponding to that x. This yields column 5 in Table 20.2. For example, to $x = 350$ there corresponds the cumulative frequency 37, and this tells us that there are 37 sample values which are smaller than or equal to 350. Division by the size n of the sample yields the **cumulative relative frequency** in column 6. For example, from column 6 we see that 76% of the sample values are smaller than or equal to 380.

Tallying is subject to error and is difficult to check. A better method is to put

Table 20.1
Sample of 100 Values of the Splitting Tensile Strength (lb/in.^{2}) of Concrete Cylinders

320	380	340	410	380	340	360	350	320	370
350	340	350	360	370	350	380	370	300	420
370	390	390	440	330	390	330	360	400	370
320	350	360	340	340	350	350	390	380	340
400	360	350	390	400	350	360	340	370	420
420	400	350	370	330	320	390	380	400	370
390	330	360	380	350	330	360	300	360	360
360	390	350	370	370	350	390	370	370	340
370	400	360	350	380	380	360	340	330	370
340	360	390	400	370	410	360	400	340	360

D. L. IVEY, Splitting tensile tests on structural lightweight aggregate concrete. Texas Transportation Institute, College Station, Texas, 1965

Table 20.2
Frequency Table of the Sample in Table 16

1 Tensile Strength x [lb/in.^{2}]	2 Absolute Frequency Tallies	3 	4 Relative Frequency	5 Cumulative Frequency	6 Cumulative Relative Frequency
300	II	2	0.02	2	0.02
310		0	0.00	2	0.02
320	IIII	4	0.04	6	0.06
330	𝍸 I	6	0.06	12	0.12
340	𝍸 𝍸 I	11	0.11	23	0.23
350	𝍸 𝍸 IIII	14	0.14	37	0.37
360	𝍸 𝍸 𝍸 I	16	0.16	53	0.53
370	𝍸 𝍸 𝍸	15	0.15	68	0.68
380	𝍸 III	8	0.08	76	0.76
390	𝍸 𝍸	10	0.10	86	0.86
400	𝍸 III	8	0.08	94	0.94
410	II	2	0.02	96	0.96
420	III	3	0.03	99	0.99
430		0	0.00	99	0.99
440	I	1	0.01	100	1.00

the data on individual cards, sort the cards into piles and count the number of cards in each pile. This can be done by hand or in a completely mechanistic fashion.

If a certain numerical value does not occur in a sample, its frequency is 0. If all the n values of the sample are numerically equal, then this number has the frequency n and, therefore, the relative frequency $n/n = 1$. Since these are the two extreme possible cases, we have

Theorem 1. (Relative frequency)
The relative frequency is at least equal to 0 *and at most equal to* 1.

Suppose that a given sample of size n consists of m *numerically different* values

$$x_1, \quad x_2, \quad \cdots, \quad x_m \qquad (m \leqq n)$$

with corresponding relative frequencies

$$\tilde{f}_1, \quad \tilde{f}_2, \quad \cdots, \quad \tilde{f}_m.$$

Then we may introduce the function[1]

$$\textbf{(1)} \qquad \tilde{f}(x) = \begin{cases} \tilde{f}_j & \text{when } x = x_j \qquad (j = 1, 2, \cdots, m) \\ 0 & \text{for any value } x \text{ not appearing in the sample.} \end{cases}$$

[1]We write $\tilde{f}$ because we want to reserve the simple symbol f for the theoretical counterpart of the frequency function. f will appear much more often in our text.

This function is called the **frequency function of the sample.** It shows how the values of the sample are distributed. We therefore say that it determines the **frequency distribution** of the sample.

For example, in Table 20.2 the values of the frequency function are shown in column 4, and we see that $\tilde{f}(300) = 0.02$, $\tilde{f}(310) = 0$, $\tilde{f}(320) = 0.04$, and so on.

The sum of all the frequencies in a sample of size n must equal n. (Why?) This yields

Theorem 2. (Sum of relative frequencies)

The sum of all the relative frequencies in a sample equals 1, *that is,*

$$\sum_{j=1}^{m} \tilde{f}(x_j) = \tilde{f}(x_1) + \tilde{f}(x_2) + \cdots + \tilde{f}(x_m) = 1.$$

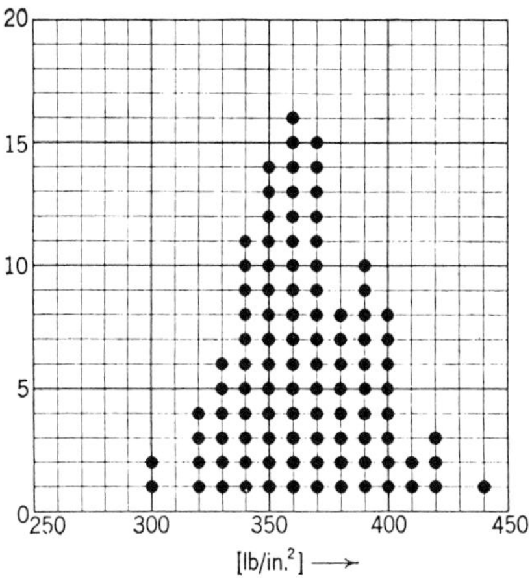

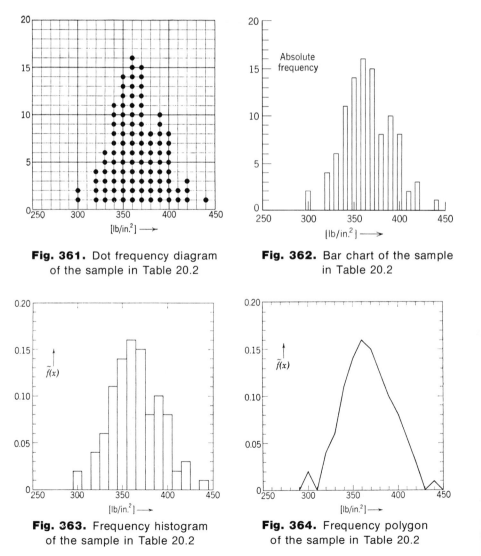

Fig. 361. Dot frequency diagram of the sample in Table 20.2

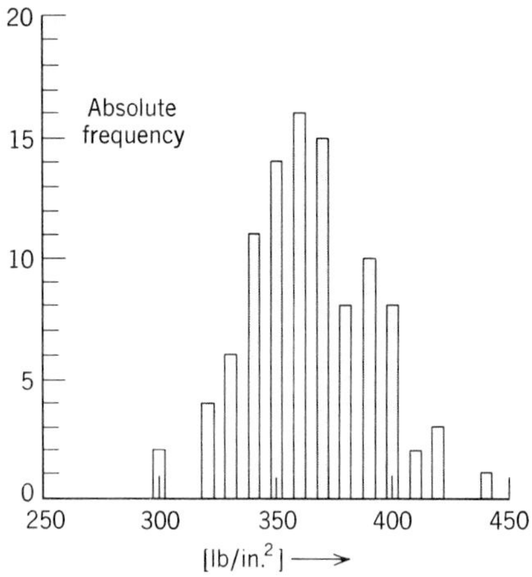

Fig. 362. Bar chart of the sample in Table 20.2

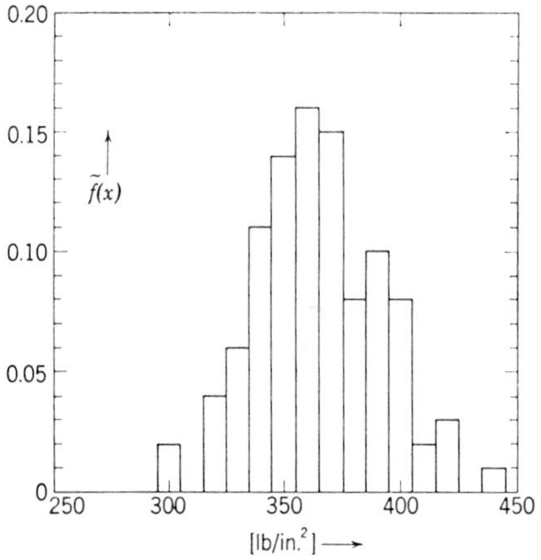

Fig. 363. Frequency histogram of the sample in Table 20.2

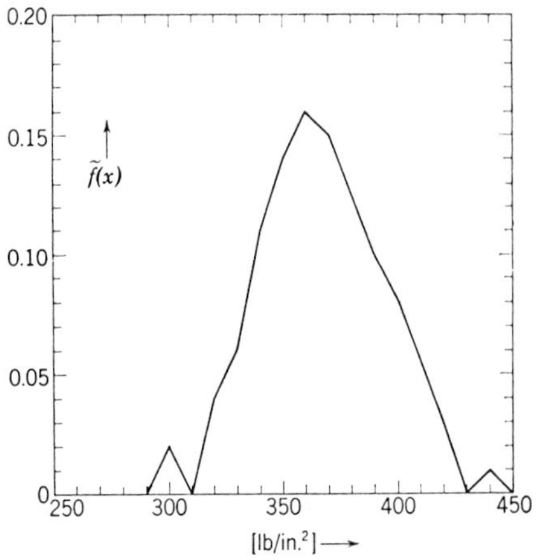

Fig. 364. Frequency polygon of the sample in Table 20.2

Graphical representations of samples are illustrated by Figs. 361 to 364. In Fig. 363 the area of each rectangle is equal to the corresponding relative frequency. Hence the ordinate should be labeled "*relative frequency per unit interval.*" Since in the present case the rectangles are equally wide, those values on the ordinate are proportional to $\tilde{f}(x)$ and we may label the ordinate in terms of $\tilde{f}(x)$. However, this would no longer be true if the rectangles were of different width. In Fig. 364 the situation is similar.

We now introduce the function

$$\tilde{F}(x) = \textit{sum of the relative frequencies of all the values that are smaller than } x \textit{ or equal to } x.$$

This function is called the **cumulative frequency function of the sample** or the **sample distribution function.** An example is shown in Fig. 365.

$\tilde{F}(x)$ is a *step function* (piecewise constant function) having jumps of magnitude $\tilde{f}(x)$ precisely at those x at which $\tilde{f}(x) \neq 0$. The first jump is at the smallest sample value and the last at the largest. Afterwards $\tilde{F}(x) = 1$.

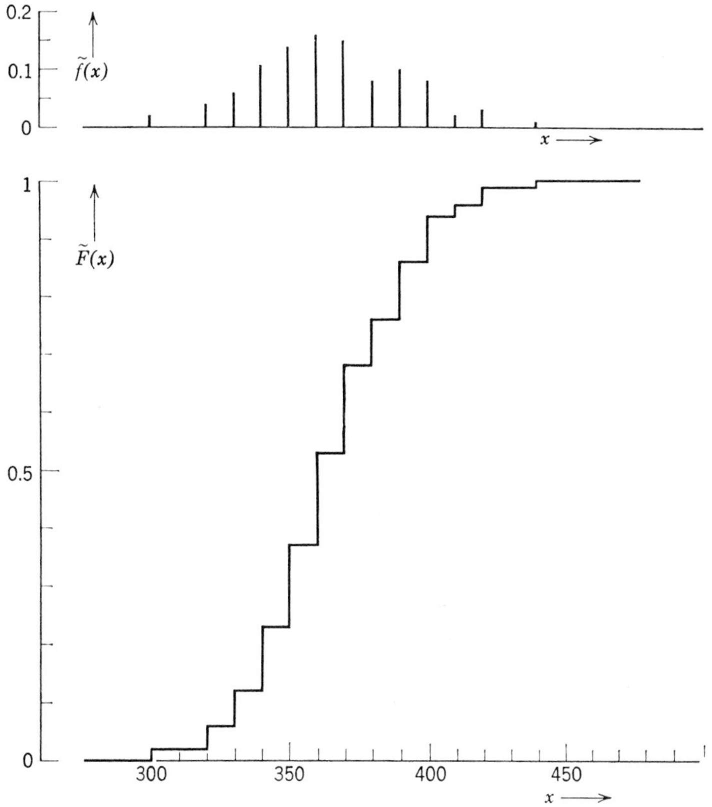

Fig. 365. Frequency function $\tilde{f}(x)$ and cumulative frequency function $\tilde{F}(x)$ of the sample in Table 20.2

The relation between $\tilde{f}(x)$ and $\tilde{F}(x)$ is

$$\tilde{F}(x) = \sum_{t \leqq x} \tilde{f}(t) \tag{2}$$

where $t \leqq x$ means that for an x we have to form the sum of all those $\tilde{f}(t)$ for which t is less than x or equal to x.

If a sample consists of too many numerically different sample values, then its tabular and graphical representations are unnecessarily complicated, but may be simplified by the process of **grouping,** as follows.

A sample being given, we choose an interval I that contains all the sample values. We subdivide I into subintervals, which are called **class intervals.** The midpoints of these intervals are called **class midpoints** or **class marks.** The sample values in each such interval are said to form a **class.** Their number is called the corresponding **class frequency.** Division by the sample size n gives the **relative class frequency.** This frequency, considered as a function $\tilde{f}(x)$ of the class marks is called the **frequency function of the grouped sample,** and the corresponding cumulative relative class frequency, considered as a function $\tilde{F}(x)$ of the class marks, is called the **distribution function of the grouped sample.** An example is shown in Tables 20.3 and 20.4.

The fewer classes we choose, the simpler the distribution of the grouped sample becomes but the more information we lose, because the original sample values no longer appear explicitly. Grouping should be done so that only unessential details are eliminated. Unnecessary complications in the later use of

Table 20.3
Strength of 50 Lots of Cotton (lb required to break a skein)

114	118	86	107	87	94	82	81	98	84
120	126	98	89	114	83	94	106	96	111
123	110	83	118	83	96	96	74	91	81
102	107	103	80	109	71	96	91	86	129
130	104	86	121	96	96	127	94	102	87

Table 20.4
Frequency Table of the Sample in Table 20.3 (Grouped)

Class Interval	Class Mark x	Absolute Frequency Tallies		$\tilde{f}(x)$	$\tilde{F}(x)$
65– 75	70	𝍷𝍷	2	0.04	0.04
75– 85	80	𝍸 𝍷𝍷𝍷	8	0.16	0.20
85– 95	90	𝍸 𝍸 𝍷	11	0.22	0.42
95–105	100	𝍸 𝍸 𝍷𝍷	12	0.24	0.66
105–115	110	𝍸 𝍷𝍷𝍷	8	0.16	0.82
115–125	120	𝍸	5	0.10	0.92
125–135	130	𝍷𝍷𝍷𝍷	4	0.08	1.00
		Sum	50	1.00	

a grouped sample are avoided by obeying the following rules.

1. All the class intervals should have the same length.
2. The class intervals should be chosen so that the class marks correspond to simple numbers (numbers with few nonzero digits).
3. If a sample value x_j coincides with the common endpoint of two class intervals, we take it into that class interval which extends from x_j to the right.

Problems for Sec. 20.2

In each case make a frequency table of the given sample and represent the sample by a dot frequency diagram, a bar chart, and a histogram.

1. Resistance [ohms] of resistors

99 100 102 101 98 103 100 102 99 101
100 100 99 101 100 102 99 101 98 100

2. Numbers that turned up on a die

6 2 4 1 2 4 3 3 2 1 6 5 6 3 4

3. Release time [sec] of a relay

1.3 1.4 1.1 1.5 1.4 1.3 1.2 1.4 1.5 1.3
1.2 1.3 1.5 1.4 1.4 1.6 1.3 1.5 1.1 1.4

4. Carbon content [%] of coal

87 86 85 87 86 87 86 81 77 85
86 84 83 83 82 84 83 79 82 73

5. Tensile strength [kg/mm²] of sheet steel

44 43 41 41 44 44 43 44 42 45 43 43 44 45 46
42 45 41 44 44 43 44 46 41 43 45 45 42 44 44

6. Number of sheets of paper over and under the desired number of 100 sheets per package in a packaging process

0 −1 0 0 1 1 2 0 1 0

7. Miles per gallon of gasoline required by six cars of the same make

15.0 15.5 14.5 15.0 15.5 15.0

8. Weight of filled bags (in grams) in an automatic filling process

200 203 199 198 201 200 201 201

9. Waiting time (in minutes, rounded) of a commuter for a train in a certain subway

3 4 1 0 2 2 3 1 5 3

10. Graph the cumulative frequency function of the sample in Prob. 3.

11. Graph the bar chart, the histogram, and the frequency polygon of the grouped sample in Table 20.4.

12. Graph the histogram of the following sample of lifetimes [hours] of light bulbs.

Lifetime	Absolute Frequency	Lifetime	Absolute Frequency	Lifetime	Absolute Frequency
950–1050	4	1350–1450	51	1750–1850	20
1050–1150	9	1450–1550	58	1850–1950	9
1150–1250	19	1550–1650	53	1950–2050	3
1250–1350	36	1650–1750	37	2050–2150	1

13. Group the sample in Table 20.1, using class intervals with midpoints 300, 320, 340, $\cdots$. Make up the corresponding frequency table. Graph the histogram and compare it with Fig. 363. Graph the cumulative frequency function.

14. Group the sample shown in Table 20.3, using class intervals with midpoints 75, 85, 95, $\cdots$. Make up the corresponding frequency table. Graph the histogram and compare it with that in Prob. 11.

15. The smallest of 1500 measurements was 10.8 cm and the largest was 11.9 cm. Suggest class intervals for grouping these data.

20.3 Sample Mean and Sample Variance

The frequency function (or equally well the distribution function) characterizes a given sample in detail. From this function we may compute measures for certain properties of the sample, such as the average size of the sample values, the "spread," the "asymmetry," etc. In the present section we shall consider the two most important such quantities, which are called the sample mean and the sample variance.

The *mean value of a sample* $x_1, x_2, \cdots, x_n$ or, briefly, **sample mean,** is denoted by $\bar{x}$ and is defined by the formula

$$\text{(1)} \qquad \bar{x} = \frac{1}{n}\sum_{j=1}^{n} x_j = \frac{1}{n}(x_1 + x_2 + \cdots + x_n).$$

It is the sum of all the sample values, divided by the size n of the sample. Obviously, it measures the average size of the sample values, and sometimes the term *average* is used for $\bar{x}$.

The *variance of a sample* $x_1, x_2, \cdots, x_n$ or, briefly, **sample variance,** is denoted by s^2 and is defined by the formula

$$\text{(2)} \qquad \begin{aligned} s^2 &= \frac{1}{n-1}\sum_{j=1}^{n}(x_j - \bar{x})^2 \\ &= \frac{1}{n-1}[(x_1 - \bar{x})^2 + (x_2 - \bar{x})^2 + \cdots + (x_n - \bar{x})^2]. \end{aligned}$$

It is the sum of the squares of the deviations of the sample values from the mean $\bar{x}$, divided by $n - 1$. It measures the spread or dispersion of the sample values and is positive, except for the rare case when all the sample values are equal (and are then equal to $\bar{x}$). The positive square root of the sample variance s^2 is called the **standard deviation** of the sample and is denoted by s.

Example 1. Sample mean and sample variance
Ten randomly selected nails had the lengths (in.)

$$0.80 \quad 0.81 \quad 0.81 \quad 0.82 \quad 0.81 \quad 0.82 \quad 0.80 \quad 0.82 \quad 0.81 \quad 0.81.$$

From (1) we see that the sample mean is

$$\bar{x} = \tfrac{1}{10}(0.80 + 0.81 + 0.81 + 0.82 + \cdots + 0.81) = 0.811 \text{ [in.]}.$$

Applying (2), we thus obtain the sample variance

$$s^2 = \tfrac{1}{9}[(0.800 - 0.811)^2 + \cdots + (0.810 - 0.811)^2] = 0.000\,054 \text{ [in.}^2\text{]}.$$

This calculation becomes simpler if we take equal sample values together. Then

$$\bar{x} = \tfrac{1}{10}(2 \cdot 0.80 + 5 \cdot 0.81 + 3 \cdot 0.82) = 0.811.$$

In the parentheses we have the sum of the three *numerically different* sample values $x_1 = 0.80$, $x_2 = 0.81$, $x_3 = 0.82$, each multiplied by its frequency. Similarly,

$$s^2 = \tfrac{1}{9}[2(0.800 - 0.811)^2 + 5(0.810 - 0.811)^2 + 3(0.820 - 0.811)^2] = 0.000\,054. \quad \blacksquare$$

The example illustrates how we may compute $\bar{x}$ and s^2 by the use of the frequency function $\tilde{f}(x)$ of the sample. If a sample of n values contains precisely m *numerically different* sample values

$$x_1, \quad x_2, \quad \cdots, \quad x_m \qquad (m \leqq n),$$

the corresponding relative frequencies are

$$\tilde{f}(x_1), \quad \tilde{f}(x_2), \quad \cdots, \quad \tilde{f}(x_m).$$

Hence the corresponding frequencies needed in the computation are

$$n\tilde{f}(x_1), \quad n\tilde{f}(x_2), \quad \cdots, \quad n\tilde{f}(x_m),$$

and (1) and (2) take the form

$$\textbf{(3)} \qquad \bar{x} = \frac{1}{n}\sum_{j=1}^{m} x_j n\tilde{f}(x_j)$$

and

$$\textbf{(4)} \qquad s^2 = \frac{1}{n-1}\sum_{j=1}^{m} (x_j - \bar{x})^2 n\tilde{f}(x_j).$$

Note that in (1) and (2) we sum over *all* the sample values, whereas now we sum over the *numerically different* sample values. The absolute frequencies $n\tilde{f}(x_j)$ are integers, while the relative frequencies $\tilde{f}(x_j)$ may be messy, for example if $n = 23$ or $n = 84$, etc.

Our formulas for s^2 are impractical because $x_j - \bar{x}$ may be small in absolute value compared with the sample values, and this may cause a loss of significant digits (undetected in automatic calculation). We shall derive the formula for s^2 which is used in computations. Inserting

$$(x_j - \bar{x})^2 = x_j^2 - 2x_j\bar{x} + \bar{x}^2$$

into (2) and decomposing into three sums, we have

$$\sum (x_j - \bar{x})^2 = \sum x_j^2 - 2\bar{x} \sum x_j + \sum \bar{x}^2.$$

The last sum equals $n\bar{x}^2$. Replacing $\bar{x}$ according to (1), we obtain

$$-2\bar{x} \sum x_j = -\frac{2}{n}\Big(\sum x_j\Big)^2 \qquad \text{and} \qquad n\bar{x}^2 = \frac{1}{n}\Big(\sum x_j\Big)^2.$$

Taking these two expressions together, we finally have

$$\text{(5)} \qquad s^2 = \frac{1}{n-1}\left[\sum_{j=1}^{n} x_j^2 - \frac{1}{n}\Big(\sum_{j=1}^{n} x_j\Big)^2\right].$$

Similarly, (4) may be transformed to

$$\text{(6)} \qquad s^2 = \frac{1}{n-1}\left[\sum_{j=1}^{m} x_j^2 n\tilde{f}(x_j) - \frac{1}{n}\Big(\sum_{j=1}^{m} x_j n\tilde{f}(x_j)\Big)^2\right].$$

For instance, in Example 1 we obtain from (3) and (6) (cf. Table 20.5) the values $\bar{x} = 8.11/10 = 0.811$ and

$$s^2 = \frac{1}{9}\left(6.5777 - \frac{8.11^2}{10}\right) = \frac{0.00049}{9} = 0.000\,054,$$

as before.

Table 20.5
Computation of Mean and Variance in Example 1 by the Use of (3) and (6)

x_j	$10\tilde{f}(x_j)$	$x_j \cdot 10\tilde{f}(x_j)$	x_j^2	$x_j^2 \cdot 10\tilde{f}(x_j)$
0.80	2	1.60	0.6400	1.2800
0.81	5	4.05	0.6561	3.2805
0.82	3	2.46	0.6724	2.0172
	Sum	8.11		6.5777

Problems for Sec. 20.3

1. Compute the mean and variance of the sample in Prob. 2, Sec. 20.2.
2. Compute the mean and variance of the sample in Prob. 4, Sec. 20.2.
3. Graph a histogram of the sample 2, 1, 4, 5 and guess $\bar{x}$ and s by inspecting the histogram. Then calculate $\bar{x}$, s^2, and s.
4. Show that $\bar{x}$ lies between the smallest and the largest sample values.

5. **(Range of a sample)** The difference between the largest and the smallest values in a sample is called the *range of the sample.* Find the range of the sample in Example 1.
6. **(Percentile, median)** The pth **percentile** of a sample is a number Q_p such that at least $p\%$ of the sample values are smaller than or equal to Q_p and also at least $(100 - p)\%$ of those values are larger than or equal to Q_p. If there is more than one such number (in which case there will be an interval of them), the pth percentile is defined as the average of the numbers (midpoint of that interval). In particular, Q_{50} is called the **middle quartile** or **median** and is denoted by $\tilde{x}$. Find $\tilde{x}$ for the sample in Table 20.2.
7. The percentiles Q_{25} and Q_{75} of a sample are called the **lower** and **upper quartiles** of the sample, and $Q_{75} - Q_{25}$, which is a measure for the spread, is called the **interquartile range.** Find Q_{25}, Q_{75}, and $Q_{75} - Q_{25}$ for the sample in Table 20.2.
8. Same tasks as in Probs. 6 and 7 for the sample in Table 20.3.
9. **(Mode)** A *mode* of a sample is a sample value that occurs most frequently in the sample. Find the mean, median, and mode of the following sample. Comment.

Value of Stocks Owned	100	1000	100,000
Frequency	100	90	20

10. **(Working origin)** If $x_j = x_j^* + c$, where $j = 1, \cdots, n$ and c is any constant, show that

$$\bar{x} = c + \bar{x}^* \quad \left(\bar{x}^* = \frac{1}{n}\sum_{j=1}^{n} x_j^*\right), \quad s^2 = s^{*2},$$

where s^{*2} is the variance of the x_j^*'s. (In practice, c is chosen so that the x_j^*'s are small in absolute value. Geometrically this is a shift of the origin and is called the *method of working origin.*)
11. Apply the method of working origin to the sample in Example 1.
12. **(Full coding)** If $x_j = c_1 x_j^* + c_2$, where $j = 1, \cdots, n$ and c_1 and c_2 are constants, show that

$$\bar{x} = c_1\bar{x}^* + c_2, \qquad s^2 = c_1^2 s^{*2},$$

where $\bar{x}^*$ and s^{*2} have the same meaning as in Prob. 10. (This is called the *method of full coding.* Obviously it is valuable in nonelectronic calculations, for instance for quick checks.)
13. Apply full coding to the sample in Example 1.
14. If a sample is being grouped, its mean will change, in general. Show that the change cannot exceed $l/2$ where l is the length of each class interval.
15. Compute the mean and the variance of the ungrouped sample in Table 20.3 (Sec. 20.2) and the grouped sample in Table 20.4 and compare the results.

20.4 Random Experiments, Outcomes, Events

Statistical experiments or observations yield samples from which we want to draw conclusions about the corresponding population. Before we can do so we must first develop mathematical models of populations, using the theory of mathematical probability. This theory, therefore, supplies an important basis in mathematical statistics, and we shall consider it to the extent needed. In this section we shall introduce some fundamental concepts.

A *random experiment* or *random observation,* briefly **experiment** or **observation,** is a process that has the following properties.

1. It is performed according to a set of rules that determines the performance completely.
2. It can be repeated arbitrarily often.
3. The result of each performance depends on "chance" (that is, on influences which we cannot control) and can therefore not be uniquely predicted.

The result of a single performance of the experiment is called the **outcome** of that trial.

Examples are games of chance such as rolling a die or flipping a coin, technical experiments such as the random selection and inspection of 10 screws from a box containing 100 screws or the determination of the yield of a chemical process under various conditions, and other experiments such as the random selection of 20 persons from a group and the determination of their blood pressures or their opinions about a certain movie.

The set of all possible outcomes of an experiment is called the **sample space** of the experiment and will be denoted by S. Each outcome is called an *element* or *point* of S. A sample space is said to be *finite* or *infinite,* depending on whether it consists of finitely many or infinitely many elements, respectively.

For instance, with the random experiment of rolling a die we may associate the sample space

$$S = \{1, 2, 3, 4, 5, 6\}$$

consisting of six elements corresponding to the six faces the die can turn up.

In connection with industrial production we may draw an item to find out whether it is defective or nondefective. Then S consists of the two elements D (defective) and N (nondefective), which may also be characterized by numbers, for example 0 (defective) and 1 (nondefective). If in this case we distinguish between several types of defects, we have a sample space consisting of more than two points.

The sample space of the experiment of measuring the strength of cotton (cf. Table 20.3 in Sec. 20.2) is infinite, because the outcome may be any positive number within a certain range.

In most practical problems we are not so much interested in the individual outcomes but in whether an outcome belongs (or does not belong) to a certain set of outcomes. Clearly, each such set A is a subset of the sample space S. It is called an **event.**

Since an outcome is a subset of S, it is an event, but a rather special one, sometimes called an *elementary event*. Similarly, the entire space S is another special event.

Example 1

If we draw 2 gaskets from a set of 5 gaskets (numbered from 1 to 5), the sample space consists of the 10 possible outcomes

$$1, 2 \quad 1, 3 \quad 1, 4 \quad 1, 5 \quad 2, 3 \quad 2, 4 \quad 2, 5 \quad 3, 4 \quad 3, 5 \quad 4, 5,$$

but we may be interested in the number of defective gaskets we get in the draws and thus distinguish among the 3 events

A: No defective gasket, *B: 1 defective gasket,* *C: 2 defective gaskets.*

Assuming that 3 gaskets, say 1, 2, 3, are defective, we see that

A occurs, if we draw	4, 5					
B occurs, if we draw	1, 4;	1, 5;	2, 4;	2, 5;	3, 4,	or 3, 5
C occurs, if we draw	1, 2;	1, 3;	or 2, 3.			

∎

A sample space S and the events of an experiment can be represented graphically by a **Venn diagram,** as follows. Suppose that the set of points inside the rectangle in Fig. 366 represents S. Then the interior of a closed curve inside the rectangle represents an event which we denote by E. The set of all the elements (outcomes) not in E is called the **complement** *of E in S* and is denoted[2] by E^c.

For example, in the experiment of rolling a die the complement of the event

E: The die turns up an even number

is the event

E^c: The die turns up an odd number.

An event containing no element is called the **impossible event** or *empty event* and is denoted by $\emptyset$.

Let A and B be any two events in an experiment. Then the event consisting of all the elements of the sample space S contained in A or B, or both, is called the **union** of A and B and is denoted by

$$A \cup B.$$

The event consisting of all the elements in S contained in both A and B is called the **intersection** of A and B and is denoted by

$$A \cap B.$$

Figure 367 illustrates how to represent these two events by a Venn diagram. If A and B have no element in common, then $A \cap B = \emptyset$, and A and B are called **mutually exclusive events** or **disjoint events.**

For instance, in Example 1, $B \cap C = \emptyset$ and $B \cup C$: *1 or 2 defective gaskets.*

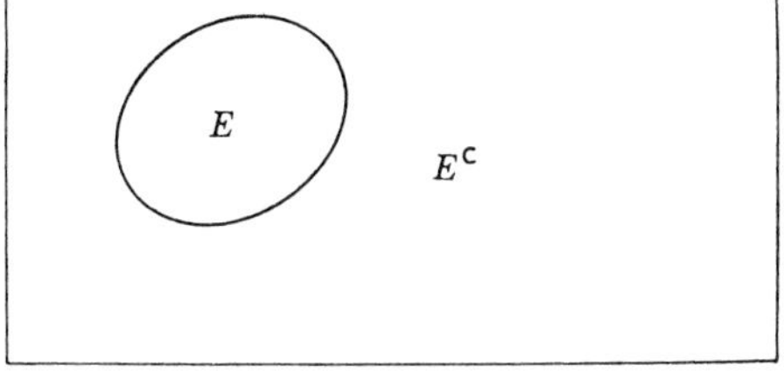

Fig. 366. Venn diagram representing a sample space S and events E and E^c

[2] Or by $\overline{E}$, but we shall not use this notation because it is used in set theory for another purpose (to denote the closure of a set).

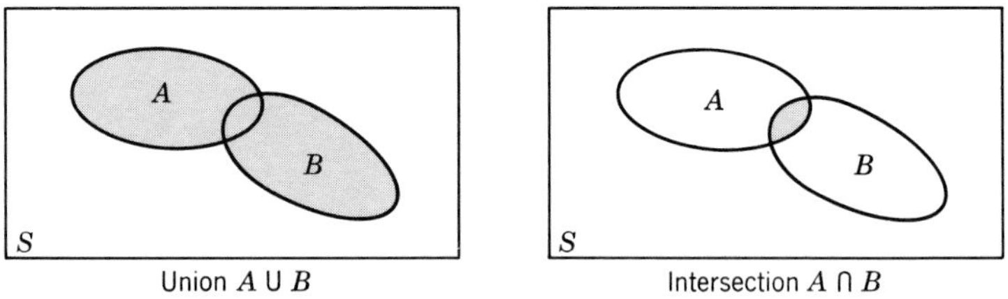

Fig. 367. Venn diagrams representing two events A and B in a sample space S and their union $A \cup B$ (shaded) and intersection $A \cap B$ (shaded)

Example 2

In the random experiment of rolling a die, the events

A: *The die turns up a number not smaller than* 4

B: *The die turns up a number divisible by* 3

have the union $A \cup B = \{3, 4, 5, 6\}$ and the intersection $A \cap B = \{6\}$ (cf. Fig. 368). ▮

If all the elements of an event A are also contained in an event B, then A is called a **subevent** of B, and we write

$$A \subset B \qquad \text{or} \qquad B \supset A.$$

Clearly, if $A \subset B$, then if B occurs, A necessarily occurs. For instance, the event $D = \{4, 6\}$ is a subevent of the event $E = \{2, 4, 6\}$ that the die turns up an even number.

Let $A_1, \cdots, A_m$ be several events in a sample space S. Then the event consisting of all elements contained in one or more of these m events is called the **union** of $A_1, \cdots, A_m$ and is denoted by

$$A_1 \cup A_2 \cup \cdots \cup A_m, \qquad \text{or more briefly} \qquad \bigcup_{j=1}^{m} A_j.$$

The event consisting of all elements contained in all those m events is called the **intersection** of $A_1, \cdots, A_m$ and is denoted by

$$A_1 \cap A_2 \cap \cdots \cap A_m, \qquad \text{or more briefly} \qquad \bigcap_{j=1}^{m} A_j.$$

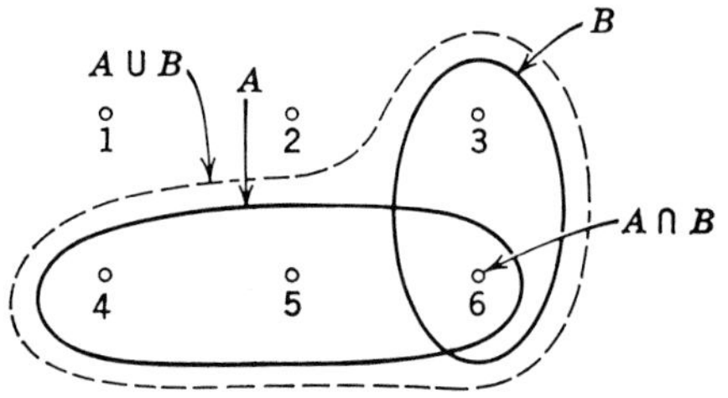

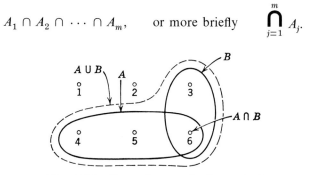

Fig. 368. Venn diagram in Example 2

More generally, let $A_1, A_2, \cdots, A_m, \cdots$ be infinitely many elements in S. Then the **union**

$$A_1 \cup A_2 \cup \cdots, \qquad \text{or more briefly} \qquad \bigcup_{j=1}^{\infty} A_j$$

is defined to be the event consisting of all the elements contained in at least one of those events, and the **intersection**

$$A_1 \cap A_2 \cap \cdots, \qquad \text{or more briefly} \qquad \bigcap_{j=1}^{\infty} A_j$$

is defined to be the event consisting of all the elements contained in all those events.

If events $A_1, A_2, \cdots, A_m \cdots$ are such that the occurrence of any of them makes the simultaneous occurrence of any other of them impossible, then we have $A_j \cap A_k = \phi$ for any j and $k \neq j$, and the events are called **mutually exclusive events** or **disjoint events.**

For instance, the events A, B, and C in Example 1 are mutually exclusive.

Suppose that we perform a random experiment n times and obtain a sample consisting of n values. Let A and B be events whose relative frequencies in those n trials are $\tilde{f}(A)$ and $\tilde{f}(B)$, respectively. Then the event $A \cup B$ has the relative frequency

$$\tilde{f}(A \cup B) = \tilde{f}(A) + \tilde{f}(B) - \tilde{f}(A \cap B). \tag{1}$$

If A and B are mutually exclusive, then $\tilde{f}(A \cap B) = 0$, and

$$\tilde{f}(A \cup B) = \tilde{f}(A) + \tilde{f}(B). \tag{2}$$

These formulas are rather obvious from the Venn diagram in Fig. 367; the formal proof is left to the reader (Prob. 5).

Problems for Sec. 20.4

1. Graph a sample space for the random experiment of tossing two coins.

2. A pair of dice is rolled once. Construct a sample space for this experiment and list its elements. On your diagram, circle and mark the events:

A: Faces are equal.

B: Sum of faces exceeds 7.

C: Sum of faces equals 5.

3. Define the sample space of the experiment of recording the lifetime of each of three electronic components.

4. In the experiment of drilling a hole into a plate and measuring its diameter, find the complement of E: The diameter of the hole is at least 2.9 in. and at most 3.1 in.

5. Prove (1).
6. A box of 20 fountain pens contains 10 nondefective pens, 8 with type A defects, 5 with type B defects, and 3 with both types of defects. Suppose that 1 pen is drawn at random. Graph a Venn diagram of the corresponding sample space S showing the events E_A of getting a type A defective, E_B of getting a type B defective, $E_A \cap E_B$, $E_A \cap E_B{}^{\mathrm{C}}$, $E_A{}^{\mathrm{C}} \cap E_B$, $E_A{}^{\mathrm{C}} \cap E_B{}^{\mathrm{C}}$, $E_A \cup E_B$, $E_A{}^{\mathrm{C}} \cup E_A$, $E_A \cup E_B{}^{\mathrm{C}}$, and $E_A{}^{\mathrm{C}} \cup E_B{}^{\mathrm{C}}$. How many elements (outcomes) are in each event?
7. Using Venn diagrams, graph and check the rules

$$A \cup (B \cap C) = (A \cup B) \cap (A \cup C)$$

$$A \cap (B \cup C) = (A \cap B) \cup (A \cap C)$$

8. **(De Morgan's laws)** Using Venn diagrams, graph and check *De Morgan's laws*

$$(A \cup B)^{\mathrm{C}} = A^{\mathrm{C}} \cap B^{\mathrm{C}}$$

$$(A \cap B)^{\mathrm{C}} = A^{\mathrm{C}} \cup B^{\mathrm{C}}$$

9. Show that, by the definition of a complement,

$$(A^{\mathrm{C}})^{\mathrm{C}} = A, \quad S^{\mathrm{C}} = \varnothing, \quad \varnothing^{\mathrm{C}} = S, \quad A \cup A^{\mathrm{C}} = S, \quad A \cap A^{\mathrm{C}} = \varnothing.$$

Here, A is any subset of a sample space S.

10. Using a Venn diagram, show that $A \subset B$ if and only if $A \cup B = B$. Find a condition for $A \subset B$ in terms of the intersection $A \cap B$.

20.5 Probability

Experience shows that most random experiments exhibit *statistical regularity* or *stability of relative frequencies;* that is, in several long sequences of such an experiment the corresponding relative frequencies of an event are almost equal. Typical examples are shown in Table 20.6 and Fig. 369. In Fig. 369 the percentage of boys seems to fluctuate less and less as the number of births increases. The percentage of defective items will show a similar behavior if the conditions of production are reasonably constant, and other examples can easily be found.

Since most random experiments exhibit statistical regularity, we may assert that for any event E in such an experiment there is a number $P(E)$ such that the relative frequency of E in a great number of performances of the experiment is approximately equal to $P(E)$.

Table 20.6
Coin Tossing

Experiments by	Number of Throws	Number of Heads	Relative Frequency of Heads
BUFFON	4,040	2,048	0.5069
K. PEARSON	12,000	6,019	0.5016
K. PEARSON	24,000	12,012	0.5005

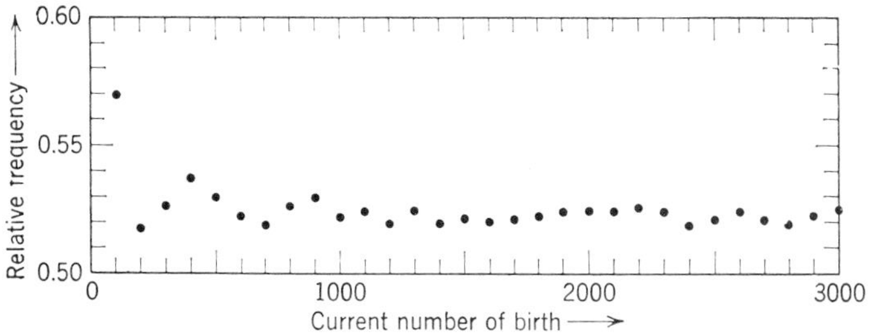

Fig. 369. Relative frequency of the event "birth of a boy." (Data from Graz, Austria, 1962)

For this reason we now postulate the existence of a number $P(E)$ which is called the **probability** *of the event E in that random experiment*. Note that this number is not an absolute property of E but refers to a certain sample space S, that is, to a certain random experiment.

The statement "E has the probability $P(E)$" then means that if we perform the experiment very often, it is practically certain that the relative frequency $\tilde{f}(E)$ is approximately equal to $P(E)$. (Here the term "approximately equal" must be made precise. This we have to postpone until Sec. 20.10.)

The probability thus introduced is the counterpart of the empirical relative frequency. It is therefore natural to require that it should have certain basic properties which the relative frequency has. These properties are suggested by Theorems 1 and 2, Sec. 20.2, and formula (2), Sec. 20.4, and may be formulated as so-called

Axioms of mathematical probability

1. *If E is any event in a sample space S, then*

$$0 \leqq P(E) \leqq 1. \tag{1}$$

2. *To the entire sample space S there corresponds*

$$P(S) = 1. \tag{2}$$

3. *If A and B are mutually exclusive events* (cf. Sec. 20.4), *then*

$$P(A \cup B) = P(A) + P(B). \tag{3}$$

If the sample space is infinite, we must replace Axiom 3 by

3*. *If $E_1, E_2, \cdots$ are mutually exclusive events, then*

$$P(E_1 \cup E_2 \cup \cdots) = P(E_1) + P(E_2) + \cdots . \tag{3*}$$

From Axiom 3 we obtain by induction the following

Theorem 1 (Addition rule for mutually exclusive events)
If $E_1, \cdots, E_m$ are mutually exclusive events, then

$$P(E_1 \cup E_2 \cup \cdots \cup E_m) = P(E_1) + P(E_2) + \cdots + P(E_m). \tag{4}$$

The reader may prove the following analogue of (1), Sec. 20.4.

Theorem 2 (Addition rule for arbitrary events)
If A and B are any events in a sample space S, then

$$P(A \cup B) = P(A) + P(B) - P(A \cap B). \tag{5}$$

Furthermore, an event E and its complement E^c (Sec. 20.4) are mutually exclusive, and $E \cup E^c = S$. Using Axioms 3 and 2, we thus have

$$P(E \cup E^c) = P(E) + P(E^c) = 1.$$

This yields

Theorem 3 (Complementation rule)
The probabilities of an event E and its complement E^c in a sample space S are related by the formula

$$P(E) = 1 - P(E^c). \tag{6}$$

This formula may be used if the computation of $P(E^c)$ is simpler than that of $P(E)$. An application will be shown in Example 2.

How can we assign probabilities to the various events in a sample space S?

If S is finite, consisting of k elements, and the nature of the experiment shows that these k outcomes are equally likely to occur, then we may assign the same probability to each outcome, and because of Axiom 2 this probability must equal $1/k$. In this case the computation of probabilities of events reduces to that of counting the elements that make up the events.

Example 1. Fair die
We consider a **fair die,** that is, a die of homogeneous material and strictly cubical form, and roll it once. In this experiment, $S = \{1, 2, 3, 4, 5, 6\}$. We thus have $P(1) = 1/6$, $P(2) = 1/6$, $\cdots$, $P(6) = 1/6$. From this and Theorem 1 we see that, for instance, the event

A: *An even number turns up*

has the probability $P(A) = P(2) + P(4) + P(6) = 1/2$, the event

B: *A number greater than* 4 *turns up*

has the probability $P(B) = P(5) + P(6) = 1/3$, etc. Formulas for more complicated cases and further examples will be included in the next section.

Example 2. Coin tossing
Five coins are tossed simultaneously. Find the probability of the event A: *At least* 1 *head turns up*. Since each coin can turn up heads or tails, the sample space consists of $2^5 = 32$ elements. Assuming that the coins are fair, we may assign the same probability (1/32) to each outcome. Then the event A^c (*no heads turn up*) consists of only 1 outcome. Hence $P(A^c) = 1/32$, and the answer is $P(A) = 1 - P(A^c) = 31/32$. ∎

If the nature of the experiment does not show that the finitely many outcomes are equally likely to occur, or if the sample space is not finite, we must assign probabilities by using relative frequencies observed in long sequences of trials in such a fashion that the axioms of probability are satisfied.

In this way we only obtain approximate values, but this does not matter. It resembles the situation in classical physics, where we postulate that each body has a certain mass but cannot determine that mass exactly. This is no handicap for developing the theory.

If we are in doubt whether we assigned the probabilities in the right fashion, we may apply a statistical test which we shall consider later (in Sec. 20.18).

Often it is required to find the probability of an event B if it is known that an event A has occurred. This probability is called the **conditional probability** *of* B *given* A and is denoted by $P(B \mid A)$. In this case A serves as a new (reduced) sample space, and that probability is the fraction of $P(A)$ which corresponds to $A \cap B$. Thus

$$P(B \mid A) = \frac{P(A \cap B)}{P(A)} \qquad [P(A) \neq 0)]. \tag{7}$$

Similarly, the *conditional probability of* A *given* B is

$$P(A \mid B) = \frac{P(A \cap B)}{P(B)} \qquad [P(B) \neq 0]. \tag{8}$$

Solving (7) and (8) for $P(A \cap B)$, we obtain

Theorem 4 (Multiplication rule)

If A *and* B *are events in a sample space* S *and* $P(A) \neq 0$, $P(B) \neq 0$, *then*

$$P(A \cap B) = P(A)P(B \mid A) = P(B)P(A \mid B). \tag{9}$$

If the events A and B are such that

$$P(A \cap B) = P(A)P(B), \tag{10}$$

they are called **independent events.** Assuming $P(A) \neq 0$, $P(B) \neq 0$, we see from (7)–(9) that in this case

$$P(A \mid B) = P(A), \qquad P(B \mid A) = P(B),$$

which means that the probability of A does not depend on the occurrence or nonoccurrence of B, and conversely. This justifies the terminology.

Similarly, m events $A_1, \cdots, A_m$ are said to be **independent** if for any k events $A_{j_1}, A_{j_2}, \cdots, A_{j_k}$ (where $1 \leqq j_1 < j_2 < \cdots < j_k \leqq m$ and $k = 2, 3, \cdots, m$)

$$P(A_{j_1} \cap A_{j_2} \cap \cdots \cap A_{j_k}) = P(A_{j_1})P(A_{j_2}) \cdots P(A_{j_k}). \tag{11}$$

To prepare for an illustrative example, we first note that there are two ways of drawing objects for obtaining a sample from a given set of objects, briefly referred to as **sampling from a population,** as follows.

1. **Sampling with replacement** means that the object which was drawn at random is placed back to the given set and the set is mixed thoroughly. Then we draw the next object at random.
2. **Sampling without replacement** means that the object which was drawn is put aside.

Example 3. Sampling with and without replacement

A box contains 10 screws, 3 of which are defective. Two screws are drawn at random. Find the probability that none of the two screws is defective. We consider the events

A: *First drawn screw nondefective.*

B: *Second drawn screw nondefective.*

Clearly, $P(A) = \frac{7}{10}$ because 7 of the 10 screws are nondefective and we sample at random, so that each screw has the same probability ($\frac{1}{10}$) of being picked. If we sample with replacement, the situation before the second drawing is the same as at the beginning, and $P(B) = \frac{7}{10}$. The events are independent, and the answer is

$$P(A \cap B) = P(A)P(B) = 0.7 \cdot 0.7 = 0.49 = 49\%.$$

If we sample without replacement, then $P(A) = \frac{7}{10}$, as before. If A has occurred, then there are 9 screws left in the box, 3 of which are defective. Thus $P(B \mid A) = \frac{6}{9} = \frac{2}{3}$, and Theorem 4 yields the answer

$$P(A \cap B) = \tfrac{7}{10} \cdot \tfrac{2}{3} \approx 47\%.$$

Problems for Sec. 20.5

1. What is the probability of obtaining at least 1 head in tossing 5 fair coins?
2. In throwing three fair dice, find the probability of the event E that at least two of the three numbers obtained are different.
3. Three screws are drawn at random from a lot of 100 screws, 10 of which are defective. Find the probability of the event that all 3 screws drawn are non-defective, assuming that we draw (*a*) with replacement, (*b*) without replacement.
4. Three urns contain 5 chips each, numbered from 1 to 5, and 1 chip is drawn from each urn. Find the probability of the event E that the sum of the numbers on the drawn chips is greater than 3.
5. A batch of 100 iron rods consists of 25 oversized rods, 25 undersized rods and 50 rods of the desired length. If two rods are drawn at random without replacement, what is the probability of obtaining (*a*) two rods of the desired length, (*b*) one of the desired length, (*c*) none of the desired length, (*d*) two undersized rods?
6. A process of producing spark plugs has been operated for a long period of time and the fraction of defective items has remained constant at 2%. Every half hour, two plugs just produced are inspected. What is the probability of obtaining (*a*) no defectives, (*b*) 1 defective, (*c*) 2 defectives? What is the sum of the probabilities in (*a*), (*b*), (*c*)?
7. A motor drives an electric generator. During a 30-day period, the motor needs repair with probability 5% and the generator needs repair with probability 6%. What is the probability that during a given period, the entire apparatus will need repair?
8. An automatic pressure control apparatus contains 6 electronic tubes. The apparatus will not work unless all tubes are operative. If the probability of failure of

each tube during some interval of time is 0.05, what is the corresponding probability of failure of the apparatus?

9. A box of 100 gaskets contains 10 gaskets with type A defects, 5 gaskets with type B defects, and 2 gaskets with both types of defects. Suppose that a gasket drawn is known to have a type A defect. What is the probability that it also has a type B defect?

10. Find the probability that in throwing 2 fair dice the sum of the faces exceeds 9, given that one of the faces is a 5.

11. Given $P(A^c) = 0.2$, $P(B) = 0.5$ and $P(A \cap B^c) = 0.4$, find $P(B \mid A \cup B^c)$. (*Hint.* Use a Venn diagram.)

12. Prove Theorem 2.

13. Prove Theorem 1.

14. Extending Theorem 4, show that

$$P(A \cap B \cap C) = P(A)P(B \mid A)P(C \mid A \cap B).$$

15. Show that if B is a subset of A, then $P(B) \leqq P(A)$.

20.6 Permutations and Combinations

From the last section we know that in the case of a finite sample space S consisting of k equally likely outcomes, each outcome has probability $1/k$, and the probability of an event A is obtained by counting the number of outcomes of which A consists. Clearly, if that number is m, then $P(A) = m/k$. For the counting of outcomes the subsequent formulas will often be useful.

Suppose that there are given n things (*elements* or *objects*). We may lay them out in a row *in any order.* Each such arrangement is called a **permutation** of those things.

Theorem 1 (Permutations)

The number of permutations of n different things taken all at a time is

$$n! = 1 \cdot 2 \cdot 3 \cdots n \qquad \text{(read "}n\text{ factorial")}. \tag{1}$$

In fact, there are n possibilities for filling the first place in the row; then $n - 1$ objects are still available for filling the second place, etc. In a similar fashion we obtain

Theorem 2 (Permutations)

If n given things can be divided into c classes such that things belonging to the same class are alike while things belonging to different classes are different, then the number of permutations of these things taken all at a time is

$$\frac{n!}{n_1!\, n_2! \cdots n_c!} \qquad (n_1 + n_2 + \cdots + n_c = n) \tag{2}$$

where n_j is the number of things in the jth class.

A **permutation of n things taken k at a time** is a permutation containing only k of the n given things. Two such permutations consisting of the same k elements, in a different order, are different, by definition. For example, there are 6 different permutations of the 3 letters *a, b, c*, taken 2 letters at a time, namely *ab, ac, bc, ba, ca, cb.*

A **permutation of n things taken k at a time with repetitions** is an arrangement obtained by putting any given thing in the first position, any given thing, including a repetition of the one just used, in the second, and continuing until k positions are filled. For example, there are $3^2 = 9$ different such permutations of *a, b, c*, taken 2 letters at a time, namely, the preceding 6 permutations and *aa, bb, cc*. The reader may prove the following theorem (cf. Prob. 9).

Theorem 3 (Permutations)

The number of different permutations of n different things taken k at a time without repetitions is

$$n(n-1)(n-2)\cdots(n-k+1) = \frac{n!}{(n-k)!} \tag{3a}$$

and with repetitions is

$$n^k. \tag{3b}$$

In the case of a permutation not only are the things themselves important, but so also is their order. A **combination** of given things is any selection of one or more of the things *without regard to order.* There are two types of combinations, as follows.

The number of **combinations of n different things, k at a time, without repetitions** is the number of sets that can be made up from the n things, each set containing k different things and no two sets containing exactly the same k things.

The number of **combinations of n different things, k at a time, with repetitions** is the number of sets that can be made up of k things chosen from the given n, each being used as often as desired.

For example, there are 3 combinations of the 3 letters *a, b, c*, taken 2 letters at a time, without repetitions, namely, *ab, ac, bc*, and 6 such combinations with repetitions, namely, *ab, ac, bc, aa, bb, cc*.

Theorem 4 (Combinations)

The number of different combinations of n different things, k at a time, without repetitions, is

$$\binom{n}{k} = \frac{n!}{k!\,(n-k)!} = \frac{n(n-1)\cdots(n-k+1)}{1\cdot 2\cdots k}, \tag{4a}$$

and the number of those combinations with repetitions is

$$\binom{n+k-1}{k}. \tag{4b}$$

The statement involving (4a) follows from the first part of Theorem 3 by noting that there are $k!$ permutations of k things from the given n things which differ by the order of the elements (cf. Theorem 1), but there is only a single combination of those k things of the type characterized in the first statement of Theorem 4. The last statement of Theorem 4 can be proved by induction (cf. Prob. 10).

Example 1. Illustration of Theorems 1 and 2

If there are 10 different screws in a box which are needed in a certain order for assembling a certain product, and these screws are drawn at random from the box, the probability P of picking them in the desired order is very small, namely (cf. Theorem 1),

$$P = 1/10! = 1/3628\,800 \approx 0.000\,03\%.$$

If the box contains 6 alike right-handed screws and 4 alike left-handed screws and the 6 right-handed ones are needed first and the 4 left-handed ones are needed afterwards, the probability P that random drawing yields the screws in the desired order is (cf. Theorem 2)

$$P = 6!\,4!/10! = 1/210 \approx 0.5\%.$$

Example 2. Illustration of Theorem 3

In a coded telegram the letters are arranged in groups of 5 letters, called *words*. From (3b) we see that there are

$$26^5 = 11\,881\,376$$

different such words. From (3a) it follows that there are

$$26!/(26-5)! = 26 \cdot 25 \cdot 24 \cdot 23 \cdot 22 = 7\,893\,600$$

different such words which contain each letter at most once.

Example 3. Illustration of Theorem 4

The number of samples of 5 light bulbs which can be selected from a lot of 500 bulbs is [cf. (4a)]

$$\binom{500}{5} = \frac{500!}{5!\,495!} = \frac{500 \cdot 499 \cdot 498 \cdot 497 \cdot 496}{1 \cdot 2 \cdot 3 \cdot 4 \cdot 5} = 255\,244\,687\,600. \qquad \blacksquare$$

Let us include some remarks about the **factorial function.** We define

$$0! = 1 \tag{5}$$

and may compute further values by the relation

$$(n+1)! = (n+1)n!. \tag{6}$$

For large n the function is very large (cf. Table A4 in Appendix 4). A convenient approximation for large n is the **Stirling formula**[3]

$$n! \sim \sqrt{2\pi n}\left(\frac{n}{e}\right)^n \qquad (e = 2.718 \cdots), \tag{7}$$

where $\sim$ means that the ratio of the two sides of (7) approaches 1 as n approaches infinity.

[3]JAMES STIRLING (1692–1770), English mathematician.

The **binomial coefficients** are defined by the formula

$$\binom{a}{k} = \frac{a(a-1)(a-2)\cdots(a-k+1)}{k!} \qquad (k \geqq 0, \text{ integral}). \tag{8}$$

The numerator has k factors. Furthermore, we define

$$\binom{a}{0} = 1, \qquad \text{in particular} \qquad \binom{0}{0} = 1. \tag{9}$$

For integral $a = n$ we obtain from (8)

$$\binom{n}{k} = \binom{n}{n-k} \qquad (n \geqq 0, 0 \leqq k \leqq n). \tag{10}$$

Binomial coefficients may be computed recursively, because

$$\binom{a}{k} + \binom{a}{k+1} = \binom{a+1}{k+1} \qquad (k \geqq 0, \text{ integral}). \tag{11}$$

Formula (8) also yields

$$\binom{-m}{k} = (-1)^k \binom{m+k-1}{k} \qquad (k \geqq 0, \text{ integral}, m > 0). \tag{12}$$

There are numerous further relations; we mention

$$\sum_{s=0}^{n-1} \binom{k+s}{k} = \binom{n+k}{k+1} \qquad (k \geqq 0, n \geqq 1, \text{ both integral}) \tag{13}$$

and

$$\sum_{k=0}^{r} \binom{p}{k}\binom{q}{r-k} = \binom{p+q}{r}. \tag{14}$$

Problems for Sec. 20.6

1. List all permutations of four digits 1, 2, 3, 4, taken all at a time.
2. List all permutations of five letters a, e, i, o, u, taken three at a time.
3. In how many ways can a committee of three be chosen from ten persons?
4. How many different license plates showing 5 symbols, namely, 2 letters followed by 3 digits, could be made?
5. Find the number of samples of 3 objects which can be drawn from a lot of 100 objects.
6. An urn contains 2 black, 3 white, and 4 red balls. We draw 1 ball at random and put it aside. Then we draw the next ball, and so on. Find the probability of drawing at first the 2 black balls, then the 3 white ones, and finally the red ones.

7. If 6 different inks are available, in how many ways can we select two colors for a printing job? Four colors?

8. Of a lot of 10 items, 2 are defective. (*a*) Find the number of different samples of 4. Find the number of samples of 4 containing (*b*) no defectives, (*c*) 1 defective, (*d*) 2 defectives.

9. Prove Theorem 3.

10. Prove the last statement of Theorem 4. *Hint.* Use (13).

11. Using (7), compute approximate values of 4! and 8! and determine the absolute and relative errors.

12. For the same number of different items taken 4 at a time without repetition, how will the number of combinations be related to the number of permutations?

13. Derive (11) from (8).

14. **(Binomial theorem)** By the binomial theorem,

$$(a + b)^n = \sum_{k=0}^{n} \binom{n}{k} a^k b^{n-k},$$

so that $a^k b^{n-k}$ has the coefficient $\binom{n}{k}$. Can you conclude this from Theorem 4 or is this merely a coincidence?

15. Prove (14) by applying the binomial theorem (Prob. 14) to

$$(1 + b)^p(1 + b)^q = (1 + b)^{p+q}.$$

20.7 Random Variables. Discrete and Continuous Distributions

If we roll two dice, we know that the sum X of the two numbers which turn up must be an integer between 2 and 12, but we cannot predict which value of X will occur in the next trial, and we may say that X depends on "chance." Similarly, if we want to draw 5 bolts from a lot of bolts and measure their diameters, we cannot predict how many will be defective, that is, will not meet given requirements; hence $X =$ *number of defectives* is again a function which depends on "chance." The lifetime X of a light bulb to be drawn at random from a lot of bulbs also depends on "chance," and so does the content X of a bottle of lemonade filled by a filling machine and selected at random from a given lot.

Roughly speaking, a **random variable** X (also called **stochastic variable** or **variate**) is a function whose values are real numbers and depend on "chance"; more precisely, it is a function X which has the following properties:

1. *X is defined on the sample space S of the experiment, and its values are real numbers.*

2. *Let a be any real number, and let I be any interval. Then the set of all outcomes in S for which $X = a$ has a well-defined probability, and the same is true for the set of all outcomes in S for which the values of X are in I. These probabilities are in agreement with the axioms in Sec. 20.5.*

Although this definition is quite general and includes many functions, we shall see that in practice the number of important types of random variables and corresponding "probability distributions" is rather small.

If we perform a random experiment and the event corresponding to a number a occurs, then we say that in this trial the random variable X corresponding to that experiment has **assumed** the value a. We also say that we have **observed** the value $X = a$. Instead of "the event corresponding to a number a," we say, more briefly, "the event $X = a$." The corresponding probability is denoted by $P(X = a)$. Similarly, the probability of the event

$$X \text{ assumes any value in the interval } a < X < b$$

is denoted by $P(a < X < b)$. The probability of the event

$$X \leqq c \text{ (}X \text{ assumes any value smaller than } c \text{ or equal to } c\text{)}$$

is denoted by $P(X \leqq c)$, and the probability of the event

$$X > c \text{ (}X \text{ assumes any value greater than } c\text{)}$$

is denoted by $P(X > c)$.

The last two events are mutually exclusive. From Axiom 3 in Sec. 20.5 we thus obtain

$$P(X \leqq c) + P(X > c) = P(-\infty < X < \infty).$$

From Axiom 2 we see that the right side equals 1, because $-\infty < X < \infty$ corresponds to the whole sample space. This yields the important formula

$$\text{(1)} \qquad P(X > c) = 1 - P(X \leqq c) \qquad (c \text{ arbitrary}).$$

For example, if X is the number that turns up in rolling a fair die, $P(X = 1) = \frac{1}{6}$, $P(X = 2) = \frac{1}{6}$, etc., $P(1 < X < 2) = 0$, $P(1 \leqq X \leqq 2) = \frac{1}{3}$, $P(0 \leqq X \leqq 3.2) = \frac{1}{2}$, $P(X > 4) = \frac{1}{3}$, $P(X \leqq 0.5) = 0$, etc.

In most practical cases the random variables are either *discrete* or *continuous*. These two important notions will now be considered one after the other.

A random variable X and the corresponding distribution are said to be **discrete**, if X has the following properties.

1. The number of values for which X has a probability different from 0 is finite or at most countably infinite.

2. If an interval $a < X \leqq b$ does not contain such a value, then $P(a < X \leqq b) = 0$.

Let

$$x_1, \quad x_2, \quad x_3, \quad \cdots$$

be the values for which X has a positive probability, and let

$$p_1, \quad p_2, \quad p_3, \quad \cdots$$

be the corresponding probabilities. Then $P(X = x_1) = p_1$, etc., We now introduce the function

(2)
$$f(x) = \begin{cases} p_j & \text{when } x = x_j \qquad (j = 1, 2, \cdots) \\ 0 & \text{otherwise.} \end{cases}$$

$f(x)$ is called the **probability function** of X.

Since $P(S) = 1$ (cf. Axiom 2 in Sec. 20.5), we must have

(3)
$$\sum_{j=1}^{\infty} f(x_j) = 1.$$

If we know the probability function of a discrete random variable X, then we may readily compute the probability $P(a < X \leqq b)$ corresponding to any interval $a < X \leqq b$. In fact,

(4)
$$P(a < X \leqq b) = \sum_{a<x_j \leqq b} f(x_j) = \sum_{a<x_j \leqq b} p_j.$$

This is the sum of all the probabilities $f(x_j) = p_j$ for which the x_j lie in that interval. For a closed, open, or infinite interval the situation is quite similar. We express this by saying that the probability function $f(x)$ determines the **probability distribution** or, briefly, the **distribution** of the random variable X in a unique fashion.

If X is any random variable, not necessarily discrete, then for any real number x there exists the probability $P(X \leqq x)$ corresponding to

$$X \leqq x \text{ (}X \textit{ assumes any value smaller than } x \textit{ or equal to } x\text{).}$$

Clearly, $P(X \leqq x)$ depends on the choice of x; it is a function of x, which is called the **distribution function**[4] of X and is denoted by $F(x)$. Thus

(5)
$$F(x) = P(X \leqq x).$$

Since for any a and $b > a$ we have

$$P(a < X \leqq b) = P(X \leqq b) - P(X \leqq a),$$

it follows that

(6)
$$P(a < X \leqq b) = F(b) - F(a).$$

This shows that the distribution function determines the distribution of X uniquely and that it can be used for computing probabilities.

Suppose that X is a discrete variable. Then we may represent the distribution

[4] We note that some authors call $F(x)$ the *cumulative distribution function,* in particular those who use the term *distribution function* for the probability function $f(x)$.

function $F(x)$ in terms of the probability function $f(x)$. In fact, by inserting (4) (with $a = -\infty$ and $b = x$) we have

(7) $$F(x) = \sum_{x_j \leqq x} f(x_j)$$

where the right-hand side is the sum of all those $f(x_j)$ for which $x_j \leqq x$. Simple examples are shown in Figs. 370 and 371. In both figures $f(x)$ is represented by a bar chart. In Fig. 370 we have $f(x) = \frac{1}{6}$ when $x = 1, 2, \cdots, 6$ and 0 otherwise. In Fig. 371 the function $f(x)$ has the values

x	2	3	4	5	6	7	8	9	10	11	12
$f(x)$	$\frac{1}{36}$	$\frac{2}{36}$	$\frac{3}{36}$	$\frac{4}{36}$	$\frac{5}{36}$	$\frac{6}{36}$	$\frac{5}{36}$	$\frac{4}{36}$	$\frac{3}{36}$	$\frac{2}{36}$	$\frac{1}{36}$

because there are $6 \cdot 6 = 36$ equally likely outcomes, each of which thus has probability $\frac{1}{36}$; we have $X = 2$ for just one outcome, namely, (1, 1) (where the first number refers to the first die and the second number to the other die); we have $X = 3$ in the case of the two outcomes (1, 2) and (2, 1); $X = 4$ in the case of the three outcomes (1, 3), (2, 2), and (3, 1), etc.

The values $x_1, x_2, x_3, \cdots$ for which a discrete random variable X has positive probability (and only these values) are called the **possible values** of X. In each interval that contains no possible value the distribution function $F(x)$ is constant. Hence $F(x)$ is a **step function** (piecewise constant function) which has an upward jump of magnitude $p_j = P(X = x_j)$ at $x = x_j$ and is constant between two subsequent possible values. Figs. 370 and 371 are illustrative examples.

We shall now define and consider continuous random variables. A random variable X and the corresponding distribution are said to be *of continuous type* or, briefly, **continuous** if the corresponding distribution function $F(x) = P(X \leqq x)$ can be represented by an integral in the form[5]

(8) $$F(x) = \int_{-\infty}^{x} f(v)\, dv$$

where the integrand is continuous, possibly except for at most finitely many values of v, and is nonnegative. The integrand f is called the *probability density* or, briefly, the **density** of the distribution. Differentiating (8) we see that

$$F'(x) = f(x)$$

for every x at which $f(x)$ is continuous. In this sense *the density is the derivative of the distribution function.*

[5] $F(x)$ is continuous, but continuity of $F(x)$ does not imply the existence of a representation of the form (8). Since continuous distribution functions that cannot be represented in the form (8) are rare in practice, the widely accepted terms "continuous random variable" and "continuous distribution" should not be confusing.

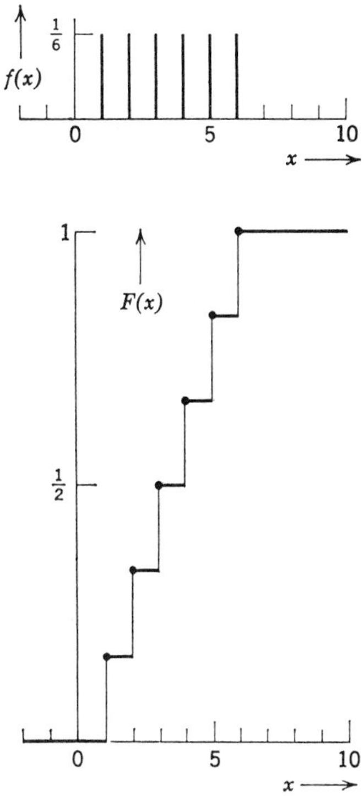

Fig. 370. Probability function $f(x)$ and distribution function $F(x)$ of the random variable X = *number obtained in rolling a fair die once*

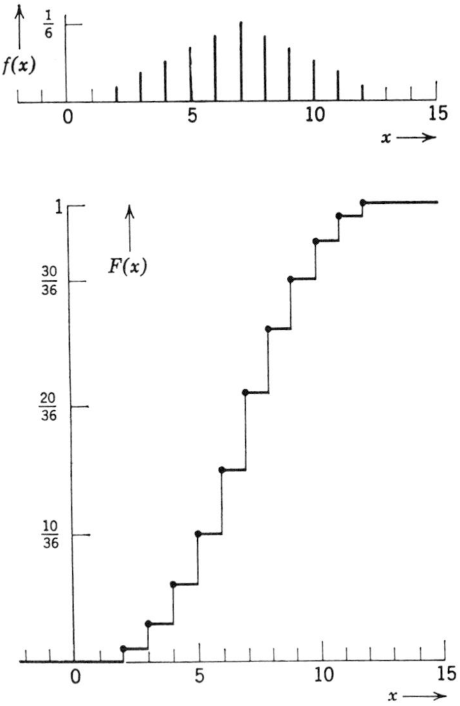

Fig. 371. Probability function $f(x)$ and distribution function $F(x)$ of the random variable X = *sum of the two numbers obtained in rolling two fair dice once*

From (8) and Axiom 2, Sec. 20.5, we also have

$$\int_{-\infty}^{\infty} f(v)\,dv = 1. \tag{9}$$

Furthermore, from (6) and (8) we obtain the formula

$$P(a < X \leqq b) = F(b) - F(a) = \int_a^b f(v)\,dv. \tag{10}$$

Hence this probability equals the area under the curve of the density $f(x)$ between $x = a$ and $x = b$, as shown in Fig. 372.

Obviously, for any fixed a and b ($>a$) the probabilities corresponding to the intervals $a < X \leqq b$, $a < X < b$, $a \leqq X < b$, and $a \leqq X \leqq b$ are all the same. This is different from the situation in the case of a discrete distribution.

Examples of continuous distributions are included in the next problem set and in later sections.

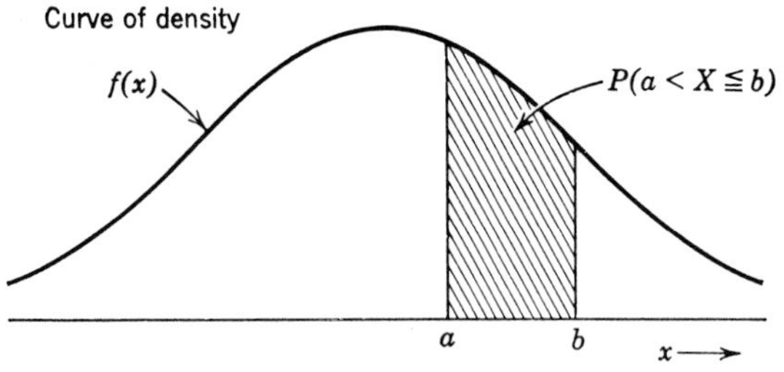

Fig. 372. Example illustrating formula (10)

Problems for Sec. 20.7

1. Graph the probability function $f(x) = x^2/14$ $(x = 1, 2, 3)$ and the distribution function.
2. If X has the probability function $f(2) = \frac{1}{2}, f(3) = \frac{1}{4}, f(4) = f(5) = \frac{1}{8}$, what is the probability that X will assume any value less than 4?
3. Let X be the number of years before a particular type of machine will need replacement. Assume that X has the probability function $f(1) = 0.3, f(2) = 0.4, f(3) = 0.2, f(4) = 0.1$. Graph f and F.
4. Graph the density $f(x) = \frac{1}{4}$ $(1 < x < 5)$ and the distribution function.
5. The daily demand for gasoline in gallons at a certain gas station is a random variable X. Assume that X has the density $f(x) = k$ if $2000 < x < 6000$ and 0 otherwise. Find k. Find and graph the distribution function $F(x)$.
6. Let $f(x) = ce^{-x}$ when $x > 0$, $f(x) = 0$ when $x < 0$. Find c. Graph f and F.
7. Find and graph the probability function $f(x)$ of the random variable $X =$ *sum of the* 3 *numbers obtained in rolling* 3 *fair dice*.
8. Let X [millimeters] be the thickness of washers a machine turns out. Assume that X has the density $f(x) = kx$ if $1.9 < x < 2.1$ and 0 otherwise. Find k. What is the probability that a washer will have thickness between 1.95 mm and 2.05 mm?
9. Show that the random variable $X =$ *number of times a fair coin is tossed until a head appears* has the probability function $f(x) = 2^{-x}$ $(x = 1, 2, \cdots)$. Show that $f(x)$ satisfies (3).
10. Let $f(x) = kx^2$ when $0 \leqq x \leqq 1$ and 0 otherwise. Find k. Find the number c such that $P(X \leqq c) = 72.9\%$.
11. Find the probability that none of 3 bulbs in a signal-light will have to be replaced during the first 1200 hours of operation if the lifetime X of a bulb is a random variable with the density

$$f(x) = 6[0.25 - (x - 1.5)^2] \qquad \text{when } 1 \leqq x \leqq 2$$

and $f(x) = 0$ otherwise, where x is measured in multiples of 1000 hours.

12. Let X be the ratio of sales to profits of some firm. Assume that X has the distribution function $F(x) = 0$ if $x < 2$, $F(x) = (x^2 - 4)/5$ if $2 \leqq x < 3$, $F(x) = 1$ if $x \geqq 3$. Find and graph the density. What is the probability that X is between 2.5 (40% profit) and 5 (20% profit)?
13. Let X be a random variable that can assume every real value. What are the complements of the events $X \leqq b, X < b, X \geqq c, X > c, b \leqq X \leqq c, b < X \leqq c$?
14. A box contains 4 right-handed and 6 left-handed screws. Two screws are drawn at

random without replacement. Let X be the number of left-handed screws drawn. Find the probabilities $P(X = 0)$, $P(X = 1)$, $P(X = 2)$, $P(1 < X < 2)$, $P(X \leqq 1)$, $P(X \geqq 1)$, $P(X > 1)$, and $P(0.5 < X < 10)$.

15. Show that $b < c$ implies $P(X \leqq b) \leqq P(X \leqq c)$.

20.8 Mean and Variance of a Distribution

The *mean value* or **mean** of a distribution is denoted by μ and is defined by

(1)

$$\text{(a)}\quad \mu = \sum_j x_j f(x_j) \qquad \text{(discrete distribution)},$$

$$\text{(b)}\quad \mu = \int_{-\infty}^{\infty} x f(x)\, dx \qquad \text{(continuous distribution)}.$$

In (1a) the function $f(x)$ is the probability function of the random variable X considered, and we sum over all possible values (cf. Sec. 20.7). In (1b) the function $f(x)$ is the density of X. The mean is also known as the *mathematical expectation of* X and is sometimes denoted by $E(X)$. By definition it is assumed that the series in (1a) converges absolutely and the integral of $|x| f(x)$ from $-\infty$ to ∞ exists. If this does not hold, we say that the distribution does not have a mean; this case will rarely occur in engineering applications.

A distribution is said to be **symmetric** with respect to a number $x = c$ if for every real x,

$$f(c + x) = f(c - x). \tag{2}$$

The reader may prove (cf. Prob. 1)

Theorem 1 (Mean of a symmetric distribution)
If a distribution is symmetric with respect to $x = c$ and has a mean μ, then $\mu = c$.

The **variance** of a distribution is denoted by σ^2 and is defined by the formula

(3)

$$\text{(a)}\quad \sigma^2 = \sum_j (x_j - \mu)^2 f(x_j) \qquad \text{(discrete distribution)},$$

$$\text{(b)}\quad \sigma^2 = \int_{-\infty}^{\infty} (x - \mu)^2 f(x)\, dx \qquad \text{(continuous distribution)},$$

where, by definition, it is assumed that the series in (3a) converges absolutely and the integral in (3b) exists.

In the case of a discrete distribution with $f(x) = 1$ at a point and $f = 0$ otherwise, we have $\sigma^2 = 0$. This case is of no practical interest. In any other case

$$\sigma^2 > 0. \tag{4}$$

The positive square root of the variance is called the **standard deviation** and is denoted by σ.

Roughly speaking, the variance is a measure of the spread or dispersion of the values which the corresponding random variable X can assume.

Example 1. Mean and variance

The random variable

$$X = \textit{number of heads in a single toss of a fair coin}$$

has the possible values $X = 0$ and $X = 1$ with probabilities $P(X = 0) = 1/2$ and $P(X = 1) = 1/2$. From (1a) we thus obtain the mean value $\mu = 0 \cdot \frac{1}{2} + 1 \cdot \frac{1}{2} = \frac{1}{2}$, and (3a) yields

$$\sigma^2 = (0 - \tfrac{1}{2})^2 \cdot \tfrac{1}{2} + (1 - \tfrac{1}{2})^2 \cdot \tfrac{1}{2} = \tfrac{1}{4}.$$

Example 2. Uniform distribution

The distribution with the density

$$f(x) = \frac{1}{b - a} \qquad \text{when} \qquad a < x < b$$

and $f = 0$ otherwise is called the *uniform* or *rectangular distribution* on the interval $a < x < b$. From Theorem 1 or from (1b) we find that $\mu = (a + b)/2$, and (3b) yields the variance

$$\sigma^2 = \int_a^b \left(x - \frac{a + b}{2}\right)^2 \frac{1}{b - a}\, dx = \frac{(b - a)^2}{12}.$$

Figure 373 shows special cases illustrating that σ^2 measures the spread. ▮

Theorem 2 (Linear transformation)

If a random variable X has mean μ and variance σ^2, the random variable $X^* = c_1 X + c_2$ $(c_1 \neq 0)$ has the mean

$$\mu^* = c_1 \mu + c_2 \tag{5}$$

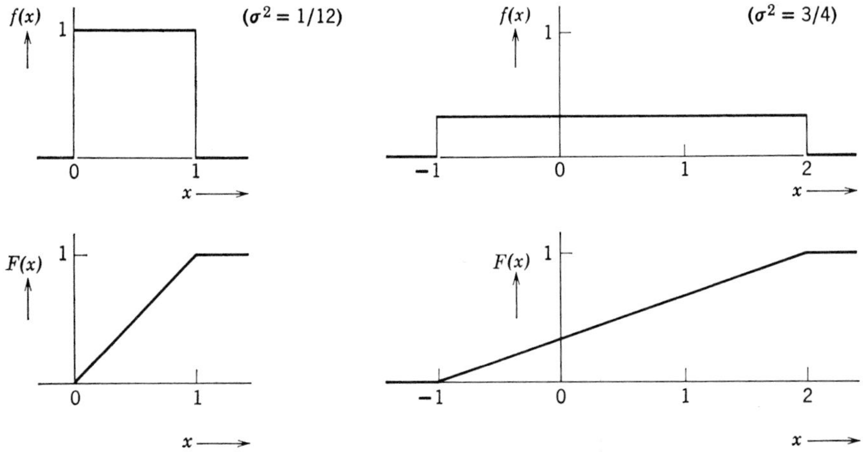

Fig. 373. Uniform distributions having the same mean (0.5) but different variances σ^2

and the variance

$$\sigma^{*2} = c_1^2\sigma^2. \tag{6}$$

Proof. We prove (5) in the continuous case, first assuming that $c_1 > 0$. For corresponding x and $x^* = c_1x + c_2$ the densities $f(x)$ of X and $f^*(x^*)$ of X^* satisfy the relation $f^*(x^*) = f(x)/c_1$, because to a small interval of length Δx on the X-axis there corresponds the probability $f(x)\,\Delta x$ (approximately) and this must equal $f^*(x^*)\,\Delta x^*$ where $\Delta x^* = c_1\,\Delta x$ is the length of the corresponding interval on the X^*-axis. Since $dx^*/dx = c_1$, $dx^* = c_1\,dx$, we thus have $f^*(x^*)\,dx^* = f(x)\,dx$. Hence

$$\mu^* = \int_{-\infty}^{\infty} x^*f^*(x^*)\,dx^* = \int_{-\infty}^{\infty} (c_1x + c_2)f(x)\,dx$$

$$= c_1\int_{-\infty}^{\infty} xf(x)\,dx + c_2\int_{-\infty}^{\infty} f(x)\,dx.$$

The last integral equals 1, cf. (9), Sec. 20.7, and formula (5) is proved. Since

$$x^* - \mu^* = (c_1x + c_2) - (c_1\mu + c_2) = c_1x - c_1\mu,$$

the definition of the variance yields

$$\sigma^{*2} = \int_{-\infty}^{\infty} (x^* - \mu^*)^2 f^*(x^*)\,dx^* = \int_{-\infty}^{\infty} (c_1x - c_1\mu)^2 f(x)\,dx = c_1^2\sigma^2.$$

If $c_1 < 0$, the results remain the same, because we get two additional minus signs, one from changing the direction of integration in x (note that $x^* = -\infty$ corresponds to $x = \infty$) and the other one from $f^*(x^*) = f(x)/(-c_1)$; here $-c_1 > 0$ is needed since densities are nonnegative.

For a discrete distribution, the proof of Theorem 2 is similar. ∎

From (5) and (6) we readily obtain

Theorem 3 (Standardized variable)

If a random variable X has mean μ and variance σ^2, then the corresponding variable $Z = (X - \mu)/\sigma$ has the mean 0 and the variance 1.

Z is called the **standardized variable** corresponding to X.

If X is any random variable and $g(X)$ is any continuous function defined for all real X, then the number

(7)

$$\text{(a)}\quad E(g(X)) = \sum_j g(x_j)f(x_j) \qquad (X \text{ discrete})$$

$$\text{(b)}\quad E(g(X)) = \int_{-\infty}^{\infty} g(x)f(x)\,dx \qquad (X \text{ continuous})$$

is called the **mathematical expectation** of $g(X)$. Here f is the probability function or the density, respectively.

Taking $g(X) = X^k$ $(k = 1, 2, \cdots)$ in (7), we obtain

$$\textbf{(8)} \qquad E(X^k) = \sum_j x_j^k f(x_j) \qquad \text{and} \qquad E(X^k) = \int_{-\infty}^{\infty} x^k f(x)\, dx,$$

respectively. $E(X^k)$ is called the ***k*th moment** of X. Taking $g(X) = (X - \mu)^k$ in (7), we have

$$\textbf{(9)} \qquad E([X - \mu]^k) = \sum_j (x_j - \mu)^k f(x_j) \qquad \text{and} \qquad \int_{-\infty}^{\infty} (x - \mu)^k f(x)\, dx,$$

respectively. This expression is called the ***k*th central moment** of X. The reader may show that

$$\textbf{(10)} \qquad E(1) = 1$$

$$\textbf{(11)} \qquad \mu = E(X)$$

$$\textbf{(12)} \qquad \sigma^2 = E([X - \mu]^2).$$

Problems for Sec. 20.8

1. Prove Theorem 1.

2. Find the mean and the variance of the distribution having the density $f(x) = \frac{1}{2}e^{-|x|}$.

3. Let X have the density $f(x) = 0.5x$ if $0 \leqq x \leqq 2$ and 0 otherwise. Show that X has the mean 4/3 and the variance 2/9.

4. Find the mean and the variance of $Y = -2X + 5$, where X is the random variable in Prob. 3.

5. Find the standardized random variable corresponding to X in Prob. 3.

6. Prove Theorem 2 in the discrete case.

7. Derive Theorem 3 from (5) and (6).

8. Assume that the mileage (in thousands of miles) which car owners get with a certain kind of tires is a random variable X having the density

$$f(x) = \theta e^{-\theta x} \qquad \text{if } x > 0$$

and 0 otherwise. Here, θ (>0) is a parameter. (*a*) What mileage can a car owner expect to get with one of these tires? (*b*) Let $\theta = 0.05$. Find the probability that the tire will last at least 30 000 miles.

9. Let X [cm] be the diameter of bolts in a production. Assume that X has the density $f(x) = k(x - 0.9)(1.1 - x)$ if $0.9 < x < 1.1$ and 0 otherwise. Determine k, graph $f(x)$ and find μ and σ^2.

10. Suppose that in Prob. 9, a bolt is regarded as being defective if its diameter deviates from 1.00 cm by more than 0.06 cm. What percentage of defective bolts should we then expect?

11. A small filling station is supplied with gasoline every Saturday afternoon. Assume that its volume X of sales in thousands of gallons has the probability density $f(x) = 6x(1 - x)$ if $0 \leqq x \leqq 1$ and 0 otherwise. Determine the mean and the variance.

12. What capacity must the tank in Prob. 11 have in order that the probability that the tank will be emptied in a given week be 10%?

13. Prove (10) and (11).

14. Prove (12).

15. Show that $E(X - \mu) = 0$.

16. Show that $\sigma^2 = E(X^2) - \mu^2$.

17. Let X have the density $f(x) = 2x$ if $0 < x < 1$ and 0 otherwise. Find all the moments. Calculate σ^2 by the formula in Prob. 16.

18. Show that $\sigma^2 = E(X[X - 1]) + \mu - \mu^2$.

19. Show that $E(ag(X) + bh(X)) = aE(g(X)) + bE(h(X))$; [$a$ and b are constants].

20. $E([X - c]^2)$ is called the *second moment of X with respect to c*. If it exists, show that it is minimum when $c = \mu$.

21. Find the moments of the uniform distribution on the interval $0 \leqq x \leqq 1$.

22. **(Skewness)** The number

$$\gamma = \frac{1}{\sigma^3} E([X - \mu]^3)$$

is called the *skewness* of X. Justify this term by showing that if X is symmetric with respect to μ and the third central moment exists, this moment is zero.

23. Find the skewness of the distribution with density $f(x) = xe^{-x}$ when $x > 0$ and $f(x) = 0$ otherwise. Graph $f(x)$.

24. Find the skewness of the distribution with density $f(x) = 2(1 - x)$ when $0 < x < 1$, $f(x) = 0$ otherwise.

25. **(Moment generating function)** The so-called *moment generating function* of a discrete or continuous random variable X is given by the formulas

$$G(t) = E(e^{tX}) = \sum_j e^{tx_j} f(x_j) \qquad \text{and} \qquad G(t) = \int_{-\infty}^{\infty} e^{tx} f(x)\, dx,$$

respectively. Assuming that differentiation under the summation sign and the integral sign is permissible, show that $E(X^k) = G^{(k)}(0)$, in particular $\mu = G'(0)$, where $G^{(k)}(t)$ is the kth derivative of G with respect to t.

20.9 Binomial, Poisson, and Hypergeometric Distributions

We shall now consider special discrete distributions which are particularly important in statistics. We start with the binomial distribution, which is obtained if we are interested in the number of times an event A occurs in n independent performances of an experiment, assuming that A has probability $P(A) = p$ in a single trial. Then $q = 1 - p$ is the probability that in a single trial the event A does not occur. We assume that the experiment is performed n times and consider the random variable

$$X = \textit{number of times A occurs.}$$

Then X can assume the values 0, 1, $\cdots$, n, and we want to determine the corresponding probabilities. For this purpose we consider any of these values, say, $X = x$, which means that in x of the n trials A occurs and in $n - x$ trials it does not occur. This may look as follows:

$$\underbrace{AA \cdots A}_{x \text{ times}} \; \underbrace{BB \cdots B}_{n-x \text{ times}}. \tag{1}$$

Here $B = A^{C}$, that is, A does not occur. We assume that the trials are *independent,* that is, do not influence each other. Then, since $P(A) = p$ and $P(B) = q$, we see that to (1) there corresponds the probability

$$\underbrace{pp \cdots p}_{x \text{ times}} \; \underbrace{qq \cdots q}_{n-x \text{ times}} = p^x q^{n-x}.$$

Clearly, (1) is merely *one* order of arranging x A's and $n - x$ B's, and the probability $P(X = x)$ thus equals $p^x q^{n-x}$ times the number of different arrangements of x A's and $n - x$ B's, as follows from Theorem 1 in Sec. 20.5. We may number the n trials from 1 to n and pick x of these numbers corresponding to those trials in which A happens. Since the order in which we pick the x numbers does not matter, we see from (4a) in Sec. 20.6 that we can pick those x numbers from the n numbers in $\binom{n}{x}$ different ways. Hence the probability $P(X = x)$ corresponding to $X = x$ equals

$$f(x) = \binom{n}{x} p^x q^{n-x} \qquad (x = 0, 1, \cdots, n) \tag{2}$$

and $f(x) = 0$ for any other value of x. This is the probability that in n independent trials an event A occurs precisely x times where p is the probability of A in a single trial and $q = 1 - p$. The distribution determined by the probability function (2) is called the **binomial distribution** or *Bernoulli distribution.* The occurrence of A is called *success,* and the nonoccurrence is called *failure.* p is called the *probability of success in a single trial.* Figure 374 shows illustrative examples of $f(x)$.

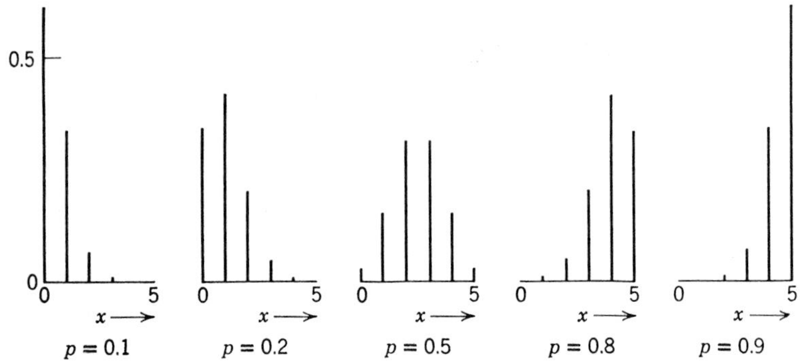

Fig. 374. Probability function (2) of the binomial distribution for $n = 5$ and various values of p

The binomial distribution has the mean (cf. Prob. 21)

$$\mu = np \tag{3}$$

and the variance (cf. Prob. 22)

$$\sigma^2 = npq. \tag{4}$$

Note that when $p = 0.5$, the distribution is symmetric with respect to μ.

A table of the binomial distribution is included in Appendix 4. For more extended tables see Ref. [J10] in Appendix 1.

The distribution with the probability function

$$f(x) = \frac{\mu^x}{x!}e^{-\mu} \qquad (x = 0, 1, \cdots) \tag{5}$$

is called the **Poisson distribution,** named after S. D. Poisson (cf. Sec. 18.4). Figure 375 shows (5) for some values of μ. It can be proved that this distribution may be obtained as a limiting case of the binomial distribution, if we let $p \to 0$ and $n \to \infty$ so that the mean $\mu = np$ approaches a finite value. The Poisson distribution has the mean μ and the variance (cf. Prob. 23)

$$\sigma^2 = \mu. \tag{6}$$

This distribution provides probabilities, e.g., of given numbers of cars passing an intersection per unit interval of time, of given numbers of defects per unit length of wire or unit area of paper or textile, and so on.

The binomial distribution is important in **sampling with replacement** (cf. Example 3, Sec. 20.5), as follows. Suppose that a box contains N things, for example, screws, M of which are defective. If we want to draw a screw at random, the probability of obtaining a defective screw is

$$p = \frac{M}{N}.$$

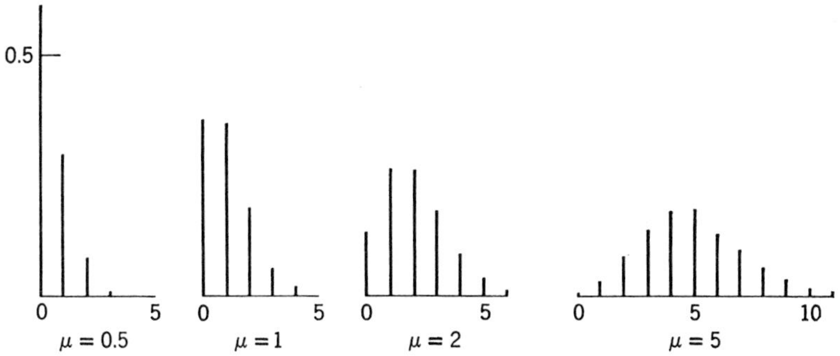

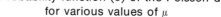

Fig. 375. Probability function (5) of the Poisson distribution for various values of μ

Hence in drawing a sample of n screws *with replacement,* the probability that precisely x of the n screws are defective is [cf. (2)]

$$f(x) = \binom{n}{x}\left(\frac{M}{N}\right)^x\left(1 - \frac{M}{N}\right)^{n-x} \qquad (x = 0, 1, \cdots, n). \tag{7}$$

In **sampling without replacement** that probability is

$$f(x) = \frac{\binom{M}{x}\binom{N-M}{n-x}}{\binom{N}{n}} \qquad (x = 0, 1, \cdots, n). \tag{8}$$

This distribution with the probability function (8) is called the **hypergeometric distribution.**[6]

To verify our statement we first note that, by (4a) in Sec. 20.6, there are

(a) $\binom{N}{n}$ different ways of picking n things from N,

(b) $\binom{M}{x}$ different ways of picking x defectives from M,

(c) $\binom{N-M}{n-x}$ different ways of picking $n - x$ nondefectives from $N - M$,

and each way in (b) combined with each way in (c) gives the total number of mutually exclusive ways of obtaining x defectives in n drawings without replacement. Since (a) is the total number of outcomes and we draw at random, each such way has the probability $1\Big/\binom{N}{n}$. From this, (8) follows.

The hypergeometric distribution has the mean (cf. Prob. 25)

$$\mu = n\frac{M}{N} \tag{9}$$

and the variance

$$\sigma^2 = \frac{nM(N - M)(N - n)}{N^2(N - 1)}. \tag{10}$$

Example 1. Sampling with and without replacement

We want to draw random samples of 2 gaskets from a box containing 10 gaskets, 3 of which are defective. Find the probability function of the random variable

$$X = \textit{number of defectives in the sample.}$$

[6] Because the moment generating function (cf. Prob. 25, Sec. 20.8) of this distribution can be represented in terms of the hypergeometric function.

We have $N = 10$, $M = 3$, $N - M = 7$, $n = 2$. For sampling with replacement, formula (7) yields

$$f(x) = \binom{2}{x}\left(\frac{3}{10}\right)^x\left(\frac{7}{10}\right)^{2-x}, \qquad f(0) = 0.49, \qquad f(1) = 0.42, \qquad f(2) = 0.09.$$

For sampling without replacement we have to use (8), finding

$$f(x) = \binom{3}{x}\binom{7}{2-x}\bigg/\binom{10}{2}, \qquad f(0) = f(1) = \frac{21}{45} \approx 0.47, \qquad f(2) = \frac{3}{45} \approx 0.07.$$

▮

If N, M, and $N - M$ are large compared with n, then it does not matter too much whether we sample with or without replacement, and in this case the hypergeometric distribution may be approximated by the binomial distribution (with $p = M/N$), *which is somewhat simpler.*

Hence in sampling from an indefinitely large population (**"infinite population"**) *we may use the binomial distribution, regardless of whether we sample with or without replacement.*

Problems for Sec. 20.9

1. Find formulas for $f(x)$, μ, and σ^2 in the case of the binomial distribution with $p = 0.5$.

2. Five fair coins are tossed simultaneously. Find the probability function of the random variable X = *number of heads* and compute the probabilities of obtaining no heads, precisely 1 head, at least 1 head, and not more than 4 heads.

3. If, in an experiment, positive and negative values are equally likely, what is the probability of obtaining at most 1 negative value in 9 trials?

4. If the probability of hitting a target is 10% and 10 shots are fired independently, what is the probability that the target will be hit at least once?

5. Let $p = 1\%$ be the probability that a certain type of light bulb will fail in a 24-hour test. Find the probability that a sign consisting of 10 such bulbs will burn 24 hours with no bulb failures.

6. Suppose that 3% of bolts made by a machine are defective, the defectives occurring at random during production. If the bolts are packaged 50 per box, what is the Poisson approximation of the probability that a given box will contain x defectives?

7. Suppose that in the production of 50-ohm radio resistors, nondefective items are those that have a resistance between 45 and 55 ohms and the probability of a resistor's being defective is 0.2%. The resistors are sold in lots of 100, with the guarantee that all resistors are nondefective. What is the probability that a given lot will violate this guarantee? (Use the Poisson distribution.)

8. Suppose that a telephone switchboard handles 300 calls on the average during a rush hour, and that the board can make at most 10 connections per minute. Using the Poisson distribution, estimate the probability that the board will be overtaxed during a given minute.

9. Classical experiments by E. Rutherford and H. Geiger in 1910 have shown that the number of alpha particles emitted per second in a radioactive process is a random variable X having a Poisson distribution. If X has mean 0.5, what is the probability of observing 2 or more particles during any given second?

10. Suppose that a certain type of magnetic tape contains, on the average, 2 defects per 100 meters. What is the probability that a roll of tape 300 meters long will contain (*a*) x defects, (*b*) no defects?

11. Let X be the number of cars per minute passing a certain point of some road between 8 A.M. and 10 A.M. on a Sunday. Assume that X has a Poisson distribution with mean 5. Find the probability of observing 3 or fewer cars during any given minute.

12. A carton contains 20 fuses, 5 of which are defective. Find the probability that, if a sample of 3 fuses is chosen from the carton by random drawing without replacement, x fuses in the sample will be defective.

13. A distributor sells rubber bands in packages of 100 and guarantees that at most 10% are defective. A consumer controls each package by drawing 10 bands without replacement. If the sample contains no defective rubber bands, he accepts the package. Otherwise he rejects it. Find the probability that in this process the consumer rejects a package which contains 10 defective bands (so that it still satisfies the guarantee).

14. In Problem 13 find the probability that any given package is accepted although it contains 20 defective rubber bands.

15. A process of manufacturing screws is checked every hour by inspecting n screws selected at random from that hour's production. If one or more screws are defective, the process is halted and carefully examined. How large should n be if the manufacturer wants the probability to be about 95% that the process will be halted when 10% of the screws being produced are defective? (Assume independence of the quality of any item of that of the other items.)

16. **(Multinomial distribution)** Suppose a trial can result in precisely one of k mutually exclusive events $A_1, \cdots, A_k$ with probabilities $p_1, \cdots, p_k$, respectively, where $p_1 + \cdots + p_k = 1$. Suppose that n independent trials are performed. Show that the probability of getting x_1 A_1's, $\cdots$, x_k A_k's is

$$f(x_1, \cdots, x_k) = \frac{n!}{x_1! \cdots x_k!} p_1^{x_1} \cdots p_k^{x_k}$$

where $0 \leqq x_j \leqq n$, $j = 1, \cdots, k$, and $x_1 + \cdots + x_k = n$. The distribution having this probability function is called the *multinomial distribution*.

17. Suppose that in a production of electrical resistors the probability of producing a resistor of resistance $R < 198$ ohms or $R > 201$ ohms is 3% and 5%, respectively. Find the probability that a random sample of 20 resistors contains precisely x_1 resistors with $R < 198$ ohms and precisely x_2 resistors with $R > 201$ ohms.

18. Show that the distribution function of the Poisson distribution satisfies $F(\infty) = 1$.

19. Let X have a binomial distribution with $p = 0.5$. Find the probability that X will assume a value in the interval $\mu - \sigma < X < \mu + \sigma$, assuming that (*a*) $n = 1$, (*b*) $n = 2$, (*c*) $n = 3$, (*d*) $n = 4$, and (*e*) $n = 5$.

20. Using the binomial theorem, show that the binomial distribution has the moment generating function (cf. Prob. 25, Sec. 20.8)

$$G(t) = \sum_{x=0}^{n} e^{tx} \binom{n}{x} p^x q^{n-x} = \sum_{x=0}^{n} \binom{n}{x} (pe^t)^x q^{n-x} = (pe^t + q)^n.$$

21. Using Prob. 20 and $p + q = 1$, prove (3).

22. Prove (4).

23. Show that the Poisson distribution has the moment generating function

$$G(t) = e^{-\mu} e^{\mu e^t}$$

and prove (6).

24. Show that $E([X-\mu]^3) = E(X^3) - 3\mu E(X^2) + 2\mu^3$. Using this and Prob. 23, show that the Poisson distribution has the skewness $\gamma = 1/\sqrt{\mu}$ and conclude that for large μ the distribution is almost symmetric (cf. Fig. 375).

25. Prove (9).

20.10 Normal Distribution

The continuous distribution having the density

$$f(x) = \frac{1}{\sigma\sqrt{2\pi}} e^{-\frac{1}{2}\left(\frac{x-\mu}{\sigma}\right)^2} \qquad (\sigma > 0) \tag{1}$$

is called the **normal distribution** or *Gauss distribution.* A random variable having this distribution is said to be **normal** or *normally distributed.* This distribution is very important, because many random variables of practical interest are normal or approximately normal or can be transformed into normal random variables in a relatively simple fashion. Furthermore, the normal distribution is a useful approximation of more complicated distributions. It also appears in the mathematical proofs of various statistical tests.

In (1), μ is the mean and σ is the standard deviation of the distribution. The curve of $f(x)$ is called the *bell-shaped curve.* It is symmetric with respect to μ. Figure 376 shows $f(x)$ for $\mu = 0$. For $\mu > 0$ ($\mu < 0$) the curves have the same shape, but are shifted $|\mu|$ units to the right (to the left). The smaller σ^2 is, the higher is the peak at $x = 0$ in Fig. 376 and the steeper are the descents on both sides. This agrees with the meaning of the variance.

From (1) we see that the normal distribution has the distribution function

$$F(x) = \frac{1}{\sigma\sqrt{2\pi}} \int_{-\infty}^{x} e^{-\frac{1}{2}\left(\frac{v-\mu}{\sigma}\right)^2} dv. \tag{2}$$

From this and (10) in Sec. 20.7 we obtain

$$P(a < X \leqq b) = F(b) - F(a) = \frac{1}{\sigma\sqrt{2\pi}} \int_{a}^{b} e^{\frac{1}{2}\left(\frac{v-\mu}{\sigma}\right)^2} dv. \tag{3}$$

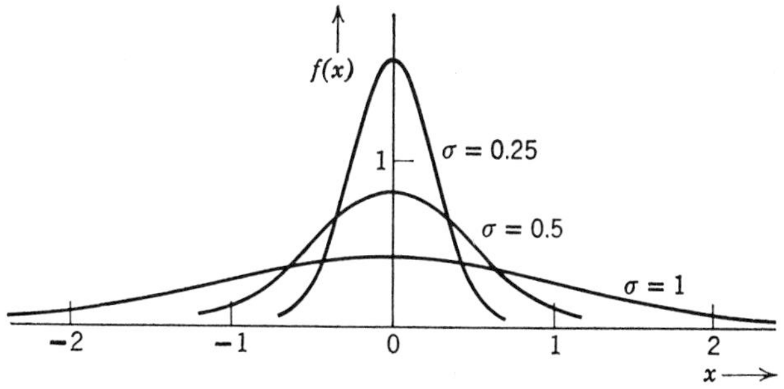

Fig. 376. Density (1) of the normal distribution with $\mu = 0$ for various values of σ

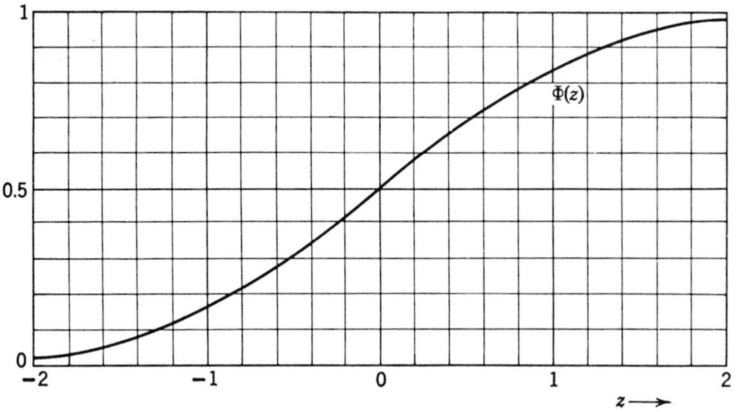

Fig. 377. Distribution function Φ(z) of the normal distribution with mean 0 and variance 1

The integral in (2) cannot be evaluated by elementary methods, but can be represented in terms of the integral (Fig. 377)

$$\textbf{(4)} \qquad \Phi(z) = \frac{1}{\sqrt{2\pi}} \int_{-\infty}^{z} e^{-u^2/2}\, du,$$

which is the distribution function of the normal distribution with mean 0 and variance 1 and has been tabulated (Table A8 in Appendix 4). In fact, if we set $(v - \mu)/\sigma = u$, then $du/dv = 1/\sigma$, $dv = \sigma\, du$, and we now have to integrate from $-\infty$ to $z = (x - \mu)/\sigma$. From (2) we thus obtain

$$F(x) = \frac{1}{\sigma\sqrt{2\pi}} \int_{-\infty}^{(x-\mu)/\sigma} e^{-u^2/2}\, \sigma\, du.$$

σ drops out, and the expression on the right equals (4) where $z = (x - \mu)/\sigma$, that is,

$$\textbf{(5)} \qquad F(x) = \Phi\left(\frac{x - \mu}{\sigma}\right).$$

From this important formula and (3) we obtain

$$\textbf{(6)} \qquad P(a < X \leqq b) = F(b) - F(a) = \Phi\left(\frac{b - \mu}{\sigma}\right) - \Phi\left(\frac{a - \mu}{\sigma}\right).$$

In particular, when $a = \mu - \sigma$ and $b = \mu + \sigma$, the right-hand side equals $\Phi(1) - \Phi(-1)$; to $a = \mu - 2\sigma$ and $b = \mu + 2\sigma$ there corresponds the value $\Phi(2) - \Phi(-2)$, etc. Using Table A8 in Appendix 4 we thus find (Fig. 378)

$$\textbf{(7)} \qquad \begin{aligned} &\text{(a)} \quad P(\mu - \sigma < X \leqq \mu + \sigma) \approx 68\%, \\ &\text{(b)} \quad P(\mu - 2\sigma < X \leqq \mu + 2\sigma) \approx 95.5\%, \\ &\text{(c)} \quad P(\mu - 3\sigma < X \leqq \mu + 3\sigma) \approx 99.7\%. \end{aligned}$$

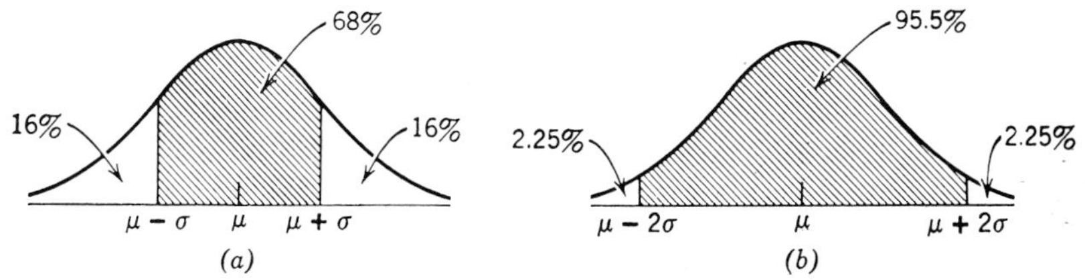

Fig. 378. Illustration of formula (7)

Hence we may expect that a large number of observed values of a normal random variable X will be distributed as follows:

(a) *About* $\frac{2}{3}$ *of the values will lie between* $\mu - \sigma$ *and* $\mu + \sigma$.

(b) *About* 95% *of the values will lie between* $\mu - 2\sigma$ *and* $\mu + 2\sigma$.

(c) *About* $99\frac{3}{4}$% *of the values will lie between* $\mu - 3\sigma$ *and* $\mu + 3\sigma$.

This may also be expressed as follows.

A value that deviates more than σ from μ will occur about once in 3 trials. A value that deviates more than 2σ or 3σ from μ will occur about once in 20 or 400 trials, respectively. Practically speaking, this means that all the values will lie between $\mu - 3\sigma$ and $\mu + 3\sigma$; these two numbers are called **three-sigma limits.**

In a similar fashion we obtain

$$\textbf{(8)} \qquad \begin{aligned} &\text{(a)} \quad P(\mu - 1.96\sigma < X \leqq \mu + 1.96\sigma) = 95\%, \\ &\text{(b)} \quad P(\mu - 2.58\sigma < X \leqq \mu + 2.58\sigma) = 99\%, \\ &\text{(c)} \quad P(\mu - 3.29\sigma < X \leqq \mu + 3.29\sigma) = 99.9\%. \end{aligned}$$

The following typical examples should help the reader to understand the practical use of Tables A8 and A9 in Appendix 4.

Example 1

Determine the probabilities

$$\text{(a) } P(X \leqq 2.44), \quad \text{(b) } P(X \leqq -1.16), \quad \text{(c) } P(X \geqq 1), \quad \text{(d) } P(2 \leqq X \leqq 10)$$

where X is assumed to be normal with mean 0 and variance 1.

Since $\mu = 0$ and $\sigma^2 = 1$, we may obtain the desired values directly from Table A8, finding

(a) 0.9927, (b) 0.1230, (c) $1 - P(X \leqq 1) = 1 - 0.8413 = 0.1587$ [cf. (6), Sec. 20.5],

(d) $\Phi(10) = 1.0000$ (why?), $\Phi(2) = 0.9772$, $\Phi(10) - \Phi(2) = 0.0228$.

Example 2

Determine the probabilities in the previous example, assuming that X is normal with mean 0.8 and variance 4.

From (6) and Table A8 we obtain

$$\text{(a) } F(2.44) = \Phi\left(\frac{2.44 - 0.80}{2}\right) = \Phi(0.82) = 0.7939$$

$$\text{(b) } F(-1.16) = \Phi(-0.98) = 0.1635$$

$$\text{(c) } 1 - P(X \leqq 1) = 1 - F(1) = 1 - \Phi(0.1) = 0.4602$$

$$\text{(d) } F(10) - F(2) = \Phi(4.6) - \Phi(0.6) = 1 - 0.7257 = 0.2743.$$

Example 3

Let X be normal with mean 0 and variance 1. Determine the constant c such that

(a) $P(X \geqq c) = 10\%$, (b) $P(X \leqq c) = 5\%$,

(c) $P(0 \leqq X \leqq c) = 45\%$, (d) $P(-c \leqq X \leqq c) = 99\%$.

From Table A9 in Appendix 4 we obtain

(a) $1 - P(X \leqq c) = 1 - \Phi(c) = 0.1,\ \Phi(c) = 0.9,\ c = 1.282$

(b) $c = -1.645$

(c) $\Phi(c) - \Phi(0) = \Phi(c) - 0.5 = 0.45,\ \Phi(c) = 0.95,\ c = 1.645$

(d) $c = 2.576$.

Example 4

Let X be normal with mean -2 and variance 0.25. Determine c such that

(a) $P(X \geqq c) = 0.2$ (b) $P(-c \leqq X \leqq -1) = 0.5$

(c) $P(-2 - c \leqq X \leqq -2 + c) = 0.9$ (d) $P(-2 - c \leqq X \leqq -2 + c) = 99.6\%$.

Using Table A9 in Appendix 4, we obtain the following results.

(a) $1 - P(X \leqq c) = 1 - \Phi\left(\dfrac{c+2}{0.5}\right) = 0.2,$

$\Phi(2c + 4) = 0.8,\ 2c + 4 = 0.842,\ c = -1.579$

(b) $\Phi\left(\dfrac{-1+2}{0.5}\right) - \Phi\left(\dfrac{-c+2}{0.5}\right) = 0.9772 - \Phi(4 - 2c) = 0.5,$

$\Phi(4 - 2c) = 0.4772,\ 4 - 2c = -0.057,\ c = 2.03$

(c) $\Phi\left(\dfrac{-2+c+2}{0.5}\right) - \Phi\left(\dfrac{-2-c+2}{0.5}\right)$

$= \Phi(2c) - \Phi(-2c) = 0.9,\ 2c = 1.645,\ c = 0.823$

(d) $\Phi(2c) - \Phi(-2c) = 99.6\%,\ 2c = 2.878,\ c = 1.439$.

Example 5

Suppose that in producing iron plates the plates are required to have a certain thickness, and the machine work is done by a shaper. In any case industrial products will differ slightly from each other because the properties of the material and the behavior of the machines and tools used show slight random variations caused by small disturbances which we cannot predict. We may therefore regard the thickness X [mm] of the plates as a random variable. We assume that for a certain setting the variable X is normal with mean $\mu = 10$ mm and standard deviation $\sigma = 0.02$ mm. We want to determine the percentage of defective plates to be expected, assuming that defective plates are (*a*) plates thinner than 9.97 mm, (*b*) plates thicker than 10.05 mm, (*c*) plates that deviate more than 0.03 mm from 10 mm. (*d*) How should we choose numbers $10 - c$ and $10 + c$ in order that the expected percentage of defectives be not greater than 5%? (*e*) How does the percentage of defectives in part (*d*) change if μ is shifted from 10 mm to 10.01 mm?

Using Table A8 in Appendix 4 and (6), we obtain the following solutions.

(a) $P(X \leqq 9.97) = \Phi\left(\dfrac{9.97 - 10.00}{0.02}\right) = \Phi(-1.5) = 0.0668 \approx 6.7\%$

(b) $P(X \geqq 10.05) = 1 - P(X \leqq 10.05) = 1 - \Phi\left(\dfrac{10.05 - 10.00}{0.02}\right)$

$= 1 - \Phi(2.5) = 1 - 0.9938 \approx 0.6\%$

(c) $P(9.97 \leqq X \leqq 10.03) = \Phi\left(\frac{10.03 - 10.00}{0.02}\right) - \Phi\left(\frac{9.97 - 10.00}{0.02}\right)$

$= \Phi(1.5) - \Phi(-1.5) = 0.8664.$ *Answer* $1 - 0.8664 \approx 13\%$.

(d) From (8a) we obtain $c = 1.96\sigma = 0.039$. *Answer* 9.961 mm and 10.039 mm.

(e) $P(9.961 \leqq X \leqq 10.039) = \Phi\left(\frac{10.039 - 10.010}{0.02}\right) - \Phi\left(\frac{9.961 - 10.010}{0.02}\right)$

$= \Phi(1.45) - \Phi(-2.45) = 0.9265 - 0.0071 \approx 92\%.$ *Answer* 8%.

We see that this slight change of the adjustment of the tool bit causes a considerable increase of the percentage of defectives. ∎

Under a linear transformation, a normal random variable transforms into a normal random variable. In fact, using (5), the reader may prove the following theorem.

Theorem 1

If X is normal with mean μ and variance σ^2, then $X^* = c_1X + c_2$ ($c_1 \neq 0$) is normal with mean $\mu^* = c_1\mu + c_2$ and variance $\sigma^{*2} = c_1^{\,2}\sigma^2$.

The normal distribution may also be used for approximating the binomial distribution when n is large, so that the binomial coefficients and powers in the probability function

$$f(x) = \binom{n}{x} p^x q^{n-x} \qquad (x = 0, 1, \cdots, n) \tag{9}$$

(cf. Sec. 20.9) of the binomial distribution become very inconvenient. In fact, the following important theorem holds true.

Theorem 2 (Limit theorem of De Moivre and Laplace)

For large n,

$$f(x) \sim f^*(x) \qquad (x = 0, 1, \cdots, n),$$

where f is given by (9),

$$f^*(x) = \frac{1}{\sqrt{2\pi}\sqrt{npq}} e^{-z^2/2}, \qquad z = \frac{x - np}{\sqrt{npq}} \tag{10}$$

is the density of the normal distribution with mean $\mu = np$ and variance $\sigma^2 = npq$ (the mean and variance of the binomial distribution), and the symbol $\sim$ (read **asymptotically equal**) *means that the ratio of both sides approaches* 1 *as n approaches ∞. Furthermore, for any nonnegative integers a and b ($> a$),*

$$P(a \leqq X \leqq b) = \sum_{x=a}^{b} \binom{n}{x} p^x q^{n-x} \sim \Phi(\beta) - \Phi(\alpha),$$
$$\alpha = \frac{a - np - 0.5}{\sqrt{npq}}, \qquad \beta = \frac{b - np + 0.5}{\sqrt{npq}}. \tag{11}$$

A proof of this theorem can be found in Ref. [H4], cf. Appendix 1. The proof shows that the term 0.5 in α and β is a correction caused by the change from a discrete to a continuous distribution.

Problems for Sec. 20.10

1. Show that the curve of (1) has two points of inflection corresponding to $x = \mu \pm \sigma$.

2. Show that $\Phi(-z) = 1 - \Phi(z)$.

3. Let X be normal with mean 10 and variance 4. Find $P(X > 12)$, $P(X < 10)$, $P(X < 11)$, and $P(9 < X < 13)$.

4. Let X be normal with mean 50 and variance 9. Determine c such that $P(X < c) = 5\%$, $P(X > c) = 1\%$, and $P(50 - c < X < 50 + c) = 50\%$.

5. In Example 5, part (*c*), what value should σ have in order that the percentage of defective plates reduces to 1%?

6. What percentage of defective iron plates can we expect in question (*c*) of Example 5 if we use a better shaper so that $\sigma = 0.01$ mm?

7. A manufacturer produces airmail envelopes whose weight is normal with mean $\mu = 1.95$ grams and standard deviation $\sigma = 0.05$ grams. The envelopes are sold in lots of 1000. How many envelopes in a lot will be heavier than 2 grams?

8. In Prob. 7, how many envelopes weighing 2.05 grams or more can be expected in a given package of 100 envelopes?

9. What is the probability of obtaining at least 2048 heads if a coin is tossed 4040 times and heads and tails are equally likely? (Cf. Table 20.6 in Sec. 20.5.)

10. If the lifetime X of a certain kind of automobile battery is normally distributed with a mean of 3 years and a standard deviation of 1 year, and the manufacturer wishes to guarantee the battery for 2 years, what percentage of the batteries will he have to replace under the guarantee?

11. The breaking strength X [kg] of a certain type of plastic blocks is normally distributed with a mean of 1000 kg and a standard deviation of 75 kg. What is the maximum load such that we can expect no more than 5% of the blocks to break?

12. Specifications for a certain job call for bolts with a diameter of 0.260 ± 0.005 cm. If the diameters of the bolts made by some manufacturer are normally distributed with $\mu = 0.259$ and $\sigma = 0.003$, what percentage of these bolts will meet specifications?

13. How does the answer to Prob. 12 change if $\sigma = 0.003$ is changed to $\sigma = 0.030$ (lower quality of production)?

14. A manufacturer knows from experience that the resistance of resistors he produces is normal with mean $\mu = 100$ ohms and standard deviation $\sigma = 2$ ohms. What percentage of the resistors will have resistance between 98 ohms and 102 ohms? Between 95 ohms and 105 ohms?

15. A producer sells electric bulbs in cartons of 1000 bulbs. Using (11), find the probability that any given carton contains not more than 1% defective bulbs, assuming the production process to be a Bernoulli experiment with $p = 1\%$ (= probability that any given bulb will be defective).

16. If X is normal with mean μ and variance σ^2, what distribution does $-X$ have?

17. Considering $\Phi^2(\infty)$ and introducing polar coordinates in the double integral, prove

$$\Phi(\infty) = \frac{1}{\sqrt{2\pi}} \int_{-\infty}^{\infty} e^{-u^2/2}\, du = 1. \tag{12}$$

18. Using (12) and integration by parts, show that σ in (1) is the standard deviation of the normal distribution.

19. Prove Theorem 1.

20. (Bernoulli's law of large numbers) In a random experiment let an event A have probability p $(0 < p < 1)$, and let X be the number of times A happens in n independent trials. Show that for any given $\epsilon > 0$,

$$P\left(\left|\frac{X}{n} - p\right| < \epsilon\right) \to 1 \qquad \text{as} \quad n \to \infty.$$

20.11 Distributions of Several Random Variables

If in a random experiment we observe a single quantity, we have to associate with that experiment a single random variable, call it X. From Sec. 20.7 we remember that the corresponding distribution function $F(x) = P(X \leqq x)$ determines the distribution completely, because for each interval $a < X \leqq b$,

$$P(a < X \leqq b) = F(b) - F(a).$$

If in a random experiment we observe two quantities, we have to associate with the experiment two random variables, say, X and Y. For example, X may correspond to the Rockwell hardness and Y to the carbon content of steel, if we measure these two quantities. Each performance of the experiment yields a pair of numbers $X = x$, $Y = y$, briefly (x, y), which may be plotted as a point in the XY-plane. We may now consider a rectangle $a_1 < X \leqq b_1$, $a_2 < Y \leqq b_2$ (Fig. 379). If for each such rectangle we know the corresponding probability

$$P(a_1 < X \leqq b_1, a_2 < Y \leqq b_2),$$

then we say that the **two-dimensional probability distribution** of the random variables X and Y or of the **two-dimensional random variable** (X, Y) is known. The function

$$F(x, y) = P(X \leqq x, Y \leqq y) \tag{1}$$

is called the **distribution function** of that distribution or of (X, Y). It determines the distribution uniquely, because (cf. Prob. 1)

$$\begin{aligned} P(a_1 < X \leqq b_1, a_2 < Y \leqq b_2) \\ = F(b_1, b_2) - F(a_1, b_2) - F(b_1, a_2) + F(a_1, a_2). \end{aligned} \tag{2}$$

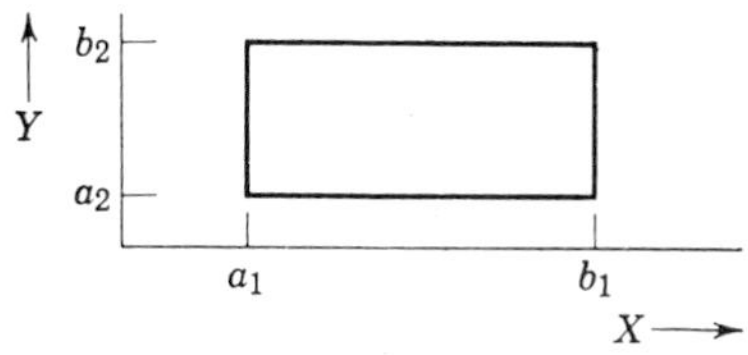

Fig. 379. Notion of a two-dimensional distribution

The distribution and the variable (X, Y) are said to be **discrete** if (X, Y) has the following properties.

(X, Y) can assume only finitely many or at most countably infinitely many pairs of values (x, y), the corresponding probabilities being positive. To each bounded domain[7] containing no such pairs there corresponds the probability 0.

Let (x_i, y_j) be any such pair and let $P(X = x_i, Y = y_j) = p_{ij}$ (where we admit that p_{ij} may be 0 for certain pairs of subscripts i, j). The function

(3) $f(x, y) = p_{ij}$ when $x = x_i,\ y = y_j$ and $f(x, y) = 0$ otherwise

is called the **probability function** of (X, Y); here $i = 1, 2, \cdots$ and $j = 1, 2, \cdots$ independently. The analogue of (7), Sec. 20.7, is

$$F(x, y) = \sum_{x_i \leqq x} \sum_{y_j \leqq y} f(x_i, y_j), \tag{4}$$

and instead of formula (3) in Sec. 20.7 we now have the condition

$$\sum_i \sum_j f(x_i, y_j) = 1. \tag{5}$$

For example, if we toss a dime and a nickel once and consider

$X =$ *number of heads the dime turns up,*

$Y =$ *number of heads the nickel turns up,*

then X and Y can have the values 0 or 1, and the probability function is

$$f(0, 0) = f(1, 0) = f(0, 1) = f(1, 1) = \tfrac{1}{4}, \qquad f(x, y) = 0 \text{ otherwise.}$$

(X, Y) and its distribution are said to be **continuous** if the corresponding distribution function $F(x, y)$ can be represented by a double integral

$$F(x, y) = \int_{-\infty}^{y} \int_{-\infty}^{x} f(x^*, y^*)\, dx^*\, dy^* \tag{6}$$

where $f(x, y)$ is defined, nonnegative, and bounded in the entire plane, possibly except for finitely many continuously differentiable curves. $f(x, y)$ is called the **probability density** of the distribution. It follows that

$$P(a_1 < X \leqq b_1, a_2 < Y \leqq b_2) = \int_{a_2}^{b_2} \int_{a_1}^{b_1} f(x, y)\, dx\, dy. \tag{7}$$

For example (cf. Fig. 380)

(8) $f(x, y) = 1/k$ when (x, y) is in R, $\quad f(x, y) = 0$ otherwise

[7] That is, a domain which can be enclosed in a circle of sufficiently large radius.

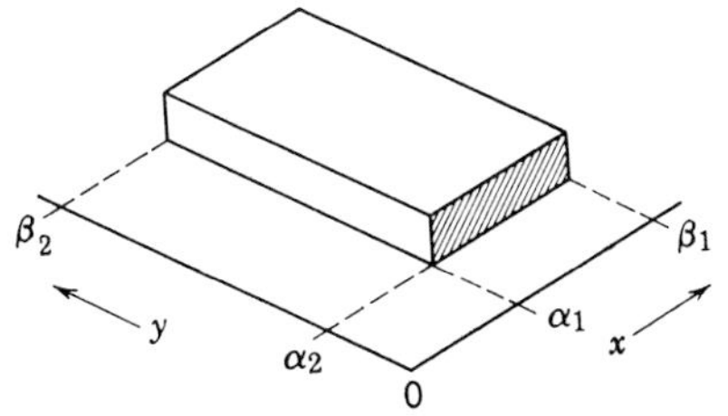

Fig. 380. Probability density function (8) of the uniform distribution

defines the so-called *uniform distribution in the rectangle R;* here k is the area of R, that is, $k = (\beta_1 - \alpha_1)(\beta_2 - \alpha_2)$. The distribution function is shown in Fig. 381.

In the case of a *discrete* random variable (X, Y) with probability function $f(x, y)$ we may ask for the probability $P(X = x,\ Y \text{ arbitrary})$ that X assume a value x, while Y may assume any value, which we ignore. This probability is a function of x, say $f_1(x)$, and we have the formula

(9) $$f_1(x) = P(X = x,\ Y \text{ arbitrary}) = \sum_y f(x, y)$$

where we sum all the values of $f(x, y)$ which are not 0 for that x. Clearly, $f_1(x)$ is the probability function of the probability distribution of a single random variable. This distribution is called the **marginal distribution** *of X with respect to the given two-dimensional distribution.* It has the distribution function

(10) $$F_1(x) = P(X \leqq x,\ Y \text{ arbitrary}) = \sum_{x^* \leqq x} f_1(x^*).$$

Similarly, the probability function

(11) $$f_2(y) = P(X \text{ arbitrary},\ Y = y) = \sum_x f(x, y)$$

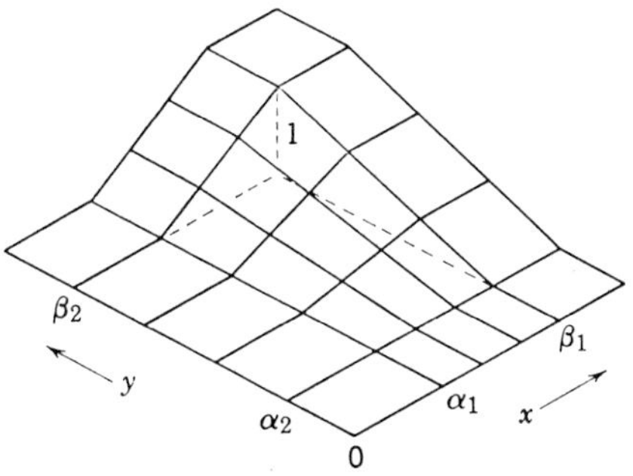

Fig. 381. Distribution function of the uniform distribution defined by (8)

determines the so-called **marginal distribution** *of Y with respect to the given two-dimensional distribution.* In (11) we sum all the values of $f(x, y)$ which are not 0 for the corresponding y. The distribution function of this distribution is

$$F_2(y) = P(X \text{ arbitrary}, Y \leqq y) = \sum_{y^* \leqq y} f_2(y^*). \tag{12}$$

Obviously, both marginal distributions of a discrete random variable (X, Y) are discrete.

Table 20.7 shows an illustrative example. X is the number of queens and Y is the number of kings or aces obtained in drawing 3 cards with replacement from a bridge deck. Since the 52 cards of the deck contain 4 queens, 4 kings, and 4 aces, the probability of obtaining a queen if we draw a single card is $\frac{4}{52} = \frac{1}{13}$ and a king or ace is obtained with probability $\frac{8}{52} = \frac{2}{13}$. Hence to this random experiment there corresponds the probability function

$$f(x, y) = \frac{3!}{x!\, y!\, (3 - x - y)!} \left(\frac{1}{13}\right)^x \left(\frac{2}{13}\right)^y \left(\frac{10}{13}\right)^{3-x-y} \qquad (x + y \leqq 3)$$

and $f(x, y) = 0$ otherwise. Table 20.7 shows the values of $f(x, y)$, $f_1(x)$, and $f_2(y)$.

Similarly, in the case of a *continuous* random variable (X, Y) with density $f(x, y)$ we may consider

$$(X \leqq x,\ Y \text{ arbitrary}) \qquad \text{or} \qquad (X \leqq x,\ -\infty < Y < \infty).$$

Obviously, the corresponding probability is

$$F_1(x) = P(X \leqq x, -\infty < Y < \infty) = \int_{-\infty}^{x} \left(\int_{-\infty}^{\infty} f(x^*, y)\, dy \right) dx^*.$$

Table 20.7
Values of the Probability Functions $f(x, y)$, $f_1(x)$, and $f_2(y)$ in Drawing 3 Cards with Replacement from a Bridge Deck where X is the Number of Queens Drawn and Y is the Number of Kings or Aces Drawn

x \ y	0	1	2	3	$f_1(x)$
0	$\frac{1000}{2197}$	$\frac{600}{2197}$	$\frac{120}{2197}$	$\frac{8}{2197}$	$\frac{1728}{2197}$
1	$\frac{300}{2197}$	$\frac{120}{2197}$	$\frac{12}{2197}$	0	$\frac{432}{2197}$
2	$\frac{30}{2197}$	$\frac{6}{2197}$	0	0	$\frac{36}{2197}$
3	$\frac{1}{2197}$	0	0	0	$\frac{1}{2197}$
$f_2(y)$	$\frac{1331}{2197}$	$\frac{726}{2197}$	$\frac{132}{2197}$	$\frac{8}{2197}$	

Setting

$$f_1(x) = \int_{-\infty}^{\infty} f(x, y)\, dy \tag{13}$$

we may write

$$F_1(x) = \int_{-\infty}^{x} f_1(x^*)\, dx^*. \tag{14}$$

$f_1(x)$ is called the *density* and $F_1(x)$ the *distribution function* of the **marginal distribution** *of X with respect to the given continuous distribution.* The function

$$f_2(y) = \int_{-\infty}^{\infty} f(x, y)\, dx \tag{15}$$

is called the *density* and

$$F_2(y) = \int_{-\infty}^{y} f_2(y^*)\, dy^* = \int_{-\infty}^{y} \int_{-\infty}^{\infty} f(x, y^*)\, dx\, dy^* \tag{16}$$

is called the *distribution function* of the **marginal distribution** *of Y with respect to the given two-dimensional distribution.* We see that both marginal distributions of a continuous distribution are continuous.

The two random variables X and Y of a two-dimensional (X, Y)-distribution with distribution function $F(x, y)$ are said to be **independent** if

$$F(x, y) = F_1(x)F_2(y) \tag{17}$$

holds for all (x, y). Otherwise these variables are said to be **dependent.**

Suppose that X and Y are either both discrete or both continuous. Then X and Y are independent if and only if the corresponding probability functions or densities $f_1(x)$ and $f_2(y)$ satisfy

$$f(x, y) = f_1(x)f_2(y) \tag{18}$$

for all (x, y); cf. Prob. 2. For example, the variables in Table 20.7 are dependent. The variables X = *number of heads on a dime,* Y = *number of heads on a nickel* in tossing a dime and a nickel once may assume the values 0 or 1 and are independent.

The notions of independence and dependence may be extended to the n random variables of an n-dimensional $(X_1, \cdots, X_n)$-distribution with distribution function

$$F(x_1, \cdots, x_n) = P(X_1 \leqq x_1, \cdots, X_n \leqq x_n).$$

These random variables are said to be **independent** if for all $(x_1, \cdots, x_n)$,

$$F(x_1, \cdots, x_n) = F_1(x_1)F_2(x_2) \cdots F_n(x_n), \tag{19}$$

where $F_j(x_j)$ is the distribution function of the marginal distribution of X_j, that is,

$$F_j(x_j) = P(X_j \leqq x_j, X_k \text{ arbitrary}, k \neq j).$$

Otherwise these random variables are said to be **dependent.**

Let (X, Y) be a random variable with probability function or density $f(x, y)$ and distribution function $F(x, y)$, and let $g(x, y)$ be any continuous function which is defined for all (x, y) and is not constant. Then $Z = g(X, Y)$ is a random variable, too. For example, if we roll two dice and X is the number that the first die turns up whereas Y is the number that the second die turns up, then $Z = X + Y$ is the sum of those two numbers (cf. Fig. 371 in Sec. 20.7).

If $(X_1, \cdots, X_n)$ is an n-dimensional random variable and $g(x_1, \cdots, x_n)$ is a continuous function which is defined for all $(x_1, \cdots x_n)$ and is not constant, then $Z = g(X_1, \cdots, X_n)$ is a random variable, too.

In the case of a *discrete* random variable (X, Y) we may obtain the probability function $f(z)$ of $Z = g(X, Y)$ by summing all $f(x, y)$ for which $g(x, y)$ equals the value of z considered; thus

$$\textbf{(20)} \qquad f(z) = P(Z = z) = \sum_{g(x,y)=z}\sum f(x, y).$$

The distribution function of Z is

$$\textbf{(21)} \qquad F(z) = P(Z \leqq z) = \sum_{g(x,y)\leqq z}\sum f(x, y),$$

where we sum all values of $f(x, y)$ for which $g(x, y) \leqq z$.

In the case of a *continuous* random variable (X, Y) we similarly have

$$\textbf{(22)} \qquad F(z) = P(Z \leqq z) = \int_{g(x,y)\leqq z}\int f(x, y)\, dx\, dy,$$

where for each z we integrate over the region $g(x, y) \leqq z$ in the xy-plane.

The number

$$\textbf{(23)} \qquad E(g(X, Y)) = \begin{cases} \displaystyle\sum_x \sum_y g(x, y) f(x, y) & [(X, Y) \text{ discrete}] \\ \displaystyle\int_{-\infty}^{\infty}\int_{-\infty}^{\infty} g(x, y) f(x, y)\, dx\, dy & [(X, Y) \text{ continuous}] \end{cases}$$

is called the **mathematical expectation** or, briefly, the **expectation of** $g(X, Y)$. Here it is assumed that the double series converges absolutely and the integral of $|g(x, y)| f(x, y)$ over the xy-plane exists. The formula

$$(24) \qquad E(ag(X, Y) + bh(X, Y)) = aE(g(X, Y)) + bE(h(X, Y))$$

can be proved in a fashion similar to that in Prob. 19, Sec. 20.8. An important special case is $E(X + Y) = E(X) + E(Y)$, and by induction we have the following result.

Theorem 1 (Addition of means)
The mean (expectation) of a sum of random variables equals the sum of the means (expectations), that is

$$E(X_1 + X_2 + \cdots + X_n) = E(X_1) + E(X_2) + \cdots + E(X_n). \tag{25}$$

Furthermore we readily obtain

Theorem 2 (Multiplication of means)
The mean (expectation) of the product of ***independent*** *random variables equals the product of the means (expectations), that is,*

$$E(X_1X_2 \cdots X_n) = E(X_1)E(X_2) \cdots E(X_n). \tag{26}$$

Proof. If X and Y are independent random variables (both discrete or both continuous), then $E(XY) = E(X)E(Y)$. In fact, in the discrete case we have

$$E(XY) = \sum_x \sum_y xyf(x, y) = \sum_x xf_1(x) \sum_y yf_2(y) = E(X)E(Y),$$

and in the continuous case the proof of the relation is similar. Extension to n independent random variables gives (26), and Theorem 2 is proved. ∎

We shall now discuss the addition of variances. Let $Z = X + Y$ and let μ and σ^2 denote the mean and variance of Z. We first have (cf. Prob. 16 in Sec. 20.8)

$$\sigma^2 = E([Z - \mu]^2) = E(Z^2) - [E(Z)]^2.$$

From (24) we see that the first term on the right equals

$$E(Z^2) = E(X^2 + 2XY + Y^2) = E(X^2) + 2E(XY) + E(Y^2),$$

and for the second term on the right, we obtain from Theorem 1

$$[E(Z)]^2 = [E(X) + E(Y)]^2 = [E(X)]^2 + 2E(X)E(Y) + [E(Y)]^2.$$

By substituting these expressions in the formula for σ^2 we have

$$\begin{aligned}\sigma^2 = E(X^2) - [E(X)]^2 + E(Y^2) - [E(Y)]^2 \\ + 2[E(XY) - E(X)E(Y)].\end{aligned}$$

From Prob. 16, Sec. 20.8, we see that the expression in the first line on the right is the sum of the variances of X and Y, which we denote by σ_1^2 and σ_2^2, respectively. The quantity

$$\sigma_{XY} = E(XY) - E(X)E(Y) \tag{27}$$

is called the **covariance** of X and Y. Consequently, our result is

(28) $$\sigma^2 = \sigma_1{}^2 + \sigma_2{}^2 + 2\sigma_{XY}.$$

If X and Y are independent, then $E(XY) = E(X)E(Y)$; hence $\sigma_{XY} = 0$, and

(29) $$\sigma^2 = \sigma_1{}^2 + \sigma_2{}^2.$$

Extension to more than two variables yields

Theorem 3 (Addition of variances)
The variance of the sum of ***independent*** *random variables equals the sum of the variances of these variables.*

Problems for Sec. 20.11

1. Prove (2).

2. Prove the statement involving (18).

3. Let $f(x, y) = 2$ when $x > 0, y > 0, x + y < 1$ and 0 otherwise. Graph f and the corresponding distribution function.

4. Find the marginal distributions of the distribution in Figs. 380 and 381 on p. 887.

5. A four-gear assembly is put together with spacers between the gears. The mean thickness of the gears is 5.020 cm with a standard deviation of 0.003 cm. The mean thickness of the spacers is 0.040 cm with a standard deviation of 0.002 cm. Find the mean and standard deviation of the assembled units consisting of 4 randomly selected gears and 3 randomly selected spacers.

6. Let X [cm] and Y [cm] be the diameter of a pin and hole, respectively. Suppose that (X, Y) has the density

$$f(x, y) = 2500 \qquad \text{if} \qquad 0.99 < x < 1.01,\ 1.00 < y < 1.02$$

and 0 otherwise. (*a*) Find the marginal distributions. (*b*) What is the probability that a pin chosen at random will fit a hole whose diameter is 1.00?

7. Find $P(X > Y)$ when (X, Y) has the density $f(x, y) = e^{-(x+y)}$ if $x \geqq 0, y \geqq 0$ and 0 otherwise.

8. Find the densities of the marginal distributions in Prob. 7.

9. An electronic device consists of two components. Let X and Y [months] be the length of time until failure of the first and second component, respectively. Assume that (X, Y) has the probability density

$$f(x, y) = 0.01e^{-0.1(x+y)} \qquad \text{if } x > 0 \text{ and } y > 0$$

and 0 otherwise. (*a*) Are X and Y dependent or independent? (*b*) Find the densities of the marginal distributions. (*c*) What is the probability that the first component has a lifetime of 10 months or longer?

10. Let (X, Y) have the probability function $f(0, 0) = f(1, 1) = 3/8$, $f(0, 1) = f(1, 0) = 1/8$. Are X and Y independent?

11. Give an example of two different discrete distributions that have the same marginal distributions.

12. Show that the random variables with the densities $f(x, y) = x + y$ and $g(x, y) = (x + \frac{1}{2})(y + \frac{1}{2})$ when $0 \leqq x \leqq 1, 0 \leqq y \leqq 1$, have the same marginal distributions.

13. Let (X, Y) have the density $f(x, y) = k$ when $x^2 + y^2 < 1$ and 0 otherwise. Determine k. Find the densities of the marginal distributions. Find $P(X^2 + Y^2 < 1/2)$.

14. Using Theorem 1, obtain the formula for the mean μ of the binomial distribution.

15. Using Theorem 3, obtain the formula for the variance σ^2 of the binomial distribution.

20.12 Random Sampling. Random Numbers

The preceding sections (20.3–20.11) were devoted to probability theory and the remaining sections (20.12–20.20) deal with statistics. Probability theory helps to create mathematical models of populations, and the statistical methods to be discussed yield relations between the theory and the observable reality, conclusions about populations by means of samples (*statistical inference;* cf. Sec. 20.1).

So far it was sufficient to know that a sample from a population is a selection taken from a population (for examples, see Sec. 20.1), but from now on we have to define this notion in a precise fashion. In fact, in order to obtain meaningful information about populations from samples, a sample must be a **random selection;** that is, each element of the population must have a known probability of being taken into the sample. This condition must be satisfied (at least approximately), otherwise the methods to be discussed may yield completely meaningless and misleading results.

If the sample space is infinite, the sample values will be **independent,** that is, the results of the n performances of a random experiment made for obtaining n sample values will not influence each other. This certainly applies to samples from a normal population. If the sample space is finite, the sample values will still be independent if we sample with replacement; they will be *practically* independent if we sample without replacement and keep the size of the sample small compared with the size of the population (for instance, samples of 5 or 10 values from a population of 1000 values). However, if we sample without replacement and take large samples from a finite population, the dependence will matter considerably.

It is not so easy to satisfy the requirement that a sample be a random selection, because there are many and subtle factors of various types that can bias results of sampling. For example, if an interested purchaser wants to draw and inspect a sample of 10 items from a lot of 80 items before he decides whether to purchase the lot, how should he select physically those 10 items so that he can be reasonably sure that all possible $\binom{80}{10}$ samples of size 10 would be equally probable?

Methods to solve this problem have been developed, and we shall now describe such a procedure, which is frequently used.

We number the items of that lot from 1 to 80. Then we select 10 items using Table A10 in Appendix 4, which contains *random numbers* made up of sets of

random digits, as follows. We first select a line number from 0 to 99 at random. This can be done by tossing a fair coin 7 times denoting heads by 1 and tails by 0, thus generating a 7-place binary number which will represent 0, 1, $\cdots$, 127 with equal probabilities. We use this number if it is 0, 1, $\cdots$, or 99. Otherwise we ignore it and repeat this process. Then we select a column number from 0 to 9 by generating a 4-place binary number in a similar fashion. Suppose that we obtained 0011010 ($= 26$) and 0111 ($= 7$), respectively. In row 26 and column 7 of Table A10 we find 44973. We keep the first 2 digits, that is, 44. We move down the column, starting with 44973 and keeping only the first 2 digits. In this way we find

44 44 83 91 55 etc.

We omit numbers which are greater than 80 or occur for the second time and continue until 10 numbers are obtained. This yields

44 55 53 03 52 61 67 78 39 54.

The 10 items having these numbers represent the desired selection.

For a larger table of random digits see Ref. [J15], Appendix 1.

However, for large samples such tables may become cumbersome. For this reason, procedures for generating numbers with similar properties have been developed and are available in a so-called **random number generator** in many computer languages and subroutine libraries. Cf. also Refs. [H7] and [H11] in Appendix 1.

Problems for Sec. 20.12

1. Suppose that in the example explained in the text we would start from row 83 and column 2 of Table A10 in Appendix 4, and move upward. What items would then be included in a sample of size 10?
2. Using Table A10 in Appendix 4, select a sample of size 20 from a lot of 250 given items.
3. How can fair dice be used in connection with random selection?
4. Consider a random variable Y having the uniform density $f(y) = 1$ if $0 < y < 1$, and 0 otherwise. We can easily simulate Y (that is, sample the values of Y) with the use of random digits. For instance, to obtain 20 values rounded to 2 decimals, consider any of the 10 columns in Table A10, start with some randomly chosen row and move downward, taking the first two digits of the five digits given in the column and put a decimal point to the left of the first digit. Suppose your choice was column 3 and row 36. Show that you get the following sample and graph a dot frequency diagram of it.

   0.89 0.40 0.67 0.86 0.87 0.86 0.06 0.20 0.38 0.12

   0.68 0.50 0.53 0.10 0.08 0.90 0.19 0.85 0.53 0.98

5. Random digits can also be used to simulate *any* continuous random variable X. For this end we graph the distribution function of X, use random digits to obtain values of the variable Y described in Prob. 4, plot these values on the vertical axis and read off the corresponding values of X. Illustrate this procedure for a normal

random variable X with mean 0 and variance 1, using the sample in Prob. 4. Graph a histogram of the sample of the 20 x-values, using the class marks $-2, -1, 0, 1, 2$.

6. The simulation technique described in Prob. 5 applies also to *discrete* random variables. Explain how you would proceed if X is the sum of the two numbers obtained in rolling two fair dice (cf. Fig. 371 in Sec. 20.7).

20.13 Estimation of Parameters

Quantities appearing in distributions, such as p in the binomial distribution and μ and σ in the normal distribution, are called **parameters.**

A **point estimate** of a parameter is a number (point on the real line), which is computed from a given sample and serves as an approximation of the unknown exact value of the parameter. An **interval estimate** is an interval ("*confidence interval*") obtained from a sample; such estimates will be considered in the next section. Estimation of parameters is an important practical problem.

As an approximation of the mean μ of a population we may take the mean $\bar{x}$ of a corresponding sample. This gives the estimate $\hat{\mu} = \bar{x}$ for μ, that is

$$(1) \qquad \hat{\mu} = \bar{x} = \frac{1}{n}(x_1 + \cdots + x_n) \qquad (n = \text{size of the sample}).$$

Similarly, an estimate $\hat{\sigma}^2$ for the variance of a population is the variance s^2 of a corresponding sample, that is,

$$(2) \qquad \hat{\sigma}^2 = s^2 = \frac{1}{n-1}\sum_{j=1}^{n}(x_j - \bar{x})^2.$$

Clearly, (1) and (2) are estimates of parameters for distributions in which μ or σ^2 appear explicitly as parameters, such as the normal and Poisson distributions. For the binomial distribution, $p = \mu/n$ [cf. (3) in Sec. 20.9]. In this case, $x_j = 1$ in (1) if the event A whose probability is p occurred in the jth trial, and $x_j = 0$ if A did not occur in that trial. From (1) we thus obtain for p the estimate

$$(3) \qquad \hat{p} = \frac{\bar{x}}{n}.$$

We mention that (1) is a very special case of the so-called **method of moments.** In this method the parameters to be estimated are expressed in terms of the moments of the distribution (cf. Sec. 20.8). In the resulting formulas those moments are replaced by the corresponding moments of the sample. This yields the desired estimates. Here the ***k*th moment of a sample** $x_1, \cdots, x_n$ is

$$m_k = \frac{1}{n}\sum_{j=1}^{n} x_j^{\,k}.$$

Another method for obtaining estimates is the so-called **maximum likelihood method** of R. A. Fisher (Messenger Math. **41,** 1912, 155–160). To explain it we consider a discrete (or continuous) random variable X whose probability function (or density) $f(x)$ depends on a single parameter θ and take a corresponding sample of n independent values $x_1, \cdots, x_n$. Then in the discrete case the probability that a sample of size n consists precisely of those n values is

$$l = f(x_1)f(x_2) \cdots f(x_n). \tag{4}$$

In the continuous case the probability that the sample consists of values in the small intervals $x_i \leqq x \leqq x_i + \Delta x$ $(i = 1, 2, \cdots, n)$ is

$$f(x_1)\,\Delta x f(x_2)\,\Delta x \cdots f(x_n)\,\Delta x = l(\Delta x)^n. \tag{5}$$

Since $f(x_i)$ depends on θ, the function l depends on $x_1, \cdots, x_n$ and θ. We imagine $x_1, \cdots, x_n$ to be given and fixed. Then l is a function of θ, which is called the **likelihood function.** The basic idea of the maximum likelihood method is very simple, as follows. We choose that approximation for the unknown value of θ for which l is as large as possible. If l is a differentiable function of θ, a necessary condition for l to have a maximum (not at the boundary) is

$$\frac{\partial l}{\partial \theta} = 0. \tag{6}$$

(We write a partial derivative, because l depends also on $x_1, \cdots, x_n$.) A solution of (6) depending on $x_1, \cdots, x_n$ is called a *maximum likelihood estimate* for θ. We may replace (6) by

$$\frac{\partial \ln l}{\partial \theta} = 0, \tag{7}$$

because $f(x) \geqq 0$, a maximum of f is in general positive, and $\ln l$ is a monotone increasing function of l. This often simplifies calculations.

If the distribution of X involves r parameters $\theta_1, \cdots, \theta_r$, then instead of (6) we have the r conditions $\partial l/\partial\theta_1 = 0, \cdots, \partial l/\partial\theta_r = 0$, and instead of (7) we have

$$\frac{\partial \ln l}{\partial \theta_1} = 0, \cdots, \frac{\partial \ln l}{\partial \theta_r} = 0. \tag{8}$$

Example 1. Normal distribution

Find maximum likelihood estimates for μ and σ in the case of the normal distribution. From (1), Sec. 20.10, and (4) we obtain

$$l = \left(\frac{1}{\sqrt{2\pi}}\right)^n \left(\frac{1}{\sigma}\right)^n e^{-h} \qquad \text{where} \qquad h = \frac{1}{2\sigma^2}\sum_{i=1}^{n}(x_i - \mu)^2.$$

Taking logarithms, we have

$$\ln l = -n \ln \sqrt{2\pi} - n \ln \sigma - h.$$

The first equation in (8) is $\partial \ln l/\partial\mu = 0$, written out

$$\frac{\partial \ln l}{\partial \mu} = -\frac{\partial h}{\partial \mu} = \frac{1}{\sigma^2}\sum_{i=1}^{n}(x_i - \mu) = 0, \qquad \text{hence} \qquad \sum_{i=1}^{n} x_i - n\mu = 0.$$

The solution is the desired estimate $\widehat{\mu}$ for μ; we find

$$\widehat{\mu} = \frac{1}{n}\sum_{i=1}^{n} x_i = \bar{x}.$$

The second equation in (8) is $\partial \ln l/\partial\sigma = 0$, written out

$$\frac{\partial \ln l}{\partial \sigma} = -\frac{n}{\sigma} - \frac{\partial h}{\partial \sigma} = -\frac{n}{\sigma} + \frac{1}{\sigma^3}\sum_{i=1}^{n}(x_i - \mu)^2 = 0.$$

Replacing μ by $\widehat{\mu}$ and solving for σ^2, we obtain the estimate

$$\widetilde{\sigma}^2 = \frac{1}{n}\sum_{i=1}^{n}(x_i - \bar{x})^2,$$

which we shall use in Sec. 20.18. Note that this differs from (2). We cannot discuss criteria for the goodness of estimates, but we want to mention that for small n, formula (2) is preferable. ▌

There are **graphical methods** for obtaining estimates. As an important case, let us consider the normal distribution and obtain estimates for μ and σ from a sample $x_1, \cdots, x_n$. From (5) in Sec. 20.10 we know that the distribution function of the normal distribution is

$$F(x) = \Phi(z), \qquad z = \frac{x - \mu}{\sigma}.$$

Fig. 377 in Sec. 20.10 shows that the corresponding curve is s-shaped and intersects the 50%-line at $z = 0$, hence at $x - \mu = 0$, that is, $x = \mu$. Furthermore, from Table A9 in Appendix 4 we see that $\Phi = 84\%$ when $z = 1$ (approximately), hence $(x - \mu)/\sigma = 1$, that is, $x = \mu + \sigma$. This suggests the following method.

Graph the distribution function of that sample and fit an s-shaped curve C by eye; let x_0 and x_1 denote the abscissas of the points of intersection of C and the 50% and 84% lines, respectively. Then x_0 is an estimate for μ, and $x_1 - x_0$ is an estimate for σ.

This method can be improved. We may "stretch" the vertical scale so that afterwards the s-shaped curve of $F(x)$ appears as a straight line. This new paper is called **normal probability paper.** The advantage is obvious, because we now have to fit a straight line, which is easier than fitting an s-shaped curve. Figure 382 shows an example, where $\bar{x} \approx 365$ (computed 364.7) and $s \approx 28$ (computed $\sqrt{720.1} \approx 26.8$). See next page.

If the sample values are equally spaced (as in Table 20.2), work may be saved by plotting merely the values of the cumulative relative frequencies as points on the probability paper and fitting a straight line as best we can "by eye." From Fig. 382 it is obvious that if we would plot each value above the corresponding

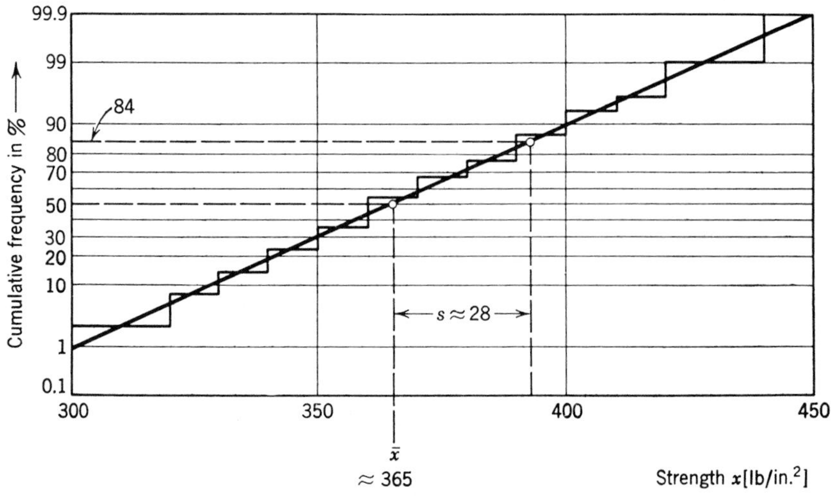

Fig. 382. Distribution function of the sample in Table 20.2 graphed on probability paper

x-value, we would make a systematic error because the line would lie higher than that in Fig. 382. To compensate for this effect, we plot each value above the corresponding x-value augmented by $\frac{1}{2}$ the distance between consecutive x-values. (In Table 20.2 this distance is $\frac{10}{2} = 5$.) This yields Fig. 383.

Similarly, if the sample is grouped, we graph the cumulative relative frequencies as points vertically above the right endpoints of the class intervals (not above the class marks). Then we fit a straight line to these points "by eye."

Normal probability paper can also be used for obtaining a rough check on whether a sample may have come from a normal population. A test for this problem will be considered in Sec. 20.18.

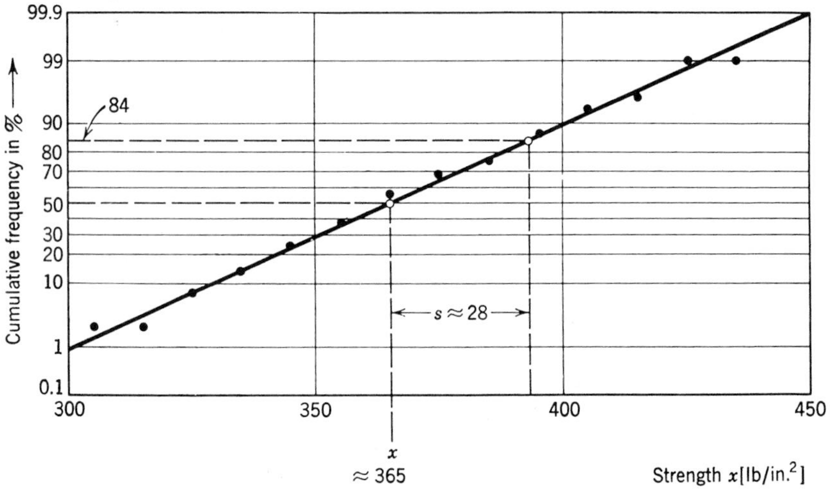

Fig. 383. Cumulative relative frequencies of the sample in Table 20.2 graphed on probability paper

Problems for Sec. 20.13

1. Find the maximum likelihood estimate for the parameter μ of a normal distribution with known variance $\sigma^2 = \sigma_0{}^2$.
2. Apply the maximum likelihood method to the normal distribution with $\mu = 0$.
3. Find the maximum likelihood estimate of θ in the density $f(x) = \theta e^{-\theta x}$ when $x \geqq 0$ and $f(x) = 0$ when $x < 0$.
4. In Prob. 3, find the mean μ, substitute it in $f(x)$, find the maximum likelihood estimate of μ, and show that it is identical with the estimate for μ which can be obtained from that for θ in Prob. 3.
5. Derive a maximum likelihood estimate for the parameter p in the binomial distribution.
6. Apply the maximum likelihood method to the Poisson distribution.

Using probability paper, estimate μ and σ from the given sample, assuming that it comes from a normally distributed population.

7. The sample in Example 1, Sec. 20.3.
8. The (ungrouped) sample in Table 20.3, Sec. 20.2.
9. The grouped sample in Table 20.4, Sec. 20.2.
10. The sample in Prob. 1, Sec. 20.2.

20.14 Confidence Intervals

The last section was devoted to point estimates of parameters, and we shall now discuss **interval estimates,** starting with a general motivation.

Whenever we use mathematical approximation formulas, we should try to find out how much the approximate value can at most deviate from the unknown true value. For example, in the case of numerical integration methods there exist "error formulas" from which we can compute the maximum possible error (that is, the difference between the approximate and the true value). Suppose that in a certain case we obtain 2.47 as an approximate value of a given integral and ± 0.02 as the maximum possible deviation from the unknown exact value. Then we are sure that the values $2.47 - 0.02 = 2.45$ and $2.47 + 0.02 = 2.49$ "*include*" the unknown exact value, that is, 2.45 is smaller than or equal to that value and 2.49 is larger than or equal to that value.

In estimating a parameter θ, the corresponding problem would be the determination of two numerical quantities that depend on the sample values and include the unknown value of the parameter with certainty. However, we already know that from a sample we cannot draw conclusions about the corresponding population that are 100% certain. So we have to be more modest and modify our problem, as follows.

Choose a probability γ close to 1 (for example, $\gamma = 95\%, 99\%$, or the like). Then determine two quantities Θ_1 and Θ_2 such that the probability that Θ_1 and Θ_2 include the exact unknown value of the parameter θ is equal to γ.

Here the idea is that we replace the impossible requirement "with certainty" by the attainable requirement "with a preassigned probability close to 1."

Numerical values of those two quantities should be computed from a given sample $x_1, \cdots, x_n$. The n sample values may be regarded as observed values of n random variables $X_1, \cdots, X_n$. Then Θ_1 and Θ_2 are functions of these random variables and therefore random variables, too. Our above requirement may thus be written

$$P(\Theta_1 \leqq \theta \leqq \Theta_2) = \gamma.$$

If we know such functions Θ_1 and Θ_2 and a sample is given, we may compute a numerical value θ_1 of Θ_1 and a numerical value θ_2 of Θ_2. The interval with endpoints θ_1 and θ_2 is called a **confidence interval**[8] or *interval estimate* for the unknown parameter θ, and we shall denote it by

$$\text{CONF } \{\theta_1 \leqq \theta \leqq \theta_2\}.$$

The values θ_1 and θ_2 are called **lower** and **upper confidence limits** for θ. The number γ is called the **confidence level.** One chooses $\gamma = 95\%$, 99%, sometimes 99.9%.

Clearly, if we intend to obtain a sample and determine a corresponding confidence interval, then γ is the probability of getting an interval that will include the unknown exact value of the parameter.

For example, if we choose $\gamma = 95\%$, then we can expect that *about* 95% of the samples that we may obtain will yield confidence intervals that do include the value θ, whereas the remaining 5% do not. Hence the statement "the confidence interval includes θ" will be correct in *about* 19 out of 20 cases, while in the remaining case it will be false.

Choosing $\gamma = 99\%$ instead of 95%, we may expect that statement to be correct even in *about* 99 out of 100 cases. But we shall see later that the intervals corresponding to $\gamma = 99\%$ are longer than those corresponding to $\gamma = 95\%$. This is the disadvantage of increasing γ.

What value of γ should we choose in a concrete case? This is not a mathematical question, but one that must be answered from the viewpoint of the application, taking into account what risk of making a false statement we can afford.

It is clear that the uncertainty involved in the present method as well as in the methods yet to be discussed comes from the sampling process, so that the statistician should be prepared for his share of mistakes. However, he is no worse off than a judge or a banker, who is subject to the laws of chance. Quite the contrary, he has the advantage that he can *measure* his chances of making a mistake.

We shall now consider methods for obtaining confidence intervals for the mean (Tables 20.8, 20.9) and the variance (Table 20.10) of the **normal distribution.** The corresponding theory will be presented in the last part of this section.

[8]The modern theory and terminology of confidence intervals were developed by J. Neyman (*Annals Math. Stat.* **6,** 1935, 111–116).

In mathematics, $\theta_1 \leqq \theta \leqq \theta_2$ means that θ lies between θ_1 and θ_2, and, to avoid misunderstandings, it seems worthwhile to characterize a confidence interval by a special symbol, such as CONF.

Table 20.8
Determination of a Confidence Interval for the Mean μ of a Normal Distribution with Known Variance σ^2

1st step. Choose a confidence level γ (95%, 99% or the like).

2nd step. Determine the corresponding c:

γ	0.90	0.95	0.99	0.999
c	1.645	1.960	2.576	3.291

3rd step. Compute the mean $\bar{x}$ of the sample $x_1, \cdots, x_n$.

4th step. Compute $k = c\sigma/\sqrt{n}$. The confidence interval for μ is

(1) $$\text{CONF}\,\{\bar{x} - k \leqq \mu \leqq \bar{x} + k\}.$$

Example 1. Confidence interval for the mean of the normal distribution with known variance

Determine a 95%-confidence interval for the mean of a normal distribution with variance $\sigma^2 = 9$, using a sample of $n = 100$ values with mean $\bar{x} = 5$.

1st step. $\gamma = 0.95$ is required.

2nd step. The corresponding c equals 1.960.

3rd step. $\bar{x} = 5$ is given.

4th step. We need $k = 1.960 \cdot 3/\sqrt{100} = 0.588$. Hence $\bar{x} - k = 4.412$, $\bar{x} + k = 5.588$, and the confidence interval is

$$\text{CONF}\,\{4.412 \leqq \mu \leqq 5.588\}.$$

Sometimes this is written $\mu = 5 \pm 0.588$, but we shall not use this notation, which may be misleading.

Example 2. Determination of sample size necessary for obtaining a confidence interval of prescribed length

How large must n in the last example be, if we want to obtain a 95%-confidence interval of length $L = 0.4$?

The interval (1) has the length $L = 2k = 2c\sigma/\sqrt{n}$. Solving for n, we obtain

$$n = (2c\sigma/L)^2.$$

In the present case the answer is $n = (2 \cdot 1.960 \cdot 3/0.4)^2 \approx 870$.

Figure 384 on the next page shows that L decreases as n increases. The shorter the confidence interval should be, the larger the sample size n must be chosen. ∎

Table 20.9 shows how to determine a confidence interval for the mean μ of a normal distribution with unknown variance σ^2. The steps are similar to those in Table 20.8, but k in Table 20.9 is different from that in Table 20.8. Moreover, c depends on n and must be determined from Table A11 in Appendix 4, which contains values z corresponding to given values of the distribution function

$$F(z) = K_m \int_{-\infty}^{z} \left(1 + \frac{u^2}{m}\right)^{-(m+1)/2} du$$

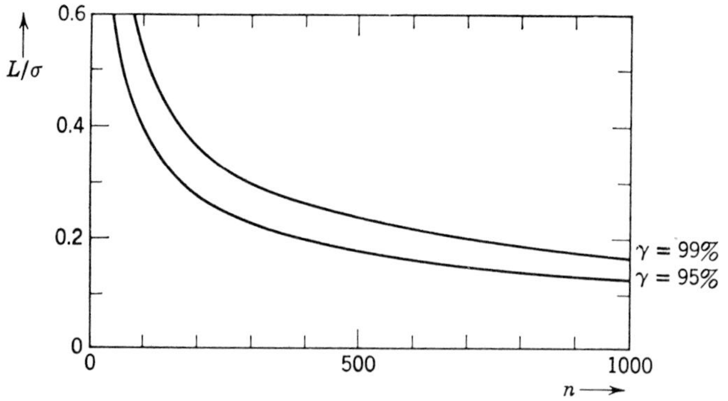

Fig. 384. Length of the confidence interval (1) (measured in multiples of σ) as a function of the sample size n

of the so-called **t-distribution** of Student (pseudonym for W. S. Gosset) where $K_m = \Gamma(\frac{1}{2}m + \frac{1}{2})/[\sqrt{m\pi}\,\Gamma(\frac{1}{2}m)]$, and $\Gamma(\alpha)$ is the gamma function [cf. (24) in Appendix 3]. $m(= 1, 2, \cdots)$ is a parameter which is called the *number of degrees of freedom* of the distribution.

Example 3. Confidence interval for the mean of the normal distribution with unknown variance

Using the sample in Table 20.2, Sec. 20.2, determine a 99%-confidence interval for the mean μ of the corresponding population, assuming that the population is normal. (This asumption will be justified in Sec. 20.18.)

1st step. $\gamma = 0.99$ is required.

2nd step. Since $n = 100$, we obtain $c = 2.63$.

3rd step. Calculation gives $\bar{x} = 364.70$ and $s = \sqrt{720.1} = 26.83$.

4th step. We find $k = 26.83 \cdot 2.63/10 = 7.06$. Hence the confidence interval is

$$\text{CONF}\ \{357.64 \leqq \mu \leqq 371.76\}.$$

Table 20.9
Determination of a Confidence Interval for the Mean μ of a Normal Distribution with Unknown Variance σ^2

1st step. Choose a confidence level γ (95%, 99% or the like).
2nd step. Determine the solution c of the equation

$$F(c) = \tfrac{1}{2}(1 + \gamma) \tag{2}$$

from the table of the t-distribution with $n - 1$ degrees of freedom (Table A11 in Appendix 4; n = sample size).
3rd step. Compute the mean $\bar{x}$ and the variance s^2 of the sample $x_1, \cdots, x_n$.
4th step. Compute $k = sc/\sqrt{n}$. The confidence interval is

$$\text{CONF}\ \{\bar{x} - k \leqq \mu \leqq \bar{x} + k\}. \tag{3}$$

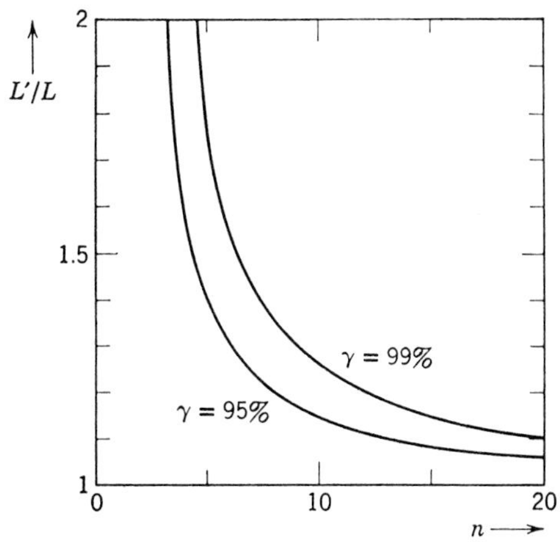

Fig. 385. Ratio of the lengths L' and L of the confidence intervals (3) and (1) as a function of the sample size n for equal s and σ

For comparison, if σ were known and equal to 26.83, Table 20.8 would give the value $k = 2.576 \cdot 26.83/\sqrt{100} = 6.91$, and CONF $\{357.79 \leqq \mu \leqq 371.61\}$. This differs but little from the preceding result because n is large. For smaller n the difference would be considerable, as Fig. 385 illustrates. ∎

The steps in Table 20.10 for obtaining a confidence interval for the variance of a normal distribution are similar to those in Tables 20.8 and 20.9, but we now have to determine two numbers c_1 and c_2. Both numbers are obtained from Table A12 in Appendix 4, which contains values z corresponding to given values of the distribution function $F(z) = 0$ when $z < 0$ and

$$F(z) = C_m \int_0^z e^{-u/2} u^{(m-2)/2} \, du \qquad \text{when} \qquad z \geqq 0.$$

Table 20.10
Determination of a Confidence Interval for the Variance σ^2 of the Normal Distribution, Whose Mean Need Not Be Known

1st step. Choose a confidence level γ (95%, 99% or the like).
2nd step. Determine solutions c_1 and c_2 of the equations

$$F(c_1) = \tfrac{1}{2}(1 - \gamma), \qquad F(c_2) = \tfrac{1}{2}(1 + \gamma) \tag{4}$$

from the table of the chi-square distribution with $n - 1$ degrees of freedom (Table A12 in Appendix 4; n = size of sample).
3rd step. Compute $(n - 1)s^2$ where s^2 is the variance of the sample $x_1, \cdots, x_n$.
4th step. Compute $k_1 = (n - 1)s^2/c_1$ and $k_2 = (n - 1)s^2/c_2$. The confidence interval is

$$\text{CONF}\,\{k_2 \leqq \sigma^2 \leqq k_1\}. \tag{5}$$

This is the distribution function of the so-called **χ^2-distribution** (*chi-square distribution*); here $C_m = 1/[2^{m/2}\Gamma(\frac{1}{2}m)]$ and m $(= 1, 2, \cdots)$ is a parameter which is called the *number of degrees of freedom* of the distribution.

Example 4. Confidence interval for the variance of the normal distribution
Using the sample in Table 20.2, Sec. 20.2, determine a 95%-confidence interval for the variance of the corresponding population.

1st step. $\gamma = 0.95$ is required.

2nd step. Since $n = 100$, we find $c_1 = 73.4$ and $c_2 = 128$.

3rd step. From Table 20.2 in Sec. 20.2 we compute $99s^2 = 71{,}291$.

4th step. The confidence interval is

$$\text{CONF}\ \{556 \leqq \sigma^2 \leqq 972\}. \qquad \blacksquare$$

Other distributions. Confidence intervals for the mean and variance of other distributions may be obtained by using the previous methods and *sufficiently large samples. Practically* speaking, if the sample indicates that the skewness of the unknown distribution is small, one should take samples of size $n = 20$ at least for obtaining confidence intervals for μ and samples of size $n = 50$ at least for obtaining confidence intervals for σ^2. The reason for this method will be explained at the end of this section.

Theoretical basis of the methods in Tables 20.8–20.10. We shall now discuss the theory which justifies our methods for obtaining confidence intervals, using the following simple but very important idea.

So far we have regarded the values $x_1, \cdots, x_n$ of a sample as n observed values of a single random variable X. We may equally well regard these n values as single observations of n random variables $X_1, \cdots, X_n$ which have the same distribution (the distribution of X) and are independent because the sample values are assumed to be independent.

For deriving (1) in Table 20.8 we need

Theorem 1 (Sum of independent normal random variables)
Suppose that $X_1, X_2, \cdots, X_n$ are independent normal random variables with means $\mu_1, \mu_2, \cdots, \mu_n$ and variances ${\sigma_1}^2, {\sigma_2}^2, \cdots, {\sigma_n}^2$, respectively. Then the random variable

$$X = X_1 + X_2 + \cdots + X_n$$

is normal with the mean

$$\mu = \mu_1 + \mu_2 + \cdots + \mu_n$$

and the variance

$$\sigma^2 = {\sigma_1}^2 + {\sigma_2}^2 + \cdots + {\sigma_n}^2.$$

The statements about μ and σ follow directly from Theorems 1 and 3 in Sec. 20.11. The proof that X is normal can be found in Ref. [H12], cf. Appendix 1.

From this theorem, Theorem 1 in Sec. 20.10, and Theorem 3 in Sec. 20.8 we obtain

Theorem 2

If $X_1, \cdots, X_n$ are independent normal random variables each of which has mean μ and variance σ^2, then the random variable

$$\bar{X} = \frac{1}{n}(X_1 + \cdots + X_n) \tag{6}$$

is normal with the mean μ and the variance σ^2/n, and the random variable

$$Z = \sqrt{n}\,\frac{\bar{X} - \mu}{\sigma} \tag{7}$$

is normal with the mean 0 and the variance 1.

Let us derive (1). From the general motivation at the beginning of this section we remember that our goal is to find two random variables Θ_1 and Θ_2 such that

$$P(\Theta_1 \leqq \mu \leqq \Theta_2) = \gamma, \tag{8}$$

where γ is chosen, and the sample gives observed values θ_1 of Θ_1 and θ_2 of Θ_2, which then yield a confidence interval CONF $\{\theta_1 \leqq \mu \leqq \theta_2\}$. In the present case this can be done as follows. We choose a number γ between 0 and 1 and determine c from Table A9, Appendix 4, such that $P(-c \leqq Z \leqq c) = \gamma$. (When $\gamma = 0.90$, etc. we obtain the values c in Table 20.8.) The inequality $-c \leqq Z \leqq c$ with Z given by (7) is

$$-c \leqq \sqrt{n}(\bar{X} - \mu)/\sigma \leqq c$$

and can be transformed into an inequality for μ. In fact, multiplication by $\sigma/\sqrt{n}$ yields $-k \leqq \bar{X} - \mu \leqq k$ where $k = c\sigma/\sqrt{n}$. Multiplying by -1 and adding $\bar{X}$, we get

$$\bar{X} + k \geqq \mu \geqq \bar{X} - k. \tag{9}$$

Thus $P(-c \leqq Z \leqq c) = \gamma$ is equivalent to $P(\bar{X} - k \leqq \mu \leqq \bar{X} + k) = \gamma$. This is of the form (8) with $\Theta_1 = \bar{X} - k$ and $\Theta_2 = \bar{X} + k$. Under our assumptions it means that with probability γ the random variables $\bar{X} - k$ and $\bar{X} + k$ will assume values that include the unknown mean μ. Regarding the sample values $x_1, \cdots, x_n$ in Table 20.8 as observed values of n independent normal random variables $X_1, \cdots, X_n$, we see that the sample mean $\bar{x}$ is an observed value of (6), and by inserting this value into (9) we obtain (1).

For deriving (3) in Table 20.9 we need

Theorem 3

Let $X_1, \cdots, X_n$ be independent normal random variables with the same mean μ and the same variance σ^2. Then the random variable

(10) $$T = \sqrt{n}\,\frac{\bar{X} - \mu}{S}$$

where $\bar{X}$ *is given by* (6) *and*

(11) $$S^2 = \frac{1}{n-1}\sum_{j=1}^{n}(X_j - \bar{X})^2$$

has a t-distribution (*cf. p.* 902) *with* $n - 1$ *degrees of freedom.*

The proof can be found in Ref. [H12], cf. Appendix 1.

The derivation of (3) is similar to that of (1). We choose a number γ between 0 and 1 and determine a number c from Table A11, Appendix 4, with $n - 1$ degrees of freedom such that

(12) $$P(-c \leqq T \leqq c) = F(c) - F(-c) = \gamma.$$

Since the t-distribution is symmetric, we have $F(-c) = 1 - F(c)$, and (12) assumes the form (2). Transforming $-c \leqq T \leqq c$ in (12) as before, we get

(13) $$\bar{X} - K \leqq \mu \leqq \bar{X} + K \qquad \text{where} \qquad K = cS/\sqrt{n},$$

and (12) becomes $P(\bar{X} - K \leqq \mu \leqq \bar{X} + K) = \gamma$. By inserting the observed values $\bar{x}$ of $\bar{X}$ and s^2 of S^2 into (13) we obtain (3).

For deriving (5) in Table 20.10 we need

Theorem 4

Under the assumptions in Theorem 3 *the random variable*

(14) $$Y = (n-1)\frac{S^2}{\sigma^2}$$

where S^2 *is given by* (11), *has a chi-square distribution* (*cf. p.* 904) *with* $n - 1$ *degrees of freedom.*

The proof can be found in Ref. [H12], cf. Appendix 1.

The derivation of (5) is similar to that of (1) and (3). We choose a number γ between 0 and 1 and determine c_1 and c_2 from Table A12, Appendix 4, such that [cf. (4)]

$$P(Y \leqq c_1) = F(c_1) = \tfrac{1}{2}(1 - \gamma), \qquad P(Y \leqq c_2) = F(c_2) = \tfrac{1}{2}(1 + \gamma).$$

Subtraction yields

$$P(c_1 \leqq Y \leqq c_2) = P(Y \leqq c_2) - P(Y \leqq c_1) = \gamma.$$

Transforming $c_1 \leqq Y \leqq c_2$ with Y given by (14) into an inequality for σ^2, we obtain

$$\frac{n-1}{c_2} S^2 \leqq \sigma^2 \leqq \frac{n-1}{c_1} S^2.$$

By inserting the observed value s^2 of S^2 we obtain (5).

Confidence intervals for the mean and variance of other distributions can be obtained by the methods in Tables 20.8 and 20.10, but we must use large samples. This follows from the basic

Theorem 5 (Central limit theorem)

Let $X_1, \cdots, X_n, \cdots$ be independent random variables that have the same distribution function and therefore the same mean μ and the same variance σ^2. Let $Y_n = X_1 + \cdots + X_n$. Then the random variable

$$Z_n = \frac{Y_n - n\mu}{\sigma\sqrt{n}} \tag{15}$$

is **asymptotically normal** *with the mean* 0 *and the variance* 1, *that is, the distribution function $F_n(x)$ of Z_n satisfies*

$$\lim_{n\to\infty} F_n(x) = \Phi(x) = \frac{1}{\sqrt{2\pi}} \int_{-\infty}^{x} e^{-u^2/2}\, du.$$

A proof can be found in Ref. [H4], cf. Appendix 1.

We know that if $X_1, \cdots, X_n$ are independent random variables with the same mean μ and the same variance σ^2, then their sum $X = X_1 + \cdots + X_n$ has the following properties.

(A) X has the mean $n\mu$ and the variance $n\sigma^2$ (cf. Theorems 1 and 3 in Sec. 20.11).

(B) If those variables are normal, then X is normal (cf. Theorem 1).

If those variables are not normal, then (B) fails to hold, but if n is large, then X is approximately normal (cf. Theorem 5) and this justifies the application of methods for the normal distribution to other distributions, but in such a case we have to use large samples.

Problems for Sec. 20.14

1. Find a 99%-confidence interval for the mean μ of a normal population with variance $\sigma^2 = 1.69$, using a sample of size 36 with mean 18.4.
2. Obtain a 95%-confidence interval for the mean μ of a normal population with variance $\sigma^2 = 16$ from Fig. 384, using a sample of size 300 with mean 87.
3. Find a 95%-confidence interval for the mean μ of a normal population with standard deviation 2.5, using the sample 16, 12, 19, 10, 15.
4. What sample size would be needed to produce a 95%-confidence interval (1) of length (*a*) 2σ, (*b*) σ?

Assuming that the populations from which the following samples are taken are normal, determine a 95%-confidence interval for the mean μ of the population.

5. A sample of diameters of 10 ball bearings with mean 4.37 cm and standard deviation 0.157 cm.

6. Density [grams per cm^3] of coke 1.40, 1.45, 1.39, 1.44, 1.38.

7. Nitrogenium content [%] of steel 0.74, 0.75, 0.73, 0.75, 0.74, 0.72.

8. What sample size should we use in Prob. 5 if we want to obtain a confidence interval of length 0.1?

9. The specific heat of iron was measured 41 times at a temperature of 25°C. The sample had the mean 0.106 [cal/g · °C] and the standard deviation 0.002 [cal/g · °C]. What can we assert with a probability of 99% about the possible size of the error if we use that sample mean to estimate the true specific heat of iron?

10. Find a 95%-confidence interval for the percentage of cars on a certain highway which have poorly adjusted brakes, using a random sample of 500 cars stopped at a roadblock on that highway, 87 of which had poorly adjusted brakes.

Assuming that the populations from which the following samples are taken are normal, determine a 99%-confidence interval for the variance σ^2 of the population.

11. Rockwell hardness of tool bits 64.9, 64.1, 63.8, 64.0.

12. The sample in Prob. 7.

13. A sample of size $n = 128$ with variance $s^2 = 1.921$.

14. Why are interval estimates in most cases more useful than point estimates?

15. If X is normal with mean 12 and variance 25, what distributions do $-X$, $2X$, and $4X - 1$ have?

16. If the weight X of bags of cement is normally distributed with a mean of 40 kg and a standard deviation of 2 kg, how many bags can a delivery truck carry so that the probability of the total load exceeding 2000 kg will be 5%?

17. A machine fills boxes weighing Y lb with X lb of salt, where X and Y are normal with mean 100 lb and 5 lb and standard deviation 1 lb and 0.5 lb, respectively. What percent of filled boxes weighing between 104 lb and 106 lb are to be expected?

18. What is the distribution of the thickness of the core of a transformer consisting of 50 layers of sheet metal and 49 insulating paper layers if the thickness X of a single metal sheet is normal with mean 0.5 mm and standard deviation 0.05 mm and the thickness Y of a paper layer is normal with mean 0.05 mm and standard deviation 0.02 mm?

19. Using Theorem 1, find out in what fraction of random matches of holes and bolts the bolts would fit, assuming that the diameter D_1 of the holes is normal with mean $\mu_1 = 1$ in. and standard deviation $\sigma_1 = 0.01$ in. and the diameter D_2 of the bolts is normal with mean $\mu_2 = 0.99$ in. and standard deviation $\sigma_2 = 0.01$ in.

20. If X_1 and X_2 are independent normal random variables with mean 6 and -2 and variance 4 and 6, respectively, what distribution does $4X_1 - X_2$ have?

20.15 Testing of Hypotheses, Decisions

A statistical **hypothesis** is an assumption about the distribution of a random variable, for example, that a certain distribution has mean 20.3, etc. A statistical **test** of a hypothesis is a procedure in which a sample is used to find out whether we may **"not reject" ("accept")** the hypothesis, that is, act as though it is true, or whether we should **"reject"** it, that is, act as though it is false.

These tests are applied quite frequently, and we may ask why they are important. Often we have to make decisions in situations where chance variation plays a role. If we have a choice between, say, two possibilities, the decision between them might be based on the result of some statistical test.

For example, if we want to use a certain lathe for producing bolts whose diameter should lie between given limits and we allow at most 2% defective bolts, we may take a sample of 100 bolts produced on that lathe and use it for testing the hypothesis $\sigma^2 = \sigma_0^2$ that the variance σ^2 of the corresponding population has a certain value σ_0^2 which we choose such that we can expect to obtain not more than 2% defectives. A meaningful *alternative* in this case is $\sigma^2 > \sigma_0^2$. Depending on the result of the test, we either do not reject the hypothesis $\sigma^2 = \sigma_0^2$ (and then use that lathe) or we reject it, assert that $\sigma^2 > \sigma_0^2$, and use another lathe. In the latter case we say that the test indicates a **significant deviation** of σ^2 from σ_0^2, that is, a deviation which is not merely caused by the unavoidable influence of chance factors but by the lack of precision of the lathe.

In other cases we may want to compare two things, for example, two different drugs, two methods of performing a certain work, the accuracy of two methods of measurement, the quality of products produced by two different tools, etc. Depending on the result of a suitable test we decide to use one of the two drugs, to introduce the better method of working, etc.

Typical sources for hypotheses are as follows.

1. The hypothesis may come from a quality requirement. (Experience about attainable quality may be gained by producing a large number of items with special care.)
2. The hypothesis is based on values known from previous experience.
3. The hypothesis results from a theory which one wants to verify.
4. The hypothesis is a pure guess caused by occasional observations.

Let us start with a simple introductory example.

Example 1. Test of a hypothesis

The birth of a single child may be regarded as a random experiment with two possible outcomes, namely, *B: Birth of a boy* and *G: Birth of a girl.* Intuitively we should feel that both outcomes are about equally likely. However, in the literature it is often claimed that births of boys are somewhat more frequent than births of girls. On the basis of this situation we want to test the hypothesis that the two outcomes B and G have the same probability. If we let p denote the probability of the outcome B, our hypothesis to be tested is $p = 50\% = 0.5$. Because of those claims we choose the alternative $p > 0.5$.

For the test we use a sample of $n = 3000$ babies born in 1962 in Graz, Austria; 1578 of these babies were boys.

If the hypothesis is true, we expect that in a sample of $n = 3000$ births *about* 1500 are boys. If the alternative holds, then we expect more than 1500 boys, on the average. Hence if the number of boys actually observed is much larger than 1500, we can use this as an indication that the hypothesis may be false, and we reject it.

To perform our test, we proceed as follows. We first determine a critical value c. Because of the alternative, c will be greater than 1500. (A method for determining c will be given below.) Then, if the observed number of boys is greater than c, we reject the hypothesis. If that number is not greater than c, we do not reject the hypothesis.

The basic question now is how we should choose c, that is, where we should draw the line between small random deviations and large significant deviations. Different people may have different opinions, and to answer the question we must use mathematical arguments. In the present case these are very simple, as will be seen in what follows.

We determine c such that if the hypothesis is true, the probability of observing more than c boys in a sample of 3000 single births is a very small number, call it α. It is customary to choose $\alpha = 1\%$ or 5%. Choosing $\alpha = 1\%$ (or 5%) we risk about once in 100 cases (in 20 cases, respectively) rejecting a hypothesis even though it is true. We shall return to this point later. Let us choose $\alpha = 1\%$ and consider the random variable

$$X = \textit{number of boys in } 3000 \textit{ births.}$$

Assuming that the hypothesis is true, we obtain the critical value c from the equation

$$P(X > c)_{p=0.5} = \alpha = 0.01. \tag{1}$$

(That assumption is indicated by the subscript $p = 0.5$.) If the observed value 1578 is greater than c, we reject the hypothesis. If $1578 \leqq c$, we do not reject the hypothesis.

To determine c from (1) we must know the distribution of X. For our purpose the binomial distribution is a sufficiently accurate model. Hence if the hypothesis is true, X has a binomial distribution with $p = 0.5$ and $n = 3000$. This distribution can be approximated by the normal distribution with mean $\mu = np = 1500$ and variance $\sigma^2 = npq = 750$; cf. Sec. 20.10. (For the sake of simplicity we shall disregard the term 0.5 in (11), Sec. 20.10.) The density is shown in Fig. 386. Using (1), we thus obtain

$$P(X > c) = 1 - P(X \leqq c) \approx 1 - \Phi\left(\frac{c - 1500}{\sqrt{750}}\right) = 0.01.$$

Table A9 in Appendix 4 yields $(c - 1500)/\sqrt{750} = 2.326$. Hence $c = 1564$. Since $1578 > c$, we reject the hypothesis and assert that $p > 0.5$. This completes the test.

For a sample of 300 values we would obtain from (1) with X = *number of boys in* 300 *births* the critical value $c = 170$, and a sample in which 158 (the same percentage as in the large sample) are boys would give $158 < c$, so that the hypothesis would not be rejected. This is interesting, because it illustrates that the usefulness of the test increases with increasing sample size n. We should choose

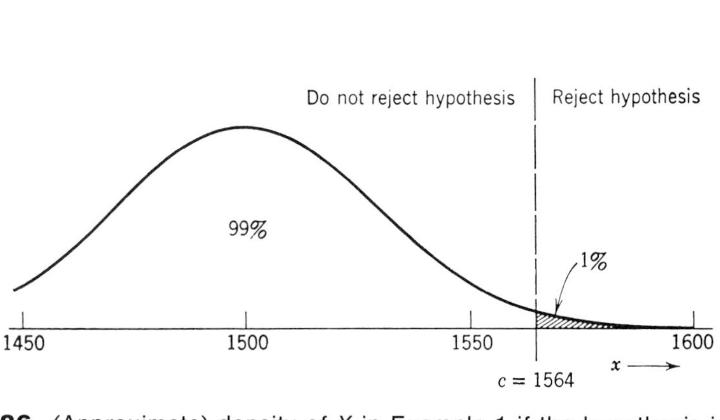

Fig. 386. (Approximate) density of X in Example 1 if the hypothesis is true. Critical value $c = 1564$

n so large that the test will yield information about the question which is of interest in a concrete case. On the other hand, n should not be larger than necessary, for reasons of time and money. In most cases an economical choice of n can be found from a small preliminary experiment. ∎

The hypothesis to be tested is sometimes called the *null hypothesis,* and a counter-assumption (like $p > 0.5$ in Example 1) is called an *alternative hypothesis* or, briefly, an **alternative.** The number α (or $100\alpha\%$) is called the **significance level** of the test. c is called the *critical value.* The region containing the values for which we reject the hypothesis is called the **rejection region** or **critical region.** The region of values for which we do not reject the hypothesis is called the **acceptance region.** A frequent choice for α is 5%.

Let θ be an unknown parameter in a distribution, and suppose that we want to test the hypothesis $\theta = \theta_0$. There are three main types of alternatives, namely

(2) $$\theta > \theta_0$$

(3) $$\theta < \theta_0$$

(4) $$\theta \neq \theta_0.$$

(2) and (3) are called **one-sided alternatives,** and (4) is called a **two-sided alternative.** (2) is of the type considered in Example 1 (where $\theta_0 = p = 0.5$ and $\theta = p > 0.5$); c lies to the right of θ_0, and the rejection region extends from c to ∞ (Fig. 387, upper part). The test is called a **right-sided test.** In the case (3) the number c lies to the left of θ_0, the rejection region extends from c to $-\infty$ (Fig. 371, middle part), and the test is called a **left-sided test.** Both types of tests are called **one-sided tests.** In the case (4) we have two critical values c_1 and c_2 ($> c_1$), the rejection region extends from c_1 to $-\infty$ and from c_2 to ∞, and the test is called a **two-sided test.**

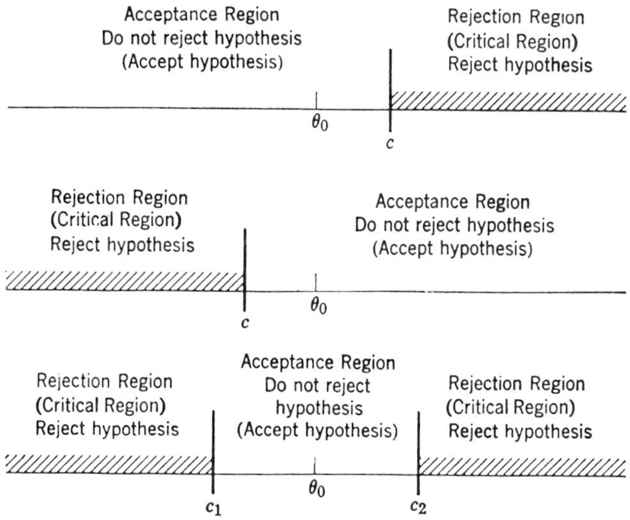

Fig. 387. Test in the case of alternative (2) (upper part of the figure), alternative (3) (middle part), and alternative (4)

All three types of hypotheses are of practical importance. For example, (3) may appear in connection with testing strength of material. θ_0 may then be the required strength, and the alternative characterizes an undesirable weakness. The case that the material may be stronger than required is acceptable, of course, and does therefore not need special attention. (4) may be important, for example, in connection with the diameter of an axle shaft. θ_0 then is the required diameter, and slimmer axle-shafts are as bad as thicker ones, so that one has to watch for undesirable deviations from θ_0 in both directions.

We shall now consider the **risks of making false decisions** in a test of a hypothesis $\theta = \theta_0$ against an alternative which, for the sake of simplicity, is a single number θ_1. Let $\theta_1 > \theta_0$ so that we have a right-sided test, the consideration for a left-sided or a two-sided test being similar. From a given sample $x_1, \cdots, x_n$ we compute a value $\widehat{\theta} = g(x_1, \cdots, x_n)$. If $\widehat{\theta} > c$, the hypothesis is rejected (as in Example 1). If $\widehat{\theta} \leqq c$, the hypothesis is not rejected. $\widehat{\theta}$ can be regarded as an observed value of the random variable

$$\widehat{\Theta} = g(X_1, \cdots, X_n)$$

because x_j may be regarded as an observed value of X_j, where $j = 1, \cdots, n$. In this test there are two possibilities of making an error, as follows.

Type I error (cf. Table 20.11). The hypothesis is true, but is rejected because Θ assumes a value $\widehat{\theta} > c$. Obviously, the probability of making such an error equals

$$P(\widehat{\Theta} > c)_{\theta=\theta_0} = \alpha, \tag{5}$$

the significance level of the test.

Type II error (cf. Table 20.11). The hypothesis is false, but is not rejected because $\widehat{\Theta}$ assumes a value $\widehat{\theta} \leqq c$. The probability of making such an error is denoted by β; thus

$$P(\widehat{\Theta} \leqq c)_{\theta=\theta_1} = \beta. \tag{6}$$

$\eta = 1 - \beta$ is called the **power** of the test. Obviously, this is the probability of avoiding a Type II error.

Table 20.11
Type I and Type II Errors in Testing a Hypothesis $\theta = \theta_0$ Against an Alternative $\theta = \theta_1$

		Unknown Truth $\theta = \theta_0$	$\theta = \theta_1$
Not rejected (Accepted)	$\theta = \theta_0$	True decision $P = 1 - \alpha$	Type II error $P = \beta$
	$\theta = \theta_1$	Type I error $P = \alpha$	True decision $P = 1 - \beta$

Formulas (5) and (6) show that both α and β depend on c, and we would like to choose c so that these probabilities of making errors are as small as possible. But Fig. 388 indicates that these are conflicting requirements, since to let α decrease we must shift c to the right, but then β increases. In practice we first choose α (5%, sometimes 1%), then determine c, and finally compute β. If β is large so that the power $\eta = 1 - \beta$ is small, we should repeat the test, choosing a larger sample, for reasons that will appear shortly.

If the alternative is not a single number but is of the form (2)–(4), then β becomes a function of θ. This function $\beta(\theta)$ is called the **operating characteristic** (OC) of the test and its curve the **OC curve.** Clearly, in this case $\eta = 1 - \beta$ also depends on θ, and this function $\eta(\theta)$ is called the **power function** of the test.

Of course, from a test that leads to the acceptance of a certain hypothesis θ_0 it does *not* follow that this is the only possible hypothesis or the best possible hypothesis. Hence the terms **"not reject"** or **"fail to reject"** are perhaps better than the term **"accept."**

The following examples will explain tests of practically important hypotheses.

Example 2. Test for the mean of the normal distribution with known variance Let X be a normal random variable with variance $\sigma^2 = 9$. Using a sample of size $n = 10$ with mean $\bar{x}$, test the hypothesis $\mu = \mu_0 = 24$ against the three types of alternatives, namely,

$$(a)\ \mu > \mu_0 \qquad (b)\ \mu < \mu_0 \qquad (c)\ \mu \neq \mu_0.$$

We choose the significance level $\alpha = 0.05$. An estimate of the mean will be obtained from

$$\bar{X} = \frac{1}{n}(X_1 + \cdots + X_n).$$

If the hypothesis is true, $\bar{X}$ is normal with mean $\mu = 24$ and variance $\sigma^2/n = 0.9$, cf. Theorem 2, Sec. 20.14. Hence we may obtain the critical value c from Table A9 in Appendix 4.

Case (a). We determine c from $P(\bar{X} > c)_{\mu=24} = \alpha = 0.05$, that is,

$$P(\bar{X} \leqq c)_{\mu=24} = \Phi\left(\frac{c - 24}{\sqrt{0.9}}\right) = 1 - \alpha = 0.95.$$

Table A9, Appendix 4, yields $(c - 24)/\sqrt{0.9} = 1.645$, and $c = 25.56$, which is greater than μ_0, as in the upper part of Fig. 387. If $\bar{x} \leqq 25.56$, the hypothesis is not rejected. If $\bar{x} > 25.56$, it is rejected. The power of the test is (cf. Fig. 389)

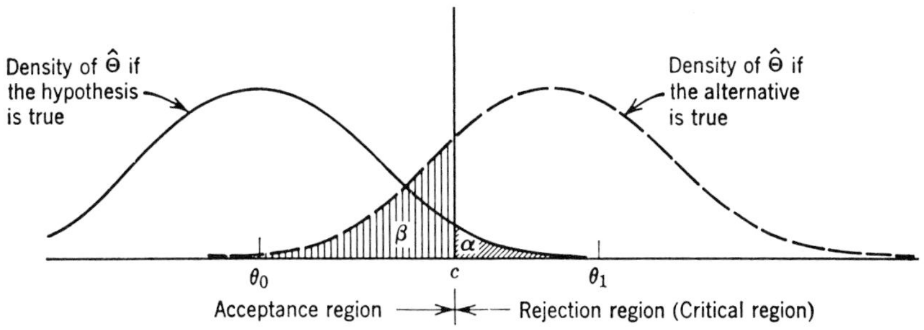

Fig. 388. Illustration of Type I and II errors in testing a hypothesis $\theta = \theta_0$ against an alternative $\theta = \theta_1$ $(> \theta_0)$

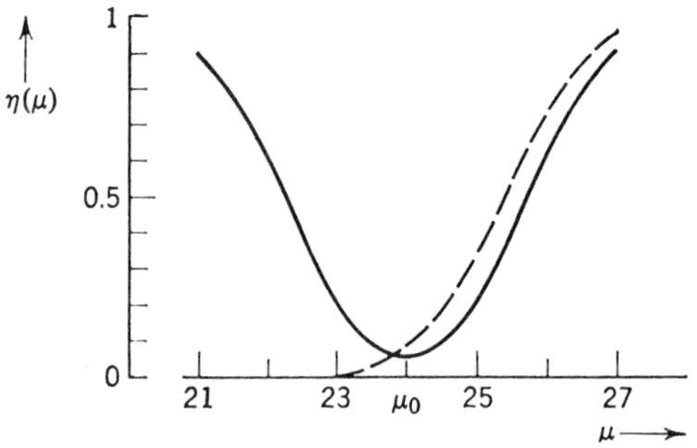

Fig. 389. Power η (μ) in Example 2, case (a) (dashed) and case (c)

$$\begin{aligned}\eta(\mu) &= P(\bar{X} > 25.56)_\mu = 1 - P(\bar{X} \leqq 25.56)_\mu \\ &= 1 - \Phi\left(\frac{25.56 - \mu}{\sqrt{0.9}}\right) = 1 - \Phi(26.94 - 1.05\mu).\end{aligned} \tag{7}$$

***Case* (*b*).** The critical value c is obtained from the equation

$$P(\bar{X} \leqq c)_{\mu=24} = \Phi\left(\frac{c - 24}{\sqrt{0.9}}\right) = \alpha = 0.05.$$

Table A9, Appendix 4, yields $c = 24 - 1.56 = 22.44$. If $\bar{x} \geqq 22.44$, we do not reject the hypothesis. If $\bar{x} < 22.44$, we reject it. The power of the test is

$$\eta(\mu) = P(\bar{X} \leqq 22.44)_\mu = \Phi\left(\frac{22.44 - \mu}{\sqrt{0.9}}\right) = \Phi(23.65 - 1.05\mu). \tag{8}$$

***Case* (*c*).** Since the normal distribution is symmetric, we choose c_1 and c_2 equidistant from $\mu = 24$, say, $c_1 = 24 - k$ and $c_2 = 24 + k$, and determine k from

$$P(24 - k \leqq \bar{X} \leqq 24 + k)_{\mu=24} = \Phi\left(\frac{k}{\sqrt{0.9}}\right) - \Phi\left(-\frac{k}{\sqrt{0.9}}\right) = 1 - \alpha = 0.95.$$

Using Table A9, Appendix 4, we obtain $k/\sqrt{0.9} = 1.960$, $k = 1.86$. Hence $c_1 = 24 - 1.86 = 22.14$ and $c_2 = 24 + 1.86 = 25.86$. If $\bar{x}$ is not smaller than c_1 and not greater than c_2, we do not reject the hypothesis. Otherwise we reject it. The power of the test is (cf. Fig. 389)

$$\begin{aligned}\eta(\mu) &= P(\bar{X} < 22.14)_\mu + P(\bar{X} > 25.86)_\mu = P(\bar{X} < 22.14)_\mu + 1 - P(\bar{X} \leqq 25.86)_\mu \\ &= 1 + \Phi\left(\frac{22.14 - \mu}{\sqrt{0.9}}\right) - \Phi\left(\frac{25.86 - \mu}{\sqrt{0.9}}\right) \\ &= 1 + \Phi(23.34 - 1.05\mu) - \Phi(27.26 - 1.05\mu).\end{aligned} \tag{9}$$

Consequently, the operating characteristic $\beta(\mu) = 1 - \eta(\mu)$ (see before) is (cf. Fig. 390)

$$\beta(\mu) = \Phi(27.26 - 1.05\mu) - \Phi(23.34 - 1.05\mu).$$

If we take a larger sample, say, of size $n = 100$ (instead of 10), then $\sigma^2/n = 0.09$ (instead of 0.9), and the critical values are $c_1 = 23.41$ and $c_2 = 24.59$, as can be verified by calculation. Then the operating characteristic of the test is

$$\beta(\mu) = \Phi\left(\frac{24.59 - \mu}{\sqrt{0.09}}\right) - \Phi\left(\frac{23.41 - \mu}{\sqrt{0.09}}\right) = \Phi(81.97 - 3.33\mu) - \Phi(78.03 - 3.33\mu).$$

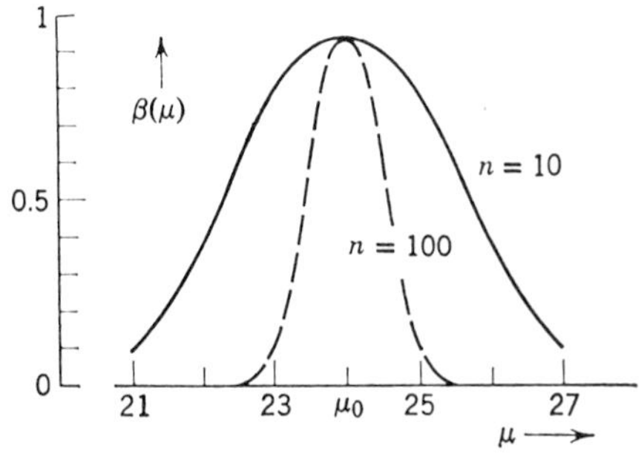

Fig. 390. Curves of the operating characteristic (OC curves) in Example 2, case (c), for two different sample sizes n

Figure 390 shows that the corresponding OC curve is steeper than that for $n = 10$. This means that the increase of n has led to an improvement of the test. In any practical case, n is chosen as small as possible but so large that the test brings out deviations between μ and μ_0 that are of practical interest. For instance, if deviations of ± 2 units are of interest, we see from Fig. 390 that $n = 10$ is much too small because when $\mu = 24 - 2 = 22$ or $\mu = 24 + 2 = 26$, then β is almost 50%. On the other hand, we see that $n = 100$ will be sufficient for that purpose.

Example 3. Test for the mean of the normal distribution with unknown variance
The tensile strength of a sample of $n = 16$ manila ropes (diameter 3 in.) was measured. The sample mean was $\bar{x} = 4482$ kg, and the sample standard deviation was $s = 115$ kg (N. C. Wiley, 41st Annual Meeting of the Amer. Soc. for Test. Materials). Assuming that the tensile strength is a normal random variable, test the hypothesis $\mu_0 = 4500$ kg against the alternative $\mu_1 = 4400$ kg. Here μ_0 may be a value given by the manufacturer, while μ_1 may result from previous experience.

We choose the significance level $\alpha = 5\%$. If the hypothesis is true, it follows from Theorem 3 in Sec. 20.14, that the random variable

$$T = \sqrt{n}\,\frac{\bar{X} - \mu_0}{S} = 4\,\frac{\bar{X} - 4500}{S}$$

has a t-distribution with $n - 1 = 15$ degrees of freedom. The critical value c is obtained from the equation

$$P(T < c)_{\mu_0} = \alpha = 0.05.$$

Table A11 in Appendix 4 yields $c = -1.75$. As an observed value of T we obtain from the sample $t = 4(4482 - 4500)/115 = -0.626$. We see that $t > c$ and do not reject the hypothesis. For obtaining numerical values of the power of the test, we would need tables called noncentral Student t-tables; we shall not discuss this question here.

Example 4. Test for the variance of the normal distribution
Using a sample of size $n = 15$ and sample variance $s^2 = 13$ from a normal population, test the hypothesis $\sigma^2 = {\sigma_0}^2 = 10$ against the alternative $\sigma^2 = {\sigma_1}^2 = 20$.

We choose the significance level $\alpha = 5\%$. If the hypothesis is true, then

$$Y = (n - 1)\frac{S^2}{{\sigma_0}^2} = 14\frac{S^2}{10} = 1.4S^2$$

has a chi-square distribution with $n - 1 = 14$ degrees of freedom, cf. Theorem 4, Sec. 20.14. From

$$P(Y > c) = \alpha = 0.05, \qquad \text{that is,} \qquad P(Y \leqq c) = 0.95,$$

and Table A12, Appendix 4, with 14 degrees of freedom we obtain $c = 23.68$. This is the critical value of Y. Hence to $S^2 = {\sigma_0}^2 Y/(n - 1) = 0.714Y$ there corresponds the critical value $c^* = 0.714 \cdot 23.68 = 16.91$. Since $s^2 < c^*$, we do not reject the hypothesis.

If the alternative is true, the variable

$$Y_1 = 14\frac{S^2}{\sigma_1^{\,2}} = 0.7S^2$$

has a chi-square distribution with 14 degrees of freedom. Hence our test has the power

$$\eta = P(S^2 > c^*)_{\sigma^2=20} = P(Y_1 > 0.7c^*)_{\sigma^2=20} = 1 - P(Y_1 \leqq 11.84)_{\sigma^2=20} \approx 62\%$$

and we see that the Type II risk is very large, namely, 38%. To make that risk smaller, we must increase the sample size.

Example 5. Comparison of the means of two normal distributions

Using a sample $x_1, \cdots, x_{n_1}$ from a normal distribution with unknown mean μ_1 and a sample $y_1, \cdots, y_{n_2}$ from another normal distribution with unknown mean μ_2, we want to test the hypothesis that the means are equal, that is, $\mu_1 = \mu_2$, against an alternative, say, $\mu_1 > \mu_2$. The variances need not be known, but are assumed to be equal.[9] Two cases are of practical importance:

Case A. *The samples have the same size. Furthermore, each value of the first sample corresponds to precisely one value of the other,* because corresponding values result from the same person or thing **(paired comparison);** for example, two measurements of the same thing by two different methods or two measurements from the two eyes of the same animal, and, more generally, they may result from pairs of *similar* individuals or things, for example, identical twins, pairs of used front tires from the same car, etc. Then we should form the differences of corresponding values and test the hypothesis that the population corresponding to the differences has mean 0, using the method in Example 3. If we have a choice, that method is better than the following one.

Case B. *The two samples are independent and not necessarily of the same size.* Then we may proceed, as follows. Suppose that the alternative is $\mu_1 > \mu_2$. We choose a significance level α. We compute the sample means $\bar{x}$ and $\bar{y}$ and $(n_1 - 1)s_1^{\,2}$, $(n_2 - 1)s_2^{\,2}$, where $s_1^{\,2}$ and $s_2^{\,2}$ are the sample variances. Using Table A11, Appendix 4, with $n_1 + n_2 - 2$ degrees of freedom, we determine c from

$$P(T \leqq c) = 1 - \alpha. \tag{10}$$

We finally compute

$$t_0 = \sqrt{\frac{n_1 n_2(n_1 + n_2 - 2)}{n_1 + n_2}}\;\frac{\bar{x} - \bar{y}}{\sqrt{(n_1 - 1)s_1^{\,2} + (n_2 - 1)s_2^{\,2}}}. \tag{11}$$

It can be shown that this is an observed value of a random variable which has a t-distribution with $n_1 + n_2 - 2$ degrees of freedom, provided the hypothesis is true. If $t_0 \leqq c$, the hypothesis is not rejected. If $t_0 > c$, it is rejected.

If the alternative is $\mu_1 \neq \mu_2$, then (10) must be replaced by

$$P(T \leqq c_1) = 0.5\alpha, \qquad P(T \leqq c_2) = 1 - 0.5\alpha. \tag{10*}$$

Note that for samples of equal size $n_1 = n_2 = n$, formula (11) reduces to

$$t_0 = \sqrt{n}\,\frac{\bar{x} - \bar{y}}{\sqrt{s_1^{\,2} + s_2^{\,2}}}. \tag{12}$$

To illustrate the numerical work, let us consider the two samples

105	108	86	103	103	107	124	105

and

89	92	84	97	103	107	111	97

[9] If the test in the next example shows that the variances differ significantly, then choose two not too small samples of the same size $n_1 = n_2 = n$ (> 30, say), use the fact that (12) is an observed value of an approximately normal random variable with mean 0 and variance 1, and proceed as in Example 2.

showing the relative output of tin plate workers under two different working conditions (J. J. B. WORTH, J. Indust. Eng. **9**, 1958, 249–253). Assuming that the corresponding populations are normal and have the same variance, let us test the hypothesis $\mu_1 = \mu_2$ against the alternative $\mu_1 \neq \mu_2$. (Equality of variances will be tested in the next example.) We find $\bar{x} = 105.125$, $\bar{y} = 97.500$, $s_1{}^2 = 106.125$, $s_2{}^2 = 84.000$. We choose the significance level $\alpha = 5\%$. From (10*) with $0.5\alpha = 2.5\%$, $1 - 0.5\alpha = 97.5\%$ and Table A11 in Appendix 4 with 14 degrees of freedom we obtain $c_1 = -2.15$ and $c_2 = 2.15$. Formula (12) with $n = 8$ gives the value $t_0 = \sqrt{8} \cdot 7.625/\sqrt{190.125} = 1.56$. Since $c_1 \leqq t_0 \leqq c_2$, we do not reject the hypothesis $\mu_1 = \mu_2$ that under both conditions the mean output is the same.

Case A applies to the example, because the two first sample values correspond to a certain type of work, the next two were obtained in another kind of work, etc. So we may use the differences

$$16 \quad 16 \quad 2 \quad 6 \quad 0 \quad 0 \quad 13 \quad 8$$

of corresponding sample values and the method in Example 3 to test the hypothesis $\mu = 0$ where μ is the mean of the population corresponding to the differences. As a logical alternative we take $\mu \neq 0$. The sample mean is $\bar{d} = 7.625$, and the sample variance is $s^2 = 45.696$. Hence $t = \sqrt{8}\,(7.625 - 0)/\sqrt{45.696} = 3.19$. From $P(T \leqq c_1) = 2.5\%$, $P(T \leqq c_2) = 97.5\%$ and Table A11 in Appendix 4 with $n - 1 = 7$ degrees of freedom we obtain $c_1 = -2.37$, $c_2 = 2.37$ and reject the hypothesis because $t = 3.19$ does not lie between c_1 and c_2. Hence our present test, in which we used more information (but the same samples), shows that the difference in output is significant.

Example 6. Comparison of the variances of two normal distributions

Using the two samples in the last example, test the hypothesis $\sigma_1{}^2 = \sigma_2{}^2$; assume that the corresponding populations are normal and the nature of the experiment suggests the alternative $\sigma_1{}^2 > \sigma_2{}^2$. We find $s_1{}^2 = 106.125$, $s_2{}^2 = 84.000$. We choose the significance level $\alpha = 5\%$. Using $P(V \leqq c) = 1 - \alpha = 95\%$ and Table A13 in Appendix 4, with $(n_1 - 1, n_2 - 1) = (7, 7)$ degrees of freedom, we determine $c = 3.79$. We finally compute $v_0 = s_1{}^2/s_2{}^2 = 1.26$. Since $v_0 \leqq c$, we do not reject the hypothesis. If $v_0 > c$, we would reject it.

This test is justified by the fact that v_0 is an observed value of a random variable which has a so-called ***F*-distribution** with $(n_1 - 1, n_2 - 1)$ degrees of freedom, provided the hypothesis is true. (Proof in Ref. [H4], cf. Appendix 1.) The F-distribution with (m, n) degrees of freedom was introduced by R. A. Fisher and has the distribution function $F(z) = 0$ when $z < 0$ and

$$F(z) = K_{mn} \int_0^z t^{(m-2)/2}(mt + n)^{-(m+n)/2}\, dt \qquad (z \geqq 0) \tag{13}$$

where $K_{mn} = m^{m/2}n^{n/2}\Gamma(\tfrac{1}{2}m + \tfrac{1}{2}n)/\Gamma(\tfrac{1}{2}m)\Gamma(\tfrac{1}{2}n)$.

Problems for Sec. 20.15

1. Test $\mu = 0$ against $\mu > 0$, assuming normality and using the sample 1, −1, 1, 3, −8, 6, 0 (deviations of the azimuth (in multiples of 0.01 radians) in the 143rd revolution of the satellite Telstar).

2. Using the sample in Example 1, Sec. 20.3, test the hypothesis $\mu = 0.80$ in. (the length indicated on the box) against the alternative $\mu \neq 0.80$ in. (Assume normality.)

3. Using the data of Buffon in Table 20.6, Sec. 20.5, test the hypothesis that the coin is fair, that is, that heads and tails have the same probability of occurrence, against the alternative that heads are more likely than tails.

4. Test the hypothesis that there is no significant difference between two methods of measuring the starch content of potatoes, assuming normality and using the sample (differences of corresponding values from 16 potatoes, measured in multiples of 0.1%)

$$2 \quad 0 \quad 0 \quad 1 \quad 2 \quad 2 \quad 3 \quad -3 \quad 1 \quad 2 \quad 3 \quad 0 \quad -1 \quad 1 \quad -2 \quad 1.$$

5. A firm sells oil in cans containing 500 g oil per can and is interested to know whether the mean weight differs significantly from 500 g at the 5%-level, in which case the filling machine has to be adjusted. Set up a hypothesis and an alternative and perform the test, assuming normality and using a sample of 10 fillings having a mean of 496 g and a standard deviation of 5 g.

6. If a sample of 50 tires of a certain kind has a mean life of 32 000 km and a standard deviation of 4000 km, can the manufacturer claim that the true mean life of such tires is greater than 30 000 km? Set up and test a corresponding hypothesis at a 5%-level, assuming normality.

7. Graph the OC curves in Example 2, cases (*a*) and (*b*).

8. Show that for a normal distribution the two types of errors in a test of a hypothesis H_0: $\mu = \mu_0$ against an alternative H_1: $\mu = \mu_1$ can be made as small as one pleases (not zero) by taking the sample sufficiently large.

9. Three specimens of high quality concrete had compressive strength 357, 359, 413 (in kilogram/centimeter^2), and for three specimens of ordinary concrete the values were 346, 358, 302. Test for equality of the population means, $\mu_1 = \mu_2$, against the alternative $\mu_1 > \mu_2$. (Assume normality and equality of variances.)

10. Assuming normality and equal variance and using independent samples with $n_1 = 9$, $\bar{x} = 12$, $s_1 = 2$, $n_2 = 9$, $\bar{y} = 15$, $s_2 = 2$, test H_0: $\mu_1 = \mu_2$ against $\mu_1 \neq \mu_2$; choose $\alpha = 5\%$.

11. Construct two simple samples such that the hypothesis of equal population means is not rejected although the difference between the sample means is large.

12. For checking the reliability of a certain electrical method for measuring temperatures, two samples were obtained. Assuming normality and equality of variances of the corresponding populations, test the hypothesis that the population means are equal. The samples are [°C]

106.9	106.3	107.0	106.0	104.9
106.5	106.7	106.8	106.1	105.6

13. If a standard drug cures about 70% of patients with a certain disease and a new drug cured 148 of the first 200 patients on whom it was tried, can we conclude that the new drug is better?

14. Using samples of sizes 10 and 5 with variances $s_1^2 = 50$ and $s_2^2 = 20$ and assuming normality of the corresponding populations, test the hypothesis H_0: $\sigma_1^1 = \sigma_2^2$ against the alternative $\sigma_1^2 > \sigma_2^2$. Choose $\alpha = 5\%$.

15. Using two samples (weight of dust in certain tubes, measured in mg) obtained in an experiment by the British Coal Utilisation Research Association and assuming normality of the corresponding populations, find whether the variances of the populations differ significantly from each other. The samples are

75	20	70	70	85	90	100	40	35	65	90	35
20	35	55	50	65	40						

20.16 Quality Control

No production process is so perfect that all the products are completely alike. There is always a small variation that is caused by a great number of small, uncontrollable factors and must therefore be regarded as a chance variation. It is important to make sure that the products have required values (for example, length, strength, or whatever property may be of importance in a particular case). For this purpose one makes a test of the hypothesis that the products have the required property, say $\mu = \mu_0$ where μ_0 is a required value. If this is done after an entire lot has been produced (for example, a lot of 100,000 screws), the test will tell us how good or how bad the products are, but it is obviously too late to alter undesirable results. It is much better to test during the production run. This is done at regular intervals of time (for example, every 30 minutes or every hour) and is called **quality control.** Each time a sample of the same size is taken, in practice, 3 to 10 items. If the hypothesis is rejected, we stop the process of production and look for the trouble that causes the deviation.

If we stop the production process even though it is progressing properly, we make a Type I error. If we do not stop the process even though something is not in order, we make a Type II error (cf. Sec. 20.15).

The result of each test is marked in graphical form on what is called a **control chart.** This was proposed by W. A. Shewhart in 1924 and makes quality control particularly effective.

Control chart for the mean. An illustration and example of a control chart is given in the upper part of Fig. 391 on p. 921. This control chart for the mean shows the **lower control limit** LCL, the **center control line** CL and the **upper control limit** UCL. The two **control limits** correspond to the critical values c_1 and c_2 in Case (*c*) of Example 2 in Sec. 20.15. As soon as a sample mean falls outside the range between the control limits, we reject the hypothesis and assert that the process of production is "out of control," that is, we assert that there has been a shift in process level. Action is called for whenever a point exceeds the limits.

If we choose control limits that are too loose, we shall not detect process shifts. On the other hand, if we choose control limits that are too tight, we shall be unable to run the process because of a frequent search for nonexistent trouble. The usual significance level is $\alpha = 1\%$. From Theorem 2 in Sec. 20.14 and Table A9 in Appendix 4 we see that in the case of the normal distribution the corresponding control limits for the mean are

$$\text{(1)} \qquad \text{LCL} = \mu_0 - 2.58\frac{\sigma}{\sqrt{n}} \qquad \text{and} \qquad \text{UCL} = \mu_0 + 2.58\frac{\sigma}{\sqrt{n}}.$$

Here σ is assumed to be known. If σ is unknown, we may calculate the standard deviations of the first 20 or 30 samples and take their arithmetic mean as an approximation of σ. The broken line connecting the means in Fig. 391 is merely to display the results effectively.

Control chart for the variance. In addition to the mean, one often controls the variance, the standard deviation, or the range. To set up a control chart for the variance in the case of a normal distribution, we may employ the method in Example 4 of Sec. 20.15 for determining control limits. It is customary to use only one control limit, namely, an upper control limit. From Example 4 of Sec. 20.15 we see that this limit is

$$\text{(2)} \qquad \text{UCL} = \frac{\sigma^2 c}{n - 1}$$

Table 20.12
Twelve Samples of 5 Values Each (Diameter of Small Cylinders, Measured in Millimeters)

Sample Number	Sample Values					$\bar{x}$	s	R
1	4.06	4.08	4.08	4.08	4.10	4.080	0.014	0.04
2	4.10	4.10	4.12	4.12	4.12	4.112	0.011	0.02
3	4.06	4.06	4.08	4.10	4.12	4.084	0.026	0.06
4	4.06	4.08	4.08	4.10	4.12	4.088	0.023	0.06
5	4.08	4.10	4.12	4.12	4.12	4.108	0.018	0.04
6	4.08	4.10	4.10	4.10	4.12	4.100	0.014	0.04
7	4.06	4.08	4.08	4.10	4.12	4.088	0.023	0.06
8	4.08	4.08	4.10	4.10	4.12	4.096	0.017	0.04
9	4.06	4.08	4.10	4.12	4.14	4.100	0.032	0.08
10	4.06	4.08	4.10	4.12	4.16	4.104	0.038	0.10
11	4.12	4.14	4.14	4.14	4.16	4.140	0.014	0.04
12	4.14	4.14	4.16	4.16	4.16	4.152	0.011	0.02

where c is obtained from the equation

$$P(Y > c) = \alpha, \qquad \text{that is,} \qquad P(Y \leqq c) = 1 - \alpha$$

and the table of the chi-square distribution (Table A12 in Appendix 4) with $n - 1$ degrees of freedom; here α (5% or 1%, say) is the probability that an observed value s^2 of S^2 in a sample is greater than the upper control limit.

If we wanted a control chart for the variance with both an upper control limit UCL and a lower control limit LCL, these limits would be

$$\text{(3)} \qquad \text{LCL} = \frac{\sigma^2 c_1}{n - 1} \qquad \text{and} \qquad \text{UCL} = \frac{\sigma^2 c_2}{n - 1}$$

where c_1 and c_2 are obtained from the equations

$$\text{(4)} \qquad P(Y \leqq c_1) = \frac{\alpha}{2} \qquad \text{and} \qquad P(Y \leqq c_2) = 1 - \frac{\alpha}{2}$$

and Table A12 in Appendix 4 with $n - 1$ degrees of freedom.

Control chart for the standard deviation. Similarly, to set up a control chart for the standard deviation, we need an upper control limit

$$\text{(5)} \qquad \text{UCL} = \frac{\sigma \sqrt{c}}{\sqrt{n - 1}}$$

obtained from (2). For example, in Table 20.12 we have $n = 5$. Assuming that the corresponding population is normal with standard deviation $\sigma = 0.02$ and

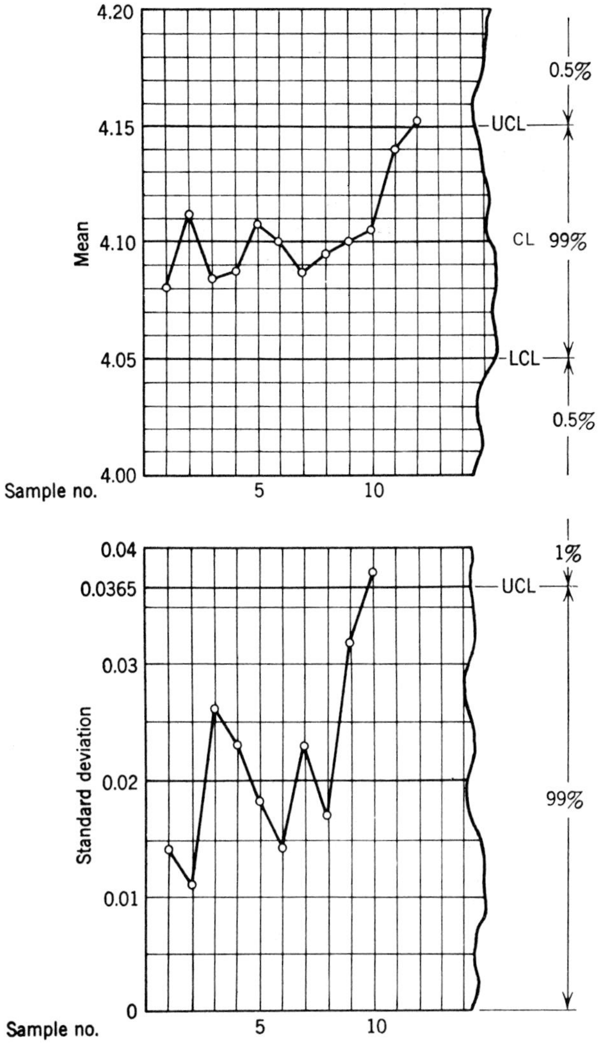

Fig. 391. Control charts for the mean (upper part of figure) and the standard deviation in the case of the sample in Table 20.12

choosing $\alpha = 1\%$, we obtain from the equation

$$P(Y \leqq c) = 1 - \alpha = 99\%$$

and Table A12 in Appendix 4 with 4 degrees of freedom the critical value $c = 13.28$ and from (5) the corresponding value

$$\text{UCL} = \frac{0.02\sqrt{13.28}}{\sqrt{4}} = 0.0365,$$

which is shown in the lower part of Fig. 391.

A control chart for the standard deviation with both an upper and a lower control limit is obtained from (3).

Control chart for the range. If we control σ^2 or σ, we must compute s^2 or s, respectively. Since this may be difficult for mathematically untrained persons one likes to replace the control of the variance or standard deviation by that of the range R (= largest sample value minus smallest sample value). It can be shown that in the case of the normal distribution, the standard deviation σ is proportional to the expectation of the random variable R^* for which R is an observed value, say, $\sigma = \lambda_n E(R^*)$, where the factor of proportionality λ_n depends on the sample size n and has the values

n	2	3	4	5	6	7	8	9	10
$\lambda_n = \sigma/E(R^*)$	0.89	0.59	0.49	0.43	0.40	0.37	0.35	0.34	0.32

n	12	14	16	18	20	30	40	50
$\lambda_n = \sigma/E(R^*)$	0.31	0.29	0.28	0.28	0.27	0.25	0.23	0.22

Since R depends on two sample values only, it gives less information about a sample than s does. Clearly, the larger the sample size n is, the more information we shall lose in using R instead of s. A practical rule is to use s when n is larger than ten.

Note that the estimate of the standard deviation based on the range is an easy approximate check on the computation of s. It may be useful in cases when somebody else (a computing center, for example) did the numerical work for the statistical investigator. Then misunderstandings and mistakes are quite common, and quick checks are therefore of increasing importance.

Problems for Sec. 20.16

1. Suppose a machine for filling cans with lubricating oil is set so that it will generate fillings which form a normal population with mean 1 gal and standard deviation 0.02 gal. Set up a control chart of the type shown in Fig. 391 for controlling the mean (that is, find LCL and UCL), assuming that the sample size is 4.
2. **(Three-sigma control chart)** Show that in Prob. 1, the requirement of the significance level $\alpha = 0.3\%$ leads to LCL $= \mu - 3\sigma/\sqrt{n}$ and UCL $= \mu + 3\sigma/\sqrt{n}$, and find the corresponding numerical values.
3. Ten samples of size 2 were taken from a production lot of bolts. The values (length in millimeters) are

Sample Nr.	1	2	3	4	5	6	7	8	9	10
Length	27.4	27.4	27.5	27.3	27.9	27.6	27.6	27.8	27.5	27.3
	27.6	27.4	27.7	27.4	27.5	27.5	27.4	27.3	27.4	27.7

Assuming that the population is normal with mean 27.5 and variance 0.024 and using (1), set up a control chart for the mean and graph the sample means on the chart.

4. Graph the means of the following 10 samples (thickness of washers, coded values)

on a control chart for means, assuming that the population is normal with mean 5 and standard deviation 1.55.

Time	8:00	8:30	9:00	9:30	10:00	10:30	11:00	11:30	12:00	12:30
	3	3	5	7	7	4	5	6	5	5
Sample	4	6	2	5	3	4	6	4	5	2
values	8	6	5	4	6	3	4	6	6	5
	4	8	6	4	5	6	6	4	4	3

5. Graph the ranges of the samples in Prob. 4 on a control chart for ranges.

6. Since the presence of a point outside control limits for the mean indicates trouble, how often would we be making the mistake of looking for nonexistent trouble if we used (*a*) 1-sigma limits, (*b*) 2-sigma limits? (Assume normality.)

7. How does the meaning of the control limits (1) change if we apply a control chart with these limits in the case of a population which is not normal?

8. What LCL and UCL should we use instead of (1) if we use the sum $x_1 + \cdots + x_n$ of the sample values instead of $\bar{x}$? Determine these limits in the case of Fig. 391.

9. (Number of defectives) Determine the LCL, CL and UCL in the case of a control chart for the number of defectives, using the binomial distribution, as follows. (*a*) Find formulas such that LCL and UCL correspond to $\mu \pm 3\sigma$. (*b*) Find numerical values if one takes samples of size 10 in a mass production of values of a certain type and knows from past experience that the percentage of defectives is about 0.07, provided the process is under statistical control. Note that p is so small that the formula for LCL will yield a negative number. In such a case one regards LCL as being 0 and uses only UCL.

10. (Number of defects per unit) A so-called *c-chart* or *defects-per-unit chart* is used for the control of the number X of defects per unit (for instance, the number of defects per 10 meters of paper, the number of missing rivets in an airplane wing, etc.) (*a*) Set up formulas for CL and LCL, UCL corresponding to $\mu \pm 3\sigma$, assuming that X has a Poisson distribution. (*b*) Compute CL, LCL and UCL in a control process of the number of imperfections in sheet glass; assume that this number is 2.5 per sheet on the average when the process is under statistical control.

20.17 Acceptance Sampling

Acceptance sampling is applied in mass production when a *producer* is supplying to a *consumer* lots of N items. In such a situation the decision to accept or reject an individual lot must be made. This decision is often based on the result of inspecting a sample of size n from the lot and determining the number of **defective items,** briefly called **defectives,** that is, items which do not meet the specifications (size, color, strength, or whatever may be important). If the number of defective items x in the sample is not greater than a specified number $c\ (< n)$, the lot is accepted. If $x > c$, the lot is rejected. c is called the *allowable number of defectives* or the **acceptance number.** It is clear that the producer and the consumer must agree on a certain **sampling plan,** that is, on a certain sample size n and an allowable number c. Such a plan is called a **single sampling plan** because it is based on a single sample. *Double sampling plans,* in which two samples are used, will be mentioned later.

Let A be the event that a lot is accepted. It is clear that the corresponding probability $P(A)$ depends not only on n and c but also on the number of defectives in the lot. Let M denote this number, let the random variable X be the number of defectives in a sample, and suppose that we sample without replacement. Then (cf. Sec. 20.9)

$$P(A) = P(X \leqq c) = \sum_{x=0}^{c} \binom{M}{x}\binom{N-M}{n-x}\Big/\binom{N}{n}. \tag{1}$$

If $M = 0$ (no defectives in the lot), then X must assume the value 0, and

$$P(A) = \binom{0}{0}\binom{N}{n}\Big/\binom{N}{n} = 1.$$

For fixed n and c and increasing M the probability $P(A)$ decreases. If $M = N$ (all items of the lot defective), then X must assume the value n, and we have $P(A) = P(X \leqq c) = 0$ because $c < n$.

The ratio $\theta = M/N$ is called the **fraction defective** in the lot. Note that $M = N\theta$, and (1) may be written

$$P(A; \theta) = \sum_{x=0}^{c} \binom{N\theta}{x}\binom{N-N\theta}{n-x}\Big/\binom{N}{n}. \tag{2}$$

Since θ can have one of the $N + 1$ values $0, 1/N, 2/N, \cdots, N/N$, the probability $P(A)$ is defined for these values only. For fixed n and c we may plot $P(A)$ as a function of θ. These are $N + 1$ points. Through these points we may then draw a smooth curve, which is called the **operating characteristic curve** (OC curve) of the sampling plan considered.

Example 1

Suppose that certain tool bits are packaged 20 to a box, and the following single sampling plan is used. A sample of 2 tool bits is drawn, and the corresponding box is accepted if and only if both bits in the sample are good. In this case, $N = 20$, $n = 2$, $c = 0$, and (2) takes the form

$$P(A; \theta) = \binom{20\theta}{0}\binom{20-20\theta}{2}\Big/\binom{20}{2} = \frac{(20-20\theta)(19-20\theta)}{380}.$$

The numerical values are

θ	0.00	0.05	0.10	0.15	0.20	$\cdots$
$P(A; \theta)$	1.00	0.90	0.81	0.72	0.63	$\cdots$

The OC curve is shown in Fig. 392. ∎

In most practical cases θ will be small (less than 10%). In many cases the lot size N will be very large (1000, 10,000 etc.), so that we may approximate the hypergeometric distribution in (1) and (2) by the binomial distribution with $p = \theta$. Then if n is such that $n\theta$ is moderate (say, less than 20), we may

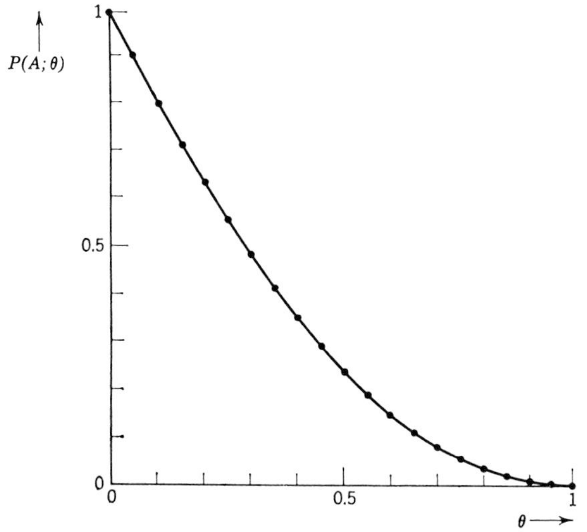

Fig. 392. OC curve of the single sampling plan with $n = 2$ and $c = 0$ for lots of size $N = 20$

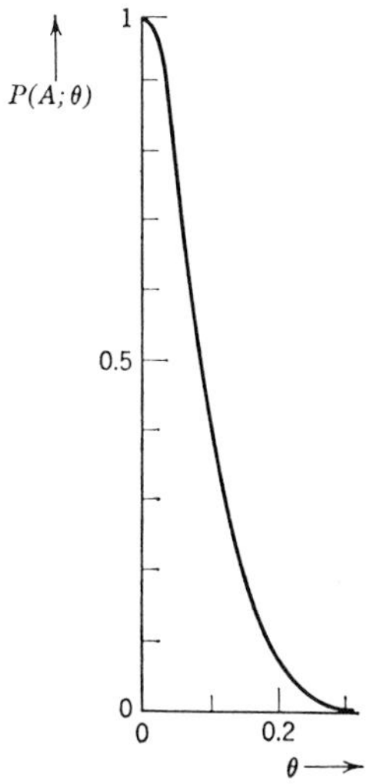

Fig. 393. OC curve in Example 2

approximate that distribution by the Poisson distribution with mean $\mu = n\theta$. From (2) we then have

$$P(A; \theta) \sim e^{-\mu} \sum_{x=0}^{c} \frac{\mu^x}{x!} \qquad (\mu = n\theta). \tag{3}$$

Example 2

Suppose that for large lots the following single sampling plan is used. A sample of size $n = 20$ is taken. If it contains not more than 1 defective, the lot is accepted. If the sample contains 2 or more defectives, the lot is rejected. In this plan, we obtain from (3)

$$P(A; \theta) \sim e^{-20\theta}(1 + 20\theta).$$

The corresponding OC curve is shown in Fig. 393. ∎

We shall now discuss the two types of possible errors in acceptance sampling and the related problem of choosing n and c. In acceptance sampling the producer and the consumer have different interests. The producer may require the probability of rejecting a "good" or "acceptable" lot to be a small number, call it α. The consumer (buyer) may demand the probability of accepting a "bad" or "unacceptable" lot to be a small number β. More precisely, suppose that the two parties agree that a lot for which θ does not exceed a certain number θ_0 is an *acceptable lot* while a lot for which θ is greater than or equal to a certain number θ_1 is an *unacceptable lot*. Then α is the probability of rejecting a lot with $\theta \leqq \theta_0$ and is called **producer's risk.** This corresponds to a Type I error in testing a hypothesis (Sec. 20.15). β is the probability of accepting a lot with $\theta \geqq \theta_1$ and is called **consumer's risk.** This corresponds to a Type II error in Sec. 20.15. Figure 394 shows an illustrative example. θ_0 is called the **acceptable quality level** (AQL), and θ_1 is called the **lot tolerance per cent defective** (LTPD) or the **rejectable quality level** (RQL). A lot with $\theta_0 < \theta < \theta_1$ may be called an *indifferent lot.*

From Fig. 394 we see that the points $(\theta_0, 1 - \alpha)$ and (θ_1, β) lie on the OC

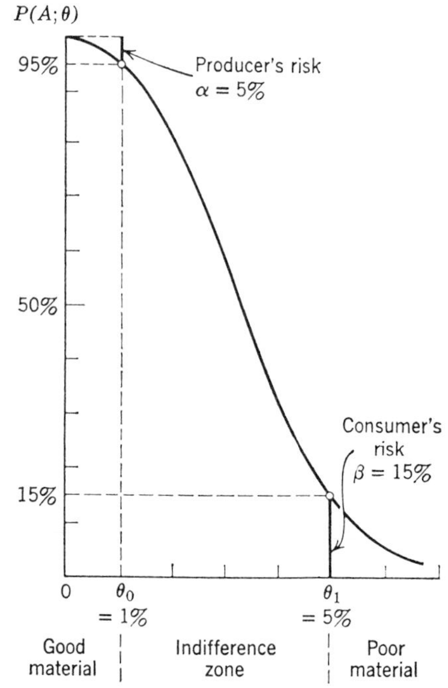

Fig. 394. OC curve, producer's and consumer's risk

curve. It can be shown that for large lots we can choose θ_0, θ_1 $(> \theta_0)$, α, β and then determine n and c such that the OC curve runs very close to those prescribed points. Sampling plans for specified α, β, θ_0, and θ_1 have been published; cf. Ref. [J6], Appendix 1.

There is a close relationship between sampling inspection and testing a hypothesis, as follows:

Sampling Inspection	**Hypothesis Testing**
Acceptable quality level (AQL) $\theta = \theta_0$	Hypothesis $\theta = \theta_0$
Lot tolerance percent defectives (LTPD) $\theta = \theta_1$	Alternative $\theta = \theta_1$
Allowable number of defectives c	Critical value c
Producer's risk α of rejecting a lot with $\theta \leqq \theta_0$	Probability α of making a Type I error (significance level)
Consumer's risk β of accepting a lot with $\theta \geqq \theta_1$	Probability β of making a Type II error

The sampling procedure by itself does not protect the consumer sufficiently well. In fact, if the producer is permitted to resubmit a rejected lot without telling that the lot has already been rejected, then even bad lots will eventually be accepted. To protect the consumer against this and other possibilities, the producer may agree with the consumer that a rejected lot is **rectified,** that is, is inspected 100%, item by item, and all defective items in the lot are removed and replaced by nondefective items.[10] Suppose that a plant produces $100\theta\%$ defec-

[10] Of course, rectification is impossible if the inspection is destructive, or is not worthwhile if it is too expensive, compared with the value of the lot. The rejected lot may then be sold at a cut-rate price or scrapped.

tive items and rejected lots are rectified. Then K lots of size N contain KN items, $KN\theta$ of which are defective. $KP(A; \theta)$ of the lots are accepted; these contain a total of $KPN\theta$ defective items. The rejected and rectified lots contain no defective items. Hence after the rectification the fraction defective in the K lots equals $KPN\theta/KN = \theta P(A; \theta)$. This function of θ is called the **average outgoing quality** (AOQ) and is denoted by AOQ(θ). Thus

(4) $$\text{AOQ}(\theta) = \theta P(A; \theta).$$

A sampling plan being given, this function and its graph, the *average outgoing quality curve* (AOQ curve), can readily be obtained from $P(A; \theta)$ and the OC curve. An example is shown in Fig. 395.

Clearly AOQ(0) = 0. Also, AOQ(1) = 0 because $P(A; 1) = 0$. From this and AOQ $\geqq 0$ we conclude that this function must have a maximum at some $\theta = \theta^*$. The corresponding value AOQ(θ^*) is called the **average outgoing quality limit** (AOQL). This is the worst average quality which may be expected to be accepted under the rectifying procedure.

It turns out that several single sampling plans may correspond to the same AOQL; see Ref. [J6] in Appendix 1. Hence if the AOQL is all the consumer cares about, the producer has some freedom in choosing a sampling plan and may select a plan which minimizes the amount of sampling, that is, the number of inspected items per lot. This number is

$$nP(A; \theta) + N(1 - P(A; \theta))$$

where the first term corresponds to the accepted lots and the last term to the rejected and rectified lots; in fact, rectification requires the inspection of all N items of the lot, and $1 - P(A; \theta)$ is the probability of rejecting a lot.

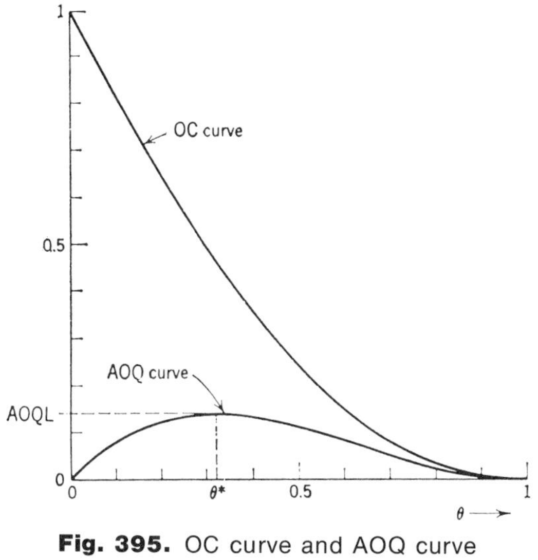

Fig. 395. OC curve and AOQ curve for the sampling plan in Fig. 392

We mention that inspection work may be saved by using a **double sampling plan,** in which the sample of size n is broken up into two samples of sizes n_1 and n_2 (where $n_1 + n_2 = n$). If the lot is very good or very bad, it may then be possible to decide about acceptance or rejectance, using one sample only, so that the other sample will be necessary in the case of a lot of intermediate quality only. Ref. [J6] contains double sampling plans using rectifying inspection of the following type (where x_1 and x_2 are the numbers of defectives in those two samples).

1. If $x_1 \leqq c_1$, accept the lot. If $x_1 > c_2$, reject the lot.
2. If $c_1 < x_1 \leqq c_2$, use the second sample, too. If $x_1 + x_2 \leqq c_2$, accept the lot. If $x_1 + x_2 > c_2$, reject the lot.

Problems for Sec. 20.17

1. Find the binomial approximation of the hypergeometric distribution in Example 1 and compare the approximate and the accurate values.

2. In Example 1, what are the producer's and consumer's risks if the AQL is 0.1 and the RQL is 0.6?

3. A purchaser checks electronic tubes by a single sampling plan which uses a sample size of 40 and an acceptance number of 1. Use Table A7 in Appendix 4 to compute the probability of acceptance of lots containing the following percentages of defective tubes: $\frac{1}{4}$%, $\frac{1}{2}$%, 1%, 2%, 5%, 10%. Graph the OC curve.

4. Large lots of batteries for pocket calculators are inspected according to the following plan. $n = 30$ batteries are randomly drawn from a lot and tested. If this sample contains at most $c = 1$ defective battery, the lot is accepted. Otherwise it is rejected. Graph the OC curve of the plan, using the Poisson approximation.

5. Graph the AOQ curve in Prob. 4. Determine the AOQL, assuming that rectification is applied.

6. Do the work required in Prob. 4 when $n = 50$ and $c = 0$.

7. Graph the OC curve and the AOQ curve for the single sampling plan for large lots with $n = 5$ and $c = 0$.

8. Same task as in Prob. 7 when $n = 4$ and $c = 1$.

9. Find θ_0 in Prob. 7 from the curve if the producer's risk is 5%. Find θ_1 in Prob. 7 if the consumer's risk is 10%.

10. Graph and compare sampling plans with $c = 0$ and increasing values of n, say, $n = 2, 3, 4$. (Use the binomial distribution.)

11. Find the risks in the single sampling plan with $n = 5$ and $c = 0$, assuming that the AQL is $\theta_0 = 1\%$ and the RQL is $\theta_1 = 15\%$.

12. Graph the OC curve and the AOQ curve for the single sampling plan for large lots with $n = 5$ and $c = 0$, and find the AOQL.

13. If in a single sampling plan for large lots of gaskets, the sample size is 100 and we want the AQL to be 5% and the producer's risk 2%, what acceptance number c should we choose? (Use the normal approximation.)

14. What is the consumer's risk in Prob. 13 if we want the RQL to be 12%?

15. Suppose in a single sampling plan for inspecting screws we have sample size 40 and $c = 3$, and we want the producer's risk to be 5%. What is the AQL?

20.18 Goodness of Fit. χ^2-Test

Using a sample $x_1, \cdots, x_n$, we want to test the hypothesis that a certain function $F(x)$ is the distribution function of the population from which the sample was taken. Clearly, the sample distribution function $\tilde{F}(x)$ is an approximation of $F(x)$, and if it approximates $F(x)$ "sufficiently well," we shall not reject the hypothesis that $F(x)$ is the distribution function of that population. If $\tilde{F}(x)$ deviates "too much" from $F(x)$, we shall reject the hypothesis.

To decide in this fashion, we have to know how much $\tilde{F}(x)$ can differ from $F(x)$ if the hypothesis is true. Hence we first have to introduce a measure for the deviation of $\tilde{F}(x)$ from $F(x)$, and we have to know the probability distribution of this measure under the assumption that the hypothesis is true. Then we

Table 20.13
Chi-square Test for the Hypothesis That $F(x)$ is the Distribution Function of a Population from Which a Sample $x_1, \cdots, x_n$ is Taken

1st step. Subdivide the x-axis into K intervals $I_1, I_2, \cdots, I_K$ such that each interval contains at least 5 values of the given sample $x_1, \cdots, x_n$. Determine the number b_j of sample values in the interval I_j $(j = 1, \cdots, K)$. If a sample value lies at a common boundary point of two intervals, add 0.5 to each of the two corresponding b_j.

2nd step. Using $F(x)$, compute the probability p_j that the random variable X under consideration assumes any value in the interval I_j $(j = 1, \cdots, K)$. Compute

$$e_j = np_j.$$

(This is the number of sample values theoretically expected in I_j if the hypothesis is true.)

3rd step. Compute the deviation

$$\text{(1)} \qquad \chi_0^2 = \sum_{j=1}^{K} \frac{(b_j - e_j)^2}{e_j}.$$

4th step. Choose a significance level (5%, 1%, or the like).

5th step. Determine the solution c of the equation

$$P(\chi^2 \leqq c) = 1 - \alpha$$

from the table of the chi-square distribution with $K - 1$ degrees of freedom (Table A12, Appendix 4). If r parameters of $F(x)$ are unknown and their maximum likelihood estimates (Sec. 20.13) are used, then use $K - r - 1$ degrees of freedom (instead of $K - 1$). If $\chi_0^2 \leqq c$, do not reject the hypothesis. If $\chi_0^2 > c$, reject the hypothesis.

proceed as follows. We determine a number c such that if the hypothesis is true, a deviation greater than c has a small preassigned probability. If, nevertheless, a deviation larger than c occurs, we have reason to doubt that the hypothesis is true and we reject it. On the other hand, if the deviation does not exceed c, so that $\tilde{F}(x)$ approximates $F(x)$ sufficiently well, we do not reject the hypothesis. Of course, if we do not reject the hypothesis, this means that we have insufficient evidence to reject it, and does not exclude the possibility that there are other functions which would not be rejected in the test. In this respect the situation is quite similar to that in Sec. 20.15.

Table 20.13 shows a test of this type, which was introduced by R. A. Fisher. This test is justified by the fact that if the hypothesis is true, then χ_0^2 is an observed value of a random variable whose distribution function approaches that of the chi-square distribution with $K - 1$ degrees of freedom (or $K - r - 1$ degrees of freedom if r parameters are estimated) as n approaches infinity. The requirement that at least 5 sample values should lie in each interval in Table 20.13 results from the fact that for finite n that random variable has only approximately a chi-square distribution. A proof can be found in Ref. [H4], cf. Appendix 1. If the sample is so small that the requirement cannot be satisfied, one may continue with the test, but use the result with great care.

Example 1. Test of normality

Test whether the population from which the sample in Table 20.2, Sec. 20.2, was taken is normal. The maximum likelihood estimates for μ and σ^2 are $\hat{\mu} = \bar{x} = 364.7$ and $\tilde{\sigma}^2 = 712.9$. The calculation in Table 20.14 yields $\chi_0^2 = 2.790$. We choose $\alpha = 5\%$. Since $K = 10$ and we estimated $r = 2$ parameters, we have to use Table A12, Appendix 4, with $K - r - 1 = 7$ degrees of freedom, finding $c = 14.07$ as the solution of $P(\chi^2 \leqq c) = 95\%$. Since $\chi_0^2 < c$, we do not reject the hypothesis that the population is normal.

Table 20.14
Calculations in Example 1

x_j	$\dfrac{x_j - 364.7}{26.7}$	$\Phi\left(\dfrac{x_j - 364.7}{26.7}\right)$	$e_j = 100p_j$	b_j	Terms in (1)
$-\infty \cdots 325$	$-\infty \cdots -1.49$	$0.0000 \cdots 0.0681$	6.81	6	0.096
$325 \cdots 335$	$-1.49 \cdots -1.11$	$0.0681 \cdots 0.1335$	6.54	6	0.045
$335 \cdots 345$	$-1.11 \cdots -0.74$	$0.1335 \cdots 0.2296$	9.61	11	0.201
$345 \cdots 355$	$-0.74 \cdots -0.36$	$0.2296 \cdots 0.3594$	12.98	14	0.080
$355 \cdots 365$	$-0.36 \cdots 0.00$	$0.3594 \cdots 0.5000$	14.06	16	0.268
$365 \cdots 375$	$0.00 \cdots 0.39$	$0.5000 \cdots 0.6517$	15.17	15	0.002
$375 \cdots 385$	$0.39 \cdots 0.76$	$0.6517 \cdots 0.7764$	12.47	8	1.602
$385 \cdots 395$	$0.76 \cdots 1.13$	$0.7764 \cdots 0.8708$	9.44	10	0.033
$395 \cdots 405$	$1.13 \cdots 1.51$	$0.8708 \cdots 0.9345$	6.37	8	0.417
$405 \cdots \infty$	$1.51 \cdots \infty$	$0.9345 \cdots 1.0000$	6.55	6	0.046

$\chi_0^2 = 2.790$

Problems for Sec. 20.18

1. In one of G. Mendel's classical experiments the result was 355 yellow peas and 123 green peas. Is this in agreement with Mendel's theory according to which in the present case the ratio yellow peas:green peas should be 3:1?
2. Between 1 P.M. and 2 P.M. on five consecutive days (Monday through Friday) a certain service station had 82, 50, 56, 80 and 52 customers, respectively. Test the hypothesis that the expected number of customers during that hour is the same on those days. (Use $\alpha = 5\%$.)

3. Three samples of 200 rivets each were taken from a large production of each of three machines. The numbers of defective rivets in the samples were 3, 8 and 4. Is this difference significant? (Use $\alpha = 5\%$.)

4. A coin was tossed 50 times. Head was obtained 27 times. Using this result, test the hypothesis that heads and tails have the same probability.

5. In the sample in Prob. 4, what is the minimum number of heads (greater than 25) that would lead to the rejection of the hypothesis if $\alpha = 5\%$ is used?

6. The following sample consists of values of the mass of cotton material (measured in gram/m²). Test the hypothesis that the corresponding population is normal.

Class mark	96	98	100	102	104	106	108	110	112
Absolute class frequency	1	0	1	2	8	19	28	30	41

Class mark	114	116	118	120	122	124	126	128	130
Absolute class frequency	66	50	27	8	5	3	0	1	1

7. Was the sample in Table 20.4, Sec. 20.2, taken from a normal population?

8. Test for normality of the polulation from which the following sample was taken. x [kg/mm²] is the tensile strength of steel sheets of 0.3 mm thickness.

x	Frequency	x	Frequency
< 42.0	15	43.5–44.0	22.5
42.0–42.5	11	44.0–44.5	19.5
42.5–43.0	15	44.5–45.0	12
43.0–43.5	14	> 45.0	19

9. Using the given sample, test that the corresponding population has a Poisson distribution. x is the number of alpha particles per 7.5-second intervals observed by E. Rutherford and H. Geiger in one of their classical experiments in 1910, and $\tilde{f}(x)$ is the frequency (= number of time periods during which exactly x particles were observed).

x	$\tilde{f}(x)$	x	$\tilde{f}(x)$	x	$\tilde{f}(x)$
0	57	5	408	10	10
1	203	6	273	11	4
2	383	7	139	12	2
3	525	8	45	$\geqq 13$	0
4	532	9	27		

10. Test the hypothesis that the population from which the given sample of 1000 sheets of paper was taken has a Poisson distribution. x is the number of dark spots on a sheet and $\tilde{f}(x)$ is the observed frequency (number of sheets with x spots). Use (*a*) $\alpha = 5\%$, (*b*) $\alpha = 1\%$.

x	$\tilde{f}(x)$
0	419
1	352
2	154
3	56
4	19
$\geqq 5$	0

20.19 Nonparametric Tests

The tests in Sec. 20.15 refer to normal populations. In many applications the populations will not be normal. Then we may apply a so-called **nonparametric test** or **distribution-free test,** which is based on order statistics and therefore is valid for any *continuous* distribution. These test are simple. However, for a normal distribution the tests in Sec. 20.15 yield better results. To illustrate the basic idea of a nonparametric test, let us consider two typical examples.

Example 1. Sign test for the median

A **median** is a solution $x = \tilde{\mu}$ of the equation $F(x) = 0.5$, where F is the distribution function. Using the sample of differences in Example 5, Sec. 20.15, that is,

$$16 \quad 16 \quad 2 \quad 6 \quad 0 \quad 0 \quad 13 \quad 8,$$

we shall test the hypothesis $\tilde{\mu} = 0$, that is, the outputs corresponding to the two different working conditions do not differ significantly. We choose the alternative $\tilde{\mu} > 0$ and the significance level $\alpha = 5\%$. If the hypothesis is true, the probability p of a positive difference is the same as that of a negative difference. Hence in this case, $p = 0.5$, and the random variable

$$X = \textit{number of positive values among } n \textit{ values}$$

has a binomial distribution with $p = 0.5$. Our sample has 8 values. We omit the values 0 which do not contribute to the decision. Then 6 values are left, all of which are positive. Since

$$P(X = 6) = \binom{6}{6}(0.5)^6(0.5)^0 = 0.0156 = 1.56\% < \alpha$$

we reject the hypothesis.

If 1 of those 6 values were negative, we would have

$$P(X \geqq 5) = \binom{6}{5}(0.5)^5 \cdot 0.5 + \binom{6}{6}(0.5)^6 = 10.9\%$$

and we would not reject the hypothesis.

Example 2. Test for arbitrary trend

A certain machine is used for cutting lengths of wire. Five successive pieces had the lengths

$$29 \quad 31 \quad 28 \quad 30 \quad 32.$$

Using this sample, test the hypothesis that there is *no trend,* that is, the machine does not have the tendency of producing longer and longer pieces. Assume that the type of machine suggests the alternative that there is *positive trend,* that is, there is the tendency that successive pieces get longer.

We count the number of *transpositions* in the sample, that is, the number of times a larger value precedes a smaller value:

29 precedes 28	(1 transposition),
31 precedes 28 and 30	(2 transpositions).

The remaining three sample values follow in ascending order. Hence in the sample there are $1 + 2 = 3$ transpositions.

We now consider the random variable

$$T = \textit{number of transpositions.}$$

If the hypothesis is true (no trend), then each of the $5! = 120$ permutations of 5 elements

1 2 3 4 5 has the same probability (1/120). We arrange these permutations according to their number of transpositions:

$T=0$	$T=1$	$T=2$	$T=3$	
1 2 3 4 5	1 2 3 5 4	1 2 4 5 3	1 2 5 4 3	
	1 2 4 3 5	1 2 5 3 4	1 3 4 5 2	
	1 3 2 4 5	1 3 2 5 4	1 3 5 2 4	
	2 1 3 4 5	1 3 4 2 5	1 4 2 5 3	
		1 4 2 3 5	1 4 3 2 5	
		2 1 3 5 4	1 5 2 3 4	
		2 1 4 3 5	2 1 4 5 3	
		2 3 1 4 5	2 1 5 3 4	etc.
		3 1 2 4 5	2 3 1 5 4	
			2 3 4 1 5	
			2 4 1 3 5	
			3 1 2 5 4	
			3 1 4 2 5	
			3 2 1 4 5	
			4 1 2 3 5	

From this we obtain

$$P(T \leqq 3) = \tfrac{1}{120} + \tfrac{4}{120} + \tfrac{9}{120} + \tfrac{15}{120} = \tfrac{29}{120} = 24\%.$$

Hence we do not reject the hypothesis.

Values of the distribution function of T in the case of no trend are shown in Table A14, Appendix 4. Our method and those values refer to *continuous* distributions. Theoretically we may then expect that all the values of a sample are different. Practically, some sample values may still be equal, because of rounding off. If m values are equal, add $m(m-1)/4$ (= mean value of the transpositions in the case of the permutations of m elements), that is, $\frac{1}{2}$ for each pair of equal values, $\frac{3}{2}$ for each triple, etc.

Problems for Sec. 20.19

1. Apply the sign test to the sample (data from the satellite Telstar) in Prob. 1, Sec. 20.15.

2. Under what condition can we use the sign test as a test for the mean of a continuous distribution?

3. Apply the sign test to Prob. 4, Sec. 20.15. (The necessary values of the binomial distribution are not contained in Table A6, Appendix 4, but must be computed.)

4. Each of 10 patients were given two different sedatives A and B. The following table shows the effect (increase of sleeping time, measured in hours).

A	1.9	0.8	1.1	0.1	−0.1	4.4	5.5	1.6	4.6	3.4
B	0.7	−1.6	−0.2	−1.2	−0.1	3.4	3.7	0.8	0.0	2.0
Difference	1.2	2.4	1.3	1.3	0.0	1.0	1.8	0.8	4.6	1.4

Using the sign test, find out whether the difference is significant.

5. Assuming that the populations corresponding to the samples in Prob. 4 are normal, apply the test explained in Example 3, Sec. 20.15.

6. Test whether a thermostatic switch is properly set to 20°C against the alternative that its setting is too low. Use a sample of 8 values, 7 of which are less than 20°C and 1 is greater than 20°C.

7. Make a table similar to that in Example 2 when $n = 4$.

8. Test the hypothesis that for a certain type of voltmeter, readings are independent of temperature T[°C] against the alternative that they tend to increase with T. Use a sample of values obtained by applying a constant voltage:

Temperature T[°C]	10	20	30	40	50
Reading V[volts]	99.8	101.0	100.4	100.8	101.5

9. Apply the test in Example 2 to the following data (x = disulphide content of a certain type of wool, measured in percent of the content in unreduced fibres; y = saturation water content of the wool, measured in percent).

x	10	15	30	40	50	55	80	100
y	50	46	43	42	36	39	37	33

10. Does the amount of fertilizer increase the yield of wheat X [kg/plot]? Use a sample of values ordered according to increasing amounts of fertilizer:

15.2 16.8 13.2 16.6 17.2 17.5 17.3 18.1

20.20 Pairs of Measurements. Fitting Straight Lines

We shall now discuss experiments in which we observe or measure two quantities simultaneously. In practice we may distinguish between two types of experiments, as follows.

1. In **correlation analysis** both quantities are random variables and we are interested in relations between them. (We shall not discuss this branch of statistics.)
2. In **regression analysis** one of the two variables, call it x, can be regarded as an ordinary variable, that is, can be measured without appreciable error. The other variable, Y, is a random variable. x is called the *independent* (sometimes the *controlled*) *variable,* and one is interested in the dependence of Y on x. Typical examples are the dependence of the blood pressure Y on the age x of a person or, as we shall now say, the regression of Y on x, the regression of the gain of weight Y of certain animals on the daily ration of food x, the regression of the heat conductivity Y of cork on the specific weight x of the cork, etc.

In the experiment the experimenter first selects n values $x_1, \cdots, x_n$ of x and then observes Y at those values of x, so that he obtains a sample of the form $(x_1, y_1), (x_2, y_2), \cdots, (x_n, y_n)$. In regression analysis the mean μ of Y is assumed to depend on x, that is, is a function $\mu = \mu(x)$ in the ordinary sense. The curve of

$\mu(x)$ is called the *regression curve of Y on x*. In the present section we shall discuss the simplest case, when $\mu(x)$ is a linear function, $\mu(x) = \alpha + \beta x$. Then we may want to plot the sample values as n points in the xY-plane, fit a straight line through them, and use this line for estimating $\mu(x)$ for given values of x, so that we know what values of Y we can expect if we choose certain values of x. If the points are scattered, fitting "by eye" becomes unreliable and we need a mathematical method for fitting lines that yields a unique result depending only on the points. A widely used procedure of this type is the **method of least squares** developed by Gauss. In our present situation it may be formulated as follows.

The straight line should be fitted through the given points so that the sum of the squares of the distances of those points from the straight line is minimum, where the distance is measured in the vertical direction (the y-direction).

General assumption (A1)

The x-values $x_1, \cdots, x_n$ of our sample $(x_1, y_1), \cdots, (x_n, y_n)$ are not all equal.

Consider a sample $(x_1, y_1), \cdots, (x_n, y_n)$ of size n. The vertical distance (distance measured in the y-direction) of a sample value (x_j, y_j) from a straight line $y = a + bx$ is $|y_j - a - bx_j|$; cf. Fig. 396. Hence the sum of the squares of these distances is

$$q = \sum_{j=1}^{n} (y_j - a - bx_j)^2. \tag{1}$$

In the method of least squares we choose a and b such that q is minimum. q depends on a and b, and a necessary condition for q to be minimum is

$$\frac{\partial q}{\partial a} = 0 \qquad \text{and} \qquad \frac{\partial q}{\partial b} = 0. \tag{2}$$

We shall see that from this condition we obtain the formula

$$y - \bar{y} = b(x - \bar{x}) \tag{3}$$

where

$$\bar{x} = \frac{1}{n}(x_1 + \cdots + x_n) \qquad \text{and} \qquad \bar{y} = \frac{1}{n}(y_1 + \cdots + y_n). \tag{4}$$

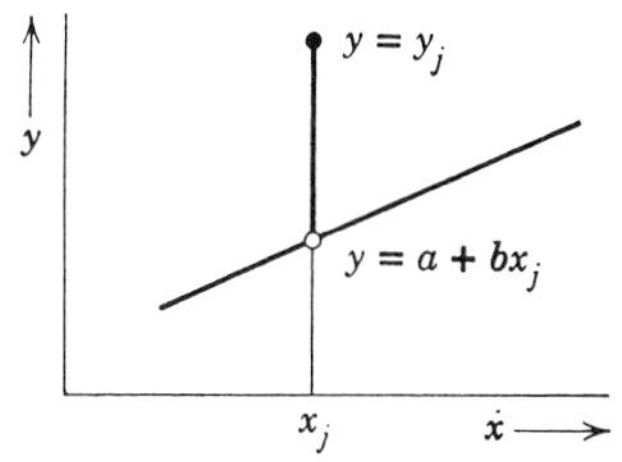

Fig. 396. Vertical distance of a point (x_j, y_j) from a line $y = a + bx$

(3) is called the **regression line** *of the y-values of the sample on the x-values of the sample. Its slope b is called the* **regression coefficient** *of y on x,* and we shall see that

$$\textbf{(5)} \qquad b = \frac{s_{xy}}{s_1^{\,2}}.$$

Here,

$$\textbf{(6)} \qquad s_1^{\,2} = \frac{1}{n-1}\sum_{j=1}^{n}(x_j - \bar{x})^2 = \frac{1}{n-1}\left[\sum_{j=1}^{n} x_j^{\,2} - \frac{1}{n}\left(\sum_{j=1}^{n} x_j\right)^2\right]$$

and

$$\textbf{(7)} \quad s_{xy} = \frac{1}{n-1}\sum_{j=1}^{n}(x_j - \bar{x})(y_j - \bar{y}) = \frac{1}{n-1}\left[\sum_{j=1}^{n} x_j y_j - \frac{1}{n}\left(\sum_{i=1}^{n} x_i\right)\left(\sum_{j=1}^{n} y_j\right)\right].$$

s_{xy} is called the **covariance** of the sample. Obviously, the regression line (3) passes through the point $(\bar{x}, \bar{y})$.

To derive (3), we use (1) and (2), finding

$$\frac{\partial q}{\partial a} = -2\sum (y_j - a - bx_j) = 0$$

$$\frac{\partial q}{\partial b} = -2\sum x_j(y_j - a - bx_j) = 0$$

(where we sum over j from 1 to n). Thus

$$na + b\sum x_j = \sum y_j$$

$$a\sum x_j + b\sum x_j^{\,2} = \sum x_j y_j.$$

Because of Assumption (A1), the determinant

$$n\sum x_j^{\,2} - \left(\sum x_j\right)^2 = n(n-1)s_1^{\,2}$$

[cf. (6)] of this system of linear equations is not zero, and the system has a unique solution [cf. (4), (6), (7)]

$$(8) \qquad a = \bar{y} - b\bar{x}, \qquad b = \frac{n\sum x_j y_j - \sum x_i \sum y_j}{n(n-1)s_1^{\,2}}.$$

This yields (3) with b given by (5)–(7).

Desk calculations can be simplified by *coding,* that is, by setting

$$(9) \qquad x_j = c_1 x_j^{\,*} + l_1, \qquad y_j = c_2 y_j^{\,*} + l_2$$

and choosing the constants c_1, c_2, l_1, l_2 such that the transformed values $x_j^{\,*}$ and

y_j^* are as simple as possible. We first compute the values $\bar{x}^*$, $\bar{y}^*$, s_1^{*2}, s_{xy}^* corresponding to the transformed values and then

$$\begin{aligned} \bar{x} &= c_1\bar{x}^* + l_1, & \bar{y} &= c_2\bar{y}^* + l_2 \\ s_1^2 &= c_1^2 s_1^{*2}, & s_{xy} &= c_1 c_2 s_{xy}^*. \end{aligned} \tag{10}$$

Example 1. Regression line

The decrease of volume y [%] of leather for certain fixed values of high pressure x [atmospheres] was measured. The results are shown in the first two columns of Table 20.15. Find the regression line of y on x. We see that $n = 4$ and obtain the values $\bar{x} = 28{,}000/4 = 7000$, $\bar{y} = 19.0/4 = 4.75$,

$$s_1^2 = \frac{1}{3}\left(216\,000\,000 - \frac{28\,000^2}{4}\right) = \frac{20\,000\,000}{3}$$

$$s_{xy} = \frac{1}{3}\left(148\,400 - \frac{28\,000 \cdot 19}{4}\right) = \frac{15\,400}{3}.$$

Hence $b = 15\,400/20\,000\,000 = 0.000\,77$, and the regression line is

$$y - 4.75 = 0.000\,77(x - 7000) \qquad \text{or} \qquad y = 0.000\,77x - 0.64.$$

∎

We now make the following two assumptions.

Assumption (A2)

For each fixed x the random variable Y is normal with mean

$$\mu(x) = \alpha + \beta x \tag{11}$$

and variance σ^2 where the latter is independent of x.

Assumption (A3)

The n performances of the experiment by which we obtain a sample $(x_1, y_1), \cdots, (x_n, y_n)$ are independent (cf. p. 893).

β in (11) is called the **regression coefficient** *of the population,* because it can be shown that under Assumptions (A1)–(A3) the maximum likelihood estimate of β is the sample regression coefficient b in (5).

Table 20.15
Regression of the Decrease of Volume y [%] of Leather on the Pressure x [Atmospheres]

Given values		Auxiliary values	
x_j	y_j	x_j^2	$x_j y_j$
4,000	2.3	16,000,000	9,200
6,000	4.1	36,000,000	24,600
8,000	5.7	64,000,000	45,600
10,000	6.9	100,000,000	69,000
28,000	19.0	216,000,000	148,400

(C. E. WEIR, Journ. Res. Nat. Bureau of Standards **45,** 1950, 468.)

Table 20.16
Determination of a Confidence Interval for β in (11) under Assumptions (A1)–(A3)

1st step. Choose a confidence level γ (95%, 99% or the like).

2nd step. Determine the solution c of the equation

$$F(c) = \tfrac{1}{2}(1 + \gamma) \tag{12}$$

from the table of the t-distribution with $n - 2$ degrees of freedom (Table A11 in Appendix 4; n = sample size).

3rd step. Using a sample $(x_1, y_1), \cdots, (x_n, y_n)$, compute $(n-1)s_1^2$ from (6), $(n-1)s_{xy}$ from (7), b from (5),

$$(n-1)s_2^2 = \sum_{j=1}^{n} y_j^2 - \frac{1}{n}\left(\sum_{j=1}^{n} y_j\right)^2 \tag{13}$$

and

$$q_0 = (n-1)(s_2^2 - b^2 s_1^2). \tag{14}$$

4th step. Compute $k = c\sqrt{q_0/(n-2)(n-1)s_1^2}$. The confidence interval is

$$\text{CONF}\,\{b - k \leqq \beta \leqq b + k\}. \tag{15}$$

Under Assumptions (A1)–(A3) we may obtain a confidence interval for β, as shown in Table 20.16.

Example 2. Confidence interval for the regression coefficient

Using the sample in Table 20.15, determine a confidence interval for β by the method in Table 20.16.

1st step. We choose $\gamma = 0.95$.

2nd step. Equation (12) takes the form $F(c) = 0.975$, and Table A11 with $n - 2 = 2$ degrees of freedom yields $c = 4.30$.

3rd step. From Example 1 we have $3s_1^2 = 20\,000\,000$ and $b = 0.000\,77$. From Table 20.15 we compute

$$3s_2^2 = 102.2 - \frac{19^2}{4} = 11.95, \qquad q_0 = 11.95 - 20\,000\,000 \cdot 0.000\,77^2 = 0.092.$$

4th step. We thus obtain $k = 4.30\sqrt{0.092/2 \cdot 20\,000\,000} = 0.000\,206$ and

$$\text{CONF}\,\{0.000\,56 \leqq \beta \leqq 0.000\,98\}.$$

Problems for Sec. 20.20

1. Fit a straight line by eye. Estimate the stopping distance of the car traveling at 35 miles per hour.

x = Speed (miles/hour)	20	30	40	50
y = Stopping distance (feet)	50	95	150	210

2. Obtain the results in Example 1 by coding; set $x_j = 2000x_j^* + 4000$, $y_j = 0.1y_j^* + 5$.

In each case find and graph the sample regression line of y on x.

3. (1, 1), (2, 1.7), (3, 3).

4. The sample in Prob. 9, Sec. 20.19.

5. Mortality y [%] of termites (*Reticulitermes lucifugus Rossi*) as a function of the concentration x [%] of chloronaphthalene

x	0.04	0.15	0.30	1.00	2.00
y	3	16	13	70	90

6. Number of revolutions x (per minute) and power y (hp) of a Diesel engine

x	400	500	600	700	750
y	580	1030	1420	1880	2100

7. Deformation x [mm] and Brinell hardness y [kg/mm^2] of a certain type of steel

x	6	9	11	13	22	26	28	33	35
y	68	67	65	53	44	40	37	34	32

In each case find a 95%-confidence interval for the regression coefficient β, using the given sample and supposing that Assumptions (A2) and (A3) are satisfied.

8. x = humidity of air (in percent), y = expansion of gelatine (in percent)

x	10	20	30	40
y	0.8	1.6	2.3	2.8

9. (1, 1), (2, 2 + p), (3, 3), where p is a constant.

10. The sample in Prob. 7.

APPENDIX

1 REFERENCES

A General References

[A1] Abramowitz, M., and I. A. Stegun, *Handbook of Mathematical Functions.* New York: Dover, 1965.

[A2] Buck, R. C., *Advanced Calculus.* 2nd ed. New York: McGraw-Hill, 1965.

[A3] Courant, R., *Differential and Integral Calculus.* 2 vols. New York: Interscience, 1964, 1956.

[A4] Courant, R., and D. Hilbert, *Methods of Mathematical Physics.* 2 vols. New York: Interscience, 1953, 1962.

[A5] Erdélyi, A., W. Magnus, F. Oberhettinger, and F. G. Tricomi, *Higher Transcendental Functions,* 3 vols. New York: McGraw-Hill, 1953, 1955.

[A6] Fletcher, A., J. C. P. Miller, L. Rosenhead, and L. J. Comrie, *An Index of Mathematical Tables.* Oxford: Blackwell, 1962.

[A7] Fulks, W., *Advanced Calculus.* 2nd ed. New York: Wiley, 1969.

[A8] Jahnke, E., F. Emde, and F. Lösch, *Tables of Higher Functions.* 7th ed. Stuttgart: Teubner, 1966.

[A9] Knopp, K., *Infinite Sequences and Series.* New York: Dover, 1956.

[A10] Kreyszig, E., *Introductory Functional Analysis With Applications.* New York: Wiley, 1978.

[A11] Magnus, W., F. Oberhettinger, and R. P. Soni, *Formulas and Theorems for the Special Functions of Mathematical Physics.* 3rd ed. New York: Springer, 1966.

[A12] Protter, M. H., and C. B. Morrey, Jr., *Modern Mathematical Analysis.* Reading, Mass.: Addison-Wesley, 1964.

[A13] Rainville, E. D., *Special Functions.* New York: Macmillan, 1960.

[A14] Thomas, G. B., *Calculus and Analytic Geometry.* 4th ed. Reading, Mass.: Addison-Wesley, 1968.

[A15] Whittaker, E. T., and G. N. Watson, *A Course of Modern Analysis.* 4th ed. Cambridge: Harvard University Press, 1927 (reprinted 1965).

B Ordinary Differential Equations (Chaps. 1–5)

[A4], [A5], [A11], [A13], [A15]

[B1] Birkhoff, G., and G.-C. Rota, *Ordinary Differential Equations.* 3rd ed. New York: Wiley, 1978.

[B2] Carslaw, H. S., and J. C. Jaeger, *Operational Methods in Applied Mathematics.* New York: Dover, 1963.

[B3] Churchill, R. V., *Operational Mathematics.* 2nd ed. New York: McGraw-Hill, 1958.

[B4] Coddington, E. A., and N. Levinson, *Theory of Ordinary Differential Equations.* New York: McGraw-Hill, 1955.

[B5] Collatz, L., *Numerical Treatment of Differential Equations.* 3rd ed. New York: Springer, 1960.

[B6] Director, S. W., *Circuit Theory.* New York: Wiley, 1975.

[B7] Erdélyi, A., W. Magnus, F. Oberhettinger, and F. Tricomi, *Tables of Integral Transforms.* 2 vols. New York: McGraw-Hill, 1954.

[B8] Fox, L., *Numerical Solution of Ordinary and Partial Differential Equations.* Reading, Mass.: Addison-Wesley, 1962.
[B9] Hahn, W., *Stability of Motion.* New York: Springer, 1967.
[B10] Hartman, P., *Ordinary Differential Equations.* New York: Wiley, 1964.
[B11] Ince, E. L., *Ordinary Differential Equations.* New York: Dover, 1956.
[B12] Kamke, E., *Differentialgleichungen, Lösungsmethoden und Lösungen. I. Gewöhnliche Differentialgleichungen.* New York: Chelsea, 1948. (This extremely useful book contains a systematic list of more than 1500 differential equations and their solutions.)
[B13] Kaplan, W., *Operational Methods for Linear Systems.* Reading, Mass.: Addison-Wesley, 1962.
[B14] McLachlan, N. W., *Ordinary Non-Linear Differential Equations in Engineering and Physical Sciences.* 2nd ed. Oxford: Clarendon, 1956.
[B15] Minorsky, N., *Nonlinear Oscillations.* New York: Van Nostrand, 1962.
[B16] Robinson, P. D., *Fourier and Laplace Transforms.* New York: Dover, 1968.
[B17] Stern, T. E., *Theory of Nonlinear Networks and Systems.* Reading, Mass.: Addison-Wesley, 1965.
[B18] Watson, G. N., *A Treatise on the Theory of Bessel Functions.* 2nd ed. Cambridge: University Press, 1944.
[B19] Widder, D. V., *The Laplace Transform.* Princeton, N.J.: Princeton University Press, 1941.

C Linear Algebra: Vectors and Matrices. Vector Calculus (Chaps. 6–9)

[C1] Bellman, R., *Introduction to Matrix Analysis.* 2nd ed. New York: McGraw-Hill, 1970.
[C2] Bodewig, E., *Matrix Calculus.* 2nd ed. Amsterdam: North-Holland Pub. 1959.
[C3] Faddeev, D. K., and V. N. Faddeeva, *Computational Methods of Linear Algebra.* San Francisco: Freeman, 1963.
[C4] Frazer, R. A., W. J. Duncan, and A. R. Collar, *Elementary Matrices.* Cambridge: University Press, 1938 (reprinted 1963).
[C5] Gantmacher, F. R., *The Theory of Matrices.* 2 vols. New York: Chelsea, 1959.
[C6] Gel'fand, I. M., *Lectures on Linear Algebra.* New York: Interscience, 1961.
[C7] Hohn, F. E., *Elementary Matrix Algebra.* 2nd ed. New York: Macmillan, 1964.
[C8] Kreyszig, E., *Introduction to Differential Geometry and Riemannian Geometry.* Toronto: University of Toronto Press, 1975.
[C9] Lamb, H., *Hydrodynamics.* 6th ed. New York: Dover, 1945.
[C10] McDuffee, C. C., *The Theory of Matrices.* New York: Chelsea, 1946.
[C11] Milne, E. A., *Vectorial Mechanics.* London: Methuen, 1948.
[C12] Nering, E. D., *Linear Algebra and Matrix Theory.* 2nd ed. New York: Wiley, 1970.
[C13] Schreier, O., and E. Spencer, *Introduction to Modern Algebra and Matrix Theory.* 2nd ed. New York: Chelsea, 1959.
[C14] Schwerdtfeger, H., *Introduction to Linear Algebra and the Theory of Matrices.* Groningen: Noordhoff, 1950.
[C15] Wilkinson, J. H., *The Algebraic Eigenvalue Problem.* Oxford: Clarendon, 1965.
[C16] Wilkinson, J. H., and C. Reinsch, *Linear Algebra.* New York: Springer, 1971.

D Fourier Series and Integrals (Chap. 10)

[D1] Carslaw, H. S., *Introduction to the Theory of Fourier's Series and Integrals.* 3rd ed. London: Macmillan, 1930.
[D2] Churchill, R. V. *Fourier Series and Boundary Value Problems.* 2nd ed. New York: McGraw-Hill, 1963.

[D3] Davis, H. F., *Fourier Series and Orthogonal Functions.* Boston: Allyn and Bacon, 1963.
[D4] Rogosinski, W., *Fourier Series.* New York: Chelsea, 1959.
[D5] Sneddon, I. N., *Fourier Transforms.* New York: McGraw-Hill, 1951.
[D6] Szegö, G., *Orthogonal Polynomials.* 3rd ed. New York: American Mathematical Society, 1967.
[D7] Titchmarsh, E. C., *Introduction to the Theory of Fourier Integrals.* 2nd ed. Oxford: Clarendon, 1948.
[D8] Tolstov, G. P., *Fourier Series.* Englewood Cliffs, N.J.: Prentice-Hall, 1962.
[D9] Zygmund, A., *Trigonometric Series.* 2 vols. 2nd ed. Cambridge: University Press, 1959.
[D10] Zygmund, A., *Trigonometrical Series.* New York: Dover, 1955.

E Partial Differential Equations (Chap. 11)

[A4], [A5], [B8], [D3]

[E1] Bateman, H., *Partial Differential Equations of Mathematical Physics.* Cambridge: University Press, 1944 (reprinted 1964).
[E2] Bergman, S., and M. Schiffer, *Kernel Functions and Elliptic Differential Equations in Mathematical Physics.* New York: Academic Press, 1953.
[E3] Duff, G. F. D., *Partial Differential Equations.* Toronto: University of Toronto Press, 1956.
[E4] Duff, G. F. D., and D. Naylor, *Differential Equations of Applied Mathematics.* New York: Wiley, 1966.
[E5] Forsythe, G. E., and W. R. Wasow, *Finite-Difference Methods for Partial Differential Equations.* New York: Wiley, 1960.
[E6] Garabedian, P. R., *Partial Differential Equations.* New York: Wiley, 1964.
[E7] Hadamard, J., *Lectures on Cauchy's Problem.* New York: Dover, 1952.
[E8] Kellog, O. D., *Foundations of Potential Theory.* New York: Dover, 1953.
[E9] Mitchell, A. R., *Computational Methods in Partial Differential Equations.* New York: Wiley, 1969.
[E10] Petrovsky, I. G., *Lectures on Partial Differential Equations.* New York: Interscience, 1954.
[E11] Lord Rayleigh, *The Theory of Sound.* 2 vols. New York: Dover, 1945.
[E12] Sagan, H., *Boundary and Eigenvalue Problems in Mathematical Physics.* New York: Wiley, 1961.
[E13] Sneddon, I. N., *Elements of Partial Differential Equations.* New York: McGraw-Hill, 1957.
[E14] Sommerfeld, A., *Partial Differential Equations in Physics.* New York: Academic Press, 1949.
[E15] Webster, A. G., *Partial Differential Equations of Mathematical Physics.* 2nd ed. New York: Hafner, 1947.

F Complex Analysis (Chaps. 12–18)

[A15]

[F1] Ahlfors, L. V., *Complex Analysis.* 2nd ed. New York: McGraw-Hill, 1966.
[F2] Bieberbach, L., *Conformal Mapping.* New York: Chelsea, 1964.
[F3] Dettman, J. W., *Applied Complex Variables.* New York: Macmillan, 1965.
[F4] Hayman, W. K., *Meromorphic Functions.* Oxford: Clarendon Press, 1975.
[F5] Hille, E., *Analytic Function Theory.* 2 vols. Boston: Ginn, 1959, 1962.
[F6] Knopp, K., *Theory of Functions.* 2 vols. New York: Dover, 1945.
[F7] Nehari, Z., *Conformal Mapping.* New York: McGraw-Hill, 1952.

[F8] Nevanlinna, R., *Analytic Functions.* New York: Springer, 1970.
[F9] Rothe, R., F. Ollendorf, and K. Pohlhausen, *Theory of Functions as Applied to Engineering Problems.* New York: Dover, 1961.
[F10] Springer, G., *Introduction to Riemann Surfaces.* Reading, Mass.: Addison-Wesley, 1957.
[F11] Titchmarsh, E. C., *The Theory of Functions.* 2nd ed. London: Oxford University Press, 1939 (reprinted 1975).

G Numerical Analysis (Chap. 19)

[B5], [B8], [C3], [C15], [C16], [E5], [E9]

[G1] Alt, F. L., *Electronic Digital Computers.* New York: Academic Press, 1958.
[G2] Buckingham, R. A., *Numerical Methods.* London: Pitman, 1957.
[G3] Forsythe, G. E., and P. C. Rosenbloom, *Numerical Analysis and Partial Differential Equations.* New York: Wiley, 1958.
[G4] Henrici, P., *Elements of Numerical Analysis.* New York: Wiley, 1964.
[G5] Hartree, D. R., *Numerical Analysis.* 2nd ed. London: Oxford University Press, 1958.
[G6] Hildebrandt, F. B., *Introduction to Numerical Analysis.* New York: McGraw-Hill, 1956.
[G7] Hull, T. E., and D. D. F. Day, *Commputers and Problem Solving.* Reading, Mass.: Addison-Wesley, 1970.
[G8] Isaacson, E., and H. B. Keller, *Analysis of Numerical Methods.* New York: Wiley, 1966.
[G9] Kopal, Z., *Numerical Analysis.* 2nd ed. London: Chapman and Hall, 1961.
[G10] McCracken, D. D., and W. S. Dorn, *Numerical Methods and* FORTRAN *Programming.* New York: Wiley, 1964.
[G11] Milne, W. E., *Numerical Calculus.* Princeton, N.J.: Princeton University Press, 1949.
[G12] National Physical Laboratory, *Modern Computing Methods.* 2nd ed. London: Her Majesty's Stationery Office, 1961.
[G13] Noble, B., *Numerical Methods.* 2 vols. New York: Interscience, 1964.
[G14] Rivlin, T. J., *An Introduction to the Approximation of Functions.* Waltham, Mass.: Blaisdell, 1969.
[G15] Sard, A., and S. Weintraub, *A Book of Splines.* New York: Wiley, 1971.
[G16] Todd, J., *Numerical Analysis.* In: Condon, E. U., and H. Odishaw, *Handbook of Physics.* 2nd ed. New York: McGraw-Hill, 1967.
[G17] Todd, J., *Survey of Numerical Analysis.* New York: McGraw-Hill, 1962.
[G18] Truitt, T. D., and A. E. Rogers, *Basics of Analog Computers.* New York: Rider, 1960.
[G19] Varga, R. S., *Matrix Iterative Analysis.* Englewood Cliffs, N.J.: Prentice-Hall, 1962.
[G20] Willers, F. A., *Practical Analysis.* New York: Dover, 1948.

H Probability and Statistics (Chap. 20)

[J2], [J6], [J8]–[J10], [J13]–[J15]

[H1] Cochran, W. G., *Sampling Techniques.* 2nd ed. New York: Wiley, 1963.
[H2] Cochran, W. G., and G. M. Cox, *Experimental Designs.* 2nd ed. New York: Wiley, 1957.
[H3] Cox, D. R., *Planning of Experiments.* New York: Wiley, 1958.
[H4] Cramér, H., *Mathematical Methods of Statistics.* Princeton: Princeton University Press, 1946.

[H5] Feller, W., *An Introduction to Probability and Its Applications*. Vol. I. 2nd ed. New York: Wiley, 1957.
[H6] Fisher, R. A., *Statistical Methods for Research Workers*. 13th ed. New York: Hafner, 1958.
[H7] Fishman, G. S., *Concepts and Methods in Discrete Event Digital Simulation*. New York: Wiley, 1973.
[H8] Fraser, D. A. S., *Nonparametric Methods in Statistics*. New York: Wiley, 1957.
[H9] Grant, E. L., *Statistical Quality Control*. 2nd ed. New York: McGraw-Hill, 1952.
[H10] Huff, D., *How to Lie with Statistics*. New York: Norton, 1954.
[H11] IRCCRAND, The Ohio State University Random Number Generator Package. Department of Statistics, The Ohio State University. Columbus, Ohio, 1974.
[H12] Kreyszig, E., *Introductory Mathematical Statistics. Principles and Methods*. New York: Wiley, 1970.
[H13] Parzen, E., *Modern Probability Theory and Its Applications*. New York: Wiley, 1960.
[H14] Rice, W. B., *Control Charts in Factory Management*. New York: Wiley, 1947.
[H15] Scheffé, H., *The Analysis of Variance*. New York: Wiley, 1959.
[H16] Wald, A., *Statistical Decision Functions*. New York: Wiley, 1950.
[H17] Wilks, S. S., *Mathematical Statistics*. New York: Wiley, 1962.

J Tables

[A6], [A8], [B7]

[J1] *Barlow's Tables of Squares, Cubes, Square Roots, Cube Roots, and Reciprocals*. 4th ed. London: Spon, 1941 (10th impression 1965).
[J2] British Association for the Advancement of Science, *Mathematical Tables*. Vol. VII. *The Probability Integral*. Cambridge: University Press, 1939 (reissued 1968).
[J3] *C.R.C. Standard Mathematical Tables*. 19th ed. Cleveland, Ohio: Chemical Rubber Co., 1971.
[J4] *C.R.C. Handbook of Tables for Mathematics*. 4th ed. Cleveland, Ohio: Chemical Rubber Co., 1975.
[J5] Davis, H. T., *Tables of the Mathematical Functions*. 3 vols. Bloomington, Ind.: Principia Press, 1962–1963.
[J6] Dodge, H. F., and H. G. Romig, *Sampling Inspection Tables*. 2nd ed. New York: Wiley, 1959.
[J7] Dwight, H. B., *Mathematical Tables*. 3rd ed. New York: Dover 1961.
[J8] Fisher, R. A., and F. Yates, *Statistical Tables for Biological, Agricultural, and Medical Research*. 6th ed. London: Oliver and Boyd, 1963.
[J9] Hald, A., *Statistical Tables and Formulas*. New York: Wiley, 1952.
[J10] Harvard Computation Laboratory, *Tables of the Cumulative Binomial Probability Distribution*. Cambridge, Mass.: Harvard University Press, 1955.
[J11] National Bureau of Standards, *Tables of the Error Function and Its Derivative*. Washington, D.C.: US Gov. Printing Office, 1954.
[J12] National Bureau of Standards, *Tables of Lagrangian Interpolation Coefficients*. New York: Columbia University Press, 1944.
[J13] National Bureau of Standards, *Tables of Normal Probability Functions*. Washington, D.C.: US Gov. Printing Office, 1953.
[J14] Pearson, E. S., and H. O. Hartley, *Biometrika Tables for Statisticians*. Vol. I. 3rd ed. Cambridge: University Press, 1966.
[J15] Rand Corporation, *A Million Random Digits with* 100,000 *Normal Deviates*. Glencoe: Free Press, 1955.

APPENDIX

2 ANSWERS TO ODD-NUMBERED PROBLEMS

SECTION 1.1, page 8

1. First order

3. Second order

5. First order

11. $y = -e^{-x} + c$

13. $y = -\frac{1}{12}\cos 2x + c_1 x + c_2$

15. $y' + y = e^{-x}$

17. $y = (x^3 - 23)/3$

19. $y = \dfrac{x}{2} + \dfrac{1}{4}\sin 2x + \dfrac{\pi}{4}$

21. $c = -1/8$

25. 0.417 mi

27. 7.42 km = 4.61 mi

29. $p(h) = p_0 \exp(-0.000\,116\,h)$

SECTION 1.2, page 11

19. Solution curves above $y = 3$ and below $y = 0$ are monotone decreasing and approach 3 and $-\infty$, respectively, as $x \to \infty$.

SECTION 1.3, page 21

3. Suppose you forgot $\tilde{c}$ in (4), wrote $\ln|y| = -x^2$, transformed this to $y = e^{-x^2}$ and afterward added a constant c. Then $y = e^{-x^2} + c$, which is not a solution of $y' = -2xy$.

5. $x^2 + y^2 = c$

7. $y = c \ln|x|$

9. $y = ce^{-ax} - b/a$

11. $y = 1 + (\ln|x| + c)^2$

13. $y = c/\cosh^2 x$

15. $I = I_0 e^{-(R/L)t}$

17. $y = \arcsin x$

19. $y = y_0 \exp(-x^2)$

21. $y = 4\ln|x|/\ln 3$

23. $r = 2\sin^2\theta$

25. 10.4 km/sec

27. $pV = c = const.$ This is the law empirically found by Boyle (1662) and Mariotte (1676).

29. $\Delta A = -kA\Delta x$ (A = amount of incident light, ΔA = absorbed light, Δx = thickness, $-k$ = constant of proportionality). Let $\Delta x \to 0$. Then $A' = -kA$. Hence $A(x) = A_0 e^{-kx}$ is the amount of light in a thick layer at depth x from the surface of incidence.

33. $I\dfrac{d\omega}{dt} = -k\omega^{1/2}$, where $k > 0$, $\omega(t) = \left(\sqrt{\omega_0} - \dfrac{k}{2I}t\right)^2$,
$t_1 = 2I\sqrt{\omega_0}(1 - 2^{-1/2})/k$, $t_2 = 2I\sqrt{\omega_0}/k$

35. $\dfrac{dV}{dt} = \dfrac{dV}{dr}\dfrac{dr}{dt} = 4\pi r^2 \dfrac{dr}{dt} = kA = 4k\pi r^2$; 5.7 months

37. $T(t) = 23 - 8e^{-kt}$, $T(1) = 19$ gives $k = \ln 2$, $t = 6.32$ min.

39. $B(h) = \pi r^2$, $r^2 = h^2\tan^2 30° = h^2/3$, $dh/dt = -12.7h^{-3/2}$,
$0.4h^{5/2} = c - 12.7t$, $c = 4 \cdot 10^4$, $t = c/12.7 = 3150$ sec ≈ 53 min

41. (a) 1.21 meter. (b) $\dot{v} = g - (\tilde{b}/m)v = -0.10(v - 98)$, so that, as $t \to \infty$,

$$v(t) = 98 + ce^{-0.10t} \to 98 \text{ [meters/sec]},$$

and the model is no longer reasonable.

43. We obtain

$$v(t) = 37.19(1 - e^{-0.1089t}),$$
$$s(t) = 37.19[t - 9.183(1 - e^{-0.1089t})],$$
$$s(2.22) = 9.2 \text{ meters.}$$

45. $y(t) = (b - acf)/(1 - cf)$, where $f(t) = \exp(b - a)kt$.

47. Hyperbolas $k^2x^2 - y^2 = const$

49. Such a normal $y = ax$ has slope $a = y/x$; slope of tangent $y' = -1/a = -x/y$, hence $yy' + x = 0$, $y^2 + x^2 = c$ (circles).

51. $y = cx$

53. $y^2 - x^2 = c$

55. $y = x + c$

SECTION 1.4, page 27

1. $y = x(\ln|x| + c)$

3. $y = x + x/(c - \ln|x|)$

5. $y = x \arcsin(x + c)$

7. $\ln(x^2 + y^2) + 2 \arctan \dfrac{y}{x} = \ln 2 + \dfrac{\pi}{2}$

9. $y = x + x(5 - x^2)^{-1/2}$

11. $y = x + \dfrac{1 + ce^{2x}}{1 - ce^{2x}}$

13. $y = x^{-1} \ln|x + c|$

15. $(y - x)^2 - 2y = c$

17. $y = ax$, $y/x = a = const$, $y' = g(a) = const$

19. $2xy^2 + xy' = y$, $y^{-2}(y\,dx - x\,dy) = 2x\,dx$, $x/y = x^2 + c$, $y = x/(x^2 + 1)$

SECTION 1.5, page 30

1. $du = 8x\,dx + 18y\,dy$

3. $du = (y\,dx + x\,dy)e^{xy}$

5. $(3x^2y - 2\sin 2x)\,dx + x^3\,dy = 0$

7. $e^{-2xy}[(1 - 2xy)\,dx - 2x^2\,dy] = 0$

9. $xy^2 = c$

11. $y/x = c$

13. $x^2 \ln|y| = c$

15. From (4b) we have $-x + k'(y) = x$, a contradiction. $y = cx$ by separation of variables.

17. No, $y = 1/(3\ln|x| + c)$

19. Yes, $\cosh x \cos y = c$

21. Yes, $\ln|x| \cos 2y = c$

23. $x^2/a^2 + y^2/b^2 = c$

25. $x^2 + y/x = c$

29. $g = ky + c_1$, $h = kx + c_2$, where c_1, c_2 and k are constants.

31. $x^3y^4 = 16$

33. Hyperbola $(x - 3)(y - 1) = 1$

35. $e^x \cos y + (x - y)^2 = \pi^2 - 1$

37. $(x + 2)^2 - (y + 1)^2 = 1$

39. $(x - 1)^2 + (y + 1)^2 = 1$

SECTION 1.6, page 32

1. $e^{2x}\cos \pi y = c$
3. $(y + 1)^2 = cx^3$
5. $x \tan xy = c$
7. $F = yx^2,\ x^3y^2 = c$
9. $F = e^x,\ e^x \sin y = c$
11. $F = \sin x,\ \sin^2 x \cos y = c$
13. $F = x,\ x^2y = -4$
15. $F = 1/xy,\ x^2y^2e^x = 18e^3$
17. $xy^2 = c$. Hence other factors are $xy^2F = xy^3$, $(xy^2)^2F = x^2y^5$, etc.
19. $\partial P/\partial y = \partial Q/\partial x + (a/x)Q$
21. $\partial Q/\partial x = \partial P/\partial y + P$

SECTION 1.7, page 37

3. $y = ce^x + x$
5. $y = cx^{-1} + x$
7. $y = ce^{4x} + x^2$
9. $y = ce^{-3x} + 0.2e^{2x} + 2$
11. $y = cx^2 + x^2e^x$
13. $y = (c + 0.5e^{x^2})x^{-2}$
15. $y = (x - 1)e^x$
17. $y = e^x x^3 - x$
25. $y(x) = r_0/f_0 + ce^{-f_0 x}$
27. $m = W/g = 4900/9.80$ kg, $dv/dt = -0.02(v - 20)$, $v(t) = 20(1 - e^{-0.02t})$, $v_\infty = 20$ meters/sec = 72 km/hr = 45 miles/hr, $t = 115$ sec, $y = 1.4$ km = 0.87 mile.

29. $v(t) = \dfrac{W - B}{k}(1 - e^{-kt/m}),\ m = \dfrac{W}{g} = \dfrac{2254}{9.80} = 230$ [kg],

$y(t) = \dfrac{W - B}{k}\left[t - \dfrac{m}{k}(1 - e^{-kt/m})\right]$, $t_{\text{crit}} = 17.3$ sec, $y_{\text{crit}} = 106$ meters.

SECTION 1.8, page 40

1. $v = e^{-x},\ u = 2e^x + c,\ y = ce^{-x} + 2$
3. $y = (3x + c)e^x$
5. $x = \frac{1}{2}y^4 + cy^2$
9. $y = 1/(1 + ce^x)$
11. $y = 1/(ce^{-x^2/2} - x^2 + 2)$
13. $y = (\frac{1}{2}xe^x + cxe^{-x})^{1/2}$
15. $u' + \left(f + \dfrac{a'}{a}\right)u = \dfrac{1}{a}(r - fb - b')$
17. The continuity of f and r implies the existence of the integrals in (4), so that (4) determines y explicitly, with c uniquely determined by $y(x_0) = y_0$.
19. $y' = -f(x)y + r(x)$. At (x_0, y_0) this equals $-f(x_0)y_0 + r(x_0)$, and $\eta - y_0 = (\xi - x_0)[r(x_0) - f(x_0)y_0]$ represents the line through (x_0, y_0) with that slope. Similarly, $\eta - y_1 = (\xi - x_0)[r(x_0) - f(x_0)y_1)]$ represents the line through (x_0, y_1) whose slope is the value of y' at (x_0, y_1) given by (1). The point of intersection of these lines has the coordinates (ξ, η) given in the enunciation of the problem, and since these coordinates do not depend on y, the statement is proved.

SECTION 1.9, page 46

5. $t = \tau_L \ln 100 = 0.009$ [sec]
7. $\tau_L = 0.025$ sec, $t = 0.02$ sec
9. $I(0) = -E_0\omega L/(R^2 + \omega^2L^2)$
15. $RQ' + Q/C = E_0,\ Q(t) = E_0C(1 - e^{-t/RC})$, $V(t) = Q(t)/C = 12(1 - e^{-0.05t})$

17. $I(t) = \omega E_0 C[1 + (\omega RC)^2]^{-1}(\cos \omega t + \omega RC \sin \omega t - e^{-t/RC})$

19. From (5*), $I = I_1 = 1 - e^{-0.01t}$ $(0 \leqq t \leqq 100)$, $I = I_2 = c_2 e^{-0.01(t-100)}$ $(100 \leqq t \leqq 200)$; $I_1(100) = I_2(100)$ yields $c_2 = 1 - e^{-1} \approx 0.63$. $I = I_3 = 1 + c_3 e^{-0.01(t-200)}$ $(200 \leqq t \leqq 300)$; $I_2(200) = I_3(200)$ yields the value $c_3 = (1 - e^{-1})e^{-1} - 1 \approx -0.77$, etc. (See the figure.)

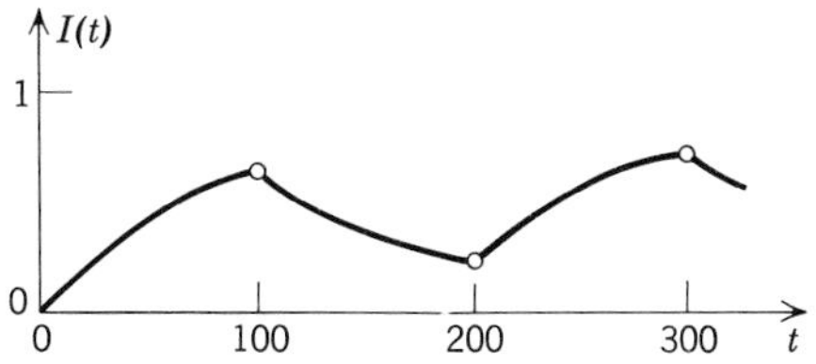

Problem 19

SECTION 1.10, page 51

1. Parallel straight lines with slope 1/3

3. Hyperbolas whose asymptotes are the coordinate axes.

5. Congruent parabolas

7. $y + 1 - c(x - 4) = 0$

9. $x^2/c^2 + y^2/(c^2 + 1) - 1 = 0$

11. $y' = 3y/x$

13. $y' = y$

15. $y' = xy/(x^2 - 1)$

17. $y = \ln |x| + c^*$

19. Ellipses $x^2 + 3y^2 = c^*$

21. $y = \sqrt{c^* - 2x}$

23. $xy = c^*$

25. Bell-shaped curves $x = c^* e^{-y^2}$

27. $2x^2 + 3y^2 = c^*$

29. $e^x \cos y = c^*$

31. $y = c^* e^{-2x}$

33. $dv = 0$ gives $dy/dx = -v_x/v_y$ and this must equal u_y/u_x (cf. Prob. 32).

35. $y' = -x/y$, $x^2 + y^2 = c^*$

39. $x^2 - y^2 = c^*$

SECTION 1.11, page 56

1. $y_n = 1 + 2\int_0^x y_{n-1}(t)\,dt$, $y_0 = 1$, $y_1 = 1 + 2x$, $y_2 = 1 + 2x + (2x)^2/2!$, etc., $y = e^{2x}$

3. $y_n = \int_0^x y_{n-1}{}^2(t)\,dt + 4x$, $y_0 = 0$, $y_1 = 4x$, $y_2 = 4x + \frac{16}{3}x^3, \cdots,$ $y = 2 \tan 2x$

5. $y_n = 1 + \int_0^x [t + y_{n-1}(t)]\,dt = 1 + \frac{1}{2}x^2 + \int_0^x y_{n-1}(t)\,dt,$
$y_0 = 1$, $y_1 = 1 + x + \frac{1}{2}x^2$, $y_2 = 1 + x + x^2 + \frac{1}{6}x^3$, etc.
Exact: $y = 2e^x - x - 1$.

9. $y_1(1) = 2$, $y_2(1) = 2.5$, $y_3(1) = 2.667$

SECTION 1.12, page 61

3. $y = cx^4$, c arbitrary

5. None

7. If they had, this would violate the uniqueness. (Why?)

9. 1/2, the largest value of $b/(1 + b^2)$, where $1 + b^2$ is the smallest K.

11. $\alpha = 1/4$

13. Yes, with $M = 1$. Not on the x-axis ($y = 0$).

15. $y = (x - 1)^2$, $y \equiv 0$. Not on the x-axis ($y = 0$).

SECTION 2.1, page 66

5. $(y_1 + y_2)'' + f(y_1 + y_2)' + g(y_1 + y_2)$
$= (y_1'' + fy_1' + gy_1) + (y_2'' + fy_2' + gy_2) = 2r$

7. $(y_1 + y_2)'' + f(y_1 + y_2)' + g(y_1 + y_2)$
$= (y_1'' + fy_1' + gy_1) + (y_2'' + fy_2' + gy_2) = r + 0 = r$

9. $y = c_1 e^{-x} + \frac{1}{2}x^2 + c_2$

11. $y = c_1 \sinh x + c_2$

13. $y = c_1 x^{-1} + c_2$

15. $y = c_1 e^t + c_2$, $c_1 + c_2 = 1$, $c_1 = 2$, $y = 2e^t - 1$,
$y(5) = 2e^5 - 1 = 296$ [meters], $\dot{y}(5) = 297$ meters/sec.

17. $y = e^x - 1$

21. $x = \frac{1}{4}e^{2y} + c_1 y + c_2$

23. $y = x^4$

SECTION 2.2, page 69

1. e^x, e^{-x}

3. e^{-2x}, e^x

5. 1, e^x

7. e^{3x}, e^{-3x}

9. $e^{-(2+i)x}$, $e^{-(2-i)x}$

11. Use Fundamental theorem 1 in Sec. 2.1.

13. (a) e^{-4x}, 1; (b) $c_1 e^{-4x} + c_2$

15. $y'' + 2y' - 3y = 0$

17. $y'' - 2y' = 0$

19. $y'' + 2y' + 5y = 0$

SECTION 2.3, page 73

1. Independent

3. Independent

5. Dependent

7. Independent

9. Dependent

11. Dependent

13. Independent

19. $c_1 e^{5x} + c_2 e^{-5x}$

21. $c_1 e^{3x} + c_2$

23. $c_1 e^{3x} + c_2 e^{-5x}$

25. $y = e^{3x} - 2e^x$

27. $y = 3 \cosh x$

29. $y = \cosh 2x$

31. $y = e^{-x/3}$

33. If $y \not\equiv 0$, $y(A) = y(B) = 0$ and y_1 satisfies (1), (5), so does $y_2 = y_1 + y$, and $y_2 \not\equiv y_1$. Conversely, if y_1 and y_2 are solutions of (1), (5), then $y = y_1 - y_2 \not\equiv 0$ is a solution of (1) and $y(A) = y(B) = 0$.

37. $y'' + \omega^2 y = 0$

39. $y'' + 6y' + 10y = 0$

SECTION 2.4, page 79

1. $y = A \cos x + B \sin x$

3. $y = e^{-2x}(A \cos 2x + B \sin 2x)$

5. $y = e^x(A \cos 3x + B \sin 3x)$

7. II, $y = A \cos \pi x + B \sin \pi x$

9. III, $y = (c_1 + c_2 x)e^{-3x}$

11. I, $y = c_1 e^{x/2} + c_2 e^{-2x}$

13. III, $y = (c_1 + c_2 x)e^{x/2}$

15. I, $y = c_1 e^{x/2} + c_2 e^{-x/4}$

17. $y'' + 2y' + 2y = 0$ **19.** $y'' - 8y' + 16y = 0$

21. $y'' - 9y = 0$ **23.** $y'' - 4y' = 0$

29. $y'' + 4y' + 5y = 0,\ y = -3e^{-2x} \sin x$

31. $y'' - 2y' + (\pi^2 + 1)y = 0,\ y = e^{x}(\cos \pi x - \sin \pi x)$

33. $y = 2e^{2x}$ **35.** $y = (2 + 2x)e^{3x}$

37. $y = e^{-2x}(\cos x - \sin x)$ **39.** $y = e^{2x} - 2e^{x}$

SECTION 2.5, page 82

1. $-16x^2 + 20x - 3,\ 0,\ 3 \sin x - 5 \cos x$

3. $0,\ 3e^{2x},\ 0,\ -3e^{-x}$ **9.** $y = c_1e^{2x} + c_2e^{-x}$

11. $y = e^{-x}(A \cos 2\pi x + B \sin 2\pi x)$ **13.** $y = c_1e^{x/2} + c_2e^{-x/3}$

SECTION 2.6, page 89

1. Independent. $1/\sqrt{2} = 0.7$ times the original frequency.

3. $1/2\pi,\ 1/\pi,\ 3/2\pi,\ 2/\pi$

5. $y = y_0 \cos \omega_0 t + (v_0/\omega_0) \sin \omega_0 t$
$= (y_0{}^2 + v_0{}^2\omega_0{}^{-2})^{1/2} \cos (\omega_0 t - \text{arc tan } (v_0/y_0\omega_0))$

9. $ml\ddot{\theta} = -mg \sin \theta \approx -mg\theta$ (the tangential component of $w = mg$), $\ddot{\theta} + \omega_0{}^2\theta = 0,\ \omega_0{}^2 = g/l$. Ans. $\sqrt{g/l}/2\pi$.

11. $T = 2\pi\sqrt{\dfrac{ml}{mg + (k_1 + k_2)l}}$

13. $m\ddot{y} = -a\gamma y$, where $m = 1$ kg, $ay = \pi \cdot 0.01^2 \cdot 2y$ meter^3 is the volume of the water which causes the restoring force $a\gamma y$ with $\gamma = 9800$ nt (= weight/meter^3). $\ddot{y} + \omega_0{}^2 y = 0,\ \omega_0{}^2 = a\gamma/m = a\gamma = 0.000628\gamma$. Frequency $\omega_0/2\pi = 0.4$ [sec^{-1}].

17. If c_1 and c_2 have the same sign.

23. The positive solutions of $\tan t = 1$, that is, $\pi/4$ (max), $5\pi/4$ (min), etc.

25. If a maximum is at t_0, the next is at $t_1 = t_0 + 2\pi/\omega^*$. Since the sine and cosine in (11) have period $2\pi/\omega^*$, the ratio is $\exp(-\alpha t_0)/\exp(-\alpha t_1) = \exp(2\pi\alpha/\omega^*)$. $\Delta = 2\pi$; $\tan t = -1$ gives $3\pi/4$ (min), $7\pi/4$ (max), etc.

27. $y = [(v_0 + \alpha y_0)t + y_0]e^{-\alpha t}$

29. $y = c_1e^{-v_0/c_1}/\alpha$, where $c_1 = v_0 + \alpha y_0$

SECTION 2.7, page 93

1. Substitution of $y_2 = uy_1,\ y_2' = uy_1' + u'y_1,\ y_2'' = uy_1'' + 2u'y_1' + u''y_1$ into (1) gives

$$(x^2y_1'' + axy_1' + by_1)u + (2x^2y_1' + axy_1)u' + x^2u''y_1 = 0.$$

Since y_1 satisfies (1), the factor of u is zero. Since $y_1' = mx^{m-1}$ and $a = 1 - 2m$ in the critical case, the remaining expression is $u' + xu'' = 0$. Thus, $u' = 1/x,\ u = \ln x,\ y_2 = x^m \ln x$.

5. $y = c_1x + c_2x^3$ **7.** $y = (c_1 + c_2 \ln x)x$

9. $y = c_1 + c_2x^4$ **11.** $y = c_1 + c_2 \ln x$

13. $2z - 3 = x$, $y = c_1(2z - 3)^{-2} + c_2(2z - 3)^{-1/2}$

15. $y = (1 - \ln x)x^2$

17. $y = x - x^3$

19. $\ddot{y} + (a - 1)\dot{y} + by = 0, x = e^t, t = \ln x$; by the chain rule, $\dot{y} = y'\dot{x} = y'x$; $\ddot{y} = y''x^2 + y'x$, $y''x^2 + y'x + (a - 1)y'x + by = 0$.

SECTION 2.8, page 100

1. $W = (\lambda_2 - \lambda_1)\exp(\lambda_1 x + \lambda_2 x)$

3. $W = qe^{2px}$

5. $W = (m_2 - m_1)x^{m_1+m_2-1}$

7. $y'' + \omega^2 y = 0$

9. $x^2y'' - 1.5xy' + y = 0$

11. $xy'' - y' = 0$

13. $y'' + 2y' + 2y = 0$

15. $y'' + \beta^2 y = 0$

17. Dependent

19. Independent

21. Dependent

23. No

25. Use Theorem 2.

SECTION 2.9, page 104

7. Dependent **9.** Dependent **11.** Dependent **13.** Independent

SECTION 2.10, page 106

3. $y = c_1e^x + c_2e^{-x} + c_3\cos 2x + c_4\sin 2x$

5. $y = c_1 + c_2\cos x + c_3\sin x$

7. $y = c_1e^x + c_2e^{-x} + c_3e^{2x} + c_4e^{-2x}$

9. $y''' - 3y'' + 3y' - y = 0$

11. $y^{\mathrm{IV}} - 6y''' + 9y'' = 0$

13. $y^{\mathrm{V}} - 5y''' + 4y' = 0$

17. $y = c_1x^{-1} + c_2x + c_3x^3$

19. $y = (c_1 + c_2e^x + c_3e^{-x})x$

SECTION 2.11, page 108

7. $y = c_1e^x + c_2e^{-x} + xe^x$

9. $y = A\cos 2x + B\sin 2x + 3x\cos 2x$

11. $y = c_1e^x + c_2e^{-x} - \cos x$

13. $y = c_1x^{1/2} + c_2x^{1/2}\ln x + \cos(x/2)$

15. $y = (c_1 + c_2x)e^{-3x} + (x\arctan x - 0.5\ln(x^2 + 1))e^{-3x}$

SECTION 2.12, page 113

1. $D(D - s)y = kx^n$, $D^{n+2}(D - s)y = kD^{n+1}[x^n] = 0$,
$y = K_0 + K_1x + \cdots + K_{n+1}x^{n+1} + ce^{sx}$

3. $x^2 + x - 2$

5. $1.5x^4 - 5x^3 + 9.5x^2 - 11x + 6$

7. $-5\cos 2x + \sin 2x$

9. $-x\cos x$

11. $2x^3 - 3x^2 + 15x - 8$

13. $A\cos x + B\sin x - x^2 - x + 2$

15. $A\cos 2x + B\sin 2x + 0.2e^{-x}$

17. $A\cos x + B\sin x - 0.5x\cos x$

19. $c_1e^{3x} + c_2e^x + 0.5xe^{3x}$

21. $ce^{-x} + x^4 - 4x^3 + 12x^2 - 24x + 24$

23. $y = 5\cos 5x - \sin 5x + 0.2x$

25. $3e^x + 2x^2$

27. $e^{-x} - e^{2x} + xe^{2x}$

29. $e^{-2x}\cos 2x + \sin x$

SECTION 2.13, page 119

1. $y = A \cos t + B \sin t - \cos 2t$
3. $y = A \cos 4t + B \sin 4t + (\cos t - \sin t)/15$
5. $y = -5 \cos 3t + 2 \sin 3t + 2 \cos t - \sin t$
7. $y = c_1 e^{-t} + c_2 e^{-2t} + \sin t - 3 \cos t$
9. $y = e^{-t}(A \cos t + B \sin t) + 6 \sin 3t - 7 \cos 3t$
11. $y = e^{-t}(A \cos 2t + B \sin 2t) + 0.1 \cos t - 0.2 \sin t$
13. $y = \cos 5t + \sin t$
15. $y = (y_0 - (1 - \omega^2)^{-1}) \cos t + v_0 \sin t + (1 - \omega^2)^{-1} \cos \omega t$
17. $y_0 = 1/(1 - \omega^2)$, $v_0 = 0$. No.
21. $y = e^{-5t} \cos 2t + \sin t$
23. $y = e^{-t} \sin t - 0.5 \sin 2t$
25. $y = \begin{cases} (1 + 2/\pi^2)(1 - \cos t) - t^2/\pi^2 & \text{if } 0 \leqq t \leqq \pi \\ -(1 + 4/\pi^2) \cos t + (2/\pi) \sin t & \text{if } t > \pi \end{cases}$

SECTION 2.14, page 124

3. $R > R_{\text{crit}} = 2\sqrt{L/C}$ is Case I, etc.
5. $S = 0$, that is, $C = 1/\omega^2 L$.
7. $I = -0.75 \cos 10t + 0.6 \sin 10t$
9. $I = e^{-t}(A \cos t + B \sin t) + 5 \cos t + 10 \sin t$
11. $I = (c_1 + c_2 t)e^{-t} + 12 \cos 2t + 9 \sin 2t$
13. $I = 60 \sin 2t - 30 \sin t$
15. $I = 5e^{-2t} \sin t$
17. $I = e^{-3t}(3 \sin 4t - 4 \cos 4t) + 4 \cos 5t$
19. $I = \begin{cases} I_1 \equiv A_1 \cos t + B_1 \sin t + 1 & \text{when } 0 < t < a, \\ I_2 \equiv A_2 \cos t + B_2 \sin t & \text{when } t > a. \end{cases}$

$I(0) = 0$, $Q(0) = 0$, $E(0) = 0$ yields $\dot{I}(0) = 0$, $A_1 = -1$, $B_1 = 0$, and $I_1 = 1 - \cos t$. From this we obtain, $I_1(a) = 1 - \cos a$, $\dot{I}_1(a) = \sin a$. Thus $I_2(a) = I_1(a) = 1 - \cos a$, $\dot{I}_2(a) = \dot{I}_1(a) - a = -a + \sin a$, that is, $A_2 \cos a + B_2 \sin a = 1 - \cos a$, $\quad -A_2 \sin a + B_2 \cos a = -a + \sin a$. Hence, $A_2 = \cos a + a \sin a - 1$, $B_2 = \sin a - a \cos a$, and

$$I = \begin{cases} 1 - \cos t & \text{when } 0 < t < a, \\ \cos (t - a) - a \sin (t - a) - \cos t & \text{when } t > a. \end{cases}$$

21. $I = \begin{cases} \frac{1}{2}(e^{-t} - \cos t + \sin t) & \text{when } 0 < t < \pi \\ -\frac{1}{2}(1 + e^{-\pi}) \cos t + \frac{1}{2}(3 - e^{-\pi}) \sin t & \text{when } t > \pi \end{cases}$
25. $I = \begin{cases} 1 - e^{-t} & \text{when } t < a \\ [(1 - a)e^{a} - 1]e^{-t} & \text{when } t > a \end{cases}$

SECTION 2.15, page 128

1. $I_p = \text{Re}[(-0.1 - 0.2i)(\cos 2t + i \sin 2t)] = -0.1 \cos 2t + 0.2 \sin 2t$
3. $I_p = (360 \sin 3t - 300 \cos 3t)/61$ 5. $I_p = (50 \sin 4t - 110 \cos 4t)/73$

7. $y_p = 2 \sin 4t$

9. $y_p = 0.1008 \sin 10t - 0.2006 \cos 10t$

SECTION 2.16, page 131

1. $y = A \cos x + B \sin x - x \cos x + \sin x \ln |\sin x| + x$

3. $y = (c_1 + c_2x + x \ln x - x)e^{2x}$

5. $y = (c_1 + c_2x - \cos x)e^{-x}$

7. $y = (c_1 + c_2x + \frac{4}{35}x^{7/2})e^{x}$

9. $y = e^{-x}(A \cos x + B \sin x - 0.5 \cos 2x/\cos x)$

11. $y = c_1x^{-1} + c_2x - 4$. Note that in (8), $r = 4/x^2$ (not 4), because the given equation must be divided by x^2 in order that it assumes the form (1) from which (8) was derived.

13. $y = c_1x + c_2x^2 + x^4/6$

15. $y = c_1x + c_2x^2 + \frac{1}{12}x^{-2}$

17. $y = c_1x^2 + c_2x^3 + \frac{1}{42}x^{-4}$

19. $y = c_1 + c_2x^2 + (2x - 2)e^{x}$

SECTION 3.1, page 136

1. $x = c_1e^{t} + c_2e^{-t}$, $y = c_1e^{t} - c_2e^{-t}$

3. $x = \frac{1}{2}(\sin t - \cos t) + c_1$, $y = \frac{1}{2}(\sin t + \cos t) + c_2$

5. $x = c_1e^{t} + c_2e^{-t} + c_3 \cos t + c_4 \sin t - t$,
$y = c_1e^{t} + c_2e^{-t} - c_3 \cos t - c_4 \sin t - 1$

7. $x = 2e^{5t} - e^{-5t}$, $y = e^{5t} + 2e^{-5t}$

9. $x = e^{5t} - \frac{5}{3}e^{2t}$, $y = e^{5t} - \frac{2}{3}e^{2t}$

13. $I_1 = -\frac{200}{3}e^{-2t} + \frac{125}{3}e^{-0.8t} + 25$, $I_2 = -\frac{100}{3}e^{-2t} + \frac{100}{3}e^{-0.8t}$

15. $I_1 = 2 - 2e^{-300t}$, $I_2 = 2 + e^{-300t}$, steady-state currents 2 amperes, 2 amperes.

SECTION 3.2, page 140

1. $\sqrt{A^2 + B^2}$

3. $y = A \cos \frac{1}{2}t + B \sin \frac{1}{2}t$; ellipses $y^2 + 4v^2 = const$

5. $y = c_1e^{t} + c_2e^{-t}$; hyperbolas $y^2 - v^2 = const$

7. Parallel straight lines $v = y + c$

9. Hyperbolas $yv = const$

SECTION 3.3, page 148

3. $x = Ae^{t}$, $y = Be^{2t}$, $y = (B/A^2)x^2$ if $A \neq 0$, arcs of parabolas and of the y-axis. Horizontal, vertical.

7. $x = Ae^{-t}$, $y = (At + B)e^{-t}$

9. $\ddot{x} = -x$, $x^2 + y^2 = const$, orientation clockwise

SECTION 4.2, page 158

1. 3 3. 1 5. ∞ 7. 1 9. ∞ 11. 3 13. 1 15. 3

17. $c_0 = 0$, $c_1 = 2$, $c_m = c_{m-1}/(m - 2)$, $y = 2x + c_2x^2\left(1 + x + \frac{x^2}{2!} + \cdots\right)$
$= c_2x^2e^{x} + 2x$

19. $y = c_1(x + x^2)$

21. $y = c_0 + c_1x + (\frac{3}{2}c_1 - c_0)x^2 + (\frac{7}{6}c_1 - c_0)x^3 + \cdots$. Setting $c_0 = A + B$ and $c_1 = A + 2B$, we obtain $y = Ae^x + Be^{2x}$. This illustrates the fact that even if the solution of an equation is a known function, the power series method may not yield it immediately in the usual form.

23. $y = c_1x + c_0(1 - x^2 - \frac{1}{3}x^4 - \frac{1}{5}x^6 - \frac{1}{7}x^8 - \cdots)$. (This is a particular case of Legendre's equation $(n = 1)$ which will be considered in the next section.)

25. $y = c_0\left(1 - \dfrac{1}{2!}x^2 - \dfrac{1}{4!}x^4 - \dfrac{3}{6!}x^6 - \dfrac{3 \cdot 5}{8!}x^8 - \cdots\right) + c_1x$

27. $(t + 1)\dot{y} - y = t + 1,\ y = c_0(1 + t) + (1 + t)\left(t - \dfrac{t^2}{2} + - \cdots\right)$

$= c_0x + x \ln x$

29. $y = c_0\left(1 + \dfrac{t^2}{2!} + \cdots\right) + c_1\left(t + \dfrac{t^3}{3!} + \cdots\right) = A \cosh x + B \sinh x$

where $A = c_0 \cosh 1 - c_1 \sinh 1$, $B = c_1 \cosh 1 - c_0 \sinh 1$

SECTION 4.3, page 163

1. $P_6(x) = \frac{1}{16}(231x^6 - 315x^4 + 105x^2 - 5)$

SECTION 4.4, page 174

1. $y_1 = \dfrac{c_0}{x^2}\left(1 - \dfrac{x^2}{2!} + \dfrac{x^4}{4!} - + \cdots\right) = c_0\dfrac{\cos x}{x^2},\ y_2 = c_0{}^*\dfrac{\sin x}{x^2}$

3. $y_1 = \dfrac{x^2}{3} - \dfrac{x^4}{2 \cdot 5} + \dfrac{x^6}{2^2 2! 7} - \dfrac{x^8}{2^3 3! 9} + - \cdots,\ y_2 = \dfrac{1}{x}$

5. $y_1 = x^{-2} \sin x^2$, $y_2 = x^{-2} \cos x^2$

9. $y_1 = 1 + \dfrac{x^2}{2^2} + \dfrac{x^4}{(2 \cdot 4)^2} + \dfrac{x^6}{(2 \cdot 4 \cdot 6)^2} + \cdots,$

$y_2 = y_1 \ln x - \dfrac{x^2}{4} - \dfrac{3x^4}{8 \cdot 16} - \cdots$

11. $y_1 = x^2$, $y_2 = x^{-2}$

13. $y_1 = x^{-3} \sinh 2x$, $y_2 = x^{-3} \cosh 2x$

15. $y_1 = e^x$, $y_2 = e^x \ln x$

23. y_1/y_2 is not a constant. This implies linear independence.

31. $y = C_1F(1, \frac{1}{2}, -\frac{1}{2}; x) + C_2x^{3/2}F(\frac{5}{2}, 2, \frac{5}{2}; x)$

33. $y = C_1F(-\frac{1}{2}, -\frac{1}{2}, \frac{1}{2}; t - 1) + C_2(t - 1)^{1/2}$

35. $y = C_1F(2, -2, -\frac{1}{2}; t - 2) + C_2(t - 2)^{3/2}F(\frac{7}{2}, -\frac{1}{2}, \frac{5}{2}; t - 2)$

SECTION 4.5, page 179

5. $J_0(1) = 1 - \dfrac{1}{2^2(1!)^2} + \dfrac{1}{2^4(2!)^2} - \dfrac{1}{2^6(3!)^2} + - \cdots$. The absolute value of each remainder R_n of this series is at most equal to the absolute value of the first term of R_n, as follows from the familiar Leibniz test (cf. also Sec. 15.4), and because of the required accuracy we must have $|R_n| < 10^{-5}$. Now

$\frac{1}{2^{2n}(n!)^2} < 10^{-5}$, $2^{2n}(n!)^2 > 10^5$ holds for $n = 4$. Hence the sum of the above explicitly written terms yields the desired approximate value. $\ln 2 = \ln(1 + 1) \approx 1 - \frac{1}{2} + \frac{1}{3} - + \cdots - \frac{1}{100{,}000}$ yields the desired approximate value, and we would need 100,000 terms.

9. Perform the product differentiations indicated in (20) and (21). Multiply the first equation by $x^{-\nu}$, the second by x^{ν}, solve for $J_{\nu-1}$ and $J_{\nu+1}$, respectively, and add.

13. $J_2(x) = 2x^{-1}J_1(x) - J_0(x)$; cf. (22) with $\nu = 1$.

17. $-2J_2(x) - J_0(x) + c$

19. $-x^{-3}J_3(x) + c$

25. $y = (A \cos x + B \sin x)/\sqrt{x}$

SECTION 4.6, page 186

1. 1.1

3. $AJ_2(x) + BY_2(x)$

5. $AJ_0(\sqrt{x}) + BY_0(\sqrt{x})$

7. $AJ_n(\lambda x) + BY_n(\lambda x)$

9. $x(AJ_1(x) + BY_1(x))$

11. $x^n(AJ_n(x^n) + BY_n(x^n))$

13. $\sqrt{x}[AJ_{1/3}(\frac{2}{3}x^{3/2}) + BY_{1/3}(\frac{2}{3}x^{3/2})]$

17. Use (12) in Sec. 4.5.

SECTION 4.7, page 189

1. $1/\sqrt{2\pi}$, $(\cos x)/\sqrt{\pi}$, $(\cos 2x)/\sqrt{\pi}$, $\cdots$

3. $\sin \pi x$, $\sin 2\pi x$, $\cdots$

5. $1/\sqrt{T}$, $\sqrt{2/T}\cos(2n\pi x/T)$ $(n = 1, 2, \cdots)$

7. $P_0/\sqrt{2}$, $\sqrt{\frac{3}{2}}P_1(x)$, $\sqrt{\frac{5}{2}}P_2(x)$, $\sqrt{\frac{7}{2}}P_3(x)$

9. Set $x = ct + k$.

SECTION 4.8, page 194

5. $\lambda = ((2n + 1)\pi/2l)^2$, $n = 0, 1, \cdots$; $y_n(x) = \sin((2n + 1)\pi x/2l)$

7. $\lambda = (n\pi/l)^2$, $n = 0, 1, \cdots$; $y_n(x) = \cos(n\pi x/l)$

9. $\lambda = n^2\pi^2$, $n = 1, 2, \cdots$; $y_n(x) = \sin(n\pi \ln |x|)$

11. $\lambda = n^2$, $n = 1, 2, \cdots$; $y_n(x) = e^{-x} \sin nx$

13. Infinitely many

15. $y'' + \lambda y = 0$, $y'(0) = 0$, $y'(\pi) = 0$

SECTION 4.9, page 196

1. $x^2 = \frac{1}{3}P_0(x) + \frac{2}{3}P_2(x)$, $x^3 = \frac{3}{5}P_1(x) + \frac{2}{5}P_3(x)$,
$x^4 = \frac{1}{5}P_0(x) + \frac{4}{7}P_2(x) + \frac{8}{35}P_4(x)$

3. $2(-P_0(x) + P_1(x) - P_2(x) + P_3(x))$

5. $P_0(x)/4 + P_1(x)/2 + 5P_2(x)/16 + \cdots$

7. $P_0(x)/2 + 5P_2(x)/8 + \cdots$

11. Differentiate the formula which defines the Hermite polynomials.

19. $\int_0^\infty e^{-x}x^k L_n(x)\,dx = \frac{1}{n!}\int_0^\infty x^k \frac{d^n}{dx^n}(x^n e^{-x})\,dx$

$= -\frac{k}{n!}\int_0^\infty x^{k-1}\frac{d^{n-1}}{dx^{n-1}}(x^n e^{-x})\,dx$

$= \cdots = (-1)^k \frac{k!}{n!}\int_0^\infty \frac{d^{n-k}}{dx^{n-k}}(x^n e^{-x})\,dx$

$= 0$ *when* $n > k$.

21. $v(\theta) = \cos n\theta$ satisfies $d^2v/d\theta^2 + n^2v = 0$. Set $x = \cos\theta$.

25. By (20), Sec. 4.5, with $\nu = 1$,

$$c_m = \frac{2}{R^2 J_1^{\,2}(\alpha_{m0})}\int_0^R xJ_0\left(\frac{\alpha_{m0}}{R}x\right)dx = \frac{2}{\alpha_{m0}{}^2 J_1^{\,2}(\alpha_{m0})}\int_0^{\alpha_{m0}} wJ_0(w)\,dw$$

$$= \frac{2}{\alpha_{m0}J_1(\alpha_{m0})}, \qquad f = 2\left(\frac{J_0(\lambda_{10}x)}{\alpha_{10}J_1(\alpha_{10})} + \frac{J_0(\lambda_{20}x)}{\alpha_{20}J_1(\alpha_{20})} + \cdots\right)$$

27. $c_m = \dfrac{2akJ_1(\alpha_{m0}a/R)}{\alpha_{m0}RJ_1^{\,2}(\alpha_{m0})}$

29. $c_m = \dfrac{4J_2(\alpha_{m0})}{\alpha_{m0}{}^2 J_1^{\,2}(\alpha_{m0})}$

31. $c_m = \dfrac{2R^2}{\alpha_{m0}J_1(\alpha_{m0})}\left[1 - \dfrac{2J_2(\alpha_{m0})}{\alpha_{m0}J_1(\alpha_{m0})}\right]$

35. $x^3 = 16\left[\dfrac{J_3(\alpha_{13}x/2)}{\alpha_{13}J_4(\alpha_{13})} + \dfrac{J_3(\alpha_{23}x/2)}{\alpha_{23}J_4(\alpha_{23})} + \cdots\right]$

SECTION 5.1, page 205

1. $k(1 - e^{-cs})/s$

3. $-e^{-s}/s + (1 - e^{-s})/s^2$

5. $1/s^2 + 2/s$

7. $a/s + b/s^2 + 2c/s^3$

9. $as/(s^2 + 4)$

11. $(\omega\cos\theta + s\sin\theta)/(s^2 + \omega^2)$

13. $\frac{1}{3}\sin 3t$

15. $a_1 + a_2t + a_3t^2/2$

17. $4\cosh 4t + \sinh 4t$

19. $\sin(2n\pi t/T)$

21. $\dfrac{1}{s+1} - \dfrac{1}{s+2}$, $e^{-t} - e^{-2t}$

23. $t^{3/2}/\Gamma(\frac{1}{2})$ since $\Gamma(\frac{5}{2}) = \frac{3}{2}\Gamma(\frac{3}{2}) = \frac{3}{4}\Gamma(\frac{1}{2})$

SECTION 5.2, page 210

13. $f' \equiv 0$, $\mathscr{L}(f') = 0 = s\mathscr{L}(f) - 0 - (1 - 0)e^{-s} - (0 - 1)e^{-2s}$. Answer: $(e^{-s} - e^{-2s})/s$

15. $(1 - e^{-s})/s^2 - e^{-2s}/s$

17. $\frac{1}{2}e^{2t} - \frac{1}{2}$

19. $\cosh t - 1$

21. $2(1 - e^{-t}) - t$

23. $2e^{3t} - 9t^2 - 6t - 2$

25. $\frac{2}{3}\sin 3t$

27. $y = -e^{-t} + 2e^{3t}$

29. $y = 2e^{2t} - e^{-4t}$

SECTION 5.3, page 218

1. $n\pi/[(s + 2)^2 + n^2\pi^2]$

3. $[A(s + \alpha) + B\beta]/[(s + \alpha)^2 + \beta^2]$

9. $e^{-t}(\cos t + 2\sin t)$

11. $e^{2t}(\frac{1}{2}t^2 + \frac{1}{24}t^4)$

13. $y = e^{-t}(2\cos 2t - \sin 2t)$

15. $y = e^{-t}\sin 3t$

17. $k(e^{-as} - e^{-bs})/s$

19. $1/s(1 + e^{-s})$

21. $1/s(1 - e^{-s})$

23. $(e^{-s} - e^{-2s})/s^2$

25. $q(t) = CV_0 e^{-t/RC}$ $(t \geqq 0)$

27. $f(t) = -1$ if $1 < t < 3$

29. $f(t) = t - 1,\ 2t - 3,\ 6 - t$ when $1 < t < 2$, $2 < t < 3$, $3 < t < 6$, and 0 otherwise

31. e^{-s}/s^2

33. $-se^{-\pi s}/(s^2 + 1)$

35. $e^{a(k-s)}/(s - k)$

37. $K\omega(1 + e^{-\pi s/\omega})/(s^2 + \omega^2)$

39. $s^{-1} - (s + 1)^{-1} - e^{-\pi s}(s^{-1} - e^{-\pi}(s + 1)^{-1})$

41. $2s^{-3} - e^{-s}(2s^{-3} + 2s^{-2} + s^{-1})$

43. $u_2(t)$

45. $e^{2(t-2)}u_2(t)$

47. $e^{3(t-1)}u_1(t)$

49. $-e^{\pi - t}\sin t$ if $t > \pi$ and 0 otherwise

51. $i' + \int_0^t i(\tau)\,d\tau = t[1 - u_a(t)] = t - (t - a)u_a(t) - au_a(t)$. *Ans.*

$$i = 1 - \cos t \qquad \text{when } 0 < t < a,$$
$$i = \cos(t - a) - a\sin(t - a) - \cos t \qquad \text{when } t > a.$$

53. $i = \frac{1}{2}(e^{-t} - \cos t + \sin t)$ when $0 < t < \pi$,
$i = -\frac{1}{2}(1 + e^{-\pi})\cos t + \frac{1}{2}(3 - e^{-\pi})\sin t$ when $t > \pi$

55. Define $\tilde{f}$ by $\tilde{f}(t) = f(t + a)$. Then $f(t) = \tilde{f}(t - a)$ and by (2),

$$\mathscr{L}\{f(t)u_a(t)\} = \mathscr{L}\{\tilde{f}(t - a)u_a(t)\} = e^{-as}\mathscr{L}\{\tilde{f}(t)\} = e^{-as}\mathscr{L}\{f(t + a)\}.$$

SECTION 5.4, page 222

1. $\dfrac{1}{(s - 1)^2}$

3. $\dfrac{6s}{(s^2 + 9)^2}$

5. $\dfrac{2s(s^2 - 3)}{(s^2 + 1)^3}$

7. $\dfrac{s^2 + 2s + 5}{(s^2 + 2s - 3)^2}$

9. $\dfrac{2(s + 2)}{(s^2 + 4s + 5)^2}$

11. $t\sin t$

13. $e^{-t}(\sin t)/t$

15. $(e^t - 1)/t$

17. $(e^{-bt} - e^{-at})/t$

SECTION 5.5, page 226

1. t

3. te^t

5. $(e^{at} - 1)/a^2 - t/a$

7. $\frac{2}{3}\sin t - \frac{1}{3}\sin 2t$

9. $\dfrac{t}{2}\sin t$

11. $e^{at}t^2/2$

13. $(1 - \cos\omega t)/\omega^2$

15. $\frac{1}{2}(\sin t - t\cos t)$

17. $\frac{1}{2}(\sin t + t\cos t)$

19. $t - \tau = \sigma$ gives $\tau = t - \sigma$, $d\tau = -d\sigma$ and

$$\int_0^t f(\tau)g(t - \tau)\,d\tau = \int_t^0 g(\sigma)f(t - \sigma)(-d\sigma).$$

23. $1/s = \mathscr{L}(1)$, $(f * 1)(t) = \int_0^t f(\tau)\,d\tau$.

25. $y = \frac{3}{8}\sin t - \frac{1}{8}\sin 3t$

27. $y = \frac{1}{2}(1 - \cos\sqrt{2}t)$ if $0 < t < 1$ and $\frac{1}{2}(\cos\sqrt{2}(t-1) - \cos\sqrt{2}t)$ if $t > 1$

29. $y = \sqrt{2}\sin\sqrt{2}t$

SECTION 5.6, page 236

3. $(e^{at} - e^{bt})/(a - b)$

5. $4\cos 4t + \sin 4t$

7. $(1 - t)e^{-t}$

9. $2e^{2t} + te^{t}$

11. $te^{2t}(\cos t - \sin t)$

13. $e^{t}(\cos t + t\sin t)$

19. $y = 2e^{2t} + e^{-t}$

21. $y = 5e^{2t} - 2t^2$

23. $y = (2t + 3)e^{-t} + \sin t$

25. $y_1 = \cos t,\ y_2 = \sin t$

27. $y_1 = e^{t} + e^{2t},\ y_2 = e^{2t}$

29. $y_1 = e^{t},\ y_2 = e^{-t},\ y_3 = e^{t} - e^{-t}$

SECTION 5.7, page 242

3. $\dfrac{\omega}{(s^2 + \omega^2)(e^{\pi s/\omega} - 1)}$

5. $\dfrac{\omega}{s^2 + \omega^2}\coth\dfrac{\pi s}{2\omega}$

7. $\dfrac{1}{s^2 + \omega^2}\left(s + \dfrac{\omega}{\sinh \pi s/2\omega}\right)$

9. $\dfrac{e^{2(1-s)\pi} - 1}{(1 - s)(1 - e^{-2\pi s})}$

11. Set $\omega = \frac{1}{2}$ in Prob. 5 and multiply by K.

13. $\left[\dfrac{\pi}{s}e^{-\pi s}(e^{-\pi s} - 1) + \dfrac{1}{s^2}(e^{-\pi s} - 1)^2\right]/(1 - e^{-2\pi s})$

15. $\left[\dfrac{1}{s^2}(1 - e^{-\pi s}) - \dfrac{\pi}{s}e^{-\pi s}\right]/(1 - e^{-2\pi s})$

17. $Ri + \dfrac{1}{C}\displaystyle\int i\,dt = v(t), \quad RI + \dfrac{I}{Cs} = \dfrac{1 - e^{-s}}{s^2} - \dfrac{e^{-s}}{s},$

$i = C(1 - e^{-t/RC})$ when $0 < t < 1$,

$i = ((C - R^{-1})e^{1/RC} - C)e^{-t/RC}$ when $t > 1$

19. $$i(t) = \begin{cases} a\cos\omega t + b\sin\omega t - ae^{-t/RC} & \text{when } 0 < t < \pi/\omega \\ -a(1 + e^{\pi/\omega RC})e^{-t/RC} & \text{when } t > \pi/\omega \end{cases}$$

where $a = \omega CK,\ b = \omega^2 RC^2 K,\ K = 1/[1 + (\omega RC)^2]$

23. $$i(t) = \begin{cases} \dfrac{V_0}{\omega^* L}e^{-\alpha t}\sin\omega^* t & \text{when } 0 < t < a \\ \dfrac{V_0}{\omega^* L}[e^{-\alpha t}\sin\omega^* t - e^{-\alpha(t-a)}\sin\{\omega^*(t - a)\}] & \text{when } t > a \end{cases}$$

where $\alpha = \dfrac{R}{2L},\ \omega^{*2} = \dfrac{1}{LC} - \alpha^2$

25. Initial current 2 amperes, initial charge 0, $i = e^{-t/5}(2\cos\frac{2}{5}t + \sin\frac{2}{5}t)$

33. $y = \dfrac{P}{K}[1 - (1 + \alpha t)e^{-\alpha t}]$

35. Instead of the term $-P$ in Problem 34 we now have $-P[1 - u_1(t)]$. For $0 < t < 1$ the solution is as in Problem 34. Denoting this solution by v_1, the solution v for $t > 1$ is

$$v = v_1 + \frac{P(t-1)}{M_1 + M_2} - \frac{P}{k(M_1 + M_2)} \sin [k(t-1)]$$
$$= \omega - \frac{P}{M_1 + M_2} + \frac{P}{k(M_1 + M_2)} [\sin kt - \sin (k(t-1))].$$

SECTION 6.2, page 254

1. 2, 0, 1, $|\mathbf{v}| = \sqrt{5}$ **3.** 1, 1, 1, $|\mathbf{v}| = \sqrt{3}$ **5.** 2, 3, 1, $|\mathbf{v}| = \sqrt{14}$
7. 0, −1, 0, $|\mathbf{v}| = 1$ **9.** 0, 0, 0, $|\mathbf{v}| = 0$ **11.** Q: (7, 4, 5)
13. Q: $(-\frac{3}{2}, \frac{5}{2}, -1)$ **15.** Q: (0, 0, 0) **17.** Q: (3, 8, −9)
19. Q: (−3, −2, −2)

SECTION 6.3, page 257

1. $-2\mathbf{i} - 4\mathbf{j} + 6\mathbf{k}$, $\frac{1}{2}\mathbf{i} + \mathbf{j} - \frac{3}{2}\mathbf{k}$, $3\mathbf{i} + 6\mathbf{j} - 9\mathbf{k}$
3. $-\mathbf{i} + \mathbf{j} - 7\mathbf{k}$, $\mathbf{i} - \mathbf{j} + 7\mathbf{k}$, $2\mathbf{i} + 6\mathbf{j} + 4\mathbf{k}$
5. $\sqrt{59}$, $\sqrt{14} - 5$ **7.** $-3\mathbf{i} + 3\mathbf{j} - 21\mathbf{k}$
9. $3\mathbf{i} - 2\mathbf{j} + \mathbf{k}$ **11.** $6\mathbf{i} + 2\mathbf{j}$
13. $\mathbf{p} = -3\mathbf{j} + 2\mathbf{k}$
15. The vector $\mathbf{r}$ from 0 to P is of the form $\mathbf{r} = k(\mathbf{a} + \mathbf{b})$. Also, $\mathbf{r} = \mathbf{a} + l(\mathbf{b} - \mathbf{a})$. Thus, $k\mathbf{a} + k\mathbf{b} = (1 - l)\mathbf{a} + l\mathbf{b}$. From this, $k = 1 - l$, $k = l$; hence $k = \frac{1}{2}$, $l = \frac{1}{2}$, and the proof is complete.

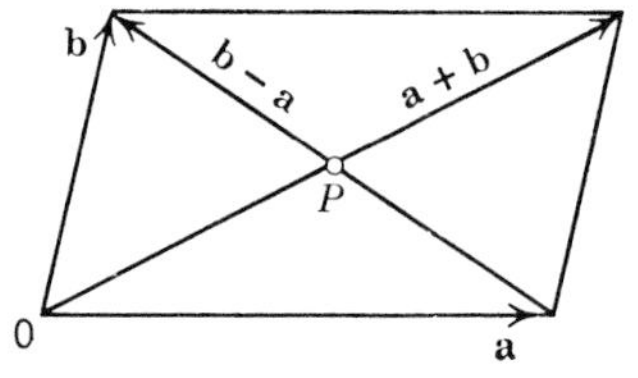

Problem 15

SECTION 6.4, page 262

1. 2; **j**, **k** **3.** 2; **i**, **j** + **k**
5. n; (1, 0, · · ·, 0), (0, 1, 0, · · ·, 0), · · ·, (0, · · ·, 0, 1)
7. 2; cos x, sin x **9.** The xy-plane
11. Assuming $\mathbf{a} = c_1\mathbf{e}_{(1)} + \cdots + c_n\mathbf{e}_{(n)} = q_1\mathbf{e}_{(1)} + \cdots + q_n\mathbf{e}_{(n)}$, we obtain $(c_1 - q_1)\mathbf{e}_{(1)} + \cdots + (c_n - q_n)\mathbf{e}_{(n)} = \mathbf{0}$, $c_j - q_j = 0$ for all j because the $\mathbf{e}_{(j)}$ are linearly independent.

SECTION 6.5, page 267

1. 5, 5 **3.** −11, −11 **5.** $\sqrt{41}$, $\sqrt{43}$ **7.** $\sqrt{18}$, $\sqrt{14} + \sqrt{26}$
9. 27 **11.** 2, 0, −2 **13.** 3 **15.** 2
17. 0 **19.** $a_1 = -8/3$
21. $\mathbf{p}_{(1)} \cdot \mathbf{d} + \mathbf{p}_{(2)} \cdot \mathbf{d} = (\mathbf{p}_{(1)} + \mathbf{p}_{(2)}) \cdot \mathbf{d}$
23. If and only if all four sides are equal.

27. The angle β of the triangle at B is the angle between the vectors $\mathbf{c} = -2\mathbf{i}$ from B to A and $\mathbf{a} = -2\mathbf{i} + 2\mathbf{j}$ from B to C. Therefore $\cos\beta = \mathbf{a}\cdot\mathbf{c}/|\mathbf{a}|\,|\mathbf{c}| = 1/\sqrt{2}$, $\beta = \pi/4$, etc.

29. $-23/26$ **31.** $\sqrt{3}/2$ **33.** -5 **35.** 2

37. $|\mathbf{a} - \mathbf{b}|^2 = (\mathbf{a} - \mathbf{b})\cdot(\mathbf{a} - \mathbf{b}) = |\mathbf{a}|^2 + |\mathbf{b}|^2 - 2|\mathbf{a}|\,|\mathbf{b}|\cos\gamma$, etc.

39. $|\mathbf{a} + \mathbf{b}|^2 = (\mathbf{a} + \mathbf{b})\cdot(\mathbf{a} + \mathbf{b}) \leqq |\mathbf{a}|^2 + 2|\mathbf{a}|\,|\mathbf{b}| + |\mathbf{b}|^2 = (|\mathbf{a}| + |\mathbf{b}|)^2$

SECTION 6.8, page 276

1. $6\mathbf{i} - 3\mathbf{j},\ -6\mathbf{i} + 3\mathbf{j}$ **3.** $3\mathbf{i} + 3\mathbf{j} + 3\mathbf{k},\ \sqrt{27},\ 1$

5. $-12\mathbf{i} + 6\mathbf{j} - 6\mathbf{k},\ 12\mathbf{i} - 6\mathbf{j} + 6\mathbf{k}$ **7.** $21\mathbf{i} + 3\mathbf{j} + 9\mathbf{k}$

9. $3\mathbf{i} - 6\mathbf{j} - 3\mathbf{k},\ -3\mathbf{i} + 6\mathbf{j} + 3\mathbf{k}$

11. $6\mathbf{i} - 3\mathbf{j}$ **15.** 2 **17.** $\sqrt{86}$ **19.** 4

21. 11 **23.** $1/2$ **25.** $\sqrt{266}\,/\,2$ **27.** $\sqrt{18}$

29. $\sqrt{26}$ **31.** Orthogonal **33.** Orthogonal **35.** Parallel

37. $\mathbf{k},\ -\mathbf{k}$ **39.** $\pm(3\mathbf{j} + 2\mathbf{k})/\sqrt{13}$

41. $\pm\mathbf{i}$ **43.** $\pm(2\mathbf{i} - \mathbf{j} - \mathbf{k})/\sqrt{6}$

45. $\mathbf{n} = 2\mathbf{i} + 3\mathbf{j} + 4\mathbf{k}$ and $\mathbf{p} = \mathbf{i} - \mathbf{j} + \mathbf{k}$ are normal vectors to the planes, and $\mathbf{v} = \mathbf{n} \times \mathbf{p} = 7\mathbf{i} + 2\mathbf{j} - 5\mathbf{k}$.

47. $\mathbf{k}$ **49.** $\mathbf{0}$

SECTION 6.9, page 280

1. 1 **3.** -1 **5.** 3 **7.** -12

9. -3 **11.** Independent **13.** Independent **15.** Dependent

17. Independent **19.** $\lambda = 5/3$ **21.** $\lambda = 5$, $\mu = 10$ **23.** 3

25. 2 **27.** 20 **29.** $7/3$ **31.** $1/6$

33. $6\mathbf{i} + 6\mathbf{j} - 7\mathbf{k},\ 9$ **35.** $6\mathbf{i} + 3\mathbf{j} - 3\mathbf{k},\ 5\mathbf{i} + 5\mathbf{j}$

37. $17\mathbf{i} - 26\mathbf{j} - 27\mathbf{k}$

39. $\mathbf{a}\cdot[\mathbf{b}\times(\mathbf{c}\times\mathbf{d})]$ equals $(\mathbf{a}\ \ \mathbf{b}\ \ [\mathbf{c}\times\mathbf{d}]) = (\mathbf{a}\times\mathbf{b})\cdot(\mathbf{c}\times\mathbf{d})$ as well as $(\mathbf{a}\cdot\mathbf{c})(\mathbf{b}\cdot\mathbf{d}) - (\mathbf{a}\cdot\mathbf{d})(\mathbf{b}\cdot\mathbf{c})$.

SECTIONS 7.1–7.3, page 289

1. $\begin{pmatrix} 0 & 1 \\ 4 & 7 \end{pmatrix}$ **3.** $\begin{pmatrix} 12 & 4 \\ -2 & 4 \end{pmatrix}$ **5.** $\begin{pmatrix} 0 & 4 \\ 1 & 7 \end{pmatrix}$

7. No **9.** $\mathbf{A}$ yes, $\mathbf{B}$ no

11. $\begin{pmatrix} 4 & 1 & 4 \\ 7 & 1 & 0 \\ 6 & -5 & 1 \end{pmatrix}$ **13.** $\begin{pmatrix} 10 & 2 & 8 \\ 13 & 3 & 0 \\ 15 & -14 & 2 \end{pmatrix}$ **15.** $\begin{pmatrix} 0 & -9 & 0 \\ -1 & 1 & -3 \\ -4 & 0 & -1 \end{pmatrix}$

17. No, no, yes, yes **21.** 3

23. Cf. Theorem 1 in Sec. 7.2. For the symmetric matrices,

$$\begin{pmatrix} 1 & 0 \\ 0 & 0 \end{pmatrix}, \begin{pmatrix} 0 & 1 \\ 1 & 0 \end{pmatrix}, \begin{pmatrix} 0 & 0 \\ 0 & 1 \end{pmatrix}.$$

25. $n(n+1)/2$. The n matrices with a single 1 in the principal diagonal and all other elements 0, and the symmetric matrices with two 1's and all other elements 0.

27. $\begin{pmatrix} -1 & 1 & 0 & 0 & -1 & -1 \\ 1 & -1 & 0 & 0 & 0 & 0 \\ 0 & 0 & 1 & -1 & 1 & 0 \end{pmatrix}$

29. $\begin{pmatrix} -1 & 1 & 1 & 0 & 0 & -1 \\ 1 & -1 & 0 & 1 & 0 & 0 \\ 0 & 0 & -1 & 0 & 1 & 0 \\ 0 & 0 & 0 & -1 & -1 & 1 \end{pmatrix}$

SECTION 7.4, page 300

1. $\mathbf{A}^T\mathbf{B} = (4 \quad -2)$, $\mathbf{B}^T\mathbf{A} = \begin{pmatrix} 4 \\ -2 \end{pmatrix}$

3. $\mathbf{C}^2 = \begin{pmatrix} 1 & -3 & -3 \\ 12 & 4 & 3 \\ 0 & -4 & 1 \end{pmatrix}$

5. $\mathbf{B}\mathbf{B}^T = \begin{pmatrix} 4 & 0 & 2 \\ 0 & 9 & -3 \\ 2 & -3 & 2 \end{pmatrix}$, $\mathbf{B}^T\mathbf{B} = \begin{pmatrix} 5 & -1 \\ -1 & 10 \end{pmatrix}$

7. $\mathbf{C}^2 = \begin{pmatrix} 1 & -3 & -3 \\ 12 & 4 & 3 \\ 0 & -4 & 1 \end{pmatrix}$, $\mathbf{C}^4 = \begin{pmatrix} -35 & -3 & -15 \\ 60 & -32 & -21 \\ -48 & -20 & -11 \end{pmatrix}$,

$$\mathbf{C}^4 - 3\mathbf{C}^2 + 2\mathbf{I} = \begin{pmatrix} -36 & 6 & -6 \\ 24 & -42 & -30 \\ -48 & -8 & -12 \end{pmatrix}$$

9. $\mathbf{A}^T\mathbf{C}\mathbf{A} = (5)$

11. $\begin{pmatrix} 1 & 0 \\ 0 & 0 \end{pmatrix}$, $\begin{pmatrix} 1 & 1 \\ 0 & 0 \end{pmatrix}$, etc.

15. Use induction and the addition formulas for the sine and cosine functions.

17. $\begin{pmatrix} y_1 \\ y_2 \end{pmatrix} = \begin{pmatrix} 1 & 3 \\ 1 & -3 \end{pmatrix} \begin{pmatrix} w_1 \\ w_2 \end{pmatrix}$

SECTION 7.5, page 307

1. $x = 1, y = 2$

3. $x = 1/7, y = -1/7$

5. $x = z, y = -z$

7. $x = 2, y = -1, z = 0$

9. $x = -1, y = 2z$

11. $x = -2, y = 0, z = 4$

13. $x = y + 1, z = 1$

15. $x = 3y + 2, z = 0$

17. $y = w - 5x + 9, z = 0$

19. $w = 2x + 1, y = 1, z = 2$

21. $w = 0, x = 3z, y = 2z + 1$

23. $I_1 = 1, I_2 = 3, I_3 = 4$ (amperes)

25. $I_1 = \dfrac{R_4E_1 + R_5(E_1 + E_2)}{(R_1 + R_2 + R_3)(R_4 + R_5) + R_4R_5}$, etc.

29. $P = 4, D = S = 5$

SECTION 7.6, page 312

1. 1 **3.** 1 **5.** 2 **7.** 2

9. 3

13. The row vectors are collinear. The row vectors are coplanar.

15. Linearly independent **17.** Linearly dependent

19. Linearly dependent **21.** Linearly dependent

23. $k = 6$ **25.** k arbitrary

SECTION 7.8, page 321

3. Use (1), with $\mathbf{A}$ replaced by $\mathbf{C}$, and set $\mathbf{C} = \mathbf{A}^{-1}$.

5. $\begin{pmatrix} 2 & 1 \\ 5 & 3 \end{pmatrix}$

7. $\begin{pmatrix} 0.2 & -0.1 & 0 \\ 0 & -0.2 & 0.1 \\ 0.1 & 0 & 0.1 \end{pmatrix}$

9. $\begin{pmatrix} 0 & 1 & 0 \\ 1 & 0 & 0 \\ 0 & 0 & 1 \end{pmatrix}$

11. $\begin{pmatrix} 0.3 & 0 & 0.1 \\ 0 & 0.5 & 0 \\ -0.1 & 0.1 & -0.1 \end{pmatrix}$

13. $\begin{aligned} x &= x^* + 2y^* + 5z^* \\ y &= \quad - y^* + 2z^* \\ z &= 2x^* + 4y^* + 11z^* \end{aligned}$

15. $\begin{aligned} x &= 19x^* + 2y^* - 9z^* \\ y &= -4x^* - y^* + 2z^* \\ z &= -2x^* \quad + z^* \end{aligned}$

17. Use (7).

19. $\mathbf{I} = \mathbf{I}^{\mathsf{T}} = (\mathbf{A}^{-1}\mathbf{A})^{\mathsf{T}} = \mathbf{A}^{\mathsf{T}}(\mathbf{A}^{-1})^{\mathsf{T}}$

SECTION 7.9, page 327

1. 14 **3.** 1 **5.** 24

7. $1 + a^2 + b^2 + c^2$ **9.** 0

SECTION 7.10, page 338

1. 90

3. $4a^2b^2c^2$

SECTION 7.11, page 343

1. 2 **3.** 1 **5.** 2 **7.** $x = 1, y = 2$

9. $x = 2, y = 4$ **11.** $x = 2, y = 1, z = -3$

15. $\begin{pmatrix} -3/13 & 5/13 \\ 2/13 & 1/13 \end{pmatrix}$ **17.** $\begin{pmatrix} -2 & 1 \\ 1.5 & -0.5 \end{pmatrix}$

19. $\begin{pmatrix} 0 & 0 & 1/a \\ 0 & 1/b & 0 \\ 1/c & 0 & 0 \end{pmatrix}$

25. Prove and use $\sum_{k=1}^{n} a_{jk}A_{jk} = \det \mathbf{A}, \quad \sum_{k=1}^{n} a_{jk}A_{ik} = 0 \ (i \neq j)$.

SECTION 7.12, page 348

1. $\mathbf{A} = \mathbf{H} + \mathbf{S}$, where $\mathbf{H} = (\mathbf{A} + \bar{\mathbf{A}}^{\mathsf{T}})/2$ is Hermitian and $\mathbf{S} = (\mathbf{A} - \bar{\mathbf{A}}^{\mathsf{T}})/2$ is skew-Hermitian.

3. $\mathbf{C}^{\mathsf{T}}$ must equal $-\mathbf{C}$, which implies a, b real.

5. $\begin{pmatrix} 6 & -2 \\ -2 & 2 \end{pmatrix}$ **7.** $\begin{pmatrix} 5 & -1 \\ -1 & 1 \end{pmatrix}$ **9.** $\begin{pmatrix} 1 & -1 & 1 \\ -1 & 1 & -1 \\ 1 & -1 & 1 \end{pmatrix}$

15. 4

17. $-|x_1|^2 + 10 \operatorname{Re} \bar{x}_1 x_2 - 2 \operatorname{Im} \bar{x}_1 x_2 + 2|x_2|^2$

19. 2 **21.** 0 **23.** $2i|x_1|^2 + 8i \operatorname{Im} \bar{x}_1 x_2$

25. $S = \alpha + i\beta = \bar{\mathbf{x}}^{\mathsf{T}}\mathbf{A}\mathbf{x}$, $\bar{S} = \alpha - i\beta = \mathbf{x}^{\mathsf{T}}\overline{\mathbf{A}\mathbf{x}} = -\mathbf{x}^{\mathsf{T}}\mathbf{A}^{\mathsf{T}}\bar{\mathbf{x}} = -(\bar{\mathbf{x}}^{\mathsf{T}}\mathbf{A}\mathbf{x})^{\mathsf{T}} = -S = -\alpha - i\beta$.

SECTION 7.13, page 354

1. No **3.** 6, 4, $\begin{pmatrix} 2 \\ 1 \end{pmatrix}$, $\begin{pmatrix} 1 \\ 1 \end{pmatrix}$ **5.** 0, any nonzero vector

7. 10, -10, $\begin{pmatrix} 2 \\ 1 \end{pmatrix}$, $\begin{pmatrix} 1 \\ -2 \end{pmatrix}$ **9.** 2.5, -2.5, $\begin{pmatrix} 2 \\ 1 \end{pmatrix}$, $\begin{pmatrix} 1 \\ -2 \end{pmatrix}$

11. a, $\begin{pmatrix} 1 \\ 0 \\ 0 \end{pmatrix}$, b, $\begin{pmatrix} 0 \\ 1 \\ 0 \end{pmatrix}$, c, $\begin{pmatrix} 0 \\ 0 \\ 1 \end{pmatrix}$ **13.** 1, $\begin{pmatrix} 0 \\ 0 \\ 1 \end{pmatrix}$, $\begin{pmatrix} 1 \\ 1 \\ 0 \end{pmatrix}$, -1, $\begin{pmatrix} 1 \\ -1 \\ 0 \end{pmatrix}$

15. 30, 25, 20

19. This follows from the fact that the zeros of a polynomial with real coefficients are real or complex conjugates in pairs.

25. From $\mathbf{A}\mathbf{x}_j = \lambda_j \mathbf{x}_j$ ($\mathbf{x}_j \neq \mathbf{0}$) and Probs. 23 and 24, $k_n\mathbf{A}^n\mathbf{x}_j = k_n\lambda_j{}^n\mathbf{x}_j$ and $k_p\mathbf{A}^p\mathbf{x}_j = k_p\lambda_j{}^p\mathbf{x}_j$ ($n \geqq 0, p \geqq 0$, integer). Adding on both sides we see that $k_n\mathbf{A}^n + k_p\mathbf{A}^p$ has the eigenvalue $k_n\lambda_j{}^n + k_p\lambda_j{}^p$. From this the statement follows.

SECTION 7.14, page 358

5. Rotation about the x_3-axis in space

7. $\lambda = \pm 1, \begin{pmatrix} 1 \\ i - i\sqrt{2} \end{pmatrix}, \begin{pmatrix} 1 \\ i + i\sqrt{2} \end{pmatrix}$

9. Let $\mathbf{A}$ be unitary. Set $\mathbf{A}^{-1} = \mathbf{B}$. Then $\mathbf{B}^{\mathsf{T}} = (\mathbf{A}^{-1})^{\mathsf{T}} = (\mathbf{A}^{\mathsf{T}})^{-1} = (\bar{\mathbf{A}}^{-1})^{-1} = \bar{\mathbf{B}}^{-1}$.

11. Let $\mathbf{Ax} = \lambda\mathbf{x}$ $(\mathbf{x} \neq \mathbf{0})$, $\mathbf{Ay} = \mu\mathbf{y}$ $(\mathbf{y} \neq \mathbf{0})$. Then $(\mathbf{Ax})^{\mathsf{T}} = \mathbf{x}^{\mathsf{T}}\mathbf{A}^{\mathsf{T}} = \mathbf{x}^{\mathsf{T}}\mathbf{A} = \lambda\mathbf{x}^{\mathsf{T}}$. Thus $\lambda\mathbf{x}^{\mathsf{T}}\mathbf{y} = \mathbf{x}^{\mathsf{T}}\mathbf{Ay} = \mathbf{x}^{\mathsf{T}}\mu\mathbf{y} = \mu\mathbf{x}^{\mathsf{T}}\mathbf{y}$. Hence, if $\lambda \neq \mu$, $\mathbf{x}^{\mathsf{T}}\mathbf{y} = 0$, which proves orthogonality.

17. trace $\widetilde{\mathbf{A}}$ = trace$[\mathbf{T}^{-1}(\mathbf{AT})]$ = trace$[(\mathbf{AT})\mathbf{T}^{-1}]$ = trace$(\mathbf{ATT}^{-1})$ = trace $\mathbf{A}$, because trace $\mathbf{AB}$ = trace $\mathbf{BA}$.

19. $\mathbf{Ax} = \lambda\mathbf{x}$ $(\mathbf{x} \neq \mathbf{0})$. Thus (*a*) $\mathbf{T}^{-1}\mathbf{Ax} = \lambda\mathbf{T}^{-1}\mathbf{x} = \lambda\mathbf{y}$. Now we have $\mathbf{T}^{-1}\mathbf{Ax} = \mathbf{T}^{-1}\mathbf{ATy} = \widetilde{\mathbf{A}}\mathbf{y}$, and (*a*) becomes $\widetilde{\mathbf{A}}\mathbf{y} = \lambda\mathbf{y}$.

23. $|a| = |b| = |c|$

25. (97 000 103 000), (94 210 105 790)

SECTION 7.15, page 365

1. Eigenvalues 3 and 2. Eigenvectors $\mathbf{x}_1 = \begin{pmatrix} 2 \\ 1 \end{pmatrix}$, $\mathbf{x}_2 = \begin{pmatrix} 1 \\ 1 \end{pmatrix}$. Answer: $\mathbf{y} = \mathbf{x}_1 e^{3t} + \mathbf{x}_2 e^{2t}$.

3. $y_1 = e^{3t}$, $y_2 = e^{3t}$

5. $y_1 = 2e^{3t}$, $y_2 = -10e^{3t}$, $y_3 = -27e^{3t}$

9. $y_1 = -2\cos\sqrt{18}t$, $y_2 = \cos\sqrt{18}t$

13. $\lambda_1 = -k$, $\lambda_2 = -3k$, $\mathbf{x}_1 = \begin{pmatrix} 1 \\ 1 \end{pmatrix}$, $\mathbf{x}_2 = \begin{pmatrix} 1 \\ -1 \end{pmatrix}$,

$$\mathbf{y} = \mathbf{x}_1 \cos\sqrt{k}\,t + \mathbf{x}_2 \sin\sqrt{3k}\,t$$

SECTION 8.1, page 369

11. $0 < y < \sqrt{3}|x|$ and $y < -\sqrt{3}|x|$

13. $T = 3y - 4y^3$

15. Parallel planes

17. Concentric spheres

19. Paraboloids of revolution

21. Ellipsoids of revolution

23. Ellipses $3x^2 + y^2 = const$

31. $x^2 + 4y^2 = const$, $x/y = const$

33. $y^2 + x^4 = const$, $y = cx^2$

35. Ellipsoids

SECTION 8.2, page 373

1. $\mathbf{u}' = \mathbf{b}$, $\mathbf{u}'' = \mathbf{0}$

3. $\mathbf{u}' = -\sin t\,\mathbf{i} + \cos t\,\mathbf{j}$, $\mathbf{u}'' = -\mathbf{u}$, $|\mathbf{u}'| = 1$

5. $\mathbf{u}' = \mathbf{i} + 2t\mathbf{j} + 3t^2\mathbf{k}$, $\mathbf{u}'' = 2\mathbf{j} + 6t\mathbf{k}$

7. $e^t\mathbf{i} - e^{-t}\mathbf{j}$, $\sqrt{e^{2t} + e^{-2t}}$, $e^t\mathbf{i} + e^{-t}\mathbf{j}$, $\sqrt{e^{2t} + e^{-2t}}$

9. $3\cos 3t\,(\mathbf{i} + \mathbf{j})$, $3\sqrt{2}\,|\cos 3t|$, $-9\sin 3t\,(\mathbf{i} + \mathbf{j})$, $9\sqrt{2}\,|\sin 3t|$

11. $-10t^4\mathbf{i} - 2t\mathbf{j} + 4t^3\mathbf{k}$

13. $-(4t^3 + 5t^4)\mathbf{i} + (4t^3 + 14t^6)\mathbf{j} - (12t^5 - 2t)\mathbf{k}$

15. $1 + 3t^2 + 12t^3$

17. $2x\mathbf{i} + 2y\mathbf{j}$, $-2y\mathbf{i} + 2x\mathbf{j}$

19. $z\mathbf{j} + y\mathbf{k},\ z\mathbf{i} + x\mathbf{k},\ y\mathbf{i} + x\mathbf{j}$

21. $2xy\mathbf{i} + z^2\mathbf{k},\ x^2\mathbf{i} + 2yz\mathbf{j},\ y^2\mathbf{j} + 2zx\mathbf{k}$

25. $(\mathbf{u} \cdot \mathbf{v})'' = \mathbf{u}'' \cdot \mathbf{v} + 2\mathbf{u}' \cdot \mathbf{v}' + \mathbf{u} \cdot \mathbf{v}''$,
$(\mathbf{u} \times \mathbf{v})'' = \mathbf{u}'' \times \mathbf{v} + 2\mathbf{u}' \times \mathbf{v}' + \mathbf{u} \times \mathbf{v}''$

SECTION 8.3, page 376

1. $\mathbf{r}(t) = t\mathbf{i} + t\mathbf{j}$

3. $\mathbf{r}(t) = 2\mathbf{i} + (1 + 2t)\mathbf{j} + t\mathbf{k}$

5. $\mathbf{r}(t) = t\mathbf{i} + t\mathbf{j} + t\mathbf{k}$

7. $\mathbf{r}(t) = (1 - t)\mathbf{i} + (5 - 3t)\mathbf{j} + (3 - 4t)\mathbf{k}$

9. $\mathbf{r}(t) = t\mathbf{i} + t\mathbf{j}$

11. $\mathbf{r}(t) = t\mathbf{i} - t\mathbf{j} + t\mathbf{k}$

13. $\mathbf{r}(t) = \cos t\,\mathbf{i} + \sin t\,\mathbf{j}$

15. $\mathbf{r}(t) = t\mathbf{i} + t^2\mathbf{j} + t^3\mathbf{k}$

17. $\mathbf{r}(t) = (-1 + \cos t)\mathbf{i} + 2 \sin t\,\mathbf{j}$

19. Circle in the xy-plane, sine curve in the yz-plane, cosine curve in the xz-plane

SECTION 8.4, page 378

1. $\sinh 1$

3. $8(\sqrt{1000} - 1)/27$

5. $\pi^2/2$

7. Start from $\mathbf{r}(t) = t\mathbf{i} + f(t)\mathbf{j}$.

11. $\sqrt{2}(e^{\pi} - 1)$

13. $8a$

SECTION 8.5, page 382

1. $\mathbf{q}(w) = (-1/\sqrt{2} + w)\mathbf{i} + (1/\sqrt{2} + w)\mathbf{j}$

3. $\mathbf{q}(w) = \mathbf{i} + w\mathbf{j} + (4\pi + 2w)\mathbf{k}$

5. $\cos \alpha = \mathbf{u} \cdot \mathbf{k} = c/\sqrt{a^2 + c^2} = const$

15. $\mathbf{r}' = \dot{\mathbf{r}}/\dot{s},\ \mathbf{r}'' = \ddot{\mathbf{r}}/\dot{s}^2 + \cdots,\ \mathbf{r}''' = \dddot{\mathbf{r}}/\dot{s}^3 + \cdots,$

$$\tau = (\mathbf{r}'\ \mathbf{r}''\ \mathbf{r}''')/\kappa^2 = \frac{(\dot{\mathbf{r}} \cdot \dot{\mathbf{r}})^3}{\dot{s}^6} \frac{(\dot{\mathbf{r}}\quad \ddot{\mathbf{r}}\quad \dddot{\mathbf{r}})}{(\dot{\mathbf{r}} \cdot \dot{\mathbf{r}})(\ddot{\mathbf{r}} \cdot \ddot{\mathbf{r}}) - (\dot{\mathbf{r}} \cdot \ddot{\mathbf{r}})^2}$$

where the dots denote terms which disappear later by applying familiar rules for simplifying determinants.

SECTION 8.6, page 386

1. $\mathbf{v} = \mathbf{i},\ |\mathbf{v}| = 1,\ \mathbf{a} = \mathbf{0}$

3. $\mathbf{v} = (2 - 2t)\mathbf{i},\ |\mathbf{v}| = |2 - 2t|,\ \mathbf{a} = -2\mathbf{i}$

5. Segment $-1 \leqq y \leqq 1$ on the y-axis, $\cos t\,\mathbf{j},\ |\cos t|,\ -\sin t\,\mathbf{j}$

7. Circle, $-2t \sin t^2\,\mathbf{i} + 2t \cos t^2\,\mathbf{j},\ 2t,$
$(-2 \sin t^2 - 4t^2 \cos t^2)\mathbf{i} + (2 \cos t^2 - 4t^2 \sin t^2)\mathbf{j}$

9. Circular helix, $\mathbf{v} = -\sin t\,\mathbf{i} + \cos t\,\mathbf{j} + \mathbf{k},\ |\mathbf{v}| = \sqrt{2},\ \mathbf{a} = -\cos t\,\mathbf{i} - \sin t\,\mathbf{j}$

11. $|\mathbf{a}| = \omega^2 R = (2\pi/2.36 \cdot 10^6)^2 \cdot 3.85 \cdot 10^8 = 0.0027$ (meter/sec^2), which is only $2.8 \cdot 10^{-4}\,g$ where g is the acceleration due to gravity.

13. $\mathbf{r}(t) = \mathbf{a}_0 t^2/2 + \mathbf{v}_0 t + \mathbf{r}_0$ ($\mathbf{a}_0$, $\mathbf{v}_0$, $\mathbf{r}_0$ constant vectors)

15. The path remains the same, but the way of moving along the path changes.

SECTION 8.7, page 390

1. $2t + 1/t$ **3.** $(g'h - gh')/h^2$ **7.** $-t(1 + t^2)^{-3/2}$

9. 2, 0 **11.** $e^{2u} \sin 2v$, $e^{2u} \cos 2v$

15. $r = \sqrt{x^2 + y^2}$, $\theta = \arc\tan(y/x)$, $r_x = x/r$, $\theta_x = -y/r^2$, $r_{xx} = y^2/r^3$, etc., $w_{xx} = x^2r^{-2}w_{rr} - 2xyr^{-3}w_{r\theta} + y^2r^{-4}w_{\theta\theta} + y^2r^{-3}w_r + 2xyr^{-4}w_\theta$, etc.

SECTION 8.8, page 396

1. $2\mathbf{i} - \mathbf{j}$ **3.** $\cos x \cosh y\,\mathbf{i} + \sin x \sinh y\,\mathbf{j}$

5. $2(x\mathbf{i} + y\mathbf{j})/(x^2 + y^2)$ **7.** $yz\mathbf{i} + xz\mathbf{j} + xy\mathbf{k}$ **9.** $2x\mathbf{i} + 2y\mathbf{j} + 2z\mathbf{k}$

11. $(yz\mathbf{i} + xz\mathbf{j} + xy\mathbf{k})e^{xyz}$ **13.** $2\mathbf{i} - 3\mathbf{j}$ **15.** $y\mathbf{i} + x\mathbf{j}$

17. $18x\mathbf{i} + 2y\mathbf{j}$ **19.** $\mathbf{i} - \mathbf{j}$ **21.** $3\mathbf{i} - \mathbf{j}$ **23.** $48\mathbf{i} + 16\mathbf{j}$

25. $\mathbf{i} + \mathbf{j} + \mathbf{k}$ **27.** $4\mathbf{i} + 4\mathbf{j} - \mathbf{k}$ **29.** $4\mathbf{i} + 4\mathbf{k}$ **31.** $x - y + z$

33. $\frac{1}{2}x^2 + y^2 + \frac{1}{2}z^2$ **35.** $\frac{1}{2}\ln(x^2 + y^2)$

37. $4, 4\sqrt{2}, 4, 0, -4, -4\sqrt{2}, -4, 0$ **39.** 3

41. 0 **43.** $8/\sqrt{5}$

SECTION 8.9, page 401

3. $c_{11} = c_{22} = c_{33} = 1$, $b_1 = -1$, $b_2 = 5$, $b_3 = 2$, all others zero

5. $c_{12} = c_{21} = c_{33} = 1$, all others zero

7. $c_{12} = c_{23} = c_{31} = 1$, all others zero

SECTION 8.10, page 405

1. 2 **3.** $2x$ **5.** $yz + zx + xy$ **7.** $-x^2$

9. No, because (1) involves coordinates.

15. 0 **19.** 272

SECTION 8.11, page 407

1. $-2\mathbf{k}$ **3.** $2y\mathbf{i} + 2z\mathbf{j} + 2x\mathbf{k}$ **5.** $y\mathbf{i} + x\mathbf{j}$

7. curl $\mathbf{v} = \mathbf{0}$, div $\mathbf{v} = 2$, compressible, $\mathbf{r} = c_1e^t\mathbf{i} + c_2e^t\mathbf{j} + c_3\mathbf{k}$

9. curl $\mathbf{v} = -3y^2\mathbf{k}$, div $\mathbf{v} = 0$, incompressible, $\mathbf{r} = (c_2{}^3t + c_1)\mathbf{i} + c_2\mathbf{j} + c_3\mathbf{k}$

17. $\mathbf{0}$, $z(x - y)\mathbf{i} + x(y - z)\mathbf{j} + y(z - x)\mathbf{k}$

19. Let a_1, a_2, a_3 and $a_1{}^*, a_2{}^*, a_3{}^*$ be the components of curl $\mathbf{v}$ with respect to right-handed Cartesian coordinates x_1, x_2, x_3 and $x_1{}^*, x_2{}^*, x_3{}^*$, respectively. Using (5′), Sec. 8.9, the chain rule for functions of several variables, and (9), Sec. 8.9, we find

$$a_1 = \frac{\partial v_3}{\partial x_2} - \frac{\partial v_2}{\partial x_3} = \sum_{m=1}^{3}\left(c_{m3}\frac{\partial v_m{}^*}{\partial x_2} - c_{m2}\frac{\partial v_m{}^*}{\partial x_3}\right)$$

$$= \sum_{m=1}^{3}\sum_{l=1}^{3}\left(c_{m3}\frac{\partial v_m{}^*}{\partial x_l{}^*}\frac{\partial x_l{}^*}{\partial x_2} - c_{m2}\frac{\partial v_m{}^*}{\partial x_l{}^*}\frac{\partial x_l{}^*}{\partial x_3}\right)$$

$$= \sum_{m=1}^{3}\sum_{l=1}^{3}(c_{m3}c_{l2} - c_{m2}c_{l3})\frac{\partial v_m{}^*}{\partial x_l{}^*}$$

$$= (c_{33}c_{22} - c_{32}c_{23})\left(\frac{\partial v_3^*}{\partial x_2^*} - \frac{\partial v_2^*}{\partial x_3^*}\right) + \cdots$$
$$= (c_{33}c_{22} - c_{32}c_{23})a_1^* + (c_{13}c_{32} - c_{12}c_{33})a_2^* + (c_{23}c_{12} - c_{22}c_{13})a_3^*.$$

Using (3) in Sec. 8.9, Lagrange's identity (Sec. 6.9), and $\mathbf{k}^* \times \mathbf{j}^* = -\mathbf{i}^*$, $\mathbf{k} \times \mathbf{j} = -\mathbf{i}$, we obtain

$$c_{33}c_{22} - c_{32}c_{23} = (\mathbf{k}^* \times \mathbf{j}^*) \cdot (\mathbf{k} \times \mathbf{j}) = \mathbf{i}^* \cdot \mathbf{i} = c_{11}, \text{ etc.}$$

Hence $a_1 = c_{11}a_1^* + c_{21}a_2^* + c_{31}a_3^*$, which is of the form (5′), Sec. 8.9. In the same way we may obtain corresponding formulas for a_2 and a_3. If the $x_1x_2x_3$-coordinates are left-handed, then $\mathbf{k} \times \mathbf{j} = +\mathbf{i}$, but then there is a minus sign in front of the determinant in (1). This proves the theorem.

SECTION 9.2, page 415

1. $\frac{5}{3}\sqrt{5}$ **3.** 4π **5.** $\frac{5}{3}$ **7.** $\ln 2$

9. $-\frac{48}{5}$ **11.** 0 **13.** $\frac{7}{5}$ **15.** 8

17. $-\frac{25}{2}$ **19.** $|W| \leqq 2$, $W = \frac{5}{6}$

SECTION 9.3, page 422

1. $\frac{2}{3}$ **3.** $\frac{1}{8}$ **5.** $\frac{67}{120}$

7. $\frac{1}{6}$ **9.** $\frac{16}{3}$ **11.** $\bar{x} = 1, \bar{y} = 2$

13. $\frac{8}{3}, \frac{4}{3}$

15. $I_x = bh^3/12$, $I_y = b^3h/4$

17. $I_x = (a + b)h^3/24$, $I_y = h(a^4 - b^4)/48(a - b)$

19. 18 **21.** $32\pi/3$ **23.** 1

25. 1

SECTION 9.4, page 427

1. 0 **3.** -1 **5.** $1/3$ **7.** 4

9. $\pi(\cosh 1 - 1)$ **11.** 0 **13.** πab

15. $9/2$ **17.** 0

19. Set $f = -ww_y$ and $g = ww_x$ in Green's theorem, where subscripts denote partial derivatives. Then $g_x - f_y = w_x^2 + w_y^2$ because $\nabla^2 w = 0$, and

$$f\,dx + g\,dy = (-ww_yx' + ww_xy')\,ds = w \text{ grad } w \cdot (y'\mathbf{i} - x'\mathbf{j})\,ds$$
$$= w(\text{grad } w) \cdot \mathbf{n}\,ds = w\frac{\partial w}{\partial n}ds$$

where primes denote derivatives with respect to s.

21. $-(e^2 - 1)/2$ **23.** 2π **25.** 1

SECTION 9.5, page 431

1. xy-plane, parallel straight lines

3. Cylinder of revolution $x^2 + y^2 = 1$, straight lines, circles

5. Elliptic cylinder $x^2 + 16y^2 = 16$, straight lines, ellipses

7. Cone of revolution $z = \sqrt{x^2 + y^2}$, circles, straight lines
9. $\mathbf{r} = u\mathbf{i} + v\mathbf{k}$
11. $\mathbf{r} = u\mathbf{i} + v\mathbf{j} + (1 - u - v)\mathbf{k}$
13. $\mathbf{r} = u\mathbf{i} + v\mathbf{j} + u^2\mathbf{k}$
15. $x^2/a^2 + y^2/b^2 + z^2/c^2 - 1 = 0$
17. $x^2/a^2 - y^2/b^2 - z = 0$
19. $(v\mathbf{i} + u\mathbf{j} - \mathbf{k})/\sqrt{1 + u^2 + v^2}$

SECTION 9.6, page 437

3. $(\mathbf{r}^* - \mathbf{r}) \cdot \operatorname{grad} f = 0$, $\mathbf{n} = \operatorname{grad} f/|\operatorname{grad} f|$
5. $x^* + y^* - z^* = 1$
7. $4x^* - z^* = 4$
9. $4x^* + 2y^* - z^* = 5$
11. $du^2 + dv^2$
13. $a^2 \cos^2 v\, du^2 + a^2\, dv^2$
15. $(1 + v^2)\, du^2 + 2uv\, du\, dv + (1 + u^2)\, dv^2$
17. $\mathbf{r}_u$ and $\mathbf{r}_v$ are tangent to those curves. Apply Theorem 1 in Sec. 6.5.
19. 2π
21. $\pi(\sqrt{125} - 1)/6$
25. $\pi\sqrt{2}$

SECTION 9.7, page 443

1. 6π
3. $(3^{5/2} - 2^{7/2} + 1)/15$
5. $(10^{3/2} - 1)/9$
7. 0
9. $(5^{3/2} - 1)/2$
13. $2\pi h$
15. $2\pi^2 ab(2a^2 + 3b^2)$
17. $2\pi^2 ab(4a^2 + 4ab + 5b^2)$

SECTION 9.9, page 452

1. 1
3. $\pi/2$
5. $\frac{1}{8}$
7. $\frac{1}{120}$
9. $2c^5/3$
11. $\pi h^5/10$
13. Take $\mathbf{u} = x\mathbf{i}, y\mathbf{j}, z\mathbf{k}$, respectively, in (3), Sec. 9.8.
19. 3
21. 0
23. 12
25. Put $f = 1$ in (9).
27. Use Prob. 26.

SECTION 9.11, page 459

1. ± 1
3. $\pm 3\pi/2$
5. Use a parametric representation of C.
7. $-18\pi\sqrt{2}$
9. $\frac{1}{3}$
11. 0
13. 0
15. 0

SECTION 9.12, page 467

1. Yes
3. No
5. Yes
7. Yes
9. $u = (x^2 - y^2 - z^2)/2$
11. $u = x + yz$
13. $u = \sin x - y^2 z$
15. $\frac{1}{2}$
17. 5
19. $e^2 - 5$

SECTION 10.1, page 470

1. $2\pi, 2\pi, \pi, \pi, 2, 2, 1, 1$
17. $\pi/2$ $(n = 0)$, 0 (n even), $1/n$ $(n = 1, 5, 9, \cdots)$, $-1/n$ $(n = 3, 7, 11, \cdots)$
19. 0

21. 0 $(n = 0)$, $2\pi/n$ $(n = 1, 3, \cdots)$, $-2\pi/n$ $(n = 2, 4, \cdots)$

23. $\pi^3/3$ $(n = 0)$, $(-1)^n 2\pi/n^2$ $(n = 1, 2, \cdots)$

25. $n[(-1)^n e^{-\pi} - 1]/(1 + n^2)$

SECTION 10.2, page 477

1. $\dfrac{1}{2} + \dfrac{2}{\pi}(\sin x + \tfrac{1}{3}\sin 3x + \tfrac{1}{5}\sin 5x + \cdots)$

3. $\dfrac{2}{\pi}(\sin x - \tfrac{2}{2}\sin 2x + \tfrac{1}{3}\sin 3x + \tfrac{1}{5}\sin 5x - \tfrac{2}{6}\sin 6x + \cdots)$

5. $2(\sin x - \tfrac{1}{2}\sin 2x + \tfrac{1}{3}\sin 3x - \tfrac{1}{4}\sin 4x + - \cdots)$

7. $2\left[\left(\dfrac{\pi^2}{1} - \dfrac{6}{1^3}\right)\sin x - \left(\dfrac{\pi^2}{2} - \dfrac{6}{2^3}\right)\sin 2x + \left(\dfrac{\pi^2}{3} - \dfrac{6}{3^3}\right)\sin 3x - + \cdots\right]$

9. $\dfrac{4}{\pi}(\cos x + \tfrac{1}{9}\cos 3x + \tfrac{1}{25}\cos 5x + \cdots)$

11. $\dfrac{4}{\pi}(\cos x - \tfrac{1}{3}\cos 3x + \tfrac{1}{5}\cos 5x - + \cdots)$

13. $\dfrac{2}{\pi}\sin x + \dfrac{1}{2}\sin 2x - \dfrac{2}{9\pi}\sin 3x - \dfrac{1}{4}\sin 4x + \dfrac{2}{25\pi}\sin 5x + \cdots$

15. $a_0 = \pi^2/6$, $a_n = (-1)^n 2/n^2$, $b_n = \{2[(-1)^n - 1]/n^3\pi\} - \{\pi(-1)^n/n\}$

SECTION 10.3, page 480

3. $\dfrac{2}{\pi} + \cos 100\pi t$

$$+ \frac{4}{\pi}\left(\frac{1}{1\cdot 3}\cos 200\pi t - \frac{1}{3\cdot 5}\cos 400\pi t + \frac{1}{5\cdot 7}\cos 600\pi t - + \cdots\right)$$

5. $\dfrac{1}{2} + \dfrac{2}{\pi}\left(\cos\dfrac{\pi t}{2} - \dfrac{1}{3}\cos\dfrac{3\pi t}{2} + \dfrac{1}{5}\cos\dfrac{5\pi t}{2} - + \cdots\right)$

7. $\dfrac{2}{\pi}(\sin \pi t - \tfrac{1}{2}\sin 2\pi t + \tfrac{1}{3}\sin 3\pi t - + \cdots)$

9. $-\dfrac{4}{\pi^2}(\cos \pi t + \tfrac{1}{9}\cos 3\pi t + \tfrac{1}{25}\cos 5\pi t + \cdots) + \dfrac{2}{\pi}(\sin \pi t + \tfrac{1}{3}\sin 3\pi t + \cdots)$

11. $\dfrac{1}{\pi}(\sin 4t - \tfrac{1}{9}\sin 12t + \tfrac{1}{25}\sin 20t - \tfrac{1}{49}\sin 28t + - \cdots)$

13. $\dfrac{1}{3} - \dfrac{4}{\pi^2}(\cos \pi t - \tfrac{1}{4}\cos 2\pi t + \tfrac{1}{9}\cos 3\pi t - \tfrac{1}{16}\cos 4\pi t + - \cdots)$

15. $a_0 = \tfrac{1}{4}(3 - e^{-2})$, $a_n = \dfrac{2k_n}{l_n}$, $b_n = \dfrac{n\pi k_n}{l_n}$ $(n = 2, 4, \cdots)$,

$b_n = \dfrac{n\pi k_n}{l_n} - \dfrac{2}{n\pi}$ $(n = 1, 3, \cdots)$, where $k_n = 1 - (-1)^n e^{-2}$,

$l_n = 4 + n^2\pi^2$

17. Set $x = \pi t$.

19. Set $x = \pi t/2$. Multiply the series by $2/\pi$. (Why?)

SECTION 10.4, page 484

3. Even: $|x^3|$, $x^2 \cos nx$, $\cosh x$. Odd: $x \cos nx$, $\sinh x$, $x|x|$

5. Odd

7. Neither even nor odd

9. Even

11. Odd

13. Neither even nor odd

15. $|f(-x)| = |-f(x)| = |f(x)|, f^2(-x) = (-1)^2 f^2(x) = f^2(x)$

21. $\cosh x + \sinh x$

23. $x^2(1 - x^2)^{-1} + x(1 - x^2)^{-1}$

25. Cf. Prob. 5 in Sec. 10.2.

27. Cf. Prob. 7 in Sec. 10.2.

29. $\frac{4}{\pi}(\sin x - \frac{1}{9}\sin 3x + \frac{1}{25}\sin 5x - + \cdots)$

31. $-\pi - \frac{4}{\pi}(\cos x + \frac{1}{9}\cos 3x + \cdots) + 2(\sin x + \frac{1}{3}\sin 3x + \cdots)$

SECTION 10.5, page 488

1. $f(t) = 1$

3. $\frac{1}{2} + \frac{2}{\pi}\left(\cos\frac{\pi t}{l} - \frac{1}{3}\cos\frac{3\pi t}{l} + \frac{1}{5}\cos\frac{5\pi t}{l} - + \cdots\right)$

5. $\frac{1}{2} + \frac{4}{\pi^2}\left(\cos\frac{\pi t}{l} + \frac{1}{9}\cos\frac{3\pi t}{l} + \frac{1}{25}\cos\frac{5\pi t}{l} + \cdots\right)$

7. $\frac{l^3}{4} + \frac{6l^3}{\pi^2}\left[\left(\frac{4}{\pi^2} - 1\right)\cos\frac{\pi t}{l} + \frac{1}{2^2}\cos\frac{2\pi t}{l} + \left(\frac{4}{3^4\pi^2} - \frac{1}{3^2}\right)\cos\frac{3\pi t}{l} + \cdots\right]$

9. $\frac{2}{\pi} - \frac{4}{\pi}\left(\frac{1}{1\cdot 3}\cos\frac{2\pi t}{l} + \frac{1}{3\cdot 5}\cos\frac{4\pi t}{l} + \frac{1}{5\cdot 7}\cos\frac{6\pi t}{l} + \cdots\right)$

11. $\frac{4}{\pi}\left(\sin\frac{\pi t}{l} + \frac{1}{3}\sin\frac{3\pi t}{l} + \frac{1}{5}\sin\frac{5\pi t}{l} + \cdots\right)$

13. $\frac{2}{\pi}\left(\sin\frac{\pi t}{l} + \frac{2}{2}\sin\frac{2\pi t}{l} + \frac{1}{3}\sin\frac{3\pi t}{l} + \frac{1}{5}\sin\frac{5\pi t}{l} + \frac{2}{6}\sin\frac{6\pi t}{l} + \cdots\right)$

15. $\frac{2l^2}{\pi}\left[\left(1 - \frac{4}{\pi^2}\right)\sin\frac{\pi t}{l} - \frac{1}{2}\sin\frac{2\pi t}{l} + \left(\frac{1}{3} - \frac{4}{3^3\pi^2}\right)\sin\frac{3\pi t}{l} - \frac{1}{4}\sin\frac{4\pi t}{l} + \cdots\right]$

SECTION 10.6, page 494

15. No

SECTION 10.7, page 497

1. $y = C_1\cos\omega t + C_2\sin\omega t + \frac{1}{\omega^2 - 1}\sin t$, the numerical values of the amplitude $A(\omega)$ of the last term being

ω	0.5	0.7	0.9	1.1	1.5	2.0	10.0
$A(\omega)$	−1.33	−0.20	−5.3	4.8	0.8	0.33	0.01

3. $y = C_1 \cos \omega t + C_2 \sin \omega t + B_1 \sin t + B_3 \sin 3t + B_5 \sin 5t$ where

ω	0.5	0.9	1.1	2.0	2.9	3.1	4.0	4.9	5.1	6.0	8.0
$B_1 = 1/(\omega^2 - 1)$	−1.33	−5.3	4.8	0.33	0.13	0.12	0.07	0.04	0.04	0.03	0.02
$B_3 = 1/9(\omega^2 - 9)$	−0.013	−0.014	−0.014	−0.02	−0.19	0.18	0.02	0.01	0.01	0.004	0.002
$B_5 = 1/25(\omega^2 - 25)$	−0.002	−0.002	−0.002	−0.002	−0.002	−0.003	−0.004	−0.04	0.04	0.004	0.001

5. $y = C_1 \cos \omega t + C_2 \sin \omega t$

$$+ \frac{\pi^2}{12\omega^2} - \frac{1}{\omega^2 - 1} \cos t + \frac{1}{4(\omega^2 - 4)} \cos 2t - + \cdots$$

7. $y = C_1 \cos \omega t + C_2 \sin \omega t$

$$+ \frac{1}{2\omega^2} - \frac{1}{1 \cdot 3(\omega^2 - 4)} \cos 2t - \frac{1}{3 \cdot 5(\omega^2 - 16)} \cos 4t - \cdots$$

9. $y = -\dfrac{K}{c} \cos t$

11. $y = \dfrac{1 - n^2}{D} a_n \cos nt + \dfrac{nc}{D} a_n \sin nt, \ D = (1 - n^2)^2 + n^2c^2$

13. $y = \displaystyle\sum_{n=1}^{\infty} \left[\frac{(-1)^n c}{n^2 D_n} \cos nt - \frac{(-1)^n (1 - n^2)}{n^3 D_n} \sin nt \right]$

where $D_n = (1 - n^2)^2 + n^2c^2$

15. $I = \displaystyle\sum_{n=1}^{\infty} (A_n \cos nt + B_n \sin nt), \ A_n = \frac{80(10 - n^2)}{\pi n^2 D_n}, \ B_n = \frac{800}{n \pi D_n}$ (n odd),

$A_n = 0, \ B_n = 0$ (n even), $D_n = (10 - n^2)^2 + 100\, n^2$

SECTION 10.8, page 500

1. $F = \dfrac{4}{\pi} \left[\sin x + \dfrac{1}{3} \sin 3x + \cdots + \dfrac{1}{N} \sin Nx \right]$ (N odd)

5. $F = \dfrac{\pi^2}{3} - 4 \left(\cos x - \dfrac{1}{4} \cos 2x + \dfrac{1}{9} \cos 3x - \cdots + \dfrac{(-1)^{N+1}}{N^2} \cos Nx \right),$

$$E^* = \frac{2\pi^5}{5} - \pi \left(\frac{2\pi^4}{9} + 16 + 1 + \frac{16}{81} + \frac{1}{16} + \cdots \right)$$

SECTION 10.9, page 506

9. $\dfrac{2}{\pi} \displaystyle\int_0^{\infty} \left[\frac{a \sin aw}{w} + \frac{\cos aw - 1}{w^2} \right] \cos xw \, dw$

11. $\dfrac{6}{\pi} \displaystyle\int_0^{\infty} \frac{2 + w^2}{4 + 5w^2 + w^4} \cos xw \, dw$

13. $\dfrac{2}{\pi} \displaystyle\int_0^{\infty} \left[\left(a^2 - \frac{2}{w^2} \right) \sin aw + \frac{2a}{w} \cos aw \right] \frac{\cos wx}{w} \, dw$

15. $f(ax) = \dfrac{1}{\pi} \displaystyle\int_0^{\infty} A(w) \cos axw \, dw = \frac{1}{\pi} \int_0^{\infty} A\left(\frac{p}{a}\right) \cos xp \, \frac{dp}{a}$ where $wa = p$

SECTION 11.1, page 510

21. $u = 3 \ln (x^2 + y^2)/\ln 4$

23. By integration, $u = f(y)$

25. $u_x = f(y)$, $u = xf(y) + g(y)$

27. $u = axy + bx + cy + k$

29. $u = ax + by + c$

31. $p_y - p = 0$, $p = A(x)e^y$, $u = \int p\, dx = f(x)e^y + g(y)$ where $f(x) = \int A\, dx$

33. $u = B(x)e^{-y} - \frac{1}{2}x^2 - xy + C(y)$

SECTION 11.3, page 520

1. The frequency is $c/2l = \sqrt{T}/2l\sqrt{\rho}$, and we see that it is an increasing function of the tension and a decreasing function of ρ and l.

3. $u = k \cos 2t \sin 2x$

5. $u = \dfrac{2k}{a(\pi - a)} \displaystyle\sum_{n=1}^{\infty} \frac{\sin na}{n^2} \cos nt \sin nx$

7. $$u = \frac{8k}{\pi^2}\left[(2 - \sqrt{2}) \cos t \sin x - \frac{2 + \sqrt{2}}{9} \cos 3t \sin 3x + \frac{2 + \sqrt{2}}{25} \cos 5t \sin 5x - \cdots\right]$$

9. $$u = 0.12\left(\cos t \sin x - \frac{1}{2^3} \cos 2t \sin 2x + \frac{1}{3^3} \cos 3t \sin 3x - + \cdots\right)$$

11. $27,960/\pi^6 \approx 0.9986$

13. $u = e^{k(x+y)}$

15. $u = x^k y^k$

17. $u = ke^{x^2+y^2+c(x-y)}$

19. $u = (c_1 e^{kx} + c_2 e^{-kx})(A \cos ky + B \sin ky)$, etc.

23. This follows immediately from the Euler formulas; note that t merely plays the role of a parameter, and the series is the Fourier sine series of the constant function 1, multiplied by $A \sin \omega t$.

25. Substitute $G_n(t)$ [Prob. 24] into $u(x, t)$ in Prob. 21 and use the given initial conditions. Then

$$G_n(0) = B_n = \frac{2}{l}\int_0^l f(x) \sin \frac{n\pi x}{l}\, dx,$$

$$\dot{G}_n(0) = \lambda_n B_n^* + \frac{2A\omega(1 - \cos n\pi)}{n\pi(\lambda_n^2 - \omega^2)} = 0.$$

SECTION 11.4, page 523

9. $u = f_1(x) + f_2(x + y)$

11. $u = xf_1(x + y) + f_2(x + y)$

15. $y'^2 - 2y' + 1 = (y' - 1)^2 = 0$, $y = x + c$, $\Psi(x, y) = x - y$, $v = x$, $z = x - y$

17. $F_n = \sin (n\pi x/l)$, $G_n = a_n \cos (cn^2\pi^2 t/l^2)$

19. $u(0, t) = 0$, $u(l, t) = 0$, $u_x(0, t) = 0$, $u_x(l, t) = 0$

SECTION 11.5, page 528

1. The solutions of the wave equation are periodic in t.

5. Since the temperatures at the ends are kept constant, the temperature will approach a steady-state (time-independent) distribution $u_I(x)$ as $t \to \infty$, and $u_I = U_1 + (U_2 - U_1)x/l$, the solution of (1) with $\partial u/\partial t = 0$, satisfying the boundary conditions.

7. $u = \sin 0.1\pi x\, e^{-1.752\pi^2 t/100}$

9. $u = \dfrac{40}{\pi^2}\left(\sin 0.1\pi x\, e^{-0.01752\pi^2 t} - \dfrac{1}{9}\sin 0.3\pi x\, e^{-0.01752(3\pi)^2 t} + - \cdots\right)$

11. $u = \dfrac{800}{\pi^3}\left(\sin 0.1\pi x\, e^{-0.01752\pi^2 t} + \dfrac{1}{3^3}\sin 0.3\pi x\, e^{-0.01752(3\pi)^2 t} + \cdots\right)$

15. $u = \dfrac{\pi^2}{3} - 4\left(\cos x\, e^{-t} - \dfrac{1}{4}\cos 2x\, e^{-4t} + \dfrac{1}{9}\cos 3x\, e^{-9t} - + \cdots\right)$

17. $u = \dfrac{\pi}{8} + \left(1 - \dfrac{2}{\pi}\right)\cos x\, e^{-t} - \dfrac{1}{\pi}\cos 2x\, e^{-4t} - \left(\dfrac{1}{3} + \dfrac{2}{9\pi}\right)\cos 3x\, e^{-9t} + \cdots$

SECTION 11.6, page 532

13. $\dfrac{1}{\sqrt{\pi}}\displaystyle\int_{(a-x)/\tau}^{(b-x)/\tau} e^{-w^2}\,dw - \dfrac{1}{\sqrt{\pi}}\int_{(a+x)/\tau}^{(b+x)/\tau} e^{-w^2}\,dw$

SECTION 11.8, page 542

1. c increases and so does the frequency.

5. $c\pi\sqrt{260}$ (corresponding eigenfunctions $F_{4,16}$, $F_{16,14}$), etc.

9. $B_{mn} = \dfrac{-8}{mn\pi^2}((-1)^m k_n + (-1)^n l_m)$

where $k_n = \begin{cases} 0 & (n \text{ even}) \\ a & (n \text{ odd}) \end{cases}$ and $l_m = \begin{cases} 0 & (m \text{ even}) \\ b & (m \text{ odd}) \end{cases}$

11. $B_{mn} = 0$ (m or n even), $B_{mn} = \dfrac{64a^2b^2}{\pi^6 m^3 n^3}$ (m, n both odd)

15. $u = \dfrac{0.64}{\pi^6}\displaystyle\sum_{\substack{m=1\\ m,n \text{ odd}}}^{\infty}\sum_{n=1}^{\infty}\dfrac{1}{m^3n^3}\cos(\pi t\sqrt{m^2+n^2})\sin m\pi x \sin n\pi y$

17. $u = k\cos \pi\sqrt{5}\, t \sin \pi x \sin 2\pi y$

SECTION 11.9, page 544

5. $a^2 u_{x^*x^*} + c^2 u_{y^*y^*}$

SECTION 11.10, page 549

1. u_2: $r = \alpha_1/\alpha_2 = 0.43565$, u_3: $r = \alpha_1/\alpha_3 = 0.27789$, $r = \alpha_2/\alpha_3 = 0.63788$

19. $u = 4k\displaystyle\sum_{m=1}^{\infty}\dfrac{J_2(\alpha_m)}{\alpha_m{}^2 J_1{}^2(\alpha_m)}\cos \alpha_m t\, J_0(\alpha_m r)$

SECTION 11.11, page 553

1. $a^2 u_{x^*x^*} + b^2 u_{y^*y^*} + c^2 u_{z^*z^*}$

19. $\dfrac{u_1 - u_0}{\ln \dfrac{r_1}{r_0}}\ln r + \dfrac{u_0 \ln r_1 - u_1 \ln r_0}{\ln \dfrac{r_1}{r_0}}$

SECTION 11.12, page 557

5. $u = 1$

7. $u = \frac{2}{3}r^2P_2(\cos\phi) + \frac{1}{3} = r^2(\cos^2\phi - \frac{1}{3}) + \frac{1}{3}$

9. $\cos 2\phi = 2\cos^2\phi - 1$, $2x^2 - 1 = \frac{4}{3}P_2(x) - \frac{1}{3}$, $u = \frac{4}{3}r^2P_2(\cos\phi) - \frac{1}{3}$

11. $u = 4r^3P_3(\cos\phi) - 2r^2P_2(\cos\phi) + rP_1(\cos\phi) - 2$

SECTION 11.13, page 561

5. $U(x, s) = \dfrac{c(s)}{x^s} + \dfrac{x}{s^2(s+1)}$, $U(0, s) = 0$, $c(s) = 0$,
$u(x, t) = x(t - 1 + e^{-t})$

9. Set $x^2/4c^2\tau = z^2$. Use z as a new variable of integration. Use $\text{erf}(\infty) = 1$.

SECTION 12.1, page 569

3. $7/29 + (26/29)i$ **5.** $7/50 - (13/25)i$ **7.** $-1/5$

9. $4x^3y - 4xy^3$ **11.** $2/25$ **13.** $(x^2 - y^2)/(x^2 + y^2)$

25. $\text{Re}(i\bar{z}) = -\text{Im}\, z$, $\text{Im}(i\bar{z}) = -\text{Re}\, z$

SECTION 12.2, page 572

1. 2 **3.** 1 **5.** $\sqrt{\dfrac{(x-1)^2 + y^2}{(x+1)^2 + y^2}}$

7. 1 **9.** $\pi/4$ **11.** $-\pi/2$

13. $5\cos\pi$

15. $\frac{1}{5}[\cos(-\text{arc}\tan\frac{3}{4}) + i\sin(-\text{arc}\tan\frac{3}{4})]$

19. Consider $c = a + ib = z_1/(z_1 + z_2)$, assuming $z_1 + z_2 \neq 0$. From (12), $|a| \leqq |c|$, $|a - 1| \leqq |c - 1|$. Thus $|a| + |a - 1| \leqq |c| + |c - 1|$. Clearly $|a| + |a - 1| \geqq 1$, and the last inequality becomes

$$1 \leqq |c| + |c - 1| = \left|\frac{z_1}{z_1 + z_2}\right| + \left|\frac{z_2}{z_1 + z_2}\right|.$$

Multiplication by $|z_1 + z_2|$ yields (10).

21. Square both sides of both inequalities.

SECTION 12.3, page 575

1. $y \geqq -1$

3. The regions to the right of the right branch and to the left of the left branch of the hyperbola $x^2 - y^2 = 1$

5. Vertical strip bounded by $x = -\pi$ and $x = \pi$

7. $(x - \frac{5}{3})^2 + y^2 = \frac{16}{9}$ **9.** The y-axis

11. Interior of the circle of radius $\frac{1}{2}$ with center at $(\frac{1}{2}, 0)$

13. The straight-line segment with endpoints z_1, z_2

SECTION 12.4, page 579

1. $11 + 13i$, $-27 + 3i$, $41 - 23i$

3. $2 - i$, $(4 - 3i)/5$, $(8 - i)/13$

5. $(1 - x)/[(1 - x)^2 + y^2]$, $y/[(1 - x)^2 + y^2]$

7. $\operatorname{Re} w > 0$ **9.** $|w| > 9$ **11.** $\operatorname{Re} w \geqq 0$

15. No, because $f(z) \to 0$ $[= f(0)]$ as $z \to 0$ along the y-axis, but $f(z) \to 1$ $[\neq f(0)]$ as $z \to 0$ along the positive x-axis.

17. Yes, because for $z \neq 0$, $|f(z)| = x^2/\sqrt{x^2 + y^2} \leqq |x| \leqq |z|$ and, therefore, $|f(z)| \to 0$ $[= f(0)]$ as $|z| \to 0$.

21. $4z(z^2 + 1)$ **23.** $(2 - 2z^2)/(z^2 + 1)^2$ **25.** $(6 - 8i)/25$

27. $-1 - 3i$

29. The quotient in (4) is $\Delta x/\Delta z$, which is 0 if $\Delta x = 0$ but 1 if $\Delta y = 0$, so that it has no limit as $\Delta z \to 0$.

SECTION 12.5, page 585

1. a **3.** $-1/z^2$ **5.** $1 - 1/z^2$ **11.** Yes

13. No **15.** Yes, except at $z = 1$

17. No **19.** No **23.** z **25.** $-iz^2/2$

27. $\cos x \cosh y - i \sin x \sinh y$

29. $u = ax^3 - 3kx^2y - 3axy^2 + ky^3$, thus $b = -3k$, $c = -3a$.

33. Let $f = u + iv$ and $|f| = c$. Then $u^2 + v^2 = c^2$. By differentiation, $uu_x + vv_x = 0$, $uu_y + vv_y = 0$. By (5), $uu_x - vu_y = 0$, $uu_y + vu_x = 0$. Hence

$$(uu_x - vu_y)^2 + (uu_y + vu_x)^2 = (u^2 + v^2)(u_x{}^2 + u_y{}^2) = c^2(u_x{}^2 + u_y{}^2) = 0.$$

If $c = 0$, then $u = v = 0$. If $c \neq 0$, then $u_x{}^2 + u_y{}^2 = 0$, i.e., $u_x = u_y = 0$ and, by (5), $v_x = 0$, $v_y = 0$. Hence, $u = const$, $v = const$.

35. Let $f = u + iv$. By (3), $f' = u_x + iv_x = 0$. Hence, $u_x = 0$, $v_x = 0$. By (4) or (5), $u_y = 0$, $v_y = 0$. Hence, $u = const$, $v = const$, $f = u + iv = const$.

SECTION 12.6, page 588

1. $\pm(1 + i)/\sqrt{2}$ **3.** $2i$, $-2i$ **5.** -1, $(1 \pm i\sqrt{3})/2$

7. i, $(\pm\sqrt{3} - i)/2$ **9.** $\pm(1 + i)/\sqrt{2}$, $\pm(-1 + i)/\sqrt{2}$

11. $\pm(\sqrt{3} + i)/2$, $\pm i$, $\pm(\sqrt{3} - i)/2$ **13.** 3, $3(-1 \pm i\sqrt{3})/2$

15. $3(\pm 1 \pm i)/\sqrt{2}$ **17.** 0 **19.** $\pm(1 + i)\sqrt{2}$

21. $\pm(2 + 3i)$ **23.** i, $-1 - i$ **25.** $3 + 2i$, $2 - i$

27. $z^4 = 4 + 4i$, $z = \pm\sqrt[8]{32}(\cos\beta + i \sin\beta)$, $\beta = \pi/16, 9\pi/16$

29. $(z^2 - z\sqrt{2} + 1)(z^2 + z\sqrt{2} + 1)$

SECTION 12.7, page 591

3. $(1 + i)/\sqrt{2}$ **5.** $e(\cos 1 + i \sin 1)$

7. $e^{2x} \cos 2y$, $e^{2x} \sin 2y$ **9.** $e^{x^2-y^2} \cos 2xy$, $e^{x^2-y^2} \sin 2xy$

11. $e^{\pi i/4}$, $e^{5\pi i/4}$, $e^{3\pi i/4}$, $e^{-\pi i/4}$ **13.** $\sqrt{8}\, e^{-\pi i/4}$

15. $\pm 2n\pi i\ (n = 0, 1, \cdots)$

17. $\ln 2 \pm (2n + 1)\pi i\ (n = 0, 1, \cdots)$

19. All z

21. $\operatorname{Re} z > 0$

23. $-e^{xy} \sin(\frac{1}{2}x^2 - \frac{1}{2}y^2)$

25. $e^z \to \infty$ when $\arg z = 0$, e^{iy} has no limit, $e^z \to 0$ when $\arg z = \pi$.

SECTION 12.8, page 594

7. $\sqrt{\cos^2 x + \sinh^2 y}$

9. $\sqrt{\dfrac{\sin^2 x + \sinh^2 y}{\cos^2 x + \sinh^2 y}}$

11. $\sin x \cos x/(\sin^2 x + \sinh^2 y)$

13. $1.175i$

15. $2.033 - 3.052i$

17. $1.960 + 3.166i$

19. $\pm 2n\pi \pm 2.29i$

21. $\pm(2n + 1)\pi i/2,\ n = 0, 1, \cdots$

23. $\pm(\pi/3)i \pm 2n\pi i,\ n = 0, 1, \cdots$

SECTION 12.9, page 597

5. $\pm 2n\pi i,\ n = 0, 1, \cdots$

7. $i(\frac{1}{2}\pi \pm 2n\pi),\ n = 0, 1, \cdots$

9. $1 + (\frac{1}{2}\pi \pm 2n\pi)i,\ n = 0, 1, \cdots$

11. $(1 \pm 2n\pi)i,\ n = 0, 1, \cdots$

13. $-i$

15. $-e$

17. $0.693 - 1.571i$

19. $1.946 + 3.142i$

21. $1 + i$

23. $\sqrt{2}e^{\pi/4}[\cos(\frac{1}{4}\pi - \ln\sqrt{2}) + i \sin(\frac{1}{4}\pi - \ln\sqrt{2})]$

25. $27[\cos(\ln 3) - i \sin(\ln 3)]$

27. $3.350 + 1.189i$

31. $\cosh w = \frac{1}{2}(e^w + e^{-w}) = z$, $(e^w)^2 - 2z\,e^w + 1 = 0$, $e^w = z + \sqrt{z^2 - 1}$

35. Immediate consequence of $\sin(w \pm 2n\pi) = \sin w$ and $\sin(\pi - w) = \sin w$.

SECTION 13.1, page 605

1. $v = -u + 2x - 2,\ v = -u - 2,\ -u,\ -u + 2,\ -u + 4$

3. $|w - 2i| \leqq 2\sqrt{2}$

5. $v = 0$ and $u \leqq 0$, $v = 2\sqrt{1 - u}$ and $u \leqq 1$, etc.

7. $u = x^2 - y^2$, $v = 2xy$. When $y = x + 1$ this becomes $u = -2x - 1$, $v = 2x(x + 1)$. Thus $x = -\frac{1}{2}(u + 1)$ and $v = \frac{1}{2}(u^2 - 1)$.

9. $u = -1$

11. $|w| \leqq 16$

13. The region between the parabolas $v^2 = 4(1 - u)$ and $v^2 = 16(4 - u)$

15. $-\pi/2 < \arg w < \pi$

17. $u = -\frac{1}{2}$

19. $v = -\frac{1}{4}$

21. $|w - \frac{1}{2}| = \frac{1}{2}$

23. The image is bounded by the circles $|w - \frac{1}{2}| = \frac{1}{2}$ and $|w - \frac{1}{4}| = \frac{1}{4}$.

27. $0 \leqq \arg w \leqq \pi/4$, $\pi/2 \leqq \arg w \leqq 3\pi/4$, $-\pi/2 \leqq \arg w \leqq -\pi/4$, $0 \leqq \arg w \leqq \pi/2$, $\pi/2 \leqq \arg w \leqq \pi$, $\pi \leqq \arg w \leqq 3\pi/2$, $-\pi/2 \leqq \arg w \leqq 0$, $0 \leqq \arg w \leqq 3\pi/4$

29. $w = z + 2 + i$

SECTION 13.2, page 610

3. No, the size is preserved, but the sense is reversed.

5. $z(t) = t + i/t$

7. $z(t) = \cosh t + i \sinh t$

9. $z(t) = 1 + 3 \cos t - 2i + 3i \sin t$

11. $z = (2n + 1)\pi/2,\ n = 0, \pm 1, \pm 2, \cdots$

13. $z = 0$

15. $z = -a/2$

SECTION 13.3, page 612

1. $z = 0$

3. $z = 0, 1$

5. $z = 0, \pm 1$

7. $z = \pm i$

9. $z = \pm i$

11. $w = -1/z$

13. $w = (az + b)/(a - bz)$

15. All translations

SECTION 13.4, page 617

1. $w_1 = iz,\ w_2 = w_1 + 4,\ w_3 = 1/w_2,\ w_4 = 5iw_3,\ w = w_4 - i$

5. $w = -2i \dfrac{1}{z + i} + 1$

7. $w = (4z - i)/(-iz - 4)$

9. $w = 1/z$

11. $w = 1/(z + 1)$

13. $w = (z - i)/(z + i)$

15. $w = (z + 1)/(z - i)$

17. $w = (z + 3)/(2z - 1)$

19. $w = az/(cz + d)$

21. All four coefficients are real, except possibly for a common complex factor.

23. $w = -(z^2 + i)/(iz^2 + 1)$

25. $w = (z^4 - i)/(-iz^4 + 1)$

SECTION 13.5, page 624

1. $e^{-1} < |w| < e,\ |\arg w| < \pi/2$

3. $1 < |w| < e,\ 0 < \arg w < 1$

5. $e^{-2} \leqq |w| \leqq e^2,\ -\pi \leqq \arg w \leqq -\pi/2$

7. $t = z^2$ maps the given region onto the strip $0 < \operatorname{Im} t < \pi$, and $w = e^t$ maps this strip onto the upper half-plane. *Ans.* $w = e^{z^2}$.

9. The region in the upper half-plane bounded by the ellipses

$$\frac{u^2}{\cosh^2 1} + \frac{v^2}{\sinh^2 1} = 1 \quad \text{and} \quad \frac{u^2}{\cosh^2 2} + \frac{v^2}{\sinh^2 2} = 1$$

11. Elliptical annulus bounded by the ellipses in Prob. 9 and cut along the positive imaginary axis

13. $\cosh z = \cos(iz) = \sin(iz + \tfrac{1}{2}\pi)$

15. $T = T_0 + \dfrac{1}{\pi}(T_1 - T_0) \arctan \dfrac{2xy}{x^2 - y^2} = T_0 + \dfrac{2}{\pi}(T_1 - T_0) \arctan \dfrac{y}{x}$

17. $T = \dfrac{100}{\pi/4} \arctan \dfrac{y}{x}$

19. $T = \dfrac{1}{\pi} \arctan \dfrac{4xy}{(x^2 + y^2)^2 - 1}$

SECTION 13.6, page 628

1. w moves once around the unit circle $|w| = 1$.
3. 4 and 5 sheets, respectively, branch point at $z = 0$
5. The interior of the ellipse $\dfrac{u^2}{(5/2)^2} + \dfrac{v^2}{(3/2)^2} = 1$. The interior of the same ellipse in the other sheet. The ring bounded by that ellipse and the ellipse

$$\frac{u^2}{(10/3)^2} + \frac{v^2}{(8/3)^2} = 1.$$

9. 2 sheets, branch point at $z = 0$
11. 3 sheets, branch point of the second order at $z = i$
13. $\pm i$, 2 sheets
15. a, b, 2 sheets
17. a, infinitely many sheets
19. None; 2 sheets, which are not connected; in fact, $\sqrt{e^z}$ represents the two functions $e^{z/2}$ and $-e^{z/2}$.

SECTION 14.1, page 634

1. $z = (1 + 2i)t,\ 0 \leqq t \leqq 1$
3. $z = 1 + i + (2 - 5i)t,\ 0 \leqq t \leqq 1$
5. $z = -2 + i + it,\ 0 \leqq t \leqq 3$
7. $z = 1 - 2i + 3e^{it},\ 0 \leqq t \leqq 2\pi$
9. $z = 3 \cos t + i \sin t,\ 0 \leqq t \leqq 2\pi$
11. $z = t + i/t,\ 1 \leqq t \leqq 4$
13. Straight line segment from 1 to $3 - i$
15. Upper half of the circle $x^2 + (y - 3)^2 = 9$
17. $y = 3x^2$ from $(-1, 3)$ to $(2, 12)$
19. $-8i$
21. $-115 - 46i$
23. 0
25. $1 + 2i$
27. (*a*) $2\pi i$, (*b*) $-2\pi i$
29. (*a*) i, (*b*) $2i$, (*c*) $2i$

SECTION 14.2, page 636

5. $2b(1 + i)$
7. $\frac{4}{3}$
9. $-1 + ie$
11. $i(\sinh 2\pi - \sinh \pi)$
13. $1 - \cosh 1$
15. $i \sin 1$
17. $5e^3$
19. 5

SECTION 14.3, page 643

3. If $z = 0$ lies exterior to C.
5. 0, yes
7. $-\pi$, no
9. 0, yes
11. 0, no
15. (*a*) $2\pi i$, (*b*) $2\pi i$, no
17. $2\pi i$, 0
19. $2\pi i$, πi
21. 0
23. $-2\pi i$, 0
25. 0

SECTION 14.4, page 647

1. $2 + 2i$ **3.** $(10 - 2i)/3$ **5.** $-2e$ **7.** $(e^{-1} - 1)/2$
9. $i \sinh \pi$ **11.** $-\frac{1}{2} \sin \pi^2$ **13.** $(\pi - \frac{1}{2} \sinh 2\pi)i$
15. $i \sin 3$ **17.** $\cos 2 - 1$ **19.** 0

SECTION 14.5, page 650

1. π **3.** 0 **5.** $\pi i/2$ **7.** $\pi/2$ **9.** $2\pi i$
11. $\pi i/2$ **13.** 0 **15.** $2\pi i$ **17.** 0 **19.** $2\pi i$

SECTION 14.6, page 653

1. $\pi i/2$ **3.** $3\pi/8$ **5.** $\pi i/8$ **7.** πi
9. $\pi(\pi + 4i)(1 - i)/32\sqrt{2}$ **11.** $2\pi i$
13. $(-1)^n 2\pi i/(2n)!$ **15.** $2\pi i$

SECTION 15.1, page 657

1. $\frac{1}{4}, \frac{2}{5}, \frac{3}{6}, \frac{4}{7}, \cdots$ **3.** $i, -\frac{1}{8}, -i/27, \frac{1}{64}, \cdots$

5. $\dfrac{i}{1+i} = \dfrac{1+i}{2}, \dfrac{-4}{2+i} = \dfrac{-8+4i}{5}, \dfrac{-9i}{3+i} = \dfrac{-9-27i}{10}, \cdots$

7. $1, i/2, -\frac{1}{2}, \frac{1}{4}, -i/8, \cdots$ **9.** Bounded, convergent, limit 0
11. Unbounded, divergent **13.** Bounded, divergent

SECTION 15.3, page 665

1. No, no, none **3.** Yes, no, $-1, 1$
5. Yes, no, $-i, -1, i, 1$ **7.** No, no, none
9. Yes, yes (limit 0) **11.** Yes, yes (limit 0)
13. Yes, no, 1, 2, 3 **15.** Yes, no, $1, i, -1, -i$

SECTION 15.4, page 667

1. Bounded, convergent, limit 0, monotone
3. Unbounded, divergent, monotone
5. Bounded, divergent, not monotone, 1, 2
7. Bounded, convergent, limit 0, not monotone
9. Bounded, convergent, limit 0, monotone when $0 \leqq c \leqq \frac{1}{2}$
11. Convergent
13. Divergent (by Theorem 3, Sec. 15.2)
15. 6 terms **17.** 5 terms **19.** 2 terms

SECTION 15.5, page 674

1. Divergent
3. Divergent (compare with the harmonic series)
5. Convergent **7.** Divergent **9.** Convergent
11. Convergent **13.** Divergent **15.** Convergent
17. Convergent **19.** 4, 8, 66

SECTION 16.1, page 687

3. 1 **5.** 3 **7.** ∞ **9.** ∞

11. $\frac{1}{4}$ **13.** ∞ **15.** $\frac{1}{6}$ **17.** $\frac{1}{9}$

SECTION 16.2, page 692

5. 5 **7.** $\frac{1}{4}$ **9.** 1

SECTION 16.4, page 699

5. $1 - \frac{(2z)^2}{2!} + \frac{(2z)^4}{4!} - + \cdots, R = \infty$

7. $1 - z + \frac{z^2}{2!} - \frac{z^3}{3!} + - \cdots, R = \infty$

9. $-1 - (z - \pi i) - \frac{1}{2!}(z - \pi i)^2 - \frac{1}{3!}(z - \pi i)^3 - \cdots, R = \infty$

11. $\frac{1}{\sqrt{2}}\left[1 + \left(z + \frac{\pi}{4}\right) - \frac{1}{2!}\left(z + \frac{\pi}{4}\right)^2 - \frac{1}{3!}\left(z + \frac{\pi}{4}\right)^3 + \cdots\right], R = \infty$

13. $-1 - (z + 1) - (z + 1)^2 - (z + 1)^3 - \cdots, R = 1$

15. $\cos^2 z = \frac{1}{2}(1 + \cos 2z) = \frac{1}{2}\left[2 - \frac{2^2}{2!}z^2 + \frac{2^4}{4!}z^4 - + \cdots\right], R = \infty$

17. $z + \frac{1}{3}z^3 + \frac{2}{15}z^5 + \cdots, R = \frac{\pi}{2}$

19. $1 - \frac{1}{3}z^2 - \frac{1}{45}z^4 - \frac{2}{945}z^6 - \cdots, R = \pi$

21. $\frac{z}{2!1} - \frac{z^3}{4!3} + \frac{z^5}{6!5} - + \cdots, R = \infty$

23. $z - \frac{z^3}{3!3} + \frac{z^5}{5!5} - \frac{z^7}{7!7} + - \cdots, R = \infty$

25. $z - \frac{z^5}{2!5} + \frac{z^9}{4!9} - \frac{z^{13}}{6!13} + - \cdots, R = \infty$

SECTION 16.5, page 702

1. $1 - z^4 + z^8 - z^{12} + - \cdots, |z| < 1$

3. $1 - z^3 + z^6 - + \cdots, |z| < 1$

5. $1 - 2z^2 + 3z^4 - 4z^6 + - \cdots, |z| < 1$

7. $1 - \frac{z^4}{2!} + \frac{z^8}{4!} - \frac{z^{12}}{6!} + - \cdots$

9. $1 + z^4 + \frac{z^8}{2!} + \frac{z^{12}}{3!} + \cdots$

11. $z^2 + z^4 + \frac{z^6}{3} + \cdots$

13. $e[1 + z + \frac{3}{2}z^2 + \frac{13}{6}z^3 + \cdots], |z| < 1$

15. $e[1 + z + z^2 + \frac{5}{6}z^3 + \cdots]$

17. $\frac{1}{1 - 3i} + \frac{3}{(1 - 3i)^2}[z - (1 + i)] + \frac{3^2}{(1 - 3i)^3}[z - (1 + i)]^2 + \cdots,$
$|z - (1 + i)| < \frac{1}{3}\sqrt{10}$

19. $\frac{i}{2}\left(1-(1+i)(z+i)+\frac{3}{4}(1+i)^2(z+i)^2-\frac{(1+i)^3}{2}(z+i)^3+\cdots\right)$, $|z+i|<\sqrt{2}$

21. $1+2\left(z-\frac{\pi}{4}\right)+2\left(z-\frac{\pi}{4}\right)^2+\frac{8}{3}\left(z-\frac{\pi}{4}\right)^3+\cdots,\ \left|z-\frac{\pi}{4}\right|<\frac{\pi}{4}$

SECTION 16.6, page 710

1. This follows from Theorem 1.
3. This follows from Theorem 1 because the radius of convergence is 4.
5. $|\sin n|z||\leqq 1$, $\Sigma 2^{-n}$ converges, apply Theorem 5.
7. $|\cos^n|z||\leqq 1$, and $\Sigma(1/n^2)$ converges.
11. Convergence follows from Theorem 1, Sec. 15.5. Let $R_n(z)$ and R_n^* be the remainders of (1) and (5), respectively. Since (5) converges, for given $\epsilon>0$ we can find an $N(\epsilon)$ such that $R_n^*<\epsilon$ for all $n>N(\epsilon)$. Since $|f_n(z)|\leqq M_n$ for all z in the region G, we have $|R_n(z)|\leqq R_n^*$ and therefore $|R_n(z)|<\epsilon$ for all $n>N(\epsilon)$ and all z in the region G. This proves that the convergence of (1) in the region G is uniform.
13. $f_n=s_n-s_{n-1}=x/(nx+1)[(n-1)x+1]$
17. The sum $s=\lim s_n=\lim x^n$ is discontinuous at $x=1$. Use Theorem 2.

SECTION 16.7, page 716

1. $\frac{1}{z^3}-\frac{1}{z^2}+\frac{1}{2z}-\frac{1}{6}+\frac{z}{24}-+\cdots,\ R=\infty$
3. $\frac{1}{z^2}-2+\frac{2}{3}z^2-\frac{4}{45}z^4+-\cdots,\ R=\infty$
5. $\frac{1}{z^2}+1+z^2+z^4+z^6+\cdots,\ R=1$
7. $\frac{3}{z^2}+\frac{9}{2}+\frac{81}{40}z^2+\cdots,\ R=\infty$
9. $\frac{1}{z^2}-\frac{2}{z}+3-4z+-\cdots,\ R=1$
11. No, because $z=0$ is a limit point of the points $z_n=2/(2n+1)\pi$ at which $\tan(1/z)$ is not analytic.
13. $\sum_{n=0}^{\infty}\binom{-4}{n}(z-1)^n,\ |z-1|<1,\ \sum_{n=0}^{\infty}\binom{-4}{n}\frac{1}{(z-1)^{n+4}},\ |z-1|>1$
15. $-\sum_{n=0}^{\infty}\left(\frac{i}{2}\right)^{n+1}(z-i)^{n-1},\ 0<|z-i|<2,\ \sum_{n=0}^{\infty}\frac{(-2i)^n}{(z-i)^{n+2}},\ |z-i|>2$
17. $(1-4z)\sum_{n=0}^{\infty}z^{4n},\ |z|<1,\ \left(\frac{4}{z^3}-\frac{1}{z^4}\right)\sum_{n=0}^{\infty}\frac{1}{z^{4n}},\ |z|>1$
19. $\sum_{n=0}^{\infty}\frac{e}{n!}(z-1)^{n-2},\ |z-1|>0$

SECTION 16.8, page 721

1. No **3.** No **5.** No
7. No, no **9.** No, no **21.** ± 1 (simple)
23. $-i$ (third order) **25.** $0, \pm 1, \pm 2, \cdots$ (third order)
27. $(2n+1)\pi i/2,\ n = 0, \pm 1, \pm 2, \cdots$ (second order)
29. $\pm 2n\pi i,\ n = 0, 1, \cdots$ (simple)
31. By assumption, $f(z) = (z-a)^n g(z)$ where $g(a) \neq 0$. Consequently, $f(z)^2 = (z-a)^{2n} g(z)^2$. Etc.
33. $-a$ (pole of third order)
35. $0, \infty$ (essential singularities)
37. 0 (pole of fifth order), ∞ (essential singularity)
39. 0 (essential singularity)

SECTION 17.1, page 726

1. 1 **3.** 0
5. $-\frac{1}{2}$ (at $z = -1$), $\frac{1}{2}$ (at $z = 1$) **7.** $\frac{1}{4}, \frac{1}{4}, -\frac{1}{4}, -\frac{1}{4}$ (at $z = 1, -1, i, -i$)
9. -1 (at $z = \pm 2n\pi i$) **11.** 1 (at $z = \pm n\pi$)
13. $-3/4,\ 3i/4,\ 3/4,\ -3i/4$ (at $z = 1, i, -1, -i$)
15. -2 (at $z = 0$) **17.** $1/17$ (at $z = -1$)
19. $2\pi i$ **21.** $2\pi i$
23. $2\pi i$ **25.** 0
27. $-\pi i$ **29.** $-2\pi i$

SECTION 17.2, page 728

1. $2\pi i$ **3.** 0 **5.** $3\pi i$ **7.** $2\pi i/3$
9. $-\pi i$ **11.** 0 **13.** $2\pi i/9$ **15.** $-16\pi i/9$
17. $-4i$ **19.** $-4i$ **21.** $-2\pi i/5$ **23.** 0
25. $-2\pi/3$ **27.** $-\pi \sin \frac{1}{2}$ **29.** 0

SECTION 17.3, page 732

1. $\dfrac{2\pi}{\sqrt{3}}$ **3.** $\dfrac{\pi}{\sqrt{k^2-1}}$ **5.** $\dfrac{\pi}{2}$ **7.** 0

9. $\dfrac{8\pi}{3}$

11. $\cos 2\theta = \dfrac{1}{2}\left(z^2 + \dfrac{1}{z^2}\right)$,

$$\int_0^{2\pi} \frac{\cos^2\theta}{26 - 10\cos 2\theta}\, d\theta = -\frac{1}{20i}\int_C \frac{(z^2+1)^2}{z(z^2-\frac{1}{5})(z^2-5)}\, dz = \frac{\pi}{20}$$

13. π
15. $2\pi/3$ **17.** $3\pi/8$ **19.** 0 **21.** $\pi/12$ **23.** $\pi/2$
25. The integrals are zero because the integrands are odd.

SECTION 17.4, page 736

3. π/e

5. $\pi e^{-s}/2$

7. $\pi e^{-1/\sqrt{2}}\,[\sin(1/\sqrt{2}) + \cos(1/\sqrt{2})]/\sqrt{2}$

SECTION 18.1, page 741

1. $u(r) = 100 \ln r/\ln 5 = 62.13 \ln r$

3. $u(r) = 100(\ln r + \ln 5)/\ln 10$

5. $(x - 1/2c)^2 + y^2 = 1/4c^2$

7. $u = c \operatorname{Re} \ln(z^2 - a^2) = c \ln|z^2 - a^2|$

9. $z = x + iy = \cosh(u + iv) = \cosh u \cos v + i \sinh u \sin v$, $x^2/\cosh^2 u + y^2/\sinh^2 u = 1$; the equipotential lines $u = const$ are confocal ellipses.

SECTION 18.2, page 747

1. $V = V_1 = K$, $Ky = const$, $Kx = const$

5. Parallel flow in the negative y-direction, $V = -i$

7. Parallel flow in the direction of $y = -x$, $V = 1 - i$

9. $V = -6xy + 3i(y^2 - x^2)$, $V_2 = 0$ on $y = x$ and $y = -x$

13. $F(z) = -[\ln(z - a)]/2\pi$

15. $F(z) = \dfrac{1}{2\pi} \ln \dfrac{z + a}{z - a}$. The streamlines are circles.

19. $V = 1 - \bar{z}^{-2} = 0$ at $z = -1$ and $z = 1$.

SECTION 18.3, page 752

3. Use that $|e^z| = e^x$ is monotone in x.

SECTION 18.4, page 755

3. $u = r \sin\theta$

5. $u = r^3 \sin 3\theta$

7. $u = 3r \sin\theta - r^3 \sin 3\theta$

9. $u = \dfrac{1}{2} + \dfrac{2}{\pi}\left(r \sin\theta + \dfrac{r^3}{3} \sin 3\theta + \dfrac{r^5}{5} \sin 5\theta + \cdots\right)$

11. $u = \left(1 + \dfrac{2}{\pi}\right) r \sin\theta - \dfrac{r^2}{2} \sin 2\theta + \left(\dfrac{1}{3} - \dfrac{2}{9\pi}\right) r^3 \sin 3\theta - \dfrac{r^4}{4} \sin 4\theta + - \cdots$

SECTION 19.1, page 761

1. 0.3818×10^{53}

3. $|d^* - d| = |\epsilon_1 - \epsilon_2| \leqq |\epsilon_1| + |\epsilon_2| \leqq \beta_1 + \beta_2 = \beta$, $a_1 = 1$, $a_1^* = 1.1$, $a_2 = 1$, $a_2^* = 0.9$ gives $|\epsilon| = |\epsilon_1| + |\epsilon_2|$.

5. $\dfrac{a_1^*}{a_2^*} = \dfrac{a_1 + \epsilon_1}{a_2 + \epsilon_2} = \dfrac{a_1 + \epsilon_1}{a_2}\left(1 - \dfrac{\epsilon_2}{a_2} + \dfrac{\epsilon_2^{\,2}}{a_2^{\,2}} - + \cdots\right) \approx \dfrac{a_1}{a_2} + \dfrac{\epsilon_1}{a_2} - \dfrac{\epsilon_2}{a_2} \cdot \dfrac{a_1}{a_2}$,

$\left(\dfrac{a_1^*}{a_2^*} - \dfrac{a_1}{a_2}\right) \Big/ \dfrac{a_1}{a_2} \approx \dfrac{\epsilon_1}{a_1} - \dfrac{\epsilon_2}{a_2}$

7. $|\epsilon_r(x_2)| = |\epsilon_r(x_1)|$ by Prob. 5 since $x_2 = 2/x_1$, and $|\epsilon_r(x_1)| \leqq 0.005/39.95$ since x_1 is rounded to 4S. Hence
$|\epsilon(x_2)| = |\epsilon_r(x_2) \cdot x_2| = |\epsilon_r(x_1) \cdot x_2| \leqq (0.005/39.95) \cdot 0.050\,06 < 0.000\,01.$

11. Rounding off a 5 in the last given place may introduce errors since no further digits are given. These errors will be more frequent if the five-place table is used.

13. $2.76065 \leqq s \leqq 2.76175$

SECTION 19.2, page 768

1. $x_1 = 1.9$ (exact 1.9)

3. The tangent to the curve of $f(x)$ at $x = 1$ intersects the x-axis at $x = 1.9$. Choose $x_0 = 1.2$, for example.

5. 1.557

7. -1.3, 1.4

11. $x_1 = -0.2$, $x_2 = -0.200\,32$, $x_3 = -0.200\,323$, $\cdots$

13. Convergence to the root near -0.2.

15. Errors 0.236 068, 0.013 932, 0.000 043, 0.000 000

17. $x \geqq \sqrt{2/(1 + 2\alpha)}$, $\alpha < 1$.

19. $f(x) = x^k - c$, $x_{n+1} = (1 - k^{-1})x_n + k^{-1}cx_n^{\,1-k}$

21. 0, 1.260

25. This follows from the intermediate value theorem of calculus.

SECTION 19.4, page 778

3. 0.257 53

5. (*a*) 0.258 27, (*b*) 0.258 27

9. $f(x) \approx 0.566\,106 + 0.496\,098x - 0.036\,022x^2$

11. $8x^2 - 15x + 9$, 36

13. 2.21921, less exact (error 1 unit of the last digit).

SECTION 19.5, page 783

7. This follows from Theorem 1.

9. $g(x) = -4x^3/\pi^3 + 3x/\pi$

SECTION 19.6, page 791

11. 0.245, $\epsilon = -0.005$

13. $\ln 2 \approx 0.693\,25$ (0.693 15 exact to 5D), $M_4^* = 0.75$, $M_4 = 24$, $0.000\,016 < \epsilon_S < 0.000\,53$ or $0.692\,72 < \ln 2 < 0.693\,24$.

15. 0.9466 (0.9461 exact to 4D)

17. 0.9458

19. $\alpha = \beta = \frac{1}{2}$, trapezoidal rule

23. 0.080, 0.320, 0.176, 0.256. Exact: $f'(0.4) = 0.256$

25. 0.52, 0.080, 0.304, 0.256 (exact: 0.256)

SECTION 19.7, page 799

3. 0.1, 0.202, 0.31008, 0.42869 etc. Exact solution: $y = e^{x^2}\int_0^x e^{-t^2}\,dt$

11. $y = \tan x$

SECTION 19.9, page 809

1. $x = 1, y = 2$

3. $x = -2, y = 3$

5. $y = 3x, z = 2x$

7. $x = y + 1, z = 1$

9. $x = 3y + 2, z = 0$

11. $w = 0, x = 2y, z = 3y + 1$

SECTION 19.10, page 813

3. $$\begin{pmatrix} 0 & 0.25 & 0.25 & 0 \\ 0 & 0.0625 & 0.0625 & 0.25 \\ 0 & 0.0625 & 0.0625 & 0.25 \\ 0 & 0.03125 & 0.03125 & 0.125 \end{pmatrix}$$

5. Exact 2, 3, 4

7. Exact 2, 2, 2

9. $\mathbf{C} = \begin{pmatrix} 0 & 0 & -1 \\ 0 & 0 & -1 \\ 0 & 0 & -1 \end{pmatrix}$, $\lambda_1 = -1, \lambda_2 = \lambda_3 = 0$, $\mathbf{I} - \mathbf{A} = \begin{pmatrix} 0 & 0 & -1 \\ 1 & 0 & 0 \\ \frac{1}{3} & \frac{2}{3} & 0 \end{pmatrix}$

$\lambda^3 + \lambda/3 + \frac{2}{3} = 0$, $\lambda_1 \approx -0.748$ (by Newton's method), hence $|\lambda_2| = |\lambda_3| = \sqrt{\frac{2}{3}/0.748} = 0.94$ because the constant term $\frac{2}{3}$ equals -1 times the product of all three eigenvalues, and λ_2 and λ_3 are conjugate.

11. $\mathbf{X}_{(1)} = \begin{pmatrix} 0.49 & -0.1 & 0.51 \\ 0 & 0.2 & 0 \\ -0.51 & 0.3 & -1.47 \end{pmatrix}$, $\mathbf{A}^{-1} = \begin{pmatrix} 0.5 & -0.1 & 0.5 \\ 0 & 0.2 & 0 \\ -0.5 & 0.3 & -1.5 \end{pmatrix}$

SECTION 19.11, page 815

1. -0.0002

5. $\mathbf{AB} = \begin{pmatrix} 1 & 0 \\ -0.1 & 1 \end{pmatrix}$, $\mathbf{BA} = \begin{pmatrix} 11 & 10 \\ -10 & -9 \end{pmatrix}$

SECTION 19.12, page 819

1. $y = -10.8 + 0.84x$

3. $y = -8.05 + 12.1x$

5. $y = 2.44 + 0.77x$

7. $y = 0.127 + 8.822\,t$. Speed 8.822 [m/sec]

9. $n = 5$, $\Sigma x_i = 13$, $\Sigma x_i^2 = 57$, $\Sigma x_i^3 = 289$, $\Sigma x_i^4 = 1569$, $\Sigma y_i = 9$, $\Sigma x_i y_i = 29$, $\Sigma x_i^2 y_i = 161$,

$$y = 2.825\,159 - 2.049\,040x + 0.377\,398x^2$$

13. $\mathbf{C} = (c_{jk})$, $c_{jk} = x_j^{\,k}$, $\mathbf{b}^\mathsf{T} = (b_0 \cdots b_m)$

15. $y^* = a^* + bx$, where $y^* = \ln y$ and $a^* = \ln b_0$

SECTION 19.13, page 824

1. 3, -3, radii 4, 4

3. $1/\sqrt{2}$, $-1/\sqrt{2}$, radii $1/\sqrt{2}$, $1/\sqrt{2}$

5. -9, -9, -9, radii 1, 2, 1

7. $\sqrt{50}$ **9.** $\sqrt{112}$ **11.** $\sqrt{247}$

15. $4 \leqq \lambda \leqq 6,\ \lambda = 5$ **17.** Circular arcs

SECTION 19.14, page 827

1. $m_0 = 1304,\ m_1 = 6412,\ m_2 = 31736,\ q = 4.9172,$
$|\epsilon| \leqq \sqrt{24.3374 - 24.1788} \approx 0.398$

3. $q = 3,\ |\epsilon| \leqq \sqrt{5}$

7. $q = \dfrac{5}{4},\ |\epsilon| \leqq \dfrac{\sqrt{11}}{4} \approx 0.83.\ q = \dfrac{14}{9},\ |\epsilon| \leqq \dfrac{\sqrt{101}}{9} \approx 1.12$

9. $\mathbf{A}$ has the eigenvalue 1 or -1 (or both).

15. $q = m_1/m_0 = -43.787/10.43 = -4.198$, which gives $\lambda_3 \approx -4.2 + 4.9 = 0.7$ (exact $\lambda_3 = 0$). "In general" means provided that we do not by chance pick $\mathbf{x}_0$ orthogonal to an (unknown) eigenvector corresponding to λ_n.

SECTION 19.15, page 837

1. Use (1).

3.
$$\operatorname{ci}(x) \sim \sin x\left(-\frac{1}{x} + \frac{2!}{x^3} - \frac{4!}{x^5} + - \cdots\right) + \cos x\left(\frac{1}{x^2} - \frac{3!}{x^4} + \frac{5!}{x^6} - + \cdots\right)$$

5. $c(x) \sim -A(x)\sin x^2 + B(x)\cos x^2$ where
$$A(x) = \frac{1}{2}\left(\frac{1}{x} - \frac{1\cdot 3}{4x^5} + \frac{1\cdot 3\cdot 5\cdot 7}{16x^9} - + \cdots\right),$$
$$B(x) = \frac{1}{2}\left(\frac{1}{2x^3} - \frac{1\cdot 3\cdot 5}{8x^7} + - \cdots\right).$$

7.
$$Q(\alpha, x) \sim x^{\alpha}e^{-x}\left[\frac{1}{x} - \frac{1-\alpha}{x^2} + - \cdots + (-1)^{n-1}\frac{(1-\alpha)(2-\alpha)\cdots(n-1-\alpha)}{x^n} + \cdots\right]$$

9. $C(x) \sim \pi^{1/2}2^{-3/2} + A(x)\sin x^2 - B(x)\cos x^2$ with A and B as in Prob. 5.

13. $y' = x^{-1}(1 - \alpha + x)y - 1$, etc.

SECTION 20.3, page 848

1. $\bar{x} = 3.47,\ s^2 = 2.98$ **3.** $\bar{x} = 3,\ s^2 = 3.3$ **5.** 0.02

7. 350, 380, 30 **9.** 10 000, 1000, 100

15. $\bar{x} = 99.2,\ s^2 = 234.7$; grouped: $\bar{x} = 99.4,\ s^2 = 254.7$

SECTION 20.4, page 853

3. The space of all ordered triples of nonnegative numbers.

5. $A \cup B$ occurs if and only if $A \cap B$ or $A \cap B^c$ or $A^c \cap B$ occurs. These three events are mutually exclusive. Let n_1, n_2, n_3 be the corresponding absolute frequencies in the sample. Then $\tilde{f}(A) = (n_1 + n_2)/n$, $\tilde{f}(B) = (n_1 + n_3)/n$, $\tilde{f}(A \cap B) = n_1/n$, $\tilde{f}(A \cup B) = (n_1 + n_2 + n_3)/n$. From this, (1) follows.

SECTION 20.5, page 858

1. $\frac{31}{32}$

3. (a) $0.9^3 = 72.9\%$. (b) $\dfrac{90}{100} \cdot \dfrac{89}{99} \cdot \dfrac{88}{98} = 72.65\%$

5. (a) 24.75%, (b) 50.5%, (c) 24.75%, (d) 6.06%

7. 10.7%

9. $P(E_B|E_A) = P(E_A \cap E_B)/P(E_A) = 0.02/0.10 = 20\%$

11. $0.4/0.9 = 0.44$

15. $A = (A \cap B) \cup (A \cap B^c) = B \cup (A \cap B^c)$ and Axiom 3 imply $P(A) = P(B) + P(A \cap B^c) \geqq P(B)$ because $P(A \cap B^c) \geqq 0$.

SECTION 20.6, page 862

3. $\binom{10}{3} = 120$ 5. $\binom{100}{3} = 161\,700$ 7. 15, 15

9. The idea of proof is the same as that of Theorem 1, but instead of filling n places we now have to fill only k places. If repetitions are permitted, we have n elements for filling each of the k places.

11. 23.5, 0.5, 2%; 39 902, 400, 1%

15. b^r has the coefficient $\binom{p+q}{r}$ on the right and $\Sigma \binom{p}{k}\binom{q}{r-k}$ on the left.

SECTION 20.7, page 868

5. $k = 1/4000$, $F(x) = 0$ if $x < 2000$, $F(x) = x/4000 - 0.5$ if $2000 \leqq x < 6000$, $F(x) = 1$ if $x \geqq 6000$

7. $f(3) = \dfrac{1}{216}, f(4) = \dfrac{3}{216}$, etc.

11. $P(X > 1200) = \int_{1.2}^{2} 6[0.25 - (x - 1.5)^2]\, dx = 0.896$. Ans. $0.896^3 = 72\%$.

13. $X > b$, $X \geqq b$, $X < c$, $X \leqq c$, $X < b$ or $X > c$, $X \leqq b$ or $X > c$.

SECTION 20.8, page 872

1. $$\mu = \int_{-\infty}^{c} tf(t)\, dt + \int_{c}^{\infty} tf(t)\, dt$$
$$= -\int_{\infty}^{0} (c - x)f(c - x)\, dx + \int_{0}^{\infty} (c + x)f(c + x)\, dx$$
$$= 2c \int_{0}^{\infty} f(c + x)\, dx = c.$$
For a discrete distribution the proof is similar.

5. $(X - 4/3)/\sqrt{2/9}$

7. Set $c_1 = 1/\sigma$ and $c_2 = -\mu/\sigma$ in (5) and (6).

9. $k = 750$, $\mu = 1$, $\sigma^2 = 0.002$

11. $\mu = 1/2$, $\sigma^2 = 1/20$

15. $E(X - \mu) = E(X) - \mu E(1) = \mu - \mu = 0$

17. $E(X^k) = 2/(k + 2)$, $\sigma^2 = 1/18$

21. $E(X^k) = 1/(k + 1)$

23. Integrate by parts, $\sigma^2 = 2$, $\gamma = 4/2\sqrt{2} = \sqrt{2}$

SECTION 20.9, page 877

1. $f(x) = \binom{n}{x} 0.5^n$, $\mu = 0.5n$, $\sigma^2 = 0.25n$

3. $\frac{5}{256}$, that is, about 2%

5. $0.99^{10} \approx 90.4\%$

7. $1 - e^{-0.2} = 1 - 0.8187 = 0.1813$
(exact $p = 0.002$, $1 - 0.998^{100} = 1 - 0.8186 = 0.1814$)

9. $f(x) = 0.5^x e^{-0.5}/x!$, $f(0) + f(1) = e^{-0.5}(1.0 + 0.5) = 0.91$.
Ans. 9%.

11. 0.265 (cf. Table A7 in Appendix 4)

13. If a package of $N = 100$ items contains precisely $M = 10$ defectives, then the probability that 10 items drawn without replacement contain no defectives is

$$f(0) = \binom{10}{0}\binom{90}{10}\Big/\binom{100}{10} = \frac{90 \cdot 89 \cdots 81}{100 \cdot 99 \cdots 91} = 33\%.$$

Ans. 67%, so the method is very poor.

15. Let X be the number of defective screws in a sample of size n. The process will be halted if $X \geqq 1$. The manufacturer wants n to be such that $P(X \geqq 1) \approx 0.95$ when $p = 0.1$, thus $P(X = 0) = q^n = 0.9^n \approx 0.95$, and $n \ln 0.9 \approx \ln 0.05$, $n = 28.4$. Ans. $n = 28$ or 29.

17. $f(x_1, x_2) = \dfrac{20!}{x_1! x_2! (20 - x_1 - x_2)!} 0.03^{x_1} 0.05^{x_2} 0.92^{20 - x_1 - x_2}$

19. (*a*) 0, (*b*) 1/2, (*c*) 3/4, (*d*) 3/8, (*e*) 5/8

23. $G(0) = 1$, $G'(t) = e^{-\mu} \exp[\mu e^t] = \mu e^t G(t)$, $G''(t) = \mu e^t[G(t) + G'(t)]$,
$E(X^2) = G''(0) = \mu + \mu^2$, $\sigma^2 = E(X^2) - \mu^2 = \mu$

25. Use $x\binom{M}{x} = M\binom{M-1}{x-1}$ and (14) in Sec. 20.6 with $p = M - 1$,
$k = x - 1$, $q = N - M$, $r - k = n - x$.

SECTION 20.10, page 884

3. 0.1587, 0.5, 0.6915, 0.6247

5. $\Phi\left(\frac{0.03}{\sigma}\right) - \Phi\left(\frac{-0.03}{\sigma}\right) = 99\%$, $0.03/\sigma = 2.576$, $\sigma = 0.012$

7. About 160

9. $\Phi\left(\frac{4040 - 2020 + 0.5}{\sqrt{1010}}\right) - \Phi\left(\frac{2048 - 2020 - 0.5}{\sqrt{1010}}\right) = 19.3\%$

11. 876 kg

13. 13.1%

15. $P = \sum_{x=0}^{10} \binom{1000}{x} 0.01^x 0.99^{1000-x}$

$$\approx \Phi\left(\frac{10 - 10 + 0.5}{\sqrt{9.9}}\right) - \Phi\left(\frac{0 - 10 - 0.5}{\sqrt{9.9}}\right) = 0.564$$

(exact 0.583)

17. $\Phi^2(\infty) = \frac{1}{2\pi}\int_{-\infty}^{\infty}\int_{-\infty}^{\infty} e^{-u^2/2}e^{-v^2/2}\,du\,dv$

$$= \frac{1}{2\pi}\int_0^{2\pi}\int_0^{\infty} e^{-r^2/2}r\,dr\,d\theta = 1$$

$(u = r\cos\theta,\ v = r\sin\theta,\ du\,dv = r\,dr\,d\theta)$

19. Set $x^* = c_1x + c_2$, then $(x - \mu)/\sigma = (x^* - \mu^*)/\sigma^*$ and, for $c_1 > 0$, $F(x^*) = P(X^* \leqq x^*) = P(X \leqq x) = \Phi((x - \mu)/\sigma) = \Phi((x^* - \mu^*)/\sigma^*)$. To include the case $c_1 < 0$, note that $X^* = -X$ is normal with mean $-\mu$ and variance σ^2 since

$$F(x^*) = P(X^* \leqq x^*) = P(X \geqq -x^*) = 1 - \Phi\left(\frac{-x^* - \mu}{\sigma}\right) = \Phi\left(\frac{x^* + \mu}{\sigma}\right).$$

SECTION 20.11, page 892

1. (X, Y) takes a value in A, B, C, or D (cf. the figure) with probability $F(b_1, b_2)$, a value in A or C with probability $F(a_1, b_2)$, a value in C or D with probability $F(b_1, a_2)$, a value in C with probability $F(a_1, a_2)$, hence a value in B with probability given by the right-hand side of (2).

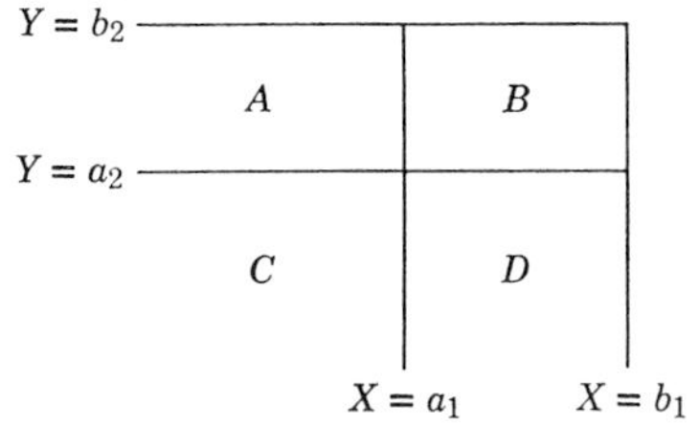

Problem 1

5. 20.200, about 0.007

7. 50%

9. (*a*) Independent, (*b*) $f_1(x) = 0.1e^{-0.1x}$ if $x > 0$, $f_2(y) = 0.1e^{-0.1y}$ if $y > 0$, (*c*) 36.8%

11. The distribution in Prob. 10 and that in the text before (6).

13. $k = 1/\pi$, $f_1(x) = \frac{2}{\pi}\sqrt{1 - x^2}$, $f_2(y) = \frac{2}{\pi}\sqrt{1 - y^2}$, 25%

SECTION 20.12, page 894

1. The items 38, 69, 02, 49, 23, 52, 73, 29, 09, 05

5. Class frequencies 1, 5, 7, 6, 1

SECTION 20.13, page 899

3. $\widehat{\theta} = n/\Sigma x_j = 1/\bar{x}$
5. $l = p^k(1-p)^{n-k}$, $k =$ number of successes in a sample of n, and $\partial \ln l/\partial p = 0$ yields $\widehat{p} = k/n$.

SECTION 20.14, page 907

1. CONF $\{17.84 \leqq \mu \leqq 18.96\}$
3. CONF $\{12.2 \leqq \mu \leqq 16.6\}$
5. CONF $\{4.25 \leqq \mu \leqq 4.49\}$
7. CONF $\{0.726 \leqq \mu \leqq 0.751\}$
9. CONF $\{0.105 \leqq \mu \leqq 0.107\}$; that error is 0.001, approximately (0.000 8 more exactly).
11. CONF $\{0.05 \leqq \sigma^2 \leqq 10\}$
13. $c_1 = (\sqrt{253} - 2.58)^2/2 = 88.8$, $c_2 = 170.9$, CONF $\{1.42 \leqq \sigma^2 \leqq 2.75\}$
15. Normal distributions, means -12, 24, 47, variances 25, 100, 400.
17. $Z = X + Y$ is normal with mean 105 and variance 1.25. Ans. $P(104 \leqq Z \leqq 106) = 63\%$.
19. $-D_2$ is normal with mean $-\mu_2$ and variance σ_2^2, cf. Theorem 1, Sec. 20.10. Hence $D = D_1 - D_2$ is normal with mean $\mu = \mu_1 - \mu_2 = 0.01$ and variance $\sigma^2 = \sigma_1^2 + \sigma_2^2 = 0.0002$. A bolt fits a hole if and only if D assumes a positive value. Ans.
$$P(D > 0) = P(0 < D < \infty) = \Phi(\infty) - \Phi\left(\frac{0 - 0.01}{\sqrt{0.0002}}\right) \approx 76\%.$$

SECTION 20.15, page 917

1. $t = \sqrt{7}(0.286 - 0)/4.31 = 0.18 < c = 1.94$ ($\alpha = 5\%$), do not reject the hypothesis.
3. If the hypothesis $p = 0.5$ is true, $X =$ *number of heads in* 4040 *trials* is approximately normal with $\mu = 2020$, $\sigma^2 = 1010$ (Sec. 20.10). $P(X \leqq c) = \Phi([c - 2020]/\sqrt{1010}) = 0.95$, $c = 2072 > 2048$, do not reject the hypothesis.
5. Alternative $\mu \neq 500$, $t = \sqrt{10}(496 - 500)/5 = -2.52 < c = -2.26$ (cf. Table A11, 9 degrees of freedom). Reject the hypothesis $\mu = 500$ g.
9. $\bar{x} = 376.3$, $\bar{y} = 335.3$, $s_1^2 = 1009.3$, $s_2^2 = 869.3$, $t_0 = 1.64 < c = 2.13$ ($\alpha = 5\%$, 4 degrees of freedom), do not reject the hypothesis.
13. Hypothesis H_0: not better. Alternative H_1: better. Under H_0 the random variable $X =$ *number of cases cured in* 200 *cases* is approximately normal with $\mu = np = 140$ and $\sigma^2 = npq = 42$, and the value $\alpha = 5\%$ gives $(c - 140)/\sqrt{42} = 1.645$, $c = 150.65 > 148$; do not reject H_0.
15. $v_0 = 679/254 = 2.67$. Choose $\alpha = 5\%$. Table A13 in Appendix 4 does not contain a value corresponding to (11, 5) degrees of freedom, but neighboring values in the table show that it must be greater than v_0, hence do not reject the hypothesis.

SECTION 20.16, page 922

1. LCL $= 1 - 2.58 \cdot 0.02/2 = 0.974$, UCL $= 1.026$
3. $2.58\sqrt{0.024}/\sqrt{2} = 0.283$, UCL $= 27.783$, LCL $= 27.217$

7. LCL and UCL will correspond to a significance level α which, in general, is not equal to 1%.

9. $\text{UCL} = np + 3\sqrt{np(1-p)} = 0.7 + 2.42 \approx 3$, $\text{CL} = np = 0.7$, $\text{LCL} = np - 3\sqrt{np(1-p)}$

SECTION 20.17, page 928

1. $(1-\theta)^2$

3. 0.9953, 0.9825, 0.9384, etc.

5. 0.028 (at $\theta = 0.054$)

7. $P(A; \theta) = \binom{N - N\theta}{5} \Big/ \binom{N}{5} \approx \theta^0(1-\theta)^{5-0} = (1-\theta)^5$,

$\text{AOQ}(\theta) \approx \theta(1-\theta)^5$

9. $\theta_0 \approx 0.01$, $\theta_1 \approx 0.37$

11. $\alpha = 5\%$, $\beta = 44\%$, which is unsatisfactorily large.

13. 9

15. 0.03

SECTION 20.18, page 930

1. $K = 2$, $n = 355 + 123 = 478$, $e_1 = 478 \cdot \frac{3}{4} = 358.5$, $e_2 = 478 \cdot \frac{1}{4} = 119.5$,

$$\chi_0^2 = \frac{(355.0 - 358.5)^2}{358.5} + \frac{(123.0 - 119.5)^2}{119.5} = 0.137.$$

We choose $\alpha = 5\%$. Then $c = 3.84$ (1 degree of freedom). Hence $\chi_0^2 < c$, that is, we assert that the deviation of the ratio in the sample from the theoretical ratio 3:1 is due to random fluctuation.

3. H_0: Number of defective screws the same for all 3 machines. We estimate $p = 15/600 = 2.5\%$. Then

$$\chi_0^2 = \tfrac{1}{5}(2^2 + 3^2 + 1^2) = 2.8 < 3.84$$

($\alpha = 5\%$, 1 degree of freedom). Ans. No.

5. $a = 32$ because $\chi_0^2 = \frac{1}{25}[(a-25)^2 + (50 - a - 25)^2] > c = 3.84$ when $a \geqq 32$ (or $a \leqq 18$).

7. $\bar{x} = 99.4$, $\tilde{\sigma} = 15.8$, $K = 5$ (endpoints $-\infty$, 85, 95, 105, 115, ∞), $\chi_0^2 = 0.7 < c = 5.99$ ($\alpha = 5\%$). Hypothesis not rejected.

9. Taking the last three rows together, we have $K - r - 1 = 9$, where $r = 1$ since the mean was estimated. $\chi_0^2 = 12.8 < c = 16.92$ ($\alpha = 5\%$). Do not reject the hypothesis.

SECTION 20.19, page 933

1. $P(X \leqq 2) = 0.5^6(1 + 6 + 15) = 34\%$. Do not reject the hypothesis $\tilde{\mu} = 0$.

3. $\alpha = 5\%$, $P(X \geqq 10) = \sum_{x=10}^{13} \binom{13}{x} 0.5^{13} = 4.6\% < \alpha$. Hypothesis rejected.

5. Hypothesis $\mu = 0$. Alternative $\mu > 0$, $\bar{x} = 1.58$, $t = \sqrt{10} \cdot 1.58/1.23 = 4.06 > c = 1.83$ ($\alpha = 5\%$). Hypothesis rejected.

9. 2 transpositions, $n = 8$, $P(T \leqq 2) \approx 0.1\%$. We reject the hypothesis and assert that there is negative trend.

SECTION 20.20, page 939

1. About 120 feet

3. $y = x - 0.1$

5. $y - 38.4 = 45.7(x - 0.70)$

7. $y - 48.89 = -1.32(x - 20.33)$

9. $2s_1^2 = 2$, $2s_{xy} = 2$, $b = 1$, $2s_2^2 = 2 + 2p^2/3$, $q_0 = 2p^2/3$,
$k = 12.7p/\sqrt{3} = 7.3p$ ($\gamma = 95\%$), CONF $\{1 - 7.3p \leqq \beta \leqq 1 + 7.3p\}$

APPENDIX 3 SOME FORMULAS FOR SPECIAL FUNCTIONS

Tables of numerical values see Appendix 4.

Exponential function e^x (Fig. 397)

$$e = 2.71828\ 18284\ 59045\ 23536\ 02874\ 71353$$

$$(1) \qquad e^x e^y = e^{x+y}, \qquad e^x/e^y = e^{x-y}, \qquad (e^x)^y = e^{xy}$$

Natural logarithm (Fig. 398)

$$(2) \quad \ln(xy) = \ln x + \ln y, \qquad \ln(x/y) = \ln x - \ln y, \qquad \ln(x^a) = a \ln x$$

$\ln x$ is the inverse of e^x, and $e^{\ln x} = x$, $e^{-\ln x} = e^{\ln(1/x)} = 1/x$.
Logarithm of base ten $\log_{10} x$ or simply $\log x$

$$(3) \quad \log x = M \ln x, \qquad M = \log e = 0.43429\ 44819\ 03251\ 82765\ 11289\ 18917$$

$$(4) \quad \ln x = \frac{1}{M} \log x, \qquad \frac{1}{M} = 2.30258\ 50929\ 94045\ 68401\ 79914\ 54684$$

$\log x$ is the inverse of 10^x, and $10^{\log x} = x$, $10^{-\log x} = 1/x$.

Sine and cosine functions (Figs. 399, 400). In calculus, angles are measured in radians, so that $\sin x$ and $\cos x$ have period 2π. $\sin x$ is odd, $\sin(-x) = -\sin x$, and $\cos x$ is even, $\cos(-x) = \cos x$.

$$1^\circ = 0.01745\ 32925\ 19943 \text{ radian}$$

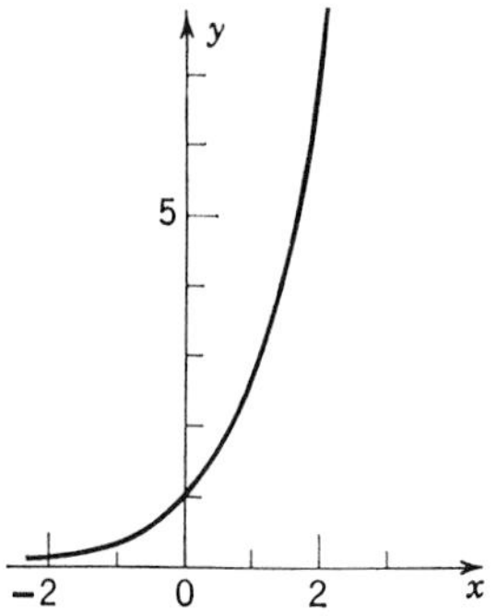

Fig. 397. Exponential function e^x

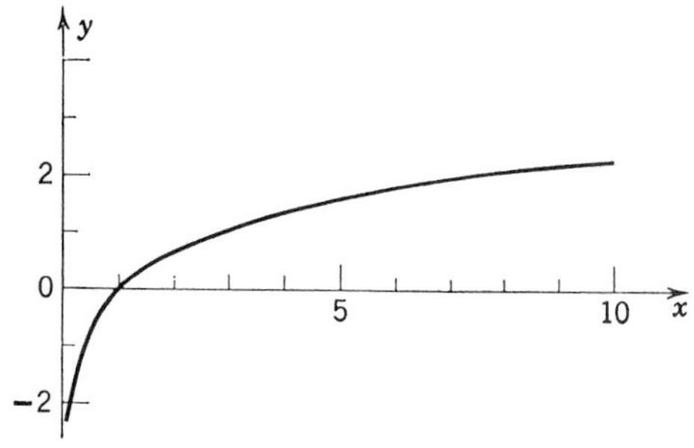

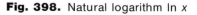

Fig. 398. Natural logarithm $\ln x$

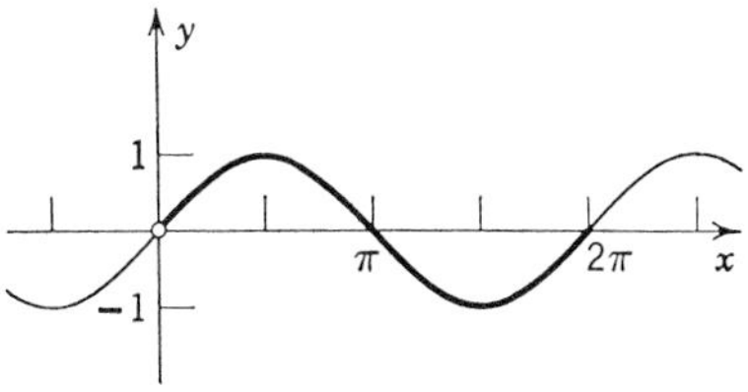

Fig. 399. sin x

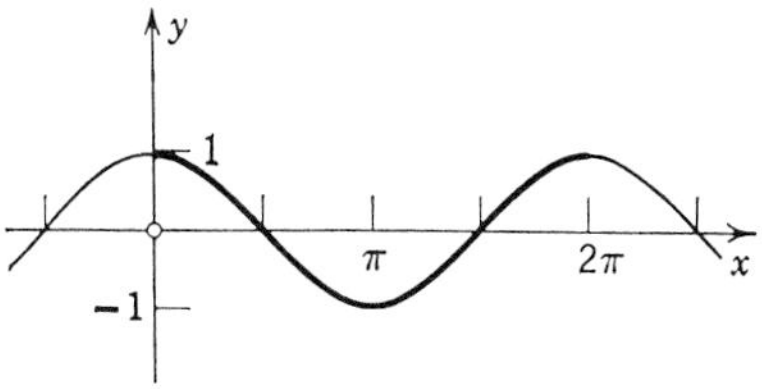

Fig. 400. cos x

$$1 \text{ radian} = 57° \ 17' \ 44.80625'' = 57.29577\,95131°$$

(5) $$\sin^2 x + \cos^2 x = 1$$

(6) $$\begin{cases} \sin(x+y) = \sin x \cos y + \cos x \sin y \\ \sin(x-y) = \sin x \cos y - \cos x \sin y \\ \cos(x+y) = \cos x \cos y - \sin x \sin y \\ \cos(x-y) = \cos x \cos y + \sin x \sin y \end{cases}$$

(7) $$\sin 2x = 2 \sin x \cos x, \qquad \cos 2x = \cos^2 x - \sin^2 x$$

(8) $$\begin{cases} \sin x = \cos\left(x - \frac{\pi}{2}\right) = \cos\left(\frac{\pi}{2} - x\right) \\ \cos x = \sin\left(x + \frac{\pi}{2}\right) = \sin\left(\frac{\pi}{2} - x\right) \end{cases}$$

(9) $$\sin(\pi - x) = \sin x, \qquad \cos(\pi - x) = -\cos x$$

(10) $$\cos^2 x = \tfrac{1}{2}(1 + \cos 2x), \qquad \sin^2 x = \tfrac{1}{2}(1 - \cos 2x)$$

(11) $$\begin{cases} \sin x \sin y = \tfrac{1}{2}[-\cos(x+y) + \cos(x-y)] \\ \cos x \cos y = \tfrac{1}{2}[\cos(x+y) + \cos(x-y)] \\ \sin x \cos y = \tfrac{1}{2}[\sin(x+y) + \sin(x-y)] \end{cases}$$

(12) $$\begin{cases} \sin u + \sin v = 2 \sin \frac{u+v}{2} \cos \frac{u-v}{2} \\ \cos u + \cos v = 2 \cos \frac{u+v}{2} \cos \frac{u-v}{2} \\ \cos v - \cos u = 2 \sin \frac{u+v}{2} \sin \frac{u-v}{2} \end{cases}$$

(13) $$A \cos x + B \sin x = \sqrt{A^2 + B^2} \cos(x \pm \delta), \qquad \tan \delta = \frac{\sin \delta}{\cos \delta} = \mp \frac{B}{A}$$

(14) $A \cos x + B \sin x = \sqrt{A^2 + B^2} \sin(x \pm \delta), \qquad \tan \delta = \dfrac{\sin \delta}{\cos \delta} = \pm \dfrac{A}{B}$

Tangent, cotangent, secant, cosecant (Figs. 401, 402)

(15) $\tan x = \dfrac{\sin x}{\cos x}, \quad \cot x = \dfrac{\cos x}{\sin x}, \quad \sec x = \dfrac{1}{\cos x}, \quad \csc x = \dfrac{1}{\sin x}$

(16) $\tan(x + y) = \dfrac{\tan x + \tan y}{1 - \tan x \tan y}, \quad \tan(x - y) = \dfrac{\tan x - \tan y}{1 + \tan x \tan y}$

Hyperbolic functions (hyperbolic sine sinh x, etc.; Figs. 403, 404)

(17) $\sinh x = \tfrac{1}{2}(e^x - e^{-x}), \qquad \cosh x = \tfrac{1}{2}(e^x + e^{-x})$

(18) $\tanh x = \dfrac{\sinh x}{\cosh x}, \qquad \coth x = \dfrac{\cosh x}{\sinh x}$

(19) $\cosh x + \sinh x = e^x, \qquad \cosh x - \sinh x = e^{-x}$

(20) $\cosh^2 x - \sinh^2 x = 1$

(21) $\sinh^2 x = \tfrac{1}{2}(\cosh 2x - 1), \qquad \cosh^2 x = \tfrac{1}{2}(\cosh 2x + 1)$

(22) $$\begin{cases} \sinh(x \pm y) = \sinh x \cosh y \pm \cosh x \sinh y \\ \cosh(x \pm y) = \cosh x \cosh y \pm \sinh x \sinh y \end{cases}$$

(23) $$\tanh(x \pm y) = \frac{\tanh x \pm \tanh y}{1 \pm \tanh x \tanh y}$$

Gamma function (cf. Fig. 405 and Table A3 in Appendix 4). The gamma function $\Gamma(\alpha)$ is defined by the integral

(24) $$\Gamma(\alpha) = \int_0^\infty e^{-t} t^{\alpha - 1}\, dt \qquad (\alpha > 0)$$

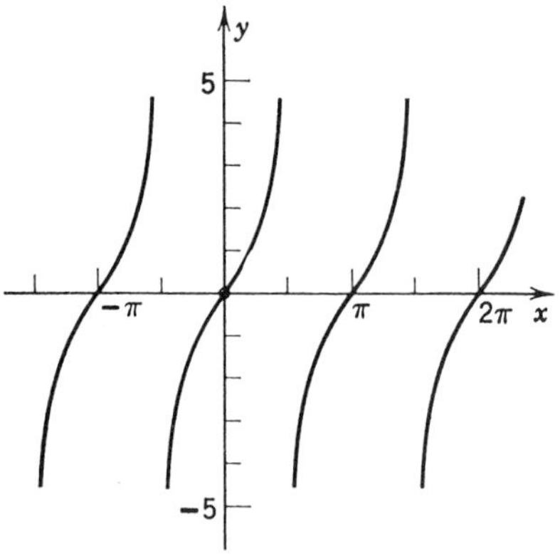

Fig. 401. tan x

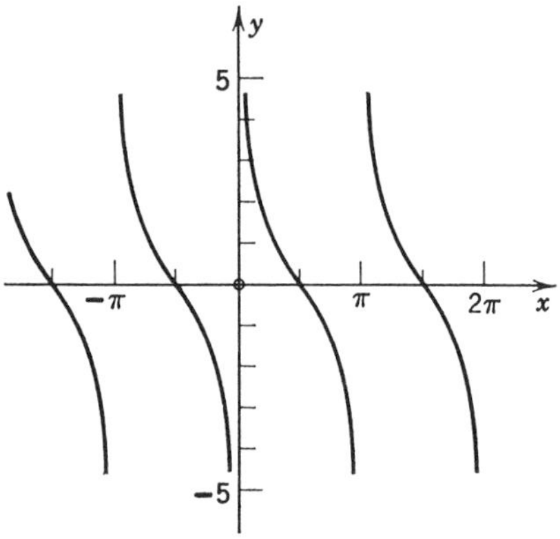

Fig. 402. cot x

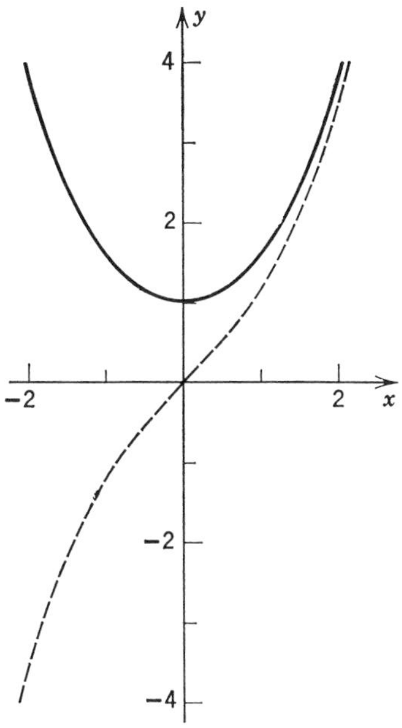

Fig. 403. sinh *x* (dashed) and cosh *x*

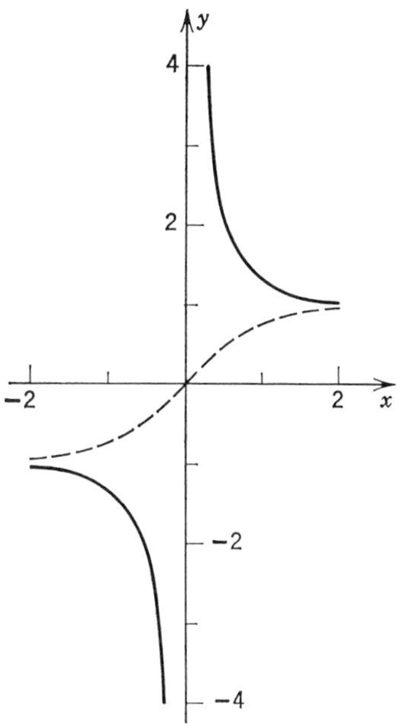

Fig. 404. tanh *x* (dashed) and coth *x*

which is meaningful only if $\alpha > 0$ (or, if we consider complex α, for those α whose real part is positive). Integration by parts gives the important *functional relation of the gamma function,*

$$\Gamma(\alpha + 1) = \alpha\Gamma(\alpha). \tag{25}$$

From (24) we readily have $\Gamma(1) = 1$; hence if α is a positive integer, say k, then by repeated application of (25) we obtain

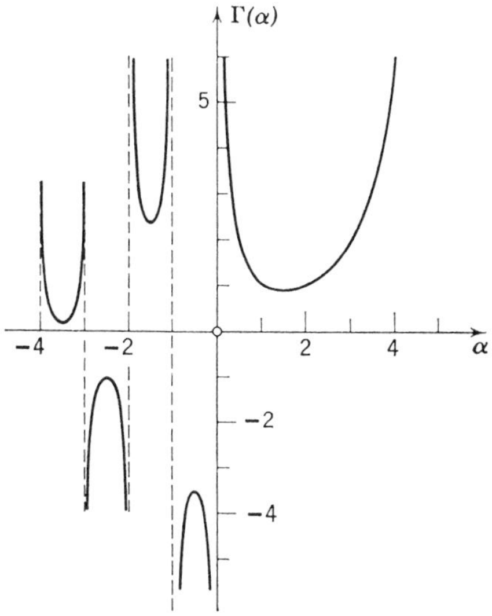

Fig. 405. Gamma function

(26) $$\Gamma(k+1) = k! \qquad (k = 0, 1, \cdots).$$

This shows that *the gamma function can be regarded as a generalization of the elementary factorial function.* [Sometimes the notation $(\alpha - 1)!$ for $\Gamma(\alpha)$ is used, even for noninteger values of α, and the gamma function is also known as the **factorial function.**]

By repeated application of (25) we obtain

$$\Gamma(\alpha) = \frac{\Gamma(\alpha+1)}{\alpha} = \frac{\Gamma(\alpha+2)}{\alpha(\alpha+1)} = \cdots = \frac{\Gamma(\alpha+k+1)}{\alpha(\alpha+1)(\alpha+2)\cdots(\alpha+k)},$$

and we may use this relation

(27) $$\Gamma(\alpha) = \frac{\Gamma(\alpha+k+1)}{\alpha(\alpha+1)\cdots(\alpha+k)} \qquad (\alpha \neq 0, -1, -2, \cdots)$$

for defining the gamma function for negative α ($\neq -1, -2, \cdots$), choosing for k the smallest integer such that $\alpha + k + 1 > 0$. *Together with* (1), *this then gives a definition of* $\Gamma(\alpha)$ *for all* α *not equal to zero or a negative integer* (Fig. 405).

It can be shown that the gamma function may also be represented as the limit of a product, namely, by the formula

(28) $$\Gamma(\alpha) = \lim_{n\to\infty} \frac{n!\, n^{\alpha}}{\alpha(\alpha+1)(\alpha+2)\cdots(\alpha+n)} \qquad (\alpha \neq 0, -1, \cdots).$$

From (27) or (28) we see that, for complex α, the gamma function $\Gamma(\alpha)$ is a meromorphic function which has simple poles at $\alpha = 0, -1, -2, \cdots$.

An approximation of the gamma function for large positive α is given by the **Stirling formula**

(29) $$\Gamma(\alpha+1) \approx \sqrt{2\pi\alpha}\left(\frac{\alpha}{e}\right)^{\alpha}$$

where e is the base of the natural logarithm. We finally mention the special value

(30) $$\Gamma(\tfrac{1}{2}) = \sqrt{\pi}.$$

Incomplete gamma functions

(31) $$P(\alpha, x) = \int_0^x e^{-t} t^{\alpha-1}\, dt, \qquad Q(\alpha, x) = \int_x^\infty e^{-t} t^{\alpha-1}\, dt \qquad (\alpha > 0)$$

(32) $$\Gamma(\alpha) = P(\alpha, x) + Q(\alpha, x)$$

Beta function

(33) $$B(x, y) = \int_0^1 t^{x-1}(1-t)^{y-1}\, dt \qquad (x > 0, y > 0)$$

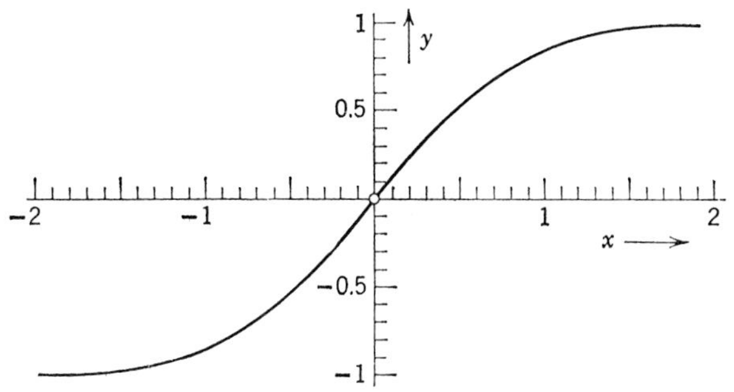

Fig. 406. Error function

Representation in terms of gamma functions:

(34)
$$B(x, y) = \frac{\Gamma(x)\Gamma(y)}{\Gamma(x + y)}$$

Error function (cf. Fig. 406 and Table A5 in Appendix 4)

(35)
$$\operatorname{erf} x = \frac{2}{\sqrt{\pi}} \int_0^x e^{-t^2}\, dt$$

(36)
$$\operatorname{erf} x = \frac{2}{\sqrt{\pi}} \left(x - \frac{x^3}{1!\,3} + \frac{x^5}{2!\,5} - \frac{x^7}{3!\,7} + - \cdots \right)$$

$\operatorname{erf}(\infty) = 1$, *complementary error function*

(37)
$$\operatorname{erfc} x = 1 - \operatorname{erf} x = \frac{2}{\sqrt{\pi}} \int_x^\infty e^{-t^2}\, dt$$

Fresnel integrals[1] (Fig. 407)

(38)
$$C(x) = \int_0^x \cos(t^2)\, dt, \qquad S(x) = \int_0^x \sin(t^2)\, dt$$

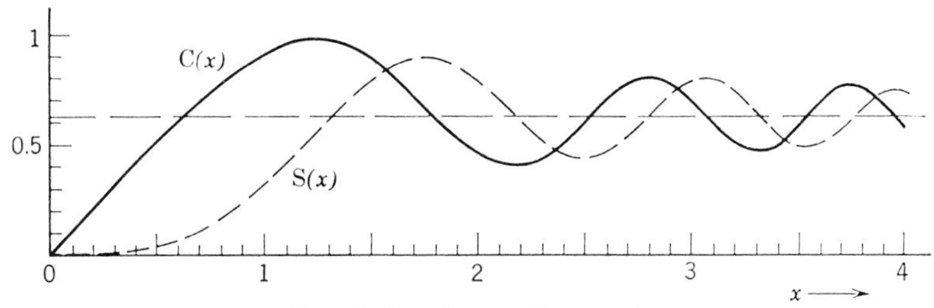

Fig. 407. Fresnel integrals

[1]AUGUSTIN FRESNEL (1788–1827), French physicist and mathematician. For tables see Refs. [A6], [A8]; cf. also *Acta mathematica*, vol. 85 (1951), p. 180, and vol. 89 (1953), p. 130.

$C(\infty) = \sqrt{\pi/8}$, $S(\infty) = \sqrt{\pi/8}$, *complementary functions*

$$c(x) = \sqrt{\frac{\pi}{8}} - C(x) = \int_x^\infty \cos(t^2)\,dt \tag{39}$$
$$s(x) = \sqrt{\frac{\pi}{8}} - S(x) = \int_x^\infty \sin(t^2)\,dt$$

Sine integral (cf. Fig. 408 and Table A5 in Appendix 4)

$$\mathrm{Si}(x) = \int_0^x \frac{\sin t}{t}\,dt \tag{40}$$

$\mathrm{Si}(\infty) = \pi/2$, *complementary function*

$$\mathrm{si}(x) = \frac{\pi}{2} - \mathrm{Si}(x) = \int_x^\infty \frac{\sin t}{t}\,dt \tag{41}$$

Cosine integral (cf. Table A5 in Appendix 4)

$$\mathrm{ci}(x) = \int_x^\infty \frac{\cos t}{t}\,dt \qquad (x > 0) \tag{42}$$

Exponential integral (cf. Sec. 19.15)

$$\mathrm{Ei}(x) = \int_x^\infty \frac{e^{-t}}{t}\,dt \qquad (x > 0) \tag{43}$$

Logarithmic integral

$$\mathrm{li}(x) = \int_0^x \frac{dt}{\ln t} \tag{44}$$

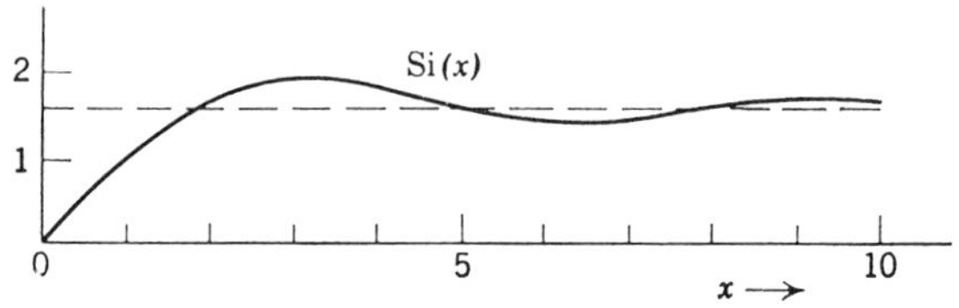

Fig. 408. Sine integral

APPENDIX

4 TABLES

Tables of Laplace Transforms in Sec. 5.8.

Table A1
Some Elementary Functions

x	$\sin x$	$\cos x$	$\tan x$	e^x	$\sinh x$	$\cosh x$
0.0	0.00000	1.00000	0.00000	1.00000	0.00000	1.00000
0.1	0.09983	0.99500	0.10033	1.10517	0.10017	1.00500
0.2	0.19867	0.98007	0.20271	1.22140	0.20134	1.02007
0.3	0.29552	0.95534	0.30934	1.34986	0.30452	1.04534
0.4	0.38942	0.92106	0.42279	1.49182	0.41075	1.08107
0.5	0.47943	0.87758	0.54630	1.64872	0.52110	1.12763
0.6	0.56464	0.82534	0.68414	1.82212	0.63665	1.18547
0.7	0.64422	0.76484	0.84229	2.01375	0.75858	1.25517
0.8	0.71736	0.69671	1.02964	2.22554	0.88811	1.33743
0.9	0.78333	0.62161	1.26016	2.45960	1.02652	1.43309
1.0	0.84147	0.54030	1.55741	2.71828	1.17520	1.54308
1.1	0.89121	0.45360	1.96476	3.00417	1.33565	1.66852
1.2	0.93204	0.36236	2.57215	3.32012	1.50946	1.81066
1.3	0.96356	0.26750	3.60210	3.66930	1.69838	1.97091
1.4	0.98545	0.16997	5.79788	4.05520	1.90430	2.15090
1.5	0.99749	0.07074	14.10142	4.48169	2.12928	2.35241
1.6	0.99957	−0.02920	−34.23253	4.95303	2.37557	2.57746
1.7	0.99166	−0.12884	−7.69660	5.47395	2.64563	2.82832
1.8	0.97385	−0.22720	−4.28626	6.04965	2.94217	3.10747
1.9	0.94630	−0.32329	−2.92710	6.68589	3.26816	3.41773
2.0	0.90930	−0.41615	−2.18504	7.38906	3.62686	3.76220

x	$\ln x$	x	$\ln x$	x	$\ln x$	x	$\ln x$
1.0	0.00000	2.0	0.69315	3.0	1.09861	5	1.60944
1.1	0.09531	2.1	0.74194	3.1	1.13140	7	1.94591
1.2	0.18232	2.2	0.78846	3.2	1.16315	11	2.39790
1.3	0.26236	2.3	0.83291	3.3	1.19392	13	2.56495
1.4	0.33647	2.4	0.87547	3.4	1.22378	17	2.83321
1.5	0.40547	2.5	0.91629	3.5	1.25276	19	2.94444
1.6	0.47000	2.6	0.95551	3.6	1.28093	23	3.13549
1.7	0.53063	2.7	0.99325	3.7	1.30833	29	3.36730
1.8	0.58779	2.8	1.02962	3.8	1.33500	31	3.43399
1.9	0.64185	2.9	1.06471	3.9	1.36098	37	3.61092

Table A1 (*continued*)

$\frac{y}{x}$	$\arctan\frac{y}{x}$	$\frac{y}{x}$	$\arctan\frac{y}{x}$	$\frac{y}{x}$	$\arctan\frac{y}{x}$	$\frac{y}{x}$	$\arctan\frac{y}{x}$
0.0	0.00000	1.0	0.78540	2.0	1.10715	4.0	1.32582
0.1	0.09967	1.1	0.83298	2.2	1.14417	4.5	1.35213
0.2	0.19740	1.2	0.87606	2.4	1.17601	5.0	1.37340
0.3	0.29146	1.3	0.91510	2.6	1.20362	5.5	1.39094
0.4	0.38051	1.4	0.95055	2.8	1.22777	6.0	1.40565
0.5	0.46365	1.5	0.98279	3.0	1.24905	7.0	1.42890
0.6	0.54042	1.6	1.01220	3.2	1.26791	8.0	1.44644
0.7	0.61073	1.7	1.03907	3.4	1.28474	9.0	1.46014
0.8	0.67474	1.8	1.06370	3.6	1.29985	10.0	1.47113
0.9	0.73282	1.9	1.08632	3.8	1.31347	11.0	1.48014

Table A2
Bessel Functions

For more extensive tables see Ref. [A8] in Appendix 1.

x	$J_0(x)$	$J_1(x)$	x	$J_0(x)$	$J_1(x)$	x	$J_0(x)$	$J_1(x)$
0.0	1.0000	0.0000	3.0	−0.2601	0.3391	6.0	0.1506	−0.2767
0.1	0.9975	0.0499	3.1	−0.2921	0.3009	6.1	0.1773	−0.2559
0.2	0.9900	0.0995	3.2	−0.3202	0.2613	6.2	0.2017	−0.2329
0.3	0.9776	0.1483	3.3	−0.3443	0.2207	6.3	0.2238	−0.2081
0.4	0.9604	0.1960	3.4	−0.3643	0.1792	6.4	0.2433	−0.1816
0.5	0.9385	0.2423	3.5	−0.3801	0.1374	6.5	0.2601	−0.1538
0.6	0.9120	0.2867	3.6	−0.3918	0.0955	6.6	0.2740	−0.1250
0.7	0.8812	0.3290	3.7	−0.3992	0.0538	6.7	0.2851	−0.0953
0.8	0.8463	0.3688	3.8	−0.4026	0.0128	6.8	0.2931	−0.0652
0.9	0.8075	0.4059	3.9	−0.4018	−0.0272	6.9	0.2981	−0.0349
1.0	0.7652	0.4401	4.0	−0.3971	−0.0660	7.0	0.3001	−0.0047
1.1	0.7196	0.4709	4.1	−0.3887	−0.1033	7.1	0.2991	0.0252
1.2	0.6711	0.4983	4.2	−0.3766	−0.1386	7.2	0.2951	0.0543
1.3	0.6201	0.5220	4.3	−0.3610	−0.1719	7.3	0.2882	0.0826
1.4	0.5669	0.5419	4.4	−0.3423	−0.2028	7.4	0.2786	0.1096
1.5	0.5118	0.5579	4.5	−0.3205	−0.2311	7.5	0.2663	0.1352
1.6	0.4554	0.5699	4.6	−0.2961	−0.2566	7.6	0.2516	0.1592
1.7	0.3980	0.5778	4.7	−0.2693	−0.2791	7.7	0.2346	0.1813
1.8	0.3400	0.5815	4.8	−0.2404	−0.2985	7.8	0.2154	0.2014
1.9	0.2818	0.5812	4.9	−0.2097	−0.3147	7.9	0.1944	0.2192
2.0	0.2239	0.5767	5.0	−0.1776	−0.3276	8.0	0.1717	0.2346
2.1	0.1666	0.5683	5.1	−0.1443	−0.3371	8.1	0.1475	0.2476
2.2	0.1104	0.5560	5.2	−0.1103	−0.3432	8.2	0.1222	0.2580
2.3	0.0555	0.5399	5.3	−0.0758	−0.3460	8.3	0.0960	0.2657
2.4	0.0025	0.5202	5.4	−0.0412	−0.3453	8.4	0.0692	0.2708
2.5	−0.0484	0.4971	5.5	−0.0068	−0.3414	8.5	0.0419	0.2731
2.6	−0.0968	0.4708	5.6	0.0270	−0.3343	8.6	0.0146	0.2728
2.7	−0.1424	0.4416	5.7	0.0599	−0.3241	8.7	−0.0125	0.2697
2.8	−0.1850	0.4097	5.8	0.0917	−0.3110	8.8	−0.0392	0.2641
2.9	−0.2243	0.3754	5.9	0.1220	−0.2951	8.9	−0.0653	0.2559

$J_0(x) = 0$ for $x = 2.405, 5.520, 8.654, 11.792, 14.931, \cdots$

$J_1(x) = 0$ for $x = 0, 3.832, 7.016, 10.173, 13.324, \cdots$

Table A2 (*continued*)

x	$Y_0(x)$	$Y_1(x)$	x	$Y_0(x)$	$Y_1(x)$	x	$Y_0(x)$	$Y_1(x)$
0.0	$(-\infty)$	$(-\infty)$	2.5	0.498	0.146	5.0	−0.309	0.148
0.5	−0.445	−1.471	3.0	0.377	0.325	5.5	−0.339	−0.024
1.0	0.088	−0.781	3.5	0.189	0.410	6.0	−0.288	−0.175
1.5	0.382	−0.412	4.0	−0.017	0.398	6.5	−0.173	−0.274
2.0	0.510	−0.107	4.5	−0.195	0.301	7.0	−0.026	−0.303

Table A3
Gamma Function (cf. (24) in Appendix 3)

α	$\Gamma(\alpha)$	α	$\Gamma(\alpha)$	α	$\Gamma(\alpha)$	α	$\Gamma(\alpha)$	α	$\Gamma(\alpha)$
1.00	1.000 000	1.20	0.918 169	1.40	0.887 264	1.60	0.893 515	1.80	0.931 384
1.02	0.988 844	1.22	0.913 106	1.42	0.886 356	1.62	0.895 924	1.82	0.936 845
1.04	0.978 438	1.24	0.908 521	1.44	0.885 805	1.64	0.898 642	1.84	0.942 612
1.06	0.968 744	1.26	0.904 397	1.46	0.885 604	1.66	0.901 668	1.86	0.948 687
1.08	0.959 725	1.28	0.900 718	1.48	0.885 747	1.68	0.905 001	1.88	0.955 071
1.10	0.951 351	1.30	0.897 471	1.50	0.886 227	1.70	0.908 639	1.90	0.961 766
1.12	0.943 590	1.32	0.894 640	1.52	0.887 039	1.72	0.912 581	1.92	0.968 774
1.14	0.936 416	1.34	0.892 216	1.54	0.888 178	1.74	0.916 826	1.94	0.976 099
1.16	0.929 803	1.36	0.890 185	1.56	0.889 639	1.76	0.921 375	1.96	0.983 743
1.18	0.923 728	1.38	0.888 537	1.58	0.891 420	1.78	0.926 227	1.98	0.991 708
1.20	0.918 169	1.40	0.887 264	1.60	0.893 515	1.80	0.931 384	2.00	1.000 000

Table A4
Factorial Function

n	$n!$	$\log(n!)$	n	$n!$	$\log(n!)$	n	$n!$	$\log(n!)$
1	1	0.000 000	6	720	2.857 332	11	39 916 800	7.601 156
2	2	0.301 030	7	5 040	3.702 431	12	479 001 600	8.680 337
3	6	0.778 151	8	40 320	4.605 521	13	6 227 020 800	9.794 280
4	24	1.380 211	9	362 880	5.559 763	14	87 178 291 200	10.940 408
5	120	2.079 181	10	3 628 800	6.559 763	15	1 307 674 368 000	12.116 500

Table A5
Error Function, Sine and Cosine Integrals (cf. (35), (40), (42) in Appendix 3)

x	erf x	Si(x)	ci(x)	x	erf x	Si(x)	ci(x)
0.0	0.0000	0.0000	∞	2.0	0.9953	1.6054	−0.4230
0.2	0.2227	0.1996	1.0422	2.2	0.9981	1.6876	−0.3751
0.4	0.4284	0.3965	0.3788	2.4	0.9993	1.7525	−0.3173
0.6	0.6039	0.5881	0.0223	2.6	0.9998	1.8004	−0.2533
0.8	0.7421	0.7721	−0.1983	2.8	0.9999	1.8321	−0.1865
1.0	0.8427	0.9461	−0.3374	3.0	1.0000	1.8487	−0.1196
1.2	0.9103	1.1080	−0.4205	3.2	1.0000	1.8514	−0.0553
1.4	0.9523	1.2562	−0.4620	3.4	1.0000	1.8419	0.0045
1.6	0.9763	1.3892	−0.4717	3.6	1.0000	1.8219	0.0580
1.8	0.9891	1.5058	−0.4568	3.8	1.0000	1.7934	0.1038
2.0	0.9953	1.6054	−0.4230	4.0	1.0000	1.7582	0.1410

Table A6
Binomial Distribution

Probability function $f(x)$ (cf. (2), Sec. 20.9) and distribution function $F(x)$

n	x	$p = 0.1$ $f(x)$	$p = 0.1$ $F(x)$	$p = 0.2$ $f(x)$	$p = 0.2$ $F(x)$	$p = 0.3$ $f(x)$	$p = 0.3$ $F(x)$	$p = 0.4$ $f(x)$	$p = 0.4$ $F(x)$	$p = 0.5$ $f(x)$	$p = 0.5$ $F(x)$
		0.		0.		0.		0.		0.	
1	0	9000	0.9000	8000	0.8000	7000	0.7000	6000	0.6000	5000	0.5000
	1	1000	1.0000	2000	1.0000	3000	1.0000	4000	1.0000	5000	1.0000
	0	8100	0.8100	6400	0.6400	4900	0.4900	3600	0.3600	2500	0.2500
2	1	1800	0.9900	3200	0.9600	4200	0.9100	4800	0.8400	5000	0.7500
	2	0100	1.0000	0400	1.0000	0900	1.0000	1600	1.0000	2500	1.0000
	0	7290	0.7290	5120	0.5120	3430	0.3430	2160	0.2160	1250	0.1250
	1	2430	0.9720	3840	0.8960	4410	0.7840	4320	0.6480	3750	0.5000
3	2	0270	0.9990	0960	0.9920	1890	0.9730	2880	0.9360	3750	0.8750
	3	0010	1.0000	0080	1.0000	0270	1.0000	0640	1.0000	1250	1.0000
	0	6561	0.6561	4096	0.4096	2401	0.2401	1296	0.1296	0625	0.0625
	1	2916	0.9477	4096	0.8192	4116	0.6517	3456	0.4752	2500	0.3125
4	2	0486	0.9963	1536	0.9728	2646	0.9163	3456	0.8208	3750	0.6875
	3	0036	0.9999	0256	0.9984	0756	0.9919	1536	0.9744	2500	0.9375
	4	0001	1.0000	0016	1.0000	0081	1.0000	0256	1.0000	0625	1.0000
	0	5905	0.5905	3277	0.3277	1681	0.1681	0778	0.0778	0313	0.0313
	1	3281	0.9185	4096	0.7373	3602	0.5282	2592	0.3370	1563	0.1875
	2	0729	0.9914	2048	0.9421	3087	0.8369	3456	0.6826	3125	0.5000
5	3	0081	0.9995	0512	0.9933	1323	0.9692	2304	0.9130	3125	0.8125
	4	0005	1.0000	0064	0.9997	0284	0.9976	0768	0.9898	1563	0.9688
	5	0000	1.0000	0003	1.0000	0024	1.0000	0102	1.0000	0313	1.0000
	0	5314	0.5314	2621	0.2621	1176	0.1176	0467	0.0467	0156	0.0156
	1	3543	0.8857	3932	0.6554	3025	0.4202	1866	0.2333	0938	0.1094
	2	0984	0.9841	2458	0.9011	3241	0.7443	3110	0.5443	2344	0.3438
6	3	0146	0.9987	0819	0.9830	1852	0.9295	2765	0.8208	3125	0.6563
	4	0012	0.9999	0154	0.9984	0595	0.9891	1382	0.9590	2344	0.8906
	5	0001	1.0000	0015	0.9999	0102	0.9993	0369	0.9959	0938	0.9844
	6	0000	1.0000	0001	1.0000	0007	1.0000	0041	1.0000	0156	1.0000
	0	4783	0.4783	2097	0.2097	0824	0.0824	0280	0.0280	0078	0.0078
	1	3720	0.8503	3670	0.5767	2471	0.3294	1306	0.1586	0547	0.0625
	2	1240	0.9743	2753	0.8520	3177	0.6471	2613	0.4199	1641	0.2266
	3	0230	0.9973	1147	0.9667	2269	0.8740	2903	0.7102	2734	0.5000
7	4	0026	0.9998	0287	0.9953	0972	0.9712	1935	0.9037	2734	0.7734
	5	0002	1.0000	0043	0.9996	0250	0.9962	0774	0.9812	1641	0.9375
	6	0000	1.0000	0004	1.0000	0036	0.9998	0172	0.9984	0547	0.9922
	7	0000	1.0000	0000	1.0000	0002	1.0000	0016	1.0000	0078	1.0000
	0	4305	0.4305	1678	0.1678	0576	0.0576	0168	0.0168	0039	0.0039
	1	3826	0.8131	3355	0.5033	1977	0.2553	0896	0.1064	0313	0.0352
	2	1488	0.9619	2936	0.7969	2965	0.5518	2090	0.3154	1094	0.1445
	3	0331	0.9950	1468	0.9437	2541	0.8059	2787	0.5941	2188	0.3633
8	4	0046	0.9996	0459	0.9896	1361	0.9420	2322	0.8263	2734	0.6367
	5	0004	1.0000	0092	0.9988	0467	0.9887	1239	0.9502	2188	0.8555
	6	0000	1.0000	0011	0.9999	0100	0.9987	0413	0.9915	1094	0.9648
	7	0000	1.0000	0001	1.0000	0012	0.9999	0079	0.9993	0313	0.9961
	8	0000	1.0000	0000	1.0000	0001	1.0000	0007	1.0000	0039	1.0000

Table A7
Poisson Distribution

Probability function $f(x)$ (cf. (5), Sec. 20.9) and distribution function $F(x)$

x	$\mu = 0.1$		$\mu = 0.2$		$\mu = 0.3$		$\mu = 0.4$		$\mu = 0.5$	
	$f(x)$	$F(x)$	$f(x)$	$F(x)$	$f(x)$	$F(x)$	$f(x)$	$F(x)$	$f(x)$	$F(x)$
	0.		**0.**		**0.**		**0.**		**0.**	
0	9048	0.9048	8187	0.8187	7408	0.7408	6703	0.6703	6065	0.6065
1	0905	0.9953	1637	0.9825	2222	0.9631	2681	0.9384	3033	0.9098
2	0045	0.9998	0164	0.9989	0333	0.9964	0536	0.9921	0758	0.9856
3	0002	1.0000	0011	0.9999	0033	0.9997	0072	0.9992	0126	0.9982
4	0000	1.0000	0001	1.0000	0003	1.0000	0007	0.9999	0016	0.9998
5							0001	1.0000	0002	1.0000

x	$\mu = 0.6$		$\mu = 0.7$		$\mu = 0.8$		$\mu = 0.9$		$\mu = 1$	
	$f(x)$	$F(x)$	$f(x)$	$F(x)$	$f(x)$	$F(x)$	$f(x)$	$F(x)$	$f(x)$	$F(x)$
	0.		**0.**		**0.**		**0.**		**0.**	
0	5488	0.5488	4966	0.4966	4493	0.4493	4066	0.4066	3679	0.3679
1	3293	0.8781	3476	0.8442	3595	0.8088	3659	0.7725	3679	0.7358
2	0988	0.9769	1217	0.9659	1438	0.9526	1647	0.9371	1839	0.9197
3	0198	0.9966	0284	0.9942	0383	0.9909	0494	0.9865	0613	0.9810
4	0030	0.9996	0050	0.9992	0077	0.9986	0111	0.9977	0153	0.9963
5	0004	1.0000	0007	0.9999	0012	0.9998	0020	0.9997	0031	0.9994
6			0001	1.0000	0002	1.0000	0003	1.0000	0005	0.9999
7									0001	1.0000

x	$\mu = 1.5$		$\mu = 2$		$\mu = 3$		$\mu = 4$		$\mu = 5$	
	$f(x)$	$F(x)$	$f(x)$	$F(x)$	$f(x)$	$F(x)$	$f(x)$	$F(x)$	$f(x)$	$F(x)$
	0.		**0.**		**0.**		**0.**		**0.**	
0	2231	0.2231	1353	0.1353	0498	0.0498	0183	0.0183	0067	0.0067
1	3347	0.5578	2707	0.4060	1494	0.1991	0733	0.0916	0337	0.0404
2	2510	0.8088	2707	0.6767	2240	0.4232	1465	0.2381	0842	0.1247
3	1255	0.9344	1804	0.8571	2240	0.6472	1954	0.4335	1404	0.2650
4	0471	0.9814	0902	0.9473	1680	0.8153	1954	0.6288	1755	0.4405
5	0141	0.9955	0361	0.9834	1008	0.9161	1563	0.7851	1755	0.6160
6	0035	0.9991	0120	0.9955	0504	0.9665	1042	0.8893	1462	0.7622
7	0008	0.9998	0034	0.9989	0216	0.9881	0595	0.9489	1044	0.8666
8	0001	1.0000	0009	0.9998	0081	0.9962	0298	0.9786	0653	0.9319
9			0002	1.0000	0027	0.9989	0132	0.9919	0363	0.9682
10					0008	0.9997	0053	0.9972	0181	0.9863
11					0002	0.9999	0019	0.9991	0082	0.9945
12					0001	1.0000	0006	0.9997	0034	0.9980
13							0002	0.9999	0013	0.9993
14							0001	1.0000	0005	0.9998
15									0002	0.9999
16									0000	1.0000

Table A8
Normal Distribution

Values of the distribution function $\Phi(z)$ (cf. (4), Sec. 20.10)
$\Phi(-z) = 1 - \Phi(z)$, $\Phi(0) = 0.5000$

z	$\Phi(z)$	z	$\Phi(z)$	z	$\Phi(z)$	z	$\Phi(z)$	z	$\Phi(z)$	z	$\Phi(z)$
	0.		**0.**		**0.**		**0.**		**0.**		**0.**
0.01	5040	0.51	6950	1.01	8438	1.51	9345	2.01	9778	2.51	9940
0.02	5080	0.52	6985	1.02	8461	1.52	9357	2.02	9783	2.52	9941
0.03	5120	0.53	7019	1.03	8485	1.53	9370	2.03	9788	2.53	9943
0.04	5160	0.54	7054	1.04	8508	1.54	9382	2.04	9793	2.54	9945
0.05	5199	0.55	7088	1.05	8531	1.55	9394	2.05	9798	2.55	9946
0.06	5239	0.56	7123	1.06	8554	1.56	9406	2.06	9803	2.56	9948
0.07	5279	0.57	7157	1.07	8577	1.57	9418	2.07	9808	2.57	9949
0.08	5319	0.58	7190	1.08	8599	1.58	9429	2.08	9812	2.58	9951
0.09	5359	0.59	7224	1.09	8621	1.59	9441	2.09	9817	2.59	9952
0.10	5398	0.60	7257	1.10	8643	1.60	9452	2.10	9821	2.60	9953
0.11	5438	0.61	7291	1.11	8665	1.61	9463	2.11	9826	2.61	9955
0.12	5478	0.62	7324	1.12	8686	1.62	9474	2.12	9830	2.62	9956
0.13	5517	0.63	7357	1.13	8708	1.63	9484	2.13	9834	2.63	9957
0.14	5557	0.64	7389	1.14	8729	1.64	9495	2.14	9838	2.64	9959
0.15	5596	0.65	7422	1.15	8749	1.65	9505	2.15	9842	2.65	9960
0.16	5636	0.66	7454	1.16	8770	1.66	9515	2.16	9846	2.66	9961
0.17	5675	0.67	7486	1.17	8790	1.67	9525	2.17	9850	2.67	9962
0.18	5714	0.68	7517	1.18	8810	1.68	9535	2.18	9854	2.68	9963
0.19	5753	0.69	7549	1.19	8830	1.69	9545	2.19	9857	2.69	9964
0.20	5793	0.70	7580	1.20	8849	1.70	9554	2.20	9861	2.70	9965
0.21	5832	0.71	7611	1.21	8869	1.71	9564	2.21	9864	2.71	9966
0.22	5871	0.72	7642	1.22	8888	1.72	9573	2.22	9868	2.72	9967
0.23	5910	0.73	7673	1.23	8907	1.73	9582	2.23	9871	2.73	9968
0.24	5948	0.74	7704	1.24	8925	1.74	9591	2.24	9875	2.74	9969
0.25	5987	0.75	7734	1.25	8944	1.75	9599	2.25	9878	2.75	9970
0.26	6026	0.76	7764	1.26	8962	1.76	9608	2.26	9881	2.76	9971
0.27	6064	0.77	7794	1.27	8980	1.77	9616	2.27	9884	2.77	9972
0.28	6103	0.78	7823	1.28	8997	1.78	9625	2.28	9887	2.78	9973
0.29	6141	0.79	7852	1.29	9015	1.79	9633	2.29	9890	2.79	9974
0.30	6179	0.80	7881	1.30	9032	1.80	9641	2.30	9893	2.80	9974
0.31	6217	0.81	7910	1.31	9049	1.81	9649	2.31	9896	2.81	9975
0.32	6255	0.82	7939	1.32	9066	1.82	9656	2.32	9898	2.82	9976
0.33	6293	0.83	7967	1.33	9082	1.83	9664	2.33	9901	2.83	9977
0.34	6331	0.84	7995	1.34	9099	1.84	9671	2.34	9904	2.84	9977
0.35	6368	0.85	8023	1.35	9115	1.85	9678	2.35	9906	2.85	9978
0.36	6406	0.86	8051	1.36	9131	1.86	9686	2.36	9909	2.86	9979
0.37	6443	0.87	8078	1.37	9147	1.87	9693	2.37	9911	2.87	9979
0.38	6480	0.88	8106	1.38	9162	1.88	9699	2.38	9913	2.88	9980
0.39	6517	0.89	8133	1.39	9177	1.89	9706	2.39	9916	2.89	9981
0.40	6554	0.90	8159	1.40	9192	1.90	9713	2.40	9918	2.90	9981
0.41	6591	0.91	8186	1.41	9207	1.91	9719	2.41	9920	2.91	9982
0.42	6628	0.92	8212	1.42	9222	1.92	9726	2.42	9922	2.92	9982
0.43	6664	0.93	8238	1.43	9236	1.93	9732	2.43	9925	2.93	9983
0.44	6700	0.94	8264	1.44	9251	1.94	9738	2.44	9927	2.94	9984
0.45	6736	0.95	8289	1.45	9265	1.95	9744	2.45	9929	2.95	9984
0.46	6772	0.96	8315	1.46	9279	1.96	9750	2.46	9931	2.96	9985
0.47	6808	0.97	8340	1.47	9292	1.97	9756	2.47	9932	2.97	9985
0.48	6844	0.98	8365	1.48	9306	1.98	9761	2.48	9934	2.98	9986
0.49	6879	0.99	8389	1.49	9319	1.99	9767	2.49	9936	2.99	9986
0.50	6915	1.00	8413	1.50	9332	2.00	9772	2.50	9938	3.00	9987

Table A9
Normal Distribution

Values of z for given values of $\Phi(z)$ (cf. (4), Sec. 20.10) and $D(z) = \Phi(z) - \Phi(-z)$
Example: $z = 0.279$ when $\Phi(z) = 61\%$; $z = 0.860$ when $D(z) = 61\%$.

%	$z(\Phi)$	$z(D)$	%	$z(\Phi)$	$z(D)$	%	$z(\Phi)$	$z(D)$
1	−2.326	0.013	41	−0.228	0.539	81	0.878	1.311
2	−2.054	0.025	42	−0.202	0.553	82	0.915	1.341
3	−1.881	0.038	43	−0.176	0.568	83	0.954	1.372
4	−1.751	0.050	44	−0.151	0.583	84	0.994	1.405
5	−1.645	0.063	45	−0.126	0.598	85	1.036	1.440
6	−1.555	0.075	46	−0.100	0.613	86	1.080	1.476
7	−1.476	0.088	47	−0.075	0.628	87	1.126	1.514
8	−1.405	0.100	48	−0.050	0.643	88	1.175	1.555
9	−1.341	0.113	49	−0.025	0.659	89	1.227	1.598
10	−1.282	0.126	50	0.000	0.674	90	1.282	1.645
11	−1.227	0.138	51	0.025	0.690	91	1.341	1.695
12	−1.175	0.151	52	0.050	0.706	92	1.405	1.751
13	−1.126	0.164	53	0.075	0.722	93	1.476	1.812
14	−1.080	0.176	54	0.100	0.739	94	1.555	1.881
15	−1.036	0.189	55	0.126	0.755	95	1.645	1.960
16	−0.994	0.202	56	0.151	0.772	96	1.751	2.054
17	−0.954	0.215	57	0.176	0.789	97	1.881	2.170
18	−0.915	0.228	58	0.202	0.806	98	2.054	2.326
19	−0.878	0.240	59	0.228	0.824	99	2.326	2.576
20	−0.842	0.253	60	0.253	0.842			
21	−0.806	0.266	61	0.279	0.860	99.1	2.366	2.612
22	−0.772	0.279	62	0.305	0.878	99.2	2.409	2.652
23	−0.739	0.292	63	0.332	0.896	99.3	2.457	2.697
24	−0.706	0.305	64	0.358	0.915	99.4	2.512	2.748
25	−0.674	0.319	65	0.385	0.935	99.5	2.576	2.807
26	−0.643	0.332	66	0.412	0.954	99.6	2.652	2.878
27	−0.613	0.345	67	0.440	0.974	99.7	2.748	2.968
28	−0.583	0.358	68	0.468	0.994	99.8	2.878	3.090
29	−0.553	0.372	69	0.496	1.015	99.9	3.090	3.291
30	−0.524	0.385	70	0.524	1.036			
31	−0.496	0.399	71	0.553	1.058	99.91	3.121	3.320
32	−0.468	0.412	72	0.583	1.080	99.92	3.156	3.353
33	−0.440	0.426	73	0.613	1.103	99.93	3.195	3.390
34	−0.412	0.440	74	0.643	1.126	99.94	3.239	3.432
35	−0.385	0.454	75	0.674	1.150	99.95	3.291	3.481
36	−0.358	0.468	76	0.706	1.175	99.96	3.353	3.540
37	−0.332	0.482	77	0.739	1.200	99.97	3.432	3.615
38	−0.305	0.496	78	0.772	1.227	99.98	3.540	3.719
39	−0.279	0.510	79	0.806	1.254	99.99	3.719	3.891
40	−0.253	0.524	80	0.842	1.282			

Table A10.
Random Digits

Cf. Sec. 20.12.

Row No.	Column Number 0	1	2	3	4	5	6	7	8	9
0	87331	82442	28104	26432	83640	17323	68764	84728	37995	96106
1	33628	17364	01409	87803	65641	33433	48944	64299	79066	31777
2	54680	13427	72496	16967	16195	96593	55040	53729	62035	66717
3	51199	49794	49407	10774	98140	83891	37195	24066	61140	65144
4	78702	98067	61313	91661	59861	54437	77739	19892	54817	88645
5	55672	16014	24892	13089	00410	81458	76156	28189	40595	21500
6	18880	58497	03862	32368	59320	24807	63392	79793	63043	09425
7	10242	62548	62330	05703	33535	49128	66298	16193	55301	01306
8	54993	17182	94618	23228	83895	73251	68199	64639	83178	70521
9	22686	50885	16006	04041	08077	33065	35237	02502	94755	72062
10	42349	03145	15770	70665	53291	32288	41568	66079	98705	31029
11	18093	09553	39428	75464	71329	86344	80729	40916	18860	51780
12	11535	03924	84252	74795	40193	84597	42497	21918	91384	84721
13	35066	73848	65351	53270	67341	70177	92373	17604	42204	60476
14	57477	22809	73558	96182	96779	01604	25748	59553	64876	94611
15	48647	33850	52956	45410	88212	05120	99391	32276	55961	41775
16	86857	81154	22223	74950	53296	67767	55866	49061	66937	81818
17	20182	36907	94644	99122	09774	29189	27212	79000	50217	71077
18	83687	31231	01133	41432	54542	60204	81618	09586	34481	87683
19	81315	12390	46074	47810	90171	36313	95440	77583	28506	38808
20	87026	52826	58341	76549	04105	66191	12914	55348	07907	06978
21	34301	76733	07251	90524	21931	83695	41340	53581	64582	60210
22	70734	24337	32674	49508	49751	90489	63202	24380	77943	09942
23	94710	31527	73445	32839	68176	53580	85250	53243	03350	00128
24	76462	16987	07775	43162	11777	16810	75158	13894	88945	15539
25	14348	28403	79245	69023	34196	46398	05964	64715	11330	17515
26	74618	89317	30146	25606	94507	98104	04239	44973	37636	88866
27	99442	19200	85406	45358	86253	60638	38858	44964	54103	57287
28	26869	44399	89452	06652	31271	00647	46551	83050	92058	83814
29	80988	08149	50499	98584	28385	63680	44638	91864	96002	87802
30	07511	79047	89289	17774	67194	37362	85684	55505	97809	67056
31	49779	12138	05048	03535	27502	63308	10218	53296	48687	61340
32	47938	55945	24003	19635	17471	65997	85906	98694	56420	78357
33	15604	06626	14360	79542	13512	87595	08542	03800	35443	52823
34	12307	27726	21864	00045	16075	03770	86978	52718	02693	09096
35	02450	28053	66134	99445	91316	25727	89399	85272	67148	78358
36	57623	54382	35236	89244	27245	90500	75430	96762	71968	65838
37	91762	78849	93105	40481	99431	03304	21079	86459	21287	76566
38	87373	31137	31128	67050	34309	44914	80711	61738	61498	24288
39	67094	41485	54149	86088	10192	21174	39948	67268	29938	32476
40	94456	66747	76922	87627	71834	57688	04878	78348	68970	60048
41	68359	75292	27710	86889	81678	79798	58360	39175	75667	65782
42	52393	31404	32584	06837	79762	13168	76055	54833	22841	98889
43	59565	91254	11847	20672	37625	41454	86861	55824	79793	74575
44	48185	11066	20162	38230	16043	48409	47421	21195	98008	57305
45	19230	12187	86659	12971	52204	76546	63272	19312	81662	96557
46	84327	21942	81727	68735	89190	58491	55329	96875	19465	89687
47	77430	71210	00591	50124	12030	50280	12358	76174	48353	09682
48	12462	19108	70512	53926	25595	97085	03833	59806	12351	64253
49	11684	06644	57816	10078	45021	47751	38285	73520	08434	65627

Table A10
Random Digits (*continued*)

Row No.	0	1	2	3	4	5	6	7	8	9
50	12896	36576	68686	08462	65652	76571	70891	09007	04581	01684
51	59090	05111	27587	90349	30789	50304	70650	06646	70126	15284
52	42486	67483	65282	19037	80588	73076	41820	46651	40442	40718
53	88662	03928	03249	85910	97533	88643	29829	21557	47328	36724
54	69403	03626	92678	53460	15465	83516	54012	80509	55976	46115
55	56434	70543	38696	98502	32092	95505	62091	39549	30117	98209
56	58227	62694	42837	29183	11393	68463	25150	86338	95620	39836
57	41272	94927	15413	40505	33123	63218	72940	98349	57249	40170
58	36819	01162	30425	15546	16065	68459	35776	64276	92868	07372
59	31700	66711	26115	55755	33584	18091	38709	57276	74660	90392
60	69855	63699	36839	90531	97125	87875	62824	03889	12538	24740
61	44322	17569	45439	41455	34324	90902	07978	26268	04279	76816
62	62226	36661	87011	66267	78777	78044	40819	49496	39814	73867
63	27284	19737	98741	72531	52741	26699	98755	19657	08665	16818
64	88341	21652	94743	77268	79525	44769	66583	30621	90534	62050
65	53266	18783	51903	56711	38060	69513	61963	80470	88018	86510
66	50527	49330	24839	42529	03944	95219	88724	37247	84166	23023
67	15655	07852	77206	35944	71446	30573	19405	57824	23576	23301
68	62057	22206	03314	83465	57466	10465	19891	32308	01900	67484
69	41769	56091	19892	96253	92808	45785	52774	49674	68103	65032
70	25993	72416	44473	41299	93095	17338	69802	98548	02429	85238
71	22842	57871	04470	37373	34516	04042	04078	35336	34393	97573
72	55704	31982	05234	22664	22181	40358	28089	15790	33340	18852
73	94258	18706	09437	96041	90052	80862	20420	24323	11635	91677
74	74145	20453	29657	98868	56695	53483	87449	35060	98942	62697
75	88881	12673	73961	89884	73247	97670	69570	88888	58560	72580
76	01508	56780	52223	35632	73347	71317	46541	88023	36656	76332
77	92069	43000	23233	06058	82527	25250	27555	20426	60361	63525
78	53366	35249	02117	68620	39388	69795	73215	01846	16983	78560
79	88057	54097	49511	74867	32192	90071	04147	46094	63519	07199
80	85492	82238	02668	91854	86149	28590	77853	81035	45561	16032
81	39453	62123	69611	53017	34964	09786	24614	49514	01056	18700
82	82627	98111	93870	56969	69566	62662	07353	84838	14570	14508
83	61142	51743	38209	31474	96095	15163	54380	77849	20465	03142
84	12031	32528	61311	53730	89032	16124	58844	35386	45521	59368
85	31313	59838	29147	76882	74328	09955	63673	96651	53264	29871
86	50767	41056	97409	44376	62219	35439	70102	99248	71179	26052
87	30522	95699	84966	26554	24768	72247	84993	85375	92518	16334
88	74176	19870	89874	64799	03792	57006	57225	36677	46825	14087
89	17114	93248	37065	91346	04657	93763	92210	43676	44944	75798
90	53005	11825	64608	87587	05742	31914	55044	41818	29667	77424
91	31985	81539	79942	49471	46200	27639	94099	42085	79231	03932
92	63499	60508	77522	15624	15088	78519	52279	79214	43623	69166
93	30506	42444	99047	66010	91657	37160	37408	85714	21420	80996
94	78248	16841	92357	10130	68990	38307	61022	56806	81016	38511
95	64996	84789	50185	32200	64382	29752	11876	00664	54547	62597
96	11963	13157	09136	01769	30117	71486	80111	09161	08371	71749
97	44335	91450	43456	90449	18338	19787	31339	60473	06606	89788
98	42277	11868	44520	01113	11341	11743	97949	49718	99176	42006
99	77562	18863	58515	90166	78508	14864	19111	57183	85808	59385

Table A11
***t*-Distribution**

Values of z for given values of the distribution function $F(z)$ (cf. p. 902)
Example: For 9 degrees of freedom, $z = 1.83$ when $F(z) = 0.95$.

	Number of Degrees of Freedom									
$F(z)$	1	2	3	4	5	6	7	8	9	10
0.5	0.00	0.00	0.00	0.00	0.00	0.00	0.00	0.00	0.00	0.00
0.6	0.33	0.29	0.28	0.27	0.27	0.27	0.26	0.26	0.26	0.26
0.7	0.73	0.62	0.58	0.57	0.56	0.55	0.55	0.55	0.54	0.54
0.8	1.38	1.06	0.98	0.94	0.92	0.91	0.90	0.89	0.88	0.88
0.9	3.08	1.89	1.64	1.53	1.48	1.44	1.42	1.40	1.38	1.37
0.95	6.31	2.92	2.35	2.13	2.02	1.94	1.90	1.86	1.83	1.81
0.975	12.7	4.30	3.18	2.78	2.57	2.45	2.37	2.31	2.26	2.23
0.99	31.8	6.97	4.54	3.75	3.37	3.14	3.00	2.90	2.82	2.76
0.995	63.7	9.93	5.84	4.60	4.03	3.71	3.50	3.36	3.25	3.17
0.999	318.3	22.3	10.2	7.17	5.89	5.21	4.79	4.50	4.30	4.14

	Number of Degrees of Freedom									
$F(z)$	11	12	13	14	15	16	17	18	19	20
0.5	0.00	0.00	0.00	0.00	0.00	0.00	0.00	0.00	0.00	0.00
0.6	0.26	0.26	0.26	0.26	0.26	0.26	0.26	0.26	0.26	0.26
0.7	0.54	0.54	0.54	0.54	0.54	0.54	0.53	0.53	0.53	0.53
0.8	0.88	0.87	0.87	0.87	0.87	0.87	0.86	0.86	0.86	0.86
0.9	1.36	1.36	1.35	1.35	1.34	1.34	1.33	1.33	1.33	1.33
0.95	1.80	1.78	1.77	1.76	1.75	1.75	1.74	1.73	1.73	1.73
0.975	2.20	2.18	2.16	2.15	2.13	2.12	2.11	2.10	2.09	2.09
0.99	2.72	2.68	2.65	2.62	2.60	2.58	2.57	2.55	2.54	2.53
0.995	3.11	3.06	3.01	2.98	2.95	2.92	2.90	2.88	2.86	2.85
0.999	4.03	3.93	3.85	3.79	3.73	3.69	3.65	3.61	3.58	3.55

	Number of Degrees of Freedom									
$F(z)$	22	24	26	28	30	40	50	100	200	∞
0.5	0.00	0.00	0.00	0.00	0.00	0.00	0.00	0.00	0.00	0.00
0.6	0.26	0.26	0.26	0.26	0.26	0.26	0.26	0.25	0.25	0.25
0.7	0.53	0.53	0.53	0.53	0.53	0.53	0.53	0.53	0.53	0.52
0.8	0.86	0.86	0.86	0.86	0.85	0.85	0.85	0.85	0.84	0.84
0.9	1.32	1.32	1.32	1.31	1.31	1.30	1.30	1.29	1.29	1.28
0.95	1.72	1.71	1.71	1.70	1.70	1.68	1.68	1.66	1.65	1.65
0.975	2.07	2.06	2.06	2.05	2.04	2.02	2.01	1.98	1.97	1.96
0.99	2.51	2.49	2.48	2.47	2.46	2.42	2.40	2.37	2.35	2.33
0.995	2.82	2.80	2.78	2.76	2.75	2.70	2.68	2.63	2.60	2.58
0.999	3.51	3.47	3.44	3.41	3.39	3.31	3.26	3.17	3.13	3.09

Table A12
Chi-square Distribution

Values of z for given values of the distribution function $F(z)$ (cf. p. 904)
Example: For 3 degrees of freedom, $z = 11.34$ when $F(z) = 0.99$.

	Number of Degrees of Freedom									
$F(z)$	1	2	3	4	5	6	7	8	9	10
0.005	0.00	0.01	0.07	0.21	0.41	0.68	0.99	1.34	1.73	2.16
0.01	0.00	0.02	0.11	0.30	0.55	0.87	1.24	1.65	2.09	2.56
0.025	0.00	0.05	0.22	0.48	0.83	1.24	1.69	2.18	2.70	3.25
0.05	0.00	0.10	0.35	0.71	1.15	1.64	2.17	2.73	3.33	3.94
0.95	3.84	5.99	7.81	9.49	11.07	12.59	14.07	15.51	16.92	18.31
0.975	5.02	7.38	9.35	11.14	12.83	14.45	16.01	17.53	19.02	20.48
0.99	6.63	9.21	11.34	13.28	15.09	16.81	18.48	20.09	21.67	23.21
0.995	7.88	10.60	12.84	14.86	16.75	18.55	20.28	21.96	23.59	25.19

	Number of Degrees of Freedom									
$F(z)$	11	12	13	14	15	16	17	18	19	20
0.005	2.60	3.07	3.57	4.07	4.60	5.14	5.70	6.26	6.84	7.43
0.01	3.05	3.57	4.11	4.66	5.23	5.81	6.41	7.01	7.63	8.26
0.025	3.82	4.40	5.01	5.63	6.26	6.91	7.56	8.23	8.91	9.59
0.05	4.57	5.23	5.89	6.57	7.26	7.96	8.67	9.39	10.12	10.85
0.95	19.68	21.03	22.36	23.68	25.00	26.30	27.59	28.87	30.14	31.41
0.975	21.92	23.34	24.74	26.12	27.49	28.85	30.19	31.53	32.85	34.17
0.99	24.73	26.22	27.69	29.14	30.58	32.00	33.41	34.81	36.19	37.57
0.995	26.76	28.30	29.82	31.32	32.80	34.27	35.72	37.16	38.58	40.00

	Number of Degrees of Freedom									
$F(z)$	21	22	23	24	25	26	27	28	29	30
0.005	8.0	8.6	9.3	9.9	10.5	11.2	11.8	12.5	13.1	13.8
0.01	8.9	9.5	10.2	10.9	11.5	12.2	12.9	13.6	14.3	15.0
0.025	10.3	11.0	11.7	12.4	13.1	13.8	14.6	15.3	16.0	16.8
0.05	11.6	12.3	13.1	13.8	14.6	15.4	16.2	16.9	17.7	18.5
0.95	32.7	33.9	35.2	36.4	37.7	38.9	40.1	41.3	42.6	43.8
0.975	35.5	36.8	38.1	39.4	40.6	41.9	43.2	44.5	45.7	47.0
0.99	38.9	40.3	41.6	43.0	44.3	45.6	47.0	48.3	49.6	50.9
0.995	41.4	42.8	44.2	45.6	46.9	48.3	49.6	51.0	52.3	53.7

	Number of Degrees of Freedom							
$F(z)$	40	50	60	70	80	90	100	> 100 (Approximation)
0.005	20.7	28.0	35.5	43.3	51.2	59.2	67.3	$\frac{1}{2}(h - 2.58)^2$
0.01	22.2	29.7	37.5	45.4	53.5	61.8	70.1	$\frac{1}{2}(h - 2.33)^2$
0.025	24.4	32.4	40.5	48.8	57.2	65.6	74.2	$\frac{1}{2}(h - 1.96)^2$
0.05	26.5	34.8	43.2	51.7	60.4	69.1	77.9	$\frac{1}{2}(h - 1.64)^2$
0.95	55.8	67.5	79.1	90.5	101.9	113.1	124.3	$\frac{1}{2}(h + 1.64)^2$
0.975	59.3	71.4	83.3	95.0	106.6	118.1	129.6	$\frac{1}{2}(h + 1.96)^2$
0.99	63.7	76.2	88.4	100.4	112.3	124.1	135.8	$\frac{1}{2}(h + 2.33)^2$
0.995	66.8	79.5	92.0	104.2	116.3	128.3	140.2	$\frac{1}{2}(h + 2.58)^2$

In the last column, $h = \sqrt{2m - 1}$ where m is the number of degrees of freedom.

Table A13
F-Distribution with (*m, n*) Degrees of Freedom

Values of z for which the distribution fuction $F(z)$ (cf. (13), Sec. 20.15) has the value **0.95**
Example: For (7, 4) degrees of freedom, $z = 6.09$ when $F(z) = 0.95$.

n	$m = 1$	$m = 2$	$m = 3$	$m = 4$	$m = 5$	$m = 6$	$m = 7$	$m = 8$	$m = 9$
1	161	200	216	225	230	234	237	239	241
2	18.5	19.0	19.2	19.2	19.3	19.3	19.4	19.4	19.4
3	10.1	9.55	9.28	9.12	9.01	8.94	8.89	8.85	8.81
4	7.71	6.94	6.59	6.39	6.26	6.16	6.09	6.04	6.00
5	6.61	5.79	5.41	5.19	5.05	4.95	4.88	4.82	4.77
6	5.99	5.14	4.76	4.53	4.39	4.28	4.21	4.15	4.10
7	5.59	4.74	4.35	4.12	3.97	3.87	3.79	3.73	3.68
8	5.32	4.46	4.07	3.84	3.69	3.58	3.50	3.44	3.39
9	5.12	4.26	3.86	3.63	3.48	3.37	3.29	3.23	3.18
10	4.96	4.10	3.71	3.48	3.33	3.22	3.14	3.07	3.02
11	4.84	3.98	3.59	3.36	3.20	3.09	3.01	2.95	2.90
12	4.75	3.89	3.49	3.26	3.11	3.00	2.91	2.85	2.80
13	4.67	3.81	3.41	3.18	3.03	2.92	2.83	2.77	2.71
14	4.60	3.74	3.34	3.11	2.96	2.85	2.76	2.70	2.65
15	4.54	3.68	3.29	3.06	2.90	2.79	2.71	2.64	2.59
16	4.49	3.63	3.24	3.01	2.85	2.74	2.66	2.59	2.54
17	4.45	3.59	3.20	2.96	2.81	2.70	2.61	2.55	2.49
18	4.41	3.55	3.16	2.93	2.77	2.66	2.58	2.51	2.46
19	4.38	3.52	3.13	2.90	2.74	2.63	2.54	2.48	2.42
20	4.35	3.49	3.10	2.87	2.71	2.60	2.51	2.45	2.39
22	4.30	3.44	3.05	2.82	2.66	2.55	2.46	2.40	2.34
24	4.26	3.40	3.01	2.78	2.62	2.51	2.42	2.36	2.30
26	4.23	3.37	2.98	2.74	2.59	2.47	2.39	2.32	2.27
28	4.20	3.34	2.95	2.71	2.56	2.45	2.36	2.29	2.24
30	4.17	3.32	2.92	2.69	2.53	2.42	2.33	2.27	2.21
32	4.15	3.30	2.90	2.67	2.51	2.40	2.31	2.24	2.19
34	4.13	3.28	2.88	2.65	2.49	2.38	2.29	2.23	2.17
36	4.11	3.26	2.87	2.63	2.48	2.36	2.28	2.21	2.15
38	4.10	3.24	2.85	2.62	2.46	2.35	2.26	2.19	2.14
40	4.08	3.23	2.84	2.61	2.45	2.34	2.25	2.18	2.12
50	4.03	3.18	2.79	2.56	2.40	2.29	2.20	2.13	2.07
60	4.00	3.15	2.76	2.53	2.37	2.25	2.17	2.10	2.04
70	3.98	3.13	2.74	2.50	2.35	2.23	2.14	2.07	2.02
80	3.96	3.11	2.72	2.49	2.33	2.21	2.13	2.06	2.00
90	3.95	3.10	2.71	2.47	2.32	2.20	2.11	2.04	1.99
100	3.94	3.09	2.70	2.46	2.31	2.19	2.10	2.03	1.97
150	3.90	3.06	2.66	2.43	2.27	2.16	2.07	2.00	1.94
200	3.89	3.04	2.65	2.42	2.26	2.14	2.06	1.98	1.93
1000	3.85	3.00	2.61	2.38	2.22	2.11	2.02	1.95	1.89
∞	3.84	3.00	2.60	2.37	2.21	2.10	2.01	1.94	1.88

Table A13
F-Distribution with (*m*, *n*) Degrees of Freedom (*continued*)

Values of z for which the distribution function $F(z)$ (cf. (13), Sec. 20.15) has the value **0.95**

n	$m = 10$	$m = 15$	$m = 20$	$m = 30$	$m = 40$	$m = 50$	$m = 100$	∞
1	242	246	248	250	251	252	253	254
2	19.4	19.4	19.4	19.5	19.5	19.5	19.5	19.5
3	8.79	8.70	8.66	8.62	8.59	8.58	8.55	8.53
4	5.96	5.86	5.80	5.75	5.72	5.70	5.66	5.63
5	4.74	4.62	4.56	4.50	4.46	4.44	4.41	4.37
6	4.06	3.94	3.87	3.81	3.77	3.75	3.71	3.67
7	3.64	3.51	3.44	3.38	3.34	3.32	3.27	3.23
8	3.35	3.22	3.15	3.08	3.04	3.02	2.97	2.93
9	3.14	3.01	2.94	2.86	2.83	2.80	2.76	2.71
10	2.98	2.85	2.77	2.70	2.66	2.64	2.59	2.54
11	2.85	2.72	2.65	2.57	2.53	2.51	2.46	2.40
12	2.75	2.62	2.54	2.47	2.43	2.40	2.35	2.30
13	2.67	2.53	2.46	2.38	2.34	2.31	2.26	2.21
14	2.60	2.46	2.39	2.31	2.27	2.24	2.19	2.13
15	2.54	2.40	2.33	2.25	2.20	2.18	2.12	2.07
16	2.49	2.35	2.28	2.19	2.15	2.12	2.07	2.01
17	2.45	2.31	2.23	2.15	2.10	2.08	2.02	1.96
18	2.41	2.27	2.19	2.11	2.06	2.04	1.98	1.92
19	2.38	2.23	2.16	2.07	2.03	2.00	1.94	1.88
20	2.35	2.20	2.12	2.04	1.99	1.97	1.91	1.84
22	2.30	2.15	2.07	1.98	1.94	1.91	1.85	1.78
24	2.25	2.11	2.03	1.94	1.89	1.86	1.80	1.73
26	2.22	2.07	1.99	1.90	1.85	1.82	1.76	1.69
28	2.19	2.04	1.96	1.87	1.82	1.79	1.73	1.65
30	2.16	2.01	1.93	1.84	1.79	1.76	1.70	1.62
32	2.14	1.99	1.91	1.82	1.77	1.74	1.67	1.59
34	2.12	1.97	1.89	1.80	1.75	1.71	1.65	1.57
36	2.11	1.95	1.87	1.78	1.73	1.69	1.62	1.55
38	2.09	1.94	1.85	1.76	1.71	1.68	1.61	1.53
40	2.08	1.92	1.84	1.74	1.69	1.66	1.59	1.51
50	2.03	1.87	1.78	1.69	1.63	1.60	1.52	1.44
60	1.99	1.84	1.75	1.65	1.59	1.56	1.48	1.39
70	1.97	1.81	1.72	1.62	1.57	1.53	1.45	1.35
80	1.95	1.79	1.70	1.60	1.54	1.51	1.43	1.32
90	1.94	1.78	1.69	1.59	1.53	1.49	1.41	1.30
100	1.93	1.77	1.68	1.57	1.52	1.48	1.39	1.28
150	1.89	1.73	1.64	1.53	1.48	1.44	1.34	1.22
200	1.88	1.72	1.62	1.52	1.46	1.41	1.32	1.19
1000	1.84	1.68	1.58	1.47	1.41	1.36	1.26	1.08
∞	1.83	1.67	1.57	1.46	1.39	1.35	1.24	1.00

Table A13
F-Distribution with (*m, n*) Degrees of Freedom (*continued*)

Values of z for which the distribution function $F(z)$ (cf. (13), Sec. 20.15) has the value **0.**

n	$m = 1$	$m = 2$	$m = 3$	$m = 4$	$m = 5$	$m = 6$	$m = 7$	$m = 8$	$m = 9$
1	4052	4999	5403	5625	5764	5859	5928	5982	6022
2	98.5	99.0	99.2	99.3	99.3	99.3	99.4	99.4	99.4
3	34.1	30.8	29.5	28.7	28.2	27.9	27.7	27.5	27.3
4	21.2	18.0	16.7	16.0	15.5	15.2	15.0	14.8	14.7
5	16.3	13.3	12.1	11.4	11.0	10.7	10.5	10.3	10.2
6	13.7	10.9	9.78	9.15	8.75	8.47	8.26	8.10	7.98
7	12.2	9.55	8.45	7.85	7.46	7.19	6.99	6.84	6.72
8	11.3	8.65	7.59	7.01	6.63	6.37	6.18	6.03	5.91
9	10.6	8.02	6.99	6.42	6.06	5.80	5.61	5.47	5.35
10	10.0	7.56	6.55	5.99	5.64	5.39	5.20	5.06	4.94
11	9.65	7.21	6.22	5.67	5.32	5.07	4.89	4.74	4.63
12	9.33	6.93	5.95	5.41	5.06	4.82	4.64	4.50	4.39
13	9.07	6.70	5.74	5.21	4.86	4.62	4.44	4.30	4.19
14	8.86	6.51	5.56	5.04	4.70	4.46	4.28	4.14	4.03
15	8.68	6.36	5.42	4.89	4.56	4.32	4.14	4.00	3.89
16	8.53	6.23	5.29	4.77	4.44	4.20	4.03	3.89	3.78
17	8.40	6.11	5.18	4.67	4.34	4.10	3.93	3.79	3.68
18	8.29	6.01	5.09	4.58	4.25	4.01	3.84	3.71	3.60
19	8.18	5.93	5.01	4.50	4.17	3.94	3.77	3.63	3.52
20	8.10	5.85	4.94	4.43	4.10	3.87	3.70	3.56	3.46
22	7.95	5.72	4.82	4.31	3.99	3.76	3.59	3.45	3.35
24	7.82	5.61	4.72	4.22	3.90	3.67	3.50	3.36	3.26
26	7.72	5.53	4.64	4.14	3.82	3.59	3.42	3.29	3.18
28	7.64	5.45	4.57	4.07	3.75	3.53	3.36	3.23	3.12
30	7.56	5.39	4.51	4.02	3.70	3.47	3.30	3.17	3.07
32	7.50	5.34	4.46	3.97	3.65	3.43	3.26	3.13	3.02
34	7.44	5.29	4.42	3.93	3.61	3.39	3.22	3.09	2.98
36	7.40	5.25	4.38	3.89	3.57	3.35	3.18	3.05	2.95
38	7.35	5.21	4.34	3.86	3.54	3.32	3.15	3.02	2.92
40	7.31	5.18	4.31	3.83	3.51	3.29	3.12	2.99	2.89
50	7.17	5.06	4.20	3.72	3.41	3.19	3.02	2.89	2.79
60	7.08	4.98	4.13	3.65	3.34	3.12	2.95	2.82	2.72
70	7.01	4.92	4.08	3.60	3.29	3.07	2.91	2.78	2.67
80	6.96	4.88	4.04	3.56	3.26	3.04	2.87	2.74	2.64
90	6.93	4.85	4.01	3.54	3.23	3.01	2.84	2.72	2.61
100	6.90	4.82	3.98	3.51	3.21	2.99	2.82	2.69	2.59
150	6.81	4.75	3.92	3.45	3.14	2.92	2.76	2.63	2.53
200	6.76	4.71	3.88	3.41	3.11	2.89	2.73	2.60	2.50
1000	6.66	4.63	3.80	3.34	3.04	2.82	2.66	2.53	2.43
∞	6.63	4.61	3.78	3.32	3.02	2.80	2.64	2.51	2.41

Table A13
F-Distribution with (*m*, *n*) Degrees of Freedom (*continued*)

Values of z for which the distribution function $F(z)$ (cf. (13), Sec. 20.15) has the value **0.99**

n	$m = 10$	$m = 15$	$m = 20$	$m = 30$	$m = 40$	$m = 50$	$m = 100$	∞
1	6056	6157	6209	6261	6287	6300	6330	6366
2	99.4	99.4	99.4	99.5	99.5	99.5	99.5	99.5
3	27.2	26.9	26.7	26.5	26.4	26.4	26.2	26.1
4	14.5	14.2	14.0	13.8	13.7	13.7	13.6	13.5
5	10.1	9.72	9.55	9.38	9.29	9.24	9.13	9.02
6	7.87	7.56	7.40	7.23	7.14	7.09	6.99	6.88
7	6.62	6.31	6.16	5.99	5.91	5.86	5.75	5.65
8	5.81	5.52	5.36	5.20	5.12	5.07	4.96	4.86
9	5.26	4.96	4.81	4.65	4.57	4.52	4.42	4.31
10	4.85	4.56	4.41	4.25	4.17	4.12	4.01	3.91
11	4.54	4.25	4.10	3.94	3.86	3.81	3.71	3.60
12	4.30	4.01	3.86	3.70	3.62	3.57	3.47	3.36
13	4.10	3.82	3.66	3.51	3.43	3.38	3.27	3.17
14	3.94	3.66	3.51	3.35	3.27	3.22	3.11	3.00
15	3.80	3.52	3.37	3.21	3.13	3.08	2.98	2.87
16	3.69	3.41	3.26	3.10	3.02	2.97	2.86	2.75
17	3.59	3.31	3.16	3.00	2.92	2.87	2.76	2.65
18	3.51	3.23	3.08	2.92	2.84	2.78	2.68	2.57
19	3.43	3.15	3.00	2.84	2.76	2.71	2.60	2.49
20	3.37	3.09	2.94	2.78	2.69	2.64	2.54	2.42
22	3.26	2.98	2.83	2.67	2.58	2.53	2.42	2.31
24	3.17	2.89	2.74	2.58	2.49	2.44	2.33	2.21
26	3.09	2.82	2.66	2.50	2.42	2.36	2.25	2.13
28	3.03	2.75	2.60	2.44	2.35	2.30	2.19	2.06
30	2.98	2.70	2.55	2.39	2.30	2.25	2.13	2.01
32	2.93	2.66	2.50	2.34	2.25	2.20	2.08	1.96
34	2.89	2.62	2.46	2.30	2.21	2.16	2.04	1.91
36	2.86	2.58	2.43	2.26	2.17	2.12	2.00	1.87
38	2.83	2.55	2.40	2.23	2.14	2.09	1.97	1.84
40	2.80	2.52	2.37	2.20	2.11	2.06	1.94	1.80
50	2.70	2.42	2.27	2.10	2.01	1.95	1.82	1.68
60	2.63	2.35	2.20	2.03	1.94	1.88	1.75	1.60
70	2.59	2.31	2.15	1.98	1.89	1.83	1.70	1.54
80	2.55	2.27	2.12	1.94	1.85	1.79	1.66	1.49
90	2.52	2.24	2.09	1.92	1.82	1.76	1.62	1.46
100	2.50	2.22	2.07	1.89	1.80	1.73	1.60	1.43
150	2.44	2.16	2.00	1.83	1.73	1.66	1.52	1.33
200	2.41	2.13	1.97	1.79	1.69	1.63	1.48	1.28
1000	2.34	2.06	1.90	1.72	1.61	1.54	1.38	1.11
∞	2.32	2.04	1.88	1.70	1.59	1.52	1.36	1.00

Table A14
Distribution Function $F(x) = P(T \leq x)$ of the Random Variable T in Section 20.19

When $n = 3$, $F(2) = 1 - 0.167 = 0.833$.
When $n = 4$, $F(3) = 1 - 0.375 = 0.625$, $F(4) = 1 - 0.167 = 0.833$, etc.

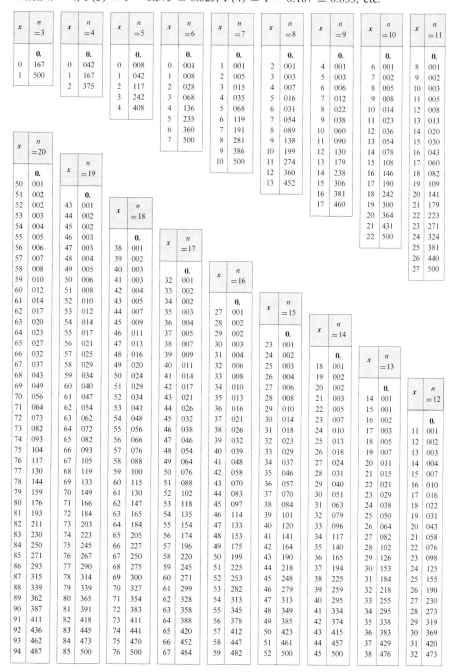

x	$n = 3$
	0.
0	167
1	500

x	$n = 4$
	0.
0	042
1	167
2	375

x	$n = 5$
	0.
0	008
1	042
2	117
3	242
4	408

x	$n = 6$
	0.
0	001
1	008
2	028
3	068
4	136
5	235
6	360
7	500

x	$n = 7$
	0.
1	001
2	005
3	015
4	035
5	068
6	119
7	191
8	281
9	386
10	500

x	$n = 8$
	0.
2	001
3	003
4	007
5	016
6	031
7	054
8	089
9	138
10	199
11	274
12	360
13	452

x	$n = 9$
	0.
4	001
5	003
6	006
7	012
8	022
9	038
10	060
11	090
12	130
13	179
14	238
15	306
16	381
17	460

x	$n = 10$
	0.
6	001
7	002
8	005
9	008
10	014
11	023
12	036
13	054
14	078
15	108
16	146
17	190
18	242
19	300
20	364
21	431
22	500

x	$n = 11$
	0.
8	001
9	002
10	003
11	005
12	008
13	013
14	020
15	030
16	043
17	060
18	082
19	109
20	141
21	179
22	223
23	271
24	324
25	381
26	440
27	500

x	$n = 20$
	0.
50	001
51	002
52	002
53	003
54	004
55	005
56	006
57	007
58	008
59	010
60	012
61	014
62	017
63	020
64	023
65	027
66	032
67	037
68	043
69	049
70	056
71	064
72	073
73	082
74	093
75	104
76	117
77	130
78	144
79	159
80	176
81	193
82	211
83	230
84	250
85	271
86	293
87	315
88	339
89	362
90	387
91	411
92	436
93	462
94	487

x	$n = 19$
	0.
43	001
44	002
45	002
46	003
47	003
48	004
49	005
50	006
51	008
52	010
53	012
54	014
55	017
56	021
57	025
58	029
59	034
60	040
61	047
62	054
63	062
64	072
65	082
66	093
67	105
68	119
69	133
70	149
71	166
72	184
73	203
74	223
75	245
76	267
77	290
78	314
79	339
80	365
81	391
82	418
83	445
84	473
85	500

x	$n = 18$
	0.
38	001
39	002
40	003
41	003
42	004
43	005
44	007
45	009
46	011
47	013
48	016
49	020
50	024
51	029
52	034
53	041
54	048
55	056
56	066
57	076
58	088
59	100
60	115
61	130
62	147
63	165
64	184
65	205
66	227
67	250
68	275
69	300
70	327
71	354
72	383
73	411
74	441
75	470
76	500

x	$n = 17$
	0.
32	001
33	002
34	002
35	003
36	004
37	005
38	007
39	009
40	011
41	014
42	017
43	021
44	026
45	032
46	038
47	046
48	054
49	064
50	076
51	088
52	102
53	118
54	135
55	154
56	174
57	196
58	220
59	245
60	271
61	299
62	328
63	358
64	388
65	420
66	452
67	484

x	$n = 16$
	0.
27	001
28	002
29	002
30	003
31	004
32	006
33	008
34	010
35	013
36	016
37	021
38	026
39	032
40	039
41	048
42	058
43	070
44	083
45	097
46	114
47	133
48	153
49	175
50	199
51	225
52	253
53	282
54	313
55	345
56	378
57	412
58	447
59	482

x	$n = 15$
	0.
23	001
24	002
25	003
26	004
27	006
28	008
29	010
30	014
31	018
32	023
33	029
34	037
35	046
36	057
37	070
38	084
39	101
40	120
41	141
42	164
43	190
44	218
45	248
46	279
47	313
48	349
49	385
50	423
51	461
52	500

x	$n = 14$
	0.
18	001
19	002
20	002
21	003
22	005
23	007
24	010
25	013
26	018
27	024
28	031
29	040
30	051
31	063
32	079
33	096
34	117
35	140
36	165
37	194
38	225
39	259
40	295
41	334
42	374
43	415
44	457
45	500

x	$n = 13$
	0.
14	001
15	001
16	002
17	003
18	005
19	007
20	011
21	015
22	021
23	029
24	038
25	050
26	064
27	082
28	102
29	126
30	153
31	184
32	218
33	255
34	295
35	338
36	383
37	429
38	476

x	$n = 12$
	0.
11	001
12	002
13	003
14	004
15	007
16	010
17	016
18	022
19	031
20	043
21	058
22	076
23	098
24	125
25	155
26	190
27	230
28	273
29	319
30	369
31	420
32	473

INDEX

Page numbers A1, A2, A3, . . . refer to Appendix 1 to Appendix 4 at the end of the book.

B

C

T

U

V

STOCKTON - BILLINGHAM
LEARNING CENTRE
COLLEGE OF F.E.

SOME CONSTANTS

$$e = 2.718\,281\,828 = 1/0.367\,879\,441$$
$$e^2 = 7.389\,056\,099 = 1/0.135\,335\,283$$
$$\sqrt{e} = 1.648\,721\,271 = 1/0.606\,530\,660$$

$$\pi = 3.141\,592\,654 = 1/0.318\,309\,886$$
$$\pi^2 = 9.869\,604\,401 = 1/0.101\,321\,184$$
$$\sqrt{\pi} = 1.772\,453\,851 = 1/0.564\,189\,584$$

$$\sqrt{2\pi} = 2.506\,628\,275 = 1/0.398\,942\,280$$
$$\sqrt{\pi/2} = 1.253\,314\,137 = 1/0.797\,884\,561$$
$$\sqrt[3]{\pi} = 1.464\,591\,888 = 1/0.682\,784\,063$$

$$e^{\pi} = 23.140\,692\,633 = 1/0.043\,213\,918$$
$$e^{\pi/2} = 4.810\,477\,381 = 1/0.207\,879\,576$$
$$e^{\pi/4} = 2.193\,280\,051 = 1/0.455\,938\,128$$

$$\log_{10}\pi = 0.497\,149\,873$$
$$\log_e \pi = 1.144\,729\,886$$

$$\sqrt{2} = 1.414\,213\,562 = 1/0.707\,106\,781$$
$$\sqrt{3} = 1.732\,050\,808 = 1/0.577\,350\,269$$
$$\sqrt{10} = 3.162\,277\,660 = 1/0.316\,227\,766$$

$$\sqrt[3]{2} = 1.259\,921\,050 = 1/0.793\,700\,526$$
$$\sqrt[3]{10} = 2.154\,434\,690 = 1/0.464\,158\,883$$

$$\log_e 2 = 0.693\,147\,181$$
$$\log_e 3 = 1.098\,612\,289$$
$$\log_e 10 = 2.302\,585\,093 = 1/0.434\,294\,482$$

$$1^\circ = 0.017\,453\,293 \text{ rad}$$
$$1 \text{ rad} = 57.295\,779\,513^\circ$$
$$= 57^\circ 17' 44.806''$$